Lecture Notes in Computer Science 13698

More information about this series at https://link.springer.com/bookseries/558

Shai Avidan · Gabriel Brostow ·
Moustapha Cissé · Giovanni Maria Farinella ·
Tal Hassner (Eds.)

Computer Vision – ECCV 2022

17th European Conference
Tel Aviv, Israel, October 23–27, 2022
Proceedings, Part XXXVIII

Editors
Shai Avidan
Tel Aviv University
Tel Aviv, Israel

Gabriel Brostow
University College London
London, UK

Moustapha Cissé
Google AI
Accra, Ghana

Giovanni Maria Farinella
University of Catania
Catania, Italy

Tal Hassner
Facebook (United States)
Menlo Park, CA, USA

ISSN 0302-9743 ISSN 1611-3349 (electronic)
Lecture Notes in Computer Science
ISBN 978-3-031-19838-0 ISBN 978-3-031-19839-7 (eBook)
https://doi.org/10.1007/978-3-031-19839-7

This Springer imprint is published by the registered company Springer Nature Switzerland AG
The registered company address is: Gewerbestrasse 11, 6330 Cham, Switzerland

Foreword

Organizing the European Conference on Computer Vision (ECCV 2022) in Tel-Aviv during a global pandemic was no easy feat. The uncertainty level was extremely high, and decisions had to be postponed to the last minute. Still, we managed to plan things just in time for ECCV 2022 to be held in person. Participation in physical events is crucial to stimulating collaborations and nurturing the culture of the Computer Vision community.

There were many people who worked hard to ensure attendees enjoyed the best science at the 16th edition of ECCV. We are grateful to the Program Chairs Gabriel Brostow and Tal Hassner, who went above and beyond to ensure the ECCV reviewing process ran smoothly. The scientific program includes dozens of workshops and tutorials in addition to the main conference and we would like to thank Leonid Karlinsky and Tomer Michaeli for their hard work. Finally, special thanks to the web chairs Lorenzo Baraldi and Kosta Derpanis, who put in extra hours to transfer information fast and efficiently to the ECCV community.

We would like to express gratitude to our generous sponsors and the Industry Chairs, Dimosthenis Karatzas and Chen Sagiv, who oversaw industry relations and proposed new ways for academia-industry collaboration and technology transfer. It's great to see so much industrial interest in what we're doing!

Authors' draft versions of the papers appeared online with open access on both the Computer Vision Foundation (CVF) and the European Computer Vision Association (ECVA) websites as with previous ECCVs. Springer, the publisher of the proceedings, has arranged for archival publication. The final version of the papers is hosted by SpringerLink, with active references and supplementary materials. It benefits all potential readers that we offer both a free and citeable version for all researchers, as well as an authoritative, citeable version for SpringerLink readers. Our thanks go to Ronan Nugent from Springer, who helped us negotiate this agreement. Last but not least, we wish to thank Eric Mortensen, our publication chair, whose expertise made the process smooth.

October 2022

Rita Cucchiara
Jiří Matas
Amnon Shashua
Lihi Zelnik-Manor

Preface

Welcome to the proceedings of the European Conference on Computer Vision (ECCV 2022). This was a hybrid edition of ECCV as we made our way out of the COVID-19 pandemic. The conference received 5804 valid paper submissions, compared to 5150 submissions to ECCV 2020 (a 12.7% increase) and 2439 in ECCV 2018. 1645 submissions were accepted for publication (28%) and, of those, 157 (2.7% overall) as orals.

846 of the submissions were desk-rejected for various reasons. Many of them because they revealed author identity, thus violating the double-blind policy. This violation came in many forms: some had author names with the title, others added acknowledgments to specific grants, yet others had links to their github account where their name was visible. Tampering with the LaTeX template was another reason for automatic desk rejection.

ECCV 2022 used the traditional CMT system to manage the entire double-blind reviewing process. Authors did not know the names of the reviewers and vice versa. Each paper received at least 3 reviews (except 6 papers that received only 2 reviews), totalling more than 15,000 reviews.

Handling the review process at this scale was a significant challenge. To ensure that each submission received as fair and high-quality reviews as possible, we recruited more than 4719 reviewers (in the end, 4719 reviewers did at least one review). Similarly we recruited more than 276 area chairs (eventually, only 276 area chairs handled a batch of papers). The area chairs were selected based on their technical expertise and reputation, largely among people who served as area chairs in previous top computer vision and machine learning conferences (ECCV, ICCV, CVPR, NeurIPS, etc.).

Reviewers were similarly invited from previous conferences, and also from the pool of authors. We also encouraged experienced area chairs to suggest additional chairs and reviewers in the initial phase of recruiting. The median reviewer load was five papers per reviewer, while the average load was about four papers, because of the emergency reviewers. The area chair load was 35 papers, on average.

Conflicts of interest between authors, area chairs, and reviewers were handled largely automatically by the CMT platform, with some manual help from the Program Chairs. Reviewers were allowed to describe themselves as senior reviewer (load of 8 papers to review) or junior reviewers (load of 4 papers). Papers were matched to area chairs based on a subject-area affinity score computed in CMT and an affinity score computed by the Toronto Paper Matching System (TPMS). TPMS is based on the paper's full text. An area chair handling each submission would bid for preferred expert reviewers, and we balanced load and prevented conflicts.

The assignment of submissions to area chairs was relatively smooth, as was the assignment of submissions to reviewers. A small percentage of reviewers were not happy with their assignments in terms of subjects and self-reported expertise. This is an area for improvement, although it's interesting that many of these cases were reviewers handpicked by AC's. We made a later round of reviewer recruiting, targeted at the list of authors of papers submitted to the conference, and had an excellent response which helped provide enough emergency reviewers. In the end, all but six papers received at least 3 reviews.

The challenges of the reviewing process are in line with past experiences at ECCV 2020. As the community grows, and the number of submissions increases, it becomes ever more challenging to recruit enough reviewers and ensure a high enough quality of reviews. Enlisting authors by default as reviewers might be one step to address this challenge.

Authors were given a week to rebut the initial reviews, and address reviewers' concerns. Each rebuttal was limited to a single pdf page with a fixed template.

The Area Chairs then led discussions with the reviewers on the merits of each submission. The goal was to reach consensus, but, ultimately, it was up to the Area Chair to make a decision. The decision was then discussed with a buddy Area Chair to make sure decisions were fair and informative. The entire process was conducted virtually with no in-person meetings taking place.

The Program Chairs were informed in cases where the Area Chairs overturned a decisive consensus reached by the reviewers, and pushed for the meta-reviews to contain details that explained the reasoning for such decisions. Obviously these were the most contentious cases, where reviewer inexperience was the most common reported factor.

Once the list of accepted papers was finalized and released, we went through the laborious process of plagiarism (including self-plagiarism) detection. A total of 4 accepted papers were rejected because of that.

Finally, we would like to thank our Technical Program Chair, Pavel Lifshits, who did tremendous work behind the scenes, and we thank the tireless CMT team.

October 2022

Gabriel Brostow
Giovanni Maria Farinella
Moustapha Cissé
Shai Avidan
Tal Hassner

Organization

General Chairs

Rita Cucchiara	University of Modena and Reggio Emilia, Italy
Jiří Matas	Czech Technical University in Prague, Czech Republic
Amnon Shashua	Hebrew University of Jerusalem, Israel
Lihi Zelnik-Manor	Technion – Israel Institute of Technology, Israel

Program Chairs

Shai Avidan	Tel-Aviv University, Israel
Gabriel Brostow	University College London, UK
Moustapha Cissé	Google AI, Ghana
Giovanni Maria Farinella	University of Catania, Italy
Tal Hassner	Facebook AI, USA

Program Technical Chair

Pavel Lifshits	Technion – Israel Institute of Technology, Israel

Workshops Chairs

Leonid Karlinsky	IBM Research, Israel
Tomer Michaeli	Technion – Israel Institute of Technology, Israel
Ko Nishino	Kyoto University, Japan

Tutorial Chairs

Thomas Pock	Graz University of Technology, Austria
Natalia Neverova	Facebook AI Research, UK

Demo Chair

Bohyung Han	Seoul National University, Korea

Social and Student Activities Chairs

Tatiana Tommasi	Italian Institute of Technology, Italy
Sagie Benaim	University of Copenhagen, Denmark

Diversity and Inclusion Chairs

Xi Yin	Facebook AI Research, USA
Bryan Russell	Adobe, USA

Communications Chairs

Lorenzo Baraldi	University of Modena and Reggio Emilia, Italy
Kosta Derpanis	York University & Samsung AI Centre Toronto, Canada

Industrial Liaison Chairs

Dimosthenis Karatzas	Universitat Autònoma de Barcelona, Spain
Chen Sagiv	SagivTech, Israel

Finance Chair

Gerard Medioni	University of Southern California & Amazon, USA

Publication Chair

Eric Mortensen	MiCROTEC, USA

Area Chairs

Lourdes Agapito	University College London, UK
Zeynep Akata	University of Tübingen, Germany
Naveed Akhtar	University of Western Australia, Australia
Karteek Alahari	Inria Grenoble Rhône-Alpes, France
Alexandre Alahi	École polytechnique fédérale de Lausanne, Switzerland
Pablo Arbelaez	Universidad de Los Andes, Columbia
Antonis A. Argyros	University of Crete & Foundation for Research and Technology-Hellas, Crete
Yuki M. Asano	University of Amsterdam, The Netherlands
Kalle Åström	Lund University, Sweden
Hadar Averbuch-Elor	Cornell University, USA

Hossein Azizpour	KTH Royal Institute of Technology, Sweden
Vineeth N. Balasubramanian	Indian Institute of Technology, Hyderabad, India
Lamberto Ballan	University of Padova, Italy
Adrien Bartoli	Université Clermont Auvergne, France
Horst Bischof	Graz University of Technology, Austria
Matthew B. Blaschko	KU Leuven, Belgium
Federica Bogo	Meta Reality Labs Research, Switzerland
Katherine Bouman	California Institute of Technology, USA
Edmond Boyer	Inria Grenoble Rhône-Alpes, France
Michael S. Brown	York University, Canada
Vittorio Caggiano	Meta AI Research, USA
Neill Campbell	University of Bath, UK
Octavia Camps	Northeastern University, USA
Duygu Ceylan	Adobe Research, USA
Ayan Chakrabarti	Google Research, USA
Tat-Jen Cham	Nanyang Technological University, Singapore
Antoni Chan	City University of Hong Kong, Hong Kong, China
Manmohan Chandraker	NEC Labs America, USA
Xinlei Chen	Facebook AI Research, USA
Xilin Chen	Institute of Computing Technology, Chinese Academy of Sciences, China
Dongdong Chen	Microsoft Cloud AI, USA
Chen Chen	University of Central Florida, USA
Ondrej Chum	Vision Recognition Group, Czech Technical University in Prague, Czech Republic
John Collomosse	Adobe Research & University of Surrey, UK
Camille Couprie	Facebook, France
David Crandall	Indiana University, USA
Daniel Cremers	Technical University of Munich, Germany
Marco Cristani	University of Verona, Italy
Canton Cristian	Facebook AI Research, USA
Dengxin Dai	ETH Zurich, Switzerland
Dima Damen	University of Bristol, UK
Kostas Daniilidis	University of Pennsylvania, USA
Trevor Darrell	University of California, Berkeley, USA
Andrew Davison	Imperial College London, UK
Tali Dekel	Weizmann Institute of Science, Israel
Alessio Del Bue	Istituto Italiano di Tecnologia, Italy
Weihong Deng	Beijing University of Posts and Telecommunications, China
Konstantinos Derpanis	Ryerson University, Canada
Carl Doersch	DeepMind, UK

Matthijs Douze	Facebook AI Research, USA
Mohamed Elhoseiny	King Abdullah University of Science and Technology, Saudi Arabia
Sergio Escalera	University of Barcelona, Spain
Yi Fang	New York University, USA
Ryan Farrell	Brigham Young University, USA
Alireza Fathi	Google, USA
Christoph Feichtenhofer	Facebook AI Research, USA
Basura Fernando	Agency for Science, Technology and Research (A*STAR), Singapore
Vittorio Ferrari	Google Research, Switzerland
Andrew W. Fitzgibbon	Graphcore, UK
David J. Fleet	University of Toronto, Canada
David Forsyth	University of Illinois at Urbana-Champaign, USA
David Fouhey	University of Michigan, USA
Katerina Fragkiadaki	Carnegie Mellon University, USA
Friedrich Fraundorfer	Graz University of Technology, Austria
Oren Freifeld	Ben-Gurion University, Israel
Thomas Funkhouser	Google Research & Princeton University, USA
Yasutaka Furukawa	Simon Fraser University, Canada
Fabio Galasso	Sapienza University of Rome, Italy
Jürgen Gall	University of Bonn, Germany
Chuang Gan	Massachusetts Institute of Technology, USA
Zhe Gan	Microsoft, USA
Animesh Garg	University of Toronto, Vector Institute, Nvidia, Canada
Efstratios Gavves	University of Amsterdam, The Netherlands
Peter Gehler	Amazon, Germany
Theo Gevers	University of Amsterdam, The Netherlands
Bernard Ghanem	King Abdullah University of Science and Technology, Saudi Arabia
Ross B. Girshick	Facebook AI Research, USA
Georgia Gkioxari	Facebook AI Research, USA
Albert Gordo	Facebook, USA
Stephen Gould	Australian National University, Australia
Venu Madhav Govindu	Indian Institute of Science, India
Kristen Grauman	Facebook AI Research & UT Austin, USA
Abhinav Gupta	Carnegie Mellon University & Facebook AI Research, USA
Mohit Gupta	University of Wisconsin-Madison, USA
Hu Han	Institute of Computing Technology, Chinese Academy of Sciences, China

Bohyung Han	Seoul National University, Korea
Tian Han	Stevens Institute of Technology, USA
Emily Hand	University of Nevada, Reno, USA
Bharath Hariharan	Cornell University, USA
Ran He	Institute of Automation, Chinese Academy of Sciences, China
Otmar Hilliges	ETH Zurich, Switzerland
Adrian Hilton	University of Surrey, UK
Minh Hoai	Stony Brook University, USA
Yedid Hoshen	Hebrew University of Jerusalem, Israel
Timothy Hospedales	University of Edinburgh, UK
Gang Hua	Wormpex AI Research, USA
Di Huang	Beihang University, China
Jing Huang	Facebook, USA
Jia-Bin Huang	Facebook, USA
Nathan Jacobs	Washington University in St. Louis, USA
C.V. Jawahar	International Institute of Information Technology, Hyderabad, India
Herve Jegou	Facebook AI Research, France
Neel Joshi	Microsoft Research, USA
Armand Joulin	Facebook AI Research, France
Frederic Jurie	University of Caen Normandie, France
Fredrik Kahl	Chalmers University of Technology, Sweden
Yannis Kalantidis	NAVER LABS Europe, France
Evangelos Kalogerakis	University of Massachusetts, Amherst, USA
Sing Bing Kang	Zillow Group, USA
Yosi Keller	Bar Ilan University, Israel
Margret Keuper	University of Mannheim, Germany
Tae-Kyun Kim	Imperial College London, UK
Benjamin Kimia	Brown University, USA
Alexander Kirillov	Facebook AI Research, USA
Kris Kitani	Carnegie Mellon University, USA
Iasonas Kokkinos	Snap Inc. & University College London, UK
Vladlen Koltun	Apple, USA
Nikos Komodakis	University of Crete, Crete
Piotr Koniusz	Australian National University, Australia
Philipp Kraehenbuehl	University of Texas at Austin, USA
Dilip Krishnan	Google, USA
Ajay Kumar	Hong Kong Polytechnic University, Hong Kong, China
Junseok Kwon	Chung-Ang University, Korea
Jean-Francois Lalonde	Université Laval, Canada

Ivan Laptev	Inria Paris, France
Laura Leal-Taixé	Technical University of Munich, Germany
Erik Learned-Miller	University of Massachusetts, Amherst, USA
Gim Hee Lee	National University of Singapore, Singapore
Seungyong Lee	Pohang University of Science and Technology, Korea
Zhen Lei	Institute of Automation, Chinese Academy of Sciences, China
Bastian Leibe	RWTH Aachen University, Germany
Hongdong Li	Australian National University, Australia
Fuxin Li	Oregon State University, USA
Bo Li	University of Illinois at Urbana-Champaign, USA
Yin Li	University of Wisconsin-Madison, USA
Ser-Nam Lim	Meta AI Research, USA
Joseph Lim	University of Southern California, USA
Stephen Lin	Microsoft Research Asia, China
Dahua Lin	The Chinese University of Hong Kong, Hong Kong, China
Si Liu	Beihang University, China
Xiaoming Liu	Michigan State University, USA
Ce Liu	Microsoft, USA
Zicheng Liu	Microsoft, USA
Yanxi Liu	Pennsylvania State University, USA
Feng Liu	Portland State University, USA
Yebin Liu	Tsinghua University, China
Chen Change Loy	Nanyang Technological University, Singapore
Huchuan Lu	Dalian University of Technology, China
Cewu Lu	Shanghai Jiao Tong University, China
Oisin Mac Aodha	University of Edinburgh, UK
Dhruv Mahajan	Facebook, USA
Subhransu Maji	University of Massachusetts, Amherst, USA
Atsuto Maki	KTH Royal Institute of Technology, Sweden
Arun Mallya	NVIDIA, USA
R. Manmatha	Amazon, USA
Iacopo Masi	Sapienza University of Rome, Italy
Dimitris N. Metaxas	Rutgers University, USA
Ajmal Mian	University of Western Australia, Australia
Christian Micheloni	University of Udine, Italy
Krystian Mikolajczyk	Imperial College London, UK
Anurag Mittal	Indian Institute of Technology, Madras, India
Philippos Mordohai	Stevens Institute of Technology, USA
Greg Mori	Simon Fraser University & Borealis AI, Canada

Vittorio Murino	Istituto Italiano di Tecnologia, Italy
P. J. Narayanan	International Institute of Information Technology, Hyderabad, India
Ram Nevatia	University of Southern California, USA
Natalia Neverova	Facebook AI Research, UK
Richard Newcombe	Facebook, USA
Cuong V. Nguyen	Florida International University, USA
Bingbing Ni	Shanghai Jiao Tong University, China
Juan Carlos Niebles	Salesforce & Stanford University, USA
Ko Nishino	Kyoto University, Japan
Jean-Marc Odobez	Idiap Research Institute, École polytechnique fédérale de Lausanne, Switzerland
Francesca Odone	University of Genova, Italy
Takayuki Okatani	Tohoku University & RIKEN Center for Advanced Intelligence Project, Japan
Manohar Paluri	Facebook, USA
Guan Pang	Facebook, USA
Maja Pantic	Imperial College London, UK
Sylvain Paris	Adobe Research, USA
Jaesik Park	Pohang University of Science and Technology, Korea
Hyun Soo Park	The University of Minnesota, USA
Omkar M. Parkhi	Facebook, USA
Deepak Pathak	Carnegie Mellon University, USA
Georgios Pavlakos	University of California, Berkeley, USA
Marcello Pelillo	University of Venice, Italy
Marc Pollefeys	ETH Zurich & Microsoft, Switzerland
Jean Ponce	Inria, France
Gerard Pons-Moll	University of Tübingen, Germany
Fatih Porikli	Qualcomm, USA
Victor Adrian Prisacariu	University of Oxford, UK
Petia Radeva	University of Barcelona, Spain
Ravi Ramamoorthi	University of California, San Diego, USA
Deva Ramanan	Carnegie Mellon University, USA
Vignesh Ramanathan	Facebook, USA
Nalini Ratha	State University of New York at Buffalo, USA
Tammy Riklin Raviv	Ben-Gurion University, Israel
Tobias Ritschel	University College London, UK
Emanuele Rodola	Sapienza University of Rome, Italy
Amit K. Roy-Chowdhury	University of California, Riverside, USA
Michael Rubinstein	Google, USA
Olga Russakovsky	Princeton University, USA

Mathieu Salzmann	École polytechnique fédérale de Lausanne, Switzerland
Dimitris Samaras	Stony Brook University, USA
Aswin Sankaranarayanan	Carnegie Mellon University, USA
Imari Sato	National Institute of Informatics, Japan
Yoichi Sato	University of Tokyo, Japan
Shin'ichi Satoh	National Institute of Informatics, Japan
Walter Scheirer	University of Notre Dame, USA
Bernt Schiele	Max Planck Institute for Informatics, Germany
Konrad Schindler	ETH Zurich, Switzerland
Cordelia Schmid	Inria & Google, France
Alexander Schwing	University of Illinois at Urbana-Champaign, USA
Nicu Sebe	University of Trento, Italy
Greg Shakhnarovich	Toyota Technological Institute at Chicago, USA
Eli Shechtman	Adobe Research, USA
Humphrey Shi	University of Oregon & University of Illinois at Urbana-Champaign & Picsart AI Research, USA
Jianbo Shi	University of Pennsylvania, USA
Roy Shilkrot	Massachusetts Institute of Technology, USA
Mike Zheng Shou	National University of Singapore, Singapore
Kaleem Siddiqi	McGill University, Canada
Richa Singh	Indian Institute of Technology Jodhpur, India
Greg Slabaugh	Queen Mary University of London, UK
Cees Snoek	University of Amsterdam, The Netherlands
Yale Song	Facebook AI Research, USA
Yi-Zhe Song	University of Surrey, UK
Bjorn Stenger	Rakuten Institute of Technology
Abby Stylianou	Saint Louis University, USA
Akihiro Sugimoto	National Institute of Informatics, Japan
Chen Sun	Brown University, USA
Deqing Sun	Google, USA
Kalyan Sunkavalli	Adobe Research, USA
Ying Tai	Tencent YouTu Lab, China
Ayellet Tal	Technion – Israel Institute of Technology, Israel
Ping Tan	Simon Fraser University, Canada
Siyu Tang	ETH Zurich, Switzerland
Chi-Keung Tang	Hong Kong University of Science and Technology, Hong Kong, China
Radu Timofte	University of Würzburg, Germany & ETH Zurich, Switzerland
Federico Tombari	Google, Switzerland & Technical University of Munich, Germany

James Tompkin	Brown University, USA
Lorenzo Torresani	Dartmouth College, USA
Alexander Toshev	Apple, USA
Du Tran	Facebook AI Research, USA
Anh T. Tran	VinAI, Vietnam
Zhuowen Tu	University of California, San Diego, USA
Georgios Tzimiropoulos	Queen Mary University of London, UK
Jasper Uijlings	Google Research, Switzerland
Jan C. van Gemert	Delft University of Technology, The Netherlands
Gul Varol	Ecole des Ponts ParisTech, France
Nuno Vasconcelos	University of California, San Diego, USA
Mayank Vatsa	Indian Institute of Technology Jodhpur, India
Ashok Veeraraghavan	Rice University, USA
Jakob Verbeek	Facebook AI Research, France
Carl Vondrick	Columbia University, USA
Ruiping Wang	Institute of Computing Technology, Chinese Academy of Sciences, China
Xinchao Wang	National University of Singapore, Singapore
Liwei Wang	The Chinese University of Hong Kong, Hong Kong, China
Chaohui Wang	Université Paris-Est, France
Xiaolong Wang	University of California, San Diego, USA
Christian Wolf	NAVER LABS Europe, France
Tao Xiang	University of Surrey, UK
Saining Xie	Facebook AI Research, USA
Cihang Xie	University of California, Santa Cruz, USA
Zeki Yalniz	Facebook, USA
Ming-Hsuan Yang	University of California, Merced, USA
Angela Yao	National University of Singapore, Singapore
Shaodi You	University of Amsterdam, The Netherlands
Stella X. Yu	University of California, Berkeley, USA
Junsong Yuan	State University of New York at Buffalo, USA
Stefanos Zafeiriou	Imperial College London, UK
Amir Zamir	École polytechnique fédérale de Lausanne, Switzerland
Lei Zhang	Alibaba & Hong Kong Polytechnic University, Hong Kong, China
Lei Zhang	International Digital Economy Academy (IDEA), China
Pengchuan Zhang	Meta AI, USA
Bolei Zhou	University of California, Los Angeles, USA
Yuke Zhu	University of Texas at Austin, USA

Todd Zickler	Harvard University, USA
Wangmeng Zuo	Harbin Institute of Technology, China

Technical Program Committee

Davide Abati
Soroush Abbasi Koohpayegani
Amos L. Abbott
Rameen Abdal
Rabab Abdelfattah
Sahar Abdelnabi
Hassan Abu Alhaija
Abulikemu Abuduweili
Ron Abutbul
Hanno Ackermann
Aikaterini Adam
Kamil Adamczewski
Ehsan Adeli
Vida Adeli
Donald Adjeroh
Arman Afrasiyabi
Akshay Agarwal
Sameer Agarwal
Abhinav Agarwalla
Vaibhav Aggarwal
Sara Aghajanzadeh
Susmit Agrawal
Antonio Agudo
Touqeer Ahmad
Sk Miraj Ahmed
Chaitanya Ahuja
Nilesh A. Ahuja
Abhishek Aich
Shubhra Aich
Noam Aigerman
Arash Akbarinia
Peri Akiva
Derya Akkaynak
Emre Aksan
Arjun R. Akula
Yuval Alaluf
Stephan Alaniz
Paul Albert
Cenek Albl
Filippo Aleotti
Konstantinos P. Alexandridis
Motasem Alfarra
Mohsen Ali
Thiemo Alldieck
Hadi Alzayer
Liang An
Shan An
Yi An
Zhulin An
Dongsheng An
Jie An
Xiang An
Saket Anand
Cosmin Ancuti
Juan Andrade-Cetto
Alexander Andreopoulos
Bjoern Andres
Jerone T. A. Andrews
Shivangi Aneja
Anelia Angelova
Dragomir Anguelov
Rushil Anirudh
Oron Anschel
Rao Muhammad Anwer
Djamila Aouada
Evlampios Apostolidis
Srikar Appalaraju
Nikita Araslanov
Andre Araujo
Eric Arazo
Dawit Mureja Argaw
Anurag Arnab
Aditya Arora
Chetan Arora
Sunpreet S. Arora
Alexey Artemov
Muhammad Asad
Kumar Ashutosh
Sinem Aslan
Vishal Asnani
Mahmoud Assran
Amir Atapour-Abarghouei
Nikos Athanasiou
Ali Athar
ShahRukh Athar
Sara Atito
Souhaib Attaiki
Matan Atzmon
Mathieu Aubry
Nicolas Audebert
Tristan T. Aumentado-Armstrong
Melinos Averkiou
Yannis Avrithis
Stephane Ayache
Mehmet Aygün
Seyed Mehdi Ayyoubzadeh
Hossein Azizpour
George Azzopardi
Mallikarjun B. R.
Yunhao Ba
Abhishek Badki
Seung-Hwan Bae
Seung-Hwan Baek
Seungryul Baek
Piyush Nitin Bagad
Shai Bagon
Gaetan Bahl
Shikhar Bahl
Sherwin Bahmani
Haoran Bai
Lei Bai
Jiawang Bai
Haoyue Bai
Jinbin Bai
Xiang Bai
Xuyang Bai

Yang Bai
Yuanchao Bai
Ziqian Bai
Sungyong Baik
Kevin Bailly
Max Bain
Federico Baldassarre
Wele Gedara Chaminda Bandara
Biplab Banerjee
Pratyay Banerjee
Sandipan Banerjee
Jihwan Bang
Antyanta Bangunharcana
Aayush Bansal
Ankan Bansal
Siddhant Bansal
Wentao Bao
Zhipeng Bao
Amir Bar
Manel Baradad Jurjo
Lorenzo Baraldi
Danny Barash
Daniel Barath
Connelly Barnes
Ioan Andrei Bârsan
Steven Basart
Dina Bashkirova
Chaim Baskin
Peyman Bateni
Anil Batra
Sebastiano Battiato
Ardhendu Behera
Harkirat Behl
Jens Behley
Vasileios Belagiannis
Boulbaba Ben Amor
Emanuel Ben Baruch
Abdessamad Ben Hamza
Gil Ben-Artzi
Assia Benbihi
Fabian Benitez-Quiroz
Guy Ben-Yosef
Philipp Benz
Alexander W. Bergman
Urs Bergmann
Jesus Bermudez-Cameo
Stefano Berretti
Gedas Bertasius
Zachary Bessinger
Petra Bevandić
Matthew Beveridge
Lucas Beyer
Yash Bhalgat
Suvaansh Bhambri
Samarth Bharadwaj
Gaurav Bharaj
Aparna Bharati
Bharat Lal Bhatnagar
Uttaran Bhattacharya
Apratim Bhattacharyya
Brojeshwar Bhowmick
Ankan Kumar Bhunia
Ayan Kumar Bhunia
Qi Bi
Sai Bi
Michael Bi Mi
Gui-Bin Bian
Jia-Wang Bian
Shaojun Bian
Pia Bideau
Mario Bijelic
Hakan Bilen
Guillaume-Alexandre Bilodeau
Alexander Binder
Tolga Birdal
Vighnesh N. Birodkar
Sandika Biswas
Andreas Blattmann
Janusz Bobulski
Giuseppe Boccignone
Vishnu Boddeti
Navaneeth Bodla
Moritz Böhle
Aleksei Bokhovkin
Sam Bond-Taylor
Vivek Boominathan
Shubhankar Borse
Mark Boss
Andrea Bottino
Adnane Boukhayma
Fadi Boutros
Nicolas C. Boutry
Richard S. Bowen
Ivaylo Boyadzhiev
Aidan Boyd
Yuri Boykov
Aljaz Bozic
Behzad Bozorgtabar
Eric Brachmann
Samarth Brahmbhatt
Gustav Bredell
Francois Bremond
Joel Brogan
Andrew Brown
Thomas Brox
Marcus A. Brubaker
Robert-Jan Bruintjes
Yuqi Bu
Anders G. Buch
Himanshu Buckchash
Mateusz Buda
Ignas Budvytis
José M. Buenaposada
Marcel C. Bühler
Tu Bui
Adrian Bulat
Hannah Bull
Evgeny Burnaev
Andrei Bursuc
Benjamin Busam
Sergey N. Buzykanov
Wonmin Byeon
Fabian Caba
Martin Cadik
Guanyu Cai
Minjie Cai
Qing Cai
Zhongang Cai
Qi Cai
Yancheng Cai
Shen Cai
Han Cai
Jiarui Cai

Bowen Cai
Mu Cai
Qin Cai
Ruojin Cai
Weidong Cai
Weiwei Cai
Yi Cai
Yujun Cai
Zhiping Cai
Akin Caliskan
Lilian Calvet
Baris Can Cam
Necati Cihan Camgoz
Tommaso Campari
Dylan Campbell
Ziang Cao
Ang Cao
Xu Cao
Zhiwen Cao
Shengcao Cao
Song Cao
Weipeng Cao
Xiangyong Cao
Xiaochun Cao
Yue Cao
Yunhao Cao
Zhangjie Cao
Jiale Cao
Yang Cao
Jiajiong Cao
Jie Cao
Jinkun Cao
Lele Cao
Yulong Cao
Zhiguo Cao
Chen Cao
Razvan Caramalau
Marlène Careil
Gustavo Carneiro
Joao Carreira
Dan Casas
Paola Cascante-Bonilla
Angela Castillo
Francisco M. Castro
Pedro Castro
Luca Cavalli
George J. Cazenavette
Oya Celiktutan
Hakan Cevikalp
Sri Harsha C. H.
Sungmin Cha
Geonho Cha
Menglei Chai
Lucy Chai
Yuning Chai
Zenghao Chai
Anirban Chakraborty
Deep Chakraborty
Rudrasis Chakraborty
Souradeep Chakraborty
Kelvin C. K. Chan
Chee Seng Chan
Paramanand Chandramouli
Arjun Chandrasekaran
Kenneth Chaney
Dongliang Chang
Huiwen Chang
Peng Chang
Xiaojun Chang
Jia-Ren Chang
Hyung Jin Chang
Hyun Sung Chang
Ju Yong Chang
Li-Jen Chang
Qi Chang
Wei-Yi Chang
Yi Chang
Nadine Chang
Hanqing Chao
Pradyumna Chari
Dibyadip Chatterjee
Chiranjoy Chattopadhyay
Siddhartha Chaudhuri
Zhengping Che
Gal Chechik
Lianggangxu Chen
Qi Alfred Chen
Brian Chen
Bor-Chun Chen
Bo-Hao Chen
Bohong Chen
Bin Chen
Ziliang Chen
Cheng Chen
Chen Chen
Chaofeng Chen
Xi Chen
Haoyu Chen
Xuanhong Chen
Wei Chen
Qiang Chen
Shi Chen
Xianyu Chen
Chang Chen
Changhuai Chen
Hao Chen
Jie Chen
Jianbo Chen
Jingjing Chen
Jun Chen
Kejiang Chen
Mingcai Chen
Nenglun Chen
Qifeng Chen
Ruoyu Chen
Shu-Yu Chen
Weidong Chen
Weijie Chen
Weikai Chen
Xiang Chen
Xiuyi Chen
Xingyu Chen
Yaofo Chen
Yueting Chen
Yu Chen
Yunjin Chen
Yuntao Chen
Yun Chen
Zhenfang Chen
Zhuangzhuang Chen
Chu-Song Chen
Xiangyu Chen
Zhuo Chen
Chaoqi Chen
Shizhe Chen

Xiaotong Chen
Xiaozhi Chen
Dian Chen
Defang Chen
Dingfan Chen
Ding-Jie Chen
Ee Heng Chen
Tao Chen
Yixin Chen
Wei-Ting Chen
Lin Chen
Guang Chen
Guangyi Chen
Guanying Chen
Guangyao Chen
Hwann-Tzong Chen
Junwen Chen
Jiacheng Chen
Jianxu Chen
Hui Chen
Kai Chen
Kan Chen
Kevin Chen
Kuan-Wen Chen
Weihua Chen
Zhang Chen
Liang-Chieh Chen
Lele Chen
Liang Chen
Fanglin Chen
Zehui Chen
Minghui Chen
Minghao Chen
Xiaokang Chen
Qian Chen
Jun-Cheng Chen
Qi Chen
Qingcai Chen
Richard J. Chen
Runnan Chen
Rui Chen
Shuo Chen
Sentao Chen
Shaoyu Chen
Shixing Chen
Shuai Chen
Shuya Chen
Sizhe Chen
Simin Chen
Shaoxiang Chen
Zitian Chen
Tianlong Chen
Tianshui Chen
Min-Hung Chen
Xiangning Chen
Xin Chen
Xinghao Chen
Xuejin Chen
Xu Chen
Xuxi Chen
Yunlu Chen
Yanbei Chen
Yuxiao Chen
Yun-Chun Chen
Yi-Ting Chen
Yi-Wen Chen
Yinbo Chen
Yiran Chen
Yuanhong Chen
Yubei Chen
Yuefeng Chen
Yuhua Chen
Yukang Chen
Zerui Chen
Zhaoyu Chen
Zhen Chen
Zhenyu Chen
Zhi Chen
Zhiwei Chen
Zhixiang Chen
Long Chen
Bowen Cheng
Jun Cheng
Yi Cheng
Jingchun Cheng
Lechao Cheng
Xi Cheng
Yuan Cheng
Ho Kei Cheng
Kevin Ho Man Cheng
Jiacheng Cheng
Kelvin B. Cheng
Li Cheng
Mengjun Cheng
Zhen Cheng
Qingrong Cheng
Tianheng Cheng
Harry Cheng
Yihua Cheng
Yu Cheng
Ziheng Cheng
Soon Yau Cheong
Anoop Cherian
Manuela Chessa
Zhixiang Chi
Naoki Chiba
Julian Chibane
Kashyap Chitta
Tai-Yin Chiu
Hsu-kuang Chiu
Wei-Chen Chiu
Sungmin Cho
Donghyeon Cho
Hyeon Cho
Yooshin Cho
Gyusang Cho
Jang Hyun Cho
Seungju Cho
Nam Ik Cho
Sunghyun Cho
Hanbyel Cho
Jaesung Choe
Jooyoung Choi
Chiho Choi
Changwoon Choi
Jongwon Choi
Myungsub Choi
Dooseop Choi
Jonghyun Choi
Jinwoo Choi
Jun Won Choi
Min-Kook Choi
Hongsuk Choi
Janghoon Choi
Yoon-Ho Choi

Yukyung Choi
Jaegul Choo
Ayush Chopra
Siddharth Choudhary
Subhabrata Choudhury
Vasileios Choutas
Ka-Ho Chow
Pinaki Nath Chowdhury
Sammy Christen
Anders Christensen
Grigorios Chrysos
Hang Chu
Wen-Hsuan Chu
Peng Chu
Qi Chu
Ruihang Chu
Wei-Ta Chu
Yung-Yu Chuang
Sanghyuk Chun
Se Young Chun
Antonio Cinà
Ramazan Gokberk Cinbis
Javier Civera
Albert Clapés
Ronald Clark
Brian S. Clipp
Felipe Codevilla
Daniel Coelho de Castro
Niv Cohen
Forrester Cole
Maxwell D. Collins
Robert T. Collins
Marc Comino Trinidad
Runmin Cong
Wenyan Cong
Maxime Cordy
Marcella Cornia
Enric Corona
Huseyin Coskun
Luca Cosmo
Dragos Costea
Davide Cozzolino
Arun C. S. Kumar
Aiyu Cui
Qiongjie Cui
Quan Cui
Shuhao Cui
Yiming Cui
Ying Cui
Zijun Cui
Jiali Cui
Jiequan Cui
Yawen Cui
Zhen Cui
Zhaopeng Cui
Jack Culpepper
Xiaodong Cun
Ross Cutler
Adam Czajka
Ali Dabouei
Konstantinos M. Dafnis
Manuel Dahnert
Tao Dai
Yuchao Dai
Bo Dai
Mengyu Dai
Hang Dai
Haixing Dai
Peng Dai
Pingyang Dai
Qi Dai
Qiyu Dai
Yutong Dai
Naser Damer
Zhiyuan Dang
Mohamed Daoudi
Ayan Das
Abir Das
Debasmit Das
Deepayan Das
Partha Das
Sagnik Das
Soumi Das
Srijan Das
Swagatam Das
Avijit Dasgupta
Jim Davis
Adrian K. Davison
Homa Davoudi
Laura Daza
Matthias De Lange
Shalini De Mello
Marco De Nadai
Christophe De Vleeschouwer
Alp Dener
Boyang Deng
Congyue Deng
Bailin Deng
Yong Deng
Ye Deng
Zhuo Deng
Zhijie Deng
Xiaoming Deng
Jiankang Deng
Jinhong Deng
Jingjing Deng
Liang-Jian Deng
Siqi Deng
Xiang Deng
Xueqing Deng
Zhongying Deng
Karan Desai
Jean-Emmanuel Deschaud
Aniket Anand Deshmukh
Neel Dey
Helisa Dhamo
Prithviraj Dhar
Amaya Dharmasiri
Yan Di
Xing Di
Ousmane A. Dia
Haiwen Diao
Xiaolei Diao
Gonçalo José Dias Pais
Abdallah Dib
Anastasios Dimou
Changxing Ding
Henghui Ding
Guodong Ding
Yaqing Ding
Shuangrui Ding
Yuhang Ding
Yikang Ding
Shouhong Ding

Haisong Ding
Hui Ding
Jiahao Ding
Jian Ding
Jian-Jiun Ding
Shuxiao Ding
Tianyu Ding
Wenhao Ding
Yuqi Ding
Yi Ding
Yuzhen Ding
Zhengming Ding
Tan Minh Dinh
Vu Dinh
Christos Diou
Mandar Dixit
Bao Gia Doan
Khoa D. Doan
Dzung Anh Doan
Debi Prosad Dogra
Nehal Doiphode
Chengdong Dong
Bowen Dong
Zhenxing Dong
Hang Dong
Xiaoyi Dong
Haoye Dong
Jiangxin Dong
Shichao Dong
Xuan Dong
Zhen Dong
Shuting Dong
Jing Dong
Li Dong
Ming Dong
Nanqing Dong
Qiulei Dong
Runpei Dong
Siyan Dong
Tian Dong
Wei Dong
Xiaomeng Dong
Xin Dong
Xingbo Dong
Yuan Dong
Samuel Dooley
Gianfranco Doretto
Michael Dorkenwald
Keval Doshi
Zhaopeng Dou
Xiaotian Dou
Hazel Doughty
Ahmad Droby
Iddo Drori
Jie Du
Yong Du
Dawei Du
Dong Du
Ruoyi Du
Yuntao Du
Xuefeng Du
Yilun Du
Yuming Du
Radhika Dua
Haodong Duan
Jiafei Duan
Kaiwen Duan
Peiqi Duan
Ye Duan
Haoran Duan
Jiali Duan
Amanda Duarte
Abhimanyu Dubey
Shiv Ram Dubey
Florian Dubost
Lukasz Dudziak
Shivam Duggal
Justin M. Dulay
Matteo Dunnhofer
Chi Nhan Duong
Thibaut Durand
Mihai Dusmanu
Ujjal Kr Dutta
Debidatta Dwibedi
Isht Dwivedi
Sai Kumar Dwivedi
Takeharu Eda
Mark Edmonds
Alexei A. Efros
Thibaud Ehret
Max Ehrlich
Mahsa Ehsanpour
Iván Eichhardt
Farshad Einabadi
Marvin Eisenberger
Hazim Kemal Ekenel
Mohamed El Banani
Ismail Elezi
Moshe Eliasof
Alaa El-Nouby
Ian Endres
Francis Engelmann
Deniz Engin
Chanho Eom
Dave Epstein
Maria C. Escobar
Victor A. Escorcia
Carlos Esteves
Sungmin Eum
Bernard J. E. Evans
Ivan Evtimov
Fevziye Irem Eyiokur Yaman
Matteo Fabbri
Sébastien Fabbro
Gabriele Facciolo
Masud Fahim
Bin Fan
Hehe Fan
Deng-Ping Fan
Aoxiang Fan
Chen-Chen Fan
Qi Fan
Zhaoxin Fan
Haoqi Fan
Heng Fan
Hongyi Fan
Linxi Fan
Baojie Fan
Jiayuan Fan
Lei Fan
Quanfu Fan
Yonghui Fan
Yingruo Fan
Zhiwen Fan

Zicong Fan
Sean Fanello
Jiansheng Fang
Chaowei Fang
Yuming Fang
Jianwu Fang
Jin Fang
Qi Fang
Shancheng Fang
Tian Fang
Xianyong Fang
Gongfan Fang
Zhen Fang
Hui Fang
Jiemin Fang
Le Fang
Pengfei Fang
Xiaolin Fang
Yuxin Fang
Zhaoyuan Fang
Ammarah Farooq
Azade Farshad
Zhengcong Fei
Michael Felsberg
Wei Feng
Chen Feng
Fan Feng
Andrew Feng
Xin Feng
Zheyun Feng
Ruicheng Feng
Mingtao Feng
Qianyu Feng
Shangbin Feng
Chun-Mei Feng
Zunlei Feng
Zhiyong Feng
Martin Fergie
Mustansar Fiaz
Marco Fiorucci
Michael Firman
Hamed Firooz
Volker Fischer
Corneliu O. Florea
Georgios Floros
Wolfgang Foerstner
Gianni Franchi
Jean-Sebastien Franco
Simone Frintrop
Anna Fruehstueck
Changhong Fu
Chaoyou Fu
Cheng-Yang Fu
Chi-Wing Fu
Deqing Fu
Huan Fu
Jun Fu
Kexue Fu
Ying Fu
Jianlong Fu
Jingjing Fu
Qichen Fu
Tsu-Jui Fu
Xueyang Fu
Yang Fu
Yanwei Fu
Yonggan Fu
Wolfgang Fuhl
Yasuhisa Fujii
Kent Fujiwara
Marco Fumero
Takuya Funatomi
Isabel Funke
Dario Fuoli
Antonino Furnari
Matheus A. Gadelha
Akshay Gadi Patil
Adrian Galdran
Guillermo Gallego
Silvano Galliani
Orazio Gallo
Leonardo Galteri
Matteo Gamba
Yiming Gan
Sujoy Ganguly
Harald Ganster
Boyan Gao
Changxin Gao
Daiheng Gao
Difei Gao
Chen Gao
Fei Gao
Lin Gao
Wei Gao
Yiming Gao
Junyu Gao
Guangyu Ryan Gao
Haichang Gao
Hongchang Gao
Jialin Gao
Jin Gao
Jun Gao
Katelyn Gao
Mingchen Gao
Mingfei Gao
Pan Gao
Shangqian Gao
Shanghua Gao
Xitong Gao
Yunhe Gao
Zhanning Gao
Elena Garces
Nuno Cruz Garcia
Noa Garcia
Guillermo Garcia-Hernando
Isha Garg
Rahul Garg
Sourav Garg
Quentin Garrido
Stefano Gasperini
Kent Gauen
Chandan Gautam
Shivam Gautam
Paul Gay
Chunjiang Ge
Shiming Ge
Wenhang Ge
Yanhao Ge
Zheng Ge
Songwei Ge
Weifeng Ge
Yixiao Ge
Yuying Ge
Shijie Geng

Zhengyang Geng
Kyle A. Genova
Georgios Georgakis
Markos Georgopoulos
Marcel Geppert
Shabnam Ghadar
Mina Ghadimi Atigh
Deepti Ghadiyaram
Maani Ghaffari Jadidi
Sedigh Ghamari
Zahra Gharaee
Michaël Gharbi
Golnaz Ghiasi
Reza Ghoddoosian
Soumya Suvra Ghosal
Adhiraj Ghosh
Arthita Ghosh
Pallabi Ghosh
Soumyadeep Ghosh
Andrew Gilbert
Igor Gilitschenski
Jhony H. Giraldo
Andreu Girbau Xalabarder
Rohit Girdhar
Sharath Girish
Xavier Giro-i-Nieto
Raja Giryes
Thomas Gittings
Nikolaos Gkanatsios
Ioannis Gkioulekas
Abhiram Gnanasambandam
Aurele T. Gnanha
Clement L. J. C. Godard
Arushi Goel
Vidit Goel
Shubham Goel
Zan Gojcic
Aaron K. Gokaslan
Tejas Gokhale
S. Alireza Golestaneh
Thiago L. Gomes
Nuno Goncalves
Boqing Gong
Chen Gong
Yuanhao Gong
Guoqiang Gong
Jingyu Gong
Rui Gong
Yu Gong
Mingming Gong
Neil Zhenqiang Gong
Xun Gong
Yunye Gong
Yihong Gong
Cristina I. González
Nithin Gopalakrishnan Nair
Gaurav Goswami
Jianping Gou
Shreyank N. Gowda
Ankit Goyal
Helmut Grabner
Patrick L. Grady
Ben Graham
Eric Granger
Douglas R. Gray
Matej Grcić
David Griffiths
Jinjin Gu
Yun Gu
Shuyang Gu
Jianyang Gu
Fuqiang Gu
Jiatao Gu
Jindong Gu
Jiaqi Gu
Jinwei Gu
Jiaxin Gu
Geonmo Gu
Xiao Gu
Xinqian Gu
Xiuye Gu
Yuming Gu
Zhangxuan Gu
Dayan Guan
Junfeng Guan
Qingji Guan
Tianrui Guan
Shanyan Guan
Denis A. Gudovskiy
Ricardo Guerrero
Pierre-Louis Guhur
Jie Gui
Liangyan Gui
Liangke Gui
Benoit Guillard
Erhan Gundogdu
Manuel Günther
Jingcai Guo
Yuanfang Guo
Junfeng Guo
Chenqi Guo
Dan Guo
Hongji Guo
Jia Guo
Jie Guo
Minghao Guo
Shi Guo
Yanhui Guo
Yangyang Guo
Yuan-Chen Guo
Yilu Guo
Yiluan Guo
Yong Guo
Guangyu Guo
Haiyun Guo
Jinyang Guo
Jianyuan Guo
Pengsheng Guo
Pengfei Guo
Shuxuan Guo
Song Guo
Tianyu Guo
Qing Guo
Qiushan Guo
Wen Guo
Xiefan Guo
Xiaohu Guo
Xiaoqing Guo
Yufei Guo
Yuhui Guo
Yuliang Guo
Yunhui Guo
Yanwen Guo

Akshita Gupta
Ankush Gupta
Kamal Gupta
Kartik Gupta
Ritwik Gupta
Rohit Gupta
Siddharth Gururani
Fredrik K. Gustafsson
Abner Guzman Rivera
Vladimir Guzov
Matthew A. Gwilliam
Jung-Woo Ha
Marc Habermann
Isma Hadji
Christian Haene
Martin Hahner
Levente Hajder
Alexandros Haliassos
Emanuela Haller
Bumsub Ham
Abdullah J. Hamdi
Shreyas Hampali
Dongyoon Han
Chunrui Han
Dong-Jun Han
Dong-Sig Han
Guangxing Han
Zhizhong Han
Ruize Han
Jiaming Han
Jin Han
Ligong Han
Xian-Hua Han
Xiaoguang Han
Yizeng Han
Zhi Han
Zhenjun Han
Zhongyi Han
Jungong Han
Junlin Han
Kai Han
Kun Han
Sungwon Han
Songfang Han
Wei Han
Xiao Han
Xintong Han
Xinzhe Han
Yahong Han
Yan Han
Zongbo Han
Nicolai Hani
Rana Hanocka
Niklas Hanselmann
Nicklas A. Hansen
Hong Hanyu
Fusheng Hao
Yanbin Hao
Shijie Hao
Udith Haputhanthri
Mehrtash Harandi
Josh Harguess
Adam Harley
David M. Hart
Atsushi Hashimoto
Ali Hassani
Mohammed Hassanin
Yana Hasson
Joakim Bruslund Haurum
Bo He
Kun He
Chen He
Xin He
Fazhi He
Gaoqi He
Hao He
Haoyu He
Jiangpeng He
Hongliang He
Qian He
Xiangteng He
Xuming He
Yannan He
Yuhang He
Yang He
Xiangyu He
Nanjun He
Pan He
Sen He
Shengfeng He
Songtao He
Tao He
Tong He
Wei He
Xuehai He
Xiaoxiao He
Ying He
Yisheng He
Ziwen He
Peter Hedman
Felix Heide
Yacov Hel-Or
Paul Henderson
Philipp Henzler
Byeongho Heo
Jae-Pil Heo
Miran Heo
Sachini A. Herath
Stephane Herbin
Pedro Hermosilla Casajus
Monica Hernandez
Charles Herrmann
Roei Herzig
Mauricio Hess-Flores
Carlos Hinojosa
Tobias Hinz
Tsubasa Hirakawa
Chih-Hui Ho
Lam Si Tung Ho
Jennifer Hobbs
Derek Hoiem
Yannick Hold-Geoffroy
Aleksander Holynski
Cheeun Hong
Fa-Ting Hong
Hanbin Hong
Guan Zhe Hong
Danfeng Hong
Lanqing Hong
Xiaopeng Hong
Xin Hong
Jie Hong
Seungbum Hong
Cheng-Yao Hong
Seunghoon Hong

Yi Hong
Yuan Hong
Yuchen Hong
Anthony Hoogs
Maxwell C. Horton
Kazuhiro Hotta
Qibin Hou
Tingbo Hou
Junhui Hou
Ji Hou
Qiqi Hou
Rui Hou
Ruibing Hou
Zhi Hou
Henry Howard-Jenkins
Lukas Hoyer
Wei-Lin Hsiao
Chiou-Ting Hsu
Anthony Hu
Brian Hu
Yusong Hu
Hexiang Hu
Haoji Hu
Di Hu
Hengtong Hu
Haigen Hu
Lianyu Hu
Hanzhe Hu
Jie Hu
Junlin Hu
Shizhe Hu
Jian Hu
Zhiming Hu
Juhua Hu
Peng Hu
Ping Hu
Ronghang Hu
MengShun Hu
Tao Hu
Vincent Tao Hu
Xiaoling Hu
Xinting Hu
Xiaolin Hu
Xuefeng Hu
Xiaowei Hu
Yang Hu
Yueyu Hu
Zeyu Hu
Zhongyun Hu
Binh-Son Hua
Guoliang Hua
Yi Hua
Linzhi Huang
Qiusheng Huang
Bo Huang
Chen Huang
Hsin-Ping Huang
Ye Huang
Shuangping Huang
Zeng Huang
Buzhen Huang
Cong Huang
Heng Huang
Hao Huang
Qidong Huang
Huaibo Huang
Chaoqin Huang
Feihu Huang
Jiahui Huang
Jingjia Huang
Kun Huang
Lei Huang
Sheng Huang
Shuaiyi Huang
Siyu Huang
Xiaoshui Huang
Xiaoyang Huang
Yan Huang
Yihao Huang
Ying Huang
Ziling Huang
Xiaoke Huang
Yifei Huang
Haiyang Huang
Zhewei Huang
Jin Huang
Haibin Huang
Jiaxing Huang
Junjie Huang
Keli Huang
Lang Huang
Lin Huang
Luojie Huang
Mingzhen Huang
Shijia Huang
Shengyu Huang
Siyuan Huang
He Huang
Xiuyu Huang
Lianghua Huang
Yue Huang
Yaping Huang
Yuge Huang
Zehao Huang
Zeyi Huang
Zhiqi Huang
Zhongzhan Huang
Zilong Huang
Ziyuan Huang
Tianrui Hui
Zhuo Hui
Le Hui
Jing Huo
Junhwa Hur
Shehzeen S. Hussain
Chuong Minh Huynh
Seunghyun Hwang
Jaehui Hwang
Jyh-Jing Hwang
Sukjun Hwang
Soonmin Hwang
Wonjun Hwang
Rakib Hyder
Sangeek Hyun
Sarah Ibrahimi
Tomoki Ichikawa
Yerlan Idelbayev
A. S. M. Iftekhar
Masaaki Iiyama
Satoshi Ikehata
Sunghoon Im
Atul N. Ingle
Eldar Insafutdinov
Yani A. Ioannou
Radu Tudor Ionescu

Umar Iqbal
Go Irie
Muhammad Zubair Irshad
Ahmet Iscen
Berivan Isik
Ashraful Islam
Md Amirul Islam
Syed Islam
Mariko Isogawa
Vamsi Krishna K. Ithapu
Boris Ivanovic
Darshan Iyer
Sarah Jabbour
Ayush Jain
Nishant Jain
Samyak Jain
Vidit Jain
Vineet Jain
Priyank Jaini
Tomas Jakab
Mohammad A. A. K. Jalwana
Muhammad Abdullah Jamal
Hadi Jamali-Rad
Stuart James
Varun Jampani
Young Kyun Jang
YeongJun Jang
Yunseok Jang
Ronnachai Jaroensri
Bhavan Jasani
Krishna Murthy Jatavallabhula
Mojan Javaheripi
Syed A. Javed
Guillaume Jeanneret
Pranav Jeevan
Herve Jegou
Rohit Jena
Tomas Jenicek
Porter Jenkins
Simon Jenni
Hae-Gon Jeon
Sangryul Jeon
Boseung Jeong
Yoonwoo Jeong
Seong-Gyun Jeong
Jisoo Jeong
Allan D. Jepson
Ankit Jha
Sumit K. Jha
I-Hong Jhuo
Ge-Peng Ji
Chaonan Ji
Deyi Ji
Jingwei Ji
Wei Ji
Zhong Ji
Jiayi Ji
Pengliang Ji
Hui Ji
Mingi Ji
Xiaopeng Ji
Yuzhu Ji
Baoxiong Jia
Songhao Jia
Dan Jia
Shan Jia
Xiaojun Jia
Xiuyi Jia
Xu Jia
Menglin Jia
Wenqi Jia
Boyuan Jiang
Wenhao Jiang
Huaizu Jiang
Hanwen Jiang
Haiyong Jiang
Hao Jiang
Huajie Jiang
Huiqin Jiang
Haojun Jiang
Haobo Jiang
Junjun Jiang
Xingyu Jiang
Yangbangyan Jiang
Yu Jiang
Jianmin Jiang
Jiaxi Jiang
Jing Jiang
Kui Jiang
Li Jiang
Liming Jiang
Chiyu Jiang
Meirui Jiang
Chen Jiang
Peng Jiang
Tai-Xiang Jiang
Wen Jiang
Xinyang Jiang
Yifan Jiang
Yuming Jiang
Yingying Jiang
Zeren Jiang
ZhengKai Jiang
Zhenyu Jiang
Shuming Jiao
Jianbo Jiao
Licheng Jiao
Dongkwon Jin
Yeying Jin
Cheng Jin
Linyi Jin
Qing Jin
Taisong Jin
Xiao Jin
Xin Jin
Sheng Jin
Kyong Hwan Jin
Ruibing Jin
SouYoung Jin
Yueming Jin
Chenchen Jing
Longlong Jing
Taotao Jing
Yongcheng Jing
Younghyun Jo
Joakim Johnander
Jeff Johnson
Michael J. Jones
R. Kenny Jones
Rico Jonschkowski
Ameya Joshi
Sunghun Joung

Felix Juefei-Xu
Claudio R. Jung
Steffen Jung
Hari Chandana K.
Rahul Vigneswaran K.
Prajwal K. R.
Abhishek Kadian
Jhony Kaesemodel Pontes
Kumara Kahatapitiya
Anmol Kalia
Sinan Kalkan
Tarun Kalluri
Jaewon Kam
Sandesh Kamath
Meina Kan
Menelaos Kanakis
Takuhiro Kaneko
Di Kang
Guoliang Kang
Hao Kang
Jaeyeon Kang
Kyoungkook Kang
Li-Wei Kang
MinGuk Kang
Suk-Ju Kang
Zhao Kang
Yash Mukund Kant
Yueying Kao
Aupendu Kar
Konstantinos Karantzalos
Sezer Karaoglu
Navid Kardan
Sanjay Kariyappa
Leonid Karlinsky
Animesh Karnewar
Shyamgopal Karthik
Hirak J. Kashyap
Marc A. Kastner
Hirokatsu Kataoka
Angelos Katharopoulos
Hiroharu Kato
Kai Katsumata
Manuel Kaufmann
Chaitanya Kaul
Prakhar Kaushik
Yuki Kawana
Lei Ke
Lipeng Ke
Tsung-Wei Ke
Wei Ke
Petr Kellnhofer
Aniruddha Kembhavi
John Kender
Corentin Kervadec
Leonid Keselman
Daniel Keysers
Nima Khademi Kalantari
Taras Khakhulin
Samir Khaki
Muhammad Haris Khan
Qadeer Khan
Salman Khan
Subash Khanal
Vaishnavi M. Khindkar
Rawal Khirodkar
Saeed Khorram
Pirazh Khorramshahi
Kourosh Khoshelham
Ansh Khurana
Benjamin Kiefer
Jae Myung Kim
Junho Kim
Boah Kim
Hyeonseong Kim
Dong-Jin Kim
Dongwan Kim
Donghyun Kim
Doyeon Kim
Yonghyun Kim
Hyung-Il Kim
Hyunwoo Kim
Hyeongwoo Kim
Hyo Jin Kim
Hyunwoo J. Kim
Taehoon Kim
Jaeha Kim
Jiwon Kim
Jung Uk Kim
Kangyeol Kim
Eunji Kim
Daeha Kim
Dongwon Kim
Kunhee Kim
Kyungmin Kim
Junsik Kim
Min H. Kim
Namil Kim
Kookhoi Kim
Sanghyun Kim
Seongyeop Kim
Seungryong Kim
Saehoon Kim
Euyoung Kim
Guisik Kim
Sungyeon Kim
Sunnie S. Y. Kim
Taehun Kim
Tae Oh Kim
Won Hwa Kim
Seungwook Kim
YoungBin Kim
Youngeun Kim
Akisato Kimura
Furkan Osman Kınlı
Zsolt Kira
Hedvig Kjellström
Florian Kleber
Jan P. Klopp
Florian Kluger
Laurent Kneip
Byungsoo Ko
Muhammed Kocabas
A. Sophia Koepke
Kevin Koeser
Nick Kolkin
Nikos Kolotouros
Wai-Kin Adams Kong
Deying Kong
Caihua Kong
Youyong Kong
Shuyu Kong
Shu Kong
Tao Kong
Yajing Kong
Yu Kong

Zishang Kong
Theodora Kontogianni
Anton S. Konushin
Julian F. P. Kooij
Bruno Korbar
Giorgos Kordopatis-Zilos
Jari Korhonen
Adam Kortylewski
Denis Korzhenkov
Divya Kothandaraman
Suraj Kothawade
Iuliia Kotseruba
Satwik Kottur
Shashank Kotyan
Alexandros Kouris
Petros Koutras
Anna Kreshuk
Ranjay Krishna
Dilip Krishnan
Andrey Kuehlkamp
Hilde Kuehne
Jason Kuen
David Kügler
Arjan Kuijper
Anna Kukleva
Sumith Kulal
Viveka Kulharia
Akshay R. Kulkarni
Nilesh Kulkarni
Dominik Kulon
Abhinav Kumar
Akash Kumar
Suryansh Kumar
B. V. K. Vijaya Kumar
Pulkit Kumar
Ratnesh Kumar
Sateesh Kumar
Satish Kumar
Vijay Kumar B. G.
Nupur Kumari
Sudhakar Kumawat
Jogendra Nath Kundu
Hsien-Kai Kuo
Meng-Yu Jennifer Kuo
Vinod Kumar Kurmi
Yusuke Kurose
Keerthy Kusumam
Alina Kuznetsova
Henry Kvinge
Ho Man Kwan
Hyeokjun Kweon
Heeseung Kwon
Gihyun Kwon
Myung-Joon Kwon
Taesung Kwon
YoungJoong Kwon
Christos Kyrkou
Jorma Laaksonen
Yann Labbe
Zorah Laehner
Florent Lafarge
Hamid Laga
Manuel Lagunas
Shenqi Lai
Jian-Huang Lai
Zihang Lai
Mohamed I. Lakhal
Mohit Lamba
Meng Lan
Loic Landrieu
Zhiqiang Lang
Natalie Lang
Dong Lao
Yizhen Lao
Yingjie Lao
Issam Hadj Laradji
Gustav Larsson
Viktor Larsson
Zakaria Laskar
Stéphane Lathuilière
Chun Pong Lau
Rynson W. H. Lau
Hei Law
Justin Lazarow
Verica Lazova
Eric-Tuan Le
Hieu Le
Trung-Nghia Le
Mathias Lechner
Byeong-Uk Lee
Chen-Yu Lee
Che-Rung Lee
Chul Lee
Hong Joo Lee
Dongsoo Lee
Jiyoung Lee
Eugene Eu Tzuan Lee
Daeun Lee
Saehyung Lee
Jewook Lee
Hyungtae Lee
Hyunmin Lee
Jungbeom Lee
Joon-Young Lee
Jong-Seok Lee
Joonseok Lee
Junha Lee
Kibok Lee
Byung-Kwan Lee
Jangwon Lee
Jinho Lee
Jongmin Lee
Seunghyun Lee
Sohyun Lee
Minsik Lee
Dogyoon Lee
Seungmin Lee
Min Jun Lee
Sangho Lee
Sangmin Lee
Seungeun Lee
Seon-Ho Lee
Sungmin Lee
Sungho Lee
Sangyoun Lee
Vincent C. S. S. Lee
Jaeseong Lee
Yong Jae Lee
Chenyang Lei
Chenyi Lei
Jiahui Lei
Xinyu Lei
Yinjie Lei
Jiaxu Leng
Luziwei Leng

Jan E. Lenssen
Vincent Lepetit
Thomas Leung
María Leyva-Vallina
Xin Li
Yikang Li
Baoxin Li
Bin Li
Bing Li
Bowen Li
Changlin Li
Chao Li
Chongyi Li
Guanyue Li
Shuai Li
Jin Li
Dingquan Li
Dongxu Li
Yiting Li
Gang Li
Dian Li
Guohao Li
Haoang Li
Haoliang Li
Haoran Li
Hengduo Li
Huafeng Li
Xiaoming Li
Hanao Li
Hongwei Li
Ziqiang Li
Jisheng Li
Jiacheng Li
Jia Li
Jiachen Li
Jiahao Li
Jianwei Li
Jiazhi Li
Jie Li
Jing Li
Jingjing Li
Jingtao Li
Jun Li
Junxuan Li
Kai Li
Kailin Li
Kenneth Li
Kun Li
Kunpeng Li
Aoxue Li
Chenglong Li
Chenglin Li
Changsheng Li
Zhichao Li
Qiang Li
Yanyu Li
Zuoyue Li
Xiang Li
Xuelong Li
Fangda Li
Ailin Li
Liang Li
Chun-Guang Li
Daiqing Li
Dong Li
Guanbin Li
Guorong Li
Haifeng Li
Jianan Li
Jianing Li
Jiaxin Li
Ke Li
Lei Li
Lincheng Li
Liulei Li
Lujun Li
Linjie Li
Lin Li
Pengyu Li
Ping Li
Qiufu Li
Qingyong Li
Rui Li
Siyuan Li
Wei Li
Wenbin Li
Xiangyang Li
Xinyu Li
Xiujun Li
Xiu Li
Xu Li
Ya-Li Li
Yao Li
Yongjie Li
Yijun Li
Yiming Li
Yuezun Li
Yu Li
Yunheng Li
Yuqi Li
Zhe Li
Zeming Li
Zhen Li
Zhengqin Li
Zhimin Li
Jiefeng Li
Jinpeng Li
Chengze Li
Jianwu Li
Lerenhan Li
Shan Li
Suichan Li
Xiangtai Li
Yanjie Li
Yandong Li
Zhuoling Li
Zhenqiang Li
Manyi Li
Maosen Li
Ji Li
Minjun Li
Mingrui Li
Mengtian Li
Junyi Li
Nianyi Li
Bo Li
Xiao Li
Peihua Li
Peike Li
Peizhao Li
Peiliang Li
Qi Li
Ren Li
Runze Li
Shile Li

Sheng Li
Shigang Li
Shiyu Li
Shuang Li
Shasha Li
Shichao Li
Tianye Li
Yuexiang Li
Wei-Hong Li
Wanhua Li
Weihao Li
Weiming Li
Weixin Li
Wenbo Li
Wenshuo Li
Weijian Li
Yunan Li
Xirong Li
Xianhang Li
Xiaoyu Li
Xueqian Li
Xuanlin Li
Xianzhi Li
Yunqiang Li
Yanjing Li
Yansheng Li
Yawei Li
Yi Li
Yong Li
Yong-Lu Li
Yuhang Li
Yu-Jhe Li
Yuxi Li
Yunsheng Li
Yanwei Li
Zechao Li
Zejian Li
Zeju Li
Zekun Li
Zhaowen Li
Zheng Li
Zhenyu Li
Zhiheng Li
Zhi Li
Zhong Li
Zhuowei Li
Zhuowan Li
Zhuohang Li
Zizhang Li
Chen Li
Yuan-Fang Li
Dongze Lian
Xiaochen Lian
Zhouhui Lian
Long Lian
Qing Lian
Jin Lianbao
Jinxiu S. Liang
Dingkang Liang
Jiahao Liang
Jianming Liang
Jingyun Liang
Kevin J. Liang
Kaizhao Liang
Chen Liang
Jie Liang
Senwei Liang
Ding Liang
Jiajun Liang
Jian Liang
Kongming Liang
Siyuan Liang
Yuanzhi Liang
Zhengfa Liang
Mingfu Liang
Xiaodan Liang
Xuefeng Liang
Yuxuan Liang
Kang Liao
Liang Liao
Hong-Yuan Mark Liao
Wentong Liao
Haofu Liao
Yue Liao
Minghui Liao
Shengcai Liao
Ting-Hsuan Liao
Xin Liao
Yinghong Liao
Teck Yian Lim
Che-Tsung Lin
Chung-Ching Lin
Chen-Hsuan Lin
Cheng Lin
Chuming Lin
Chunyu Lin
Dahua Lin
Wei Lin
Zheng Lin
Huaijia Lin
Jason Lin
Jierui Lin
Jiaying Lin
Jie Lin
Kai-En Lin
Kevin Lin
Guangfeng Lin
Jiehong Lin
Feng Lin
Hang Lin
Kwan-Yee Lin
Ke Lin
Luojun Lin
Qinghong Lin
Xiangbo Lin
Yi Lin
Zudi Lin
Shijie Lin
Yiqun Lin
Tzu-Heng Lin
Ming Lin
Shaohui Lin
SongNan Lin
Ji Lin
Tsung-Yu Lin
Xudong Lin
Yancong Lin
Yen-Chen Lin
Yiming Lin
Yuewei Lin
Zhiqiu Lin
Zinan Lin
Zhe Lin
David B. Lindell
Zhixin Ling

Zhan Ling
Alexander Liniger
Venice Erin B. Liong
Joey Litalien
Or Litany
Roee Litman
Ron Litman
Jim Little
Dor Litvak
Shaoteng Liu
Shuaicheng Liu
Andrew Liu
Xian Liu
Shaohui Liu
Bei Liu
Bo Liu
Yong Liu
Ming Liu
Yanbin Liu
Chenxi Liu
Daqi Liu
Di Liu
Difan Liu
Dong Liu
Dongfang Liu
Daizong Liu
Xiao Liu
Fangyi Liu
Fengbei Liu
Fenglin Liu
Bin Liu
Yuang Liu
Ao Liu
Hong Liu
Hongfu Liu
Huidong Liu
Ziyi Liu
Feng Liu
Hao Liu
Jie Liu
Jialun Liu
Jiang Liu
Jing Liu
Jingya Liu
Jiaming Liu
Jun Liu
Juncheng Liu
Jiawei Liu
Hongyu Liu
Chuanbin Liu
Haotian Liu
Lingqiao Liu
Chang Liu
Han Liu
Liu Liu
Min Liu
Yingqi Liu
Aishan Liu
Bingyu Liu
Benlin Liu
Boxiao Liu
Chenchen Liu
Chuanjian Liu
Daqing Liu
Huan Liu
Haozhe Liu
Jiaheng Liu
Wei Liu
Jingzhou Liu
Jiyuan Liu
Lingbo Liu
Nian Liu
Peiye Liu
Qiankun Liu
Shenglan Liu
Shilong Liu
Wen Liu
Wenyu Liu
Weifeng Liu
Wu Liu
Xiaolong Liu
Yang Liu
Yanwei Liu
Yingcheng Liu
Yongfei Liu
Yihao Liu
Yu Liu
Yunze Liu
Ze Liu
Zhenhua Liu
Zhenguang Liu
Lin Liu
Lihao Liu
Pengju Liu
Xinhai Liu
Yunfei Liu
Meng Liu
Minghua Liu
Mingyuan Liu
Miao Liu
Peirong Liu
Ping Liu
Qingjie Liu
Ruoshi Liu
Risheng Liu
Songtao Liu
Xing Liu
Shikun Liu
Shuming Liu
Sheng Liu
Songhua Liu
Tongliang Liu
Weibo Liu
Weide Liu
Weizhe Liu
Wenxi Liu
Weiyang Liu
Xin Liu
Xiaobin Liu
Xudong Liu
Xiaoyi Liu
Xihui Liu
Xinchen Liu
Xingtong Liu
Xinpeng Liu
Xinyu Liu
Xianpeng Liu
Xu Liu
Xingyu Liu
Yongtuo Liu
Yahui Liu
Yangxin Liu
Yaoyao Liu
Yaojie Liu
Yuliang Liu

Yongcheng Liu
Yuan Liu
Yufan Liu
Yu-Lun Liu
Yun Liu
Yunfan Liu
Yuanzhong Liu
Zhuoran Liu
Zhen Liu
Zheng Liu
Zhijian Liu
Zhisong Liu
Ziquan Liu
Ziyu Liu
Zhihua Liu
Zechun Liu
Zhaoyang Liu
Zhengzhe Liu
Stephan Liwicki
Shao-Yuan Lo
Sylvain Lobry
Suhas Lohit
Vishnu Suresh Lokhande
Vincenzo Lomonaco
Chengjiang Long
Guodong Long
Fuchen Long
Shangbang Long
Yang Long
Zijun Long
Vasco Lopes
Antonio M. Lopez
Roberto Javier Lopez-Sastre
Tobias Lorenz
Javier Lorenzo-Navarro
Yujing Lou
Qian Lou
Xiankai Lu
Changsheng Lu
Huimin Lu
Yongxi Lu
Hao Lu
Hong Lu
Jiasen Lu
Juwei Lu
Fan Lu
Guangming Lu
Jiwen Lu
Shun Lu
Tao Lu
Xiaonan Lu
Yang Lu
Yao Lu
Yongchun Lu
Zhiwu Lu
Cheng Lu
Liying Lu
Guo Lu
Xuequan Lu
Yanye Lu
Yantao Lu
Yuhang Lu
Fujun Luan
Jonathon Luiten
Jovita Lukasik
Alan Lukezic
Jonathan Samuel Lumentut
Mayank Lunayach
Ao Luo
Canjie Luo
Chong Luo
Xu Luo
Grace Luo
Jun Luo
Katie Z. Luo
Tao Luo
Cheng Luo
Fangzhou Luo
Gen Luo
Lei Luo
Sihui Luo
Weixin Luo
Yan Luo
Xiaoyan Luo
Yong Luo
Yadan Luo
Hao Luo
Ruotian Luo
Mi Luo
Tiange Luo
Wenjie Luo
Wenhan Luo
Xiao Luo
Zhiming Luo
Zhipeng Luo
Zhengyi Luo
Diogo C. Luvizon
Zhaoyang Lv
Gengyu Lyu
Lingjuan Lyu
Jun Lyu
Yuanyuan Lyu
Youwei Lyu
Yueming Lyu
Bingpeng Ma
Chao Ma
Chongyang Ma
Congbo Ma
Chih-Yao Ma
Fan Ma
Lin Ma
Haoyu Ma
Hengbo Ma
Jianqi Ma
Jiawei Ma
Jiayi Ma
Kede Ma
Kai Ma
Lingni Ma
Lei Ma
Xu Ma
Ning Ma
Benteng Ma
Cheng Ma
Andy J. Ma
Long Ma
Zhanyu Ma
Zhiheng Ma
Qianli Ma
Shiqiang Ma
Sizhuo Ma
Shiqing Ma
Xiaolong Ma
Xinzhu Ma

Gautam B. Machiraju
Spandan Madan
Mathew Magimai-Doss
Luca Magri
Behrooz Mahasseni
Upal Mahbub
Siddharth Mahendran
Paridhi Maheshwari
Rishabh Maheshwary
Mohammed Mahmoud
Shishira R. R. Maiya
Sylwia Majchrowska
Arjun Majumdar
Puspita Majumdar
Orchid Majumder
Sagnik Majumder
Ilya Makarov
Farkhod F. Makhmudkhujaev
Yasushi Makihara
Ankur Mali
Mateusz Malinowski
Utkarsh Mall
Srikanth Malla
Clement Mallet
Dimitrios Mallis
Yunze Man
Dipu Manandhar
Massimiliano Mancini
Murari Mandal
Raunak Manekar
Karttikeya Mangalam
Puneet Mangla
Fabian Manhardt
Sivabalan Manivasagam
Fahim Mannan
Chengzhi Mao
Hanzi Mao
Jiayuan Mao
Junhua Mao
Zhiyuan Mao
Jiageng Mao
Yunyao Mao
Zhendong Mao
Alberto Marchisio
Diego Marcos
Riccardo Marin
Aram Markosyan
Renaud Marlet
Ricardo Marques
Miquel Martí i Rabadán
Diego Martin Arroyo
Niki Martinel
Brais Martinez
Julieta Martinez
Marc Masana
Tomohiro Mashita
Timothée Masquelier
Minesh Mathew
Tetsu Matsukawa
Marwan Mattar
Bruce A. Maxwell
Christoph Mayer
Mantas Mazeika
Pratik Mazumder
Scott McCloskey
Steven McDonagh
Ishit Mehta
Jie Mei
Kangfu Mei
Jieru Mei
Xiaoguang Mei
Givi Meishvili
Luke Melas-Kyriazi
Iaroslav Melekhov
Andres Mendez-Vazquez
Heydi Mendez-Vazquez
Matias Mendieta
Ricardo A. Mendoza-León
Chenlin Meng
Depu Meng
Rang Meng
Zibo Meng
Qingjie Meng
Qier Meng
Yanda Meng
Zihang Meng
Thomas Mensink
Fabian Mentzer
Christopher Metzler
Gregory P. Meyer
Vasileios Mezaris
Liang Mi
Lu Mi
Bo Miao
Changtao Miao
Zichen Miao
Qiguang Miao
Xin Miao
Zhongqi Miao
Frank Michel
Simone Milani
Ben Mildenhall
Roy V. Miles
Juhong Min
Kyle Min
Hyun-Seok Min
Weiqing Min
Yuecong Min
Zhixiang Min
Qi Ming
David Minnen
Aymen Mir
Deepak Mishra
Anand Mishra
Shlok K. Mishra
Niluthpol Mithun
Gaurav Mittal
Trisha Mittal
Daisuke Miyazaki
Kaichun Mo
Hong Mo
Zhipeng Mo
Davide Modolo
Abduallah A. Mohamed
Mohamed Afham Mohamed Aflal
Ron Mokady
Pavlo Molchanov
Davide Moltisanti
Liliane Momeni
Gianluca Monaci
Pascal Monasse
Ajoy Mondal
Tom Monnier

Aron Monszpart
Gyeongsik Moon
Suhong Moon
Taesup Moon
Sean Moran
Daniel Moreira
Pietro Morerio
Alexandre Morgand
Lia Morra
Ali Mosleh
Inbar Mosseri
Sayed Mohammad Mostafavi Isfahani
Saman Motamed
Ramy A. Mounir
Fangzhou Mu
Jiteng Mu
Norman Mu
Yasuhiro Mukaigawa
Ryan Mukherjee
Tanmoy Mukherjee
Yusuke Mukuta
Ravi Teja Mullapudi
Lea Müller
Matthias Müller
Martin Mundt
Nils Murrugarra-Llerena
Damien Muselet
Armin Mustafa
Muhammad Ferjad Naeem
Sauradip Nag
Hajime Nagahara
Pravin Nagar
Rajendra Nagar
Naveen Shankar Nagaraja
Varun Nagaraja
Tushar Nagarajan
Seungjun Nah
Gaku Nakano
Yuta Nakashima
Giljoo Nam
Seonghyeon Nam
Liangliang Nan
Yuesong Nan
Yeshwanth Napolean
Dinesh Reddy Narapureddy
Medhini Narasimhan
Supreeth Narasimhaswamy
Sriram Narayanan
Erickson R. Nascimento
Varun Nasery
K. L. Navaneet
Pablo Navarrete Michelini
Shant Navasardyan
Shah Nawaz
Nihal Nayak
Farhood Negin
Lukáš Neumann
Alejandro Newell
Evonne Ng
Kam Woh Ng
Tony Ng
Anh Nguyen
Tuan Anh Nguyen
Cuong Cao Nguyen
Ngoc Cuong Nguyen
Thanh Nguyen
Khoi Nguyen
Phi Le Nguyen
Phong Ha Nguyen
Tam Nguyen
Truong Nguyen
Anh Tuan Nguyen
Rang Nguyen
Thao Thi Phuong Nguyen
Van Nguyen Nguyen
Zhen-Liang Ni
Yao Ni
Shijie Nie
Xuecheng Nie
Yongwei Nie
Weizhi Nie
Ying Nie
Yinyu Nie
Kshitij N. Nikhal
Simon Niklaus
Xuefei Ning
Jifeng Ning
Yotam Nitzan
Di Niu
Shuaicheng Niu
Li Niu
Wei Niu
Yulei Niu
Zhenxing Niu
Albert No
Shohei Nobuhara
Nicoletta Noceti
Junhyug Noh
Sotiris Nousias
Slawomir Nowaczyk
Ewa M. Nowara
Valsamis Ntouskos
Gilberto Ochoa-Ruiz
Ferda Ofli
Jihyong Oh
Sangyun Oh
Youngtaek Oh
Hiroki Ohashi
Takahiro Okabe
Kemal Oksuz
Fumio Okura
Daniel Olmeda Reino
Matthew Olson
Carl Olsson
Roy Or-El
Alessandro Ortis
Guillermo Ortiz-Jimenez
Magnus Oskarsson
Ahmed A. A. Osman
Martin R. Oswald
Mayu Otani
Naima Otberdout
Cheng Ouyang
Jiahong Ouyang
Wanli Ouyang
Andrew Owens
Poojan B. Oza
Mete Ozay
A. Cengiz Oztireli
Gautam Pai
Tomas Pajdla
Umapada Pal

Simone Palazzo
Luca Palmieri
Bowen Pan
Hao Pan
Lili Pan
Tai-Yu Pan
Liang Pan
Chengwei Pan
Yingwei Pan
Xuran Pan
Jinshan Pan
Xinyu Pan
Liyuan Pan
Xingang Pan
Xingjia Pan
Zhihong Pan
Zizheng Pan
Priyadarshini Panda
Rameswar Panda
Rohit Pandey
Kaiyue Pang
Bo Pang
Guansong Pang
Jiangmiao Pang
Meng Pang
Tianyu Pang
Ziqi Pang
Omiros Pantazis
Andreas Panteli
Maja Pantic
Marina Paolanti
Joao P. Papa
Samuele Papa
Mike Papadakis
Dim P. Papadopoulos
George Papandreou
Constantin Pape
Toufiq Parag
Chethan Parameshwara
Shaifali Parashar
Alejandro Pardo
Rishubh Parihar
Sarah Parisot
JaeYoo Park
Gyeong-Moon Park
Hyojin Park
Hyoungseob Park
Jongchan Park
Jae Sung Park
Kiru Park
Chunghyun Park
Kwanyong Park
Sunghyun Park
Sungrae Park
Seongsik Park
Sanghyun Park
Sungjune Park
Taesung Park
Gaurav Parmar
Paritosh Parmar
Alvaro Parra
Despoina Paschalidou
Or Patashnik
Shivansh Patel
Pushpak Pati
Prashant W. Patil
Vaishakh Patil
Suvam Patra
Jay Patravali
Badri Narayana Patro
Angshuman Paul
Sudipta Paul
Rémi Pautrat
Nick E. Pears
Adithya Pediredla
Wenjie Pei
Shmuel Peleg
Latha Pemula
Bo Peng
Houwen Peng
Yue Peng
Liangzu Peng
Baoyun Peng
Jun Peng
Pai Peng
Sida Peng
Xi Peng
Yuxin Peng
Songyou Peng
Wei Peng
Weiqi Peng
Wen-Hsiao Peng
Pramuditha Perera
Juan C. Perez
Eduardo Pérez Pellitero
Juan-Manuel Perez-Rua
Federico Pernici
Marco Pesavento
Stavros Petridis
Ilya A. Petrov
Vladan Petrovic
Mathis Petrovich
Suzanne Petryk
Hieu Pham
Quang Pham
Khoi Pham
Tung Pham
Huy Phan
Stephen Phillips
Cheng Perng Phoo
David Picard
Marco Piccirilli
Georg Pichler
A. J. Piergiovanni
Vipin Pillai
Silvia L. Pintea
Giovanni Pintore
Robinson Piramuthu
Fiora Pirri
Theodoros Pissas
Fabio Pizzati
Benjamin Planche
Bryan Plummer
Matteo Poggi
Ashwini Pokle
Georgy E. Ponimatkin
Adrian Popescu
Stefan Popov
Nikola Popović
Ronald Poppe
Angelo Porrello
Michael Potter
Charalambos Poullis
Hadi Pouransari
Omid Poursaeed

Shraman Pramanick
Mantini Pranav
Dilip K. Prasad
Meghshyam Prasad
B. H. Pawan Prasad
Shitala Prasad
Prateek Prasanna
Ekta Prashnani
Derek S. Prijatelj
Luke Y. Prince
Véronique Prinet
Victor Adrian Prisacariu
James Pritts
Thomas Probst
Sergey Prokudin
Rita Pucci
Chi-Man Pun
Matthew Purri
Haozhi Qi
Lu Qi
Lei Qi
Xianbiao Qi
Yonggang Qi
Yuankai Qi
Siyuan Qi
Guocheng Qian
Hangwei Qian
Qi Qian
Deheng Qian
Shengsheng Qian
Wen Qian
Rui Qian
Yiming Qian
Shengju Qian
Shengyi Qian
Xuelin Qian
Zhenxing Qian
Nan Qiao
Xiaotian Qiao
Jing Qin
Can Qin
Siyang Qin
Hongwei Qin
Jie Qin
Minghai Qin
Yipeng Qin
Yongqiang Qin
Wenda Qin
Xuebin Qin
Yuzhe Qin
Yao Qin
Zhenyue Qin
Zhiwu Qing
Heqian Qiu
Jiayan Qiu
Jielin Qiu
Yue Qiu
Jiaxiong Qiu
Zhongxi Qiu
Shi Qiu
Zhaofan Qiu
Zhongnan Qu
Yanyun Qu
Kha Gia Quach
Yuhui Quan
Ruijie Quan
Mike Rabbat
Rahul Shekhar Rade
Filip Radenovic
Gorjan Radevski
Bogdan Raducanu
Francesco Ragusa
Shafin Rahman
Md Mahfuzur Rahman Siddiquee
Hossein Rahmani
Kiran Raja
Sivaramakrishnan Rajaraman
Jathushan Rajasegaran
Adnan Siraj Rakin
Michaël Ramamonjisoa
Chirag A. Raman
Shanmuganathan Raman
Vignesh Ramanathan
Vasili Ramanishka
Vikram V. Ramaswamy
Merey Ramazanova
Jason Rambach
Sai Saketh Rambhatla
Clément Rambour
Ashwin Ramesh Babu
Adín Ramírez Rivera
Arianna Rampini
Haoxi Ran
Aakanksha Rana
Aayush Jung Bahadur Rana
Kanchana N. Ranasinghe
Aneesh Rangnekar
Samrudhdhi B. Rangrej
Harsh Rangwani
Viresh Ranjan
Anyi Rao
Yongming Rao
Carolina Raposo
Michalis Raptis
Amir Rasouli
Vivek Rathod
Adepu Ravi Sankar
Avinash Ravichandran
Bharadwaj Ravichandran
Dripta S. Raychaudhuri
Adria Recasens
Simon Reiß
Davis Rempe
Daxuan Ren
Jiawei Ren
Jimmy Ren
Sucheng Ren
Dayong Ren
Zhile Ren
Dongwei Ren
Qibing Ren
Pengfei Ren
Zhenwen Ren
Xuqian Ren
Yixuan Ren
Zhongzheng Ren
Ambareesh Revanur
Hamed Rezazadegan Tavakoli
Rafael S. Rezende
Wonjong Rhee
Alexander Richard

Christian Richardt
Stephan R. Richter
Benjamin Riggan
Dominik Rivoir
Mamshad Nayeem Rizve
Joshua D. Robinson
Joseph Robinson
Chris Rockwell
Ranga Rodrigo
Andres C. Rodriguez
Carlos Rodriguez-Pardo
Marcus Rohrbach
Gemma Roig
Yu Rong
David A. Ross
Mohammad Rostami
Edward Rosten
Karsten Roth
Anirban Roy
Debaditya Roy
Shuvendu Roy
Ahana Roy Choudhury
Aruni Roy Chowdhury
Denys Rozumnyi
Shulan Ruan
Wenjie Ruan
Patrick Ruhkamp
Danila Rukhovich
Anian Ruoss
Chris Russell
Dan Ruta
Dawid Damian Rymarczyk
DongHun Ryu
Hyeonggon Ryu
Kwonyoung Ryu
Balasubramanian S.
Alexandre Sablayrolles
Mohammad Sabokrou
Arka Sadhu
Aniruddha Saha
Oindrila Saha
Pritish Sahu
Aneeshan Sain
Nirat Saini
Saurabh Saini
Takeshi Saitoh
Christos Sakaridis
Fumihiko Sakaue
Dimitrios Sakkos
Ken Sakurada
Parikshit V. Sakurikar
Rohit Saluja
Nermin Samet
Leo Sampaio Ferraz Ribeiro
Jorge Sanchez
Enrique Sanchez
Shengtian Sang
Anush Sankaran
Soubhik Sanyal
Nikolaos Sarafianos
Vishwanath Saragadam
István Sárándi
Saquib Sarfraz
Mert Bulent Sariyildiz
Anindya Sarkar
Pritam Sarkar
Paul-Edouard Sarlin
Hiroshi Sasaki
Takami Sato
Torsten Sattler
Ravi Kumar Satzoda
Axel Sauer
Stefano Savian
Artem Savkin
Manolis Savva
Gerald Schaefer
Simone Schaub-Meyer
Yoni Schirris
Samuel Schulter
Katja Schwarz
Jesse Scott
Sinisa Segvic
Constantin Marc Seibold
Lorenzo Seidenari
Matan Sela
Fadime Sener
Paul Hongsuck Seo
Kwanggyoon Seo
Hongje Seong
Dario Serez
Francesco Setti
Bryan Seybold
Mohamad Shahbazi
Shima Shahfar
Xinxin Shan
Caifeng Shan
Dandan Shan
Shawn Shan
Wei Shang
Jinghuan Shang
Jiaxiang Shang
Lei Shang
Sukrit Shankar
Ken Shao
Rui Shao
Jie Shao
Mingwen Shao
Aashish Sharma
Gaurav Sharma
Vivek Sharma
Abhishek Sharma
Yoli Shavit
Shashank Shekhar
Sumit Shekhar
Zhijie Shen
Fengyi Shen
Furao Shen
Jialie Shen
Jingjing Shen
Ziyi Shen
Linlin Shen
Guangyu Shen
Biluo Shen
Falong Shen
Jiajun Shen
Qiu Shen
Qiuhong Shen
Shuai Shen
Wang Shen
Yiqing Shen
Yunhang Shen
Siqi Shen
Bin Shen
Tianwei Shen

Xi Shen
Yilin Shen
Yuming Shen
Yucong Shen
Zhiqiang Shen
Lu Sheng
Yichen Sheng
Shivanand Venkanna Sheshappanavar
Shelly Sheynin
Baifeng Shi
Ruoxi Shi
Botian Shi
Hailin Shi
Jia Shi
Jing Shi
Shaoshuai Shi
Baoguang Shi
Boxin Shi
Hengcan Shi
Tianyang Shi
Xiaodan Shi
Yongjie Shi
Zhensheng Shi
Yinghuan Shi
Weiqi Shi
Wu Shi
Xuepeng Shi
Xiaoshuang Shi
Yujiao Shi
Zenglin Shi
Zhenmei Shi
Takashi Shibata
Meng-Li Shih
Yichang Shih
Hyunjung Shim
Dongseok Shim
Soshi Shimada
Inkyu Shin
Jinwoo Shin
Seungjoo Shin
Seungjae Shin
Koichi Shinoda
Suprosanna Shit
Palaiahnakote Shivakumara
Eli Shlizerman
Gaurav Shrivastava
Xiao Shu
Xiangbo Shu
Xiujun Shu
Yang Shu
Tianmin Shu
Jun Shu
Zhixin Shu
Bing Shuai
Maria Shugrina
Ivan Shugurov
Satya Narayan Shukla
Pranjay Shyam
Jianlou Si
Yawar Siddiqui
Alberto Signoroni
Pedro Silva
Jae-Young Sim
Oriane Siméoni
Martin Simon
Andrea Simonelli
Abhishek Singh
Ashish Singh
Dinesh Singh
Gurkirt Singh
Krishna Kumar Singh
Mannat Singh
Pravendra Singh
Rajat Vikram Singh
Utkarsh Singhal
Dipika Singhania
Vasu Singla
Harsh Sinha
Sudipta Sinha
Josef Sivic
Elena Sizikova
Geri Skenderi
Ivan Skorokhodov
Dmitriy Smirnov
Cameron Y. Smith
James S. Smith
Patrick Snape
Mattia Soldan
Hyeongseok Son
Sanghyun Son
Chuanbiao Song
Chen Song
Chunfeng Song
Dan Song
Dongjin Song
Hwanjun Song
Guoxian Song
Jiaming Song
Jie Song
Liangchen Song
Ran Song
Luchuan Song
Xibin Song
Li Song
Fenglong Song
Guoli Song
Guanglu Song
Zhenbo Song
Lin Song
Xinhang Song
Yang Song
Yibing Song
Rajiv Soundararajan
Hossein Souri
Cristovao Sousa
Riccardo Spezialetti
Leonidas Spinoulas
Michael W. Spratling
Deepak Sridhar
Srinath Sridhar
Gaurang Sriramanan
Vinkle Kumar Srivastav
Themos Stafylakis
Serban Stan
Anastasis Stathopoulos
Markus Steinberger
Jan Steinbrener
Sinisa Stekovic
Alexandros Stergiou
Gleb Sterkin
Rainer Stiefelhagen
Pierre Stock

Ombretta Strafforello
Julian Straub
Yannick Strümpler
Joerg Stueckler
Hang Su
Weijie Su
Jong-Chyi Su
Bing Su
Haisheng Su
Jinming Su
Yiyang Su
Yukun Su
Yuxin Su
Zhuo Su
Zhaoqi Su
Xiu Su
Yu-Chuan Su
Zhixun Su
Arulkumar Subramaniam
Akshayvarun Subramanya
A. Subramanyam
Swathikiran Sudhakaran
Yusuke Sugano
Masanori Suganuma
Yumin Suh
Yang Sui
Baochen Sun
Cheng Sun
Long Sun
Guolei Sun
Haoliang Sun
Haomiao Sun
He Sun
Hanqing Sun
Hao Sun
Lichao Sun
Jiachen Sun
Jiaming Sun
Jian Sun
Jin Sun
Jennifer J. Sun
Tiancheng Sun
Libo Sun
Peize Sun
Qianru Sun
Shanlin Sun
Yu Sun
Zhun Sun
Che Sun
Lin Sun
Tao Sun
Yiyou Sun
Chunyi Sun
Chong Sun
Weiwei Sun
Weixuan Sun
Xiuyu Sun
Yanan Sun
Zeren Sun
Zhaodong Sun
Zhiqing Sun
Minhyuk Sung
Jinli Suo
Simon Suo
Abhijit Suprem
Anshuman Suri
Saksham Suri
Joshua M. Susskind
Roman Suvorov
Gurumurthy Swaminathan
Robin Swanson
Paul Swoboda
Tabish A. Syed
Richard Szeliski
Fariborz Taherkhani
Yu-Wing Tai
Keita Takahashi
Walter Talbott
Gary Tam
Masato Tamura
Feitong Tan
Fuwen Tan
Shuhan Tan
Andong Tan
Bin Tan
Cheng Tan
Jianchao Tan
Lei Tan
Mingxing Tan
Xin Tan
Zichang Tan
Zhentao Tan
Kenichiro Tanaka
Masayuki Tanaka
Yushun Tang
Hao Tang
Jingqun Tang
Jinhui Tang
Kaihua Tang
Luming Tang
Lv Tang
Sheyang Tang
Shitao Tang
Siliang Tang
Shixiang Tang
Yansong Tang
Keke Tang
Chang Tang
Chenwei Tang
Jie Tang
Junshu Tang
Ming Tang
Peng Tang
Xu Tang
Yao Tang
Chen Tang
Fan Tang
Haoran Tang
Shengeng Tang
Yehui Tang
Zhipeng Tang
Ugo Tanielian
Chaofan Tao
Jiale Tao
Junli Tao
Renshuai Tao
An Tao
Guanhong Tao
Zhiqiang Tao
Makarand Tapaswi
Jean-Philippe G. Tarel
Juan J. Tarrio
Enzo Tartaglione
Keisuke Tateno
Zachary Teed

Ajinkya B. Tejankar
Bugra Tekin
Purva Tendulkar
Damien Teney
Minggui Teng
Chris Tensmeyer
Andrew Beng Jin Teoh
Philipp Terhörst
Kartik Thakral
Nupur Thakur
Kevin Thandiackal
Spyridon Thermos
Diego Thomas
William Thong
Yuesong Tian
Guanzhong Tian
Lin Tian
Shiqi Tian
Kai Tian
Meng Tian
Tai-Peng Tian
Zhuotao Tian
Shangxuan Tian
Tian Tian
Yapeng Tian
Yu Tian
Yuxin Tian
Leslie Ching Ow Tiong
Praveen Tirupattur
Garvita Tiwari
George Toderici
Antoine Toisoul
Aysim Toker
Tatiana Tommasi
Zhan Tong
Alessio Tonioni
Alessandro Torcinovich
Fabio Tosi
Matteo Toso
Hugo Touvron
Quan Hung Tran
Son Tran
Hung Tran
Ngoc-Trung Tran
Vinh Tran
Phong Tran
Giovanni Trappolini
Edith Tretschk
Subarna Tripathi
Shubhendu Trivedi
Eduard Trulls
Prune Truong
Thanh-Dat Truong
Tomasz Trzcinski
Sam Tsai
Yi-Hsuan Tsai
Ethan Tseng
Yu-Chee Tseng
Shahar Tsiper
Stavros Tsogkas
Shikui Tu
Zhigang Tu
Zhengzhong Tu
Richard Tucker
Sergey Tulyakov
Cigdem Turan
Daniyar Turmukhambetov
Victor G. Turrisi da Costa
Bartlomiej Twardowski
Christopher D. Twigg
Radim Tylecek
Mostofa Rafid Uddin
Md. Zasim Uddin
Kohei Uehara
Nicolas Ugrinovic
Youngjung Uh
Norimichi Ukita
Anwaar Ulhaq
Devesh Upadhyay
Paul Upchurch
Yoshitaka Ushiku
Yuzuko Utsumi
Mikaela Angelina Uy
Mohit Vaishnav
Pratik Vaishnavi
Jeya Maria Jose Valanarasu
Matias A. Valdenegro Toro
Diego Valsesia
Wouter Van Gansbeke
Nanne van Noord
Simon Vandenhende
Farshid Varno
Cristina Vasconcelos
Francisco Vasconcelos
Alex Vasilescu
Subeesh Vasu
Arun Balajee Vasudevan
Kanav Vats
Vaibhav S. Vavilala
Sagar Vaze
Javier Vazquez-Corral
Andrea Vedaldi
Olga Veksler
Andreas Velten
Sai H. Vemprala
Raviteja Vemulapalli
Shashanka Venkataramanan
Dor Verbin
Luisa Verdoliva
Manisha Verma
Yashaswi Verma
Constantin Vertan
Eli Verwimp
Deepak Vijaykeerthy
Pablo Villanueva
Ruben Villegas
Markus Vincze
Vibhav Vineet
Minh P. Vo
Huy V. Vo
Duc Minh Vo
Tomas Vojir
Igor Vozniak
Nicholas Vretos
Vibashan VS
Tuan-Anh Vu
Thang Vu
Mårten Wadenbäck
Neal Wadhwa
Aaron T. Walsman
Steven Walton
Jin Wan
Alvin Wan
Jia Wan

Jun Wan
Xiaoyue Wan
Fang Wan
Guowei Wan
Renjie Wan
Zhiqiang Wan
Ziyu Wan
Bastian Wandt
Dongdong Wang
Limin Wang
Haiyang Wang
Xiaobing Wang
Angtian Wang
Angelina Wang
Bing Wang
Bo Wang
Boyu Wang
Binghui Wang
Chen Wang
Chien-Yi Wang
Congli Wang
Qi Wang
Chengrui Wang
Rui Wang
Yiqun Wang
Cong Wang
Wenjing Wang
Dongkai Wang
Di Wang
Xiaogang Wang
Kai Wang
Zhizhong Wang
Fangjinhua Wang
Feng Wang
Hang Wang
Gaoang Wang
Guoqing Wang
Guangcong Wang
Guangzhi Wang
Hanqing Wang
Hao Wang
Haohan Wang
Haoran Wang
Hong Wang
Haotao Wang
Hu Wang
Huan Wang
Hua Wang
Hui-Po Wang
Hengli Wang
Hanyu Wang
Hongxing Wang
Jingwen Wang
Jialiang Wang
Jian Wang
Jianyi Wang
Jiashun Wang
Jiahao Wang
Tsun-Hsuan Wang
Xiaoqian Wang
Jinqiao Wang
Jun Wang
Jianzong Wang
Kaihong Wang
Ke Wang
Lei Wang
Lingjing Wang
Linnan Wang
Lin Wang
Liansheng Wang
Mengjiao Wang
Manning Wang
Nannan Wang
Peihao Wang
Jiayun Wang
Pu Wang
Qiang Wang
Qiufeng Wang
Qilong Wang
Qiangchang Wang
Qin Wang
Qing Wang
Ruocheng Wang
Ruibin Wang
Ruisheng Wang
Ruizhe Wang
Runqi Wang
Runzhong Wang
Wenxuan Wang
Sen Wang
Shangfei Wang
Shaofei Wang
Shijie Wang
Shiqi Wang
Zhibo Wang
Song Wang
Xinjiang Wang
Tai Wang
Tao Wang
Teng Wang
Xiang Wang
Tianren Wang
Tiantian Wang
Tianyi Wang
Fengjiao Wang
Wei Wang
Miaohui Wang
Suchen Wang
Siyue Wang
Yaoming Wang
Xiao Wang
Ze Wang
Biao Wang
Chaofei Wang
Dong Wang
Gu Wang
Guangrun Wang
Guangming Wang
Guo-Hua Wang
Haoqing Wang
Hesheng Wang
Huafeng Wang
Jinghua Wang
Jingdong Wang
Jingjing Wang
Jingya Wang
Jingkang Wang
Jiakai Wang
Junke Wang
Kuo Wang
Lichen Wang
Lizhi Wang
Longguang Wang
Mang Wang
Mei Wang

Min Wang
Peng-Shuai Wang
Run Wang
Shaoru Wang
Shuhui Wang
Tan Wang
Tiancai Wang
Tianqi Wang
Wenhai Wang
Wenzhe Wang
Xiaobo Wang
Xiudong Wang
Xu Wang
Yajie Wang
Yan Wang
Yuan-Gen Wang
Yingqian Wang
Yizhi Wang
Yulin Wang
Yu Wang
Yujie Wang
Yunhe Wang
Yuxi Wang
Yaowei Wang
Yiwei Wang
Zezheng Wang
Hongzhi Wang
Zhiqiang Wang
Ziteng Wang
Ziwei Wang
Zheng Wang
Zhenyu Wang
Binglu Wang
Zhongdao Wang
Ce Wang
Weining Wang
Weiyao Wang
Wenbin Wang
Wenguan Wang
Guangting Wang
Haolin Wang
Haiyan Wang
Huiyu Wang
Naiyan Wang
Jingbo Wang
Jinpeng Wang
Jiaqi Wang
Liyuan Wang
Lizhen Wang
Ning Wang
Wenqian Wang
Sheng-Yu Wang
Weimin Wang
Xiaohan Wang
Yifan Wang
Yi Wang
Yongtao Wang
Yizhou Wang
Zhuo Wang
Zhe Wang
Xudong Wang
Xiaofang Wang
Xinggang Wang
Xiaosen Wang
Xiaosong Wang
Xiaoyang Wang
Lijun Wang
Xinlong Wang
Xuan Wang
Xue Wang
Yangang Wang
Yaohui Wang
Yu-Chiang Frank Wang
Yida Wang
Yilin Wang
Yi Ru Wang
Yali Wang
Yinglong Wang
Yufu Wang
Yujiang Wang
Yuwang Wang
Yuting Wang
Yang Wang
Yu-Xiong Wang
Yixu Wang
Ziqi Wang
Zhicheng Wang
Zeyu Wang
Zhaowen Wang
Zhenyi Wang
Zhenzhi Wang
Zhijie Wang
Zhiyong Wang
Zhongling Wang
Zhuowei Wang
Zian Wang
Zifu Wang
Zihao Wang
Zirui Wang
Ziyan Wang
Wenxiao Wang
Zhen Wang
Zhepeng Wang
Zi Wang
Zihao W. Wang
Steven L. Waslander
Olivia Watkins
Daniel Watson
Silvan Weder
Dongyoon Wee
Dongming Wei
Tianyi Wei
Jia Wei
Dong Wei
Fangyun Wei
Longhui Wei
Mingqiang Wei
Xinyue Wei
Chen Wei
Donglai Wei
Pengxu Wei
Xing Wei
Xiu-Shen Wei
Wenqi Wei
Guoqiang Wei
Wei Wei
XingKui Wei
Xian Wei
Xingxing Wei
Yake Wei
Yuxiang Wei
Yi Wei
Luca Weihs
Michael Weinmann
Martin Weinmann

Congcong Wen
Chuan Wen
Jie Wen
Sijia Wen
Song Wen
Chao Wen
Xiang Wen
Zeyi Wen
Xin Wen
Yilin Wen
Yijia Weng
Shuchen Weng
Junwu Weng
Wenming Weng
Renliang Weng
Zhenyu Weng
Xinshuo Weng
Nicholas J. Westlake
Gordon Wetzstein
Lena M. Widin Klasén
Rick Wildes
Bryan M. Williams
Williem Williem
Ole Winther
Scott Wisdom
Alex Wong
Chau-Wai Wong
Kwan-Yee K. Wong
Yongkang Wong
Scott Workman
Marcel Worring
Michael Wray
Safwan Wshah
Xiang Wu
Aming Wu
Chongruo Wu
Cho-Ying Wu
Chunpeng Wu
Chenyan Wu
Ziyi Wu
Fuxiang Wu
Gang Wu
Haiping Wu
Huisi Wu
Jane Wu
Jialian Wu
Jing Wu
Jinjian Wu
Jianlong Wu
Xian Wu
Lifang Wu
Lifan Wu
Minye Wu
Qianyi Wu
Rongliang Wu
Rui Wu
Shiqian Wu
Shuzhe Wu
Shangzhe Wu
Tsung-Han Wu
Tz-Ying Wu
Ting-Wei Wu
Jiannan Wu
Zhiliang Wu
Yu Wu
Chenyun Wu
Dayan Wu
Dongxian Wu
Fei Wu
Hefeng Wu
Jianxin Wu
Weibin Wu
Wenxuan Wu
Wenhao Wu
Xiao Wu
Yicheng Wu
Yuanwei Wu
Yu-Huan Wu
Zhenxin Wu
Zhenyu Wu
Wei Wu
Peng Wu
Xiaohe Wu
Xindi Wu
Xinxing Wu
Xinyi Wu
Xingjiao Wu
Xiongwei Wu
Yangzheng Wu
Yanzhao Wu
Yawen Wu
Yong Wu
Yi Wu
Ying Nian Wu
Zhenyao Wu
Zhonghua Wu
Zongze Wu
Zuxuan Wu
Stefanie Wuhrer
Teng Xi
Jianing Xi
Fei Xia
Haifeng Xia
Menghan Xia
Yuanqing Xia
Zhihua Xia
Xiaobo Xia
Weihao Xia
Shihong Xia
Yan Xia
Yong Xia
Zhaoyang Xia
Zhihao Xia
Chuhua Xian
Yongqin Xian
Wangmeng Xiang
Fanbo Xiang
Tiange Xiang
Tao Xiang
Liuyu Xiang
Xiaoyu Xiang
Zhiyu Xiang
Aoran Xiao
Chunxia Xiao
Fanyi Xiao
Jimin Xiao
Jun Xiao
Taihong Xiao
Anqi Xiao
Junfei Xiao
Jing Xiao
Liang Xiao
Yang Xiao
Yuting Xiao
Yijun Xiao

Yao Xiao
Zeyu Xiao
Zhisheng Xiao
Zihao Xiao
Binhui Xie
Christopher Xie
Haozhe Xie
Jin Xie
Guo-Sen Xie
Hongtao Xie
Ming-Kun Xie
Tingting Xie
Chaohao Xie
Weicheng Xie
Xudong Xie
Jiyang Xie
Xiaohua Xie
Yuan Xie
Zhenyu Xie
Ning Xie
Xianghui Xie
Xiufeng Xie
You Xie
Yutong Xie
Fuyong Xing
Yifan Xing
Zhen Xing
Yuanjun Xiong
Jinhui Xiong
Weihua Xiong
Hongkai Xiong
Zhitong Xiong
Yuanhao Xiong
Yunyang Xiong
Yuwen Xiong
Zhiwei Xiong
Yuliang Xiu
An Xu
Chang Xu
Chenliang Xu
Chengming Xu
Chenshu Xu
Xiang Xu
Huijuan Xu
Zhe Xu
Jie Xu
Jingyi Xu
Jiarui Xu
Yinghao Xu
Kele Xu
Ke Xu
Li Xu
Linchuan Xu
Linning Xu
Mengde Xu
Mengmeng Frost Xu
Min Xu
Mingye Xu
Jun Xu
Ning Xu
Peng Xu
Runsheng Xu
Sheng Xu
Wenqiang Xu
Xiaogang Xu
Renzhe Xu
Kaidi Xu
Yi Xu
Chi Xu
Qiuling Xu
Baobei Xu
Feng Xu
Haohang Xu
Haofei Xu
Lan Xu
Mingze Xu
Songcen Xu
Weipeng Xu
Wenjia Xu
Wenju Xu
Xiangyu Xu
Xin Xu
Yinshuang Xu
Yixing Xu
Yuting Xu
Yanyu Xu
Zhenbo Xu
Zhiliang Xu
Zhiyuan Xu
Xiaohao Xu
Yanwu Xu
Yan Xu
Yiran Xu
Yifan Xu
Yufei Xu
Yong Xu
Zichuan Xu
Zenglin Xu
Zexiang Xu
Zhan Xu
Zheng Xu
Zhiwei Xu
Ziyue Xu
Shiyu Xuan
Hanyu Xuan
Fei Xue
Jianru Xue
Mingfu Xue
Qinghan Xue
Tianfan Xue
Chao Xue
Chuhui Xue
Nan Xue
Zhou Xue
Xiangyang Xue
Yuan Xue
Abhay Yadav
Ravindra Yadav
Kota Yamaguchi
Toshihiko Yamasaki
Kohei Yamashita
Chaochao Yan
Feng Yan
Kun Yan
Qingsen Yan
Qixin Yan
Rui Yan
Siming Yan
Xinchen Yan
Yaping Yan
Bin Yan
Qingan Yan
Shen Yan
Shipeng Yan
Xu Yan

Yan Yan
Yichao Yan
Zhaoyi Yan
Zike Yan
Zhiqiang Yan
Hongliang Yan
Zizheng Yan
Jiewen Yang
Anqi Joyce Yang
Shan Yang
Anqi Yang
Antoine Yang
Bo Yang
Baoyao Yang
Chenhongyi Yang
Dingkang Yang
De-Nian Yang
Dong Yang
David Yang
Fan Yang
Fengyu Yang
Fengting Yang
Fei Yang
Gengshan Yang
Heng Yang
Han Yang
Huan Yang
Yibo Yang
Jiancheng Yang
Jihan Yang
Jiawei Yang
Jiayu Yang
Jie Yang
Jinfa Yang
Jingkang Yang
Jinyu Yang
Cheng-Fu Yang
Ji Yang
Jianyu Yang
Kailun Yang
Tian Yang
Luyu Yang
Liang Yang
Li Yang
Michael Ying Yang
Yang Yang
Muli Yang
Le Yang
Qiushi Yang
Ren Yang
Ruihan Yang
Shuang Yang
Siyuan Yang
Su Yang
Shiqi Yang
Taojiannan Yang
Tianyu Yang
Lei Yang
Wanzhao Yang
Shuai Yang
William Yang
Wei Yang
Xiaofeng Yang
Xiaoshan Yang
Xin Yang
Xuan Yang
Xu Yang
Xingyi Yang
Xitong Yang
Jing Yang
Yanchao Yang
Wenming Yang
Yujiu Yang
Herb Yang
Jianfei Yang
Jinhui Yang
Chuanguang Yang
Guanglei Yang
Haitao Yang
Kewei Yang
Linlin Yang
Lijin Yang
Longrong Yang
Meng Yang
MingKun Yang
Sibei Yang
Shicai Yang
Tong Yang
Wen Yang
Xi Yang
Xiaolong Yang
Xue Yang
Yubin Yang
Ze Yang
Ziyi Yang
Yi Yang
Linjie Yang
Yuzhe Yang
Yiding Yang
Zhenpei Yang
Zhaohui Yang
Zhengyuan Yang
Zhibo Yang
Zongxin Yang
Hantao Yao
Mingde Yao
Rui Yao
Taiping Yao
Ting Yao
Cong Yao
Qingsong Yao
Quanming Yao
Xu Yao
Yuan Yao
Yao Yao
Yazhou Yao
Jiawen Yao
Shunyu Yao
Pew-Thian Yap
Sudhir Yarram
Rajeev Yasarla
Peng Ye
Botao Ye
Mao Ye
Fei Ye
Hanrong Ye
Jingwen Ye
Jinwei Ye
Jiarong Ye
Mang Ye
Meng Ye
Qi Ye
Qian Ye
Qixiang Ye
Junjie Ye

Sheng Ye
Nanyang Ye
Yufei Ye
Xiaoqing Ye
Ruolin Ye
Yousef Yeganeh
Chun-Hsiao Yeh
Raymond A. Yeh
Yu-Ying Yeh
Kai Yi
Chang Yi
Renjiao Yi
Xinping Yi
Peng Yi
Alper Yilmaz
Junho Yim
Hui Yin
Bangjie Yin
Jia-Li Yin
Miao Yin
Wenzhe Yin
Xuwang Yin
Ming Yin
Yu Yin
Aoxiong Yin
Kangxue Yin
Tianwei Yin
Wei Yin
Xianghua Ying
Rio Yokota
Tatsuya Yokota
Naoto Yokoya
Ryo Yonetani
Ki Yoon Yoo
Jinsu Yoo
Sunjae Yoon
Jae Shin Yoon
Jihun Yoon
Sung-Hoon Yoon
Ryota Yoshihashi
Yusuke Yoshiyasu
Chenyu You
Haoran You
Haoxuan You
Yang You
Quanzeng You
Tackgeun You
Kaichao You
Shan You
Xinge You
Yurong You
Baosheng Yu
Bei Yu
Haichao Yu
Hao Yu
Chaohui Yu
Fisher Yu
Jin-Gang Yu
Jiyang Yu
Jason J. Yu
Jiashuo Yu
Hong-Xing Yu
Lei Yu
Mulin Yu
Ning Yu
Peilin Yu
Qi Yu
Qian Yu
Rui Yu
Shuzhi Yu
Gang Yu
Tan Yu
Weijiang Yu
Xin Yu
Bingyao Yu
Ye Yu
Hanchao Yu
Yingchen Yu
Tao Yu
Xiaotian Yu
Qing Yu
Houjian Yu
Changqian Yu
Jing Yu
Jun Yu
Shujian Yu
Xiang Yu
Zhaofei Yu
Zhenbo Yu
Yinfeng Yu
Zhuoran Yu
Zitong Yu
Bo Yuan
Jiangbo Yuan
Liangzhe Yuan
Weihao Yuan
Jianbo Yuan
Xiaoyun Yuan
Ye Yuan
Li Yuan
Geng Yuan
Jialin Yuan
Maoxun Yuan
Peng Yuan
Xin Yuan
Yuan Yuan
Yuhui Yuan
Yixuan Yuan
Zheng Yuan
Mehmet Kerim Yücel
Kaiyu Yue
Haixiao Yue
Heeseung Yun
Sangdoo Yun
Tian Yun
Mahmut Yurt
Ekim Yurtsever
Ahmet Yüzügüler
Edouard Yvinec
Eloi Zablocki
Christopher Zach
Muhammad Zaigham Zaheer
Pierluigi Zama Ramirez
Yuhang Zang
Pietro Zanuttigh
Alexey Zaytsev
Bernhard Zeisl
Haitian Zeng
Pengpeng Zeng
Jiabei Zeng
Runhao Zeng
Wei Zeng
Yawen Zeng
Yi Zeng

Hua Zhang
Jiaxin Zhang
Zhenyu Zhang
Jian Zhang
Jianfeng Zhang
Jiao Zhang
Jiakai Zhang
Lefei Zhang
Le Zhang
Mi Zhang
Min Zhang
Ning Zhang
Pan Zhang
Pu Zhang
Qing Zhang
Renrui Zhang
Shifeng Zhang
Shuo Zhang
Shaoxiong Zhang
Weizhong Zhang
Xi Zhang
Xiaomei Zhang
Xinyu Zhang
Yin Zhang
Zicheng Zhang
Zihao Zhang
Ziqi Zhang
Zhaoxiang Zhang
Zhen Zhang
Zhipeng Zhang
Zhixing Zhang
Zhizheng Zhang
Jiawei Zhang
Zhong Zhang
Pingping Zhang
Yixin Zhang
Kui Zhang
Lingzhi Zhang
Huaiwen Zhang
Quanshi Zhang
Zhoutong Zhang
Yuhang Zhang
Yuting Zhang
Zhang Zhang
Ziming Zhang
Zhizhong Zhang
Qilong Zhangli
Bingyin Zhao
Bin Zhao
Chenglong Zhao
Lei Zhao
Feng Zhao
Gangming Zhao
Haiyan Zhao
Hao Zhao
Handong Zhao
Hengshuang Zhao
Yinan Zhao
Jiaojiao Zhao
Jiaqi Zhao
Jing Zhao
Kaili Zhao
Haojie Zhao
Yucheng Zhao
Longjiao Zhao
Long Zhao
Qingsong Zhao
Qingyu Zhao
Rui Zhao
Rui-Wei Zhao
Sicheng Zhao
Shuang Zhao
Siyan Zhao
Zelin Zhao
Shiyu Zhao
Wang Zhao
Tiesong Zhao
Qian Zhao
Wangbo Zhao
Xi-Le Zhao
Xu Zhao
Yajie Zhao
Yang Zhao
Ying Zhao
Yin Zhao
Yizhou Zhao
Yunhan Zhao
Yuyang Zhao
Yue Zhao
Yuzhi Zhao
Bowen Zhao
Pu Zhao
Bingchen Zhao
Borui Zhao
Fuqiang Zhao
Hanbin Zhao
Jian Zhao
Mingyang Zhao
Na Zhao
Rongchang Zhao
Ruiqi Zhao
Shuai Zhao
Wenda Zhao
Wenliang Zhao
Xiangyun Zhao
Yifan Zhao
Yaping Zhao
Zhou Zhao
He Zhao
Jie Zhao
Xibin Zhao
Xiaoqi Zhao
Zhengyu Zhao
Jin Zhe
Chuanxia Zheng
Huan Zheng
Hao Zheng
Jia Zheng
Jian-Qing Zheng
Shuai Zheng
Meng Zheng
Mingkai Zheng
Qian Zheng
Qi Zheng
Wu Zheng
Yinqiang Zheng
Yufeng Zheng
Yutong Zheng
Yalin Zheng
Yu Zheng
Feng Zheng
Zhaoheng Zheng
Haitian Zheng
Kang Zheng
Bolun Zheng

Haiyong Zheng
Mingwu Zheng
Sipeng Zheng
Tu Zheng
Wenzhao Zheng
Xiawu Zheng
Yinglin Zheng
Zhuo Zheng
Zilong Zheng
Kecheng Zheng
Zerong Zheng
Shuaifeng Zhi
Tiancheng Zhi
Jia-Xing Zhong
Yiwu Zhong
Fangwei Zhong
Zhihang Zhong
Yaoyao Zhong
Yiran Zhong
Zhun Zhong
Zichun Zhong
Bo Zhou
Boyao Zhou
Brady Zhou
Mo Zhou
Chunluan Zhou
Dingfu Zhou
Fan Zhou
Jingkai Zhou
Honglu Zhou
Jiaming Zhou
Jiahuan Zhou
Jun Zhou
Kaiyang Zhou
Keyang Zhou
Kuangqi Zhou
Lei Zhou
Lihua Zhou
Man Zhou
Mingyi Zhou
Mingyuan Zhou
Ning Zhou
Peng Zhou
Penghao Zhou
Qianyi Zhou
Shuigeng Zhou
Shangchen Zhou
Huayi Zhou
Zhize Zhou
Sanping Zhou
Qin Zhou
Tao Zhou
Wenbo Zhou
Xiangdong Zhou
Xiao-Yun Zhou
Xiao Zhou
Yang Zhou
Yipin Zhou
Zhenyu Zhou
Hao Zhou
Chu Zhou
Daquan Zhou
Da-Wei Zhou
Hang Zhou
Kang Zhou
Qianyu Zhou
Sheng Zhou
Wenhui Zhou
Xingyi Zhou
Yan-Jie Zhou
Yiyi Zhou
Yu Zhou
Yuan Zhou
Yuqian Zhou
Yuxuan Zhou
Zixiang Zhou
Wengang Zhou
Shuchang Zhou
Tianfei Zhou
Yichao Zhou
Alex Zhu
Chenchen Zhu
Deyao Zhu
Xiatian Zhu
Guibo Zhu
Haidong Zhu
Hao Zhu
Hongzi Zhu
Rui Zhu
Jing Zhu
Jianke Zhu
Junchen Zhu
Lei Zhu
Lingyu Zhu
Luyang Zhu
Menglong Zhu
Peihao Zhu
Hui Zhu
Xiaofeng Zhu
Tyler (Lixuan) Zhu
Wentao Zhu
Xiangyu Zhu
Xinqi Zhu
Xinxin Zhu
Xinliang Zhu
Yangguang Zhu
Yichen Zhu
Yixin Zhu
Yanjun Zhu
Yousong Zhu
Yuhao Zhu
Ye Zhu
Feng Zhu
Zhen Zhu
Fangrui Zhu
Jinjing Zhu
Linchao Zhu
Pengfei Zhu
Sijie Zhu
Xiaobin Zhu
Xiaoguang Zhu
Zezhou Zhu
Zhenyao Zhu
Kai Zhu
Pengkai Zhu
Bingbing Zhuang
Chengyuan Zhuang
Liansheng Zhuang
Peiye Zhuang
Yixin Zhuang
Yihong Zhuang
Junbao Zhuo
Andrea Ziani
Bartosz Zieliński
Primo Zingaretti

Nikolaos Zioulis
Andrew Zisserman
Yael Ziv
Liu Ziyin
Xingxing Zou
Danping Zou
Qi Zou
Shihao Zou
Xueyan Zou
Yang Zou
Yuliang Zou
Zihang Zou
Chuhang Zou
Dongqing Zou
Xu Zou
Zhiming Zou
Maria A. Zuluaga
Xinxin Zuo
Zhiwen Zuo
Reyer Zwiggelaar

Contents – Part XXXVIII

Talisman: Targeted Active Learning for Object Detection with Rare Classes and Slices Using Submodular Mutual Information

Suraj Kothawade[1(✉)], Saikat Ghosh[1], Sumit Shekhar[2], Yu Xiang[1], and Rishabh Iyer[1]

[1] University of Texas at Dallas, Richardson, USA
{suraj.kothawade,saikat.ghosh,yu.xiang,rishabh.iyer}@utdallas.edu
[2] Adobe Research, Bengaluru, India
sushekha@adobe.com

Abstract. Deep neural networks based object detectors have shown great success in a variety of domains like autonomous vehicles, biomedical imaging, *etc.*, however their success depends on the availability of a large amount of data from the domain of interest. While deep models perform well in terms of overall accuracy, they often struggle in performance on rare yet critical data slices. For *e.g.*, detecting objects in rare data slices like "motorcycles at night" or "bicycles at night" for self-driving applications. Active learning (AL) is a paradigm to incrementally and adaptively build training datasets with a human in the loop. However, current AL based acquisition functions are not well-equipped to mine rare slices of data from large real-world datasets, since they are based on uncertainty scores or global descriptors of the image. We propose TALISMAN, a novel framework for **T**argeted **A**ctive **L**earning for object detect**I**on with rare slices using **S**ubmodular **M**utu**A**l i**N**formation. Our method uses the submodular mutual information functions instantiated using features of the region of interest (RoI) to efficiently target and acquire images with rare slices. We evaluate our framework on the standard PASCAL VOC07+12 [8] and BDD100K [31], a real-world large-scale driving dataset. We observe that TALISMAN consistently outperforms a wide range of AL methods by ≈ 5%- - 14% in terms of average precision on rare slices, and ≈ 2%–4% in terms of mAP. The code for TALISMAN is available here: https://github.com/surajkothawade/talisman.

Keywords: Targeted active learning · Object detection · Class imbalance · Rare slices · Submodular mutual information

1 Introduction

Deep learning approaches for object detection have made a lot of progress, with accuracies improving consistently over the years. As a result, object detection

Supplementary Information The online version contains supplementary material available at https://doi.org/10.1007/978-3-031-19839-7_1.

S. Avidan et al. (Eds.): ECCV 2022, LNCS 13698, pp. 1–16, 2022.
https://doi.org/10.1007/978-3-031-19839-7_1

technology is extensively being used and deployed in applications like self-driving cars and medical imaging, and is approaching human performance. One critical aspect, though in high-stake applications like self-driving cars and medical imaging, is that the cost of failure is very high. Even a single mistake in the detection and specifically a false-negative (*e.g.*, missing a pedestrian on a highway or a motorcycle at night) can result in a major and potentially fatal accident[1].

An important aspect in such problems is that there are a number of rare yet critical slices of objects and scenarios. Because many of these rare slices are severely under-represented in the data, deep learning based object detectors often perform poorly in such scenarios. Some examples of such data slices are "motorcycles at night", "pedestrians on a highway", and "bicycles at night". Figure 1 shows the distribution of slices in the BDD100K [31] dataset. As is evident, these slices are very rare – for instance, the number of motorcycles at night, is 0.094% of the number of cars in the dataset.

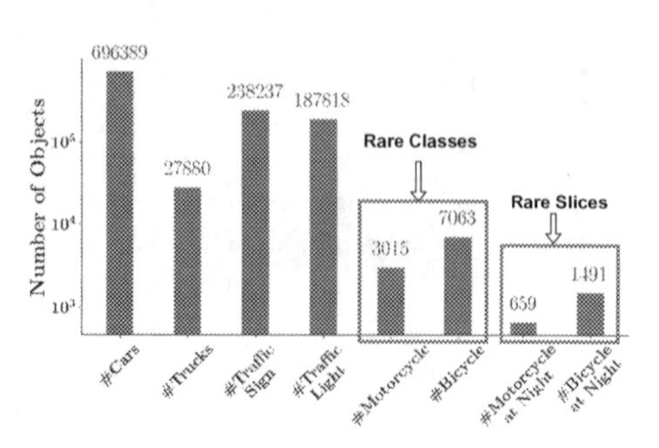

Fig. 1. Problem Statement: Rare classes and Rare slices in BDD100K [31]. Motorcycle and bicycle classes have the least number of objects, thereby making them *rare classes*, on which the model performs the worst in terms of average precision (AP). Further, motorcycle/bicycle objects at night are *rarer*, thereby making them *rare slices* on which the model performs the worst.

This causes a more pronounced issue in the limited data setting. To understand the effect of this imbalance, we trained a Faster-RCNN Model [23] on a small subset of BDD100K (5% of the dataset) and we noticed a significant difference in mAP between "cars" class (around 55% mAP) and "motorcycle" (around 9% mAP). This gap is even more pronounced for rare slices. Active learning based data sampling is an increasingly popular paradigm for training deep learning classifiers and object detectors [6,10,25,29] because such approaches significantly reduce the amount of labeled data required to achieve a certain desired accuracy. On the other hand, current active learning based paradigms are heavily dependent on aspects like uncertainty and diversity, and often miss rare slices of data. This is because such slices, though critical for the end task, are a small fraction of the full dataset, and play a negligible role in the overall accuracy.

1.1 Our Contributions

In this paper, we propose TALISMAN, a novel active learning framework for object detection, which (a) provides a mechanism to encode the similarity between an

[1] An unfortunate example of this is the self-driving car crash with Uber: https://www.theverge.com/2019/11/6/20951385/uber-self-driving-crash-death-reason-ntsb-dcouments where the self-driving car did not detect a pedestrian on a highway at night, resulting in a fatal accident.

unlabeled image and a small query set of targeted examples (e.g., images with "motorcycles at night" RoIs), and (b) mines these examples in a scalable manner from a large unlabeled set using the recently proposed submodular mutual information functions. We also provide an approach where we can mine examples based on multiple such rare slices. Similar to standard active learning, TALISMAN is an interactive human-in-the-loop approach where images are chosen iteratively and provided to a human for labeling. However, the key difference is that TALISMAN does the selection by targeting rare slices using only a few exemplars. The overview of targeted selection using TALISMAN is shown in Fig. 3.

Empirically, we demonstrate the utility of TALISMAN on a diverse set of rare slices that occur in the real-world. Specifically, we see that TALISMAN outperforms the best baseline by significant margins on different rare slices (c.f. Fig. 2).

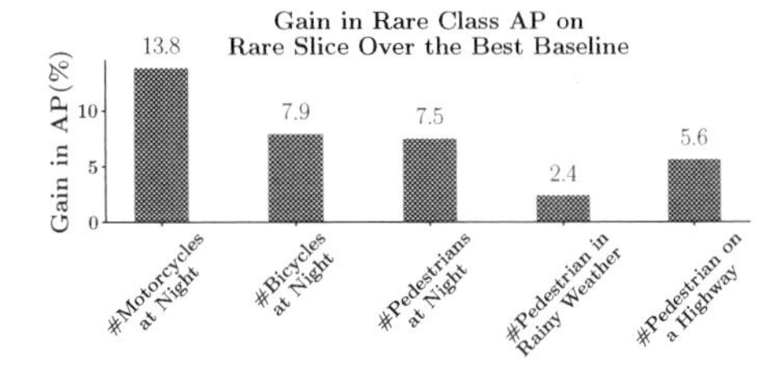

Fig. 2. Efficiency of TALISMAN over the best-performing baseline on a variety of rare slices in BDD100K

2 Related Work

A number of recent works have studied deep active learning for image classification [2,5,16,19,26,27,30]. The most common approach for active learning is to select the most uncertain examples. These include approaches like ENTROPY [27], LEAST CONFIDENCE [29], and MARGIN [24]. One challenge of this approach is that all the samples within a batch can be potentially similar, even though they are uncertain. Hence, a number of recent works have ensured that we select examples that are both uncertain and diverse. Examples include BADGE [2], FASS [30], BATCH-BALD [18], CORESET [26], and so on.

Recently, researchers have started applying active learning to the problem of object detection. [6] proposed an uncertainty sampling based approach for active object detection, while [25] proposed a 'query-by-committee' paradigm to select the most uncertain items for object detection. Recently [10], studied several scoring functions for active learning, including entropy based functions, coreset based functions, and so on. [15] proposed an active learning approach based on the localization of the detections, and studied the role of two metrics called "localization tightness" and "localization stability" as uncertainty measures. [7] studied active learning in the setting of users providing weak supervision (i.e., just suggesting the label and a rough location as opposed to drawing bounding boxes around the objects). All these approaches have shown significant labeling cost reductions and gains in accuracy compared to random sampling. However, the major limitation with these approaches (which are mostly variations of uncertainty) is that they focus on the overall accuracy, and do not necessarily try to select instances specific to certain rare yet critical data slices. To overcome these limitations, we provide a generalized paradigm for active learning in object detection, where we can target specific rare data slices.

A related thread of research is the use of the recently proposed submodular information measures [13] for data selection and active learning. [17] extended the work of [13] and proposed a general family of *parameterized submodular information measures* for guided summarization and data subset selection. [19] use the submodular information measures for active learning in the image classification setting to address realistic scenarios like imbalance, redundancy, and out-of-distribution data. Finally, [20] use the submodular information measures for personalized speech recognition. To our knowledge, this is the first work which proposes an active learning framework for object detection capable of handling rare slices of data.

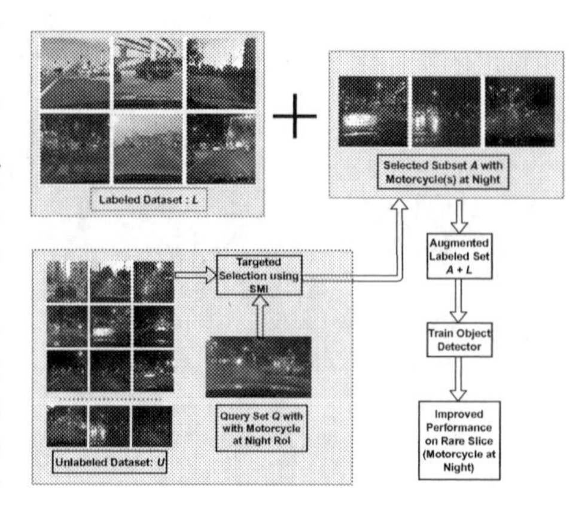

Fig. 3. Targeted Selection using TALISMAN for one round of targeted active learning. Motorcycles at night is a rare slice in the labeled data. We mine images from the unlabeled set that semantically similar to the RoIs in the query set by using the submodular mutual information (SMI) functions. These images are then labeled and added to the labeled data to improve performance on the rare slice.

3 Background

In this section, we discuss different submodular functions and their mutual information instantiations.

3.1 Submodular Functions

Submodular functions are an appealing class of functions for data subset selection in real-world applications due to their diminishing returns property and their ability to model properties of a good subset, such as diversity, representation and coverage [3,4,14,28]. Consider an unlabeled set $\mathcal{U} = \{1, 2, 3, \cdots, n\}$ and a set function $f : 2^{\mathcal{U}} \rightarrow \mathbb{R}$. Formally, f is defined to be submodular [9] if for $x \in \mathcal{U}$, $f(\mathcal{A} \cup x) - f(\mathcal{A}) \geq f(\mathcal{B} \cup x) - f(\mathcal{B}), \forall \mathcal{A} \subseteq \mathcal{B} \subseteq \mathcal{U}$ and $x \notin \mathcal{B}$. For data subset selection and active learning, a number of recent approaches [16,30] use f as an acquisition function to obtain a real-valued score for $f(\mathcal{A})$, where $\mathcal{A} \subseteq \mathcal{U}$. Given a budget $\mathcal{B}$ (the number of elements to select at every round of subset selection of batch active learning), the optimization problem is: $\max_{\mathcal{A}:|\mathcal{A}| \leq \mathcal{B}} f(\mathcal{A})$. Two examples of submodular functions that we use in this work are Facility Location (FL) and Graph Cut (GC) functions (see Table 1(a)). They are instantiated by using a similarity matrix S, that stores the similarity scores S_{ij} between any two data points i, j. The submodular functions admit a constant factor approximation $1 - \frac{1}{e}$ [22] for cardinality constraint maximization. Importantly, submodular maximization can be done in *near-linear time* using variants of greedy algorithms [21].

3.2 Submodular Mutual Information (SMI)

While submodular functions are a good choice of functions for standard active learning, in this work, we want to not only select the most informative and diverse set of points, but also select points which are *similar* to a specific target slice (typically only a few examples from a rare slice). The Submodular mutual information (SMI) functions capture this second property and are defined as $I_f(\mathcal{A}; \mathcal{Q}) = f(\mathcal{A}) + f(\mathcal{Q}) - f(\mathcal{A} \cup \mathcal{Q})$, where $\mathcal{Q}$ is a query or target set (e.g., a few sample images of "motorcycles at night"). Intuitively, maximizing the SMI functions ensure that we obtain *diverse* subsets that are *relevant* to a query set $\mathcal{Q}$. We discuss the details of the SMI functions used in our work in the next section.

3.3 Specific SMI Functions Used in TALISMAN

We adapt the mutual information variants of Facility Location (FL) and Graph Cut (GC) functions [17] for targeted active learning.

Table 1. Instantiations of different submodular functions.

(a) Instantiations of Submodular functions.

SF	$f(\mathcal{A})$
FL	$\sum_{i \in \mathcal{U}} \max_{j \in \mathcal{A}} S_{ij}$
GC	$\sum_{i \in \mathcal{A}, j \in \mathcal{U}} S_{ij} - \sum_{i,j \in \mathcal{A}} S_{ij}$

(b) Instantiations of SMI functions.

SMI	$I_f(\mathcal{A}; \mathcal{Q})$
FLMI	$\sum_{i \in \mathcal{Q}} \max_{j \in \mathcal{A}} S_{ij} + \sum_{i \in \mathcal{A}} \max_{j \in \mathcal{Q}} S_{ij}$
GCMI	$2 \sum_{i \in \mathcal{A}} \sum_{j \in \mathcal{Q}} S_{ij}$

Facility Location: The FL function models representation (i.e., it picks the most representative points or "centroids"). The FL based SMI function called FLMI can be written as $I_f(\mathcal{A}, \mathcal{Q}) = \sum_{i \in \mathcal{Q}} \max_{j \in \mathcal{A}} S_{ij} + \sum_{i \in \mathcal{A}} \max_{j \in \mathcal{Q}} S_{ij}$ [17]. This function models representation as well as query relevance.

Graph Cut: The GC function models diversity and representation, and has modeling properties similar to FL. The SMI variant of GC is defined as GCMI, which maximizes the pairwise similarity between the query set and the unlabeled set. The GCMI function can be written as $I_f(\mathcal{A}; \mathcal{Q}) = 2 \sum_{i \in \mathcal{A}} \sum_{j \in \mathcal{Q}} S_{ij}$.

Table 1(a) and (b) demonstrate the SMI functions we will use in this work and the corresponding submodular functions instantiating them. Note that in [13,17], a number of other SMI functions and instantiations have been proposed. However, keeping scalability to large datasets in mind (see Sect. 4.4), we only focus on these two.

4 TALISMAN: Our Targeted Active Learning Framework for Object Detection

4.1 TALISMAN Framework

In this section, we present TALISMAN, our targeted active learning framework for object detection. We show that TALISMAN can efficiently target any imbalanced scenario with rare classes or rare slices. We summarize our method in Algorithm 1, and illustrate it in Fig. 4. The core idea of our framework lies within instantiating the SMI functions such that they can mine for images from the unlabeled set which contain proposals semantically similar to the region of interests (RoIs) in the query set. The query set contains exemplars of the rare slice that we want to target.

Algorithm 1. TALISMAN: Targeted AL Framework for Object Detection (Illustration in Fig. 4)

Require: Initial labeled set of data points: $\mathcal{L}$, large unlabeled dataset: $\mathcal{U}$, small query set $\mathcal{Q}$, object detection model $\mathcal{M}$, batch size: B, number of selection rounds: N.
1: **for** selection round $i = 1 : N$ **do**
2: Train model $\mathcal{M}$ on the current labeled set $\mathcal{L}$ and obtain parameters θ_i
3: Compute $S \in \mathbb{R}^{|\mathcal{Q}| \times |\mathcal{U}|}$ such that: $S_{qu} \leftarrow$ TARGETEDSIM$(\mathcal{M}_{\theta_i}, \mathcal{I}_q, \mathcal{I}_u)$, $\forall q \in \mathcal{Q}, \forall u \in \mathcal{U}$ {Algorithm 2}
4: Instantiate a submodular function f based on S.
5: $\mathcal{A}_i \leftarrow \text{argmax}_{\mathcal{A} \subseteq \mathcal{U}, |\mathcal{A}| \leq B} I_f(\mathcal{A}; \mathcal{Q})$ {Greedy maximization of SMI function to select a subset $\mathcal{A}$}
6: Get labels $L(\mathcal{A}_i)$ for batch $\mathcal{A}_i$ and $\mathcal{L} \leftarrow \mathcal{L} \cup L(\mathcal{A}_i)$, $\mathcal{U} \leftarrow \mathcal{U} - \mathcal{A}_i$
7: **end for**
8: **Return** trained model $\mathcal{M}$ and parameters θ.

We start with training an object detection model $\mathcal{M}$ on an initial labeled set $\mathcal{L}$. Using $\mathcal{M}$, we compute embeddings of the query set $\mathcal{Q}$ and the unlabeled set $\mathcal{U}$. Next, we compute pairwise cosine similarity scores $S_{qu}, \forall q \in \mathcal{Q}, \forall u \in \mathcal{U}$ to obtain a similarity matrix $S \in \mathbb{R}^{|\mathcal{Q}| \times |\mathcal{U}|}$. We discuss the details of computing S_{qu} for a single query image q and a single unlabeled image u in Sect. 4.2. Using the similarity matrix S, we instantiate the SMI function $I_f(\mathcal{A}; \mathcal{Q})$ as discussed in Sect. 3 (note that both the SMI functions we consider in this work are similarity based functions). Finally,

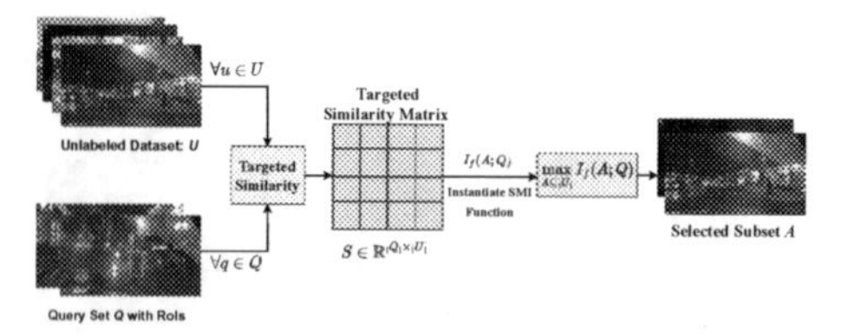

Fig. 4. Architecture of TALISMAN during one round of targeted active learning. We illustrate the targeted similarity computation in Fig. 5.

we acquire a subset $\mathcal{A}$ that contains regions that are semantically similar to the RoI in $\mathcal{Q}$ by maximizing the SMI function $I_f(\mathcal{A};\mathcal{Q})$:

$$\max_{\mathcal{A}\subseteq\mathcal{U},|\mathcal{A}|\leq B} I_f(\mathcal{A};\mathcal{Q}). \quad (1)$$

Since this function is submodular (i.e. $I_f(\mathcal{A};\mathcal{Q})$ is submodular in $\mathcal{A}$ for a fixed query set $\mathcal{Q}$), we use a greedy algorithm [22] (to solve Eq. (1) and Line 5 in Algorithm 1) which ensures a $1-\frac{1}{e}$ approximation guarantee of the optimal solution.

4.2 Targeted Similarity Computation

We summarize our method for targeted similarity computation in Algorithm 2 and illustrate it in Fig. 5. For simplicity, consider a single query image $\mathcal{I}_q \in \mathcal{Q}$ with T RoIs (targets) indicating a rare slice, and an unlabeled image $\mathcal{I}_u \in \mathcal{U}$ with P region proposals obtained using a region proposal network (RPN). Using $\mathcal{M}$ that is trained on $\mathcal{L}$, we compute the embedding of the RoIs in $\mathcal{I}_q$ to obtain $\mathcal{E}_q \in \mathbb{R}^{T\times D}$, and for the proposals of $\mathcal{I}_u$ to obtain $\mathcal{E}_u \in \mathbb{R}^{P\times D}$. Here, D denotes the dimensionality of each feature vector representing a RoI or region proposal. We use the embeddings $\mathcal{E}_q$ and $\mathcal{E}_u$ to represent $\mathcal{I}_q$ and $\mathcal{I}_u$ respectively. We use these embeddings to compute the targeted similarity (see Algorithm 2).

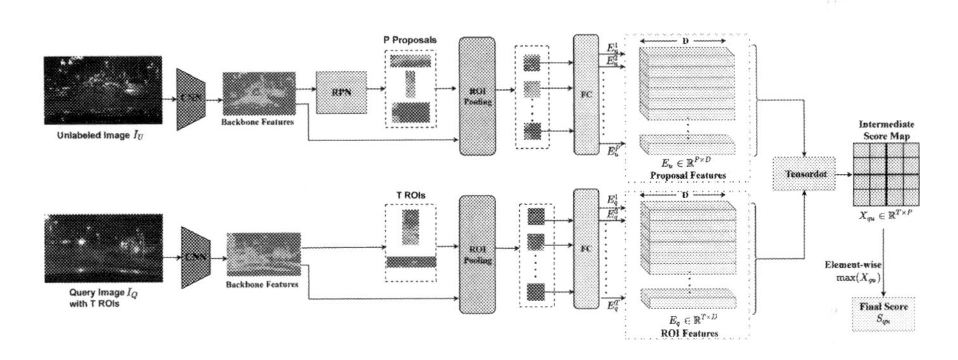

Fig. 5. TARGETEDSIM: Targeted Similarity computation in TALISMAN.

In order to compute cosine similarity between $\mathcal{E}_q$ and $\mathcal{E}_u$ efficiently, we L2-normalize along the feature dimension of length D. This enables us to highly parallelize the similarity computation via off-the-shelf GPU enabled dot product[2] implementations. Next, we compute the dot product along the feature dimension to obtain pairwise similarities between T RoIs in $\mathcal{I}_q$ and the P proposals in $\mathcal{I}_u$ which gives us RoI-proposal score map $\mathcal{X}_{qu} \in \mathbb{R}^{T\times P}$. Finally, we assign the similarity score S_{qu} between $\mathcal{I}_q$ and $\mathcal{I}_u$ by computing the *element-wise* maximum of $\mathcal{X}_{qu}$, which entails the best matching proposal of the P region proposals to some query RoI in the T RoIs.

[2] See torch.tensordot.

Algorithm 2. TARGETEDSIM: Targeted Similarity Matching (Illustration in Fig. 5)

Require: Local feature extraction model F_θ, $\mathcal{I}_q \in \mathcal{Q}$ with T RoIs and $\mathcal{I}_u \in \mathcal{U}$ with P region proposals.
1: $\mathcal{E}_q \leftarrow F_\theta(\mathcal{I}_q)$ $\{\mathcal{E}_q \in \mathbb{R}^{T\times D}\}$
2: $\mathcal{E}_u \leftarrow F_\theta(\mathcal{I}_u)$ $\{\mathcal{E}_u \in \mathbb{R}^{P\times D}\}$
3: $\mathcal{X}_{qu} \leftarrow$ COSINE_SIMILARITY$(\mathcal{E}_q, \mathcal{E}_u)$ $\{\mathcal{X}_{qu} \in \mathbb{R}^{T\times P}$. Compute Cosine similarity along the feature dimension$\}$
4: $S_{qu} \leftarrow \max(\mathcal{X}_{qu})$ {Element-wise Max, S_{qu} represents the score between the best matching proposal $j \in P$ to some query RoI $i \in T$}
5: **Return** Similarity score S_{qu}

4.3 Using TALISMAN to Mine Rare Slices

A critical input to TALISMAN (Algorithm 1) is the query set $\mathcal{Q}$. The query set consists of a specific target slice, which could be a rare class (e.g. "motorcycles") or a rare slice ("motorcycles at night"). In our experiments, we study the role of TALISMAN for both scenarios. For our setting to be realistic, we need to ensure that $\mathcal{Q}$ is tiny – since these are rare slices, we cannot assume that we have access to numerous of these rare examples. For this reason, we set $\mathcal{Q}$ to be between 2 and 5 examples in our experiments. It is worth noting that since the SMI functions naturally model relevance to the query set and diversity within the selected subset, they pick a diverse set of data points which are relevant to the query set $\mathcal{Q}$.

4.4 Scalability of TALISMAN

A key factor in the efficiency of TALISMAN is the choice of SMI functions FLMI and GCMI. The memory and time complexity of computing the similarity kernel for both these functions is only $|\mathcal{Q}| \times |\mathcal{U}|$ – since $\mathcal{Q}$ is a tiny held-out set of the examples from the rare slice (of size 2 to 5), the time complexity of creating and storing the FLMI and GCMI functions is only $O(\mathcal{U})$. For the greedy algorithm [22], we use memoization [12]. This ensures that the complexity of computing the gains for both FLMI and GCMI functions is in fact $O(|\mathcal{Q}|$, which is a constant, so the amortized complexity using the lazy greedy algorithm is $|\mathcal{U}| \log |\mathcal{U}|$. We can also use the lazier than lazy greedy algorithm [21], which ensures that the worst case complexity of the greedy algorithm is only $|\mathcal{U}|$. As a result, both FLMI and GCMI can be optimized in linear time (with respect to the size of the unlabeled set), thereby ensuring that TALISMAN can scale to very large datasets.

5 Experimental Results

In this section, we empirically evaluate the effectiveness of TALISMAN for a wide range of real-world scenarios where the dataset has one or more rare classes or rare slices. We do so by comparing the performance of TALISMAN instantiated SMI functions (Table 1(b)) with existing active learning approaches using a wide variety of metrics, namely the mean average precision (mAP), average precision

(AP) of the rare slice, and the number of data points selected that belong to the rare slice. We summarize all notations used in Appendix. A.

5.1 Experimental Setup

We apply TALISMAN for object detection tasks on two diverse public datasets: 1) the standard PASCAL VOC07+12 (VOC07+12) [8] and 2) BDD100K [31], a large scale driving dataset. VOC07+12 has 16,551 images in the training set and 4,952 images in the test set which come from the test set of VOC07. BDD100K consists of 70K images in the training set, 10K images in the validation set and 20K images in the test set. Since the labels for the test set are not publicly available, we use the validation set for evaluation. For active learning (AL), we split the training set into the labeled set $\mathcal{L}$ and unlabeled set $\mathcal{U}$.

Since the problem of targeted active learning is more about sampling *objects* semantically similar to a region of interest, we create the initial seed set for AL by randomly sampling images such that an *object-level* budget is satisfied for each class. This allows us to simulate multiple scenarios with rare classes or rare slices. In the following sections, we provide the individual splits for $\mathcal{L}, \mathcal{U}$, and $\mathcal{Q}$ in each rare class or rare slice scenario.

In all the AL experiments discussed below, we use a common training procedure and hyperparameters to ensure fair comparison across all acquisition functions. We use standard data augmentation techniques like random flips followed by normalization. For all experiments on both datasets, we train a Faster RCNN model [23] based on a ResNet50 backbone [11]. In each round of AL, we reinitialize the model parameters, and train the model for 150 epochs using SGD with momentum. The initial learning rate is set to 0.001 with a step size of 3, the momentum and weight decay are set to 0.9 and 0.0005 respectively. For comparing multiple acquisition functions in the AL loop, we start with an identical model that is trained on the initial labeled set $\mathcal{L}$. All the experiments were run 5$\times$ on a V100 GPU and the error bars (std deviation) are reported.

5.2 Baselines in All Scenarios

We compare TALISMAN instantiated SMI functions with multiple AL baselines: namely ENTROPY [10,27], Targeted Entropy (T-ENTROPY), Least Confidence (LEAST-CONF) [29], MARGIN [24], FASS [30], CORESET [26], BADGE [2] and RANDOM sampling. Below, we discuss the details for each baseline:

Entropy [10,27]**:** We compute the entropy for each region proposal of a specific class by using the probability scores generated by the model $\mathcal{M}$. This entropy is computed as follows:

$$\mathcal{H}(R_c) = -R_c \log R_c - (1 - R_c) \log(1 - R_c), \tag{2}$$

where R_c represents the probability for class c at the region proposal R. We set the number of region proposals $P = 300$ in our experiments. We compute the final entropy score s of an unlabeled image $\mathcal{I}_u$ by taking the maximum across all C classes for each proposal followed by an average across all proposals as follows:

$$s = \underset{R}{avg} \max_{c \in C} \mathcal{H}(R_c). \tag{3}$$

Targeted Entropy: In order to encourage ENTROPY sampling to select more points relevant to the query set, we make ENTROPY *target-aware* by following a two-step process. First, we select the top-K data points with maximum entropy. Next, we compute a $|\mathcal{Q}| \times K$ similarity matrix using these top-K uncertain data points and the query set $\mathcal{Q}$ (as done in lines 3–5 of Algorithm 1). Finally, we select the top-B samples from these top-K samples that have a region semantically similar to some RoI in $\mathcal{Q}$. We refer to this method as Targeted Entropy (T-ENTROPY). We set $K > B$ in our experiments so that T-ENTROPY has enough samples to choose from.

Least Confidence [29]**:** Least confidence (LEAST-CONF) is another intuitive acquisition function based on uncertainty. We compute the LEAST-CONF score s of each data point by averaging over the minimum predicted class probability of P region proposals as follows:

$$s = \underset{R}{avg} \min_{c \in C} R_c \tag{4}$$

Intuitively, we select the bottom B data points that have the smallest predicted class probability scored by s.

Margin [24]**:** For MARGIN sampling, we score each data point by averaging over the difference between the top two predicted class probabilities of P region proposals as follows:

$$s = \underset{R}{avg} \min_{c1, c2 \in C} R_{c1} - R_{c2} \tag{5}$$

We use the score s to select B data points that have the least difference in the probability score of the first and the second most probable labels.

FASS [30]**:** In order to encourage uncertainty and diversity using FASS, we first select top $K \times B$ uncertain data points using ENTROPY. Finally, we select top-B data points using the facility location (FL) submodular function. We use FL since it performs the best in [30].

Coreset [26]**:** CORESET is a diversity based approach that selects core-sets such that the geometric arrangement of the superset is maintained. The core-sets are acquired using a greedy k-center clustering approach. In our experiments, we use the features of the last convolutional layer to represent each data point.

Badge [2]**:** BADGE proposes to select diverse and uncertain data points that have a high gradient magnitude. The gradients are computed using hypothesized labels and distanced from previously selected data points using K-MEANS++ [1]. In our experiments, we use the gradients from the penultimate convolutional layer of the ResNet50 backbone of the Faster RCNN model.

Random: For RANDOM, we select B data points randomly.

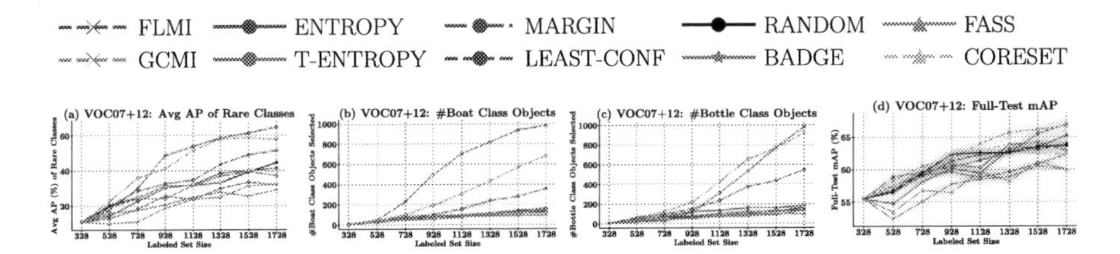

Fig. 6. Active Learning with rare classes on VOC07+12. Plot (a) shows the average AP of the rare classes, plots (b–c) show the number of boat and bottle objects selected respectively, plot (d) shows the mAP on the VOC07+12 test set. We observe that the SMI functions (FLMI, GCMI) outperform other baselines by ≈ 8%–10% average AP of the rare classes.

5.3 Rare Classes

Dataset Setting: We conduct the experiments for the rare classes scenario on the VOC07+12 dataset. In particular, we create the initial labeled set, $\mathcal{L}$ which simulates the rare classes by creating a class imbalance at an object level. Let $\mathcal{C}_i^{\mathcal{L}}$ be the number of objects from a rare (infrequent) class i and $\mathcal{B}_j^{\mathcal{L}}$ be the number of objects from a frequent class j. The initial labeled set $\mathcal{L}$ is created such that the imbalance ratio between $\mathcal{C}_i^{\mathcal{L}}$ and $\mathcal{B}_j^{\mathcal{L}}$ is *at least* ρ, *i.e.*, $\rho \leq (\mathcal{B}_j^{\mathcal{L}}/\mathcal{C}_i^{\mathcal{L}})$. All the remaining data points are used in the unlabeled set $\mathcal{U}$. In our experiments, we choose two classes to be rare from VOC07+12: 'boat' and 'bottle'. We do so due to two reasons: 1) they are by default the most uncommon objects in VOC, thereby making them the natural choice, and 2) they are comparatively smaller objects than other classes like 'sofa', 'chair', 'train', *etc.*. We use a small query set $\mathcal{Q}$ containing 5 randomly chosen data points representing the rare classes (RoIs). We construct the initial labeled set by setting $\rho = 10$, $|\mathcal{C}^{\mathcal{L}}| = 20$ and $|\mathcal{B}^{\mathcal{L}}| = 2858$. This gives us an initial labeled seed set of size $|\mathcal{L}| = 1143$ images. Note that the imbalance ratio is not exact because objects of some classes are predominantly present in most images, thereby increasing the size of $|\mathcal{B}^{\mathcal{L}}|$.

Results: In Fig. 6, we compare the performance of TALISMAN on the rare classes scenario in VOC07+12 [8]. We observe that TALISMAN significantly outperforms all state-of-the art uncertainty based methods (ENTROPY, LEAST-CONF, and MARGIN) by ≈ 8%–10% (Fig. 6(a)) in terms of average precision (AP) on the rare classes and by ≈ 2%–3% in terms of mAP (Fig. 6(d)). This improvement is performance is because the TALISMAN instantiated functions (GCMI and FLQMI) are able to select more data points that contain regions with objects belonging to the rare classes (see Fig. 6(c)). Interestingly, the TALISMAN functions were also able to give a fair treatment to multiple rare classes at the same time by selecting significant number of objects belonging to both the rare classes ('boat' and 'bottle', see Fig. 6(b, c)). This suggests that TALISMAN is able to select diverse data points by appropriately targeting regions containing rare class objects in the query image.

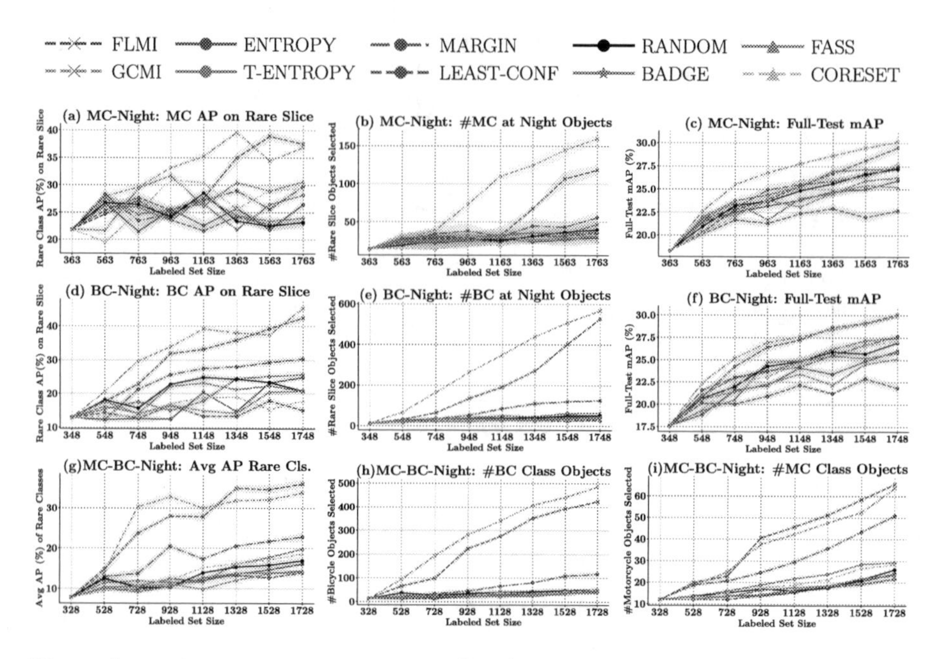

Fig. 7. Active Learning with Motorcycle (**MC**) at Night (top row) and Bicycle (**BC**) at Night (bottom row) rare slices on BDD100K. Left side plots (a, d, g) show the AP of the rare class on the rare slice of data, center plots (b, e) show the number of objects selected that belong to the rare slice, and right side plots (c, f) show the mAP on the full test set of BDD100K. We observe that the SMI functions (FLMI, GCMI) outperform other baselines by $\approx 5\%$– 14% AP of the rare class on the rare slice. In (h, i), we show that TALISMAN selects more objects from multiple rare slices in comparison to the existing methods.

5.4 Rare Slices

Dataset Setting: We chose BDD100K [31] since it is a realistic, large, and challenging dataset that allows us to evaluate the performance of TALISMAN on datasets with naturally occurring rare slices. Since we want to evaluate rare slices, the procedure to simulate the initial labeled set and the evaluation is slightly different from the rare classes experiment in the above section. In the following sections, we discuss experiments where the initial labeled set $\mathcal{L}$ has a rare slice made of a *class* and an *attribute*. For instance, *motorcyles* (class) at *night* (attribute), *pedestrians* (class) in *rainy weather* (attribute), *etc.*. Let $|\mathcal{O}_c^A|$ be the number of objects in $\mathcal{L}$ that belong to class c and attribute A. Concretely, we create a balanced initial labeled set $\mathcal{L}$, such that each class c contributes an equal number of objects, *i.e.*$|\mathcal{O}_{c1}| = |\mathcal{O}_{c2}|, \forall c1, c2$. Let i be the class involved in the rare slice. We simulate the rare slice by creating an imbalance in $\mathcal{O}_i$ based on an attribute A such that the ratio between the number of objects of class i with attribute A and the ones without attribute A (denoted by $\tilde{\mathbb{A}}$) is *at least* ρ, *i.e.*$\rho \leq (|\mathcal{O}_i^{\tilde{\mathbb{A}}}|/|\mathcal{O}_i^A|)$. In all the rare slice experiments, we start with an initial

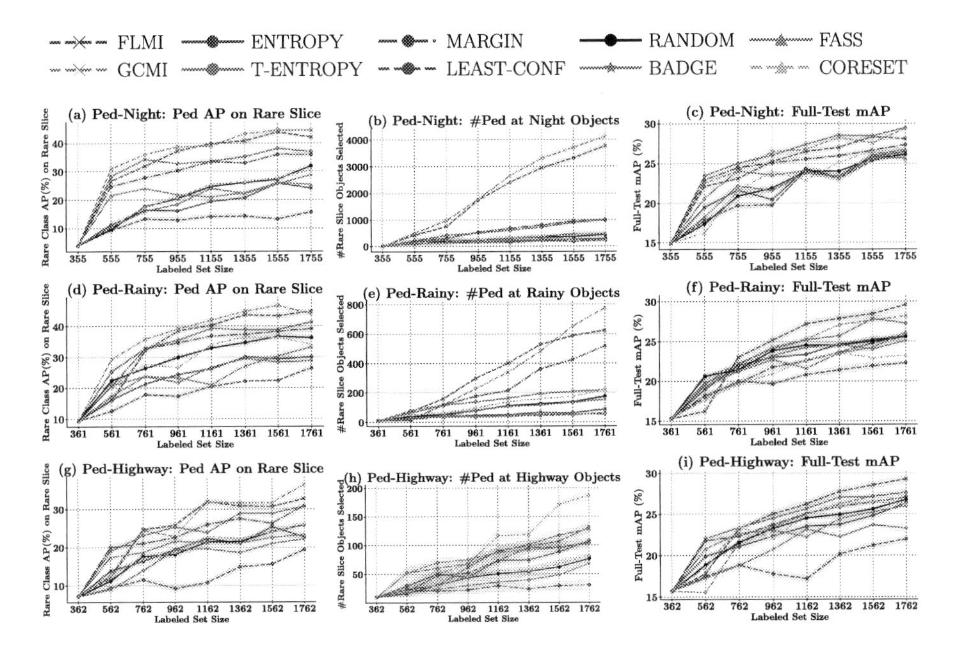

Fig. 8. AL with Pedestrian (**Ped**) at Nighttime (top row), Pedestrian in Rainy Weather (middle row), and Pedestrian on a Highway (bottom row) rare slices on BDD100K. Left side plots (a, d, g) show the AP of the rare class on the rare slice of data, center plots (b, e, h) show the number of objects selected that belong to the rare slice, and right side plots (c, f, i) show the mAP on the full test set of BDD100K. We observe that the SMI functions (FLMI, GCMI) outperform other baselines by ≈ 5%–10% AP of the pedestrian class on the rare slice.

labeled set by setting $\rho = 10, |\mathcal{O}_i^{\tilde{\mathbb{A}}}| = 100$, and $|\mathcal{O}_i^{A}| = 10$. For all other classes j, we randomly pick objects such that $|\mathcal{O}_j| = 110$. Note that we use a small query set in all experiments (≈ 3–5 images). The exact number of images in $\mathcal{L}$ and $\mathcal{Q}$ for each experiment is given in Appendix. B. For evaluation, we compare the performance of TALISMAN using three metrics: 1) *Rare Class Rare Slice AP*: the average precision (AP) of the 'rare class' (*e.g.motorcycle*) on the 'rare slice' (*e.g.night*), 2) *# Rare Slice Objects*: the number of objects selected that belong to the rare slice, and 3) *Overall Test mAP*: the mAP on the complete test set.

Motorcycle *or* Bicycle at Night Rare Slice Results: We show the results for the 'motorcycle at night' and 'bicycle at night' rare slices in Fig. 7(top and middle row). We observe that the TALISMAN outperforms other baselines by ≈ 5%–14% AP of the rare class on the rare slice (see Fig. 7(a, d)), and by ≈ 2%–4% (see Fig. 7(c, f)) in terms of mAP on the full test set. The gain in AP of the rare class and mAP increases in the later rounds of active learning, since the embedding representation of the model improves. Specifically, GCMI outperforms all methods since it models query-relevance well by selecting many rare class objects that belong to the rare slice (see Fig. 7(b, e)). In Fig. 9, we qualitatively show an example false negative motorcycle at night fixed using TALISMAN.

Motorcycle *and* Bicycle at Night Rare Slice Results: We show the results for a scenario with multiple rare slices: 'motorcycle and bicycle at night' (Fig. 7(bottom row)). Importantly, we observe that Talisman selects more number of objects from both the rare slices in comparison to existing methods (see Fig. 7(h, i)). This is critical in real-world scenarios, since there are often cases with multiple co-occuring rare slices.

Pedestrian at Night *or* Rainy *or* Highway Rare Slice Results: To study the robustness of Talisman in diverse real-world scenarios, we evaluate its performance for the 'pedestrian' rare class on multiple attributes - 1) 'night', 2) 'rainy', and 3) 'highway' (see Fig. 8). We observe consistent performance of both the Talisman instantiated functions (Gcmi and Flmi) across all scenarios. Concretely, we show that our framework can robustly find more pedestrians than any other baseline *across all rare slices* (see Fig. 8(b, e, h)), which leads to a performance gain of $\approx$ 5%–10% AP over existing baselines for the pedestrian class on the rare slice. This reinforces the need for a framework like Talisman for improving the performance of object detectors on such rare slices.

Fig. 9. A false negative motorcycle at night (left) fixed to a true positive detection (right) using Talisman.

6 Conclusion

In this paper, we present a targeted active learning framework Talisman that enables improving the performance of object detection models on rare classes and slices. We showed the utility of our framework across a variety of real-world scenarios with one or more rare classes and slices on the PASCAL VOC07+12 and BDD100K driving dataset, and observe a $\approx$ 5%–14% gain compared to the existing baselines. Moreover, Talisman can select objects belonging to *multiple* co-occuring rare slices and simultaneously improve their performance, which is critical for modern object detectors. The main limitation of our work is the requirement of a reasonable feature embedding for computing similarity.

Acknowledgments.. This work is supported by the National Science Foundation under Grant No. IIS-2106937, a startup grant from UT Dallas, and by a Google, Adobe, and Amazon research award, and an Adobe data science award.

References

1. Arthur, D., Vassilvitskii, S.: k-means++: the advantages of careful seeding. In: SODA 2007: Proceedings of the eighteenth annual ACM-SIAM symposium on Discrete algorithms, pp. 1027–1035. Society for Industrial and Applied Mathematics, Philadelphia (2007)
2. Ash, J.T., Zhang, C., Krishnamurthy, A., Langford, J., Agarwal, A.: Deep batch active learning by diverse, uncertain gradient lower bounds. arXiv preprint arXiv:1906.03671 (2019)
3. Bach, F.: Learning with submodular functions: a convex optimization perspective. arXiv preprint arXiv:1111.6453 (2011)
4. Bach, F.: Submodular functions: from discrete to continuous domains. Math. Program. **175**(1), 419–459 (2019)
5. Beck, N., Sivasubramanian, D., Dani, A., Ramakrishnan, G., Iyer, R.: Effective evaluation of deep active learning on image classification tasks. arXiv preprint arXiv:2106.15324 (2021)
6. Brust, C.A., Käding, C., Denzler, J.: Active learning for deep object detection. arXiv preprint arXiv:1809.09875 (2018)
7. Desai, S.V., Chandra, A.L., Guo, W., Ninomiya, S., Balasubramanian, V.N.: An adaptive supervision framework for active learning in object detection. arXiv preprint arXiv:1908.02454 (2019)
8. Everingham, M., Van Gool, L., Williams, C.K., Winn, J., Zisserman, A.: The pascal visual object classes (VOC) challenge. Int. J. Comput. Vision **88**(2), 303–338 (2010)
9. Fujishige, S.: Submodular functions and optimization. Elsevier (2005)
10. Haussmann, E., et al.: Scalable active learning for object detection. In: 2020 IEEE Intelligent Vehicles Symposium (IV), pp. 1430–1435. IEEE (2020)
11. He, K., Zhang, X., Ren, S., Sun, J.: Deep residual learning for image recognition. In: Proceedings of the IEEE Conference on Computer Vision and Pattern Recognition, pp. 770–778 (2016)
12. Iyer, R., Bilmes, J.: A memoization framework for scaling submodular optimization to large scale problems. In: The 22nd International Conference on Artificial Intelligence and Statistics, pp. 2340–2349. PMLR (2019)
13. Iyer, R., Khargoankar, N., Bilmes, J., Asanani, H.: Submodular combinatorial information measures with applications in machine learning. In: Algorithmic Learning Theory, pp. 722–754. PMLR (2021)
14. Iyer, R.K.: Submodular optimization and machine learning: Theoretical results, unifying and scalable algorithms, and applications. Ph.D. thesis (2015)
15. Kao, C.-C., Lee, T.-Y., Sen, P., Liu, M.-Y.: Localization-aware active learning for object detection. In: Jawahar, C.V., Li, H., Mori, G., Schindler, K. (eds.) ACCV 2018. LNCS, vol. 11366, pp. 506–522. Springer, Cham (2019). https://doi.org/10.1007/978-3-030-20876-9_32
16. Kaushal, V., Iyer, R., Kothawade, S., Mahadev, R., Doctor, K., Ramakrishnan, G.: Learning from less data: a unified data subset selection and active learning framework for computer vision. In: 2019 IEEE Winter Conference on Applications of Computer Vision (WACV), pp. 1289–1299. IEEE (2019)
17. Kaushal, V., Kothawade, S., Ramakrishnan, G., Bilmes, J., Iyer, R.: Prism: a unified framework of parameterized submodular information measures for targeted data subset selection and summarization. arXiv preprint arXiv:2103.00128 (2021)
18. Kirsch, A., Van Amersfoort, J., Gal, Y.: Batchbald: efficient and diverse batch acquisition for deep bayesian active learning. Adv. Neural. Inf. Process. Syst. **32**, 7026–7037 (2019)

19. Kothawade, S., Beck, N., Killamsetty, K., Iyer, R.: Similar: submodular information measures based active learning in realistic scenarios. arXiv preprint arXiv:2107.00717 (2021)
20. Kothyari, M., Mekala, A.R., Iyer, R., Ramakrishnan, G., Jyothi, P.: Personalizing ASR with limited data using targeted subset selection. arXiv preprint arXiv:2110.04908 (2021)
21. Mirzasoleiman, B., Badanidiyuru, A., Karbasi, A., Vondrák, J., Krause, A.: Lazier than lazy greedy. In: Proceedings of the AAAI Conference on Artificial Intelligence, vol. 29 (2015)
22. Nemhauser, G.L., Wolsey, L.A., Fisher, M.L.: An analysis of approximations for maximizing submodular set functions-i. Math. Program. **14**(1), 265–294 (1978)
23. Ren, S., He, K., Girshick, R., Sun, J.: Faster R-CNN: towards real-time object detection with region proposal networks. Adv. Neural. Inf. Process. Syst. **28**, 91–99 (2015)
24. Roth, D., Small, K.: Margin-based active learning for structured output spaces. In: Fürnkranz, J., Scheffer, T., Spiliopoulou, M. (eds.) ECML 2006. LNCS (LNAI), vol. 4212, pp. 413–424. Springer, Heidelberg (2006). https://doi.org/10.1007/11871842_40
25. Roy, S., Unmesh, A., Namboodiri, V.P.: Deep active learning for object detection. In: BMVC, vol. 362, p. 91 (2018)
26. Sener, O., Savarese, S.: Active learning for convolutional neural networks: a core-set approach. arXiv preprint arXiv:1708.00489 (2017)
27. Settles, B.: Active learning literature survey (2009)
28. Tohidi, E., Amiri, R., Coutino, M., Gesbert, D., Leus, G., Karbasi, A.: Submodularity in action: from machine learning to signal processing applications. IEEE Signal Process. Mag. **37**(5), 120–133 (2020)
29. Wang, D., Shang, Y.: A new active labeling method for deep learning. In: 2014 International Joint Conference on Neural Networks (IJCNN), pp. 112–119. IEEE (2014)
30. Wei, K., Iyer, R., Bilmes, J.: Submodularity in data subset selection and active learning. In: International Conference on Machine Learning, pp. 1954–1963. PMLR (2015)
31. Yu, F., Xian, W., Chen, Y., Liu, F., Liao, M., Madhavan, V., Darrell, T.: Bdd100k: A diverse driving video database with scalable annotation tooling. arXiv preprint arXiv:1805.04687 2(5), 6 (2018)

An Efficient Person Clustering Algorithm for Open Checkout-free Groceries

Junde Wu[1], Yu Zhang[2], Rao Fu[2], Yuanpei Liu[3], and Jing Gao[1(✉)]

[1] Purdue University, West Lafayette, IN 47907, USA
jinggao@purdue.edu
[2] Harbin Institute of Technology, Harbin 150080, China
[3] Beijing Institute of Technology, Beijing 102213, China

Abstract. Open checkout-free grocery is the grocery store where the customers never have to wait in line to check out. Developing a system like this is not trivial since it faces challenges of recognizing the dynamic and massive flow of people. In particular, a clustering method that can efficiently assign each snapshot to the corresponding customer is essential for the system. In order to address the unique challenges in the open checkout-free grocery, we propose an efficient and effective person clustering method. Specifically, we first propose a Crowded Sub-Graph (CSG) to localize the relationship among massive and continuous data streams. CSG is constructed by the proposed Pick-Link-Weight (PLW) strategy, which **picks** the nodes based on time-space information, **links** the nodes via trajectory information, and **weighs** the links by the proposed von Mises-Fisher (vMF) similarity metric. Then, to ensure that the method adapts to the dynamic and unseen person flow, we propose Graph Convolutional Network (GCN) with a simple Nearest Neighbor (NN) strategy to accurately cluster the instances of CSG. GCN is adopted to project the features into low-dimensional separable space, and NN is able to quickly produce a result in this space upon dynamic person flow. The experimental results show that the proposed method outperforms other alternative algorithms in this scenario. In practice, the whole system has been implemented and deployed in several real-world open checkout-free groceries.

Keywords: Open checkout-free groceries · Person clustering · Graph convolutional network

1 Introduction

Traditional checkout-free groceries are in small closed venues with limited commodities. Customers are required to register upon entry, which may cause privacy and security issues. Recently, a concept of open checkout-free grocery is

J. Wu and Y. Zhang—Equally contributed.

Supplementary Information The online version contains supplementary material available at https://doi.org/10.1007/978-3-031-19839-7_2.

S. Avidan et al. (Eds.): ECCV 2022, LNCS 13698, pp. 17–33, 2022.
https://doi.org/10.1007/978-3-031-19839-7_2

proposed to address the existing issues. Open checkout-free grocery allows free entry of consumers without registering customer information. Customers can walk in such grocery stores just like what they do in the traditional supermarkets while enjoying the benefit of automatic checkout. To achieve this goal, it is essential to automatically identify the customers in the grocery, which is a very challenging task in an open environment. The number of needed identifications is not accessible beforehand, and the identification process needs to work continuously and steadily for every customer. Therefore, many tools in computer vision to associate/track a person, e.g., human tracking, person re-identification, or face verification, cannot be directly applied in this scenario. Instead, a person clustering component is essential to cluster the customers based on features extracted from their video snapshots. Implementing an effective clustering algorithm in this scenario is non-trivial. Among many challenges, below we discuss the two major challenges of this problem.

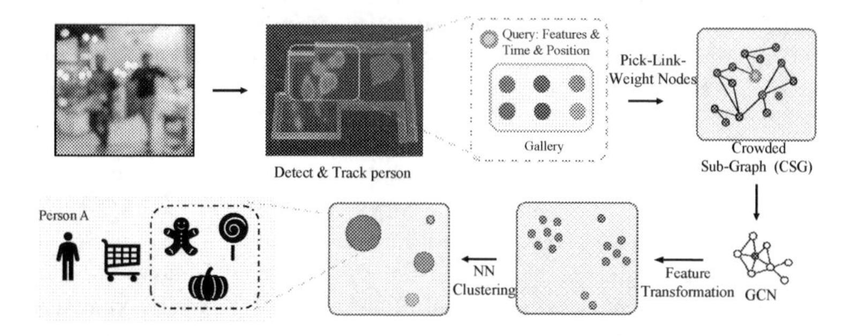

Fig. 1. Overview of the person clustering algorithm in open checkout-free groceries.

The first challenge comes from the difficulty of handling data streams [2], which are the input of the person clustering algorithm. The proposed clustering algorithm has to work over massive, unbounded sequences of data objects that are continuously generated at rapid rates. In addition, the algorithm must react immediately when a *query* comes, under an extremely harsh condition where the observed data stream is generally too large to store and too expensive to access.

The second challenge is that person clustering in this scenario is an open-world problem [5]. The data that the system needs to process is dynamic and unknown person flow. Therefore, the system is required to continuously distinguish and memorize the unseen new customer once an individual comes and could recognize the person in any other locations and views since then.

For the first challenge, the commonly applied method is to cluster locally on the summarized data (statistic summaries of data streams) [2]. In this paper, we propose a Crowded Sub-Graph (CSG) to efficiently collect the local clusters and provide the divergence of the summarized data. CSG constructs a local relationship network for each incoming *query* through the Pick-Link-Weight (PLW) strategy. Specially, we leverage time-space information to **pick** the nodes, and utilize trajectory information to **link** the nodes, and propose a novelty von Mises-Fisher (vMF) similarity metric to **weigh** the links. With the CSG, we can compare locally to accelerate the system without degrading performance.

The open-world nature of the scenario leads to dynamic changes in the distribution of person features. Thus, some complex clustering techniques may consume too much time, while some simple techniques may lead to performance degradation. In this paper, we propose Graph Convolutional Network (GCN) with a simple Nearest Neighbor (NN) strategy to accurately cluster the instances of CSG. GCN is adopted to project the features into low-dimensional separable space, and NN is able to quickly produce a result in this space upon dynamic person flow. To our knowledge, our work is the first towards a comprehensive strategy to identify a person in this data-stream & open-world environment.

In brief, the contributions of the paper are as follows.

- We are the first to define the People Clustering task in an open checkout-free grocery scenario, and we propose an effective framework to address this important problem. As the first research report of this scene, we believe it will be an essential starting point for future research and practical applications.
- We propose Crowded Sub-Graph (CSG), a local relationship graph constructed by PLW strategy. PLW strategy can model the distance of nodes through the lens of the probability distribution, so as to construct a subgraph that fairly represents the relevance of local nodes.
- Given CSG as input data, we apply Graph Convolutional Network (GCN) on it following a simple NN (Nearest Neighbor) strategy to cluster the nodes, which is able to quickly adapt to the dynamic person distribution in groceries. Experimental results show the proposed algorithm considerably outperforms best alternative methods.

2 Related Works

2.1 Data-stream Clustering

For the last decade, we have seen an increasing interest in managing the data streams [2,14]. Clustering on data streams requires a process to continuously cluster data objects within memory and time restrictions [14].

In the data abstraction step, the data structures to summarize the data are also diversely proposed for the different tasks [25], like feature vectors [40], prototype arrays [10], coreset trees [1], and data grids [20]. However, most of them are bound with particular clustering methods, which narrows their applications.

In the clustering step, many k-means variants have been presented to deal with summarized features to cluster in data streams in real-time. Bradley et al. [6] proposed Scalable k-means, which uses the CF vectors of the processed and new data objects as input to find k clusters. The ClusTree algorithm [20] proposes to use a weighted CF vector, which is kept into a hierarchical tree (R-tree family). These well-designed data-stream clustering methods, on the one hand, to date, are limited to Euclidean spaces, and on the other hand, hard to take a balance between efficiency and performance.

2.2 GCN on Clustering

Recently, with a part of the various applications of deep neural networks [16,32–34,36], Graph Convolutional Network (GCN) has shown outstanding performance on data clustering. GCN can extract high-level node representations, thus simplifying the sensitive discrimination step [35].

To apply GCN on the data which naturally has the graph structure seems straightforward, such as graph-based recommendation systems [27], point clouds classification [21], and molecular properties prediction [22], etc. However, the graph nature of some other data may not be so explicit. In this case, researchers have to construct the graph of the data. For example, [39] addressed the traffic prediction problem using STGNNs. [29] applied the GCN to text classification based on the syntactic dependency tree of a sentence. [30] proposed Instance Pivot Subgraph (IPS) to construct the sub-graph for person face features. However, despite the outstanding performance of GCN on data clustering, there is still a research gap between GCN and data-stream clustering.

3 Preliminary

3.1 Data-stream Clustering

A data stream S is a massive sequence of data objects $x_1, x_2, ..., x_K$, that is, $S = \{x_i\}_{i=1}^K$, which is potentially unbounded ($K \rightarrow \infty$). Each data object is described by a n-dimensional attribute vector $x_i = [x_i^j]_{j=1}^n$ belonging to an attribute space Ω that can be continuous, categorical, or mixed.

It is impossible to store and get access to each data object in the data stream. Developing suitable data structures to store statistic summaries of data streams is indispensable in the data-stream clustering tasks. Cluster Feature vector (CF vector) is a commonly used data structure for summarizing large amounts of data. The CF vector has three components: K, the number of data objects, LS, the linear sum of the data objects, and SS, the sum of squared data objects. The structures LS and SS are n-dimensional vectors. These three components allow to compute cluster measures, such as cluster mean μ and radius σ (Eq. (1)).

$$\mu = \frac{LS}{K}, \ \sigma = \sqrt{\left(\frac{SS}{K} - \left(\frac{LS}{K}\right)^2\right)}, \tag{1}$$

where $(\cdot)^2$ and $\sqrt{\cdot}$ represent element-wise square and square root. Obviously, the three components of the CF vector have incrementality and additivity properties, which make the CF vector widely used in clustering (More details can be found in the supplementary material, or [25]).

In the open checkout-free groceries, a complete *person record* p is represented as a set (Eq. (2))

$$p = \{\{\tilde{t}_i\}_{i=1}^K, \{z_i\}_{i=1}^K, \{v_i\}_{i=1}^K, CF[K]\}, \tag{2}$$

where $\tilde{t}_i = t_i^s + t_i^e$, t_i^s and t_i^e are the time stamps of the person appeared in the camera view and left the camera view, respectively. z_i is a two dimensional point records the plane coordinates of the camera, v_i is a two-dimensional normalized vector to denote the direction of the pedestrian's walking. It has $v_i = (\cos(\theta), \sin(\theta))$, where θ is the angle between the last straight pedestrian path in the camera and the horizontal line. The pedestrian path is got from the move path of the bounding box. $CF[K]$ represents incremented CF vectors of person features updated K times. $K \geq 1$ is the number of pieces of data incremented in p. Each coming data was a *query*, which is represented as:

$$q = \{\tilde{t}, z, v, CF[1]\}, \tag{3}$$

where $CF[1]$ represents the initial tracked person features.

4 Methodology

The complete abstracted features in open checkout-free grocery combined several different sensors, and the vision system is one of the most important parts. The visual features are obtained through a flow of object tracking [24], person detection [7], image deblurring [31], image enhancement [41] and deep learning-based person feature abstraction [3]. When the camera captures a customer, it tracks the customer until the individual is out of view. We would sample several frames from this track and use a person detection algorithm to abstract a set of images of the person. These pictures will then be sent to the pre-trained neural networks to abstract a set of visual features. These visual features, combined with the appeared time-space information and person walking track, will be sent to the clustering algorithm to be identified. We call this piece of data sent to the clustering algorithm a *query*.

The overflow of our algorithm contains two steps, which are shown in Fig. 1. In the first step, when a *query* comes, we construct a Crowded Sub-Graph for this *query*. CSG contains N nodes, and one node is the *query*, and the other N-1 nodes are *person records*. The N-1 *person records* are **picked** depending on the time-space constraint to assume a person is impossible to appear in a far place in a short time. Then we **link** the nodes depending on the person walking track and **weight** these links based on the vMF divergence of the person features. In the second step, we propose Graph Convolutional Network (GCN) with a simple Nearest Neighbor (NN) strategy to accurately cluster the instances of CSG. GCN is adopted to project the features into low-dimensional separable space, and NN is able to quickly produce a result in this space upon dynamic person flow.

4.1 Crowded Sub-Graph

In this section, we introduce Crowded Sub-Graph (CSG) to construct a graph based on the raw data of open checkout-free groceries. Constructing a CSG contains three steps: to **pick** the nodes, **link** the nodes, and **weight** the links.

In this way, CSG constructed a local sub-graph for the *query*. The association of each pair of nodes in the sub-graph can be well represented (Fig. 2).

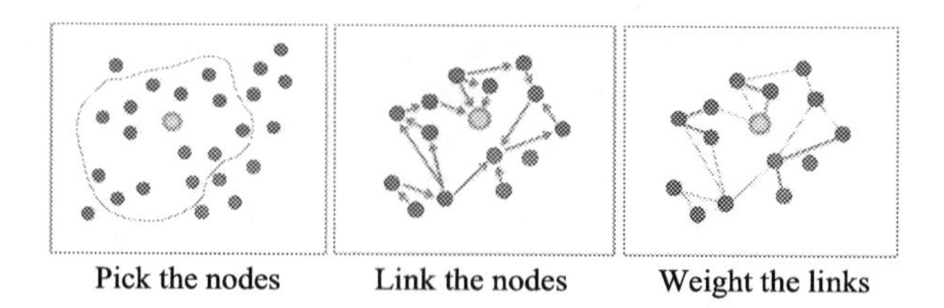

Fig. 2. Construct Crowded Sub-Graph through the proposed Pick-Link-Weight (PLW) strategy

Pick the Nodes. Our algorithm is required to react immediately once a *query* comes. However, clustering globally on the nearly unbounded data stream, i.e., clustering the *query* based on all of the recorded person in the grocery, is impossible. Fortunately, a customer would not move far within the two captures of the cameras. This allows us to cluster locally based on time and space constraints. Considering a *query* q captured by a camera at location z^q in time t^q, other *person records* that have a smaller time-space distance with the *query* q would have a higher possibility to be contained to the sub-graph. For a *person record* p with time and location sets $\{\tilde{t}_i^p\}_{i=1}^{K^p}$ and $\{z_i^p\}_{i=1}^{K^p}$, we define the time-space distance between p and q is:

$$\chi(p;q) = \min \sqrt{\frac{E(z_i^p, z^q)^2}{s^2} + (\tilde{t}_i^p - \tilde{t}^q)^2},\ i \in [1, K^p], \tag{4}$$

where E denotes Euclidean distance, s is the standard human walking speed, which is set as 3 mi per hour. Then we can collect the nodes of CSG. The number of the nodes we collect depends on the setting size of CSG.

Link the Nodes. In the graph, two linked nodes usually mean that they are related to each other to some extent. In the open checkout-free grocery scenario, we assume two nodes are linked if they are on the same trajectory. Practically, we adopt an attentive dot-product mechanism [28] for the classification of the trajectories.

As shown in Fig. 3, the recorded information of each pair of picked nodes are treated as inputs of the attention mechanism, including the position z_i and pose v_i of person relative to cameras and time stamps t_i. To predict whether Node a is on the trajectory of node b, a QKV dot-product is adopted to activate node b value matrix based on the calculated affinity map of two nodes. A binary classification head is applied after the attention to get the probability of the prediction. The binary classification head consists of a Global Average Pooling (GAP) layer and the Multilayer Perceptron (MLP) layer. For each piece of recorded information in the node, we use the position vector, pose vector, and timestamp to constitute the input embedding. Specially, we concatenate the position vector and pose vector, then add a temporal embedding on it. The temporal embedding is learned from the time stamp, followed by [8]. Formally, the sequential input embedding of trajectory predictor is represented as:

$$[z_i; v_i] + TE(t_i), i \in [1, K_1 + K_2], \tag{5}$$

where K_1 and K_2 are the numbers of recorded information of two nodes. $TE(\cdot)$ represents the temporal embedding. We apply a trajectory predictor on each pair of nodes to get the final link relationship of the sub-graph.

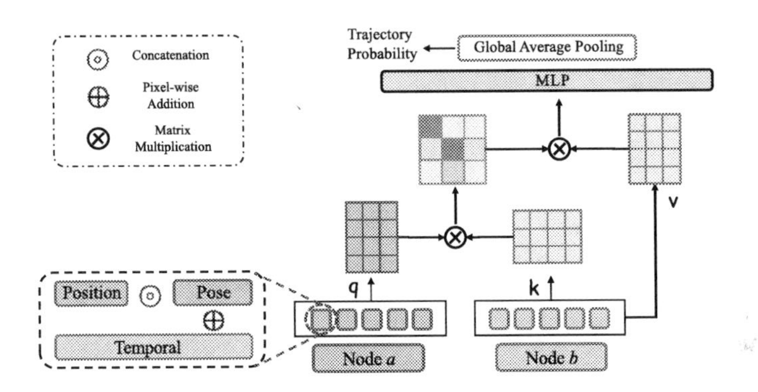

Fig. 3. Trajectory predictor which predicts whether two nodes are on the same trajectory.

Weight the Link & vMF Similarity of CF Vectors. After we **link** the nodes in the graph, we then **weight** these links by measuring how strong the linked pair is associated. Specially, we propose a similarity distance for the CF vectors based on vMF distribution and use this distance to generate the weights.

Note that each node in CSG is represented as a CF vector. Since the centroid and radius of the node can be easily computed from CF vector, previous work often represented the distance of two nodes by the discrepancy of two normal distributions. For example, a probabilistic model on the CF vector can be represented as a Gaussian model $\mathcal{N}(\mu, \sigma^2)$, where mean μ and variance σ can be easily through Equation (1). Then, we can **weight** the link of node a and node b through the distance of $\mathcal{N}(\mu_a, \sigma_a^2)$ and $\mathcal{N}(\mu_b, \sigma_b^2)$.

However, such a practice implicitly assumes the distance of the features can be fairly represented by Euclidean distance, which is invalid for the high-dimensional person features in our hand. The person features abstracted by the neural networks have intrinsic angular distribution because of the softmax loss in the neural networks [12]. Thus the probability distributions on Euclidean space, like Gaussian distribution, are invalid on this d-sphere. In this paper, we use von Mises–Fisher (vMF) distribution instead to model these clusters [4]. The von Mises–Fisher distribution is an isotropic distribution over the d-dimensional unit hypersphere, which can fairly represent the distribution high-dimensional person feature. It has mean direction μ and concentration κ, and the probability density function of it for the d-dimensional unit vector x is given by:

$$f(\mu,\kappa)=C_d(\kappa)e^{\kappa\mu^T x}, \tag{6}$$

where $f(\cdot)$ represents the probability distribution, $\kappa \geq 0$, $\| \mu \|= 1$, and the normalization constant $C_p(\kappa)$, is equal to

$$C_d(\kappa)=\frac{\kappa^{d/2-1}}{(2\pi)^{d/2}I_{d/2-1}(\kappa)}, \tag{7}$$

where I_v denotes the modified Bessel function of the first kind at order v. In vMF distribution, μ denotes the mean direction, and κ denotes the concentration. The greater the value of κ, the higher the concentration of the distribution around the mean direction μ. When it is needed, we can follow [26] to get the numerical solutions of κ and I_v.

Consider two vMF probability distributions of Node a and Node b are $f(\theta_a)$ and $f(\theta_b)$, and $f(\theta_{ab})$ denotes vMF distribution of their merged features. We **weight** the link l_{ab} by:

$$e^{-\frac{1}{2}(D_{JS}[f(\theta_a),f(\theta_{ab})]+D_{JS}[f(\theta_b),f(\theta_{ab})])}, \tag{8}$$

where D_{JS} denotes Jensen–Shannon divergence [13]:

$$D_{JS}[f(\theta_a),f(\theta_{ab})]=\frac{1}{2}(D_{KL}(f(\theta_a))\|f(\theta_{ab})+D_{KL}(f(\theta_{ab}))\|f(\theta_a)), \tag{9}$$

where D_{KL} denotes Kullback–Leibler divergence.

To get the analytic solution of the KL divergence of two vMF distribution is challenging. For the benefit of the computation and graph sparsification, we **weight** the links of similar feature distributions and set the others as zero. We consider the parameters of a distribution are close to one another, so that:

$$f(\theta_{ab})=f(\theta_a)+\sum_j \Delta\theta^j \left.\frac{\partial f}{\partial \theta^j}\right|_{\theta_a}, \tag{10}$$

where θ^j represents a small change of θ in the j direction. Then Kullback–Leibler divergence $D_{\mathrm{KL}}[v(\theta_a)\|v(\theta_b)]$ has the second order Taylor Expansion in $\theta=\theta_0$ of the form

$$D_{\mathrm{KL}}[f(\theta_a)\|f(\theta_{ab})]=\frac{1}{2}\sum_{jk}\Delta\theta^j\Delta\theta^k g_{jk}(\theta_a)+\mathrm{O}(\Delta\theta^3), \tag{11}$$

in which

$$g_{jk}(\theta)=\int_X \frac{\partial \log f(\theta)}{\partial \theta_j}\frac{\partial \log f(\theta)}{\partial \theta_k} f(\theta)\,dx. \tag{12}$$

In our case, we have the parameter $\theta = (\kappa, \mu^T)^T$. Substituting Eq. (12) for the given parameters, we can get

$$\begin{aligned} g_{\kappa,\kappa}(\kappa,\kappa) =& \tau_\kappa(\kappa,\mu),\ g_{\kappa,\mu}(\kappa,\mu) = \tau_{\kappa\mu}(\kappa)\mu^T, \\ g_{\mu,\kappa}(\mu,\kappa) =& \tau_{\kappa\mu}(\kappa)\mu,\ g_{\mu,\mu}(\mu,\mu) = \tau_\mu(\kappa)\mu\mu^T, \end{aligned} \tag{13}$$

in which

$$\begin{aligned} \tau_\kappa(\kappa,\mu) = & [e^{\kappa\mu^T x}(\kappa\mu^T - 1)]^2 + [\frac{d-2}{\kappa} - \frac{(I_{\frac{d}{2}-2}(\kappa) + I_{\frac{d}{2}}(\kappa))}{I_{\frac{d}{2}-1}(\kappa)}] \\ & + x[e^{\kappa\mu^T x}(\kappa\mu^T - 1)] + [\frac{d}{2} - \frac{2(I_{\frac{d}{2}-2}(\kappa) + I_{\frac{d}{2}}(\kappa))}{\kappa I_{\frac{d}{2}-1}(\kappa)} - 1]^2, \end{aligned} \tag{14}$$

$$\tau_{\kappa\mu}(\kappa) = \frac{(d-2)I_{\frac{d}{2}-1}(\kappa) - \kappa(I_{\frac{d}{2}-2}(\kappa) - I_{\frac{d}{2}}(\kappa))}{2(I_{\frac{d}{2}-1}(\kappa))^2}, \tag{15}$$

$$\tau_\mu(\kappa) = \kappa^2. \tag{16}$$

Let us say $f(\theta_a)$ has parameters κ and μ, and $f(\theta_{ab})$ has parameters $\bar{\kappa}$ and $\bar{\mu}$. Then, according Eq. (11), it has

$$\begin{aligned} D_{\mathrm{KL}}[f(\theta_a)\|f(\theta_{ab})] &= \frac{1}{2}\sum_{jk} \Delta\theta^j \Delta\theta^k f_{jk}(\theta_a) \\ &= \frac{1}{2}[\tau_\kappa(\bar{\kappa},\bar{\mu})(\bar{\kappa}-\kappa)^2 + 2\tau_{\kappa\mu}(\bar{\kappa})\bar{\mu}^T(\bar{\kappa}-\kappa)(\bar{\mu}-\mu) \\ &\quad + \tau_\mu(\bar{\kappa})(\bar{\mu}-\mu)\bar{\mu}\bar{\mu}^T(\bar{\mu}-\mu)]. \end{aligned} \tag{17}$$

After constructing CSG, we can get an adjacency matrix A. Non-zero values in the adjacency matrix indicate the existence of links between nodes. The values are normalized, and the sum of each row or column is equal to 1.

4.2 Graph Convolution Network

CSG is highly valuable to identify the nodes. To leverage this, we adopt a graph convolution network (GCN) to perform reasoning on CSG. Specifically, in order to adapt to the changing person features distribution caused by the open-world challenge, we use GCN to project the features into a linearly separable low dimensional space. With a simple Nearest Neighbor strategy adopted after then for the discrimination, the method can achieve high precision clustering results.

The input of the GCN is the original node feature matrix together with the adjacency matrix A. The output is the projected node feature matrix with the same number of input features but with lower dimensions. Each graph convolution layer is inputted by the last node feature matrix together with the adjacency matrix and outputs the next node feature matrix. Formally, for the lth layer, we have the following equation:

$$\mathbf{X}^{l+1} = \sigma([\mathbf{X}^l | \mathbf{\Lambda}^{-\frac{1}{2}}\mathbf{A}\Lambda^{-\frac{1}{2}}\mathbf{X}^l]\mathbf{W}_l), \tag{18}$$

where $\mathbf{X}^l \in \mathbb{R}^{n \times d_l}$ is the node feature matrix inputted to the lth layer. n represents the number of nodes, and d^l represents the feature's dimension of the lth layer. $\mathbf{\Lambda}$ is a diagonal matrix with $\mathbf{\Lambda}_{ii} = \sum_j \mathbf{A}_{ij}$. $\sigma(\cdot)$ is an nonlinear activation function. $\mathbf{W}_l \in \mathbb{R}^{2d_l \times d_{l+1}}$ is a layer-specific trainable weight matrix for the convolution layer. $\mathbf{X}^{l+1} \in \mathbb{R}^{n \times d_{l+1}}$ is the output node feature matrix of the layer.

GCN is supervised to project the nodes of the same person to be closer and others to be more distant. Toward that end, We adopt triplet loss [23] to our case. Denote the outputted features matrix as $\mathbf{O} \in \mathbb{R}^{n \times d_{out}}$, and each feature for one node as $o \in \mathbb{R}^{d_{out}}$. Given a crowded graph containing h people, and each person corresponding c node in the graph, we denote the feature of kth node of ith person as o_i^k, then our loss is given by:

$$\mathcal{L}_{GCN} = \sum_{i=1}^{h} \sum_{k=1}^{c_i} \max(log \sum_{r=1, r \neq k}^{c_i} e^{D(o_i^k, o_i^r)} + log \sum_{j=1, j \neq i}^{h} \sum_{s=1}^{c_j} e^{m - D(o_i^k, o_j^s)}, 0), \tag{19}$$

where m is the least margin that the projected feature is closer to the same class than the different classes. D is the distance measure, which is implemented by cosine similarity here.

In the inference stage, we use a simple Nearest Neighbor (NN) clustering strategy on the projected features. The NN has two types, the NN-A (Assign a new *query* to the existing nodes) and NN-M (Merge existing nodes). We set different distance thresholds ξ_A and ξ_M for the NN-A and NN-M, respectively. And the distance measure is set to be as same as that in the training stage. NN-A will assign the new *query* to the nearest *person records* if their distance is lower than ξ_A. If this distance is alternatively higher than the threshold ξ_A, the *query* will be left as an outlier, which represents a new *person record.* The outliers will be processed as nodes when being contained in the other CSG.

NN-M merges two *person records* together if their distance is lower than ξ_M, which is used to make up the early stage division problem (predict one customer as multiple people). In this data-stream clustering process, with the continuous arriving of the queries, a node has the chance to be actives in many different sub-graphs. This process helps to gradually attach the global information, then correct the previous errors caused by over-conservative clustering strategy.

5 Experiments

5.1 Data and Evaluation Metric

It should be noted that all the customers we collected information from assigned the informed agreement before going into the grocery. In addition, all data points used in the clustering stage are obfuscated high-dimensional features, which do not contain any customers' personal information.

Table 1. Ablation study setup. 'vMF' and 'Cos' represent applying vMF similarity and Cosine similarity in the corresponding element, respectively. The 'NN-A' represents Nearest Neighbor clustering which Assigns a *query* to existing nodes. The 'NN-M' represents Nearest Neighbor clustering which Merges existing nodes

Model \ Element	Time-Space&Track	CSG	GCN	NN-A	NN-M
Baseline	-	-	-	Cos	-
TS	√	-	-	Cos	-
TS-M	√	-	-	Cos	Cos
TS-M-vMF	√	-	-	vMF	vMF
Cos-GCN	√	Cos	√	Cos	Cos
CSG-GCN	√	vMF	√	Cos	Cos

Table 2. Ablation study results. P : BCubed Precision, R : BCubed Recall, $F = \frac{2PR}{P+R}$, T : Time (Seconds per one thousand quires)

	DaiCOFG				IseCOFG			
	P	R	F	T	P	R	F	T
Baseline	84.46	73.28	78.47	46.61	86.62	84.21	85.40	26.78
TS	91.25	80.16	85.35	**8.27**	93.31	84.02	88.42	**7.32**
TS-M	91.22	84.39	87.67	9.13	96.02	89.30	92.54	7.95
TS-M-vMF	95.75	93.70	94.71	11.06	97.47	94.51	95.97	8.72
Cos-GCN	96.83	96.37	96.60	11.81	97.73	96.91	97.32	8.96
CSG-GCN	**98.77**	**98.24**	**98.50**	13.68	**99.07**	**98.75**	**98.91**	10.54

Table 3. Comparison with SOTA. P : BCubed Precision, R : BCubed Recall, $F = \frac{2PR}{P+R}$, T : Time (Seconds per one thousand quires, omitted if $> 50\,\text{s/ptq}$)

	DaiCOFG				IseCOFG			
	P	R	F	T	P	R	F	T
Scalable-kmeans	84.49	79.36	81.84	16.68	86.63	84.36	85.48	16.15
ClusTree	91.81	85.73	88.67	14.00	93.41	90.64	91.87	14.00
ClusterGCN	94.55	90.75	92.61	–	94.12	91.46	92.77	–
AffinityGCN	97.05	96.82	96.93	–	98.47	97.83	98.14	–
GraphSaint	95.84	92.34	94.05	46.62	96.16	95.67	95.91	28.79
IPS-GCN	96.32	94.17	95.23	**11.49**	97.15	96.30	96.72	**9.69**
GLCN	96.86	94.96	95.90	12.37	97.82	96.51	97.16	10.37
CSG-GCN	**98.77**	**98.24**	**98.50**	13.68	**99.07**	**98.75**	**98.91**	10.54

To evaluate our method in different scenes, we establish two datasets collected from a large grocery and a smaller grocery, respectively. The recurring identities across training and test set are removed to avoid bias. The dataset we collected from a large grocery is called DaiCOFG. DaiCOFG contains $362,300$ snapshots with $10,176$ identities for training, in which $125,378$ snapshots are labeled, and each identity contains at least one snapshot labeled. DaiCOFG contains $250,710$ labeled snapshots with $7,406$ identities for testing. The snapshots are taken by 186 cameras deployed at the key spots of the grocery. The dataset we collected from the smaller grocery is called IseCOFG. IseCOFG contains $78,630$ snapshots with $4,116$ identities for training, in that $21,648$ snapshots are labeled, each identity contains at least one snapshot labeled. IseCOFG contains $54,606$ snapshots with $2,773$ people for testing. The snapshots are taken by 76 cameras in the grocery.

To evaluate the performance of the proposed algorithm, we adopt the mainstream BCubed evaluation metrics [11,30]. Denote ground truth label and predicted label as a y and a y' respectively, the pairwise correctness is represented as:

$$Correct(i,j) = \begin{cases} 1 & y_i = y_j \ and \ y'_i = y'_j \\ 0 & otherwise \end{cases}, \tag{20}$$

If the ith *query* and the jth *query* belong to the same customer during clustering and labeling, we can get $Correct(i,j) = 1$. The BCubed Precision P and BCubed Recall R are respectively defined as:

$$P = \mathbb{E}_i[\mathbb{E}_{j:y'_i=y'_j}[Correct(i,j)]], \ R = \mathbb{E}_i[\mathbb{E}_{j:y_i=y_j}[Correct(i,j)]], \tag{21}$$

When taking both precision and recall into consideration, BCubed F-measure is defined as $F = \frac{2PR}{P+R}$. To evaluate the algorithms' speed, we record the seconds per one thousand quires (s/ptq) as the metric in the comparisons.

5.2 Experiment Setting

The variables in **link** operation are pre-trained on the training set of the selected dataset of the experiment. In the training stage, it is supervised by binary cross-entropy loss function with a mini-batch of 64 for 80 epochs using ADAM algorithm [18]. The learning rate is set to 0.01. We set the number of convolution layers in our GCN to 3. The number of units in the graph convolution network's hidden layer is set to 256, 128, and 64. The number of the nearby nodes is set as 256. We train GCN for a maximum of 120 epochs (training iterations) using an ADAM algorithm with a learning rate 0.01, and stop training if the validation loss does not decrease for 10 consecutive epochs, as suggested in work [19]. All the network weights θ are initialized using Glorot initialization [15]. The thresholds ξ_A and ξ_M are set as 0.91 and 0.88, respectively.

The experiments are run on the server cluster with 16 CPU: Intel Xeon Gold 5120, 256GB memory, and 8 GPU: NVIDIA Tesla P40.

5.3 Ablation Study

To show the advantages of the proposed components, we do comprehensive ablation studies. The setup and results of the ablation study are shown in Tables 1 and 2.

As shown in Table 2, the raw NN-A strategy based only on the features' cosine similarity (Baseline) takes a long time and gets a lower recall. Applying time-space constraint and track information (TS) helps to narrow the search range since we only compare the nodes under time-space constraints and linked through track information, which speeds up the algorithm a lot with 6.88% and 3.02% F1 score improvement on DaiCOFG and IseCOFG respectively. Applying NN-M strategy (TS-M) actually turns the method to a k-means liked algorithm. It can be seen that, NN-M strategy helps to significantly improve the recall with a slight degradation of the precision. Further applying vMF-based divergence in NN-A and NN-M (TS-M-vMF) significantly improves both the precision and the recall. Also, the processing time increased by the extra consumption. Applying GCN (Cos-GCN) helps to improve the recall, due to the low-dimensional features can be better discriminated in the early stage. Further replacing cosine similarity by vMF-based divergence to construct the graph (Cos-GCN→CSG-GCN) turns it to the proposed algorithm. It can be seen that vMF weight strategy improves 1.90% and 1.59% F1 score on DaiCOFG and IseCOFG, respectively.

From the comparison of TS-M/Cos-GCN with TS-M-vMF/CSG-GCN, it can be seen that modeling the person features by vMF distribution significantly outperforms cosine similarity based pointwise comparison. Comparing TS-M with Cos-GCN, the combination of proposed CSG and GCN shows better precision and recall even without vMF distribution modeling. This shows the effectiveness of the proposed CSG when facing the dynamic and unseen person flow. In general, from Table 2, we can see that the proposed algorithm CSG-GCN gets both high performance and processes in the tolerable time on the dataset.

5.4 Comparison with Alternative Methods

In this part, to verify the effectiveness of the proposed method, we compare the proposed method with a wide range of alternative clustering methods, including non-learning-based and learning-based methods. Since those methods are not designed for our scene, and most of the learning-based methods are oriented to closed data sets, we have to adapt some methods for the comparison.

Most previous methods toward data-stream clustering are not learning-based. We compare our method with scalable k-means [6] and ClusTree [20], which are two commonly used methods in the data-stream clustering tasks. Scalable k-means employs different mechanisms to identify objects that need to be retained in memory. It stores data objects in a buffer in the main memory. By utilizing CF vectors, it discards objects that were previously statistically summarized into the buffer. When the block is full, an extended version of k-means is executed over the stored data. ClusTree algorithm [20] proposed to use a weighted CF vector, which is kept into a hierarchical tree (R-tree family). ClusTree provides strategies for dealing with time constraints for anytime clustering, that is, the possibility of interrupting the process of inserting new objects in the tree at any moment.

Recently, GCN has been proved an efficient method for clustering tasks [37]. However, few works applied GCN to the data-stream clustering. We adapted ClusterGCN [9], AffinityGCN [37], GraphSaint [38], IPS (Instance Pivot Subgraph)-GCN [30] and GLCN [17] for the comparison. The results are shown in Table 3. [9,30,37,38] cluster the features by applying GCN on the constructed subgraph. Their methods are applied to construct the subgraph for each coming *query* and then trained by our own strategy. [17] proposed to learn a graph for GCN clustering based on Euclidean distance, which is called Graph Learning Convolutional Network (GLCN). To adapt it to our data-stream scenario, we apply it on CSG linked nodes to cluster each sub-graph locally.

Since the algorithm takes more time when the grocery becomes larger, we record the algorithms' speed when the grocery is full. ClusTree is a time-adaptable method that fits our scenario well. However, when we simulate the fast stream in the open checkout-free groceries (e.g., 14s/ptq), ClusTree gets lower Precision and recall.

In addition, through Table 3, we can see that learning-based methods are much more efficient than traditional methods. They generally achieve better performance by projecting the features to the low-dimensional linear subspace. [37] achieves competitive overall performance due to the global-aware clustering, but the time consumption of the strategy is intolerable in this scene. IPS based GCN [17,30] are more efficient comparing with the others. To take our *query* as pivot for IPS construction, the sub-graph represents the local correlation of nodes just like the proposed method, and gains a balanced time and performance improvement. However, they still measure the node distance by cosine similarity, which ignores the distribution nature of the nodes.

As shown in Table 3, the proposed method surpasses the other clustering methods by a large margin and achieves state-of-the-art performance on the datasets. It also outperforms CSG-linked GLCN by a 2.60% and 1.75% F1 score on DaiCOFG and IseCOFG, respectively, with comparable time consumption, indicating the vMF-based *weight* strategy works even better than the learning-based strategy. In practice, the proposed method is thus the most applicable algorithm for the complicated open check-out free grocery scenario.

6 Conclusion and Future Work

This paper proposed a real-time Person Clustering method, namely CSG-GCN, for Open checkout-free groceries under data streams and the unknown number of persons. The proposed method fully utilizes the human time-space information and makes a variance-considered comparison on the spherical summarized data to improve the method's speed and accuracy. And the experimental results show the effectiveness and advantages of the method. Future research will focus on applying the technology to more real-world open checkout-free groceries.

References

1. Ackermann, M.R., Märtens, M., Raupach, C., Swierkot, K., Lammersen, C., Sohler, C.: StreamKM++ a clustering algorithm for data streams. J. Exp. Algorithmics (JEA) **17**, 1–2 (2012)
2. Aggarwal, C.C.: Data Streams: Models and Algorithms. Advances in Database Systems, vol. 31. Springer Science & Business Media, New York (2007). https://doi.org/10.1007/978-0-387-47534-9
3. Almasawa, M.O., Elrefaei, L.A., Moria, K.: A survey on deep learning-based person re-identification systems. IEEE Access **7**, 175228–175247 (2019)
4. Banerjee, A., Dhillon, I.S., Ghosh, J., Sra, S., Ridgeway, G.: Clustering on the unit hypersphere using von Mises-Fisher distributions. J. Mach. Learn. Res. **6**(9), 1–38 (2005)
5. Bendale, A., Boult, T.: Towards open world recognition. In: Proceedings of the IEEE Conference on Computer Vision and Pattern Recognition, pp. 1893–1902 (2015)
6. Bradley, P.S., et al.: Scaling clustering algorithms to large databases. Microsoft research report (1998)
7. Braun, M., Krebs, S., Flohr, F., Gavrila, D.M.: Eurocity persons: A novel benchmark for person detection in traffic scenes. IEEE Trans. Pattern Anal. Mach. Intell. **41**(8), 1844–1861 (2019). https://doi.org/10.1109/TPAMI.2019.2897684
8. Carion, N., Massa, F., Synnaeve, G., Usunier, N., Kirillov, A., Zagoruyko, S.: End-to-end object detection with transformers. In: Vedaldi, A., Bischof, H., Brox, T., Frahm, J.-M. (eds.) ECCV 2020. LNCS, vol. 12346, pp. 213–229. Springer, Cham (2020). https://doi.org/10.1007/978-3-030-58452-8_13
9. Chiang, W.L., Liu, X., Si, S., Li, Y., Bengio, S., Hsieh, C.J.: Cluster-GCN: an efficient algorithm for training deep and large graph convolutional networks. In: Proceedings of the 25th ACM SIGKDD International Conference on Knowledge Discovery & Data Mining, pp. 257–266 (2019)
10. Domingos, P., Hulten, G.: A general method for scaling up machine learning algorithms and its application to clustering. In: In proceedings of the Eighteenth International Conference on Machine Learning. Citeseer (2001)
11. Amigó, E., Gonzalo, J., Artiles, J., Verdejo, F.: A comparison of extrinsic clustering evaluation metrics based on formal constraints. Inf. Retriev. **12**, 613 (2009)
12. Fan, X., Jiang, W., Luo, H., Fei, M.: Spherereid: deep hypersphere manifold embedding for person re-identification. J. Vis. Commun. Image Represent. **60**, 51–58 (2019)
13. Fuglede, B., Topsoe, F.: Jensen-Shannon divergence and Hilbert space embedding. In: International Symposium on Information Theory, 2004. ISIT 2004. Proceedings, p. 31. IEEE (2004)
14. Gama, J.: Knowledge Discovery From Data Streams. CRC Press, Boca Raton (2010)
15. Glorot, X., Bengio, Y.: Understanding the difficulty of training deep feedforward neural networks. In: Proceedings of the Thirteenth International Conference on Artificial Intelligence and Statistics, pp. 249–256. JMLR Workshop and Conference Proceedings (2010)
16. Ji, W., et al.: Learning calibrated medical image segmentation via multi-rater agreement modeling. In: Proceedings of the IEEE/CVF Conference on Computer Vision and Pattern Recognition, pp. 12341–12351 (2021)

17. Jiang, B., Zhang, Z., Lin, D., Tang, J., Luo, B.: Semi-supervised learning with graph learning-convolutional networks. In: Proceedings of the IEEE/CVF Conference on Computer Vision and Pattern Recognition (CVPR), June 2019
18. Kingma, D.P., Ba, J.: Adam: a method for stochastic optimization. arXiv preprint arXiv:1412.6980 (2014)
19. Kipf, T.N., Welling, M.: Semi-supervised classification with graph convolutional networks. arXiv preprint arXiv:1609.02907 (2016)
20. Kranen, P., Assent, I., Baldauf, C., Seidl, T.: The clustree: indexing micro-clusters for anytime stream mining. Knowl. Inf. Syst. **29**(2), 249–272 (2011)
21. Mohammadi, S.S., Wang, Y., Bue, A.D.: Pointview-GCN: 3d shape classification with multi-view point clouds. In: 2021 IEEE International Conference on Image Processing (ICIP), pp. 3103–3107 (2021). https://doi.org/10.1109/ICIP42928.2021.9506426
22. Ryu, S., Kwon, Y., Kim, W.Y.: A Bayesian graph convolutional network for reliable prediction of molecular properties with uncertainty quantification. Chem. Sci. **10**(36), 8438–8446 (2019)
23. Schroff, F., Kalenichenko, D., Philbin, J.: Facenet: A unified embedding for face recognition and clustering. In: Proceedings of the IEEE Conference on Computer Vision and Pattern Recognition, pp. 815–823 (2015)
24. Shen, J., Liu, Y., Dong, X., Lu, X., Khan, F.S., Hoi, S.C.: Distilled Siamese networks for visual tracking. IEEE Trans. Pattern Anal. Mach. Intell. (2021)
25. Silva, J.A., Faria, E.R., Barros, R.C., Hruschka, E.R., Carvalho, A.C., Gama, J.: Data stream clustering: a survey. ACM Comput. Surv. (CSUR) **46**(1), 1–31 (2013)
26. Sra, S.: A short note on parameter approximation for von Mises-Fisher distributions: and a fast implementation of i s (x). Comput. Statist. **27**(1), 177–190 (2012)
27. Tang, H., Zhao, G., Bu, X., Qian, X.: Dynamic evolution of multi-graph based collaborative filtering for recommendation systems. Knowl. Based Syst. **228**, 107251 (2021)
28. Vaswani, A., et al.: Attention is all you need. In: Advances in Neural Information Processing Systems, pp. 5998–6008 (2017)
29. Veličković, P., Cucurull, G., Casanova, A., Romero, A., Lio, P., Bengio, Y.: Graph attention networks. arXiv preprint arXiv:1710.10903 (2017)
30. Wang, Z., Zheng, L., Li, Y., Wang, S.: Linkage based face clustering via graph convolution network. In: Proceedings of the IEEE/CVF Conference on Computer Vision and Pattern Recognition, pp. 1117–1125 (2019)
31. Wu, J., Di, X.: Integrating neural networks into the blind deblurring framework to compete with the end-to-end learning-based methods. IEEE Trans. Image Process. **29**, 6841–6851 (2020)
32. Wu, J., et al.: Learning self-calibrated optic disc and cup segmentation from multi-rater annotations. arXiv preprint arXiv:2206.05092 (2022)
33. Wu, J., et al.: Seatrans: learning segmentation-assisted diagnosis model via transforme. arXiv preprint arXiv:2206.05763 (2022)
34. Wu, J., Fang, H., Wu, B., Yang, D., Yang, Y., Xu, Y.: Opinions vary? Diagnosis first! arXiv preprint arXiv:2202.06505 (2022)
35. Wu, J., Fu, R.: Universal, transferable and targeted adversarial attacks. arXiv preprint arXiv:2109.07217 (2019)
36. Wu, J., et al.: Leveraging undiagnosed data for glaucoma classification with teacher-student learning. In: Martel, A.L., et al. (eds.) MICCAI 2020. LNCS, vol. 12261, pp. 731–740. Springer, Cham (2020). https://doi.org/10.1007/978-3-030-59710-8_71

37. Yang, L., Zhan, X., Chen, D., Yan, J., Loy, C.C., Lin, D.: Learning to cluster faces on an affinity graph. In: Proceedings of the IEEE/CVF Conference on Computer Vision and Pattern Recognition, pp. 2298–2306 (2019)
38. Zeng, H., Zhou, H., Srivastava, A., Kannan, R., Prasanna, V.: GraphSaint: graph sampling based inductive learning method. arXiv preprint arXiv:1907.04931 (2019)
39. Zhang, J., Shi, X., Xie, J., Ma, H., King, I., Yeung, D.Y.: GAAN: gated attention networks for learning on large and spatiotemporal graphs. arXiv preprint arXiv:1803.07294 (2018)
40. Zhang, T., Ramakrishnan, R., Livny, M.: Birch: an efficient data clustering method for very large databases. ACM SIGMOD Rec. **25**(2), 103–114 (1996)
41. Zhang, Y., Di, X., Zhang, B., Ji, R., Wang, C.: Better than reference in low-light image enhancement: conditional re-enhancement network. IEEE Trans. Image Process. **31**, 759–772 (2022). https://doi.org/10.1109/TIP.2021.3135473

POP: Mining POtential Performance of New Fashion Products via Webly Cross-modal Query Expansion

Christian Joppi[1], Geri Skenderi[2](✉), and Marco Cristani[1,2]

[1] Humatics srl, Verona, Italy
{Christian.Joppi,Marco.Cristani}@sys-datgroup.com
[2] University of Verona, Verona, Italy
{Geri.Skender,Marco.Cristani}i@univr.it

Abstract. We propose a data-centric pipeline able to generate exogenous observation data for the New Fashion Product Performance Forecasting (NFPPF) problem, i.e., predicting the performance of a brand-new clothing probe with no available past observations. Our pipeline manufactures the missing past starting from a single, available image of the clothing probe. It starts by expanding textual tags associated with the image, querying related fashionable or unfashionable images uploaded on the web at a specific time in the past. A binary classifier is robustly trained on these web images by confident learning, to learn what was fashionable in the past and how much the probe image conforms to this notion of fashionability. This compliance produces the POtential Performance (POP) time series, indicating how performing the probe could have been if it were available earlier. POP proves to be highly predictive for the probe's future performance, ameliorating the sales forecasts of all state-of-the-art models on the recent VISUELLE fast-fashion dataset. We also show that POP reflects the ground-truth popularity of new styles (ensembles of clothing items) on the Fashion Forward benchmark, demonstrating that our webly-learned signal is a truthful expression of popularity, accessible by everyone and generalizable to any time of analysis. Forecasting code, data and the POP time series are available at: https://github.com/HumaticsLAB/POP-Mining-POtential-Performance.

Keywords: Computer vision for fashion · Data-centric artificial intelligence · Time series forecasting

1 Introduction

Forecasting the performance of a new clothing item is a crucial challenge for fashion companies [7,11]. A good forecast in terms of predicted sales or product

C. Joppi and G. Skenderi—Equal contribution.

Supplementary Information The online version contains supplementary material available at https://doi.org/10.1007/978-3-031-19839-7_3.

S. Avidan et al. (Eds.): ECCV 2022, LNCS 13698, pp. 34–50, 2022.
https://doi.org/10.1007/978-3-031-19839-7_3

popularity can greatly help optimize the supply chain [30] and minimize losses on multiple levels. Unfortunately, standard forecasting approaches require observations from the past to provide a forecast for the same product in the future [1,14] and this information is typically available for evergreen products only (Fig. 1a). In other cases, judgemental forecasts [14] from fashion professionals[1] are the only ones that can help. Starting from photos or realistic renderings, which we call *probe* images, they perform comparisons with trends as they surface and then infer the probe's success [30]. In this paper, we try to model this line of reasoning and create a data-centric [25] pipeline that is able to extract a highly predictive signal from the web, dubbed "POtential Performance" (POP), which can be fed into any NFFPF forecasting model as an additional variable and lead to more accurate forecasts. (see Fig. 1b).

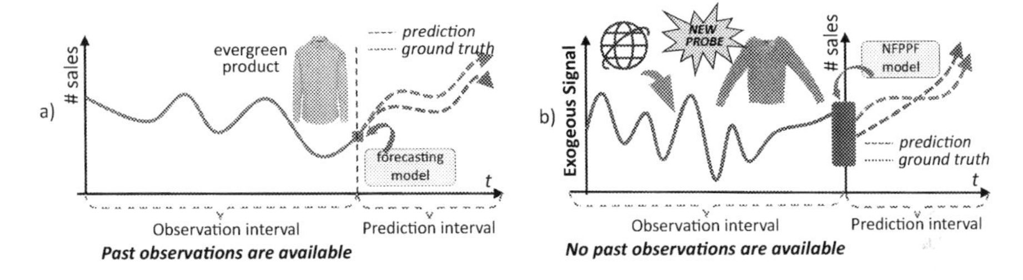

Fig. 1. a) A standard forecasting setup, where an evergreen item has past observations to exploit, *e.g.*, # sales; b) New Fashion Product Performance Forecasting (NFPPF) problem, where no past observations are available and exogenous data must be considered. Here we propose POP, the POtential Performance series, which is webly learned. The signals in b) appear on the same scale purely for visualization purposes, in reality this might not be the case.

Our cross-modal, query expansion based pipeline is sketched in Fig. 2. The input is a single probe image of the product to be analyzed, or a photorealistic rendering[2]. The pipeline first extracts textual tags from the probe automatically or by directly considering the associated technical sheet. The tag set is expanded with *positive* and *negative* tags that are used to perform a *time-dependent* query online, i.e., collecting images of "fashionable" and "unfashionable" items related to the tags, which have been uploaded during some specified K_{past} intervals in the past. These images are used to *confidently learn* [27] a binary classifier that captures what is fashionable VS unfashionable in that interval. This learning procedure prunes noisy images from both the positive and negative classes, resulting in a robust model. Subsequently, pruned positive images are projected into an embedding space by the learned model and compared with the (also

[1] A commercial example is Trendstop https://www.trendstop.com/ and its "Trend Platform Membership" service.

[2] Several such tools are available, for instance https://www.tg3ds.com/3d-fashion-design-tools.

projected) initial probe image, providing the K_{past}-long POP signal. The POP signal indicates how popular the probe could have been over time if it were available earlier in the past.

Our approach should be cast in the field of *data-centric artificial intelligence* (DCAI) [25], since it automates the creation of high quality training data that can be used to improve any forecasting model which accommodates multivariate time series forecasting. POP has been tested on diverse state-of-the-art NFPPF algorithms that predict sales curves of new products on the recent VISUELLE fast-fashion dataset [34], providing superior performance when compared to other types of training signals. It has also been customized to deal with fashion styles (*i.e.,* ensembles of clothing items) on the Fashion Forward benchmark [1]. Fashion Forward (FF) calculates a popularity time series for an automatically extracted style based on the dataset properties and then applies standard forecasting algorithms. We substitute their popularity series with POP, reaching similar predictions despite relying only on an exogenous input. Surprisingly, on the Dresses partition of FF, we reach the absolute best, suggesting that POP can foresee the success of a potentially new fashion style. Summarizing, the contributions of this work are threefold:

1. The first data-centric strategy tailored to forecasting, used to create an exogenous observation signal which improves forecasts of the performance (number of sold items, popularity) of brand new clothing items with non-existent pasts.
2. A webly-learned method to freely collect information about fashion trends without relying on private or costly repositories.
3. Best overall results on all the tested NFPPF tasks.

The rest of the paper is organized as follows: related literature is analyzed in Sect. 2; the proposed approach is detailed in Sect. 3; experiments are reported in Sect. 4, and finally; concluding remarks are drawn in Sect. 5.

2 Related Literature

NFPPF Problem. The NFPPF problem has been deeply investigated in the fields of quantitative fashion design [3,16,29], marketing and social sciences [12,32], but is relatively new in the computer vision community. In both [9,33], the main idea is that new products will sell comparably to similar, older products; this similarity is exploited in [33] via textual tags only, while in [9] an autoregressive RNN model takes past sales, textual product attributes, and the product image as input, to forecast the item sales. The work in [34] focuses on the additional direction of checking the past to look for predictive exogenous signals. In particular, the authors exploit Google Trends, querying textual attributes related to the probe and embed the resulting trend into a Transformer-based [37] architecture, which considers images, text and other metadata. The authors also rendered accessible the first publicly available dataset for NFPPF, VISUELLE. In our paper we follow the idea of looking back to web data, but use images as the main representation of online fashionability, obtaining a richer

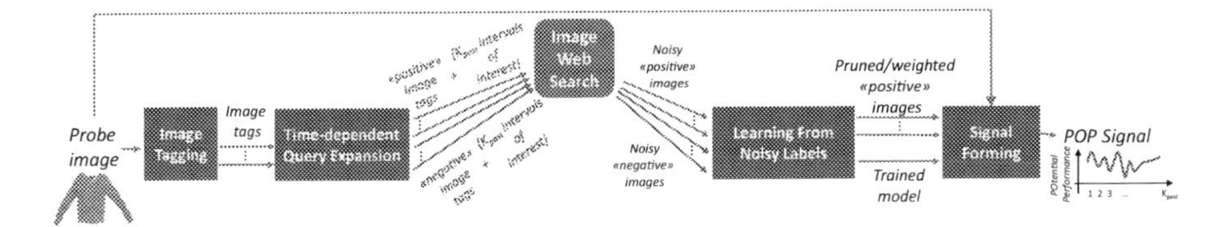

Fig. 2. Schematic pipeline of our approach; we start with a probe image and obtain the POtential Performance (POP) signal at the end. Along this pipeline, we sequentially process information in different modalities, thereby creating a *cross-modal signal.*

exogenous signal. Predicting the success of new fashion styles has never been taken into account, with past works [1,20,22] focusing on the standard forecasting setup.

Data-centric AI. Data-Centric AI [25] (DCAI) shifts the attention from the models to the data used to train and evaluate them. It is a topic whose importance is constantly growing in many AI communities [2,26,28][3], with important effects on CV & ML. In general, DCAI investigates methodologies for accelerating open-source dataset creation from lower-quality resources. Consequently, it is tightly coupled with learning on noisy data, which aims at producing consistent and low noise data samples, or removing labeling noise and inconsistencies from existing data [27,35,38]. Our methodology is data-centric, since it automates the creation of training data from a large amount of web resources, while removing labeling noise. Notably, it represents a novelty in the DCAI panorama, since it creates *temporally-dependent* training data, i.e., time series, as it is required by NFPPF and in general by forecasting tasks.

3 Methodology

The goal of our approach is to produce an exogenous variable that can aid a forecasting model in predicting the future performance of a product (sales, popularity). The input to our approach is the probe image $\mathbf{z}^{(t)}$, where $\mathbf{z}$ represents the new clothing item and t the *observation time*, which is the date from when we begin to look into the past. The output is the POP signal $S_{\mathbf{z}}^{(t)} = s_{\mathbf{z}}^{(t-K_{past})}, \dots, s_{\mathbf{z}}^{(t-k)}, \dots, s_{\mathbf{z}}^{(t-1)}$, defined for K_{past} time steps preceding t, where $k = 1, \dots, K_{past}$ and $s_{\mathbf{z}}^{(t-k)} \in \mathbb{R}$. In this paper, we describe the observation times in terms of weeks and set $K_{past} = 52$. This translates to looking one year prior to the observation time t, as typically done in fashion market analysis [36]. The next sections will sequentially detail the general pipeline of our approach, depicted in Fig. 2.

[3] https://datacentricai.org/.

3.1 Image Tagging

The first operation is the extraction of textual tags $\{a_{\mathbf{z}}^{(j)}\}_{j=1,\ldots,J}$ associated to $\mathbf{z}$. These tags should represent the clothing item with sufficient generality, capturing at least categorical information (*e.g.* "long sleeve") and a dominant color (*e.g.* "yellow"). Empirically, we found these tags to work well while being easily obtainable. Category and color can be automatically extracted with high accuracy [19] or are usually contained in the technical data sheet accompanying the product, which is what we exploit in this work, as shown in Sect. 4.

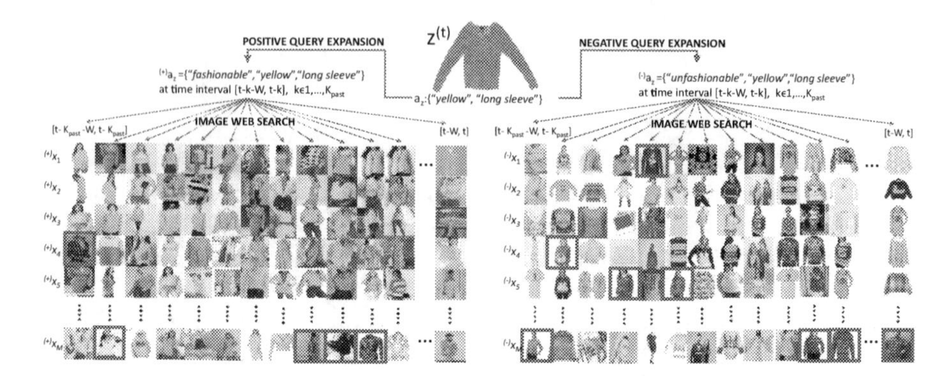

Fig. 3. Pipeline insights on *Time-dependent Query Expansion* (Sect. 3.2), *Image Web Search* (Sect. 3.3) and *Learning From Noisy Labels* (Sect. 3.4) steps. This figure reports a real world excerpt of the download and processing of $N = 2600$ images ($N = 2(M \times K_{past})$, $M = 25$, $K_{past} = 52$). (Color figure online)

3.2 Time-dependent Query Expansion

The second operation (detailed in Fig. 3 on a real example) performs two different textual query expansions, generating *positive expansions*, $\{a_{\mathbf{z}}^{(j)}\}_{j=1,\ldots,J} \cup J^{(+)}$ where the additional $J^{(+)}$ tags indicate attractive clothing items, and conversely for *negative expansions*. In this paper, we found the tags $J^{(+)}$ = "fashionable" and $J^{(-)}$ = "unfashionable" to be the most effective for positive and negative expansions, respectively. Alternatives as "best seller" and "unattractive" were considered, returning similar results.

Each expansion, either positive or negative, is associated to a particular $k = 1, \ldots, K_{past}$ for the time interval $[t-k-W, t-k]$, where W is a temporal window we wish to consider for the image search, also expressed in weeks. In our experiments we set $W = 4$, which translates to having a sliding window of size 4 and stride of 1 over the temporal axis. This allows the pool of downloaded images to disclose what are newly indexed items in relation to previous time steps, developing a temporal locality in the data pool. The precise value of W was chosen after an empirical evaluation over the range $1, \ldots, 12$.

3.3 Image Web Search

A given expanded textual query along with a time interval is fed into a web API request to gather M representative fashionable and unfashionable images ${}^{(+)}\{\mathbf{x}_i\}_{i=1,\dots,M}^{(t-k)}$; ${}^{(-)}\{\mathbf{x}_i\}_{i=1,\dots,M}^{(t-k)}$ that have been uploaded in the interval $[t-k-W, t-k]$, for $k = 1, \dots, K_{past}$. In particular, we adopt Google Image search, selecting the first $M = 25$ images returned, assuming the ordering of Google Images perfectly mirrors a genuine image relevance [17]. After the image web search phase, $M \times K_{past}$ fashionable and unfashionable images are collected respectively (as shown in Fig. 3). These images are then used to train a binary classifier θ, aimed at distinguishing fashionable from unfashionable images. Webly learning and supervision based on Google Images has been considered before in computer vision, especially for image classification and object detection [6,10,18]. POP goes one step further, merging visual and textual search while adding a time-dependent query expansion to create more discriminative image sets. Nevertheless, the labels assigned to the images from the query expansions might be noisy, therefore we apply a confident learning method.

3.4 Learning from Noisy Labels

In the following, we adapt the confident learning (CL) methodology specifically for our binary problem. For a broader overview, readers may refer to [27]. Let $\mathbf{X} = \{\mathbf{x}_i, \tilde{y}_i\}_{1\dots N}$ be our set of $N = 2(M \times K_{past})$ images with associated observed noisy binary labels $\tilde{y}_i \in \{$"fashionable", "unfashionable"$\}$. CL assumes that a true, latent label $y_i^* \in \{$"fashionable", "unfashionable"$\}$ exists for every sample. CL requires two inputs: 1) the out-of-sample $N \times 2$ matrix $\hat{\mathbf{P}}$ of predicted probabilities where $\hat{\mathbf{P}}_{i,h} = \hat{p}(\tilde{y}_i = h; \mathbf{x}_i, \theta)$ with θ a generic (binary) classifier initially trained on $\mathbf{X}$; 2) the set of noisy labels $\{\tilde{y}_i\}$. Subsequently, a robust 2×2 confusion matrix, called the *confident joint* matrix $\mathbf{C}_{\tilde{y},y^*}$, is computed[4]:

$$\begin{aligned} \mathbf{C}_{\tilde{y},y^*}(h,l) &= |\hat{\mathbf{X}}_{\tilde{y}=h,y^*=l}|, \text{with} \\ \hat{\mathbf{X}}_{\tilde{y}=h,y^*=l} &= \left\{ \mathbf{x} \in \mathbf{X}_{\tilde{y}=h} : \hat{p}(\tilde{y}=l; \mathbf{x}, \theta) \geq t_l \right\} \end{aligned} \tag{1}$$

where t_l is a threshold that represents the expected self confidence value for each class:

$$t_l = \frac{1}{|\mathbf{X}_{\tilde{y}=l}|} \sum_{x \in \mathbf{X}_{\tilde{y}=l}} \hat{p}(\tilde{y}=l; x, \theta) \tag{2}$$

In practice, $\mathbf{C}_{\tilde{y},y^*}$ counts only those elements which have been confidently classified in a particular class, where the term "confident" means with a probability that is higher than the average probability of an element belonging to that class. In simpler words, if samples labeled as belonging to class h tend to have higher probabilities because the model is over-confident about class h, then t_h will be

[4] We drop the index i for clarity.

proportionally larger. It also worth noting that Eq. 1 corresponds to a simplified version of the general building procedure of the confident joint matrix $\mathbf{C}_{\tilde{y},y^*}$ of [27], which nonetheless in our case is perfectly acceptable since we deal with binary classification and no *label collision* may happen, *i.e.,* the fact that a noisy label can correspond to a more than a single alternative class.

On this robust confusion matrix, we estimate label errors from the off diagonal elements of $\mathbf{C}_{\tilde{y},y^*}(h,l)$. Wrongly labeled images are therefore pruned (indicated by the red boxes in Fig. 3), obtaining the cleaned fashionable and unfashionable images ${}^{(+)}\{x_i'\}_{i=1,\dots,M'^{(t-k)}}^{(t-k)};{}^{(-)}\{x_i'\}_{i=1,\dots,M''^{(t-k)}}^{(t-k)}$, where $M'^{(t-k)}$ and $M''^{(t-k)}$ indicate that we can have a different number of positive and negative images, respectively, related to each $t-k$ time step, due to the noisy sample elimination. The classifier is retrained on the cleaned data, obtaining a robust trained model θ'. This procedure is data-centric and model agnostic; the specific θ used in this work is described in Sect. 4.

3.5 Signal Forming

The POP signal $S_{\mathbf{z}}^{(t)} = s_{\mathbf{z}}^{(t-K_{past})}, \dots, s_{\mathbf{z}}^{(t-k)}, \dots, s_{\mathbf{z}}^{(t-1)}$, is computed by considering the cleaned fashionable images ${}^{(+)}\{\mathbf{x}_i'\}_{i=1,\dots,M'^{(t-k)}}^{(t-k)}$, the robust model θ', and the image $\mathbf{z}$, as follows:

$$s_{\mathbf{z}}^{(t-k)} = \frac{1}{M'^{(t-k)}} \sum_{i=1}^{M'^{(t-k)}} \frac{\langle \theta' \left({}^{(+)}\mathbf{x}_i'^{(t-k)} \right) \cdot \theta' (\mathbf{z}) \rangle}{\| \theta' \left({}^{(+)}\mathbf{x}_i'^{(t-k)} \right) \| \| \theta' (\mathbf{z}) \|} \tag{3}$$

where $\theta'(\mathbf{z})$ indicates the extracted features of $\mathbf{z}$ from θ', and $\langle \cdot \rangle$ indicates the dot product. In other words, the signal value $s_{\mathbf{z}}^{(t-k)}$ is the average cosine similarity between the embedding of the probe image $\mathbf{z}$ and each fashionable image $\mathbf{x}'_{i(t-k)}$ from the $M'^{(t-k)}$ downloaded images. An assessment of alternative signal forming options is shown in the supplementary material.

4 Experiments

In line with the general requirements of DCAI [25], we show how our automatically manufactured time series helps a forecasting model ψ achieve better results on a given task γ. The main idea behind our approach is that by knowing POP, the forecasting model can gain a context on the past which otherwise would be missing and therefore improve. To demonstrate this, we perform extensive evaluation on two tasks (and different forecasting models): *new fashion product sales curve prediction* [9,33,34], and *style popularity forecasting* [1]. We show ablative studies on the first task and an impressive outcome on the second.

The binary classifier θ for learning on noisy data (Sect. 3.4) is based on a ResNet50 [13], pre-trained on ImageNet [8], with two additional fully connected layers. During the confident learning procedure, we fine-tune its last convolutional block and fully connected layers for 50 epochs with a batch size of 64, using CE loss, following a 5-fold cross validation protocol. AdamW [21] is used as optimizer, with a learning rate of $1e-4$. The forecasting neural network models are all trained for 200 epochs with a batch size of 128 and L2 loss, using the AdaFactor [31] optimizer. The experiments are performed on two NVIDIA 3090 RTX GPUs.

4.1 Task 1: New Fashion Product Sales Curve Prediction

The output of a sales curve forecasting model for a probe clothing item $\mathbf{z}$ is a time series $O_{\mathbf{z}}^{(st)} = o_{\mathbf{z}}^{(st+1)}, \ldots, o_{\mathbf{z}}^{(st+k)}, \ldots, o_{\mathbf{z}}^{(st+K_{fut})}$ that indicates how many pieces of z will be sold starting at a particular time step st (typically the start of the season), for the next K_{fut} time steps.

We run our first set of experiments on the VISUELLE dataset [34]. For each available product, multi-modal information is provided: i) images, ii) text tags, iii) Google Trends, iv) sales curves. The evaluation protocol follows that of VISUELLE, simulating how a fast-fashion company deals with new products on two particular moments: the *first order setup* and the *release setup*. The former takes place when the company decides which products and how many pieces to order by looking at probe images. The latter is right before the season, and is useful to obtain an accurate forecast in order to plan stock replenishment. These two setups use 28 and 52 week long exogenous signals (originally Google Trends [34]), respectively.

Note that for the sake of fairness, we do not alter the training setup or models from [34], keeping the cardinality and the type of the training data fixed and *substituting only the Google Trends with our POP signal.* All the models are trained considering the 12-week long sales signals, whilst the evaluation is done on a 6-week horizon. This is shown to give the best predictions while simulating politics of real fashion companies [34].

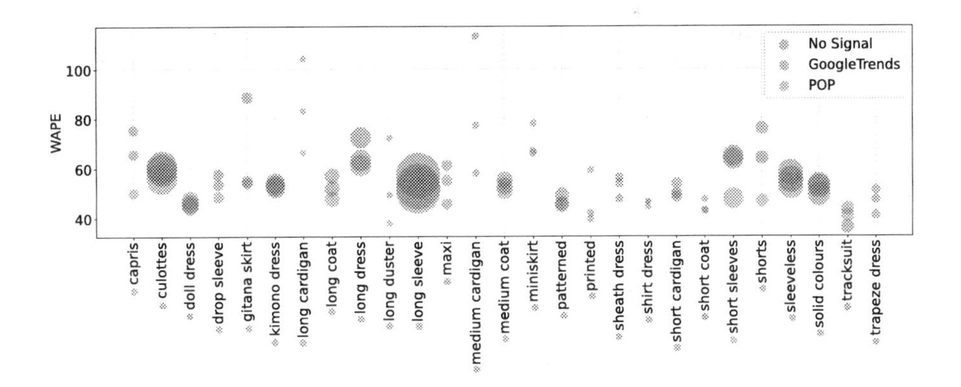

Fig. 4. Forecasting WAPE results per clothing category; the larger the blob, the higher the # of items in that category; the color below each category name indicates the type of training setup which gives the best WAPE.

We consider 5 algorithms (from oldest to newest): *Gradient Boosting* for forecasting [15], Concat Multi-Modal RNN [9] (*Concat MM RNN* in the tables), Residual Multi-Modal RNN [9] (*Residual MM RNN*), Cross-Attention RNN [9] (*X-Attention RNN*) and GTM Transformer [34] (*GTM Transf.*) We consider the Weighted Absolute Percentage Error (WAPE) as primary evaluation metric and also compute the Mean Absolute Error (MAE) to demonstrate the error on an absolute scale [34]:

Table 1. Results on VISUELLE with the *first order setup*; "W" stands for WAPE, "M" for MAE. Lower is better for all metrics.

First order setup (K_{best} = *28 weeks*)															
Exogenous Signal	*Gradient Boosting* [15] **2020**			*Concat MM RNN* [9] **2020**			*Residual MM RNN* [9] **2020**			*X-Attention RNN* [9] **2020**			*GTM Transformer* [34] **2021**		
	W	M	ERP	W	M	ERP	W	M	ERP	W	M	ERP	W	M	ERP
No signal	64.10	35.02	0.43	63.31	34.41	0.42	64.26	34.92	0.44	59.49	32.33	0.38	56.62	30.93	0.37
GoogleTrends	64.29	35.12	0.43	64.11	34.84	0.43	68.11	37.02	0.47	58.70	31.90	0.38	56.83	31.05	0.35
POPSignal	**63.75**	**34.83**	**0.42**	**58.09**	**31.73**	**0.39**	**58.88**	**32.16**	**0.39**	**57.78**	**31.56**	**0.38**	**53.41**	**29.18**	**0.32**

Table 2. Results on VISUELLE with the *release setup*; "W" stands for WAPE, "M" for MAE. Lower is better for all metrics.

Release setup (K_{best} = *52 weeks*)															
Exogenous Signal	*Gradient Boosting* [15] **2020**			*Concat MM RNN* [9] **2020**			*Residual MM RNN* [9] **2020**			*X-Attention RNN* [9] **2020**			*GTM* ***Transformer*** [34] **2021**		
	W	M	ERP	W	M	ERP	W	M	ERP	W	M	ERP	W	M	ERP
NoSignal	64.10	35.02	0.43	63.31	34.41	0.42	64.26	34.92	0.44	59.49	32.33	0.38	56.62	30.93	0.37
GoogleTrends	63.52	34.70	0.42	65.87	35.80	0.44	68.46	37.21	0.48	59.02	32.08	0.38	55.24	30.18	0.33
POPSignal	**63.38**	**34.62**	**0.42**	**57.43**	**31.37**	**0.36**	**58.38**	**31.89**	**0.39**	**57.36**	**31.33**	**0.36**	**52.39**	**28.62**	**0.29**

Note that the WAPE is not bounded by 100. Finally, we measure the similarity of the slope of the predicted curve with the ground truth using the *Edit distance with Real Penalty* (ERP) [5]. This metric counts the number of edit operations (insert, delete, replace) that are necessary to transform one series into the other. Because we are dealing with continuous values, a threshold ϵ=0.03 is used to decide if values are considered different and have to be edited.

The results are shown in Table 1 for the *first order setup* and in Table 2 for the *release setup*. As reference, we also report results *without* any exogenous series, to show the net value of these indicators. For all the algorithms and both setups, adding POP to the model boosts the performances over all the metrics, reaching the absolute best when coupled with GTM Transformer. On average, in the *first order setup*, we improve the WAPE by 3.42% over the Google Trends and by 3.21% over not using any exogenous signals. In the *release setup* we improve by 2.85% over the Google Trends and by 4.23% over not using exogenous signals. These results demonstrate how our data-centric approach can provide optimal forecasts by creating a highly-predictive signal of past popularity that is image-based, unlike Google Trends. The forecasts are performed on 497 products over different stores, meaning that these improvements can provide a large impact on the supply chain operations.

In Fig. 4 we show the WAPE *per clothing category*. We mostly perform better than the other training alternatives, yet some particular categories display limitations of our approach. These limitations arise due to the fact that the Image Tagging phase is assumed as flawless, since we rely on the technical sheet accompanying the probe image to extract the tags. The results per category (Fig. 4) display how possibly mislabeled categories, or categories labeled in a general manner ("solid colours","doll dress") may lead to misleading web images. As visible in Fig. 5, the related images from the web, both fashionable and not, are completely useless, since the tag of the category itself is misleading. In such cases, a robust automated category extraction could potentially lead to better results.

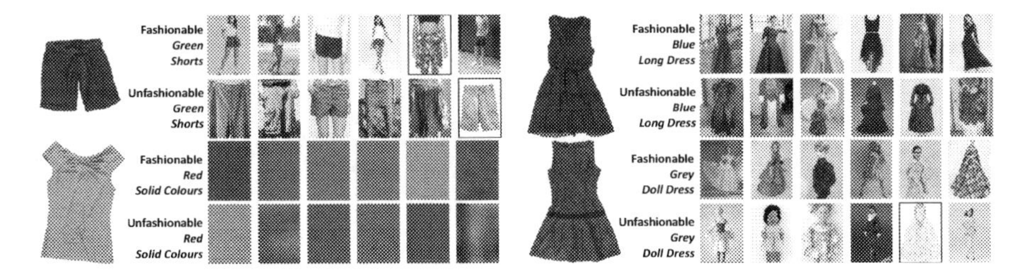

Fig. 5. Examples of VISUELLE items (seasons SS17 and SS18 on the left, SS19 and AI19 on the right, respectively) and the correspondent fashionable/unfashionable images from the web. Some web images can be misleading, due to the questionable category names of the VISUELLE dataset ("solid colours", "doll dress").

Ablation Studies. In the following, we focus on alternative versions of our proposed pipeline, ablating the specific modules illustrated in Fig. 2. Table 3 contains all the results.

Time dependent query expansion

- *No expansion:* Images are queried with the original tags collected in the Image Tagging phase, without generating positive or negative expansions. This is equivalent to querying only with "color + category". The learning step is impacted directly, since no positive or negative classes are available for learning, therefore we use our backbone model to extract image features. For each image $\mathbf{z}^{(\mathbf{t})}$, the web images $\{\mathbf{x}_i\}_{i=1,\ldots,M}^{(t-k)}$ that have been uploaded in the interval $[t-k-W, t-k]$, for $k=1,\ldots,K_{past}$ are collected. The signal forming Eq. 3 changes accordingly, using all the M downloaded images;

- *Misaligned past:* We modify the query expansions by looking one year earlier than the "correct" past. Given the observation time t of the probe $\mathbf{z}^{(t)}$, instead of looking backwards from $t-1$ weeks to $t-K_{past}$, we go from $t-1-K_{past}$ to $t-2 \cdot K_{past}$.

With respect to all the alternative versions in this study, the *No expansion* ablation gives the worst result. POP provides an improvement of 0.73% and 1.06% WAPE for the *first order setup* and *release setup*, respectively. The *Misaligned past* yields slightly better results, but still performs worse than POP by 0.63% and 0.22% WAPE for the *first order setup* and *release setup*, respectively. This confirms that fashion has an evolution that changes year after year that we have to take into account.

Table 3. Alternative versions of our pipeline (Fig. 2) on both the *release* and *first order* setups; "W" stands for WAPE, "M" for MAE. Lower is better for all metrics.

Time dependent query expansion				
	Release setup		*First order setup*	
Strategy	W	M	W	M
No expansion	53.12	29.02	54.47	29.77
Misaligned past	53.02	28.96	53.63	29.30
Learning with noisy labels				
	Release setup		*First order setup*	
Strategy	W	M	W	M
No learning	53.03	28.97	53.83	29.41
No robust learning	52.81	28.85	53.59	29.28
Symmetric cross entropy [38]	52.63	28.75	53.58	29.27
SELFIE [35]	52.56	28.71	53.51	29.23
POP	**52.39**	**28.62**	**53.41**	**29.18**

Learning from noisy data

- *No learning*: A predefined image classification network is used to compute the distance among embeddings of the probe image with the positive, downloaded images. This is equivalent to ablating the "Learning from Noisy Data" phase of Fig. 2. It will highlight the importance of dealing with distances among embeddings which are specifically learned against distances coming from a general purpose network. We utilise the backbone of our binary classifier, specified in the introduction of Sect. 4;
- *No robust learning*: All of the downloaded positive and negative images are used to learn our binary classifier without pruning noisy data by confident learning;

- *Symmetric cross entropy* [38]: SCE is a robust classification loss; it adds to the standard cross entropy loss a *reverse cross entropy* term which assumes the predicted labels as ground truth, and the original labels as possibly faulty. In practice, it penalizes noisy labels, without removing any associated training data;
- *SELFIE* [35]: the key idea is to correct the label of noisy *refurnishable* samples with high precision, with the help of clean data which is defined as those samples within a mini-batch creating a small loss. Repeated training runs (dubbed "restarts") allow to use more training data, *i.e.*, noisy samples which have been corrected in their labels. In particular, we use 3 restarts, after which 1.1% of both fashionable and unfashionable items have been removed from the training data.

The results in Table 3 show slightly different performances, promoting the general idea of learning from webly data. *No learning* gives the worse performance, indicating that a fine tuning on the web data is beneficial (53.03 and 53.83 WAPE); when learning is done on the web data, there is some increase (52.81 and 53.59 WAPE); when learning is robust to label noise, with SCE, performances are better (52.63 and 53.58 WAPE); removing some outliers with SELFIE gives a further help (52.56 and 53.51 WAPE). Confident learning remains the best solution, with 52.39 and 53.41 WAPE, while removing 0.8% and 1.1% of fashionable and unfashionable items respectively, from the 14,500,200 images mined using our cross-modal pipeline (Fig. 6).

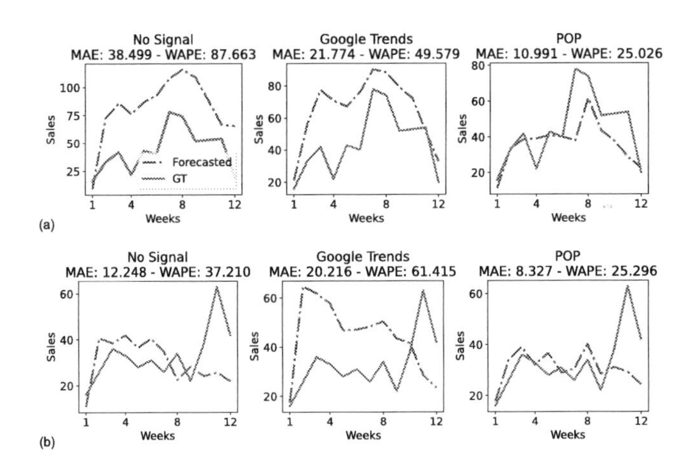

Fig. 6. Qualitative results for the sales forecast of two different products on VISUELLE, considering all 12 time-steps. In all cases, using POP provides better forecasts. In the bottom row (bottom right plot), we show a forecasting failure case, where the product is discounted in its final week of sales.

4.2 Task 2: Popularity Prediction of Fashion Styles

The style popularity prediction task [1] is different from product sales forecasting in that it considers a popularity signal y based on multiple clothing items. In the literature [1,23], style is defined as a latent property of a set of clothing images that share some common visual features. Concretely, in Fashion Forward (FF) [1], Non-negative Matrix factorization is applied to extract K styles from the attribute extraction features [19] of all the product images. Formally, let $\mathbf{A} \in \mathbb{R}^{M \times N}$ indicate the confidence that each of the M visual attributes is contained in each of the N images. $\mathbf{A}$ can be factorized into two matrices with non-negative entries:

$$\mathbf{A} \approx \mathbf{W}\mathbf{H}, \mathbf{W} \in \mathbb{R}^{M \times K} \quad \text{and} \quad \mathbf{H} \in \mathbb{R}^{K \times N} \tag{4}$$

where $\mathbf{W}$ represents the confidence that each attribute is part of a style and $\mathbf{H}$ represents the confidence that each style is associated to an image. The popularity signal y for a style k is built by considering the interactions in the Amazon Reviews dataset [24] of all the items $\{\mathbf{z}\} \in A$ at time t, weighted by their style membership $H(k, z)$. For a detailed explanation, we refer to [1].

Table 4. Average results over all Fashion Forward [1] dataset partitions and specific results for the Dresses partition, where POP outperforms even the original GT style popularity time series (*Oracle*).

Global average												
Signals	Mean		Last		Drift		AR		ARIMA		SES	
	MAE	MAPE	MAE	MAPE	MAE	MAPE	MAE	MAPE	MAE	MAPE	MAE	MAPE
Oracle	*0.136*	*0.170*	*0.093*	*0.114*	*0.174*	*0.222*	*0.271*	*0.403*	*0.136*	*0.167*	*0.094*	*0.116*
GoogleTrends	0.846	1.000	0.846	1.000	0.846	1.000	0.846	1.000	0.846	1.000	0.846	1.000
POP	0.152	0.192	0.116	0.144	0.182	0.229	0.281	0.418	0.235	0.293	0.125	0.156
Dresses												
Signals	Mean		Last		Drift		AR		ARIMA		SES	
	MAE	MAPE	MAE	MAPE	MAE	MAPE	MAE	MAPE	MAE	MAPE	MAE	MAPE
Oracle	*0.155*	*0.197*	*0.130*	*0.158*	*0.203*	*0.263*	*0.307*	*0.409*	*0.173*	*0.209*	*0.129*	*0.157*
GoogleTrends	0.849	1.000	0.849	1.000	0.849	1.000	0.849	1.000	0.849	1.000	0.849	1.000
POP	**0.119**	**0.157**	**0.108**	**0.127**	**0.173**	**0.216**	**0.229**	**0.334**	**0.162**	**0.193**	**0.109**	**0.130**

To extend this problem to a NFPPF setup, we have to imagine we are evaluating the performance of a brand new style that does not have a past. The purpose of POP becomes replacing the original style popularity series. This means that POP has to be modified to deal with a style and not with a single clothing item, where two challenges are presented: 1) To verify how similar POP is to the ground truth popularity signal and; 2) To check if POP is highly predictive of the future popularity.

To deal with the first challenge, we consider for each style k the 2 textual attributes [19] w_1, w_2 (extracted from $\mathbf{W}$) with the highest confidence scores and

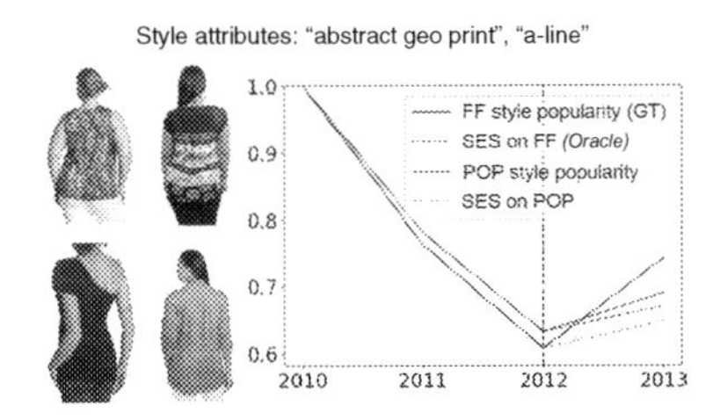

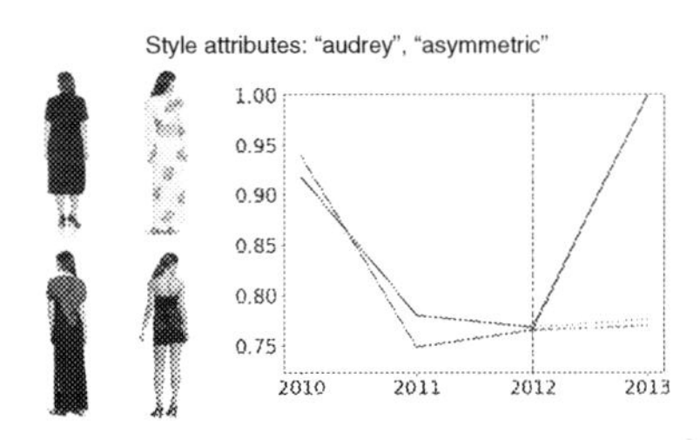

Fig. 7. Qualitative results on the forecasting of two different styles from FF, represented by their respective "style-defining" images and top (automatically extracted) attributes. In both cases, POP and the ground-truth (GT) style popularity from Fashion Forward are substantially similar. The plot on the right shows a forecasting failure case, which holds for both POP and the GT).

use them for the time dependent query expansion. FF provides the only dataset for style forecasting where both images and product metadata are available. The task is to predict a popularity score on a yearly basis. The data ranges from [2008 – 2013], but since Google Images returns little to no images for queries before 2010, we use the range [2010 – 2013] in our experiments. We set $K_{past} = 208$, meaning that we investigate 4 years back. In this way we can create a weekly series for each year and use the average as the value representing the popularity for that year. As probe image to create our POP signal, we consider the top 10 images $\{\mathbf{z}\}$ that represent a style (based on their membership weight $H(k, z)$). Each image will lead to one POP signal, which we average together to obtain the *POP style signal*. This process is repeated for all the dataset partitions presented in FF. To deal with the second challenge, we adopt the best performing statistical forecasting techniques from Fashion Forward and feed them the style POP signal described above. For more details on the forecasting techniques, we refer the reader to [4,14]:

1. **Naive methods.** These methods infer by utilizing general information from the training data. *Mean* forecasts the future as the mean of past observations, while *Last* as the last observed value. *Drift* is the same as *Last*, but the forecasts change over time based on the global trend of the series;
2. **Auto Regressive and Moving Average methods.** These methods forecast using a linear combination of *some* past observations in a regression framework. The most famous and representative method of this class is the *AutoRegressive Integrated Moving Average*(ARIMA) model.
3. **Simple Exponential Smoothing.** Stands for simple exponential smoothing, is a weighted average of previous observations where the weights decrease exponentially as we go further in the past.

Following the protocol of [1], all models are trained on all but the last timestep, which is used for testing. We utilise the mean absolute percentage error (MAPE) and the mean absolute error (MAE) to evaluate the forecasting accuracy on the last timestep of the signal stemming from FF. To provide an additional comparison, we show additional results using Google Trends as the substitute popularity time series [34]. Note that to obtain fair and comparable results, all the signals are rescaled in the range [0,1] using min-max normalization. The results are shown in Table 4, where *Oracle* refers to the original ground-truth style popularity series given as input to the forecasting models (Fig. 7).

POP proves to be a natural substitute to the GT style popularity time series and it allows for optimal forecasts, providing better results than the GT signal itself for the *Dresses* partition. On the other hand, Google Trends are not able to convey such similarities, partially because searching only for the popularity of textual tags might not provide a series that is as predictive as ours.

5 Conclusion

Metaphorically, our approach performs a kind of "time travel": it sends a fashion probe image in the past, before its launch in the market. It then models the popularity from that past point forward by relying on highly ranked web images, queried by using general textual tags related to the probe. The probe similarity with the past is then shown to be a good exogenous indicator for future performance. This pipeline provides a new, effective and data-centric scheme for NFPPF problems.

Acknowledgements. This work was supported by the Italian MIUR through the project "Dipartimenti di Eccellenza 2018–2022".

References

1. Al-Halah, Z., Stiefelhagen, R., Grauman, K.: Fashion forward: Forecasting visual style in fashion. In: ICCV (2017)
2. Anik, A.I., Bunt, A.: Data-centric explanations: Explaining training data of machine learning systems to promote transparency. In: Proceedings of the 2021 CHI Conference on Human Factors in Computing Systems (2021)
3. Arvan, M., Fahimnia, B., Reisi, M., Siemsen, E.: Integrating human judgement into quantitative forecasting methods: A review. Omega 86 (2019)
4. Box, G., Jenkins, G., Reinsel, G., Ljung, G.: Time Series Analysis: Forecasting and Control. John Wiley & Sons (2015)
5. Chen, L., Ng, R.: On the marriage of LP-norms and edit distance. In: Proceedings of the Thirtieth international conference on Very Large Data Bases, vol. 30 (2004)
6. Chen, X., Gupta, A.: Webly supervised learning of convolutional networks. In: 2015 IEEE International Conference on Computer Vision (ICCV), pp. 1431–1439. IEEE Computer Society, Los Alamitos, CA, USA (2015). https://doi.org/10.1109/ICCV.2015.168, https://doi.ieeecomputersociety.org/10.1109/ICCV.2015.168
7. Cheng, W.H., Song, S., Chen, C.Y., Hidayati, S.C., Liu, J.: Fashion meets computer vision: a survey. ACM Comput. Surv. (CSUR) **54**(4), 1–41 (2021)
8. Deng, J., et al.: ImageNet: a large-scale hierarchical image database. In: 2009 IEEE Conference on Computer Vision and Pattern Recognition (2009). https://doi.org/10.1109/CVPR.2009.5206848

9. Ekambaram, V., Manglik, K., Mukherjee, S., Sajja, S.S.K., Dwivedi, S., Raykar, V.: Attention based multi-modal new product sales time-series forecasting. In: Proceedings of the 26th ACM SIGKDD International Conference on Knowledge Discovery & Data Mining. ACM, Virtual Event CA USA, August 2020. https://doi.org/10.1145/3394486.3403362, https://dl.acm.org/doi/10.1145/3394486.3403362
10. Fergus, R., Fei-Fei, L., Perona, P., Zisserman, A.: Learning object categories from Google's image search. In: Tenth IEEE International Conference on Computer Vision (ICCV 2005) Volume 1, vol. 2, pp. 1816–1823 (2005). https://doi.org/10.1109/ICCV.2005.142
11. Fildes, R., Ma, S., Kolassa, S.: Retail forecasting: research and practice. Int. J. Forecast. **38**, 1283–1318 (2019)
12. Garcia, C.C.: Fashion forecasting: an overview from material culture to industry. J. Fashion Mark. Manage. Int. J. **26**, 436–451 (2021)
13. He, K., Zhang, X., Ren, S., Sun, J.: Deep residual learning for image recognition (2015)
14. Hyndman, R., Athanasopoulos, G.: Forecasting: Principles and Practice, 2nd edn. OTexts, Australia (2018)
15. Ilic, I., Görgülü, B., Cevik, M., Baydoğan, M.G.: Explainable boosted linear regression for time series forecasting. Pattern Recogn. **120**, 108144 (2021)
16. Jeon, Y., Jin, S., Kim, B., Han, K.: FashionQ: an interactive tool for analyzing fashion style trend with quantitative criteria. In: Extended Abstracts of the 2020 CHI Conference on Human Factors in Computing Systems (2020)
17. Jing, Y., Baluja, S.: VisualRank: applying pageRank to large-scale image search. IEEE Trans. Pattern Anal. Mach. Intell. **30**, 1877–1890 (2008)
18. Li, J., et al.: Learning from large-scale noisy web data with ubiquitous reweighting for image classification. IEEE Trans. Pattern Anal. Mach. Intell. **43**(5), 1808–1814 (2021). https://doi.org/10.1109/TPAMI.2019.2961910
19. Liu, Z., Luo, P., Qiu, S., Wang, X., Tang, X.: DeepFashion: powering robust clothes recognition and retrieval with rich annotations. In: Proceedings of IEEE Conference on Computer Vision and Pattern Recognition (CVPR), June 2016
20. Lo, L., Liu, C., Lin, R., Wu, B., Shuai, H., Cheng, W.: Dressing for Attention: Outfit Based Fashion Popularity Prediction. In: 2019 IEEE International Conference on Image Processing (ICIP), September 2019. https://doi.org/10.1109/ICIP.2019.8803461, ISSN: 2381-8549
21. Loshchilov, I., Hutter, F.: Decoupled weight decay regularization. In: International Conference on Learning Representations (2018)
22. Ma, Y., Ding, Y., Yang, X., Liao, L., Wong, W.K., Chua, T.S.: Knowledge enhanced neural fashion trend forecasting. In: Proceedings of the 2020 International Conference on Multimedia Retrieval. ACM, Dublin Ireland, June 2020. https://doi.org/10.1145/3372278.3390677, https://dl.acm.org/doi/10.1145/3372278.3390677
23. Ma, Y., Ding, Y., Yang, X., Liao, L., Wong, W.K., Chua, T.S.: Knowledge enhanced neural fashion trend forecasting. In: Proceedings of the 2020 International Conference on Multimedia Retrieval. ICMR 2020, Association for Computing Machinery, New York, NY, USA (2020). https://doi.org/10.1145/3372278.3390677
24. McAuley, J., Targett, C., Shi, Q., van den Hengel, A.: Image-based recommendations on styles and substitutes. In: Proceedings of the 38th International ACM SIGIR Conference on Research and Development in Information Retrieval, pp. 43–52. SIGIR 2015, Association for Computing Machinery, New York, NY, USA (2015). https://doi.org/10.1145/2766462.2767755
25. Motamedi, M., Sakharnykh, N., Kaldewey, T.: A data-centric approach for training deep neural networks with less data. arXiv preprint arXiv:2110.03613 (2021)

26. Ng, A.: A chat with Andrew on MLOps: From model-centric to data-centric AI, May 2021. https://www.youtube.com/watch?v=06-AZXmwHjo
27. Northcutt, C., Jiang, L., Chuang, I.: Confident learning: estimating uncertainty in dataset labels. J. Artif. Intell. Res. **70**, 1373–1411 (2021)
28. Northcutt, C.G., ChipBrain, M., Athalye, A., Mueller, J.: Pervasive label errors in test sets destabilize machine learning benchmarks. stat 1050 (2021)
29. Ren, S., Chan, H.L., Ram, P.: A comparative study on fashion demand forecasting models with multiple sources of uncertainty. Ann. Oper. Res. **257**(1), 335–355 (2017)
30. Ren, S., Chan, H.L., Siqin, T.: Demand forecasting in retail operations for fashionable products: methods, practices, and real case study. Ann. Oper. Res. **291**(1), 761–777 (2020)
31. Shazeer, N., Stern, M.: Adafactor: Adaptive learning rates with sublinear memory cost. In: Dy, J., Krause, A. (eds.) Proceedings of the 35th International Conference on Machine Learning. Proceedings of Machine Learning Research, vol. 80. PMLR, 10–15 July 2018. https://proceedings.mlr.press/v80/shazeer18a.html
32. Silva, E.S., Hassani, H., Madsen, D.Ø., Gee, L.: Googling fashion: forecasting fashion consumer behaviour using google trends. Soc. Sci. **8**(4), 111 (2019)
33. Singh, P.K., Gupta, Y., Jha, N., Rajan, A.: Fashion retail: forecasting demand for new items. In: 26th ACM SIGKDD Conference on Knowledge Discovery and Data Mining, June 2019. http://arxiv.org/abs/1907.01960
34. Skenderi, G., Joppi, C., Denitto, M., Cristani, M.: Well googled is half done: multimodal forecasting of new fashion product sales with image-based google trends. arXiv preprint arXiv:2109.09824 (2021)
35. Song, H., Kim, M., Lee, J.G.: Selfie: Refurbishing unclean samples for robust deep learning. In: International Conference on Machine Learning. PMLR (2019)
36. Sorger, R., Udale, J.: The Fundamentals of Fashion Design. Bloomsbury Publishing, London (2017)
37. Vaswani, A., et al.: Attention is all you need (2017)
38. Wang, Y., Ma, X., Chen, Z., Luo, Y., Yi, J., Bailey, J.: Symmetric cross entropy for robust learning with noisy labels. In: Proceedings of the IEEE/CVF International Conference on Computer Vision (2019)

Pose Forecasting in Industrial Human-Robot Collaboration

Alessio Sampieri[1], Guido Maria D'Amely di Melendugno[1](✉), Andrea Avogaro[2], Federico Cunico[2], Francesco Setti[2], Geri Skenderi[2], Marco Cristani[2], and Fabio Galasso[1]

[1] Sapienza University of Rome, Rome, Italy
{sampieri,damely,galasso}@di.uniroma1.it
[2] University of Verona, Verona, Italy
{andrea.avogaro,federico.cunico,francesco.setti,geri.skenderi,marco.cristani}@univr.it

Abstract. Pushing back the frontiers of collaborative robots in industrial environments, we propose a new Separable-Sparse Graph Convolutional Network (SeS-GCN) for pose forecasting. For the first time, SeS-GCN bottlenecks the interaction of the spatial, temporal and channel-wise dimensions in GCNs, and it learns sparse adjacency matrices by a teacher-student framework. Compared to the state-of-the-art, it only uses 1.72% of the parameters and it is ~4 times faster, while still performing comparably in forecasting accuracy on Human3.6M at 1 s in the future, which enables cobots to be aware of human operators. As a second contribution, we present a new benchmark of Cobots and Humans in Industrial COllaboration (CHICO). CHICO includes multi-view videos, 3D poses and trajectories of 20 human operators and cobots, engaging in 7 realistic industrial actions. Additionally, it reports 226 genuine collisions, taking place during the human-cobot interaction. We test SeS-GCN on CHICO for two important perception tasks in robotics: human pose forecasting, where it reaches an average error of 85.3 mm (MPJPE) at 1 sec in the future with a run time of 2.3 ms, and collision detection, by comparing the forecasted human motion with the known cobot motion, obtaining an F1-score of 0.64.

Keywords: Human pose forecasting · Graph convolutional networks · Human-robot collaboration in industry

1 Introduction

Collaborative robots (cobots) and modern Human Robot Collaboration (HRC) depart from the traditional separation of functions of industrial robots [36], because of the shared workspace [32]. Additionally cobots and humans perform

---

Supplementary Information The online version contains supplementary material available at https://doi.org/10.1007/978-3-031-19839-7_4.

S. Avidan et al. (Eds.): ECCV 2022, LNCS 13698, pp. 51–69, 2022.
https://doi.org/10.1007/978-3-031-19839-7_4

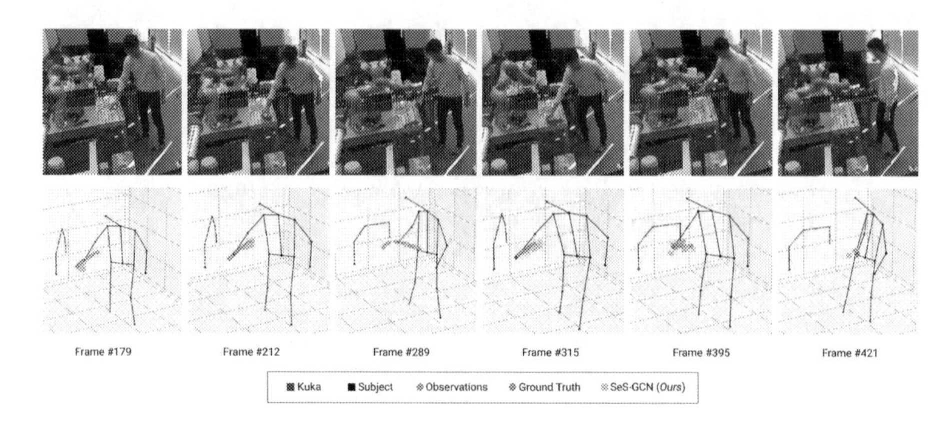

Fig. 1. A collision example from our CHICO dataset. On the top row some frames of the *Lightweight pick and place* action captured by one of the three cameras. On the bottom row, operator + robot skeletons. The forecasting takes an observation sequence (in yellow, here pictured for the right wrist only), and performs a prediction (cyan) which is compared with the ground truth (green). On frame 395 is it easy to see the robot hitting the operator, which is retracting, as it is evident in frame 421. See how the predictions by SeS-GCN follow closely the GT, except during the collision. At collision time, due to the impact, the abrupt change of the arm motion produces uncertain predictions, as it shows from the very irregular predicted trajectory. (Color figure online)

actions concurrently and they will therefore physically engage in contact. While there is a clear advantage in increased productivity [62] (improved by as much as 85% [63]) due to the minimization of idle times, there are challenges in the workplace safety [24]: it is not about whether there will be contact, but rather about understanding its consequences [53].

The pioneering work of Shah et al. [63] has already shown that, in order to seamlessly and efficiently interact with human co-workers, cobots need to abide by two collaborative principles: (1) Making decisions on-the-fly, and (2) Considering the consequences of their decision on their teammates. The first calls for promptly and accurately detecting the human motion in the workspace. The second principle implies that cobots need to anticipate pose trajectories of their human co-workers and predict future collisions.

Motivated by these problems, the first contribution of our work is a novel Separable-Sparse Graph Convolutional Neural Network (SeS-GCN) for human pose forecasting. Pose forecasting requires understanding of the complex spatio-temporal joint dynamics of the human body and recent trends have highlighted the promises of modelling body kinematics within a single GCN framework [15,17,42,44,50,69,74]. We have designed SeS-GCN with performance and efficiency in mind, by bringing together, for the first time, three main modelling principles: depthwise-separable graph convolutions [37], space-time separable graph adjacency matrices [65], and sparse graph adjacency matrices [64]. In SeS-GCN, *separable* stands for limiting the interplay of joints with others (space), at different frames (time) and per channel (depth-wise). Within the GCN, different channels, frames and joints still interact by means of multi-hop messages For the first time, sparsity is achieved by a teacher-student framework. The reduced interaction and sparsity results in comparable or less parameters than all GCN-based baselines [37,64,65], while improving performance by at least 2.7%. Compared to the state-of-the-art (SoA) [49], SeS-GCN is lightweight, only using 1.72% of its parameters, it is ~4 times faster, while remaining competitive with just 1.5% larger error on Human 3.6M [31] when predicting 1 sec in the future. The model is described in detail in Sect. 3, experiments and ablation studies are illustrated in Sect. 5.

We also introduce the very first benchmark of Cobots and Humans in Industrial COllaboration (CHICO, an excerpt in Fig. 1). CHICO includes multi-view videos, 3D poses and trajectories of the joints of 20 human operators, in close collaboration with a robotic arm *KUKA LBR iiwa* within a shared workspace. The dataset features 7 realistic industrial actions, taken at a real industrial assembly line with a marker-less setup. The goal of CHICO is to endow cobots with perceptive awareness to enable human-cobot collaboration with contact. Towards this frontier, CHICO proposes to benchmark two key tasks: human pose forecasting and collision detection. Cobots currently detect collisions by mechanical-only events (transmission of contact wrenches, control torques, sensitive skins). This ensures safety but it harms the human-cobot interaction, because collisions break the motion of both, which reduces productivity, and may be annoying to the human operator. CHICO features 240 1-minute video recordings, from which two separate sets of test sequences are selected: one for estimating the accuracy in pose forecasting, so cobots may be aware of the next future (1.0 sec); and one with 226 genuine collisions, so cobots may foresee them and possibly re-plan. The dataset is detailed in Sect. 4, experiments are illustrated in Sect. 6.

When tested on CHICO, the proposed SeS-GCN outperforms all baselines and reaches an error of 85.3 mm (MPJPE) at 1.00 sec, with a negligible run time of 2.3 ms (as reported in Table 5). Additionally, the forecast human motion is used to detect human-cobot collision, by checking whether the predicted trajectory of the human body intersects that of the cobot. This is also encouraging, as SeS-GCN allows to reach an F1-score of 0.64. Both aspects contribute to a cobot awareness of the future, which is instrumental for HRC in industrial applications.

2 Related Work

Human Pose Forecasting. Human pose forecasting is a recent field which has some intersection with human action anticipation in computer vision [42] and HRC [18]. Previous studies exploited Temporal Convolutional Networks (TCNs) [2,22,41,59] and Recurrent Neural Networks (RNNs) [20,23,33]. Both architectures are naturally suited to model the temporal dimension. Recent works have expanded the range of available methods by using Variational Auto-Encoders [8], specific and model-agnostic layers that implicitly model the spatial structure of the human skeleton [1], or Transformer Networks [9].

Table 1. Comparison between the state-of-the-art datasets and the proposed CHICO; *unk* stands for "unknown".

	Quantitative details							Rec.Scene	Actions type		Tasks			Markerless
	#Classes	#Subj.	Avg Rec.Time	#Joints	FPS	AspectRatio	#Sensors		Industr.	HRC	ActionRecog.	PoseForec.	Coll.Det.	
Human3.6M [31]	15	11	100.49 s	32	25	Normalized	15	mo-capstudio				✓		
AMASS [48]	11265	344	12.89 s	Variable	Variable	Original	variable	mo-capstudio				✓		
3DPW [51]	47	7	28.33 s	18	60	Original	18	outdoorlocations				✓		
ExPI [25]	16	4	*unk*	18	25	Original	88	mo-capstudio				✓		
CHI3D [19]	8	6	*unk*	*unk*	*unk*	Original	14	mo-capstudio				✓		
InHARD [16]	14	16	< 8 s	17	120	Original	20	assemblyline	✓	✓	✓			
CHICO (ours)	7	20	55 s	15	25	Original	3	assemblyline	✓	✓		✓	✓	✓

Pose Forecasting Using Graph Convolutional Networks (GCN). Most recent research uses GCNs [17,44,49,65,74]. In [49], the authors have mixed GCN for modelling the joint-joint interaction with Transformer Networks for the temporal patterns. Others [44,65,74] have adopted GCNs to model the space-time body kinematics, devising, in the case of [17], hierarchical architectures to model coarse-to-fine body dynamics.

We identify three main research directions for improving efficiency in GCNs: **i.** space-time separable GCNs [65], which factorizes the spatial joint-joint and temporal patterns of the adjacency matrix; **ii.** depth-wise separable graph convolutions [30], which has been explored by [3] in the spectral domain; and **iii.** sparse GCNs [64], which iteratively prunes the terms of the adjacency matrix of a GCN. Notably, all three techniques also yield better performance than the plain GCN. Here, for the first, we bring together these three aspects into an end-to-end space-time-depthwise-separable and sparse GCN. The three techniques are complementary to improve both efficiency and performance, but their integration requires some structural changes (*e.g.*, adopting teacher-student architectures for sparsifying), as we describe in Sect. 3.

Human Robot Collaboration (HRC). HRC is the study of collaborative processes where human and robot agents work together to achieve shared goals [4,11]. Computer vision studies on HRC are mostly related to pose estimation [10,21,40] to locate the articulated human body in the scene.

In [12,34,57], methodologies for robot motion planning and collision avoidance are proposed; their study perspective is opposite to ours, since we focus on the human operator. In this regard, the works of [5,14,35,45] model the operators' whereabouts through detection algorithms which approximate human shapes using simple bounding boxes. Approaches that predict the human motion during collaborative tasks are in [66,73] using RNNs and in [68] using Guassian processes. Other work [38] models the upper body and the human right hand (which they call the Human End Effector) by considering the robot-human handover phase. As motion prediction engine, DCT-RNN-GCN [49] is considered, against which we compare in the experiments.

Datasets for Pose Forecasting. Human pose forecasting datasets cover a wide spectrum of scenarios, see Table 1 for a comparative analysis. Human3.6M [31] considers everyday actions such as conversing, eating, greeting and smoking. Data were acquired using a 3D marker-based motion capture system, composed of 10 high-speed infrared cameras. AMASS [48] is a collection of 15 datasets where daily actions were captured by an optical marker-based motion capture. Human3.6M and AMASS are standard benchmarks for human pose forecasting, with some overlap in the type of actions they deal with. The 3DPW dataset [51] focuses on outdoor actions, captured with a moving camera and 17 Inertial Measurement Units (IMU), embedded on a special suit for motion capturing [60]. The recent ExPI dataset [25] contains 16 different dance actions performed by professional dancers, for a total of 115 sequences, and it is aimed at motion prediction. ExPI has been acquired with 68 synchronised and calibrated color cameras and a motion capture system with 20 mocap cameras. Finally, the CHI3D dataset [19] reports 3D data taken from MOCAP systems to study human interactions.

None of these datasets answer our research needs, i.e., a benchmark taken by a sparing, energy-efficient markerless system, focused on the industrial HRC scenario, where forecasting may be really useful for anticipating collisions between the humans and robots. In fact, the only dataset relating to industrial applications is InHARD [16]. Therein, humans are asked to perform an assembly task while wearing inertial sensors on each limb. The dataset is designed for human action recognition, and it involves 16 individuals performing 13 different actions each, for a total of 4800 action samples over more than 2 million frames. Despite showcasing a collaborative robot, in this dataset the robot is mostly static, making it unsuitable for collision forecasting.

3 Methodology

We build an accurate, memory efficient and fast GCN by bridging three diverse research directions: **i.** Space-time separable adjacency matrices; **ii.** Depth-wise separable graph convolutions; **iii.** Sparse adjacency matrices. This results in an all-Separable and Sparse GCN encoder for the human body kinematics, which we dub SeS-GCN, from which the future frames are forecast by a Temporal Convolutional Network (TCN).

3.1 Background

Problem Formalization. Pose forecasting is formulated as observing the 3D coordinates $\boldsymbol{x}_{v,t}$ of V joints across T frames and predicting their location in the K future frames. For convenience of notation, we gather the coordinates from all joints at frame t into the matrix $X_t = [\boldsymbol{x}_{v,t}]_{v=1}^{V} \in \mathbb{R}^{3\times V}$. Then we define the tensors $\mathcal{X}_{in} = [X_1, X_2..., X_T]$ and $\mathcal{X}_{out} = [X_{T+1}, X_{T+2}..., X_{T+K}]$ that contain all observed input and target frames, respectively.

We consider a graph $\mathcal{G} = (\mathcal{V}, \mathcal{E})$ to encode the body kinematics, with all joints at all observed frames as the node set $\mathcal{V} = \{\boldsymbol{v}_{i,t}\}_{i=1,t=1}^{V,T}$, and edges $(\boldsymbol{v}_{i,t}, \boldsymbol{v}_{j,s}) \in \mathcal{E}$ that connect joints i, j at frames t, s.

Graph Convolutional Networks (GCN). A GCN is a layered architecture:

$$\mathcal{X}^{(l+1)} = \sigma\left(A^{(l)}\mathcal{X}^{(l)}W^{(l)}\right) \tag{1}$$

The input to a GCN layer l is the tensor $\mathcal{X}^{(l)} \in \mathbb{R}^{C^{(l)} \times V \times T}$ which maintains the correspondence to the V body joints and the T observed frames, but increases the depth of features to $C^{(l)}$ channels. $\mathcal{X}^{(1)} = \mathcal{X}_{in}$ is the input tensor at the first layer, with $C^{(1)} = 3$ channels given by the 3D coordinates. $A^{(l)} \in \mathbb{R}^{VT \times VT}$ is the adjacency matrix relating pairs of VT joints from all frames. Following most recent literature [17,49,64,65], $A^{(l)}$ is learnt. $W^{(l)} \in \mathbb{R}^{C^{(l)} \times 1 \times 1}$ are the learnable weights of the graph convolutions. σ is a the non-linear PReLU activation function.

3.2 Separable and Sparse Graph Convolutional Networks (SeS-GCN)

We build SeS-GCN by integrating the three mentioned modelling dimensions: **i.** separating spatial and temporal interaction terms in the adjacency matrix of a GCN; **ii.** separating the graph convolutions depth-wise; **iii.** sparsifying the adjacency matrices of the GCN.

Separating Space-Time. STS-GCN [65] has factored the adjacency matrix $A^{(l)}$ of the GCN, at each layer l, into the product of two terms $A_s^{(l)} \in \mathbb{R}^{V \times V \times T}$ and $A_t^{(l)} \in \mathbb{R}^{T \times T \times V}$, respectively responsible for the temporal-temporal and joint-joint relations. The GCN formulation becomes:

$$\mathcal{X}^{(l+1)} = \sigma\left(A_s^{(l)} A_t^{(l)} \mathcal{X}^{(l)} W^{(l)}\right) \tag{2}$$

Eq. (2) bottlenecks the interplay of joints across different frames, implicitly placing more emphasis on the interaction of joints on the same frame ($A_s^{(l)}$) and on the temporal pattern of each joint ($A_t^{(l)}$). This reduces the memory-footprint of a GCN by approx. 4x while improving its performance (cf. Sect. 5.1). Note that this differs from alternating spatial and temporal modules, as it is done in [71] and [7], respectively for trajectory forecasting and action recognition.

Separating Depth-Wise. Inspired by depth-wise convolutions [13,30], the approach in [37] has introduced depth-wise graph convolutions for image classification, followed by [3] which resorted to a spectral formulation of depth-wise graph convolutions for graph classification. Here we consider depth-wise graph convolutions for pose forecasting. The depth-wise formulation bottlenecks the interplay of space and time (operated by the adjacency matrix $A^{(l)}$) with the channels of the graph convolution $W^{(l)}$. The resulting all-separable model which we dub STS-DW-GCN is formulated as such:

$$\mathcal{H}^{(l)} = \gamma\left(A_s^{(l)} A_t^{(l)} \mathcal{X}^{(l)} W_{DW}^{(l)}\right) \tag{3a}$$

$$\mathcal{X}^{(l+1)} = \sigma\left(\mathcal{H}^{(l)} W_{\text{MLP}}^{(l)}\right) \tag{3b}$$

Adding the depth-wise graph convolution splits the GCN of layer l into two terms. The first, Eq. (3a), focuses on space-time interaction and limits the channel cross-talk by the use of $W_{DW}^{(l)} \in \mathbb{R}^{\frac{C^{(l)}}{\alpha} \times 1 \times 1}$, with $1 \leq \alpha \leq C^{(l)}$ setting the number of convolutional groups ($\alpha = C^{(l)}$ is the plain single-group depth-wise convolution). The second, Eq. (3b), models the intra-channel communication just. This may be understood as a plain (MLP) 1D-convolution with $W_{\text{MLP}}^{(l)} \in \mathbb{R}^{C^{(l)} \times 1 \times 1}$ which re-maps features from $C^{(l)}$ to $C^{(l+1)}$. γ is the ReLU6 non-linear activation function. Overall, this does not significantly reduce the number of parameters, but it deepens the GCN without over-smoothing [58], which improves performance (see Sect. 5.1 for details).

Sparsifying the GCN. Sparsification has been used to improve the efficiency (memory and, in some cases, runtime) of neural networks since the seminal pruning work of [39]. [64] has sparsified GCNs for trajectory forecasting. This consists in learning masks $\mathcal{M}$ which selectively erase certain parameters in the adjacency matrix of the GCN. Here we integrate sparsification with the all-separable GCN design, which yields our proposed SeS-GCN for human pose forecasting:

$$\mathcal{H}^{(l)} = \gamma \left((\mathcal{M}_s^{(l)} \odot A_s^{(l)})(\mathcal{M}_t^{(l)} \odot A_t^{(l)}) \mathcal{X}^{(l)} W_{DW}^{(l)} \right) \tag{4a}$$

$$\mathcal{X}^{(l+1)} = \sigma \left(\mathcal{H}^{(l)} W_{\text{MLP}}^{(l)} \right) \tag{4b}$$

$\odot$ is the element-wise product and $\mathcal{M}_{\{s,t\}}^{(l)}$ are binary masks. Both at training and inference, [64] generates masks, it uses those to zero certain coefficients of the adjacency matrix A, and it adopts the resulting GCN for trajectory forecasting. By contrast, we adopt a teacher-student framework during training. The teacher learns the masks, and the student only considers the spared coefficients in A. At inference, our proposed SeS-GCN only consists of the student, which simply adopts the learnt sparse A_s and A_t. Compared to [64], the approach of SeS-GCN is more robust at training, it yields fewer model parameters at inference (~30% less for both A_s and A_t), and it reaches a better performance, as it is detailed in Sect. 5.1.

3.3 Decoder Forecasting

Given the space-time representation, as encoded by the SeS-GCN, the future frames are then decoded by using a temporal convolutional network (TCN) [2,22,41,65]. The TCN remaps the temporal dimension to match the sought output number of predicted frames. This part of the model is not considered for improvement because it is already efficient and it performs satisfactorily.

4 The CHICO dataset

In this section the CHICO dataset is detailed by describing the acquisition scenario and devices, the cobot and the performed actions. We release RGB videos, skeletons and calibration parameters[1].

[1] Code and dataset are available at: https://github.com/AlessioSam/CHICO-PoseForecasting.

The Scenario. We are in a smart-factory environment, with a single human operator standing in front of a 0.9 m $\times$ 0.6 m workbench and a cobot at its end (see Fig. 1). The human operator has some free space to turn towards some equipment and carry out certain assembly, loading and unloading actions [54]. In particular, light plastic pieces and heavy tiles, a hammer, abrasive sponges are available. The detailed setups for each action are reported graphically in the additional material. A total of 20 human operators have been hired for this study. They attended a course on how to operate with the cobot and signed an informed consent form prior to the recordings.

The Collaborative Robot. A 7 degrees-of-freedom Kuka LBR iiwa 14 R820 collaborates with the human operator during the data acquisition process. Weighing in at 29.5 kg and with the ability to handle a payload up to 14 kg, it is widely used in modern production lines. More details on the cobot can be found in the supplementary material.

The Acquisition Setup. The acquisition system is based on three RGB HD cameras providing three different viewpoints of the same workplace: two frontal-lateral, one rear view. The frame rate 25 Hz. The videos were first checked for erroneous or spurious frames, then we used Voxelpose [67] to extract 3D human pose for each frame. Extrinsic parameters of each camera are estimated w.r.t. the robot's reference frame by means of a calibration chessboard of 1$\times$1m, and temporal alignment is guaranteed by synchronization of all the components with an Internet Time Server. In our environment, Voxelpose estimates a joint positioning accuracy in terms of Mean Per Joint Position Error (MPJPE) of 24.99mm using three cameras, which is enough for our purposes, as an ideal compromise between portability of the system and accuracy. We confirm these numbers in two ways: the first is by checking that human-cobot collisions were detected with 100% F1 score (we have a collision when the minimum distance between the human limbs and the robotic links is below a predefined threshold). Secondly, we show that the new CHICO dataset does not suffer from a trivial zero velocity solution [52], *i.e.* results achieved by a zero velocity model underperform the current SoA in equal proportion as for the large-scale established Human3.6M.

Actions. The 7 types of actions of CHICO are inspired from ordinary work sessions in an HRC disassembly line as described in the review work of [29]. Each action is repeated over a time interval of $\sim$1 min on average. Each action is associated to a goal that the human operator has to achieve by a given time limit, which requires them to move with a certain velocity. Each action consists of repeated interactions with the robot (*e.g.*, robot place, human picks) which, due to the limited space, lead to some *unconstrained collisions*[2] which we label accordingly. Globally, from the 7 actions $\times$ 20 operators, we collect 226 different collisions. On the additional material, an excerpt of the videos with collisions are available. In the following, each action is shortly described.

[2] Unconstrained collisions is a term coming from [26], indicating a situation in which only the robot and human are directly involved into the collision.

- ***Lightweight pick and place*** (*Light P&P*). The human operator is required to move small objects of approximately 50 g from a loading bay to a delivery location within a given time slot. The bay and the delivery location are at the opposite sides of the workbench. Meanwhile, the robot loads on of this bay so that the human operator has to pass close to the robotic arm. In many cases the distance between the limbs and the robotic arm is few centimeters.
- ***Heavyweight pick and place*** (*Heavy P&P*). The setup of this action is the same as before, but the objects to be moved are floor tiles weighing 0.75 kg. This means that the actions have to be carried out with two hands.
- ***Surface polishing*** (*Polishing*). This action was inspired by [47], where the human operator polishes the border of a 40 by 60cm tile with some abrasive sponge, and the robot mimics a visual quality inspection.
- ***Precision pick and place*** (*Prec. P&P*). The robot places four plastic pieces in the four corners of a 30×30 cm table in the center of the workbench, and the human has to remove them and put on a bay, before the robot repeats the same unloading.
- ***Random pick and place*** (*Rnd. P&P*). Same as the previous action, except for the plastic pieces which were continuously placed by the robot randomly on the central 30×30cm table, and the human operator has to remove them.
- ***High shelf lifting*** (*High lift*). The goal was to pick light plastic pieces (50 g each) on a sideway bay filled by the robot, putting them on a shelf located at 1.70m, at the opposite side of the workbench. Due to the geometry of the workspace, the arms of the human operator were required to pass above or below the moving robotic arm. In this way, close distances between the human arm and forearm and the robotic links were realized.
- ***Hammering*** (*Hammer*). The operator hits with a hammer a metallic tide held by the robot. In this case, the interest was to check how much the collision detection is robust to an action where the human arm is colliding close to the robotic arm (that is, on the metallic tile) without properly colliding *with the robotic arm.*

5 Experiments on Human3.6M

We benchmark the proposed SeS-GCN model on the large and established Human3.6M [31]. In Sect. 5.1, we analyze the design choices corresponding to the models discussed in Sect. 3, then we compare with the state-of-the-art in Sect. 5.2.

Human3.6M [31] is an established dataset for pose forecasting, consisting of 15 daily life actions (e.g. Walking, Eating, Sitting Down). From the original skeleton of 32 joints, 22 are sampled as the task, representing the body kinematics. A total of 3.6 million poses are captured at 25 fps. In line with the literature [17,49,52], subjects 1, 6, 7, 8, 9 are used for training, subject 11 for validation, and subject 5 for testing.

Metric. The prediction error is quantified via the MPJPE error metric [31,50], which considers the displacement of the predicted 3D coordinates w.r.t. the ground truth, in millimeters, at a specific future frame t:

$$L_{\text{MPJPE}} = \frac{1}{V} \sum_{v=1}^{V} ||\hat{\boldsymbol{x}}_{vt} - \boldsymbol{x}_{vt}||_2. \tag{5}$$

5.1 Modelling Choices of SeS-GCN

We review and quantify the impact of the modelling choices of SeS-GCN:

Efficient GCN Baselines. In Table 2, we first validate the three difference modelling approaches to efficient GCNs, namely space-time separable STS-GCN [65], depth-wise separable graph convolutions DW-GCN [37], and Sparse-GCN [64]. STS-GCN yields the lowest MPJPE error of 117.0 mm at a 1 sec forecasting horizon (2.4% better than DW-GCN, 4.8% better than Sparse-GCN) with the fewest parameters, 57.6k (ca. x4 less). We build therefore on this approach.

Deeper GCNs. It is a long standing belief that Deep Neural Networks (DNN) owe their performance to depth [27,46,70,72]. However, deeper models require more parameters and have a longer processing time. Additionally, deeper GCNs may suffer from over-smoothing [58]. Seeking both better accuracy and efficiency, we consider three pathways for improvement: (1) add GCN layers; (2) add MLP layers between layers of GCNs; (3) adopt depth-wise graph convolutions, which also add MLP layers between GCN ones (cf. Sect. 3.2).

As shown in Table 2, there is a slight improvement in performance with 5 STS-GCN layers (MPJPE of 115.9 mm), but deeper models underperform. Adding MLP layers between the GCN ones (depth of 5+5) also decreases performance (MPJPE of 125.2). By contrast, adding depth by depth-wise separable graph convolutions (STS-DW-GCN of depth 5+5) reduces the error to 114.8 mm. This may be explained by the virtues of the increased depth in combination

Table 2. MPJPE error (millimeters) for long-range predictions (25 frames) on Human3.6M [31] and numbers of parameters. Best figures overall are reported in bold, while underlined figures represent the best in each block. The proposed model has comparable or less parameters than the GCN-based baselines [30,64,65] and it outperforms the best of them [65] by 2.6%.

	Depth	MPJPE	Parameters (K)	DW-separable	ST-separable	Sparse	w/ MLP layers	Teacher-student
GCN	4	123.2	222.7					
DW-GCN [37]	4+4	119.8	223.2	✓			✓	
STS-GCN[a] [65]	4	117.0	**57.6**		✓			
Sparse-GCN [64]	4	122.7	257.9			✓		
STS-GCN	5	115.9	68.6		✓			
STS-GCN	6	116.1	79.9		✓			
STS-GCN w/ MLP	5+5	125.2	101.4		✓		✓	
STS-DW-GCN	5+5	114.8	70.0	✓	✓		✓	
STS-DW-Sparse-GCN	5+5	115.7	122.4	✓	✓	✓	✓	
SeS-GCN (proposed)	5+5	**113.9**	58.6	✓	✓	✓	✓	✓

[a]Results for STS-GCN differ from [65], due to revision by the authors, cf. https://github.com/FraLuca/STSGCN

with the limiting cross-talk of joint-time-channels, which existing literature confirms [13,37,65]. We note that space-time and depth-wise channel separability is complementary. Altogether, this performance is beyond the STS-GCN performance (114.8 Vs. 117.0 mm), at a slight increase of the parameter count ($70k$ Vs. $57.6k$).

Sparsifying GCNs and the Proposed SeS-GCN. Finally, we target to improve efficiency by model compression. Trends have reduced the size of models by reducing the parameter precision [61], by pruning and sparsifying some of the parameters [56], or by constructing teacher-student frameworks, whereby a smaller student model is paired with a larger teacher to reach its same performance [28,43]. Note that the last technique is the current go-to choice in deploying very large networks such as Transformers [6].

We start off by compressing the model with sparse adjacency matrices by the approach of Sparse-GCN [64]. They iteratively optimize the learnt parameters and the masks to select some (the selection occurs by a network branch, also at inference, cf. Sect. 3.2). As illustrated in Table 2, the approach of [64] does not make a viable direction (STS-DW-Sparse-GCN), since the error increases to 115.7 mm and the parameter count to $122.4k$.

Reminiscent of teacher-student models, in the proposed SeS-GCN we first train a teacher STS-DW-GCN, then use its learnt parameters to sparsify the affinity matrices of a student STS-DW-GCN, which is then trained from scratch. SeS-GCN achieves a competitive parameter count and the lowest MPJPE error of 113.9 mm, being comparable with the current SoA [49] and using only 1.72% of its parameters ($58.6k$ Vs. $3.4M$).

5.2 Comparison with the State-of-the-art (SoA)

In Table 3, we evaluate the proposed SeS-GCN against three most recent techniques, over a short time horizon (10 frames, 400 ms) and a long time horizon (25 frames, 1000 ms). The first, DCT-RNN-GCN [49], the current SoA, uses DCT encoding, motion attention and RNNs and, differently from other models, demands more frames as input (50 vs. 10). The other two, MSR-GCN [17] and STS-GCN [65] adopt GCN-only frameworks, the former adopts a multi-scale approach, the latter acts a separation between spatial and temporal encoding.

Both on Short- and long-term predictions, at the 400 and 1000 ms horizons, the proposed SeS-GCN outperforms other techniques [49,65] and it is within a 1.5% error w.r.t. the current SoA [49], while only using 1.72% parameters and being $\sim$4 times faster than [49].

6 Experiments on CHICO

We benchmark on CHICO the SoA and the proposed SeS-GCN model. The two HRC tasks of human pose forecasting and collision detection are discussed in Sects. 6.1 and 6.2 respectively.

Table 3. MPJPE error in mm for short-term (400 ms, 10 frames) and long-term (1000 ms, 25 frames) predictions of 3D joint positions on Human3.6M. The proposed model achieves competitive performance with the SoA [49], while adopting 1.72% of its parameters and running ∼4 times faster, cf. Table 5. Results are discussed in Sect. 5.2.

	Walking		Eating		Smoking		Discussion		Directions		Greeting		Phoning		Posing	
Time Horizon (msec)	*400*	*1000*	*400*	*1000*	*400*	*1000*	*400*	*1000*	*400*	*1000*	*400*	*1000*	*400*	*1000*	*400*	*1000*
DCT-RNN-GCN [49]	**39.8**	**58.1**	**36.2**	**75.5**	**36.4**	**69.5**	**65.4**	119.8	56.5	106.5	**78.1**	138.8	**49.2**	105.0	**75.8**	178.2
MSR-GCN [17]	45.2	63.0	40.4	77.1	38.1	71.6	69.7	117.5	**53.8**	**100.5**	93.3	147.2	51.2	**104.3**	85.0	174.3
STS-GCN[a] [65]	51.0	70.2	43.3	82.6	42.3	76.1	71.9	118.9	63.2	109.6	86.4	**136.1**	53.8	108.3	84.7	178.4
SeS-GCN (proposed)	48.8	67.3	41.7	78.1	40.8	73.7	70.6	**116.7**	60.3	106.9	83.8	137.2	52.6	106.7	82.6	**173.5**

	Purchases		Sitting		Sitting down		Taking photo		Waiting		Walking dog		Walking together		Average	
Time Horizon (msec)	*400*	*1000*	*400*	*1000*	*400*	*1000*	*400*	*1000*	*400*	*1000*	*400*	*1000*	*400*	*1000*	*400*	*1000*
DCT-RNN-GCN [49]	**73.9**	**134.2**	**56.0**	**115.9**	**72.0**	**143.6**	**51.5**	**115.9**	**54.9**	108.2	**86.3**	**146.9**	**41.9**	**64.9**	**58.3**	**112.1**
MSR-GCN [17]	79.6	139.1	57.8	120.0	76.8	155.4	56.3	121.8	59.2	**106.2**	93.3	148.2	43.8	65.9	62.9	114.1
STS-GCN[a] [65]	83.1	141.0	60.8	121.4	79.4	148.4	59.4	126.3	62.0	113.6	97.3	151.5	49.1	72.5	65.8	117.0
SeS-GCN (proposed)	82.2	139.1	59.9	117.5	78.1	146.0	57.7	121.2	58.5	107.5	94.0	147.7	48.3	70.8	64.0	113.9

[a]Results for STS-GCN differ from [65], due to revision by the authors, cf. https://github.com/FraLuca/STSGCN

6.1 Pose Forecasting Benchmark

Here we describe the evaluation protocol proposed for CHICO and report comparative evaluation of pose forecasting techniques.

Evaluation Protocol. We create the train/validation/test split by assigning 2 subjects to the validation (subjects 0 and 4), 4 to the test set (subjects 2, 3, 18 and 19), and the remaining 14 to the training set. For short-range prediction experiments, abiding the setup of Human3.6M [31], we consider 10 frames as observation time and 10 or 25 frames as forecasting horizon. Differently from all reported techniques, DCT-RNN-GCN requires 50 input frames.

We adopt the same Mean Per Joint Position Error (MPJPE) [31] as Human3.6M, in Eq. (5), which also defines the training loss for the evaluated techniques.

None of the motion sequences for pose forecasting contains collisions. In fact, the objective is to train and test the "correct" collaborative human behavior, and not the human retractions and the pauses due to the collisions[3].

Table 4. MPJPE error in mm for short-term (400 ms, 10 frames) and long-term (1000 ms, 25 frames) prediction of 3D joint positions on CHICO dataset. The average error is 7.9% lower than the other models in the short-term and 2.4% lower in the long-term prediction. See Sect. 6.1 for a discussion.

	Hammer		High Lift		Prec. P&P		Rnd. P&P		Polishing		Heavy P&P		Light P&P		Average	
Time horizon (ms)	*400*	*1000*	*400*	*1000*	*400*	*1000*	*400*	*1000*	*400*	*1000*	*400*	*1000*	*400*	*1000*	*400*	*1000*
DCT-RNN-GCN [49]	41.1	**39.0**	69.4	128.8	50.6	83.3	52.7	88.2	42.1	76.0	64.1	121.5	62.1	104.2	54.6	91.6
MSR-GCN [17]	41.6	39.7	67.8	130.2	50.2	81.3	53.4	90.3	41.1	73.2	62.7	118.2	61.5	101.9	54.1	90.7
STS-GCN [65]	46.6	52.1	64.2	116.4	48.3	79.5	52.0	87.9	42.1	73.9	60.6	106.5	57.2	95.2	53.0	87.4
SeS-GCN (proposed)	**40.9**	49.3	**62.1**	**116.3**	**46.0**	**77.4**	**48.4**	**84.8**	**38.8**	**72.4**	**56.1**	**104.4**	**56.2**	**92.2**	**48.8**	**85.3**

[3] After the collisions, the robot stops for 1 s, during which the human operator usually stands still, waiting for the robot to resume operations.

Comparative Evaluation. In Table 4, we compare pose forecasting techniques from the SoA and the proposed SeS-GCN. On the short-term predictions the best performance is that of SeS-GCN, reaching an MPJPE error of 48.8 mm, which is 7.9% better than the second best STS-GCN [65].

On the longer-term predictions, the best performance (MPJPE error of 85.3 mm) is also detained by SeS-GCN, which is 2.4% better than the second best STS-GCN [65]. The proposed model outperforms all techniques on all actions except *Hammer*, a briefly repeating action which may differ for single hits. We argue that DCT-RNN-GCN [49] may get an advantage from using 50 input frames (all other methods use 10 frames)

For a graphical illustration, Fig. 2 shows a distribution of the error per joint calculated over all the actions, for the horizons 400 (*left*) and 1000 ms (*right*). In both cases the error gets larger as we get closer to the extrema of the kinematic skeleton, since those joints move the most. The slightly larger error at the right hands (70.03 and 125.76 mm, respectively) matches that subjects are right-handed (but some actions are operated with both hands).

For a sanity check of results, we have also evaluated the performance of a trivial zero velocity model. [52] has found that keeping the last observed positions may be a surprisingly strong (trivial) baseline. For CHICO, the zero velocity model scores an MPJPE of 110.6 at 25-frames, worse than the 85.3 mm score of SeS-GCN. This is in line with the large-scale dataset Human3.6M [31], where the performance of the trivial model is 153.3 mm.

6.2 Collision Detection Experiments

Evaluation Protocol. We consider a collision to occur when any body limb of the subject gets too close to any part of the cobot, i.e. within a distance threshold, for at least one frame. In particular, a collision refers to the proximity between the cobot and the human in the forecast portion of the trajectory. The (Euclidean) distance threshold is set to 13 cm.

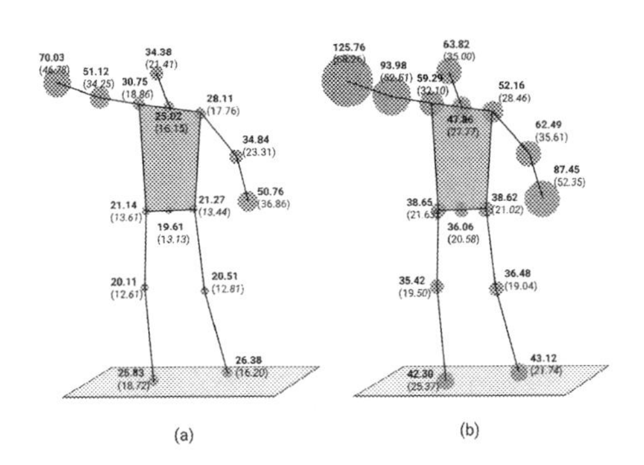

Fig. 2. Average MPJPE distribution for all actions in CHICO on different joints for (a) short-term (0.40 s) and (b) long-term (1.00 s) predictions. The radius of the blob gives the spatial error with the same scale of the skeleton.

The motion of the cobot is scripted beforehand, thus known. The motion of the human subjects in the next 1000 ms needs to be forecast, starting from the observation of 400 ms. The train/validation/test sets sample sequences of 10+25 frames with stride of 10.

Evaluation of Collision Detection. For the evaluation of collision, following [55], both the cobot arm parts and the human body limbs are approximated by cylinders. The diameters for the cobot are fixed to 8 cm. Those of the body limbs are taken from a human atlas.

In Table 5, we report precision, recall and F_1 scores for the detection of collisions on the motion of 2 test subjects, which contains 21 collisions. The top performer in pose forecasting, our proposed SeS-GCN, also yields the largest F_1 score of 0.64. The lower performing MSR-GCN [17] yields poor collision detection capabilities, with an F_1 score of 0.31.

Table 5. Evaluation of collision detection performance achieved by competing pose forecasting techniques, with indication of inference run time. See discussion in Sect. 6.2.

Time horizon (ms)	***1000***			
Metrics	*Prec*	*Recall*	F_1	*Inference time (s)*
DCT-RNN-GCN [49]	0.63	0.58	0.56	9.1×10^{-3}
MSR-GCN [17]	0.63	0.30	0.31	25.2×10^{-3}
STS-GCN [65]	0.68	0.61	0.63	$\mathbf{2.3 \times 10^{-3}}$
SeS-GCN (proposed)	0.84	0.54	**0.64**	$\mathbf{2.3 \times 10^{-3}}$

7 Conclusions

Towards the goal of forecasting the human motion during human-robot collaboration in industrial (HRC) environments, we have proposed the novel SeS-GCN model, which integrates three most recent modelling methodologies for accuracy and efficiency: space-time separable GCNs, depth-wise separable graph convolutions and sparse GCNs. Also, we have contributed a new CHICO dataset, acquired at real assembly line, the first providing a benchmark of the two fundamental HRC tasks of human pose forecasting and collision detection. Featuring an MPJPE error of 85.3 mm at 1 sec in the future with a negligible run time of 2.3 ms, SeS-GCN and CHICO unleash great potential for perception algorithms and their application in robotics.

Acknowledgements. This work was supported by the Italian MIUR through the project "Dipartimenti di Eccellenza 2018–2022", and partially funded by DsTech S.r.l.

References

1. Aksan, E., Kaufmann, M., Hilliges, O.: Structured prediction helps 3d human motion modelling. In: Proceedings of the IEEE/CVF International Conference on Computer Vision (ICCV) (October 2019)
2. Bai, S., Kolter, J.Z., Koltun, V.: An empirical evaluation of generic convolutional and recurrent networks for sequence modeling. arXiv:abs/1803.01271 (2018)
3. Balcilar, M., Renton, G., Héroux, P., Gaüzère, B., Adam, S., Honeine, P.: Spectral-designed depthwise separable graph neural networks. In: Proceedings of Thirty-seventh International Conference on Machine Learning (ICML 2020)-Workshop on Graph Representation Learning and Beyond (GRL+ 2020) (2020)
4. Bauer, A., Wollherr, D., Buss, M.: Human-robot collaboration: a survey. Int. J. Humanoid Rob. **5**(01), 47–66 (2008)
5. Beltran, E.P., Diwa, A.A.S., Gales, B.T.B., Perez, C.E., Saguisag, C.A.A., Serrano, K.K.D.: Fuzzy logic-based risk estimation for safe collaborative robots. In: 2018 IEEE 10th International Conference on Humanoid, Nanotechnology, Information Technology, Communication and Control, Environment and Management (HNICEM), pp. 1–5 (2018)
6. Benesova, K., Svec, A., Suppa, M.: Cost-effective deployment of BERT models in serverless environment (2021)
7. Bertasius, G., Wang, H., Torresani, L.: Is space-time attention all you need for video understanding? In: Proceedings of the International Conference on Machine Learning (ICML) (2021)
8. Bütepage, J., Kjellström, H., Kragic, D.: Anticipating many futures: Online human motion prediction and synthesis for human-robot collaboration. arXiv:abs/1702.08212 (2017)
9. Cai, Y., et al.: Learning progressive joint propagation for human motion prediction. In: Vedaldi, A., Bischof, H., Brox, T., Frahm, J.-M. (eds.) ECCV 2020. LNCS, vol. 12352, pp. 226–242. Springer, Cham (2020). https://doi.org/10.1007/978-3-030-58571-6_14
10. Cao, Z., Hidalgo Martinez, G., Simon, T., Wei, S., Sheikh, Y.A.: OpenPose: real-time multi-person 2d pose estimation using part affinity fields. In: IEEE Transactions on Pattern Analysis and Machine Intelligence (2019)
11. Castro, A., Silva, F., Santos, V.: Trends of human-robot collaboration in industry contexts: handover, learning, and metrics. Sensors **21**(12), 4113 (2021)
12. Chen, J.H., Song, K.T.: Collision-free motion planning for human-robot collaborative safety under cartesian constraint. In: IEEE International Conference on Robotics and Automation, pp. 4348–4354 (2018)
13. Chollet, F.: Xception: deep learning with depthwise separable convolutions. In: 2017 IEEE Conference on Computer Vision and Pattern Recognition (CVPR), pp. 1800–1807 (2017)
14. Costanzo, M., De Maria, G., Lettera, G., Natale, C.: A multimodal approach to human safety in collaborative robotic workcells. IEEE Trans. Autom. Sci. Eng. **19**, 1–15 (2021)
15. Cui, Q., Sun, H., Yang, F.: Learning dynamic relationships for 3d human motion prediction. In: 2020 IEEE/CVF Conference on Computer Vision and Pattern Recognition (CVPR), pp. 6518–6526 (2020)
16. Dallel, M., Havard, V., Baudry, D., Savatier, X.: Inhard - industrial human action recognition dataset in the context of industrial collaborative robotics. In: 2020 IEEE International Conference on Human-Machine Systems (ICHMS) (2020)

17. Dang, L., Nie, Y., Long, C., Zhang, Q., Li, G.: MSR-GCN: Multi-scale residual graph convolution networks for human motion prediction. In: Proceedings of the IEEE/CVF International Conference on Computer Vision (ICCV) (2021)
18. Duarte, N.F., Raković, M., Tasevski, J., Coco, M.I., Billard, A., Santos-Victor, J.: Action anticipation: reading the intentions of humans and robots. IEEE Robot. Autom. Lett. **3**(4), 4132–4139 (2018)
19. Fieraru, M., Zanfir, M., Oneata, E., Popa, A.I., Olaru, V., Sminchisescu, C.: Three-dimensional reconstruction of human interactions. In: Proceedings of the IEEE/CVF Conference on Computer Vision and Pattern Recognition, pp. 7214–7223 (2020)
20. Fragkiadaki, K., Levine, S., Felsen, P., Malik, J.: Recurrent network models for human dynamics. In: 2015 IEEE International Conference on Computer Vision (ICCV), pp. 4346–4354 (2015)
21. Garcia-Esteban, J.A., Piardi, L., Leitao, P., Curto, B., Moreno, V.: An interaction strategy for safe human Co-working with industrial collaborative robots. In: Proceedings of 2021 4th IEEE International Conference on Industrial Cyber-Physical Systems ICPS 2021, pp. 585–590 (2021)
22. Gehring, J., Auli, M., Grangier, D., Yarats, D., Dauphin, Y.N.: Convolutional sequence to sequence learning. In: The International Conference on Machine Learning (ICML) (2017)
23. Gopalakrishnan, A., Mali, A., Kifer, D., Giles, L., Ororbia, A.G.: A neural temporal model for human motion prediction. In: 2019 IEEE/CVF Conference on Computer Vision and Pattern Recognition (CVPR), pp. 12108–12117 (2019)
24. Gualtieri, L., Palomba, I., Wehrle, E.J., Vidoni, R.: The opportunities and challenges of SME manufacturing automation: safety and ergonomics in human–robot collaboration. In: Matt, D.T., Modrák, V., Zsifkovits, H. (eds.) Industry 4.0 for SMEs, pp. 105–144. Springer, Cham (2020). https://doi.org/10.1007/978-3-030-25425-4_4
25. Guo, W., Bie, X., Alameda-Pineda, X., Moreno-Noguer, F.: Multi-person extreme motion prediction with cross-interaction attention. arXiv preprint arXiv:2105.08825 (2021)
26. Haddadin, S., Albu-Schaffer, A., Frommberger, M., Rossmann, J., Hirzinger, G.: The "DLR crash report": Towards a standard crash-testing protocol for robot safety-part i: Results. In: 2009 IEEE International Conference on Robotics and Automation, pp. 272–279. IEEE (2009)
27. He, K., Zhang, X., Ren, S., Sun, J.: Deep residual learning for image recognition. In: 2016 IEEE Conference on Computer Vision and Pattern Recognition (CVPR), pp. 770–778 (2016)
28. Hinton, G., Dean, J., Vinyals, O.: Distilling the knowledge in a neural network. In: NIPS, pp. 1–9 (2014)
29. Hjorth, S., Chrysostomou, D.: Human-robot collaboration in industrial environments: a literature review on non-destructive disassembly. Robot. Comput. Integr. Manuf. **73**, 102–208 (2022)
30. Howard, A.G., et al.: MobileNets: efficient convolutional neural networks for mobile vision applications (2017)
31. Ionescu, C., Papava, D., Olaru, V., Sminchisescu, C.: Human3.6m: large scale datasets and predictive methods for 3d human sensing in natural environments. IEEE Trans. Pattern Anal. Mach. Intell. **36**(7), 1325–1369 (2014)
32. ISO: ISO/TS 15066:2016. Robots and robotic devices - Collaborative robots (2021). https://www.iso.org/obp/ui/#iso:std:iso:ts:15066:ed-1:v1:en

33. Jain, A., Zamir, A.R., Savarese, S., Saxena, A.: Structural-RNN: deep learning on spatio-temporal graphs. In: 2016 IEEE Conference on Computer Vision and Pattern Recognition (CVPR), pp. 5308–5317 (2016)
34. Kanazawa, A., Kinugawa, J., Kosuge, K.: Adaptive motion planning for a collaborative robot based on prediction uncertainty to enhance human safety and work efficiency. IEEE Trans. Robot. **35**(4), 817–832 (2019)
35. Kang, S., Kim, M., Kim, K.: Safety monitoring for human robot collaborative workspaces. In: International Conference on Control, Automation and System, 2019-October (ICCAS), pp. 1192–1194 (2019)
36. Knudsen, M., Kaivo-oja, J.: Collaborative robots: frontiers of current literature. J. Intell. Syst. Theory App. **3**, 13–20 (2020)
37. Lai, G., Liu, H., Yang, Y.: Learning graph convolution filters from data manifold (2018)
38. Laplaza, J., Pumarola, A., Moreno-Noguer, F., Sanfeliu, A.: Attention deep learning based model for predicting the 3d human body pose using the robot human handover phases. In: 2021 30th IEEE International Conference on Robot & Human Interactive Communication (RO-MAN), pp. 161–166. IEEE (2021)
39. LeCun, V., Denker, J., Solla, S.: Optimal brain damage. In: Advances in Neural Information Processing Systems (1989)
40. Lemmerz, K., Glogowski, P., Kleineberg, P., Hypki, A., Kuhlenkötter, B.: A hybrid collaborative operation for human-robot interaction supported by machine learning. In: International Conference on Human System Interaction, HSI 2019-June, pp. 69–75 (2019)
41. Li, C., Zhang, Z., Sun Lee, W., Hee Lee, G.: Convolutional sequence to sequence model for human dynamics. In: The IEEE Conference on Computer Vision and Pattern Recognition (CVPR) (2018)
42. Li, M., Chen, S., Zhao, Y., Zhang, Y., Wang, Y., Tian, Q.: Dynamic multiscale graph neural networks for 3d skeleton based human motion prediction. In: 2020 IEEE/CVF Conference on Computer Vision and Pattern Recognition (CVPR), pp. 211–220 (2020)
43. Li, M., Lin, J., Ding, Y., Liu, Z., Zhu, J.Y., Han, S.: Gan compression: efficient architectures for interactive conditional GANs. In: 2020 IEEE/CVF Conference on Computer Vision and Pattern Recognition (CVPR), pp. 5283–5293 (2020)
44. Li, X., Li, D.: GPFS: a graph-based human pose forecasting system for smart home with online learning. ACM Trans. Sen. Netw. **17**(3), 1–9 (2021)
45. Lim, J., et al.: Designing path of collision avoidance for mobile manipulator in worker safety monitoring system using reinforcement learning. In: ISR 2021–2021 IEEE International Conference on Intelligence and Safety for Robotics, pp. 94–97 (2021)
46. Liu, S., Deng, W.: Very deep convolutional neural network based image classification using small training sample size. In: 2015 3rd IAPR Asian Conference on Pattern Recognition (ACPR), pp. 730–734 (2015)
47. Magrini, E., Ferraguti, F., Ronga, A.J., Pini, F., De Luca, A., Leali, F.: Human-robot coexistence and interaction in open industrial cells. Robot. Comput. Integr. Manuf. **61**, 101846 (2020)
48. Mahmood, N., Ghorbani, N., Troje, N.F., Pons-Moll, G., Black, M.J.: AMASS: Archive of motion capture as surface shapes. In: International Conference on Computer Vision (2019)
49. Mao, W., Liu, M., Salzmann, M.: History repeats itself: human motion prediction via motion attention. In: Vedaldi, A., Bischof, H., Brox, T., Frahm, J.-M. (eds.)

ECCV 2020. LNCS, vol. 12359, pp. 474–489. Springer, Cham (2020). https://doi.org/10.1007/978-3-030-58568-6_28
50. Mao, W., Liu, M., Salzmann, M., Li, H.: Learning trajectory dependencies for human motion prediction. In: The IEEE International Conference on Computer Vision (ICCV) (2019)
51. von Marcard, T., Henschel, R., Black, M.J., Rosenhahn, B., Pons-Moll, G.: Recovering accurate 3D human pose in the wild using IMUS and a moving camera. In: Ferrari, V., Hebert, M., Sminchisescu, C., Weiss, Y. (eds.) ECCV 2018. LNCS, vol. 11214, pp. 614–631. Springer, Cham (2018). https://doi.org/10.1007/978-3-030-01249-6_37
52. Martinez, J., Black, M.J., Romero, J.: On human motion prediction using recurrent neural networks. In: The IEEE Conference on Computer Vision and Pattern Recognition (CVPR) (2017)
53. Matthias, B., Reisinger, T.: Example application of ISO/TS 15066 to a collaborative assembly scenario. In: 47th International Symposium on Robotics ISR 2016 2016, pp. 88–92 (2016)
54. Michalos, G., Makris, S., Tsarouchi, P., Guasch, T., Kontovrakis, D., Chryssolouris, G.: Design considerations for safe human-robot collaborative workplaces. Proc. CIrP **37**, 248–253 (2015)
55. Minelli, M., et al.: Integrating model predictive control and dynamic waypoints generation for motion planning in surgical scenario. In: IEEE/RSJ International Conference on Intelligent Robots and Systems (IROS), pp. 3157–3163 (2020)
56. Molchanov, P., Tyree, S., Karras, T., Aila, T., Kautz, J.: Pruning convolutional neural networks for resource efficient inference (2017)
57. Nascimento, H., Mujica, M., Benoussaad, M.: Collision avoidance in human-robot interaction using kinect vision system combined with robot's model and data. In: IEEE International Conference on Intelligent Robotics and Systems, pp. 10293–10298 (2020)
58. Oono, K., Suzuki, T.: Graph neural networks exponentially lose expressive power for node classification. In: International Conference on Learning Representations (2020)
59. Pavllo, D., Feichtenhofer, C., Grangier, D., Auli, M.: 3d human pose estimation in video with temporal convolutions and semi-supervised training. In: Conference on Computer Vision and Pattern Recognition (CVPR) (2019)
60. Ramon, J.A.C., Herias, F.A.C., Torres, F.: Safe human-robot interaction based on dynamic sphere-swept line bounding volumes. Robot. Comput. Integr. Manuf. **27**(1), 177–185 (2011)
61. Rastegari, M., Ordonez, V., Redmon, J., Farhadi, A.: XNOR-Net: Imagenet classification using binary convolutional neural networks (2016)
62. Rodriguez-Guerra, D., Sorrosal, G., Cabanes, I., Calleja, C.: Human-robot interaction review: challenges and solutions for modern industrial environments. IEEE Access **9**, 108557–108578 (2021)
63. Shah, J., Wiken, J., Breazeal, C., Williams, B.: Improved human-robot team performance using Chaski, a human-inspired plan execution system. In: HRI 2011 - Proceedings of 6th ACM/IEEE International Conference on Human-Robot Interaction, pp. 29–36 (2011)
64. Shi, L., Wang, L., Long, C., Zhou, S., Zhou, M., Niu, Z., Hua, G.: Sparse graph convolution network for pedestrian trajectory prediction. In: Proceedings of the IEEE/CVF Conference on Computer Vision and Pattern Recognition (2021)

65. Sofianos, T., Sampieri, A., Franco, L., Galasso, F.: Space-time-separable graph convolutional network for pose forecasting. In: Proceedings of the IEEE/CVF International Conference on Computer Vision (ICCV) (2021)
66. Torkar, C., Yahyanejad, S., Pichler, H., Hofbaur, M., Rinner, B.: RNN-based human pose prediction for human-robot interaction. In: Proceedings of the ARW & OAGM Workshop 2019, pp. 76–80 (2019)
67. Tu, H., Wang, C., Zeng, W.: VoxelPose: towards multi-camera 3d human pose estimation in wild environment. In: Vedaldi, A., Bischof, H., Brox, T., Frahm, J.-M. (eds.) ECCV 2020. LNCS, vol. 12346, pp. 197–212. Springer, Cham (2020). https://doi.org/10.1007/978-3-030-58452-8_12
68. Vianello, L., Mouret, J.B., Dalin, E., Aubry, A., Ivaldi, S.: Human posture prediction during physical human-robot interaction. IEEE Robot. Autom. Lett. **6**, 6046–6053 (2021)
69. Wang, C., Wang, Y., Huang, Z., Chen, Z.: Simple baseline for single human motion forecasting. In: Proceedings of the IEEE/CVF International Conference on Computer Vision (ICCV) Workshops, pp. 2260–2265 (2021)
70. Xie, S., Girshick, R., Dollár, P., Tu, Z., He, K.: Aggregated residual transformations for deep neural networks. In: 2017 IEEE Conference on Computer Vision and Pattern Recognition (CVPR), pp. 5987–5995 (2017)
71. Yu, C., Ma, X., Ren, J., Zhao, H., Yi, S.: Spatio-temporal graph transformer networks for pedestrian trajectory prediction. In: Vedaldi, A., Bischof, H., Brox, T., Frahm, J.-M. (eds.) ECCV 2020. LNCS, vol. 12357, pp. 507–523. Springer, Cham (2020). https://doi.org/10.1007/978-3-030-58610-2_30
72. Zeiler, M.D., Fergus, R.: Visualizing and understanding convolutional networks. In: Fleet, D., Pajdla, T., Schiele, B., Tuytelaars, T. (eds.) ECCV 2014. LNCS, vol. 8689, pp. 818–833. Springer, Cham (2014). https://doi.org/10.1007/978-3-319-10590-1_53
73. Zhang, J., Liu, H., Chang, Q., Wang, L., Gao, R.X.: Recurrent neural network for motion trajectory prediction in human-robot collaborative assembly. CIRP Ann. **69**(1), 9–12 (2020)
74. Zhao, Y., Dou, Y.: Pose-forecasting aided human video prediction with graph convolutional networks. IEEE Access **8**, 147256–147264 (2020)

Actor-Centered Representations for Action Localization in Streaming Videos

Sathyanarayanan Aakur[1(✉)] and Sudeep Sarkar[2]

[1] Oklahoma State University, Stillwater, OK 74074, USA
saakurn@okstate.edu
[2] University of South Florida, Tampa, FL 33620, USA
sarkar@usf.edu

Abstract. Event perception tasks such as recognizing and localizing actions in streaming videos are essential for scaling to real-world application contexts. We tackle the problem of learning *actor-centered* representations through the notion of *continual hierarchical predictive learning* to *localize* actions in streaming videos *without* the need for training labels and outlines for the objects in the video. We propose a framework driven by the notion of hierarchical predictive learning to construct *actor-centered* features by attention-based contextualization. The key idea is that predictable features or objects do not attract attention and hence do not contribute to the action of interest. Experiments on three benchmark datasets show that the approach can learn robust representations for localizing actions *using only one epoch of training*, i.e., a single pass through the streaming video. We show that the proposed approach outperforms unsupervised and weakly supervised baselines while offering competitive performance to fully supervised approaches. Additionally, we extend the model to multi-actor settings to recognize group activities while localizing the multiple, plausible actors. We also show that it generalizes to out-of-domain data with limited performance degradation.

1 Introduction

Understanding events in videos requires understanding beyond recognition, such as localizing the actor, understanding their future behavior from current and past observations, and building robust representations at the event and actor levels. While many recent works have focused on action recognition [1,18,21] and action localization [6,8,20], significant progress has primarily been driven by the use of large-scale, annotated training data. While self-supervised learning [7,42] has reduced the need for labeled data for *recognition*, there is still a dependency on large amounts of manual annotations for *localization*.

Supplementary Information The online version contains supplementary material available at https://doi.org/10.1007/978-3-031-19839-7_5.

S. Avidan et al. (Eds.): ECCV 2022, LNCS 13698, pp. 70–87, 2022.
https://doi.org/10.1007/978-3-031-19839-7_5

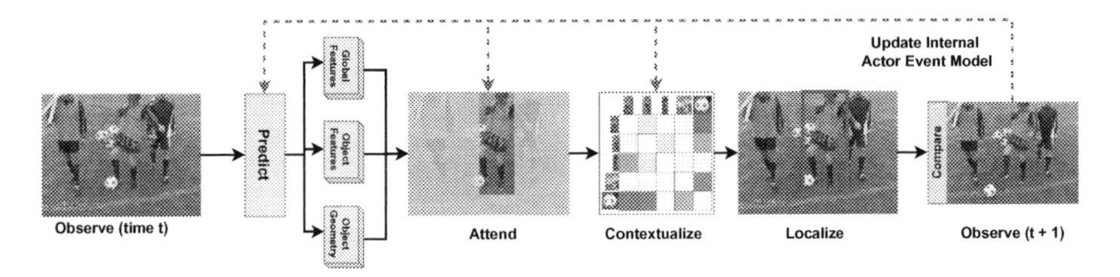

Fig. 1. Our goal is to learn actor-centered representations for actor localization in streaming videos *without explicit annotations*. Given a frame at time t, we follow the sequence of: observe, predict, attend based on prediction error, contextualize actor representations, and localize. The internal event models are constantly updated based on the prediction errors with the observation at time $t + 1$.

We consider the problem of learning *actor-centered* representations to localize actions in *streaming videos* i.e., needing a single-pass through video for training (single epoch) and without training labels and outlines. We do not need multiple training epochs to build the representations. We define an actor-centered representation as a compositional structure of the scene that encodes the properties (location, geometry, and relational cues) of the *dominant* actor contributing to the action of interest. For example, in Fig. 1, there are many actors (three players, a soccer ball, etc.) in the scene, but only one *dominant actor* (the player in the middle) is involved in the action "*kicking ball*". Hence, an actor-centered representation would encode the appearance and geometry of the player in the middle and *contextualize* their features concerning the other objects in the scene. Such representations allow us to capture action-specific contextual cues in a generalizable representation.

We build actor-centered event representations by contextualizing the actor's features with environment-level (or scene) at both a perceptual level (such as color, texture, and movement) and a conceptual level (such as actor-environment interactions and action goals). Computationally, we model this process by following a sequence of operations given by *observe, predict, compare, attend, contextualize*, and *localize*. Figure 1 illustrates this process. The *key idea* is that predictable features or objects do not attract attention and hence do not contribute to the action of interest. We introduce the idea of hierarchical prediction that enables the framework to select objects of interest and maintain context in prediction to localize the action by navigating spurious motion patterns such as camera motion and background clutter. This hierarchical prediction differs from prior versions of predictive learning for action recognition [2,3], which do not consider the actor-centered features such as appearance, geometry, and their evolution with respect to the scene.

The **contributions** are four-fold: (i) we introduce the idea of *hierarchical* predictive learning to learn actor-centered representations to localize actions in *streaming videos* in an *unsupervised manner*, (ii) introduce a novel, attention-driven formulation for learning robust, actor-centered event features for action localization and recognition, (iii) demonstrate that the proposed approach can be trivially extended to multi-actor group activity recognition and localization, and (iv) show that the use of actor-centered feature representations helps learn robust features that can generalize to data from outside the training domain *without finetuning for both localization and group activity recognition.*

2 Related Work

Action localization has largely been tackled through *supervised* learning approaches [9,12,14,35,36,39,41,43,45], which aim to simultaneously generate bounding box proposals and labels learned from annotated training data. The common pipeline uses convolutional neural networks (both 2D and 3D [40]) to extract features from RGB images, optionally the optical flow images, and generate bounding box proposals to localize objects in the video sequence. A linking algorithm (Viterbi or actor linking [6]) is used to extract action tubes from the generated bounding boxes. Annotated training data is used to train recognition and bounding box regression modules.

Weakly supervised [6,21,32] reduce the dependency on training data by negating the need for spatial-temporal annotations and using either attention-based pooling [21,32] or appearance-based linking from generic object detection-based proposals [6]. They typically require video-level label annotations that are used to learn representations for recognition and use object-level labels and characteristics to select bounding box proposals from pre-trained object detection models. Hence, they may be constrained to localizing actions specific to classes from the detection models.

Unsupervised approaches [3,37] do not require annotations for labels or bounding boxes. Soomro *et al.* [37] use pre-trained object detection models to generate proposals and score each with a "humanness" score that ranks the likelihood of belonging to an action class and uses a knapsack-based algorithm to discover action classes to self-label videos. Aakur *et al.* [3] use a predictive learning-based approach (PredLearn for brevity) to create spatial-temporal attention maps which are used to localize objects of interest. Closely related to our approach, PredLearn anticipates the future spatial feature using a motion-weighted loss function at the feature level. However, it does not enforce consistency in actor-specific features such as geometry or contextualized representations to help reject the background clutter and maintain context in prediction. Additionally, we localize multiple actors together with learning robust features.

3 Actor-Centered Action Localization

Problem Formulation. In our setup, we consider the problem of localizing the *dominant* action a_i at each time instant t in a *streaming video.* Each video can contain multiple objects in the scene with one *dominant* action performed by one or more *actors.* The key challenge is ignore clutter and identify the object(s) of interest (i.e., the *actors*) *without any supervision* while building robust representations that capture the motion and relational dynamics of the event. Figure 2 illustrates the proposed action localization framework. We begin with perceptual features extracted from a convolutional neural network [34] and progressively refine these scene-level features with context from event-level dynamics (Sect. 3.2) and actor-centered context (Sect. 3.3) through the notion of hierarchical predictive learning (Sect. 3.4), to jointly model both the evolution of the action and the actors in a unified framework.

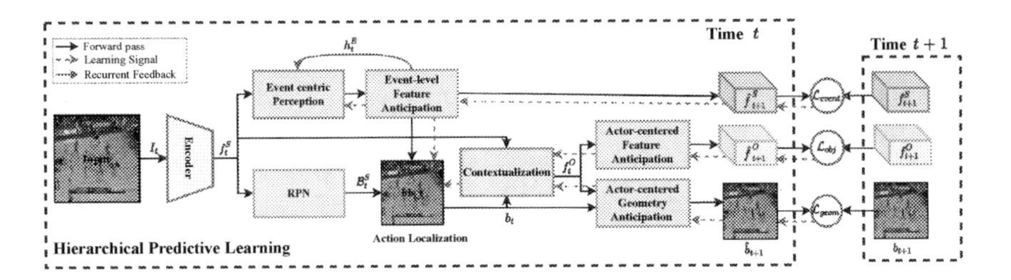

Fig. 2. Overview of our approach. Given a sequence of frames in *streaming fashion*, our model constructs an actor-centered representation using the notion of hierarchical predictive learning. A prediction-driven attention map is used to localize the action.

3.1 Extracting Perceptual Features

First, we extract a global, scene-level representation of the given visual sequence. This representation includes both perceptual features and identifying regions of likely interest representing objects. While our approach is general enough to handle different object proposal approaches (see Sect. 4), we use a pre-trained convolutional neural network to extract the scene-level representation (f_t^S) and use a Single-Short Object Detector (SSD) [23] layer to generate region-proposals $(\mathcal{B}_t^S)$, where objects are likely to exist. Following prior work in [3], we make the SSD class-agnostic by considering *all* bounding boxes returned (at an "objectness" threshold of 0.01) regardless of the predicted class and their corresponding confidence scores. This allows us to remove biases towards certain actors, such as human actors, and help handle any visual variations (such as pose and occlusion) that can cause missed detection.

3.2 Event-Centric Perception

The second step in the proposed model is to construct a feature representation of the current scene (at time t) influenced by the observed event's spatial-temporal dynamics. While CNN features provide an efficient spatial representation, it does not consider the contextual knowledge provided by temporal transitions and spatial interactions among the scene's entities. This process requires modeling a stable event representation h_t^E and an attention mechanism α_t^S that uses this global representation to jointly perceive and anticipate the spatial and temporal dynamics of the event. Formally, we define the event-centric perception model as a prediction function that maximizes the probability $P(\hat{f}_{t+1}^S|W_p, \alpha_t^S, h_t^E, f_t^S)$, where $\hat{f}_{t+1}^S$ is the anticipated features at time $t+1$ conditioned on an internal event representation h_t^E, a temporally-weighted, spatial attention function α_t^S and the current observed features f_t^S. W_p is the set of learnable parameters in this module. There are two steps in constructing the event-centric perceptual features - (i) learning an efficient global, event-level representation and (ii) using the learned event representation to drive the perception in a recurrent manner. First, we first create the event-centric scene representation by weighting the CNN feature f_t^S of the frame at time t by an attention vector (α_t^S), influenced by the spatial temporal dynamics of the current event. Hence, the event-centric representation is given by $f_t^E = \alpha_t^S \odot f_t^S$, where $\alpha_t^S = f_a(f_t^S, h_{t-1}^E)$ and $f_a(\cdot)$ is a learned attention function [4]. Second, we use a hierarchical stack of Long Short Term Memory (LSTM) networks [10] to construct the internal event representation. We take a continual predictive learning approach, inspired by [2,3], to learn an efficient global representation of the event that captures the relevant spatial-temporal patterns to provide context for event-based perception. The hierarchical LSTM stack is used as a spatial-temporal decoder network. It takes a sequence of event-centric image features as input and propagates its prediction up the stack. The output of the top-most LSTM is taken as the anticipated features ($\hat{f}_{t+1}^S$) at the next time step. Hence, the hierarchical LSTM stack acts as a generative model that learns and uses a stable event representation to anticipate the scene's spatial and temporal evolution. Formally, this is represented as

$$\hat{f}_{t+1}^{\ell}, h_t^{\ell} = LSTM(\hat{f}_{t+1}^{\ell-1}, W_{\ell}, h_{t-1}^{\ell}) \tag{1}$$

$$\hat{f}_{t+1}^{0}, h_t^{0} = LSTM(f_t^E, W_0, h_{t-1}^{0}) \tag{2}$$

where $\hat{f}_{t+1}^{\ell-1}$ refers to the predicted features at the ℓth LSTM in the stack and W^{ℓ} refers to the weights associated with the LSTM at the ℓth layer; Eq. 2 shows the initialization for the bottom-most 0th-level LSTM.

Note that $\hat{f}_{t+1}^{\ell}$ for the top most LSTM network is taken as the prediction for time $t+1$ and the corresponding hidden state is taken as the event representation such that $h_t^E = h_t^{\ell}$. The memory is not shared within the stack and hence allows each level of the stack to model the spatial-temporal dynamics at different granularity, with the ℓth-level LSTM influenced by the lower-level LSTMs.

Event-Level Prediction. The event-centric perception module is trained in a predictive learning approach with the training objective given by

$$\mathcal{L}_{event} = \|f_{t+1}^S - f_t^S\|_2 \odot \|f_{t+1}^S - \hat{f}_{t+1}^S\|_2 \tag{3}$$

where the first term represents the weighted difference between the *features* at consecutive time steps t and $t+1$ and the resulting value $\mathcal{L}_{event}$ represents a weighted $L-2$ norm of the predicted and expected value that penalizes incorrect predictions at spatial locations with maximal change at the *feature level*. Hence, $\mathcal{L}_{event}$ is the prediction-based drive, a measure of the effectiveness of the learned event representation at a coarse spatial quantization. This predictive learning process forms the bottom level of the hierarchy and helps learn event-level dynamics for modeling the action in the scene. While effective, as shown in PredLearn [3], it is not enough to handle complex scenes and multiple actors in the scene. There is a need to model the actor-environment interactions for effective visual understanding.

3.3 Contextualization: Actor-Centered Features

The next step is to construct actor-centered representations that contextualize the event-level dynamics with the actor-environment interactions. We consider a feature representation of a scene to be *actor-centered* if the resulting representation can (i) reject clutter in the scene, (ii) reduce the impact of background or spurious motion patterns, and (iii) *contextualize* the actor's motion dynamics with the rest of the scene or environment. This representation is analogous to a posterior-weighted spatial representation that highlights areas of interest while suppressing spatially irrelevant features. In our framework, the posterior is obtained by updating the prior, captured by the prediction-based error from Eq. 3, with the current observation (f_t^S). The contextualized representation is obtained by computing the dot-product attention [25] between the posterior-weighted representation and the actual representation and is defined as

$$f_t^O = GAP(softmax(f_t^S \odot \mathcal{F}_t^S) \odot \mathcal{F}_t^S) \tag{4}$$

where GAP refers to the Global Average Pooling function [22] and $\mathcal{F}_t^S$ is a contextualized feature representation conditioned by the posterior probability provided by the spatial-temporal prediction loss $\mathcal{L}_{Event}$. We compute this function as $\mathcal{F}_t^S{=}softmax(\mathcal{L}_{Event}) \odot f_t^S$, which intuitively provides a representation that rejects clutter by scaling down the spatial regions that do not contribute to the prediction uncertainty. These areas typically involve background scenes or actors whose actions are more predictable and less likely to be of interest. This formulation helps preserve the scene's spatial-temporal structure by summing out any trivial motion-based changes, making it more robust to spurious motion patterns in the input, such as those induced by background noise and small camera motion. Empirically, this formulation results in a more robust video-level representation that can generalize across domains (see Sect. 4.2).

3.4 Hierarchical Predictive Learning

We use the notion of *continual, hierarchical predictive* learning to train the model end-to-end without needing labels and outlines of the objects in the video. This approach aims to model the dynamics of the observed event at different levels of granularity, moving beyond just scene-level dynamics [3] or temporal dynamics [2]. To this end, we create a hierarchy of predictions that are performed at every time step that models the event-level and actor-level dynamics in the event. At the lowest level is the event-level prediction (Sect. 3.2). At the next level is the prediction of the actor's dynamics within the scene's context. At the top level is the prediction of the actor's visual properties. The goal is to anticipate the actor's location and geometry in the context of event-level and actor-level dynamics. Each level of the stack influences the prediction of the upper level and hence forms a hierarchy of predictions that capture the inherent dynamics within the event. The actor-level predictions (levels 2 and 3) are conditioned on the event-level representation by constructing a global representation given by $\hat{f}^E_t = GAP(\alpha^S_t \odot \hat{f}^S_t)$, where $\hat{f}^S_t$ refers to the anticipated spatial features (from the previous prediction step) and α^S_t refers to the spatial attention constructed (conditioned on the current observation) at time t. This formulation of the global representation forces the model to learn spatially relevant features that are important *across time steps* and hence helps ensure that the event representation is robust by acting as temporal smoothing.

Computationally, we learn two LSTM-based prediction models that use this global representation ($\hat{f}^E_t$) to anticipate the actors' dynamics in terms of contextualized features and geometry. One LSTM anticipates the *changes* in the actor's geometry rather than directly predicting the BB location, which allows the predictor to focus on the *evolution* of geometry. The other LSTM anticipates the actor-centered representations ($\hat{f}^O_{t+1}$) at time $t+1$. Hence, the goal of these two LSTMs is to minimize the actor-centered prediction errors defined as

$$\mathcal{L}_{object} = \|f^O_{t+1} - \hat{f}^O_{t+1}\|_2 + \mathcal{D}_{bb}(b_{t+1}, \hat{b}_{t+1}) + \mathcal{D}_g(b_{t+1}, \hat{b}_{t+1}) \tag{5}$$

where $\mathcal{D}_{bb}(b_{t+1}, \hat{b}_{t+1})$ is the distance between the predicted bounding box and the actual observed bounding centers; $\mathcal{D}_g(b_{t+1}, \hat{b}_{t+1}) = (\sqrt{w} - \sqrt{\hat{w}})^2 + (\sqrt{h} - \sqrt{\hat{h}})^2$, where ($\hat{h}$ and $\hat{w}$) and (h, w) are the predicted and actual height and widths of the bounding box bb_{t+1}, respectively. Hence, the entire framework is trained end-to-end using the overall objective function given by

$$\mathcal{L}_{total} = \lambda_1 \frac{1}{n_f} \sum_{i=1}^{w_f} \sum_{j=1}^{h_f} \mathcal{L}_{event} + \lambda_2 \mathcal{L}_{object} \tag{6}$$

where λ_1 and λ_2 are modulating factors to balance the trade-off between predicting the event-level and object-level prediction errors. Both losses directly penalize the event-level representation (h^E_t), the spatial attention (α^S_t) and the contextualization module ($\mathcal{F}^S_t$. Hence it adds an implicit regularization to prevent overfitting since the model's parameters are updated *continuously per frame*. The resulting spatial-temporal loss $\mathcal{L}_{event}$ can then considered to be reflective of the *predictability* of both the actor *and* scene. Hence, spatial locations with a higher error indicate the location of the actor [3,11]. Note that the entire process is unsupervised, there are no labels or bounding box annotations needed for training since the predictions at time t are compared to observations at time $t+1$ to provide supervision to progressively refine the representations.

3.5 Attention-Based Action Localization

The final step in the proposed approach is using attention to localize the actor (the object of interest) in the given video. We create an attention-like representation using the prediction-based error $\mathcal{L}_{event}$ to identify areas of interest. The input to the localization process consists of (i) initial regions of interests generated based on spatial features $\mathcal{B}_t^S$ (from Sect. 3.1), (ii) the spatial-temporal prediction error $\mathcal{L}_{event}$ (from Sect. 3.2), (iii) number of attention "grids" to consider K, and (iv) the total number of bounding box predictions per frame t. We first construct an attention-like representation by running the spatial-temporal prediction error through a *softmax* function to produce an attention map of shape $c_x \times c_y$ where c_x and c_y are spatial dimensions of the observed feature maps, with each point corresponding to a "grid" in the frame (following notation from YOLO [29]). The softmax operation magnifies areas of high errors while suppressing areas of low prediction errors.

We consider areas of high prediction error to be regions of interest. However, we allow the attention map to be split between multiple objects and consider the top K grids (sorted based on prediction error) to select bounding box localization. Following the notation from YOLO-based object detection models [29], we define a binary function $\mathbb{1}(\cdot)$ that returns *True* if a bounding box proposal's center falls within the "grid" $e_{i,j}$ and *False* otherwise. This allows us to select objects that are most likely to contribute to the grid's prediction error. Note that this is different from [3], where each bounding box is assigned an energy term based on distance from a prior position and the magnitude of the prediction error, which does not allow them to attend to multiple objects. This is further explored in Sect. 4, where the use of hierarchical prediction allows the model to attend to multiple objects simultaneously for multi-actor localization.

Implementation Details. In our experiments, we use a VGG-16 network [34], pre-trained on ImageNet [31], as the backbone network for training a Single Shot Multbox Detector (SSD) [23] to extract frame-level representations and generate localization proposals. The SSD is trained on MS-COCO with input re-sizes to 512×512. We use the output of the max-pooling layer after the fifth convolutional layer as f_t^S. We use the SSD as a class-agnostic region proposal network by taking the bounding box proposals without any predicted classes or associated probabilities. The number of layers ℓ in the hierarchical prediction network (in Sect. 3.2) as 3 and set the dimensions of the hidden state at each layer to 512. We set the number of attention grids $K = 5$ and the number of localization per frame $N = 10$ (Sect. 3.5). We train with adaptive learning [2], with an initial learning rate of 1×10^{-10} and scaling factors $\Delta_t^- = 0.1$ and $\Delta_t^+ = 0.01$.

4 Experimental Evaluation

4.1 Data, Metrics and Baselines

We use three standard benchmark datasets (UCF Sports [30], JHMDB [15], and THUMOS'13 [17]) to evaluate the proposed approach for action localization. We also evaluate on the Collective Activity dataset [5] to demonstrate and evaluate our approach on multi-actor action localization. **UCF Sports** [30] contains 10 classes characterizing sports-based actions such as weight-lifting and diving. We use the official splits containing 103 videos for training and 47 videos for testing, as defined in [19] for evaluation. **JHMDB** [15] has 21 action classes from 928 trimmed videos, each annotated with human joints and bounding box for every frame. It offers several significant challenges for unsupervised action localization, such as camera motion that causes significant occlusions and background objects that act as distractions. We report all results as the average across all three splits. **THUMOS'13** [17] (or the *UCF-101–24* dataset) is a subset of the UCF-101 [38] dataset, consisting of 24 classes and 3,207 videos. It is one of the most challenging action localization datasets with complex motion, background clutter and high intra-class variability. Following prior works [21,37], we report results on the first split. **Collective Activities** [5] is a group activity dataset where the goal is to recognize the activity performed by multiple actors such as talking, queueing, and walking. It comprises 44 short video sequences with 5 group activities, with every 10 frames annotated with bounding boxes of actors involved in the group activity. This dataset offers a unique challenge in localizing

Table 1. Comparison with state-of-the-art approaches on three common benchmark datasets - UCF Sports, JHMDB and THUMOS'13. We report the video-level mAP at different overlap thresholds. * refers to the use of class-specific object proposals.

Approach	Supervision		UCF Sports		JHMDB		THUMOS'13	
	Spatial	Label	σ=0.2	σ=0.5	σ=0.2	σ=0.5	σ=0.2	σ=0.5
Tube CNN [12]	✓	✓	0.47	–	–	0.77	0.47	0.41
Action Tubelets [14]	✓	✓	0.53	0.27	–	–	0.48	–
Action Tubes [9]	✓	✓	0.56	0.49	0.55	0.45	–	–
MRSTL [48]	✓	✓	–	–	–	0.37	–	0.68
MENET [24]	✓	✓	–	–	–	**0.82**	–	**0.84**
HISAN [27]	✓	✓	–	–	–	0.77	–	0.73
ACAR-Net [26]	✓	✓	–	–	–	–	–	0.84
ALSTM [32]	✗	✓	–	–	–	–	0.06	–
VideoLSTM [21]	✗	✓	–	–	–	–	0.37	–
Actor Supervision [6]	✗	✓	-	**0.48**	–	**0.36**	**0.46**	–
Soomro *et al* [37]	✗	✗	0.46*	0.30*	**0.43***	**0.22***	0.21*	0.06*
PredLearn [3] (k=k_{gt})	✗	✗	0.55	0.32	0.30	0.10	0.31	0.10
AC-HPL (Ours, k=k_{gt})	✗	✗	**0.70**	**0.59**	**0.43**	0.15	**0.38**	**0.20**

all actors ($>= 1$) involved in the action while learning robust features that can capture the dynamics of each actor in the context of their collective activity. We follow prior works [8,13,28,44,46] and use 1/3 of the video sequences for testing and the rest for training. We report results for both recognition and localization.

Label Prediction and Metrics. Due to the unsupervised nature of learning representations, we use *k-means* clustering to obtain class labels. The frame-level features are max-pooled to obtain video-level features. Following prior work [3, 16,47], we use the Hungarian method to map from predicted clusters to the ground-truth labels. We set the number of clusters to the number of classes in the ground-truth for comparison with state-of-the-art. For *action localization*, we report the mean average precision (mAP) metric at different overlap thresholds for a fair comparison with prior works [21,37].

Baselines. We compare against several fully supervised baselines such as MRSTL [48], MENET [24], HISAN [27], ACAR-Net [26], tube convolution networks [12], motion-based action tublets [14] and action tubes [9] and weakly supervised such as ALSTM [32], VideoLSTM [21] and Actor Supervision [6]. We also evaluate our approach against unsupervised action localization approaches such as Soomro *et al.* [37] and the closely related predictive learning approach [3], which we term as PredLearn. Note, we compare against PredLearn when the number of clusters is set to the ground-truth clusters ($k = k_{gt}$).

Table 2. Generalization capability when evaluated on out-of-domain test samples *without finetuning*. PredLearn refers to [3] and AC-HPL refers to our approach.

Test Data →	UCF Sports		JHMDB		THUMOS'13	
Train Data ↓	AC-HPL	PredLearn	AC-HPL	PredLearn	AC-HPL	PredLearn
	$\sigma = 0.5$		$\sigma = 0.2$		$\sigma = 0.2$	
UCF Sports	**0.59**	0.32	**0.39**	0.19	**0.38**	0.20
JHMDB	**0.48**	0.23	**0.43**	0.30	**0.35**	0.26
THUMOS'13	**0.50**	0.27	**0.40**	0.24	**0.38**	0.31

4.2 Quantitative Analysis

We first present the quantitative results of the proposed in Table 1, where we compare against different baseline approaches. We report the mean average precision (mAP) scores over the most commonly reported overlap thresholds of 0.2 and 0.5. The approaches are ordered by the amount of supervision required for training. The models at the top require *strong* supervision in terms of spatial annotations such as bounding boxes *and* video-level labels to localize and classify the action. The models in the middle are *weakly* supervised and hence only

Table 3. Evaluation of our approach on the **Collective Activities** dataset for (a) multi-actor group activity recognition and (b) multi-actor group activity localization.

Approach	Supervision?		Acc.
	Label	Box	
LRCN [28]	✓	✓	64.0
VGG-16 [28]	✓	✗	68.3
VGG-16 [28]	✓	✓	71.2
Hierarchical LSTM [13]	✓	✓	81.1
CERN [33]	✓	✓	84.8
stagNET [28]	✓	✓	89.1
ARG [46]	✓	✓	91.0
Action Transformer [8]	✓	✓	92.8
GroupFormer [20]	✓	✓	96.3
AC-HPL (k-means)*	✗	✗	72.2
AC-HPL (Finetuned)*	✓	✗	**80.2**

(a)

Approach	Avg. IOU	Recall	mAP	
			0.2	0.5
PredLearn	0.18	0.172	0.378	0.011
AC-HPL (K=1)	0.205	0.181	0.442	0.017
AC-HPL (K=5)	0.271	0.284	0.545	0.068
AC-HPL (K=10)	0.342	0.396	0.551	0.14
AC-HPL (K=25)	0.472	0.634	0.723	0.449

* denotes features are trained for only 1 epoch on the target dataset.

(b)

require video-level labels for training. The approaches at the bottom require no training annotations. It can be seen that our approach outperforms all baselines, including fully supervised models, on the *UCF Sports* dataset, even at higher thresholds. Interestingly, we significantly outperform the closely related PredLearn by a significant margin ($\approx 15\%$ in absolute mAP).

On datasets with significantly higher complexity, such as JHMDB and THUMOS'13, we see consistent improvements over the other unsupervised models such as PredLearn and Soomro *et al.*'s action discovery approach, that use bounding box proposals from *class-specific* proposals and hence are restricted to objects (humans) that are present in the pre-trained object detection models. On the other hand, we use *class-agnostic* proposals and are not restricted to any object class. Also, it is interesting to note that hierarchical predictive learning and actor-centered feature representations help overcome the challenges posed by occlusions and clutter, as indicated by the significant gains over PredLearn at higher thresholds on JHMDB and THUMOS'13.

Generalization to Novel Domains. In addition to evaluating the proposed approach in traditional settings, we also assess its ability to generalize to *novel* domains. To be specific, we check its generalization capability by training on one dataset and testing its performance on a different dataset *without finetuning.* We begin by evaluating the approach on the generalization task by training the model on the training data from one of the three standard benchmarks (UCF Sports, JHMDB - Split 1, and THUMOS'13) and evaluating on the others. While the three datasets have similar actions, they have varying amounts of data, camera motion, and occlusions, which provide a challenging benchmark for evaluating generalization performance. We also report the closely related PredLearn approach's performance, which does not use hierarchical prediction and actor-centered representations. Table 2 summarizes the results. It can be seen that the proposed approach generalizes well across datasets, *regardless of the training data size*. For example, UCF Sports has a very small number of training data (103) and classes (10). However, the model can transfer well to other datasets with more classes (21 for JHMDB and 24 for THUMOS'13). It is to be noted that the model is *not finetuned* on any data in the target domain yet performs as well as weakly supervised models such as VideoLSTM (0.37 mAP@0.2 on THUMOS'13), *which was trained on the data*. Similarly, the use of actor-centered representations allows for better generalization compared to PredLearn, which has a poorer recognition performance due to the lack of *contextualized* feature representations.

4.3 Multi-Actor Group Activity Localization

Our approach can be naturally extended to multi-actor group activity recognition and localization using the Collective Activities dataset [5]. The goal is to recognize the collective or group activity performed by the majority of the actors in the scene. While majority of the prior works [8,20,28,33,46] have focused on *recognition*, there have not been efforts for multi-actor *localization*. Using the prediction errors outlined in Sect. 3.5, we can increase the number of attention *grids* (K_{attn}) and use the resulting attention points to localize multiple actors in a given scene. We show that our approach can attend to multiple actors and learn robust representations for simultaneous localization and recognition in multi-actor videos. Table 3(a) summarizes the performance for *recognition*,

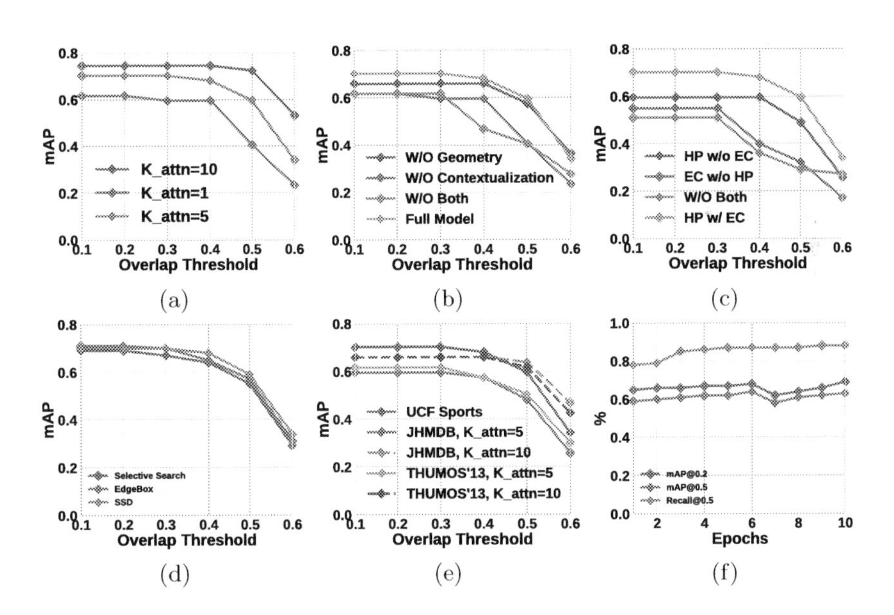

Fig. 3. Ablation experiments on UCF Sports to evaluate the effect of (a) number of attention "grids", (b) actor-centered prediction, and (c) event-centric perception, (d) choice of region proposal, (e) out-of-domain data, and (e) number of training epochs.

while Table 3(b) summarizes the performance for *localization*. For recognition, we evaluate two versions of the proposed approach to generate the labels - a completely unsupervised version with k-means for prediction and a fine-tuned version where the features are categorized into classes using a 2-layer feedforward neural network. As can be seen, without extensive training and annotations such as bounding boxes, we can achieve a recognition of 80.2%, which is remarkable considering that other state-of-the-art approaches require large amounts of training annotations and epochs. We only need one epoch of training to finetune the features to the multi-actor setting and do not need any annotations for learning. For localization, we report the average IOU of all bounding boxes produced by the approach and the mAP at 0.2 IOU and 0.5 IOU. To evaluate the upper bound of the approach, we also compute recall by considering groundtruth bounding boxes with at least one attention point as a true positive (TP). As can be seen from Table 3(b), as the number of attention points (K_{attn}) increases, we achieve a better IOU, recall, and mAP. Note that the closely related PredLearn is not able to handle multiple actor localization since their attention focuses only on the dominant *actor* and considers other actors as clutter. Nevertheless, these results and the qualitative visualizations in Fig. 4 show that the approach can perform multi-actor localization without any bells and whistles while not being explicitly trained for the task.

4.4 Ablation Studies

In Fig. 3(a) we show the effect of the changing the number of attention "*grids*" (K_{attn}) (Sect. 3.5). As the K_{attn} increases, the localization performance also increases and allows the model to keep track of the object of interest even if there are other potential actors. We also evaluate the effect of the different terms in the actor-centered prediction loss (Eq. 5). As can be seen from Fig. 4(b), the use of both geometry prediction and contextualized feature prediction help improve the performance significantly, with the use of contextualized prediction providing a greater jump in performance. In Fig. 4(c), we present the effect of using event-centric perceptual features (Sect. 3.2) on the framework with and without hierarchical prediction. It can be that the use of both improves the performance, especially at higher thresholds, indicating that the use of hierarchical prediction with event-centric features helps attend to areas of interest.

Effect of Region Proposal Methods. To evaluate the effect of object detection modules on the proposed approach, we try other bounding box proposals from *untrained* approaches such as EdgeBox and Selective Search. We present the results on UCF Sports below in Fig. 4(d). As can be seen, our approach is not dependent on SSD as the object proposal mechanism and is able to use any region proposal mechanism as its input. We do not use any class labels from SSD and make it class-agnostic to ensure that we do not have any assumptions about the actor or domain semantics, unlike supervised or weakly-supervised approaches that use object detectors as part of the action proposals.

Effect of Out-of-Domain Data. We also evaluated the localization performance of the approach when trained with out-of-domain data. Figure 3(e) shows that the real performance drop is at higher overlap thresholds when testing on data outside of the training domain. However, increasing K_{attn} helps alleviate this issue and even outperforms models trained in the same domain.

Effect of Multiple Training Epochs. Although our approach is designed to work with one epoch of training, we also evaluate the impact of multi-epoch training on the UCF Sports dataset and present the results in Fig. 3(f). It can be seen that increasing the number of epochs allows the model to learn better features for recognition while the localization is improved as well.

4.5 Qualitative Analysis

We qualitatively analyze our approach and visualize some interesting instances in Fig. 4. We show two specific groups of examples - (i) a comparison with PredLearn [3] to highlight the importance of using actor-centered features and hierarchical prediction beyond numbers presented in Sect. 4, and (ii) some failure modes of the approach to identify possible ways to mitigate them. In rows 1 and 4, it can be seen that although there are other objects in the scene whose motion is unpredictable, the use of multiple attention grids and actor-centered prediction helps the model to maintain focus on the actor. Row 3 shows that the model can overcome the challenges posed by camera motion *and* object deformation to maintain context in prediction, whereas PredLearn (without hierarchical prediction) is influenced by the camera motion and loses track of the object. We also visualize some of the failure modes of the proposed model in the final row in Fig. 4. In particular, we would like to highlight two areas that lead to failure. First, consider the sequence on the left. The model's attention is initially on the wrong player in the scene and continues to attend to areas of the same player, which we attribute to the actor-centered prediction. Although the object is well localized, it is not the *labeled* object of interest to which the model does not have access. The second failure mode, highlighted in the sequence on the right, is bounding box selection. Although the attention is on the object for most frames, *class-agnostic* proposals returns poorer bounding box fit. The mAP score does not take these factors into account.

5 Conclusion

We showed that we can learn actor-centered representations to localize actions with just a single-pass through video for training (single epoch) and without training labels and outlines. We do not need multiple training epochs to build the representations. This makes the approach useful for many real-world context where storing the video raises privacy or high storage cost concerns. Our solution was a hierarchical predictive learning framework that continuously predicts and learns from errors at different granularities. The resulting spatial-temporal error

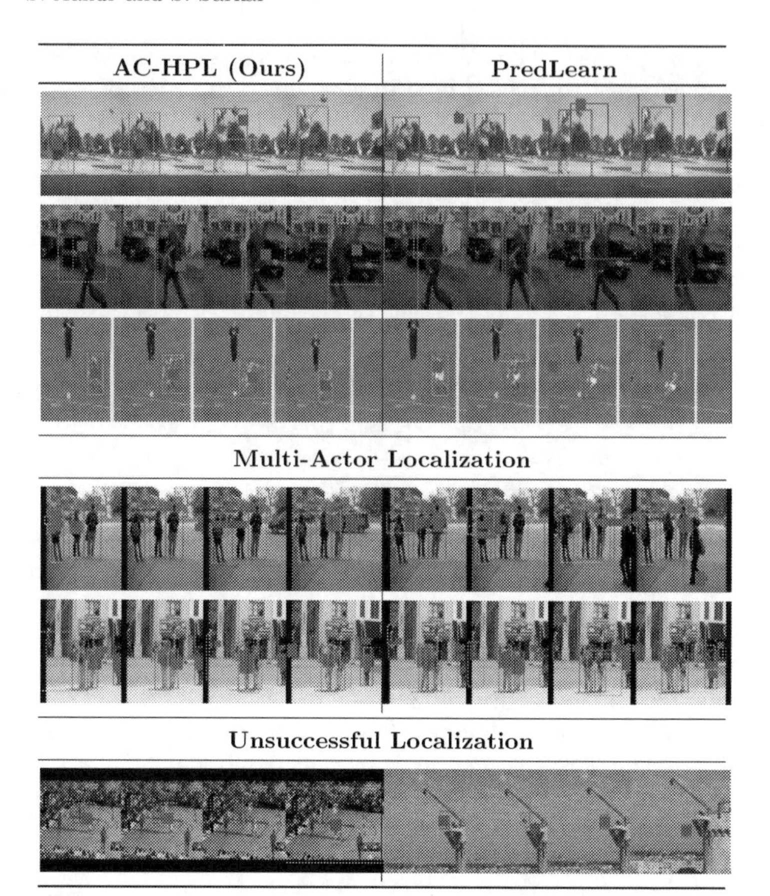

Fig. 4. Qualitative Examples. *Top:* handling camera motion and background motion to maintain context in localization. *Middle*: multi-actor localizations from AC-HPL. *Bottom*: Unsuccessful localizations from AC-HPL. *Visualization Legend:* Red BB: Grountruth, Green BB: Predictions, Blue Squares: Attention Locations. (Color figure online)

localized the action. The model leverages a novel actor-centered representation to learn robust features that mitigate the effect of camera motion and background clutter. We showed that we can beat SOTA on unsupervised action localization and multi-actor group activity localization while generalizing to novel domains *without finetuning.*

Acknowledgements. This research was supported in part by the US National Science Foundation grants CNS 1513126, IIS 1956050, IIS 2143150, and IIS 1955230.

References

1. Aakur, S., de Souza, F.D., Sarkar, S.: Going deeper with semantics: exploiting semantic contextualization for interpretation of human activity in videos. In: IEEE Winter Conference on Applications of Computer Vision (WACV). IEEE (2019)
2. Aakur, S.N., Sarkar, S.: A perceptual prediction framework for self supervised event segmentation. In: The IEEE Conference on Computer Vision and Pattern Recognition (CVPR) (2019)
3. Aakur, S.N., Sarkar, S.: Action localization through continual predictive learning. arXiv preprint arXiv:2003.12185 (2020)
4. Bahdanau, D., Cho, K., Bengio, Y.: Neural machine translation by jointly learning to align and translate. arXiv preprint arXiv:1409.0473 (2014)
5. Choi, W., Shahid, K., Savarese, S.: What are they doing?: collective activity classification using spatio-temporal relationship among people. In: 2009 IEEE 12th International Conference on Computer Vision Workshops, pp. 1282–1289. IEEE (2009)
6. Escorcia, V., Dao, C.D., Jain, M., Ghanem, B., Snoek, C.: Guess where? actor-supervision for spatiotemporal action localization. Comput. Vis. Image Underst. **192**, 102886 (2020)
7. Gan, C., Gong, B., Liu, K., Su, H., Guibas, L.J.: Geometry guided convolutional neural networks for self-supervised video representation learning. In: Proceedings of the IEEE Conference on Computer Vision and Pattern Recognition, pp. 5589–5597 (2018)
8. Gavrilyuk, K., Sanford, R., Javan, M., Snoek, C.G.: Actor-transformers for group activity recognition. In: Proceedings of the IEEE/CVF Conference on Computer Vision and Pattern Recognition, pp. 839–848 (2020)
9. Gkioxari, G., Malik, J.: Finding action tubes. In: Proceedings of the IEEE Conference on Computer Vision and Pattern Recognition, pp. 759–768 (2015)
10. Hochreiter, S., Schmidhuber, J.: Long short-term memory. Neural Comput. **9**(8), 1735–1780 (1997)
11. Horstmann, G., Herwig, A.: Surprise attracts the eyes and binds the gaze. Psychon. Bull. Rev. **22**(3), 743–749 (2015)
12. Hou, R., Chen, C., Shah, M.: Tube convolutional neural network (t-CNN) for action detection in videos. In: Proceedings of the IEEE International Conference on Computer Vision (ICCV), pp. 5822–5831 (2017)
13. Ibrahim, M.S., Muralidharan, S., Deng, Z., Vahdat, A., Mori, G.: A hierarchical deep temporal model for group activity recognition. In: Proceedings of the IEEE Conference on Computer Vision and Pattern Recognition, pp. 1971–1980 (2016)
14. Jain, M., Van Gemert, J., Jégou, H., Bouthemy, P., Snoek, C.G.: Tubelets: unsupervised action proposals from spatiotemporal super-voxels. Int. J. Comput. Vision **124**(3), 287–311 (2017)
15. Jhuang, H., Gall, J., Zuffi, S., Schmid, C., Black, M.J.: Towards understanding action recognition. In: Proceedings of the IEEE International Conference on Computer Vision, pp. 3192–3199 (2013)
16. Ji, X., Henriques, J.F., Vedaldi, A.: Invariant information clustering for unsupervised image classification and segmentation. In: Proceedings of the IEEE International Conference on Computer Vision, pp. 9865–9874 (2019)
17. Jiang, Y.G., et al.: Thumos challenge: action recognition with a large number of classes (2014)

18. Kuehne, H., Arslan, A., Serre, T.: The language of actions: recovering the syntax and semantics of goal-directed human activities. In: IEEE Conference on Computer Vision and Pattern Recognition (CVPR), pp. 780–787 (2014)
19. Lan, T., Wang, Y., Mori, G.: Discriminative figure-centric models for joint action localization and recognition. In: 2011 International Conference on Computer Vision, pp. 2003–2010. IEEE (2011)
20. Li, S., et al.: Groupformer: group activity recognition with clustered spatial-temporal transformer. In: Proceedings of the IEEE/CVF International Conference on Computer Vision, pp. 13668–13677 (2021)
21. Li, Z., Gavrilyuk, K., Gavves, E., Jain, M., Snoek, C.G.: Videolstm convolves, attends and flows for action recognition. Comput. Vis. Image Underst. **166**, 41–50 (2018)
22. Lin, M., Chen, Q., Yan, S.: Network in network. arXiv preprint arXiv:1312.4400 (2013)
23. Liu, W., et al.: SSD: single shot multibox detector. In: Leibe, B., Matas, J., Sebe, N., Welling, M. (eds.) ECCV 2016. LNCS, vol. 9905, pp. 21–37. Springer, Cham (2016). https://doi.org/10.1007/978-3-319-46448-0_2
24. Liu, Y., Tu, Z., Lin, L., Xie, X., Qin, Q.: Real-time spatio-temporal action localization via learning motion representation. In: Proceedings of the Asian Conference on Computer Vision (2020)
25. Luong, M.T., Pham, H., Manning, C.D.: Effective approaches to attention-based neural machine translation. arXiv preprint arXiv:1508.04025 (2015)
26. Pan, J., Chen, S., Shou, M.Z., Liu, Y., Shao, J., Li, H.: Actor-context-actor relation network for spatio-temporal action localization. In: Proceedings of the IEEE/CVF Conference on Computer Vision and Pattern Recognition, pp. 464–474 (2021)
27. Pramono, R.R.A., Chen, Y.T., Fang, W.H.: Hierarchical self-attention network for action localization in videos. In: Proceedings of the IEEE/CVF International Conference on Computer Vision, pp. 61–70 (2019)
28. Qi, M., Qin, J., Li, A., Wang, Y., Luo, J., Van Gool, L.: stagNet: an attentive semantic RNN for group activity recognition. In: Ferrari, V., Hebert, M., Sminchisescu, C., Weiss, Y. (eds.) ECCV 2018. LNCS, vol. 11214, pp. 104–120. Springer, Cham (2018). https://doi.org/10.1007/978-3-030-01249-6_7
29. Redmon, J., Farhadi, A.: Yolo9000: better, faster, stronger. In: Proceedings of the IEEE Conference on Computer Vision and Pattern Recognition, pp. 7263–7271 (2017)
30. Rodriguez, M.D., Ahmed, J., Shah, M.: Action MACH a spatio-temporal maximum average correlation height filter for action recognition. In: 2008 IEEE Conference on Computer Vision and Pattern Recognition, pp. 1–8. IEEE (2008)
31. Russakovsky, O., et al.: ImageNet large scale visual recognition challenge. Int. J. Comput. Vis. (IJCV) **115**(3), 211–252 (2015). https://doi.org/10.1007/s11263-015-0816-y
32. Sharma, S., Kiros, R., Salakhutdinov, R.: Action recognition using visual attention. In: Neural Information Processing Systems: Time Series Workshop (2015)
33. Shu, T., Todorovic, S., Zhu, S.C.: CERN: confidence-energy recurrent network for group activity recognition. In: Proceedings of the IEEE Conference on Computer Vision and Pattern Recognition, pp. 5523–5531 (2017)
34. Simonyan, K., Zisserman, A.: Very deep convolutional networks for large-scale image recognition. arXiv preprint arXiv:1409.1556 (2014)
35. Soomro, K., Idrees, H., Shah, M.: Action localization in videos through context walk. In: Proceedings of the IEEE International Conference on Computer Vision, pp. 3280–3288 (2015)

36. Soomro, K., Idrees, H., Shah, M.: Predicting the where and what of actors and actions through online action localization. In: Proceedings of the IEEE Conference on Computer Vision and Pattern Recognition, pp. 2648–2657 (2016)
37. Soomro, K., Shah, M.: Unsupervised action discovery and localization in videos. In: Proceedings of the IEEE International Conference on Computer Vision, pp. 696–705 (2017)
38. Soomro, K., Zamir, A.R., Shah, M.: UCF101: a dataset of 101 human actions classes from videos in the wild. arXiv preprint arXiv:1212.0402 (2012)
39. Tian, Y., Sukthankar, R., Shah, M.: Spatiotemporal deformable part models for action detection. In: Proceedings of the IEEE Conference on Computer Vision and Pattern Recognition, pp. 2642–2649 (2013)
40. Tran, D., Bourdev, L., Fergus, R., Torresani, L., Paluri, M.: Learning spatiotemporal features with 3d convolutional networks. In: Proceedings of the IEEE International Conference on Computer Vision, pp. 4489–4497 (2015)
41. Tran, D., Yuan, J.: Max-margin structured output regression for spatio-temporal action localization. In: Advances in neural information processing systems, pp. 350–358 (2012)
42. Wang, J., Jiao, J., Bao, L., He, S., Liu, Y., Liu, W.: Self-supervised spatio-temporal representation learning for videos by predicting motion and appearance statistics. In: Proceedings of the IEEE Conference on Computer Vision and Pattern Recognition, pp. 4006–4015 (2019)
43. Wang, L., Qiao, Yu., Tang, X.: Video action detection with relational dynamic-poselets. In: Fleet, D., Pajdla, T., Schiele, B., Tuytelaars, T. (eds.) ECCV 2014. LNCS, vol. 8693, pp. 565–580. Springer, Cham (2014). https://doi.org/10.1007/978-3-319-10602-1_37
44. Wang, M., Ni, B., Yang, X.: Recurrent modeling of interaction context for collective activity recognition. In: Proceedings of the IEEE Conference on Computer Vision and Pattern Recognition, pp. 3048–3056 (2017)
45. Weinzaepfel, P., Harchaoui, Z., Schmid, C.: Learning to track for spatio-temporal action localization. In: Proceedings of the IEEE international conference on computer vision, pp. 3164–3172 (2015)
46. Wu, J., Wang, L., Wang, L., Guo, J., Wu, G.: Learning actor relation graphs for group activity recognition. In: Proceedings of the IEEE/CVF Conference on Computer Vision and Pattern Recognition, pp. 9964–9974 (2019)
47. Xie, J., Girshick, R., Farhadi, A.: Unsupervised deep embedding for clustering analysis. In: International Conference on Machine Learning (ICML), pp. 478–487 (2016)
48. Zhang, D., He, L., Tu, Z., Zhang, S., Han, F., Yang, B.: Learning motion representation for real-time spatio-temporal action localization. Pattern Recogn. **103**, 107312 (2020)

Bandwidth-Aware Adaptive Codec for DNN Inference Offloading in IoT

Xiufeng Xie[1](✉), Ning Zhou[2], Wentao Zhu[2], and Ji Liu[1]

[1] Kwai Inc, New York, USA
{xiufengxie,jiliu}@kuaishou.com
[2] Amazon, Seattle, USA
{ningzhou,wentaozhu}@amazon.com

Abstract. The lightweight nature of IoT devices makes it challenging to run deep neural networks (DNNs) locally for applications like augmented reality. Recent advances in IoT communication like LTE-M have significantly boosted the link bandwidth, enabling IoT devices to stream visual data to edge servers running DNNs for inference. However, uncompressed visual data can still easily overload the IoT link, and the wireless spectrum is shared by numerous IoT devices, causing unstable link bandwidth. Mainstream codecs can reduce the traffic but at the cost of severe inference accuracy drops. Recent works on differentiable JPEG train the codec to tackle the damage to inference accuracy. But they rely on heuristic configurations in the loss function to balance the rate-accuracy tradeoff, providing no guarantee to meet the IoT bandwidth constraint. This paper presents AutoJPEG, a bandwidth-aware adaptive compression solution that learns the JPEG encoding parameters to optimize the DNN inference accuracy under bandwidth constraints. We model the compressed image size as a closed-form function of encoding parameters by analyzing the JPEG codec workflow. Furthermore, we formulate a constrained optimization framework to minimize the original DNN loss while ensuring the image size strictly meets the bandwidth constraint. Our evaluation validates AutoJPEG on various DNN models and datasets. In our experiments, AutoJPEG outperforms the mainstream codecs (like JPEG and WebP) and the state-of-the-art solutions that optimize the image codec for DNN inference.

Keywords: DNN-friendly image compression · IoT · Edge computing · Inference offloading · Admm optimizer · Bandwidth constraint

1 Introduction

AI applications are becoming ubiquitous in the Internet-of-Things (IoT) era. For example, wearable items like augmented reality (AR) glasses can employ deep neural networks (DNNs) to understand the environment and show the users an

S. Avidan et al. (Eds.): ECCV 2022, LNCS 13698, pp. 88–104, 2022.
https://doi.org/10.1007/978-3-031-19839-7_6

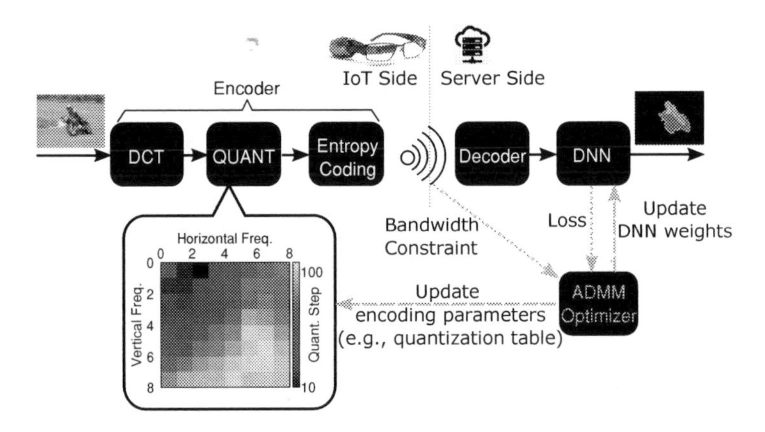

Fig. 1. Given the IoT uplink bandwidth constraint, AutoJPEG uses an alternating direction method of multipliers (ADMM) optimizer to learn the DNN model weights and the encoding parameters jointly.

augmented world. However, IoT devices like AR glasses are generally resource-constrained with limited computing power, memory, and battery capacity, thus cannot afford to run a DNN of any size. Recent advances [13–15] in IoT communication like 5G and Long Term Evolution for Machines (LTE-M) have made it possible to transport the captured data from the IoT device to an edge computing server running DNN inference. Due to the limited wireless bandwidth and huge IoT device population in the future, the IoT device needs to compress large data footprints such as images or videos. Unfortunately, the compression artifacts of mainstream codecs (e.g., JPEG) usually cause severe inference accuracy drop because they target human perception rather than DNNs.

This paper presents AutoJPEG, a bandwidth-aware data compression solution for DNN inference offloading. As outlined in Fig. 1, AutoJPEG jointly learns the DNN weights and encoding parameters like the quantization table of discrete cosine transform (DCT) coefficients in an end-to-end fashion. The core of AutoJPEG is a constrained optimization framework to minimize the DNN loss with the image size as the constraint, enabling it to balance the tradeoff between the inference accuracy and bandwidth cost. The offline optimization and deployment (e.g., wireless broadcast) of encoding parameters to the IoT devices happen only once before the online inference, adding no extra end-to-end latency. The IoT device encodes its captured images using the optimized encoding parameters. It then sends compressed images to the server that runs the optimized DNN for inference. Considering the dynamic IoT bandwidth, we optimize multiple sets of encoding parameters in advance, each for a particular bandwidth constraint. Then the encoder adapts between these sets following the current bandwidth.

We built AutoJPEG on top of the prevalent JPEG codec as a testbed of our optimization framework, which can be easily extended to improve codecs with a similar architecture like MPEG. Our implementation is compatible with JPEG since it only trains the configurable encoding parameters in the JPEG standard. Our evaluation in Sect. 4 includes semantic segmentation (large image, complicated DNN) and classification (small image, simple DNN). In our experiments, AutoJPEG outperforms recent DNN-friendly compression works [36], DNN inference offloading works [20], and mainstream codecs like JPEG and WebP, in terms of both inference accuracy and compression ratio.

The main contributions of this work can be summarized as follows:

- To the best of our knowledge, AutoJPEG is the first bandwidth-aware data compression solution optimized for inference offloading in IoT networks.
- We dig deep into a representative codec workflow to obtain the closed-form expressions of the compressed image size & recovered image data and use smoothed estimators to make them compatible with gradient descent.
- The crux of this work is a constrained optimization framework designed to learn the encoding parameters under strict bandwidth constraint – a contribution from the methodological aspect as prior works are typically a collection of heuristic designs with no guarantee to meet the constraint.

2 Related Work

Running DNNs on resource-constrained mobile devices like smartphones is challenging. DNN weight pruning [9,9–11,23,23,24,27,28,35,42,42] and weight quantization [4,5,17,22,29,41] can reduce the computation load with little or no impact on the inference accuracy. Another line of work [16,20] splits the DNN model at a "bottleneck" to form a lightweight head model deployed at the mobile devices and a heavier tail model at the server. However, the above solutions cannot migrate to IoT scenarios because most IoT devices cannot even afford to run a lightweight DNN.

Recent works investigated the DNN-friendly JPEG codec where the compressed images are inputs of DNN inference rather than human vision. DeepN-JPEG [25] assumes the standard deviation of a DCT coefficient determines its contribution to DNN learning, thus using a heuristic function to map the standard deviations of DCT coefficients to a JPEG quantization table. GRACE [36] uses the gradient *w.r.t.* the loss function to measure a DNN's perceptual sensitivity to different DCT frequency bands and then optimizes the JPEG quantization table following the frequency-domain sensitivity. Making JPEG trainable [2,26,32,40] is another related research direction, which incorporates the encoder and DNN in an end-to-end differentiable training framework. A comprehensive work [26] investigates the tradeoff between rate, distortion, and inference accuracy in such a differentiable setup by formulating a loss function as a weighted sum of loss components corresponding to the three targets.

AutoJPEG differs from the above works in the following aspects. 1) most existing works roughly estimate the compressed image size, whereas our work digs into the details of the encoding algorithms (like run-length encoding) and the data formats to estimate a more accurate image size. 2) As shown in Fig. 2, existing works typically use a weighted sum with configurable weight coefficients

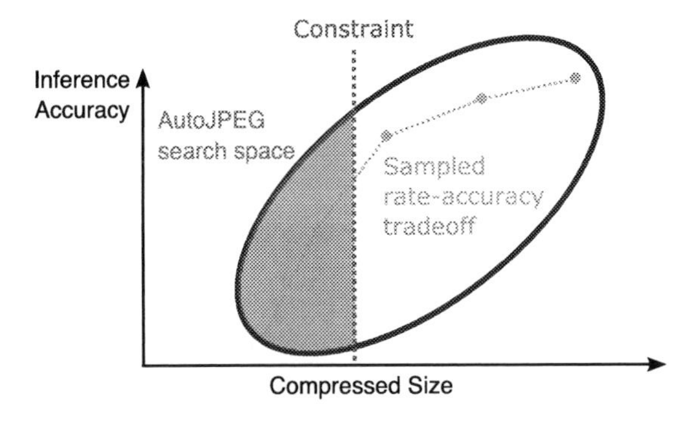

Fig. 2. AutoJPEG searches for the optimal encoding parameters under the bandwidth constraint, while existing works coarsely sample the rate-accuracy relation by tuning the weight coefficients in the weighted-sum loss. (The area in the black oval is the space of encoding parameters allowed by JPEG.)

to combine the original loss and the penalty of image size into one loss function, which fails to precisely control the image size to fit the bandwidth constraint. In contrast, we formulate a constrained optimization problem, guaranteeing the compressed image size meets the given IoT bandwidth constraint.

3 AutoJPEG Design

In this section, we first mathematically model the JPEG codec from the viewpoint of differentiable and non-differentiable operations (Sect. 3.1), then discuss how AutoJPEG makes JPEG codec trainable (Sect. 3.3), and elaborate on AutoJPEG's ADMM-based optimization (Sect. 3.4) framework.

3.1 Mathematical Modeling of JPEG Codec Workflow

***(1)* RGB-to-YUV Conversion (Differentiable).** First, the encoder converts the 3-channel RGB image $\boldsymbol{x} = (\boldsymbol{r}, \boldsymbol{g}, \boldsymbol{b})^\top$ to the YUV color space as $\boldsymbol{x}_\tau = (\boldsymbol{y}, \boldsymbol{u}, \boldsymbol{v})^\top$. The YUV color space provides more room for compression by concentrating salient information to the $\boldsymbol{y}$ channel so that the other two less informative channels $\boldsymbol{u}$ and $\boldsymbol{v}$ allow more compression. The parameter $\boldsymbol{\omega} = (w_r, w_g, w_b)^\top$ uniquely defines the RGB-to-YUV conversion as shown in Eq. (1).

$$\begin{bmatrix} \boldsymbol{y} \\ \boldsymbol{u} \\ \boldsymbol{v} \end{bmatrix} = \begin{bmatrix} w_r & w_g & w_b \\ -\frac{1}{2}\frac{w_r}{1-w_b} & -\frac{1}{2}\frac{w_g}{1-w_b} & \frac{1}{2} \\ \frac{1}{2} & -\frac{1}{2}\frac{w_g}{1-w_r} & -\frac{1}{2}\frac{w_b}{1-w_r} \end{bmatrix} \begin{bmatrix} \boldsymbol{r} \\ \boldsymbol{g} \\ \boldsymbol{b} \end{bmatrix} \tag{1}$$

Eq. (1) is a linear transformation and thus differentiable. It can be written as $\boldsymbol{x}_\tau = \Phi_\omega \boldsymbol{x}$, where Φ_ω is the 3×3 matrix that defines the transform. Conventional encoders choose $\boldsymbol{\omega}$ based on the color sensitivity of the human vision system [6], e.g., JPEG uses $(w_r, w_g, w_b) = (0.299, 0.587.0.114)$ following [18]. Overall, $\boldsymbol{\omega}$ follows the constraint $\boldsymbol{\omega}^\top \mathbf{1} = 1, \boldsymbol{\omega} \geq 0$. The decoder can recover the RGB image from the YUV image by the inverse transformation $\boldsymbol{x} = \Phi_\omega^{-1} \boldsymbol{x}_\tau$.

(2) **DCT (Differentiable).** The encoder then uses the 2D Discrete Cosine Transform (DCT) to convert the YUV image $\boldsymbol{x}_\tau$ to the DCT coefficients $\boldsymbol{X} = \mathrm{DCT}(\boldsymbol{x}_\tau) = (\mathrm{DCT}(\boldsymbol{y}), \mathrm{DCT}(\boldsymbol{u}), \mathrm{DCT}(\boldsymbol{v}))$. DCT is a linear transformation and differentiable. The decoder can recover the image from the DCT coefficients by the Inverse Discrete Cosine Transform (IDCT), which is also differentiable.

(3) **Quantization (Non-Differentiable).** Next, the encoder quantizes each channel of the DCT coefficients $\boldsymbol{X}$ by a quantization table. For simplicity, we denote $\boldsymbol{X} = (X_1, \ldots, X_N)$ as all DCT coefficients across 3 channels, and $\boldsymbol{Q} = (q_1, \ldots, q_N)$ contains all elements of the 3 quantization tables, where q_i is the *quantization step* on the i-th DCT coefficient. The quantized result is denoted as $\boldsymbol{X}_Q := \mathcal{R}(\boldsymbol{X}/\boldsymbol{Q})$, where $\mathcal{R}(\cdot)$ denotes the round function $\lfloor \cdot \rceil$ and $\mathcal{R}(\boldsymbol{X}/\boldsymbol{Q}) = (\mathcal{R}(X_1/q_1), \ldots, \mathcal{R}(X_N/q_N))$. Quantization is non-differentiable. Later, the decoder can only recover a lossy version of the original DCT coefficients by dequantization as $\hat{\boldsymbol{X}}_Q = \boldsymbol{Q} \cdot \boldsymbol{X}_Q$, and dequantization is differentiable.

(4) **Lossless Encoding (Transparent).** The encoder then vectorizes the quantized DCT coefficients $\boldsymbol{X}_Q$ by running a zigzag scan from low to high frequency, which forms groups of consecutive zeros. Run-length encoding (RLE) leverages such consecutive zeros for compression: the encoder only stores the non-zero data points and the count of consecutive 0s after each non-zero data point (*i.e.*, the *skip length*). We elaborate on the image size analysis in Sect. 3.2. The JPEG encoder then employs entropy coding to reduce the image size further. Both RLE and entropy coding are lossless, which means the gradient can pass through them in backward propagation.

(5) **Decoding (Differentiable).** The decoder recovers the image from the compressed data by reversing the above encoding steps: entropy decoding (transparent), RLE decoding (transparent), dequantization (differentiable), IDCT (differentiable), and YUV-to-RGB conversion (differentiable). The recovered image $\hat{\boldsymbol{x}}$ is not exactly the same as the original image $\boldsymbol{x}$. Using the recovered image $\hat{\boldsymbol{x}}$ as the DNN's input, the inference accuracy is generally lower than using the original image due to the compression artifacts.

Summary. Let $\mathcal{N}(\boldsymbol{Q}, \boldsymbol{\omega}, \cdot)$ denote the end-to-end JPEG encoding & decoding process, it can be summarized as:

$$\mathcal{N}(\boldsymbol{Q}, \boldsymbol{\omega}, \boldsymbol{x}) := \mathrm{IDCT}\left(\Phi_\omega^{-1}\left(\boldsymbol{Q} \cdot \mathcal{R}\left(\frac{\mathrm{DCT}(\Phi_\omega \boldsymbol{x})}{\boldsymbol{Q}}\right)\right)\right) \tag{2}$$

Since RLE and entropy coding are lossless, their encoding and decoding cancel out with each other in Eq. (2). $\mathcal{N}(\boldsymbol{Q}, \boldsymbol{\omega}, \cdot)$ is non-differentiable.

3.2 Modeling Compressed Image Size of JPEG Encoder

Let $\mathcal{H}$ denote the size of image $\boldsymbol{x}$ after JPEG compression, we want to model it as a function $\mathcal{H}(\boldsymbol{Q}, \boldsymbol{\omega}, \boldsymbol{x})$ of the quantization table $\boldsymbol{Q}$, RGB-to-YUV parameters $\boldsymbol{\omega}$, and the original image $\boldsymbol{x}$. Since the effect of the entropy coding on the image size is always marginal and difficult to model, we use the total number of bits after RLE as an estimation, which is an upper bound of the actual size. If $\mathcal{H}$ is below the image size constraint, the actual size strictly satisfies the constraint. $\mathcal{H}$ consists of two parts: *(i)* the number of bits ($\mathcal{V}$) carrying the value of quantized data points $\boldsymbol{X_Q}$ and *(ii)* the number of bits ($\mathcal{M}$) carrying the supporting data associated with each non-zero data point, including the bit lengths for variable-length binary encoding, the sign bit (DCT can yield negative values), and the skip lengths for RLE.

$$\mathcal{H}(\boldsymbol{Q}, \boldsymbol{\omega}, \boldsymbol{x}) = \mathcal{V}(\boldsymbol{Q}, \boldsymbol{\omega}, \boldsymbol{x}) + \mathcal{M}(\boldsymbol{Q}, \boldsymbol{\omega}, \boldsymbol{x}) \tag{3}$$

(i) Bits Carrying the Datapoint Values (Non-Differentiable). We look into $\mathcal{V}$ first. Note that a data point with value 0 consumes 0 bit in RLE. Given a quantized DCT data point with value $\mathcal{R}(X_j/q_j)$, Eq. (4) shows the number of bits b_j to represent this data point. We can further remove the round function $\mathcal{R}(\cdot)$ without affecting the value of b_j:

$$b_j = 1 + \left\lfloor \log_2 \left(\mathcal{R}\left(\frac{|X_j|}{q_j}\right) + 0.5 \right) \right\rfloor = 1 + \left\lfloor \log_2 \left(\frac{|X_j|}{q_j} + 0.5 \right) \right\rfloor \tag{4}$$

By summing the sizes of all quantized data points, we have:

$$\mathcal{V} = \sum_{j=1}^{N} b_j = N + \sum_{j=1}^{N} \left\lfloor \log_2 \left(\frac{|X_j|}{q_j} + 0.5 \right) \right\rfloor \tag{5}$$

(ii) Bits Carrying the Supporting Data (Non-Differentiable). Then we look into $\mathcal{M}$. For each non-zero data point, the JPEG encoder spends 4 bits to represent its binary data length, and it also uses 4 bit to represent the skip length. Since only the non-zero data point needs the sign bit in RLE, we also count it as 1 bit here. Therefore, for each non-zero data point X_j, besides the data size b_j, the encoder needs an extra 9 bits. As a result, the total supporting data size $\mathcal{M}$ is the scaled L0 norm of the quantized data $\mathcal{R}(\boldsymbol{X}/\boldsymbol{Q})$:

$$\mathcal{M} = 9 \left\| \mathcal{R}(\boldsymbol{X}/\boldsymbol{Q}) \right\|_0 \tag{6}$$

3.3 AutoJPEG Makes JPEG Codec End-to-End Trainable

The compression artifacts harming the DNN inference accuracy come from the quantization and are indirectly affected by the RGB-to-YUV conversion. Therefore, we optimize the quantization table $\boldsymbol{Q}$ and the RGB-to-YUV parameters $\boldsymbol{\omega}$ to make the compression artifacts mostly "invisible" to the DNN.

As shown in Eq. (7), AutoJPEG jointly optimizes the encoding parameters ($\boldsymbol{Q}$ and $\boldsymbol{\omega}$) and DNN weights $\mathcal{W}$. The object is to minimize the DNN loss function $\ell(\cdot)$ under a given image size constraint $\mathcal{C} := \mathcal{H}_{\text{raw}}S$, where $S \in (0,1]$ is the compression ratio, $\mathcal{H}_{\text{raw}}$ is the uncompressed image size, $\boldsymbol{x}^{(i)}$ is the i-th training sample, $\mathcal{X} = \{x^{(1)}, \ldots, x^{(M)}\}$ is the dataset with M samples, $\mathcal{N}(\boldsymbol{Q}, \boldsymbol{\omega}, \boldsymbol{x}^{(i)})$ from Eq. (2) is the recovered image, $\mathcal{H}(\boldsymbol{Q}, \boldsymbol{\omega}, \boldsymbol{x}^{(i)})$ from Eq. (3) is the compressed image size in bits.

$$\min_{\mathcal{W},\boldsymbol{Q},\boldsymbol{\omega}} \quad \sum_{i=1}^{M} \ell\left(\mathcal{W}, \mathcal{N}(\boldsymbol{Q}, \boldsymbol{\omega}, \boldsymbol{x}^{(i)})\right) \tag{7a}$$

$$\text{s.t.} \quad \mathcal{H}(\boldsymbol{Q}, \boldsymbol{\omega}, \boldsymbol{x}^{(i)}) \leq \mathcal{C}, \quad \forall \boldsymbol{x}^{(i)} \in \mathcal{X} \tag{7b}$$

$$\boldsymbol{\omega}^{\top}\mathbf{1} = 1, \boldsymbol{\omega} \geq 0 \tag{7c}$$

Solving the problem in Eq.(7) is difficult because both $\mathcal{N}(\boldsymbol{Q}, \boldsymbol{\omega}, \boldsymbol{x}^{(i)})$ and $\mathcal{H}(\boldsymbol{Q}, \boldsymbol{\omega}, \boldsymbol{x}^{(i)})$ are non-differentiable. In what follows, we discuss how to design the smoothed estimators $\hat{\mathcal{N}}(\boldsymbol{Q}, \boldsymbol{\omega}, \boldsymbol{x}^{(i)})$ and $\hat{\mathcal{H}}(\boldsymbol{Q}, \boldsymbol{\omega}, \boldsymbol{x}^{(i)})$ to enable gradient descent.

Smoothed Estimator of the Image Size. The image size $\mathcal{H}(\boldsymbol{Q}, \boldsymbol{\omega}, \boldsymbol{x}^{(i)})$ is the sum of $\mathcal{V}$ and $\mathcal{M}$. However, both $\mathcal{V}$ and $\mathcal{M}$ are non-differentiable, so we define their smoothed estimators $\hat{\mathcal{V}}$ and $\hat{\mathcal{M}}$ to enable gradient descent.

$$\hat{\mathcal{H}}(\boldsymbol{Q}, \boldsymbol{\omega}, \boldsymbol{x}^{(i)}) = \hat{\mathcal{V}}(\boldsymbol{Q}, \boldsymbol{\omega}, \boldsymbol{x}^{(i)}) + \hat{\mathcal{M}}(\boldsymbol{Q}, \boldsymbol{\omega}, \boldsymbol{x}^{(i)}) \tag{8}$$

The representation of $\mathcal{V}$ in Eq. (5) can be rewritten as:

$$\mathcal{V} = 2N + \sum_{j=1}^{N} \left\lfloor \log_2 \left(\max\left(\frac{X_j^{(i)}}{q_j} + \frac{1}{2}, \frac{1}{2} \right) \right) \right\rfloor + \sum_{j=1}^{N} \left\lfloor \log_2 \left(\max\left(-\frac{X_j^{(i)}}{q_j} + \frac{1}{2}, \frac{1}{2} \right) \right) \right\rfloor \tag{9}$$

The floor function $\lfloor \cdot \rfloor$ in the above equation is piece-wise constant, which is non-differentiable at integer inputs while having 0 gradient elsewhere, and thus is incompatible[1] with gradient descent. We use the straight through estimator (STE) to solve this problem and define the smoothed estimator of $\mathcal{V}$ as:

$$\hat{\mathcal{V}}(\boldsymbol{Q}, \boldsymbol{\omega}, \boldsymbol{x}) := \begin{cases} \mathcal{V}(\boldsymbol{Q}, \boldsymbol{\omega}, \boldsymbol{x}), & \text{if in forward pass} \\ \mathcal{V}^d(\boldsymbol{Q}, \boldsymbol{\omega}, \boldsymbol{x}), & \text{otherwise} \end{cases}$$

$\hat{\mathcal{V}}$ equals $\mathcal{V}$ in the forward pass, while $\mathcal{V}^d$ is simply Eq. (9) without the floor functions $\lfloor \cdot \rfloor$ so that gradients can pass through in the backward propagation.

$\mathcal{M}$ in Eq. (6) is essentially an L0 regularization for sparsity, and the common practice is to relax it to the L1 norm. We also remove the non-differentiable

[1] Although the max function is also non-differentiable, gradients can still pass through the max function during backward propagation (just like ReLU).

round function $\mathcal{R}(\cdot)$ following the idea of STE. The smoothed estimator $\hat{\mathcal{M}}$ is:

$$\hat{\mathcal{M}}(\boldsymbol{Q}, \boldsymbol{\omega}, \boldsymbol{x}) := \begin{cases} \mathcal{M}(\boldsymbol{Q}, \boldsymbol{\omega}, \boldsymbol{x}), & \text{if in forward pass} \\ 9 \left\| (\boldsymbol{X}^{(i)}/\boldsymbol{Q} \right\|_1, & \text{otherwise} \end{cases}$$

Smoothed Estimator of the Recovered Image. According to Eq. (2), the non-differentiable part of $\mathcal{N}(\boldsymbol{Q}, \boldsymbol{\omega}, \boldsymbol{x}^{(i)})$, the DNN input recovered from the compressed image, is the round function $\mathcal{R}(\cdot)$, but we cannot apply STE to remove $\mathcal{R}(x)$. For the quantization and recovery process, the recovered value $Y_j = \mathcal{R}(X_j/q_j) \cdot q_j$ with quantization step q_j, if we simply ignore the round function, then we have $Y_j = (X_j/q_j)q_j = X_j$, where q_j is cancelled out. Since this work is about learning the optimal q_j, we cannot let q_j disappear.

$\mathcal{R}(\cdot)$ can be written as a piece-wise constant function $\mathcal{R}(x) = \mathcal{A}(x - 0.5 - K) + K$, if $K \leq x < K + 1$. Different piece of the function has different constant integer K whose value equals $\lfloor x \rfloor$. $\mathcal{A}(\cdot)$ is an unit step function which is non-differentiable at 0 and has 0 gradient elsewhere.

$$\mathcal{A}(x) := \begin{cases} 1, & x \geq 0 \\ 0, & x < 0 \end{cases} \qquad \hat{\mathcal{A}}(x) := \begin{cases} \mathcal{A}(x), & \text{if in forward pass} \\ \frac{1}{2}\left(\tanh(Tx) + 1\right), & \text{otherwise} \end{cases}$$

We use a smoothed estimator $\hat{\mathcal{A}}(x)$ of the step function $\mathcal{A}(x)$ to enable gradient descent, where T controls the steepness of the smoothed step function. This work uses $T = 5$ to balance the approximation accuracy and training stability as discussed in [38]. By replacing $\mathcal{A}(\cdot)$ in function $\mathcal{R}(\cdot)$ with $\hat{\mathcal{A}}(\cdot)$, we have $\hat{\mathcal{R}}(\cdot)$, a smoothed estimator of $\mathcal{R}(\cdot)$. Then the DNN input recovered from the compressed image can be estimated as $\hat{\mathcal{N}}(\boldsymbol{Q}, \boldsymbol{\omega}, \boldsymbol{x})$ in Eq. (10), which allows gradients to pass through during backward propagation.

$$\hat{\mathcal{N}}(\boldsymbol{Q}, \boldsymbol{\omega}, \boldsymbol{x}) = \mathrm{IDCT}\left(\Phi_{\omega}^{-1}\left(\boldsymbol{Q} \cdot \hat{\mathcal{R}}\left(\frac{\mathrm{DCT}(\Phi_{\omega}\boldsymbol{x})}{\boldsymbol{Q}}\right)\right)\right) \tag{10}$$

By replacing the non-differentiable operators in Eq. (7) with the smoothed estimators in Eqs. (8) and (10), we have a new optimization problem:

$$\min_{\mathcal{W}, \boldsymbol{Q}, \boldsymbol{\omega}} \quad \sum_{i=1}^{M} \ell\left(\mathcal{W}, \hat{\mathcal{N}}(\boldsymbol{Q}, \boldsymbol{\omega}, \boldsymbol{x}^{(i)})\right) \tag{11a}$$

$$\text{s.t.} \quad \hat{\mathcal{H}}(\boldsymbol{Q}, \boldsymbol{\omega}, \boldsymbol{x}^{(i)}) \leq \mathcal{C}, \ \forall \boldsymbol{x}^{(i)} \in \mathcal{X} \tag{11b}$$

$$\boldsymbol{\omega}^{\top}\mathbf{1} = 1, \boldsymbol{\omega} \geq 0 \tag{11c}$$

3.4 Solving the Optimization Problem Using ADMM

In what follows, we discuss how AutoJPEG solves the constrained optimization problem in Eq. (11). First, we use the augmented Lagrangian method to convert

Algorithm 1: Codec & DNN Joint Optimization Under Size Constraint

Input : Image size constraint $\mathcal{C}$,
Pretrained DNN weights $\mathcal{W}^I$,
BT.601 RGB-to-YUV parameter $\boldsymbol{\omega}^I$

Output : Quantization table $\boldsymbol{Q}^*$,
RGB-to-YUV parameters $\boldsymbol{\omega}^*$,
DNN weights $\mathcal{W}^*$

Initialize $\mathcal{W} = \mathcal{W}^I, \boldsymbol{Q} = J_{3,8,8}, \boldsymbol{\omega} = \boldsymbol{\omega}^I$

while $\boldsymbol{x}^{(t)} \in \mathcal{X}$ *and* $\mathcal{H}\left(\boldsymbol{Q}, \boldsymbol{\omega}, \boldsymbol{x}^{(t)}\right) > \mathcal{H}_c$ **do**

  Update the primary variable $\mathcal{W}$ by SGD following Eq.(13);
  Update the primary variable $\boldsymbol{Q}$ by SGD following Eq.(14);
  Update the primary variable $\boldsymbol{\omega}$ by PGD following Eq.(15);
  Update the dual variable z following Eq.(16);

end

$\boldsymbol{Q}^* = \boldsymbol{Q}$, $\boldsymbol{\omega}^* = \boldsymbol{\omega}$, $\mathcal{W}^* = \mathcal{W}$

the problem to its equivalent minimax problem:

$$\min_{\mathcal{W},\boldsymbol{Q},\boldsymbol{\omega}} \max_{z\geq 0} \quad \sum_{i=1}^{M} \left(\ell(\mathcal{W}, \hat{\mathcal{N}}(\boldsymbol{Q}, \boldsymbol{\omega}, \boldsymbol{x}^{(i)})) + \zeta(\boldsymbol{Q}, \boldsymbol{\omega}, \boldsymbol{x}^{(i)}, z)\right) \tag{12a}$$

$$\text{s.t.} \quad \boldsymbol{\omega}^\top \mathbf{1} = 1, \boldsymbol{\omega} \geq 0 \tag{12b}$$

where $\zeta(\boldsymbol{Q}, \boldsymbol{\omega}, \boldsymbol{x}^{(i)}, z)$ is the augmented Lagrangian:

$$\zeta(\boldsymbol{Q}, \boldsymbol{\omega}, \boldsymbol{x}^{(i)}, z) := \frac{\rho_z}{2}\left[\hat{\mathcal{H}}(\boldsymbol{Q}, \boldsymbol{\omega}, \boldsymbol{x}^{(i)}) - \mathcal{C}\right]_+^2 + z\left(\hat{\mathcal{H}}(\boldsymbol{Q}, \boldsymbol{\omega}, \boldsymbol{x}^{(i)}) - \mathcal{C}\right)$$

We use $[\cdot]_+$ to denote the non-negative clamp $\max(\cdot, 0)$. ρ_z is the learning rate of the dual variable z when using ADMM to solve the problem.

Following Algorithm 1, in the t-th iteration, we first optimize the DNN weights $\mathcal{W}$ by stochastic gradient descent (SGD) with learning rate $\rho_\mathcal{W}$:

$$\mathcal{W}^{(t+1)} = \mathcal{W}^{(t)} - \rho_\mathcal{W} \nabla_{\mathcal{W}^{(t)}} \mathcal{L}^{(t)} \tag{13}$$

$\mathcal{L}^{(t)}$ is the loss $\sum_{i=1}^{M}(\ell(\mathcal{W}^{(t)}, \hat{\mathcal{N}}(\boldsymbol{Q}^{(t)}, \boldsymbol{\omega}^{(t)}, \boldsymbol{x}^{(t)})) + \zeta(\boldsymbol{Q}^{(t)}, \boldsymbol{\omega}^{(t)}, \boldsymbol{x}^{(t)}, z^{(t)}))$ from Eq. (12a) in the t-th iteration. We then use SGD to optimize the quantization table $\boldsymbol{Q}$ with learning rate ρ_q:

$$\boldsymbol{Q}^{(t+1)} = \boldsymbol{Q}^{(t)} - \rho_q \nabla_{\boldsymbol{Q}^{(t)}} \mathcal{L}^{(t)} \tag{14}$$

The constraint $\boldsymbol{\omega}^\top \mathbf{1} = 1, \boldsymbol{\omega} \geq 0$ is an unit 3-simplex, hence the variable $\boldsymbol{\omega}$ can be optimized by projected gradient descent (PGD) with learning rate ρ_ω:

$$\boldsymbol{\omega}^{(t+1)} = \mathcal{P}(\boldsymbol{\omega}^{(t)} - \rho_\omega \nabla_{\boldsymbol{\omega}^{(t)}} \mathcal{L}^{(t)}) \tag{15}$$

where $\mathcal{P}(\cdot)$ is the projection to an unit 3-simplex.

Finally, the optimizer performs the dual update by updating the dual variable z using project gradient ascent with learning rate ρ_z, where the non-negative clamp $[\cdot]_+$ projects the gradient ascent result to the space where $z \geq 0$:

$$z^{(t+1)} = \left[z^{(t)} + \rho_z \left(\hat{\mathcal{H}}(\boldsymbol{Q}^{(t)}, \boldsymbol{\omega}^{(t)}, \boldsymbol{x}^{(t)}) - \mathcal{C}\right)\right]_+ \tag{16}$$

The initial quantization table $\boldsymbol{Q}$ is a $3 \times 8 \times 8$ all-one matrix $J_{3,8,8}$ that barely compresses the image. Meanwhile, the initial RGB-to-YUV parameters $\boldsymbol{\omega}^I = (0.299, 0.587, 0.114)$ comes from BT.601 standard, and the initial DNN weights $\mathcal{W}^I$ comes from a model pre-trained by uncompressed training dataset.

4 Evaluation

4.1 Experiment Setup

Benchmarks. We compare AutoJPEG with a recent work GRACE [36] that optimizes quantization tables following the DNN's perceptual sensitivity. Neurosurgeon [20] is another benchmark, which partitions the DNN into two parts, one running on the client and the other on the server. Our benchmarks also include the mainstream codecs like JPEG [34], PNG [31], WebP [7], and H.264 [30].

Datasets. Since semantic segmentation typically requires high-quality images that may overload the IoT link, it is an ideal application of AutoJPEG. Our evaluation in Sect. 4.2 tests semantic segmentation on the Cityscapes dataset [3] with 2048×1024 resolution and lossless PNG format. We also evaluate AutoJPEG in classification tasks on CIFAR10 [21] dataset with small 32×32 images in Sect. 4.3. Such scenarios can be found in low-cost IoT devices.

DNN Models. AutoJPEG aims to achieve less inference accuracy loss than other encoders. Therefore, we test commonly used models instead of the latest models with state-of-the-art accuracy. For image classification, we use the ResNet models including ResNet20, ResNet32, ResNet56, ResNet110 [8] and VGG models including VGG11 and VGG13 [33]. For semantic segmentation, we use the dialated ResNet models [39] DRN-D-38 and DRN-D-22, following the evaluation setup of GRACE, the vital benchmark algorithm.

Hardware and Training Details. We run the experiments on a machine with a 2.1 GHz Intel Xeon Gold 5218R CPU, 120 GB memory, and 4 Nvidia Geforce RTX 2080Ti GPUs. The ADMM-based optimization framework and DNNs are based on PyTorch. The joint optimization finishes when the compressed image size satisfies the given constraint so there is no preset total epoch number. The initial model weights are pretrained on uncompressed datasets so we set the learning rate of the model weights to a small 10^{-4} for finetuning.

4.2 AutoJPEG in Semantic Segmentation

Figure 3 shows AutoJPEG's superior performance to existing solutions when compressing the input image for semantic segmentation model. The target DNN model is DRN-D-38, and the dataset is Cityscapes. We observe that AutoJPEG achieves the best balance between the inference accuracy and compression ratio.

Comparison with Neurosurgeon. Neurosurgeon has poor performance because its effectiveness relies on the DNN structure. It only works for DNNs with "bottleneck" layers whose output tensor is small. However, for DNNs to perform complicated tasks, the output tensor of every hidden layer can be bulky. Here the output data size of any layer in DRN-D-38 is larger than the input image, so the Neurosurgeon fails to reduce the data size. We further improve it as Neurosurgeon-Prune by using the DNN pruning solution from [37] to carve a "bottleneck" layer if the DNN does not have any, and thus the output tensor size of the partitioning point is reduced. Figure 3 shows that Neurosurgeon-Prune's performance is still far behind AutoJPEG. It only slightly compresses the data size but lowers the mIoU by more than 2%, because the channel-wise pruning is too coarse-grained compared to the spectral quantization in AutoJPEG.

Comparison with GRACE. We then perform a detailed comparison between AutoJPEG and GRACE, the state-of-the-art solution on DNN-friendly image compression. We run the experiments under multiple compression levels for both algorithms to profile their tradeoff between the image size and inference accuracy. As shown in Figs. 4a and b, AutoJPEG outperforms GRACE on both models by achieving higher mIoU at smaller image size. For example, from Fig. 4b, we observe that AutoJPEG achieves a 71.24% mIoU at image size of 408KB, while GRACE has a 0.24% lower mIoU at a larger image size of 418KB.

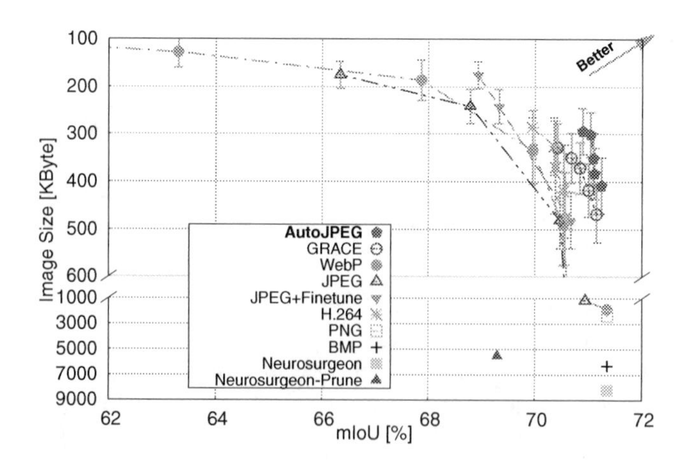

Fig. 3. AutoJPEG achieves a better balance between the DNN inference accuracy (mIoU) and compressed image size than existing solutions.

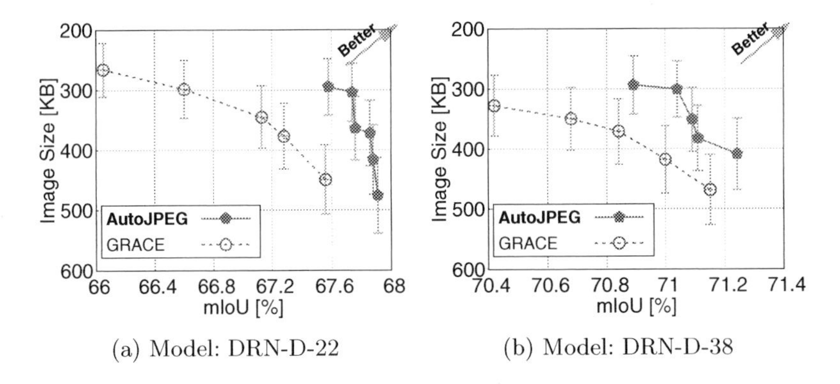

(a) Model: DRN-D-22 (b) Model: DRN-D-38

Fig. 4. Our solution AutoJPEG vs. GRACE.

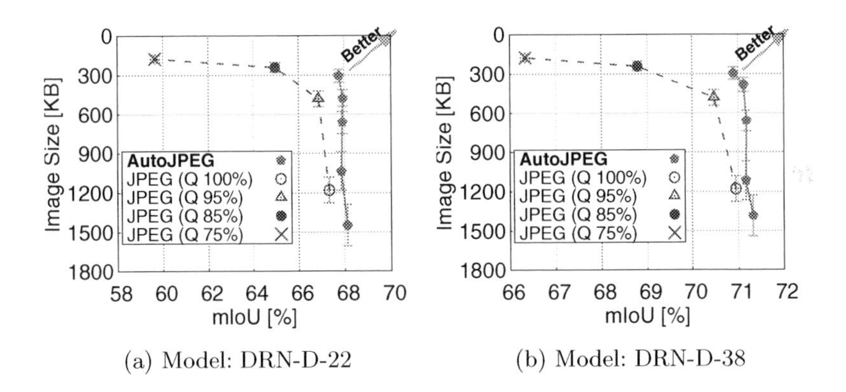

(a) Model: DRN-D-22 (b) Model: DRN-D-38

Fig. 5. Our solution AutoJPEG vs. JPEG.

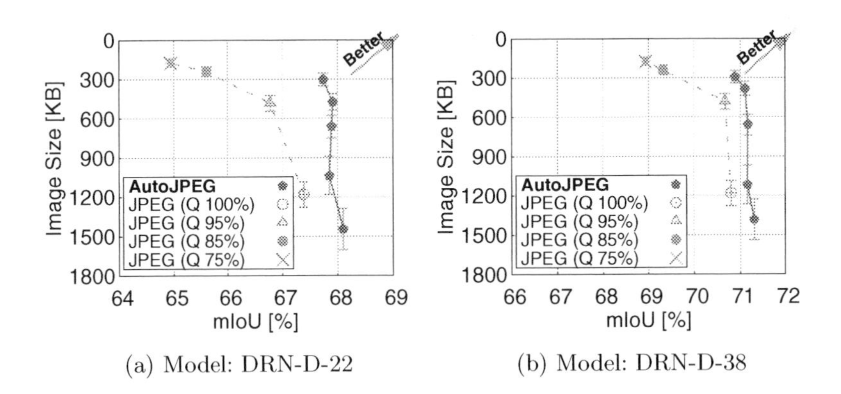

(a) Model: DRN-D-22 (b) Model: DRN-D-38

Fig. 6. Our solution AutoJPEG vs. JPEG+FT.

Comparison with JPEG and WebP. In this experiment, we first compare AutoJPEG with the JPEG encoder. We use multiple JPEG quality levels (75%, 85%, 95%, and 100%) since image size depends on the quality level ranging from 1% to 100% [34]. As shown in Fig. 5, AutoJPEG achieves higher mIoU than JPEG with similar image size. When the image size is small, JPEG suffers from a significant accuracy drop, while AutoJPEG's accuracy only reduces slightly. For instance, in Fig. 5a, the 85% JPEG quality causes a 7.43% mIoU loss compared to the uncompressed image. AutoJPEG with a similar size achieves a high mIoU with only 0.26% accuracy loss. When the image size is large, AutoJPEG still outperforms JPEG. For instance, the 100% JPEG quality yields a large image size of 1180KB but causes a 0.69% mIoU loss, while AutoJPEG has a slight 0.15% mIoU loss at a smaller image size of 1036KB. Finetuning the DNN by the JPEG-encoded image (JPEG+FT) can improve the inference accuracy, but the accuracy remains far below AutoJPEG as shown in Fig. 6. We further compare AutoJPEG with the WebP encoder of quality 80%, 85%, 90%, 100%, and lossless. As shown in Fig. 7, AutoJPEG also outperforms WebP in our experiments.

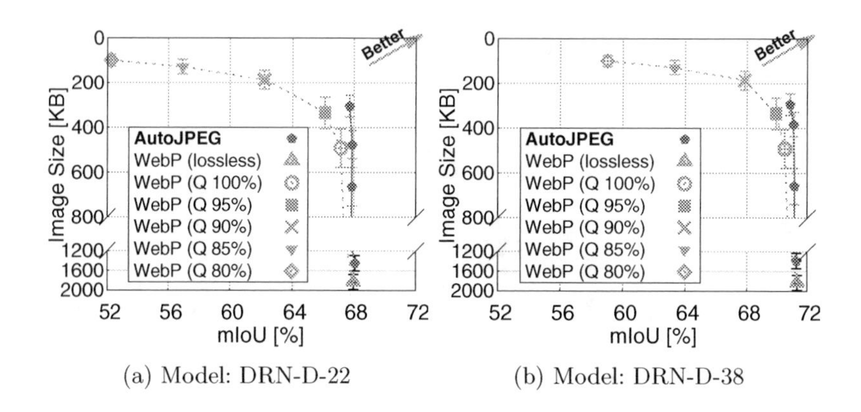

(a) Model: DRN-D-22 (b) Model: DRN-D-38

Fig. 7. Our solution AutoJPEG vs. WebP.

Table 1. Ablation study for components of AutoJPEG's joint optimization.

DNN tuning	YUV tuning	mIoU (%)	Avg. Size (KB)
✓	✓	71.24	408
✓	×	71.17	419
×	✓	70.94	404
×	×	70.75	402

Ablation Study. We validate AutoJPEG's components separately in this experiment. The results in Table 1 show that enabling both DNN tuning and YUV tuning yields the highest mIoU. Without YUV tuning (using JPEG's default YUV color space), the mIoU drops by 0.07% even when the image size is slightly larger. We also evaluate AutoJPEG without DNN tuning (use the original pre-trained DNN for inference), as some users may not want to update the DNNs already deployed on servers. In this case, the mIoU is slightly lower than when the DNN tuning is on but still outperforms using JPEG with 95% quality.

4.3 AutoJPEG in Image Classification

We further evaluate AutoJPEG in image classification tasks on CIFAR10. The experiment setup follows Sect. 4.1, and the benchmark is the widely used 75% quality JPEG. Figure 8 shows that AutoJPEG outperforms JPEG (no matter with or without finetuning the DNN by the JPEG training set) on all tested DNN models, with significantly higher Top-1 classification accuracy and smaller image size. For instance, AutoJPEG reduces the image size by 13% while improving the Top-1 classification accuracy by 4.73% on ResNet56 even when the JPEG benchmark uses DNN finetuned by JPEG-encoded dataset for inference.

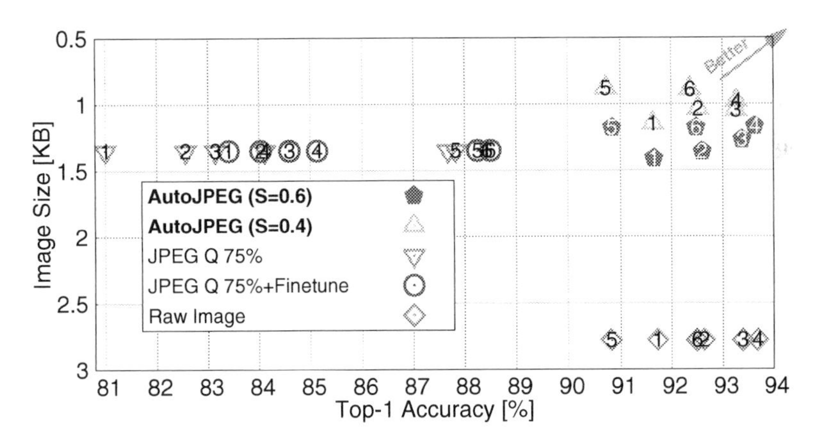

Fig. 8. Comparison of classification accuracy (Top-1) and compressed image size targeting different models, label 1, 2, 3, 4, 5, 6 correspond to ResNet20, ResNet32, ResNet56, ResNet110, VGG11, VGG13, respectively.

5 Limitations and Future Works

Our constrained optimization algorithm uses STE [1] to approximate the non-differentiable functions. A more rigorous approach to handle such non-differentiable functions is proximal gradient descent [12], which guarantees the same convergence rate as gradient descent. However, finding the proximity operator is challenging, and we leave it to future work. Meanwhile, our end-to-end framework enables joint optimization of the DNN model weights quantization (like QAT [19]) and DNN input image quantization, which is a future direction we plan to explore. Finally, inference offloading faces a general challenge that is out of the scope of this work: communication latency. A low-latency link is required to exploit inference offloading in delay-sensitive applications like AR.

6 Conclusion

This paper presents AutoJPEG, a DNN-friendly image compression solution tailored for DNN inference offloading in IoT networks. By harnessing ADMM to jointly optimize the encoding parameters and DNN model weights under a given image size constraint, AutoJPEG achieves significant IoT link bandwidth saving while preserving high inference accuracy. It has low complexity and is compatible with the JPEG codec, making deployment easy. In our evaluation of semantic segmentation and image classification tasks, AutoJPEG demonstrates superior performance to recent DNN-friendly data compression algorithms, DNN splitting algorithms, and mainstream codecs.

References

1. Bengio, Y., Léonard, N., Courville, A.: Estimating or Propagating Gradients Through Stochastic Neurons for Conditional Computation. arXiv preprint arXiv:1308.3432 (2013)
2. Choi, J., Han, B.: Task-aware quantization network for JPEG image compression. In: Vedaldi, A., Bischof, H., Brox, T., Frahm, J.-M. (eds.) ECCV 2020. LNCS, vol. 12365, pp. 309–324. Springer, Cham (2020). https://doi.org/10.1007/978-3-030-58565-5_19
3. Cordts, M., et al.: The cityscapes dataset for semantic urban scene understanding. In: Proceedings of the IEEE Conference on Computer Vision and Pattern Recognition, pp. 3213–3223 (2016). https://www.cityscapes-dataset.com/
4. Courbariaux, M., Bengio, Y., David, J.P.: BinaryConnect: training deep neural networks with binary weights during propagations. In: Advances in Neural Information Processing Systems (2015)
5. Courbariaux, M., Hubara, I., Soudry, D., El-Yaniv, R., Bengio, Y.: Binarized neural networks: training deep neural networks with weights and activations constrained to + 1 or -1. arXiv preprint arXiv:1602.02830 (2016)
6. Fairman, H., Brill, M., Hemmendinger, H.: How the cie 1931 color-matching functions were derived from wright-guild data. Color. Res. Appl. **22**(1), 11–23 (1997)
7. Google: an image format for the web (2021). https://developers.google.com/speed/webp/
8. He, K., Zhang, X., Ren, S., Sun, J.: Deep Residual Learning for Image Recognition. In: Proceedings of the IEEE Conference on Computer Vision and Pattern Recognition, pp. 770–778 (2016)
9. He, Y., Kang, G., Dong, X., Fu, Y., Yang, Y.: Soft filter pruning for accelerating deep convolutional neural networks. In: Proceedings of the Twenty-Seventh International Joint Conference on Artificial Intelligence, IJCAI-18, pp. 2234–2240. International Joint Conferences on Artificial Intelligence Organization (2018). https://doi.org/10.24963/ijcai.2018/309
10. He, Y., Lin, J., Liu, Z., Wang, H., Li, L.-J., Han, S.: AMC: AutoML for model compression and acceleration on mobile devices. In: Ferrari, V., Hebert, M., Sminchisescu, C., Weiss, Y. (eds.) ECCV 2018. LNCS, vol. 11211, pp. 815–832. Springer, Cham (2018). https://doi.org/10.1007/978-3-030-01234-2_48
11. He, Y., Zhang, X., Sun, J.: Channel pruning for accelerating very deep neural networks. In: Proceedings of the IEEE International Conference on Computer Vision, pp. 1389–1397 (2017)

12. Hiriart-Urruty, J.B., Lemaréchal, C.: Convex analysis and minimization algorithms I: fundamentals, vol. 305. Springer Science & Business Media (2013). https://doi.org/10.1007/978-3-662-02796-7
13. Hoglund, A., et al.: Overview of 3GPP release 14 further enhanced MTC. IEEE Commun. Stand. Mag. **2**(2), 84–89 (2018)
14. Hoglund, A., et al.: Overview of 3GPP release 14 enhanced NB-IoT. IEEE Network **31**(6), 16–22 (2017)
15. Hoymann, C., et al.: LTE release 14 outlook. IEEE Commun. Mag. **54**(6), 44–49 (2016)
16. Hu, C., Bao, W., Wang, D., Liu, F.: Dynamic adaptive dnn surgery for inference acceleration on the edge. In: IEEE INFOCOM 2019-IEEE Conference on Computer Communications, pp. 1423–1431. IEEE (2019)
17. Hubara, I., Courbariaux, M., Soudry, D., El-Yaniv, R., Bengio, Y.: Binarized neural networks. In: Advances in Neural Information Processing Systems (2016)
18. ITUR: BT 601: studio encoding parameters of digital television for standard 4: 3 and wide-screen 16: 9 aspect ratios. ITU-R Rec. BT 656 (1995)
19. Jacob, B., et al.: Quantization and training of neural networks for efficient integer-arithmetic-only inference. In: Proceedings of the IEEE Conference on Computer Vision and Pattern Recognition, pp. 2704–2713 (2018)
20. Kang, Y., et al.: Neurosurgeon: collaborative intelligence between the cloud and mobile edge. ACM SIGARCH Comput. Archit. News **45**(1), 615–629 (2017)
21. Krizhevsky, A., et al.: Learning Multiple Layers of Features from Tiny Images. Tech. rep., University of Toronto (2009). https://www.cs.toronto.edu/~kriz/cifar-10-python.tar.gz
22. Li, F., Zhang, B., Liu, B.: Ternary Weight Networks. arXiv preprint arXiv:1605.04711 (2016)
23. Li, H., Kadav, A., Durdanovic, I., Samet, H., Graf, P.H.: Pruning filters for efficient convNets. In: International Conference on Learning Representations (2017)
24. Lin, M., et al.: HRank: filter pruning using high-rank feature map. In: Proceedings of the IEEE/CVF Conference on Computer Vision and Pattern Recognition, pp. 1529–1538 (2020)
25. Liu, Z., et al.: DeepN-JPEG: a deep neural network favorable JPEG-based image compression framework. In: Proceedings of the 55th Annual Design Automation Conference, pp. 1–6 (2018)
26. Luo, X., Talebi, H., Yang, F., Elad, M., Milanfar, P.: The rate-distortion-accuracy tradeoff: JPEG case study. arXiv preprint arXiv:2008.00605 (2020)
27. Ma, X., et al.: PCONV: the missing but desirable sparsity in dnn weight pruning for real-time execution on mobile devices. In: Proceedings of the AAAI Conference on Artificial Intelligence, vol. 34, pp. 5117–5124 (2020)
28. Peng, H., Wu, J., Chen, S., Huang, J.: Collaborative channel pruning for deep networks. In: International Conference on Machine Learning, pp. 5113–5122 (2019)
29. Rastegari, M., Ordonez, V., Redmon, J., Farhadi, A.: XNOR-Net: imagenet classification using binary convolutional neural networks. In: Leibe, B., Matas, J., Sebe, N., Welling, M. (eds.) ECCV 2016. LNCS, vol. 9908, pp. 525–542. Springer, Cham (2016). https://doi.org/10.1007/978-3-319-46493-0_32
30. Richardson, I.E.: H. 264 and MPEG-4 video compression: video coding for next-generation multimedia. John Wiley & Sons (2004)
31. Roelofs, G., Koman, R.: PNG: the definitive guide. O'Reilly & Associates, Inc. (1999)
32. Shin, R., Song, D.: JPEG-resistant adversarial images. In: NIPS 2017 Workshop on Machine Learning and Computer Security, vol. 1 (2017)

33. Simonyan, K., Zisserman, A.: Very deep convolutional networks for large-scale image recognition. arXiv preprint arXiv:1409.1556 (2014)
34. Wallace, G.K.: The JPEG still picture compression standard. In: IEEE Transactions on Consumer Electronics (1992)
35. Wang, Y., et al.: Pruning from scratch. In: AAAI Conference on Artificial Intelligence (2020)
36. Xie, X., Kim, K.H.: Source compression with bounded dnn perception loss for IoT edge computer vision. In: The 25th Annual International Conference on Mobile Computing and Networking, pp. 1–16 (2019)
37. Yang, H., Zhu, Y., Liu, J.: ECC: platform-independent energy-constrained deep neural network compression via a bilinear regression model. In: Proceedings of the IEEE Conference on Computer Vision and Pattern Recognition, pp. 11206–11215 (2019)
38. Yang, J., et al.: Quantization networks. In: Proceedings of the IEEE/CVF Conference on Computer Vision and Pattern Recognition, pp. 7308–7316 (2019)
39. Yu, F., Koltun, V., Funkhouser, T.: Dilated residual networks. In: Proceedings of the IEEE Conference on Computer Vision and Pattern Recognition, pp. 472–480 (2017)
40. Zhang, C., Karjauv, A., Benz, P., Kweon, I.S.: Towards robust data hiding against (jpeg) compression: a pseudo-differentiable deep learning approach. arXiv preprint arXiv:2101.00973 (2020)
41. Zhu, C., Han, S., Mao, H., Dally, W.J.: Trained ternary quantization. arXiv preprint arXiv:1612.01064 (2016)
42. Zhuang, Z., et al.: Discrimination-aware channel pruning for deep neural networks. In: Advances in Neural Information Processing Systems, pp. 875–886 (2018)

Domain Knowledge-Informed Self-supervised Representations for Workout Form Assessment

Paritosh Parmar[1,2(✉)], Amol Gharat[2], and Helge Rhodin[1]

[1] University of British Columbia, Vancouver, Canada
paritosh.parmar@alumni.unlv.edu
[2] FlexAI Inc., Seattle, USA

Abstract. Maintaining proper form while exercising is important for preventing injuries and maximizing muscle mass gains. Detecting errors in workout form naturally requires estimating human's body pose. However, off-the-shelf pose estimators struggle to perform well on the videos recorded in gym scenarios due to factors such as camera angles, occlusion from gym equipment, illumination, and clothing. To aggravate the problem, the errors to be detected in the workouts are very subtle. To that end, we propose to learn exercise-oriented image and video representations from unlabeled samples such that a small dataset annotated by experts suffices for supervised error detection. In particular, our domain knowledge-informed self-supervised approaches (pose contrastive learning and motion disentangling) exploit the harmonic motion of the exercise actions, and capitalize on the large variances in camera angles, clothes, and illumination to learn powerful representations. To facilitate our self-supervised pretraining, and supervised finetuning, we curated a new exercise dataset, *Fitness-AQA* (https://github.com/ParitoshParmar/Fitness-AQA), comprising of three exercises: BackSquat, BarbellRow, and OverheadPress. It has been annotated by expert trainers for multiple crucial and typically occurring exercise errors. Experimental results show that our self-supervised representations outperform off-the-shelf 2D- and 3D-pose estimators and several other baselines. We also show that our approaches can be applied to other domains/tasks such as pose estimation and dive quality assessment.

1 Introduction

Detecting errors in gym exercise execution and providing feedback on it is crucial for preventing injuries and maximizing muscle gain. However, feedback from personal trainers is a costly option and hence used only sparingly—typically only a few days a month, just enough to learn the basic form. We believe that

Supplementary Information The online version contains supplementary material available at https://doi.org/10.1007/978-3-031-19839-7_7.

S. Avidan et al. (Eds.): ECCV 2022, LNCS 13698, pp. 105–123, 2022.
https://doi.org/10.1007/978-3-031-19839-7_7

an automated computer vision-based workout form assessment (*e.g.*, in the form of an app) would provide a cheap and viable substitute for personal trainers to continuously monitor users' workout form when their trainers are not around. Such an option would also be helpful to the socio-economically disadvantaged demographic who cannot afford or have access to personal trainers.

While fitness apps have recently become popular, the existing apps only allow the users to make workout plans—they do not provide a functionality to assess the workout form of the users. To detect errors in the workout videos, it is important to analyze the posture of the person. Academic research in workout form assessment so far has been limited to simple, controlled conditions [3,27], where posture can be reliably estimated using off-the-shelf (OTS) pose estimators [2,19,22]. Ours, on the other hand, is the first work to tackle the problem of workout form assessment distinctly in complex, real-world gym scenarios, where, people generally record themselves using ubiquitous cellphone cameras that they place somewhere in the vicinity; which results in large variances in terms of camera angles, alongside clothing styles, lighting, and occlusions due to gym equipment (barbells, dumbbells, racks). These environmental factors combined with the subtle nature of workout errors (refer to Fig. 1) and the convoluted, uncommon poses that people go through while exercising, cause major challenges for OTS pose estimators (refer to Fig. 1), and consequently, workout form errors cannot be reliably detected from pose. To mitigate this in the absence of workout datasets labeled for human body pose, we propose to replace the error-prone pose estimators with our more robust domain knowledge-informed self-

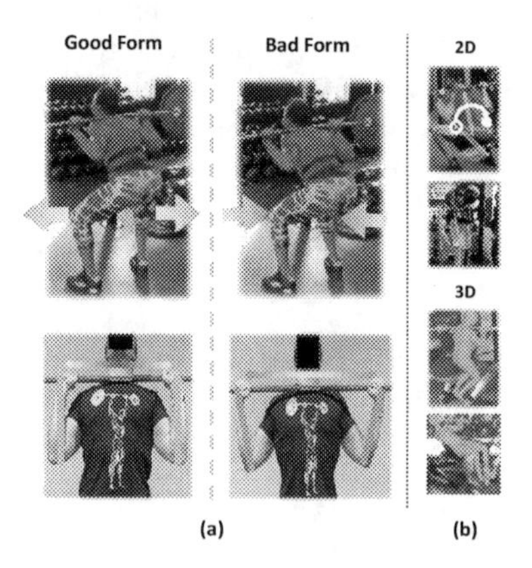

Fig. 1. Concept. (a) Errors of small magnitude generally occurring in workout form: Good column shows correct posture/execution (knees should be outwards), while the Bad column shows erroneous form during exercising. (b) Examples of failures of off-the-shelf 2D- and 3D-pose estimators in real-world gym scenarios (compare the discrepancies in pose estimation with the magnitude of the errors to be detected). We tackle the problem of detecting errors in workout form. To do so more accurately, we replace the error-prone pose estimators with our more robust fitness domain-oriented representations learnt using self-supervision.

supervised representations that are sensitive to pose and motion, learned from unlabeled videos—helps in avoiding annotation efforts. Towards those ends, our contributions are as follows:

1. **Novel self-supervised approaches that leverage domain knowledge.** We initiate the work in the direction of domain knowledge-informed self-supervised representation learning by developing two contrastive learning-based approaches that capitalize on the harmonic motion of workout actions and the large variance in unlabeled gym videos to learn robust fitness domain-oriented representations (Sect. 3). Our domain knowledge-informed self-supervised representations outperform 2D- and 3D-pose estimators [2,22,27], and various general self-supervised approaches [1,5,17,18] on the task of workout form assessment on existing and our newly introduced datasets. This indicates that future work on representation learning would benefit from using domain knowledge in designing self-supervised methods, especially when tackling problems involving real-world data.
2. **Workout form assessment dataset.** To facilitate our self-supervised approaches, as well as the subsequent supervised workout form error detection, we collected the largest, first-of-its-kind, in-the-wild, fine-grained fitness assessment dataset, covering three different exercises (Sect. 4) and a small labeled subset for evaluation. We show that this in-the-wild dataset provides a significantly more challenging benchmark than the existing ones recorded in controlled conditions.

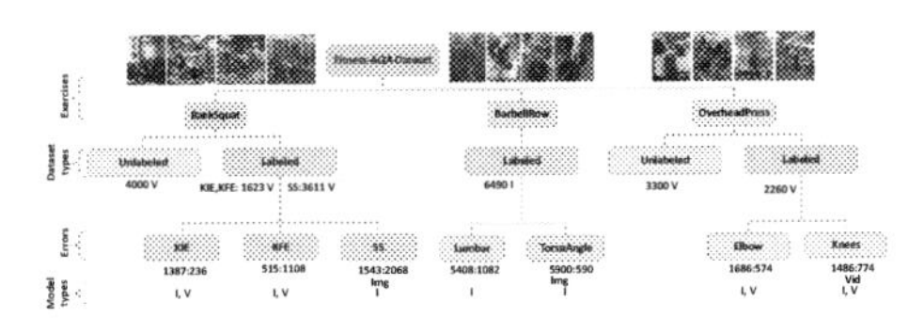

Fig. 2. Fitness-AQA dataset hierarchy. Numbers below the dataset type indicate dataset size; and those under the errors indicate the ratio of non-erroneous:erroneous samples. I, V indicate if the error detection is static image- or multiframe (video)-based.

2 Related Work

Action Quality Assessment (AQA)/Skills Assessment (SA). Our work can be classified under AQA/SA, which involve the computer vision-based quantification of the quality of movements and actions. Works in AQA/SA have mainly been focused on domains like physiotherapy [8,24,31,41,45], Olympic sports [4,30,34,44,49,51], various types of skills [7,25,32,48]. However, workout form assessment, especially, in real-world conditions, has not received much attention.

Approaches in AQA can be organized into 1) human pose features-based [28,35]; 2) image and video features-based [33,34]. Pose-based approaches use OTS pose estimators to extract 2D or 3D coordinate positions of various human body joints. These approaches have the disadvantage that poor estimation of the pose can adversely affect the final output. This is especially prevalent in non-daily action classes like fitness and sports domains. This can be mitigated, for example, by annotating domain-specific datasets [4], but that requires a considerable amount of manual annotation efforts, financial resources, and 3D annotations can only be obtained in controlled conditions. Therefore, we propose to learn domain-oriented pose-sensitive representations from unlabeled videos, which can be finetuned using only a small labeled dataset.

Closest to ours is the work on backsquat assessment by Ogata *et al.* [27]. However, a) they used OTS pose estimators, whereas we develop self-supervised approaches to learn more powerful representations; b) being dependent on OTS pose estimators, their approach is limited only to simple, controlled environments, whereas our approach is applicable to complex, real-world scenarios (Sect. 5); and c) their dataset contains only single exercise and was collected in simpler conditions and a single human, whereas our dataset contains three exercises and was collected in real-world gym scenarios and numerous humans (further differences discussed in Sect. 4).

SSL. Earlier work in this area include those of autoencoders [12], which learn low-dimensional representations by reconstructing the input. Le *et al.* [23] propose a way to learn hierarchical representations from unlabeled videos using unsupervised learning, which was also considered as a feature extractor in an earlier AQA work [35], but was found to perform worse than an OTS pose estimator. Recently, Chen *et al.* [5] proposed a simple siamese approach to learn representations that obtain competitive results on various benchmarks. Various general SSL works also propose to leverage properties of videos. Misra *et al.* [26] and Xu *et al.* [50] propose to exploit temporal order of frames and clips. Predicting the amount of rotation in images and videos was used as a pretext task by Gidaris *et al.* [9] and Jing *et al.* [18]. Wang *et al.* [46] leveraged motion and appearance statistics to learn self-supervised video representations. Benaim *et al.* [1] and Wang *et al.* [47] used video speed prediction as the pretext task. In addition to video speed prediction, Jenni *et al.* [17] proposed to use wider range of temporal transformations for pretext task. In contrast, we developed domain knowledge-informed SSL approaches that we show outperform general SSL approaches. A few works propose to leverage time-contrast to learn representations using self-supervision [14,15,42]. However, these temporal models either consider a single-view or a single subject. Our pose contrastive approach, on the other hand, simultaneously exploits cross-view and cross-subject information to learn more meaningful representations.

Another work proposes to disentangle pose and appearance from multiple views with a geometry-aware representation [37]. However, this approach is not tailored for exercise analysis, and requires calibrated multi-view datasets.

Inspired by this method, we develop a variant—our pose and appearance disentangling baseline—applicable to our dataset.

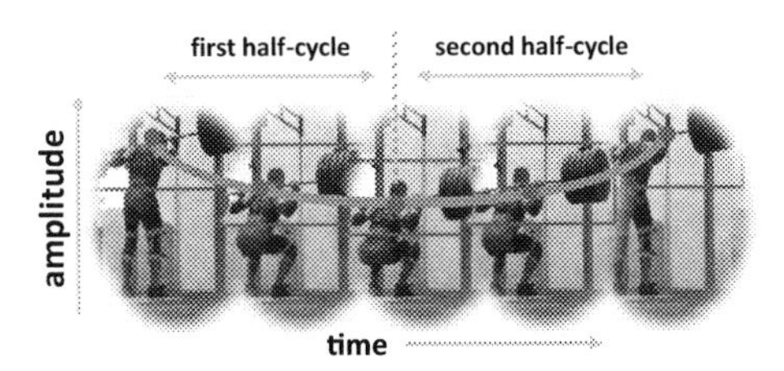

Fig. 3. Barbell trajectory. Red bounding boxes (bboxes) - barbell object detected; Red dots: the center of bboxes; Blue curve: the parabolic trajectory of the barbell traced out. (Color figure online)

3 Method

In this section, we present our self-supervised approaches for learning image and video representations. Subsequently, an error detection network is trained to map these self-supervised representations to workout form error probabilities. Note that in the following, we have presented our approaches using BackSquat as an exemplary exercise, but our methods are applicable to other exercises.

Preliminary: *Synchronizing videos*. Our methods build upon quasi-synchronized videos. In some datasets, such as Human3.6M [16], synchronized videos recorded from multiple angles is already available using special setups, which allows unsupervised learning, *e.g.*, as done in [37]. However, we are not using any kind of such special setups. Thus, we quasi-synchronize the videos using the following method. Given a collection of videos of (different) people performing the same exercise, we detect the barbell/weight over time to get a motion trajectory, which when plotted against time traces an approximately parabolic curve as shown in Fig. 3. These trajectories are then amplitude-normalized. Object size, resolutions do not affect the normalization, as we are using the center of the bounding box; and the vertical movement of the barbell can be reliably recorded from various viewpoints (unless extreme, like top-view of the scene—unrealistic, anyway). Now, we leverage the following property to synchronize the videos: for a given elevation of the object (or equivalently, the amplitude of the trajectory), the people doing the same exercise would be in approximately the same pose. This holds across different subjects, different video instances, and across different views/camera angles, which allows us to synchronize videos of different subjects in different environments/scenes.

3.1 Self-supervised Pose Contrastive Learning

Objective. Given the synchronized video samples of the same exercise (*e.g.*, BackSquat), in this approach, we aim to learn richer human pose information using self-supervised contrastive learning. In contrastive learning, same or similar samples are pulled together, while dissimilar samples are pushed apart [6]. In our case, we hypothesize that we can extend contrastive learning to learn human pose-sensitive representations. Particularly, we propose a self-supervised pretext task, which aims to pull together images (frames of videos) containing humans in similar poses, while pushing apart images with humans in dissimilar poses as shown in Fig. 4. Note that, this approach operates on single frame-triplets (not videos or clips) at a time.

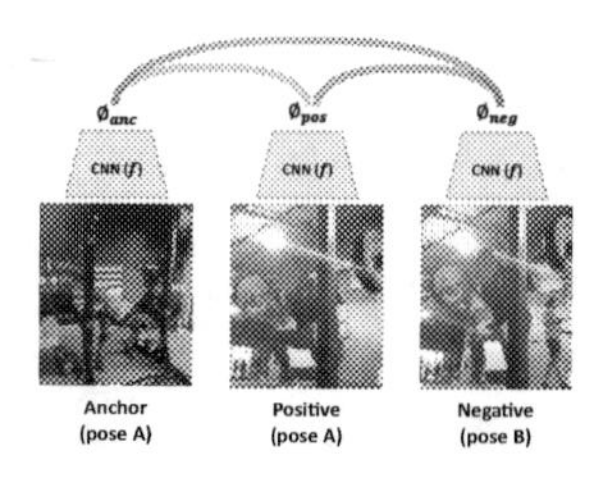

Fig. 4. Cross-View Cross-Subject Pose Contrastive learning (CVCSPC). Red lines indicate repulsion, while the green line indicates attraction in the representation space. (Color figure online)

Constructing Triplets for Contrastive Learning. Once we have the normalized barbell trajectories, for any given anchor input, I_{anc}, we retrieve the corresponding positive input frames with similar object elevation, I_{pos}, and the negative input frames with a difference in object elevation of more than a threshold value (δ), I_{neg}, from across video instances; and subsequently build triplets of $\{I_{\text{anc}}, I_{\text{pos}}, I_{\text{neg}}\}$. Such triplets provide a cross-view, cross-subject, cross-video-instance self-supervisory signal that has not yet been leveraged by the existing computer vision approaches to learn pose sensitive representations. These triplets also offer strong, in-built data augmentations. A recent work [40] observed that background augmentation can help increase the robustness of self-supervised learning. Our method not only provides such background augmentation, but also provides foreground augmentation in terms of appearance (clothing, body type, gender, etc.). We term our approach Cross-View Cross-Subject Pose Contrastive learning (CVCSPC).

Contrastive Learning. We use the constructed triplet, $\{I_{\text{anc}}, I_{\text{pos}}, I_{\text{neg}}\}$, to learn good representations through self-supervised contrastive learning. Let f represent a 2D-convolutional neural network (CNN) backbone, which when applied to $I_{\text{anc}}, I_{\text{pos}}, I_{\text{neg}}$, yields $\phi_{\text{anc}}, \phi_{\text{pos}}, \phi_{\text{neg}}$, respectively. In contrastive learning, ϕ_{anc} and ϕ_{pos} are forced to be similar, *i.e.*, $\phi_{\text{anc}} \approx \phi_{\text{pos}}$, while ϕ_{anc} and ϕ_{neg}

are forced to be dissimilar, *i.e.*, $\phi_{\text{anc}} \neq \phi_{\text{neg}}$, as illustrated in Fig. 4. Following [43], we optimize the parameters of f during the self-supervised training, by minimizing the distance ratio loss [13],

$$\mathcal{L} = -\log \frac{e^{-||\phi_{\text{anc}}-\phi_{\text{pos}}||_2}}{e^{-||\phi_{\text{anc}}-\phi_{\text{pos}}||_2} + e^{-||\phi_{\text{anc}}-\phi_{\text{neg}}||_2}}. \tag{1}$$

3.2 Self-supervised Motion Disentangling

Motion cues can be useful in detecting many workout form errors. Different from our pose-contrastive approach, this approach uses motion information to detect anomalies in workout form. In the following, we first present the preliminary information, before describing our method.

Preliminaries

- **Useful property 1: Harmonic motion**. Workout actions have a desirable property of exhibiting harmonic motion. For example, during benchpress (an exercise targeting the chest muscles), the person would be lifting the barbell above their chest and then bringing it down to the starting point; or during squats, the person would be squatting down (first half-cycle in Fig. 3) and then getting up (second half-cycle in Fig. 3).
- **Useful property 2: Bias in temporal location of form-errors**. People are more likely to make errors (anomalous motions) when lifting up the weights (one half-cycle of the harmonic motion, as in Fig. 3), rather than lowering the weights (another half-cycle of the harmonic motion).
- **Global motion.** The actual, regular motion of the workout action. For example, in Backsquat, the person squatting down and getting up.
- **Local motion.** The small-scale, fine-grained, irregular motion of the body parts (ref. Fig. 5). For example, in Backsquat, the knees abnormally going inward/outward or forward. So, while the global motion refers to regularities in motion patterns, local motion would cover anomalies in motion patterns.

Objective. Our goal is to learn self-supervised representations that are sensitive to local (anomalous) motions. The above discussed properties can provide a very useful, freely available signal that has not yet been exploited for this task by the existing computer vision approaches. We design a contrastive learning-based self-supervised approach to disentangle the local motion from the global motion.

Accentuating the Local Motion. Temporally reversing any one of the half-cycles would, in general, make both half-cycles identical in terms of the global motion, while they would still differ in terms of the local motion. In other words, contrasting the two half-cycles after temporally reversing any one of them, helps accentuate the anomalous local motion, as shown in Fig. 5.

Constructing Triplets for Contrastive Learning. The first half-cycle serves as the anchor; an augmented copy of the anchor serves as the positive input. The second half-cycle serves as the negative input. As discussed previously, we

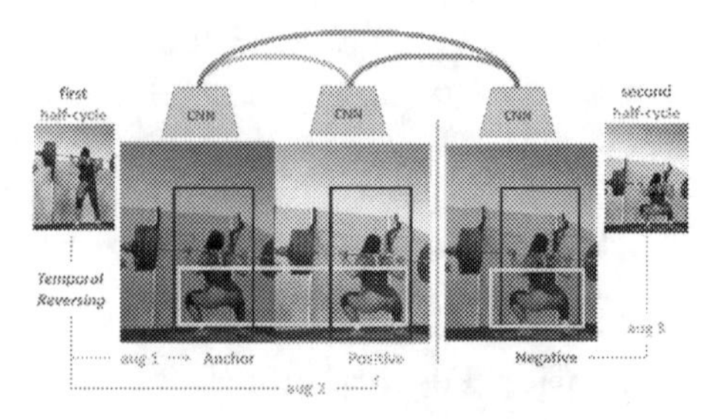

Fig. 5. Motion Disentangling (MD) approach. *Please view in AdobeReader to play the embedded animation for better explanation.* **Black** boxes: global motion (getting up, here); **Yellow** boxes: local motions (the knees rotating inwards under the influence of heavy training weight); aug: augmentations. Here we have applied very weak augmentation (only color augmentation) for representative purpose—to better illustrate the concept. However, in practice, we apply much stronger augmentations. Red lines indicate repulsion, while green indicates attraction in the representation space. (Color figure online)

randomly temporally-reverse either the {anchor, positive} pair or the {negative} input to make the global motion of all three identical. In practice, we randomly and independently applied the following augmentations on the triplets: image horizontal flipping, partial image masking, image translation, image rotation, image blurring, image zooming, color channel swapping, temporal shifting.

Contrastive Learning. We use a 3DCNN as the backbone for this model, and Eq. 1 as the loss function for this self-supervision task. Through contrastive learning, the 3DCNN learns to capture the previously discussed local, anomalous motions that are accentuated in our specially created triplets.

Anomalous motions maybe harmful or they can be beneficial. For example, knees buckling inwards during squatting is harmful, while knees going outwards is not. Therefore, during the finetuning phase, we aim to calibrate representations learnt using self-supervision to distinguish between harmful irregularities and harmless variations.

4 Fitness-AQA Dataset

Since exercise or workout assessment is an emerging field, there is a shortage of dedicated video datasets. To the best of our knowledge, the Waseda backsquat dataset by Ogata *et al.* [27] is the only publicly available such dataset. However, this dataset has shortcomings such as: it contains samples from a single human subject; the human subject is deliberately faking exercise errors; no kind of exercising weights, such as barbells and dumbbells, are used; the videos do not include realistic occlusions.

Dataset Collection. To fill the void of real-world datasets, we collected the largest exercise assessment dataset from video sharing sites such as Instagram and

YouTube. We considered the following three exercises: 1) BackSquat; 2) BarbellRow; and 3) Overhead (shoulder) Press. In addition to the labeled data, we also collected an unlabeled dataset to learn human pose focused representations in self-supervised ways (discussed in Sect. 3). The purpose of the labeled dataset is to finetune our models to do actual error detection and quantify the performance of our models. We have provided statistics and illustrated the full hierarchy of our Fitness-AQA dataset in Fig. 2. Illustrations of exercise errors are provided in the supplementary material.

Annotations by Expert Trainers. We employed two professional gym trainers to annotate our dataset for error labels. Due to this, even very subtle errors are caught and annotated accordingly. Errors range from very subtle to very severe.

Unique properties of our dataset:

- **Real-world videos**. Unlike the existing dataset [27], we collected our dataset from actual real-world videos in actual gyms recorded by the people without any scripts. Due to this, the videos are naturally recorded from a wide range of azimuthal angles, inclination angles, and distances. Our samples were automatically processed to contain a single repetition.
- **People making errors under the impact of actual weights**. In the existing dataset [27], people are instructed to make deliberate exercise mistakes without being under the influence of actual weights. Our dataset, on the other hand, captures cases where people are naturally making mistakes (without any instructions), under the influence of actually heavy weights. Due to this, we believe that there is no bias towards exaggerated errors, and contains natural, subtler error cases.
- **Occlusions**. Having captured in actual gyms, human subjects are partially occluded by barbell weights, weight racks or other equipment like benches.
- **Various types of clothing, background, illumination**. Since we did not hire any specific group of people to collect the dataset, the samples in our dataset are likely to come from numerous unique individuals, which results in a large number of clothing styles, and colors; different gyms (in terms of the room arrangement, and the background); other people in the background; and lighting conditions.
- **Unusual poses**. Exercise actions result in much more convoluted human body positions than those covered in the existing pose estimation datasets.

5 Experiments

To validate our contributions, we compared our features against various baselines and off-the-shelf pose estimators in simple (Case Study 1) and complex conditions (Case Study 2), showing significant improvements in the latter case.

We took a two-step approach towards detecting errors in exercising videos. Our models were first trained on the unlabeled datasets using self-supervision, and then used as feature extractors on the supervised datasets. For imbalanced datasets, we used class weights (in cross-entropy loss) inversely proportional to the class size. Note that the labeled dataset contains only the exercise error as ground-truth annotation and no information related to human pose. As such, our models did not use any pose-related ground-truth.

For the motion disentangling model, since the temporal model is already baked in it, we simply finetuned the model end-to-end on the labeled dataset for error detection. We used 32 frames for all types of errors.

For all 2DCNN-based approaches, we a learnt ResNet1D temporal model [27] that aggregates frame-level features for supervised error prediction on our labeled dataset. We used about 200 frames during error detection. Finetuning end-to-end on such a long sequence is not recommended [33,49,52]. Therefore, in this case, the 2DCNN backbone is not finetuned unless specified otherwise.

Implementation Details. We used ResNet-18 [11] as the backbone CNN unless specified otherwise. We used custom YOLOv3 [36] to detect barbells/weights; and normalized the amplitudes of the trajectories to -180 to 180 (simply for a resemblance to a circle). Specifications regarding each approach are as follows:

- **Pose Contrastive approach (CVCSPC)**. We used a threshold gap of 30 between anchor/positive and negative inputs. We initialized our backbone CNN with ImageNet weights. We used ADAM optimizer [21] with an initial learning rate of 1e-4 and optimized for 100 epochs with a batch size of 25.
- **Motion Disentanglement approach (MD)**. We used R(2+1)D-18 [10] as our backbone CNN. We sampled 16 frames from each half-cycle. We randomly applied strong augmentations. We initialized our backbone CNN with Kinetics [20] pretrained weights. We optimized our models using ADAM optimizer with an initial learning rate of 1e-4 for 20 epochs with a batch size of 5.

Further details provided in the supplementary material.

5.1 Case Study 1: Simple Conditions

The Waseda Squat dataset [27] provides an excellent labeled dataset for evaluating exercise errors in controlled conditions. The publicly available portion of this dataset contains samples from a single human subject. This dataset was not captured in a gym-like setting, but rather in home, and office-like settings. Each sample contains multiple squat repetitions. Note that the publicly available train/val/test split is different from that used in the original paper. Using this dataset, we experimented detecting the following errors: knees inward error

Table 1. Performance comparison on Waseda Squat dataset.

Feature extraction	Modality	Accuracies ↑					
		KIE	CVRB	CCRB	SS	KFE	Avg
HMR-TDM [27]	3D Pose	89.80	**98.65**	93.05	**87.30**	83.58	89.08
Ours CVCSPC	Image	**95.92**	91.89	**94.44**	77.77	**89.55**	**89.92**

(KIE); convex rounded back (spine) (CVRB); concave rounded back (spine) (CCRB); shallow squat (SS); knees forward error (KFE). To do so, we trained classifiers to distinguish between each of these error classes and good squat class (samples belonging to this class did not contain any errors). In this experiment, we compared features from our CVCSPC method (self-supervisedly trained on our unlabeled BackSquat dataset) against the Temporal Distances Matrices (TDM) derived from HMR pose estimator [19]. HMR-TDM features were made available by Ogata *et al.* [27]. During feature extraction, we resized the input images to 320×320 pixels, and considered the center 224×224 pixel crop. We did not consider our MD model because this dataset has multiple repetitions in each sample, and the sequence length is 300 frames, which is about 9 times longer than our MD model sequence length (32 frames). And, consequently, if we temporally downsample the sequence, it would lose a lot of information.

The results are summarized in Table 1, where we report accuracies. We found that our model outperformed existing methods [27] on three types of errors: KIE, CCRB, and KFE; with the performances being notably better on KIE and KFE errors. Even though not consistently across all the errors, our self-supervisedly learnt features outperformed HMR-TDM features on overall average performance. Note that large performance gap is not expected on this dataset, as OTS pose estimators work quite well in these simpler conditions.

5.2 Case Study 2: In-the-Wild Conditions

Next, we considered evaluating our approach on more complex datasets. For that, we considered our labeled datasets, which we introduced in Sect. 4, where we also discussed the reasons that make our new in-the-wild dataset more challenging. Unless mentioned otherwise, we divided the datasets into train-, validation-, and test-splits of 70%, 15%, and 15%, respectively.

Baselines. We compared our self-supervised feature extractors with the following models and features:

- ImageNet: ImageNet [39] pretrained ResNet-18 [11]
- Kinetics: Kinetics [20] pretrained R(2+1)D-18 [10]
- SPIN-TDM: Temporal Distance Matrices (TDM) [27] constructed from the output of SPIN [22] (3D joint positions)
- OpenPose-TDM: Temporal Distance Matrices [27] constructed from the output of OpenPose [2] (2D joint positions). Originally, TDM was proposed for 3D joint positions, but we also experiment with constructing TDMs from 2D joint positions.
- SimSiam: ImageNet pretrained model adapted/trained to our dataset using a general self-supervised image representation learning approach: SimSiam [5].
- Ours PAD: Inspired from [29,37], we developed an autoencoder-based approach that learns to disentangle pose and appearance of the human. Pose vector is then used for error-detection. We term this pose and appearance disentangling approach Ours PAD. We initialized the encoder with ImageNet weights. We have elaborated on this baseline in the supplementary material.

- VideoSpeed-1: Kinetics pretrained model adapted/trained to our dataset using the pretext task of predicting speed of videos [1]. We considered the following speeds: 1x (normal), 2x (faster), 3x, 4x (fastest).
- VideoSpeed-2: same as VideoSpeed-1, but for 1x speed, we sampled frames uniformly from entire sequence. For higher speeds, it would create the effect of repeating the sequences. So, it can equivalently be considered as counting the exercise repetitions.
- VideoRot: Kinetics pretrained model adapted/trained to our dataset using the pretext task of predicting rotation amount of videos [18]. Rotation amount is selected randomly from $\{0°, 90°, 180°, 270°\}$.
- TemporalXform: Kinetics pretrained model adapted to our dataset using the pretext task of predicting various temporal transforms [17].
- Ours TemporalXform-1: We developed a contrastive learning-based approach in which the negative input is more temporally shifted than the positive input relative to the anchor. We initialized with Kinetics pretrained model.
- Ours TemporalXform-2: We developed another contrastive learning-based approach in which the negative is more temporally distorted than the positive input. We initialized with Kinetics pretrained model.

Performance Metric. Since this dataset is imbalanced, we report the F1-score, instead of the accuracy.

Dataset: Fitness-AQA BackSquat

Knees Inward and Knees Forward Errors. First, we evaluated all the approaches on knees inward (KIE) and forward (KFE) errors. The results are summarized in Table 2. Additionally, here, we also considered a single-view, single-subject version of our cross-view, cross-subject pose-contrastive approach. In this version, anchor, positive, and negative inputs all belonged to the same video instance. We applied strong augmentations (rotation, translation, masking image regions, color channel order changing, zooming, blurring) during training this model. We refer to this approach as Vanilla-PC. We observed the following. 1) Training both image- and video-based self-supervision methods on our dataset helped in improving over their respective base models (ImageNet pretrained model and Kinetics pretrained model). 2) Our Vanilla Pose Contrastive learning improved the performance even more than our PAD. However, off-the-shelf pose estimator, OpenPose, still worked better than this model. 3) By contrast, our full pose-contrastive model, CVCSPC outperformed all the models on KIE; for completeness, we also computed OpenPose baseline with our hyperparameter settings referred to as OpenPose*. 4) CVCSPC performing better than Vanilla PC also reinforced the importance of considering our cross-view and cross-subject conditions during pose contrastive learning. 5) Our MD model performed the best and second best on KFE and KIE, respectively. TemporalXform performed the best among general video self-supervised approaches. 6) Our domain knowledge-informed self-supervised approaches outperformed general self-supervised approaches, indicating the importance of using domain knowledge in designing self-supervised approaches. 7) Our contrastive learning-based

Table 2. Performance comparison on Knees Inward and Knees Forward errors on our BackSquat dataset.

Feature extraction model	Modality	F-score ↑	
		KIE	KFE
OpenPose-TDM [2,27]	2D Pose	0.4143	0.8123
OpenPose-TDM* [2,27]	2D Pose	0.3186	0.7968
SPIN-TDM [22,27]	3D Pose	0.2878	0.7761
ImageNet [39]	Image	0.1923	0.7725
SimSiam [5]	Image	0.2270	0.7868
Ours PAD	Image	0.3180	0.7784
Ours Vanilla PC	Image	0.4118	0.7965
Ours CVCSPC	Image	**0.5195**	0.8286
Kinetics [20]	Video	0.2970	0.8184
VideoSpeed-1 [1]	Video	0.3095	0.8155
VideoSpeed-2	Video	0.3617	0.8000
VideoRot [18]	Video	0.3333	0.8138
TemporalXform [17]	Video	0.3414	0.8319
Ours TemporalXform-1	Video	0.3457	0.8097
Ours TemporalXform-2	Video	0.2286	0.8184
Ours MD	Video	0.4186	**0.8338**
Ours MD + CVCSPC	Image, Video	**0.5263**	**0.8468**

Table 3. Performance comparison on detecting Shallow Squat error.

Feature extraction model	Modality	F-score ↑
OpenPose-TDM [2,27]	2D Pose	0.8340
SimSiam [5]	Image	0.8286
Ours CVCSPC	Image	**0.8694**

approaches (CVCSPC and MD) worked better than our reconstruction-based approach (PAD). Furthermore, ensemble of our contrastive approaches outperformed all the models. Attention visualizations presented in the supplementary material.

Note that in all the subsequent experiments, we selected only the best performing methods for further evaluation.

Shallow Squat Error. We further considered evaluating and comparing approaches on another squat error—shallow squat error. Since shallow depth error is a static type of error, image models (2DCNN-based) are more suitable, where errors are detected in singular images, as opposed to in a stack of video frames. Using a 3DCNN for detecting single frame-based errors does not make sense. Therefore, we have not considered our MD approach for single frame-based errors. Single image detection also made end-to-end learning more feasible, so we finetuned our models end-to-end. The results are summarized in Table 3. We observed that OpenPose worked better than SimSiam. Our self-supervised learning performed the best, showing the importance of learning task-oriented representations, and its utility even in end-to-end finetuning scenarios.

Dataset: Fitness-AQA OverheadPress. Further, we evaluated and compared approaches on a different exercise—OverheadPress. The results are summarized in Table 4. We observed that video-based approaches worked better than image-based approaches on this exercise. Both of our proposed approaches outperformed the off-the-shelf pose estimator.

Table 4. Performance comparison on detecting Elbow and Knees errors in OverheadPress exercise.

Feature extraction model	Modality	F-score ↑	
		Elbow Err.	Knees Err.
OpenPose-TDM [2,27]	2D Pose	0.4265	0.7131
SimSiam [5]	Image	0.4145	0.5301
Ours CVCSPC	Image	0.4522	0.7203
TemporalXform [17]	Video	0.4138	0.8416
Ours MD	Video	**0.4552**	**0.8452**

Table 5. Cross-exercise transfer performance. Detecting Lumbar and Torso-Angle errors in BarbellRow exercise.

Feature extraction model	Modality	F-score ↑	
		Lumbar Err.	Torso Err.
OpenPose-TDM [2,27] (SQ→BR)	2D Pose	0.5422	0.4060
SimSiam [5] (SQ→BR)	Image	0.5934	0.4543
Ours CVCSPC (SQ→BR)	Image	**0.6057**	**0.4800**
Ours CVCSPC (OHP→BR)	Image	0.5760	0.4675
Ours CVCSPC (SQ+OHP→BR)	Image	**0.6338**	**0.5261**

5.3 Cross-Exercise Transfer

It is common to not have enough labeled data for each exercise. In such cases, it would be useful to transfer models from an exercise with abundant data over to exercises with limited data. So, in this experiment, we first transferred our model trained on BackSquat (SQ) exercise to BarbellRow exercise, where we detected two kinds of errors: Lumbar and TorsoAngle errors. Since these errors are static errors, we considered transferring our CVCSPC model. Note that in this experiment we used only a small amount of training data (details in the Supplementary Material). The results are presented in Table 5. We observed that models pretrained using our proposed self-supervised approach performed better than baselines even when finetuned to a different exercise action. We also transferred from Overhead Press (OHP), & noted improvements. Lastly, we also tried the ensemble of our SQ & OHP transferred models, which worked the best.

5.4 Applications to Other Domains

Pose Estimation. We conducted a novel pose retrieval experiment where we retrieved images based on query poses using our pose-contrastive embeddings.

From the results shown in Fig. 6, it can be seen that compared to SimSiam embeddings, ours are much better at encoding pose information, even with camera angle variation. We believe that our representations can be decoded into actual 2D/3D joint positions, by using a small pose-annotated dataset. We will explore this further in future research.

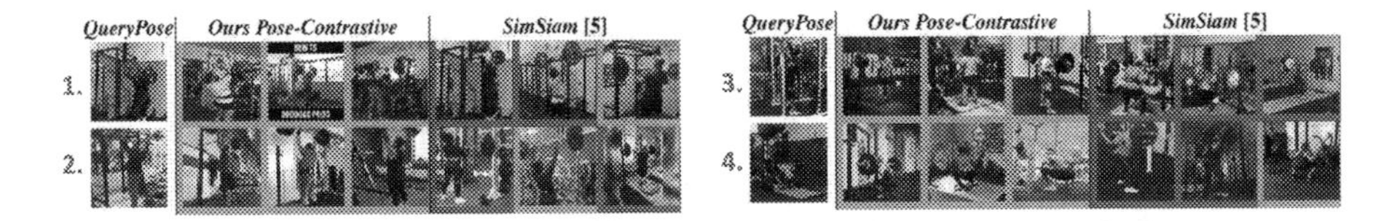

Fig. 6. Results of **pose-based retrieval** experiment.

Dive Quality Assessment. *While we use symmetry to simplify problems, our methods are generalizable, e.g.*, we applied our motion disentangling method for assessing the quality of Olympic dives on MTL-AQA dataset [34]. Global motion & local motions here refer to the motion of the dive-classes & the errors in them, respectively. To disentangle local motion, we match-contrast dives from the same dive-class from the same diving events so that the background remains same. We used supervised dive-classification pretraining as the baseline. Performance metric is Spearman's rank correlation (higher is better). We found significant improvement after incorporating our motion disentangling approach as shown in Table 6, even surpassing previous self-supervised state-of-the-art [38].

Table 6. Motion disentangling for Dive quality assessment.

Model	SSL SoTA [38]	Ours baseline	Ours MD
Sp. Corr.	0.7700	0.5665	**0.7763**

6 Conclusion

In this paper, we addressed the problem of assessing the workout form in real-world gym scenarios, where we showed that pose-features from off-the-shelf pose estimators cannot be reliably used for detecting subtle errors in workout form, as these pose estimators struggle to perform well due to unusual poses, occlusions, illumination, and clothing styles. We tackled the problem by replacing these noisy pose features with our more robust image and video representations learnt from unlabeled videos using domain knowledge-informed self-supervised approaches. Using self-supervision helped in avoiding the cost of annotating poses. Mapping of our self-supervised representations to workout form error probabilities was learnt using a much smaller labeled dataset. We also introduced a novel dataset, Fitness-AQA, containing actual, unscripted exercise samples from real-world gyms. Experimentally, we found that while our self-supervised features performed comparably in simpler conditions, they outperformed off-the-shelf pose estimators and various baselines in complex real-world conditions on multiple exercises. We also showed that pose information is encoded in our representations; and our motion disentangling approach can be used to assess quality of motion in other domains.

References

1. Benaim, S., et al.: SpeedNet: learning the speediness in videos. In: Proceedings of the IEEE/CVF Conference on Computer Vision and Pattern Recognition, pp. 9922–9931 (2020)
2. Cao, Z., Hidalgo, G., Simon, T., Wei, S.E., Sheikh, Y.: OpenPose: realtime multi-person 2D pose estimation using part affinity fields. IEEE Trans. Pattern Anal. Mach. Intell. **43**(1), 172–186 (2019)
3. Chen, S., Yang, R.R.: Pose trainer: correcting exercise posture using pose estimation. arXiv preprint arXiv:2006.11718 (2020)
4. Chen, X., Pang, A., Yang, W., Ma, Y., Xu, L., Yu, J.: SportsCap: monocular 3D human motion capture and fine-grained understanding in challenging sports videos. arXiv preprint arXiv:2104.11452 (2021)
5. Chen, X., He, K.: Exploring simple siamese representation learning. In: Proceedings of the IEEE/CVF Conference on Computer Vision and Pattern Recognition, pp. 15750–15758 (2021)
6. Chopra, S., Hadsell, R., LeCun, Y.: Learning a similarity metric discriminatively, with application to face verification. In: 2005 IEEE Computer Society Conference on Computer Vision and Pattern Recognition (CVPR 2005), vol. 1, pp. 539–546. IEEE (2005)
7. Doughty, H., Mayol-Cuevas, W., Damen, D.: The pros and cons: rank-aware temporal attention for skill determination in long videos. In: Proceedings of the IEEE/CVF Conference on Computer Vision and Pattern Recognition, pp. 7862–7871 (2019)
8. Du, C., Graham, S., Depp, C., Nguyen, T.: Assessing physical rehabilitation exercises using graph convolutional network with self-supervised regularization. In: 2021 43rd Annual International Conference of the IEEE Engineering in Medicine & Biology Society (EMBC), pp. 281–285. IEEE (2021)

9. Gidaris, S., Singh, P., Komodakis, N.: Unsupervised representation learning by predicting image rotations. arXiv preprint arXiv:1803.07728 (2018)
10. Hara, K., Kataoka, H., Satoh, Y.: Can spatiotemporal 3D CNNs retrace the history of 2D CNNs and imagenet? In: Proceedings of the IEEE Conference on Computer Vision and Pattern Recognition, pp. 6546–6555 (2018)
11. He, K., Zhang, X., Ren, S., Sun, J.: Deep residual learning for image recognition. In: Proceedings of the IEEE Conference on Computer Vision and Pattern Recognition, pp. 770–778 (2016)
12. Hinton, G.E., Salakhutdinov, R.R.: Reducing the dimensionality of data with neural networks. Science **313**(5786), 504–507 (2006)
13. Hoffer, E., Ailon, N.: Deep metric learning using triplet network. In: Feragen, A., Pelillo, M., Loog, M. (eds.) SIMBAD 2015. LNCS, vol. 9370, pp. 84–92. Springer, Cham (2015). https://doi.org/10.1007/978-3-319-24261-3_7
14. Honari, S., Constantin, V., Rhodin, H., Salzmann, M., Fua, P.: Unsupervised learning on monocular videos for 3D human pose estimation. arXiv preprint arXiv:2012.01511 (2020)
15. Hyvarinen, A., Morioka, H.: Unsupervised feature extraction by time-contrastive learning and nonlinear ICA. Adv. Neural. Inf. Process. Syst. **29**, 3765–3773 (2016)
16. Ionescu, C., Papava, D., Olaru, V., Sminchisescu, C.: Human3.6m: large scale datasets and predictive methods for 3D human sensing in natural environments. IEEE Trans. Pattern Anal. Mach. Intell. **36**(7), 1325–1339 (2014)
17. Jenni, S., Meishvili, G., Favaro, P.: Video representation learning by recognizing temporal transformations. In: Vedaldi, A., Bischof, H., Brox, T., Frahm, J.-M. (eds.) ECCV 2020. LNCS, vol. 12373, pp. 425–442. Springer, Cham (2020). https://doi.org/10.1007/978-3-030-58604-1_26
18. Jing, L., Yang, X., Liu, J., Tian, Y.: Self-supervised spatiotemporal feature learning via video rotation prediction. arXiv preprint arXiv:1811.11387 (2018)
19. Kanazawa, A., Black, M.J., Jacobs, D.W., Malik, J.: End-to-end recovery of human shape and pose. In: Computer Vision and Pattern Regognition (CVPR) (2018)
20. Kay, W., et al.: The kinetics human action video dataset. arXiv preprint arXiv:1705.06950 (2017)
21. Kingma, D.P., Ba, J.: Adam: a method for stochastic optimization. arXiv preprint arXiv:1412.6980 (2014)
22. Kolotouros, N., Pavlakos, G., Black, M.J., Daniilidis, K.: Learning to reconstruct 3D human pose and shape via model-fitting in the loop. In: ICCV (2019)
23. Le, Q.V., Zou, W.Y., Yeung, S.Y., Ng, A.Y.: Learning hierarchical invariant spatio-temporal features for action recognition with independent subspace analysis. In: CVPR 2011, pp. 3361–3368. IEEE (2011)
24. Li, J., Bhat, A., Barmaki, R.: Improving the movement synchrony estimation with action quality assessment in children play therapy. In: Proceedings of the 2021 International Conference on Multimodal Interaction, pp. 397–406 (2021)
25. Liu, D., et al.: Towards unified surgical skill assessment. In: Proceedings of the IEEE/CVF Conference on Computer Vision and Pattern Recognition, pp. 9522–9531 (2021)
26. Misra, I., Zitnick, C.L., Hebert, M.: Shuffle and learn: unsupervised learning using temporal order verification. In: Leibe, B., Matas, J., Sebe, N., Welling, M. (eds.) ECCV 2016. LNCS, vol. 9905, pp. 527–544. Springer, Cham (2016). https://doi.org/10.1007/978-3-319-46448-0_32
27. Ogata, R., Simo-Serra, E., Iizuka, S., Ishikawa, H.: Temporal distance matrices for squat classification. In: Proceedings of the IEEE/CVF Conference on Computer Vision and Pattern Recognition Workshops (2019)

28. Pan, J.H., Gao, J., Zheng, W.S.: Action assessment by joint relation graphs. In: Proceedings of the IEEE/CVF International Conference on Computer Vision (ICCV), October 2019
29. Park, T., et al.: Swapping autoencoder for deep image manipulation. Adv. Neural. Inf. Process. Syst. **33**, 7198–7211 (2020)
30. Parmar, P., Morris, B.: Action quality assessment across multiple actions. In: 2019 IEEE Winter Conference on Applications of Computer Vision (WACV), pp. 1468–1476. IEEE (2019)
31. Parmar, P., Morris, B.T.: Measuring the quality of exercises. In: 2016 38th Annual International Conference of the IEEE Engineering in Medicine and Biology Society (EMBC), pp. 2241–2244. IEEE (2016)
32. Parmar, P., Reddy, J., Morris, B.: Piano skills assessment. arXiv preprint arXiv:2101.04884 (2021)
33. Parmar, P., Tran Morris, B.: Learning to score olympic events. In: Proceedings of the IEEE Conference on Computer Vision and Pattern Recognition Workshops, pp. 20–28 (2017)
34. Parmar, P., Tran Morris, B.: What and how well you performed? A multitask learning approach to action quality assessment. In: Proceedings of the IEEE Conference on Computer Vision and Pattern Recognition, pp. 304–313 (2019)
35. Pirsiavash, H., Vondrick, C., Torralba, A.: Assessing the quality of actions. In: Fleet, D., Pajdla, T., Schiele, B., Tuytelaars, T. (eds.) ECCV 2014. LNCS, vol. 8694, pp. 556–571. Springer, Cham (2014). https://doi.org/10.1007/978-3-319-10599-4_36
36. Redmon, J., Farhadi, A.: Yolov3: an incremental improvement. arXiv preprint arXiv:1804.02767 (2018)
37. Rhodin, H., Salzmann, M., Fua, P.: Unsupervised geometry-aware representation for 3D human pose estimation. In: Proceedings of the European Conference on Computer Vision (ECCV), pp. 750–767 (2018)
38. Roditakis, K., Makris, A., Argyros, A.: Towards improved and interpretable action quality assessment with self-supervised alignment. In: The 14th PErvasive Technologies Related to Assistive Environments Conference, pp. 507–513 (2021)
39. Russakovsky, O., et al.: Imagenet large scale visual recognition challenge. Int. J. Comput. Vision **115**(3), 211–252 (2015)
40. Ryali, C.K., Schwab, D.J., Morcos, A.S.: Characterizing and improving the robustness of self-supervised learning through background augmentations. arXiv preprint arXiv:2103.12719 (2021)
41. Sardari, F., Paiement, A., Hannuna, S., Mirmehdi, M.: VI-Net-view-invariant quality of human movement assessment. Sensors **20**(18), 5258 (2020)
42. Sermanet, P., et al.: Time-contrastive networks: self-supervised learning from video. In: 2018 IEEE International Conference on Robotics and Automation (ICRA), pp. 1134–1141. IEEE (2018)
43. Sigurdsson, G.A., Gupta, A., Schmid, C., Farhadi, A., Alahari, K.: Actor and observer: joint modeling of first and third-person videos. In: Proceedings of the IEEE Conference on Computer Vision and Pattern Recognition, pp. 7396–7404 (2018)
44. Tang, Y., et al.: Uncertainty-aware score distribution learning for action quality assessment. In: Proceedings of the IEEE/CVF Conference on Computer Vision and Pattern Recognition, pp. 9839–9848 (2020)
45. Tao, L., et al.: A comparative study of pose representation and dynamics modelling for online motion quality assessment. Comput. Vis. Image Underst. **148**, 136–152 (2016)

46. Wang, J., Jiao, J., Bao, L., He, S., Liu, Y., Liu, W.: Self-supervised spatio-temporal representation learning for videos by predicting motion and appearance statistics. In: CVPR, pp. 4006–4015 (2019)
47. Wang, J., Jiao, J., Liu, Y.-H.: Self-supervised video representation learning by pace prediction. In: Vedaldi, A., Bischof, H., Brox, T., Frahm, J.-M. (eds.) ECCV 2020. LNCS, vol. 12362, pp. 504–521. Springer, Cham (2020). https://doi.org/10.1007/978-3-030-58520-4_30
48. Wang, T., Wang, Y., Li, M.: Towards accurate and interpretable surgical skill assessment: a video-based method incorporating recognized surgical gestures and skill levels. In: Martel, A.L., et al. (eds.) MICCAI 2020. LNCS, vol. 12263, pp. 668–678. Springer, Cham (2020). https://doi.org/10.1007/978-3-030-59716-0_64
49. Xu, C., Fu, Y., Zhang, B., Chen, Z., Jiang, Y.G., Xue, X.: Learning to score figure skating sport videos. IEEE Trans. Circuits Syst. Video Technol. **30**(12), 4578–4590 (2019)
50. Xu, D., Xiao, J., Zhao, Z., Shao, J., Xie, D., Zhuang, Y.: Self-supervised spatiotemporal learning via video clip order prediction. In: Computer Vision and Pattern Recognition (CVPR) (2019)
51. Yu, X., Rao, Y., Zhao, W., Lu, J., Zhou, J.: Group-aware contrastive regression for action quality assessment. In: Proceedings of the IEEE/CVF International Conference on Computer Vision, pp. 7919–7928 (2021)
52. Zeng, L.A., et al.: Hybrid dynamic-static context-aware attention network for action assessment in long videos. In: Proceedings of the 28th ACM International Conference on Multimedia, pp. 2526–2534 (2020)

Responsive Listening Head Generation: A Benchmark Dataset and Baseline

Mohan Zhou[1], Yalong Bai[2], Wei Zhang[2], Ting Yao[2], Tiejun Zhao[1](✉), and Tao Mei[2]

[1] Harbin Institute of Technology, Harbin, China
mhzhou99@outlook.com, tjzhao@hit.edu.cn
[2] JD Explore Academy, Beijing, China
ylbai@outlook.com, wzhang.cu@gmail.com, tingyao.ustc@gmail.com, tmei@jd.com

Abstract. We present a new listening head generation benchmark, for synthesizing responsive feedbacks of a listener (*e.g.*, nod, smile) during a face-to-face conversation. As the indispensable complement to talking heads generation, listening head generation has seldomly been studied in literature. Automatically synthesizing listening behavior that actively responds to a talking head, is critical to applications such as digital human, virtual agents and social robots. In this work, we propose a novel dataset "ViCo", highlighting the listening head generation during a face-to-face conversation. A total number of 92 identities (67 speakers and 76 listeners) are involved in ViCo, featuring 483 clips in a paired "speaking-listening" pattern, where listeners show three listening styles based on their attitudes: positive, neutral, negative. Different from traditional speech-to-gesture or talking-head generation, listening head generation takes as input both the audio and visual signals from the speaker, and gives non-verbal feedbacks (*e.g.*, head motions, facial expressions) in a real-time manner. Our dataset supports a wide range of applications such as human-to-human interaction, video-to-video translation, cross-modal understanding and generation. To encourage further research, we also release a listening head generation baseline, conditioning on different listening attitudes. Code & ViCo dataset: https://project.mhzhou.com/vico.

Keywords: Listening head generation · Video synthesis

M. Zhou—This work was done at JD Explore Academy.
M. Zhou and Y. Bai—Equal contribution.

Supplementary Information The online version contains supplementary material available at https://doi.org/10.1007/978-3-031-19839-7_8.

S. Avidan et al. (Eds.): ECCV 2022, LNCS 13698, pp. 124–142, 2022.
https://doi.org/10.1007/978-3-031-19839-7_8

1 Introduction

Communication [5,24,34,42,52,54] is one of the most common activities that everybody engages in their daily lives. During a face-to-face communication [29], two persons shift their roles in turn between the speaker and listener, to effectively exchange information. The speaker verbally transmits information to the listener, while the listener provides real-time feedbacks to the speaker mostly through non-verbal behaviors such as *affirmative nod*, *smiling*, *head shake*.

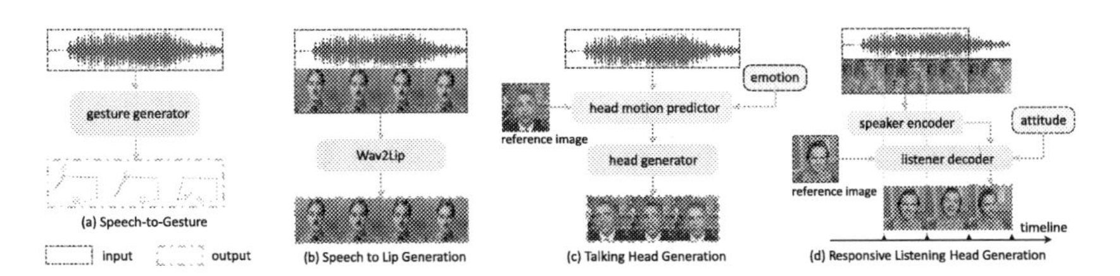

Fig. 1. Illustrations of three related tasks and our proposed responsive listening head generation. (a) Speech-to-gesture translation: generates plausible gestures to go along with the given speech. (b) Speech to lip generation: produces lip-synchronization in talking-head video. (c) Talking head generation: synthesizes talking face video conditioned on the identity of the speaker, audio speech, and/or the speaker emotion. (d) Our proposed responsive listening head synthesizes videos in responding to the speaker video stream

Although static images, repeated frames, or pre-scripted animations are often used to synthesize listeners in practice, they are often rigid and not realistic enough to respond to the speaker appropriately. According to studies in social psychology and anthropology, listening is a function-specific [18] and conditioned behavior [3], where learnable patterns can be inferred from training data. First, common patterns of listeners are observed to express their viewpoints, symmetrical and cyclic motions were employed to signal 'yes', 'no' or equivalents; narrow linear movements occurred in phase with stressed syllables in the other's speech; wide, linear movements occurred during pauses in the other's speech. Even the duration of eye blinks of the listener is perceived as communicative signals in human face-to-face interaction [23]. Second, these patterns in listener motions are mainly affected by two signals: the attitude of listener [21], and signals from the speaker [10,16,35]. Different attitudes of the listener results in diverse facial expressions, e.g., attitude of *agree* is meant by a *nod* and *accept*, attitude of *disbelieve* is represented by the combination of *head tilt* and *frown*. Meanwhile, listening behavior is heavily affected by speaker motion and audio signals. For example, the flow of movement of listener may be rhythmically coordinated with the speech and motions by the speaker [28]. These psychological and ethological studies motivate us to propose a data-driven method for modeling listening behaviors for face-to-face communication.

There have been extensive research efforts on speaker-centric synthesis. As shown in Fig. 1, speech to gesture generation [17] learns a mapping between the audio signal and speaker's pose. Speech to lip generation [45] aims to refine the lip-synchronization of a given video input. Talking-head synthesis [11,56,62] tries to generate a vivid talking video of a specific speaker with facial animations from a still image and a clip of audio. However, these works only focus on the speaking role, while ignore the indispensable counterpart of listener. Notably during a face-to-face conversation, listening behavior is even more important, as proper feedbacks to the speaker (*e.g.*, *nod*, *smile*, *eye contact*, *etc.*) are vital for a successful communication [37,49–51]. Through real-time feedbacks, listeners show how they are engaged (*e.g.*, *interested*, *understand*, *agree*, *etc.*) to the speech, such that conversation gets more accessible for both participates.

In this work, we propose a new task to highlight listener-centric generation. Specifically, listening-head generation aims to synthesize a video of listening head, conditioning on the corresponding talking-head video of the speaker and the identity information of the listener, as shown in Fig. 1d. Proper reactions of the listener are expected to coordinate with the input talking video. This task is critical to a wide range of applications including virtual anchors, digital influencers, customer representatives, digital avatar in Metaverse, wherever involves interactive communication.

To address this, we construct a high-quality speaker-listener dataset, named ViCo, by capturing the high-definition video data from public conversations between two persons containing frontal faces on the same screen. The data strictly follows the principle that a video clip contains only uniquely identified listener and speaker, and requires that the listener has responsive non-verbal feedback to the speaker. After data cleaning, we further annotate the listener with three different attitudes: positive, neutral and negative. In total, our ViCo dataset contains 483 video clips of 76 listeners responding to 67 speakers. Compared to speaker-centric datasets such as MEAD [56] and VoxCeleb2 [12], ViCo highlights the listener role, making an indispensable couterpart to those speaker-centric ones. Compared to SEMAINE [36] (human interacts with a limited artificial agent) and MAHNOB Laughter [44] (people watching movies), ViCo features real persons in real conversations, such that natural reactions between genuine humans during a conversation make a key difference.

Together with the dataset, we propose a listening-head generation baseline method. We are aware that previous speaker-centric tasks are usually modeled in an idiosyncratic way (different speakers are modeled independently). However, listening behavior patterns are typically well coordinated with the speaker video. Thus we decouple the identity features from the listener and focus on learning the general motion patterns of responsive listening behaviors. We model listening head generation as a video-to-video translation task, by designing a sequence-to-sequence architecture to sequentially decode the listener's head motion and expression. Through quantitative evaluation and user study, we show our baseline is able to automatically capture the salient moments of speaker video and responds properly with clear motions and expressions.

2 Related Works

Active Listener. Active listening is an effective communication skill that not only means focusing fully on the speaker but also actively showing the non-verbal signals of listening with attitude. Usually, the active listener would mirror some facial expressions used by the speaker or shows more eye contact with the speaker. Active listening have shown its positive effects in many areas, such as teaching [26], medical consultations [15], team management [39], *etc.* In this paper, we aim to generate an active listener that could provide responsive feedback, the listener would understand the speaker's verbal and non-verbal signals first and then give proper feedback to the speaker.

Speaker-Centered Video Synthesis. Given time-varying signals and a reference still image of the speaker, the talking head synthesis task aims to generate a vivid clip for the speaker with the time-varying signals matched. Based on the different types of time-varying signals, we can group these tasks into two groups: 1) audio-driven talking head synthesis [11,45,60], 2) video-driven talking head synthesis [2,58]. The goal of the former one is to generate a video of the speaker that matches the audio. And the latter one is to generate videos of speakers with expressions similar to those in the video. This differs from our task: the "listener" is forced to perceive the speaker's visual and audio signals and make an active response. Our task does not focus on only a single person or transfers face expression and slight head movements from another person. There are two roles in our task: listener and speaker, and the listener should actively respond to the speaker with non-verbal signals.

Listening Behaviors Modeling. Many applications and research papers have focused on speaking, while the "listener modeling" is seldomly explored. Gillies *et al.* [16] first propose the data-driven method that can generate an animated character that can respond to speaker's voice. This lacks the supervision of speaker visual signals, which is incomplete for responsive listener modeling. And this method can not be applied to realistic head synthesis. Heylen *et al.* [20] further studied the relationship between listener and speaker audio/visual signals from a cognitive technologies view. SEMAINE [36] records the conversation between a human and a limited artificial listener. MAHNOB Laughter database [44] focuses on studying laughter's behaviors when watching funny video clips. Apart from these related work, ALICO [8] corpus about active listener analysis is the most relevant dataset with our proposed task. However, it has not been made public and also not constructed from the real scene conversations. Moreover, the main objective of ALICO is for psychology analysis, the data mode of that dataset is vastly different from the audio-video corpora in computer vision area. In the past few years, the social AI intelligence [27,40] has been introduced to model the nonverbal social signals in triadic or multi-party interactions. Joo *et al.* [27] concerned with the overall posture and head movement of a person, and Oertel *et al.* [40] aims to mine listening motion rules for robotics controlling. Both

Table 1. Comparison with other listener-related datasets

Dataset	Public	Environment	Style	Interact with Real
Gillies *et al.* [16]	✗	Lab	Simulated	✗
SEMAINE [36]	✓	Lab	Simulated	✓
Heylen *et al.* [20]	✗	Lab	Simulated	✓
MAHNOB Laughter [44]	✓	Lab	Realistic	✗
ALICO [8]	✗	Lab	Realistic	✓
Ours	✓	Wild	Realistic	✓

related works only deal with the speaking status and ignore the speaker's content. What's more, they rarely care about two-person interactions nor pay attention to model the face in detail, which is also different from our task. A detailed comparison to exising listener-related datasets is shown in Table 1.

As far as we know, this is the first time to introduce the learning-based listening head generation task in computer vision area. In this work, we propose a formulation of responsive listening head generation and construct a public ViCo dataset for this task. Meanwhile, a baseline method is proposed for listening head synthesis by perceiving both speaker's audio/visual signals and preset attitude.

3 Task Overview

We present Responsive Listening Head Generation, a new task that challenges vision systems to generate listening heads actively responding to the speaker's face or/and audio in real-time. In particular, we need to understand the head motion, facial expression, including eye blinks, mouth movements, *etc.*, of the input speaker video frame, and simultaneously understand the speaker's voice, then synchronously generate the active listening face video conditioned by the given attitude.

Given an input video sequence $\mathcal{V}_t^s = \{v_1^s, \cdots, v_t^s\}$ of a speaker head in time stamps ranging from $\{1, ..., t\}$, and an corresponding audio signal sequence $\mathcal{A}_t^s = \{a_1, \cdots, a_t\}$ of the speaker, listening head generation aims to generate a listener's head v_{t+1}^l of the next time stamp:

$$v_{t+1}^l = \mathbf{G}(\mathcal{V}_t^s, \mathcal{A}_t^s, v_1^l, e), \tag{1}$$

where v_1^l is the reference head of the listener, e denotes the attitude of the listener. The whole generated listener video $\mathcal{V}_{t+1}^l$ can be denoted as the concatenation of $\{v_2^l, \cdots, v_{t+1}^l\}$.

Listening Attitude Definition. During conversation, after perceiving the signals from the speaker, the listener usually reacts with an active, responsive *attitude*, including epistemic attitudes (*e.g.*, agree, disagree) and affective attitudes (*e.g.*, like, dislike). In this work, we group the attitudes into three categories: positive, negative and neutral. Positive attitude consists of *agree*, *like*, *interested*. Conversely, negative attitude consists of *disagree*, *dislike*, *disbelieve*, *not interested*. In general, attitude potentially guides the listener's behavior and consequently affects the conversation. Also, different attitude results in different facial expressions and behaviors of the listener [21], e.g. a smile appears as the most appropriate signal for *like*, a combination of smile and raise eyebrows could be a possibility for *interested*, *disagree* can be meant by a head shake, *dislike* is represented by a frown and tension of the lips, *etc.* A listener example with different attitudes is illustrated in Fig. 2.

Fig. 2. During a conversation, different attitudes of the listener could show different pose and expression patterns

Feature Extraction. In this work, we extract the energy feature, temporal domain feature, and frequency domain feature of the input audio; and model the facial expression and head poses using 3DMM [6] coefficients.

For the audio, we extract the Mel-frequency cepstral coefficients (MFCC) feature with the corresponding MFCC Delta and Delta-Delta feature. Besides, the energy, loudness and zero-crossing rate (ZCR) are also embedded into audio features s_i for each audio clip a_i. The audio feature extracted from $\mathcal{A}_t^s$ can be denoted as $\mathcal{S}_t^s = \{s_1, \cdots, s_t\}$.

We leverage the state-of-the-art deep learning-based 3D face reconstruction model [14] for the videos to get the 3DMM [6] coefficients. Specially, for each image, we can get the reconstruction coefficients $\{\alpha, \beta, \delta, p, \gamma\}$ which denote the identity, expression, texture [9,43], pose and lighting [46], respectively. Further, we distinguished the 3D reconstruction coefficients into two parts: $\mathcal{I} = (\alpha, \delta, \gamma)$ to represent relatively fixed, identity-dependent features, and $m = (\beta, p)$ to represent relatively dynamic, identity-independent features. The identity-independent feature extracted from speaker videos can be denoted as $\mathcal{M}_t^s = \{m_1^s, \cdots, m_t^s\}$, where $m_i^s \in \mathbb{R}^{1\times C_v}$ is the expression and pose feature of 3D reconstruction coefficients for the i-th frame v_i^s, where $C_v = |\beta| + |p|$.

Task Definition. To ignore identity-dependent features and learn general listener patterns that can be adapted to multiple listener identities, we use only the head motion and facial expression feature m for responsive listening head generation model training, and then adapt the identity-dependent features $\mathcal{I}$ of different listener identities for visualization and evaluation. Thus our listening head synthesis task can be formulated as:

$$\begin{aligned} m_{t+1}^l &= \mathbf{G}_m(\mathcal{M}_t^s, \mathcal{S}_t^s, m_1^l, e), \\ v_{t+1}^l &= \mathbf{G}_v(m_{t+1}^l, \mathcal{I}^l, v_1^l), \end{aligned} \tag{2}$$

where m_{t+1}^l is the dynamic feature predicted for listener's head, and $\mathcal{I}^l$ denotes the identity-dependent features of the given listener. In a real implementation, we use T frame of speaker's audio and video for responsive listening head generation model training.

The 3D face rendering technology $\mathbf{G}_v$ has been well studied in many recent related works [30,60]. Moreover, the face rendering models are usually identity-specific, so one may need to train the rendering model separately for each identity for better performance. To highlight the properties of the interactive digital human synthesis task, and decouple the critical factor in this task, our proposed responsive listening head synthesis model primarily focuses on the motion-related and identity-independent 3D facial coefficients prediction task $\mathbf{G}_m$, and use the pretrained rendering model [47] for simplified visualization.

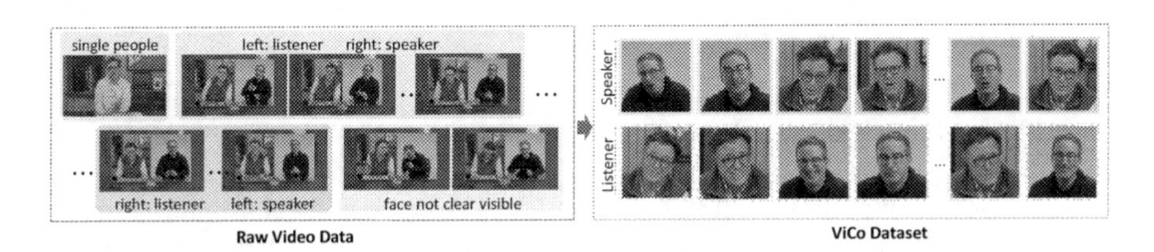

Fig. 3. In ViCo, valid clips are selected in accordance with the standards that 1) both the speaker and listener behaviors are clearly visible, and 2) listeners are responsively engaged to the conversation. The facial regions of listener-speaker pairs are further cropped for constructing our ViCo dataset (right)

Table 2. Statistics of ViCo dataset. #ID indicates the number of identities. The same person/identity can play different roles with multiple attitudes

Attitude	#Videos	#Speaker	#Listener	#ID	#Clips	Duration
Positive	42	53	62	81	226	49 min 18 s
Neutral	35	38	48	63	134	27 min 7 s
Negative	11	11	9	18	123	18 min 57 s
Total	50	67	76	92	483	95 min 22 s

4 Dataset Construction

We construct a dataset for responsive listening-head generation by capturing conversational video clips from YouTube containing two people's frontal faces. A *valid* video clip is required to meet the following conditions:

- The screen contains only two people, and one of them is speaking while the other is listening carefully.
- The frontal faces of both people are clearly visible. The facial expression is natural and stable.
- The listener actively responds to the speaker in a dynamic and real-time manner.

The annotators were asked to accurately record the start and end time of each *valid* clip, label the position of the speaker (left or right of the screen) and identify the attitude of the listener in the video. Cross-validation was applied among at least three annotators for each candidate clip for quality control. For each valid clip, we use the MTCNN [61] to detect the face regions in each frame, and then crop and resize the detected face regions to 384×384 resolution image sequence for model training and evaluation, as shown in Fig. 3.

Table 2 shows the statistical information of our annotated responsive listening head generation dataset ViCo. The proposed dataset contains rich samples of 483 video clips. We normalize all videos to 30 FPS, forming more than 0.1 million frames in total. Moreover, our dataset has following properties:

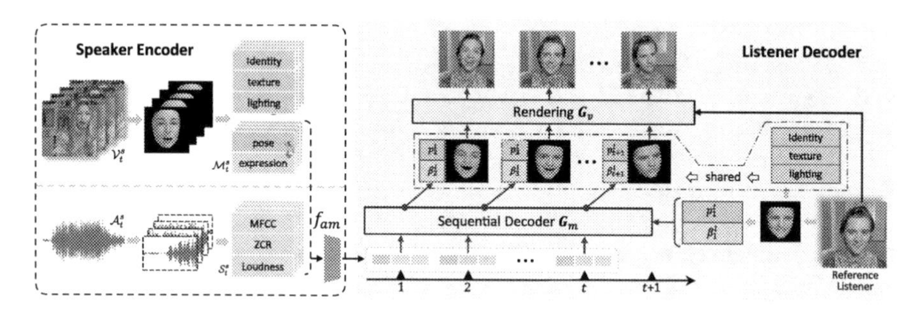

Fig. 4. The overall pipeline of our responsive listening head generation baseline. The speaker encoder aims to encode the head motion, facial expression and audio features. Starting from the fused feature from reference listener image, the listener decoder receives signals from speaker encoder in temporal order, and predicts the head motion and facial expression features. These features are adapted to reconstruct the 3DMM coefficients with the reference listener's identity-dependent features, and then fed to a neural renderer to generate realistic listening video

High Quality. All raw videos are of high resolutions (1920×1080), so that the subtle differences between different attitudes and changing moods are well preserved. And audios are in 44.1 kHz/16bit such that the speech-related features can be well preserved, too.

High Diversity. Our dataset contains various scenarios, including news interviews, entertainment interviews, TED discussions, variety shows, *etc.* These diverse scenarios provide rich semantic information and various listener patterns in different situations. The video clips length ranges from 1 to 71 seconds.

Realtime and Interactivity. Different from the existing talking head video datasets [1,12,32,56,59] which aim to generate head or face in synchronization with the audio signals, our dataset focus on the face-to-face *response*. These responses are generated by jointly understanding the speaker's audio, facial, and head motion signals, then adapting to different listener heads. It matters about mutual interaction rather than a monologue.

5 Responsive Listening Head Generation

Based on ViCo dataset, we propose a responsive listening head generation baseline. The overview of our approach is illustrated in Fig. 4.

5.1 Model Architecture

According to the psychological knowledge, an active listener tends to respond based on speaker's audio [28] and visual signals [10,16,35] comprehensively. And at a given moment, the listener receives information from the speaker of that moment as well as information from history and adopts a certain attitude to present actions in response to the speaker. Thus, the goal of our model is to estimate the conditioned probability $P(\mathcal{M}_{t+1}^l | \mathcal{M}_t^s, \mathcal{S}_t^s, m_1^l, e)$, where the $\mathcal{M}_t^s$ and $\mathcal{S}_t^s$ are time-varying signals that the listener should respond to, and the reference listener feature m_1^l and attitude e constrain the pattern of the entire generated sequence.

Inspired by the sequence-to-sequence model [53], a multi-layer sequential decoder module $\mathbf{G}_m$ is applied for modeling the time-sequential information of conversation. Unlike talking-head generation [11,56,60,62], which accepts an entire input of audio and then processes it using a bidirectional LSTM or attention layer; in our scenario, the model $\mathbf{G}_m$ receives the streaming input of the speaker where future information is not available.

For the speaker feature encoder, at each time step t, we first extract the audio feature s_t and the speaker's head and facial expression representation m_t^s, then apply non-linear feature transformations following a multi-modal feature fusion function f_{am} to get the encoded feature of speaker. The representation of reference listener m_1^l and attitude e can be embeded as the initial state h_1 for the sequential motion decoder. At each time step t, taking the speaker's fused feature $f_{am}(s_t, m_t^s)$ as input, $\mathbf{G}_m$ in Eq. 2 is functioned as updating current state h_{t+1} and generating the listener motion m_{t+1}^l, which contains two feature vectors, *i.e.* β_{t+1}^l for the expression and p_{t+1}^l for the head rotation and translation. Our responsive listening head generator supports an arbitrary length of speaker input. The procedure can be formulated as:

$$\beta_{t+1}^l, p_{t+1}^l = \mathbf{G}_m(h_t, f_{am}(s_t, m_t^s)). \tag{3}$$

For optimization, with the ground truth listener patterns denoted as $\hat{\mathcal{M}}_T^l = [\hat{m}_2^l, \hat{m}_3^l, \cdots, \hat{m}_T^l]$, we drop the last prediction m_{T+1}^l due to the lack of supervision signals and use L_2 distance to optimize the training procedure:

$$\mathcal{L}_{gen} = \sum_{t=2}^{T} \|\beta_t^l - \hat{\beta}_t^l\|_2 + \|p_t^l - \hat{p}_t^l\|_2. \tag{4}$$

Moreover, a motion constraint loss $\mathcal{L}_{mot}$ is applied to guarantee the inter-frame continuity across $\hat{\mathcal{M}}_T^l$ is similar to the predicted $\mathcal{M}_T^l$:

$$\mathcal{L}_{mot} = \sum_{t=2}^{T} w_1\|\mu(\beta_t^l) - \mu(\hat{\beta}_t^l)\|_2 + w_2\|\mu(p_t^l) - \mu(\hat{p}_t^l)\|_2, \tag{5}$$

where $\mu(\cdot)$ measures the inter-frame changes of current frame and its adjacent previous frame, i.e., $\mu(\beta_t^l) = \beta_t^l - \beta_{t-1}^l$, w_1 and w_2 is a weight to balance the motion constraint loss and generation loss. The final loss function of our proposed listening head motion generation baseline can be formulated as:

$$\mathcal{L}_{total} = \mathcal{L}_{gen} + \mathcal{L}_{mot}. \tag{6}$$

By optimizing $\mathcal{L}_{total}$, our model can generate attitude conditioned responsive listening head for a given speaker video and audio.

5.2 Implementation Details

To verify that our model can learn a generic listening pattern rather than conditioning on any particular individual, we divide the ViCo dataset ($\mathcal{D}$) into three parts: i) training set $\mathcal{D}_{train}$ for learning listener patterns, ii) test set $\mathcal{D}_{test}$ for validating our model on in-domain data, and iii) out-of-domain (OOD) test set $\mathcal{D}_{ood}$ for evaluating the generalization and transferability. In this case, all identities in $\mathcal{D}_{test}$ have appeared in $\mathcal{D}_{train}$, while identities in the $\mathcal{D}_{ood}$ do not overlap with those in $\mathcal{D}_{train}$.

We extract 45-dimensional acoustic features for audios, including 14-dim MFCC, 28-dim MFCC-Delta, energy, ZCR and loudness. There are multiple choices to implement $\mathbf{G}_m$, such as standard sequential model like LSTM [22], GRU [13], or a Transformer [55] decoder with sliding window [4]. Here we adopt

LSTM for our baseline, since it has been widely used in many similar applications such as motion generation [48], and achieve stable state-of-the-art performance when training on small corpus [38]. Our listening head generation model is trained with AdamW [33] optimizer with a learning rate of 1×10^{-3} (decayed exponentially by 0.8 every 30 epochs), $\beta_1 = 0.9$ and $\beta_2 = 0.999$, for 300 epochs. For all experiments, we set hyper-parameter w_1 to 0.1 and w_2 to 1×10^{-4}.

5.3 Experimental Results

Quantitative Results. Since we use a detached renderer rather than an end-to-end pipeline, we can divide the assessment into two sides: the performance of listener generator $\mathbf{G}_m$ and the visual effects of renderer $\mathbf{G}_v$. For the former one, we use use L_1 distance between the generated features and the ground-truth features (FD) to ensure the predicted fine-grained head and expression coefficients similar to the ground-truth. And for the latter one, we select the Fréchet Inception Distance (FID) [19], Structural SIMilary (SSIM) [57], Peak Signal-to-Noise Ratio (PSNR) and Cumulative Probability of Blur Detection (CPBD) [7] to evaluate the visual effects of renderer. The high-level metric on $\mathbf{G}_m$ can help us analyze the model, and the low-level metrics on $\mathbf{G}_v$ provide a baseline for the successors.

In Table 3, we report the FD of the 3D facial coefficients across different listening head generation methods, including: 1) "Random": generate frames from reference image but injecting small perturbations in a normal distribution to mimic random head motion. 2) "Simulation": simulating natural listening behavior by repeating the motion patterns sampled from $\mathcal{D}_{train}$. 3) "Simulation*": repeating the natural listening motion patterns sampled from $\mathcal{D}_{train}$ with the corresponding attitude. 4) "Ours": our proposed responsive listening head generation method. The random permutations is the worst and not able to present a listener. Our method can reach the best performance which demonstrates the superiority of our algorithm over traditional non-parametric listeners. Also, we provide the evaluation results of zero-shot 3D face rendering [47] in Table 4, as a basic criterion for evaluation of future realistic face rendering research work.

Qualitative Results. Further, we visualize the results of those generations to analyze the differences between different configurations intuitively. Given $\mathcal{D}_{test}$ and $\mathcal{D}_{ood}$, we generate listener videos with a given attitude. We randomly select two sequences from each set and then down-sample to six frames to qualitatively visualize the generated results and the ground truth video. The results are shown in Fig. 5, we can find that our model is generally able to capture listener patterns (*e.g.*, *eye*, *mouth*, and *head motion*, *etc.*), which may differ from the ground-truth while still making sense.

From these two groups, we observe that the "Random" patterns behave very confusing and messy. The "Simulation" patterns depend heavily on whether we can randomize to a given attitude, and as long as this fails, the result is bad. The "Simulation*" performs a little bit better, while its motion is also limited by the

Table 3. The Feature Distance (×100) of different listening head generation methods. Each cell in the table represent the feature distance of `angle/expression/translation` coefficients respectively. Lower is better

Attitude	Motion	Random		Simulation		Simulation*		Ours	
		$\mathcal{D}_{test}$	$\mathcal{D}_{ood}$	$\mathcal{D}_{test}$	$\mathcal{D}_{ood}$	$\mathcal{D}_{test}$	$\mathcal{D}_{ood}$	$\mathcal{D}_{test}$	$\mathcal{D}_{ood}$
Positive	Angle	17.92	17.99	9.86	10.48	9.79	11.57	**6.79**	**9.72**
	Exp	44.71	44.86	27.08	27.81	30.00	30.27	**15.37**	**24.89**
	Trans	19.74	20.14	16.25	12.06	9.07	13.82	**6.48**	**9.51**
Neutral	Angle	17.85	17.78	10.94	9.47	14.18	8.94	**8.79**	**6.33**
	Exp	44.26	44.29	27.37	29.50	26.44	27.56	**13.61**	**23.51**
	Trans	19.98	20.17	8.47	12.27	11.53	9.40	**6.68**	**8.95**
Negative	Angle	19.53	18.70	17.86	9.66	13.24	18.75	**12.45**	**8.54**
	Exp	45.68	44.62	29.57	28.78	31.62	27.04	**16.98**	**18.99**
	Trans	19.69	20.92	8.06	10.69	24.42	11.09	**6.35**	**5.81**
Average	Angle	18.04	18.11	10.81	9.91	11.24	12.58	**7.79**	**8.23**
	Exp	44.67	44.60	27.37	28.66	29.20	28.46	**15.04**	**22.83**
	Trans	19.80	20.36	13.52	11.76	11.00	11.55	**6.52**	**8.32**

Table 4. Quantitative Results of Renderer $\mathbf{G}_v$ on $\mathcal{D}_{test}$ and $\mathcal{D}_{ood}$

Attitude	FID↓		SSIM↑		PSNR↑		CPBD↑	
	$\mathcal{D}_{test}$	$\mathcal{D}_{ood}$	$\mathcal{D}_{test}$	$\mathcal{D}_{ood}$	$\mathcal{D}_{test}$	$\mathcal{D}_{ood}$	$\mathcal{D}_{test}$	$\mathcal{D}_{ood}$
Positive	29.736	27.865	0.565	0.496	17.075	15.421	0.122	0.121
Neutral	36.551	27.366	0.686	0.544	21.220	17.703	0.106	0.113
Negative	46.277	28.406	0.610	0.528	16.870	16.709	0.219	0.211
Average	30.529	24.962	0.601	0.521	18.149	16.558	0.126	0.142

size and diversity of dataset, and intuitively, it cannot respond to the speaker in a dynamic manner. Our results are more visually plausible than others with head motions and expression changes.

We also provide the comparison results of listening head generation under different attitudes in Fig. 6. Obviously, the facial expression and head motions under different attitudes are expressive and distinguishable.

Ablation Studies. We conduct an ablation study on the impact of different speaker signals for listening head modeling on $\mathcal{D}_{test}$. As Table 5 shown, the listening head driven by audio-only inputs prefers expression modeling but performs badly in head motion (angle, trans), while the model with visual-only inputs would capture the motion or sightline changes of the speaker during conversation and provide reasonable listening head action response. Removing these two modalities and simply mirror the speaker causes the worst listener. And with

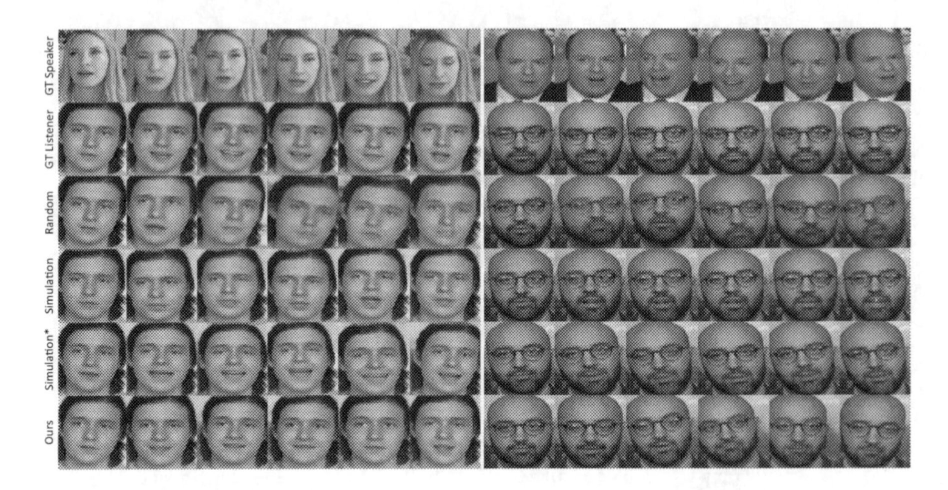

Fig. 5. Qualitative comparison of listening head generation methods conditioned by the same reference frame (the first column of each group) and the same attitude (left: positive listener in $\mathcal{D}_{test}$, right: neutral listener in $\mathcal{D}_{ood}$). Our method can generate various, vivid and responsive listening motions for the given speaker video stream

these two joint input signals, our model exhibits the best performance. That is, *only when we look and hear, can we act as better responsive listeners.*

Besides, we also report the performance of listener generation by giving wrong audio inputs and correct visual inputs in Table 5. It shows that Audio Only < Wrong Audio < Visual Only for expression FD, while Visual Only ≈ Wrong Audio < Audio Only for head motion FD (lower is better). This also reveals that expression modeling depends more on audio inputs and head motion modeling is more corresponding to visual inputs.

Table 5. The averaged Feature Distance (×100) of listener generations across all attitudes on $\mathcal{D}_{test}$.

Method	Angle	Exp.	Trans.
Audio only	8.69	17.19	8.49
Visual only	8.06	18.85	7.54
Wrong Audio	8.15	18.02	7.60
Mirroring	9.99	26.48	11.39
Ours	**7.79**	**15.04**	**6.52**

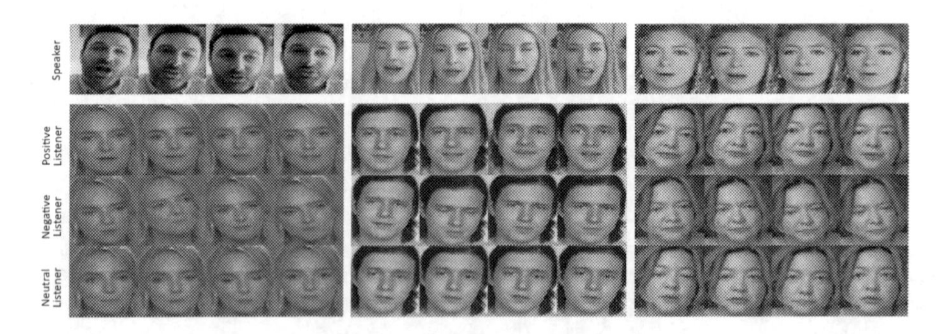

Fig. 6. Comparative results of our listening heads generation method conditioned by the same reference image and speaker video but different attitudes

Fig. 7. Diverse visual patterns can be observed from the generated listening-heads. Left: diverse expressions can be generated corresponding to different attitudes. Right: motion patterns including *shaking head*, *nod*, *glare*, *looking askance*, *pressing lip* and *focusing*

Diversity of Generations. Figure 7 illustrates the diversity of visual patterns learned by our method. The single images on the left side show our model can generate different expressions while the six groups of images on the right demonstrate the head motions and eye contacts can be modeled by our method.

Runtime Complexity. With a given speaker's streaming video, it takes 52.5 ms to fit the 3DMM coefficients, and generate the next step listener's motion in 0.0372 ms, then render the motion to RGB images in 29.3 ms. The per-frame delay of this process on one Tesla-V100 GPU is 81.84 ms without any optimization strategy such as ONNX or TensorRT. The FPS of generated listening head videos can reach 12, and can be further improved to 19 with pipeline parallelism. The bottleneck of our proposed baseline method is the efficiency of 3D face reconstruction (fitting). Some video post-processing methods, such as video frame interpolation [25,31,41] can be used for real-time interaction.

User Study. Since responsive listening head modeling is a user-oriented task, it is essential to conduct user studies for evaluation. For $\mathcal{D}_{test}$ and $\mathcal{D}_{ood}$, we had 10 volunteers doing two double-blind tests: 1) **Preference Test (PT).** Given the shuffled tuple of ⟨ground-truth listening head, generated listening head⟩, along with the ground-truth attitude, speaker's audio and speaker's video, the volunteers were asked to pick the best listening head. 2) **Attitude Matching Test (AttMatch).** Given the randomly shuffled listening heads generations conditioned by three different attitudes for each identity, the volunteers were asked to identify one attitude for each listening head.

For the first user-study, each volunteer was asked to check 30 samples in a preference test. As shown in Table 6a, fpr $\mathcal{D}_{test}$ and $\mathcal{D}_{ood}$, volunteers voted that nearly 43.8% and 44.7% of the generated heads can get equal or even better rating than the ground-truth heads respectively, which verified that our model could generate responsive listeners consistent with subjective human perceptions, and even can be reached confused as real ones.

Our second user-study results are shown in Table 6b. Each volunteer was given 90 generated listening clips under three different attitudes. For each attitude, we calculate the mean precision and precision variance across all volunteers

Table 6. (a) The result (mean/variance) of preference test. "Equal": the generated results and the ground-truth are visually equivalent. "PD Better": the generated results are more in line with human perception. (b) Mean precision and precision variance of attitude matching test

(a) Preference Test

PT	$\mathcal{D}_{test}$	$\mathcal{D}_{ood}$
Equal	31.3/17.1	13.9/6.2
GT Better	56.2/14.9	65.3/6.1
PD Better	12.5/9.3	20.8/4.6

(a) Attitude Matching Test

AttMatch	$\mathcal{D}_{test}$	$\mathcal{D}_{ood}$
Positive	66.7/15.7	69.0/7.1
Negative	63.9/16.4	50.0/12.8
Neutral	69.4/9.2	53.6/3.9

for analysis. The results show that both in $\mathcal{D}_{test}$ and $\mathcal{D}_{ood}$, our model is capable of generating listening head yielding to the required attitude. The precision for negative and neutral attitude is slightly decreased in $\mathcal{D}_{ood}$, which might be caused by the unbalanced attitude distributions.

6 Conclusion

In this paper, we define the responsive listening head generation task. It aims to generate a responsive video clip for a listener with the understanding of the speaker's facial signals and voices. Further, the high-quality responsive listener dataset (ViCo) is contributed for addressing this problem. The responsive listener generation baseline can synthesis active listeners, which are more consistent with human perception. We expect that ViCo could benefit the face-to-face communication modeling in computer vision area and facilitate the applications in more scenarios, such as intelligence assistance, virtual human, *etc.*

Ethical Impact. The ViCo dataset will be released only for research purposes under restricted licenses. The responsive listening patterns are identity-independent, which reduces the abuse of facial data. The only potential social harm is "fake content". However, different from talking head synthesis, responsive listening can hardly harm information fidelity.

Acknowledgment. This work was supported by the National Key R&D Program of China under Grant No. 2020AAA0108600.

References

1. Afouras, T., Chung, J.S., Zisserman, A.: Lrs3-ted: a large-scale dataset for visual speech recognition. arXiv preprint arXiv:1809.00496 (2018)
2. Bansal, A., Ma, S., Ramanan, D., Sheikh, Y.: Recycle-gan: unsupervised video retargeting. In: Proceedings of the European Conference on Computer Vision (ECCV), pp. 119–135 (2018)
3. Barker, L.L.: Listening behavior (1971)
4. Beltagy, I., Peters, M.E., Cohan, A.: Longformer: the long-document transformer. arXiv preprint arXiv:2004.05150 (2020)
5. Berger, C.R.: Interpersonal communication: theoretical perspectives, future prospects. J. Commun. **55**, 415–477 (2005)
6. Blanz, V., Vetter, T.: A morphable model for the synthesis of 3d faces. In: Proceedings of the 26th Annual Conference on Computer Graphics and Interactive Techniques, pp. 187–194 (1999)
7. Bohr, P., Gargote, R., Vhorkate, R., Yawle, R., Bairagi, V.: A no reference image blur detection using cumulative probability blur detection (cpbd) metric. Int. J. Sci. Modern Eng. **1**(5) (2013)
8. Buschmeier, H., et al.: Alico: a multimodal corpus for the study of active listening. In: LREC 2014, Ninth International Conference on Language Resources and Evaluation, Reykjavik, Iceland,, 26–31 May 2014, pp. 3638–3643 (2014)
9. Cao, C., Weng, Y., Zhou, S., Tong, Y., Zhou, K.: Facewarehouse: a 3d facial expression database for visual computing. IEEE Trans. Vis. Comput. Graph. **20**(3), 413–425 (2013)
10. Cassel, N.N.W.W.: Elements of face-to-face conversation for embodied conversational agents, embodied conversational agents (2000)
11. Chung, J.S., Jamaludin, A., Zisserman, A.: You said that? arXiv preprint arXiv:1705.02966 (2017)
12. Chung, J.S., Nagrani, A., Zisserman, A.: Voxceleb2: deep speaker recognition. arXiv preprint arXiv:1806.05622 (2018)
13. Chung, J., Gulcehre, C., Cho, K., Bengio, Y.: Empirical evaluation of gated recurrent neural networks on sequence modeling. arXiv preprint arXiv:1412.3555 (2014)
14. Deng, Y., Yang, J., Xu, S., Chen, D., Jia, Y., Tong, X.: Accurate 3D face reconstruction with weakly-supervised learning: from single image to image set. In: IEEE Computer Vision and Pattern Recognition Workshops (2019)
15. Fassaert, T., van Dulmen, S., Schellevis, F., Bensing, J.: Active listening in medical consultations: development of the active listening observation scale (alos-global). Patient Educ. Counsel. **68**(3), 258–264 (2007)
16. Gillies, M., Pan, X., Slater, M., Shawe-Taylor, J.: Responsive listening behavior. Comput. Anim. Virt. Worlds **19**(5), 579–589 (2008)
17. Ginosar, S., Bar, A., Kohavi, G., Chan, C., Owens, A., Malik, J.: Learning individual styles of conversational gesture. In: Proceedings of the IEEE/CVF Conference on Computer Vision and Pattern Recognition, pp. 3497–3506 (2019)
18. Hadar, U., Steiner, T.J., Rose, F.C.: Head movement during listening turns in conversation. J. Nonverbal Behav. **9**(4), 214–228 (1985)
19. Heusel, M., Ramsauer, H., Unterthiner, T., Nessler, B., Hochreiter, S.: Gans trained by a two time-scale update rule converge to a local nash equilibrium. Adv. Neural Inf. Processi. Syst. **30** (2017)
20. Heylen, D., Bevacqua, E., Pelachaud, C., Poggi, I., Gratch, J., Schröder, M.: Generating listening behaviour. In: Emotion-Oriented Systems, pp. 321–347. Springer, Heidleberg (2011). https://doi.org/10.1007/978-3-642-15184-2_17

21. Heylen, D., Bevacqua, E., Tellier, M., Pelachaud, C.: Searching for prototypical facial feedback signals. In: Pelachaud, C., Martin, J.-C., André, E., Chollet, G., Karpouzis, K., Pelé, D. (eds.) IVA 2007. LNCS (LNAI), vol. 4722, pp. 147–153. Springer, Heidelberg (2007). https://doi.org/10.1007/978-3-540-74997-4_14
22. Hochreiter, S., Schmidhuber, J.: Long short-term memory. Neural Comput. **9**(8), 1735–1780 (1997)
23. Hömke, P., Holler, J., Levinson, S.C.: Eye blinks are perceived as communicative signals in human face-to-face interaction. PloS One **13**(12), e0208030 (2018)
24. Honeycutt, J.M., Ford, S.G.: Mental imagery and intrapersonal communication: a review of research on imagined interactions (iis) and current developments. Ann. Int. Commun. Assoc. **25**(1), 315–345 (2001)
25. Huang, Z., Zhang, T., Heng, W., Shi, B., Zhou, S.: Real-time intermediate flow estimation for video frame interpolation. In: Proceedings of the European Conference on Computer Vision (ECCV) (2022)
26. Jalongo, M.R.: Promoting active listening in the classroom. Childhood Educ. **72**(1), 13–18 (1995)
27. Joo, H., Simon, T., Cikara, M., Sheikh, Y.: Towards social artificial intelligence: nonverbal social signal prediction in a triadic interaction. In: Proceedings of the IEEE/CVF Conference on Computer Vision and Pattern Recognition, pp. 10873–10883 (2019)
28. Kendon, A.: Movement coordination in social interaction: some examples described. Acta Psychologica **32**, 101–125 (1970)
29. Kendon, A., Harris, R.M., Key, M.R.: Organization of behavior in face-to-face interaction. Walter de Gruyter (2011)
30. Kim, H.: Deep video portraits. ACM Trans. Graph. (TOG) **37**(4), 1–14 (2018)
31. Kong, L., et al.: Ifrnet: Intermediate feature refine network for efficient frame interpolation. In: Proceedings of the IEEE/CVF Conference on Computer Vision and Pattern Recognition, pp. 1969–1978 (2022)
32. Li, L., et al.: Write-a-speaker: text-based emotional and rhythmic talking-head generation. In: Proceedings of the AAAI Conference on Artificial Intelligence, vol. 35, pp. 1911–1920 (2021)
33. Loshchilov, I., Hutter, F.: Decoupled weight decay regularization. arXiv preprint arXiv:1711.05101 (2017)
34. Luhmann, N.: What is communication? Commun. Theory **2**(3), 251–259 (1992)
35. Maatman, R.M., Gratch, J., Marsella, S.: Natural behavior of a listening agent. In: Panayiotopoulos, T., Gratch, J., Aylett, R., Ballin, D., Olivier, P., Rist, T. (eds.) IVA 2005. LNCS (LNAI), vol. 3661, pp. 25–36. Springer, Heidelberg (2005). https://doi.org/10.1007/11550617_3
36. McKeown, G., Valstar, M., Cowie, R., Pantic, M., Schroder, M.: The semaine database: annotated multimodal records of emotionally colored conversations between a person and a limited agent. IEEE Trans. Affect. Comput. **3**(1), 5–17 (2011)
37. McNaughton, D., Hamlin, D., McCarthy, J., Head-Reeves, D., Schreiner, M.: Learning to listen: teaching an active listening strategy to preservice education professionals. Topics Early Childhood Spec. Educ. **27**(4), 223–231 (2008)
38. Melis, G., Kočiskỳ, T., Blunsom, P.: Mogrifier lstm. arXiv preprint arXiv:1909.01792 (2019)
39. Mineyama, S., Tsutsumi, A., Takao, S., Nishiuchi, K., Kawakami, N.: Supervisors' attitudes and skills for active listening with regard to working conditions and psychological stress reactions among subordinate workers. J. Occup. Health **49**(2), 81–87 (2007)

40. Oertel, C., Jonell, P., Kontogiorgos, D., Mora, K.F., Odobez, J.M., Gustafson, J.: Towards an engagement-aware attentive artificial listener for multi-party interactions. Front. Rob. AI **189** (2021)
41. Park, J., Lee, C., Kim, C.S.: Asymmetric bilateral motion estimation for video frame interpolation. In: Proceedings of the IEEE/CVF International Conference on Computer Vision, pp. 14539–14548 (2021)
42. Parker, J., Coiera, E.: Improving clinical communication: a view from psychology. J. Am. Med. Inf. Assoc. **7**(5), 453–461 (2000)
43. Paysan, P., Knothe, R., Amberg, B., Romdhani, S., Vetter, T.: A 3D face model for pose and illumination invariant face recognition. In: 2009 Sixth IEEE International Conference on Advanced Video and Signal Based Surveillance, pp. 296–301. IEEE (2009)
44. Petridis, S., Martinez, B., Pantic, M.: The mahnob laughter database. Image Vision Comput. **31**(2), 186–202 (2013)
45. Prajwal, K., Mukhopadhyay, R., Namboodiri, V.P., Jawahar, C.: A lip sync expert is all you need for speech to lip generation in the wild. In: Proceedings of the 28th ACM International Conference on Multimedia, pp. 484–492 (2020)
46. Ramamoorthi, R., Hanrahan, P.: An efficient representation for irradiance environment maps. In: Proceedings of the 28th Annual Conference on Computer Graphics and Interactive Techniques, pp. 497–500 (2001)
47. Ren, Y., Li, G., Chen, Y., Li, T.H., Liu, S.: Pirenderer: controllable portrait image generation via semantic neural rendering. In: Proceedings of the IEEE/CVF International Conference on Computer Vision, pp. 13759–13768 (2021)
48. Richard, A., Zollhöfer, M., Wen, Y., De la Torre, F., Sheikh, Y.: Meshtalk: 3D face animation from speech using cross-modality disentanglement. In: Proceedings of the IEEE/CVF International Conference on Computer Vision, pp. 1173–1182 (2021)
49. Robertson, K.: Active listening: more than just paying attention. Aust. Family Phys. **34**(12) (2005)
50. Rogers, C.R., Farson, R.E.: Active listening (1957)
51. Rost, M., Wilson, J.: Active Listening. Routledge, Abingdon (2013)
52. Stacks, D.W., Salwen, M.B.: An Integrated Approach to Communication Theory and Research. Routledge, Abingdon (2014)
53. Sutskever, I., Vinyals, O., Le, Q.V.: Sequence to sequence learning with neural networks. Adv. Neural Inf. Process. Syst. **27**, 3104–3112 (2014)
54. Tomasello, M.: Origins of Human Communication. MIT press, London (2010)
55. Vaswani, A., et al.: Attention is all you need. Adv. Neural Inf. Process. Syst. **30** (2017)
56. Wang, K., et al.: MEAD: a large-scale audio-visual dataset for emotional talking-face generation. In: Vedaldi, A., Bischof, H., Brox, T., Frahm, J.-M. (eds.) ECCV 2020. LNCS, vol. 12366, pp. 700–717. Springer, Cham (2020). https://doi.org/10.1007/978-3-030-58589-1_42
57. Wang, Z., Bovik, A.C., Sheikh, H.R., Simoncelli, E.P.: Image quality assessment: from error visibility to structural similarity. IEEE Trans. Image Process. **13**(4), 600–612 (2004)
58. Wu, W., Zhang, Y., Li, C., Qian, C., Loy, C.C.: Reenactgan: learning to reenact faces via boundary transfer. In: Proceedings of the European Conference on Computer Vision (ECCV), pp. 603–619 (2018)
59. Zhang, C., Ni, S., Fan, Z., Li, H., Zeng, M., Budagavi, M., Guo, X.: 3d talking face with personalized pose dynamics. IEEE Trans. Vis. Comput. Graph. (2021)

60. Zhang, C., et al.: Facial: synthesizing dynamic talking face with implicit attribute learning. In: Proceedings of the IEEE/CVF International Conference on Computer Vision, pp. 3867–3876 (2021)
61. Zhang, K., Zhang, Z., Li, Z., Qiao, Y.: Joint face detection and alignment using multitask cascaded convolutional networks. IEEE Signal Process. Lett. **23**(10), 1499–1503 (2016)
62. Zhu, H., Luo, M.D., Wang, R., Zheng, A.H., He, R.: Deep audio-visual learning: a survey. Int. J. Autom. Comput., 1–26 (2021)

Towards Scale-Aware, Robust, and Generalizable Unsupervised Monocular Depth Estimation by Integrating IMU Motion Dynamics

Sen Zhang(✉), Jing Zhang, and Dacheng Tao

The University of Sydney, Sydney, Australia
szha2609@uni.sydney.edu.au, {jing.zhang1,dacheng.tao}@sydney.edu.au

Abstract. Unsupervised monocular depth and ego-motion estimation has drawn extensive research attention in recent years. Although current methods have reached a high up-to-scale accuracy, they usually fail to learn the true scale metric due to the inherent scale ambiguity from training with monocular sequences. In this work, we tackle this problem and propose DynaDepth, a novel scale-aware framework that integrates information from vision and IMU motion dynamics. Specifically, we first propose an IMU photometric loss and a cross-sensor photometric consistency loss to provide dense supervision and absolute scales. To fully exploit the complementary information from both sensors, we further drive a differentiable camera-centric extended Kalman filter (EKF) to update the IMU preintegrated motions when observing visual measurements. In addition, the EKF formulation enables learning an ego-motion uncertainty measure, which is non-trivial for unsupervised methods. By leveraging IMU during training, DynaDepth not only learns an absolute scale, but also provides a better generalization ability and robustness against vision degradation such as illumination change and moving objects. We validate the effectiveness of DynaDepth by conducting extensive experiments and simulations on the KITTI and Make3D datasets (Code https://github.com/SenZHANG-GitHub/ekf-imu-depth).

Keywords: Unsupervised monocular depth estimation · Differentiable camera-centric EKF · Visual-inertial SLAM · Ego-motion uncertainty

1 Introduction

Monocular depth estimation is a fundamental computer vision task which plays an essential role in many real-world applications such as autonomous driving,

Supplementary Information The online version contains supplementary material available at https://doi.org/10.1007/978-3-031-19839-7_9.

S. Avidan et al. (Eds.): ECCV 2022, LNCS 13698, pp. 143–160, 2022.
https://doi.org/10.1007/978-3-031-19839-7_9

robot navigation, and virtual reality [20,36,44]. Classical geometric methods resolve this problem by leveraging the geometric relationship between temporally contiguous frames and formulating depth prediction as an optimization problem [8,9,29]. While geometric methods have achieved good performance, they are sensitive to either textureless regions or illumination changes. The computational cost for dense depth prediction also limits their practical use. Recently deep learning techniques have reformed this research field by training networks to predict depth directly from monocular images and designing proper losses based on ground-truth depth labels or geometric depth clues from visual data. While supervised learning methods achieve the best performance [1,7,11,24,45], the labour cost for collecting ground-truth labels prohibits their use in real-world. To address this issue, unsupervised monocular depth estimation has drawn a lot of research attention [14,48], which leverages the photometric error from backwarping.

Although unsupervised monocular depth learning has made great progress in recent years, there still exist several fundamental problems that may obstruct its usage in real-world. First, current methods suffer from the scale ambiguity problem since the backwarping process is equivalent up to an arbitrary scaling factor w.r.t. depth and translation. While current methods are usually evaluated by re-scaling each prediction map using the median ratio between the ground-truth depth and the prediction, it is difficult to obtain such median ratios in practice. Secondly, it is well-known that the photometric error is sensitive to illumination change and moving objects, which violate the underlying assumption of the backwarping projection. In addition, though uncertainty has been introduced for the photometric error map under the unsupervised learning framework [21,41], it remains non-trivial to learn an uncertainty measure for the predicted ego-motion, which could further benefit the development of a robust and trustworthy system.

In this work, we tackle the above-mentioned problems and propose DynaDepth, a novel scale-aware monocular depth and ego-motion prediction method that explicitly integrates IMU *motion dynamics* into the vision-based system under a camera-centric extended Kalman filter (EKF) framework. Modern sensor suites on vehicles that collect data for training neural networks usually contain multiple sensors beyond cameras. IMU presents a commonly-deployed one which is advantageous in that (1) it is robust to the scenarios when vision fails such as in illumination-changing and textureless regions, (2) the absolute scale metric can be recovered by inquiring the IMU motion dynamics, and (3) it does not suffer from the visual domain gap, leading to a better generalization ability across datasets. While integrating IMU information has dramatically improved the performance of classical geometric odometry and simultaneous localization and mapping (SLAM) systems [22,28,31], its potential in the regime of unsupervised monocular depth learning is much less explored, which is the focus of this work.

Specifically, we propose a scale-aware IMU photometric loss which is constructed by performing backwarping using ego-motion integrated from IMU measurements, which provides dense supervision by using the appearance-based

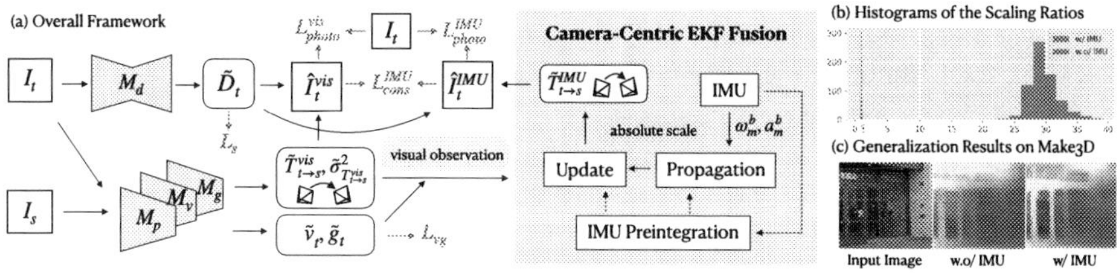

Fig. 1. (a) The overall framework of DynaDepth. $\hat{I}_t^{vis}$ and $\hat{I}_t^{IMU}$ denote the reconstructed target frames from the source frame I_s. Detailed notations of other terms are given in Sect. 3. (b) Histograms of the scaling ratios between the medians of depth predictions and the ground-truth. (c) Generalization results on Make3D using models trained on KITTI with (w/) and without (w.o/) IMU.

photometric loss instead of naively constraining the ego-motion predicted by networks. To accelerate the training process, the IMU preintegration technique [10,26] is adopted to avoid redundant computation. To correct the errors that result from illumination change and moving objects, we further propose a cross-sensor photometric consistency loss between the synthesized target views using network-predicted and IMU-integrated ego-motions, respectively. Unlike classical visual-inertial SLAM systems that accumulate the gravity and the velocity estimates from initial frames, these two metrics are unknown for the image triplet used in unsupervised depth estimation methods. To address this issue, DynaDepth trains two extra lightweight networks that take two consecutive frames as input and predict the camera-centric gravity and velocity during training.

Considering that IMU and camera present two independent sensing modalities that complement each other, we further derive a differentiable camera-centric EKF framework for DynaDepth to fully exploit the potential of both sensors. When observing new ego-motion predictions from visual data, DynaDepth updates the preintegrated IMU terms based on the propagated IMU error states and the covariances of visual predictions. The benefit is two-fold. First, IMU is known to suffer from inherent noises, which could be corrected by the relatively accurate visual predictions. Second, fusing with IMU under the proposed EKF framework not only introduces scale-awareness, but also provides an elegant way to learn an uncertainty measure for the predicted ego-motion, which can be beneficial for recently emerging research methods that incorporate deep learning into classical SLAM systems to achieve the synergy of learning, geometry, and optimization.

Our overall framework is shown in Fig. 1. In summary, our contributions are:

- We propose an IMU photometric loss and a cross-sensor photometric consistency loss to provide dense supervision and absolute scales
- We derive a differentiable camera-centric EKF framework for sensor fusion.
- We show that DynaDepth benefits (1) the learning of the absolute scale, (2) the generalization ability, (3) the robustness against vision degradation

such as illumination change and moving objects, and (4) the learning of an ego-motion uncertainty measure, which are also supported by our extensive experiments and simulations on the KITTI and Make3D datasets.

2 Related Work

2.1 Unsupervised Monocular Depth Estimation

Unsupervised monocular depth estimation has drawn extensive research attention recently [14,27,48], which uses the photometric loss by backwarping adjacent images. Recent works improve the performance by introducing multiple tasks [19,32,42], designing more complex networks and losses [15,18,38,49], and constructing the photometric loss on learnt features [35]. However, monocular methods suffer from the scale ambiguity problem. DynaDepth tackles this problem by integrating IMU dynamics, which not only provides absolute scale, but also achieves state-of-the-art accuracy even if only lightweight networks are adopted.

2.2 Scale-Aware Depth Learning

Though supervised depth learning methods [1,7,11] can predict depths with absolute scale, the cost of collecting ground-truth data limits its practical use. To relieve the scale problem, local reprojected depth consistency loss has been proposed to ensure the scale consistency of the predictions [2,43,47]. However, the absolute scale is not guaranteed in these methods. Similar to DynaDepth, there exist methods that resort to other sensors than monocular camera, such as stereo camera that allows a scale-aware left-right consistency loss [13,14,46], and GPS that provides velocities to constrain the ego-motion network [3,15]. In comparison with these methods, using IMU is beneficial in that (1) IMU provides better generalizability since it does suffer from the visual domain gap, and (2) unlike GPS that cannot be used indoors and cameras that fail in texture-less, dynamic and illumination changing scenes, IMU is more robust to the environments.

2.3 Visual-Inertial SLAM Systems

The fusion of vision and IMU has achieved great success in classical visual-inertial SLAM systems [22,28,31], yet this topic is much less explored in learning-based depth and ego-motion estimation. Though recently IMU has been introduced into both supervised [4,5] and unsupervised [16,34,40] odometry learning, most methods extract IMU features implicitly, while we explicitly utilize IMU dynamics to derive explicit supervisory signals. Li et al. [23] and Wagstaff et al. [37] similarly use EKF for odometry learning. Ours differs in that we do not require ground-truth information [23] or an initialization step [37] to align the velocities and gravities, but learn these quantities using networks. Instead of expressing the error states in the IMU frame, we further derive a camera-centric EKF framework to facilitate the training process. In addition, compared with odometry methods that do not consider the requirements for depth estimation, we specifically design the losses to provide dense depth supervision for monocular depth estimation.

3 Methodology

We present the technical details of DynaDepth in this section. We first revisit the preliminaries of IMU motion dynamics. Then we give the details of camera-centric IMU preintegration and the two IMU-related losses, i.e., the scale-aware IMU photometric loss and the cross-sensor photometric consistency loss. Finally, we present the differentiable camera-centric EKF framework which fuses IMU and camera predictions based on their uncertainties and complements the limitations of each other. A discussion on the connection between DynaDepth and classical visual-inertial SLAM algorithms is also given to provide further insights.

3.1 IMU Motion Dynamics

Let $\{\boldsymbol{w}_m^b, \boldsymbol{a}_m^b\}$ and $\{\boldsymbol{w}^b, \boldsymbol{a}^w\}$ denote the IMU measurements and the underlying vehicle angular and acceleration. The superscript b and w denote the vector is expressed in the body (IMU) frame or the world frame, respectively. Then we have $\boldsymbol{w}_m^b = \boldsymbol{w}^b + \boldsymbol{b}^g + \boldsymbol{n}^g$ and $\boldsymbol{a}_m^b = \boldsymbol{R}_{bw}(\boldsymbol{a}^w + \boldsymbol{g}^w) + \boldsymbol{b}^a + \boldsymbol{n}^a$, where $\boldsymbol{g}^w$ is the gravity in the world frame and $\boldsymbol{R}_{bw}$ is the rotation matrix from the world frame to the body frame [17]. $\{\boldsymbol{b}^g, \boldsymbol{b}^a\}$ and $\{\boldsymbol{n}^g, \boldsymbol{n}^a\}$ denote the Gaussian bias and random walk of the gyroscope and the accelerometer, respectively. Let $\{\boldsymbol{p}_{wb_t}, \boldsymbol{q}_{wb_t}\}$ and $\boldsymbol{v}_t^w$ denote the translation and rotation from the body frame to the world frame, and the velocity expressed in the world frame at time t, where $\boldsymbol{q}_{wb_t}$ denotes the quaternion. The first-order derivatives of $\{\boldsymbol{p}, \boldsymbol{v}, \boldsymbol{q}\}$ read: $\dot{\boldsymbol{p}_{wb_t}} = \boldsymbol{v}_t^w$, $\dot{\boldsymbol{v}_t^w} = \boldsymbol{a}_t^w$, and $\dot{\boldsymbol{q}_{wb_t}} = \boldsymbol{q}_{wb_t} \otimes [0, \frac{1}{2}\boldsymbol{w}^{b_t}]^T$, where $\otimes$ denotes the quaternion multiplication. Then the continuous IMU motion dynamics from time i to j can be derived as:

$$\boldsymbol{p}_{wb_j} = \boldsymbol{p}_{wb_i} + \boldsymbol{v}_i^w \Delta t + \iint_{t\in[i,j]} (\boldsymbol{R}_{wb_t}\boldsymbol{a}^{b_t} - \boldsymbol{g}^w)\mathrm{d}t^2, \tag{1}$$

$$\boldsymbol{v}_j^w = \boldsymbol{v}_i^w + \int_{t\in[i,j]} (\boldsymbol{R}_{wb_t}\boldsymbol{a}^{b_t} - \boldsymbol{g}^w)\mathrm{d}t, \tag{2}$$

$$\boldsymbol{q}_{wb_j} = \int_{t\in[i,j]} \boldsymbol{q}_{wb_t} \otimes [0, \frac{1}{2}\boldsymbol{w}^{b_t}]^T \mathrm{d}t, \tag{3}$$

where Δt is the time gap between i and j. For the discrete cases, we use the averages of $\{\boldsymbol{w}, \boldsymbol{a}\}$ within the time interval to approximate the integrals.

3.2 The DynaDepth Framework

DynaDepth aims at jointly training a scale-aware depth network $\mathcal{M}_d$ and an ego-motion network $\mathcal{M}_p$ by fusing IMU and camera information. The overall framework is shown in Fig. 1. Given IMU measurements between two consecutive images, we first recover the camera-centric ego-motion $\{\boldsymbol{R}_{c_k \check{c}_{k+1}}, \boldsymbol{p}_{c_k \check{c}_{k+1}}\}$ with absolute scale using IMU motion dynamics, and train two network modules $\{\mathcal{M}_g, \mathcal{M}_v\}$ to predict the camera-centric gravity and velocity. Then a scare-aware IMU photometric loss and a cross-sensor photometric consistency loss are built based on the ego-motion from IMU. To complement IMU and camera with each other, DynaDepth further integrates a camera-centric EKF module, leading to an updated ego-motion $\{\boldsymbol{R}_{c_k \hat{c}_{k+1}}, \boldsymbol{p}_{c_k \hat{c}_{k+1}}\}$ for the two IMU-related losses.

IMU Preintegration. IMU usually collects data at a much higher frequency than camera, i.e., between two image frames there exist multiple IMU records. Since the training losses are defined on ego-motions at the camera frequency, naive use of the IMU motion dynamics requires recalculating the integrals at each training step, which could be computationally expensive. IMU preintegration presents a commonly-used technique to avoid the online integral computation [10,26], which preintegrates the relative pose increment from the IMU records by leveraging the multiplicative property of rotation, i.e., $\boldsymbol{q}_{wb_t} = \boldsymbol{q}_{wb_i} \otimes \boldsymbol{q}_{b_i b_t}$. Then the integration operations can be put into three preintegration terms which only rely on the IMU measurements and can be precomputed beforehand: (1) $\boldsymbol{\alpha}_{b_i b_j} = \int \int_{t \in [i,j]} (\boldsymbol{R}_{b_i b_t} \boldsymbol{a}^{b_t}) \mathrm{d}t^2$, (2) $\boldsymbol{\beta}_{b_i b_j} = \int_{t \in [i,j]} (\boldsymbol{R}_{b_i b_t} \boldsymbol{a}^{b_t}) \mathrm{d}t$, and (3) $\boldsymbol{q}_{b_i b_j} = \int_{t \in [i,j]} \boldsymbol{q}_{b_i b_t} \otimes [0, \frac{1}{2}\boldsymbol{w}^{b_t}]^T \mathrm{d}t$. Since IMU preintegration is performed in the IMU body frame while the network predicts ego-motions in the camera fame, we thus establish the discrete camera-centric IMU preintegrated ego-motion as:

$$\boldsymbol{R}_{c_k \check{c}_{k+1}} = \boldsymbol{R}_{cb} \mathcal{F}^{-1}(\boldsymbol{q}_{b_k b_{k+1}}) \boldsymbol{R}_{bc}, \tag{4}$$

$$\boldsymbol{p}_{c_k \check{c}_{k+1}} = \boldsymbol{R}_{cb} \boldsymbol{\alpha}_{b_k b_{k+1}} + \boldsymbol{R}_{c_k \check{c}_{k+1}} \boldsymbol{R}_{cb} \boldsymbol{p}_{bc} - \boldsymbol{R}_{cb} \boldsymbol{p}_{bc} + \boldsymbol{v}^{\tilde{c}_k} \Delta t_k - \frac{1}{2} \boldsymbol{g}^{\tilde{c}_k} \Delta t_k^2, \tag{5}$$

where $\mathcal{F}$ denotes the transformation from rotation matrix to quaternion. $\{\boldsymbol{R}_{cb}, \boldsymbol{p}_{cb}\}$ and $\{\boldsymbol{R}_{bc}, \boldsymbol{p}_{bc}\}$ are the extrinsics between the IMU and the camera frames. Of note is the estimation of $\boldsymbol{v}^{\tilde{c}_k}$ and $\boldsymbol{g}^{\tilde{c}_k}$, which are the velocity and the gravity vectors expressed in the camera frame at time k.

Classical visual-inertial SLAM systems jointly optimize the velocity and the gravity vectors, and accumulate their estimates from previous steps. A complicated initialization step is usually required to achieve good performance. For unsupervised learning where the training units are randomly sampled short-range clips, it is difficult to apply the aforementioned initialization and accumulation. To address this issue, we propose to predict these two quantities directly from images as well during training, using two extra network modules $\{\mathcal{M}_v, \mathcal{M}_g\}$.

IMU Photometric Loss. State-of-the-art visual-inertial SLAM systems usually utilize IMU preintegrated ego-motions by constructing the residues between the IMU preintegrated terms and the system estimates to be optimized. However, naively formulating the training loss as these residues on IMU preintegration terms can only provide sparse supervision for the ego-motion network and thus is inefficient in terms of the entire unsupervised learning system. In this work, we propose an IMU photometric loss L_{photo}^{IMU} to tackle this problem which provides dense supervisory signals for both the depth and the ego-motion networks. Given an image $\boldsymbol{I}$ and its consecutive neighbours $\{\boldsymbol{I}_{-1}, \boldsymbol{I}_1\}$, L_{photo}^{IMU} reads:

$$L_{photo}^{IMU} = \frac{1}{N}\sum_{i=1}^{N} \min_{\delta \in \{-1,1\}} \mathcal{L}(\boldsymbol{I}(\boldsymbol{y}_i), \boldsymbol{I}_\delta(\psi(\boldsymbol{K}\hat{\boldsymbol{R}}_\delta \boldsymbol{K}^{-1}\boldsymbol{y}_i + \frac{\boldsymbol{K}\hat{\boldsymbol{p}}_\delta}{\tilde{z}_i}))), \tag{6}$$

$$\mathcal{L}(\boldsymbol{I}, \boldsymbol{I}_\delta) = \alpha\frac{1 - SSIM(\boldsymbol{I}, \boldsymbol{I}_\delta)}{2} + (1-\alpha)||\boldsymbol{I} - \boldsymbol{I}_\delta||_1, \tag{7}$$

where $\boldsymbol{K}$ and N are the camera intrinsics and the number of utilized pixels, $\boldsymbol{y}_i$ and $\tilde{z}_i$ are the pixel coordinate in image $\boldsymbol{I}$ and its depth predicted by $\mathcal{M}_d$, $\boldsymbol{I}(\boldsymbol{y}_i)$ is the pixel intensity at $\boldsymbol{y}_i$, and $\psi(\cdot)$ denotes the depth normalization function. $\{\hat{\boldsymbol{R}}_\delta, \hat{\boldsymbol{p}}_\delta\}$ denotes the ego-motion estimate from image $\boldsymbol{I}$ to $\boldsymbol{I}_\delta$, which is obtained by fusing the IMU preintegrated ego-motion and the ones predicted by $\mathcal{M}_p$ under our camera-centric EKF framework. $SSIM(\cdot)$ denotes the structural similarity index [39]. We also adopt the per-pixel minimum trick proposed in [14].

Cross-Sensor Photometric Consistency Loss. In addition to L_{photo}^{IMU}, we further propose a cross-sensor photometric consistency loss L_{photo}^{cons} to align the ego-motions from IMU preintegration and $\mathcal{M}_p$. Instead of directly comparing the ego-motions, we use the photometric error between the backwarped images, which provides denser supervisory signals for both $\mathcal{M}_d$ and $\mathcal{M}_p$:

$$L_{photo}^{cons} = \frac{1}{N}\sum_{i=1}^{N} \min_{\delta \in \{-1,1\}} \mathcal{L}(\boldsymbol{I}_\delta(\psi(\boldsymbol{K}\tilde{\boldsymbol{R}}_\delta \boldsymbol{K}^{-1}\boldsymbol{y}_i + \frac{\boldsymbol{K}\tilde{\boldsymbol{p}}_\delta}{\tilde{z}_i})), \boldsymbol{I}_\delta(\psi(\boldsymbol{K}\hat{\boldsymbol{R}}_\delta \boldsymbol{K}^{-1}\boldsymbol{y}_i + \frac{\boldsymbol{K}\hat{\boldsymbol{p}}_\delta}{\tilde{z}_i}))), \tag{8}$$

where $\{\tilde{\boldsymbol{R}}_\delta, \tilde{\boldsymbol{p}}_\delta\}$ are the ego-motion predicted by $\mathcal{M}_p$.

Remark: Of note is that using L_{photo}^{cons} actually increases the tolerance for illumination change and moving objects which may violate the underlying assumption of the photometric loss between consecutive frames. Since we are comparing two backwarped views in L_{photo}^{cons}, the errors incurred by the corner cases will be exhibited equally in both backwarped views. In this sense, L_{photo}^{cons} remains valid, and minimizing L_{photo}^{cons} helps to align $\{\tilde{\boldsymbol{R}}_\delta, \tilde{\boldsymbol{p}}_\delta\}$ and $\{\hat{\boldsymbol{R}}_\delta, \hat{\boldsymbol{p}}_\delta\}$ under such cases.

The Camera-Centric EKF Fusion. To fully exploit the complementary IMU and camera sensors, we propose to fuse ego-motions from both sensors under a camera-centric EKF framework. Different from previous methods that integrate EKF into deep learning-based frameworks to deal with IMU data [23,25], ours differs in that we do not require ground-truth ego-motion and velocities to obtain the aligned velocities and gravities for each IMU frame, but propose $\{\mathcal{M}_v, \mathcal{M}_g\}$ to predict these quantities. In addition, instead of expressing the error states in the IMU body frame, we derive the camera-centric EKF propagation and update processes to facilitate the training process which takes camera images as input.

EKF Propagation: Let c_k denote the camera frame at time t_k, and $\{b_t\}$ denote the IMU frames between t_k and time t_{k+1} when we receive the next visual measurement. We then propagate the IMU information according to the state transition model: $\boldsymbol{x}_t = f(\boldsymbol{x}_{t-1}, \boldsymbol{u}_t) + \boldsymbol{w}_t$, where $\boldsymbol{u}_t$ is the IMU record at time t, $\boldsymbol{w}_t$ is the noise term, and $\boldsymbol{x}_t = [\boldsymbol{\phi}^T_{c_k b_t}, \boldsymbol{p}^T_{c_k b_t}, \boldsymbol{v}^{c_k T}, \boldsymbol{g}^{c_k T}, \boldsymbol{b}_w^{b_t T}, \boldsymbol{b}_a^{b_t T}]^T$ is the state vector expressed in the camera frame c_k except for $\{\boldsymbol{b}_w, \boldsymbol{b}_a\}$. $\boldsymbol{\phi}_{c_k b_t}$ denotes the so(3) Lie algebra of the rotation matrix $\boldsymbol{R}_{c_k b_t}$ s.t. $\boldsymbol{R}_{c_k b_t} = exp([\boldsymbol{\phi}_{c_k b_t}]^\wedge)$, where $[\cdot]^\wedge$ denotes the operation from a so(3) vector to the corresponding skew symmetric matrix. To facilitate the derivation of the propagation process, we further separate the state into the nominal states denoted by $\bar{(\cdot)}$, and the error states $\delta\boldsymbol{x}_{b_t} = [\delta\boldsymbol{\phi}^T_{c_k b_t}, \delta\boldsymbol{p}^T_{c_k b_t}, \delta\boldsymbol{v}^{c_k T}, \delta\boldsymbol{g}^{c_k T}, \delta\boldsymbol{b}_w^{b_t T}, \delta\boldsymbol{b}_a^{b_t T}]^T$, such that:

$$\boldsymbol{R}_{c_k b_t} = \bar{\boldsymbol{R}}_{c_k b_t} exp([\delta\boldsymbol{\phi}_{c_k b_t}]^\wedge), \quad \boldsymbol{p}_{c_k b_t} = \bar{\boldsymbol{p}}_{c_k b_t} + \delta\boldsymbol{p}_{c_k b_t}, \tag{9}$$

$$\boldsymbol{v}^{c_k} = \bar{\boldsymbol{v}}^{c_k} + \delta\boldsymbol{v}^{c_k}, \quad \boldsymbol{g}^{c_k} = \bar{\boldsymbol{g}}^{c_k} + \delta\boldsymbol{g}^{c_k}, \tag{10}$$

$$\boldsymbol{b}_w^{b_t} = \bar{\boldsymbol{b}_w}^{b_t} + \delta\boldsymbol{b}_w^{b_t}, \quad \boldsymbol{b}_a^{b_t} = \bar{\boldsymbol{b}_a}^{b_t} + \delta\boldsymbol{b}_a^{b_t}. \tag{11}$$

The nominal states can be computed using the preintegration terms, while the error states are used for propagating the covariances. It is noteworthy that the state transition model of $\delta\boldsymbol{x}_{b_t}$ is non-linear, which prevents a naive use of the Kalman filter. EKF addresses this problem and performs propagation by linearizing the state transition model at each time step using the first-order Taylor approximation. Therefore, let $\dot{(\cdot)}$ denote the derivative w.r.t. time t, we derive the continuous-time propagation model for the error states as: $\delta\dot{\boldsymbol{x}}_{b_t} = \boldsymbol{F}\delta\boldsymbol{x}_{b_t} + \boldsymbol{G}\boldsymbol{n}$. Detailed derivations are given in the Supplementary material, and $\boldsymbol{F}$ and $\boldsymbol{G}$ read:

$$\boldsymbol{F} = \begin{bmatrix} -[\bar{\boldsymbol{w}}^{b_t}]^\wedge & 0 & 0 & 0 & -\boldsymbol{I}_3 & 0 \\ 0 & 0 & \boldsymbol{I}_3 & 0 & 0 & 0 \\ -\bar{\boldsymbol{R}}_{c_k b_t}[\bar{\boldsymbol{R}}^T_{c_k b_t}\bar{\boldsymbol{g}}^{c_k} + \bar{\boldsymbol{a}}^{b_t}]^\wedge & 0 & 0 & -\boldsymbol{I}_3 & 0 & -\bar{\boldsymbol{R}}_{c_k b_t} \\ 0 & 0 & 0 & 0 & 0 & 0 \\ 0 & 0 & 0 & 0 & 0 & 0 \\ 0 & 0 & 0 & 0 & 0 & 0 \end{bmatrix}, \boldsymbol{G} = \begin{bmatrix} -\boldsymbol{I}_3 & 0 & 0 & 0 \\ 0 & 0 & 0 & 0 \\ 0 & 0 & -\bar{\boldsymbol{R}}_{c_k b_t} & 0 \\ 0 & 0 & 0 & 0 \\ 0 & \boldsymbol{I}_3 & 0 & 0 \\ 0 & 0 & 0 & \boldsymbol{I}_3 \end{bmatrix} \tag{12}$$

where $\bar{\boldsymbol{w}}^{b_t} = \boldsymbol{w}_m^{b_t} - \bar{\boldsymbol{b}_w}^{b_t}$ and $\bar{\boldsymbol{a}}^{b_t} = \boldsymbol{a}_m^{b_t} - \bar{\boldsymbol{R}}^T_{c_k b_t}\bar{\boldsymbol{g}}_{c_k} - \bar{\boldsymbol{b}_a}^{b_t}$. Given the continuous error propagation model and the initial condition $\boldsymbol{\Phi}_{t_\tau, t_\tau} = \boldsymbol{I}_{18}$, the discrete state-transition matrix $\boldsymbol{\Phi}_{(t_{\tau+1}, t_\tau)}$ can be found by solving $\dot{\boldsymbol{\Phi}}_{(t_{\tau+1}, t_\tau)} = \boldsymbol{F}_{t_{\tau+1}}\boldsymbol{\Phi}_{(t_{\tau+1}, t_\tau)}$:

$$\boldsymbol{\Phi}_{t_{\tau+1}, t_\tau} = exp(\int_{t_\tau}^{t_{\tau+1}} \boldsymbol{F}(s)\mathrm{d}s) \approx \boldsymbol{I}_{18} + \boldsymbol{F}\delta t + \frac{1}{2}\boldsymbol{F}^2\delta t^2, \quad \delta t = t_{\tau+1} - t_\tau. \tag{13}$$

Let $\check{\boldsymbol{P}}$ and $\hat{\boldsymbol{P}}$ denote the prior and posterior covariance estimates during propagation and after an update given new observations. Then we have

$$\check{\boldsymbol{P}}_{t_{\tau+1}} = \boldsymbol{\Phi}_{t_{\tau+1},t_\tau} \check{\boldsymbol{P}}_{t_\tau} \boldsymbol{\Phi}^T_{t_{\tau+1},t_\tau} + \boldsymbol{Q}_{t_\tau}, \tag{14}$$

$$\boldsymbol{Q}_{t_\tau} = \int_{t_\tau}^{t_{\tau+1}} \boldsymbol{\Phi}_{s,t_\tau} \boldsymbol{G}\boldsymbol{Q}\boldsymbol{G}^T \boldsymbol{\Phi}^T_{s,t_\tau} \mathrm{d}s \approx \boldsymbol{\Phi}_{t_{\tau+1},t_\tau} \boldsymbol{G}\boldsymbol{Q}\boldsymbol{G}^T \boldsymbol{\Phi}^T_{t_{\tau+1},t_\tau} \delta t, \tag{15}$$

where $\boldsymbol{Q} = \mathcal{D}([\sigma_w^2 \boldsymbol{I}_3, \sigma_{b_w}^2 \boldsymbol{I}_3, \sigma_a^2 \boldsymbol{I}_3, \sigma_{b_a}^2 \boldsymbol{I}_3])$. $\mathcal{D}$ is the diagonalization function.

EKF Update: In general, given an observation measurement $\boldsymbol{\xi}_{k+1}$ and its corresponding covariance $\boldsymbol{\Gamma}_{k+1}$ from the camera sensor at time t_{k+1}, we assume the following observation model: $\boldsymbol{\xi}_{k+1} = h(\boldsymbol{x}_{k+1}) + \boldsymbol{n}_r$, $\boldsymbol{n}_r \sim N(0, \boldsymbol{\Gamma}_{k+1})$.

Let $\boldsymbol{H}_{k+1} = \frac{\partial h(\boldsymbol{x}_{k+1})}{\partial \delta \boldsymbol{x}_{k+1}}$. Then the EKF update applies as following:

$$\boldsymbol{K}_{k+1} = \check{\boldsymbol{P}}_{k+1} \boldsymbol{H}^T_{k+1} (\boldsymbol{H}_{k+1} \check{\boldsymbol{P}}_{k+1} \boldsymbol{H}^T_{k+1} + \boldsymbol{\Gamma}_{k+1})^{-1}, \tag{16}$$

$$\hat{\boldsymbol{P}}_{k+1} = (\boldsymbol{I}_{18} - \boldsymbol{K}_{k+1} \boldsymbol{H}_{k+1}) \check{\boldsymbol{P}}_{k+1}, \tag{17}$$

$$\delta \hat{\boldsymbol{x}}_{k+1} = \boldsymbol{K}_{k+1} (\boldsymbol{\xi}_{k+1} - h(\check{\boldsymbol{x}}_{k+1})). \tag{18}$$

In DynaDepth, the observation measurement is defined as the ego-motion predicted by $\mathcal{M}_p$, i.e., $\boldsymbol{\xi}_{k+1} = [\tilde{\boldsymbol{\phi}}^T_{c_k c_{k+1}}, \tilde{\boldsymbol{p}}^T_{c_k c_{k+1}}]^T$. Of note is that the covariances $\boldsymbol{\Gamma}_{k+1}$ of $\{\tilde{\boldsymbol{\phi}}^T_{c_k c_{k+1}}, \tilde{\boldsymbol{p}}^T_{c_k c_{k+1}}\}$ are also predicted by the ego-motion network $\mathcal{M}_p$. To finish the camera-centric EKF update step, we derive $h(\check{\boldsymbol{x}}_{k+1})$ and $\boldsymbol{H}_{k+1}$ as:

$$h(\check{\boldsymbol{x}}_{k+1}) = \begin{bmatrix} \bar{\boldsymbol{\phi}}_{c_k c_{k+1}} \\ \bar{\boldsymbol{R}}_{c_k b_{k+1}} \boldsymbol{p}_{bc} + \bar{\boldsymbol{p}}_{c_k b_{k+1}} \end{bmatrix}, \boldsymbol{H}_{k+1} = \begin{bmatrix} J_l(-\bar{\boldsymbol{\phi}}_{c_k c_{k+1}})^{-1} \boldsymbol{R}_{cb} & 0 & 0\,0\,0\,0 \\ -\bar{\boldsymbol{R}}_{c_k b_{k+1}} [\boldsymbol{p}_{bc}]^\wedge & \boldsymbol{I}_3 & 0\,0\,0\,0 \end{bmatrix}. \tag{19}$$

After obtaining the updated error states $\delta \hat{\boldsymbol{x}}_{k+1}$, we add $\delta \hat{\boldsymbol{x}}_{k+1}$ back to the accumulated nominal states to get the corrected ego-motion. In detail, $\delta \hat{\boldsymbol{x}}_{k+1}$ is obtained by inserting Eq. (19) into Eq. (16–18), which can be inserted into Eq. (9) to get the updated $\{\hat{\boldsymbol{\phi}}_{c_k b_{k+1}}, \hat{\boldsymbol{p}}_{c_k b_{k+1}}\}$. Then by projecting $\{\hat{\boldsymbol{\phi}}_{c_k b_{k+1}}, \hat{\boldsymbol{p}}_{c_k b_{k+1}}\}$ using the camera intrinsics, we obtain the corrected ego-motion $\{\hat{\boldsymbol{\phi}}_{c_k b_{k+1}}, \hat{\boldsymbol{p}}_{c_k b_{k+1}}\}$ that fuses IMU and camera information based on their covariances as confidence indicators, which are used to compute L^{IMU}_{photo} and L^{cons}_{photo}.

Finally, in addition to $\{L^{IMU}_{photo}, L^{cons}_{photo}\}$, the total training loss L_{total} in DynaDepth also includes the vision-based photometric loss L^{vis}_{photo} and the disparity smoothness loss L_s as proposed in monodepth2 [14] to leverage the visual clues. We also consider the weak L2-norm loss L_{vg} for the velocity and gravity predictions from $\mathcal{M}_v$ and $\mathcal{M}_g$. In summary, L_{total} reads:

$$L_{total} = L^{vis}_{photo} + \lambda_1 L_s + \lambda_2 L^{IMU}_{photo} + \lambda_3 L^{cons}_{photo} + \lambda_4 L_{vg}, \tag{20}$$

where $\{\lambda_1, \lambda_2, \lambda_3, \lambda_4\}$ denote the loss weights which are determined empirically.

Remark: Alhough we have witnessed a paradigm shift from EKF to optimization in classical visual-inertial SLAM systems in recent years [22,28,31], we argue that in the setting of unsupervised depth estimation, EKF provides a better choice than optimization. The major problem of EKF is its limited ability to handle long-term data because of the Markov assumption between updates, the first-order approximation for the non-linear state-transition and observation models, and the memory consumption for storing the covariances. However, in our setting, short-term image clips are usually used as the basic training unit, which indicates that the Markov property and the linearization in EKF will approximately hold within the short time intervals. In addition, only the ego-motions predicted by $\mathcal{M}_p$ are used as the visual measurements, which is memory-efficient.

On the other hand, by using EKF, we are able to correct the IMU preintegrated ego-motions and update $\{L_{photo}^{IMU}, L_{photo}^{cons}\}$ accordingly when observing new visual measurements. Compared with formulating the commonly-used optimization objective, i.e., the residues of the IMU preintegration terms, as the training losses, our proposed L_{photo}^{IMU} and L_{photo}^{cons} provide denser supervision for both $\mathcal{M}_d$ and $\mathcal{M}_p$. From another perspective, EKF essentially can be regarded as weighting the ego-motions from IMU and vision based on their covariances, and thus naturally provides a framework for estimating the uncertainty of the ego-motion predicted by $\mathcal{M}_p$, which is non-trivial for the unsupervised learning frameworks.

Table 1. Per-image rescaled depth evaluation on KITTI using the Eigen split. The best and the second best results are shown in **bold** and underline. † denotes our reproduced results. Results are rescaled using the median ground-truth from Lidar. The means and standard errors of the scaling ratios are reported in Scale.

Methods	Year	Scale	Error↓				Accuracy↑		
			AbsRel	SqRel	RMSE	$RMSE_{log}$	$\sigma < 1.25$	$\sigma < 1.25^2$	$\sigma < 1.25^3$
Monodepth2 R18 [14]	ICCV 2019	NA	0.112	0.851	4.754	0.190	0.881	0.960	0.981
Monodepth2 R50† [14]	ICCV 2019	29.128 ± 0.084	0.111	0.806	4.642	0.189	0.882	**0.962**	**0.982**
PackNet-SfM [15]	CVPR 2020	NA	0.111	0.785	**4.601**	0.189	0.878	0.960	**0.982**
Johnston R18 [18]	CVPR 2020	NA	0.111	0.941	4.817	0.189	**0.885**	0.961	0.981
R-MSFM6 [49]	ICCV 2021	NA	0.112	0.806	4.704	0.191	0.878	0.960	0.981
G2S R50 [3]	ICRA 2021	1.031 ± 0.073	0.112	0.894	4.852	0.192	0.877	0.958	0.981
ScaleInvariant R18 [38]	ICCV 2021	NA	0.109	0.779	4.641	**0.186**	0.883	**0.962**	**0.982**
DynaDepth R18	2022	1.021 ± **0.069**	0.111	0.806	4.777	0.190	0.878	0.960	**0.982**
DynaDepth R50	2022	**1.013** ± 0.071	**0.108**	**0.761**	4.608	0.187	0.883	**0.962**	**0.982**

4 Experiment

We evaluate the effectiveness of DynaDepth on KITTI [12] and test the generalization ability on Make3D [33]. In addition, we perform extensive ablation studies on our proposed IMU losses, the EKF framework, the learnt ego-motion uncertainty, and the robustness against illumination change and moving objects.

4.1 Implementation

DynaDepth is implemented in pytorch [30]. We adopt the monodepth2 [14] network structures for $\{\mathcal{M}_d, \mathcal{M}_p\}$, except that we increase the output dimension of $\mathcal{M}_p$ from 6 to 12 to include the uncertainty predictions. $\{\mathcal{M}_g, \mathcal{M}_v\}$ share the same network structure as $\mathcal{M}_p$ except that the output dimensions are both set to 3. $\{\lambda_1, \lambda_2, \lambda_3, \lambda_4\}$ are set to $\{0.001, 0.5, 0.01, 0.001\}$. We train all networks for 30 epochs using an initial learning rate 1e-4, which is reduced to 1e-5 after the first 15 epochs. The training process takes $1 \sim 2$ days on a single NVIDIA V100 GPU. The source codes and the trained models will be released.

4.2 Scale-Aware Depth Estimation on KITTI

We use the Eigen split [6] for depth evaluation. In addition to the removal of static frames as proposed in [48], we discard images without the corresponding IMU records, leading to 38,102 image-and-IMU triplets for training and 4,238 for validation. WLOG, we use the image resolution 640×192 and cap the depth predictions at 80m, following the common practice in [3,14,15,18,38].

We compare DynaDepth with state-of-the-art monocular depth estimation methods in Table 1, which rescale the results using the ratio of the median depth between the ground-truth and the prediction. For a fair comparison, we only present results achieved with image resolution 640×192 and an encoder with moderate size, i.e., ResNet18 (R18) or ResNet50 (R50). In addition to standard depth evaluation metrics [7], we report the means and standard errors of the rescaling factors to demonstrate the scale-awareness ability. DynaDepth achieves the best up-to-scale performance w.r.t. four metrics and achieves the second best for the other three metrics. Of note is that DynaDepth also achieves a nearly perfect absolute scale. In terms of scale-awareness, even our R18 version outperforms G2S R50 [3], which uses a heavier encoder. For better illustration, we also show the scaling ratio histograms with and without IMU in Fig. 1(b).

We then report the unscaled results in Table 2, and compare with PackNet-SfM [15] and G2S [3], which use the GPS information to construct velocity constraints. Without rescaling, Monodepth2 [14] fails completely as expected. In this case, DynaDepth achieves the best performance w.r.t. all metrics, setting a new benchmark of unscaled depth evaluation for monocular methods.

Table 2. Unscaled depth evaluation on KITTI using the Eigen split. † denotes our reproduced results. The best results are shown in **bold**.

Methods	Year	Error↓				Accuracy↑		
		AbsRel	SqRel	RMSE	RMSE$_{log}$	$\sigma < 1.25$	$\sigma < 1.25^2$	$\sigma < 1.25^3$
Monodepth2 R50† [14]	ICCV 2019	0.966	15.039	19.145	3.404	0.000	0.000	0.000
PackNet-SfM [15]	CVPR 2020	0.111	0.829	4.788	0.199	0.864	0.954	0.980
G2S R50 [3]	ICRA 2021	**0.109**	0.860	4.855	0.198	0.865	0.954	0.980
DynaDepth R50	2022	**0.109**	**0.787**	**4.705**	**0.195**	**0.869**	**0.958**	**0.981**

4.3 Generalizability on Make3D

We further test the generalizability of DynaDepth on Make3D [33] using models trained on KITTI [12]. The test images are centre-cropped to a 2×1 ratio for a fair comparison with previous methods [14]. A qualitative example is given in Fig. 1(c), where the model without IMU fails in the glass and shadow areas, while our model achieves a distinguishable prediction. Quantitative results are reported in Table 3. A reasonably good scaling ratio has been achieved for DynaDepth, indicating that the scale-awareness learnt by DynaDepth can be well generalized to unseen datasets. Surprisingly, we found that DynaDepth that only uses the gyroscope and accelerator IMU information (w.o/ L_{vg}) achieves the best generalization results. The reason can be two-fold. First, our full model may overfit to the KITTI dataset due to the increased modeling capacity. Second, the performance degradation can be due to the domain gap of the visual data, since both $\mathcal{M}_v$ and $\mathcal{M}_g$ take images as input. This also explains the scale loss of G2S in this case. We further show that DynaDepth w.o/ L_{vg} significantly outperforms the stereo version of Monodepth2, which can also be explained by the visual domain gap, especially the different camera intrinsics used in their left-right consistency loss. Our generalizability experiment justifies the advantages of using IMU to provide scale information, which will not be affected by the visual domain gap and varied camera parameters, leading to improved generalization performance. In addition, it is also shown that the use of EKF in training significantly improves the generalization ability, possibly thanks to the EKF fusion framework that takes the uncertainty into account and integrates the generalizable IMU motion dynamics and the domain-specific vision information in a more reasonable way.

Table 3. Generalization results on Make3D. * denotes unscaled results while the others present per-image rescaled results. The best results are shown in **bold**. M, S, GPS, and IMU in Type denote whether monocular, stereo, GPS and IMU information are used for training the model on KITTI. – means item not available.

Methods	L_{vg}	EKF	Type	Scale	Error↓				Accuracy↑		
					Abs_{rel}	Sq_{rel}	RMSE	$RMSE_{log}$	$\sigma < 1.25$	$\sigma < 1.25^2$	$\sigma < 1.25^3$
Zhou [48]	–	–	M	–	0.383	5.321	10.470	0.478	–	–	–
Monodepth2 [14]	–	–	M	–	0.322	3.589	7.417	0.163	–	–	–
G2S [3]	–	–	M+GPS	2.81 ± 0.85	–	–	–	–	–	–	–
DynaDepth			M+IMU	1.37 ± 0.27	0.316	3.006	7.218	0.164	0.522	0.797	0.914
DynaDepth		✓	M+IMU	$\mathbf{1.26} \pm 0.27$	**0.313**	**2.878**	**7.133**	**0.162**	**0.527**	**0.800**	**0.916**
DynaDepth (full)	✓	✓	M+IMU	$1.45 \pm \mathbf{0.26}$	0.334	3.311	7.463	0.169	0.497	0.779	0.908
Monodepth2* [14]	–	–	M+S	–	0.374	3.792	8.238	**0.201**	–	–	–
DynaDepth*			M+IMU	–	0.360	3.461	8.833	0.226	0.295	0.594	0.794
DynaDepth*		✓	M+IMU	–	**0.337**	**3.135**	**8.217**	**0.201**	**0.384**	**0.671**	**0.845**
DynaDepth* (full)	✓	✓	M+IMU	–	0.378	3.655	9.034	0.240	0.261	0.550	0.758

4.4 Ablation Studies

We conduct ablation studies on KITTI to investigate the effects of the proposed IMU-related losses, the EKF fusion framework, and the learnt ego-motion uncertainty. In addition, we design simulated experiment to demonstrate the robustness of DynaDepth against vision degradation such as illumination change and moving objects. WLOG, we use ResNet18 as the encoder for all ablation studies.

The Effects of the IMU-related Losses and the EKF Fusion Framework. We report the ablation results of the IMU-related losses and the EKF fusion framework in Table 4. First, L_{photo}^{IMU} presents the main contributor to learning the scale. However, only a rough scale is learnt using L_{photo}^{IMU} only. And the up-to-scale accuracy is also not as good as the other models. L_{photo}^{cons} provides better up-to-scale accuracy, but using L_{photo}^{cons} alone is not enough to learn the absolute scale due to the relatively weak supervision. Instead, combining L_{photo}^{IMU} and L_{photo}^{cons} together boosts the performance of both the scale-awareness and the accuracy. The use of L_{vg} further enhances the evaluation results. Nevertheless, as shown in Sect. 4.3, L_{vg} may lead to overfitting to current dataset and harm the generalizability, due to its dependence on visual data that suffers from the visual domain gap between different datasets. On the other hand, EKF improves the up-to-scale accuracy w.r.t. almost all metrics, while decreasing the learnt scale information a little bit. Since the scale information comes from IMU, and the visual data contributes most to the up-to-scale accuracy, EKF achieves a good balance between the two sensors. Moreover, as shown in Table 3, the use of EKF leads to the best generalization results w.r.t. both the scale and the accuracy.

Table 4. Ablation results of the IMU-related losses and the EKF fusion framework on KITTI. The best results are shown in **bold**.

EKF	L_{photo}^{IMU}	L_{photo}^{cons}	L_{vg}	Scale	Error↓				Accuracy↑		
					AbsRel	SqRel	RMSE	$RMSE_{log}$	$\sigma < 1.25$	$\sigma < 1.25^2$	$\sigma < 1.25^3$
✓	✓			1.130 ± 0.099	0.115	0.804	4.806	0.193	0.871	0.959	**0.982**
✓		✓		4.271 ± 0.089	0.114	0.832	4.780	0.192	0.876	0.959	0.981
✓	✓	✓		1.076 ± 0.095	0.113	**0.794**	**4.760**	0.191	0.874	**0.960**	**0.982**
✓	✓	✓	✓	**1.021 ± 0.069**	**0.111**	0.806	4.777	**0.190**	**0.878**	**0.960**	**0.982**
	✓	✓		**0.968** ± 0.098	0.115	0.839	4.898	0.194	0.869	0.958	0.981
✓	✓	✓		1.076 ± **0.095**	**0.113**	**0.794**	**4.760**	**0.191**	**0.874**	**0.960**	**0.982**
	✓	✓	✓	**1.013 ± 0.069**	0.112	0.808	**4.751**	0.191	0.877	**0.960**	**0.982**
✓	✓	✓	✓	1.021 ± **0.069**	**0.111**	**0.806**	4.777	**0.190**	**0.878**	**0.960**	**0.982**

The Robustness Against Vision Degradation. We then examine the robustness of DynaDepth against illumination change and moving objects, two major cases that violate the underlying assumption of the photometric loss. We simulate the illumination change by randomly alternating image contrast within a range 0.5. The moving objects are simulated by randomly inserting three 150×150 black squares. In contrast to data augmentation, we perform the perturbation for each image independently, rather than applying the same perturbation to all images in a triplet. Results are given in Table 5. Under illumination change, the accuracy of Monodepth2 degrades as expected, while DynaDepth rescues the accuracy to a certain degree and maintains the correct absolute scales. EKF improves almost all metrics in this case, and using both EKF and L_{vg} achieves the best scale and AbsRel. However, the model without L_{vg} obtains the best performance on most metrics. The reason may be the dependence of L_{vg} on the visual data, which is more sensitive to image qualities. When there exist moving objects, Monodepth2 fails completely. Using DynaDepth without EKF and L_{vg} improves the up-to-scale accuracy a little bit, but the results are still far from expected. Using EKF significantly improves the up-to-scale results, while it is still hard to learn the scale given the difficulty of the task. In this case, using L_{vg} is shown to provide strong scale supervision and achieve a good scale result.

The Learnt Ego-Motion Uncertainty. We illustrate the training progress of the ego-motion uncertainty in Fig. 2. We report the averaged covariance as the uncertainty measure. The learnt uncertainty exhibits a similar pattern as the depth error (AbsRel), meaning that the model becomes more certain about its predictions as the training continues. Of note is that only indirect supervision is provided, which justifies the effectiveness of our fusion framework. In addition, DynaDepth R50 achieves a lower uncertainty than R18, indicating that a larger model capacity also contributes to the prediction confidence, yet such difference can hardly be seen w.r.t. AbsRel. Table 6 presents another interesting observation. In KITTI, the axis-z denotes the forward direction. Since most test images correspond to driving forward, the magnitude of t_z is significantly larger than $\{t_x, t_y\}$. Accordingly, DynaDepth shows a high confidence on t_z, while large

Table 5. Ablation results of the robustness against vision degradation on the simulated data from KITTI. The best results are shown in **bold**. IC and MO denote the two investigated vision degradation types, i.e., illumination change and moving objects. – means item not available. † denotes our reproduced results.

Methods	EKF	L_{vg}	Type	Scale	Error↓				Accuracy↑		
					AbsRel	SqRel	RMSE	$RMSE_{log}$	$\sigma < 1.25$	$\sigma < 1.25^2$	$\sigma < 1.25^3$
Monodepth2† [14]	–	–	IC	27.701 ± 0.096	0.127	0.976	5.019	0.220	0.855	0.946	0.972
DynaDepth			IC	1.036 ± 0.099	0.124	**0.858**	4.915	0.226	0.852	0.950	0.977
DynaDepth	✓		IC	0.946 ± 0.089	0.123	0.925	**4.866**	**0.196**	**0.863**	**0.957**	**0.981**
DynaDepth	✓	✓	IC	$\mathbf{1.019 \pm 0.074}$	**0.121**	0.906	4.950	0.217	0.859	0.954	0.978
Monodepth2† [14]	–	–	MO	0.291 ± 0.176	0.257	2.493	8.670	0.398	0.584	0.801	0.897
DynaDepth			MO	0.083 ± 0.225	0.169	1.290	6.030	0.278	0.763	0.915	0.960
DynaDepth	✓		MO	0.087 ± 0.119	0.126	**0.861**	5.312	**0.210**	0.840	0.948	**0.979**
DynaDepth	✓	✓	MO	$\mathbf{0.956 \pm 0.084}$	**0.125**	0.926	**4.954**	0.214	**0.852**	**0.949**	0.976

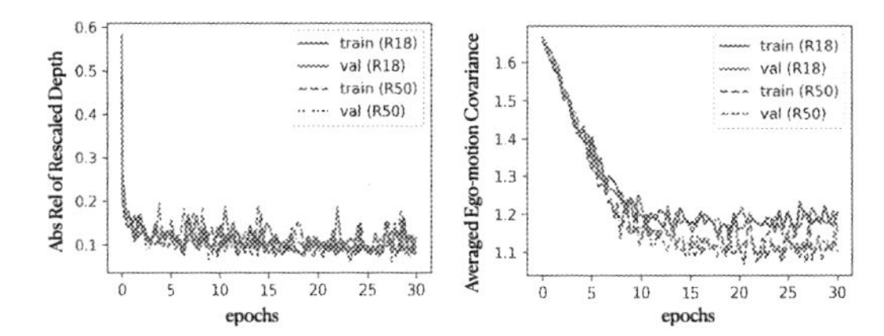

Fig. 2. The training processes w.r.t. AbsRel (left) and the averaged ego-motion covariance (right).

Table 6. The averaged magnitude $|\bar{t}|$ and the variance $\bar{\sigma}_t^2$ of the translation predictions along each axis.

	Axis-x	Axis-y	Axis-z
$\lvert\bar{t}\rvert$	0.017	0.018	0.811
$\bar{\sigma}_t^2$	7.559	5.222	0.105

variances are observed for $\{t_x, t_y\}$, potentially due to the difficulty to distinguish the noises from the small amount of translations along axis-x and axis-y.

5 Conclusion

In this paper, we propose DynaDepth, a scale-aware, robust, and generalizable monocular depth estimation framework using IMU motion dynamics. Specifically, we propose an IMU photometric loss and a cross-sensor photometric consistency loss to provide dense supervision and absolution scales. In addition, we derive a camera-centric EKF framework for the sensor fusion, which also provides an ego-motion uncertainty measure under the setting of unsupervised learning. Extensive experiments support that DynaDepth is advantageous w.r.t. learning absolute scales, the generalizability, and the robustness against vision degradation.

Acknowledgment. This work is supported by ARC FL-170100117, DP-180103424, IC-190100031, and LE-200100049.

References

1. Bhat, S.F., Alhashim, I., Wonka, P.: AdaBins: depth estimation using adaptive bins. In: Proceedings of the IEEE/CVF Conference on Computer Vision and Pattern Recognition, pp. 4009–4018 (2021)
2. Bian, J., et al.: Unsupervised scale-consistent depth and ego-motion learning from monocular video. In: Advances in Neural Information Processing Systems 32 (2019)
3. Chawla, H., Varma, A., Arani, E., Zonooz, B.: Multimodal scale consistency and awareness for monocular self-supervised depth estimation. In: 2021 IEEE International Conference on Robotics and Automation (ICRA), pp. 5140–5146. IEEE (2021)
4. Chen, C., et al.: Selective sensor fusion for neural visual-inertial odometry. In: Proceedings of the IEEE/CVF Conference on Computer Vision and Pattern Recognition, pp. 10542–10551 (2019)
5. Clark, R., Wang, S., Wen, H., Markham, A., Trigoni, N.: VINet: visual-inertial odometry as a sequence-to-sequence learning problem. In: Proceedings of the AAAI Conference on Artificial Intelligence, vol. 31 (2017)

6. Eigen, D., Fergus, R.: Predicting depth, surface normals and semantic labels with a common multi-scale convolutional architecture. In: Proceedings of the IEEE International Conference on Computer Vision, pp. 2650–2658 (2015)
7. Eigen, D., Puhrsch, C., Fergus, R.: Depth map prediction from a single image using a multi-scale deep network. In: Advances in Neural Information Processing Systems 27 (2014)
8. Engel, J., Koltun, V., Cremers, D.: Direct sparse odometry. IEEE Trans. Pattern Anal. Mach. Intell. **40**(3), 611–625 (2017)
9. Engel, J., Schöps, T., Cremers, D.: LSD-SLAM: large-scale direct monocular SLAM. In: Fleet, D., Pajdla, T., Schiele, B., Tuytelaars, T. (eds.) ECCV 2014. LNCS, vol. 8690, pp. 834–849. Springer, Cham (2014). https://doi.org/10.1007/978-3-319-10605-2_54
10. Forster, C., Carlone, L., Dellaert, F., Scaramuzza, D.: IMU preintegration on manifold for efficient visual-inertial maximum-a-posteriori estimation. Georgia Institute of Technology (2015)
11. Fu, H., Gong, M., Wang, C., Batmanghelich, K., Tao, D.: Deep ordinal regression network for monocular depth estimation. In: Proceedings of the IEEE Conference on Computer Vision and Pattern Recognition, pp. 2002–2011 (2018)
12. Geiger, A., Lenz, P., Stiller, C., Urtasun, R.: Vision meets robotics: the KITTI dataset. Int. J. Robot. Res. **32**(11), 1231–1237 (2013)
13. Godard, C., Mac Aodha, O., Brostow, G.J.: Unsupervised monocular depth estimation with left-right consistency. In: Proceedings of the IEEE Conference on Computer Vision and Pattern Recognition, pp. 270–279 (2017)
14. Godard, C., Mac Aodha, O., Firman, M., Brostow, G.J.: Digging into self-supervised monocular depth estimation. In: Proceedings of the IEEE/CVF International Conference on Computer Vision, pp. 3828–3838 (2019)
15. Guizilini, V., Ambrus, R., Pillai, S., Raventos, A., Gaidon, A.: 3D packing for self-supervised monocular depth estimation. In: Proceedings of the IEEE/CVF Conference on Computer Vision and Pattern Recognition, pp. 2485–2494 (2020)
16. Han, L., Lin, Y., Du, G., Lian, S.: DeepVIO: self-supervised deep learning of monocular visual inertial odometry using 3D geometric constraints. In: 2019 IEEE/RSJ International Conference on Intelligent Robots and Systems (IROS), pp. 6906–6913. IEEE (2019)
17. Huang, G.: Visual-inertial navigation: a concise review. In: 2019 IEEE International Conference on Robotics and Automation (ICRA), pp. 9572–9582. IEEE (2019)
18. Johnston, A., Carneiro, G.: Self-supervised monocular trained depth estimation using self-attention and discrete disparity volume. In: Proceedings of the IEEE/CVF Conference on Computer Vision and Pattern Recognition, pp. 4756–4765 (2020)
19. Jung, H., Park, E., Yoo, S.: Fine-grained semantics-aware representation enhancement for self-supervised monocular depth estimation. In: Proceedings of the IEEE/CVF International Conference on Computer Vision, pp. 12642–12652 (2021)
20. Khan, F., Salahuddin, S., Javidnia, H.: Deep learning-based monocular depth estimation methods-a state-of-the-art review. Sensors **20**(8), 2272 (2020)
21. Klodt, M., Vedaldi, A.: Supervising the new with the old: learning SFM from SFM. In: Ferrari, V., Hebert, M., Sminchisescu, C., Weiss, Y. (eds.) ECCV 2018. LNCS, vol. 11214, pp. 713–728. Springer, Cham (2018). https://doi.org/10.1007/978-3-030-01249-6_43
22. Leutenegger, S., Lynen, S., Bosse, M., Siegwart, R., Furgale, P.: Keyframe-based visual-inertial odometry using nonlinear optimization. Int. J. Robot. Res. **34**(3), 314–334 (2015)

23. Li, C., Waslander, S.L.: Towards end-to-end learning of visual inertial odometry with an EKF. In: 2020 17th Conference on Computer and Robot Vision (CRV), pp. 190–197. IEEE (2020)
24. Liu, F., Shen, C., Lin, G., Reid, I.: Learning depth from single monocular images using deep convolutional neural fields. IEEE Trans. Pattern Anal. Mach. Intell. **38**(10), 2024–2039 (2015)
25. Liu, W., et al.: TLIO: tight learned inertial odometry. IEEE Robot. Autom. Lett. **5**(4), 5653–5660 (2020)
26. Lupton, T., Sukkarieh, S.: Visual-inertial-aided navigation for high-dynamic motion in built environments without initial conditions. IEEE Trans. Robot. **28**(1), 61–76 (2011)
27. Mahjourian, R., Wicke, M., Angelova, A.: Unsupervised learning of depth and ego-motion from monocular video using 3D geometric constraints. In: Proceedings of the IEEE Conference on Computer Vision and Pattern Recognition, pp. 5667–5675 (2018)
28. Mourikis, A.I., Roumeliotis, S.I., et al.: A multi-state constraint Kalman filter for vision-aided inertial navigation. In: 2007 IEEE International Conference on Robotics and Automation (ICRA), vol. 2, p. 6 (2007)
29. Mur-Artal, R., Montiel, J.M.M., Tardos, J.D.: ORB-SLAM: a versatile and accurate monocular SLAM system. IEEE Trans. Robot. **31**(5), 1147–1163 (2015)
30. Paszke, A., et al.: PyTorch: an imperative style, high-performance deep learning library. In: Advances in Neural Information Processing Systems 32 (2019)
31. Qin, T., Li, P., Shen, S.: VINS-Mono: a robust and versatile monocular visual-inertial state estimator. IEEE Trans. Robot. **34**(4), 1004–1020 (2018)
32. Ranjan, A., et al.: Competitive collaboration: joint unsupervised learning of depth, camera motion, optical flow and motion segmentation. In: Proceedings of the IEEE/CVF Conference on Computer Vision and Pattern Recognition, pp. 12240–12249 (2019)
33. Saxena, A., Sun, M., Ng, A.Y.: Make3D: learning 3D scene structure from a single still image. IEEE Trans. Pattern Anal. Mach. Intell. **31**(5), 824–840 (2008)
34. Shamwell, E.J., Lindgren, K., Leung, S., Nothwang, W.D.: Unsupervised deep visual-inertial odometry with online error correction for RGB-D imagery. IEEE Trans. Pattern Anal. Mach. Intell. **42**(10), 2478–2493 (2019)
35. Shu, C., Yu, K., Duan, Z., Yang, K.: Feature-metric loss for self-supervised learning of depth and egomotion. In: Vedaldi, A., Bischof, H., Brox, T., Frahm, J.-M. (eds.) ECCV 2020. LNCS, vol. 12364, pp. 572–588. Springer, Cham (2020). https://doi.org/10.1007/978-3-030-58529-7_34
36. Taketomi, T., Uchiyama, H., Ikeda, S.: Visual SLAM algorithms: a survey from 2010 to 2016. IPSJ Trans. Comput. Vis. Appl. **9**(1), 1–11 (2017)
37. Wagstaff, B., Wise, E., Kelly, J.: A self-supervised, differentiable Kalman filter for uncertainty-aware visual-inertial odometry. In: IEEE/ASME International Conference on Advanced Intelligent Mechatronics (2022)
38. Wang, L., Wang, Y., Wang, L., Zhan, Y., Wang, Y., Lu, H.: Can scale-consistent monocular depth be learned in a self-supervised scale-invariant manner? In: Proceedings of the IEEE/CVF International Conference on Computer Vision, pp. 12727–12736 (2021)
39. Wang, Z., Bovik, A.C., Sheikh, H.R., Simoncelli, E.P.: Image quality assessment: from error visibility to structural similarity. IEEE Trans. Image Proces. **13**(4), 600–612 (2004)

40. Wei, P., Hua, G., Huang, W., Meng, F., Liu, H.: Unsupervised monocular visual-inertial odometry network. In: Proceedings of the Twenty-Ninth International Conference on International Joint Conferences on Artificial Intelligence, pp. 2347–2354 (2021)
41. Yang, N., Stumberg, L.v., Wang, R., Cremers, D.: D3VO: deep depth, deep pose and deep uncertainty for monocular visual odometry. In: Proceedings of the IEEE/CVF Conference on Computer Vision and Pattern Recognition, pp. 1281–1292 (2020)
42. Yin, Z., Shi, J.: Geonet: Unsupervised learning of dense depth, optical flow and camera pose. In: Proceedings of the IEEE Conference on Computer Vision and Pattern Recognition, pp. 1983–1992 (2018)
43. Zhan, H., Weerasekera, C.S., Bian, J.W., Reid, I.: Visual odometry revisited: What should be learnt? In: 2020 IEEE International Conference on Robotics and Automation (ICRA), pp. 4203–4210. IEEE (2020)
44. Zhang, J., Tao, D.: Empowering things with intelligence: a survey of the progress, challenges, and opportunities in artificial intelligence of things. IEEE Internet Things J. **8**(10), 7789–7817 (2020)
45. Zhang, S., Zhang, J., Tao, D.: Information-theoretic odometry learning. arXiv preprint arXiv:2203.05724 (2022)
46. Zhang, S., Zhang, J., Tao, D.: Towards scale consistent monocular visual odometry by learning from the virtual world. arXiv preprint arXiv:2203.05712 (2022)
47. Zhao, W., Liu, S., Shu, Y., Liu, Y.J.: Towards better generalization: joint depth-pose learning without PoseNet. In: Proceedings of the IEEE/CVF Conference on Computer Vision and Pattern Recognition, pp. 9151–9161 (2020)
48. Zhou, T., Brown, M., Snavely, N., Lowe, D.G.: Unsupervised learning of depth and ego-motion from video. In: Proceedings of the IEEE Conference on Computer Vision and Pattern Recognition, pp. 1851–1858 (2017)
49. Zhou, Z., Fan, X., Shi, P., Xin, Y.: R-MSFM: recurrent multi-scale feature modulation for monocular depth estimating. In: Proceedings of the IEEE/CVF International Conference on Computer Vision, pp. 12777–12786 (2021)

TIPS: Text-Induced Pose Synthesis

Prasun Roy[1(✉)], Subhankar Ghosh[1], Saumik Bhattacharya[2], Umapada Pal[3], and Michael Blumenstein[1]

[1] University of Technology Sydney, Ultimo, Australia
{prasun.roy,subhankar.ghosh}@student.uts.edu.au, michael.blumenstein@uts.edu.au
[2] Indian Institute of Technology Kharagpur, Kharagpur, India
saumik@ece.iitkgp.ac.in
[3] Indian Statistical Institute Kolkata, Kolkata, India
umapada@isical.ac.in
https://prasunroy.github.io/tips

Abstract. In computer vision, human pose synthesis and transfer deal with probabilistic image generation of a person in a previously unseen pose from an already available observation of that person. Though researchers have recently proposed several methods to achieve this task, most of these techniques derive the target pose directly from the desired target image on a specific dataset, making the underlying process challenging to apply in real-world scenarios as the generation of the target image is the actual aim. In this paper, we first present the shortcomings of current pose transfer algorithms and then propose a novel text-based pose transfer technique to address those issues. We divide the problem into three independent stages: (a) text to pose representation, (b) pose refinement, and (c) pose rendering. To the best of our knowledge, this is one of the first attempts to develop a text-based pose transfer framework where we also introduce a new dataset DF-PASS, by adding descriptive pose annotations for the images of the DeepFashion dataset. The proposed method generates promising results with significant qualitative and quantitative scores in our experiments.

Keywords: Text-guided generation · Pose transfer · GAN · DeepFashion

1 Introduction

Generating novel views of a given object is a challenging yet necessary task for many computer vision applications. Pose transfer is a subclass of the view synthesis problem where the goal is to estimate an unseen view (*target* image) of a person with a particular pose from a given observation (*source* image)

Supplementary Information The online version contains supplementary material available at https://doi.org/10.1007/978-3-031-19839-7_10.

S. Avidan et al. (Eds.): ECCV 2022, LNCS 13698, pp. 161–178, 2022.
https://doi.org/10.1007/978-3-031-19839-7_10

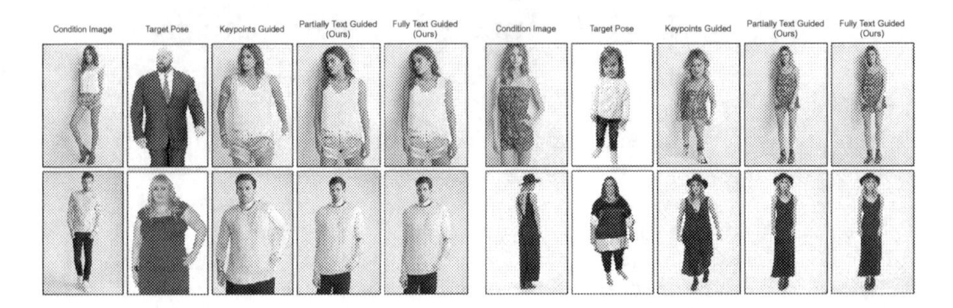

Fig. 1. Overview of the proposed approach. Keypoint-guided methods tend to produce structurally inconsistent images when the physical appearance of the target pose reference significantly differs from the condition image. The proposed text-guided technique successfully addresses this issue while retaining the ability to generate visually decent results close to the keypoint-guided baseline.

of that person. As there can be significant differences between the source and target images, the pose transfer pipeline requires a very accurate generative algorithm to infer both the visible and occluded body parts in the target image. The method also needs to preserve the person's general appearance, including facial expression, skin color, attire, and background. In particular, the goal is to generate a target person image I_B for a specific pose P_B from an input source image I_A of that person having an observed pose P_A. A human pose P is usually expressed by a set of body-joint locations (*keypoints*), denoted as K. As the location of the keypoints can vary significantly from person to person, two different sets of keypoints K and K' may represent the same pose P.

As initial solutions, researchers have introduced coarse to fine generation schemes [28,29] by splitting the problem into separate sub-tasks for handling background, foreground, and pose separately. The architectural complexity of such an approach is later streamlined with a unified pipeline by utilizing deformable GANs [42], variational U-Net [9], and progressive attention transfer [55]. Although the state-of-the-art (SOTA) algorithms have produced visually compelling results, a common yet noticeable flaw is present in these techniques. For training and evaluation of the models, SOTA algorithms extract keypoints K_B directly from I_B to represent P_B and use it as one of the inputs. However, I_B should not be ideally known to users, and such an over-simplified training process creates a dilemma. One way to circumvent the problem is training the model to adapt to a target pose P_B, represented by keypoints K'_B, which is extracted from the image I'_B of some other person. However, as the models are trained using K_B directly, they adapt poorly to any other set of keypoints $K'_B \neq K_B$ representing the same pose P_B. In Fig. 1, we have shown the limitation of the existing keypoint-based models. We use the existing keypoint-guided pose transfer algorithm PATN [55] as a baseline in our experiments. The keypoint-based models try to follow the body structure of the target reference rather than the general pose. Thus, they fail occasionally in the absence of the target image

I_B to provide the keypoints. On the other hand, the proposed algorithm is not biased toward the target image as it exclusively works on the textual description of the target pose.

In this paper, we propose a novel pose-transfer pipeline guided by the textual description of the pose. Initially, we estimate the target keypoint set K_B from the textual description T_B of the target pose P_B. The estimated keypoint set K_B is then used to generate the pose-transferred image $\tilde{I}_B$. As the estimation of K_B is directly conditioned on T_B, we do not need the target image I_B for estimating the target pose P_B, and the training is free from any bias. The main contributions of our work are as follows.

- We propose a pose transfer pipeline that takes the source image and a textual description of the target pose to generate the target image. To the best of our knowledge, this is one of the first attempts to design a pose transfer algorithm based on textual descriptions of the pose.
- We introduce a new dataset DF-PASS derived from the DeepFashion dataset. The proposed dataset contains a human-annotated text description of the pose for 40488 images of the DeepFashion dataset.
- We extensively explore different perceptual metrics to analyze the performance of the proposed technique and introduce a new metric (GCR) for evaluating the gender consistency in the generated images.
- Most importantly, the algorithm is designed not to require the target image at the time of inference, making it more suitable for real-world applications than the existing pose transfer algorithms.

2 Related Work

Novel view synthesis is an intriguing problem in computer vision. Recently, Generative Adversarial Networks (GANs) [10] have been explored extensively for perceptually realistic image generation [10,17,20,21,31,36]. Conditional generative models [16,31,41,54] have become popular in different fields of computer vision, such as inpainting [46], super-resolution [8,18] etc. Pose transfer can be viewed as a sub-category of the conditional generation task where a target image is generated from a source image by conditioning on the target pose. Thus, with the progress of conditional generative models, pose transfer algorithms have significantly enhanced performance in the last decade. Initial multi-stage approaches divide the complex task into relatively simpler sub-problems. In [50], Zhao *et al.* adopt a coarse to fine approach to generate multi-view images of a person from a single observation. Ma *et al.* [28,29] introduce a multi-stage framework to generate the final pose-transferred image from a single source image. Balakrishnan *et al.* [3] propose a method of pose transfer by segmenting and generating the foreground and background individually. Wang *et al.* [44] introduce a characteristic preserving generative network with a geometric matching module. The coarse to fine generation technique is further improved by incorporating the idea of disentanglement [29] where the generative model is designed as a multi-branch network to handle foreground, background, and pose separately. In [34],

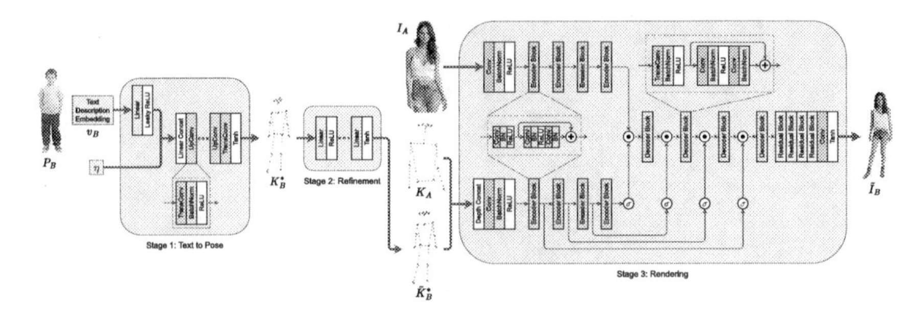

Fig. 2. Architecture of the proposed pipeline. The workflow is divided into three stages. In stage 1, we estimate a spatial representation K_B^* for the target pose P_B from the corresponding text description embedding v_B. In stage 2, we regressively refine the initial estimation of the facial keypoints to obtain the refined target keypoints $\tilde{K}_B^*$. Finally, in stage 3, we render the target image $\tilde{I}_B$ by conditioning the pose transfer on the source image I_A having the keypoints K_A corresponding to the source pose P_A.

the authors propose a pose conditioned bidirectional generator in an unsupervised multi-level generation strategy. In [55], the authors introduce a progressive attention transfer technique to transfer the pose gradually. Li *et al.* [22] propose a method to progressively select important regions of an image using pose-guided non-local attention with a long-range dependency. Researchers have also investigated 3D appearance flow [23], pose flow [51], and surface-based modeling [11,33] for pose transfer. In [47], the authors first approximate a 3D mesh from a single image, and then the 3D mesh is used to transfer the pose. Siarohin *et al.* [42] propose a nearest neighbour loss for pose transfer using deformable GANs. In [6,53], the authors generate talking-face dynamics from a single face image and a target audio signal.

Text-based image generations are also an intriguing topic in computer vision. In [37], the authors propose a GAN-based architecture for synthesizing the images. Qiao *et al.* [35] have incorporated redescription of textual descriptions for image synthesis. Recently, text-based approaches are also explored for generating human pose [4] and appearance [49]. In [24], the authors use a Variational Autoencoder (VAE) to generate human actions from text descriptions. In [4], the authors generate 3D human meshes from text using a recurrent GAN and SMPL [27] model. In [52], the authors propose a text-guided method for generating human images by selecting a pose from a set of eight basic poses, followed by controlling the appearance attributes of the selected basic pose. However, text-based visual generation techniques are limited in the literature, and text-guided pose transfer is not well-explored previously to the best of our knowledge.

3 Methodology

The proposed technique is divided into three independent sequential stages, each specific to a particular task. In the first stage, we derive an initial estimation of the target pose from the corresponding text description embedding. This coarse pose is then refined through regression at the next step. Finally, pose transfer is performed by conditioning the transformation on the appearance of the source image. We show our integrated generation pipeline in Fig. 2.

3.1 Text to Keypoints Generation

For a given source image I_A, our algorithm aims to generate the pose-transferred image I_B where the target pose P_B is described by textual description T_B. At first, we encode T_B into an embedded vector v_B either by many-hot encoding or using a pre-trained NLP model such as BERT [7], FastText [2], or Word2Vec [30]. We first aim to estimate the keypoint set K_B from v_B using a generative model to guide the pose transfer process in a later stage. To train such a generative model, we represent the keypoints $k_j \in \mathbb{R}^{m \times n}$ where $k_j \in K; \forall j$ and the domain of both I_A and I_B is $\mathbb{R}^{m \times n}$. As a slight spatial variation of k_j does not change the pose P_B, it is better to represent it with a Gaussian distribution $\mathcal{N}(k_j, \sigma_j)$; $\forall j$ for mitigating the high sparsity in the data. Although for different k_j, the invariance of the pose is valid for different amounts of spatial perturbations, we can assume $\sigma_j = \sigma$, a constant, $\forall j$, if σ_j is small. Such representation of keypoints is often referred to as heatmaps.

Taking motivation from [49], we design a generative adversarial network to estimate the target keypoint set K_B from the text embedding v_B. In our generator G_T, we first project v_B into a 128-dimensional latent space ϕ_B using a linear layer with leaky ReLU activation. To allow some structural variations in the generated poses, we sample a 128-dimensional noise vector $\eta \sim N(\mathbf{0}, I)$, where I is a 128×128 identity matrix. Both ϕ_B and η are linearly concatenated and passed through 4 up-convolution blocks. At each block, we perform a transposed convolution followed by batch normalization [15] and ReLU activation [32]. The four transposed convolutions use 256, 128, 64, and 32 filters, respectively. We produce the final output from G_T by passing the output of the last up-convolution block through another transposed convolution layer with 18 filters and *tanh* activation. The final generator output $G_T(v_B, \eta)$ has a spatial dimension of $64 \times 64 \times 18$, where each channel represents one of the 18 keypoints k_j, $j \in \{1, 2, \ldots, 18\}$. In our discriminator (*critic*) D_T, we first perform 4 successive convolutions, each followed by leaky ReLU activation, on the 18-channel heatmap. The four convolutions use 32, 64, 128, and 256 filters, respectively. The output of the last convolution layer is concatenated with 16 copies of ϕ_B arranged in a 4×4 tile. The concatenated feature map is then passed through a point convolution layer with 256 filters and leaky ReLU activation. We estimate the final scalar output from D_T by passing the feature map through another convolution layer with a single filter. We mathematically define the objective function for D_T as

$$L_D = -\mathbb{E}_{(x,v_B)\sim p_t, \eta\sim p_\eta}[D_T(x, v_B) - D_T(G_T(\eta, v_B), v_B)] \tag{1}$$

where $(x, v_B) \sim p_t$ is the heatmap and text embedding pair sampled from the training set, $\eta \sim p_\eta$ is the noise vector sampled from a Gaussian distribution, and $G_T(\eta, v_B)$ is the generated heatmap for the given text embedding

v_B. Researchers [12] have shown that the WGAN training is more stable if D_T is Lipschitz continuous, which mitigates the undesired behavior due to gradient clipping. To enforce the Lipschitz constraint, we compute gradient penalty as

$$\mathcal{C}_T = \mathbb{E}_{(\tilde{x}, v_B) \sim p_{\tilde{x}, v_B}} [(\|\nabla_{\tilde{x}, v_B} D_T(\tilde{x}, v_B)\|_2 - 1)^2] \tag{2}$$

where $\|.\|_2$ indicates the l_2 norm and $\tilde{x}$ is an interpolated sample between a real sample x and a generated sample $G_T(\eta, v_B)$, i.e., $\tilde{x} = \alpha x + (1 - \alpha) G_T(\eta, v_B)$, where α is a random number, selected from a uniform distribution between 0 and 1. Equation 2 enforces the Lipschitz constraint by restricting the gradient magnitude to 1. We define the overall objective of D_T by combining Eqs. 1 and 2 as

$$L_{D_T} = L_D + \lambda \mathcal{C}_T \tag{3}$$

where λ is a regularization constant. We keep $\lambda = 10$ in all of our experiments. We mathematically define the objective function for G_T as

$$\begin{aligned} L_{G_T} = &- \mathbb{E}_{\eta \sim p_\eta, v_B \sim p_{v_B}} [D_T(G_T(\eta, v_B), v_B)] \\ &- \mathbb{E}_{\eta \sim p_\eta, v_B^1, v_B^2 \sim p_{v_B}} \left[D_T(G_T(\eta, \frac{v_B^1 + v_B^2}{2}), \frac{v_B^1 + v_B^2}{2}) \right] \end{aligned} \tag{4}$$

where $v_B^1, v_B^2 \sim p_{v_B}$ are text encodings sampled from the training set. The second term in Eq. 4 helps the generator learn from the interpolated text encodings, which are not originally present in the training set.

We estimate the target keypoint set K_B^* from the 18-channel heatmap generated from G_T by computing the maximum activation ψ_j^{max}, $j \in \{1, 2, \ldots, 18\}$ for every channel. The spatial location of the maximum activation for the j-th channel determines the coordinates of the j-th keypoint if $\psi_j^{max} \geq 0.2$. Otherwise, the j-th keypoint is considered occluded if $\psi_j^{max} < 0.2$.

3.2 Facial Keypoints Refinement

While G_T produces a reasonable estimate of the target keypoints from the corresponding textual description, the estimation K_B^* is often noisy. The spatial perturbation is most prominent for the facial keypoints (nose, two eyes, and two ears) due to their proximity. Slight positional variations for other keypoints generally do not drastically affect the pose representation. Therefore, we refine the initial estimate of the facial keypoints by regression using a linear fully-connected network N_R (RefineNet). At first, the five facial keypoints k_i^f, $i \in \{1, 2, \ldots, 5\}$ are translated by $(k_i^f - k_n)$ where k_n is the spatial location of the nose. In this way, we align the nose with the origin of the coordinate system. Then, we normalize the translated facial keypoints such that the scaled keypoints k_i^s are within a square of span ± 1 and the scaled nose is at the origin (0, 0). Next, we flatten the coordinates of the five normalized keypoints to a 10-dimensional vector v_f and pass it through three linear fully-connected layers, where each layer has 128 nodes and ReLU activation. The final output layer of the network consists of 10 nodes and *tanh* activation. While training, we augment k_i^s with small amounts of random 2D spatial perturbations and try to predict the original values of k_i^s. We optimize the parameters of N_R by minimizing the mean squared error (MSE) between the actual and the predicted coordinates. Finally, we denormalize and retranslate the predicted facial keypoints. The refined set of keypoints $\tilde{K}_B^*$ is obtained by updating the coordinates of the facial keypoints of K_B^* with the predictions from RefineNet.

3.3 Pose Rendering

To render the final pose-transferred image $\tilde{I}_B$, we first extract the keypoints K_A from the source image I_A using a pre-trained Human Pose Estimator (HPE) [5]. However, we may also estimate the keypoints K_A^* from the embedding vector v_A for the text description T_A of the source pose P_A. If we compute the keypoints K_A^* from T_A, then the refinement is also applied on K_A^* to obtain the refined source keypoints $\tilde{K}_A^*$. Thus, depending on the source keypoints selection, we propose two slightly different variants of the method – (a) partially text-guided, where we use HPE to extract K_A and (b) fully text-guided, where we estimate $\tilde{K}_A^*$ using G_T followed by N_R. However, in both cases, the pose rendering step works similarly. For simplicity, we discuss the rendering network using K_A as the notation for the source keypoints. We represent the keypoints K_A and $\tilde{K}_B^*$ as multi-channel heatmaps H_A and $\tilde{H}_B$, respectively, where each channel of a heatmap corresponds to one particular keypoint.

We adopt an attention-guided conditional GAN architecture [38,39] for the target pose rendering. We take I_A, H_A, and $\tilde{H}_B$ as inputs for our generator network G_S, which produces the final rendered image output $\tilde{I}_B$ as an estimate for the target image I_B. The discriminator network D_S utilizes a PatchGAN [16] to evaluate the quality of the generated image by taking a channel-wise concatenation between I_A and either I_B or $\tilde{I}_B$. In G_S, we have two downstream branches for separately encoding the condition image I_A and the channel-wise concatenated heatmaps $(H_A, \tilde{H}_B)$. After mapping both inputs to a 256×256 feature space by convolution (kernel size $= 3 \times 3$, stride $= 1$, padding $= 1$, bias $= 0$), batch normalization, and ReLU activation, we pass the feature maps through four consecutive encoder blocks. Each block encodes the input feature space by reducing the dimension to half but doubling the number of filters. Each encoder block features a sequence of convolution (kernel size $= 4 \times 4$, stride $= 2$, padding $= 1$, bias $= 0$), batch normalization, ReLU activation, and a basic residual block [13]. We combine the encoded feature maps and pass the merged feature space through an upstream branch with four consecutive decoder blocks. Each block decodes the feature space by doubling the dimension but reducing the number of filters by half. Each decoder block features a sequence of transposed convolution (kernel size $= 4 \times 4$, stride $= 2$, padding $= 1$, bias $= 0$), batch normalization, ReLU activation, and a basic residual block. We use attention links between encoding and decoding paths at every resolution level to retain coarse and fine attributes in the generated image. Mathematically, for the lowest resolution level, $L = 4$,

$$I_3^{\delta} = \delta_4(I_4^{\pi^i} \odot \sigma(H_4^{\pi^h}))$$

and for the higher resolution levels, $L = \{1, 2, 3\}$,

$$I_{L-1}^{\delta} = \delta_L(I_L^{\pi^i} \odot \sigma(H_L^{\pi^h}))$$

where, at the resolution level L, I_L^{δ} denotes the output of the decoding block δ_L, $I_L^{\pi^i}$ denotes the output of the image encoding block π_L^i, $H_L^{\pi^h}$ denotes the output of the pose encoding block π_L^h, σ is an element-wise *sigmoid* activation function, and $\odot$ is an element-wise product. Finally, we pass the resulting feature maps through four consecutive basic residual blocks followed by a point-wise convolution (kernel size $= 1 \times 1$, stride $= 1$, padding $= 0$, bias $= 0$) with *tanh* activation to map the feature space into a $256 \times 256 \times 3$ normalized image $\tilde{I}_B$.

The optimization objective of G_S consists of three loss components – a pixel-wise l_1 loss $\mathcal{L}_{l_1}^{G_S}$, a discrimination loss $\mathcal{L}_{GAN}^{G_S}$ by D_S, and a perceptual loss $\mathcal{L}_{P_\rho}^{G_S}$ computed using a pre-trained VGG-19 network [43]. We measure the pixel-wise l_1 loss as $\mathcal{L}_{l_1}^{G_S} = \|\tilde{I}_B - I_B\|_1$, where $\|.\|_1$ denotes the l_1 norm or the mean absolute error. We compute the discrimination loss as

$$\mathcal{L}_{GAN}^{G_S} = \mathcal{L}_{BCE}(D_S(I_A, \tilde{I}_B), 1) \quad (5)$$

where $\mathcal{L}_{BCE}$ denotes the binary cross-entropy loss. Finally, we estimate the perceptual loss as

$$\mathcal{L}_{P_\rho}^{G_S} = \frac{1}{h_\rho w_\rho c_\rho} \sum_{x=1}^{h_\rho} \sum_{y=1}^{w_\rho} \sum_{z=1}^{c_\rho} \|q_\rho(\tilde{I}_B) - q_\rho(I_B)\|_1 \quad (6)$$

where q_ρ is the output of dimension $(h_\rho \times w_\rho \times c_\rho)$ from the ρ-th layer of a pre-trained VGG-19 network. We add two perceptual loss terms for $\rho = 4$ and $\rho = 9$ to the objective function. So, in our method, the overall optimization objective for G_S is given by

$$\mathcal{L}_{G_S} = \lambda_1 \mathcal{L}_{l_1}^{G_S} + \lambda_2 \mathcal{L}_{GAN}^{G_S} + \lambda_3(\mathcal{L}_{P_4}^{G_S} + \mathcal{L}_{P_9}^{G_S}) \quad (7)$$

where λ_1, λ_2, and λ_3 denote the weighing parameters for respective loss terms. We keep $\lambda_1 = 5$, $\lambda_2 = 1$, and $\lambda_3 = 5$ in our experiments. Lastly, we define the optimization objective for D_S as

$$\mathcal{L}_{D_S} = \frac{1}{2}\left[\mathcal{L}_{BCE}(D_S(I_A, I_B), 1) + \mathcal{L}_{BCE}(D_S(I_A, \tilde{I}_B), 0)\right] \quad (8)$$

4 Dataset and Training

As this is one of the earliest attempts to perform a text-guided pose transfer, we introduce a new dataset called *DeepFashion Pose Annotations and Semantics* (DF-PASS) to compensate for the lack of similar public datasets. DF-PASS

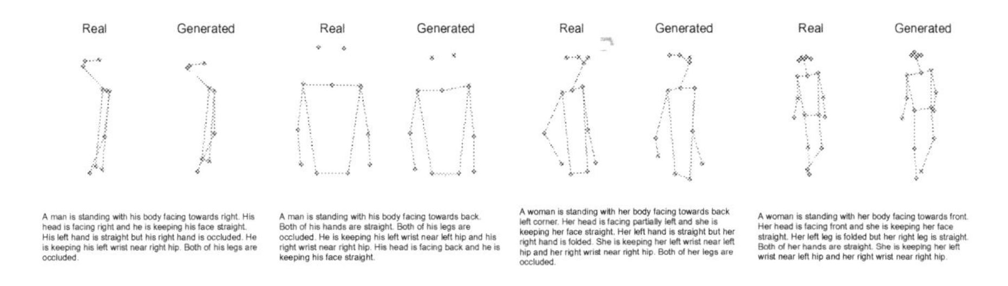

Fig. 3. Qualitative results of text to pose generation using G_T.

contains a human-annotated textual description of the pose for 40488 images of the DeepFashion dataset [26]. Each text annotation contains (1) the person's gender (e.g. 'man', 'woman' etc.); (2) visibility flags of the body keypoints (e.g. 'his left eye is visible', 'her right ear is occluded' etc.); (3) head and face orientations (e.g. 'her head is facing partially left', 'he is keeping his face straight' etc.); (4) body orientation (e.g. 'facing towards front', 'facing towards right' etc.); (5) hand and wrist positioning (e.g. 'his right hand is folded', 'she is keeping her left wrist near left hip' etc.); (6) leg positioning (e.g. 'both of his legs are straight', 'her right leg is folded' etc.). We recruit five in-house annotators to acquire the text descriptions, which two independent verifiers have validated. Each annotator describes a pose during data acquisition by selecting options from a set of possible attribute states. In this way, we have collected many-hot embedding vectors alongside the text descriptions. We use 37344 samples for training and 3144 samples for testing out of 40488 annotated samples following the same data split provided by [55].

In stage 1, the text to pose conversion network uses the stochastic Adam optimizer [19] to train both G_T and D_T. We keep learning rate $\eta_1 = 1e^{-4}$, $\beta_1 = 0$, $\beta_2 = 0.9$, $\epsilon = 1e^{-8}$, and weight decay $= 0$ for the optimizer. While training, we update G_T once after every five updates of D_T. In stage 2, we train the facial keypoints refinement network N_R using stochastic gradient descent keeping learning rate $\eta_2 = 1e^{-2}$. In stage 3, the pose rendering network also uses the Adam optimizer to train both G_S and D_S. In this case, we keep learning rate $\eta_3 = 1e^{-3}$, $\beta_1 = 0.5$, $\beta_2 = 0.999$, $\epsilon = 1e^{-8}$, and weight decay $= 0$. Before training, the parameters of G_T, D_T, G_S, and D_S are initialized by sampling from a normal distribution of 0 mean and 0.02 standard deviation.

5 Results

In Fig. 3, we demonstrate the output K_B^* of the text to keypoints generator G_T. The textual descriptions used for estimating the respective K_B^* are also shown in the figure. It can be observed that the estimated keypoints K_B^* capture the pose P_B and closely resembles K_B. However, a precise observation may reveal that the facial keypoints of K_B^* significantly differ from K_B. In Fig. 4, the advantage of regressive refinement of keypoints K_B^* is shown. As depicted in

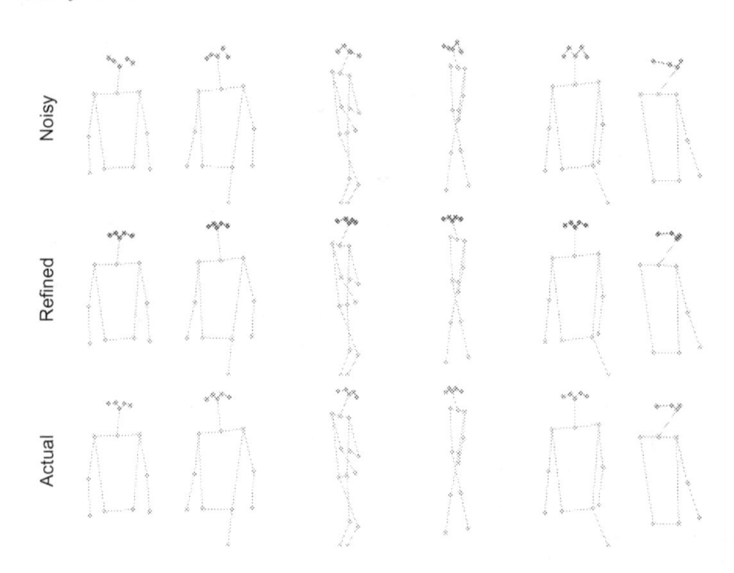

Fig. 4. Qualitative results of regressive refinement using N_R.

the figure, the refinement network aims to rectify only the facial keypoints. In Fig. 5, we demonstrate that when K_B is selected from the DeepFashion dataset, the existing PATN algorithm performs satisfactorily. However, when K_B comes from real-world samples (out of the DeepFashion dataset), PATN fails to generate the pose-transferred images while maintaining structural consistency. In the case of the proposed algorithm, as the pose description does not require structural information of the target image, the generated pose-transferred images are consistent with the respective source images. Our proposed algorithm performs well irrespective of the representation of the source pose, i.e., for the partially text-guided approach, where the source pose is represented using keypoints, and for the fully text-guided approach, where the source pose is described using text.

5.1 Evaluation

As the proposed method has three major steps – text to keypoint generation, refinement of the generated keypoints, and generation of the pose-transferred image, it is important to analyze each step qualitatively and quantitatively.

Metrics: Quantifying the generated image quality is a challenging problem. However, researchers [9,28,42,55] have used a few well-known quantitative metrics to judge the quality of the synthesis. This includes a Structural Similarity Index (SSIM) [45], Inception Score (IS) [40], Detection Score (DS) [25], and PCKh [1]. We also evaluate the Learned Perceptual Image Patch Similarity (LPIPS) [48] metric as a more modern replacement of the SSIM for perceptual image quality assessment. In our evaluation, we calculate LPIPS using both VGG19 [43] and SqueezeNet [14] backbones. As we are dealing with human poses

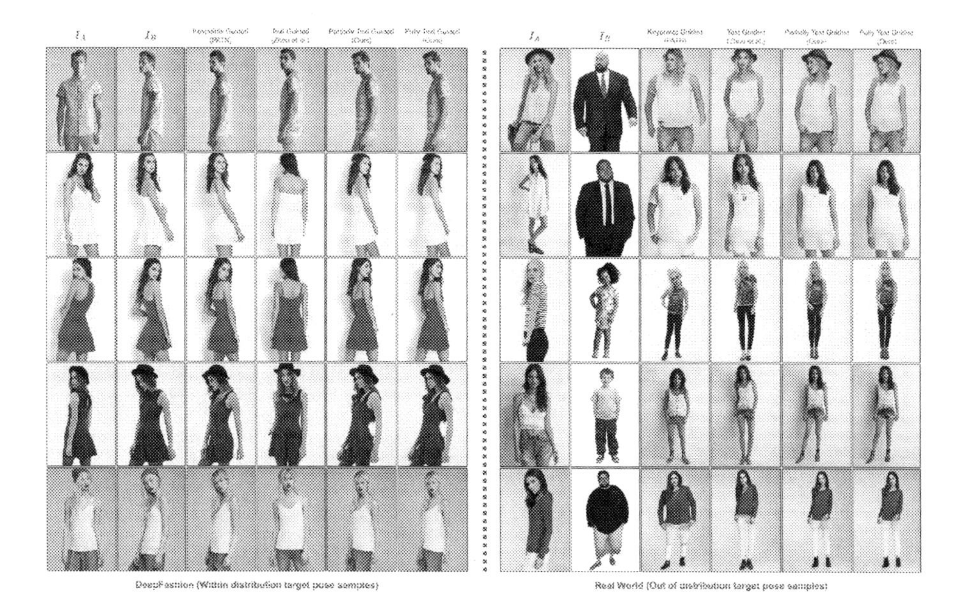

Fig. 5. Qualitative results of different pose transfer algorithms.

and evaluating the generation quality, we propose a novel metric, named Gender Consistency Rate (GCR), that evaluates whether the generated image $\tilde{I}_B$ can be identified to be of the same gender as the source image I_A by a pre-trained classifier. GCR serves two purposes: first, it ensures that the gender-specific features are present in the generated image, and second, it ensures that the generated target image is consistent with the source image. To calculate GCR, we remove the last layer of the VGG19 network and add a single neuron with sigmoid activation to design a binary classifier and train it with the image samples from the DeepFashion dataset with label 0 for males and label 1 for females. The pre-trained network achieves a test accuracy of 0.995. We use this pre-trained model to compute the gender recognition rate for the generated images.

In Table 1, we evaluate our proposed algorithm on the DeepFashion dataset. The keypoints-guided baseline [55] performs well for *within distribution* target poses from DeepFashion. However, the proposed text-guided approach performs satisfactorily, as reflected in SSIM, IS, DS, and LPIPS scores. As PCKh uses keypoint coordinates, our method achieves a low PCKh score compared to the keypoint-based method, which uses precise keypoints for the target image generation. For evaluating *out of distribution* target poses, we select 50 pairs of source and target images from DeepFashion; however, we estimate the target keypoints from real-world images (outside DeepFashion) having similar poses as the original target images. As shown in Table 2, in such a case, the proposed technique achieves significantly higher SSIM and PCKh values, indicating much better structural generation for real-world pose references.

User Study: It is known that quantitative metrics do not always reflect the perceptual quality of images well [42,55] and a quantifiable metric for evaluating image quality is still an open problem in computer vision. Therefore, we also perform an opinion-based user assessment to judge the realness of the generated images. Following the similar protocol as [28,42,55], the observer needs to provide an instant decision whether an image is real or fake. We create a subset of 260 real and 260 generated images with 10 images of each type used as a practice set. During the test, 20 random images (10 real + 10 fake) are drawn from the remaining images and shown to the examiner. We compute the **R2G** (the fraction of real images identified as generated) and **G2R** (the fraction of generated images identified as real) scores from the user submissions. Our method achieves a mean G2R score of 0.6968 for submissions by 156 individual volunteers.

Table 1. Performance of pose transfer algorithms on DeepFashion.

Pose generation algorithm	SSIM	IS	DS	PCKh	GCR	LPIPS (VGG)	LPIPS (SqzNet)
Partially text guided (Ours)	0.549	3.269	0.950	0.53	0.963	0.402	0.290
Fully text guided (Ours)	0.549	3.296	0.950	0.53	0.963	0.402	0.289
Zhou et al. [52]	0.373	2.320	0.864	0.62	0.979	0.310	0.215
PATN [55]	0.773	3.209	0.976	0.96	0.983	0.299	0.170
Real data	1.000	3.790	0.948	1.00	0.995	0.000	0.000

Table 2. Performance of pose transfer algorithms for real-world targets.

Pose generation algorithm	SSIM	IS	DS	PCKh	GCR	LPIPS (VGG)	LPIPS (SqzNet)
Partially text guided (Ours)	0.696	2.093	0.990	0.84	1.000	0.262	0.155
Fully text guided (Ours)	0.695	2.171	0.991	0.85	1.000	0.263	0.157
Zhou et al. [52]	0.615	2.891	0.931	0.52	1.000	0.271	0.182
PATN [55]	0.677	2.779	0.996	0.64	1.000	0.294	0.183
Real data	1.000	2.431	0.984	1.00	1.000	0.000	0.000

5.2 Ablation

We perform exhaustive ablation experiments to understand the effectiveness of different architectural components of the proposed pipeline. As shown in Table 3, refinement helps to improve SSIM, IS, and GCR scores in both partially and fully text-based approaches. Though the improvement in terms of metric values may look incremental, as shown in Fig. 6, the qualitative improvement due to the refinement operation is remarkable. Facial features play an essential role in the overall human appearance. Thus, the use of refinement is highly desirable in the pipeline. We report the rest of the ablation results with a partially text-based scheme while keeping the refinement operation intact in the pipeline.

We also explore several text embedding techniques and their effects on the generation pipeline. As shown in Table 4, the encoding methods like FastText [2] and Word2Vec [30] perform closely to BERT [7]. Thus, we can conclude that our method is robust to standard text embedding algorithms.

We also observe the effect of multi-resolution attention used in G_S. In the case of single-scale attention, we only take point-wise multiplication of the channels of the final pose encoder and the final image encoder and skip all the following supervision of the pose encoders at higher resolution levels. As shown in Table 5, multi-scale attention significantly improves majority of the evaluation metrics.

Table 3. Effects of source encoding and regressive refinement.

Source encoding	Refinement	SSIM	IS	DS	PCKh	GCR	LPIPS (VGG)	LPIPS (SqzNet)
Keypoints	✘	0.545	3.221	0.952	0.53	0.960	0.404	0.290
Keypoints	✔	0.549	3.269	0.950	0.53	0.963	0.402	0.290
Text embedding	✘	0.545	3.261	0.952	0.53	0.960	0.404	0.290
Text embedding	✔	0.549	3.296	0.950	0.53	0.963	0.402	0.289
Real data		1.000	3.790	0.948	1.00	0.995	0.000	0.000

Table 4. Effects of different text encoding methods.

Text embedding	SSIM	IS	DS	PCKh	GCR	LPIPS (VGG)	LPIPS (SqzNet)
Multi-hot	0.558	3.228	0.953	0.60	0.970	0.388	0.274
BERT [7]	0.549	3.269	0.950	0.53	0.963	0.402	0.290
FastText [2]	0.548	3.275	0.949	0.52	0.968	0.399	0.285
Word2Vec [30]	0.550	3.251	0.949	0.52	0.973	0.401	0.289
Real data	1.000	3.790	0.948	1.00	0.995	0.000	0.000

Table 5. Effects of multi-resolution attention.

Pose transfer method	SSIM	IS	DS	PCKh	GCR	LPIPS (VGG)	LPIPS (SqzNet)
Single-scale attention guided	0.540	3.170	0.921	0.54	0.954	0.415	0.298
Multi-scale attention guided	0.549	3.269	0.950	0.53	0.963	0.402	0.290
Real data	1.000	3.790	0.948	1.00	0.995	0.000	0.000

Fig. 6. Qualitative results by the proposed pipeline using the partially text-guided generator with and without refinement.

Fig. 7. Failure cases of the proposed framework.

6 Limitations

To the best of our knowledge, this is one of the earliest attempts to transfer pose using textual supervision. As shown in Sect. 5, although the results produced by the proposed approach are often at par with the existing keypoint-based baseline, it fails to perform well in some instances. When the textual description is brief and lacks a fine-grained description of the pose, the generator G_T fails to interpret the pose correctly. Some of the failed cases produced by our algorithm are shown in Fig. 7.

7 Conclusion

In this paper, we have shown that the existing keypoint-based approaches for human pose transfer suffer from a significant flaw that occasionally prevents these techniques from being useful in real-world situations when the target pose reference is unavailable. Thus, we propose a novel text-guided pose transfer pipeline to mitigate the dependency on the target pose reference. To perform the task, first, we have designed a *text to keypoints* generator for estimating the keypoints from a text description of the target pose. Next, we use a linear *refinement* network to regressively obtain a refined spatial estimation of the keypoints representing the target pose. Lastly, we *render* the target pose by conditioning a multi-resolution attention-based generator on the appearance of the source image. Due to the lack of similar public datasets, we have also introduced a new dataset DF-PASS, by extending the DeepFashion dataset with human annotations for poses.

Acknowledgment. This work was partially supported by the Technology Innovation Hub, Indian Statistical Institute Kolkata, India. The ISI-UTS Joint Research Cluster (JRC) partly funded the project.

References

1. Andriluka, M., Pishchulin, L., Gehler, P., Schiele, B.: 2D human pose estimation: new benchmark and state of the art analysis. In: The IEEE Conference on Computer Vision and Pattern Recognition (CVPR) (2014)
2. Athiwaratkun, B., Wilson, A.G., Anandkumar, A.: Probabilistic FastText for multi-sense word embeddings. arXiv preprint arXiv:1806.02901 (2018)
3. Balakrishnan, G., Zhao, A., Dalca, A.V., Durand, F., Guttag, J.: Synthesizing images of humans in unseen poses. In: The IEEE Conference on Computer Vision and Pattern Recognition (CVPR) (2018)
4. Briq, R., Kochar, P., Gall, J.: Towards better adversarial synthesis of human images from text. arXiv preprint arXiv:2107.01869 (2021)
5. Cao, Z., Simon, T., Wei, S.E., Sheikh, Y.: Realtime multi-person 2D pose estimation using part affinity fields. In: The IEEE Conference on Computer Vision and Pattern Recognition (CVPR) (2017)
6. Chen, L., Maddox, R.K., Duan, Z., Xu, C.: Hierarchical cross-modal talking face generation with dynamic pixel-wise loss. In: The IEEE Conference on Computer Vision and Pattern Recognition (CVPR) (2019)
7. Devlin, J., Chang, M.W., Lee, K., Toutanova, K.: BERT: pre-training of deep bidirectional transformers for language understanding. arXiv preprint arXiv:1810.04805 (2018)
8. Dong, C., Loy, C.C., He, K., Tang, X.: Image super-resolution using deep convolutional networks. IEEE Trans. Pattern Analy. Mach. Intell. (TPAMI) **38**, 295–307 (2015)
9. Esser, P., Sutter, E., Ommer, B.: A variational U-Net for conditional appearance and shape generation. In: The IEEE Conference on Computer Vision and Pattern Recognition (CVPR) (2018)
10. Goodfellow, I., et al.: Generative adversarial nets. In: The Conference on Neural Information Processing Systems (NeurIPS) (2014)
11. Güler, R.A., Neverova, N., Kokkinos, I.: DensePose: dense human pose estimation in the wild. In: The IEEE Conference on Computer Vision and Pattern Recognition (CVPR) (2018)

12. Gulrajani, I., Ahmed, F., Arjovsky, M., Dumoulin, V., Courville, A.: Improved training of Wasserstein GANs. arXiv preprint arXiv:1704.00028 (2017)
13. He, K., Zhang, X., Ren, S., Sun, J.: Deep residual learning for image recognition. In: The IEEE Conference on Computer Vision and Pattern Recognition (CVPR) (2016)
14. Iandola, F.N., Han, S., Moskewicz, M.W., Ashraf, K., Dally, W.J., Keutzer, K.: SqueezeNet: AlexNet-level accuracy with 50x fewer parameters and <0.5 MB model size. arXiv preprint arXiv:1602.07360 (2016)
15. Ioffe, S., Szegedy, C.: Batch Normalization: accelerating deep network training by reducing internal covariate shift. In: The International Conference on Machine Learning (ICML) (2015)
16. Isola, P., Zhu, J.Y., Zhou, T., Efros, A.A.: Image-to-Image translation with conditional adversarial networks. In: The IEEE Conference on Computer Vision and Pattern Recognition (CVPR) (2017)
17. Johnson, J., Alahi, A., Fei-Fei, L.: Perceptual losses for real-time style transfer and super-resolution. In: The European Conference on Computer Vision (ECCV) (2016)
18. Kim, J., Kwon Lee, J., Mu Lee, K.: Accurate image super-resolution using very deep convolutional networks. In: The IEEE Conference on Computer Vision and Pattern Recognition (CVPR) (2016)
19. Kingma, D.P., Ba, J.: Adam: a method for stochastic optimization. In: The International Conference on Learning Representations (ICLR) (2015)
20. Lassner, C., Pons-Moll, G., Gehler, P.V.: A generative model of people in clothing. In: The IEEE International Conference on Computer Vision (ICCV) (2017)
21. Ledig, C., et al.: Photo-realistic single image super-resolution using a generative adversarial network. In: The IEEE Conference on Computer Vision and Pattern Recognition (CVPR) (2017)
22. Li, K., Zhang, J., Liu, Y., Lai, Y.K., Dai, Q.: PoNA: pose-guided non-local attention for human pose transfer. IEEE Trans. Image Process. (TIP) **29**, 9584–9599 (2020)
23. Li, Y., Huang, C., Loy, C.C.: Dense intrinsic appearance flow for human pose transfer. In: The IEEE Conference on Computer Vision and Pattern Recognition (CVPR) (2019)
24. Li, Y., Min, M., Shen, D., Carlson, D., Carin, L.: Video generation from text. In: The AAAI Conference on Artificial Intelligence (2018)
25. Liu, W., et al.: SSD: single shot MultiBox detector. In: Leibe, B., Matas, J., Sebe, N., Welling, M. (eds.) ECCV 2016. LNCS, vol. 9905, pp. 21–37. Springer, Cham (2016). https://doi.org/10.1007/978-3-319-46448-0_2
26. Liu, Z., Luo, P., Qiu, S., Wang, X., Tang, X.: DeepFashion: powering robust clothes recognition and retrieval with rich annotations. In: The IEEE Conference on Computer Vision and Pattern Recognition (CVPR) (2016)
27. Loper, M., Mahmood, N., Romero, J., Pons-Moll, G., Black, M.J.: SMPL: a skinned multi-person linear model. ACM Trans. Graph. (TOG) **34**, 1–16 (2015)
28. Ma, L., Jia, X., Sun, Q., Schiele, B., Tuytelaars, T., Van Gool, L.: Pose guided person image generation. In: The Conference on Neural Information Processing Systems (NeurIPS) (2017)
29. Ma, L., Sun, Q., Georgoulis, S., Van Gool, L., Schiele, B., Fritz, M.: Disentangled person image generation. In: The IEEE Conference on Computer Vision and Pattern Recognition (CVPR) (2018)
30. Mikolov, T., Chen, K., Corrado, G., Dean, J.: Efficient estimation of word representations in vector space. arXiv preprint arXiv:1301.3781 (2013)

31. Mirza, M., Osindero, S.: Conditional generative adversarial nets. arXiv preprint arXiv:1411.1784 (2014)
32. Nair, V., Hinton, G.E.: Rectified linear units improve Restricted Boltzmann Machines. In: The International Conference on Machine Learning (ICML) (2010)
33. Neverova, N., Alp Güler, R., Kokkinos, I.: Dense pose transfer. In: Ferrari, V., Hebert, M., Sminchisescu, C., Weiss, Y. (eds.) ECCV 2018. LNCS, vol. 11207, pp. 128–143. Springer, Cham (2018). https://doi.org/10.1007/978-3-030-01219-9_8
34. Pumarola, A., Agudo, A., Sanfeliu, A., Moreno-Noguer, F.: Unsupervised person image synthesis in arbitrary poses. In: The IEEE Conference on Computer Vision and Pattern Recognition (CVPR) (2018)
35. Qiao, T., Zhang, J., Xu, D., Tao, D.: MirrorGAN: learning text-to-image generation by redescription. In: The IEEE Conference on Computer Vision and Pattern Recognition (CVPR) (2019)
36. Radford, A., Metz, L., Chintala, S.: Unsupervised representation learning with deep convolutional generative adversarial networks. In: The International Conference on Learning Representations (ICLR) (2016)
37. Reed, S., Akata, Z., Yan, X., Logeswaran, L., Schiele, B., Lee, H.: Generative adversarial text to image synthesis. In: The International Conference on Machine Learning (ICML) (2016)
38. Roy, P., Bhattacharya, S., Ghosh, S., Pal, U.: Multi-scale attention guided pose transfer. arXiv preprint arXiv:2202.06777 (2022)
39. Roy, P., Ghosh, S., Bhattacharya, S., Pal, U., Blumenstein, M.: Scene aware person image generation through global contextual conditioning. In: The International Conference on Pattern Recognition (ICPR) (2022)
40. Salimans, T., Goodfellow, I.J., Zaremba, W., Cheung, V., Radford, A., Chen, X.: Improved techniques for training GANs. In: The Conference on Neural Information Processing Systems (NeurIPS) (2016)
41. Sangkloy, P., Lu, J., Fang, C., Yu, F., Hays, J.: Scribbler: controlling deep image synthesis with sketch and color. In: The IEEE Conference on Computer Vision and Pattern Recognition (CVPR) (2017)
42. Siarohin, A., Sangineto, E., Lathuilière, S., Sebe, N.: Deformable GANs for pose-based human image generation. In: The IEEE Conference on Computer Vision and Pattern Recognition (CVPR) (2018)
43. Simonyan, K., Zisserman, A.: Very deep convolutional networks for large-scale image recognition. In: The International Conference on Learning Representations (ICLR) (2015)
44. Wang, B., Zheng, H., Liang, X., Chen, Y., Lin, L., Yang, M.: Toward characteristic-preserving image-based virtual try-on network. In: Ferrari, V., Hebert, M., Sminchisescu, C., Weiss, Y. (eds.) ECCV 2018. LNCS, vol. 11217, pp. 607–623. Springer, Cham (2018). https://doi.org/10.1007/978-3-030-01261-8_36
45. Wang, Z., Bovik, A.C., Sheikh, H.R., Simoncelli, E.P.: Image quality assessment: from error visibility to structural similarity. IEEE Trans. Image Process. (TIP) **13**, 600–612 (2004)
46. Yeh, R.A., Chen, C., Yian Lim, T., Schwing, A.G., Hasegawa-Johnson, M., Do, M.N.: Semantic image inpainting with deep generative models. In: The IEEE Conference on Computer Vision and Pattern Recognition (CVPR) (2017)
47. Zanfir, M., Popa, A.I., Zanfir, A., Sminchisescu, C.: Human appearance transfer. In: The IEEE Conference on Computer Vision and Pattern Recognition (CVPR) (2018)

48. Zhang, R., Isola, P., Efros, A.A., Shechtman, E., Wang, O.: The unreasonable effectiveness of deep features as a perceptual metric. In: The IEEE Conference on Computer Vision and Pattern Recognition (CVPR) (2018)
49. Zhang, Y., Briq, R., Tanke, J., Gall, J.: Adversarial synthesis of human pose from text. In: The DAGM German Conference on Pattern Recognition (GCPR) (2020)
50. Zhao, B., Wu, X., Cheng, Z.Q., Liu, H., Jie, Z., Feng, J.: Multi-view image generation from a single-view. In: The ACM International Conference on Multimedia (MM) (2018)
51. Zheng, H., Chen, L., Xu, C., Luo, J.: Pose flow learning from person images for pose guided synthesis. IEEE Trans. Image Process. (TIP) **30**, 1898–1909 (2020)
52. Zhou, X., Huang, S., Li, B., Li, Y., Li, J., Zhang, Z.: Text guided person image synthesis. In: The IEEE Conference on Computer Vision and Pattern Recognition (CVPR) (2019)
53. Zhou, Y., Han, X., Shechtman, E., Echevarria, J., Kalogerakis, E., Li, D.: MakeItTalk: speaker-aware talking-head animation. ACM Trans. Graph. (TOG) **39**, 1–15 (2020)
54. Zhu, J.Y., Park, T., Isola, P., Efros, A.A.: Unpaired image-to-image translation using cycle-consistent adversarial networks. In: The IEEE International Conference on Computer Vision (ICCV) (2017)
55. Zhu, Z., Huang, T., Shi, B., Yu, M., Wang, B., Bai, X.: Progressive pose attention transfer for person image generation. In: The IEEE Conference on Computer Vision and Pattern Recognition (CVPR) (2019)

Addressing Heterogeneity in Federated Learning via Distributional Transformation

Haolin Yuan[1(✉)], Bo Hui[1], Yuchen Yang[1], Philippe Burlina[1,2], Neil Zhenqiang Gong[3], and Yinzhi Cao[1]

[1] Department of Computer Science, Johns Hopkins University, Baltimore, USA
{hyuan4,bo.hui,yc.yang,yinzhi.cao}@jhu.edu
[2] Johns Hopkins University Applied Physics Laboratory (JHU/APL), Laurel, USA
Philippe.Burlina@jhuapl.edu
[3] Duke University, Durham, USA
neil.gong@duke.edu

Abstract. Federated learning (FL) allows multiple clients to collaboratively train a deep learning model. One major challenge of FL is when data distribution is heterogeneous, i.e., differs from one client to another. Existing personalized FL algorithms are only applicable to narrow cases, e.g., one or two data classes per client, and therefore they do not satisfactorily address FL under varying levels of data heterogeneity. In this paper, we propose a novel framework, called DisTrans, to improve FL performance (i.e., model accuracy) via train and test-time distributional transformations along with a double-input-channel model structure. DisTrans works by optimizing distributional offsets and models for each FL client to shift their data distribution, and aggregates these offsets at the FL server to further improve performance in case of distributional heterogeneity. Our evaluation on multiple benchmark datasets shows that DisTrans outperforms state-of-the-art FL methods and data augmentation methods under various settings and different degrees of client distributional heterogeneity (e.g., for CelebA and 100% heterogeneity DisTrans has accuracy of 80.4% vs. 72.1% or lower for other SOTA approaches).

1 Introduction

Federated learning [17,29,34,47] (FL) is an emerging distributed machine learning (ML) framework that enables clients to learn models together with the help of a central server. In FL, each client learns a local model that is sent to the FL server for aggregation, and subsequently the FL server returns the aggregated model to the client. The process is repeated until convergence. One emerging and unsolved FL challenge is that the data distribution at each client can be heterogeneous. For example, for FL based skin diagnostics, the skin disease distribution for each hospital/client can vary significantly. In another use case of smartphone face verification, data distributions collected at each mobile device

H. Yuan, B. Hui, Y. Yang—Equal contributions to the paper.

S. Avidan et al. (Eds.): ECCV 2022, LNCS 13698, pp. 179–195, 2022.
https://doi.org/10.1007/978-3-031-19839-7_11

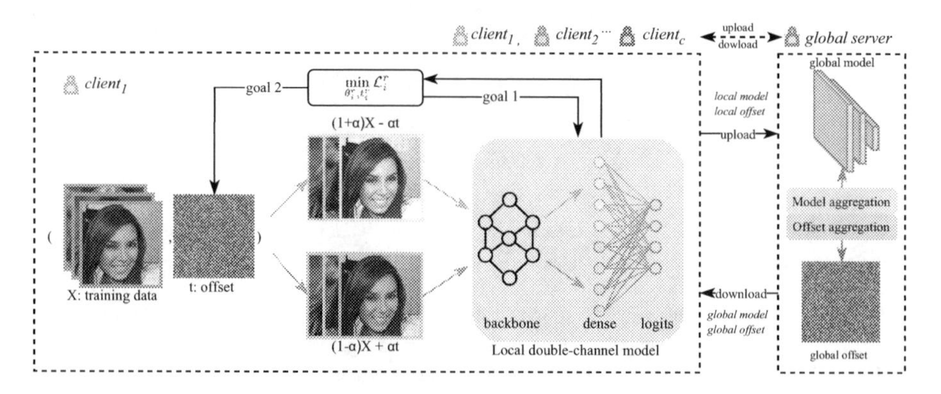

Fig. 1. The pipelines of DisTrans. Each client jointly optimizes the offset and model in local training phase, then uploads both to the central server for aggregation. The aggregated model and offset are sent back to clients for next-round.

can vary from one client to another. Such distributional heterogeneity often leads to suboptimal accuracy of the final FL model.

There are two types of approaches to learn FL models under data heterogeneity: (i) improving FL's training process and (ii) improving clients' local data. Unfortunately, neither improves FL under varied levels of data heterogeneity. On one hand, existing FL methods [2,13,34], especially personalized FLs [25,27], learn a model (or even multiple models) using customized loss functions or model architectures based on heterogeneity level. However, existing personalized FL algorithms are designed for highly heterogeneous distribution. FedAwS [49] can only train FL models when local client's data has one positive label. The performance of pfedMe [42] and pfedHN [38] degrades to even 5% to 18% lower accuracy than FedAvg [34], when the data distribution is between heterogeneity and homogeneity.

On the other hand, traditional centralized machine learning also rely on data transformations, i.e., data augmentation, [7,8,26,30,45,51,52] to improve model's performance. Such transformations could be used for a pre-processing of all the training data or an addition to the existing training set. Until very recently, data transformations are also used during test time [14,20,37,39,41] to improve learning models, e.g., adversarial robustness [37]. However, it remains unclear whether and how data transformation can improve FL particularly under different client heterogeneity. The major challenge is how to tailor transformations for each client with different data distributions.

In this paper, we propose the *first* FL distributional transformation framework, called DisTrans, to address this heterogeneity challenge by altering local data distributions via a client-specific data shift applied both on train and test/inference data. Our distributional transformation alters each client's data distribution so that such distribution becomes less heterogeneous and thus the local models can be better aggregated at the server. Specifically, DisTrans performs a so-called *joint optimization*, at each client, to train the local model and generate an offset that is added to the local data. That is, an DisTrans's client alternately performs two steps in each round: 1) optimizing the personalized *offset* to transform the local data via distribution shifts and 2) optimizing a local model to fit its offsetted local data. After client-side optimization, the FL server aggregates both the personalized offsets and the local models from all the clients and sends the aggregated global model and offset back to each client. During testing, each client adds its personalized offset to each testing input before using the global model to predict its label.

DisTrans is designed with a special network architecture, called a double-input-channel model, to accommodate client-side offsets. This double-input-channel model has a backbone network shared by both channels, a dense layer accepting outputs from two channels in parallel, and a logits layer that merges channel-related outputs from the dense layer. This double architecture allows the offset to be added to an (training or testing) input in one channel but subtracted from the input in the other. Such addition and subtraction better preserves the information in the original training and testing data because the original data can be recovered from the data with offset in the two channels.

We perform extensive evaluation of DisTrans using five different image datasets and compare it against state-of-the-art (SOTA) methods. Our evaluation shows that DisTrans outperforms SOTA FL methods across various distributional settings of the clients' local data by 1%–10% with respect to testing accuracy. Moreover, our evaluation shows that DisTrans achieves 1%–7% higher testing accuracy than other data transformation/augmentation approaches, i.e., mixup [50] and AdvProp [45]. The code for DisTrans is made available under (https://github.com/hyhmia/DisTrans).

2 Related Work

Existing federated learning (FL) studies focus on improving accuracy [34,38,42,49], convergence [6,11,16,31,36,44], communication cost [3,12,21–23,33,40,48], security and privacy [4,5,9,35], or others [10,15,19,46]. Our work focuses on FL accuracy.

Personalized Federated Learning. Prior studies [38,42,49] have attempted to address personalization, i.e., to make a model better fit a client's local training data. For instance, FedAwS [49] investigates FL problems where each local model only has access to the positive data associated with only a single class and imposes a geometric regularizer at the server after each round to encourage classes to spread out in the embedding space. pFedMe [42] formulates a new bi-level optimization problem and uses Moreau envelopes to regularize each client loss function and to decouple personalized model optimization from the global model learning. pFedHN [38] utilizes a hypernetwork model as the global model to generate weights for each local model. MOON [28] uses contrastive learning to maximize the agreement between local and global model.

Data Transformation. Data transformation applies label-preserving transformations to images and is a standard technique to improve model accuracy in centralized learning. Most of the recent data transformation methods

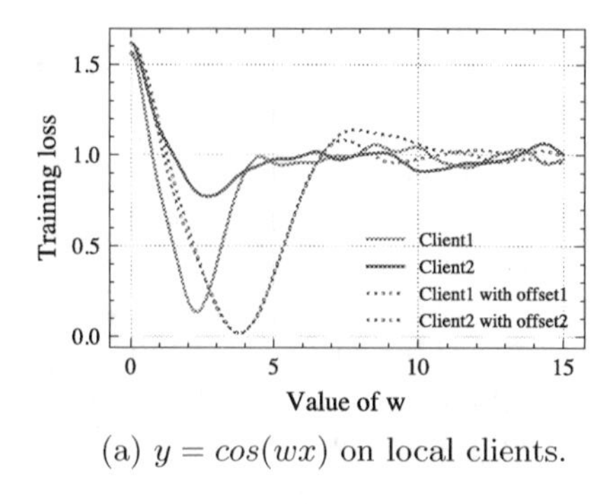

(a) $y = cos(wx)$ on local clients.

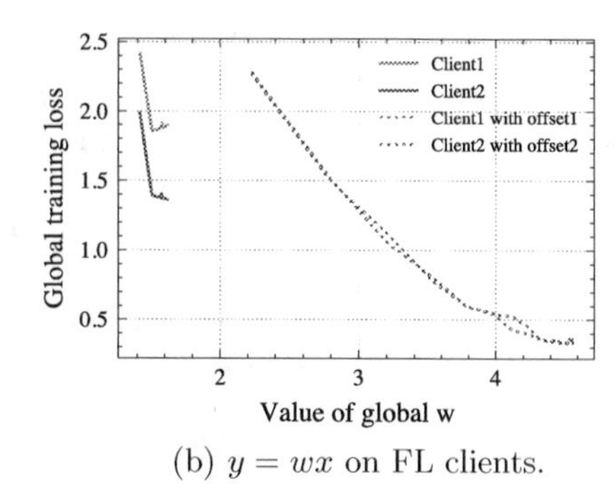

(b) $y = wx$ on FL clients.

Fig. 2. Training loss with respect to optimal weight w on two clients' local training data with and w/o offset. We observe that offsets can make the training loss against weight more consistent on local clients and help FL model converge.

[7,8,26,30,45,51,52] focus on transforming datasets during the training phase. For instance, mixup [50] transforms the training data by mixing up the features and their corresponding labels; and AdvProp [45] transforms the training data by adding adversarial examples. Additionally, transforming data at testing time [14,20,37,39,41] has received increased attention. The basic test-time transformations use multiple data augmentations [14,41] at test time to classify one image and get the averaged results. Pérez et.al [37] aims to enhance adversarial robustness via test-time transformation. As a comparison, DisTrans is the first to utilize test-time transformation to improve federated learning accuracy under data heterogeneity.

3 Motivation

DisTrans's intuition is to transform each client's training and testing data with offsets to improve FL under heterogeneous data. That is, DisTrans transforms the client-side data distribution so that the learned local models are less heterogeneous and can be better aggregated. To better illustrate this intuition, we describe two simple learning problems as motivating examples. Specifically, we show that well-optimized and selected offsets can (i) align two learning problems at different FL clients and (ii) help the aggregated model converge.

Local Non-convex Learning Problems. We consider a non-convex learning problem, i.e., $f(x) = cos(wx)$ where $w \in \mathbb{R}$, at two local clients with heterogeneous data. The local data is generated via $x, y \in \mathbb{R}$ with $y = cos(w^{true}_{clientk}x) + \epsilon_{clientk}$, where x is drawn i.i.d from Gaussian distribution and $\epsilon_{clientk}$ is Gaussian noise with mean value as 0. The offsets are $px + q$ where p is a fixed value at both clients and q is chosen via brute force search. Figure 2a shows the squared training loss with and without offsets. The difference between the training losses of two learning models are reduced, thus making two clients consistent.

Linear Regression Problems with An Aggregation Server. We train two local linear models, i.e., $f(x) = wx$ with the model parameter $w \in \mathbb{R}^2$, aggregate

Algorithm 1. Pseudo-code of DisTrans

Input: Number of clients C, local training dataset D_i for client i, number of rounds R, batch size B, number of epochs E, and learning rates η and η_p for model and offset t, respectively
Output: Offset t_i for client i and global model θ
1: Server initializes global model θ^0 and offset t_i^0 for each client i
2: **for** $r = 0$ to $R-1$ **do**
3: Server sends θ^r and t_i^r to client i
4: **for** $i = 0$ to $C-1$ **do**
5: $\theta_i^r \leftarrow \theta^r$ // Initialize local model θ_i^r for client i
6: **for** $e = 0$ to $E-1$ **do**
7: **for** each mini-batch D_m from D_i **do**
8: $t_i^r \leftarrow SGD(\nabla_{t_i^r}\mathcal{L}_i^r, t_i^r, \eta_t)$ // Update offset t_i^r
9: $x_t \leftarrow ((1-\alpha)x + \alpha t_i^r,\ (1+\alpha)x - \alpha t_i^r)$ // Combine t_i^r with each $x \in D_m$
10: $\theta_i^r \leftarrow SGD(\nabla_{\theta_i^r}\mathcal{L}_i^r, \theta_i^r, \eta)$ // Update local model
11: **end for**
12: **end for**
13: Client i sends θ_i^r and t_i^r to server
14: **end for**
15: Server updates global model: $\theta^{r+1} \leftarrow \frac{1}{C}\sum_{i\in[C]}\theta_i^r$
16: Server updates offset t_i^{r+1} for each client i via Offset Aggregation
17: **end for**

the parameters at a server following FL, and then repeat the two steps following FL until convergence. The local training data is heterogeneous and generated as $y = w_{clientk}^{true}x + \epsilon_{clientk}$, where each of the two dimensions of x is drawn i.i.d from normal distribution and $\epsilon_{clientk}$ is a Gaussian noise. The offset is the same as the non-convex learning problem. We fix p and optimize q and w via SGD at each client to minimize learning loss respectively. Figure 2b shows the squared training loss with respect to the optimal w (sum value of two dimensions) with and without the offsets. Clearly, when offsets are not present, the aggregated model does not converge, resulting in a set of sub-optimal weights. Instead, the aggregated model converges with a small training loss with the presence of offsets, confirming our intuition.

4 Method

In this section, we present our proposed method in detail. DisTrans aims to learn a single shared global model for the clients. Algorithm 1 shows the pseudo-code of DisTrans. In each round, each client learns a local model and an offset, which are sent to the central server. The server aggregates the clients' local models and local offsets, and sends them back to the clients. Based on the intuition presented above, we propose a joint optimization method to learn a local model and offset for a client in each round. Figure 1 illustrates our joint optimization that each client performs.

Notations. We assume C clients and denote by D_i the local training dataset for client i, where $i = 1, 2, \cdots, C$ and $|D_i| = n_i$. We consider $z = (x, y)$ a training sample, where $x \in \mathbb{R}^m$ denotes the training input and y the label of the training input. We also denote by D_{ti} the offsetted local training dataset for client i, x_t an offsetted training input, and $z_t = (x_t, y)$ a training sample offsetted with offset. We denote by θ the global model.

4.1 Double-Input-Channel Model Architecture

DisTrans uses a double-input-channel neural network architecture (see Fig. 1) for a local/global model. Our architecture has a shared backbone network, a dense layer concatenating two channels' outputs, and a logits layer merging outputs from the dense layer. Specifically, these two channels shift the local data distribution in two different ways using the same offset t. Formally, Eq. 1 shows our two linear shifts:

$$x_t = ((1-\alpha)x + \alpha t,\ (1+\alpha)x - \alpha t)), \tag{1}$$

where the first channel adds the offset t to the input x with a coefficient α (i.e., $(1-\alpha)x + \alpha t$ is the input for the first channel) and the second subtracts t from x with the same α (i.e., $(1+\alpha)x - \alpha t$ is the input to the second channel). Unless otherwise mentioned, our default setting for α is 0.3 in our experiments.

4.2 Joint Optimization

In our joint optimization, each client aims to achieve the following two goals:

- *Goal 1.* Optimizing *offset* to shift local data distribution to better fit with local model.
- *Goal 2.* Optimizing local model to fit with offsetted local data distribution.

We formulate the two goals as an optimization problem. Specifically, client i aims to solve the following optimization problem in round r:

$$\min_{\theta_i^r, t_i^r} \mathcal{L}_i^r = \frac{1}{n} \sum_{z_t \in D_{ti}} l(\theta_i^r, z_t), \tag{2}$$

where θ_i^r is the local model of client i, t_i^r is the offset of client i, and $\mathcal{L}_i^r$ is the loss function of client i in round r. We choose cross entropy as loss term in our implementation. Solving t_i^r in Eq. 2 while fixing θ_i^r achieves Goal 1; and solving θ_i^r in Eq. 2 while fixing t_i^r achieves Goal 2. Therefore, we initialize θ_i^r as the global model θ^r and alternately optimize t_i^r and θ_i^r for each mini-batch. Algorithm 1 illustrates our pseudo-code.

4.3 Model and Offset Aggregation

The server aggregates both the local models and the offsets from the clients. The model aggregation follows the traditional FL, e.g., the server computes the mean of the clients' local models as the global model like FedAvg [34]. Our offset aggregation leverages the class distribution at each client. Next, we first introduce a metric to measure *distributional heterogeneity* and then our offset aggregation method based on the metric.

Distributional Heterogeneity. We define distributional heterogeneity to characterize the class heterogeneity among the clients. Formally, we denote distributional heterogeneity as DH and define it as follows:

$$DH = 1 - \frac{\sum_{j\in[1,N]} c_j}{N \times C}, \tag{3}$$

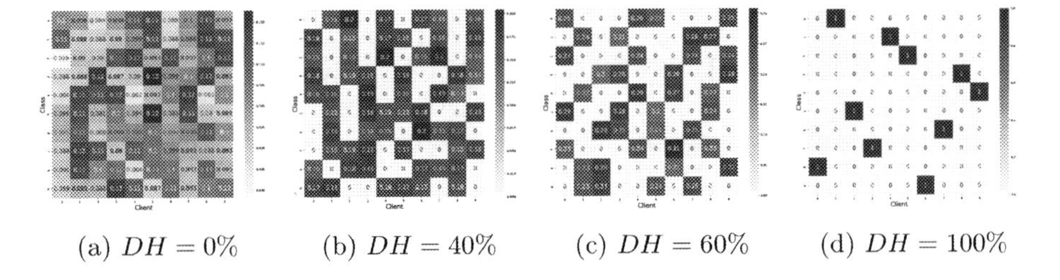

(a) $DH = 0\%$ (b) $DH = 40\%$ (c) $DH = 60\%$ (d) $DH = 100\%$

Fig. 3. Different distributional heterogeneity levels on CIFAR-10.

where N is the total number of classes, C is the total number of clients, and c_j is defined as follows:

$$c_j = \begin{cases} 0, & \text{if only one client has data from class j}, \\ k, & \text{if } k > 1 \text{ clients have data from class j}. \end{cases} \tag{4}$$

Our defined DH has a value between 0 and 100%. In particular, $DH = 0\%$ means that each client has data from the C classes, e.g., the clients' local data are i.i.d., while $DH = 100\%$ means that each class of data belongs to only one client, i.e., an extreme non-i.i.d. setting. Figure 3 shows examples of different levels of distributional heterogeneity visualized by heatmaps for clients' local data in our experiments on CIFAR-10. We list clients on the x-axis and classes on the y-axis; and each cell is the fraction of the data from the corresponding class that are on the corresponding client.

4.4 Offset Aggregation Methods

DisTrans aggregates clients' offsets based on distributional heterogeneity. Intuitively, when the distributional heterogeneity is very large, the offset of one client may not be informative for the offset of another client, as their data distributions are substantially different. Therefore, we aggregate clients' offsets only if the distributional heterogeneity is smaller than a threshold (we set the threshold to be 50% in experiments).

Suppose the distributional heterogeneity of an FL system is smaller than the threshold. One naive way to aggregate the clients' offsets is to compute their average as a global offset, which is sent back to all clients. However, such naive aggregation method uses the same global offset for all clients, which achieves sub-optimal accuracy as shown in our experiments. Therefore, we propose a neural network based aggregation method, which produces different aggregated offsets for the clients. Specifically, the server maintains a neural network, which takes a client-specific embedding vector $e \in \mathbb{R}^{1 \times N}$ and a client's offset as input and outputs an aggregated offset for the client, where an entry e_i of the embedding vector is the fraction of the training data in class i that are on the client.

The server learns the offset aggregation network by treating it as a regression problem during the FL training process. Specifically, in each round of DisTrans, the server collects a set of pairs $(t_i,\ t_i')$, where t_i is the offset from client i in the current round, t_i' is the aggregated offset the server outputs for client i in the previous round, and $i = 1, 2, \cdots, C$. The server learns the offset aggregation network by minimizing the ℓ_2 distance between t_i and t_i', i.e., $\min \sum_{i=1}^{C} ||t_i - t_i'||_2$, using Stochastic Gradient Descent (SGD).

5 Experiments

Hyperparameters. Our model's architecture is the double-input-channel model as shown in Fig. 1. Our default α value is 0.3, number of epochs $E = 1$, and the learning rates for the model and offset optimization are 5e-3 and 1e-3 respectively. Our neural network based offset aggregator's architecture is a single-input-channel generator with four convolutional layers.

Datasets and Model Architectures. We use six different datasets in the experiment to show the generality of DISTRANS. (i) The BioID [1] dataset contains 1521 gray level images with the frontal view of 23 people's face and eye positions. We keep 20 people's images in a descending order and central-crop the images into 256×256. (ii) The CelebA [32] dataset contains 202,599 face images of 10,177 unique, unnamed celebrities. Due to computation resource limit, we choose images of 50 identities in descending order, central-crop them to 178×178, and then resize to 128×128. (iii) The CH-MNIST [18] dataset contains eight classes of 5,000 histology tiles images (64 × 64) from patients with colorectal cancer, (iv) The CIFAR-10 [24] dataset contains 60,000 32×32 color images in 10 different classes, we resize them to 64×64, (v)The CIFAR-100 [24] dataset contains 60,000 32×32 color images in 100 different classes, and (vi) Caltech-UCSD Birds-200-2011 [43] (referred as Bird-200. The Bird-200 dataset contains 11,788 image from 200 bird species. Due to computation resource limit, we resize them to 128×128.

Here are the model architectures for each dataset. We use LeNet as the backbone for BioID, AlexNet for CelebA, CH-MNIST and CIFAR-100, ResNet18 and ResNet50 for CIFAR-100, and ResNet18 for Bird200.

Local Data Distribution. Our local data distribution ranges from entirely i.i.d. to extreme non-i.i.d., i.e., with distributional heterogeneity value ranging from 0% to 100%. Our data splitting method follows SOTA approach [38]. Specifically, we first assign a specific number of classes u out of total classes N for each client. Then, we sample $s_{i,c} \in (0.4, 0.6)$ for each client i and a selected class c, and then assign the client with $\frac{s_{i,c}}{\sum_n s_{n,c}}$ of the samples for the class c. We repeat the same process for each client.

5.1 Results Under Different Data Distributions

We evaluate DISTRANS's accuracy with different data distributions and compare with SOTA personalized FL works.

Extreme non-i.i.d. The extreme non-i.i.d. setting, following prior work [49], is a setup where each client only has one class (called positive labels), thus being

Table 1. DisTrans vs. SOTA under different data distribution. — means that the approach is not applicable under that setting, and DH means distributional heterogeneity (0%: i.i.d. and 100%: extreme non-i.i.d.). We did not evaluate the datasets of BioID and CelebA under other distributional settings due to the relative small number of images per class.

Dataset	# clients	DH	DisTrans (ours)	FedAvg	pFedMe	pFedHN	MOON	FedAwS
CH-MNIST	8	0% (i.i.d.)	**0.908**	0.891	0.778	0.702	0.887	—
		50%	**0.907**	0.892	0.834	0.871	0.894	—
		100%	**0.946**	0.908	0.908	0.641	0.910	0.942
CIFAR-10	10	0% (i.i.d.)	**0.829**	0.809	0.520	0.652	0.789	—
		40%	**0.819**	0.782	0.523	0.721	0.809	—
		60%	**0.846**	0.751	0.673	0.785	0.798	—
		80%	**0.891**	0.702	0.736	0.869	0.794	—
		100%	**0.860**	0.726	0.751	0.629	0.813	0.829
CIFAR-100	10	0% (i.i.d.)	**0.533**	0.531	0.020	0.354	0.532	—
		40%	**0.586**	0.538	0.018	0.492	0.564	—
		60%	**0.646**	0.523	0.017	0.604	0.628	—
		80%	**0.734**	0.461	0.013	0.669	0.709	—
		100%	**0.834**	0.524	0.015	0.469	0.820	—
Bird-200	10	0% (i.i.d.)	**0.556**	0.518	0.018	0.053	0.523	—
		40%	**0.548**	0.521	0.015	0.064	0.528	—
		60%	**0.542**	0.528	0.012	0.086	0.532	—
		80%	**0.565**	0.524	0.010	0.125	0.550	—
		100%	**0.641**	0.549	0.014	0.309	0.621	—
BioID	20	100%	**0.988**	0.911	0.902	0.932	0.961	0.983
CelebA	50	100%	**0.804**	0.639	0.527	0.545	0.497	0.721

disjointed from each other. The distributional heterogeneity value is thus 100%. We single out this setting, because the evaluation metrics are different from other settings given that each client only has positive images. That is, the same amount of negative images (i.e., randomly-selected images from other classes) are introduced in the testing dataset just like prior work [49].

The rows with 100% distributional heterogeneity values in Table 1 show the model's accuracy of DisTrans and the comparison with SOTA works. As shown in those results, DisTrans outperforms all prior works with five different datasets with an improvement ranging from 0.4% to 7.7%. FedAwS is clearly the SOTA, which always performs next to DisTrans, because it is designed for this extreme setting. Due to the negative test images, pFedHN performs poor since the server assigns each client model weights that are trained on only positive images according to its mechanism. FedAvg performs better than we expect because the features of negative examples are aggregated from other clients. We did not evaluate CIFAR-100 or Bird-200 under positive labels scenario (FedAws), since the number of classes per client does not satisfy positive labels setting when # clients equals to 10 for them. Instead, each client is assigned 10 or 20 disjoint classes as the extreme non-i.i.d. case.

Table 2. Comparison with data transformation for CH-MNIST dataset.

Method	Distributional heterogeneity				
	0%	25%	50%	75%	100%
FedAvg	0.891	0.893	0.892	0.847	0.908
DisTrans	**0.908**	**0.904**	**0.907**	**0.905**	**0.946**
mixup	0.896	0.895	0.882	0.839	0.901
AdvProp	0.879	0.880	0.877	0.859	0.919

Other Distributional Settings. Other settings include distributional heterogeneity values ranging from 0% (i.i.d.) to 80%. The evaluation also follows prior FL works [34,42], i.e., each client evaluates testing data with the same classes as its training data. Table 1 also shows the accuracy of DisTrans and four other SOTA works (FedAvg, pFedMe, MOON, and pFedHN). DisTrans outperforms STOA works in every data distribution for all datasets. Note that we do not evaluate FedAwS in these settings because its design is only applicable to the extreme non-i.i.d. setting.

5.2 Comparing with Data Transformation

We compare DisTrans with two state-of-the-art, popular data transformation (augmentation) methods, mixup [50] and AdvProp [45]. The former, i.e., mixup, augments training data with virtual training data based on existing data samples and one hot encoding of the label. The latter, i.e., AdvProp, augments training data with its adversarial counterpart. We add both data transformation methods for local training data at each client of FedAvg.

The comparison results are shown in Table 2. DisTrans appears to outperform both mixup and AdvProp in different data distributions from i.i.d. to non-i.i.d. There are two major reasons. First, DisTrans shifts local training and testing data distribution to fit the global model, but existing data transformation only improves training data. Second, DisTrans aggregates the offset based on data distributions, but neither data transformation approaches did so. Another thing worth noting is that mixup improves FedAvg under an i.i.d. setting, but AdvProp improves FedAvg under a non-i.i.d. setting. On one hand, that is likely because virtual examples under a non-i.i.d. setting may introduce further distributional discrepancies, while adversarial examples may help each local model better know the boundary. On the other hand, the distribution is the same under an i.i.d. setting and so does the virtual examples, but different adversarial examples may explore different boundaries at different clients.

5.3 Ablation Studies

Single vs. Double-Input Channel. We compare the performance of single vs. double-input-channel models to demonstrate the necessity in using the double-input-channel model. Table 3 shows the model's accuracy on three datasets with

Table 3. Ablation study on model structures. We adopt different distributional heterogeneity values according to the number of classes in the dataset, i.e., 0%, 25%, 50%, 75%, and 100% for CH-MNIST (8 classes) and 0%, 40%, 60%, 80%, and 100% for CIFAR-10 (10 classes) and Bird-200 (200 classes).

Dataset	Structure	Distributional heterogeneity				
		0%	40%/25%	60%/50%	80%/75%	100%
CH-MNIST	single	0.874	0.871	0.872	0.874	0.889
	double	**0.908**	**0.904**	**0.907**	**0.905**	**0.946**
CIFAR-10	single	0.775	0.802	0.785	0.796	0.811
	double	**0.829**	**0.819**	**0.846**	**0.891**	**0.860**
Bird-200	single	0.569	0.512	0.497	0.501	0.505
	double	**0.556**	**0.548**	**0.542**	**0.565**	**0.641**

Table 4. Ablation study on aggregation methods. We adopt different distributional heterogeneity values according to the number of classes in the dataset, i.e., 0%, 25%, 50%, 75%, and 100% for CH-MNIST (8 classes) and 0%, 40%, 60%, 80%, and 100% for CIFAR-10 (10 classes) and Bird-200 (200 classes).

Dataset	Aggregation	Distributional heterogeneity				
		0%	40%/25%	60%/50%	80%/75%	100%
CH-MNIST	no agg	0.868	0.887	0.907	0.905	0.946
	avg agg	0.903	0.902	0.907	0.865	0.899
	nn agg	0.908	0.904	0.905	0.887	0.921
	nn+no (default)	**0.908**	**0.904**	**0.907**	**0.905**	**0.946**
CIFAR-10	no agg	0.767	0.789	0.846	0.891	0.860
	avg agg	0.811	0.814	0.813	0.702	0.798
	nn agg	0.829	0.819	0.799	0.743	0.839
	nn+no (default)	**0.829**	**0.819**	**0.846**	**0.891**	**0.860**
Bird-200	no agg	0.501	0.522	0.542	0.565	0.641
	avg agg	0.526	0.529	0.525	0.515	0.489
	nn agg	0.556	0.548	0.551	0.532	0.513
	nn+no (default)	**0.556**	**0.548**	**0.551**	**0.565**	**0.641**

different distributional heterogeneity values. As shown, the double-input-channel model always outperforms the single-input-channel with around 3%–9% accuracy improvement on three different datasets.

Different Offset Aggregations. We compare different offset aggregation methods, i.e., no aggregation, average aggregation and neural network (NN) based aggregation, on three datasets with various distributional heterogeneity values. Table 4 shows the comparison results. No aggregation performs best when the distributional heterogeneity is greater than 50%, and NN aggregation performs the best when the distributional heterogeneity is smaller than 50%. Average aggregation always performs worse than the other two. This motivates the design of DisTrans in adopting no aggregation for greater than 50% distributional heterogeneity and NN aggregation for less than 50% distributional heterogeneity, which is the "nn+no (default)" row in Table 4.

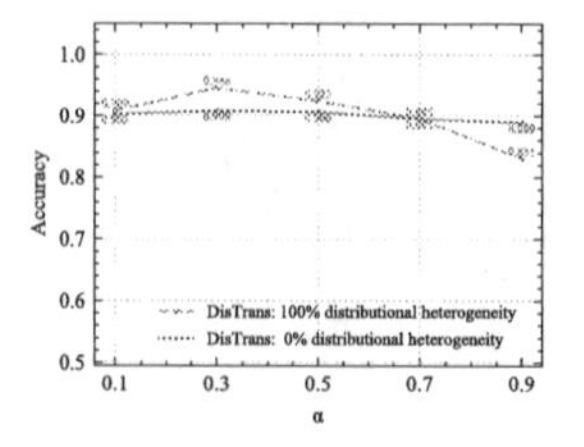

Fig. 4. Accuracy vs. α for CH-MNIST.

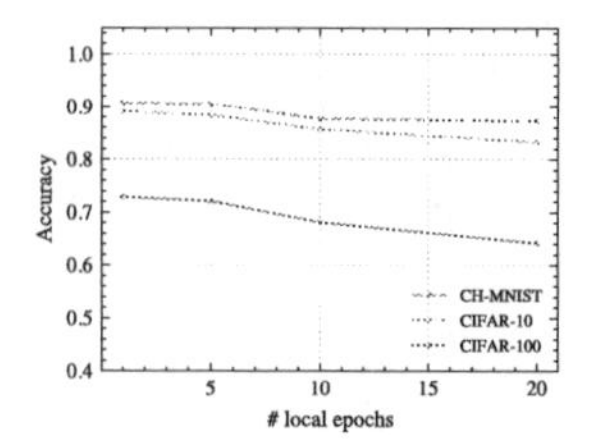

Fig. 5. Accuracy vs. # of local epochs.

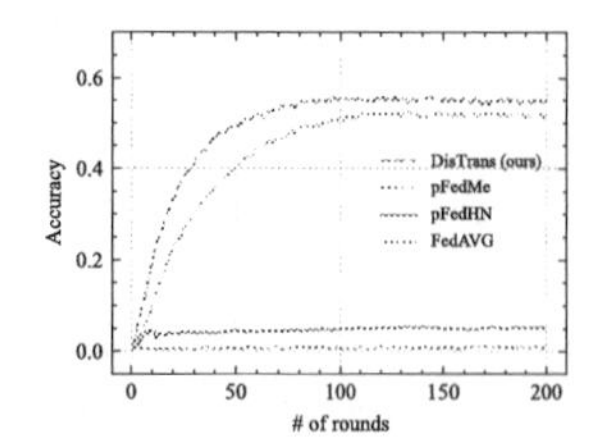

Fig. 6. Accuracy vs. # of rounds for Birds-200.

Different α Values. We evaluate top-1 accuracy of DisTrans with different α values, and 0% and 100% distributional heterogeneity to justify why we choose 0.3 as α. Figure 4 shows the results. The accuracy with 100% distributional heterogeneity is more sensitive to α than that with 0%. In both data distributions, the accuracy is the highest when α equals 0.3. The reason is as follows. When α is small, the offset is too weak to shift the distribution. When the α is large, the offset is too strong in overriding the original data distribution.

Different Local Epochs. We study different local training epochs for each round. Figure 5 shows the accuracy for CH-MNIST, CIFAR-100, and Bird-200 with epochs from 1, 5, 10, to 20. The accuracy is the highest with the local epoch as 1, and decreases when the epoch increases. The reason is too much local training makes offsets become overfitted to local data.

Convergence Rate. We study the convergence rate of three SOTA works and DisTrans in terms of communication rounds between the server and local clients. Figure 6 shows the number of communication rounds as the x-axis and the model's accuracy as the y-axis for the Birds-200 dataset under the i.i.d. setting (i.e., 0% distributional heterogeneity). There are two things worth noting. First, as shown, the convergence rate of DisTrans is similar to that of FedAvg, which needs approximately 100 rounds. Second, the accuracy of DisTrans is constantly better than FedAvg for each communication between client and server.

5.4 Scalability

We study the scalability of DisTrans using two datasets CH-MNIST and CIFAR-100 as the number of FL clients increases. First, we test the number of clients from 8, 16, 24, to 40 using 50% distributional heterogeneity for CH-MNIST. The third column in Table 5 shows the accuracy of four different works including DisTrans as the number of clients increases. The fourth to eighth columns in Table 5 show the total number of rounds to reach certain accuracy. Generally, the convergence needs more rounds with more clients, which aligns with the previous work [34]. Second, we show the testing accuracy in Table 6 for CIFAR-100 (with ResNet18) of 50, 100, and 500 clients. Each client has 10 classes and the sample rate is 0.2 [28]. DISTRANS outperforms SOTA by 8.8%–30.7%. Note that the accuracy of DISTRANS drops by 8.4% while SOTA drops by 10.8% to 27.6% for 500 clients.

Table 5. Best accuracy and the number of rounds to achieve it vs. different number of clients using the CH-MNIST dataset when reaching listed accuracy, e.g., DISTRANS needs 5 rounds to achieve a 0.800 accuracy with 8 clients. (—: the approach cannot reach the accuracy under that setting.)

	# clients	Best accuracy	# of rounds to achieve				
			>0.700	>0.800	>0.850	>0.870	>0.890
DISTRANS (ours)	8	**0.907**	**2**	**5**	**10**	36	**63**
	16	**0.898**	8	15	25	51	**123**
	24	**0.897**	12	26	**37**	**68**	**154**
	40	**0.895**	14	29	**40**	**87**	**192**
FedAVG	8	0.892	3	**5**	15	**24**	78
	16	0.883	7	**12**	**23**	**48**	—
	24	0.880	10	**21**	39	74	—
	40	0.878	19	34	44	100	—
pFedMe	8	0.834	690	779	—	—	—
	16	0.805	844	1,225	—	—	—
	24	0.725	1,859	—	—	—	—
	40	0.719	3,071	—	—	—	—
pFedHN	8	0.871	3	8	23	117	—
	16	0.817	**6**	29	—	—	—
	24	0.816	**4**	28	—	—	—
	40	0.832	**5**	**27**	—	—	—

Table 6. Best accuracy vs. number of clients on CIFAR-100.

	DISTRANS			pFedHN			pFedHN-pc			MOON		
#Client	50	100	500	50	100	500	50	100	500	50	100	500
Accuracy	**0.729**	**0.681**	**0.645**	0.614	0.538	0.338	0.623	0.541	0.372	0.615	0.593	0.507

Table 7. Communication overhead of weights and offset for 64×64 RGB images.

	Baseline (in bytes)	DISTRANS (in bytes)		Δ overhead
	single-input-channel weight	double-input-channel weight	offset	
LeNet	4,346,447	4,350,415	49,280	1.225%
AlexNet	244,449,263	244,489,263	49,280	0.036%
ResNet18	44,805,709	44,826,189	49,280	0.155%
ResNet50	94,326,992	94,408,912	49,280	0.139%

5.5 Communication Overhead

We study the communication overhead by calculating the Δ bytes brought by our double-input-channel weights and offset in each communication round. The comparison baseline used is FedAvg with conventional single-input-channel model. Table 7 shows results for different backbone neural network architectures: The overhead is between 0.036% to 1.225% with an average value of 0.389% for different network architectures because the double-input-channel model only introduces one additional layer and the offsets are small.

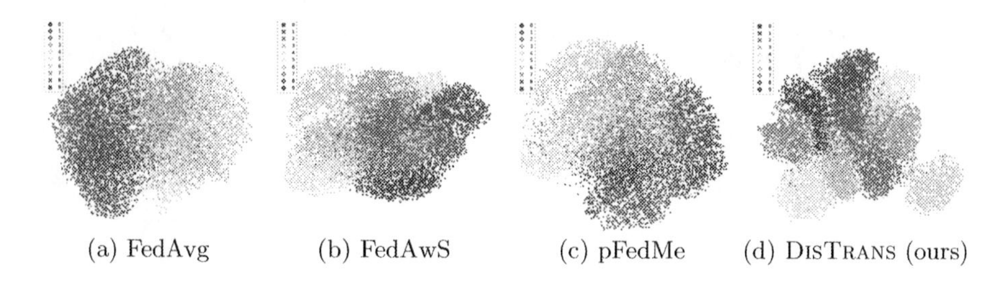

(a) FedAvg (b) FedAwS (c) pFedMe (d) DisTrans (ours)

Fig. 7. UMAP visualization of embedded feature representations in the global model for test images in CIFAR-10. DisTrans learns better feature representations than FedAvg, FedAwS, and pFedMe.

5.6 Prediction Visualization

We perform an experiment on CIFAR-10 using ten FL clients where each client has data for only one class, and visualize hidden feature representations using Uniform Manifold Approximation and Projection (UMAP) in Fig. 7. The model trained using FedAvg learns poor features, which are mixed and indistinguishable. The feature representations of FedAwS and pFedMe also highly overlap. By contrast, the feature representations of DisTrans are well separated in Fig. 7d as a result of shifting the local data distributions via personalized offsets.

6 Conclusion

FL often needs to contend with client-side local training data with different distributions with high heterogeneity. This paper advances a novel approach, DisTrans, based on distributional transformation, that jointly optimizes local model and data with a personalized offset and then aggregates both at a central server. We perform an empirical evaluation of DisTrans using five different datasets, which shows that DisTrans outperforms SOTA FL and data augmentation methods, under different degrees of data distributional heterogeneity ranging from extreme non-i.i.d. to i.i.d.

Acknowledgment. This work was supported in part by Johns Hopkins University Institute for Assured Autonomy (IAA) with grants 80052272 and 80052273, and National Science Foundation (NSF) under grants CNS-21-31859, CNS-21-12562, and CNS-18-54001. The views and conclusions contained herein are those of the authors and should not be interpreted as necessarily representing the official policies or endorsements, either expressed or implied, of NSF or JHU-IAA.

References

1. Bioid face dataset. https://www.bioid.com/facedb/
2. Bonawitz, K., et al.: Towards federated learning at scale: system design. arXiv preprint arXiv:1902.01046 (2019)
3. Caldas, S., Konečny, J., McMahan, H.B., Talwalkar, A.: Expanding the reach of federated learning by reducing client resource requirements. arXiv preprint arXiv:1812.07210 (2018)
4. Cao, X., Fang, M., Liu, J., Gong, N.Z.: Fltrust: byzantine-robust federated learning via trust bootstrapping. arXiv preprint arXiv:2012.13995 (2020)
5. Cao, X., Jia, J., Gong, N.Z.: Provably secure federated learning against malicious clients. In: Proceedings of the AAAI Conference on Artificial Intelligence, vol. 35, pp. 6885–6893 (2021)
6. Chen, M., Poor, H.V., Saad, W., Cui, S.: Convergence time optimization for federated learning over wireless networks. IEEE Trans. Wirel. Commun. **20**(4), 2457–2471 (2021). https://doi.org/10.1109/TWC.2020.3042530
7. Cubuk, E.D., Zoph, B., Mane, D., Vasudevan, V., Le, Q.V.: Autoaugment: learning augmentation strategies from data. In: Proceedings of the IEEE/CVF Conference on Computer Vision and Pattern Recognition (CVPR), June 2019
8. Cubuk, E.D., Zoph, B., Shlens, J., Le, Q.: Randaugment: practical automated data augmentation with a reduced search space. In: Larochelle, H., Ranzato, M., Hadsell, R., Balcan, M.F., Lin, H. (eds.) Advances in Neural Information Processing Systems, vol. 33, pp. 18613–18624. Curran Associates, Inc. (2020). https://proceedings.neurips.cc/paper/2020/file/d85b63ef0ccb114d0a3bb7b7d808028f-Paper.pdf
9. Fang, M., Cao, X., Jia, J., Gong, N.: Local model poisoning attacks to byzantine-robust federated learning. In: 29th USENIX Security Symposium (USENIX Security 20), pp. 1605–1622 (2020)
10. Guo, P., Wang, P., Zhou, J., Jiang, S., Patel, V.M.: Multi-institutional collaborations for improving deep learning-based magnetic resonance image reconstruction using federated learning. In: Proceedings of the IEEE/CVF Conference on Computer Vision and Pattern Recognition (CVPR), pp. 2423–2432 (June 2021)
11. Haddadpour, F., Mahdavi, M.: On the convergence of local descent methods in federated learning. arXiv preprint arXiv:1910.14425 (2019)
12. Hamer, J., Mohri, M., Suresh, A.T.: Fedboost: a communication-efficient algorithm for federated learning. In: International Conference on Machine Learning, pp. 3973–3983. PMLR (2020)
13. Hard, A., et al.: Federated learning for mobile keyboard prediction. arXiv preprint arXiv:1811.03604 (2018)
14. He, K., Zhang, X., Ren, S., Sun, J.: Deep residual learning for image recognition. In: Proceedings of the IEEE Conference on Computer Vision and Pattern Recognition (CVPR), June 2016

15. Hsu, T.M.H., Qi, H., Brown, M.: Federated visual classification with real-world data distribution (2020)
16. Jin, Y., Jiao, L., Qian, Z., Zhang, S., Lu, S., Wang, X.: Resource-efficient and convergence-preserving online participant selection in federated learning. In: 2020 IEEE 40th International Conference on Distributed Computing Systems (ICDCS), pp. 606–616 (2020). https://doi.org/10.1109/ICDCS47774.2020.00049
17. Kairouz, P., et al.: Advances and open problems in federated learning. arXiv preprint arXiv:1912.04977 (2019)
18. Kather, J.N., et al.: Multi-class texture analysis in colorectal cancer histology. Sci. Rep. **6**(1), 1–11 (2016)
19. Kim, H., Park, J., Bennis, M., Kim, S.L.: Blockchained on-device federated learning. IEEE Commun. Lett. **24**(6), 1279–1283 (2020). https://doi.org/10.1109/LCOMM.2019.2921755
20. Kim, I., Kim, Y., Kim, S.: Learning loss for test-time augmentation. In: Proceedings of Advances in Neural Information Processing Systems (2020)
21. Konečnỳ, J., McMahan, H.B., Ramage, D., Richtárik, P.: Federated optimization: distributed machine learning for on-device intelligence. arXiv preprint arXiv:1610.02527 (2016)
22. Konečnỳ, J., McMahan, H.B., Yu, F.X., Richtárik, P., Suresh, A.T., Bacon, D.: Federated learning: strategies for improving communication efficiency. arXiv preprint arXiv:1610.05492 (2016)
23. Konečnỳ J., McMahan, H.B., Ramage, D., Richtárik, P.: Federated optimization: distributed machine learning for on-device intelligence (2016)
24. Krizhevsky, A.: Learning multiple layers of features from tiny images. Technical report (2009)
25. Laguel, Y., Pillutla, K., Malick, J., Harchaoui, Z.: Device heterogeneity in federated learning: A superquantile approach. arXiv preprint arXiv:2002.11223 (2020)
26. Lemley, J., Bazrafkan, S., Corcoran, P.: Smart augmentation learning an optimal data augmentation strategy. IEEE Access **5**, 5858–5869 (2017). https://doi.org/10.1109/ACCESS.2017.2696121
27. Li, D., Wang, J.: FEDMD: heterogenous federated learning via model distillation. arXiv preprint arXiv:1910.03581 (2019)
28. Li, Q., He, B., Song, D.: Model-contrastive federated learning. In: Proceedings of the IEEE/CVF Conference on Computer Vision and Pattern Recognition (2021)
29. Li, T., Sahu, A.K., Talwalkar, A., Smith, V.: Federated learning: challenges, methods, and future directions. IEEE Signal Process. Mag. **37**(3), 50–60 (2020). https://doi.org/10.1109/MSP.2020.2975749
30. Li, Y., et al.: Shape-texture debiased neural network training. CoRR abs/2010.05981 (2020). https://arxiv.org/abs/2010.05981
31. Liu, W., Chen, L., Chen, Y., Zhang, W.: Accelerating federated learning via momentum gradient descent. IEEE Trans. Parallel Distrib. Syst. **31**(8), 1754–1766 (2020). https://doi.org/10.1109/TPDS.2020.2975189
32. Liu, Z., Luo, P., Wang, X., Tang, X.: Deep learning face attributes in the wild. In: Proceedings of International Conference on Computer Vision (ICCV), December 2015
33. Luo, B., Li, X., Wang, S., Huang, J., Tassiulas, L.: Cost-effective federated learning design. In: IEEE INFOCOM 2021-IEEE Conference on Computer Communications, pp. 1–10. IEEE (2021)
34. McMahan, B., Moore, E., Ramage, D., Hampson, S., Arcas, B.A.: Communication-efficient learning of deep networks from decentralized data. In: Artificial Intelligence and Statistics, pp. 1273–1282. PMLR (2017)

35. Nasr, M., Shokri, R., Houmansadr, A.: Comprehensive privacy analysis of deep learning: passive and active white-box inference attacks against centralized and federated learning. In: 2019 IEEE Symposium on Security and Privacy (SP), pp. 739–753. IEEE (2019)
36. Nguyen, H.T., Sehwag, V., Hosseinalipour, S., Brinton, C.G., Chiang, M., Vincent Poor, H.: Fast-convergent federated learning. IEEE J. Sel. Areas Commun. **39**(1), 201–218 (2021). https://doi.org/10.1109/JSAC.2020.3036952
37. Pérez, J.C., et al.: Enhancing adversarial robustness via test-time transformation ensembling. In: Proceedings of the IEEE/CVF International Conference on Computer Vision (ICCV) (2021)
38. Shamsian, A., Navon, A., Fetaya, E., Chechik, G.: Personalized federated learning using hypernetworks. In: Proceedings of the 38th International Conference on Machine Learning (ICML), PMLR 139 (2021)
39. Shanmugam, D., Blalock, D.W., Balakrishnan, G., Guttag, J.V.: Better aggregation in test-time augmentation. In: Proceedings of International Conference on Computer Vision (ICCV) (2021)
40. Suresh, A.T., Felix, X.Y., Kumar, S., McMahan, H.B.: Distributed mean estimation with limited communication. In: International Conference on Machine Learning, pp. 3329–3337. PMLR (2017)
41. Szegedy, C., et al.: Going deeper with convolutions. In: Proceedings of the IEEE Conference on Computer Vision and Pattern Recognition, pp. 1–9 (2015)
42. T Dinh, C., Tran, N., Nguyen, T.D.: Personalized federated learning with moreau envelopes. In: Advances in Neural Information Processing Systems, vol. 33 (2020)
43. Wah, C., Branson, S., Welinder, P., Perona, P., Belongie, S.: The Caltech-UCSD Birds-200-2011 Dataset. Technical report CNS-TR-2011-001, California Institute of Technology (2011)
44. Wang, J., Xu, Z., Garrett, Z., Charles, Z., Liu, L., Joshi, G.: Local adaptivity in federated learning: convergence and consistency (2021)
45. Xie, C., Tan, M., Gong, B., Wang, J., Yuille, A.L., Le, Q.V.: Adversarial examples improve image recognition. In: Proceedings of the IEEE/CVF Conference on Computer Vision and Pattern Recognition (CVPR), June 2020
46. Xu, J., Glicksberg, B.S., Su, C., Walker, P., Bian, J., Wang, F.: Federated learning for healthcare informatics. J. Healthcare Inform. Res. **5**(1), 1–19 (2021)
47. Yang, Q., Liu, Y., Chen, T., Tong, Y.: Federated machine learning: concept and applications. ACM Trans. Intell. Syst. TechnoL. (TIST) **10**(2), 1–19 (2019)
48. Yao, X., Huang, T., Wu, C., Zhang, R., Sun, L.: Towards faster and better federated learning: a feature fusion approach. In: 2019 IEEE International Conference on Image Processing (ICIP), pp. 175–179 (2019). https://doi.org/10.1109/ICIP.2019.8803001
49. Yu, F., Rawat, A.S., Menon, A., Kumar, S.: Federated learning with only positive labels. In: International Conference on Machine Learning, pp. 10946–10956. PMLR (2020)
50. Zhang, H., Cisse, M., Dauphin, Y.N., Lopez-Paz, D.: mixup: Beyond empirical risk minimization. In: International Conference on Learning Representations (2018)
51. Zhang, H., Moustapha Cisse, Yann N. Dauphin, D.L.P.: mixup: beyond empirical risk minimization. International Conference on Learning Representations (ICLR) (2018).https://openreview.net/forum?id=r1Ddp1-Rb
52. Zhang, X., Wang, Q., Zhang, J., Zhong, Z.: Adversarial autoaugment (2019)

Where in the World Is This Image? Transformer-Based Geo-localization in the Wild

Shraman Pramanick[1(✉)], Ewa M. Nowara[1], Joshua Gleason[2], Carlos D. Castillo[1], and Rama Chellappa[1]

[1] Johns Hopkins University, Baltimore, USA
{spraman3,carlosdc,rchella4}@jhu.edu
[2] University of Maryland, College Park, USA
gleason@umd.edu

Abstract. Predicting the geographic location (geo-localization) from a single ground-level RGB image taken anywhere in the world is a very challenging problem. The challenges include huge diversity of images due to different environmental scenarios, drastic variation in the appearance of the same location depending on the time of the day, weather, season, and more importantly, the prediction is made from a single image possibly having only a few geo-locating cues. For these reasons, most existing works are restricted to specific cities, imagery, or worldwide landmarks. In this work, we focus on developing an efficient solution to planet-scale single-image geo-localization. To this end, we propose TransLocator, a unified dual-branch transformer network that attends to tiny details over the entire image and produces robust feature representation under extreme appearance variations. TransLocator takes an RGB image and its semantic segmentation map as inputs, interacts between its two parallel branches after each transformer layer and simultaneously performs geo-localization and scene recognition in a multi-task fashion. We evaluate TransLocator on four benchmark datasets - Im2GPS, Im2GPS3k, YFCC4k, YFCC26k and obtain 5.5%, 14.1%, 4.9%, 9.9% continent-level accuracy improvement over the state-of-the-art. TransLocator is also validated on real-world test images and found to be more effective than previous methods.

Keywords: Geo-location estimation · Vision transformer · Multi-task learning · Semantic segmentation

1 Introduction

Can we determine the location of a scene given a single ground-level RGB image? For famous and characteristic scenes, such as the Eiffel Tower, it is trivial because the landmark is so distinctive of Paris. Moreover, there is a lot of images captured in such prominent locations under different viewing angles, at different times of

Supplementary Information The online version contains supplementary material available at https://doi.org/10.1007/978-3-031-19839-7_12.

S. Avidan et al. (Eds.): ECCV 2022, LNCS 13698, pp. 196–215, 2022.
https://doi.org/10.1007/978-3-031-19839-7_12

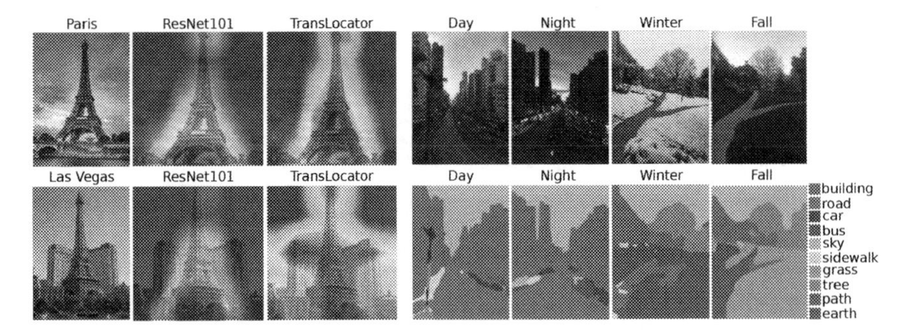

Fig. 1. *First three columns:* **Utilizing global context -** TransLocator focuses on the entire image to correctly geo-locate the Eiffel Tower and its replica as visualized with Grad-CAM [45] class activation maps. *Last four columns:* **Robustness to appearance variations -** Semantic segmentation maps are robust to extreme appearance variations.

the day, and even in different weather or lighting conditions. However, some scenes, especially places outside cities and tourist attractions, may not have characteristic landmarks and it is not so obvious to locate where they were snapped. This is the case for the vast majority of the places in the world. Moreover, such places are less popular and are less photographed. As a result, there are very few images from such locations and the existing ones do not capture a diversity of viewing angles, time of the day, or weather conditions, making it much harder to geo-locate. Because of the complexity of this problem, most existing geo-localization approaches have been constrained to small parts of the world [8,16,47,55]. Recently, convolutional neural networks (CNNs) trained with large datasets have significantly improved the performance of geo-localization methods and enabled extending the task to the scale of the entire world [17,18,34,63]. However, planet-scale unconstrained geo-localization is still a very challenging problem and existing state-of-the-art methods struggle to geo-locate images taken anywhere in the world.

In contrast to many other vision applications, single-image geo-localization often depends on fine-grained visual cues present in small regions of an image. In Fig. 1, consider the photo of the Eiffel Tower in Paris and its replica in Las Vegas. Even though these two images seem to come from the same location, the buildings and vegetation in the background play a decisive role in distinguishing them. Similarly, in the case of most other images, the global context spanned over the entire image is more important than individual foreground objects in geo-localization. Recently, a few studies [12,39] comparing vision transformer (ViT) with CNNs have revealed that the early aggregation of global information using self-attention enables transformers to build long-range dependencies within an image. Moreover, higher layers of ViT maintain spatial location information better than CNNs. Hence, we argue that transformer-based networks are more effective than CNNs for geo-localization because they focus on detailed visual cues present in the entire image.

Another challenge of single-image geo-localization is the drastic appearance variation of the exact location under different daytime or weather conditions. Semantic segmentation offers a solution to this problem by generating robust representations in such extreme variations [47]. For example, consider the drastic disparity of the RGB images of same locations in day and night or winter and fall in Fig. 1. In contrast to the RGB images, the semantic segmentation maps remain almost unchanged. Furthermore, semantic segmentation provides auxiliary information about the objects present in the image. This additional information can be a valuable pre-processing step since it enables the model to learn which objects occur more frequently in which geographic locations. For example, as soon as the semantic map detects mountains, the model immediately eliminates all flat regions, thus, reducing the complexity of the problem.

Planet-scale geo-localization deals with a diverse set of input images caused by different environmental settings (e.g., outdoors vs indoors), which entails different features to distinguish between them. For example, to geo-locate outdoor images, features such as the architecture of buildings or the type of vegetation are important. In contrast, for indoor images, the shape and style of furniture may be helpful. To address such variations, Muller et al. [34] proposed to train different networks for different environmental settings. Though such an approach produces good results, they are cost-prohibitive and are not generalizable to a higher number of environmental scenarios. In contrast, we propose a unified multi-task framework for simultaneous geo-localization and scene recognition applied to images from all environmental settings.

This work addresses the challenges of planet-scale single-image geo-localization by designing a novel dual-branch transformer architecture, TransLocator. We treat the problem as a classification task [34,63] by subdividing the earth's surface into a high number of geo-cells and assigning each image to one geo-cell. TransLocator takes an RGB image and its corresponding semantic segmentation map as input, divides them into non-overlapping patches, flattens the patches, and feeds them into two parallel transformer encoder modules to simultaneously predict the geo-cell and recognize the environmental scene in a multi-task framework. The two parallel transformer branches interact after every layer, ensuring an efficient fusion strategy. The resulting features learned by TransLocator are robust under appearance variation and focus on tiny details over the entire image.

In summary, our contributions are three-fold. (*i*) We propose TransLocator - a unified solution to planet-scale single-image geo-localization with a dual-branch transformer network. TransLocator is able to distinguish between similar images from different locations by precisely attending to tiny visual cues. (*ii*) We propose a simple yet efficient fusion of two transformer branches, which helps TransLocator to learn robust features under extreme appearance variation. (*iii*) We achieve state-of-the-art performance on four datasets with a significant improvement of 5.5%, 14.1%, 4.9%, 9.9% continent-level geolocational accuracy on Im2GPS [17], Im2GPS3k [18], YFCC4k [59], and YFCC26k [51], respectively. We also qualitatively evaluate the effectiveness of the proposed method on real-world images.

2 Related Works

2.1 Single-image Geo-Localization

Small-Scale Approaches: Planet-scale single-image geo-localization is difficult due to several challenges, including the large variety of images due to different environmental scenarios and drastic differences in the appearance of same location based on the weather, time of day, or season. For this reason, many existing approaches are limited to geo-locating images captured in very specific and constrained locations. For example, many approaches have been restricted to geo-locating an image within a single city [3,16,47], such as Pittsburgh [56], San Francisco [8], or Tokyo [55]. Far fewer approaches have attempted geo-localization in natural, non-urban environments. Some have limited the problem to only very specific natural environments, such as beaches [5,61], deserts [57], or mountains [2,44].

Cross-view Approaches: The challenge of large-scale image geo-localization with few landmarks and limited training data has led some researchers to propose cross-view approaches, which match a query ground-level RGB image with a large reference dataset of aerial or satellite images [21,28,41,53,54,67,78,79]. However, these approaches require access to a relevant reference dataset which is not available for many locations.

Planet-scale Approaches: Only a few approaches have attempted to geo-locate images on a scale of an entire world without any restrictions. Im2GPS [17] was the first work to geo-locate images taken anywhere on the earth using hand-crafted features and nearest neighbor search. The same authors have later improved their approach [18] by refining the search with multi-class support vector machines. PlaNet [63] was the first deep learning approach for unconstrained geo-localization which significantly outperformed the two Im2GPS approaches [17,18]. Vo et al. [59] combined the Im2GPS and PlaNet approaches by using the deep-learning-based features to match a query image with a nearest neighbors search. The approach by Muller et al. [34] achieved the current state-of-the-art performance for unconstrained geo-localization. They fine-tuned three separate ResNet101 networks, each trained only on images of outdoor natural, outdoor urban, or indoor scenes. Detailed surveys of existing work on geo-localization can be found at [4,32].

2.2 Vision Transformer

Following the success of transformers [58] in machine translation, convolution-free networks that rely only on self-attentive transformer layers have gone viral in computer vision. In particular, Vision Transformer (ViT) [13] was the first to apply a pure transformer architecture on non-overlapping image patches and surpassed CNNs for image classification. Inspired by ViT, transformers have been widely adopted for various computer vision tasks, such as object detection [6,14,26,80], image segmentation [10,48,64,74], video understanding [49,62,65, 69], low-shot learning [12,71], image super resolution [9,66], 3D classification [33,73], and multimodal learning [1,20,23,25,27,31,68].

We are the first to address the ill-posed planet-scale single-image geo-localization problem by designing a novel dual-branch transformer architecture. Unlike [34], we use only a single network for all kinds of input images, allowing our model to take advantage of the similar features from different scenes while maintaining its ability to learn scene-specific features. Moreover, our approach is robust under extreme appearance variations and generalizes to challenging real-world images.

3 Proposed System - TransLocator

This section presents our proposed multimodal multi-task system, TransLocator for geo-location estimation and scene recognition. Following the literature [34,46,63], we treat geo-localization as a classification problem by subdividing the earth into several geographical cells[3] containing a similar number of images. As the size and the number of geo-cells poses a trade-off between system performance and classification difficulty, we use multiple partitioning schemes which allow the system to learn geographical features at different scales, leading to a more discriminative classifier. Furthermore, we incorporate semantic representations of RGB images to learn robust feature representation across different daytime and weather conditions. Finally, we predict the geo-cell and environmental scenario of the input image by exploiting contextual information in a multi-task fashion (Fig. 2).

3.1 Global Context with Vision Transformer

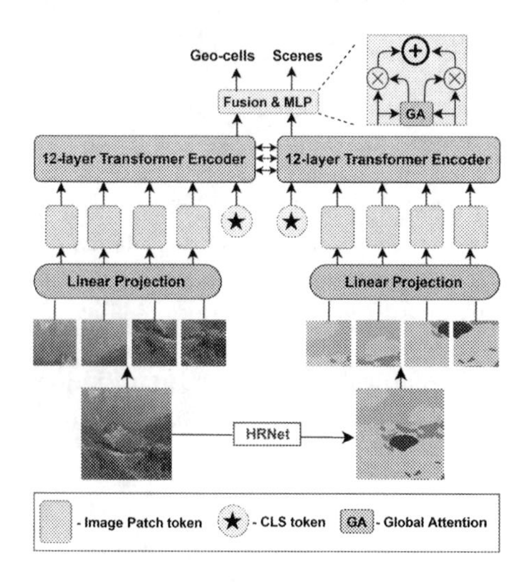

Fig. 2. Overview of the proposed model - TransLocator. A dual-branch transformer network is trained using RGB images and corresponding semantic segmentation maps in a multi-task learning framework to simultaneously predict the environmental scene and the geographic location (geo-cell) of the image.

We use a vision transformer (ViT) [13] model as the backbone of our architecture. Transformers have traditionally been used for sequential data, such as text or audio [11, 29,36,58,70]. In order to extend transformers to vision problems, an image is split into a sequence of 2-D patches which are then flattened and fed into the stacked transformer encoders through a trainable linear projection layer. An additional classification token (`CLS`) is added to the sequence, as in the original BERT approach [11]. Moreover, positional embeddings are added to each token to preserve the order of the patches in the original image. A transformer encoder is made up with a sequence of blocks containing multi-headed self-attention (`MSA`) with a feed-forward network (`FFN`). `FFN` contains two multi-layer perceptron (`MLP`) layers with GELU non-linearity applied after the first layer. Layer normalization (`LN`) is applied before every block, and residual connections after every block. The output of k-th block $x^{(k)}$ in the transformer encoder can be expressed as

$$x^{(0)} = [x_{cls}^{(0)} || x_{patch}^{(0)}] + x_{pos} \tag{1}$$

[3] Following [34], we use the $S2$ *geometry library* to generate the geo-cells.

$$y^{(k)} = x^{(k-1)} + \texttt{MSA}(\texttt{LN}(x^{(k-1)})) \tag{2}$$

$$x^{(k)} = y^{(k)} + \texttt{FFN}(\texttt{LN}(y^{(k)})) \tag{3}$$

where $x_{cls}^{(i)} \in \mathbb{R}^{1\times C}$ is the CLS token, $x_{patch}^{(i)} \in \mathbb{R}^{N\times C}$ is the patch tokens at i-th layer, and $x_{pos} \in \mathbb{R}^{(1+N)\times C}$ is the positional embeddings. N and C is the patch-sequence length and embedding dimension, respectively. $||$ denotes concatenation.

The self-attention in ViT is a mechanism to aggregate information from all spatial locations and is structurally very different from the fixed-sized receptive field of CNNs. Every layer of ViT can access the whole image and, thus, learn the global structure more effectively. This particular attribute of ViT plays a vital role in our classification system. In agreement with Raghu et al. [39], we empirically establish how the global receptive field of ViT can help to attend to small but essential visual cues, which CNNs often neglect. We provide a detailed comparative evaluation of this phenomenon in Sect. 5.3.

3.2 Semantic Segmentation for Robustness to Appearance Variation

In order to improve the network's ability to generalize to scenes captured in different daytime and weather conditions, we train a dual-branch vision transformer on the RGB images and their corresponding semantic segmentation maps. As shown in Fig. 1, compared to RGB features, the semantic layout of a scene is generally more robust to drastic appearance variations. We use HRNet [60] pre-trained on ADE20K and scene parsing datasets [76,77] to obtain high resolution semantic maps. HRNet assigns each pixel in the RGB image to one of the pre-defined 150 object classes, such as *sky*, *tree*, *sea*, *building*, *grass*, *pier*, etc.

Multimodal Feature Fusion (MFF): Our proposed framework contains two parallel transformer branches, one for the RGB image and the other for the corresponding semantic map. Since the two branches carry complementary information of the same input, effective fusion is the key for learning multimodal feature representations. We propose a simple and computationally light fusion scheme, where we sum the CLS tokens of each branch after every transformer encoder layer. At each layer, the CLS token is considered as an abstract global feature representation. This strategy is as effective as concatenating all feature tokens, but avoids quadratic complexity. Once the CLS tokens are fused, the information will be passed back to patch tokens at the later transformer encoder layers. More formally, ${}^{(i)}x^{(k)}$, the token sequence at k-th layer of a branch i can be expressed as

$$ {}^{(i)}x^{(k)} = \Big[g\big(\sum_{j\in\{\text{rgb, seg}\}} f({}^{(j)}x_{\text{cls}}^{(k)})\big) || {}^{(i)}x_{patch}^{(k)} \Big] \tag{4}$$

where $f(.)$ and $g(.)$ are the projection and back-projection functions used to align the dimensions.

Since the relative importance of the two branches depends upon the structure of the input image, we attentively fuse the CLS tokens from the last layer of each branch. Motivated by [15,37,38], we design our attention module with two major parts – modality attention generation and weighted concatenation. In the first part, a sequence of dense layers followed by a softmax layer is used to generate the attention scores $w_{mm} = [w_{rgb}, w_{seg}]$ for the two branches. In the second part, the CLS tokens from the last transformer layer are weighted using their respective attention scores and concatenated together as follows

$$f_{rgb} = (1 + w_{rgb})^{rgb} x_{cls}^{(last)} \tag{5}$$

$$f_{seg} = (1 + w_{seg})^{seg} x_{cls}^{(last)} \tag{6}$$

$$f_{mm} = [f_{rgb} || f_{seg}] \tag{7}$$

We use residual connections to improve the gradient flow. The final multimodal representation, f_{mm}, is fed into fully-connected classifier heads.

3.3 Single Model with Multi-Task Learning

Different features are essential for various environmental settings, such as indoor and outdoor urban or natural scenes. Hence, geo-localization can benefit from contextual knowledge about the surroundings, and this information can reduce the complexity of the data space, thus simplifying the classification problem. Muller et al. [34] addressed this issue and approached the problem by fine-tuning three individual networks separately on natural, urban, and indoor images. However, one immediate drawback of their approach is that using multiple separate networks is cost-prohibitive and limits the number of scene kinds. In addition, the separately trained networks can not share the learned features, which likely have semantic similarities across different scene kinds, which effectively reduces the size and potency of the training set.

We address these two drawbacks by training a single network with a multi-task learning objective for simultaneous geo-localization and scene recognition. Following [34], we use a ResNet-152 pre-trained on *Places2* dataset[4] [75] to label the training images with corresponding 365 different scene categories. Additionally, based on the provided scene hierarchy[5], we label each image with a coarser 16 and 3 different scene labels. In multi-task learning, adding complementary tasks has been proven to improve the results of the main task [7,40,72]. We follow a multi-task learning approach known as *hard parameter sharing* [43] which shares the parameters of hidden layers for all tasks and uses task-specific classifier heads. This strategy reduces overfitting because learning the same weights for multiple tasks forces the model to learn a generalized representation useful for the different tasks. More specifically, we only add a classifier head on top of the fused multimodal features and train the system end-to-end for both tasks.

3.4 Training Objective

Since there is a trade-off between the classification difficulty and the prediction accuracy caused by the number and size of the geo-cells, we use partitioning of

[4] *Places2 ResNet152 model:* https://github.com/CSAILVision/places365.

[5] http://places2.csail.mit.edu/download.html.

the earth's surface at three different resolutions, which we call *coarse*, *middle* and *fine* geo-cells[6]. We feed the final multimodal feature representation f_{mm} into four parallel classifier heads for the final classification: three for geo-cell prediction and one for scene recognition. We use a cross-entropy loss for each head. Our overall loss function can be summarized as follows:

$$\mathcal{L}_{total} = (1 - \alpha - \beta)\mathcal{L}_{geo}^{coarse} + \alpha\mathcal{L}_{geo}^{middle} + \beta\mathcal{L}_{geo}^{fine} + \gamma\mathcal{L}_{scene} \quad (8)$$

4 Experiments

We conduct extensive experiments to show the effectiveness of our proposed method. In this section, we describe the datasets, the evaluation metrics, the baseline methods, and the detailed experimental settings.

4.1 Datasets

We use publicly available RGB image datasets with corresponding ground truth GPS (latitude, longitude) tags for training, validation, and testing. We trained our model on the MediaEval Placing Task 2016 dataset (MP-16) [24] containing 4.72M geo-tagged images sourced from Flickr. Like Vo et al. [59], during training we excluded images taken by the same authors in our validation or test sets. We validated and tested our model on two randomly sampled subsets of images from the Yahoo Flickr Creative Commons 100 Million dataset (YFCC100M) [52], referred to as YFCC26k [51] and YFCC4k [59] containing 25,600 and 4536 images, respectively. Since the images of MP-16, YFCC26k and YFCC4k were sourced without any scene and user restrictions, these datasets contain images of landmarks and landscapes, but also ambiguous images with little to no geographical cues, such as photographs of food and portraits of people.

We have additionally tested our model on two smaller datasets commonly used for geo-localization – Im2GPS [17] and Im2GPS3k [18]. Im2GPS contains 237 manually selected geo-localizable images. In contrast to the previous three datasets, Im2GPS is specially designed to evaluate geo-localization systems and contains images from popular landmarks and famous tourist locations. Im2GPS3k is an extended version of Im2GPS. Im2GPS3k contains 2997 images with geo-tags. Unlike Im2GPS, this dataset was not manually filtered and hence, it is a slightly more challenging test compared to Im2GPS.

4.2 Baselines

We compare our method to several existing geo-estimation methods, including **Im2GPS** [17], **[L]kNN** [59], **MvMF** [22], **Planet** [63], **CPlanet** [46], and **ISNs** [34]. ISNs reports the state-of-the-art results on Im2GPS and Im2GPS3k. More details about the baselines are provided in the supplementary materials. Since neither of the baselines reported their performance on all considered test sets, we re-implement the missing ones. We also provide a detailed ablation study of our method by removing one component at a time from TransLocator in Table 2.

[6] Details on geo-cell partitioning can be found in the supplementary material.

4.3 Evaluation Metrics

We evaluate the performance of our approach using geolocational accuracy at multiple error levels, i.e. the percentage of images correctly localized within a predefined distance from the ground truth GPS coordinates [34,46,63]. Formally, if the predicted and ground truth coordinates are (lat_{pred}, lon_{pred}) and (lat_{gt}, lon_{gt}), the geo-locational accuracy a_r at scale r (in km) is defined as follows for a set of N samples:

$$a_r \equiv \frac{1}{N} \sum_{i=1}^{N} u(GCD((lat_{pred}^{(i)}, lon_{pred}^{(i)}), (lat_{gt}^{(i)}, lon_{gt}^{(i)})) - r) \quad (9)$$

where GCD is the great circle distance and $u(x) = \begin{cases} 1, \text{if } x < 0 \\ 0, \text{otherwise} \end{cases}$ is an indicator function whether the distance is smaller than the tolerated radius r. We report the results at 1 km, 25 km, 200 km, 750 km, and 2500 km ranges from the ground truth, which correspond to the scale of the same street, city, region, country, and continent, respectively [63].

Following Muller et al. [34], we assign the predicted geo-cell a GPS tag by using the mean locations of all training images in that cell. Since the models are trained in a classification framework and evaluated using a distance metric, we empirically verified a strong linear relationship between classification and geo-locational accuracy. The average Pearson correlation coefficient between these two metrics on Im2GPS and Im2GPS3k test sets for TransLocator is 0.981, which implies the validity of the classification framework for geo-localization (see Fig. 3). More details regarding the strong correlation between these two metrics can be found in the supplementary material.

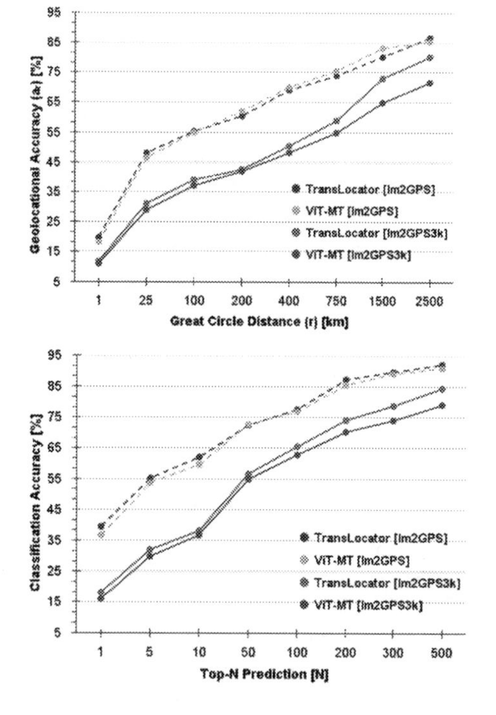

Fig. 3. Strong linear relationship between the geolocational and classification accuracy metrics. The positive correlation enables us to treat geo-localization as a classification problem.

4.4 Implementation Details

We use ViT-B/16 [13] as the backbone of both RGB and segmentation channels. While training a single-channel (RGB only) multi-task transformer network, which we

Table 1. Geolocational accuracy of the proposed systems compared to several baselines across four datasets and five scales. The methods re-implemented by [46] are denoted by dagger(†), and the methods re-implemented by us are denoted by double-dagger(‡). Δ indicates improvement over state-of-the-art achieved by TransLocator.

Dataset	Method	Distance (a_r [%] @ km)				
		Street 1 km	City 25 km	Region 200 km	Country 750 km	Continent 2500 km
Im2GPS [18]	Human [59]	–	–	3.8	13.9	39.3
	[L]kNN, $\sigma = 4$ [59]	14.4	33.3	47.7	61.6	73.4
	MvMF [22]	8.4	32.6	39.4	57.2	80.2
	PlaNet [63]	8.4	24.5	37.6	53.6	71.3
	CPlaNet [46]	16.5	37.1	46.4	62.0	78.5
	ISNs (M, f, S_3) [34]	16.5	42.2	51.9	66.2	81.0
	ISNs (M, f*, S_3) [34]	16.9	43.0	51.9	66.7	80.2
	ViT-MT	18.2	46.4	62.1	74.5	85.2
	TransLocator	**19.9**	**48.1**	**64.6**	**75.6**	**86.7**
	$\Delta_{\texttt{Ours - ISNs}}$	3.0 ↑	5.1 ↑	12.7 ↑	8.9 ↑	5.5 ↑
Im2GPS3k[17]	[L]kNN, $\sigma = 4$ [59]	7.2	19.4	26.9	38.9	55.9
	PlaNet† [63]	8.5	24.8	34.3	48.4	64.6
	CPlaNet [46]	10.2	26.5	34.6	48.6	64.6
	ISNs (M, f, S_3) [34]	10.1	27.2	36.2	49.3	65.6
	ISNs (M,f*,S_3) [34]	10.5	28.0	36.6	49.7	66.0
	ViT-MT	11.0	29.0	42.6	54.8	71.6
	TransLocator	**11.8**	**31.1**	**46.7**	**58.9**	**80.1**
	$\Delta_{\texttt{Ours - ISNs}}$	1.3 ↑	3.1 ↑	6.1 ↑	9.2 ↑	14.1 ↑
YFCC4k[51]	[L]kNN, $\sigma = 4$ [59]	2.3	5.7	11.0	23.5	42.0
	PlaNet† [63]	5.6	14.3	22.2	36.4	55.8
	CPlaNet [46]	7.9	14.8	21.9	36.4	55.5
	ISNs (M, f, S_3)‡ [34]	6.5	16.2	23.8	37.4	55.0
	ISNs (M,f*,S_3)‡ [34]	6.7	16.5	24.2	37.5	54.9
	ViT-MT	8.1	18.0	26.2	40.0	59.9
	TransLocator	**8.4**	**18.6**	**27.0**	**41.1**	**60.4**
	$\Delta_{\texttt{Ours - CPlanet}}$	0.5 ↑	3.8 ↑	5.1 ↑	4.7 ↑	4.9 ↑
	$\Delta_{\texttt{Ours - ISNs}}$	1.7 ↑	2.1 ↑	2.8 ↑	3.6 ↑	5.5 ↑
YFCC26k [59]	PlaNet‡ [63]	4.4	11.0	16.9	28.5	47.7
	ISNs (M, f, S_3)‡ [34]	5.3	12.1	18.8	31.8	50.6
	ISNs (M, f*, S_3)‡ [34]	5.3	12.3	19.0	31.9	50.7
	ViT-MT	6.9	17.3	27.5	40.5	59.5
	TransLocator	**7.2**	**17.8**	**28.0**	**41.3**	**60.6**
	$\Delta_{\texttt{Ours - ISNs}}$	1.9 ↑	5.5 ↑	9.0 ↑	9.4 ↑	9.9 ↑

denote as ViT-MT, we use the pre-trained weights on the large ImageNet21K [42] dataset containing 14 million images and 21 thousand classes. We fine-tuned it for the geo-localization task. Both channels of the dual-branch TransLocator system are initialized with the weights of the ViT-MT backbone trained for 10 epochs. The weights of the classifier heads are randomly initialized with a zero-mean Gaussian distribution with a standard deviation of 0.02. Following the standard ViT literature, we linearly project the non-overlapping patches of 16×16 pixels for both channels, and we add the `CLS` token and the positional embeddings.

Since the training set contains images in various resolutions and scales, we use extensive data augmentation, as detailed in the supplementary materials. We implement the methods in Pytorch [35] framework. Our ViT-MT and TransLocator took 6/10 days to train on 20 NVIDIA RTX 3090 GPUs, respectively, with 24 GB dedicated memory in each GPU. We train both systems using a AdamW [30] optimizer with an base learning rate of 0.1, a momentum of 0.9, and a weight decay of 0.0001. We train the network for a total of 40 epochs with a batch size of 256. During testing, we convert the *fine* geo-cells to corresponding GPS coordinates. Other necessary hyper-parameters and data augmentation details are given in the supplementary materials.

5 Results, Discussions and Analysis

In this section, we compare the performance of TransLocator system with different baselines, and conduct a detailed ablation study to demonstrate the importance of different components in our system. Furthermore, we visualize the interpretability of TransLocator using Grad-CAM [45] and perform an error analysis.

Table 2. Ablation Study on the Im2GPS and Im2GPS3K datasets. *Seg* denotes the segmentation branch of TransLocator, MFF represents multimodal feature fusion, and *Scene* denotes the multi-task learning framework.

Dataset	Method	Distance (a_r [%] @ km)				
		Street 1 km	City 25 km	Region 200 km	Country 750 km	Continent 2500 km
Im2GPS[18]	ResNet101	14.3	41.4	51.9	64.1	78.9
	EfficientNet-B4	15.4	42.7	52.8	64.8	79.5
	ViT base	16.9	43.4	54.5	67.8	80.7
	+ Seg	17.6	44.8	58.9	70.0	83.3
	+ Seg + MFF	19.0	47.2	62.7	73.5	85.7
	+ Seg + MFF + Scene	**19.9**	**48.1**	**64.6**	**75.6**	**86.7**
Im2GPS3k [18]	ResNet101	9.0	25.1	32.8	46.1	63.5
	EfficientNet-B4	9.2	26.8	32.7	47.0	63.9
	ViT base	9.9	28.0	37.8	54.2	70.7
	+ Seg	10.5	29.1	42.5	55.8	73.6
	+ Seg + MFF	11.1	30.2	45.0	56.8	78.1
	+ Seg + MFF + Scene	**11.8**	**31.1**	**46.7**	**58.9**	**80.1**

5.1 Comparison with Baselines

Table 1 presents the performance of our proposed TransLocator system and baseline methods on all four evaluation datasets. The reported baselines have a similar number of training images and geographic classes, and hence, we can directly compare the results of our system with them. Since the Im2GPS and

Table 3. Effect of number of scenes on TransLocator. Fine-grained scene information helps TransLocator to achieve superior performance.

Dataset	Method	#Scenes	Distance (a_r [%] @ km)				
			Street 1 km	City 25 km	Region 200 km	Country 750 km	Continent 2500 km
Im2GPS [17]	TransLocator	3	18.4	46.3	55.6	68.6	84.0
		16	19.0	47.1	56.5	69.7	85.4
		365	**19.9**	**48.1**	**57.4**	**70.9**	**86.5**
	$\Delta_{\text{Scenes}_{365}-\text{Scenes}_3}$		1.5 ↑	1.8 ↑	1.8 ↑	2.3 ↑	2.5 ↑
Im2GPS3k [18]	TransLocator	3	10.8	29.9	41.0	56.8	78.7
		16	11.6	30.5	42.1	57.6	79.4
		365	**11.8**	**31.1**	**42.6**	**58.9**	**80.1**
	$\Delta_{\text{Scenes}_{365}-\text{Scenes}_3}$		1.0 ↑	1.2 ↑	1.6 ↑	2.1 ↑	1.4 ↑

Im2GPS3k dataset mainly contains images of landmarks and popular tourist locations, the systems yield high accuracy on these two datasets. On Im2GPS, even the earlier methods like PlaNet and MvMF surpass human performance considerably. CPlaNet, a combinational geo-partitioning approach, brings a substantial improvement of 8.1% in street-level accuracy over PlaNet. A similar trend of results is seen in the case of Im2GPS3k. On this dataset, CPlaNet beats PlaNet by 1.7% street-level accuracy. For other distance scales, the results improve proportionally. The Individual Scene Networks (ISNs) report the state-of-the-art result on both of these datasets, achieving 16.5% and 10.1% street-level accuracy on Im2GPS and Im2GPS3k, respectively. The ensemble of hierarchical classifications (denoted by f*) improves their results. However, note that none of these methods use semantic maps in their framework. Thus, we first implement the single-branch multi-task ViT-MT model. Interestingly, even this model produces a significant improvement over ISNs for all scales on both datasets, which can be explained by the global context used by ViT architecture. Our final dual-branch TransLocator system improves on top of ViT-MT. Overall, we push the current state-of-the-art by **3.0%** and **1.3%** street-level and **5.5%** and **14.1%** continent-level accuracy on Im2GPS and Im2GPS3k, respectively.

The YFCC4k and YFCC26k datasets contain more challenging samples which have little to no geo-locating cues. However, these large datasets of unconstrained real-world images examine the generalizability of the systems. On YFCC4k, CPlaNet produces the best street-level accuracy among baselines, while ISNs beats CPlaNet at the continent level. Our proposed TransLocator outperforms both CPlaNet and ISNs in every distance scale, improving the state-of-the-art by **1.7%** and **5.5%** street-level and continent-level accuracy. The YFCC26k dataset is even more challenging than YFCC4k; the best baseline produces only 5.3% street-level accuracy. However, our proposed TransLocator system yields an impressive 7.2% street-level accuracy on YFCC26k, which proves the appreciable generalizability of TransLocator.

5.2 Ablation Study

Role of Vision Transformer: We conduct a detailed ablation study to understand the contributions of different components proposed in TransLocator architecture. We start by comparing a base ViT-B/16 encoder with two conv networks - ResNet101 [19] and EfficientNet-B4 [50]. We re-train these three systems using only RGB images of the MP-16 dataset. As shown in Table 2, the base ViT architecture produces consistent improvements over the conv models on Im2GPS and Im2GPS3k, which confirms the effectiveness of the larger receptive field of ViT for geo-localization.

Role of Segmentation Maps: We then add the segmentation branch to base-ViT by concatenating the `CLS` tokens from the last layers. This method yields an improvement of 0.7% over base-ViT on Im2GPS. However, this kind of fusion is not optimal because the two channels do not interact in-between. Hence, we then incorporate our proposed multimodal feature fusion (MFF), which sums `CLS` tokens after each transformer layer. MFF improves on the base-ViT by 2.1% and 1.2% street-level accuracy on Im2GPS and Im2GPS3k, suggesting that the two branches learn complementary and robust features for different images.

Role of Multi-task Learning: Next, we incorporate an additional classifier head for scene recognition, which significantly improves the performance. Moreover, we evaluate the effect of coarse- and fine-grained contextual knowledge by varying the number of scene categories. As shown in Table 3, a higher number of scenes improves street-level accuracy by 1–1.5% across the two datasets.

5.3 Interpretability of TransLocator

In this section, we comprehend the interpretability of TransLocator by generating visual explanations using Grad-CAM [45]. First, we focus on similar-looking images coming from different portions of the world. As shown in Fig. 4, these images[7] can be discriminated only by close attention to the global context.

[7] Collected from the Internet under creative commons license.

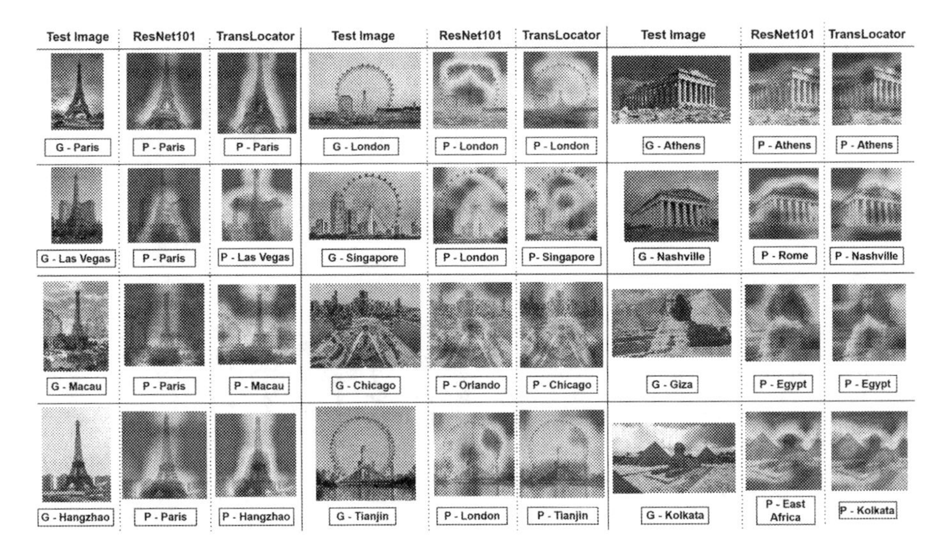

Fig. 4. Qualitative comparison of TransLocator and ResNet101 on images with similar landmarks but from different geographic locations. Unlike ResNet101, TransLocator focuses both on the foreground and background and is able to correctly geo-locate the very similar looking images. G and P denotes ground truth and predicted location.

For example, the famous Eiffel Tower in Paris closely resembles the towers in Las Vegas, Macau, and Hangzhou. However, there are differences in the background pixels between these images, such as different characteristic buildings and vegetation. While the ResNet101 architecture fails to locate the image from Las Vegas, our proposed TransLocator network correctly discriminates it from the Eiffel Tower in Paris by adequately focusing on the background structure. The similar superior performance of TransLocator is also shown for other images containing Ferris Wheel, Parthenon-like, and pyramid-like constructions in different geographic locations.

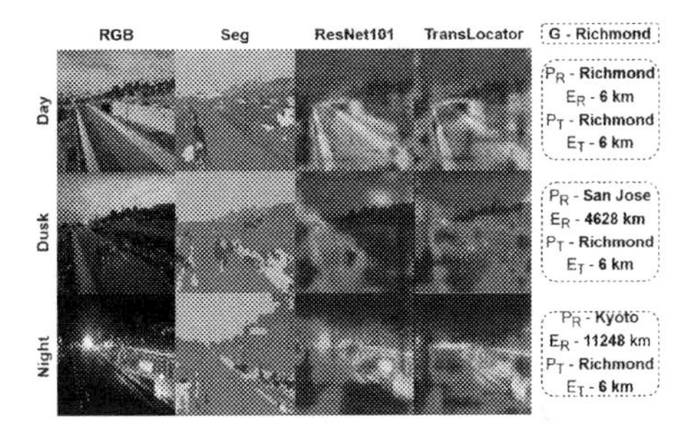

Fig. 5. Qualitative comparison of TransLocator and ResNet101 on images with same location but under challenging appearance variations. Unlike ResNet101, TransLocator attends to similar regions in each image and locates all three images correctly. G denotes ground truth, P_R, E_R, P_T and E_T denotes predicted location and prediction error by ResNet101 and TransLocator, respectively.

Next, we investigate the case of drastic appearance variation in the same location. Figure 5 shows three images[8]

[8] Collected from https://www.virginiadot.org under creative commons license.

taken in Richmond, Virginia by a highway surveillance camera in the morning, dusk, and night. Though the RGB image varies with the change in the daytime, the corresponding semantic segmentation maps remain similar, and thus TransLocator can learn robust multimodal features. In contrast to ResNet101, which suffers from such appearance variation, TransLocator attends to similar regions in each image and locates all three images correctly.

5.4 Error Analysis

Although TransLocator achieves better quantitative and qualitative results than baselines, there are still some open geo-localization problems that TransLocator does not solve. TransLocator can not locate images without geo-locating cues. For example, a photo of a cherry blossom tree can come anywhere from Tokyo, Paris, Washington DC, Dublin, etc. Hence TransLocator can not locate such an image. Similarly, pictures of sea beaches, deserts, or pictures of foods with no background cues can never precisely be geo-located. A few samples of such incorrectly located images are illustrated in Fig. 6.

Fig. 6. Limitations of our method: images without salient geo-locating cues cannot be geo-located correctly. *G* and *P* denotes ground truth and predicted location.

6 Conclusion

Planet-scale single-image geo-localization is a highly challenging problem. These challenges include images with a large diversity in various environmental scenarios and appearance variation due to daytime, season, or weather changes. Hence, most existing approaches limit geo-localization in the scale of landmarks, a specific area, or an environmental scenario. Some approaches propose to use separate systems for different environments. In this paper, we address this challenging problem by proposing TransLocator, a unified dual-branch transformer network that attends to tiny details over the entire image and produces robust feature representation under extreme appearance variations. TransLocator takes an RGB image with its semantic segmentation map as input, interacts between its two parallel channels after each transformer layer, and concatenates the learned RGB and semantic representations using global attention. We train TransLocator in a unified multi-task framework for simultaneous geo-localization and scene recognition, and thus, our system can be applied to images from all environmental settings. Extensive experiments with TransLocator on four benchmark datasets - Im2GPS [17], Im2GPS3k [18], YFCC4k [59] and YFCC26k [51] shows a significant improvement of 5.5%, 14.1%, 4.9%, 9.9% continent-level accuracy over current state-of-the-art. We also obtain better qualitative results when we test TransLocator on challenging real-world images.

Acknowledgement. This research is partially supported by an ARO MURI Grant No. W911NF-17-1-0304.

References

1. Akbari, H., et al.: VATT: transformers for multimodal self-supervised learning from raw video, audio and text. In: Advances in Neural Information Processing Systems, vol. 34, pp. 24206–24221 (2021)
2. Baatz, G., Saurer, O., Köser, K., Pollefeys, M.: Large scale visual geo-localization of images in mountainous terrain. In: Fitzgibbon, A., Lazebnik, S., Perona, P., Sato, Y., Schmid, C. (eds.) ECCV 2012. LNCS, vol. 7573, pp. 517–530. Springer, Heidelberg (2012). https://doi.org/10.1007/978-3-642-33709-3_37
3. Berton, G., Masone, C., Caputo, B.: Rethinking visual geo-localization for large-scale applications. In: Proceedings of the IEEE/CVF Conference on Computer Vision and Pattern Recognition, pp. 4878–4888 (2022)
4. Brejcha, J., Čadík, M.: State-of-the-art in visual geo-localization. Pattern Anal. Appl. **20**(3), 613–637 (2017)
5. Cao, L., Smith, J.R., Wen, Z., Yin, Z., Jin, X., Han, J.: Bluefinder: estimate where a beach photo was taken. In: Proceedings of the 21st International Conference on World Wide Web, pp. 469–470 (2012)
6. Carion, N., Massa, F., Synnaeve, G., Usunier, N., Kirillov, A., Zagoruyko, S.: End-to-end object detection with transformers. In: Vedaldi, A., Bischof, H., Brox, T., Frahm, J.-M. (eds.) ECCV 2020. LNCS, vol. 12346, pp. 213–229. Springer, Cham (2020). https://doi.org/10.1007/978-3-030-58452-8_13
7. Caruana, R.: Multitask learning. Mach. Learn. **28**(1), 41–75 (1997)
8. Chen, D.M., et al.: City-scale landmark identification on mobile devices. In: Proceedings of the IEEE/CVF Conference on Computer Vision and Pattern Recognition, pp. 737–744 (2011)
9. Chen, H., et al.: Pre-trained image processing transformer. In: Proceedings of the IEEE/CVF Conference on Computer Vision and Pattern Recognition, pp. 12299–12310 (2021)
10. Cheng, B., Misra, I., Schwing, A.G., Kirillov, A., Girdhar, R.: Masked-attention mask transformer for universal image segmentation. In: Proceedings of the IEEE/CVF Conference on Computer Vision and Pattern Recognition, pp. 1290–1299 (2022)
11. Devlin, J., Chang, M.W., Lee, K., Toutanova, K.: BERT: pre-training of deep bidirectional transformers for language understanding. In: Proceedings of the 2019 Conference of the North American Chapter of the Association for Computational Linguistics: Human Language Technologies, Volume 1 (Long and Short Papers), pp. 4171–4186. Association for Computational Linguistics, Minneapolis, Minnesota, Jun 2019. 10.18653/v1/N19-1423, https://aclanthology.org/N19-1423
12. Doersch, C., Gupta, A., Zisserman, A.: Crosstransformers: spatially-aware few-shot transfer. In: Advances in Neural Information Processing Systems, vol. 33, pp. 21981–21993 (2020)
13. Dosovitskiy, A., et al.: An image is worth 16x16 words: transformers for image recognition at scale. In: International Conference on Learning Representations (2020)

14. Fang, Y., Liao, B., Wang, X., Fang, J., Qi, J., Wu, R., Niu, J., Liu, W.: You only look at one sequence: Rethinking transformer in vision through object detection. Advances in Neural Information Processing Systems 34 (2021)
15. Gu, Y., Yang, K., Fu, S., Chen, S., Li, X., Marsic, I.: Hybrid attention based multimodal network for spoken language classification. In: ACL, vol. 2018, p. 2379 (2018)
16. Hausler, S., Garg, S., Xu, M., Milford, M., Fischer, T.: Patch-netvlad: multi-scale fusion of locally-global descriptors for place recognition. In: Proceedings of the IEEE/CVF Conference on Computer Vision and Pattern Recognition, pp. 14141–14152 (2021)
17. Hays, J., Efros, A.A.: Im2gps: estimating geographic information from a single image. In: Proceedings of the IEEE/CVF Conference on Computer Vision and Pattern Recognition, pp. 1–8. IEEE (2008)
18. Hays, J., Efros, A.A.: Large-scale image geolocalization. In: Choi, J., Friedland, G. (eds.) Multimodal Location Estimation of Videos and Images, pp. 41–62. Springer, Cham (2015). https://doi.org/10.1007/978-3-319-09861-6_3
19. He, K., Zhang, X., Ren, S., Sun, J.: Deep residual learning for image recognition. In: Proceedings of the IEEE/CVF Conference on Computer Vision and Pattern Recognition, pp. 770–778 (2016)
20. Hu, R., Singh, A.: Unit: multimodal multitask learning with a unified transformer. In: Proceedings of the IEEE/CVF International Conference on Computer Vision, pp. 1439–1449 (2021)
21. Hu, S., Feng, M., Nguyen, R.M., Lee, G.H.: Cvm-net: Cross-view matching network for image-based ground-to-aerial geo-localization. In: Proceedings of the IEEE/CVF Conference on Computer Vision and Pattern Recognition, pp. 7258–7267 (2018)
22. Izbicki, M., Papalexakis, E.E., Tsotras, V.J.: Exploiting the earth's spherical geometry to geolocate images. In: Brefeld, U., Fromont, E., Hotho, A., Knobbe, A., Maathuis, M., Robardet, C. (eds.) ECML PKDD 2019. LNCS (LNAI), vol. 11907, pp. 3–19. Springer, Cham (2020). https://doi.org/10.1007/978-3-030-46147-8_1
23. Kant, Y., et al.: Spatially aware multimodal transformers for TextVQA. In: Vedaldi, A., Bischof, H., Brox, T., Frahm, J.-M. (eds.) ECCV 2020. LNCS, vol. 12354, pp. 715–732. Springer, Cham (2020). https://doi.org/10.1007/978-3-030-58545-7_41
24. Larson, M., Soleymani, M., Gravier, G., Ionescu, B., Jones, G.J.: The benchmarking initiative for multimedia evaluation: mediaeval 2016. IEEE Multimedia **24**(1), 93–96 (2017)
25. Lei, J., Li, L., Zhou, L., Gan, Z., Berg, T.L., Bansal, M., Liu, J.: Less is more: clipbert for video-and-language learning via sparse sampling. In: Proceedings of the IEEE/CVF Conference on Computer Vision and Pattern Recognition, pp. 7331–7341 (2021)
26. Li, L.H., et al.: Grounded language-image pre-training. In: Proceedings of the IEEE/CVF Conference on Computer Vision and Pattern Recognition, pp. 10965–10975 (2022)
27. Li, X., et al.: Oscar: object-semantics aligned pre-training for vision-language tasks. In: Vedaldi, A., Bischof, H., Brox, T., Frahm, J.-M. (eds.) ECCV 2020. LNCS, vol. 12375, pp. 121–137. Springer, Cham (2020). https://doi.org/10.1007/978-3-030-58577-8_8
28. Lin, T.Y., Belongie, S., Hays, J.: Cross-view image geolocalization. In: Proceedings of the IEEE/CVF Conference on Computer Vision and Pattern Recognition, pp. 891–898 (2013)

29. Liu, Y., et al.: Roberta: a robustly optimized Bert pretraining approach. arXiv preprint arXiv:1907.11692 (2019)
30. Loshchilov, I., Hutter, F.: Decoupled weight decay regularization. In: International Conference on Learning Representations (2018)
31. Lu, J., Batra, D., Parikh, D., Lee, S.: Vilbert: pretraining task-agnostic visiolinguistic representations for vision-and-language tasks. In: Advances in Neural Information Processing Systems, vol. 32 (2019)
32. Masone, C., Caputo, B.: A survey on deep visual place recognition. IEEE Access **9**, 19516–19547 (2021)
33. Misra, I., Girdhar, R., Joulin, A.: An end-to-end transformer model for 3D object detection. In: Proceedings of the IEEE/CVF International Conference on Computer Vision, pp. 2906–2917 (2021)
34. Müller-Budack, E., Pustu-Iren, K., Ewerth, R.: Geolocation estimation of photos using a hierarchical model and scene classification. In: Ferrari, V., Hebert, M., Sminchisescu, C., Weiss, Y. (eds.) ECCV 2018. LNCS, vol. 11216, pp. 575–592. Springer, Cham (2018). https://doi.org/10.1007/978-3-030-01258-8_35
35. Paszke, A., et al.: Pytorch: an imperative style, high-performance deep learning library. In: Advances in Neural Information Processing Systems, vol. 32 (2019)
36. Peters, M., Neumann, M., Iyyer, M., Gardner, M., Clark, C., Lee, K., Zettlemoyer, L.: Deep contextualized word representations. In: Proceedings of the 2018 Conference of the North American Chapter of the Association for Computational Linguistics: Human Language Technologies, Volume 1 (Long Papers), pp. 2227–2237 (2018)
37. Pramanick, S., Roy, A., Patel, V.M.: Multimodal learning using optimal transport for sarcasm and humor detection. In: Proceedings of the IEEE/CVF Winter Conference on Applications of Computer Vision, pp. 3930–3940 (2022)
38. Pramanick, S., Sharma, S., Dimitrov, D., Akhtar, M.S., Nakov, P., Chakraborty, T.: Momenta: a multimodal framework for detecting harmful memes and their targets. In: Findings of the Association for Computational Linguistics: EMNLP 2021, pp. 4439–4455 (2021)
39. Raghu, M., Unterthiner, T., Kornblith, S., Zhang, C., Dosovitskiy, A.: Do vision transformers see like convolutional neural networks? In: Advances in Neural Information Processing Systems, vol. 34 (2021)
40. Ranjan, R., Patel, V.M., Chellappa, R.: Hyperface: a deep multi-task learning framework for face detection, landmark localization, pose estimation, and gender recognition. IEEE Trans. Pattern Anal. Mach. Intell. **41**(1), 121–135 (2017)
41. Regmi, K., Shah, M.: Bridging the domain gap for ground-to-aerial image matching. In: Proceedings of the IEEE/CVF International Conference on Computer Vision, pp. 470–479 (2019)
42. Ridnik, T., Ben-Baruch, E., Noy, A., Zelnik-Manor, L.: Imagenet-21k pretraining for the masses. arXiv preprint arXiv:2104.10972 (2021)
43. Ruder, S.: An overview of multi-task learning in deep neural networks. arXiv preprint arXiv:1706.05098 (2017)
44. Saurer, O., Baatz, G., Köser, K., Pollefeys, M., et al.: Image based geo-localization in the alps. Int. J. Comput. Vision **116**(3), 213–225 (2016)
45. Selvaraju, R.R., Cogswell, M., Das, A., Vedantam, R., Parikh, D., Batra, D.: Grad-cam: visual explanations from deep networks via gradient-based localization. In: Proceedings of the IEEE/CVF International Conference on Computer Vision, pp. 618–626 (2017)

46. Seo, P.H., Weyand, T., Sim, J., Han, B.: CPlaNet: enhancing image geolocalization by combinatorial partitioning of maps. In: Ferrari, V., Hebert, M., Sminchisescu, C., Weiss, Y. (eds.) ECCV 2018. LNCS, vol. 11214, pp. 544–560. Springer, Cham (2018). https://doi.org/10.1007/978-3-030-01249-6_33
47. Seymour, Z., Sikka, K., Chiu, H.P., Samarasekera, S., Kumar, R.: Semantically-aware attentive neural embeddings for image-based visual localization. arXiv preprint arXiv:1812.03402 (2018)
48. Strudel, R., Garcia, R., Laptev, I., Schmid, C.: Segmenter: transformer for semantic segmentation. In: Proceedings of the IEEE/CVF International Conference on Computer Vision, pp. 7262–7272 (2021)
49. Sun, C., Myers, A., Vondrick, C., Murphy, K., Schmid, C.: Videobert: a joint model for video and language representation learning. In: Proceedings of the IEEE/CVF International Conference on Computer Vision. pp. 7464–7473 (2019)
50. Tan, M., Le, Q.: Efficientnet: rethinking model scaling for convolutional neural networks. In: International Conference on Machine Learning, pp. 6105–6114. PMLR (2019)
51. Theiner, J., Müller-Budack, E., Ewerth, R.: Interpretable semantic photo geolocation. In: Proceedings of the IEEE/CVF Winter Conference on Applications of Computer Vision, pp. 750–760 (2022)
52. Thomee, B., et al.: Yfcc100m: the new data in multimedia research. Commun. ACM **59**(2), 64–73 (2016)
53. Tian, Y., Chen, C., Shah, M.: Cross-view image matching for geo-localization in urban environments. In: Proceedings of the IEEE/CVF Conference on Computer Vision and Pattern Recognition, pp. 3608–3616 (2017)
54. Toker, A., Zhou, Q., Maximov, M., Leal-Taixé, L.: Coming down to earth: Satellite-to-street view synthesis for geo-localization. In: Proceedings of the IEEE/CVF Conference on Computer Vision and Pattern Recognition, pp. 6488–6497 (2021)
55. Torii, A., Arandjelovic, R., Sivic, J., Okutomi, M., Pajdla, T.: 24/7 place recognition by view synthesis. In: Proceedings of the IEEE/CVF Conference on Computer Vision and Pattern Recognition, pp. 1808–1817 (2015)
56. Torii, A., Sivic, J., Pajdla, T., Okutomi, M.: Visual place recognition with repetitive structures. In: Proceedings of the IEEE/CVF Conference on Computer Vision and Pattern Recognition, pp. 883–890 (2013)
57. Tzeng, E., Zhai, A., Clements, M., Townshend, R., Zakhor, A.: User-driven geolocation of untagged desert imagery using digital elevation models. In: Proceedings of the IEEE/CVF Conference on Computer Vision and Pattern Recognition Workshops, pp. 237–244 (2013)
58. Vaswani, A., et al.: Attention is all you need. In: Advances in Neural Information Processing Systems, pp. 5998–6008 (2017)
59. Vo, N., Jacobs, N., Hays, J.: Revisiting im2gps in the deep learning era. In: Proceedings of the IEEE/CVF International Conference on Computer Vision, pp. 2621–2630 (2017)
60. Wang, J., et al.: Deep high-resolution representation learning for visual recognition. IEEE Trans. Pattern Anal. Mach. Intell. **43**(10), 3349–3364 (2020)
61. Wang, Y., Cao, L.: Discovering latent clusters from geotagged beach images. In: Li, S., et al. (eds.) MMM 2013. LNCS, vol. 7733, pp. 133–142. Springer, Heidelberg (2013). https://doi.org/10.1007/978-3-642-35728-2_13
62. Wang, Y., et al.: End-to-end video instance segmentation with transformers. In: Proceedings of the IEEE/CVF Conference on Computer Vision and Pattern Recognition, pp. 8741–8750 (2021)

63. Weyand, T., Kostrikov, I., Philbin, J.: PlaNet - photo geolocation with convolutional neural networks. In: Leibe, B., Matas, J., Sebe, N., Welling, M. (eds.) ECCV 2016. LNCS, vol. 9912, pp. 37–55. Springer, Cham (2016). https://doi.org/10.1007/978-3-319-46484-8_3
64. Xie, E., Wang, W., Yu, Z., Anandkumar, A., Alvarez, J.M., Luo, P.: Segformer: simple and efficient design for semantic segmentation with transformers. In: Advances in Neural Information Processing Systems, vol. 34 (2021)
65. Xu, H., et al.: Videoclip: contrastive pre-training for zero-shot video-text understanding. In: Proceedings of the 2021 Conference on Empirical Methods in Natural Language Processing, pp. 6787–6800 (2021)
66. Yang, F., Yang, H., Fu, J., Lu, H., Guo, B.: Learning texture transformer network for image super-resolution. In: Proceedings of the IEEE/CVF Conference on Computer Vision and Pattern Recognition, pp. 5791–5800 (2020)
67. Yang, H., Lu, X., Zhu, Y.: Cross-view geo-localization with layer-to-layer transformer. Adv. Neural. Inf. Process. Syst. **34**, 29009–29020 (2021)
68. Yang, J., et al.: Unified contrastive learning in image-text-label space. In: Proceedings of the IEEE/CVF Conference on Computer Vision and Pattern Recognition, pp. 19163–19173 (2022)
69. Yang, L., Fan, Y., Xu, N.: Video instance segmentation. In: Proceedings of the IEEE/CVF International Conference on Computer Vision, pp. 5188–5197 (2019)
70. Yang, Z., Dai, Z., Yang, Y., Carbonell, J., Salakhutdinov, R.R., Le, Q.V.: XLNet: generalized autoregressive pretraining for language understanding. In: Advances in Neural Information Processing Systems, vol. 32 (2019)
71. Ye, H.J., Hu, H., Zhan, D.C., Sha, F.: Few-shot learning via embedding adaptation with set-to-set functions. In: Proceedings of the IEEE/CVF Conference on Computer Vision and Pattern Recognition, pp. 8808–8817 (2020)
72. Zhang, Y., Yang, Q.: A survey on multi-task learning. IEEE Trans. Knowl. Data Eng. (2021)
73. Zhao, H., Jiang, L., Jia, J., Torr, P.H., Koltun, V.: Point transformer. In: Proceedings of the IEEE/CVF International Conference on Computer Vision, pp. 16259–16268 (2021)
74. Zheng, S., et al.: Rethinking semantic segmentation from a sequence-to-sequence perspective with transformers. In: Proceedings of the IEEE/CVF Conference on Computer Vision and Pattern Recognition, pp. 6881–6890 (2021)
75. Zhou, B., Lapedriza, A., Khosla, A., Oliva, A., Torralba, A.: Places: a 10 million image database for scene recognition. IEEE Trans. Pattern Anal. Mach. Intell. **40**(6), 1452–1464 (2017)
76. Zhou, B., Zhao, H., Puig, X., Fidler, S., Barriuso, A., Torralba, A.: Scene parsing through ade20k dataset. In: Proceedings of the IEEE/CVF Conference on Computer Vision and Pattern Recognition, pp. 633–641 (2017)
77. Zhou, B., et al.: Semantic understanding of scenes through the ade20k dataset. Int. J. Comput. Vision **127**(3), 302–321 (2019)
78. Zhu, S., Shah, M., Chen, C.: Transgeo: transformer is all you need for cross-view image geo-localization. In: Proceedings of the IEEE/CVF Conference on Computer Vision and Pattern Recognition, pp. 1162–1171 (2022)
79. Zhu, S., Yang, T., Chen, C.: Vigor: cross-view image geo-localization beyond one-to-one retrieval. In: Proceedings of the IEEE/CVF Conference on Computer Vision and Pattern Recognition (CVPR), pp. 3640–3649, June 2021
80. Zhu, X., Su, W., Lu, L., Li, B., Wang, X., Dai, J.: Deformable detr: deformable transformers for end-to-end object detection. In: International Conference on Learning Representations (2020)

Colorization for *in situ* Marine Plankton Images

Guannan Guo[1,2], Qi Lin[3], Tao Chen[1,2], Zhenghui Feng[4], Zheng Wang[1,2], and Jianping Li[1,2](✉)

[1] Shenzhen Institute of Advanced Technology, Chinese Academy of Sciences, Shenzhen, China
jp.li@siat.ac.cn
[2] University of Chinese Academy of Sciences, Beijing, China
[3] Xiamen University, Xiamen, China
[4] Harbin Institute of Technology, Shenzhen, China

Abstract. Underwater imaging with red-NIR light illumination can avoid phototropic aggregation-induced observational deviation of marine plankton abundance under white light illumination, but this will lead to the loss of critical color information in the collected grayscale images, which is non-preferable to subsequent human and machine recognition. We present a novel deep networks-based vision system IsPlanktonCLR for automatic colorization of *in situ* marine plankton images. IsPlanktonCLR uses a reference module to generate self-guidance from a customized palette, which is obtained by clustering *in situ* plankton image colors. With this self-guidance, a parallel colorization module restores input grayscale images into their true color counterparts. Additionally, a new metric for image colorization evaluation is proposed, which can objectively reflect the color dissimilarity between comparative images. Experiments and comparisons with state-of-the-art approaches are presented to show that our method achieves a substantial improvement over previous methods on color restoration of scientific plankton image data.

Keywords: Image colorization · Deep learning · Underwater imaging · *in situ* observation · Marine plankton

1 Introduction

In situ imaging of marine plankton has been demonstrated very promising for scientific research to understand marine ecosystems and also become appealing for modern ocean management [25,36]. Limited by working principle and device performance, most early underwater plankton cameras can only capture grayscale images [5]. With recent technology development, some dark-field underwater cameras have been enabled for color imaging [6,13,21,33]. The color images captured by them have been shown to improve a machine classifier's accuracy than

Supplementary Information The online version contains supplementary material available at https://doi.org/10.1007/978-3-031-19839-7_13.

S. Avidan et al. (Eds.): ECCV 2022, LNCS 13698, pp. 216–232, 2022.
https://doi.org/10.1007/978-3-031-19839-7_13

that achieved on grayscale ones [21]. However, color imaging requires white light illumination, which causes phototropic aggregation of zooplankton frequently, thus resulting in plankton abundance measurement bias and great concern for observation accuracy [38,39].

Since most zooplankton are insensitive to longer wavelengths [12], *in situ* cameras can use red-NIR light for illumination to avoid phototropic aggregation [9,32]. But using red-NIR lighting will make an underwater camera acquire just grayscale images. If such grayscale images can be colorized with high-fidelity using deep learning techniques [1], it would not only enrich the image data with extra color information to facilitate subsequent human and machine recognition, but can also completely resolve the concern on plankton abundance measurement error associated with white light illumination. Moreover, a grayscale image not only has higher spatial resolution than its color counterpart with the same pixel size and number, its file size is also much smaller than the color version. This is beneficial to reduce the resource stress of image data processing, storage, and transmission for achieving sustainable ocean observation.

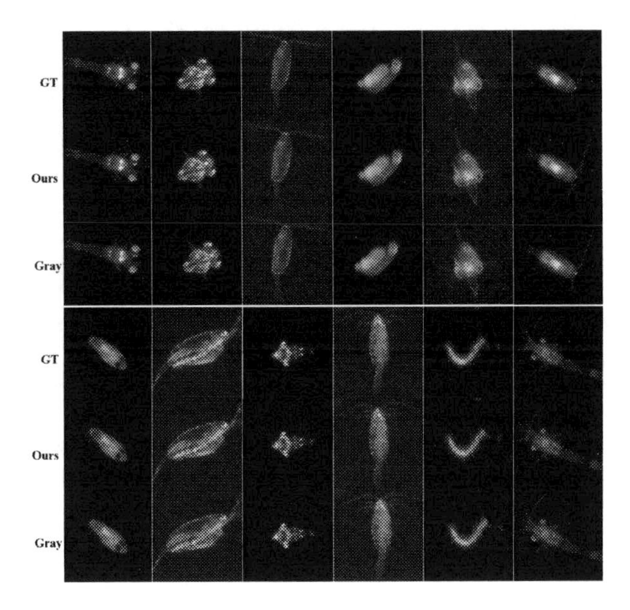

Fig. 1. Representative examples of *in situ* marine plankton grayscale images, their colorizations by IsPlanktonCLR, and the ground truth. (Color figure online)

The demand for *in situ* plankton grayscale image colorization obviously corresponds to an image restoration problem. However, existing colorization algorithms are mainly developed for colorizing natural scene images captured in the real world [4,7,8,17,30,40,41], which corresponds essentially to an image enhancement task. As the colorization of these images is more in pursuit of rationality, comfort and diversity of human visual perception, the same target can be artificially painted into multiple colors. For example, a blue T-shirt can be colorized into a green, or a yellow or a red one. This task is an ill-posed problem, in which the deep colorization networks are difficult to establish deterministic mapping between the input and the output, and hence unable to meet higher demands in colorization accuracy for scientific imaging applications. On the other hand, many restoration-oriented algorithms need user guidance to ensure the colorization effect [11,15]. However, these guidance is either given by human in advance or obtained through human-computer interaction, which is not conducive to their applications in long-term and automated ocean observation activities.

On this regard, we treat it as a color classification problem and propose a self-guided automatic deep colorization algorithm for color restoration of *in situ* plankton grayscale images. The network is named IsPlanktonCLR, whose idea and architecture is illustrated in Fig. 2. In the design of this network, we firstly customize a reference palette to reduce the number of colors in the searching space, so as to achieve satisfying colorization effect with better efficiency. Then, to further ensure colorization accuracy and avoid color averaging effect, we combine the advantages of both user-guided and big-data driven algorithms to improve the model performance of color restoration in a self-guided and automatic way. This is achieved through a parallel network architecture consisting of a primary module for image colorization and an additional reference module for providing guidance from the customized palette. Using this method, we successfully achieve satisfying colorization effect on the *in situ* marine plankton image data as shown in Fig. 1.

The feasibility of IsPlanktonCLR is based on the premise that the coloration of marine plankton is relatively monotonous in their *in situ* images. By investigating human visual perception of a large number of images, we realize that most plankton only show one or two families of colors. This allows us to group the plankton images based on their color families, and use this information to label them as references for guiding the colorization.

We are also aware that the obvious imbalance in plankton image dataset leads to the imbalance in the color quantities. For example, there are a lot of yellowish colors in the dataset, while the reddish colors are less common. Direct use of imbalanced data to train the model will cause dominant color effect, which is a known problem associated with many deep colorization models [44]. To resolve this issue, we use data augmentation and loss reweighting to enforce the model learning more on rare colors.

In addition, we notice an obvious lack of objective and quantitative colorization evaluation metrics for restoration-oriented scientific image colorization. Although PSNR, SSIM and other metrics are often used for image restoration evaluation, they are known to be less effective for colorization evaluation [7,17,29,37,40,47], and are often inconsistent with human perception. Therefore, we propose a new metric Color Dissimilarity (CDSIM) to better characterize the color restoration accuracy of an output image relative to its ground truth. CDSIM is obtained by calculating the Euclidean distance between color feature vectors extracted from two comparative images. We demonstrate its effectiveness on both plankton and natural scene images.

To summarize, the contribution of this work includes:

- A new idea in automatic grayscale image colorization for scientific domain imagery is provided, which improves the colorization accuracy and efficiency by referencing the colorization model with a simplified color palette.
- A customized self-guided automatic colorization network IsPlanktonCLR is designed and its colorization performance on plankton images has been verified superior to SOTA methods. To the best of our knowledge, this is the first endeavor to study scientific colorization of *in situ* marine plankton imagery.
- A new metric CDSIM is proposed for evaluation of color similarity between input and output images of a colorization model, which has been verified suitable for restoration-oriented image colorization problems.

2 Related Work

2.1 Underwater Plankton Cameras for *in situ* Plankton Color Imaging

All the underwater cameras that are good at capturing *in situ* colorful images of marine plankton have adopted dark-field imaging principle. The early Video Plankton Recorder (VPR) can only capture grayscale images [9], but it is reported to support color imaging latterly after device upgrade [33]. The Continuous Particle Imaging and Classification System (CPICS) has been deployed on the sea floor [14]. The Scripps Plankton Camera (SPC) has been deployed underwater at the shore [31], with profilers [6], and under a floating station in the Lake Greifensee [28]. The Imaging Plankton Probe (IPP) has been deployed under a moored buoy for monitoring coastal waters [21].

There is no significant difference in the imaging light path of these dark-field cameras. While they all use flashed white-light as sources, their lighting path design is different. VPR uses lateral side lighting [9], CPICS uses annular oblique lighting [13], SPC uses standard hollow-cone illumination as adopted in traditional dark-field microscopes [31], and IPP uses annular orthogonal compressed lighting [21]. VPR and SPC are not reported to be color-calibrated, while CPICS and IPP both perform white balance calibration using external reference targets [13,21]. It is worth noting that IPP has not only achieved high-quality *in situ* true color imaging of marine plankton through spatially compressed and condensed laminar white-light illumination, but also greatly inhibited the leakage of white light to the adjacent underwater environment, thus greatly reducing the phototropic aggregation of zooplankton [21]. However, in principle, this lighting design still cannot completely eliminate the white-light leakage scattered by the seawater within the imaging area. It is still suspicious whether the influence by phototaxis of zooplankton can be completely avoided.

2.2 Deep Learning-Based Image Colorization

The colorization of grayscale image has long been a very challenging problem. With the development of deep learning technology and the emergence of large-scale image datasets such as ImageNet [10] and MSCoco [24], various deep networks have been applied to the field of natural scene image colorization and achieved good results [1,20,40,41,44,45]. Based on the difference of their objectives, these deep image colorization algorithms can be roughly classified into two groups for image enhancement and restoration. The algorithms for image enhancement [3,8,17,27,34,40–42,46] aim at converting grayscale images into color images with visual comfortableness and fit of human commonsense, but pay little attention to whether the generated colors are the same as ground truth.

The deep colorization algorithms for image restoration aim to recover the original true color of scenes or objects in the grayscale images. CIC [45] transforms the colorization problem into a classification problem of 313 colors. Pixelated Semantic Colorization uses pixel-level semantic features to guide a network for colorization [47]. LetColor achieves end-to-end colorization by learning the global prior and local features of the image [19]. MemoColor [44] remembers color information of rare instances through an external storage network [20], achieving colorization with limited data. We notice some networks use paired image data for colorization learning. For example, Colorization in the dual-lens system [11] and Low-light Color Imaging [15] both use dual-camera systems to obtain grayscale-color image pairs at the same time. Then the grayscale images are colorized with the guidance from their color counterparts, so that the color information is transferred from the color images with low resolution and poor brightness to the grayscale images with higher quality. This is similar to our idea, except that their goal is to improve the resultant quality of natural scene color images, while ours is to automatically and accurately restore the true color of marine plankton in grayscale images to be taken under red-NIR lighting in seawater.

2.3 Metrics for Colorization Evaluation

Existing metrics for image colorization are mainly used to evaluate whether the color of network output images conforms to human commonsense and cognition. For example, the Colorfulness Score [16] evaluates the quality and diversity of image colorization, but it is difficult to compare the color authenticity of the output image with its ground truth. UCIQE [43] evaluates the color quality of underwater images through a linear combination of chroma, saturation and contrast, but it does not consider the spatial distribution of color. Although PSNR and SSIM are also frequently used for image restoration evaluation, they are proved insensitive to color difference between images [7,17,29,37,40,47]. FID [18] calculates the Fréchet distance between feature vectors extracted by an Inception V3 network to evaluate the overall similarity between two comparative images, but the deep features are not interpretable enough to clarify the relationship between both color images. In a word, it is difficult for previous metrics to simulate the perception of human vision, and to objectively and quantitatively evaluate the color similarity between images.

3 Methodology

As illustrated in Fig. 2, the IsPlanktonCLR network mainly consists of two parts: a customized color palette, which simplifies the color space of the plankton image data; and a deep colorization network, which is used to achieve self-guided and automated colorization of the input grayscale image.

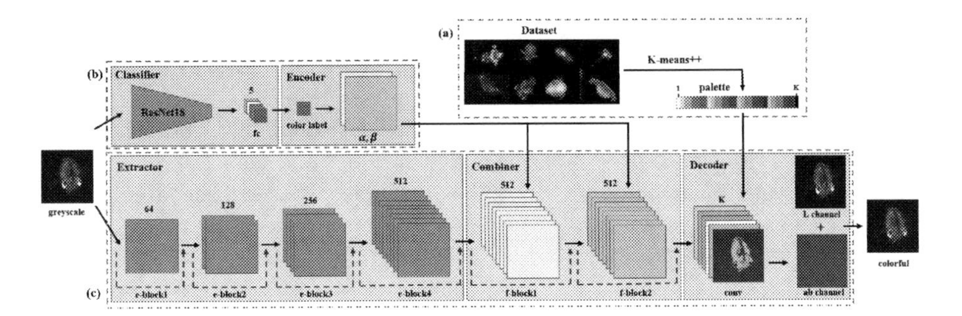

Fig. 2. Overview of IsPlanktonCLR: (a) palette customization; (b) reference module; (c) colorization module. (Color figure online)

3.1 Palette Customization

We use the color ROI images in the DYB-PlanktonNet [22] dataset to customize the reference palette. For convenience, we convert all the RGB ROIs into *Lab* color space with (a_i, b_i) denotes the ith color. Since the background of dark-field image is nearly zero, we only extract color information from the foreground pixels and integrate similar colors with a clustering algorithm based on K-means++ [2] to reduce the number of colors. To achieve this, we firstly select manually 2–3 ROIs from each plankton class of the dataset, which can represent as more colors as possible in this class. Then the foreground pixels of these ROIs are extracted by thresholding, and their (a, b) values are used to train the K-means++ clustering algorithm. After 10 iterations, the model with the smallest inertia (sum of squared distances of samples to their closest cluster center) is selected and denoted as $Model_{km}$. The palette is then customized to contain colors represented by the clustering centers of $Model_{km}$:

$$palette = \{(a_i, b_i), i \in K\}, \tag{1}$$

where K is the number of clustering centers. We select a reasonable value range of K by observing the inertia variation curve with K, and finally determine an optimized value of K by comparing the image colorization results obtained with different K.

3.2 Colorization Network

As shown in Fig. 2 (b) and (c), the IsPlanktonCLR network consists of two parallel modules. We classify the plankton images in the DYB-PlanktonNet into five color families, namely white, red, yellow, green, and blue. The reference module firstly determines the color family labels of an input grayscale image, and then the colorization module completes the colorization under the guidance of these labels.

The reference module consists of a classifier and an encoder. The classifier is built on a ResNet18 network and is responsible for giving a color family label L to the input grayscale image G. L is an integer ranging from 1 to 5, corresponding to 5 color family labels, respectively. The encoder is used to encode the color family label L and generate the reference information α and β for the colorization module. We test two encoding methods. The first one is discrete encoding, which directly encodes L with embedding to obtain the reference information R; the other is continuous encoding, whose formula is

$$R = L - 1 + P, \tag{2}$$

where P is the probability that G belongs to L. The R values obtained by the two encoding methods are fed into an 1×1 convolution layer, and finally split into α and β.

The colorization module is mainly composed of an extractor, a combiner and a decoder. The extractor is mainly responsible for extracting the feature map fm_{ex}, which consists of four e-blocks. Each e-block contains three convolutional layers with batch-normalization and ReLU. The combiner is responsible for adding α and β to fm_{ex}, which consists of two f-blocks. Each f-block is composed of convolutional and FILM layers [27]. FILM layer implements the function as expressed in Eq. (3), where fm_{ref} is a feature map with reference information.

$$fm_{ref} = \alpha fm_{ex} + \beta. \tag{3}$$

The decoder achieves color classification of each pixel by the convolutional and softmax layers. We encode the softmax output as a one-hot vector and then decode it by multiplying it with the palette vector to obtain the *ab* channel of the image. Finally, the *ab* channel is overlapped with the input L channel to form the colorized image.

3.3 Loss Function

We treat pixels containing common colors as easy examples, and pixels containing rare colors as hard examples. We use a loss function incorporating OHEM [35] and Focal-Loss [23] to solve the color imbalance problem by training the model with more weights on rare colors to avoid dominant color issue in the results. The loss function is formulated as follows:

$$loss_{pixle}\left(x, p_t\right) = -\omega_t \left(1 - p_t\right)^{\gamma} \log\left(p_t\right), \tag{4}$$

$$loss = \frac{1}{N}\sum_{i \in S} loss_{pixle}\left(i, p_t\right). \tag{5}$$

In Formula (4), $loss_{pixel}(x)$ denotes the loss of classifying pixel x, where p_t denotes the probability that x is classified correctly, and ω_t denotes the weight of the correctly classified category in calculating the loss. ω is determined by $Model_{km}$ in Sect. 3.1. We use $Model_{km}$ to quantify the image color in the entire training set to obtain information about the number of colors in each class. The higher the number of colors, the smaller the weight is assigned to pixels belong to that color class. Therefore, ω is mainly used to balance the differences between categories. γ is a modulating factor to make the weights of hard examples larger than those of easy examples for balance.

In Formula (5), S denotes the set of hard examples, and N is the number of hard examples. We first sort the color classification loss of all pixels in a batch, and then take N pixels with the largest loss to form S. This allows the model to focus only on the pixels with larger loss and ignore the remaining pixels. In addition, as the training progresses, the number of hard examples gradually decreases, so the size of N also decreases.

The reference module and the colorization module produces a loss as expressed by Formula (5), respectively. We add the two in proportion as the final training loss as follows:

$$loss_{final} = \mu loss_{ref} + \vartheta loss_{color}, \tag{6}$$

where μ and ϑ are the scaling factors for the reference module loss $loss_{ref}$ and colorization module loss $loss_{color}$, respectively.

3.4 Evaluation Metric

We propose CDSIM to evaluate the difference in color quantity and spatial distribution between the colorization results and their ground truth. The calculation of this metric is dependent on the extraction of color features from the images. The color features we use include color histogram, color coherence vector, color correlogram, and color gradients. Details of their definition and feature reduction can be referred to the Supplementary Materials.

The combination and reduction of these features can produce an 1260-dimensional vector. Then the Euclidean distance between the color feature vectors of two images is defined as CDSIM, which is expressed in Eq. (7).

$$CDSIM\left(X, Y\right) = \sqrt{\sum_{i=1}^{l}\left(x_i - y_i\right)^2}, \tag{7}$$

where X and Y are the color feature vectors of two comparative images, and l is the vector length. The smaller the CDSIM is, the better the colorization is.

4 Experiments

4.1 Dataset

We construct a dataset consisting of 2967 *in situ* marine plankton ROI images for training and testing the IsPlanktonCLR network, whose composition is shown in Table 1. There are two main sources for this dataset. The first is the DYB-PlanktonNet dataset [22], which contains ROI images of 92 classes of marine plankton and suspended particles recorded *in situ* by IPP [21]. The second is from a dataset we exclusively obtained for this study by imaging natural seawater sample with the customized dual-channel dark-field imaging apparatus as used in [26]. For acquiring the IsPlanktonCLR dataset, this apparatus is firstly modified with installment of two lenses with the same magnification, replacement of one color camera with a grayscale camera with the same pixel size and number, and addition of an 850nm NIR light source to the white light source to illuminate the plankton in the seawater sample. Then the grayscale and color cameras are synchronized to capture image pairs of the same plankton sample in real seawater, which eventually constitute the IsPlanktonCLR dataset after image registration similar to that used in [26]. The IsPlanktonCLR dataset is available at https://drive.google.com/drive/folders/1GspuXRqd_GbB2k12UWiclN3MPFoclxYn?usp=sharing.

Table 1. Composition of the *in situ* plankton image dataset for experiments.

	Training Set	Testing Set 1	Testing Set 2	Total
DYB-PlanktonNet	2117	356	0	2473
IsPlanktonCLR	344	30	60×2	494
Total	2461	386	120	2967

4.2 Comparisons with Previous Works

Figure 3 compares the colorization results of multiple marine plankton grayscale images produced by IsPlanktonCLR and several SOTA approaches. Among them, images in row (1)–(11) are results from Testing Set 1, and images in row (a)–(c) are results from Testing Set 2.

Judging from human visual perception, the colorization performance of IsPlanktonCLR is obviously better than other models. The plankton shown in row (1)–(4) are a Megalopa larvae, an amphipod, and two copepods, which are numerous and common in the dataset. The rest of the tested images contain rare colors, on which the results of IsPlanktonCLR remained excellent, but the results of other models on them are significantly degraded. Specifically, other models show varying degrees of dominant color effect in row (5)–(7), color averaging effect in row (8), wrong colorization in row (9)–(10), and poor colorization of details in row (11).

IsPlanktonCLR also achieves good colorization effect on the images from Testing Set 2, *i.e.*, row (a)–(c), while the colorization results of other models are far from the ground truth. This demonstrates very good device- and content-generalization potential of the IsPlanktonCLR network.

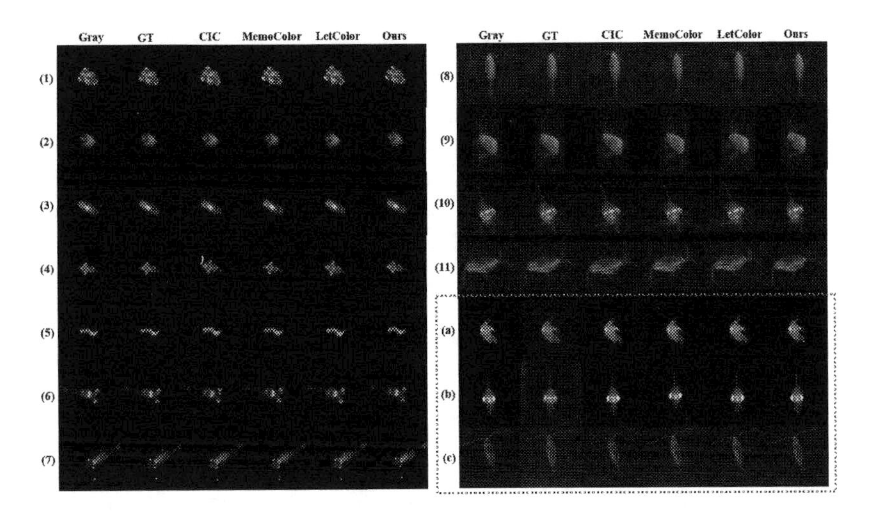

Fig. 3. Visual perception comparison of colorization performance by IsPlanktonCLR and other SOTA methods.

Table 2 compares the evaluation results between IsPlanktonCLR and other models on various numerical metrics. IsPlanktonCLR obtains the highest scores on both CDSIM and FID, which proves that the images generated by this algorithm have the highest similarity with ground truth; it performs slightly worse than other models on Colorfulness Score, because this metric mainly measures the color richness of image and is irrelevant to the colorization accuracy. Although the results of MemoColor get the best evaluation on Colorfulness Score, they contain many unreal colors. We also provide the evaluation results of PSNR and SSIM for reference only.

In order to make a fair comparison between IsPlanktonCLR and SOTA approaches, we further conduct an online survey to collect human visual evaluation from 115 volunteers (mainly composed of PhD students, marine biologists and several marine plankton experts) on the color similarity between the colorization results of four models and the ground truth. The survey questionnaire is designed to include 14 groups of marine plankton images, which can be referred to in the Supplemental Materials. The volunteers are asked to score the color similarity between each colorized image generated by one of four colorization models with its ground truth. The score is based on an 1–5 points scale with 5 for the most similar and 1 for the least similar. The average score for each model is finally tabulated in the rightmost column in Table 2, which indicates that IsPlanktonCLR still performs the best.

Table 2. Numerical comparison of colorization performance by IsPlanktonCLR and other SOTA methods under various evaluation metrics.

	CDSIM	FID	Colorfulness	UCIQE	PSNR	SSIM	Human
	↓	↓	↑	↑	↑	↑	↑
CIC	466.001	40.904	5.901	0.443	42.903	**0.997**	2.716
MemoColor	734.186	29.348	**6.789**	0.429	43.307	0.995	2.668
LetColor	384.637	29.063	6.009	**0.447**	43.905	0.983	2.920
IsPlanktonCLR	**346.434**	**24.578**	5.921	0.425	**44.269**	0.996	**3.785**

4.3 Ablation Experiments

Color Number. In order to select an appropriate number of color clusters K during palette customization, we calculate the variation of inertia with the number of clusters K as shown in Fig. 4 (a). In general, the smaller the inertia, the better the color clustering, and the larger the corresponding K. We compare the colorization results when K is taken as 32, 64, and 128, respectively, and find that there is little difference among them. Therefore, we choose $K = 32$ as the final number of clusters.

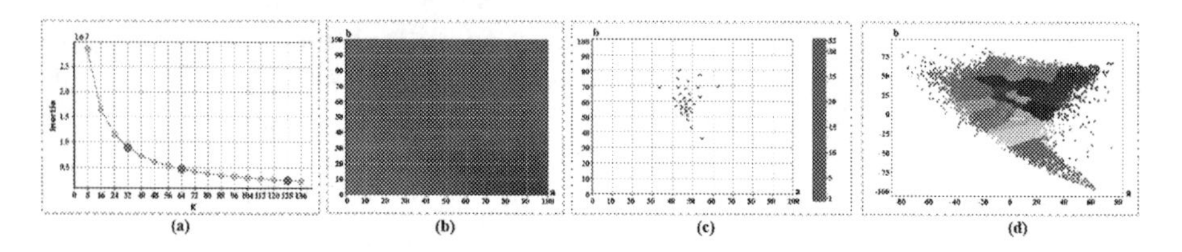

Fig. 4. (a) Inertia variation with the color cluster numbers. (b) The palette of the natural scene, (fixing $L = 50$ and normalizing ab channel values to [0,100]). (c) The customized palette of DYB-PlanktonNet dataset at $K = 32$. (d) Color clustering result of the DYB-PlanktonNet dataset with the customized palette (colors are not real but only for visualization). (Color figure online)

As can be compared in Fig. 4 (b) and (c), the customized palette of the *in situ* plankton images has significantly reduced colors over the palette of natural scenes after color clustering. This not only greatly simplifies the search space of the colorization algorithm, but also limits the color abuse by the model. Figure 4 (d) shows the 32 clustered colors of the DYB-PlanktonNet dataset are well separated.

Color Label. To validate the effectiveness of reference information provided by the reference module, the activation maps before and after the combiner in the colorization module are visualized in Fig. 5 (a). It can be seen that before the reference information is added, the network can only distinguish foreground plankton from the background, and the mean activation within the plankton is relatively homogeneous. After the addition of the reference information, the mean activation at different colors within the plankton appears significantly different. We can see the values of reddish colors in the ground truths are higher in the activation map, while the those of yellowish or greenish colors are lower. These results prove that the reference information can really help the network to distinguish different colors and provide effective guidance for colorization.

We also compare the colorization effect between discrete encoding and continuous encoding. For the majority of images, there is little difference between the two encoding methods. However, for some plankton with similar morphology and are difficult to determine color family labels, the continuous encoding has achieved slightly better colorization effects than those obtained by discrete encoding. Figure 5(b) shows the difference of colorization effect between the two encoding methods on two confusable examples. It can be seen that the colorization module will generate wrong colors with the guidance of discrete encoding when the reference module gives a wrong color family label, while the continuous coding can enable the model to recover correct colors in many positions. Taking the bottom row of Fig. 5(b) as an example, the network wrongly generates the greenish colors with discrete encoding, while it can recover correctly reddish colors with continuous encoding.

Loss Function. During training, we set the initial number of N, *i.e.*, the difficult examples, to be 0.05 of the total number of pixels, and it decays every 200 epochs with an attenuation rate of 0.5. Both the reference module and the colorization module use the same loss function as expressed in Formula (6) with coefficients μ and ϑ equal to 0.1 and 0.9, respectively.

Figure 5(c) compares the effects of IsPlanktonCLR and other models on colorization of rare colors. As the bodies of most plankton are semi-transparent or transparent, there are many white examples in our dataset while the reddish examples are rare. Except for a few species, reddish colors mainly appear in the positions of plankton eyes. As can be seen, our model overcomes the dominant color effect well and achieves accurate colorization of the decapod's eyes and the copepod's eye-spot, while other models perform much poorer. In addition, we conduct further experiment to compare the performance of our loss with the cross-entropy loss as baseline. The results are detailed in the Materials, which show that our loss can make the model converge faster than the baseline loss does under the same conditions.

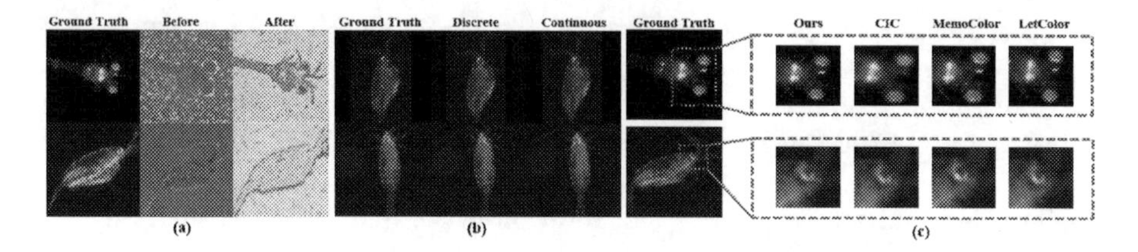

Fig. 5. (a) Examples indicating the change in mean activation before and after the addition of reference information. The colors in the activation maps are related to the mean value of activation with low values indicating cooler colors and high values indicating warmer colors. (b) Comparison of colorization effects between discrete encoding and continuous encoding. (c) Colorization effect comparison of each model on rare colors. (Color figure online)

4.4 CDSIM Metric

We firstly use a dataset of *in situ* plankton images to validate the proposed CDSIM. Figure 6 (a) shows the evaluation results of color dissimilarity obtained by different colorization models, where only the foreground pixels of plankton are used for CDSIM calculation. Intuitively, the color differences of images in each row from the ground truth become gradually more obvious from left to right, and their CDSIM scores also gradually increase, indicating that the colorization quality is getting worse. This result verifies that CDSIM can not only quantify color dissimilarity between plankton images, but also its evaluation results are consistent with the perception of human eyes.

In addition, we select some images from ImageNet [10] to further assess CDSIM on natural scene image colorization, and the results are compared with those obtained by other common metrics, as shown in Fig. 6(b). Before this test, we replace some colors in the original pictures with other colors that look still reasonable to human eyes. In Case1, some colors of the scenes are replaced with visually similar colors; while in Case2, some colors in the pictures are replaced with obviously different colors. In this test, all the pixels of a whole image are evaluated. The results show that when using CDSIM and FID, the scores of Case1 images are significantly lower than those of Case2, which is consistent with human visual perception. But the results of PSNR and SSIM are very similar in the two cases, indicating that they cannot distinguish the color difference. The evaluation results of Colorfulness are not consistent at all. This result proves that CDSIM is also suitable for objective and quantitative evaluation of natural scene image colorization.

Compared with PSNR, SSIM, FID and other metrics, CDSIM has higher computational complexity. Especially for high resolution natural scene colorful images, its computation can be intensive. However, for *in situ* marine plankton dark-field ROIs, the CDSIM computation cost is significantly reduced due to limited image resolution and even lower proportion of foreground pixels, which is completely acceptable in practice.

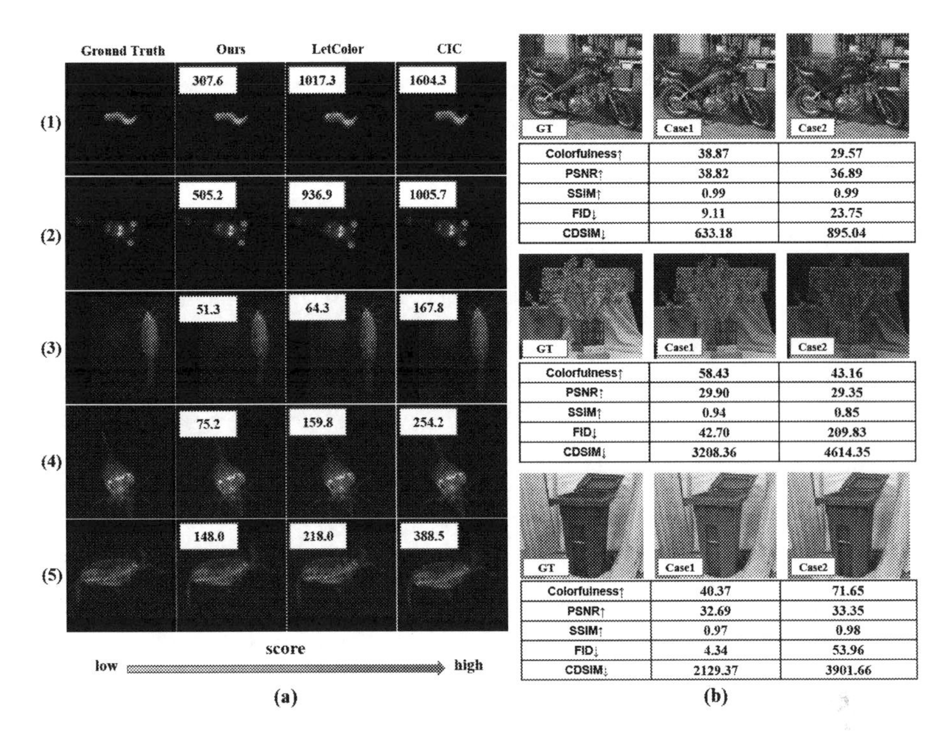

Fig. 6. (a) Comparison of visual perception and CDSIM evaluations on marine plankton images produced by various colorization models. (b) Comparison of visual perception and numerical metrics-based evaluations on artificially colorized natural scene images. (Color figure online)

5 Conclusion

We present a deep colorization model for automatic color restoration of *in situ* marine plankton grayscale images. The model achieves the state-of-the-art performances on color restoration of *in situ* marine plankton image data. This is the first endeavor, to the best of our knowledge, to apply deep colorization for marine plankton scientific imagery. We also propose a metric for comparing color dissimilarity between images, which provides a new and objective evaluation for restoration-oriented image colorization algorithms. This method is expected to inspire new design of next generation instruments or systems for achieving long-term, continuous, high-frequency, and *in situ* ocean observation.

Acknowledgement. This work was supported by International Partnership Program of Chinese Academy of Sciences No. 172644kysb20210022, Scientific Instrument Development Project of Chinese Academy of Sciences No. YJKYYQ201 90028, and Shenzhen Science and Technology Innovation Program No. JCYJ2020 0109105823170. We thank the participants for replying our online survey.

References

1. Anwar, S., Tahir, M., Li, C., Mian, A., Khan, F.S., Muzaffar, A.W.: Image colorization: a survey and dataset. arXiv preprint arXiv:2008.10774 (2020)
2. Arthur, D., Vassilvitskii, S.: k-means++: The advantages of careful seeding. Technical report, Stanford (2006)
3. Bahng, H., et al.: Coloring with words: guiding image colorization through text-based palette generation. In: Ferrari, V., Hebert, M., Sminchisescu, C., Weiss, Y. (eds.) ECCV 2018. LNCS, vol. 11216, pp. 443–459. Springer, Cham (2018). https://doi.org/10.1007/978-3-030-01258-8_27
4. Baig, M.H., Torresani, L.: Multiple hypothesis colorization and its application to image compression. Comput. Vis. Image Underst. **164**, 111–123 (2017)
5. Benfield, M.C., et al.: Rapid: research on automated plankton identification. Oceanography **20**(2), 172–187 (2007)
6. Campbell, R., Roberts, P., Jaffe, J.: The prince William sound plankton camera: a profiling in situ observatory of plankton and particulates. ICES J. Mar. Sci. **77**(4), 1440–1455 (2020)
7. Cao, Y., Zhou, Z., Zhang, W., Yu, Y.: Unsupervised diverse colorization via generative adversarial networks. In: Ceci, M., Hollmén, J., Todorovski, L., Vens, C., Džeroski, S. (eds.) ECML PKDD 2017. LNCS (LNAI), vol. 10534, pp. 151–166. Springer, Cham (2017). https://doi.org/10.1007/978-3-319-71249-9_10
8. Ci, Y., Ma, X., Wang, Z., Li, H., Luo, Z.: User-guided deep anime line art colorization with conditional adversarial networks. In: Proceedings of the 26th ACM International Conference on Multimedia, pp. 1536–1544 (2018)
9. Davis, C., Gallager, S., Berman, M., Haury, L., Strickler, J.: The video plankton recorder (VPR): design and initial results. Arch. Hydrobiol. Beih **36**, 67–81 (1992)
10. Deng, J., Dong, W., Socher, R., Li, L.J., Li, K., Fei-Fei, L.: ImageNet: a large-scale hierarchical image database. In: 2009 IEEE Conference on Computer Vision and Pattern Recognition, pp. 248–255. IEEE (2009)
11. Dong, X., Li, W.: Shoot high-quality color images using dual-lens system with monochrome and color cameras. Neurocomputing **352**, 22–32 (2019)
12. Forward, R.B.: Light and diurnal vertical migration: photobehavior and photophysiology of plankton. In: Smith, K.C. (ed.) Photochemical and photobiological reviews, pp. 157–209. Springer, Boston (1976). https://doi.org/10.1007/978-1-4684-2574-1_4
13. Gallager, S.M.: Continuous particle imaging and classification system. US Patent 10,222,688, 5 March 2019
14. Grossmann, M.M., Gallager, S.M., Mitarai, S.: Continuous monitoring of near-bottom mesoplankton communities in the east china sea during a series of typhoons. J. Oceanogr. **71**(1), 115–124 (2015)
15. Guo, P., Ma, Z.: Low-light color imaging via dual camera acquisition. In: Proceedings of the Asian Conference on Computer Vision (2020)
16. Hasler, D., Suesstrunk, S.E.: Measuring colorfulness in natural images. In: Human Vision and Electronic Imaging VIII, vol. 5007, pp. 87–95. International Society for Optics and Photonics (2003)
17. He, M., Chen, D., Liao, J., Sander, P.V., Yuan, L.: Deep exemplar-based colorization. ACM Trans. Graph. (TOG) **37**(4), 1–16 (2018)
18. Heusel, M., Ramsauer, H., Unterthiner, T., Nessler, B., Hochreiter, S.: GANs trained by a two time-scale update rule converge to a local Nash equilibrium. Adv. Neural Inf. Process. Syst. **30** (2017)

19. Iizuka, S., Simo-Serra, E., Ishikawa, H.: Let there be color! Joint end-to-end learning of global and local image priors for automatic image colorization with simultaneous classification. ACM Trans. Graph. (ToG) **35**(4), 1–11 (2016)
20. Kaiser, Ł., Nachum, O., Roy, A., Bengio, S.: Learning to remember rare events. arXiv preprint arXiv:1703.03129 (2017)
21. Li, J., et al.: Development of a buoy-borne underwater imaging system for in situ mesoplankton monitoring of coastal waters. IEEE J. Oceanic Eng. **47**(1), 88–110 (2021)
22. Li, J., Yang, Z., Chen, T.: DYB-planktonnet. IEEE Dataport (2021)
23. Lin, T.Y., Goyal, P., Girshick, R., He, K., Dollár, P.: Focal loss for dense object detection. In: Proceedings of the IEEE International Conference on Computer Vision, pp. 2980–2988 (2017)
24. Lin, T.-Y., et al.: Microsoft COCO: common objects in context. In: Fleet, D., Pajdla, T., Schiele, B., Tuytelaars, T. (eds.) ECCV 2014. LNCS, vol. 8693, pp. 740–755. Springer, Cham (2014). https://doi.org/10.1007/978-3-319-10602-1_48
25. Lombard, F., et al.: Globally consistent quantitative observations of planktonic ecosystems. Fron. Marine Sci. 196 (2019)
26. Ma, W., et al.: Super-resolution for in situ plankton images. In: Proceedings of the IEEE/CVF International Conference on Computer Vision, pp. 3683–3692 (2021)
27. Manjunatha, V., Iyyer, M., Boyd-Graber, J., Davis, L.: Learning to color from language. arXiv preprint arXiv:1804.06026 (2018)
28. Merz, E., et al.: Underwater dual-magnification imaging for automated lake plankton monitoring. Water Res. **203**, 117524 (2021)
29. Messaoud, S., Forsyth, D., Schwing, A.G.: Structural consistency and controllability for diverse colorization. In: Ferrari, V., Hebert, M., Sminchisescu, C., Weiss, Y. (eds.) ECCV 2018. LNCS, vol. 11210, pp. 603–619. Springer, Cham (2018). https://doi.org/10.1007/978-3-030-01231-1_37
30. Nazeri, K., Ng, E., Ebrahimi, M.: Image colorization using generative adversarial networks. In: Perales, F.J., Kittler, J. (eds.) AMDO 2018. LNCS, vol. 10945, pp. 85–94. Springer, Cham (2018). https://doi.org/10.1007/978-3-319-94544-6_9
31. Orenstein, E.C., et al.: The scripps plankton camera system: a framework and platform for in situ microscopy. Limnol. Oceanogr. Methods **18**(11), 681–695 (2020)
32. Picheral, M., Grisoni, J.M., Stemmann, L., Gorsky, G.: Underwater video profiler for the "in situ" study of suspended particulate matter. In: IEEE Oceanic Engineering Society. OCEANS 1998. Conference Proceedings (Cat. No. 98CH36259), vol. 1, pp. 171–173. IEEE (1998)
33. Plonus, R.M., Conradt, J., Harmer, A., Janßen, S., Floeter, J.: Automatic plankton image classification - can capsules and filters help cope with data set shift? Limnol. Oceanogr. Methods **19**(3), 176–195 (2021)
34. Sangkloy, P., Lu, J., Fang, C., Yu, F., Hays, J.: Scribbler: controlling deep image synthesis with sketch and color. In: Proceedings of the IEEE Conference on Computer Vision and Pattern Recognition, pp. 5400–5409 (2017)
35. Shrivastava, A., Gupta, A., Girshick, R.: Training region-based object detectors with online hard example mining. In: Proceedings of the IEEE Conference on Computer Vision and Pattern Recognition, pp. 761–769 (2016)
36. Steinberg, D.K., Landry, M.R.: Zooplankton and the ocean carbon cycle. Ann. Rev. Mar. Sci. **9**, 413–444 (2017)
37. Su, J.W., Chu, H.K., Huang, J.B.: Instance-aware image colorization. In: Proceedings of the IEEE/CVF Conference on Computer Vision and Pattern Recognition, pp. 7968–7977 (2020)

38. Tanaka, M., Genin, A., Endo, Y., Ivey, G.N., Yamazaki, H.: The potential role of turbulence in modulating the migration of demersal zooplankton. Limnol. Oceanogr. **66**(3), 855–864 (2021)
39. Tanaka, M., Genin, A., Lopes, R.M., Strickler, J.R., Yamazaki, H.: Biased measurements by stationary turbidity-fluorescence instruments due to phototactic zooplankton behavior. Limnol. Oceanogr. Methods **17**(9), 505–513 (2019)
40. Vitoria, P., Raad, L., Ballester, C.: ChromaGAN: adversarial picture colorization with semantic class distribution. In: Proceedings of the IEEE/CVF Winter Conference on Applications of Computer Vision, pp. 2445–2454 (2020)
41. Wu, Y., Wang, X., Li, Y., Zhang, H., Zhao, X., Shan, Y.: Towards vivid and diverse image colorization with generative color prior. In: Proceedings of the IEEE/CVF International Conference on Computer Vision, pp. 14377–14386 (2021)
42. Xu, Z., Wang, T., Fang, F., Sheng, Y., Zhang, G.: Stylization-based architecture for fast deep exemplar colorization. In: Proceedings of the IEEE/CVF Conference on Computer Vision and Pattern Recognition, pp. 9363–9372 (2020)
43. Yang, M., Sowmya, A.: An underwater color image quality evaluation metric. IEEE Trans. Image Process. **24**(12), 6062–6071 (2015)
44. Yoo, S., Bahng, H., Chung, S., Lee, J., Chang, J., Choo, J.: Coloring with limited data: few-shot colorization via memory augmented networks. In: Proceedings of the IEEE/CVF Conference on Computer Vision and Pattern Recognition, pp. 11283–11292 (2019)
45. Zhang, R., Isola, P., Efros, A.A.: Colorful image colorization. In: Leibe, B., Matas, J., Sebe, N., Welling, M. (eds.) ECCV 2016. LNCS, vol. 9907, pp. 649–666. Springer, Cham (2016). https://doi.org/10.1007/978-3-319-46487-9_40
46. Zhang, R., et al.: Real-time user-guided image colorization with learned deep priors. arXiv preprint arXiv:1705.02999 (2017)
47. Zhao, J., Han, J., Shao, L., Snoek, C.G.: Pixelated semantic colorization. Int. J. Comput. Vision **128**(4), 818–834 (2020)

Efficient Deep Visual and Inertial Odometry with Adaptive Visual Modality Selection

Mingyu Yang[(✉)], Yu Chen, and Hun-Seok Kim

University of Michigan, Ann Arbor, MI 48109, USA
{mingyuy,unchenyu,hunseok}@umich.edu

Abstract. In recent years, deep learning-based approaches for visual-inertial odometry (VIO) have shown remarkable performance outperforming traditional geometric methods. Yet, all existing methods use both the visual and inertial measurements for every pose estimation incurring potential computational redundancy. While visual data processing is much more expensive than that for the inertial measurement unit (IMU), it may not always contribute to improving the pose estimation accuracy. In this paper, we propose an adaptive deep-learning based VIO method that reduces computational redundancy by opportunistically disabling the visual modality. Specifically, we train a policy network that learns to deactivate the visual feature extractor on the fly based on the current motion state and IMU readings. A Gumbel-Softmax trick is adopted to train the policy network to make the decision process differentiable for end-to-end system training. The learned strategy is interpretable, and it shows scenario-dependent decision patterns for adaptive complexity reduction. Experiment results show that our method achieves a similar or even better performance than the full-modality baseline with up to 78.8% computational complexity reduction for KITTI dataset evaluation. The code is available at https://github.com/mingyuyng/Visual-Selective-VIO.

Keywords: Visual-inertial odometry · Deep neural networks · Long short-term memory · Gumbel-softmax · Adaptive learning

1 Introduction

Visual-inertial odometry (VIO) estimates the agent's self-motion using information collected from cameras and inertial measurement unit (IMU) sensors. With its wide applications in navigation and autonomous driving, VIO became one of the most important problems in the field of robotics and computer vision. Compared with visual odometry (VO) methods [3,9,10,30], VIO systems [24,34]

M. Yang and Y. Chen—Equally contributed.

Supplementary Information The online version contains supplementary material available at https://doi.org/10.1007/978-3-031-19839-7_14.

S. Avidan et al. (Eds.): ECCV 2022, LNCS 13698, pp. 233–250, 2022.
https://doi.org/10.1007/978-3-031-19839-7_14

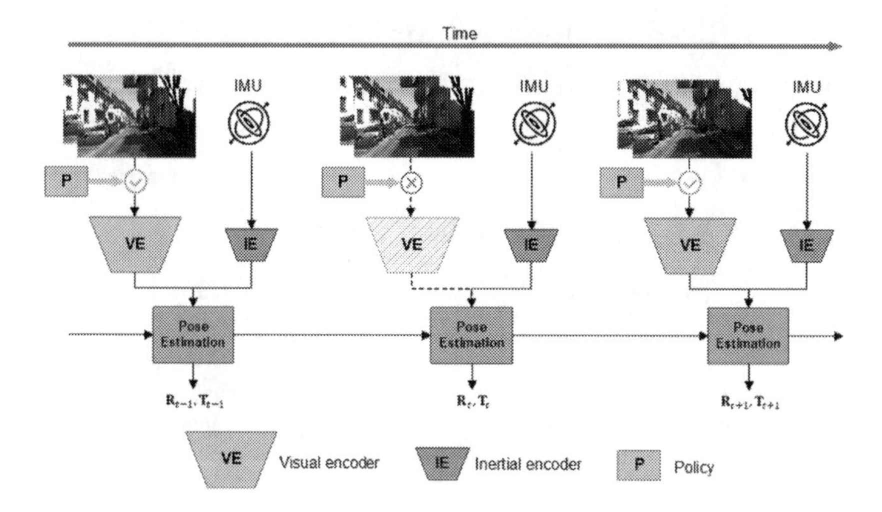

Fig. 1. An overview of our approach. For deep learning-based VIO methods, the computational cost of the visual encoder is much higher than that of the inertial encoder due to the difference in data dimension. Thus, rather than using images for every pose estimation, our method learns a policy that controls the usage of the visual encoder to avoid unnecessary image processing while maintaining a reasonable accuracy.

incorporate additional IMU measurements and thus achieve more robust performance in texture-less environments and/or in extreme lightning conditions. However, classical VIO methods (not based on deep learning) rely heavily on manual interventions for system initialization and careful parameter tuning (e.g., number of features per frame, threshold of feature matching, and keyframe selection) for each test environment. Besides, there are still significant challenges to deploying such systems with rapid calibration for fast-moving scenarios [53].

With the tremendous success of deep learning in various computer vision tasks [22,35,41], data-driven VIO methods [1,6,7,15,25,40] have drawn significant attention to the community, and they achieve competitive performance in both accuracy and robustness in challenging scenarios. Compared with classical geometric-based methods, these learning-based VIO solutions extract better features using deep neural networks (DNN). In addition, they can learn a better fusion mechanism between visual and inertial features to filter out abnormal sensor data while training on large-scale datasets. However, such learning-based methods typically have significant overhead in computation and power consumption, which is not affordable to energy-constrained mobile platforms operating with low-cost, energy-efficient cameras and IMU sensors.

Motivated by recent works that apply temporal adaptive inference to realize efficient action recognition [27,28,33,49] and fast text classification [4,16,39], we propose a new adaptive policy-based method to alleviate the high computational cost of deep learning-based VIO methods. The trained policy network opportunistically disables the visual (image) modality, as illustrated in Fig. 1, to reduce the computational overhead when the visual features do not contribute significantly to the overall pose estimation accuracy. We choose to dynamically disable the visual modality while keeping IMU always available because the image encoder is much more computationally demanding than the inertial encoder due to their modality dimension difference. Thus, skipping the image processing significantly reduces the overall computational complexity. Besides, visual information is not always necessary for an accurate pose estimation, especially when the motion state does not vary much over time. Thus, occasionally skipping unimportant image inputs does not necessarily degrade the odometry accuracy. For our method, the proposed policy uses sampling from a Bernoulli distribution parameterized by the output of a light-weight policy network. We adopt the Gumbel-Softmax trick [20] to make the decision process differentiable. The model is trained to strike a balance between accuracy and efficiency with a joint loss. Our experiments demonstrate that our method significantly reduces computation (up to 78.8%) without compromising VIO accuracy. Thus, the proposed framework is suitable for mobile platforms with limited computation resources and energy budgets. Also, our method is modal-agnostic and can be applied to any visual and inertial encoders with different structures. Moreover, the learned policy is interpretable and yields scenario-dependent decision patterns in various test sequences.

Overall, our contributions are summarized as follows:

- We propose a novel method that adaptively disables the visual modality on the fly for efficient deep learning-based VIO. To the best of our knowledge, we are the first to demonstrate such a system reducing the complexity and energy consumption of deep learning-based VIO.
- A novel policy network is jointly trained with a pose estimation network to learn a visual modality selection strategy to enable or disable a visual feature extractor based on the motion state and IMU measurements. We adopt a Gumbel-Softmax trick to make the end-to-end system differentiable.
- The proposed method is tested extensively on the KITTI Odometry dataset. Experiments show that our approach achieves up to 78.8% computation reduction without noticeable performance degradation. Furthermore, we show that the learned policy exhibits an interpretable behavior that depends on motion states and patterns.

2 Related Works

2.1 Visual-Inertial Odometry

Visual odometry (VO) is a process to estimate ego-motion from sequential camera images [32], and it is extended to visual inertial odometry (VIO) including an IMU as an additional input. The datapath of conventional schemes typically consists of the following steps: feature detection, feature matching and tracking, motion estimation, and local optimization [38]. The VO/VIO system can be integrated into a simultaneous localization and mapping (SLAM) system [30,31,34] by performing additional steps of 3D environment mapping, global optimization, and loop closure. The performance of conventional VIO/SLAM systems is largely affected by visual feature matching and tracking accuracy, and the sensor fusion strategy. Hence, identifying superior handcrafted feature descriptors [26,37], adaptive filtering [24] or nonlinear optimization [17,23] based sensor fusion schemes are key challenges of such methods.

In recent years, deep learning-based methods have achieved remarkable successes on various computer vision applications, including VIO. VINet [7] is the first end-to-end trainable deep learning-based VIO where a DNN learns pose regression from the sequence of images and IMU measurements in a supervised manner. A long short-term memory (LSTM) network is introduced in VINet to model the temporal motion correlation. Later, Chen et al. [6] propose two different masking techniques that selectively fuse the visual and inertial features. ATVIO [25] introduces an attention-based fusion function and uses an adaptive loss for pose regression. Some recent works also propose to learn the 6-DoF ego-motion through a self-supervised learning framework that does not require ground-truth annotations during training. Shamwell et al. [40] introduce VIOLearner that estimates the poses through a view-synthesis approach with multi-level error correction. DeepVIO [15] improves VIO poses by additional self-supervision of optical flow, and similarly Almalioglu et al. [1] demonstrate a self-supervised VIO based on depth estimation [13,55].

Whereas these prior works always rely on both the visual and inertial modality for each pose estimation, we propose a new framework to save the computation and power consumption overhead by opportunistically disabling the visual modality based on a learned strategy.

2.2 Adaptive Inference

An adaptive inference scheme dynamically allocates computing resources based on each task input instance to minimize the redundant computation for relatively 'easy' task inputs. Several techniques for adaptive inference have been proposed including early exiting [2,19,42], layer skipping [14,43,45], and dynamic channel pruning [18,52,54]. Recently, the idea of adaptive inference has been extend to sequential data (e.g., text [4,16,39] and videos [27,28,33,49]) that are processed by recurrent neural networks (RNNs). Our technique is closely related to adaptive video recognition first proposed in [49], which introduces a memory-augmented LSTM to select only the relevant frames for efficient action recognition by training with a policy gradient method. Similarly, AR-Net [27] learns a policy that dynamically selects more relevant image frames and also adjusts their resolutions. The training in AR-Net is simplified using the Gumbel-Softmax trick. Later, this idea was extended to adaptively selecting a proper modality [33] or patches [46]. Our approach is motivated by these prior works to apply a similar framework to adaptive computation on deep learn-based VIO for the first time. We formulate it as a discrete-time pose regression problem that produces a pose estimation for every time step.

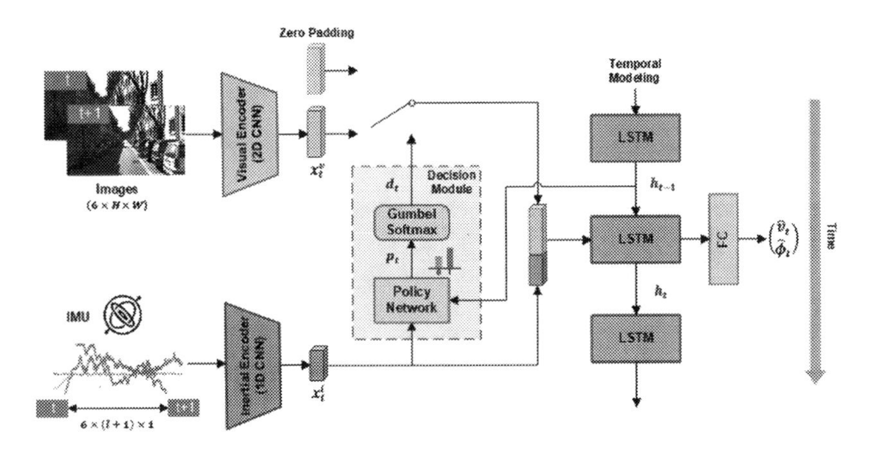

Fig. 2. Illustration of the proposed framework. At each time step, the policy network takes the inertial features and the previous hidden state vector to decide whether to use the visual modality or not. Once the policy network decides to use the visual modality, the current image is passed through the visual encoder, and the corresponding visual features are fed to the pose regression LSTM together with the inertial features for pose estimation. Otherwise, the visual encoder is disabled to save computations and LSTM input is zero padded. The decisions are sampled from a Gumbel-Softmax distribution during training to make the system end-to-end differentiable. During inference, the decision is sampled from a Bernoulli distribution controlled by the policy network.

3 Method

The inputs for VIO are the monocular video frames $\{V_i\}_{i=1}^{N}$, IMU measurements $\{I_i\}_{i=1}^{Nl}$ captured with a sampling frequency l times higher than the video frame rate, and the initial camera pose P_1. The goal of VIO is to estimate the camera poses $\{P_i\}_{i=2}^{N}$ for the entire path where $V_i \in \mathbb{R}^{3\times H\times W}$, $I_i \in \mathbb{R}^6$, and $P_i \in \mathbf{SE}(3)$. One typical way to perform VIO is to estimate the 6-DoF relative pose $T_{t\rightarrow t+1}$ that satisfies $P_t T_{t\rightarrow t+1} = P_{t+1}$ using two consecutive images $V_{t\rightarrow t+1} = \{V_t, V_{t+1}\}$ and a set of IMU measurements $I_{t\rightarrow t+1} = \{I_{tl}, \ldots, I_{(t+1)l}\}$ for the time index $t = 1, 2, \ldots, N-1$. The relative pose $T_{t\rightarrow t+1}$ can be further decomposed into a rotational vector $\phi_t \in \mathbb{R}^3$ containing Eular angles and a translational vector $v_t \in \mathbb{R}^3$. Our method learns a selection strategy that opportunistically skips the visual information $V_{t\rightarrow t+1}$ to reduce the computational complexity while maintaining the relative pose estimation accuracy.

3.1 End-to-End Neural Visual-Inertial Odometry

End-to-end neural VIO methods [6,7,25] consist of a visual feature encoder E_{visual} and an inertial feature encoder $E_{inertial}$ that extracts learned features from the input images and IMU measurements as follows:

$$x_t^v = E_{visual}(V_{t\rightarrow t+1}), \quad x_t^i = E_{inertial}(I_{t\rightarrow t+1}). \tag{1}$$

Typically, the visual feature encoder is much larger than the inertial feature encoder as the image dimension is much larger than that of the IMU measurement.

Visual feature x_t^v and inertial feature x_t^i are combined as z_t through concatenation [7] or attention modules [6,25]. For accurate pose estimation, estimated motions and states of previous frames are used together with the newly extracted features of the current frame. Because of this temporally sequential nature of the problem, an RNN is typically employed to learn the correlation within the sequence of motions. The RNN employes fully connected layers as the last step for the final 6-DoF pose regression as in:

$$(h_t, \hat{\phi}_t, \hat{v}_t) = \mathrm{RNN}(z_t, h_{t-1}), \tag{2}$$

where h_{t-1} and h_t are the hidden latent vectors of the RNN at time t and $t-1$. $\hat{\phi}_t$ and $\hat{v}_t$ denotes the estimated rotational vector and translational vector, respectively.

3.2 Deep VIO with Visual Modality Selection

The overview of our proposed method is illustrated in Fig. 2. As an adaptive method, we aim to learn a binary decision d_t to determine whether the visual modality is not necessary and can be disabled without significant pose estimation accuracy degradation. We introduce a decision module where the decision d_t is sampled from a Bernoulli distribution whose probability p_t is generated by a light-weight policy network Φ. The policy network takes the current IMU features x_t^i and the last hidden latent vector h_{t-1} that contains the history information as the input. Thus, we have

$$p_t = \Phi(h_{t-1}, x_i^t), \tag{3}$$

where $p_t \in \mathbb{R}^2$ denotes the probability of the Bernoulli distribution. To make the system end-to-end trainable, we sample the binary decision $d_t \in \{0, 1\}$ via the Gumbel-Softmax operation,

$$d_t \sim \mathrm{GUMBEL}(p_t). \tag{4}$$

The detail of training with Gumbel-Softmax is discussed in the next section. When $d_t = 1$, visual features are enabled and the combined feature is obtained by concatenation of visual features and inertial features. On the other hand, when $d_t = 0$, visual features are disabled thus we apply zero padding to replace visual features to keep the same input dimension for the following RNN. This can be expressed as:

$$z_t = \begin{cases} x_t^v \oplus x_t^i & \text{if } d_t = 1 \\ \mathbf{0} \oplus x_t^i & \text{otherwise} \end{cases}, \tag{5}$$

where $\oplus$ denotes the concatenation operation. The combined feature z_t is then fed to the RNN that produces the estimated pose outputs ($\hat{\phi}_t$ and $\hat{v}_t$) via regression as in Eq. (2). In this paper, we adopt a two-layer LSTM for the pose estimation RNN.

3.3 Training with Gumbel-Softmax

Sampled d_t that follows a Bernoulli distribution is discrete in nature and it makes the network non-differentiable. Thus, it is not trivial to train the policy network through back-propagation. One common choice is to use a score function estimator (e.g., REINFORCE [12,47]) to estimate the gradient through the 'log-derivative trick'. However, that approach often has issues with slow convergence and high variance [48] for many applications. As an alternative, we adopt the Gumbel-Softmax scheme [20] to resolve non-differentiability by sampling from a corresponding Gumbel-Softmax distribution, which is essentially a reparametrization trick for categorical distributions [21,29,36]. Though reparameterization tricks may be less general than score function estimators, they usually exhibit several advantages such as lower variance and easier implementation.

Consider a categorical distribution where the probability for the k_{th} category is p_k for $k = 1, ..., K$. Then, following the Gumbel-Max trick [20], a discrete sample $\hat{P}$ that follows the target distribution can be drawn by:

$$\hat{P} = \underset{k}{\operatorname{argmax}}(\log p_k + g_k), \quad k \in [1, 2, ..., K], \tag{6}$$

where $g_k = -\log(-\log U_k)$ is a standard Gumbel distribution with a random variable U_k sampled from a uniform distribution $U(0, 1)$. Later, the softmax function is applied to relax the argmax operation to obtain a real-valued vector $\tilde{P} \in \mathbb{R}^K$ by a differentiable function as in

$$\tilde{P}_k = \frac{\exp((\log p_k + g_k)/\tau)}{\sum_{j=1}^{K} \exp((\log p_j + g_j)/\tau)}, \quad k = 1, 2, ..., K, \tag{7}$$

where τ is a temperature parameter that controls the 'discreteness' of $\tilde{P}$. When τ goes to infinity, $\tilde{P}$ tends to be a uniformly distributed vector, whereas $\tau \approx 0$ makes $\tilde{P}$ close to a one-hot vector and indistinguishable from the discrete distribution. In our case, we only have two categories $K = 2$ since we are dealing with a binary decision. During training, we sample the policy from the target Bernoulli distribution through (6) for the forward pass whereas the continuous relaxation (7) is used for the backward pass to approximate the gradient.

3.4 Loss Function

During training, we apply the mean squared error (MSE) loss to reduce the pose estimation error given by:

$$\mathcal{L}_{pose} = \frac{1}{T-1} \sum_{t=1}^{T-1} (\|\hat{v}_t - v_t\|_2^2 + \alpha \|\hat{\phi}_t - \phi_t\|_2^2), \tag{8}$$

where T is the sequence length of training. v_t and ϕ_t denote the ground-truth translational and rotational vectors. α is a weight to balance the translational loss and rotational loss. We set $\alpha = 100$ as in the setting in prior supervised learning VO/VIO methods [6,7,25,44,51].

Besides, we apply an additional penalty factor λ to every visual encoder usage to encourage disabling visual feature computations. During the training, we calculate the averaged penalty and denote it as the efficiency loss defined by:

$$\mathcal{L}_{eff} = \frac{1}{T-1} \sum_{t=1}^{T-1} \lambda d_t. \tag{9}$$

Finally, the end-to-end system is trained with the summation of the pose estimation loss and efficiency loss (10) to strike a balance between good accuracy and computational efficiency.

$$\mathcal{L} = \mathcal{L}_{pose} + \mathcal{L}_{eff} \tag{10}$$

4 Experiments

In this section, we conduct an ablation study on the penalty factor to compare the proposed adaptive scheme with the full-modality baseline that always uses visual features. Results in this section will show that our proposed visual modality selection strategy can significantly reduces computational overhead while maintaining a similar or better accuracy compared to the full-modality baseline.

4.1 Experiment Setup

Dataset. We evaluate our approach on KITTI Odometry dataset [11], which is one of the most influential VO/VIO benchmarks. The KITTI Odometry dataset consists of 22 sequences of stereo videos, where Sequence *00–10* contain the ground-truth trajectory and Sequence *11–22* exclude the ground-truth for evaluation. Following the procedure in [6], we train our model with Sequence *00, 01, 02, 04, 06, 08, 09* and test with Sequence *05, 07*, and *10*. We exclude Sequence *03* because of the lack of the raw IMU data. The images and ground-truth poses are recorded 10 Hz and the IMU data is recorded 100 Hz. The IMU data and images are not strictly synchronized. Thus, we interpolate the raw IMU data to time-synchronize it with the images and ground-truth poses. We use the monocular images from the left camera of KITTI Odometry dataset.

Implementation Details. During training, we resize all images to 512×256 and set the training subsequence length to 11. We have 11 IMU measurements between every two consecutive images and thus the dimension of the input IMU data is 6×11. For the visual encoder, we adopt the FlowNet-S network [8] (except for the last layer) pretrained on the FlyingChairs dataset [8] for optical flow estimation. A fully connected layer is attached at the end of the network to produce a visual feature of length 512. The inertial encoder contains three 1D-convolutional layers and a fully connected layer to generate the inertial feature

Table 1. Evaluation of the full-modality baseline and our proposed method with various penalty factors λ on the KITTI dataset. Due to the stochastic nature of our policy, we test our model with 10 different random seeds and show the average performance.

Method	Trans. RMSE (m)	Rot. RMSE (°)	Visual encoder usage	GFLOPS
Full Modality	**0.0355**	0.0648	100%	77.87
$\lambda = 1 \times 10^{-5}$	0.0364	0.0505	62.89%	49.04
$\lambda = 3 \times 10^{-5}$	0.0406	**0.0495**	21.02%	16.51
$\lambda = 5 \times 10^{-5}$	0.0477	0.0529	11.37%	9.02
$\lambda = 7 \times 10^{-5}$	0.0609	0.0592	**6.85%**	**5.50**

Table 2. The relative translational & rotational error, and visual encoder usage of the baseline model and our proposed method with different penalty factors λ on Sequence 05, 07, and 10. The results are averaged over 10 tests with different seeds. The last column also shows the standard deviation to quantify the stability.

Method	Seq. 05			Seq. 07			Seq. 10			Average		
	t_{rel}	r_{rel}	Usage	t_{rel}	r_{rel}	Usage	t_{rel}	r_{rel}	Usage	t_{rel}	r_{rel}	Usage
Full modality	2.61	1.06	100%	1.83	1.35	100%	**3.11**	1.12	100%	2.52	1.18	100%
$\lambda = 1 \times 10^{-5}$	2.15	0.78	60.30%	2.25	1.19	63.35%	3.30	**0.94**	65.01%	2.57 ± 0.052	0.97 ± 0.018	62.89%
$\lambda = 3 \times 10^{-5}$	**2.01**	**0.75**	20.60%	**1.79**	**0.76**	19.79%	3.41	1.08	22.68%	$\mathbf{2.40 \pm 0.064}$	$\mathbf{0.86 \pm 0.018}$	21.02%
$\lambda = 5 \times 10^{-5}$	2.71	1.03	11.34%	2.22	1.14	10.57%	3.59	1.20	12.20%	2.84 ± 0.102	1.13 ± 0.045	11.37%
$\lambda = 7 \times 10^{-5}$	3.00	1.20	**6.83%**	2.48	1.60	**6.03%**	3.67	1.57	**7.68%**	3.05 ± 0.086	1.46 ± 0.046	**6.85%**

■ t_{rel} and r_{rel} are the average translational error (%) and average rotational error (°/100 m) obtained from various segment lengths of 100 m–800 m.

of size 256. The pose estimation network contains a two-layer LSTM each with 1024 hidden units. At each time step, the hidden state of the last LSTM layer is passed through a two-layer multi-layer perceptron (MLP) to estimate the 6-DoF pose. The policy network is designed with a light-weight three-layer MLP.

The training process consists of two stages: warm-up stage and joint-training stage. In the warm-up stage, we train the visual encoder, inertial encoder, and the pose estimation network for 40 epochs with a random policy where we have a 50% chance to use the visual encoder at each time step. The learning rate is set to 5×10^{-4} in this stage. Next, in the joint-training stage, we train all end-to-end components including the policy network for 40 epochs with a learning rate of 5×10^{-5}, and then decrease the learning rate to 1×10^{-6} for additional 20 epochs. We set the initial temperature of Gumbel-Softmax to 5 and apply exponential decaying for each epoch with a factor of -0.05. We use Adam optimization with $\alpha = 0.9$ and $\beta = 0.999$, and the batch size is set to 16. During training, we always use the visual modality for the first frame to guarantee a qualified initial pose estimation. Similarly, during inference, we always enable the visual modality for the first pose estimation before we run the policy network without intervention for the rest of the path. Although the sequence length for training is set to 11, our method can run on any length of inputs for the inference.

Metric. We calculate the root mean square error (RMSE) for the estimated translational vectors $\{\hat{v}_t\}_{t=1}^{N-1}$ and rotational vectors $\{\hat{\phi}_t\}_{t=1}^{N-1}$ of the entire path (i.e., $\sqrt{\frac{1}{N-1}\sum_{t=1}^{N-1} \|\hat{v}_t - v_t\|_2^2}$ and $\sqrt{\frac{1}{N-1}\sum_{t=1}^{N-1} \|\hat{\phi}_t - \phi_t\|_2^2}$). We also evaluate

Table 3. Comparison with two sub-optimal policies (regular skipping and random sampling) that use a similar visual encoder usage.

Method	Params	t_{rel} (%)	r_{rel} (°)	Visual encoder usage
Policy network	$\lambda = 3 \times 10^{-5}$	2.40 ± 0.064	0.86 ± 0.018	21.02%
	$\lambda = 5 \times 10^{-5}$	2.84 ± 0.102	1.13 ± 0.045	11.37%
Regular skipping	$n = 5$	3.40	0.95	20%
	$n = 8$	5.39	2.15	12.5%
Random policy	$p = 0.2$	3.11 ± 0.11	1.16 ± 0.073	20.41%
	$p = 0.125$	4.42 ± 0.239	1.2 ± 0.027	12.69%

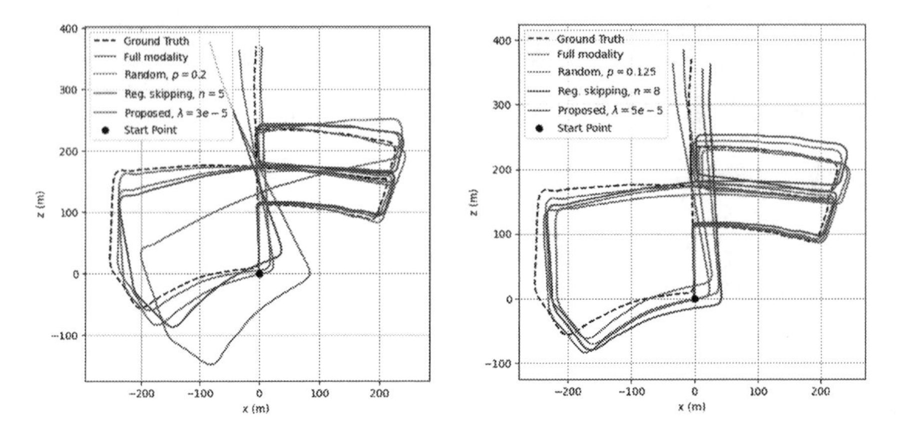

Fig. 3. Trajectories of ground-truth, full modality baseline, random and regular skipping, and proposed method on KITTI Sequence *05*.

the relative translation/rotation error denoted by t_{rel} and r_{rel} for various subsequence path lengths such as 100, 200, ..., 800 m as in [11]. To evaluate our policy network, we calculate the average usage rate of the visual modality and GFLOPS (giga floating-point operations per second).

4.2 Main Results

Ablation Study on the Penalty Factor. We first test our method on KITTI using four different penalty factors: $1 \times 10^{-5}, 3 \times 10^{-5}, 5 \times 10^{-5}$ and 7×10^{-5} to compare with the full modality baseline. For a fair comparison, we train the proposed and baseline full modality models with the same optimizer and common hyperparameters including the number of epochs and learning rate. Since our method is non-deterministic with a random sampling process, we test our model with 10 different random seeds and show the average performance. In Table 1, we present the average usage rate of the visual encoder, average GFLOPS, and average translational and rotational RMSE. It is observed that, as we gradually increase the penalty factor λ, both the usage of the visual encoder and system GFLOPS decrease as expected. In the meantime, as the visual encoder usage (and GFLOPS) drops, the translational RMSE becomes monotonically worse

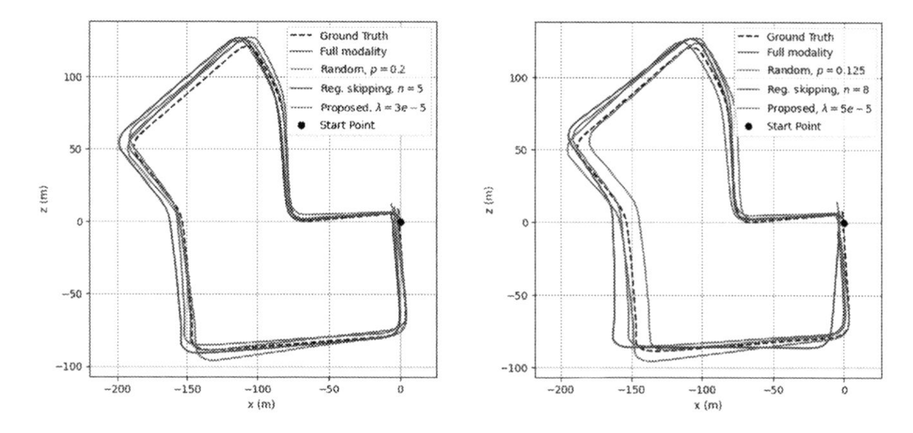

Fig. 4. Trajectories of ground-truth, full modality baseline, random and regular skipping, and proposed method on KITTI Sequence *07*.

while the rotational RMSE does not show a monotonic behavior. This indicates that visual features do not necessarily always contribute to improving rotation estimation accuracy. A particular setting of $\lambda = 3 \times 10^{-5}$ provides 78.8% reduction in GFLOPS at the cost of a relatively small 14.3% loss in translational RMSE while improving rotational RMSE by 23.6%. We also conduct evaluations of t_{rel} and r_{rel} obtained from various subsequent path lengths and report the results in Table 2. Similarly, our method achieves comparable accuracy to the fully modality baseline with $\lambda = 1 \times 10^{-5}$ and achieves an even better result with $\lambda = 3 \times 10^{-5}$ which results in 78.8% lower GFLOPS. Very aggressive policy network settings at $\lambda = 5 \times 10^{-5}$ and $\lambda = 7 \times 10^{-5}$ experience mild performance degradation. Note that the standard deviation shown in the last column remains quite small, demonstrating the stability of our proposed method.

Comparison with Sub-optimal Selection Strategies. In this section, we compare our proposed method with two sub-optimal visual modality selection strategies: regular skipping and random sampling. For regular skipping, we train the model with a fixed selection pattern where the visual encoder is enabled every n time indices. For random sampling, the visual modality is enabled with probability of p for each time index. These two methods are trained with the same number of epochs, optimizer, learning rate decaying strategy, and the other hyperparameters as in our proposed method. For each penalty factor λ applied to our method, we carefully choose a corresponding skipping rate parameter n and probability p such that all methods share a similar visual encoder usage. Table 3 shows our method significantly outperforms those two sub-optimal policies especially for t_{rel}. We also plot the path trajectories based on estimated poses from all methods on Sequence *05* in Fig. 3 and Sequence *07* in Fig. 4 for comparison. The proposed method exhibits the most reliable trajectory among all evaluated policies (Table 4).

Table 4. Comparison with prior VO/VIO works in translational & rotational error and image usage. The best performance in each block is marked in **bold**. Loop closure is excluded for ORB-SLAM2 and VINS-Mono.

	Method	Seq. 05			Seq. 07			Seq. 10		
		t_{rel} (%)	r_{rel} (°)	Usage	t_{rel} (%)	r_{rel} (°)	Usage	t_{rel} (%)	r_{rel} (°)	Usage
Geo.	ORB-SLAM2* [31]	**9.12**	**0.2**	100%	10.34	**0.3**	100%	**4.04**	**0.3**	100%
	VINS-Mono† [34]	11.6	1.26	100%	**10.0**	1.72	100%	16.5	2.34	100%
Self-Sup.	Monodepth2 * [13]	4.66	1.7	100%	4.58	2.6	100%	7.73	3.4	100%
	Zou et al.* [56]	**2.63**	**0.5**	100%	6.43	2.1	100%	5.81	1.8	100%
	VIOLearner† [40]	3.00	1.40	100%	3.60	2.06	100%	2.04	1.37	100%
	DeepVIO† [15]	2.86	2.32	100%	**2.71**	**1.66**	100%	**0.85**	**1.03**	100%
Sup.	GFS-VO* [50]	3.27	1.6	100%	3.37	2.2	100%	6.32	2.3	100%
	BeyondTracking* [51]	2.59	1.2	100%	3.07	1.8	100%	3.94	1.7	100%
	ATVIO† [25]	4.93	2.4	100%	3.78	2.59	100%	5.71	2.96	100%
	Soft Fusion† [5]	4.44	1.69	100%	2.95	1.32	100%	3.41	1.41	100%
	Hard Fusion† [5]	4.11	1.49	100%	3.44	1.86	100%	**1.51**	**0.91**	100%
	(Ours) baseline†	2.61	1.06	100%	1.83	1.35	100%	3.11	1.12	100%
	(Ours) $\lambda = 3 \times 10^{-5}$†	**2.01**	**0.75**	20.6%	**1.79**	**0.76**	19.79%	3.41	1.08	22.68%
	(Ours) $\lambda = 5 \times 10^{-5}$†	2.71	1.03	**11.34%**	2.22	1.14	**10.57%**	3.59	1.20	**12.2%**

*: Visual Odometry
†: Visual-Inertial Odometry

Comparison with Other VO/VIO Baselines. Now, we compare our method with geometric (non-learning-based) methods such as ORB-SLAM2 [31] and VINS-Mono [34] without loop closure, and also with state-of-the-art deep learning-based VO/VIO methods. Among those, deep learning-based self-supervised methods are [13,15,40,56], and supervised learning methods are [5,25,50,51]. All self-supervised methods are trained on Sequence *00-08* and tested on *09-10*. Among supervised methods, [50] and [51] are trained on Sequence *00, 02, 08, 09*. The other methods use the same training set as ours. It can be seen that although our main goal is not necessarily maximizing the odometry accuracy, our method still achieves the best performance among all the supervised methods. Compared with the state-of-the-art self-supervised methods [15,40] (which are known to outperform supervised methods in general), our method achieves a competitive performance especially for Sequence *05* and *07* that belong to their training set. This demonstrates the robustness of our policy network and also the effectiveness of our network structure and training strategy.

Interpretation of the Learned Policy. In Fig. 5, we present the visual interpretation of our learned policy evaluated on Sequence *07* with $\lambda = 5 \times 10^{-5}$. On the top left, we plot the local visual encoder usage with a color coding that represents the visual modality usage rate for a local window of 31 frames. Darker (lighter) colors represent lower (higher) usages. On the top right, we show the speed of the agent (a vehicle) at each time step where darker colors represent lower speed. An obvious correlation is observed between the visual modality usage and the speed which is also correlated with the turning angle. When the agent is moving slowly or making a turn, the policy network utilizes the visual modality less frequently. When the agent moves straight and fast, the visual encoder is activated more frequently.

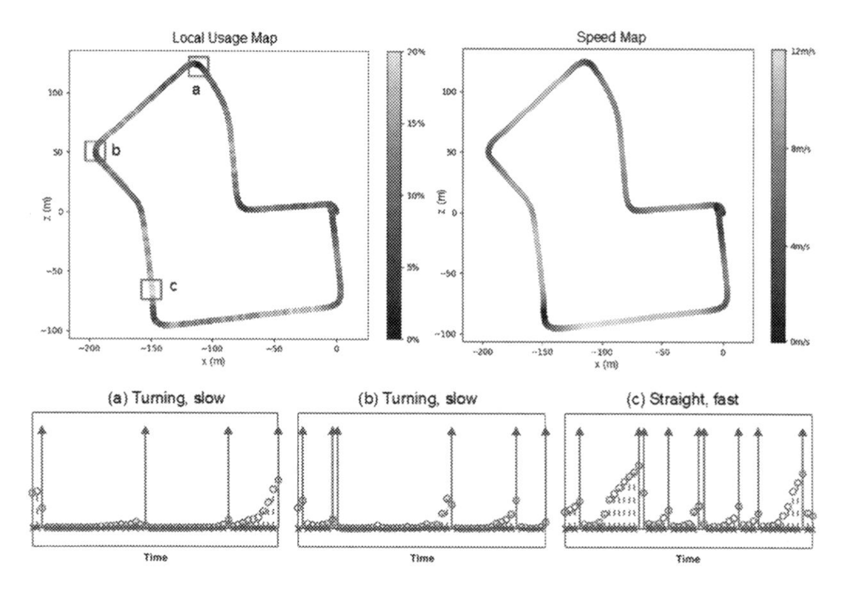

Fig. 5. Visual interpretation of the learned policy on Sequence *07* with $\lambda = 5 \times 10^{-5}$. Top left is the usage map that shows the local usage rate at each time step calculated by averaging the activation rate of the visual encoder during a local window of 31 frames. The agent vehicle speed map is shown on the top right. We selected three short segments from the path to visualize the policy network's behavior by showing the decisions d_t (blue pulses) and probabilities p_t (orange circles) on the bottom for different scenarios. Seg. a and b show low speed with turning scenarios whereas Seg. c is a high speed straight movement scenario. The policy network tends to activate the visual encoder more frequently when the agent is moving fast in straight, and decrease the usage of the visual encoder when the agent is moving slowly and making a turn. (Color figure online)

One explanation for this behavior is based on the inherent IMU's property that directly measures the angular velocity. Unlike visual feature based estimation, it is relatively easy to estimate the turning angle using IMU because it is obtained by simple first-order integration. However, estimating translation requires additional process with IMU measurements because it only measures the acceleration which is the second-order differential of translation, requiring a qualified initialization of the velocity. Thus, when the agent is moving fast, IMU-only estimation tends to make large translation errors and hence the policy network enables the visual modality more frequently to reduce the errors.

To provide more insights on the behavior of the policy network, we selected three short segments from the path (marked with red squares in Fig. 5 top left) to show the decisions d_t and corresponding probabilities to enable the visual modality (p_t, generated by the policy network) on the bottom of Fig. 5. We mark d_t using blue pulses and p_t using orange circles. The policy network exhibits a clear 'integrate-and-fire' pattern where it immediately resets the probability to ≈ 0 after the visual encoder is activated, and it keeps increasing the probability until the visual modality is enabled again. The slope of increasing p_t varies along

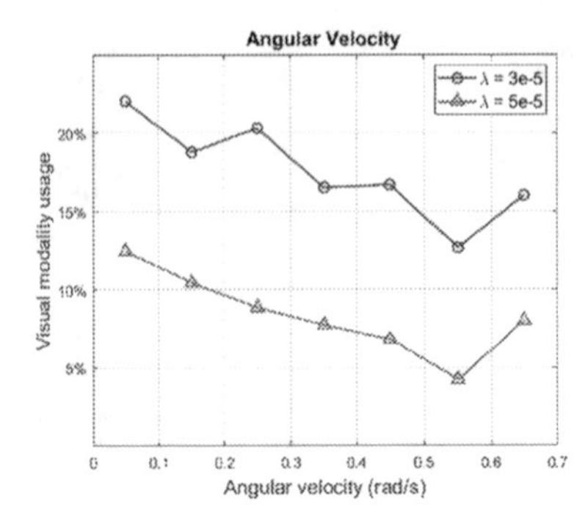

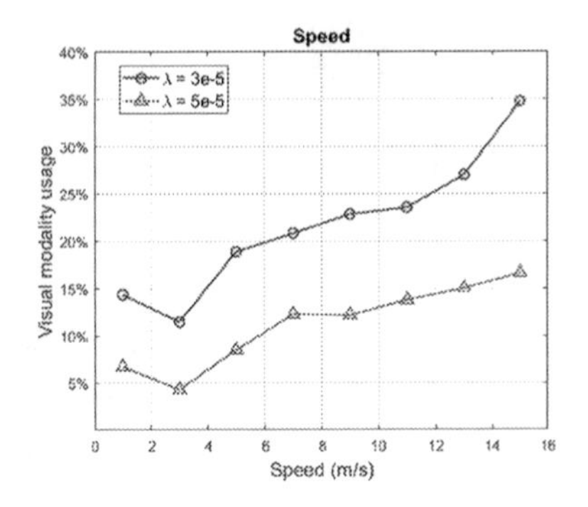

Fig. 6. The average usage rate of the visual modality for different angular velocities (left) and speeds (right) with two different penalty factors λ over the entire test set (Sequence *05, 07, 10*). The learned policy tends to use more images with lower angular velocity and higher speed.

the path. When the agent is making a sharp turn and moving slowly, p_t tends to increase slower and thus the gaps between two visual modality usages are relatively large (segments a and b). When the agent moves fast and straight, p_t surges much faster leading to smaller gaps to enable the visual encoder.

To show the general trend, we also plot the visual modality usage versus angular velocity and speed over all test paths for two different λ's in Fig. 6. We calculate the averaged visual encoder usage for the intervals of $[0, 0.1)$, $[0.1, 0.2)$, $\ldots$, $[0.6, 0.7)$ rad/s for the angular velocity and the intervals of $[0, 2)$, $[2, 4)$, $\ldots$, $[14, 16)$ m/s for the speed. It is observed that the usage is closely related to the angular velocity and speed. The usage in general tends to decrease with higher angular velocity and lower speed, although there can be occasional spots where this observation does not necessarily hold since our method is stochastic in nature.

5 Conclusion

In this paper, we propose a novel deep learning-based VIO system that reduces computation overhead and power consumption by opportunistically disabling the visual modality when the visual information is not critical to maintain the accuracy of pose estimation. To learn the selection strategy, we introduce a decision module to the neural VIO structure and end-to-end train it with the Gumbel-Softmax trick. Our experiments show that our approach provides up to 78.8% computation reduction without obvious performance degradation. Our learned strategy significantly outperform simple sub-optimal strategies. Furthermore, the learned policy is interpretable and shows scenario-dependent adaptive behaviours. Our adaptive learning strategy is model-agnostic and can be easily adopted to other deep VIO systems.

Acknowledgement. This work was supported in part by Meta Platforms, Inc. We also acknowledge Google LLC for providing GCP computing resources.

References

1. Almalioglu, Y., et al.: SelfVio: self-supervised deep monocular visual-inertial odometry and depth estimation. arXiv preprint arXiv:1911.09968 (2019)
2. Bolukbasi, T., Wang, J., Dekel, O., Saligrama, V.: Adaptive neural networks for efficient inference. In: International Conference on Machine Learning, pp. 527–536. PMLR (2017)
3. Cadena, C., et al.: Past, present, and future of simultaneous localization and mapping: toward the robust-perception age. IEEE Trans. Rob. **32**(6), 1309–1332 (2016)
4. Campos, V., Jou, B., Giró-i Nieto, X., Torres, J., Chang, S.F.: Skip RNN: learning to skip state updates in recurrent neural networks. arXiv preprint arXiv:1708.06834 (2017)
5. Chen, C., Rosa, S., Lu, C.X., Trigoni, N., Markham, A.: SelectFusion: a generic framework to selectively learn multisensory fusion. arXiv preprint arXiv:1912.13077 (2019)
6. Chen, C., et al.: Selective sensor fusion for neural visual-inertial odometry. In: Proceedings of the IEEE/CVF Conference on Computer Vision and Pattern Recognition, pp. 10542–10551 (2019)
7. Clark, R., Wang, S., Wen, H., Markham, A., Trigoni, N.: ViNet: visual-inertial odometry as a sequence-to-sequence learning problem. In: Proceedings of the AAAI Conference on Artificial Intelligence, vol. 31 (2017)
8. Dosovitskiy, A., et al.: FlowNet: learning optical flow with convolutional networks. In: Proceedings of the IEEE International Conference on Computer Vision, pp. 2758–2766 (2015)
9. Engel, J., Koltun, V., Cremers, D.: Direct sparse odometry. IEEE Trans. Pattern Anal. Mach. Intell. **40**(3), 611–625 (2017)
10. Forster, C., Pizzoli, M., Scaramuzza, D.: SVO: fast semi-direct monocular visual odometry. In: 2014 IEEE International Conference on Robotics and Automation (ICRA), pp. 15–22. IEEE (2014)
11. Geiger, A., Lenz, P., Urtasun, R.: Are we ready for autonomous driving? The Kitti vision benchmark suite. In: 2012 IEEE Conference on Computer Vision and Pattern Recognition, pp. 3354–3361. IEEE (2012)
12. Glynn, P.W.: Likelihood ratio gradient estimation for stochastic systems. Commun. ACM **33**(10), 75–84 (1990)
13. Godard, C., Mac Aodha, O., Firman, M., Brostow, G.J.: Digging into self-supervised monocular depth estimation. In: Proceedings of the IEEE/CVF International Conference on Computer Vision, pp. 3828–3838 (2019)
14. Graves, A.: Adaptive computation time for recurrent neural networks. arXiv preprint arXiv:1603.08983 (2016)
15. Han, L., Lin, Y., Du, G., Lian, S.: DeepVio: self-supervised deep learning of monocular visual inertial odometry using 3D geometric constraints. In: 2019 IEEE/RSJ International Conference on Intelligent Robots and Systems (IROS), pp. 6906–6913. IEEE (2019)
16. Hansen, C., Hansen, C., Alstrup, S., Simonsen, J.G., Lioma, C.: Neural speed reading with structural-jump-LSTM. arXiv preprint arXiv:1904.00761 (2019)
17. Hong, E., Lim, J.: Visual inertial odometry using coupled nonlinear optimization. In: 2017 IEEE/RSJ International Conference on Intelligent Robots and Systems (IROS), pp. 6879–6885. IEEE (2017)
18. Hua, W., Zhou, Y., De Sa, C.M., Zhang, Z., Suh, G.E.: Channel gating neural networks. Adv. Neural Inf. Process. Syst. **32** (2019)

19. Huang, G., Chen, D., Li, T., Wu, F., Van Der Maaten, L., Weinberger, K.Q.: Multi-scale dense networks for resource efficient image classification. arXiv preprint arXiv:1703.09844 (2017)
20. Jang, E., Gu, S., Poole, B.: Categorical reparameterization with gumbel-softmax. arXiv preprint arXiv:1611.01144 (2016)
21. Kingma, D.P., Welling, M.: Auto-encoding variational Bayes. arXiv preprint arXiv:1312.6114 (2013)
22. Krizhevsky, A., Sutskever, I., Hinton, G.E.: ImageNet classification with deep convolutional neural networks. Adv. Neural Inf. Process. Syst. **25** (2012)
23. Leutenegger, S., Lynen, S., Bosse, M., Siegwart, R., Furgale, P.: Keyframe-based visual-inertial odometry using nonlinear optimization. Int. J. Robot. Res. **34**(3), 314–334 (2015)
24. Li, M., Mourikis, A.I.: High-precision, consistent EKF-based visual-inertial odometry. Int. J. Robot. Res. **32**(6), 690–711 (2013)
25. Liu, L., Li, G., Li, T.H.: AtVio: attention guided visual-inertial odometry. In: ICASSP 2021–2021 IEEE International Conference on Acoustics, Speech and Signal Processing (ICASSP), pp. 4125–4129. IEEE (2021)
26. Lowe, D.G.: Object recognition from local scale-invariant features. In: Proceedings of the Seventh IEEE International Conference on Computer Vision, vol. 2, pp. 1150–1157. IEEE (1999)
27. Meng, Y., et al.: AR-Net: adaptive frame resolution for efficient action recognition. In: Vedaldi, A., Bischof, H., Brox, T., Frahm, J.-M. (eds.) ECCV 2020. LNCS, vol. 12352, pp. 86–104. Springer, Cham (2020). https://doi.org/10.1007/978-3-030-58571-6_6
28. Meng, Y., et al.: AdaFuse: adaptive temporal fusion network for efficient action recognition. arXiv preprint arXiv:2102.05775 (2021)
29. Mohamed, S., Rosca, M., Figurnov, M., Mnih, A.: Monte Carlo gradient estimation in machine learning. J. Mach. Learn. Res. **21**(132), 1–62 (2020)
30. Mur-Artal, R., Montiel, J.M.M., Tardos, J.D.: ORB-SLAM: a versatile and accurate monocular slam system. IEEE Trans. Rob. **31**(5), 1147–1163 (2015)
31. Mur-Artal, R., Tardós, J.D.: ORB-SLAM2: an open-source slam system for monocular, stereo, and RGB-D cameras. IEEE Trans. Rob. **33**(5), 1255–1262 (2017)
32. Nistér, D., Naroditsky, O., Bergen, J.: Visual odometry. In: Proceedings of the 2004 IEEE Computer Society Conference on Computer Vision and Pattern Recognition. CVPR 2004, vol. 1, p. I-I. IEEE (2004)
33. Panda, R., et al.: AdaMML: adaptive multi-modal learning for efficient video recognition. In: Proceedings of the IEEE/CVF International Conference on Computer Vision, pp. 7576–7585 (2021)
34. Qin, T., Li, P., Shen, S.: VINS-MONO: a robust and versatile monocular visual-inertial state estimator. IEEE Trans. Rob. **34**(4), 1004–1020 (2018)
35. Ren, S., He, K., Girshick, R., Sun, J.: Faster R-CNN: towards real-time object detection with region proposal networks. Adv. Neural Inf. Process. Syst. **28** (2015)
36. Rezende, D.J., Mohamed, S., Wierstra, D.: Stochastic backpropagation and approximate inference in deep generative models. In: International Conference on Machine Learning, pp. 1278–1286. PMLR (2014)
37. Rublee, E., Rabaud, V., Konolige, K., Bradski, G.: ORB: an efficient alternative to sift or surf. In: 2011 International Conference on Computer Vision, pp. 2564–2571. IEEE (2011)
38. Scaramuzza, D., Fraundorfer, F.: Visual odometry [tutorial]. IEEE Robot. Autom. Mag. **18**(4), 80–92 (2011)

39. Seo, M., Min, S., Farhadi, A., Hajishirzi, H.: Neural speed reading via skim-RNN. arXiv preprint arXiv:1711.02085 (2017)
40. Shamwell, E.J., Leung, S., Nothwang, W.D.: Vision-aided absolute trajectory estimation using an unsupervised deep network with online error correction. In: 2018 IEEE/RSJ International Conference on Intelligent Robots and Systems (IROS), pp. 2524–2531. IEEE (2018)
41. Simonyan, K., Zisserman, A.: Very deep convolutional networks for large-scale image recognition. arXiv preprint arXiv:1409.1556 (2014)
42. Teerapittayanon, S., McDanel, B., Kung, H.T.: BranchyNet: fast inference via early exiting from deep neural networks. In: 2016 23rd International Conference on Pattern Recognition (ICPR), pp. 2464–2469. IEEE (2016)
43. Veit, A., Belongie, S.: Convolutional networks with adaptive inference graphs. In: Ferrari, V., Hebert, M., Sminchisescu, C., Weiss, Y. (eds.) ECCV 2018. LNCS, vol. 11205, pp. 3–18. Springer, Cham (2018). https://doi.org/10.1007/978-3-030-01246-5_1
44. Wang, S., Clark, R., Wen, H., Trigoni, N.: DeepVO: towards end-to-end visual odometry with deep recurrent convolutional neural networks. In: 2017 IEEE International Conference On Robotics and Automation (ICRA), pp. 2043–2050. IEEE (2017)
45. Wang, X., Yu, F., Dou, Z.-Y., Darrell, T., Gonzalez, J.E.: SkipNet: learning dynamic routing in convolutional networks. In: Ferrari, V., Hebert, M., Sminchisescu, C., Weiss, Y. (eds.) ECCV 2018. LNCS, vol. 11217, pp. 420–436. Springer, Cham (2018). https://doi.org/10.1007/978-3-030-01261-8_25
46. Wang, Y., Chen, Z., Jiang, H., Song, S., Han, Y., Huang, G.: Adaptive focus for efficient video recognition. In: Proceedings of the IEEE/CVF International Conference on Computer Vision, pp. 16249–16258 (2021)
47. Williams, R.J.: Simple statistical gradient-following algorithms for connectionist reinforcement learning. Mach. Learn. **8**(3), 229–256 (1992)
48. Wu, Z., et al.: BlockDrop: dynamic inference paths in residual networks. In: Proceedings of the IEEE Conference on Computer Vision and Pattern Recognition, pp. 8817–8826 (2018)
49. Wu, Z., Xiong, C., Ma, C.Y., Socher, R., Davis, L.S.: AdaFrame: adaptive frame selection for fast video recognition. In: Proceedings of the IEEE/CVF Conference on Computer Vision and Pattern Recognition, pp. 1278–1287 (2019)
50. Xue, F., Wang, Q., Wang, X., Dong, W., Wang, J., Zha, H.: Guided feature selection for deep visual odometry. In: Jawahar, C.V., Li, H., Mori, G., Schindler, K. (eds.) ACCV 2018. LNCS, vol. 11366, pp. 293–308. Springer, Cham (2019). https://doi.org/10.1007/978-3-030-20876-9_19
51. Xue, F., Wang, X., Li, S., Wang, Q., Wang, J., Zha, H.: Beyond tracking: selecting memory and refining poses for deep visual odometry. In: Proceedings of the IEEE/CVF Conference on Computer Vision and Pattern Recognition, pp. 8575–8583 (2019)
52. Yang, M., Kim, H.S.: Deep joint source-channel coding for wireless image transmission with adaptive rate control. arXiv preprint arXiv:2110.04456 (2021)
53. Yang, N., Wang, R., Gao, X., Cremers, D.: Challenges in monocular visual odometry: photometric calibration, motion bias, and rolling shutter effect. IEEE Robot. Autom. Lett. **3**(4), 2878–2885 (2018)
54. Yuan, Z., Wu, B., Sun, G., Liang, Z., Zhao, S., Bi, W.: S2DNAS: transforming static CNN model for dynamic inference via neural architecture search. In: Vedaldi, A., Bischof, H., Brox, T., Frahm, J.-M. (eds.) ECCV 2020. LNCS, vol. 12347, pp. 175–192. Springer, Cham (2020). https://doi.org/10.1007/978-3-030-58536-5_11

55. Zhou, T., Brown, M., Snavely, N., Lowe, D.G.: Unsupervised learning of depth and ego-motion from video. In: Proceedings of the IEEE Conference on Computer Vision and Pattern Recognition, pp. 1851–1858 (2017)
56. Zou, Y., Ji, P., Tran, Q.-H., Huang, J.-B., Chandraker, M.: Learning monocular visual odometry via self-supervised long-term modeling. In: Vedaldi, A., Bischof, H., Brox, T., Frahm, J.-M. (eds.) ECCV 2020. LNCS, vol. 12359, pp. 710–727. Springer, Cham (2020). https://doi.org/10.1007/978-3-030-58568-6_42

A Sketch is Worth a Thousand Words: Image Retrieval with Text and Sketch

Patsorn Sangkloy[1,3](✉), Wittawat Jitkrittum[2], Diyi Yang[1], and James Hays[1]

[1] Georgia Institute of Technology, Atlanta, GA 30332, USA
diyi.yang@cc.gatech.edu, hays@gatech.edu
[2] Google Research, New York, NY 10011, USA
wittawat@google.com
[3] Phranakhon Rajabhat University, Bang Khen 10220, Bangkok, Thailand
patsorn.s@pnru.ac.th

Abstract. We address the problem of retrieving in-the-wild images with *both a sketch and a text query.* We present TASK-former (Text And SKetch transformer), an end-to-end trainable model for image retrieval using a text description and a sketch as input. We argue that both input modalities complement each other in a manner that cannot be achieved easily by either one alone. TASK-former follows the late-fusion dual-encoder approach, similar to CLIP [35], which allows efficient and scalable retrieval since the retrieval set can be indexed independently of the queries. We empirically demonstrate that using an input sketch (even a poorly drawn one) in addition to text considerably increases retrieval recall compared to traditional text-based image retrieval. To evaluate our approach, we collect 5,000 hand-drawn sketches for images in the test set of the COCO dataset. The collected sketches are available a https://janesjanes.github.io/tsbir/.

1 Introduction

Cross-modal retrieval [44] is a retrieval problem where the query and the set of retrieved objects take different forms. A representative problem of this class is Text-Based Image Retrieval (TBIR), where the goal is to retrieve relevant images from an input text query. TBIR has been studied extensively, with recent interest focused on transformer-based models [36]. A core component of a cross-modal retrieval model is its scoring function that assesses the similarity between a text description and an image. Given a text description, retrieving relevant images then amounts to finding the top k images that achieve the highest scores.

While a text description is suitable for describing qualitative attributes of an object (e.g., object color, object shape) in TBIR, it can be cumbersome when there is a need to describe multiple objects or a complex shape. Consider the query "This flower is red and wavy" in Fig. 1 as an example. This query alone can match red flowers of a wide range of shapes. Further, with multiple

Supplementary Information The online version contains supplementary material available at https://doi.org/10.1007/978-3-031-19839-7_15.

S. Avidan et al. (Eds.): ECCV 2022, LNCS 13698, pp. 251–267, 2022.
https://doi.org/10.1007/978-3-031-19839-7_15

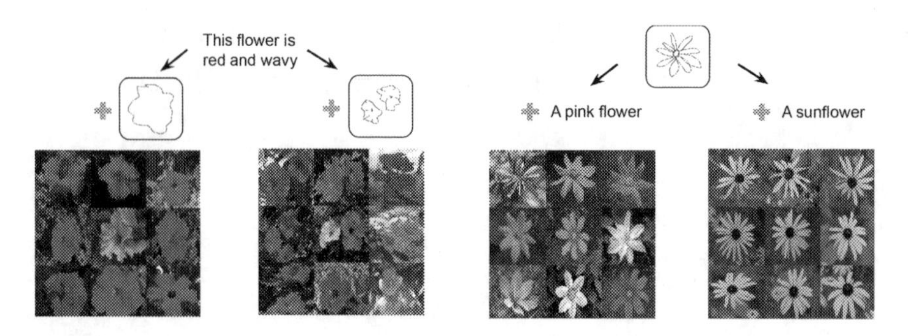

Fig. 1. We present TASK-former, a Text And SKetch transformer based method for image retrieval. We demonstrate that the presence of a sketch input, even a poorly drawn one, helps narrow down the set of retrieved images to ones that match the joint description provided by the sketch and the text query. Retrieval results are from our model trained on Flower102 dataset [29].

objects, it becomes necessary to describe their relative positions, making the task too cumbersome to be practical. These limitations naturally led to a related thread of research on Sketch-Based Image Retrieval (SBIR), where the goal is to retrieve images from an input sketch [13,31,32,37,38,47,50]. Compared to text, specifying object positions with a hand-drawn sketch is relatively easy.

Recent works on SBIR tend to focus on a specific set of retrievable images. For instance, [38] considers object based retrieval: each image has only one object centered in the image. Compared to [38], [47] allows a more free-form sketch input, but from a single category e.g., shoes. While suitable for an e-commerce product search, searching from an arbitrarily drawn sketch, also known as *in-the-wild SBIR*, was not studied in [47]. In-the-wild image retrieval is the problem we tackle in this work.

In-the wild SBIR presents two challenges. Firstly, there is semantic ambiguity in a sketch drawn by a non-artist user. To a non-artist, it takes effort to draw a sketch which sufficiently accurately represents the desired image to retrieve. While adding details to the sketch would naturally narrow down the candidate image set, to a non-artist user, the extra effort required to do so may outweigh the intended convenience of SBIR. Ideally the retrieval system should be able to extract relevant information from a poorly drawn sketch. Secondly, the few publicly available SBIR datasets contain only either images of single objects [38,47], or images describing single concepts [18]. These images are different from target images in in-the-wild image search, which may contain multiple objects with each belonging to a distinct category.

In this work, we address these two challenges in in-the-wild image retrieval by proposing to *use both a sketch and a text query as input.* The (optional) input sketch is treated as a supplement to the text query to provide more information that may be difficult to express with text (e.g., positions of multiple objects, object shapes). In particular, we do not require the input sketch to be drawn well. As will be seen in our results, when coupled with a text query, an extra input sketch (even a poorly drawn one) can help considerably narrow down the set of candidate images, leading to an increase in retrieval recall. We show that both input modalities can complement each other in a manner that cannot be easily achieved by either one alone. To illustrate this point concretely, two example queries can be found in Fig. 1 where dropping one input modality would make it difficult to retrieve the same set of images. This idea directly addresses the first challenge of in-the-wild SBIR – sketch ambiguity.

Combining two input modalities is made possible with our proposed similarity scoring model, *TASK-former*, which follows the late-fusion dual-encoder approach, and allows efficient retrieval (see Fig. 2 for the model and our training pipeline). Our proposed training objective comprises 1) an embedding loss (to learn a shared embedding space for text, sketch, and image); 2) a multi-label classification loss (to allow the model to recognize objects); and 3) a caption generation loss (to encourage a strong correspondence between the learned joint embedding and text description). Crucially, training our model only requires synthetically generated sketches, not human-drawn ones. These sketches are generated from the target images, and are further transformed by appropriate augmentation procedures (e.g., random affine transformation, dropout) to provide robustness to ambiguity in the input sketch. The ability of our model to make use of synthetically generated sketches during training addresses the second challenge of in-the-wild SBIR – lack of training data. We show that our model is robust to sketches with missing strokes (Fig. 5), and is able to generalize and operate on human-drawn input sketches (Fig. 4).

Our contributions are as follows.

1. We present TASK-former (Text And SKetch transformer), a scalable, end-to-end trainable model for image retrieval using a text description and a sketch as input.
2. We collect 5000 hand-drawn sketches for images in the test set of COCO [27], a commonly used dataset to benchmark image retrieval methods. The collected sketches are available at https://janesjanes.github.io/tsbir/.
3. We empirically demonstrate (in Sect. 4) that using an input sketch (even with a poorly drawn one) in addition to text helps increases retrieval recall, compared to the traditional TBIR where only a text description is used as input.

2 Related Work

Sketch Based Image Retrieval (SBIR) There are several works that tackle SBIR [13,31,32,38,47]. The works of [10,13,28,31,42] consider zero-shot SBIR where retrieving from unseen categories is the focus. Zhang et al. [48] tackles SBIR using only weakly labeled data. Fine-grained SBIR of a restricted set of images (e.g., from a single category) is studied in [32,33,39]. An interactive variant of SBIR was considered in [8] where images are retrieved on-the-fly as an input sketch is being drawn. They propose a form of sketch completion system for SBIR based on user's choice of image cluster to refine the retrieval results. Bhunia et al. [3] present a reinforcement learning based pipeline for SBIR and allows a retrieval

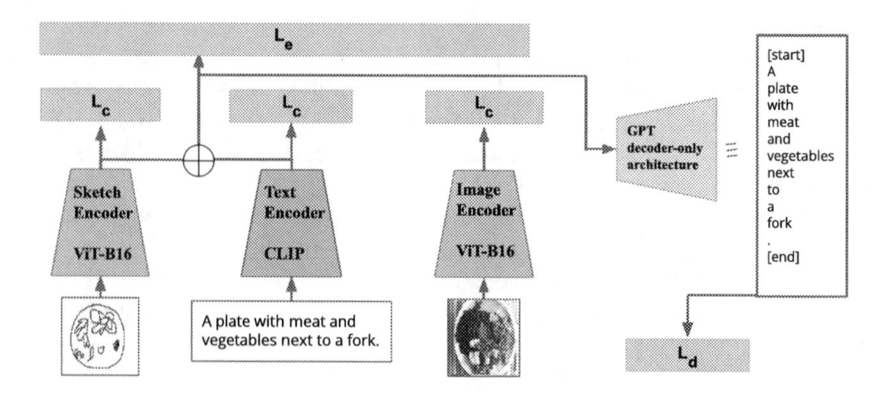

Fig. 2. Overview of our model and the training pipeline. ViT is the vision transformer [12], and ⊕ represents an element-wise addition. We extend CLIP [35] to incorporate additional sketch input from user. We add two auxiliary tasks: multi-label classification (L_c) and caption generation (L_d). Our objective losses are explained in Sect. 3.2. We encourages the network to learn a discriminative representation to simultaneously distinguish at the instance level and at the category level, both of which are crucial for a successful retrieval.

even before the sketch is completed. Being able to retrieve relevant images even with incomplete sketch information is an important aspect of SBIR. In this work, we achieved this by supplementing the (incomplete) sketch with an input text description.

Text-Based Image Retrieval (TBIR). The problem of retrieving images based on a text description is closely related to the ability to learn a good representation between these two different domains. Previous successes in TBIR relied on the cross attention mechanism (a specific form of early fusion) to learn the similarity between text and images [24,26,49]. It is generally believed that, compared to a late fusion model, an early fusion model can offer more capacity for modeling a complex scoring function owing to its non-factorized form [1,46]. Interestingly however the recently proposed CLIP [35], a late fusion model, is able to achieve comparable TBIR results on COCO to existing early fusion models. This feat is impressive and is presumably due to the use of a large proprietary dataset of text-image pairs crawled from the Internet. The representation learned by CLIP is incredibly rich, and correlates well with human perception on image-text similarity as observed in [17].

Similar to CLIP, ALIGN [20] also proposes a similar contrastive training scheme on a large-scale data. Fine-tuning from a pre-trained network in ALIGN is able to achieve state-of-the-art (SOTA) accuracy of 59.9% in retrieving a single correct image based on text description in the COCO image retrieval benchmark (karphathy split).[1]. In our work, we opt to use the pre-trained CLIP network to initialize our network for training.

[1] More recently, Microsoft team also presented T-Bletchley [41] with a similar pipeline.

Image Retrieval with Multi-modal Queries. Using text and image as queries for image retrieval has been explored extensively in the literature, for example as a textual instruction for a desired modification of the query image [6,11,16,40,43]. Jia et al. [20] demonstrate that it is possible to directly combine representations of the two input modalities (text and image in this case) with a simple element-wise addition, for the purpose of image retrieval with maximum inner product search. In our work, we take this insight and combine the representation of sketch and text inputs in the same way. Using a sketch as input which is further supplemented by additional information (such as a classification label) has also been studied in Sketch2Tag [45] and Sketch2Photo [5]. Instead of sketch, [4] explores using fine grain association between mouse trace and text description for image retrieval. Surprisingly, combining sketch and full text description as queries for image retrieval has been relatively under explore despite its potential usefulness. [9,15,21] explore the benefit of training with both sketch and text inputs for the purpose of improving image retrieval performance for each modality separately. Only [21] demonstrate examples of retrieving with both sketch and text inputs by linearly combining their score during test time. Note that unlike ours, the proposed loss function of [21] does not explicitly consider interactions between sketch and text inputs, since the goal is to simply improve retrieving with each modality separately. Further to find the similarity between a sketch and a given image, the proposed method of [21] requires that the image be first converted to a sketch via edge detection. This conversion would incur a loss of information such as the color. The proposed method also requires a conversion of image into sketch via edge detection, and would lose color information which is not ideal. While [21] is one of the most related works to our work, we are unfortunately not able to obtain the data used, or the trained model for a quantitative comparison.

3 Proposal: TASK-former

TASK-former can be considered an extension of the training pipeline proposed in CLIP [35]. We start with the same network architectures and contrastive learning as in CLIP, but modify them and add additional loss terms to allow the use of both a sketch and an input text query. Specifically, we include two additional auxiliary tasks: multi-label classification and caption generation. Our motivation is to improve the learned embedding space so as to achieve the following goals.

1. Discriminate between the positive and negative pairs;
2. Distinguish objects of different categories; and
3. Contain sufficient information to reconstruct the original text caption from embeddings of an image and its sketch.

We achieve these goals via CLIP's symmetric cross entropy, multi-label classification objective, and caption generation objective, respectively. While these three objectives appear to seek conflicting goals (e.g., the classification loss might encourage discarding class-invariant information, whereas the image captioning loss would suffer less if all describable details are kept), we observe that combining them with appropriate weights can lead to gains in performance for image retrieval as shown in Table 2.

We start by describing our model and training pipeline in Sect. 3.1, and describe the three aforementioned loss terms in Sect. 3.2.

3.1 Model and Training Pipeline

Our pipeline is summarized in Fig. 2. Each input query consists of 1) a hand-drawn sketch, and 2) a text description of desired target images. We use the same image and text encoder architecture as described in CLIP [35] in order to leverage networks that have been the pre-trained with large-scale training data. This choice is in accord with the common practice of using, for instance, an ImageNet pre-trained network for various vision tasks. We use CLIP's publicly available pre-trained model ViT-B/16, in which the image encoder is based on the Vision Transformer [12] which we found to be the best performer for our task. For the details of the network architecture of each encoder that we use, please refer to CLIP [35] and ViT [12].

There are three encoders in our pipeline for: 1) input sketch, 2) input text description, and 3) a candidate retrieved image. The output embeddings from the sketch and the text encoders are combined together and used for contrastive learning with the image embedding as the target. We explore several options for combining features in Sect. 4.1. Since images and sketches are in the same domain (visual), we use the same architecture for them (ViT-B/16 pre-trained on CLIP). We also found that the performance is much better when sharing weight parameters across the sketch and the image encoders. Yu et al. [47] also observed similar results where the Siamese network performs significantly better than heterogeneous networks for the SBIR task. It was speculated that weight sharing is advantageous because of relatively small training datasets. This explanation may also hold in our case as well given the complexity of our task. For classification, we feed each embedding to two additional fully connected layers with ReLU activation.

Additionally, we also train a captioning generator from the embeddings of sketches and images using a transformer based text decoder. This decoder is an autoregressive language model similar to GPT decoder-only architecture [36]. We use absolute positional embedding, with six stacks of decoder blocks, each with eight attention heads. More details can be found in the supplementary materials.

3.2 Objective Function

Our objective function consists of three main components: symmetric cross entropy, asymmetric loss for multi-label classification, and auxiliary caption generation.

Embedding Loss (L_e). To learn a shared embedding space for text, image and sketch, we follow the contrastive learning objective from CLIP [35], which use a form of *InfoNCE Loss* as originally proposed in [30] to learn to match the right image-text pairs as observed in the batch. This proxy task is accomplished via a symmetric cross entropy loss over all possible pairs in each batch, effectively maximizing cosine similarity of each matching pair and minimizing it for non-matching ones. We add the sketch as an additional query, and replace the text embedding in CLIP with our combined embedding constructed by summing the text and sketch embeddings.

Classification Loss (L_c). For classification, we consider this as a multi-label classification problem as each image can belong to multiple categories. We follow a common practice in multi-label classification, which frames the problem as a series of many binary classification problems. We use object annotation available in the datasets as ground truth. Specifically, we use Asymmetric Loss For Multi-Label Classification (ASL Loss), proposed in [2]. The loss is designed to help alleviate the effect of the imbalanced label distribution in multi-label classification.

Auxiliary Caption Generation (Decoder Loss, L_d). For caption generation, the decoder attempts to predict the most likely token given the accumulated embedding and the previous tokens. The decoded output tokens are then compared with the ground truth sentences via the cross entropy loss $\sum_t^T \log(p_t|p_1, \ldots, p_{t-1})$, where T is the maximum sequence length.

Our final objective is given by a weighted combination of all the loss terms. We refer to supplementary materials for our hyperparameter choices.

3.3 Sketch Generation and Data Augmentation

During training, we synthetically generate sketches using the method proposed in [25]. In general the method produces drawings similar to human sketches. The synthesized sketches are however in exact alignment with their source images, which would not be the case in sketches drawn by humans. To achieve invariance to small misalignment, we further apply a random affine transformation on the synthetic sketch as an augmentation. We also apply a similar transformation to the image (with different random seeds). Introducing this misalignment is crucial for the network to generalize to hand-drawn input sketches at test time.

To help deal with partial sketches at test time, we also randomly occlude parts of each sketch. We randomly replace black strokes with white pixel. The completion level of each synthesized sketch in the training set is between 60% to 100%

Evaluation with Sketch-Text-Image Tuples. Since our approach retrieves images with sketches and text, evaluating our approach naturally requires an annotated dataset where each record contains a hand-drawn sketch, a human-annotated text description, and a source image. SketchyCOCO [14] fits this description and is a candidate dataset. However, the dataset was constructed by having the sketches drawn first based on categories. The sketch for each category was then pasted into their supposed areas based on incomplete annotation (not all objects were annotated). As a result, the sketch in each record may poorly represent the image because 1) each category contains the same sketch, 2) the choice of which object to draw is based on categories, rather than human judgement of what is salient in each particular image. More related is the dataset mentioned in [21] which provides 1,112 matching sketch/image/text of shoes. However, the dataset is not yet available at the time of writing. Also worth mentioning is [34], which also proposes a dataset containing synchronized annotation between text description and the associated location (in a form of mouse trace) in the image. However, we argue that a line drawing sketch represents more than just the location of the objects; even a badly drawn sketch can provide information such as shape, details, or even relative scale. Owing to lack of an appropriate evaluation dataset, we construct a new benchmark by collecting hand-drawn sketches for images in the COCO image retrieval benchmark. These images are in the COCO 5k split from [22], which is widely used to evaluate text based image retrieval methods. As part of our contributions, the collected data will be made publicly available. We describe the sketch collection process in the next section.

3.4 Data Collection: Sketching from Memory

We collect sketches for COCO 5k via Amazon Mechanical Turk crow sourcing platform (AMT). We follow the split from [22], which has been widely used as benchmark for text based image retrieval. The test set contains 5000 images which are disjoint from the training split.

To solicit a sketch, we first show the participant (Turker) the target image for 15 s, before replacing it with a noise image. This is to mimic the typical scenario at deployment time where we there is no concrete reference image, but instead only a mental image of a retrieval target. The participant is then asked to draw a sketch from memory. In the instructions, we ask the participants to draw as if they are explaining the image to a friend but using a sketch. We do not put any restrictions on how the sketch has to be drawn, other than no shading. The participants are free to draw any parts of the image they think are important and distinctive enough to be included in the sketch. The sketches are collected in SVG (a vector format), and contain information of all individual strokes. In experiments, we use this information to randomly drop individual strokes to test the robustness of our approach to incomplete input sketches (see Fig. 5). As a sanity check, we also ask the participants to put one or two words describing the image in an open-ended fashion.

From the results, we manually filter out sketches that are not at all representative of the target image (e.g., empty sketch, random lines, wrong object). Our only criteria is that each sketch has to be recognizable as describing the target image (even if only remotely recognizable). Our goal is not to collect complete or perfect sketches, but to collect in-the-wild sketches, which may be poorly drawn, that can be used along side the text description to explain image.

Fig. 3. Example retrieved images from TASK-former, randomly selected from our benchmark. Each query consists of a text description (shown at the top of each block, and an input sketch (at the top left of each block). In each case, the image that forms a matching pair to the sketch is highlighted with a green border. See Sect. 3.4 for details on how we collect human drawn sketches for images in the evaluation set. (Color figure online)

4 Results and Discussion

For quantitative evaluation, we calculate Recall@K, which is the fraction of times that the target image is included in the top-K retrieved images. We evaluate on COCO [27]; a commonly used datasets for language based image retrieval. We use the same data split as proposed in [22]. Specifically, COCO's evaluation set contains 5,000 images. Each image is annotated with multiple captions, and we additionally add a sketch to each image. For COCO, we collect hand drawn sketch as described in 3.4.

Implementation Details. We use Adam [23] as the optimization method to train the model. Training hyperparameters are set in accordance with Open Clip [19] with the initial learning rate set to 10^{-5}. To provide robustness to incomplete input, for each (text query, sketch, image) training tuple, we set a 20% probability to drop either the sketch (replaced with a white image) or the text query (replaced with an empty text). We demonstrate the robustness of our model

Table 1. Recall@{1, 5, 10} of state-of-the-art methods on COCO dataset. We use our collected hand-drawn sketches for evaluation. Among ALIGN, and CLIP, only CLIP has released pre-trained model. Our TASK-former is initialized for training with this CLIP model.

Type	Method	Input	R@1	R@5	R@10
Early fusion	Uniter [7]	Text	0.529	0.799	0.880
Early fusion	OSCAR [26]	Text	0.575	0.828	0.898
Dual encoders	ALIGN [20] (fine-tuned)	Text	0.599	0.833	0.898
Dual encoders	CLIP [35] (fine-tuned)	Text	0.518	0.811	0.891
Dual encoders	**TASK-former** (ours)	(Text, Sketch)	**0.609**	**0.847**	**0.917**

to incomplete input in Sect. 4.2. Code to reproduce our results will be made publicly available.

For evaluation, our model TASK-former is initialized with the publicly released CLIP model (ViT-16) [35], a text-based image retrieval model. Table 1 compares Recall@{1, 5, 10} of our method and that of current state-of-the-art approaches on text based image retrieval: Uniter, OSCAR, and ALIGN as reported in [7,20,26], respectively. To provide a fairer comparison, we further finetune the publicly released CLIP model (ViT-16) on COCO.[2] We observe that our method is able to achieve a considerably higher recall than other methods, and CLIP in particular, owing in part to the use of a sketch as a supplemental input to text. Our method can retrieve the correct image with recall@1 of 60.9%, compared to using text alone, which achieves 51.8% on the same CLIP architecture (ViT-16).

At the time of writing, models for ALIGN and T-Bletchley have not been released. Both methods have been shown to perform better than CLIP on text based image retrieval, and without any early fusion of text and image. We note that both ALIGN and T-Bletchley can be used as part of our framework by simply replacing the pre-trained encoders for text and image.

4.1 Ablation Study

In this section, we seek to understand the effect of each of our proposed loss functions (L_e, L_c, L_d), as described in Sect. 3. Baselines used as part of this ablation study are:

- **CLIP zero shot**. Recall@K that has been reported in [35] for zero-shot image retrieval. We note that their best model for the reported results (ViT-L/14@336px) is not available publicly.

[2] Note that this baseline makes use of our best-effort implementation and training. We do not have access to the official training code and there is no reported result on using a fine-tuned CLIP on an image retrieval dataset.

Table 2. Results from our abalation study as described in Sect. 4.1. We construct variants of our proposed method by selectively including only some training losses. These are denoted by L_e, $L_e + L_c$, and $L_e + L_c + L_d$ (see Sect. 3.2). Sketch and text inputs are combined by adding their embeddings. For "Feature max" and "Feature concat", the embeddings are combined by coordinate-wise maximum, and concatenation, respectively, and the full objective is used.

Method	R@1	R@5	R@10
CLIP (zero shot)	0.378	0.624	0.722
Ours: L_e	0.493	0.748	0.836
Ours: $L_e + L_c$	0.509	0.764	0.850
Ours: $L_e + L_c + L_d$	0.527	0.778	0.862
Ours: Feature max	0.443	0.704	0.804
Ours: Feature concat	0.357	0.650	0.768
Ours (final)	**0.609**	**0.847**	**0.917**

- **Ours:** L_e (ablated TASK-former). We start adding a sketch as an additional query, as shown in Fig. 2. The only objective in this baseline is to correctly classify the correct matching pair between the query (sketch+text) and the image via *symmetric cross entropy loss.*
- **Ours:** $L_e + L_c$ (ablated TASK-former). In this baseline, we add the multi-label classification loss term in the objective.
- **Ours:** $L_e + L_c + L_d$ (ablated TASK-former). In this baseline, we add both classification loss and decoder loss

For the above three baselines, text and sketch embeddings are combined by adding them, as described in Fig. 2. We further compare two additional ways to combine sketch and text embeddings: coordinate-wise maximum, and concatenation:

- **Ours: Feature max** (ablated TASK-former). Embeddings from sketch and text are combined using element-wise max.
- **Ours: Feature concatenate** (ablated TASK-former). Embeddings from sketch and text are concatenated, and projected into the same dimension as embedding from image.

We use the full objective (i.e., $L_e + L_c + L_d$) for the above two variants.

- **Ours (final)** This is our complete model trained with the full objective ($L_e + L_c + L_d$). We augment training sketches and images with random affine transformation, randomly remove parts of each sketch (as describe in Sect. 3) and train for 50 epochs.

Except for the final model, we train each baseline for 10 epoches. For ablation study, we simplify the training by only performing simple augmentation (random cropping and flipping).

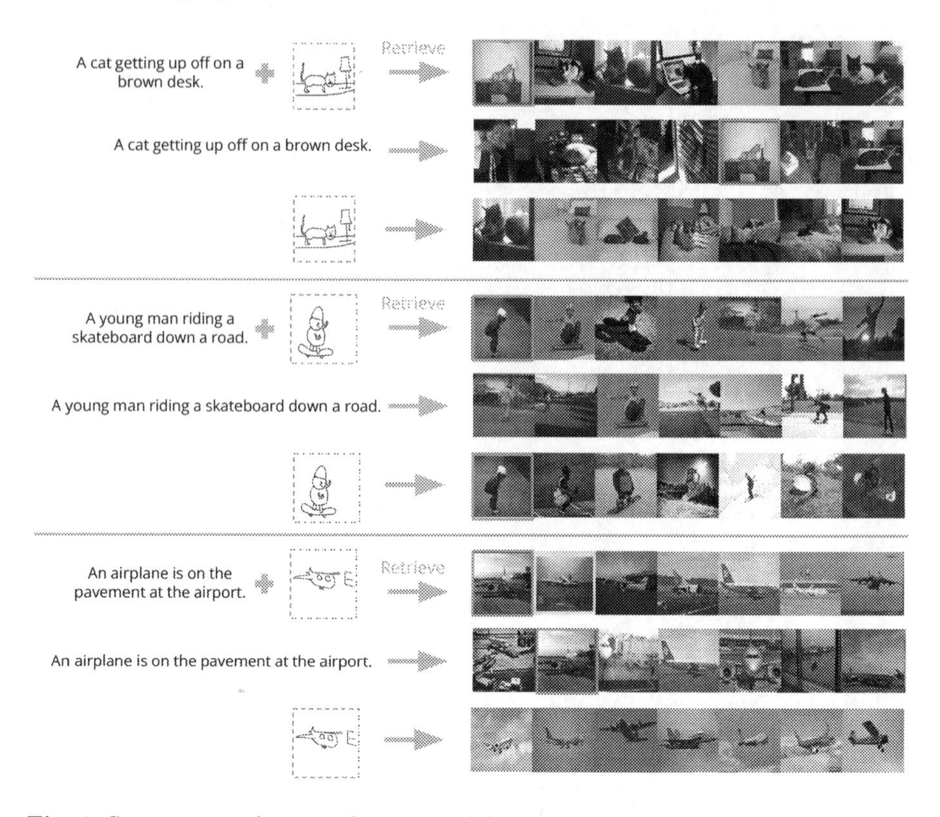

Fig. 4. Some retrieved images by our model when using only a sketch query, only a text query, and with both queries at the same time. Our model is robust to missing input: when only one input modality is present, it can still retrieve relevant image, albeit suboptimally. We observe that the presence of an input sketch clearly helps rank the most relevant image higher.

Table 2 reports recalls of each baselines. Both L_c and L_d further improve the retrieval performance compare to our baseline with only embedding loss (L_e). Surprisingly, our experiment also show that a simple element wise addition leads to the best performance compare to element wise max, and concatenation. We hypothesize that direct combination of the sketch and text embedding implicitly helps with feature alignment, compare to concatenate them.

4.2 Robustness to Missing Input

One key benefit of our pipeline is the ability to retrieve image even when missing one of the input modalities (sketch or text). We accomplish this by adding a query drop-out augmentation which replaces either the sketch or text with an empty sketch or an empty string. This ensures the network can operate even when no sketch or text description is given. Figure 4 shows the results breaking

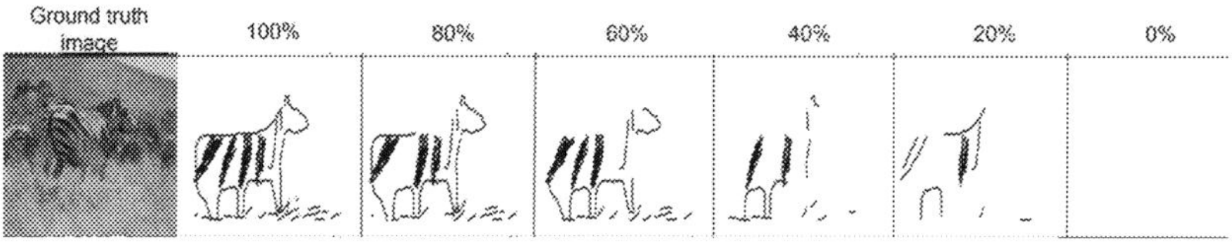

Sketch completeness	100%	80%	60%	40%	20%	0%
R@1	.615	.609	.602	.582	.515	.458
R@5	.883	.876	.870	.843	.833	.776
R@10	.947	.926	.940	.923	.916	.870

Fig. 5. Recall@{1, 5, 10} of TASK-former, averaged over 300 randomly sampled (text query, sketch, image) tuples in the COCO dataset. For each test instance, we vary the completeness of the sketch, where completeness level $x\%$ is defined as the original sketch with only $x\%$ of all strokes randomly chosen. Candidate retrievable images are the full COCO test set of 5k. The completeness 0% corresponds to using only text as input. We observe that even with a small fraction of strokes retained (e.g., 20%), there is gain in recall compared to using only text as input (i.e., 0% sketch completeness)

down into querying with sketch only, querying with text only, and querying with both modalities. We emphasize that sketch based image retrieval is particularly challenging for unconstrained, in-the-wild images with multiple objects and no fixed categories. In spite of that, our network can retrieve reasonably relevant results for sketch query alone, all without requiring any hand drawn sketch during training.

Nevertheless, we find that using the sketch or text as a sole input is often sub-optimal for image retrieval in this setup. A combination of sketch and text can provide a more complete picture of the target image, leading to an improve performance. For instance, in the first example of Fig. 4, it can be difficult to draw a *brown* desk with sketch alone, but this can be included easily with text description.

4.3 Sketch Complexity and Retrieval Performance

In this section, we investigate the effect of varying the level of sketch details on the retrieval performance. We sample 300 sketches from our collected COCO hand-drawn sketches, and randomly subsample strokes to keep only 20% to 100%. Table 5 shows the recall performance on each level of sketch complexity. We observe a large gain near the lower end of the completeness level, with a diminishing gain as the completeness level increases. In particular, this suggests that even a poorly drawn sketch helps in retrieving more relevant images. We observe that reducing the sketch completeness level from 100% to 60% only slightly decreases recall.

Table 3. Retrieval performance of our model as measured by recall@{1, 5, 10} when both a complete sketches, and text queries are inputted to the model. Words in each text query are sub-sampled to investigate the effect of incomplete input text on retrieval.

Text completeness	R@1	R@5	R@10
0%	0.099	0.195	0.257
20%	0.316	0.545	0.640
40%	0.357	0.613	0.708
60%	0.446	0.710	0.813
80%	0.530	0.807	0.884
100%	0.607	0.866	0.930

4.4 On the Effect of Text Completeness

In Sect. 4.3, we investigate the effect of an incomplete input sketch on the retrieval recall. The goal of this section is to provide an analogous analysis on the effect of an incomplete text input. We sample 300 records of from our collected COCO dataset. Each record consists of three objects: a sketch, a text description, and an image. For each text description, we vary its degree of completeness by randomly subsampling x% (for a number of values of x) of the tokens (words), while keeping its corresponding sketch intact. Table 3 shows the resulted retrieval recall for each degree of text completeness. We observe that compared to when no text query is given as an input (0%), recall@1 increases from 0.099 to 0.316 when only 20% of tokens are included. The gain in recall diminishes as more tokens are added.

5 Limitations and Future Work

Figure 3 shows the potentials of our retrieval model. However, it also reveals some examples of when our method fails to retrieve the target image. For example, in the forth row, the network is unable to retrieve the correct image (with blue jacket and black horse). This is likely due to the confusion with the color (both black and blue color are in the text description). While the network does place images with both black and blue color closer to the query embedding, it is unable to find the target image with black horse, and blue jacket within the top 10 results. This shows the current limitation of the model in understanding complicated text description.

Beyond these examples, we generally observe that the network suffers when the sketch is not representative of the target image. For example, the difference in scale and location can contribute to incorrect retrieval results (sixth row of Fig. 3). Designing a model that is more tolerant to scale mismatch will be an interesting topic for future research. On the text encoder, automatically augmenting the text query to be more concrete by leveraging text augmentation techniques also deserves further attention.

References

1. Alberti, C., Ling, J., Collins, M., Reitter, D.: Fusion of detected objects in text for visual question answering. arXiv preprint arXiv:1908.05054 (2019)
2. Ben-Baruch, E., et al.: Asymmetric loss for multi-label classification (2020)
3. Bhunia, A.K., Yang, Y., Hospedales, T.M., Xiang, T., Song, Y.Z.: Sketch less for more: on-the-fly fine-grained sketch-based image retrieval. In: Proceedings of the IEEE/CVF Conference on Computer Vision and Pattern Recognition, pp. 9779–9788 (2020)
4. Changpinyo, S., Pont-Tuset, J., Ferrari, V., Soricut, R.: Telling the what while pointing to the where: multimodal queries for image retrieval. In: Proceedings of the IEEE/CVF International Conference on Computer Vision, pp. 12136–12146 (2021)
5. Chen, T., Cheng, M.M., Tan, P., Shamir, A., Hu, S.M.: Sketch2Photo: internet image montage. ACM Trans. Graph. **28**(5), 1–10 (2009). https://doi.org/10.1145/1618452.1618470
6. Chen, Y., Bazzani, L.: Learning joint visual semantic matching embeddings for language-guided retrieval. In: Vedaldi, A., Bischof, H., Brox, T., Frahm, J.-M. (eds.) ECCV 2020. LNCS, vol. 12367, pp. 136–152. Springer, Cham (2020). https://doi.org/10.1007/978-3-030-58542-6_9
7. Chen, Y.-C., et al.: UNITER: UNiversal image-TExt representation learning. In: Vedaldi, A., Bischof, H., Brox, T., Frahm, J.-M. (eds.) ECCV 2020. LNCS, vol. 12375, pp. 104–120. Springer, Cham (2020). https://doi.org/10.1007/978-3-030-58577-8_7
8. Collomosse, J., Bui, T., Jin, H.: LiveSketch: query perturbations for guided sketch-based visual search. In: Proceedings of the IEEE/CVF Conference on Computer Vision and Pattern Recognition, pp. 2879–2887 (2019)
9. Dey, S., Dutta, A., Ghosh, S.K., Valveny, E., Lladós, J., Pal, U.: Learning cross-modal deep embeddings for multi-object image retrieval using text and sketch. In: 2018 24th International Conference on Pattern Recognition (ICPR), pp. 916–921. IEEE (2018)
10. Dey, S., Riba, P., Dutta, A., Llados, J., Song, Y.Z.: Doodle to search: practical zero-shot sketch-based image retrieval. In: The IEEE Conference on Computer Vision and Pattern Recognition (CVPR), June 2019
11. Dong, H., Wang, Z., Qiu, Q., Sapiro, G.: Using text to teach image retrieval (2020)
12. Dosovitskiy, A., et al.: An image is worth 16x16 words: transformers for image recognition at scale. In: ICLR (2021)
13. Dutta, A., Akata, Z.: Semantically tied paired cycle consistency for zero-shot sketch-based image retrieval. In: Proceedings of the IEEE/CVF Conference on Computer Vision and Pattern Recognition, pp. 5089–5098 (2019)
14. Gao, C., Liu, Q., Xu, Q., Wang, L., Liu, J., Zou, C.: SketchyCOCO: image generation from freehand scene sketches. In: Proceedings of the IEEE/CVF Conference on Computer Vision and Pattern Recognition (CVPR), June 2020
15. Han, T., Schlangen, D.: Draw and tell: multimodal descriptions outperform verbal- or sketch-only descriptions in an image retrieval task. In: Proceedings of the Eighth International Joint Conference on Natural Language Processing (Volume 2: Short Papers), pp. 361–365. Asian Federation of Natural Language Processing, Taipei, November 2017. https://aclanthology.org/I17-2061
16. Han, X., et al.: Automatic spatially-aware fashion concept discovery. In: ICCV (2017)

17. Hessel, J., Holtzman, A., Forbes, M., Bras, R.L., Choi, Y.: ClipScore: a reference-free evaluation metric for image captioning (2021)
18. Hu, R., Collomosse, J.: A performance evaluation of gradient field hog descriptor for sketch based image retrieval. Comput. Vis. Image Underst. **117**(7), 790–806 (2013). https://doi.org/10.1016/j.cviu.2013.02.005
19. Ilharco, G., et al.: Openclip (2021). https://doi.org/10.5281/zenodo.5143773
20. Jia, C., et al.: Scaling up visual and vision-language representation learning with noisy text supervision. arXiv preprint arXiv:2102.05918 (2021)
21. Song, J., Song, Y.Z., Xiang, T., Hospedales, T.M.: Fine-grained image retrieval: the text/sketch input dilemma. In: Kim, T.K., Zafeiriou, S., Brostow, G., Mikolajczykpp, K. (eds.) 45.1-45.12. BMVA Press, September 2017. https://doi.org/10.5244/C.31.45
22. Karpathy, A., Fei-Fei, L.: Deep visual-semantic alignments for generating image descriptions (2015)
23. Kingma, D.P., Ba, J.: Adam: a method for stochastic optimization. arXiv preprint arXiv:1412.6980 (2014)
24. Lee, K.H., Chen, X., Hua, G., Hu, H., He, X.: Stacked cross attention for image-text matching. arXiv preprint arXiv:1803.08024 (2018)
25. Li, M., Lin, Z., Mech, R., Yumer, E., Ramanan, D.: Photo-sketching: inferring contour drawings from images. In: WACV (2019)
26. Li, X., et al.: Oscar: object-semantics aligned pre-training for vision-language tasks. In: Vedaldi, A., Bischof, H., Brox, T., Frahm, J.-M. (eds.) ECCV 2020. LNCS, vol. 12375, pp. 121–137. Springer, Cham (2020). https://doi.org/10.1007/978-3-030-58577-8_8
27. Lin, T., et al.: Microsoft COCO: common objects in context. CoRR abs/1405.0312 (2014). http://arxiv.org/abs/1405.0312
28. Liu, Q., Xie, L., Wang, H., Yuille, A.L.: Semantic-aware knowledge preservation for zero-shot sketch-based image retrieval. In: Proceedings of the IEEE/CVF International Conference on Computer Vision, pp. 3662–3671 (2019)
29. Nilsback, M.E., Zisserman, A.: Automated flower classification over a large number of classes. In: 2008 Sixth Indian Conference on Computer Vision, Graphics and Image Processing, pp. 722–729. IEEE (2008)
30. van den Oord, A., Li, Y., Vinyals, O.: Representation learning with contrastive predictive coding. arXiv preprint arXiv:1807.03748 (2018)
31. Pandey, A., Mishra, A., Verma, V.K., Mittal, A., Murthy, H.: Stacked adversarial network for zero-shot sketch based image retrieval. In: Proceedings of the IEEE/CVF Winter Conference on Applications of Computer Vision, pp. 2540–2549 (2020)
32. Pang, K., et al.: Generalising fine-grained sketch-based image retrieval. In: Proceedings of the IEEE/CVF Conference on Computer Vision and Pattern Recognition (CVPR), June 2019
33. Pang, K., Song, Y.Z., Xiang, T., Hospedales, T.M.: Cross-domain generative learning for fine-grained sketch-based image retrieval. In: BMVC, pp. 1–12 (2017)
34. Pont-Tuset, J., Uijlings, J., Changpinyo, S., Soricut, R., Ferrari, V.: Connecting vision and language with localized narratives. In: Vedaldi, A., Bischof, H., Brox, T., Frahm, J.-M. (eds.) ECCV 2020. LNCS, vol. 12350, pp. 647–664. Springer, Cham (2020). https://doi.org/10.1007/978-3-030-58558-7_38
35. Radford, A., et al.: Learning transferable visual models from natural language supervision. In: Meila, M., Zhang, T. (eds.) Proceedings of the 38th International Conference on Machine Learning. Proceedings of Machine Learning Research, vol.

139, pp. 8748–8763. PMLR, 18–24 July 2021. https://proceedings.mlr.press/v139/radford21a.html
36. Radford, A., Wu, J., Child, R., Luan, D., Amodei, D., Sutskever, I.: Language models are unsupervised multitask learners (2019)
37. Sain, A., Bhunia, A.K., Yang, Y., Xiang, T., Song, Y.Z.: StyleMeUp: towards style-agnostic sketch-based image retrieval. In: Proceedings of the IEEE/CVF Conference on Computer Vision and Pattern Recognition, pp. 8504–8513 (2021)
38. Sangkloy, P., Burnell, N., Ham, C., Hays, J.: The Sketchy database: learning to retrieve badly drawn bunnies. ACM Trans. Graph. (Proceedings of SIGGRAPH) (2016)
39. Song, J., Yu, Q., Song, Y.Z., Xiang, T., Hospedales, T.M.: Deep spatial-semantic attention for fine-grained sketch-based image retrieval. In: Proceedings of the IEEE International Conference on Computer Vision, pp. 5551–5560 (2017)
40. Tautkute, I., Trzcinski, T., Skorupa, A., Brocki, L., Marasek, K.: DeepStyle: multimodal search engine for fashion and interior design (2019)
41. Tiwary, S.: Turing Bletchley: A Universal Image Language Representation Model by Microsoft (2021). https://www.microsoft.com/en-us/research/blog/turing-bletchley-a-universal-image-language-representation-model-by-microsoft/. Accessed 7 March 2021
42. Tursun, O., Denman, S., Sridharan, S., Goan, E., Fookes, C.: An efficient framework for zero-shot sketch-based image retrieval. arXiv preprint arXiv:2102.04016 (2021)
43. Vo, N., et al.: Composing text and image for image retrieval - an empirical odyssey. In: CVPR (2019). https://arxiv.org/abs/1812.07119
44. Wang, B., Yang, Y., Xu, X., Hanjalic, A., Shen, H.T.: Adversarial cross-modal retrieval. In: Proceedings of the 25th ACM International Conference on Multimedia, pp. 154–162 (2017)
45. Wang, C., Sun, Z., Zhang, L., Zhang, L.: Sketch2Tag: automatic hand-drawn sketch recognition. In: ACM Conference on Multimedia, January 2012. https://www.microsoft.com/en-us/research/publication/sketch2tag-automatic-hand-drawn-sketch-recognition/
46. Yang, Y., Jin, N., Lin, K., Guo, M., Cer, D.: Neural retrieval for question answering with cross-attention supervised data augmentation. arXiv preprint arXiv:2009.13815 (2020)
47. Yu, Q., Liu, F., Song, Y.Z., Xiang, T., Hospedales, T.M., Loy, C.C.: Sketch me that shoe. In: Proceedings of the IEEE Conference on Computer Vision and Pattern Recognition, pp. 799–807 (2016)
48. Zhang, H., Liu, S., Zhang, C., Ren, W., Wang, R., Cao, X.: SketchNet: sketch classification with web images. In: Proceedings of the IEEE Conference on Computer Vision and Pattern Recognition, pp. 1105–1113 (2016)
49. Zhang, Q., Lei, Z., Zhang, Z., Li, S.Z.: Context-aware attention network for image-text retrieval. In: Proceedings of the IEEE/CVF Conference on Computer Vision and Pattern Recognition, pp. 3536–3545 (2020)
50. Zhang, Z., Zhang, Y., Feng, R., Zhang, T., Fan, W.: Zero-shot sketch-based image retrieval via graph convolution network. In: Proceedings of the AAAI Conference on Artificial Intelligence, vol. 34, pp. 12943–12950 (2020)

A Cloud 3D Dataset and Application-Specific Learned Image Compression in Cloud 3D

Tianyi Liu, Sen He, Vinodh Kumaran Jayakumar, and Wei Wang[(✉)]

The University of Texas at San Antonio, San Antonio, USA
{tianyi.liu,sen.he,wei.wang}@utsa.edu,
vinodhkumaran.jayakumar@my.utsa.edu

Abstract. In Cloud 3D, such as Cloud Gaming and Cloud Virtual Reality (VR), image frames are rendered and compressed (encoded) in the cloud, and sent to the clients for users to view. For low latency and high image quality, fast, high compression rate, and high-quality image compression techniques are preferable. This paper explores computation time reduction techniques for learned image compression to make it more suitable for cloud 3D. More specifically, we employed slim (low-complexity) and application-specific AI models to reduce the computation time without degrading image quality. Our approach is based on two key insights: (1) as the frames generated by a 3D application are highly homogeneous, application-specific compression models can improve the rate-distortion performance over a general model; (2) many computer-generated frames from 3D applications are less complex than natural photos, which makes it feasible to reduce the model complexity to accelerate compression computation. We evaluated our models on six gaming image datasets. The results show that our approach has similar rate-distortion performance as a state-of-the-art learned image compression algorithm, while obtaining about 5x to 9x speedup and reducing the compression time to be less than 1 s (0.74s), bringing learned image compression closer to being viable for cloud 3D. Code is available at https://github.com/cloud-graphics-rendering/AppSpecificLIC.

Keywords: Cloud gaming · Cloud virtual reality · Learned image compression · Model simplification · Application-specific modeling · Model-task balance

1 Introduction

Image compression plays an important role in cloud 3D, including cloud gaming and cloud virtual reality (VR). Compared with local (non-cloud) 3D applications, cloud 3D needs an extra step, image/video encoding and streaming, to transmit compressed (encoded) frames of 3D applications to end users. The smaller the file size of the compressed image (i.e., higher compression rate), the lower

S. Avidan et al. (Eds.): ECCV 2022, LNCS 13698, pp. 268–284, 2022.
https://doi.org/10.1007/978-3-031-19839-7_16

network bandwidth usage and network latency will be. However, high compression rate usually either implies longer compression/decompression time or lower image quality. Therefore, fast, high compression rate, and high-quality image compression techniques are highly preferable.

Recently, learned image compression [3–5,9,10,12,14,17,19,23,27–29,41] has shown great potential for further increasing the compression rate, while maintaining similar image quality. By replacing traditional discrete cosine transform (DCT) [40] or discrete wavelet transform (DWT) [32] with deep neural networks, the latent representation after transform operation can have smaller data sizes (less spatial redundancy). Typical learned image compression frameworks [4,5,17,29] utilize cascaded auto-encoder/decoder layers to optimize the entropy encoding. Some studies [17,23] also aggregate information from all the decoder layers to get a better compression rate and reconstruction quality. Therefore, learned image compression has demonstrated better or comparable compression capability, compared to traditional methods [6,7,32,40].

Although learned image compression has made great progress, there is still a major challenge that hinders its application to cloud 3D: **The high inference latency of compression/decompression violates the timing constraint of cloud 3D**. For example, the Coarse-to-Fine model [17] takes about 5 and 7 s to compress and decompress an image with 1920 × 1080p resolution. Compression and decompression at several seconds are not tolerable for cloud 3D, as its latency is typically required to be less than a second [11].

In this work, we focus on reducing the computation time for learned image compression for cloud 3D while preserving low bitrate and high image quality. More specifically, we employed low-complexity and application-specific AI models to reduce the computation time without degrading image quality. Our approach is based on two key insights: (1) as the frames generated by a 3D application are highly homogeneous, application-specific models can improve the rate-distortion overall a general model; (2) many of the computer-generated frames from real 3D applications are less complex than natural photos, which also makes it feasible to employ less complex models to accelerate computation.

We collected six gaming image datasets to assist the development of our learned image compression method. We evaluated our simplified application-specific models on these datasets. The results show that our approach has similar rate-distortion performance as a state-of-the-art (SOTA) learned image compression algorithm, while obtaining 5x to 9x speedup and reducing the compression time to be less than 1 s, bringing learned image compression closer to being viable for cloud 3D. Our method also has a lower bitrate than BPG and VTM.

Note that, in this work, we focus on image compression than video-based encoding. Our results show that, for cloud 3D, the bitrate of our image compression solution is about 3 times better than H.264 videos (about 20 Mbps V.S. 60 Mbps), which is commonly used for cloud gaming [26]. Moreover, this image compression is similar to the I-Frames (independent frames) in videos. Using images (i.e., only I-Frames) have better tolerance over lost frames, which is common in cloud 3D, as their decoding does not depend on previous (lost) frames, and thus, potentially providing a better user experience. Finally, our conclusions can also be applied to the encoding of I-Frames for video-based cloud 3D.

In summary, the contributions of this work include:

1. Six gaming image datasets for learned image compression research on rendered images.

2. A low-complexity and application-specific approach for learned image compression for images generated/rendered by 3D applications.

3. A thorough evaluation of the proposed approach with various parameters which showcases the viability of learned image compression for cloud 3D.

The rest of this paper is organized as following: Sect. 2 discusses the related work; Sect. 3 explains the problem and presents our approach; Sect. 4 demonstrates the effectiveness of our methodologies. Section 5 discusses special design issues, and Sect. 6 concludes the paper.

2 Related Work

2.1 Traditional Image Compression

Traditional image compression frameworks employ several processes: image transformation, quantization, and entropy encoding. JPEG, first proposed by Wallace in 1992 [40], utilizes Discrete Cosine Transform (DCT) to make key information compact at the left-up corner of an image. JPEG2000 [32] was proposed later, which uses Discrete Wavelet Transform (DWT) to further improve the compression rate. Based on HEVC video encoding standard, Bellard proposed a new compression solution, BPG [6], which uses intra image prediction techniques and has been the state-of-the-art algorithm until VTM. Based on the latest video encoding standard, VVC or H.266 [7], VTM uses improved methods, such as larger coding units, more intra-prediction modes, and more transform types, to increase image compression rate. However, VTM is complex and may spend several hundred seconds in compression computation.

2.2 Learned Image Compression

As deep neural networks (DNN) have demonstrated excellent modeling and representation ability on computer vision tasks, many deep learning-based image compression schemes are proposed to improve the rate-distortion performance of image compression. Based on network types, these algorithms fall into two categories: recurrent models (RNN) [20,22,37,38] and convolutional models (CNN) [3–5,9,10,12,14,17,19,23,27–29,41].

Recurrent neural network (RNN) models typically compress images or the residual information progressively. Toderici et al. proposed a variable-rate RNN model [37], which is the first to utilize convolutional LSTM to compress images. Their approach supports end-to-end training and can generate multiple bitrates through a single model. Later, Toderici et al. and Johnston et al. introduced full-resolution RNN [38] and priming RNN [20], both of which achieve comparable and even better performance when compared with BPG in terms of image quality (MS-SSIM). Spatial RNN [22] excels BPG significantly by removing the redundant information among larger pixel ranges. However, RNN models tend to have high complexity, especially for higher ranges of bit-rates.

Being less complex, CNN methods are widely studied recently. Ballé explored an auto-encoder/decoder architecture with GDN network [4] to compress and reconstruct images. Later, a hyperprior model [5] is invented to predict the probability distribution of pixels to help with the entropy coding, which is comparable to BPG. The autoregressive entropy model [29] is based on the hyperprior model and utilizes context information to further remove spatial redundancy. EDIC model [23] and checkerboard context model [14] accelerate the decoding process for the autoregressive entropy model [29]. Coarse-to-Fine model [17] is proposed in 2020 and becomes the SOTA image compression solution. Coarse-to-Fine adds one more auto-encoder/decoder layer to the hyperprior architecture and aggregated information from all the auto-encoder/decoder layers to get better compression rate and reconstruction quality. However, the compression and decompression time of Coarse-to-Fine is usually high [34]. Slimmable network architectures [41–43] provide an effective solution for variable rate and adaptive complexity, making the AI models suitable for resource-limited devices.

Our work is inspired by these prior studies. Nonetheless, more work is still required to speed up learned image compression for real-time processing to meet cloud 3D's latency and quality requirements.

2.3 Formulation of Learned Image Compression

A basic learned image compression framework [4] follows the equations below:

$$\begin{aligned} y &= g_a(x, \alpha) \\ \hat{y} &= Q(y); b \leftarrow \xi(\hat{y}); \hat{y} \leftarrow \xi^{-1}(b) \\ \hat{x} &= g_s(\hat{y}, \beta) \end{aligned} \tag{1}$$

where g_a and g_s are analysis and synthesis transforms that are used to reduce image redundancy and reconstruct the image. α and β are optimized parameters for analysis and synthesis transforms. x and $\hat{x}$ denote the input raw image and the reconstructed image. y is the latent representation, which is quantized by Q. After quantization, the result $\hat{y}$ is imported into ξ for entropy encoding. The symbol b denotes the bitstream after image compression. For image decompression, the bitstream b is imported into ξ^{-1} for entropy decoding. The entropy encoding and decoding processes are lossless [34,35]. Hence, the quantized latent representation $\hat{y}$ can be fully recovered after these two steps. Finally, g_s reconstructs the raw image from $\hat{y}$. Since quantization is a lossy process, the reconstructed information $\hat{x}$ may be different from the original input x. The difference between x and $\hat{x}$ is the reconstruction distortion.

For quantization, round-based quantization values can be used to generate $\hat{y}$. However, learned image compression usually employs end-to-end models, and the round-based quantization method is non-differentiable, making it challenging for end-to-end training. Prior work [4] solved this problem by adding a uniform noise $u(-0.5, +0.5)$ to signal y during model training. This method creates a noisy signal $\tilde{y}$, which is used to approximate $\hat{y}$ during model training. That is, this solution makes the quantization process differentiable.

For entropy coding, prior work [4] used a basic factorized model, and later the authors proposed a more advanced trainable hyperprior model [5]. The hyperprior model added another auto-encoder/decoder layer to the baseline model, and the new layer is used for probability estimation. To model long term dependency, the Coarse-to-Fine model [17] proposed hierarchical layers of hyperpriors to conduct more comprehensive analysis, which can be formulated as,

$$\begin{aligned} &y = g_a(x, \alpha); z = h_{a1}(y, \alpha_1); w = h_{a2}(z, \alpha_2) \\ &\hat{y} = Q(y); \hat{z} = Q(z); \hat{w} = Q(w); \\ &p_{\hat{z}|\hat{w}}(\hat{z}|\hat{w}) \leftarrow h_2 \leftarrow h_{s2}(\hat{w}; \theta_{s2}) \\ &p_{\hat{y}|\hat{z}}(\hat{y}|\hat{z}) \leftarrow h_1 \leftarrow h_{s1}(\hat{z}; \theta_{s1}) \\ &\hat{x} \leftarrow g_s(\hat{y}, \beta), h_1, h_2 \end{aligned} \tag{2}$$

where g_a, g_s, x, $\hat{x}$, y, $\hat{y}$, α, β, and Q are same as the Formula 1. h_{a1} and h_{s1} are the auto-encoder and decoder in the first auxiliary layer, while h_{a2} and h_{s2} are the auto-encoder and decoder in the second auxiliary layer. After the transformation of g_a, the latent representation y (output of transformation) is imported into h_{a1}. h_{a1}'s output z is in turn imported into h_{a2} to produce w. Then, y, z, and w are further quantized into $\hat{y}$, $\hat{z}$, and $\hat{w}$. h_1 and h_2 are side information from h_{s1} and h_{s2}, which help reconstruct the original image, along with $g_s(\hat{y}, \beta)$. $p_{\hat{z}|\hat{w}}(\hat{z}|\hat{w})$ is the estimated distribution conditioned on $\hat{w}$, and $p_{\hat{y}|\hat{z}}(\hat{y}|\hat{z})$ is the estimated distribution conditioned on $\hat{z}$. $p_{\hat{z}|\hat{w}}(\hat{z}|\hat{w})$ and $p_{\hat{y}|\hat{z}}(\hat{y}|\hat{z})$ are used to do entropy coding of $\hat{y}$ and $\hat{z}$. The distribution of $\hat{w}$ uses the typical factorized model.

The typical loss function for learned image compression is a rate-distortion trade-off. For the Coarse-to-Fine model, the bitrate includes three parts: R_y, R_z, and R_w. The distortion, $D(x, \hat{x})$, is the difference between x and $\hat{x}$. The loss function can be described as,

$$\begin{aligned} L &= R + \lambda D = R_y + R_z + R_w + \lambda D(x, \hat{x}) \\ &= E[-\log_2 p_{\hat{y}|\hat{z}}(\hat{y}|\hat{z})] + E[-\log_2 p_{\hat{z}|\hat{w}}(\hat{z}|\hat{w})] + E[-\log_2 p_{\hat{w}}(\hat{w})] + \lambda D(x, \hat{x}), \end{aligned} \tag{3}$$

where λ is a trade-off parameter. Larger λ leads to better (i.e., less) distortion, and hence, higher bitrate. According to information theory, the minimum bitrate to encode signal x is the cross entropy of the real and estimated distribution of x. The $E(.)$ calculates the size of output bitstream from each layer.

3 Proposed Research

3.1 Application-Specific Learned Image Compression in Cloud 3D: A Practical Use Case

Learned image compression is a data-driven method, and the quality of the reconstructed image is highly dependent on the training dataset and the practical use case. The image quality would become poor if obvious differences are found between the training dataset and practical use case. To solve the problem of image quality drift (degradation), we propose application-specific learned image compression, which collects datasets and trains AI models for each specific use case. To make this solution more practical, we are targeting the cloud 3D system.

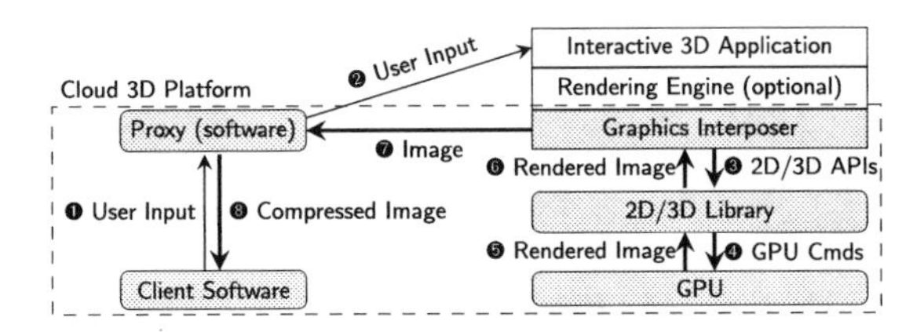

Fig. 1. A typical architecture of cloud 3D system.

Figure 1 illustrates a typical architecture of a cloud 3D system. Cloud 3D platform and 3D applications are the two main components, and the 3D applications run above the cloud 3D platform. For the cloud 3D platform, it involves server-side software (server proxy) and client-side software, and these two parts are separated by the network. On the client-side, the user interacts with the client software that is also responsible for displaying frames/images of remote 3D applications. The server-side includes several modules: proxy software, graphics interposer, 2D/3D library, and GPU. The proxy receives user inputs (step ❶), forwards inputs to corresponding 3D applications (step ❷), and transmits rendered frames from the server-side to client-side (step ❽). The graphics interposer intercepts API calls from 3D applications and invokes the real 2D/3D library in the cloud (step ❸). The GPU at the bottom of cloud 3D platform is responsible for executing rendering commands (step ❹). In cloud 3D, 3D applications are offloaded from local computers to the cloud, and the GUIs of 3D applications are transmitted to the client-side in the form of compressed images (step ❽). Therefore, image compression plays an important role in cloud 3D streaming.

Note that, an AI model is trained for each 3D application to conduct application-specific image compression/decompression. Hence, instead of downloading and installing a 3D application on local computers, an AI model (which is typically smaller than a 3D application) will be downloaded for decoding.

3.2 Cloud 3D Image Dataset

Images in cloud 3D systems are rendered by programs. To study learned image compression for cloud 3D images, datasets are collected from six 3D applications.

Dataset Collection. Pictor benchmark suite [25] is created for cloud 3D research. We used this benchmark suite to collect GUI frames of each 3D application automatically. More specifically, after rendering a frame on GPU, the pixels of the current frame will be copied from the GPU memory to CPU memory (step ❺, ❻, and ❼ in Fig. 1). Before sending the frame to client side, we compress the frame into the PNG format [1] and save it to disk (in the graphics interposer or the server proxy of Fig. 1).

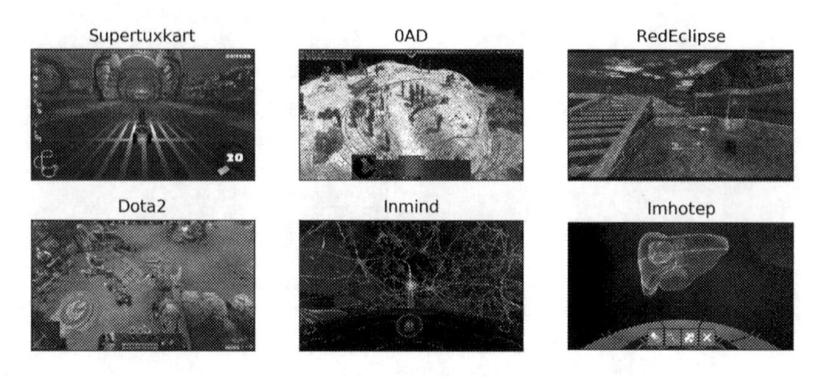

Fig. 2. A sample of cloud 3D image dataset from six 3D applications. Each 3D application has 2000 images. 1000 images are 1080p, and the other 1000 images are 720p.

Dataset Construction. We have collected six datasets from six 3D applications with two resolutions, 1920 × 1080 and 1280 × 720). Each 3D application has 2000 images. 1000 images are 1080p, and the other 1000 images are 720p. For each resolution, there are 800 images for training, 100 images for validation, and 100 images for testing. With each game having 2000 images, there are 12000 images in total. Figure 2 gives a sample of the cloud 3D image dataset. Supertuxkart [16] is an open-source racing game; 0AD [13] is an open-source real-time strategy game; Red Eclipse [33] is a first-person shooting game, and it is also open-source. Dota2 [39] is a highly popular online battle arena game. Inmind [30] and Imhotep [31] are VR applications.

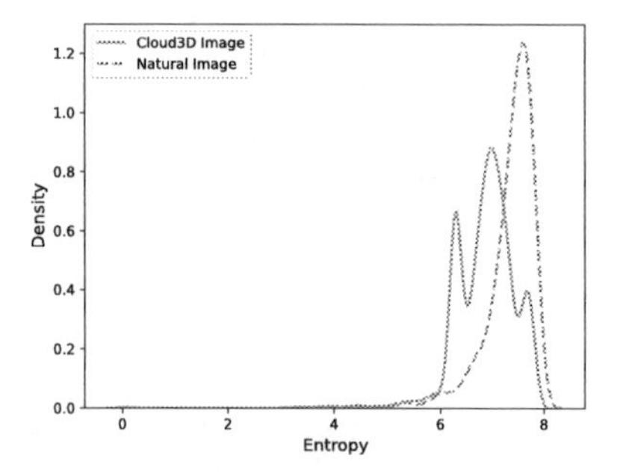

Fig. 3. Entropy density comparison of Cloud 3D images and natural images. The natural image dataset is from DIV2K_train_HR [2,18].

Dataset Analysis. Rendered images from 3D applications usually have fewer details and are relatively simpler than natural photos. To verify this intuition, we analyzed the complexity of rendered images and natural photos that are from the cloud 3D dataset and DIV2K_train_HR dataset respectively. Figure 3 provides quantitative evidence by comparing the difference between these two datasets in terms of entropy density. Entropy, $H(X) = -\sum_{i=1}^{n} p(x_i) log p(x_i)$, is a typical metric to evaluate the complexity of an image. Figure 3 shows that the majority of natural photos' entropy is on the right side of (i.e., larger than) the cloud 3D datasets, indicating that **cloud 3D images statistically have lower entropy and are simpler than natural photos**. Interestingly, Fig. 3 also shows that the entropy of cloud 3D images has a wide distribution, and one of the peak values is very close to that of natural photos, indicating that some rendered images have comparable complexity with natural photos.

3.3 A Slim Framwork: How Slim the Framework Can Be?

The SOTA learned image compression [17] has outperformed the traditional image compression solutions. However, it has complex structures, such as cascaded DNNs. These structures make image compression and decompression time-consuming. As discussed in Sect. 3.2, rendered images are simpler and have fewer details than natural photos. However, it is unclear if these simpler images can lead to more lightweight frameworks with lower bitrates to make it feasible for compressing gaming images. To further explore this feasibility, we studied the large, medium, small, xsmall models, and pruned models.

Table 1. Simplifying the Coarse-to-Fine model by continuously reducing the number(#) of channels in the main and auxiliary auto-encoder/decoders.

Model name	#of chnls in g_a/g_s	#of chnls in h_{a1}/h_{s1}	#of chnls in h_{a2}/h_{s2}
Large Com.	(3, [384, 384, 384, 384])	(384, [768, 1536, 256])	(256, [512, 256, 128])
Large Dec.	(384, [384, 384, 384, 3])	(256, [1536,1536,384])	(128, [512, 512, 256])
Medium Com.	(3, [192, 192, 192, 192])	(192, [384, 768, 256])	(256, [512, 128, 128])
Medium Dec.	(192, [192, 192, 192, 3])	(256, [768, 384, 192])	(128, [128, 256, 256])
Small Com.	(3, [96, 96, 96, 96])	(96, [192, 384, 256])	(256, [256, 128, 128])
Small Dec.	(96, [96, 96, 96, 3])	(256, [384, 192, 96])	(128, [128, 256, 256])
XSmall Com.	(3, [48, 48, 48, 48])	(48, [96, 192, 256])	(256, [256, 128, 128])
XSmall Dec.	(48, [48, 48, 48, 3])	(256, [192, 96, 48])	(128, [128, 256, 256])
Pruning Com.	Same as above	Same as above	Pruned
Pruning Dec.	Same as above	Same as above	Pruned

Large, Medium, Small, and Extra-Small Models. To reduce the computation time of the Coare-to-Fine model, we aim to obtain a slim framework by continuously reducing the number of channels in g_a, g_s, h_{a1}, h_{s1}, h_{a2} and h_{s2} (Please refer to Formula 2 and prior work [17]). Table 1 shows the details of large, medium, small, and xsmall models. "*Com" denotes image compression (analysis), while "*Dec" denotes image decompression (synthesis). Each model has separate analysis and synthesis processes. In each row of Table 1, the numbers stand for the input and output channels (chnls) of a CNN layer. E.g., in (3, [384, 384, 384, 384]), 3 refers to the RGB channels of an image, and [384, 384, 384, 384] denotes 4 convolutional layers each with 384 filters.

Framework Pruning. Coarse-to-Fine proposed hierarchical hyperprior layers to estimate the probability for entropy coding. Other studies [9,23] also proved that side information can benefit image reconstruction. Therefore, our framework kept these structures. Nonetheless, although adding one more hyperprior layer benefits probability estimation, it also increases the complexity of the model and the computation time. As we target cloud 3D, which has simpler images, we removed the last hyper-prior layer to further speed up the computation time (see Fig. 4 and the "Pruning *" rows in Table 1).

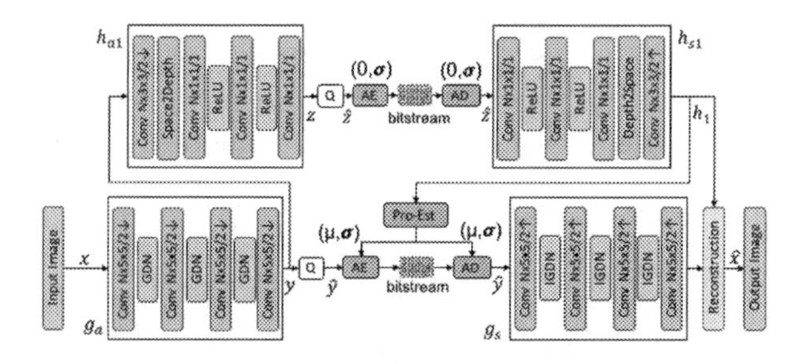

Fig. 4. A slim framework for image compression and decompression.

After pruning h_{a2} and h_{s2}, the original model becomes a slim framework. In Fig. 4, for image compression, the raw image x is first transformed by g_a, and then the output y is quantized into $\hat{y}$ for arithmetic encoding (AE) in the bottom path. In the top path, y is sent to h_{a1}. After quantization and AE, y is compressed into bitsteam. The h_{s1} in the top path generates h_1, which is imported into a probability estimation (Pro-Est) module to help with AE in the bottom path. For image decompression, the bitsteam in the top path is imported into the arithmetic decoding (AD) and further synthesized by h_{s1}. Finally, with h_1, the image is reconstructed after g_s processes $\hat{y}$. Consequently, inference and loss function can be simplified as Formula 4 and 5,

$$\begin{aligned} &y = g_a(x, \alpha); z = h_{a1}(y, \alpha_1) \\ &\hat{y} = Q(y); \hat{z} = Q(z); \\ &p_{\hat{y}|\hat{z}}(\hat{y}|\hat{z}) \leftarrow h_1 \leftarrow h_{s1}(\hat{z}; \theta_{s1}) \\ &\hat{x} \leftarrow g_s(\hat{y}, \beta), h_1 \end{aligned} \tag{4}$$

$$\begin{aligned} L &= R + \lambda D = R_y + R_z + \lambda D(x, \hat{x}) \\ &= E[-\log_2 p_{\hat{y}|\hat{z}}(\hat{y}|\hat{z})] + E[-\log_2 p_{\hat{z}}(\hat{z})] + \lambda D(x, \hat{x}) \end{aligned} \tag{5}$$

Although most of the symbols have been described in this section, please also refer to Formula 2 and 3 for more details about other symbols.

3.4 Model-Task Balance on GPU

GPUs usually have limited video memory, which makes it challenging to run large-scale AI models with high-resolution images. Splitting big images (e.g., 1080p/2K/4K/8K) into smaller tiles (e.g., 256×256) might help learned image compression. These small tiles can fit into GPU memory, allowing large AI model processing in GPU without memory error. However, processing these tiles in a sequential manner is suboptimal.

To further speed up the image processing on GPU, we employed Model-Task Balance (M-T-Balance). **The key idea is to increase the task size (size of input tile) as we switch to more lightweight learned image compression models.** More specifically, as we continuously simplify or prune the compression model, the input size can be increased accordingly. After getting a slim framework (Sect. 3.3), the AI model occupies less video memory, while the saved video memory can be utilized to accommodate larger image tiles. In this way, the computing resources in GPU can be fully utilized, which, in turn, increases the parallelism and further speeds up the compression and decompression process. In our evaluation, our final slim framework allows whole image frames to be stored in GPU memory.

4 Evaluation

Rate-distortion performance, compression/decompression speedup, and visualization quality are evaluated in our experiments. For each dataset, the model was trained with an initial λ value of 0.004. However, as a hyperparameter, λ was tuned for each application-specific model, with potential values of 0.01, 0.02, 0.04, 0.08, 0.16, 0.002, 0.001, 0.0005, 0.0002, and 0.0001. Each model was trained for 2500 epochs until the loss became stable. The speedup evaluation was conducted on a server with an 8-core Intel i7-7820x CPU, 16 GB memory, and an NVIDIA GTX1080Ti GPU with 11 GB GPU memory.

4.1 Performance Comparison

Rate-Distortion Comparison. Our method (slim framework with application-specific learned image compression) has similar or better rate-distortion performance, when compared with other algorithms. Figure 5 compares the rate-distortion performance of our approach with the Coarse-to-Fine model, BPG, and VTM. In the first five subfigures of Fig. 5, our method is very close to the performance of the SOTA model. In the last subfigure for Imhotep,

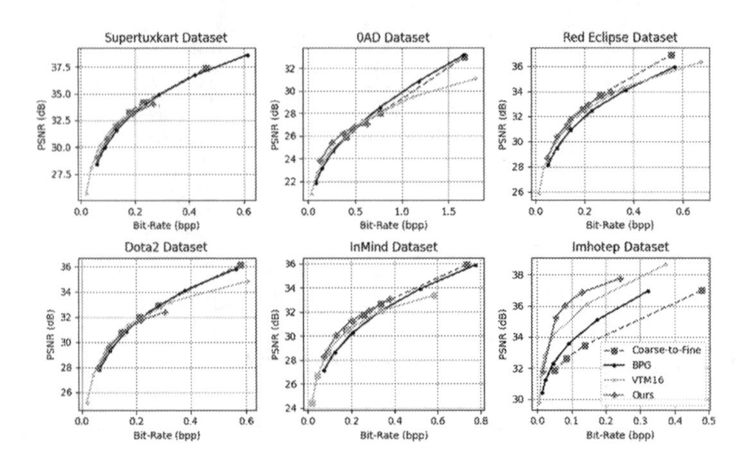

Fig. 5. Rate-Distortion curves (PSNR) of Coarse-to-Fine model [17], handcrafted BPG4:4:4 [6], VTM16-4:2:0 [36], and our method on six gaming datasets.

our method has significant performance improvement and saved about 0.3 bpp (bits per pixel) with a PSNR of 36dB. When comparing with BPG and VTM, our method has similar performance on Supertuxkart, 0AD, and Dota2 datasets. For the other three datasets, our method outperformed BPG and VTM. The Imhotep dataset is particularly interesting, as it only has a liver in the image with regular patterned background (see Fig. 2), our compressing scheme is more effectively on this dataset over other algorithms.

Computing Speedup. The goal of this work is to speed up the compression and decompression of learned image compression without harming the rate-distortion performance. The previous evaluation has shown that our method has better image quality at similar bitrates than other methods, the computing time will be further evaluated. Table 2 compares the computing time of our approach with the Coarse-to-Fine (SOTA) model, BPG, and VTM. Table 2 shows that our method outperforms the other three methods on image compression speed. More specifically, the compression time is reduced to 0.74s, which is about 9x and 156x faster than Coarse-to-Fine model and VTM, respectively. For image decompression, our method is about 4.7x faster than Coarse-to-Fine model, but slower than BPG and VTM. Further reduction of computing time will be investigated in the future (See Sect. 5 for more discussion).

Table 2. Computing time comparison (in seconds) of Coarse-to-Fine model [17], BPG [6], VTM-16 [36], and our method on six gaming image datasets.

Computing time (s)	Coarse-to-fine (SOTA)	BPG-4:4:4	VTM16-4:2:0	Ours
Compression	6.52	0.76	115.16	0.74
Decompression	7.72	0.38	0.39	1.64

4.2 Ablation Study

Application-Specific Compression. To illustrate the impact of application-specific model, we also report the performance of the application-specific models with the "large" size. Figure 6 compares the rate-distortion performance of the Coarse-to-Fine model (green curve) and the application-specific models (blue curve with square marker). Figure 6 shows that the application-specific scheme improves the rate-distortion performance of all applications significantly on both high and low bitrates. In particular, Imhotep has the largest improvement, and the reason of the large improvement is discussed in Sect. 4.1. Similarly, given the same bitrate, our application-specific method can improve PSNR by 2 to 4dB in most cases.

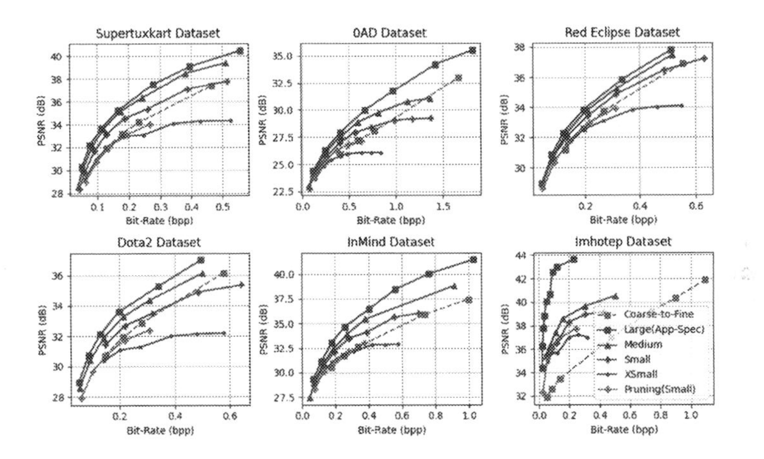

Fig. 6. Rate-Distortion curves of the Coarse-to-Fine model [17] and application-specific learned image compression (with model pruning) on six datasets. (Color figure online)

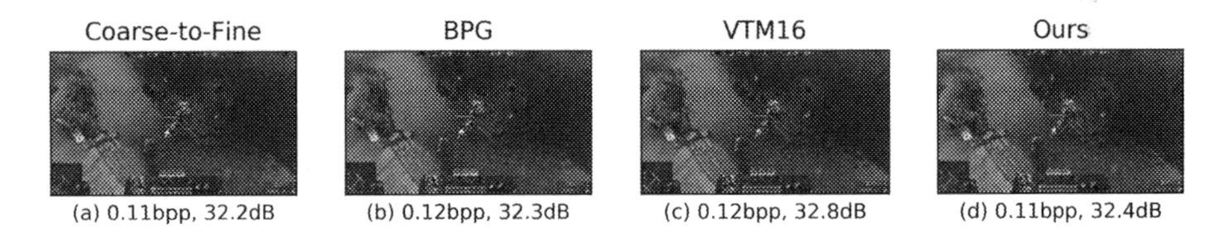

Fig. 7. Comparison on visual quality with the Coarse-to-Fine model, BPG, VTM, and our method. This sample image is 665.png under "test/dota2-1080p/" directory.

Rate-Distortion Comparison of Model Simplification and Prunning. To determine the proper size of a slim framework, an ablation study on each simplified model is also conducted. As shown in Table 1, large, medium, small, and xsmall models are trained and evaluated, and the rate-distortion performance is shown in Fig. 6. From Fig. 6, it is observed that the performance of these models gradually approaches the Coarse-to-Fine model as we keep simplifying the original model. The xsmall model has worse performance than the Coarse-to-Fine model, whereas the other three models behave well. As the small model has lower complexity, it is selected to perform hyperprior layer pruning, which achieves similar or better performance when compared with Coarse-to-Fine as shown in Fig. 6.

Computing Speedup. We also analyze the compression and decompression time for the simplified and pruned models. Their computing time is shown in Table 3. The M-T-Balance scheme is implemented on the "Small" model, and the pruning scheme is based on the "Small" model and employs the M-T-Balance. Table 3 shows that our model simplification can effectively reduce the computing time for both compression and decompression. Moreover, M-T-Balance achieved impressive speedup, and its average computing time is reduced by about 37%. At last, pruning the last hyperprior layer saved another 0.3 s.

Table 3. Computing time comparison of application-specific learned image compression with different model sizes, model-task balance (based on Small model), and model pruning (based on the small model and M-T-Balance).

Computing time (s)	Large	Medium	Small	XSmall	M-T-Balance (Small)	Pruning
Compression	4.16	2.24	1.63	1.43	0.93	0.74
Decompression	5.27	3.42	2.60	2.25	1.75	1.64

4.3 Visualization Study

To exemplify the effectiveness and image quality of our approach, visual examples of some reconstructed images are presented in Fig. 7. A relatively complex image is selected from the Dota2 dataset (as Dota2 is a popular 3D game and is welcomed by many players). In this figure, we compared our approach with the learned image compression algorithm, Coarse-to-Fine, as well as hybrid coding methods (BPG and VTM). Figure 7 shows that these images are very close, and we can barely distinguish these four images based on visual qualities. This is coherent with the quantitative results presented in Fig. 5.

In summary, our method has a lower compression time, similar or better rate-distortion performance, compared with Coarse-to-Fine, BPG, and VTM.

5 Discussion and Future Work

Model Size in Bytes. The model size of our solution is about 30.9 MB. As our models are application-specific, to use our models in cloud 3D, users need to download a model for each 3D application. As the models are small, this download should not be an obstacle, especially when considering many existing mobile games are significantly larger than our models.

Comparison with H.264 Video Encoding. Contemporary cloud gaming, such as Google's Stadia [8], primarily employed video encoding to stream the game images to the client. The most common encoding standard used is H.264 [8, 15,21,24,26]. Due to space limitation, we cannot provide a detailed comparison with H.264 video-based encoding. The results are summarized there. On average, at PSNR 30 to 35, the H.264 video bitrate for our datasets is more than 60Mbps, which is higher than our solution. That is, our solution (and learned image compression in general) has better bitrates than H.264 encoding.

The main benefit of H.264 is its faster encoding and decoding, especially with hardware accelerators. This fast computation makes H.264 more suitable for real-time encoding for cloud 3D than the general learned image compression. However, our application-specific solution can significantly reduce the compression/decompression time for learned compression. Although our solution is still slower than H.264, further optimizations (more in the next paragraph) may make learned image compression's computation time closer to H.264. With smaller bitrates (thus smaller network latency), the overall latency of learned image compression can be potentially lower than H.264 for cloud 3D.

Further Reduction of Computation Time. This paper focuses on the slim and application-specific model design for cloud 3D image compression. Therefore, we only conducted limited optimization (i.e., the Model-to-Task balance) on the implementation of our models. In the future, we plan to explore more optimization options, such as GPU code optimization, FPGA implementation, and/or ASIC (Application-specific Integrated Circuits) implementations, to eventually make learned compression feasible for cloud 3D.

6 Conclusions

This work focuses on reducing the computing time of learned image compression to make it one step closer to meeting the real-time requirement of cloud gaming and VR. We proposed application-specific compression to reduce the model complexity to speed up model computation time. Evaluations show that our approach significantly accelerated compression/decompression without degrading image quality, making learned compression potentially viable for cloud 3D.

Acknowledgments. This work was partially supported by the National Science Foundation under grants, 2221843, 2155096, 2202632, and 2215359. The views and conclusions contained herein are those of the authors and should not be interpreted as necessarily representing the official policies or endorsements, either expressed or implied of NSF. The authors would like to thank the anonymous reviewers for their insightful comments. We would also like to thank Kebin Peng for his valuable input.

References

1. libpng. https://www.libpng.org/pub/png/libpng.html. Accessed 17 Sep 2020
2. Agustsson, E., Timofte, R.: Ntire 2017 Challenge on Single Image Super-Resolution: Dataset and Study. In: Proceedings of the IEEE conference on computer vision and pattern recognition workshops, pp. 126–135 (2017)
3. Bai, Y., Liu, X., Zuo, W., Wang, Y., Ji, X.: Learning Scalable lY=-Constrained Near-Lossless Image Compression via Joint Lossy Image and Residual Compression. In: Proceedings of the IEEE/CVF Conference on Computer Vision and Pattern Recognition (CVPR), pp. 11946–11955 (2021)
4. Ballé, J., Laparra, V., Simoncelli, E.P.: End-to-end Optimized Image Compression. arXiv preprint arXiv:1611.01704 (2016)
5. Ballé, J., Minnen, D., Singh, S., Hwang, S.J., Johnston, N.: Variational Image Compression with a Scale Hyperprior. arXiv preprint arXiv:1802.01436 (2018)
6. Bellard, F.: Bpg image format. https://bellard.org/bpg/ (2017)
7. Bross, B., Chen, J., Liu, S., Wang, Y.K.: Versatile video coding editorial refinements on draft 10. JVET-Q2002-v3 (2020)
8. Carrascosa, M., Bellalta, B.: Cloud-gaming: Analysis of google stadia traffic. CoRR abs/2009.09786 (2020), https://arxiv.org/abs/2009.09786
9. Chen, T., Liu, H., Ma, Z., Shen, Q., Cao, X., Wang, Y.: End-to-end learnt image compression via non-local attention optimization and improved context modeling. IEEE Trans. Image Process. **30**, 3179–3191 (2021)
10. Cheng, Z., Sun, H., Takeuchi, M., Katto, J.: Learned Image Compression With Discretized Gaussian Mixture Likelihoods and Attention Modules. In: Proceedings of the IEEE/CVF Conference on Computer Vision and Pattern Recognition (CVPR) (2020)
11. Claypool, M., Claypool, K.: Latency and player actions in online games. Commun. ACM **49**(11), 40–45 (2006)
12. Cui, Z., Wang, J., Gao, S., Guo, T., Feng, Y., Bai, B.: Asymmetric Gained Deep Image Compression With Continuous Rate Adaptation. In: Proceedings of the IEEE/CVF Conference on Computer Vision and Pattern Recognition (CVPR), pp. 10532–10541 (2021)
13. Games, W.: 0 A.D. https://play0ad.com/. Accessed 11 Nov 2018
14. He, D., Zheng, Y., Sun, B., Wang, Y., Qin, H.: Checkerboard Context Model for Efficient Learned Image Compression. In: Proceedings of the IEEE/CVF Conference on Computer Vision and Pattern Recognition (CVPR), pp. 14771–14780 (June 2021)
15. Hegazy, M., et al.: Content-aware video encoding for cloud gaming. In: Proceedings of the 10th ACM Multimedia Systems Conference (2019)
16. Henrichs, J.: SuperTuxKart. https://supertuxkart.net/Main_Page. Accessed 11 Nov 2018

17. Hu, Y., Yang, W., Liu, J.: Coarse-to-Fine Hyper-Prior Modeling for Learned Image Compression. In: Proceedings of the AAAI Conference on Artificial Intelligence (2020)
18. Ignatov, A., Timofte, R., et al.: PIRM Challenge on Perceptual Image Enhancement on Smartphones: Report. In: European Conference on Computer Vision (ECCV) Workshops (2019)
19. Jiang, F., Tao, W., Liu, S., Ren, J., Guo, X., Zhao, D.: An End-to-End Compression Framework Based on Convolutional Neural Networks. IEEE Transactions on Circuits and Systems for Video Technology (2018)
20. Johnston, N., et al.: Improved lossy Image Compression with Priming and Spatially Adaptive Bit Rates for Recurrent Networks. In: Proceedings of the IEEE Conference on Computer Vision and Pattern Recognition (2018)
21. Lai, Z., Hu, Y.C., Cui, Y., Sun, L., Dai, N.: Furion: Engineering high-quality immersive virtual reality on today's mobile devices. In: Proceedings of the 23rd Annual International Conference on Mobile Computing and Networking (2017)
22. Lin, C., Yao, J., Chen, F., Wang, L.: A Spatial RNN Codec for End-to-End Image Compression. In: Proceedings of the IEEE/CVF Conference on Computer Vision and Pattern Recognition (CVPR) (2020)
23. Liu, J., Lu, G., Hu, Z., Xu, D.: A Unified End-to-End Framework for Efficient Deep Image Compression. arXiv preprint arXiv:2002.03370 (2020)
24. Liu, L., et al.: Cutting the cord: Designing a high-quality untethered vr system with low latency remote rendering. In: Proceedings of the 16th Annual International Conference on Mobile Systems, Applications, and Services (2018)
25. Liu, T., et al.: A Benchmarking Framework for Interactive 3D Applications in the Cloud. In: International Symposium on Microarchitecture (MICRO) (2020)
26. Meng, J., Paul, S., Hu, Y.C.: Coterie: Exploiting frame similarity to enable high-quality multiplayer vr on commodity mobile devices. In: Proceedings of the Twenty-Fifth International Conference on Architectural Support for Programming Languages and Operating Systems (2020)
27. Mentzer, F., Agustsson, E., Tschannen, M., Timofte, R., Gool, L.V.: Practical Full Resolution Learned Lossless Image Compression. In: Proceedings of the IEEE/CVF Conference on Computer Vision and Pattern Recognition (CVPR) (2019)
28. Mentzer, F., Gool, L.V., Tschannen, M.: Learning Better Lossless Compression Using Lossy Compression. In: Proceedings of the IEEE/CVF Conference on Computer Vision and Pattern Recognition (CVPR) (2020)
29. Minnen, D., Ballé, J., Toderici, G.: Joint Autoregressive and Hierarchical Priors for Learned Image Compression. arXiv preprint arXiv:1809.02736 (2018)
30. Nival: InMind VR. https://luden.io/inmind/. Accessed 22 July 2018
31. Pfeiffer, M., et al.: IMHOTEP: virtual reality framework for surgical applications. Int. J. Comput. Assisted Radiol. Surgery **13**(5) (2018)
32. Rabbani, M., Joshi, R.: An Overview of the JPEG 2000 Still Image Compression Standard. Image communication, Signal processing (2002)
33. Reeves, Q., Salzman, L.: Red Eclipse: A Free Arena Shooter Featuring Parktour. https://www.redeclipse.net/. Accessed 11 Nov 2018
34. Rippel, O., Bourdev, L.: Real-Time Adaptive Image Compression. In: Proceedings of the 34th International Conference on Machine Learning, pp. 2922–2930 (2017)
35. Rissanen, J., Langdon, G.: Universal Modeling and Coding. IEEE Transactions on Information Theory (1981)
36. Suehring, K.: VVCSoftware_VTM. https://vcgit.hhi.fraunhofer.de/jvet/VVCSoftware_VTM (2022)

37. Toderici, G., et al.: Variable Rate Image Compression with Recurrent Neural Networks. arXiv preprint arXiv:1511.06085 (2015)
38. Toderici, G., et al.: Full Resolution Image Compression With Recurrent Neural Networks. In: Proceedings of the IEEE Conference on Computer Vision and Pattern Recognition (CVPR) (2017)
39. Valve: InMind VR. https://blog.dota2.com/?l=english. Accessed 22 July 2018
40. Wallace, G.K.: The jpeg still picture compression standard. IEEE transactions on consumer electronics (1992)
41. Yang, F., Herranz, L., Cheng, Y., Mozerov, M.G.: Slimmable Compressive Autoencoders for Practical Neural Image Compression. In: Proceedings of the IEEE/CVF Conference on Computer Vision and Pattern Recognition (CVPR), pp. 4998–5007 (2021)
42. Yu, J., Huang, T.S.: Universally Slimmable Networks and Improved Training Techniques. In: Proceedings of the IEEE/CVF International Conference on Computer Vision, pp. 1803–1811 (2019)
43. Yu, J., Yang, L., Xu, N., Yang, J., Huang, T.: Slimmable Neural Networks. arXiv preprint arXiv:1812.08928 (2018)

AutoTransition: Learning to Recommend Video Transition Effects

Yaojie Shen[1,2,3], Libo Zhang[1,2], Kai Xu[3], and Xiaojie Jin[3](✉)

[1] Institute of Software, Chinese Academy of Sciences, Beijing, China
[2] University of Chinese Academy of Sciences, Beijing, China
[3] ByteDance Inc., Beijing, China
jinxiaojie@bytedance.com

Abstract. Video transition effects are widely used in video editing to connect shots for creating cohesive and visually appealing videos. However, it is challenging for non-professionals to choose best transitions due to the lack of cinematographic knowledge and design skills. In this paper, we present the premier work on performing automatic video transitions recommendation (VTR): given a sequence of raw video shots and companion audio, recommend video transitions for each pair of neighboring shots. To solve this task, we collect a large-scale video transition dataset using publicly available video templates on editing softwares. Then we formulate VTR as a multi-modal retrieval problem from vision/audio to video transitions and propose a novel multi-modal matching framework which consists of two parts. First we learn the embedding of video transitions through a video transition classification task. Then we propose a model to learn the matching correspondence from vision/audio inputs to video transitions. Specifically, the proposed model employs a multi-modal transformer to fuse vision and audio information, as well as capture the context cues in sequential transition outputs. Through both quantitative and qualitative experiments, we clearly demonstrate the effectiveness of our method. Notably, in the comprehensive user study, our method receives comparable scores compared with professional editors while improving the video editing efficiency by **300** $\times$. We hope our work serves to inspire other researchers to work on this new task. The dataset and codes are public at https://github.com/acherstyx/AutoTransition.

Keywords: Video transition effects recommendation · Multi-modal retrieval · Video editing

Y. Shen, L. Zhang, K. Xu and X. Jin—Equal contribution.

Supplementary Information The online version contains supplementary material available at https://doi.org/10.1007/978-3-031-19839-7_17.

S. Avidan et al. (Eds.): ECCV 2022, LNCS 13698, pp. 285–300, 2022.
https://doi.org/10.1007/978-3-031-19839-7_17

1 Introduction

With the advance of multimedia technology and network infrastructures, video is ubiquitous, occurring in numerous everyday activities such as education, entertainment, surveillance, etc. There is a massive amount of needs for people to edit videos and share with others. However, video editing is challenging for non-professionals since it is not only laborious but also needs a lot of cinematography and design knowledge. Some editing tools like *Adobe Premier* and *Apple Final Cut Pro* are developed to assist video editing, however their main target users are professionals while novices may find it difficult to learn. Moreover, they still lack the ability of automatic video editing, i.e., users have to manipulate videos on their own. Recently, popular video editing tools like *InShot Video Editor* and *CapCut* provide the function of creating videos in one-click. Nevertheless, since they only utilize simple strategies or fixed video templates and ignore the content of input vision/audio, the quality of generated video is unsatisfactory.

The video transition effects play an important role in video editing to join shots for creating smooth and cohesive videos. In this paper, we introduce a new task of automatic video transition recommendation (VTR) and provide a systematic solution. Specifically, VTR is defined as: given a sequence of raw video shots and the companion audio (which can either be original sound or overwritten music), recommend a sequence of video transitions for each neighboring shot. Different from conventional classification problems which choose only one most probable category as the output, VTR aims to provide a ranking of candidate transition categories so that users can choose freely in practical usage.

When working on this task, we encounter following challenges. First, there is no publicly available video transition dataset for training and evaluation. It takes enormous efforts to manually collect and annotate a large-scale video dataset. Meanwhile, due to the large complexity and diversity of video editing, the evaluation of video quality is subjective and vary from person to person. Thus during creating the dataset, it is crucial to design proper criteria for selecting video samples that are appealing to most. Besides dataset, solving VTR is also challenging. A good video transitions recommender should ensure top transitions match well with both the dynamics and contents of videos and the rhythm and theme of audio (or music). Moreover, transitions recommended at multiple video connections should be harmonious so that the final video is visually smooth and unified. Being the first effort in addressing VTR, we need to take all of above factors into consideration for delivering the optimal solution.

We start with building a video transition dataset from those video editing templates that are publicly available on video editing tools. We design comprehensive rules via trials to select high-quality video templates followed by pre-processing for refinement. Afterwards, we extract video shots and corresponding transitions (used as ground-truth in training), creating the first large-scale video transitions dataset. More details are introduced in Sect. 3.

To figure out the best way for modeling VTR, we conduct extensive experiments to compare the classification-based and retrieval-based solutions and finally demonstrate the latter performs better. In the classification-based solution, the model takes neighboring video shots as inputs and outputs the prediction of transition categories. The cross-entropy loss between predictions and ground-truth transitions is used. In the retrieval-based solution, we first pre-train a transition classification network to learn the embedding of transitions. Then we propose a multi-modal transformer to learn the fused vision/audio features in sequential video shots. A triplet margin loss is devised to minimize the distance between fused input features and pre-trained transition embedding. Similarly, in testing, according to the distance calculated between input features and transition embedding, the transition categories with smaller distances are in higher ranking position.

To conclude, our contributions are threefold:

1. We introduce a new task of automatically recommending video transition effects given video and audio inputs. We collect the first large-scale video transitions dataset to facilitate future research on this task.
2. We formulate the transition recommendation task as a multi-modal retrieval problem and propose a framework for learning the correspondence between input vision/audio and video transitions in feature space. The proposed framework is capable of fully utilizing the multi-modal input information for generating sequential transition outputs.
3. Through both quantitative and qualitative evaluation, we demonstrate that the proposed method can successfully learn the matching from vision/audio to transitions and generate reasonable recommendation results. Moreover, a user study conducted to evaluate the quality of generated videos further demonstrates the effectiveness of our method.

2 Related Work

Video Editing. Automatic video editing is challenging due to following reasons. First, the model has to fully understand the spatial-temporal context and multi-modal information in videos to obtain semantically coherent results. Second, video editing requires lots of professional knowledge to endow videos with creativity and particular aesthetic taste. Third, the evaluation of the quality of the generated video may be subjective. Recently, there are some progresses towards performing automatic video editing, each focusing on different aspects. Frey et al. [1] proposes an automatic approach to transfer the editing styles of an edited video to the new raw shots. Wang et al. [2] builds a tool for creating video montage based on the text description. Koorathota et al. [3] proposes a method to perform video editing according to a short text query by utilizing contextual and multi-modal information. Liao et al. [4] introduces a method for music-driven video montage. Several methods focus on solving video ordering and shot selection [1,2,5]. Distinguished from all tasks above, VTR is still unexplored although its equal importance in creating high-quality videos in video editing. In this work, we take the first step to close the research gap.

Video Transitions. Video transition is a widely used post-production technique for achieving smooth transitions between neighboring shots via special image/video transforms. There are various kinds of transitions including straight cuts, fades, and 3D animations among many others. To professionals, each type of transition is with a dedicated meaning to convey specific emotions, feelings or scene information to viewers, thus should be used meticulously. Moreover, when multiple video transitions are used, they should work in a harmonious way to ensure the visual unification of the final video. Due to above reasons, it is difficult for non-professionals to apply video transitions in their edit. Our work can substantially assist these people by automatically recommending reasonable video transitions on the fly.

Visual-semantic Embedding. Many recent works on video retrieval are based on the alignment of visual-semantic embedding [6,7]. Embedding techniques are employed to measure the similarity of different modalities in cross-modal video retrieval tasks, where features from different modals are mapped into a shared embedding space for better alignment [8,9]. Miech et al. [9] utilizes millions of video text pairs to learn the text-video embedding. Escorcia et al. [10] aligns the embedding of text and moments in the videos. They share the same idea of jointly aligning embeddings from two different modals. Triplet loss is initially proposed for learning the distance metric [11]. It is used for learning multi-modal embedding through deep neural networks in recent works [12,13]. By optimizing the distance directly with a soft margin, triplet loss is suitable for ranking tasks [14]. Our method also employs triplet loss to learn the distance between representations. A multi-modal transformer is used therein to learn the features of vision/audio inputs which aim to match with the pre-trained video transition embedding. Through extensive experiments, we demonstrate the effectiveness of our methods.

Multi-modal Transformer and Sequence Modeling. Our task is closely related to recent progress in modeling the spatial-temporal and multi-modal information in vision, speech and text. Transformer [15] is widely used in these tasks to encode cross-modal and spatial-temporal information [16–19]. It employs an attention mechanism to represent multi-modal information in a common latent space. In other sequential problems, Lin et al. [20] uses a modality-specific classifier and a differentiable tokenization scheme to fuse multi-modal information via transformer. Gabeur et al. [17] introduces a video encoder architecture with multi-modal transformer for video retrieval. We also use modal-specific networks to extract embeddings from vision and audio inputs. Specifically, we use SlowFast [21] and Harmonic CNN [22] to extract video and audio features respectively. Different with the vanilla transformer architecture which adopts an encoder-decoder architecture [15], recent works [23] proposes unified encoder-decoder transformer model for sequential modeling. In our method, we exploit a multi-modal transformer for learning the multi-modal representations as well as capturing the context information in sequential transitions.

3 Task Definition

We aim to solve the task of video transitions effect recommendation (VTR), the goal of which is to recommend appropriate transitions between neighboring video shots. As shown in Fig. 1a, we take a sequence of raw video shots $\{v_1, v_2, \ldots, v_n\}$ as inputs. To simplify the task, we assume that the order of the videos is already determined, all the videos are already cut and scaled to the target range, and the background audio (either original sound and/or overwritten music) is already specified. Then for a pair of video shots v_k and v_{k+1}, a transition effect is added to join them. We denote the video clip added with transition effect as $t_{k,k+1}$. Since video transitions generally mix neighboring video shots, we separate the final video after adding transitions into two parts, one is the clips added by transitions (i.e. $t_{k,k+1}$), the other is uncontaminated video shots (i.e. v'_k and v'_{k+1}). Then the output video after adding transitions can be denoted as $v' = \{v'_1, t_{1,2}, v'_2, \ldots, t_{n-1,n}, v'_n\}$. The dataset we collect contains both the start and end timestamps of each transition so that we can obtain the exact positions of video shots and the categories of transitions. Note that in the collected dataset, we can only get access to the output video v' since the original videos v are not publicly available.

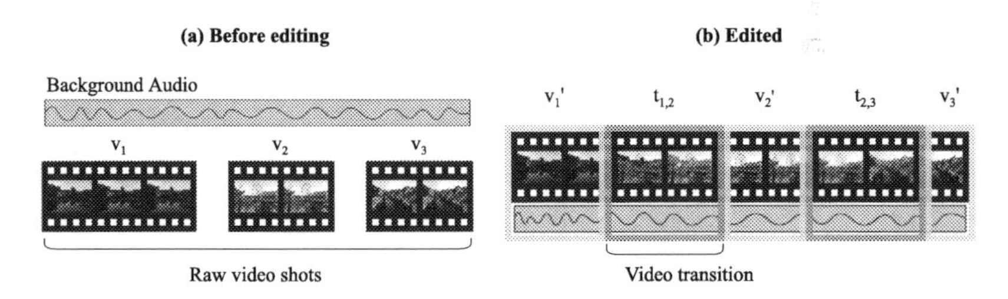

Fig. 1. The definition of the task. We take a sequence of raw video shots, for example $\{v_1, v_2, v_3\}$, and companion audio as the inputs, the task is to recommend categories of the transitions $\{t_{1,2}, t_{2,3}\}$ used in the edited video.

In our method, we use the video clip $t_{k,k+1}$ and the label of corresponding transition effect $c_{k,k+1}$ to train the transition classification network, for learning the embedding of transitions based on their visual representation. When training the transition recommendation model, we remove $t_{k,k+1}$ from the input to be consistent with the inference setting where input videos are without any transitions. The uncontaminated video shots v'_k and v'_{k+1} are used to represent the original video shots, i.e. v_k and v_{k+1} respectively. This strategy is reasonable since the duration of transition effect is short, v'_k and v'_{k+1} can serve as a good approximation to their original counterparts.

4 Video Transition Dataset

4.1 Raw Data Collection

With the development of video editing tools and platforms, a large amount of well-designed video templates are publicly available. Produced by professionals, video templates define fixed combinations of essential editing elements, including the number of video shots, the length of video, music, transitions, animations, and camera movement, etc. By simply replacing materials in templates with self videos, even novice users can create edited videos easily. Each video template also comes with an example video made by the designer using this template and his/her original videos. In our experiments, we collect video templates from these online video platforms and get the annotations related to transition effects, including each transition's category and corresponding start/end time. Though video templates may contain other special visual effects like animation and 3D movements, we only consider video transition effects in this paper.

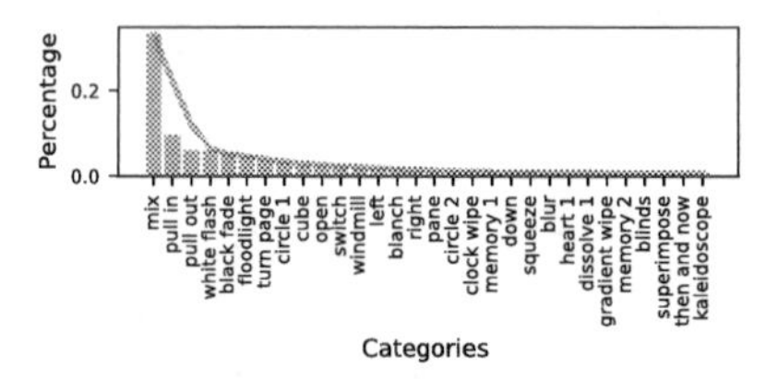

(a) The label distribution on the small dataset at the first stage of data collection.

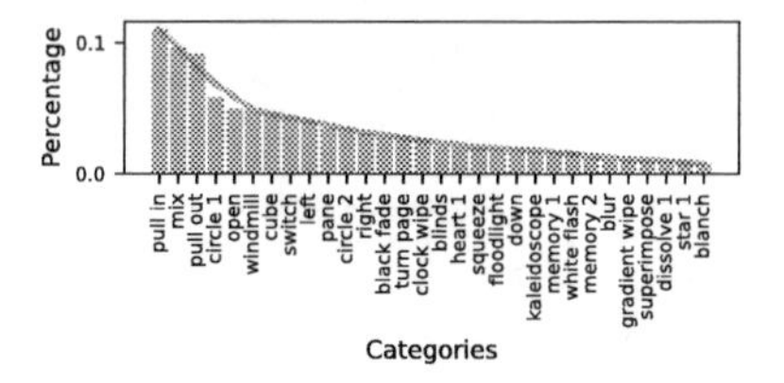

(b) The label distribution on the final dataset after filtering.

Fig. 2. Label distribution of top-30 categories in the dataset. (a) The label distribution on the small dataset at the first stage of data collection. (b) The label distribution on the final dataset after filtering.

Table 1. The statistical information of the transition dataset.

Dataset	Train set	Test set
Number of videos	29,998	5,000
Number of transitions	118,984	19,869
Transitions per video	3.966	3.973
Average video length	15.83 s	15.79 s

4.2 Data Filtering

We perform data filtering to improve the quality and diversity of the dataset. At first, we limit the maximum length of the video to 60 s, and the templates are filtered depending on the number of user likes and usage. We ignore those videos without any transition. We gather a small dataset through manual crawling and examine the overall distribution of collected samples. To guarantee that there are enough training samples for each category, we only select top-30 categories for training and testing according to the amount of samples. By statistical results, we observe that there are many duplicated transitions in a single video, and different types of transitions are distributed in a severe long-tail manner. The label distribution of the small dataset is shown in Fig. 2a.

We believe that the duplication and long-tailed distribution are harmful to the diversity of recommendation. To solve this issue, we use two additional rules to select samples: each video should contain more than two different types of transitions, and the usage times of the same transition should be no more than six. Following these rules, we acquire the final dataset which contains 34998 videos (train-29998 and test-5000) in total and 138.8K valid transitions between neighboring video shots. Table 1 shows more statistical results of the dataset. The label distribution is shown in Fig. 2b.

5 Video Transition Recommendation

5.1 Pre-training Transition Embedding

We formulate the video transition effects recommendation as a multi-modal retrieval problem. Since retrieving from vision/audio to video transitions requires the model to learn correspondence between vision/audio inputs and video transitions, the first problem we need to solve is thus how to learn a strong representation for each transition. We notice that some video transitions have similar visual effects like move up and move down. It is natural that we expect the learned embedding can also reflect these connections. To achieve this goal, we employ a video classification network to learn transition embedding based on their visual appearance. As shown in Fig. 3, we take the transition clips t as input and use the video backbone to extract visual representations. After passing through linear transform and normalization, we obtain a unit vector for each transition. Then we apply another linear transform and use the cross-entropy loss to optimize the classification objective. As expected, the embedding of the transitions are separably distributed in latent space and similar transitions stay close to each other. The visualization result of learned embeddings through t-SNE is illustrated in Fig. 5.

5.2 Multi-modal Transformer

As shown in Fig. 4, we propose a multi-modal transformer to extract representations from the raw video shots and audio. For recommending a transition $t_{k,k+1}$

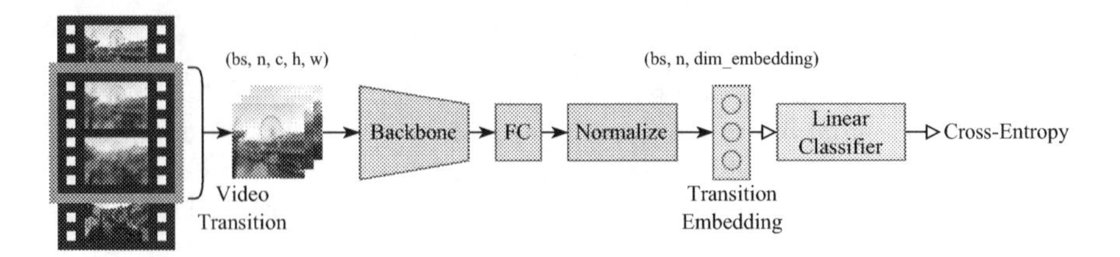

Fig. 3. Extracting transition embedding. A transition classification network is built to learn the transition embedding.

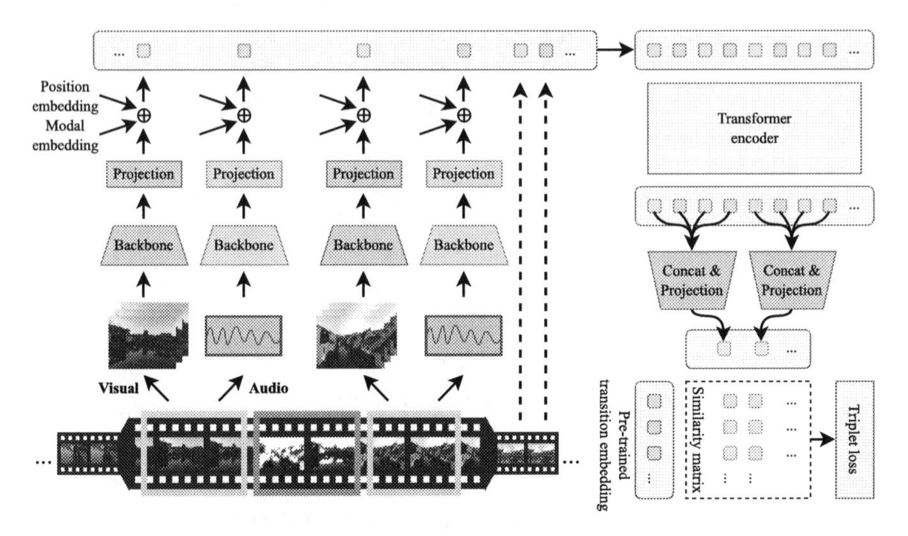

Fig. 4. Transition recommendation model for retrieving matching transitions based on vision/audio inputs. First, we use modality-specific networks to extract visual and audio features. After that, a multi-modal transformer encoder fuses the tokens from different modalities. Finally, a triplet loss is used to optimize the network end-to-end.

between video shots v_k and v_{k+1}, we take both the video frames and audio waves from the end period of video shot v_k and the start period of video shot v_{k+1} as input. Specifically, n video frames are sampled uniformly from each video shot.

After obtaining video frames and corresponding audio waves, we extract their features by feeding into visual and audio backbone respectively. Note that these backbones can be conveniently replaced by other common video or audio models. In our experiments, we use the SlowFast [21] and the Harmonic CNN [22] as video and audio backbone respectively.

In order to make full use of multi-modal information of vision/audio, we combine the visual and audio features as the input for the multi-modal transformer. By doing so, the model not only learns the matching relationship from input to transitions but also captures the context cues among sequential transitions.

Before being fed into the transformer, visual tokens and audio tokens are projected to the same dimension by independent linear transformations. Then learnable positional embedding and modal embedding are element-wisely added to these tokens. We share the positional embedding for vision/audio tokens at the same time point. Modal embedding is applied to inform the model which modality the token belongs to. After above processing, tokens from all video shots are input into transformer as a whole. In this way, the transformer is encouraged to learn the contextual relationship in sequential transitions. As demonstrated by experimental results, such a context-aware training method contributes to generate harmonious sequential outputs. The self-attention mechanism in transformer encoder can model complex mutual relationship among input tokens. From all output tokens, we concatenate those four tokens which corresponds to each transition point (for each modality of vision and audio, there are two tokens before and after the transition point) along the feature dimension and get the final representation by a projection to the same dimension of transition embedding.

5.3 Transition Recommendation

As introduced in Sect. 5.2, the multi-modal video embedding is extracted by the multi-modal transformer, which is used as the query to retrieve the transitions by matching the pre-trained transition embedding mentioned in Sect. 5.1. During retrieval, we apply learnable linear transformations to both the pre-trained transition embedding and the video embedding for achieving better alignment.

We denote $E^{trans} = \{e_1^{trans}, \ldots, e_N^{trans}\}$ as the set of pre-trained transition embedding where N is the number of categories and $N = 30$ in our experiments. $E^{video} = \{e_1^{video}, \ldots, e_V^{video}\}$ is denoted as the multi-modal input embedding in a batch. V is the number of all transition points in a batch. For each sample e_v^{video}, we calculate its matching score with every transition embedding through a similarity metric $\Phi(e_v^{video}, e_k^{trans})$. In our implementation, Φ adopts the form of dot-production due to its simplicity, i.e.

$$\Phi(e_v^{video}, e_k^{trans}) = \langle e_v^{video}, e_k^{trans} \rangle. \tag{1}$$

Eq. (1) is used to calculate the ranking loss in the following training steps.

Training. We expect that the model can learn to rank transitions by their distances with inputs in embedding space. To achieve so, we utilize the triplet margin loss to optimize the similarity between the transition embedding and multi-modal video embedding. For the embedding of each training sample e^{video} with ground truth label c, we define our training objective with triplet loss as

$$\mathcal{L}(e^{video}) = \frac{1}{N-1} \sum_{k \neq c, k \in \{1,\ldots,N\}} \mathcal{T}(e^{video}, e_c^{trans}, e_k^{trans})$$

where $\mathcal{T}$ calculates the triplet margin loss for each triplet $(e^{video}, e_c^{trans}, e_k^{trans})$:

$$\mathcal{T}(a, p, n) = max(\Phi(a, p) - \Phi(a, n) + M, 0).$$

M is the soft margin, a, p and, n are anchor, positive sample, and negative sample, respectively. Φ follows the definition of Eq. (1). In our settings, we take video embedding e^{video} as the anchor, the transition embedding with category c as the positive sample, others transitions as negative samples. The final objective is the average loss over all samples, i.e.

$$\mathcal{L}(E^{video}) = \frac{1}{V} \sum_{v \in \{1,...,V\}} \mathcal{L}(e_v^{video}). \tag{2}$$

By optimizing Eq. (2), the model encourages the similarity between the multi-modal video embedding and its ground-truth transition embedding higher than the similarity between non-matching pairs with a margin of M.

Evaluation. In evaluation, we follow Eq. (1) to measure the matching degree between multi-modal video embedding and candidate transitions embedding. For two neighboring video shots, we sort the similarities of candidate transitions in descending order and select the top one as final result.

6 Experiments

6.1 Implementation Details

Model Details. We employ SlowFast 8×8 [21] as the backbone to extract visual features. The same backbone is also used as the transition embedding network. By default, we freeze SlowFast from stage 1 to stage 3 during our experiments to save memory. We train all models on one machine with 8 NVIDIA V100 GPUs, except the experiment without freezing the SlowFast backbone, in which we use two machines and 16 GPUs in total. For audio modal, we use Harmonic CNN [22] to extract the local audio features around the transition. Then we linearly project the feature of two modals to the same dimension of d_{model} and take them as the input tokens of the multi-modal transformer. The multi-modal transformer consists of two transformer layers with $d_{model} = 2048$ and $n_{head} = 8$. Before matching, we apply linear projections to both the video embedding and pre-trained transition embedding in order to map them into a joint space. Such a projection is experimentally proved to be beneficial as shown in Table 3.

Data Preprocessing and Training. When training the transition embedding network, we uniformly sample 16 frames with the image size of 224×224 from the transition duration as the input. The batch size is set to 256. The training process is 30 epochs in total, and start with a warm up for 5 epoch to raise the learning rate from 1e-6 to the initial learning rate 1e-3, then decay by a factor of 0.1 every 10 epochs. We use the model parameters of the last epoch to generate the transition embedding.

When training the transition recommendation model, we uniformly sample 16 frames with the image size of 224×224 as the visual inputs. The local audio features are extracted using the pre-trained Harmonic CNN before training. Given a time point, the Harmonic CNN generate a feature vector with a dimension of 100 based on the audio in around one second. For the sequential inputs, we set the maximum sequence length to 8. Redundant transitions beyond the maximum length are dropped while zero tensors are padded if the length is less than 8. We use Adam optimizer in all the experiments and set the soft margin $M = 0.3$ in triplet margin loss. The initial learning rate is set to 1e-5, then decay by a factor of 0.1 every 10 epochs. The total training epoch is 30 epochs.

Metrics. For testing the fine-grained performance of our model, we evaluate the individual recommendation results in testing. The commonly used Recall@K metric ($k \in \{1, 5\}$) and Mean Rank are employed as evaluation metrics. Since in our dataset, there is only one ground-truth for each transition, Recall@K indicates the likelihood of hitting the target in top K retrieval results. Mean rank is the averaged rank of all ground-truth transitions, whose math formula is

$$\mathrm{MR} = \frac{1}{|\mathcal{S}|} \sum_{i \in \mathcal{S}} \mathrm{Rank}(i)$$

where $\mathcal{S}$ represents the set of all transitions in test.

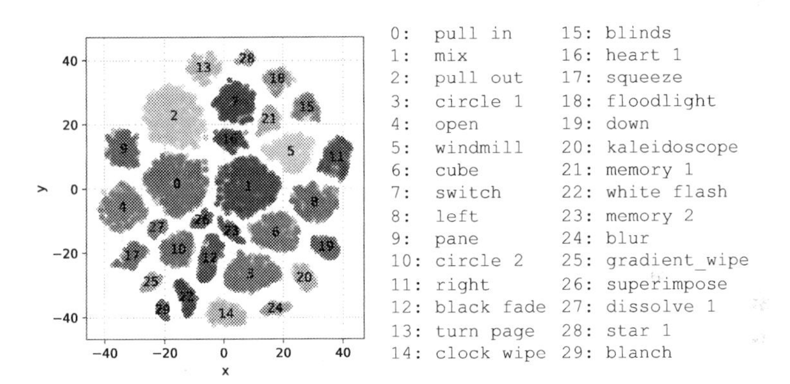

Fig. 5. t-SNE visualization of the transition embedding on the train set after pre-training. We use cosine similarity as the distance metrics when running t-SNE. We drop the outliers and randomly pick 10K samples from the t-SNE output for visualization. We can see that transitions with similar visual effects, such as "left" (8), "right" (11), and "down" (19) are close to each other in the embedding space.

6.2 Extracting Transition Embedding

A transition classification network is trained using annotated transition category in the dataset as the transition embedding network. After training, we use the pre-trained network to extract the embeddings of all transitions in the training set. Normalization is then applied for the embedding as shown in Fig. 3 to convert the embedding into a unit vector. We drop the outliers utilizing the three-sigma rule by assuming the embeddings follow a normal distribution. The embeddings after dropping are visualized in Fig. 5. The remaining embeddings for each category are averaged to generate the final transition embedding. We then demonstrate the effectiveness of the transition embedding by the experiments shown in Table 3. Notably, the transition embedding network is advantageous in extending the model to support new transition categories since retraining the recommendation model is circumvented.

6.3 Ablation Studies and Comparisons

We start by showing the advantages of leveraging contextual and multi-modal information in inputs. Then we verify the effectiveness of the pre-trained transition embedding by comparing it with random initialization. We also compare with the classification method to demonstrate the superiority of retrieval methods on this task. Due to space limit, more details of results are referred to the supplementary material.

Context and Multi-modal. In this experiment, we study the impact of contextual and multi-modal inputs on the recommendation performance, and the result is shown in Table 2. From the first and third rows of Table 2, we can see that sequential inputs introduce the context information to the model, which is helpful for modeling the temporal relations between the transitions. From the second and third rows of Table 2, visual modal inputs perform much better than audio as input, which indicates that visual content is more related to the transition effects than audio. The results in the last three rows demonstrate that the multi-modal inputs can improve the accuracy of recommendations than the single modal inputs.

Table 2. The impact of contextual and multi-modal information to the recommendation results.

Sequential (Context)	Modal		Recall@1	Recall@5	Mean Rank
	Visual	Audio			
	✓		24.12%	66.25%	5.758
✓		✓	19.39%	56.61%	7.012
✓	✓		25.40%	66.33%	5.665
✓	✓	✓	**28.06%**	**66.85%**	**5.480**

Table 3. Advantages of pre-training transition embedding from the visual effects. We demonstrate its effectiveness by comparing it with a random initialized embedding.

Transition Embedding	Projection	Recall@1	Recall@5	Mean Rank
Random initialization		25.67%	66.3%	5.646
Pre-trained transition embedding		26.24%	66.03%	5.623
Pre-trained transition embedding	✓	**28.06%**	**66.85%**	**5.480**

The Effectiveness of Pre-trained Transition Embedding. In this experiment, we study the effectiveness of our proposed pre-trained transition embedding, and the results are shown in Table 3. All the embedding is frozen during training. In the random initialization setting, we use a normalized random embedding as the replacement of the pre-trained transition embeddings. It is observed that the performance of using random embedding is worse compared to our pre-trained embedding. From the results in second and third rows in Table 3, the importance of the linear projection can be demonstrated. We conjecture the linear projection helps learning better mapping between the pre-trained transition embedding and multi-modal input embedding in a shared space (Table 4).

Table 4. Comparing with the classification method.

Methods	Recall@1	Recall@5	Mean Rank
Classification	22.27%	61.82%	6.099
Matching with pre-trained transition embedding	**28.06%**	**66.85%**	**5.480**

Comparing with the Classification Method. In this ablation study, we remove the transition embedding, replace triplet margin loss with cross-entropy, and train the recommendation model utilizing the transition category label. The comparison result is shown in Fig. 4. The classification model performs worse compared with the retrieval model. The reason is that the learned transition embeddings in retrieval model contain richer visual properties of the transitions compared with semantically meaningless one-hot vectors used in the classification model. In addition, the cross-entropy loss may impose excessive punishment for negative categories due to using one-hot ground-truth, thus neglecting the fact that there are similar transitions as the ground-truth transition and they can also be used as favorable alternatives.

6.4 User Study

Since the transition recommendation is subjective, the viewer's feeling is essential to the evaluation. Therefore, we conduct a user study to further verify its effectiveness. Specifically, we collect raw video shots from online copyright free video sources, e.g. videvo.net[1], covering various topics such as travel, life, entertainment, sports, nature, and animals. For each set of video shots, we fix their orders and assign an appropriate background music, leaving only the transitions between neighboring video shots to be added. After selecting video transitions, we use the tool of *CapCut* to connect the raw video shots by transitions, producing the final videos. We compare among following three methods of selecting video transitions.

1. **Weighted random pick.** At each transition point, select the transition category by a random sampling from a multinomial distribution. The probability of each category is its frequency in our collected video transition dataset.
2. **Professional video editor.** We ask a professional editor who has 6 years of video editing to select transitions. He is free to take as long as he wants to select the best transitions depending on his understanding of the given video/audio.
3. **Our method.** The top-1 video transition predicted by our method at each transition point is used as the best selection.

We collect 20 groups of video results in total for user study. Each group contains three videos edited using above three methods respectively. We invite overall 15 non-expert volunteers to participate in the evaluation. They are asked to choose a favorite video from each group (Q1) and rate each video on a scale of 1 to 5 (1 = poor, 5 = excellent, Q2), taking into account the general visual quality of videos and the matching degree between transitions and video/audio. Table 5 and Fig. 6 show the statistics of the results. As shown in Table 5, the videos

Table 5. The statistical results of user study. The inference time is reported as the time cost of our method.

Method	Avg. score	Avg. time (per video)
Weighted random pick	2.96	–
Professional video editor	3.80	7.5 min
Our method	3.76	1.5 s

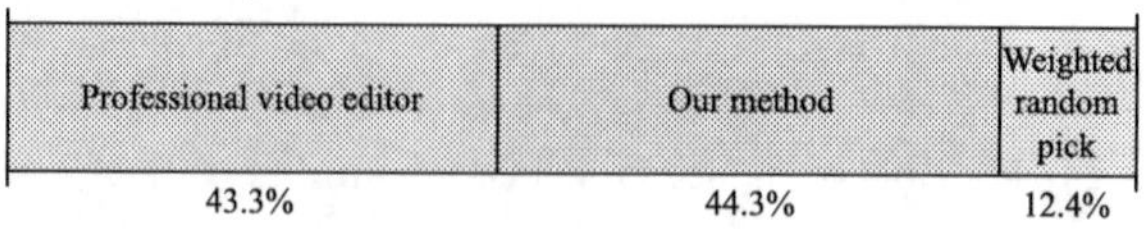

Fig. 6. The voting results for three methods in Q1.

[1] https://www.videvo.net/.

generated using random pick receive the lowest score. Our method is pretty close to the professional editing in terms of average score, but being much more efficient by drastically reducing the average processing time of each video from 7.5 min to only 1.5 s (a **300**× speedup). Interestingly, in Fig. 6 which shows the voting results of Q1, one can see that videos from our methods are slightly more appealing than that from the professional editor. Above experimental results clearly demonstrate the advantages of our method in producing high-quality video transitions recommendations.

7 Conclusions and Future Works

The recent development of online video tools and platforms creates a high demand for a user-friendly video editing experience, which asks for a computational method or artificial intelligent model to lower the barrier, improve efficiency and ensure quality for doing video editing. Therefore, we propose a new task of video transition recommendation (VTR) to automatically recommend transitions based on any visual and audio inputs. We start with building a large-scale transition dataset. Then we formulate VTR as a multi-modal retrieval problem and propose a flexible framework for addressing the task. Through extensive qualitative and quantitative evaluations, we clearly demonstrate the effectiveness of our method. We hope this work can inspire more researchers to work on VTR and bring creativity and convenience to both professionals and non-professional users. Future works include but is not limited to extending the framework to support more video editing effects like video animation, 3D movements, etc., developing more efficient models for mobile deployment and integrating with other video editing techniques to create more comprehensive video editing systems.

Acknowledgment. Yaojie Shen did this work when interning at ByteDance Inc.

References

1. Frey, N., Chi, P., Yang, W., Essa, I.: Automatic non-linear video editing transfer. arXiv preprint arXiv:2105.06988 (2021)
2. Wang, M., Yang, G.W., Hu, S.M., Yau, S.T., Shamir, A.: Write-a-video: computational video montage from themed text. ACM Trans. Graph. **38**(6), 1–177 (2019)
3. Koorathota, S., Adelman, P., Cotton, K., Sajda, P.: Editing like humans: a contextual, multimodal framework for automated video editing. In: 2021 IEEE/CVF Conference on Computer Vision and Pattern Recognition Workshops (CVPRW), IEEE Computer Society, pp. 1701–1709 (2021)
4. Liao, Z., Yu, Y., Gong, B., Cheng, L.: Audeosynth: music-driven video montage. ACM Trans. Graph. (TOG) **34**(4), 1–10 (2015)
5. Pardo, A., Caba, F., Alcázar, J.L., Thabet, A.K., Ghanem, B.: Learning to cut by watching movies. In: Proceedings of the IEEE/CVF International Conference on Computer Vision, pp. 6858–6868 (2021)
6. Hendricks, L.A., Wang, O., Shechtman, E., Sivic, J., Darrell, T., Russell, B.: Localizing moments in video with natural language. In: 2017 IEEE International Conference on Computer Vision (ICCV), IEEE Computer Society, pp. 5804–5813 (2017)

7. Akbari, H., et al.: Vatt: transformers for multimodal self-supervised learning from raw video, audio and text. In: Advances in Neural Information Processing Systems (2021)
8. Karpathy, A., Fei-Fei, L.: Deep visual-semantic alignments for generating image descriptions. IEEE Trans. Pattern Anal. Mach. Intell. **39**(4), 664–676 (2017)
9. Miech, A., Zhukov, D., Alayrac, J.B., Tapaswi, M., Laptev, I., Sivic, J.: Howto100m: learning a text-video embedding by watching hundred million narrated video clips. In: 2019 IEEE/CVF International Conference on Computer Vision (ICCV), pp. 2630–2640. IEEE (2019)
10. Escorcia, V., Soldan, M., Sivic, J., Ghanem, B., Russell, B.: Temporal localization of moments in video collections with natural language. arXiv preprint arXiv:1907.12763 (2019)
11. Schultz, M., Joachims, T.: Learning a distance metric from relative comparisons. In: Advances in Neural Information Processing Systems, vol. 16 (2003)
12. Faghri, F., Fleet, D.J., Kiros, J.R., Fidler, S.: Vse++: improving visual-semantic embeddings with hard negatives. arXiv preprint arXiv:1707.05612 (2017)
13. Schroff, F., Kalenichenko, D., Philbin, J.: Facenet: a unified embedding for face recognition and clustering. In: 2015 IEEE Conference on Computer Vision and Pattern Recognition (CVPR), IEEE Computer Society pp. 815–823 (2015)
14. Hoffer, E., Ailon, N.: Deep metric learning using triplet network. In: Feragen, A., Pelillo, M., Loog, M. (eds.) SIMBAD 2015. LNCS, vol. 9370, pp. 84–92. Springer, Cham (2015). https://doi.org/10.1007/978-3-319-24261-3_7
15. Vaswani, A., et al.: Attention is all you need. In: Advances in neural information processing systems, vol. 30 (2017)
16. Arnab, A., Dehghani, M., Heigold, G., Sun, C., Lučić, M., Schmid, C.: Vivit: a video vision transformer. arXiv preprint arXiv:2103.15691 (2021)
17. Gabeur, V., Sun, C., Alahari, K., Schmid, C.: Multi-modal transformer for video retrieval. In: Vedaldi, A., Bischof, H., Brox, T., Frahm, J.-M. (eds.) ECCV 2020. LNCS, vol. 12349, pp. 214–229. Springer, Cham (2020). https://doi.org/10.1007/978-3-030-58548-8_13
18. Lin, T., Wang, Y., Liu, X., Qiu, X.: A survey of transformers. arXiv preprint arXiv:2106.04554 (2021)
19. Hendricks, L.A., Mellor, J., Schneider, R., Alayrac, J.B., Nematzadeh, A.: Decoupling the role of data, attention, and losses in multimodal transformers. Trans. Assoc. Comput. Linguist. **9**, 570–585 (2021)
20. Lin, X., Bertasius, G., Wang, J., Chang, S.F., Parikh, D., Torresani, L.: Vx2text: end-to-end learning of video-based text generation from multimodal inputs. In: 2021 IEEE/CVF Conference on Computer Vision and Pattern Recognition (CVPR), IEEE Computer Society, pp. 7001–7011 (2021)
21. Feichtenhofer, C., Fan, H., Malik, J., He, K.: Slowfast networks for video recognition. In: 2019 IEEE/CVF International Conference on Computer Vision (ICCV), IEEE Computer Society, pp. 6201–6210 (2019)
22. Won, M., Chun, S., Nieto, O., Serrc, X.: Data-driven harmonic filters for audio representation learning. In: ICASSP 2020–2020 IEEE International Conference on Acoustics, Speech and Signal Processing (ICASSP), pp. 536–540. IEEE (2020)
23. Lei, J., Wang, L., Shen, Y., Yu, D., Berg, T.L., Bansal, M.: Mart: memory-augmented recurrent transformer for coherent video paragraph captioning. arXiv preprint arXiv:2005.05402 (2020)

Online Segmentation of LiDAR Sequences: Dataset and Algorithm

Romain Loiseau[1,2(✉)], Mathieu Aubry[1], and Loïc Landrieu[2]

[1] LIGM, Ecole des Ponts, Univ Gustave Eiffel, CNRS, 77454 Marne-la-Vallée, France
romain.loiseau@enpc.fr
[2] LASTIG, Univ Gustave Eiffel, IGN-ENSG, 94160 Saint-Mandé, France

Abstract. Roof-mounted spinning LiDAR sensors are widely used by autonomous vehicles. However, most semantic datasets and algorithms used for LiDAR sequence segmentation operate on 360° frames, causing an acquisition latency incompatible with real-time applications. To address this issue, we first introduce HelixNet, a 10 billion point dataset with fine-grained labels, timestamps, and sensor rotation information necessary to accurately assess the real-time readiness of segmentation algorithms. Second, we propose Helix4D, a compact and efficient spatio-temporal transformer architecture specifically designed for rotating LiDAR sequences. Helix4D operates on acquisition slices corresponding to a fraction of a full sensor rotation, significantly reducing the total latency. Helix4D reaches accuracy on par with the best segmentation algorithms on HelixNet and SemanticKITTI with a reduction of over 5× in terms of latency and 50× in model size. The code and data are available at: https://romainloiseau.fr/helixnet.

Keywords: LiDAR · Transformer · Autonomous driving · Real-time · Online segmentation

1 Introduction

Due to their low acquisition latency and high precision, rotating LiDAR sensors are among the most prevalent sensors for autonomous vehicles [38]. The acquired sequences of 3D points exhibit a complex structure in which the temporal and spatial dimensions are entangled through the rotation of the sensor around a reference point in motion; see Fig. 1. However, this structure is often not reflected in the formatting of open-access LiDAR datasets [3,25,28], which are discrete sequences of range images, or frames, each corresponding to a 360° degree arc around the sensor. Consequently, most LiDAR semantic segmentation methods operate on one or several such frames at the same time, in the image [12] or point cloud [42,47,50] format. However, waiting for an entire frame to be acquired

S. Avidan et al. (Eds.): ECCV 2022, LNCS 13698, pp. 301–317, 2022.
https://doi.org/10.1007/978-3-031-19839-7_18

introduces an unavoidable latency of more than 100 ms on top of the processing time, excluding applications for high-speed or urban driving. In this paper, we address this issue by introducing (i) HelixNet, the largest available LiDAR dataset, and whose fine-grained point information allows for the realistic real-time evaluation of segmentation methods, and (ii) Helix4D, a spatio-temporal transformer designed for the efficient segmentation of LiDAR sequences.

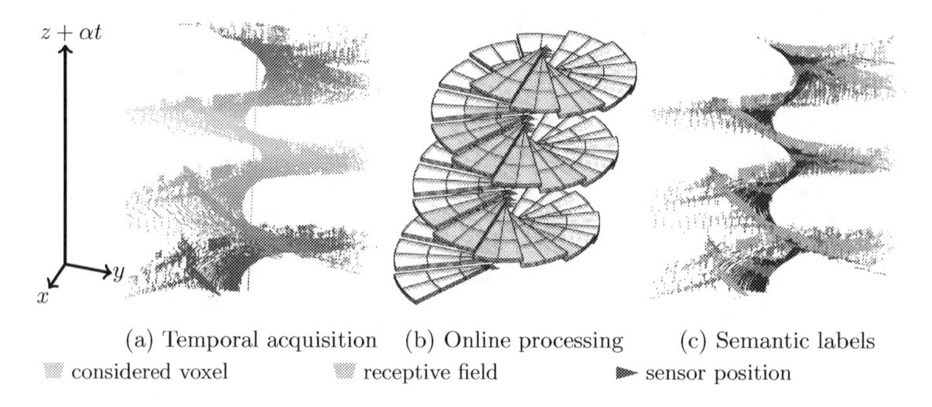

Fig. 1. Online LiDAR segmentation. The 3D point sequences of rotating LiDAR data of our proposed dataset HelixNet follow a complex helix-like structure in space and time, represented in (a) by using the vertical axis for both time and elevation. We propose an efficient spatio-temporal transformer to process angular slices of data centered on the sensor's position. The slices are partitioned into voxels, each attending other voxels from past slices to build a large spatio-temporal receptive field (b) Our proposed model can segment the LiDAR point stream (c) with state-of-the-art accuracy and in real-time.

Our dataset HelixNet, has several key advantages compared to standard datasets such as SemanticKITTI [3], see Table 1. By organizing points with respect to sensor rotation and reporting their precise release times, we can accurately benchmark the real-time readiness of leading state-of-the-art LiDAR sequence segmentation algorithms. Furthermore, the pointwise sensor orientation allows us to split the data into slices of acquisition corresponding to a fraction of the sensor's rotation. These slices can be processed sequentially by our proposed network Helix4D, resulting in a lower acquisition latency and a more realistic scenario for autonomous driving. Based on a spatio-temporal transformer designed explicitly for LiDAR sequences, Helix4D is more than 50 times smaller than the current best semantic segmentation architectures and reaches state-of-the-art performance with significantly reduced latency.

2 Related Work

We present an overview of the existing LiDAR datasets related to autonomous driving, and a summary of the recent developments in 3D semantic segmentation.

Autonomous Driving 3D Datasets. As autonomous driving becomes an increasingly realistic prospect, multiple datasets have been proposed to evaluate the performance of perception algorithms [11,33]. In addition to cameras, rotating LiDARs have become one of the most prevalent sensors mounted on autonomous vehicles due to their high accuracy, low latency, and steadily decreasing prices [38]. ApolloScape [23], DublinCity [51], and TerraMobilita/iQmulus [43] have been acquired with LiDAR setups that offer scans of urban environments with high precision and density. However, the vertical orientation of the emitters is not compatible with real-time road perception. Several prominent datasets such as NuScene [6] or the very large ONCE dataset [31] provide only object-level annotations (*i.e.* boxes).

Table 1. Embarked LiDAR datasets with semantic point annotations. With over 8.8B annotated 3D points, HelixNet is 70% larger than SemanticKITTI, and includes more diverse scenes spanning 6 different French cities. Contrary to other datasets, HelixNet arranges points with respect to the sensor rotation and contains fine-grained information about their release time.

Dataset	Labels	Frames	Classes	Span	Format
HelixNet (Ours)	8.85B	78k	9	6 cities	Sensor rotation
SemanticKITTI [3,16]	5.2B	43k	19	1 city	Frame
Rellis3D [25]	1.5B	13k	16	1 city	Frame
KITTI-360 [28]	1.0B	81k	37	1 city	Frame
A2D2 [17]	387M	41k	38	3 cities	Frames
Paris-Lille-3D [37]	143M	N/A	50	2 cities	Multi-frame
Toronto3D [41]	78M	N/A	8	1 city	Multi-frame

This paper focuses on semantic segmentation algorithms for roof-mounted rotating LiDAR sensors. In Table 1, we report several key characteristics of such datasets [3,16,17,25,28,37,41]. Our proposed dataset HelixNet is 70% larger than SemanticKITTI [3,16], and spans 6 cities and various environments. In contrast to previously released datasets, the 3D points of HelixNet are given with respect to the sensor rotation and in the order in which they are made available. This last point proves crucial for evaluating the precision and latency of segmentation algorithms in a setting that is compatible with real-time inference.

Deep Semantic Segmentation of 3D Point Clouds. The development of specific deep architectures for the semantic segmentation of 3D point clouds has led to a tremendous increase in performance [21]. The first set of methods that operate on rotating LiDAR sequences processes data in range image format [12,27,40]. Taking advantage of advances in the implementation of sparse convolutions [9, 18], a second set of methods uses fine grids in polar [47], Cartesian [8,42] or cylindrical [22,50] coordinates. A third kind of approach proposes exploiting the temporal dimension of LiDAR acquisitions by *stacking* contiguous frames [2,9]. Observing that cylindrical partitions better capture the geometry of rotating LiDAR acquisition, our proposed Helix4D builds on the idea of Cylinder3D [50] and adds a temporal component to the architecture.

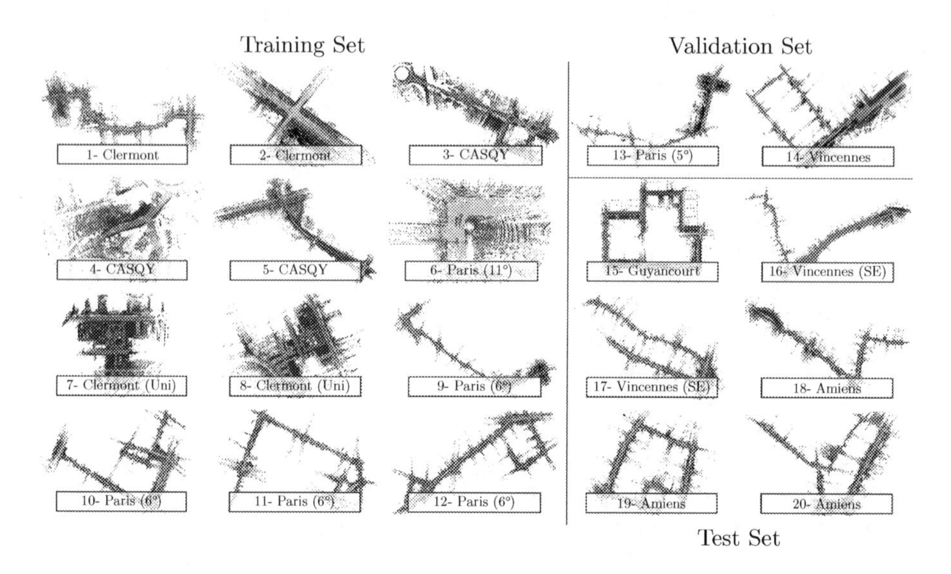

Fig. 2. Coverage from HelixNet. We split the acquisitions into 12 training, 2 validation, and 6 testing sequences. HelixNet contains diverse scenes in various urban environments from static or mobile sensors.

Due to their remarkable performance and scalability, transformers [44] have quickly been adapted from text processing to images [7,13,30,39], videos [1], or meshes [29]. Transformers are also well suited to handle unordered sets, such as 3D point clouds [20,48]. In particular, their scalability can be leveraged to achieve large receptive fields [32,34] and more discriminative features [5] than purely convolutional approaches. Transformers can also efficiently process complex temporal [26,45] and spatio-temporal [15] sequences. In the wake of hybrid convolution-transformer models [10,14,19], our proposed Helix4D combines efficient cylindrical convolutions with a simplified spatio-temporal transformer architecture operating at low resolution.

3 HelixNet: A Dataset for Online LiDAR Segmentation

We introduce HelixNet, a new large-scale and open-access LiDAR dataset intended for the evaluation of real-time semantic segmentation algorithms. In contrast to other large-scale datasets, HelixNet includes fine-grained data on sensor rotation and position, as well as point release time.

General Characteristics. As seen in Fig. 2, HelixNet contains 20 sequences of 3D points, each corresponding to 6 to 7 min of continuous acquisition, for a total of 129 min. Scanning was performed by an HDL-64E Velodyne rotating LiDAR [24] mounted on a mobile platform [35]. As shown in Fig. 3, HelixNet covers multiple cities and a wide variety of environments such as a university campus, dense historical centers, and a highway interchange. With a total of 10 billion points across 78 800 frames and 8.85 billion individual labels, HelixNet is the largest densely annotated open-access rotating LiDAR dataset by a factor of 1.7 as shown in Table 1. HelixNet follows the file format of SemanticKITTI [3], allowing researchers to evaluate existing code with minimal effort.

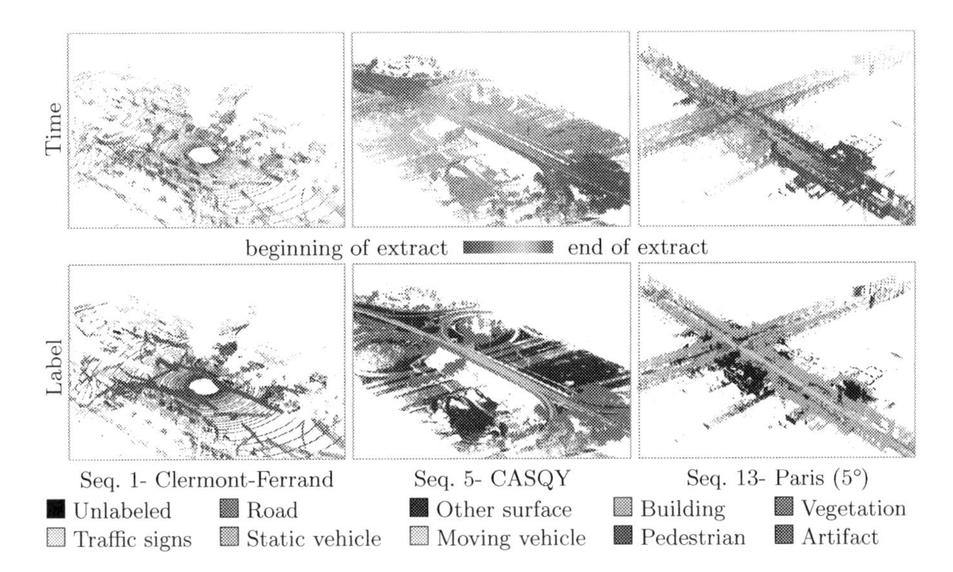

Fig. 3. Extracts from HelixNet. Our proposed dataset contains various urban scenes from motorway to pedestrian plazas and historical centers. In the first row, we represent extracts of 15 to 30s of acquisition colored according to the point release time. In the second row, we represent the point semantic labels.

We use a 9-classes nomenclature: *road* (16.4% of all points), *other surface* (22.0%), *building* (31.3%), *vegetation* (8.5%), *traffic signs* (1.6%), *static vehicle* (4.9%), *moving vehicle* (2.1%), *pedestrian* (0.9%), and *acquisition artifact* (0.05%). Points without labels correspond to either *un-annotated* (6.2%)

parts of the clouds due to their ambiguity, or point without echos (6.1%). Compared to fine-grained classes such as the ones used by SemanticKITTI [3] or Paris-Lille3D's [37], our focused nomenclature limits class imbalance and makes macro-averaged metrics more stable.

Each point is associated with the 9 following values: (1–3) Cartesian coordinates in a fixed frame of a reference (4–6) cylindrical coordinate relative to the sensor at the time of acquisition (7) intensity (8) fiber index, and (9) packet output time. As detailed in the next paragraph, the last two features are not typically available in large-scale datasets and cannot be inferred.

Sensor-Based Timing and Grouping. A rotating LiDAR consists of a set of lasers—or fibers—arranged on a rotating sensor head. The lasers send periodic pulses of light whose return times give the position of the impact points relative to the sensor. In the context of autonomous driving, these sensors are typically deployed on a moving platform and capture 3D points with centimetric accuracy. The sensor releases the data stream as a discrete temporal sequence of *packets* of 3D points. For an HDL-64E LiDAR, each packet contains 6×64 points, corresponding to around $1°$ rotation of the sensor. To represent the real-time operational setting of autonomous driving, we associate with each point the timestamp of its *packet output event*, *i.e.* the instant the packet is available and not the acquisition time of the point. The latency between the acquisition of the first point and the complete transfer of its packet is $278\,\mu$s. Although small compared to acquisition and inference times, this more rigorous timing constitutes a step towards a more realistic evaluation setting of segmentation algorithms of LiDAR sequences.

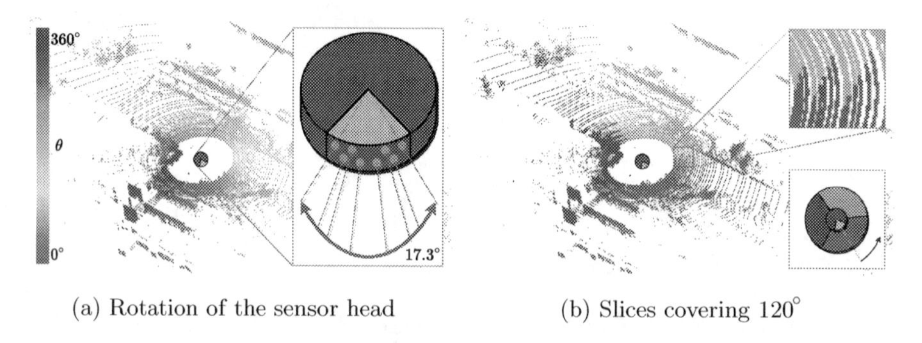

(a) Rotation of the sensor head (b) Slices covering 120°

Fig. 4. Sensor Acquisition Geometry. We represent in (a) the acquisition of a rotating sensor, which is split into ⅓ turn slices in (b) As the Laser emitters position forms an angle of over $17.3°$ around the sensor head, taking slices with respect to the sensor rotation θ results in a jagged profile.

On top of its absolute position, we associate with each individual point its cylindrical coordinates relative to the position of the sensor at the exact time of its acquisition. This differs from other datasets such as SemanticKITTI [3], which gives the relative position of all points but the absolute position of the sensor only once per frame. While sensor movement can be interpolated, the vehicle trajectory might not be linear and the sensor head rotates. For comparison, at 50 km/h, the sensor moves more than 1.4 m during each rotation.

LiDAR sequences are typically split into frames containing points that cover a 360° degree arc around the sensor. However, the acquisition geometry makes this grouping artificial. Indeed, the fibers (*i.e.* the individual lasers) do not all face the same direction: they are arranged around the sensor's heads at different angles, with a range of more than 17.3°. This means that the points within a packet are not vertically aligned but present a jagged profile as seen in Fig. 4. In order to obtain frames with *straight edges* such as those of SemanticKITTI [3], we would have to consider an acquisition over a sensor rotation of 377°, adding a further 5 ms of latency. Contrary to other datasets, HelixNet contains the index of the emitter of each point and organizes the points with respect to the angle of the sensor itself This allows us to easily build frames or frame portions that are directly consistent with the rotation *of the sensor head itself.* This is important for measuring the real latency of segmentation methods and, as described in the next section, contributes to the efficiency of our proposed network.

4 Helix4D: Fast LiDAR Segmentation with Transformers

We consider a sequence of 3D points acquired by a rotating LiDAR on a mobile platform, which we split into chronologically ordered slices of acquisition. As represented in Fig. 5, we process each slice with a U-Net architecture [36] with cylindrical convolutions [50]. At the lowest resolution, a spatio-temporal transformer network connects neighboring voxels in space and time, resulting in a large receptive field. We first describe the construction of slices, then our cylindrical U-Net, and finally the transformer module.

4.1 Temporal Slicing

Instead of processing the data frame-by-frame, we propose to split the sequence into slices covering a fixed portion of the sensor rotation, resulting in a shorter acquisition time and a lower latency. Each point i of the sequence is characterized by the angular position θ_i *of the sensor head* at its exact time of acquisition. The points are sorted in chronological acquisition order *i.e.* $\theta_i \leq \theta_j$ if $i < j$. We partition the sequence into groups of contiguous points called slices, acquired during a portion $\Delta\theta \in]0, 2\pi]$ of a full rotation of the sensor itself. Choosing $\Delta\theta = 2\pi$ corresponds to the classic frame-by-frame setting and implies an acquisition latency of 104 ms in HelixNet or SemanticKITTI [3]. A slice size of $\Delta\theta = 2\pi/5$ leads to an acquisition latency of 21 ms, which is more conducive to real-time processing of driving data.

4.2 Cylindrical U-Net

Inspired by the Cylinder3D model [50], we first discretize each slice along a fine cylindrical partition $\mathsf{grid}\,(1)$. Each point i is associated with a descriptor x_i^{point} based on its intensity, relative position with respect to the sensor in Cartesian and cylindrical coordinates, and its offset with respect to the center of its voxels in $\mathsf{grid}\,(1)$. We compute the point feature f_i^{point} by applying a shared Multi-Layer Perceptron (MLP) $\mathcal{E}^{\mathsf{point}}$ to x_i^{point} for all points i in the slice. The resulting f_i^{point} are then maxpooled with respect to the voxels of $\mathsf{grid}\,(1)$ to serve as input to a convolutional encoder $\mathcal{E}^{\mathsf{grid}}$. The network $\mathcal{E}^{\mathsf{grid}}$ is composed of sparse cylindrical convolutions [18] and strided convolutions for downsampling. $\mathcal{E}^{\mathsf{grid}}$ produces a set of L sparse feature maps $f^{\mathsf{grid}(1)}, \cdots, f^{\mathsf{grid}(L)}$ with decreasing resolutions:

$$f_i^{\mathsf{point}} = \mathcal{E}^{\mathsf{point}}\left(x_i^{\mathsf{point}}\right) \tag{1}$$

$$f^{\mathsf{grid}(1)}, \cdots, f^{\mathsf{grid}(L)} = \mathcal{E}^{\mathsf{grid}}\left(\mathrm{maxpool}\left(f^{\mathsf{point}}\right)\right) \;, \tag{2}$$

where maxpool is performed with respect to $\mathsf{grid}\,(1)$. At the lowest resolution $\mathsf{grid}\,(L)$, we apply the transformer-based module $\mathcal{T}$ presented in the next subsection to the feature map $f^{\mathsf{grid}(L)}$ to obtain the coarse cylindrical map $g^{\mathsf{grid}(L)}$:

$$g^{\mathsf{grid}(L)} = \mathcal{T}\left(f^{\mathsf{grid}(L)}\right) \;. \tag{3}$$

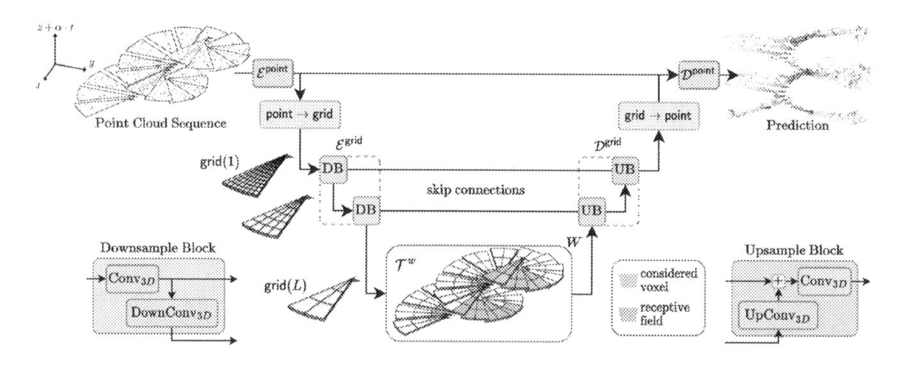

Fig. 5. Helix4D architecture. A point sequence is split into angular slices, whose points are encoded by $\mathcal{E}^{\mathsf{point}}$ and pooled along a fine-grained cylindrical partition. A convolutional encoder $\mathcal{E}^{\mathsf{grid}}$ yields feature maps at lower resolutions. We apply W consecutive spatio-temporal transformer blocks $\mathcal{T}^w$ on the coarse voxels, with attention spanning across current and past slices. The resulting features are up-sampled to full resolution with a convolutional decoder $\mathcal{D}^{\mathsf{grid}}$ using the encoder's maps at intermediate resolutions through skip connections. Finally, the grid features are allocated to the points, which are classified by $\mathcal{D}^{\mathsf{point}}$.

The decoder $\mathcal{D}^{\mathsf{grid}}$ combines cylindrical convolutions and strided transposed convolutions to map $g^{\mathsf{grid}(L)}$ to a feature map $g^{\mathsf{grid}(1)}$ at the highest resolution, and uses the maps $f^{\mathsf{grid}(L-1)}, \cdots, f^{\mathsf{grid}(1)}$ through residual skip connections. We concatenate for each point i the descriptor $g^{\mathsf{grid}(1)}(i)$ of its voxel in $\mathsf{grid}\,(1)$ and its point feature f_i^{point}. Finally, the point decoder $\mathcal{D}^{\mathsf{point}}$ associates a vector of class scores c_i^{point} with each point i:

$$g^{\mathsf{grid}(1)} = \mathcal{D}^{\mathsf{grid}}\left(g^{\mathsf{grid}(L)}, f^{\mathsf{grid}(L-1)}, \cdots, f^{\mathsf{grid}(1)}\right) \tag{4}$$

$$c_i^{\mathsf{point}} = \mathcal{D}^{\mathsf{point}}\left(\left[g^{\mathsf{grid}(1)}(i), f_i^{\mathsf{point}}\right]\right), \tag{5}$$

where $[\,\cdot\,]$ is the channelwise concatenation operator. The network is supervised by the cross-entropy and Lovász-softmax [4] losses directly on the point prediction, without class weights.

Our approach differs from Cylinder3D [50] by relying on simple $3\times3\times3$ sparse cylindrical convolutions instead of asymmetrical convolutions and dimension-based context modeling. Furthermore, we do not use voxel-wise supervision.

Our simplified architecture results in a lighter computational and memory load, but can still learn rich spatio-temporal features thanks to the addition of the transformer module described below.

4.3 Spatio-Temporal Transformer

We denote by $\mathcal{V}$ the set of non-empty voxels at the lowest resolution $\mathsf{grid}\,(L)$ *for all slices of the considered sequence.* We associate with each voxel v of $\mathcal{V}$ a feature f_v^{voxel} defined as the value of $f^{\mathsf{grid}(L)}$ at v. We remark that f^{voxel} can be ordered as a non-strictly ordered time sequence, and propose to successively apply W independent transformer blocks $\mathcal{T}^1, \cdots, \mathcal{T}^W$ whose architecture is described below. We denote by g^{voxel} the resulting spatio-temporal voxel representation:

$$g^{\mathsf{voxel}} = \mathcal{T}^W \circ \cdots \circ \mathcal{T}^1(f^{\mathsf{voxel}})\,. \tag{6}$$

We associate each voxel v of $\mathcal{V}$ with the absolute position (X_v, Y_v, Z_v) of its center, the release time T_v of its first point, and the index I_v of the sensor rotation of its corresponding slice. In order to use a sparse attention scheme, we define for each voxel v a spatio-temporal mask $M(v)$ characterized by a radius R and a set of rotation offsets $P \subset \mathbb{N}$:

$$M(v) = \{u \mid \|(X_v, Y_v, Z_v) - (X_u, Y_u, Z_u)\| < R\,,\, I_v - I_u \in P\}\,. \tag{7}$$

In the context of autonomous driving, we choose $R = 6\,\mathrm{m}$ and $P = \{0, 5, 10\}$. With a standard rotation speed of 10Hz, this corresponds to considering slices 0.5 and 1 seconds in the past along with the current one. See Fig. 6 for an illustration of the receptive field and attention maps.

Simplified Transformer Block. We now define a single transformer block $\mathcal{T}^w$ with H heads operating on a sequence of voxel features f^{voxel} of dimension D. For each head h and each voxel v, we apply the following operations:

(i) A single linear layer $\mathcal{L}^h$ generates both a key k_v^h of dimension K and a value val_v^h of dimension D/H.
(ii) For all voxels u in the mask $M(v)$, we define the compatibility score $y_{u,v}^h$ as the cross-product between keys and with a learned relative positional encoding $\text{PE}^h(u,v)$.
(iii) The cross-voxel attention $a_{u,v}^h$ is obtained with a scaled softmax.
(iv) The values val_u^h of voxels in $M(v)$ are averaged into a vector $\tilde{f}_v^h$ using their respective cross-voxel attention as weights.
(v) The vectors $\tilde{f}_v^h$ are concatenated channelwise across heads and added to the input of the block to define its output.

These operations can be summarized as follows:

$$k_v^h, \text{val}_v^h = \mathcal{L}^h\left(f_v^{\text{voxel}}\right) \tag{8}$$

$$y_{u,v}^h = \left(k_v^h\right)^\intercal \left(k_u^h + \text{PE}^h(u,v)\right) \quad \text{for } u \in M(v) \tag{9}$$

$$\left\{a_{u,v}^h\right\}_{u\in M(v)} = \text{softmax}\left(\left\{y_{u,v}^h\right\}_{u\in M(v)} / \sqrt{K}\right) \tag{10}$$

$$\tilde{f}_v^h = \sum\nolimits_{u\in M(v)} a_{u,v}^h \text{val}_u^h \tag{11}$$

$$\mathcal{T}^w(f^{\text{voxel}})_v = f_v^{\text{voxel}} + [\tilde{f}_v^1, \cdots, \tilde{f}_v^H]\,. \tag{12}$$

Our design is similar to the classical transformer architecture but uses keys as queries to save memory and computation. We also do not use feed-forward networks after averaging the values: the only learnable part of a block $\mathcal{T}^w$ is its linear layers $\mathcal{L}^h$ and its relative positional encoding PE^h.

Since g^{voxel} only requires information about the voxels of the current and past slices, it can be computed sequentially for all slices in the order in which the sensor releases them. For a given slice, the voxel map $g^{\text{grid}(L)}$ for non-empty voxels is given by the values of g^{voxel}, and set to zero otherwise. To save computation at inference time, we store in memory the keys, values, and absolute positions of the voxels in past slices with a fixed buffer of $\max(P)$ rotations. This allows us to allocate a large spatio-temporal receptive field to each voxel without supplementary computations.

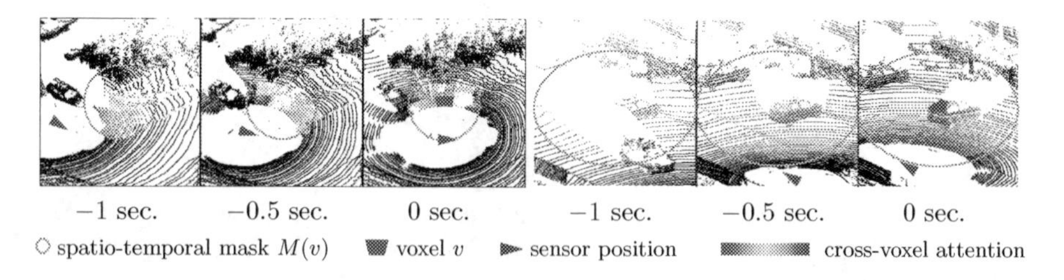

Fig. 6. Spatio-temporal attention. We represent the spatio-temporal mask and attention score of one head of the transformer for two different voxels. The network gathers information from different frame offsets P as the sensor moves.

Relative Positional Encoding. We propose to learn relative positional vectors $\text{PE}^h(u, v)$ that encode the spatio-temporal offset $(X_u, Y_u, Z_u, T_u) - (X_v, Y_v, Z_v, T_v)$ between voxels u and v for each transformer block w independently. Inspired by the work of Wu *et al.* [46], we first discretize the offsets along each dimension $d \in \{X, Y, Z, T\}$ with B_d irregular bins. For each dimension d and head h, we learn B_d weight vectors of size K. We define the functions $\text{PE}^h_d : \mathbb{R} \mapsto \mathbb{R}^K$ that map the d-dimension of an offset to the vector associated with its corresponding bin. The positional encoding between two voxels u and v is the sum of the vectors corresponding to their discretized offsets in each dimension:

$$\begin{aligned} \text{PE}^h(u, v) = \text{PE}^h_X(X_u - X_v) \ + \ \text{PE}^h_Y(Y_u - Y_v) \\ + \text{PE}^h_Z(Z_u - Z_v) \ + \ \text{PE}^h_T(T_u - T_v) \ . \end{aligned} \tag{13}$$

Relative positional encoding vectors are used directly in the calculation of the compatibility score, as given in (9). Additional details on positional encoding are given in the supplementary material.

5 Evaluating Online Semantic Segmentation

We evaluate the performance and inference time of our approach and other state-of-the-art methods in both online and frame-by-frame settings. We use our proposed dataset HelixNet and the standard SemanticKITTI dataset.

Online Evaluation Setting. We aim at evaluating the real-time readiness of rotating LiDAR semantic segmentation algorithms in the context of autonomous driving. The total latency of a model is determined by its inference speed and also the time it takes to acquire its input. Operating on full frames requires at least 104 ms of acquisition, which is incompatible with realistic autonomous driving scenarios. Instead, we propose an online evaluation setting using the slices defined in Sect. 4.1. By default, we use a slice size of a fifth turn of the sensor head: $\Delta\theta = 2\pi/5$, corresponding to 21 ms of acquisition.

Slices are processed sequentially. We define the inference latency of a segmentation method as the average time between the release of the last point of a slice and its segmentation. To meet the real-time requirement, inference must be faster than the acquisition of a slice. Slower processing would cause the classification to continuously fall behind. Although thinner slices directly reduce acquisition latency, they also make the real-time requirement more strict: as a full turn must be processed in less than 104 ms, a fifth turn must be in at most 21 ms.

Adapting SemanticKITTI. SemanticKITTI [3,16] contains 43 552 frames along 22 sequences of LiDAR scans densely annotated with 19 classes. In contrast to HelixNet, SemanticKITTI is not formatted with respect to the sensor rotation and only gives the acquisition time and sensor position once per frame. To measure the latency, we make the following approximation: (i) the fibers are assumed to be vertically aligned, meaning that the angle of the points is the same as the sensor's; (ii) we interpolate the acquisition time of points between frames from their angular positions; (iii) we use the acquisition time as release time. To obtain the absolute positions of the voxels, we assume that the sensor jumps between the positions given by the camera poses for each frame. In our open-source implementation, we provide an adapted dataloader allowing methods already running on SemanticKITTI to be evaluated in the online setting with minimal adaptation.

Adapting Competing Methods. To evaluate the semantic segmentation performance and latency of other segmentation algorithms in the online setting, we process the point clouds corresponding to each slice independently and sequentially. This approach restricts the spatial receptive field to the extent of the slices. However, as the sensor moves, it is not straightforward to add past slices whose relative positions may no longer be valid. By explicitly modeling the spatio-temporal offset between voxels, Helix4D does not suffer from this limitation.

We selected five segmentation algorithms with open-source implementations and trained models for SemanticKITTI. SalsaNeXt [12] uses range images, PolarNet [47] and panoptic PolarNet [49] a bird's eye view polar grid, SPVNAS [42] a regular grid, and Cylinder3D [50] a cylindrical grid. We do not consider methods that stack frames as their structure and resulting latency is incompatible with the online setting. When using SemanticKITTI, we evaluate the provided pre-trained models on the validation set. On HelixNet, we retrain the models from scratch using the procedure of their official repository. We removed all test-time

Table 2. Semantic segmentation results. Performance of Helix4D and competing approaches on HelixNet and on the validation set of SemanticKITTI⋆, in the frame-by-frame and online setting. We report the mean Intersection-over-Union (mIoU) and the inference time in ms. Methods meeting the real-time requirement are indicated with ✓ and those who do not with ✗. ⋆ SemanticKITTI is denoted as SK. Measuring the latency on this dataset requires making non-realistic approximations about the fiber position.

Method	Size	Full frame ●		104 ms	⅕ frame ◗		21 ms
	$\times 10^6$	HelixNet	SK⋆	Inf. (ms)	HelixNet	SK⋆	Inf. (ms)
SalsaNeXt [12]	6.7	69.4	55.8	**23** ✓	68.2	55.6	**10** ✓
PolarNet [47]	13.6	73.6	58.2	49 ✓	72.2	56.9	36 ✗
Pan. PolarNet [49]	13.7	—	64.5	50 ✓	—	60.3	44 ✗
SPVNAS [42]	10.8	73.4	64.7	73 ✓	69.9	57.8	44 ✗
Cylinder3D [50]	55.9	76.6	**66.9**	108 ✗	75.0	65.3	54 ✗
Helix4D (Ours)	**1.0**	**79.4**	66.7	45 ✓	**78.7**	**66.8**	19 ✓

Table 3. HelixNet Semantic Segmentation Scores. We report the IoU for each class of HelixNet evaluated in the online setting with slices of 72°.

Method	Road	Other surface	Building	Vegetation	Traffic signs	Static vehicle	Moving vehicle	Pedestrian	Artifact	Avg
SalsaNeXt [12]	84.4	76.1	88.7	70.7	61.4	58.6	35.1	68.5	69.7	68.2
PolarNet [47]	86.2	77.9	91.2	77.9	63.2	64.8	35.4	68.1	84.8	72.2
SPVNAS [42]	80.5	77.1	93.0	81.8	68.0	60.9	36.9	71.7	59.0	69.9
Cylinder3D [50]	85.3	78.4	93.5	83.9	66.2	63.3	35.7	77.7	90.9	75.0
Helix4D (Ours)	**87.8**	**82.5**	**94.0**	**84.4**	**68.9**	**72.3**	**46.4**	**78.8**	**93.3**	**78.7**

augmentations that resulted in prohibitive inference time. All methods are evaluated on the same workstation using a NVIDIA TESLA V100 32Go GPU.

Analysis. In Table 2, we report performance in frame-by-frame and online setting with slices of 72°, for Helix4D and competing methods, for HelixNet and SemanticKITTI. We observe that Helix4D yields state-of-the-art accuracy, with mIoU scores only matched by Cylinder3D [50]. However, Cylinder3D is 50 times larger in terms of parameters and twice slower, not meeting the real-time requirement even in the full frame setting. As reported in Table 3, distinguishing moving vehicles in HelixNet is particularly difficult. Our approach even largely outperforms Panoptic PolarNet despite this method using instance annotation as supervision, preventing us from evaluating on HelixNet. Helix4D yields significantly improved scores thanks to its larger spatio-temporal receptive fields: 14 m and 1000 ms *vs.* 8 m and 21 ms for Cylinder3D for a fifth rotation.

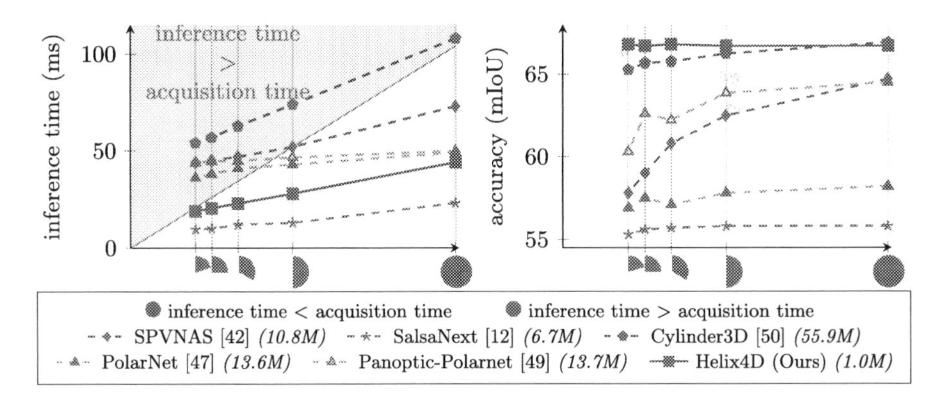

Fig. 7. Influence of slice size. We plot the processing time (left, in ms) and precision (right, in mIoU) of different methods with respect to the considered size of slices, estimated on the validation set of SemanticKITTI [3]. Methods whose inference time is slower than the acquisition time of the slice (red shaded area) do not meet the real time requirement.

Table 4. Ablation study. We report the speed and accuracy of several modification of our Helix4D on the validation set of SemanticKITTI.

Method	Size	Full Frame ●	104 ms	⅕ Frame ▶	21 ms
	$\times 10^3$	mIoU	Inf. (ms)	mIoU	Inf. (ms)
Helix4D	985	**66.7**	45 ✓	**66.8**	19 ✓
(a) Asymmetric Convolutions	1171	66.6	56 ✓	66.6	31 ✗
(b) Cylindrical U-Net	985	58.6	**22** ✓	60.2	**16** ✓
(c) Slice-by-Slice	985	62.9	29 ✓	62.6	19 ✓
(d) w. Queries	993	65.2	45 ✓	64.8	20 ✓
(e) w/o. Positional Encoding	983	64.3	41 ✓	64.1	18 ✓
(f) Helix4D Tiny	306	65.3	45 ✓	64.9	17 ✓

In the online setting, only two approaches meet the real-time requirement: SalsaNeXt [12] and Helix4D. Our approach outperforms SalsaNeXt by over 10 mIoU points in both the full frame and the on-line settings. In short, Helix4D is as accurate as the largest and slowest models with an inference speed comparable to that of the fastest and less accurate models. The total latency (acquisition plus inference time) of our model evaluated online is 40 ms (21 + 19 ms), and reaches the same performance as Cylinder3D evaluated on full frame with a latency of 212 ms (104 + 108 ms), an acceleration of more than 5 folds.

In Fig. 7, we report the inference time and mIoU for different slice sizes. Due to various overheads, the inference time appears in an affine relationship with the size of slices, making the real-time requirement stricter for smaller slices. Due to its very design, the performance of Helix4D is not affected by the slice size. In contrast, competing methods perform worse with smaller slices.

Ablation Study. We assess on SemanticKITTI the impact of different design choices by evaluating several alterations of our method, reported in Table 4.
(a) Asymmetric Convolutions: we replace the $3 \times 3 \times 3$ convolutions in our U-Net with the convolution design proposed by Cylinder3D [50]. We did not observe a significant change in performance and an increase in run-time of 50%, failing the real-time requirement for slices of 72°.
(b) Cylindrical U-Net: we replace the transformer by a $1{\times}1{\times}1$ convolution on the voxels of the lowest resolution. We observe a slight decrease in run-time and a significant drop of over 6 mIoU points. This result shows that the transformer is able to learn meaningful spatio-temporal features at low resolution.
(c) Slice-by-Slice: we restrict the mask $M(v)$ of each voxel to its current slice. This reduction in the temporal receptive field results in a drop of 4 mIoU points, without any appreciable acceleration.
(d) w. Queries: we modify our simplified transformer to associate a query for each voxel along with keys and values, and use key-queries compatibilities. This does not affect the run-time and slightly decreases the performance.

(e) w/o. Positional Encoding: we remove the relative positional encoding PE in the calculation of compatibilities in Eq. (9). This leads to a slightly decreased run time, but decreases performance by more than 2.5 points. This illustrates the advantage of explicitly modeling the spatio-temporal voxel offsets.
(f) Helix4D Tiny: we replace the learned pooling in our U-Net with maxpools and use narrower feature maps for a total of $306k$ parameters. This method only performs two points under Helix4D with a third of its parameters.

6 Conclusion

In this paper, we introduced a novel online inference setting for the semantic segmentation of sequences of rotating liDAR 3D point clouds. Our proposed large-scale dataset HelixNet contains specific sensor information that allows a rigorous evaluation of the performance and latency of segmentation methods in our online setting. We also introduced Helix4D, a transformer-based network specifically designed for online segmentation, achieving state-of-the-art results with a fraction of the latency and parameters of competing methods. We hope that our open-source dataset and implementation will encourage the evaluation of future semantic liDAR segmentation methods in more realistic settings and help to bridge the gap between academic work on 3D perception and the operational constraints of autonomous driving.

Acknowledgements. This work was supported in part by ANR project READY3D ANR-19-CE23-0007 and was granted access to the HPC resources of IDRIS under the allocation 2022-AD011012096R1 made by GENCI. The point cloud sequences of HelixNet were acquired during the Stereopolis II project [35]. HelixNet was annotated by FUTURMAP. We thank Zenodo for hosting the dataset. We thank François Darmon, Tom Monnier, Mathis Petrovich and Damien Robert for inspiring discussions and valuable feedback.

References

1. Arnab, A., Dehghani, M., Heigold, G., Sun, C., Lučić, M., Schmid, C.: ViViT: a video vision transformer. In: ICCV (2021)
2. Aygun, M., et al.: 4D panoptic LiDAR segmentation. In: CVPR (2021)
3. Behley, J., et al.: SemanticKITTI: a dataset for semantic scene understanding of LiDAR sequences. In: ICCV (2019)
4. Berman, M., Triki, A.R., Blaschko, M.B.: The Lovász-softmax loss: a tractable surrogate for the optimization of the intersection-over-union measure in neural networks. In: CVPR (2018)
5. Bhattacharyya, P., Huang, C., Czarnecki, K.: SA-Det3D: self-attention based context-aware 3D object detection. In: ICCV Workshops (2021)
6. Caesar, H., et al.: nuScenes: a multimodal dataset for autonomous driving. In: CVPR (2020)
7. Carion, N., Massa, F., Synnaeve, G., Usunier, N., Kirillov, A., Zagoruyko, S.: End-to-end object detection with transformers. In: Vedaldi, A., Bischof, H., Brox, T., Frahm, J.-M. (eds.) ECCV 2020. LNCS, vol. 12346, pp. 213–229. Springer, Cham (2020). https://doi.org/10.1007/978-3-030-58452-8_13

8. Cheng, R., Razani, R., Taghavi, E., Li, E., Liu, B.: AF2-S3Net: attentive feature fusion with adaptive feature selection for sparse semantic segmentation network. In: CVPR (2021)
9. Choy, C., Gwak, J., Savarese, S.: 4D Spatio-temporal convnets: minkowski convolutional neural networks. In: CVPR (2019)
10. Coccomini, D., Messina, N., Gennaro, C., Falchi, F.: Combining efficientNet and vision transformers for video deepfake detection. arXiv preprint arXiv:2107.02612 (2021)
11. Cordts, M., et al.: The cityscapes dataset for semantic urban scene understanding. In: CVPR (2016)
12. Cortinhal, T., Tzelepis, G., Aksoy, E.E.: SalsaNext: fast, uncertainty-aware semantic segmentation of LiDAR point clouds for autonomous driving. arXiv:2003.03653 (2020)
13. Dosovitskiy, A., et al.: An image is worth 16×16 words: transformers for image recognition at scale. In: ICLR (2021)
14. d'Ascoli, S., Touvron, H., Leavitt, M.L., Morcos, A.S., Biroli, G., Sagun, L.: Convit: improving vision transformers with soft convolutional inductive biases. In: ICML (2021)
15. Fan, H., Yang, Y., Kankanhalli, M.: Point 4D transformer networks for spatio-temporal modeling in point cloud videos. In: CVPR (2021)
16. Geiger, A., Lenz, P., Stiller, C., Urtasun, R.: Vision meets robotics: the KITTI dataset. Int. J. Robot. Res. **32**(11), 1231–1237 (2013)
17. Geyer, J., et al.: A2D2: audi autonomous driving dataset. arXiv preprint arXiv:2004.06320 (2020)
18. Graham, B., van der Maaten, L.: Submanifold sparse convolutional networks. arXiv preprint arXiv:1706.01307 (2017)
19. Guo, J., et al.: CMT: convolutional neural networks meet vision transformers. arXiv preprint arXiv:2107.06263 (2021)
20. Guo, M.H., Cai, J.X., Liu, Z.N., Mu, T.J., Martin, R.R., Hu, S.M.: PCT: point cloud transformer. Comput. Vis. Media **7**(2), 187–199 (2021)
21. Guo, Y., Wang, H., Hu, Q., Liu, H., Liu, L., Bennamoun, M.: Deep learning for 3D point clouds: a survey. Trans. Pattern Anal. Mach. Intell. **43**(12), 4338–4364 (2020)
22. Hong, F., Zhou, H., Zhu, X., Li, H., Liu, Z.: LiDAR-based panoptic segmentation via dynamic shifting network. In: CVPR (2021)
23. Huang, X., et al.: The Apolloscape dataset for autonomous driving. In: CVPR Workshop (2018)
24. Inc., V.L.: HDL-64E User's Manual. Velodyne LiDAR Inc. 345 Digital Drive, Morgan Hill, CA 95037 (2008)
25. Jiang, P., Osteen, P., Wigness, M., Saripalli, S.: Rellis-3D dataset: data, benchmarks and analysis. In: ICRA (2021)
26. Katharopoulos, A., Vyas, A., Pappas, N., Fleuret, F.: Transformers are RNNs: fast autoregressive transformers with linear attention. In: ICML (2020)
27. Liang, Z., Zhang, Z., Zhang, M., Zhao, X., Pu, S.: Rangeioudet: range image based real-time 3D object detector optimized by intersection over union. In: CVPR (2021)
28. Liao, Y., Xie, J., Geiger, A.: KITTI-360: a novel dataset and benchmarks for urban scene understanding in 2D and 3D. arXiv preprint arXiv:2109.13410 (2021)
29. Lin, K., Wang, L., Liu, Z.: End-to-end human pose and mesh reconstruction with transformers. In: CVPR (2021)
30. Liu, Z., et al.: Swin transformer: hierarchical vision transformer using shifted windows. In: ICCV (2021)

31. Mao, J., et al.: One million scenes for autonomous driving: once dataset. arXiv preprint arXiv:2106.11037 (2021)
32. Mao, J., et al.: Voxel transformer for 3D object detection. In: ICCV (2021)
33. Neuhold, G., Ollmann, T., Rota Bulo, S., Kontschieder, P.: The mapillary vistas dataset for semantic understanding of street scenes. In: CVPR (2017)
34. Pan, X., Xia, Z., Song, S., Li, L.E., Huang, G.: 3D object detection with pointformer. In: CVPR (2021)
35. Paparoditis, N., et al.: Stereopolis ii: a multi-purpose and multi-sensor 3D mobile mapping system for street visualisation and 3D metrology. Revue française de photogrammétrie et de télédétection (2012)
36. Ronneberger, O., Fischer, P., Brox, T.: U-Net: convolutional networks for biomedical image segmentation. In: Navab, N., Hornegger, J., Wells, W.M., Frangi, A.F. (eds.) MICCAI 2015. LNCS, vol. 9351, pp. 234–241. Springer, Cham (2015). https://doi.org/10.1007/978-3-319-24574-4_28
37. Roynard, X., Deschaud, J.E., Goulette, F.: Paris-Lille-3D: a large and high-quality ground-truth urban point cloud dataset for automatic segmentation and classification. Int. J. Robot. Res. **37** (2018)
38. Royo, S., Ballesta-Garcia, M.: An overview of LiDAR imaging systems for autonomous vehicles. Appl. Sci. **9**(19), 4093 (2019)
39. Strudel, R., Garcia, R., Laptev, I., Schmid, C.: Segmenter: transformer for semantic segmentation. In: ICCV (2021)
40. Sun, P., et al.: RSN: range sparse net for efficient, accurate LiDAR 3D object detection. In: CVPR (2021)
41. Tan, W., et al.: Toronto-3D: a large-scale mobile LiDAR dataset for semantic segmentation of urban roadways. In: CVPR Workshop (2020)
42. Tang, H., et al.: Searching efficient 3D architectures with sparse point-voxel convolution. In: Vedaldi, A., Bischof, H., Brox, T., Frahm, J.-M. (eds.) ECCV 2020. LNCS, vol. 12373, pp. 685–702. Springer, Cham (2020). https://doi.org/10.1007/978-3-030-58604-1_41
43. Vallet, B., Brédif, M., Serna, A., Marcotegui, B., Paparoditis, N.: Terramobilita/iqmulus urban point cloud analysis benchmark. Comput. Graph. **49**, 126–133 (2015)
44. Vaswani, A., et al.: Attention is all you need. In: NeurIPS (2017)
45. Vyas, A., Katharopoulos, A., Fleuret, F.: Fast transformers with clustered attention. In: NeurIPS (2020)
46. Wu, K., Peng, H., Chen, M., Fu, J., Chao, H.: Rethinking and improving relative position encoding for vision transformer. In: ICCV (2021)
47. Zhang, Y., et al.: PolarNet: an improved grid representation for online LiDAR point clouds semantic segmentation. In: CVPR (2020)
48. Zhao, H., Jiang, L., Jia, J., Torr, P.H., Koltun, V.: Point transformer. In: CVPR (2021)
49. Zhou, Z., Zhang, Y., Foroosh, H.: Panoptic-polarNet: proposal-free LiDAR point cloud panoptic segmentation. In: CVPR (2021)
50. Zhu, X., et al.: Cylindrical and asymmetrical 3D convolution networks for LiDAR segmentation. In: CVPR (2021)
51. Zolanvari, S., et al.: DublinCity: annotated LiDAR point cloud and its applications. In: BMVC (2019)

Open-world Semantic Segmentation for LIDAR Point Clouds

Jun Cen[1], Peng Yun[1], Shiwei Zhang[2(✉)], Junhao Cai[1], Di Luan[1], Mingqian Tang[2], Ming Liu[1], and Michael Yu Wang[1]

[1] The Hong Kong University of Science and Technology, Hong Kong, China
{jcenaa,pyun,jcaiaq,dluan}@connect.ust.hk, {eelium,mywang}@ust.hk
[2] Alibaba Group, Hangzhou, China
{zhangjin.zsw,mingqian.tmq}@alibaba-inc.com

Abstract. Current methods for LIDAR semantic segmentation are not robust enough for real-world applications, *e.g.*, autonomous driving, since it is *closed-set* and *static*. The closed-set assumption makes the network only able to output labels of trained classes, even for objects never seen before, while a static network cannot update its knowledge base according to what it has seen. Therefore, in this work, we propose the *open-world semantic segmentation* task for LIDAR point clouds, which aims to 1) identify both old and novel classes using open-set semantic segmentation, and 2) gradually incorporate novel objects into the existing knowledge base using incremental learning without forgetting old classes. For this purpose, we propose a **RE**dund**A**ncy c**L**assifier (REAL) framework to provide a general architecture for both the open-set semantic segmentation and incremental learning problems. The experimental results show that REAL can simultaneously achieves state-of-the-art performance in the open-set semantic segmentation task on the SemanticKITTI and nuScenes datasets, and alleviate the catastrophic forgetting problem with a large margin during incremental learning.

Keywords: Open-world semantic segmentation · LIDAR point clouds · Open-set semantic segmentation · Incremental learning

1 Introduction

3D LIDAR sensors play an important role in the perception system of autonomous vehicles. Semantic segmentation for LIDAR point clouds has grown very fast in recent years [10,26,42,44], benefiting from well-annotated datasets including SemanticKITTI [2,3,13] and nuScenes [6]. However, existing methods for LIDAR semantic segmentation are all *closed-set* and *static*. The closed-set network regards all inputs as categories encountered during training, so it will assign the labels of old classes to novel classes by mistake, which may have disastrous consequences in safety-sensitive applications, such as autonomous driving [5]. Meanwhile, the static network is constrained to certain scenarios, as it

code is available at: https://github.com/Jun-CEN/Open_world_3D_semantic_segmentation

S. Avidan et al. (Eds.): ECCV 2022, LNCS 13698, pp. 318–334, 2022.
https://doi.org/10.1007/978-3-031-19839-7_19

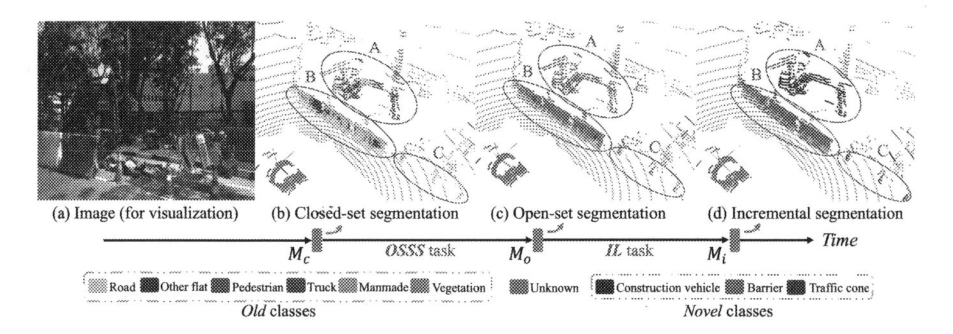

Fig. 1. Closed-set model $\mathcal{M}_c$ wrongly assigns the labels of old classes to novel objects (A: construction vehicle is classified as the manmade, truck, and even pedestrian; B: barrier is classified as the road, manmade and other flat; C: traffic cone is classified as the manmade). After open-set semantic segmentation (OSeg) task, the open-set model $\mathcal{M}_o$ can identify the novel objects and assign the label *unknown* for them. After incremental learning (IL) task, the model $\mathcal{M}_i$ can classify both old and novel classes.

cannot update itself to adapt to new environments. In addition, training from scratch to adapt to new scenes is extremely time-consuming, and the annotations of old classes are sometimes unavailable, due to privacy constrains.

To solve the *closed-set* and *static* problem, we propose the *open-world semantic segmentation* for LIDAR point clouds, which is composed of two tasks: 1) open-set semantic segmentation (OSeg) to assign the *unknown* label to novel classes as well as to assign the correct labels to old classes, and 2) incremental learning (IL) to gradually incorporate the novel classes into the knowledge base after labellers provide the labels of novel classes. Figure 1 illustrates an example of open-world semantic segmentation for LIDAR point clouds.

As we are the first to study OSeg task in the 3D LIDAR point cloud domain, we refer to the existing methods in the 2D image domain, which can be divided into two types, generative network-based methods [1,22,39] and uncertainty-based methods [12,15,19], though none of them can be directly utilized. Generative network-based methods adopt a conditional generative adversarial network (cGAN) [27] to reconstruct the input based on the closed-set prediction results, and assume the novel regions have a larger difference in appearance between the reconstructed input and original input. However, cGAN is not appropriate for reconstruction of the point cloud as all information is determined by the geometry information, *i.e.*, coordinates of points, and cGAN can only reconstruct the channel information, *i.e.*, RGB values, while keeping the geometry information, including coordinates of pixels and the shape of an image, unchanged. The uncertainty-based methods also work poorly as we find the network predicts the novel classes as old classes with high confidence scores, as shown in Fig. 3 (a).

In addition to the challenges of the OSeg task, the catastrophic forgetting of old classes in incremental learning [25] is another problem to solve. Directly fine-tuning the network using only the labels of novel classes will make the network classify everything as novel classes. Thus a method is needed to incrementally learn novel classes while keeping the performance of the old classes.

We find that the closed-set and static properties of the traditional closed-set model is due to the fixed classifier architecture, *i.e.*, one classifier corresponds to one old class. Therefore, we propose a **RE**dund**A**ncy c**L**assifier (REAL) framework to provide a dynamic classifier architecture to adapt the model to both the OSeg and IL tasks. For the OSeg task, we add several redundancy classifiers (RCs) on the basis of the original network to predict the probability of the unknown class. Then, during the IL task, several RCs are trained to classify the newly introduced classes, while the remaining RCs are still responsible for the unknown class, as shown in Fig. 2. We provide the training strategies for the OSeg and IL tasks under REAL, based on the unknown object synthesis, predictive distribution calibration, and pseudo label generation. We show the effectiveness of REAL and corresponding training strategies through our comprehensive experiments. In summary, our contributions are three-folds:

- We are the first to define the open-world semantic segmentation problem for LIDAR point clouds, which is composed of OSeg and IL tasks;
- We propose a REAL model to provide a general architecture for both the OSeg and IL tasks, as well as training strategies for each task, based on the unknown objects synthesis, predictive distribution calibration, and pseudo labels generation;
- We construct benchmark and evaluation protocols for OSeg and IL in the 3D LIDAR point cloud domain, based on the SemanticKITTI and nuScenes datasets, to measure the effectiveness of our training strategies under REAL.

2 Related Work

Closed-set LIDAR Semantic Segmentation: Semantic segmentation for LIDAR point clouds can be categorized into point-based and voxel-based methods. Typical point-based methods [16,33,38] use PointNet [29] and PointNet++ [30] to directly operate on the LIDAR point cloud. However, they have limited performance due to the varying density and large scale of the LIDAR point cloud. The other type of point-based methods convert the LIDAR point cloud to 2D grids and then apply 2D convolutional operations for semantic segmentation. SqueezeSeg [37] and its alternatives [26,35,42] convert the point cloud to a range image or bird's-eye-view. However, 2D representations inevitably lose some of the 3D topology and geometric information. Cylinder3D [44] is a voxel-based method and it tackles the sparsity and varying density problems of LIDAR point clouds through cylindrical partition and asymmetrical 3D convolutional networks. Cylinder3D achieves state-of-the-art performance on SemanticKITTI [2,3,13] and nuScenes [6], so we adopt it as the base architecture.

Incremental Learning: Neural networks tend to forget what they have learned when trained with new data, which is called catastrophic forgetting. Some researchers adopt knowledge distillation to overcome this problem. Knowledge-distillation-based methods [8,20,21,28,32,40] retain the learnt knowledge by restricting the prediction $p(\hat{y}|x,\theta)$ close to that computed with the optimal parameter of previous tasks $p(\hat{y}|x,\theta_0^*)$. A regularization term, proportional to the distance between these two conditional distributions, is added to the original loss function. In classification, the distance is commonly measured by the Kullback-Leibler (KL) divergence [21,31]. Shmelkov *et al.* [32] extended the knowledge distillation to the image-based 2D object detection problem and proposed an incremental learning object detector, ILOD. In recent years, there have been follow-up works [8,20,28,40,41] in 2D/3D object detection and semantic segmentation. The most similar method to the incremental learning part of our work is [23], where they adopted a pseudo-label generation approach to provide supervision of learned knowledge in image-based object detection.

Open-set 2D Classification: There are two trends of open-set 2D classification methods: uncertainty-based methods and generative model-based methods. Maximum softmax probability (MSP) [15] is the baseline of uncertainty-based methods, while Dan *et al.* [14] found that Maximum Logit (MaxLogit) is a better choice than the probability. MC-Dropout [12] and Ensembles [19] are used to approximate Bayesian inference [18,24], which regards the network from a probabilistic view. Meanwhile, generative-based methods, including SynthCP [39] and DUIR [22], adopt conditional GAN (cGAN) [27] to reconstruct the input, and find the novel regions by comparing the reconstructed input with the original input. However, these methods cannot adapt to the 3D LIDAR point cloud domain directly, as discussed in Sect. 1. [34,43] propose to use redundancy classifiers (RCs) to directly output the score of the unknown class, and adopt manifold mixup and a sampler based on Stochastic Gradient Langevin Dynamics (SGLD) [36] to approximate the unknown class distribution. We draw inspiration from them, and take a step further by using RCs for both OSeg and IL, as well as developing suitable training strategies for the 3D point cloud domain.

Open-world Classification and Detection: The open-world problem was first proposed by Abhijit *et al.* [4], who argued that the network should be able to deal with a dynamic category set which is practical in the real world. Therefore, they introduced the open-world classification pipeline: first identify both known and unknown images, and then gradually learn to classify unknown images when labels are given. They presented the Nearest Non-Outlier method to manage the open-world classification task. Joseph *et al.* [17] extended the open-world problem to the 2D object detection domain and Jun *et al.* [7] later adopted deep metric learning for open-world semantic segmentation for 2D images. We extend the open-world problem to the 3D LIDAR cloud point domain, and both sub-tasks including OSeg and IL for 3D LIDAR point clouds are not studied yet.

3 Open-World Semantic Segmentation

In this section, we formalise the definition of open-world semantic segmentation for LIDAR point clouds. Let the classes of the training set be called old classes and labeled by positive integers $\mathcal{K}_0 = \{1, 2, ..., C\} \subset \mathbb{N}^+$. Unlike the traditional closed-set semantic segmentation where the classes of the test set are the same as the training set, some novel classes $\mathcal{U} = \{C + 1, ...\}$ are involved in the test set in the open-world semantic segmentation problem. Let one LIDAR point cloud sample be formulated as $\mathcal{D} = \{\mathbf{P}, \mathbf{Y}\}$, where $\mathbf{P} = \{\mathbf{p}_1, \mathbf{p}_2, ..., \mathbf{p}_M\}$ is the input LIDAR point cloud composed of M points and every point $\mathbf{p}$ is represented by three coordinates $\mathbf{p} = (x, y, z)$. The label $\mathbf{Y} = \{y_1, y_2, ..., y_M\}$ contains the semantic class for every point, in which $y \in \mathcal{K}_0$ for the training data and $y \in \mathcal{K}_0 \cup \mathcal{U}$ for the test data.

Suppose we already have a model $\mathcal{M}_c$ which is trained under the closed-set condition, so its outputs are within the domain of $\mathcal{K}_0$. As discussed in Sect. 1, the open-world semantic segmentation is composed of two tasks: open-set semantic segmentation (OSeg) and incremental learning (IL). For the OSeg task, the model $\mathcal{M}_c$ will be finetuned to $\mathcal{M}_o$ so that it can assign the correct labels for the points of old classes $\mathcal{K}_0$, as well as assign the *unknown* label to the points of novel classes $\mathcal{U}$. For the IL task, the model $\mathcal{M}_o$ will be further finetuned to $\mathcal{M}_i$ when the labels of novel classes $\mathcal{K}_n$ are given, so that its knowledge base is enlarged from $\mathcal{K}_0$ to $\mathcal{K}_0 \cup \mathcal{K}_n$, where $\mathcal{K}_n = \{C + 1, ..., C + n\}$. So the classes in $\mathcal{K}_n$ change from *unknown* to *known* for the network. We follow the classical task IL setting [9,11,40] that the new given labels only contain the annotation of the novel class $\mathcal{K}_n$, while the remaining points of old classes $\mathcal{K}_0$ are not annotated. Additionally, the model after IL $\mathcal{M}_i$ still keeps the open-set property, *i.e.*, assigns the *unknown* label to the remaining novel classes $\mathcal{K}_{rn} = \{C + n + 1, ...\}$.

4 Methodology

In this section, we introduce our strategies to solve the open-world semantic segmentation problem for LIDAR point clouds. The open-world semantic segmentation is composed of two tasks: OSeg task and IL task. We first introduce the redundancy classifier framework (REAL) in Sect. 4.1, which provides a general network architecture for both the OSeg task and IL task. Then, we introduce the training strategies and inference procedures for the OSeg task and IL task in Sect. 4.2 and Sect. 4.3 respectively.

4.1 Redundancy Classifier Framework (REAL)

The overall view of REAL is shown in Fig. 2. The trained closed-set model $\mathcal{M}_c$, which can well classify old classes $\mathcal{K}_0$, is composed of a feature extractor f and normal classifiers $g_{nm} = \{g^1_{nm}, g^2_{nm}, ..., g^C_{nm}\}$. For a certain input $\mathbf{P} \in \mathbb{R}^{M\times3}$, the output of the model $\mathcal{M}_c$ is

$$\mathcal{M}_c(\mathbf{P}) = [y^{old}] = [g_{nm}(f(\mathbf{P}))] \in \mathbb{R}^{M\times C}. \tag{1}$$

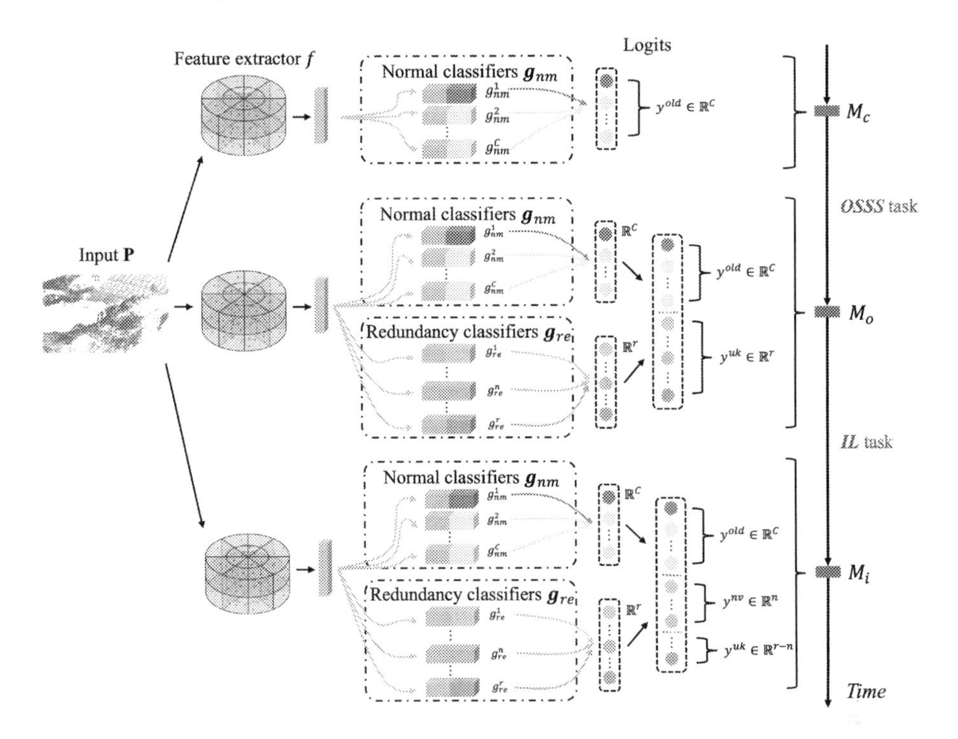

Fig. 2. Redundancy classifier framework (REAL). Closed-set model $\mathcal{M}_c$ can only output logits for old classes y^{old}. Redundancy Classifiers g_{re} are added on top of the original framework in our REAL. All g_{re} in $\mathcal{M}_o$ are used to output the scores y^{uk} for the unknown class. After the IL task, part of g_{re} are used to output logits for the newly introduced classes y^{nv}, while the remaining are still for the unknown class y^{uk}.

OSeg Task: The OSeg task is to adapt closed-set model $\mathcal{M}_c$ to open-set model $\mathcal{M}_o$ so that $\mathcal{M}_o$ can identify novel classes $\mathcal{U}$ as *unknown*. To achieve this goal, we add r redundancy classifiers (RCs) $g_{re} = \{g_{re}^1, g_{re}^2, ..., g_{re}^r\}$ on top of the original feature extractor f, as shown in Fig. 2 $\mathcal{M}_o$. All RCs in $\mathcal{M}_o$ are used to predict the scores y^{uk} for the unknown class. We let the maximum response of y^{uk} be the score of the unknown class, which is represented by class 0. In this way, the output of the open-set model $\mathcal{M}_o$ is

$$\mathcal{M}_o(\mathbf{P}) = [\max y^{uk}, y^{old}] = [\max\ g_{re}(f(\mathbf{P})), g_{nm}(f(\mathbf{P}))] \in \mathbb{R}^{M\times(1+C)}. \tag{2}$$

IL Task: The IL task is to train open-set model $\mathcal{M}_o$ to $\mathcal{M}_i$ so that newly introduced classes $\mathcal{K}_n$ change from *unknown* to *known*. $\mathcal{M}_i$ is still open-set, *i.e.*, it can classify remaining novel classes $\mathcal{K}_{rn}$ as *unknown*. In this task, among all RCs g_{re}, some of the RCs $g_{re}^{nv} = \{g_{re}^1, g_{re}^2, ..., g_{re}^n\}$ are used to classify newly introduced classes $\mathcal{K}_n$, *i.e.*, y^{nv} in Fig. 2 $\mathcal{M}_i$, and the remaining RCs $g_{re}^{uk} = \{g_{re}^{n+1}, g_{re}^{n+2}, ..., g_{re}^r\}$ are kept for the unknown class $\mathcal{K}_{rn}$, *i.e.*, y^{uk} in Fig. 2 $\mathcal{M}_i$.

In this way, the output of $\mathcal{M}_i$ can be represented as

$$\mathcal{M}_i(\mathbf{P}) = [\max y^{uk}, y^{old}, y^{nv}] = [\max \ g_{re}^{uk}(f(\mathbf{P})), g_{nm}(f(\mathbf{P})), g_{re}^{nv}(f(\mathbf{P}))]. \quad (3)$$

where $\mathcal{M}_i(\mathbf{P}) \in \mathbb{R}^{M \times (1+C+n)}$.

4.2 Open-set Semantic Segmentation (OSeg)

The OSeg task is to train the closed-set model $\mathcal{M}_c$ to the open-set model $\mathcal{M}_o$ which can identify novel classes $\mathcal{U}$ as *unknown*, as shown in Fig. 1 (c). The network architecture of $\mathcal{M}_o$ is shown in Fig. 2 $\mathcal{M}_o$. We introduce two training methods including *Unknown Object Synthesis* and *Predictive Distribution Calibration* as well as inference procedure in this section.

Unknown Object Synthesis: We synthesize pseudo unknown objects in the LIDAR point cloud to approximate the distribution of real novel objects. The synthesis process should meet two requirements: 1) the synthesized object should share some invariant basic geometry features with existing objects, such as curved and flat surfaces, so that it can be regarded as an *object* rather than noise and possibly have a similar appearance to real unknown objects; 2) the synthesis process should be as quick as possible.

We find that resizing the existing objects with a proper factor is a simple but effective way to conduct the synthesis process, as it keeps the geometric shape of an object, but the different size determines it is a new object. For instance, a car, truck, bus, and construction vehicle have similar local geometric features, such as the shape of the body and tires, but their size can be different. Therefore, we pick up objects of specific old classes $\mathcal{K}_{syn}$ with a probability p_{syn} and resize them from 0.25 to 0.5 times or 1.5 to 3 times as pseudo unknown objects, such as B in Fig. 4 (c) and (d). In this way, the input $\mathbf{P}$ is divided into two parts: $\mathbf{P} = \mathbf{P}_{syn} \cup \mathbf{P}_{nm}$, where $\mathbf{P}_{syn}$ and $\mathbf{P}_{nm}$ represent the points of synthesized objects and unchanged normal objects respectively. For the points of synthesized objects $\mathbf{P}_{syn}$, the synthesis loss $\mathcal{L}_{syn}$ is

$$\mathcal{L}_{syn} = \ell(\mathcal{M}(\mathbf{P}_{syn}), \mathbf{0}), \quad (4)$$

where ℓ is the cross-entropy loss. The ground truth labels of synthesized objects are set to be the unknown class 0, so the first term in Eq. 2 is trained to give high scores to objects never seen before.

Predictive Distribution Calibration: We find that in the closed-set prediction, the novel objects are classified as old classes with high probability, as shown in Fig. 3 (a). We intend to alleviate this problem by probability calibration, and the calibrated scores of the unknown class are shown as Fig. 3 (b).

We force every point of old classes to have the largest score on its original class, and have the second largest score on the unknown class [34]. By this design, the network is supposed to output high probability scores on the unknown class for the novel objects as they do not belong to any one of the old classes. Therefore, for the points of unchanged normal objects $\mathbf{P}_{nm}$, the calibration loss is designed as

$$\mathcal{L}_{cal} = \mathcal{L}_{cal}^{ori} + \lambda_{cal}\mathcal{L}_{cal}^{uk}, \tag{5}$$

where $\mathcal{L}_{cal}^{ori}$ and $\mathcal{L}_{cal}^{uk}$ are defined as

$$\mathcal{L}_{cal}^{ori} = \ell(\mathcal{M}(\mathbf{P}_{nm}), \mathbf{Y}_{nm}), \tag{6}$$

$$\mathcal{L}_{cal}^{uk} = \ell(\mathcal{M}(\mathbf{P}_{nm}) \setminus \mathbf{Y}_{nm}, \mathbf{0}), \tag{7}$$

where $\mathbf{Y}_{nm}$ is the ground truth of $\mathbf{P}_{nm}$. $\mathcal{M}(\mathbf{P}_{nm}) \setminus \mathbf{Y}_{nm}$ means to remove the response of the corresponding ground truth old class. $\mathcal{L}_{cal}^{ori}$ is to ensure the good closed-set prediction, while $\mathcal{L}_{cal}^{uk}$ is to make every point have the second largest probability on the unknown class.

Loss Function: The overall loss function to train the model $\mathcal{M}_c$ to $\mathcal{M}_o$ is

$$\mathcal{L}^{OSeg} = \mathcal{L}_{cal}^{OSeg} + \lambda_{syn}\mathcal{L}_{syn}^{OSeg}, \tag{8}$$

where $\mathcal{L}_{cal}^{OSeg}$ is determined by Eq. 5, Eq. 6, and Eq. 7, while $\mathcal{L}_{syn}^{OSeg}$ is determined by Eq. 4. All $\mathcal{M}$ in the related terms are $\mathcal{M}_o$ in the OSeg task.

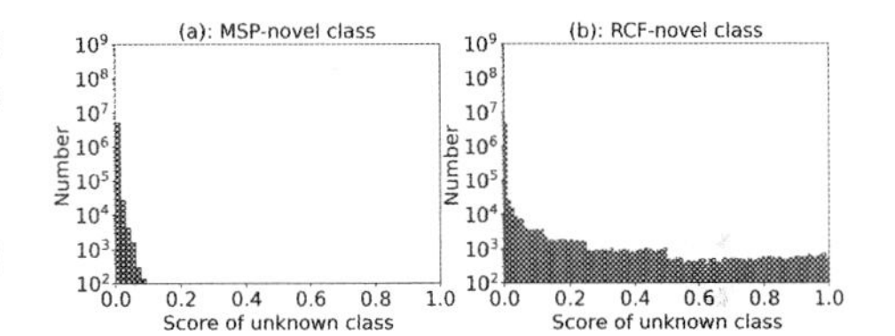

Fig. 3. Distribution of scores of the unknown class for Maximum Softmax Probability (MSP) and our REAL method. The scores of the unknown class for novel classes are low in MSP (a), meaning the closed-set prediction classifies novel classes as old classes with high confidence.

Inference: Both the closed-set and open-set performance of the fine-tuned model $\mathcal{M}_o$ will be evaluated. For the closed-set prediction, the inference result $\hat{\mathbf{Y}}_{close}$ is defined as

$$\hat{\mathbf{Y}}_{close} = \underset{i=1,2,\ldots,C}{\arg\max}\ g_{nm}(f(\mathbf{P})). \tag{9}$$

For the open-set prediction, we have to classify both old classes and the novel class, so the inference result $\hat{\mathbf{Y}}_{open}$ is defined as:

$$\hat{\mathbf{Y}}_{open} = \begin{cases} \underset{i=1,2,\ldots,C}{\arg\max}\ g_{nm}(f(\mathbf{P})) & \lambda_{conf} < \lambda_{th} \\ 0 & otherwise, \end{cases} \tag{10}$$

where $\lambda_{conf} = \max\ g_{re}(f(\mathbf{P}))$ is the confidence score of the unknown class, and λ_{th} is the threshold. The unknown class is represented by class 0.

4.3 Incremental Learning (IL)

The IL task is to train $\mathcal{M}_o$ to $\mathcal{M}_i$ when the labels of novel classes $\mathcal{K}_n$ are available. $\mathcal{M}_i$ can classify both newly introduced classes $\mathcal{K}_n$ and old classes $\mathcal{K}_0$, as well as identify remaining novel classes $\mathcal{K}_{rn}$ as *unknown*. The inference example is shown in Fig. 1 (d) and the architecture is shown in Fig. 2 $\mathcal{M}_i$.

As mentioned in Sect. 3, only the labels of introduced novel classes $\mathcal{K}_n$ are given in this task. Therefore, we divide the unchanged normal points $\mathbf{P}_{nm}$ into two parts, $\mathbf{P}_{nm}^{old}$, which belongs to old classes $\mathcal{K}_0$, and $\mathbf{P}_{nm}^{nv}$, which belongs to newly introduced classes $\mathcal{K}_n$, so that $\mathbf{P}_{nm} = \mathbf{P}_{nm}^{old} \cup \mathbf{P}_{nm}^{nv}$. The labels of points $\mathbf{P}_{nm}^{nv}$ are given as $\mathbf{Y}_{nm}^{nv}$, *e.g.*, labels of A in Fig. 4 (a), but labels of $\mathbf{P}_{nm}^{old}$ are not given, *e.g.*, gray points in Fig. 4 (a). If we only use $\mathbf{Y}_{nm}^{nv}$ to directly finetune the model, it will classify all points as the newly introduced class as there is only one kind of class in the training process. This is called the catastrophic forgetting and we use *Pseudo Label Generation* to solve this problem.

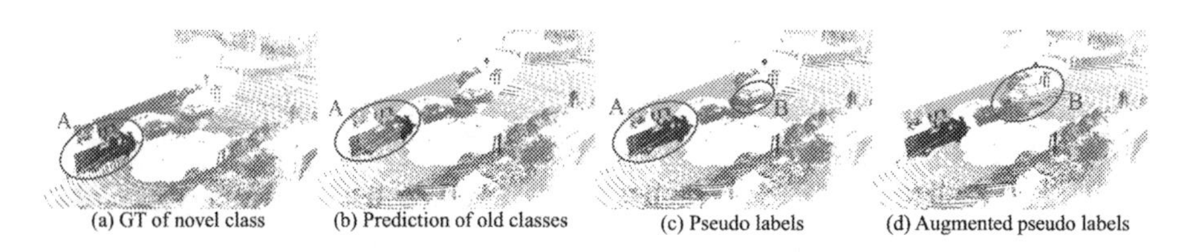

Fig. 4. Pseudo labels generating process for incremental learning. Ground truth (a) only contains the label of the novel class (A: other-vehicle). So we combine the prediction results of $\mathcal{M}_o$ (b) to generate the pseudo labels (c). Then we resize objects of old classes as the synthesized objects in (d) (B: resized car).

Pseudo Label Generation: We use model $\mathcal{M}_o$ to predict the pseudo labels $\mathbf{pY}_{nm}^{old}$ for $\mathbf{P}_{nm}^{old}$ [7,9], as shown in Fig. 4 (b). In this way, the learned knowledge of old classes is preserved in $\mathbf{pY}_{nm}^{old}$ to alleviate the catastrophic forgetting problem. Then we combine $\mathbf{pY}_{nm}^{old}$ with $\mathbf{Y}_{nm}^{nv}$ to generate the pseudo labels of the whole point cloud $\mathbf{Y}_{nm}$, such as in Fig. 4 (c).

Loss Function: Note that we keep the open-set property after IL, so the methods in OSeg task including *Unknown Object Synthesis* and *Predictive Distribution Calibration* are still used in IL task. The overall loss function to train the model $\mathcal{M}_o$ from $\mathcal{M}_i$ is

$$\mathcal{L}^{il} = \mathcal{L}_{cal}^{il} + \lambda_{syn}\mathcal{L}_{syn}^{il}, \tag{11}$$

where $\mathcal{L}_{cal}^{il}$ and $\mathcal{L}_{syn}^{il}$ are determined by Eq. 5, Eq. 6, Eq. 7, and Eq. 4. All $\mathcal{M}$ in the related terms are $\mathcal{M}_i$. Note that $\mathbf{Y}_{nm}$ in Eq. 6 and Eq. 7 are generated as

$$\mathbf{Y}_{nm} = \mathbf{pY}_{nm}^{old} \cup \mathbf{Y}_{nm}^{nv}, \tag{12}$$

where $\mathbf{Y}_{nm}^{nv}$ is the ground truth label of newly introduced classes $\mathcal{K}_n$ and $\mathbf{pY}_{nm}^{old}$ is the pseudo labels of old classes $\mathcal{K}_0$ generated by $\mathcal{M}_o$,

$$\mathbf{pY}_{nm}^{old} = \mathcal{M}_o(\mathbf{P}_{nm}^{old}). \tag{13}$$

The $\mathbf{Y}_{nm}$ in Eq. 12 contains both newly introduced classes K_n and old classes K_0, so $\mathcal{M}_i$ can learn new classes without forgetting old classes.

Inference: To evaluate the performance of IL, we only calculate the closed-set prediction results. This is because, for incremental learning we care about how well the catastrophic forgetting problem is alleviated and the new classes are learned, while the ability to classify the unknown class is already evaluated by Eq. 10 in OSeg task, although after IL the model $\mathcal{M}_i$ can still classify the unknown class $\mathcal{K}_{rn}$. The closed-set inference result $\hat{\mathbf{Y}}'_{close}$ is defined as

$$\hat{\mathbf{Y}}'_{close} = \underset{i=1,2,\dots,C+n}{\arg\max} \; [g_{nm}(f(\mathbf{P}), g_{re}^{nv}(f(\mathbf{P}))]. \tag{14}$$

Table 1. Benchmark of open-set semantic segmentation for LIDAR point clouds. Results are evaluated on the validation set.

Dataset	SemanticKITTI			nuScenes		
Methods	AUPR	AUROC	mIoU$_{\text{old}}$	AUPR	AUROC	mIoU$_{\text{old}}$
Closed-set	0	0	58.0	0	0	58.7
Upper bound	73.6	97.1	63.5	86.1	99.3	73.8
MSP	6.7	74.0	58.0	4.3	76.7	58.7
MaxLogit	7.6	70.5	58.0	8.3	79.4	58.7
MC-Dropout	7.4	74.7	58.0	14.9	82.6	58.7
REAL	**20.8**	**84.9**	57.8	**21.2**	**84.5**	56.8

5 Experiments

We conduct experiments for both tasks of the open-world semantic segmentation, including OSeg and IL tasks. We evaluate our proposed method on two large-scale datasets, SemanticKITTI and nuScenes.

5.1 Open-World Evaluation Protocol

Data Split: We set the novel classes of SemanticKITTI $\mathcal{K}_n^{sk}$ and nuScenes $\mathcal{K}_n^{ns}$ as:

$$\mathcal{K}_n^{sk} = \{\textit{other-vehicle}\}$$

$$\mathcal{K}_n^{ns} = \{barrier,\ construction\text{-}vehicle,\ traffic\text{-}cone,\ trailer\}$$

All remaining classes are included in the old class set $\mathcal{K}_0^{sk}$ and $\mathcal{K}_0^{ns}$. During training of the closed-set model $\mathcal{M}_c$ and open-set model $\mathcal{M}_o$, we set the labels of novel classes $\mathcal{K}_n^{sk}$ and $\mathcal{K}_n^{ns}$ to be void and ignore them. During incremental learning, we gradually introduce the labels of novel classes $\mathcal{K}_n^{sk}$ and $\mathcal{K}_n^{ns}$ one by one, and set the labels of old classes $\mathcal{K}_0^{sk}$ and $\mathcal{K}_0^{ns}$ to be void.

Evaluation Metrics: To evaluate the performance of the open-set semantic segmentation model $\mathcal{M}_o$, we consider both the closed-set and open-set segmentation ability. The closed-set ability is measured by $\text{mIoU}_{\textbf{close}}$, while the open-set evaluation is regarded as a binary classification problem between the known class and unknown class, which is measured by area under the ROC curve (AUROC) and area under the precision-recall curve (AUPR) [14].

To evaluate the performance of the model $\mathcal{M}_i$ after incremental learning, we calculate the performance of the old classes $\text{mIoU}_{\textbf{old}}$ and newly introduced classes $\text{mIoU}_{\textbf{novel}}$ respectively, and also the mIoU of all classes.

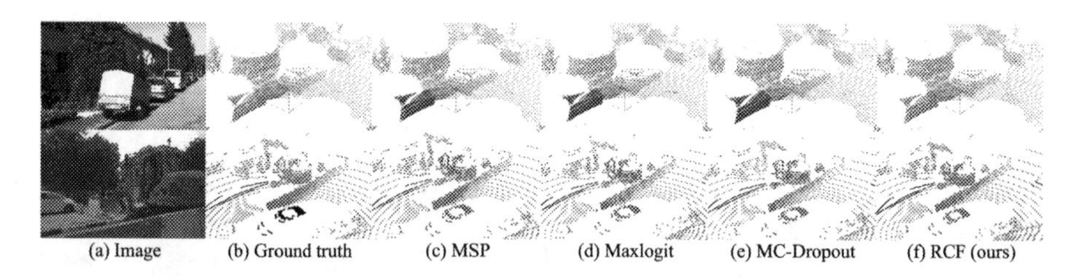

Fig. 5. Qualitative results of OSeg task. Novel classes are in pink (other-vehicle in SemanticKITTI (top), and construction-vehicle and barrier in nuScenes (bottom)). The results show that our method has a better performance in distinguishing the novel class from old classes than all the baselines. Best viewed in zoom. (Color figure online)

5.2 Open-set Semantic Segmentation (OSeg)

Implementation: We adopt Cylinder3D as the base network and train the traditional closed-set model $\mathcal{M}_c$ following the training settings in [44] using the labels of old classes $\mathcal{K}_0^{sk}$ and $\mathcal{K}_0^{ns}$. Then we add several redundancy classifiers on top of the $\mathcal{M}_0$ and finetune the model $\mathcal{M}_c$ to $\mathcal{M}_o$ based on the training strategies described in Sect. 4.2. The old classes used to synthesize novel objects $\mathcal{K}_{syn}$ are *car* for SemanticKITTI and *car, bus,* and *truck* for nuScenes. The probability of resizing these objects p_{syn} is set to 0.5. The unknown object synthesis time is 0.5–4 ms based on our experiments, which is sufficiently quick.

Baselines and Upper Bound: We refer to several methods from the open-set 2D semantic segmentation domain and implement them in our 3D LIDAR points domain as our baselines, including MSP, Maxlogit, and MC-Dropout, as discussed in Sect. 2. The upper bound is to use labels of all classes $\mathcal{K}_0 \cup \mathcal{K}_n$ to train the network and regard the softmax probability of the classes $\mathcal{K}_n$ as the confidence score.

Quantitative Results: The quantitative results of OSeg are shown in Table 1. The closed-set method does not consider the unknown class at all, so the open-set evaluation metrics are 0. Among all open-set semantic segmentation baselines, our REAL achieves remarkably better results on the open-set evaluation metrics. The closed-set mIoU_{old} shows that our method does not sacrifice the ability to classify old classes. The upper bound naturally achieves the best performance as it is conducted in a supervised manner, while the information of the unknown class is not provided for other open-set methods.

Qualitative Results: Figure 5 contains the qualitative results from SemanticKITTI and nuScenes respectively. Figure 5 top row shows that our method can identify the other-vehicle as the novel class, while all baselines consider it as the truck. In Fig. 5 bottom row, the baselines classify the construction-vehicle as the truck, pedestrian, and manmade, while our method distinguishes it as the novel object.

Ablation Experiments: We carefully conduct ablation experiments on the SemanticKITTI dataset to verify the effectiveness of our we proposed components. According to the results of Row ID 2 in Table 2, using the calibration loss alone can already outperforms all baselines in Table 1. Furthermore, the result of Row ID 3 illustrates that resizing the objects of existing classes with a proper factor is a simple but useful way to imitate novel objects. λ_{syn} and r are set to be 1 and 3 according to Fig. 6. λ_{cal} is 0.1, and it does not influence the result with a large margin based on Fig. 6.

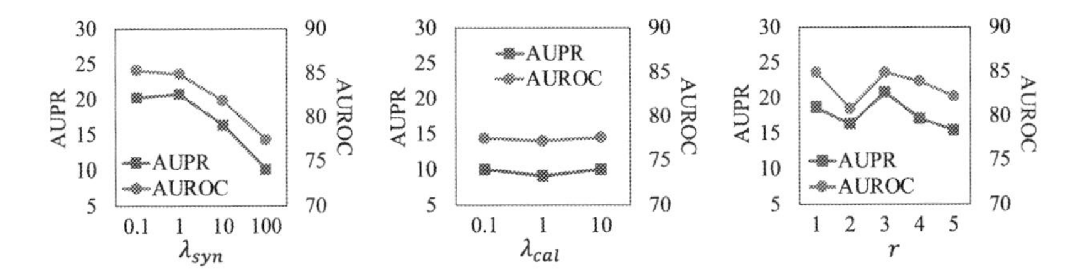

Fig. 6. Ablation experiments of coefficient λ_{syn}, λ_{cal} and number of redundancy classifiers r for OSeg task on SemanticKITTI.

Table 2. Ablation study results of $\mathcal{L}_{cal}$ and $\mathcal{L}_{syn}$ for OSeg task on SemanticKITTI.

Row ID	$\mathcal{L}_{cal}$	$\mathcal{L}_{syn}$	AUPR	AUROC	mIoU$_{\mathbf{old}}$
1	×	×	0	0	58.0
2	✓	×	10.0	77.5	**58.1**
3	✓	✓	**20.8**	**84.9**	57.8

5.3 Incremental Learning

Implementation: We adopt the training strategies described in Sect. 4.3 to finetune the model $\mathcal{M}_o$ to $\mathcal{M}_i$. The old classes used for synthesis $\mathcal{K}_{syn}$ are the same as the set during training from $\mathcal{M}_c$ to $\mathcal{M}_o$.

Baselines and Upper Bound: We adopt direct finetuning of $\mathcal{M}_o$ to $\mathcal{M}_i$ using only the labels of novel classes $\mathcal{K}_n^{sk}$ and $\mathcal{K}_n^{ns}$ to illustrate the catastrophic forgetting problem. Two methods including Feature Extraction and Learning without Forgetting (LwF) [21] using $\mathcal{K}_n^{sk}$ and $\mathcal{K}_n^{ns}$ are regarded as the baselines. The upper bound is the same as the upper bound in the open-set semantic segmentation task, which uses all labels $\mathcal{K}_0 \cup \mathcal{K}_n$ to train the network.

Quantitative Results: Table 3 and Table 4 show the IL performance of SemanticKITTI and nuScenes dataset respectively. Directly finetuning the model $\mathcal{M}_o$ to $\mathcal{M}_i$ only using labels of the novel class incurs the catastrophic forgetting problem, *i.e.*, the network classifies all points as the new class. mIoU$_{\mathbf{old}}$ becomes 0 as there is no prediction results in old classes. mIoU$_{\mathbf{novel}}$ is also close to 0 as newly introduced class only counts a little portion in the whole point cloud. In contrast, mIoU$_{\mathbf{old}}$ in our method is similar with the closed-set, meaning our method can learn the new classes one by one without forgetting the old classes. Our methods has better performance compared to two baselines, showing that

Table 3. Incremental learning results on SemanticKITTI 18 + 1 (other-vehicle) setting.

SemanticKITTI 18+1	Validation set			Test set		
Method	mIoU	mIoU$_{\mathbf{novel}}$	mIoU$_{\mathbf{old}}$	mIoU	mIoU$_{\mathbf{novel}}$	mIoU$_{\mathbf{old}}$
Closed-set	58.0	0	61.2	61.8	0	65.3
Upper bound	63.5	44.1	64.6	62.2	40.1	63.5
Finetune	0	0.5	0	0	0	0
Feature extraction	6.8	0.6	7.1	6.9	0.4	7.3
LwF	21.6	1.7	22.7	20.2	0.9	21.3
REAL	**64.3**	**51.5**	**65.0**	**61.1**	**25.3**	**63.1**

Table 4. Incremental learning results on nuScenes for 12 + 4 (barrier, construction-vehicle, traffic-cone, and trailer) setting.

nuScenes 12+4	Validation set			Test set		
Method	mIoU	$\text{mIoU}_{\text{novel}}$	mIoU_{old}	mIoU	$\text{mIoU}_{\text{novel}}$	mIoU_{old}
Closed-set	58.7	0	78.3	55.8	0	74.4
Upper bound	73.8	62.5	77.6	73.8	70.4	74.8
Finetune	0	0	0	0	0	0
Feature extraction	5.5	2.1	6.6	5.3	1.9	6.4
LwF	6.1	2.4	7.3	5.6	2.5	6.6
REAL	**74.9**	**62.2**	**79.1**	**74.2**	**71.9**	**75.0**

using the unlabeled background points $\mathbf{Y}_{nm}^{old}$ is extremely helpful to preserve the old knowledge. Compared to the upper bound, our method only needs the ground truth of newly introduced classes $\mathcal{K}_n$ and consumes much less time in training (5 epochs v.s. 35 epochs), while keeping the similar performance.

We show the performance of the model on the nuScenes dataset during IL in Fig. 7. Figure 7 (a) shows during IL the model are gradually learning novel classes while keeping the performance of old classes. Figure 7 (b) illustrates the model starts from the closed-set model and finally achieves the comparable performance with the upper bound.

5.4 Open-world Semantic Segmentation

We illustrate the whole open-world semantic segmentation system in Fig. 8. Traditional closed-set model $\mathcal{M}_c$ classifies objects of novel classes $\mathcal{K}_n$ as old classes $\mathcal{K}_0$. In Fig. 8 (c), A (construction vehicle) is classified as manmade, pedestrian, and truck; B (barrier) is classified as road and manmade; C (traffic-cone) is classified as road. Such misclassification may cause serious problems in autonomous driving. Thus we conduct the methods in Eq. 8 to finetune $\mathcal{M}_c$ to $\mathcal{M}_o$ so that this open-set model can identify these novel objects as *unknown*, as shown in pink area of Fig. 8 (d). Then, after incremental learning using the methods described

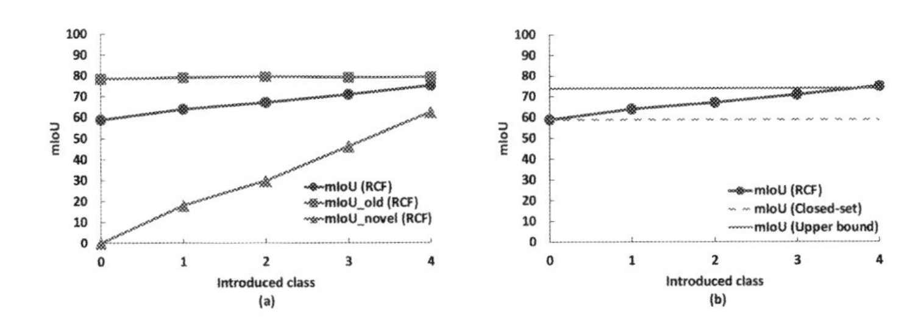

Fig. 7. Incremental learning results for nuScenes validation set. Introduced class: 1: barrier; 2: construction-vehicle; 3: traffic-cone; 4: trailer.

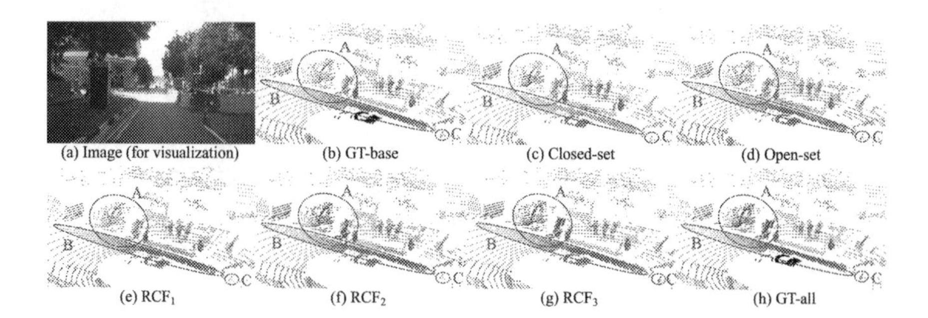

Fig. 8. Qualitative results of open-world semantic segmentation. GT: ground truth. In (b) GT-base we set the novel classes $\mathcal{K}_n$ in pink (**A**: construction-vehicle; **B**: barrier; **C**: traffic-cone). (c) Closed-set prediction classifies novel objects as old classes. (d) Open-set prediction can identify these novel objects as *unknown*. We gradually introduce the labels of barrier, construction-vehicle, and traffic-cone in (e) REAL$_1$, (f) REAL$_2$, and (g) REAL$_3$, so they can classify these novel classes one by one. (h) GT-all contains ground truth of all classes. (Color figure online)

in Eq. 11, the model can gradually classify new classes, *e.g.*, A (barrier), B (construction-vehicle), and C (traffic-cone) in Fig. 8 (e), (f), and (g). Note that after incremental learning the model can still identify unknown classes, as shown in the pink areas of Fig. 8 (e).

6 Conclusion

Traditional closed-set semantic segmentation cannot handle objects of novel classes. In this paper, we propose the open-world semantic segmentation for LIDAR point clouds, where the model can identify novel objects (open-set semantic segmentation) and then gradually learn them when labels are available (incremental learning). We propose the redundancy classifier framework (REAL) and corresponding training and inference strategies to fulfill the open-world semantic segmentation system. We hope this work can draw the attention of researchers toward this meaningful and open problem.

References

1. Baur, C., Wiestler, B., Albarqouni, S., Navab, N.: Deep autoencoding models for unsupervised anomaly segmentation in brain MR images. In: International MICCAI Brainlesion Workshop (2018)
2. Behley, J., et al.: Towards 3D LiDAR-based semantic scene understanding of 3D point cloud sequences: The SemanticKITTI Dataset. In: The International Journal on Robotics Research (2021)
3. Behley, J., et al.: A Dataset for Semantic Scene Understanding of LiDAR Sequences. In: ICCV (2019)
4. Bendale, A., Boult, T.: Towards open world recognition. In: CVPR (2015)

5. Bozhinoski, D., Di Ruscio, D., Malavolta, I., Pelliccione, P., Crnkovic, I.: Safety for mobile robotic systems: A systematic mapping study from a software engineering perspective. J. Syst. Softw. **151** 150–179 (2019)
6. Caesar, H., et al.: nuscenes: A multimodal dataset for autonomous driving. In: CVPR (2020)
7. Cen, J., Yun, P., Cai, J., Wang, M.Y., Liu, M.: Deep metric learning for open world semantic segmentation. In: ICCV (2021)
8. Cermelli, F., Mancini, M., Bulo, S.R., Ricci, E., Caputo, B.: Modeling the background for incremental learning in semantic segmentation. In: Proceedings of the IEEE Computer Society Conference on Computer Vision and Pattern Recognition, pp. 9230–9239 (2020). https://doi.org/10.1109/CVPR42600.2020.00925
9. Cermelli, F., Mancini, M., Bulo, S.R., Ricci, E., Caputo, B.: Modeling the background for incremental learning in semantic segmentation. In: CVPR (2020)
10. Cheng, R., Razani, R., Taghavi, E., Li, E., Liu, B.: 2–s3net: Attentive feature fusion with adaptive feature selection for sparse semantic segmentation network. In: CVPR (2021)
11. Delange, M., et al.: A continual learning survey: Defying forgetting in classification tasks. IEEE Trans. Pattern Analysis Mach. Intell. **44** 3366–3385 (2021)
12. Gal, Y., Ghahramani, Z.: Dropout as a bayesian approximation: Representing model uncertainty in deep learning. In: ICML (2016)
13. Geiger, A., Lenz, P., Urtasun, R.: Are We Ready for Autonomous Driving? CVPR, The KITTI Vision Benchmark Suite. In (2012)
14. Hendrycks, D., Basart, S., Mazeika, M., Mostajabi, M., Steinhardt, J., Song, D.: Scaling out-of-distribution detection for real-world settings. arXiv preprint arXiv:1911.11132 (2019)
15. Hendrycks, D., Gimpel, K.: A baseline for detecting misclassified and out-of-distribution examples in neural networks. In: ICLR (2017)
16. Hu, Q., et al.: Learning semantic segmentation of large-scale point clouds with random sampling. IEEE Trans. Pattern Analysis Mach. Intell. textbf44(11), 8338–8354 (2021)
17. Joseph, K.J., Khan, S., Khan, F.S., Balasubramanian, V.N.: Towards open world object detection. In: CVPR (2021)
18. Kendall, A., Gal, Y.: What uncertainties do we need in bayesian deep learning for computer vision? In: NeurIPS (2017)
19. Lakshminarayanan, B., Pritzel, A., Blundell, C.: Simple and scalable predictive uncertainty estimation using deep ensembles. In: NeurIPS (2017)
20. Li, D., Tasci, S., Ghosh, S., Zhu, J., Zhang, J.T., Heck, L.: Rilod: near real-time incremental learning for object detection at the edge. In: Proceedings of the 4th ACM/IEEE Symposium on Edge Computing, pp. 113–126 (2019)
21. Li, Z., Hoiem, D.: Learning without forgetting. IEEE Trans. Pattern Analysis Mach. Intell. **40**(12), 2935–2947 (2018)
22. Lis, K., Nakka, K., Fua, P., Salzmann, M.: Detecting the unexpected via image resynthesis. In: ICCV (2019)
23. Liu, L., Kuang, Z., Chen, Y., Xue, J., Yang, W., Zhang, W.: Incdet: In defense of elastic weight consolidation for incremental object detection. IEEE Transactions on Neural Networks and Learning Systems, pp. 1–14 (2020)
24. MacKay, D.J.: Bayesian neural networks and density networks. Nuclear Instruments and Methods in Physics Research Section A: Accelerators, Spectrometers, Detectors and Associated Equipment (1995)
25. McCloskey, M., Cohen, N.J.: Catastrophic interference in connectionist networks: The sequential learning problem. In: Psychology of learning and motivation (1989)

26. Milioto, A., Vizzo, I., Behley, J., Stachniss, C.: Rangenet++: Fast and accurate lidar semantic segmentation. In: IROS (2019)
27. Park, T., Liu, M.Y., Wang, T.C., Zhu, J.Y.: Semantic image synthesis with spatially-adaptive normalization. In: CVPR (2019)
28. Peng, C., Zhao, K., Lovell, B.: Faster ilod: Incremental learning for object detectors based on faster rcnn. arXiv preprint arXiv:2003.03901 (2020)
29. Qi, C.R., Su, H., Mo, K., Guibas, L.J.: Pointnet: Deep learning on point sets for 3d classification and segmentation. In: CVPR (2017)
30. Qi, C.R., Yi, L., Su, H., Guibas, L.J.: Pointnet++: Deep hierarchical feature learning on point sets in a metric space. arXiv preprint arXiv:1706.02413 (2017)
31. Rannen, A., Aljundi, R., Blaschko, M.B., Tuytelaars, T.: Encoder based lifelong learning. In: Proceedings of the IEEE International Conference on Computer Vision, pp. 1320–1328 (2017)
32. Shmelkov, K., Schmid, C., Alahari, K.: Incremental learning of object detectors without catastrophic forgetting. In: 2017 IEEE International Conference on Computer Vision (ICCV), pp. 3420–3429 (2017)
33. Thomas, H., Qi, C.R., Deschaud, J.E., Marcotegui, B., Goulette, F., Guibas, L.J.: Kpconv: Flexible and deformable convolution for point clouds. In: ICCV (2019)
34. Wang, Y., Li, B., Che, T., Zhou, K., Liu, Z., Li, D.: Energy-based open-world uncertainty modeling for confidence calibration. In: ICCV (2021)
35. Wang, Y., Shi, T., Yun, P., Tai, L., Liu, M.: Pointseg: Real-time semantic segmentation based on 3d lidar point cloud. arXiv preprint arXiv:1807.06288 (2018)
36. Welling, M., Teh, Y.W.: Bayesian learning via stochastic gradient langevin dynamics. In: ICML, pp. 681–688 (2011)
37. Wu, B., Zhou, X., Zhao, S., Yue, X., Keutzer, K.: Squeezesegv 2: Improved model structure and unsupervised domain adaptation for road-object segmentation from a lidar point cloud. In: ICRA (2019)
38. Wu, W., Qi, Z., Fuxin, L.: Pointconv: Deep convolutional networks on 3d point clouds. In: CVPR (2019)
39. Xia, Y., Zhang, Y., Liu, F., Shen, W., Yuille, A.L.: Synthesize then compare: detecting failures and anomalies for semantic segmentation. In: Vedaldi, A., Bischof, H., Brox, T., Frahm, J.-M. (eds.) ECCV 2020. LNCS, vol. 12346, pp. 145–161. Springer, Cham (2020). https://doi.org/10.1007/978-3-030-58452-8_9
40. Yun, P., Cen, J., Liu, M.: Conflicts between likelihood and knowledge distillation in task incremental learning for 3d object detection. In: 3DV (2021)
41. Yun, P., Liu, Y., Liu, M.: In defense of knowledge distillation for task incremental learning and its application in 3D object detection. IEEE Robot. Autom. Lett. **6**(2), 2012–2019 (2021). https://doi.org/10.1109/LRA.2021.3060417
42. Zhang, Y., et al.: Polarnet: An improved grid representation for online lidar point clouds semantic segmentation. In: CVPR (2020)
43. Zhou, D.W., Ye, H.J., Zhan, D.C.: Learning placeholders for open-set recognition. In: CVPR (2021)
44. Zhu, X., et al.: Cylindrical and asymmetrical 3d convolution networks for lidar segmentation. In: CVPR (2021)

KING: Generating Safety-Critical Driving Scenarios for Robust Imitation via Kinematics Gradients

Niklas Hanselmann[1,2,3(✉)], Katrin Renz[2,3], Kashyap Chitta[2,3], Apratim Bhattacharyya[2,3], and Andreas Geiger[2,3]

[1] Mercedes-Benz AG R&D, Stuttgart, Germany
[2] University of Tübingen, Tübingen, Germany
[3] Max Planck Institute for Intelligent Systems, Tübingen, Germany
niklas.hanselmann@mercedes-benz.com
https://lasnik.github.io/king/

Abstract. Simulators offer the possibility of safe, low-cost development of self-driving systems. However, current driving simulators exhibit naïve behavior models for background traffic. Hand-tuned scenarios are typically added during simulation to induce safety-critical situations. An alternative approach is to adversarially perturb the background traffic trajectories. In this paper, we study this approach to safety-critical driving scenario generation using the CARLA simulator. We use a kinematic bicycle model as a proxy to the simulator's true dynamics and observe that gradients through this proxy model are sufficient for optimizing the background traffic trajectories. Based on this finding, we propose KING, which generates safety-critical driving scenarios with a 20% higher success rate than black-box optimization. By solving the scenarios generated by KING using a privileged rule-based expert algorithm, we obtain training data for an imitation learning policy. After fine-tuning on this new data, we show that the policy becomes better at avoiding collisions. Importantly, our generated data leads to reduced collisions on both held-out scenarios generated via KING as well as traditional hand-crafted scenarios, demonstrating improved robustness.

1 Introduction

After years of steady progress, autonomous driving systems are getting closer to maturity [25]. Due to the high consequences of failure, they have to satisfy extraordinarily high standards of robustness in the face of unseen and safety-critical scenarios. However, real-world data collection and validation for these situations is dangerous and lacks the necessary scalability [29,31]. These problems can be addressed with realistic simulation. Unfortunately, current simulators such as CARLA [16] are not only insufficient in terms of visual fidelity but

Supplementary Information The online version contains supplementary material available at https://doi.org/10.1007/978-3-031-19839-7_20.

S. Avidan et al. (Eds.): ECCV 2022, LNCS 13698, pp. 335–352, 2022.
https://doi.org/10.1007/978-3-031-19839-7_20

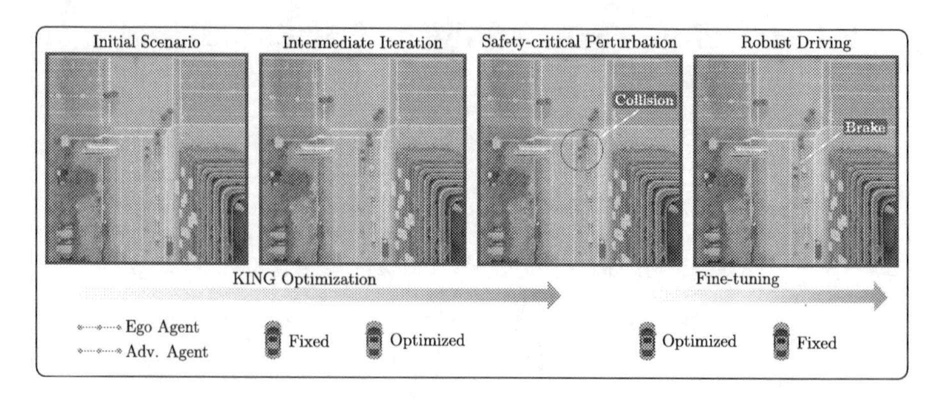

Fig. 1. Generating safety-critical scenarios for robust driving. Left: we propose KING, a novel optimization method to generate safety-critical driving scenarios which iteratively updates the initial scenario using gradients through a differentiable kinematics model and successfully induces a collision with the ego agent. Right: fine-tuning on expert behavior in safety-critical perturbations leads to a more robust agent.

also lack the necessary diversity of driving scenarios: there exists both an *appearance* and a *content gap* to the real world [26]. The content gap poses a major challenge in the adoption of driving agents trained in simulation using imitation learning IL or reinforcement learning (RL), which are often brittle to o.o.d. inputs underrepresented during training [17]. In this work, we aim to address the content gap by improving the behavior of simulated background traffic agents.

Background agents in current simulators follow naïve behavioral models, resulting in limited diversity of the emerging traffic [16,40]. Critical scenarios are often hand-crafted [1,18]. This strategy is unlikely to be successful in fully covering the long-tailed distribution of critical situations that might be encountered in the real world. Furthermore, these scenarios are often non-adaptive to the driving agent under test. A more targeted approach is to actively seek possible failure modes. To do so, existing work perturbs the trajectories of background agents in a physically plausible manner to induce failures in the driving agent [2,48]. This paradigm can be framed as a kinematically constrained adversarial attack on the driving agent, where the amount of safety-critical data generated within a given compute budget is dependent on the success rate of the attack. The prevalent approach for this task is black-box optimization BBO, since simulators are often not differentiable [19,23]. However, as we observe on the widely used CARLA simulator [16], existing attacks based on BBO (e.g. [19,20,23]) struggle to reliably induce collisions in IL-based driving agents (see Table 2).

As observed in image-space adversarial attacks, gradient-based optimization has the potential to be faster and more successful than BBO [3,13]. Moreover, there has been a trend towards end-to-end differentiability, both in simulation [5,44–46] and driving agents [6,7,10,11,36,42]. Using differentiable components enables gradient-based generation of adversarial traffic scenarios. In this paper, we answer an important question: *does the entire simulation pipeline need to be differentiable to provide useful gradients for the optimization of traffic scenarios?* We present KING, a simple and effective approach for safety-critical scenario generation. Our key idea is to use a kinematic bicycle model as proxy to a driving simulator's true dynamics, and solve for safety-critical perturbations of non-critical initial scenarios via backpropagation. The process of optimizing a non-critical scenario with KING is visualized in Fig. 1 (left). Further, we show that KING generates challenging but solvable test cases for driving systems that use both (1) a planner that acts on a bird's-eye view (BEV) grid input and (2) a camera and LiDAR-based driving agent [37]. Finally, we demonstrate that scenarios generated by KING can augment the original training distribution which has limited diversity. This leads to improved collision avoidance, as shown in Fig. 1 (right).

Contributions. (1) We propose KING, a simple procedure for generating safety-critical scenarios via backpropagation that is more reliable and requires less optimization time than BBO. (2) We show that KING generates challenging, diverse, and solvable scenarios for two different driving agents with different input modalities. (3) We use the generated scenarios to augment the CARLA simulator's non-diverse traffic, improving the robustness of an end-to-end IL-based driving agent on both our generated test cases and a benchmark containing CARLA's hand-crafted scenarios.

2 Related Work

End-to-End Driving. We are interested in stress-testing and improving end-to-end learning-based autonomous driving systems. While there are a few RL methods for this task [8,47], most work leverages IL. Some adhere closely to the end-to-end learning paradigm [6,11,12,30,35,37], directly inferring driving actions from raw sensor observations. However, others use interpretable intermediate representations [4,43,49]. In particular, BEV semantic occupancy grid representations are widely used in modern driving approaches [7,10,42,51,52]. This representation can be inferred from images [10,21,22,27,28,32,33,41]. In our study we consider two IL-based driving agents reflecting both schools of thought: (1) a planner called AIM-BEV acting on ground-truth perception represented as a BEV semantic occupancy grid, and (2) an end-to-end agent acting on camera and LiDAR observations called TransFuser [37].

Generating Safety-Critical Scenarios. Previous work on generating safety-critical scenarios relies on BBO techniques and explores a variety of search space parameterizations, such as initial velocity or position of adversarial agents [14,15,38], a high-level route graph [2] or parameterized driving policies [31]. In AdvSim [48], the search space is parameterized as a sequence of kinematic bicycle model states for each adversarial agent, with steering and acceleration actions as free parameters. We also adopt this simple and expressive parameterization for KING. Different from this line of work, we propose a gradient-based procedure to optimize over these parameters rather than resorting to BBO techniques. Concurrent work presents STRIVE [39], a framework that also generates critical scenarios via gradient-based optimization. Here, an adversarial agent is parameterized as a latent vector of a learned motion forecasting model. STRIVE focuses on a simple, privileged rule-based planner rather than end-to-end IL agents and uses a proxy of the driving agent to enable gradient-based optimization, while KING directly optimizes for collisions wrt. the actual driving agent.

3 Safety-Critical Scenario Generation for Robust Imitation

In this section, we outline our overall approach for stress-testing and improving the robustness of IL-based driving agents, which is illustrated in Fig. 2. Given a driving agent trained on a dataset $\mathcal{D}_{reg}$ of regular traffic, we propose KING, a novel gradient-based optimization procedure for automatically generating safety-critical perturbations of non-critical scenarios tailored to the agent under consideration. These scenarios serve to augment the original training distribution with limited diversity. In the following, we formally present our task settings, detail the parameterization and objective function used for scenario generation, and describe our robust training approach for IL.

Driving Agent and Regular Training. We assume that the driving policy of the agent is a neural network π_ω with parameters ω that takes in an observation $\mathbf{o}_t \in \mathbb{R}^{H_o \times W_o \times C_o}$ and goal location $\mathbf{x}_{goal} \in \mathbb{R}^2$ indicating the intended high-level route on the map, and plans a trajectory represented by four future 2D waypoints $\mathbf{w} \in \mathbb{R}^{4\times 2}$:

$$\pi_\omega\left(\mathbf{o}_t, \mathbf{x}_{goal}\right) : \mathbb{R}^{H_o \times W_o \times C_o} \times \mathbb{R}^2 \rightarrow \mathbb{R}^{4\times 2}. \tag{1}$$

Based on the predicted waypoints, the final actions $\mathbf{a}_t^0 \in [-1,1]^2$ in the form of throttle and steering commands are produced by lateral and longitudinal controllers. Currently, several state-of-the-art IL agents fall under this paradigm [9,10,37]. With this general scheme, we consider both an IL policy with an intermediate representation as well as a strictly end-to-end model in our study. The first is a planner acting on ground-truth visual abstractions which we will refer to as AIM-BEV. This is inspired by [4] and the AIM-VA model in [10], but uses a BEV intermediate representation instead of 2D semantics since the BEV is an orthographic projection of the physical 3D space which is better correlated with vehicle kinematics than the 2D image domain. Here, the observations $\mathbf{o}_t \in \mathbb{R}^{192\times 192\times 3}$ are a rasterized BEV grid encoding HD map information with channels for (1) road and (2) lanes as well as a separate channel for dynamic obstacles such as background agents (3). The grid represents the environment ahead and to each side of the agent at a resolution of 5 pixels per meter. In addition to AIM-BEV, we also stress-test the publicly available checkpoint released by the authors of TransFuser [37]. This is a recent state-of-the-art IL-based self-driving model acting on observations $\mathbf{o}_t^{rgb} \in \mathbb{R}^{256 \times 256 \times 3}$ obtained from a front-facing camera and a discretized BEV lidar-histogram with two height bins $\mathbf{o}_t^{lid} \in \mathbb{R}^{256 \times 256 \times 2}$. Both AIM-BEV and TransFuser are trained on observation-waypoint pairs $(\mathbf{o}, \mathbf{w})$ drawn from $\mathcal{D}_{reg}$. The observations are mapped to a latent representation which is input to a gated recurrent unit GRU that plans the trajectory $\mathbf{w}$ in an autoregressive fashion. For additional details, please refer to the supplementary material and the original TransFuser paper.

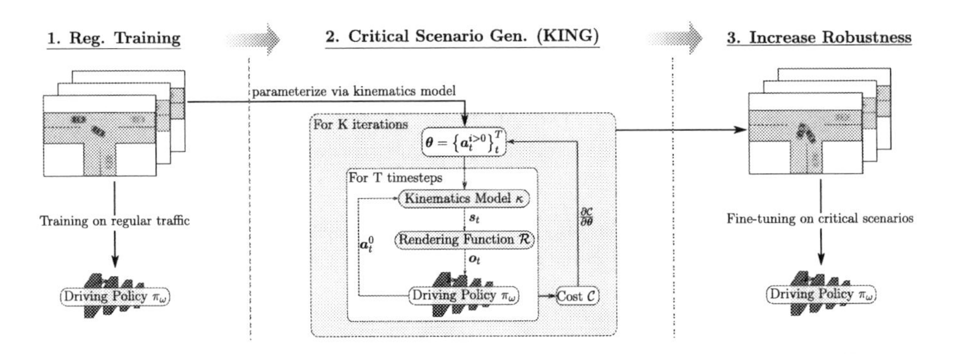

Fig. 2. Robust training pipeline. Given any agent with a driving policy π_ω trained on regular traffic data, we propose to increase its robustness under safety-critical scenarios by generating targeted augmentations. We propose KING, a gradient-based optimization procedure to obtain safety-critical perturbations of initial regular traffic scenarios. These perturbations then serve as additional training data for π_ω.

Gradient-Based Scenario Generation. To optimize for safety-critical perturbations of an initial non-critical scenario (regular traffic), we iteratively simulate the scenario with the driving agent under attack (ego agent) in a closed-loop simulation. In particular, we aim to create a collision between the ego agent and one of the background actors (adversarial agents). At each iteration, we adjust the scenario's parameters (i.e. the trajectories of adversarial agents) in order to induce such a collision. Importantly, the ego agent is able to react to the perturbations of the adversarial agents, since the attacks take place in a closed loop. Therefore, the scenarios generated are adaptive to the specific ego agent being attacked. In the following, we formally describe the simulation process and scenario generation procedure.

Let $\mathbf{x}_t^i \in \mathbb{R}^2$, $\psi_t^i \in [0, 2\pi]$ and $v_t^i \in \mathbb{R}$ be the ground-plane position, orientation and speed of the i-th agent at time t, where the index 0 indicates the ego agent. We denote the traffic state as $\mathbf{s}_t = \left\{\mathbf{x}_t^i, \psi_t^i, v_t^i\right\}_{i=0}^N$, where N is the number of agents. In slight abuse of notation, we will use $\mathbf{s}_t^i$ to refer to the state of a specific agent. We instantiate a particular scenario as a sequence of these states $\mathcal{S} = \{\mathbf{s}_t\}_{t=0}^T$, where T is a fixed simulation horizon. $\mathcal{S}$ is initialized using regular, non-critical traffic behavior as described in Sect. 4.2. To unroll the simulation forward in time, we compute the state at the next timestep $\mathbf{s}_{t+1}$ given the current state $\mathbf{s}_t$ and actions of all agents $\mathbf{a}_t = \{\mathbf{a}_t^i\}_{i=0}^{N}$ using the kinematics model κ, i.e., $\mathbf{s}_{t+1} = \kappa(\mathbf{s}_t, \mathbf{a}_t)$. We choose the bicycle model, which provides a strong prior on physically plausible motion of non-holonomic vehicles [8,34] and is differentiable, enabling backpropagation through the unrolled state sequence $\mathcal{S}$. The ego agent is reactive to the simulation and chooses its actions $\mathbf{a}_t^0$ based on observations $\mathbf{o}_t$ of the true underlying state, which are obtained through a rendering function $\mathcal{R}$, i.e., $\mathbf{o}_t = \mathcal{R}(\mathbf{s}_t, \mathcal{M})$. To render BEV semantic occupancy grids for AIM-BEV, we query a differentiable rasterizer [24] for the given current state $\mathbf{s}_t$ and HD map $\mathcal{M}$, representing other agents by their bounding polygons. To render sensor data, e.g., camera and LiDAR data for TransFuser, we query the CARLA simulator's graphics engine. Note that all components of the simulation (π, κ and $\mathcal{R}$) are differentiable for AIM-BEV but $\mathcal{R}$ is not differentiable for TransFuser.

Safety-Critical Perturbation. We perturb the sequence of states $\{\mathbf{s}_t^{i>0}\}_{t=0}^{T}$ for the N adversarial agents to induce a collision. If the ego agent collides within T timesteps, we terminate the simulation successfully. However, we would like the behavior of the adversarial agents to remain plausible. Hence, if any adversarial agent deviates from the drivable areas of the map or collides with another adversarial agent, the simulation terminates unsuccessfully. To detect collisions, we perform intersection checks between the bounding boxes of the agents. For out-of-bounds violations, we check if the adversarial agent bounding boxes enter the off-road area of the map.

As in [48], we parameterize the trajectories of adversarial agents as a sequence of states obtained by unrolling the kinematics model κ. Specifically, a safety-critical perturbation is found by optimizing the sequence of actions $\{\mathbf{a}_t^{i>0}\}_{t=1}^{T}$ for each adversarial agent. The overall search space can be written as $\boldsymbol{\theta} = \{\boldsymbol{\theta}^i\}_{i=1}^{N}$ where $\boldsymbol{\theta}^i = \{\mathbf{a}_{t=0}^i, ..., \mathbf{a}_{t=T}^i\}$, with dimensionality $N \times T \times 2$. We optimize an objective $\mathcal{C}$ which is motivated by prior work on safety-critical scenario generation [2,15,48]:

$$\boldsymbol{\theta}^* = \operatorname*{argmin}_{\boldsymbol{\theta}} \; \mathcal{C}(\mathcal{S}) \quad \text{with} \quad \mathcal{C}(\mathcal{S}) = \phi_{col}^{ego}(\mathcal{S}) + \lambda\, \phi_{col}^{adv}(\mathcal{S}) + \gamma\, \phi_{dev}^{adv}(\mathcal{S}). \tag{2}$$

We encourage collisions involving the ego agent with the cost ϕ_{col}^{ego}, which measures the ℓ_2 distance to the closest adversarial agent. Unsuccessful terminations of the simulation are discouraged via the costs ϕ_{col}^{adv} and ϕ_{dev}^{adv}, weighted using hyper-parameters λ and γ. Here, ϕ_{col}^{adv} discourages collisions between adversarial agents by penalizing violations of a safety margin and ϕ_{dev}^{adv} penalizes deviations from the drivable area. These costs are similar to those commonly used in planning [7,42,50]. For details, we refer to the supplementary material.

Note that the realism of the generated scenarios is determined by the choice of regularizing terms in $\mathcal{C}$. While additional regularization may be beneficial, we find that the three terms in Eq. (2) are sufficient to find meaningful scenarios. We remark that our goal is to discover challenging scenarios that lie in the long tail of the distribution of traffic. Therefore, the scenarios discovered by our objective are not all likely to occur frequently in daily traffic. Importantly, however, the discovered scenarios are diverse, solvable, and enable learning more robust driving behaviors as demonstrated in Sects. 4.2 and 4.4.

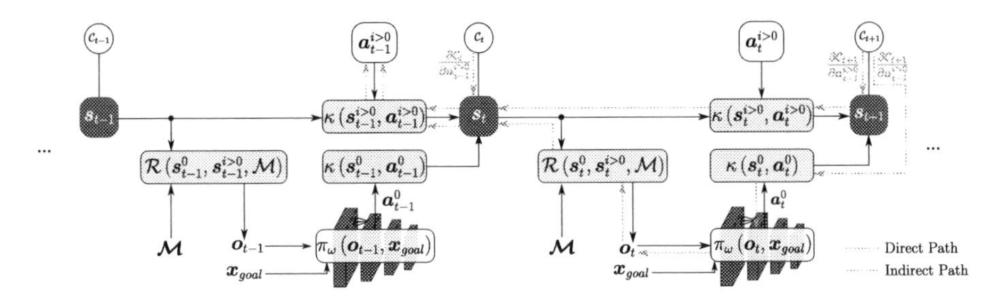

Fig. 3. Gradient paths. To unroll a simulation, we first render an observation $\mathbf{o}_t$ of the traffic state $\mathbf{s}_t$ using a rendering function $\mathcal{R}$. Both the ego agent policy π_ω and adversarial agents then take actions. The actions of the ego agent $\mathbf{a}_t^0$ depend on the observation and a goal location $\mathbf{x}_{goal}$. The actions of the adversarial agents $\mathbf{a}_t^{i>0}$ form the search space of the generation procedure. Given all actions, the next state $\mathbf{s}_{t+1}$ is computed using a differentiable kinematics model κ. Gradients from the cost at time t can then be propagated back to preceding timesteps. The derivative has components along two paths: an efficient direct path and a compute-intensive indirect path.

Kinematics Gradients. Given that the sequence of states $\mathcal{S}$ is unrolled based on the differentiable kinematics model, we can backpropagate costs at any timestep t to the set of actions $\{\mathbf{a}_{t-1}, \mathbf{a}_{t-2}, ..., \mathbf{a}_0\}$ at previous timesteps. In the full unrolled computation graph of the simulation, the true gradients of the cost at any timestep can be taken wrt. the actions in preceding timesteps by recursively applying the chain rule along two paths: a direct path through the kinematics model and an indirect path, which additionally involves the driving policy π_ω and renderer $\mathcal{R}$. This is illustrated in Fig. 3.

With KING, we propose an approximation to the true gradients, which only considers the direct path and stops gradients through the indirect path. While this introduces an error in the gradient estimation, we empirically find it to work well while leading to several advantages. Firstly, as we show in Sect. 4.2, it enables gradient-based generation in the common case where the rendering function or driving policy is non-differentiable, preventing gradients to be taken wrt. the indirect path. Secondly, even when all components are differentiable, taking gradients wrt. to the indirect path involves backpropagating through the driving policy and rendering function (dotted red arrows in Fig. 3), incurring significant computational overhead. We investigate this setting for AIM-BEV where both the driving policy and rendering function are differentiable in Sect. 4.2 and show that given a fixed computational budget, this computational overhead leads to worse overall results compared to KING. We hypothesize that utilizing gradients through both paths becomes more important as the driving policy becomes robust to attacks.

Robust Training for IL. After stress-testing the IL-based driving agents, we are further interested in improving robustness by augmenting the original training data with the generated safety-critical scenarios. To this end, we pursue a simple yet effective strategy: (1) we generate a large set of safety-critical scenarios (2) we filter these for scenarios in which a privileged rule-based expert algorithm finds a safe alternate trajectory, (3) we collect a dataset of observation-waypoint pairs $\mathcal{D}_{crit}$ for the filtered scenarios using the expert, and (4) we fine-tune the policy π_ω with the standard L_1 loss $\mathcal{L}$ on a mix of the safety-critical data $\mathcal{D}_{crit}$ and the original dataset $\mathcal{D}_{reg}$:

$$\omega^* = \underset{\omega}{\operatorname{argmin}}\ \mathbb{E}_{(\mathbf{o}_t,\mathbf{x}_{goal},\mathbf{w})\sim(\mathcal{D}_{crit}\ \cup\ \mathcal{D}_{reg})}\left[\mathcal{L}(\mathbf{w}, \pi_\omega\left(\mathbf{o}_t, \mathbf{x}_{goal}\right))\right]. \tag{3}$$

4 Experiments

We begin by presenting the research questions we aim to answer in this study.
Can gradient-based attacks outperform black-box optimization (BBO) for safety-critical scenario generation? We are interested in reducing the optimization time needed to take a set of non-critical scenario initializations and find interesting scenarios. Given the computational overhead of computing gradients and performing a backward pass, we analyze the gains that can be achieved for this task with gradient-based attacks over BBO in Sect. 4.2. In addition, as shown in Fig. 3, there are two paths for gradients through a simulator. We aim to understand the computational cost of backpropagating through each path and the corresponding gains in terms of collision rates.
Are gradient-based attacks applicable to non-differentiable simulators? While our main experiments are conducted using a differentiable simulator, in Sect. 4.3, we aim to investigate the applicability of KING to non-differentiable rendering functions, such as CARLA's camera and LiDAR sensors.
Can we improve robustness by augmenting the training distribution with critical scenarios? We are interested in the analyzing robustness of the fine-tuned IL model that uses the data augmentation strategy described in Sect. 3. In Sect. 4.4, we investigate this on both the regular benchmark (hand-crafted scenarios) and held-out safety-critical test scenarios generated by KING.

4.1 Benchmarking IL Agents on Hand-Crafted Scenarios

To gain an initial understanding of their robustness, we first benchmark the agents used in our study with hand-crafted scenarios from CARLA. As an additional benchmark that aims to maximize the traffic interactions achievable with such scenarios, we select a set of short routes through intersections involving dense traffic. We describe these benchmarks below. The results provide a reference for performance of our AIM-BEV agent and the existing TransFuser agent on these settings which are relevant for the following experiments. All our experiments are conducted using CARLA version 0.9.10.1.

Experimental Setup. AIM-BEV and TransFuser [37] are trained via supervised learning to imitate a privileged expert on data containing regular CARLA traffic. The expert is a rule-based algorithm similar to the CARLA traffic manager autopilot. We evaluate these models on two benchmarks: (1) the NEAT validation routes from [10], and (2) a set of 82 routes through intersections in CARLA's Town10 with dense traffic. The NEAT routes provide a holistic evaluation of the driving performance, but the evaluation is time-consuming. This set contains routes varying in length from 100 m–3 km with regular CARLA traffic and hand-crafted scenarios. Since several of the routes are long and contain low traffic densities, poor collision avoidance has limited impact on the final metrics. For a more focused evaluation on collisions with traffic, the Town10 intersection routes are shorter in length (80 m–100 m). In this setting, we ensure a high density of dynamic agents by spawning vehicles at every possible spawn point permitted by the CARLA simulator. Furthermore, each route is guaranteed to contain a hand-crafted scenario in which multiple vehicles enter the intersection from different directions at the same time. We selected Town10 for this benchmark as we found it to be the most challenging in preliminary experiments. On both of these benchmarks, we report the official metrics of the CARLA leaderboard, **Route Completion (RC)**, **Infraction Score (IS)** and **Driving Score (DS)**. RC is the percentage of the route completed by an agent before it gets blocked or deviates from the route. IS is a cumulative multiplicative penalty for every red light violation, stop sign violation, collision, and lane infraction. DS is the final metric, computed as the RC multiplied by the IS for each route. Each model is tested with three different evaluation seeds. In addition, we report the **collision rate (CR)**, which is the percentage of routes in which the agent was involved in a collision. Additional details regarding the driving metrics, rule-based expert, and training dataset for the driving policy are provided in the supplementary.

Results. The performance of both IL-agents as well as the rule-based expert which uses privileged information is shown in Table 1. Note that these methods have different inputs, and are not directly comparable. AIM-BEV achieves superior results in comparison to TransFuser. In particular, its significantly higher IS on the NEAT routes indicates that it is proficient at avoiding collisions when placed in sparse and non-adversarial CARLA traffic. On the Town10 intersections, AIM-BEV has a better IS than TransFuser, but we observe that the CR of both agents is similar (17.48%). This is much higher than the expert (CR=3.66%), showing that hand-crafted scenarios in dense traffic remain challenging for current IL-based methods. These hand-crafted scenarios are not adaptive to the agent, i.e., the same scenarios are applied for both AIM-BEV and TransFuser. In the following, we study the more targeted approach of actively generating safety-critical scenarios that are adaptive to the agent being attacked.

4.2 Comparison to BBO for Safety-Critical Scenario Generation

Next, we analyze the efficacy of KING for the generation of safety-critical scenarios, by comparing it with several BBO baselines for attacking AIM-BEV.

Experimental Setup. One scenario in our experimental setup involves rolling out a policy for 20 s of simulation time (80 timesteps at 4fps). We find this time horizon to be sufficient for the ego agent to traverse a route from the start location to the end location while coming in close proximity to the adversarial agents. We compare several adversarial optimization techniques on 80 such scenarios. We obtain 4 maps (Town03-Town06) from the CARLA simulator. The 4 maps have a wide variety of road layouts, including intersections, single-lane roads, multi-lane highways, exits, and roundabouts. We sample a dense set of candidate start locations and end locations for the ego agent from the set of all junctions available in these 4 maps. The 80 ego agent routes in our evaluation are obtained by uniformly sampling 20 candidate routes per CARLA town. For each of these routes, we then initialize the adversarial agents to mimic regular CARLA traffic to obtain an initial, non-critical scenario, which allows explicit control over the traffic density. We use three traffic densities in our evaluation: 1, 2 and 4 agents (additional details in supplementary). The adversarial scenarios are evaluated using the **collision rate (CR)**, which is the percentage of routes for which the adversarial scenario search yielded a collision while respecting behavioral constraints. A search is only considered successful if all adversarial agents stay on drivable parts of the map (i.e., the road) and do not collide with other adversarial agents. To evaluate convergence, we report the average **time to 50% collision rate ($t_{50\%}$)**. This measures the average computation cost (in GPU seconds) required to find a collision in 50% of the total scenarios available. Finally, we report the runtime of each technique as the average number of optimization **seconds per iteration (s/it)**. The $t_{50\%}$ and s/it metrics for KING as well as all baselines are evaluated on a single RTX 2080Ti GPU. For all

Table 1. Performance on hand-crafted scenarios. We show the mean ± std over 3 evaluations. AIM-BEV has fewer infractions than TransFuser on the NEAT validation routes. However, both agents collide in over 17% of the Town10 intersection routes.

	NEAT validation routes [10]				Town10 intersections			
Method	RC ↑	IS ↑	DS ↑	CR ↓	RC ↑	IS ↑	DS ↑	CR ↓
AIM-BEV	96.77±3.32	0.95±0.00	92.24±3.32	2.38±4.12	93.86±0.14	0.92±0.01	86.74±0.67	17.48±1.86
TransFuser [37]	99.25±1.30	0.78±0.03	77.59±2.01	11.90±4.12	93.68±2.01	0.85±0.00	80.03±0.79	17.48±0.70
Privileged Expert	99.83±0.07	1.00±0.00	99.83±0.07	0.00±0.00	94.89±0.33	0.97±0.00	92.81±0.53	3.66±0.00

methods, we use a compute budget of 180 s per route on a single GPU, leading to a total experimental budget of up to 4 GPU hours for 80 routes.

Results. We now assess the efficiency of KING compared to BBO. To this end, we report the CR, $t_{50\%}$ and s/it of our approach and several baselines in Table 2. We consider the three traffic density settings separately, as well as the overall metrics for the complete set of 80×3 scenarios. Our baselines optimize the scenario parameters via BBO. In particular, besides **Random Search** and **Bayesian Optimization**, we consider **SimBA** [19], **CMA-ES** [20] and **Bandit-TD** [23]. SimBA is a variant of Random Search that greedily maximizes the objective and CMA-ES is a state-of-the-art evolutionary algorithm. Finally, Bandit-TD computes numerical gradients by integrating priors into a finite differences approach.

KING obtains a significantly higher CR than the BBO baselines in all 3 settings, increasing the number of scenarios for which a safety-critical perturbation is found by over 20%. Among the BBO baselines, CMA-ES attains the best overall scores with respect to both CR and $t_{50\%}$. Interestingly, the best performance for BBO is often observed for $N = 2$ agents. As we increase N from 1 to 2, it becomes easier for the baselines to find one nearby agent that can be perturbed to collide with the ego agent. However, further increasing N to 4 makes it harder to maintain plausible trajectories where the adversarial agents do not collide with each other or go off-road, leading to reduced performance. As the dimensionality of the search space increases (e.g. $N = 4$), KING begins to outperform the baselines in terms of $t_{50\%}$ by a large margin.

Table 2. Critical scenario generation on CARLA. We show the CR, $t_{50\%}$ and s/it for different optimization techniques in three traffic settings, as well as the aggregated metrics. KING finds collisions in over 80% of the initializations.

	1 Agent			2 Agents			4 Agents			Overall		
Method	CR ↑	$t_{50\%}$ ↓	s/it ↓	CR ↑	$t_{50\%}$ ↓	s/it ↓	CR ↑	$t_{50\%}$ ↓	s/it ↓	CR ↑	$t_{50\%}$ ↓	s/it ↓
Random Search	62.50	**9.25**	1.30	68.75	7.38	1.35	68.75	15.22	1.48	66.67	9.66	1.38
Bayesian Optimization	63.75	11.88	1.46	68.75	10.01	1.66	63.75	22.12	2.06	65.00	14.34	1.73
SimBA [19]	60.00	14.14	1.30	71.25	14.35	1.35	61.25	19.68	1.48	64.17	15.84	1.38
CMA-ES [20]	67.50	9.34	1.31	75.00	**6.73**	1.36	62.50	9.39	1.52	68.33	8.17	1.40
Bandit-TD [23]	37.50	–	3.87	30.00	–	4.39	21.25	–	5.02	29.58	–	4.43
KING Direct + Indirect	81.25	17.63	3.17	76.25	11.58	3.22	**80.00**	13.14	3.40	79.17	14.09	3.26
KING (Ours)	**86.25**	9.98	1.78	**82.50**	6.96	1.88	78.75	**6.40**	2.03	**82.50**	**7.78**	1.90

We also compare the proposed approximation in KING against the setting where we use gradients through entire simulation, including the driving policy and renderer ("KING Direct + Indirect" in Table 2). While also reliably finding safety-critical perturbations, the computational overhead of backpropagating through the indirect path leads to worse overall results given the same computation budget. This suggests the approximation in KING is reasonable for efficiently generating safety-critical scenarios. Additional results and details regarding the hyper-parameter choices for BBO are provided in the supplementary material. Since we observe that gradients through the direct path only are sufficient, we now conduct a detailed qualitative analysis where we apply KING to attack TransFuser, which requires the use of CARLA's non-differentiable camera and LiDAR sensors for rendering.

4.3 Analysis of Safety-Critical Scenarios

In this section, we analyze the safety-critical scenarios generated by KING for both AIM-BEV and TransFuser in detail. Specifically, we show the distribution of the resulting scenarios with a traffic density of $N = 4$ agents in Fig. 4. For both driving agents, we first filter out the set of scenarios where KING is unable to find a collision ("No Collision") as well as those that are not solvable by the rule-based expert ("Not Solvable"). We cluster the remaining scenarios using k-means (similar to [39]) to obtain 6 clusters of failure modes such as cut-ins (a_1), rear-ends (a_2) and unsafe behavior in unprotected turns (e,f). From the frequency of scenarios with "No collision" in Fig. 4, we observe that both AIM-BEV and TransFuser collide in at least 80% of the scenarios. This is a significant deviation from the collision avoidance of both models in the benchmarks shown in Table 1, where they attain a CR below 20%. The large amount of collisions for TransFuser indicates that KING can achieve promising results when applied out-of-the-box to driving simulators with non-differentiable rendering functions.

We show qualitative examples in Figs. 5 and 6, and additional examples in the supplementary material. Both AIM-BEV and TransFuser frequently collide in intersections when they encounter traffic that behaves differently from the traffic observed during training. Importantly, the "Not solvable" column shows

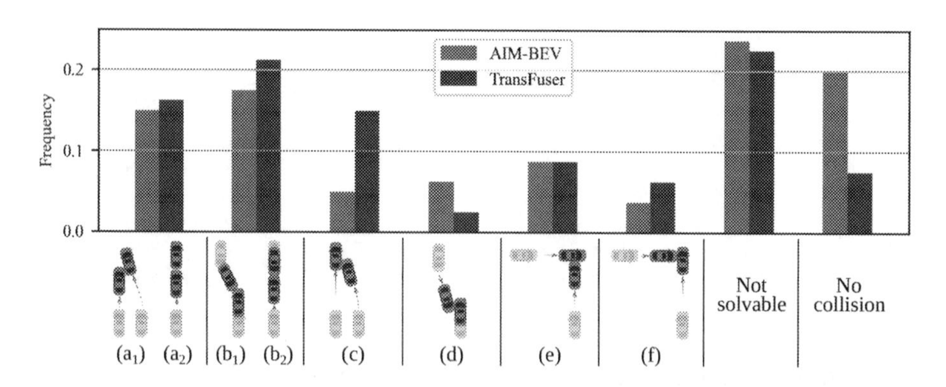

Fig. 4. Collision types. KING generates a diverse set of challenging but solvable scenarios. We group these into 6 clusters (a-f). The illustrations depict the ego agent in red and the adversarial agent in blue. The scenarios include (a) cut-ins and rear-ends caused by the ego agent, (b) head-on collisions, (c) merges, (d) side collisions with oncoming traffic, and t-bone collisions in intersections (e and f).

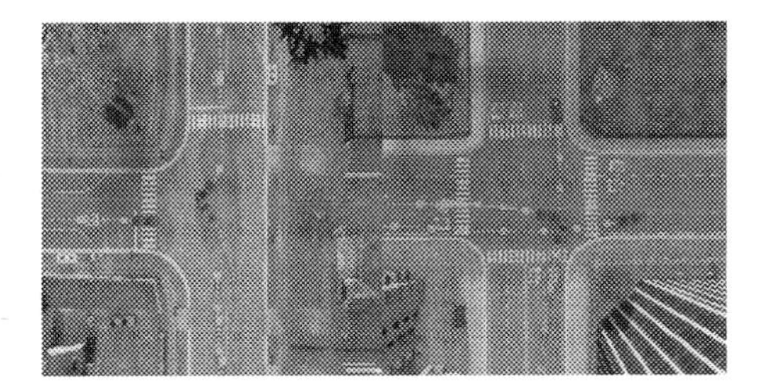

Fig. 5. KING scenarios - AIM-BEV. In intersections, AIM-BEV often fails to yield to the perturbed traffic. This leads to t-bone collisions, either by AIM-BEV (left) or the adversarial agent (right), corresponding to clusters (e) and (f) in Fig. 4.

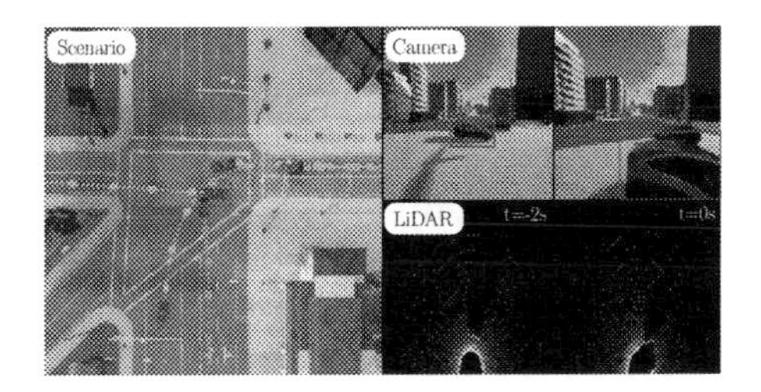

Fig. 6. KING scenario - TransFuser [37]. We show a scenario along with camera and LiDAR inputs two seconds before and at collision. TransFuser fails to slow down for an adversarial agent which stops inside the intersection (red box).

that for both agents, only around 20% of the scenarios have no feasible alternate trajectory. This leaves a large proportion of solvable scenarios in the 6 clusters shown in Fig. 4. The most frequent failure modes of both models are observed in clusters (a) and (b), which involve cut-ins, rear-ends, and head-on collisions. The rule-based expert solves these challenging scenarios by accurately forecasting the motion of the adversarial actors using privileged information. Interestingly, the failure cases are fairly evenly distributed over the 6 clusters which involve a wide variety of relative orientations between the colliding agents. The examples in Fig. 5 correspond to clusters (e) and (f). We highlight examples from clusters (a_1) and (a_2) for our experiment in Fig. 7. The high frequency and diversity of solvable scenarios generated by KING indicate its potential to augment the original training data for IL models, which we investigate next.

Table 3. Robust training for AIM-BEV. Results shown are the mean and std over 3 evaluation seeds. Fine-tuning with safety-critical scenarios reduces the CR by over 50% on other safety-critical scenarios as well as hand-crafted scenarios from CARLA.

	Held-out KING scenarios	Hand-crafted scenarios (Town10 intersections)			
Dataset	CR ↓	RC ↑	IS ↑	DS ↑	CR ↓
No Fine-tuning	100.00 ± 0.00	93.86 ± 0.14	0.92 ± 0.01	86.74 ± 0.67	17.48 ± 1.86
$\mathcal{D}_{reg}$	57.14 ± 0.00	**95.66 ± 0.51**	0.90 ± 0.00	86.85 ± 0.62	19.51 ± 0.00
$\mathcal{D}_{crit}$	**28.57 ± 0.00**	91.92 ± 0.19	**0.96 ± 0.00**	88.37 ± 0.41	**6.10 ± 0.00**
$\mathcal{D}_{crit} \cup \mathcal{D}_{reg}$	**28.57 ± 0.00**	94.42 ± 0.36	**0.96 ± 0.36**	**90.20 ± 0.00**	8.13 ± 0.70

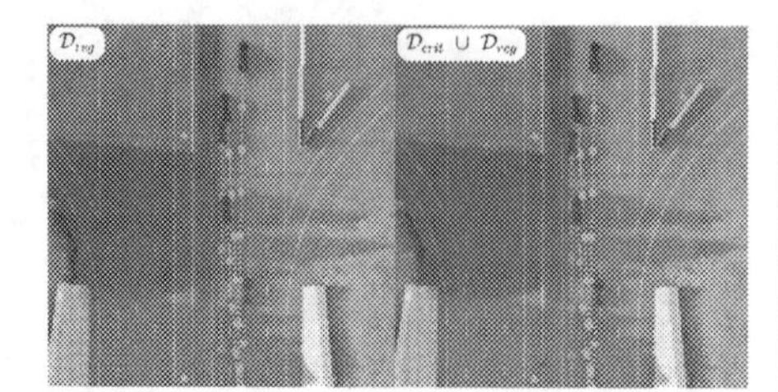

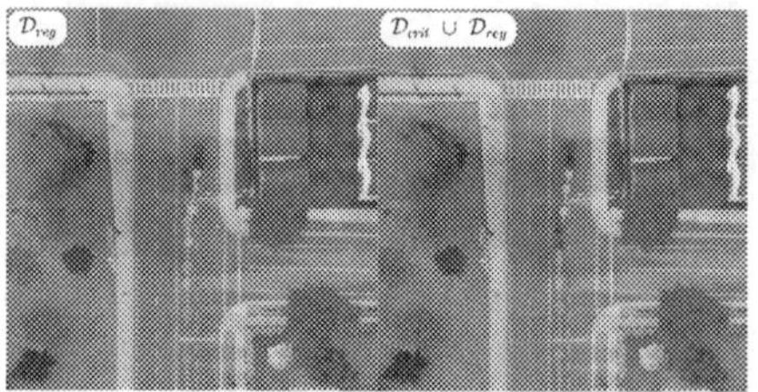

Fig. 7. Improved collision avoidance on held-out KING scenarios with AIM-BEV. Comparison of AIM-BEV before and after fine-tuning on $\mathcal{D}_{crit} \cup \mathcal{D}_{reg}$ in two KING scenarios. Ego agent in red, adversarial agent in blue. Best viewed zoomed in.

4.4 Evaluating Robustness After Fine-Tuning

Finally, we analyze the efficacy of the generated scenarios in augmenting the regular data $\mathcal{D}_{reg}$ to yield more robust driving agents. Here, we evaluate robustness both wrt. safety-critical scenarios generated by KING and to hand-crafted scenarios in the CARLA simulator (using the Town10 intersections benchmark).

Experimental Setup: The goal of this experiment is to collect training data for improving collision avoidance. To this end, we build a large set of safety-critical scenarios by attacking AIM-BEV using initializations from Town03-Town06 of CARLA with $N = 4$ agents. To ensure meaningful supervision, we filter the resulting scenarios for ones where KING finds collisions that are solvable by the expert. This results in around 300 scenarios from which we hold out 20% for evaluation. We ensure that there is no overlap between the training and evaluation during this split by preventing routes with the same ego vehicle start location from being in both splits. Additional details regarding the training data and hyper-parameters are provided in the supplementary material.

Results. We report the driving performance of AIM-BEV after fine-tuning on $\mathcal{D}_{crit} \cup \mathcal{D}_{reg}$ in Table 3. Since the trajectories of the adversarial agents are fixed after optimization via KING, some of the scenarios may be solvable by simply adopting different overall driving styles, rather than becoming more proficient at collision avoidance. To quantify this, we fine-tune each model with only the original training data $\mathcal{D}_{reg}$ as a baseline, which reduces the CR from 100% to 57.14% on the held-out KING scenarios. Additionally, we compare to fine-tuning on only the critical scenarios $\mathcal{D}_{crit}$ and the initial checkpoint from Table 1 ("No Fine-tuning"). Among the three fine-tuning strategies, using only $\mathcal{D}_{reg}$ leads to unsatisfactory results, with a CR of 19.51% on the Town10 intersections benchmark. Using only $\mathcal{D}_{crit}$ leads to a large reduction in CR on both evaluation settings. However, the model has a lower RC and only a small improvement in DS when compared to the $\mathcal{D}_{reg}$ baseline on the Town10 intersections. Finally, using the combined dataset of $\mathcal{D}_{crit} \cup \mathcal{D}_{reg}$ gives the best results. In this setting, we obtain a CR of 28.57% on the KING scenarios, which is identical to the model fine-tuned with only $\mathcal{D}_{crit}$. However, the DS of this model on Town10 is improved by over 3 points, since it reduces the CR while maintaining a similar RC to the original model. This shows that the simple strategy of fine-tuning on a mixture of regular and safety-critical data is an effective way of learning from the scenarios generated by KING.

In Fig. 7, we show qualitative driving examples of the original and fine-tuned AIM-BEV agents on held-out KING scenarios, which belong to clusters (a_1) and (a_2) from Fig. 4. While these scenarios are straightforward to handle for an expert driver, AIM-BEV fails to brake for a vehicle stopping in between two lanes and is unable to maintain a safe distance in merging maneuvers, which highlights its brittleness in o.o.d scenarios. These scenario types do not frequently emerge naturally from the CARLA simulator's background agent behavior which governs $\mathcal{D}_{reg}$. By incorporating data from $\mathcal{D}_{crit}$ during training, the driving agent can learn to handle these scenarios safely.

5 Conclusion

In this work, we proposed a novel gradient-based generation procedure for safety-critical driving scenarios, KING, which achieves significantly higher success rates compared to existing BBO-based approaches while being more efficient. The key to our success is a compute-efficient direct gradient path through a kinematic motion model to guide the adversarial scenario generation process. Our analysis indicates that KING can achieve promising results when applied out-of-the-box to arbitrary driving agents. Furthermore, we show that despite having access to privileged BEV semantic maps as inputs, state-of-the-art IL-based driving policies are surprisingly brittle to minor perturbations in the behavior of the background actors. By augmenting their training data with scenarios from KING, we are able to significantly improve their collision avoidance. Exploring the robustness of agents with different training procedures (e.g. RL) offers an interesting direction for future research.

Acknowledgments. This work was supported by the BMWi (KI Delta Learning, project numbers: 19A19013A, 19A19013O), the BMBF (Tbingen AI Center, FKZ: 01IS18039A, 01IS18039B) and the DFG (SFB 1233, TP 17, project number: 276693517). We thank the International Max Planck Research School for Intelligent Systems (IMPRS-IS) for supporting Katrin Renz and Kashyap Chitta. We also thank Aditya Prakash and Bernhard Jaeger for proofreading.

References

1. Carla autonomous driving leaderboard. https://leaderboard.carla.org/ (2020)
2. Abeysirigoonawardena, Y., Shkurti, F., Dudek, G.: Generating adversarial driving scenarios in high-fidelity simulators. In: Proceedings of the IEEE International Conference on Robotics and Automation (ICRA) (2019)
3. Andriushchenko, M., Croce, F., Flammarion, N., Hein, M.: Square attack: a query-efficient black-box adversarial attack via random search. In: Vedaldi, A., Bischof, H., Brox, T., Frahm, J.-M. (eds.) ECCV 2020. LNCS, vol. 12368, pp. 484–501. Springer, Cham (2020). https://doi.org/10.1007/978-3-030-58592-1_29

4. Behl, A., Chitta, K., Prakash, A., Ohn-Bar, E., Geiger, A.: Label efficient visual abstractions for autonomous driving. In: Proceedings IEEE International Conference on Intelligent Robots and Systems (IROS) (2020)
5. Bergamini, L., et al.: Simnet: Learning reactive self-driving simulations from real-world observations. In: Proceedings IEEE International Conference on Robotics and Automation (ICRA) (2021)
6. Bojarski, M., et al.: End to end learning for self-driving cars. arXiv.org 1604.07316 (2016)
7. Casas, S., Sadat, A., Urtasun, R.: Mp3: A unified model to map, perceive, predict and plan. In: Proceedings IEEE Conference on Computer Vision and Pattern Recognition (CVPR) (2021)
8. Chen, D., Koltun, V., Krähenbühl, P.: Learning to drive from a world on rails. In: Proceeidngs of the IEEE International Conference on Computer Vision (ICCV) (2021)
9. Chen, D., Zhou, B., Koltun, V., Krähenbühl, P.: Learning by cheating. In: Proceeidngs of the Conference on Robot Learning (CoRL) (2019)
10. Chitta, K., Prakash, A., Geiger, A.: Neat: Neural attention fields for end-to-end autonomous driving. In: Proceedings of the IEEE International Conference on Computer Vision (ICCV) (2021)
11. Codevilla, F., Miiller, M., López, A., Koltun, V., Dosovitskiy, A.: End-to-end driving via conditional imitation learning. In: Proceedings of the IEEE International Conference on Robotics and Automation (ICRA) (2018)
12. Codevilla, F., Santana, E., López, A.M., Gaidon, A.: Exploring the limitations of behavior cloning for autonomous driving. In: Proceedings of the IEEE International Conference on Computer Vision (ICCV) (2019)
13. Deng, Y., Zheng, X., Zhang, T., Chen, C., Lou, G., Kim, M.: An analysis of adversarial attacks and defenses on autonomous driving models. In: Proceedings of IEEE International Conference on Pervasive Computing and Communications (2020)
14. Ding, W., Chen, B., Li, B., Eun, K.J., Zhao, D.: Multimodal safety-critical scenarios generation for decision-making algorithms evaluation. IEEE Robot. Auto. Lett. (RA-L) **6**(2), 1551–1558 (2021)
15. Ding, W., Xu, M., Zhao, D.: Learning to collide: An adaptive safety-critical scenarios generating method. In: Proceedings of the IEEE International Confernece on Intelligent Robots and Systems (IROS) (2020)
16. Dosovitskiy, A., Ros, G., Codevilla, F., Lopez, A., Koltun, V.: CARLA: An open urban driving simulator. In: Proceedings of the Conference on Robot Learning (CoRL) (2017)
17. Filos, A., Tigas, P., McAllister, R., Rhinehart, N., Levine, S., Gal, Y.: Can autonomous vehicles identify, recover from, and adapt to distribution shifts? In: Proceedings of the International Conference on Machine learning (ICML) (2020)
18. Fremont, D.J., et al.: Scenic: A language for scenario specification and data generation (2022)
19. Guo, C., Gardner, J.R., You, Y., Wilson, A.G., Weinberger, K.Q.: Simple black-box adversarial attacks. In: Proceeidngs of the International Confernce on Machine learning (ICML) (2019)
20. Hansen, N., Ostermeier, A.: Completely derandomized self-adaptation in evolution strategies. Evol. Comput. **9**(2), 159–195 (2001)
21. Hendy, N., et al.: Fishing net: Future inference of semantic heatmaps in grids. arXiv.org 2006.09917 (2020)

22. Hu, A., et al.: FIERY: future instance prediction in bird's-eye view from surround monocular cameras. In: Proceedings of the IEEE International Conference on Computer Vision (ICCV) (2021)
23. Ilyas, A., Engstrom, L., Madry, A.: Prior convictions: Black-box adversarial attacks with bandits and priors. In: Proceedings of the International Conference on Learning Representations (ICLR) (2019)
24. Jaderberg, M., Simonyan, K., Zisserman, A., Kavukcuoglu, K.: Spatial transformer networks. In: Advances in Neural Information Processing Systems (NIPS) (2015)
25. Janai, J., Güney, F., Behl, A., Geiger, A.: Computer Vision for Autonomous Vehicles: Problems, Datasets and State of the Art, vol. 12. Foundations and Trends in Computer Graphics and Vision (2020)
26. Kar, A., et al.: Meta-sim: Learning to generate synthetic datasets. In: Proceedings of the IEEE International Conference on Computer Vision (ICCV) (2019)
27. Loukkal, A., Grandvalet, Y., Drummond, T., Li, Y.: Driving among Flatmobiles: Bird-Eye-View occupancy grids from a monocular camera for holistic trajectory planning. arXiv.org 2008.04047 (2020)
28. Mani, K., et al.: MonoLayout: Amodal scene layout from a single image. In: Proceeding of the IEEE Winter Conference on Applications of Computer Vision (WACV) (2020)
29. Norden, J., O'Kelly, M., Sinha, A.: Efficient black-box assessment of autonomous vehicle safety. arXiv.org 1912.03618 (2019)
30. Ohn-Bar, E., Prakash, A., Behl, A., Chitta, K., Geiger, A.: Learning situational driving. In: Proceedings IEEE Conference on Computer Vision and Pattern Recognition (CVPR) (2020)
31. O' Kelly, M., Sinha, A., Namkoong, H., Tedrake, R., Duchi, J.C.: Scalable end-to-end autonomous vehicle testing via rare-event simulation. In: Advances in Neural Information Processing Systems (NeurIPS) (2018)
32. Pan, B., Sun, J., Leung, H.Y.T., Andonian, A., Zhou, B.: Cross-view semantic segmentation for sensing surroundings. IEEE Robot. Autom. Lett. (RA-L) **99** 1–1 (2020)
33. Philion, J., Fidler, S.: Lift, Splat, Shoot: encoding images from arbitrary camera rigs by implicitly unprojecting to 3D. In: Vedaldi, A., Bischof, H., Brox, T., Frahm, J.-M. (eds.) ECCV 2020. LNCS, vol. 12359, pp. 194–210. Springer, Cham (2020). https://doi.org/10.1007/978-3-030-58568-6_12
34. Polack, P., Altché, F., d'Andréa Novel, B., de La Fortelle, A.: The kinematic bicycle model: A consistent model for planning feasible trajectories for autonomous vehicles? In: Proceedings of the IEEE Intelligent Vehicles Symposium (IV) (2017)
35. Pomerleau, D.: ALVINN: an autonomous land vehicle in a neural network. In: Advances in Neural Information Processing Systems (NIPS) (1988)
36. Prakash, A., Behl, A., Ohn-Bar, E., Chitta, K., Geiger, A.: Exploring data aggregation in policy learning for vision-based urban autonomous driving. In: Proceedings IEEE Conference on Computer Vision and Pattern Recognition (CVPR) (2020)
37. Prakash, A., Chitta, K., Geiger, A.: Multi-modal fusion transformer for end-to-end autonomous driving. In: Proceedings IEEE Conference on Computer Vision and Pattern Recognition (CVPR) (2021)
38. Priisalu, M., Pirinen, A., Paduraru, C., Sminchisescu, C.: Generating scenarios with diverse pedestrian behaviors for autonomous vehicle testing. In: Proceedings Conference on Robot Learning (CoRL) (2022)
39. Rempe, D., Philion, J., Guibas, L.J., Fidler, S., Litany, O.: Generating useful accident-prone driving scenarios via a learned traffic prior. In: Proceedings IEEE Conference on Computer Vision and Pattern Recognition (CVPR) (2022)

40. Richter, S.R., Hayder, Z., Koltun, V.: Playing for benchmarks. In: Proceedings of the IEEE International Conference on Computer Vision (ICCV) (2017)
41. Roddick, T., Cipolla, R.: Predicting semantic map representations from images using pyramid occupancy networks. In: Proceedings of the IEEE Conference on Computer Vision and Pattern Recognition (CVPR) (2020)
42. Sadat, A., Casas, S., Ren, M., Wu, X., Dhawan, P., Urtasun, R.: Perceive, predict, and plan: safe motion planning through interpretable semantic representations. In: Vedaldi, A., Bischof, H., Brox, T., Frahm, J.-M. (eds.) ECCV 2020. LNCS, vol. 12368, pp. 414–430. Springer, Cham (2020). https://doi.org/10.1007/978-3-030-58592-1_25
43. Sauer, A., Savinov, N., Geiger, A.: Conditional affordance learning for driving in urban environments. In: Proceedings of the Conference on Robot Learning (CoRL) (2018)
44. Scheel, O., Bergamini, L., Wolczyk, M., Osinski, B., Ondruska, P.: Urban driver: Learning to drive from real-world demonstrations using policy gradients. In: Proceedings of the Conference on Robot Learning (CoRL) (2021)
45. Ścibior, A., Lioutas, V., Reda, D., Bateni, P., Wood, F.: Imagining the road ahead: Multi-agent trajectory prediction via differentiable simulation (2021)
46. Suo, S., Regalado, S., Casas, S., Urtasun, R.: Trafficsim: Learning to simulate realistic multi-agent behaviors. In: Proceedings of the IEEE Confernce on Computer Vision and Pattern Recognition (CVPR) (2021)
47. Toromanoff, M., Wirbel, E., Moutarde, F.: End-to-end model-free reinforcement learning for urban driving using implicit affordances. In: Proceedings of the IEEE Conference on Computer Vision and Pattern Recognition (CVPR) (2020)
48. Wang, J., et al.: Advsim: Generating safety-critical scenarios for self-driving vehicles. In: Proceedings of the IEEE Conference on Computer Vision and Pattern Recognition (CVPR) (2021)
49. Xiao, Y., Codevilla, F., Pal, C., López, A.M.: Action-Based Representation Learning for Autonomous Driving. In: Proceedings Conference on Robot Learning (CoRL) (2020)
50. Zeng, W., et al.: End-to-end interpretable neural motion planner. In: Proceedings IEEE Conference on Computer Vision and Pattern Recognition (CVPR) (2019)
51. Zhang, Z., Liniger, A., Dai, D., Yu, F., Van Gool, L.: End-to-end urban driving by imitating a reinforcement learning coach. In: Proceedings of the IEEE International Conference on Computer Vision (ICCV) (2021)
52. Zhou, Y., et al.: End-to-end multi-view fusion for 3d object detection in lidar point clouds. In: Proc. Conf. on Robot Learning (CoRL) (2019)

Differentiable Raycasting for Self-Supervised Occupancy Forecasting

Tarasha Khurana[1(✉)], Peiyun Hu[2], Achal Dave[3], Jason Ziglar[2], David Held[1,2], and Deva Ramanan[1,2]

[1] Carnegie Mellon University, Pittsburgh, USA
tkhurana@cs.cmu.edu
[2] Argo AI, Pittsburgh, USA
[3] Amazon, Seattle, USA

Abstract. Motion planning for safe autonomous driving requires learning how the environment around an ego-vehicle evolves with time. Ego-centric perception of driveable regions in a scene not only changes with the motion of actors in the environment, but also with the movement of the ego-vehicle itself. Self-supervised representations proposed for large-scale planning, such as ego-centric freespace, confound these two motions, making the representation difficult to use for downstream motion planners. In this paper, we use *geometric occupancy* as a natural alternative to view-dependent representations such as freespace. Occupancy maps naturally disentagle the motion of the environment from the motion of the ego-vehicle. However, one cannot directly observe the full 3D occupancy of a scene (due to occlusion), making it difficult to use as a signal for learning. Our key insight is to use *differentiable raycasting* to "render" future occupancy predictions into future LiDAR sweep predictions, which can be compared with ground-truth sweeps for self-supervised learning. The use of differentiable raycasting allows occupancy to *emerge* as an internal representation within the forecasting network. In the absence of groundtruth occupancy, we quantitatively evaluate the forecasting of raycasted LiDAR sweeps and show improvements of upto 15 F1 points. For downstream motion planners, where emergent occupancy can be directly used to guide non-driveable regions, this representation relatively reduces the number of collisions with objects by up to 17% as compared to freespace-centric motion planners.

1 Introduction

To navigate in complex and dynamic environments such as urban cores, autonomous vehicles need to perceive actors and predict their future movements.

T. Khurana and P. Hu—Equal contribution.

Supplementary Information The online version contains supplementary material available at https://doi.org/10.1007/978-3-031-19839-7_21.

S. Avidan et al. (Eds.): ECCV 2022, LNCS 13698, pp. 353–369, 2022.
https://doi.org/10.1007/978-3-031-19839-7_21

(a) Sensor position y, Scene s

(b) New sensor position $y + \Delta y$, Scene s

(c) New sensor position $y + \Delta y$, New scene $s + \Delta s$

Fig. 1. We propose emergent occupancy as a novel self-supervised representation for motion planning. Occupancy is independent of changes in sensor pose Δ_y, which is in contrast to prior work on self-supervised learning from LiDAR [9,15,28,29] specifically, ego-centric freespace [9], which changes with (**a-b**) sensor pose motion Δ_y and (**b-c**) scene motion Δ_s. We use differentiable raycasting to naturally decouple ego motion from scene motion, allowing us to learn to forecast occupancy by self-supervision from pose-aligned LiDAR sweeps.

Such knowledge is often represented in some form of forecasted occupancy [23], which downstream motion planners rely on to produce safe trajectories. When tackling the tasks of perception and prediction, standard solutions consist of perceptual modules such as object detection, tracking, and trajectory forecasting, which require a massive amount of object track labels. Such solutions do not scale given the speed that log data is being collected by large fleets.

Freespace Versus Occupancy: To avoid the need for costly human annotations, and to enable learning at scale, self-supervised representations such as ego-centric freespace [9] have been proposed. However, such a representation couples the motion of the world with the motion of the ego-vehicle (Fig. 1). Our key innovation in this paper is to learn an ego-pose independent and explainable representation for safe motion planning, which we call *emergent occupancy*. Emergent occupancy decouples ego motion and scene motion using differentiable raycasting: we design a network that learns to "space-time complete" the future volumetric state of the world (in a world-coordinate frame) given past LiDAR observations. Consider an ego-vehicle that moves in a static scene. Here, LiDAR returns (even when aligned to a world-coordinate frame) will still *swim* along the surfaces of the fixed scene (Fig. 2). This implies that even when the world is static, most of what the ego-vehicle observes through the LiDAR sensor appears to move with complex nonlinear motion, but in fact those observations can be fully explained by static geometry and ego-motion (via raycasting). LiDAR forecasters need to implicitly predict this ego-motion of the car to produce accurate future returns. However, we argue that such prediction doesn't make sense for autonomous agents that *plan* their future motion. Importantly, our differentiable raycasting network has access to future camera ego-poses as *input*, both during training (since they are available in archival logs) and testing (since state-of-the-art planners explicitly search over candidate trajectories).

Self-Supervision: Note that ground-truth future volumetric occupancy is largely unavailable without human supervision, because the full 3D world is rarely observed; the ego-vehicle only sees a limited number of future views as recorded in a single archival log. To this end, we apply a differentiable raycaster that projects the forecasted volumetric occupancy into a LiDAR sweep, as seen by the future ego-vehicle motion in the log. We then use the difference between the raycasted sweep and actual sweep as a signal for self-supervised learning, allowing us to train models on massive amounts of unannotated logs.

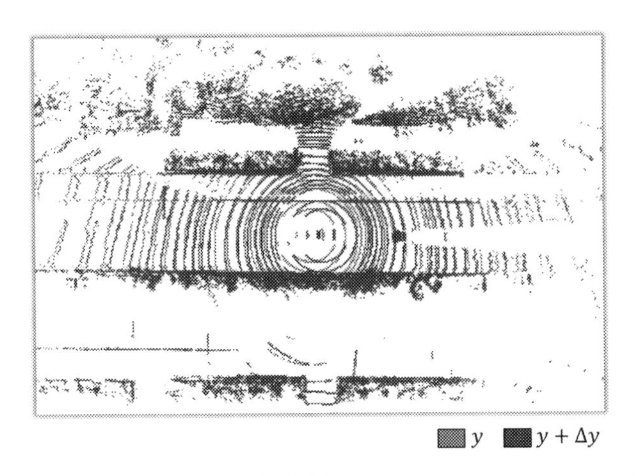

Fig. 2. We pose-align two succesive LiDAR sweeps of a static scene s to a common world coordinate-frame (using the notation of Fig. 1). Even though there is zero scene motion Δs, points appear to drift or *swim* across surfaces. This is due to the fact that points are obtained by intersecting rays from a moving sensor Δy with static scene geometry. This in turn implies that points can appear to move since they are not tied to physical locations on a surface. This apparent movement ($\Delta \tilde{s}$) is in general a complex nonlinear transformation, even when the sensor motion Δy is a simple translation (as shown above). Traditional methods for self-supervised LiDAR forecasting [9,15,28,29] require predicting the complex transformation $\Delta \tilde{s}$ which depends on the unknown Δy, while our differentiable-raycasting framework assumes Δy is an *input*, dramatically simplifying the task of the forecasting network. From a planning perspective, we argue that the future (planned) change-in-pose *should* be an input rather than an output.

Planning: Lastly, we show that such forecasted space-time occupancy can be jointly learned with space-time costmaps for end-to-end motion planning. Owing to LiDAR self-supervision, we are able to train on recent unsupervised LiDAR datasets [13] that are orders of magnitude larger than their annotated counterparts, resulting in significant improvement in accuracy for both forecasted occupancy and motion plans. Interestingly, as we increase the amount of archival training data at the cost of zero additional human annotation, object shape, tracks, and multiple futures "emerge" in the arbitrary quantities predicted by our model despite there being no direct supervision on ground-truth occupancy.

2 Related Work

Occupancy as a Scene Representation: Knowledge regarding what is around an autonomous vehicle (AV) and what will happen next is captured in different representations throughout the standard modular perception and prediction (P&P) pipeline [4,11,24,26]. Instead of separate optimization of these modules [16,25], Sadat et al. [23] propose bird's-eye view (BEV) *semantic occupancy* that is end-to-end optimizable. As an alternative to *semantic occupancy*, Hu et al. [10] propose BEV *ego-centric freespace* that can be self-supervised by raycasting on aligned LiDAR sweeps. However, the ego-centric freespace entangles motion from other actors, which is arguably more relevant for motion planning, with ego-motion. In this paper, we propose *emergent occupancy* to isolate motion of other actors. While we focus on self-supervised learning at scale, we acknowledge that for motion planning, some semantic labelling is required (e.g., state of a traffic light) which can be incorporated via semi-supervised learning.

Differentiable Raycasting: Differentiable raycasting has shown great promise in learning the underlying scene structure given samples of observations for downstream novel view synthesis [14], pose estimation [30], etc. In contrast, our application is best described as "space-time scene completion", where we learn a network to predict an explicit space-time occupancy volume. Furthermore, our approach differs from existing approaches in the following ways. We use LiDAR sequences as input and raycast LiDAR sweeps given future occupancy and sensor pose. We work with explicit volumetric representations [12] for dynamic scenes with a feed-forward network instead of test-time optimization [18].

Self-Supervision: Standard P&P solutions do not scale given how fast log data is collected by large fleets and how slow it is to curate object track labels. To enable learning on massive amount of unlabeled logs, supervision from simulation [5–8], auto labeling using multi-view constraints [20], and self-supervision have been proposed. Notably, tasks that can be naturally self-supervised by LiDAR sweeps e.g., scene flow [15] have the potential to generalize better as they can leverage more data. More recently, LiDAR self-supervision has been explored in the context of point cloud forecasting [27–29]. However, when predicting future sweeps given the history, as stated before, past approaches often tend to couple motion of the world with the motion of the ego-vehicle [27].

Motion Planning: An understanding of what is around an AV and what will happen next [25] is crucial. This is typically done in the bird's eye-view (BEV) space by building a modular P&P pipeline. Although BEV motion planning does not precisely reflect planning in the 3D world, it is widely used as the highest-resolution and computation- and memory-efficient representation [3,23,31]. However, training such modules often requires a massive amount of data. End-to-end learned planners requiring less human annotation have emerged, with end-to-end imitation learning (IL) methods showing particular promise [5,6,22]. Such methods often learn a neural network to map sensor data to either action (known as behavior cloning) or "action-ready" cost function (known as inverse optimal control) [17]. However, they are often criticized for lack of explainable intermediate representations, making them less accountable for safety-critical applications [19]. More recently, end-to-end learned but modular methods producing explainable representations, e.g., neural motion planners [3,23,31] have been proposed. However, these still require costly object track labels. Unlike them, our approach learns explainable intermediate representations that are explainable quantities for safety-critical motion planning without the need of track labels.

3 Method

Autonomous fleets provide an abundance of *aligned* sequences of LiDAR sweeps $\mathbf{x}$ and ego vehicle trajectories $\mathbf{y}$. How can we make use of such data to improve perception, prediction, and planning? In the sections to follow, we first define occupancy. Then we describe a self-supervised approach to predicting future occupancy. Finally, we describe an approach for integrating this forecasted occupancy into neural motion planners. Note that in the text that follows, we use ego-centric freespace and freespace interchangeably.

3.1 Occupancy

We define occupancy as the state of occupied space at a particular time instance. We use $\mathbf{z}$ to denote the true occupancy, which may not be directly observable due to visibility constraints. Let us write

$$\mathbf{z}[\mathbf{u}] \in \{0, 1\}, \mathbf{u} = (x, y, t), \mathbf{u} \in \mathbf{U} \tag{1}$$

to denote the occupancy of a voxel $\mathbf{u}$ in the space-time voxel grid $\mathbf{U}$, which can be *occupied* (1) or *free* (0). The spatial index of $\mathbf{u}$, i.e., (x, y) represents the spatial location from a bird's-eye view. Given a sequence of *aligned* sensor data and ego-vehicle trajectory $(\mathbf{x}, \mathbf{y})$, there may be multiple plausible occupancy states $\mathbf{z}$ that "explain" the sensor measurements. We denote this set of plausible occupancy states as $\mathbf{Z}$.

Forecasting Occupancy. Suppose we split an aligned sequence of LiDAR sweeps and ego-vehicle trajectory $(\mathbf{x}, \mathbf{y})$ into a historic pair $(\mathbf{x_1}, \mathbf{y_1})$ and a future pair $(\mathbf{x_2}, \mathbf{y_2})$. Our goal is to learn a function f that takes historical observations $(\mathbf{x_1}, \mathbf{y_1})$ as input and predicts emergent future occupancy $\hat{\mathbf{z}}_\mathbf{2}$. Formally,

$$\hat{\mathbf{z}}_\mathbf{2} = f(\mathbf{x_1}, \mathbf{y_1}), \tag{2}$$

If the true occupancy $\mathbf{z_2}$ were observable, we could directly supervise our forecaster, f. Unfortunately, in practice, we only observe LiDAR sweeps, $\mathbf{x}$. We show in the next section how to supervise f with LiDAR sweeps using differentiable raycasting techniques.

3.2 Raycasting

Given an occupancy estimate $\hat{\mathbf{z}}$, sensor origin $\mathbf{y}$ and directional unit vectors for rays $\mathbf{r}$, a differentiable raycaster $\mathcal{R}$ can raycast LiDAR sweeps $\hat{\mathbf{x}}$. We use $\hat{\mathbf{d}}$ to represent the expected distance these rays travel before hitting obstacles: $\hat{\mathbf{d}} = \mathcal{R}(\mathbf{r}; \hat{\mathbf{z}}, \mathbf{y})$. Then we can reconstruct the raycast LiDAR sweep $\hat{\mathbf{x}}$ as $\hat{\mathbf{x}} = \mathbf{y} + \hat{\mathbf{d}} * \mathbf{r}$.

3.3 Learning to Forecast Occupancy

Given the predicted occupancy $\hat{\mathbf{z}}_\mathbf{2}$ (Eq. 2), and the captured sensor pose $\mathbf{y_2}$, a differentiable raycaster $\mathcal{R}$ can take rays $\mathbf{r_2}$ as input and produce $\hat{\mathbf{d}}_\mathbf{2} = \mathcal{R}(\mathbf{r_2}; \hat{\mathbf{z}}_\mathbf{2}, \mathbf{y_2})$. Note that this formulation allows us to decouple the motion of the world captured by change in occupancy, $\hat{\mathbf{z}}_\mathbf{2}$, and the motion of the ego-vehicle captured by change in sensor origin, $\mathbf{y_2}$.

This also allows us to supervise $\hat{\mathbf{z}}_\mathbf{2}$ using a loss function that measures the difference between the raycast distance $\hat{\mathbf{d}}_\mathbf{2}$ and the ground-truth distance $\mathbf{d_2}$.

$$L_r = \text{loss}(\hat{\mathbf{d}}_\mathbf{2}, \mathbf{d_2}) \tag{3}$$

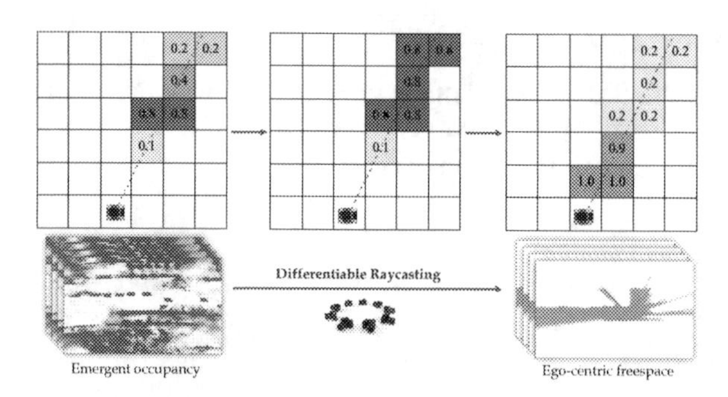

Fig. 3. Differentiable procedure for estimating ego-centric freespace from volumetric occupancy, necessary for computing the loss from (3). The left image depicts predicted emergent occupancy, on which we perform a cumulative max along the LiDAR ray from known sensor poses (middle), which is differentiable because it is essentially re-indexing. The result is then inverted to produce (soft) visible ego-centric freespace estimates. To identify BEV pixels along the LiDAR ray, we perform fast voxel traversal in 2D [1].

Loss Function: One natural loss function might be distance between the ray-cast depth and measured depth along each ray. In practice, we care most about disagreements of freespace which can inform safe motion plans. To emphasize such disagreements, we define voxels encountered along the ray as having a free versus not-free binary label, and use a binary cross-entropy loss (summed over all voxels encountered by each ray until the boundary of voxel grid, ref. Fig. 3). We adopt an encoder-decoder architecture that predicts future emergent occupancy given historical LiDAR sweeps, differentiably raycasts future LiDAR sweeps and self-supervises using archival sweeps (ref. highlighted branch of Fig. 4 (a)).

3.4 Learning to Plan

The previous section described an approach for predicting future LiDAR returns via differentiable raycasting of BEV space-time occupancy maps. We now show that such costmaps can be integrated directly into an end-to-end motion planner that makes use of space-time costmaps for scoring candidate trajectories. We follow [9], but modify their derivation to take into account emergent occupancy.

Max-Margin Planning: We learn a model g to predict a space-time cost map, $\mathbf{c_2}$, over future timestamps given past observations $(\mathbf{x_1}, \mathbf{y_1})$:

$$\mathbf{c_2} = g(\mathbf{x_1}, \mathbf{y_1}), \quad \text{where} \quad \mathbf{c_2}[\mathbf{u}] \in \mathbb{R}, \mathbf{u} \in \mathbf{U_2} \tag{4}$$

where $\mathbf{U_2}$ represents the space-time voxel grid over future timestamps. We define the cost of a trajectory as the sum of costs at its space-time way-points. The best candidate future trajectory according to the cost map is the one with the lowest cost:

$$\hat{\mathbf{y}}_2^* = \arg\min_{\hat{\mathbf{y}} \in \mathbf{Y_2}} C(\hat{\mathbf{y}}; \mathbf{c_2}) = \arg\min_{\hat{\mathbf{y}} \in \mathbf{Y_2}} \sum_{\mathbf{u} \in \hat{\mathbf{y}}} \mathbf{c_2}[\mathbf{u}] \tag{5}$$

where $\mathbf{Y_2}$ represents the set of viable future trajectories.

Loss Function: We use a max-margin loss function, where the target cost of a candidate trajectory $(\hat{\mathbf{y}})$ is equal to the cost of the expert trajectory $(\mathbf{y_2})$ plus a margin. We can write the objective as follows:

$$L_p = \left[C(\mathbf{y_2}; \mathbf{c_2}) - \left(\min_{\hat{\mathbf{y}} \in \mathbf{Y_2}} C(\hat{\mathbf{y}}; \mathbf{c_2}) - D(\hat{\mathbf{y}}, \mathbf{y_2}) \right) \right]_+ \tag{6}$$

where $[\cdot]_+ = \max(\cdot, 0)$ and D is a function that quantifies the desired margin between the cost of a candidate trajectory and the cost of an expert trajectory. A common choice for D is Euclidean distance between pairs of way-points:

$$D(\hat{\mathbf{y}}_2, \mathbf{y_2}) = ||\hat{\mathbf{y}}_2, \mathbf{y_2}||_2 \tag{7}$$

Learning cost maps that reflect such cost margins only requires expert demonstrations, which are readily available in archival log data. However, sometimes candidates trajectories that are equally distant from the expert one should bear different costs. We provide an example (right) where the red trajectory should cost more than blue in the presence of an obstacle despite both being equidistant from the expert demonstration.

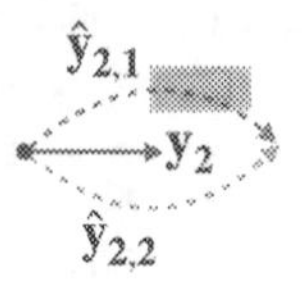

Guided Planning: To further distinguish among candidate trajectories, one could introduce extra penalty terms given additional supervision.

$$D(\hat{\mathbf{y}}_2, \mathbf{y}_2) = ||\hat{\mathbf{y}}_2, \mathbf{y}_2||_2 + \gamma\ P(\hat{\mathbf{y}}_2) \tag{8}$$

where P represents a penalty function and γ is a predefined scaling factor. Zeng et al. [31] propose to define an additional penalty such that candidate trajectories that collide with object boxes would cost an additional γ in addition to the deviation from the expert demonstration. We refer to this approach as *object-guided planning*, which is effective but costly as it requires object track labels.

More scalable alternatives to object supervision can be adopted, such as formulation of the penalty term proposed by Hu et al. [9]. Concretely, candidate trajectories that reach outside the freespace as observed by future LiDAR poses would incur an additional penalty. We refer to this as *freespace-guided planning.*

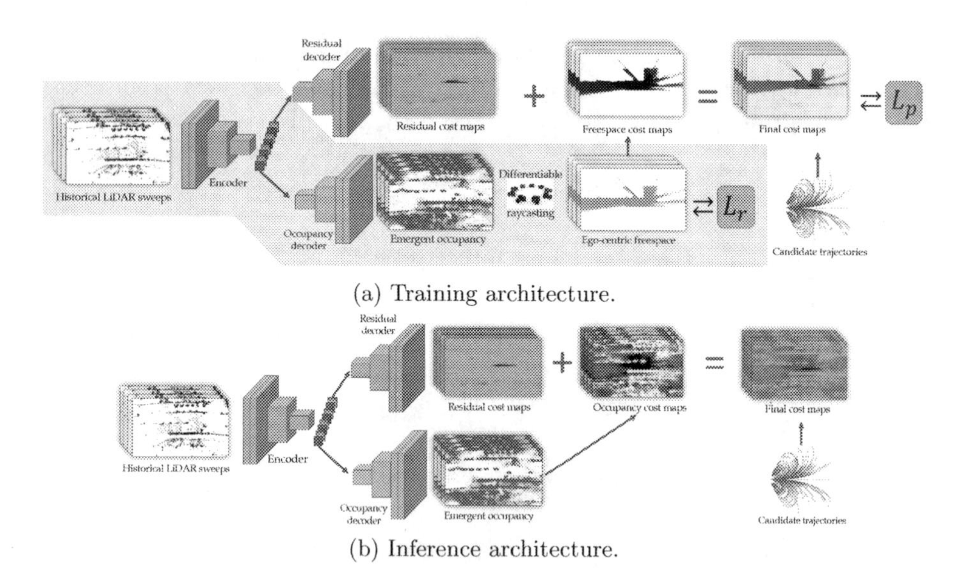

Fig. 4. Overview of our training and inference-time planning architectures. Highlighted network branch in (a) is used to learn future emergent occupancy, which is augmented by the residual branch that predicts residual cost maps, eventually used in computing a guided planning loss.

Residual Costmaps: Instead of directly predicting the cost map $\mathbf{c_2}[\mathbf{u}]$, we follow prior work [9] and predict a residual cost map $\tilde{\mathbf{c}}_\mathbf{2}[\mathbf{u}]$ that is added to the cost map from freespace estimate based on predicted emergent occupancy.

$$\mathbf{c_2}[\mathbf{u}] = \tilde{\mathbf{c}}_\mathbf{2}[\mathbf{u}] + \alpha \ \text{proj}(\hat{\mathbf{z}}_\mathbf{2}; \mathbf{y_2})[\mathbf{u}], \ \ \mathbf{u} \in \hat{\mathbf{y}}_\mathbf{2} \tag{9}$$

where α is a predefined constant and $\tilde{\mathbf{c}}_\mathbf{2}$ represents the predicted residual cost map. The operation $\text{proj}(\hat{\mathbf{z}}_\mathbf{2}; \mathbf{y_2})$ is illustrated in Fig. 3.

Multi-Task Planning. (new): In addition to the raycasting loss in Fig. 4, we add L_p as an additional planning loss. In other words, the emergent occupancy prediction architecture is augmented with another decoder branch to predict the residual cost maps while sharing the encoder features. Because of this, emergent occupancy forecasting becomes the auxiliary task for the end-to-end motion planner. We illustrate the network architecture during training in Fig. 4 (a).

Test-Time Occupancy Cost Maps. (new): At test time, to compute ego-centric freespace cost maps based on predicted emergent occupancy, for each candidate sample trajectory, one would need to perform raycasting from its way-points, which is prohibitively expensive. Fortunately, this is exactly equivalent to directly accessing emergent occupancy on the waypoints along the candidate trajectory (because of the cumulative max-operation used in deriving freespace from occupancy - see Fig. 3), as formally expressed in Eq. (10).

$$\text{proj}(\hat{\mathbf{z}}_\mathbf{2}; \hat{\mathbf{y}}_\mathbf{2})[\mathbf{u}] = \hat{\mathbf{z}}_\mathbf{2}[\mathbf{u}], \ \ \mathbf{u} \in \hat{\mathbf{y}}_\mathbf{2} \tag{10}$$

The simplified test-time architecture is illustrated in Fig. 4(b). When optimizing for future trajectories, we restrict the search space of future trajectories to the ones with a smooth transition from the past trajectory [9,31]. Please refer to the supplement for other implementation information such as detailed network architecture.

Table 1. Indirect evaluation of emergent occupancy forecasting with respect to groundtruth LiDAR sweeps. On both nuScenes and ONCE, we significantly improve forecasting accuracy across all metrics by using differentiable raycasting for decoupling the scene and ego-motion, unlike Hu *et al.* [9].

Dataset	Diff. Raycast	$\frac{\lvert \hat{d}-d \rvert}{d}$ (↓)	BCE (↓)	F1 (↑)	AP (↑)
nuScenes	- [9]	0.297	0.221	0.665	0.769
	✓	**0.242**	**0.140**	**0.777**	**0.863**
ONCE	- [9]	0.371	0.143	0.635	0.732
	✓	**0.243**	**0.097**	**0.787**	**0.827**

4 Experiments

Datasets: We evaluate occupancy forecasting and motion planning on two datasets: nuScenes [2] and ONCE [13]. nuScenes features real-world driving data with 1,000 fully annotated 15 s logs. ONCE is the largest driving dataset with 150 h of real-world data including 1 million LiDAR sweeps, collected in a range of diverse environments such as urban and suburban areas. As annotation is expensive, only a small subset of logs in ONCE are fully annotated, making it ideal for self-supervised learning. We include comparison against state-of-the-art forecasting and planning approaches on both datasets. We also construct multiple baselines for all ablative evaluation for bird's eye-view motion planning. To understand how our occupancy forecasting and motion planning performance scales to an increasing amount of training data, we randomly curate different training sets of the datasets. Since only a small subset of 8K samples in ONCE is labeled, we do this by progressively increasing the number of training samples by adding scenes from both their labeled and unlabeled-small splits, which include 8K, and 86K training samples respectively. Some of our analysis exists only on the combined labeled and unlabeled-small split which totals to 94K samples. For nuScenes, we randomly sample scenes from their official training set. For all experiments that follow, we take in a historical LiDAR stack of 2 s and forecast for the next 3 s.

4.1 Emergent Occupancy Forecasting

Metrics: Since, the groundtruth for true occupancy is unavailable, we quantitatively evaluate the LiDAR sweeps raycast from the emergent occupancy predictions. Specifically, our first evaluation computes the absolute relative error between the groundtruth distance traveled by every ray starting from the sensor origin, and the expected distance traveled by corresponding rays; where the expected distance is obtained by casting rays through the forecasted occupancy. Second, we score *every* BEV voxel traversed by a ray using its 'free' or 'not-free' state. This dense per-ray evaluation is equivalent to evaluating the per-pixel binary classification of an ego-centric freespace map with respect to its groundtruth, allowing us to compare to the baseline discussed below. We compute the dense binary cross-entropy, average precision and the F1-score. All metrics are averaged across all prediction timesteps (up to 3s).

Baseline: We re-implement the future-freespace architecture from [9] which directly forecasts ego-centric freespace. For building our architecture, we adapt this network to predict an arbitrary quantity which differentiably raycasts into ego-centric freespace given a sensor location. On training this architecture in a self-supervised manner, the arbitrary quantity *emerges* into emergent occupancy, an explainable intermediate representation for downstream motion planners.

Main Results: We compare the performance of both approaches in Table 1. Note the drastic improvement in all metrics on using differentiable raycasting to decouple the scene motion from the ego-motion of the sensor on both nuScenes

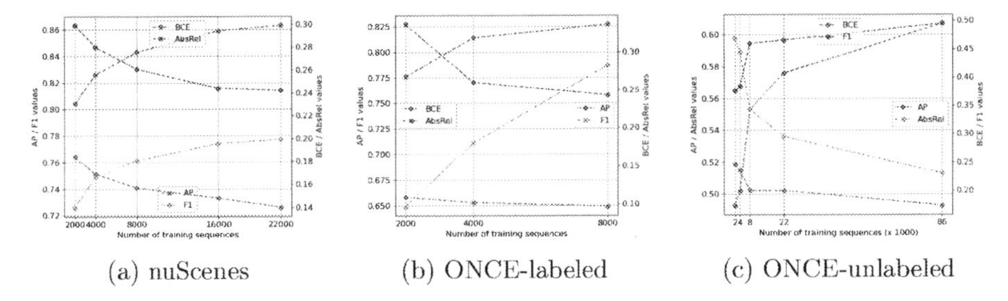

(a) nuScenes (b) ONCE-labeled (c) ONCE-unlabeled

Fig. 5. We highlight the merits of our self-supervised approach which can be given any amount of unlabeled LiDAR data to train on, in the form of posed archival LiDAR sweeps, thereby increasing the performance of emergent occupancy forecasting (evaluated using classification metrics such as average-precision and F1). Please refer to the supplement for corresponding tables.

and ONCE. With increase of up to 15% F1 points, we highlight the high-quality of our predicted occupancy and the pronounced effect of adding differentiable raycasting. Our results show that occupancy reasoning is an important intermediate task, *even* if the end-goal is simply understanding freespace: Our method, which predicts occupancy as an intermediate target, outperforms [9], which directly aims to predict freespace. Fig. 6 visualizes predicted ego-centric freespace for a single scenario in ONCE using [9] and our approach at $t = 0, 3s$ in the future. In Fig. 5, we show how adding more training samples to both datasets result in an upward trend in performance across *all* metrics. This increasing generalizability and scaling of training data comes for free with our self-supervised approach.

4.2 Motion Planning

Metrics: We follow prior works and compute three metrics for evaluating motion planning performance, including (1) L2 error; (2) point collision rate; (3) box collision rate. The L2 distance measures how close the planned trajectory follows the expert trajectory at each future timestamp. The point collision rate measures how often the planned waypoint is within the BEV boxes of other objects. The box collision rate measures how often the BEV box of the ego-vehicle intersects with BEV boxes of other objects.

Trajectory Sampling: When evaluating performance on nuScenes, we follow previous state-of-the-art approaches [9,31] and sample a combination of straight lines, circles, and clothoid curves as trajectory samples. Owing to the scene diversity in ONCE, we notice that such a sampling strategy does not capture the distribution of expert trajectories on ONCE as they range widely in their velocities and directions.

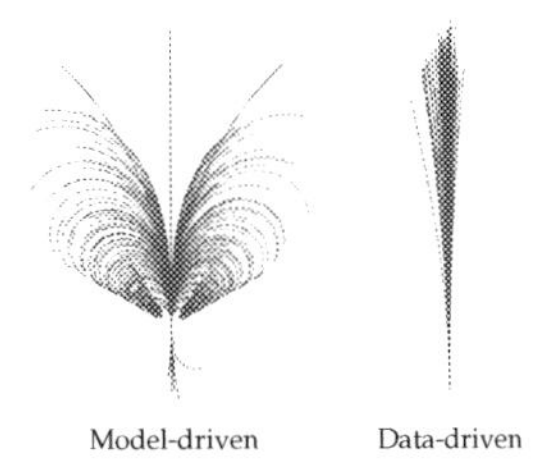

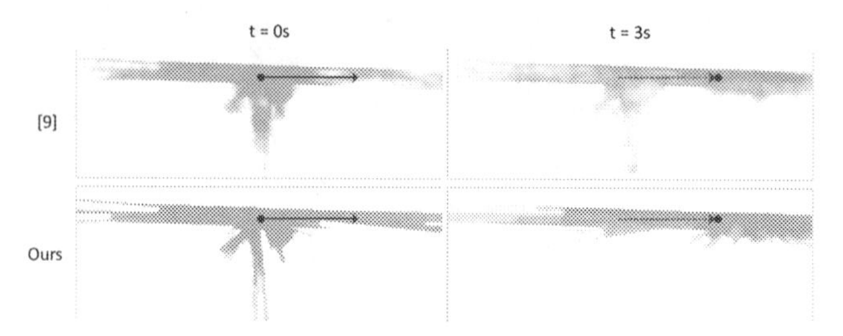

Fig. 6. Future ego-centric freespace from [9] and our model, raycasted from predicted emergent occupancy. Note how the presence of moving and parked cars on roadsides is captured well by our approach even 3s in the future.

Table 2. We compare end-to-end state-of-the-art motion planners on nuScenes-val. NMP and P3 are supervised approaches that have access to object tracking labels.

nuScenes	Box Collision (%)			L2 Error (m)		
	1s	2s	3s	1s	2s	3s
IL [21]	0.08	0.27	1.95	**0.44**	**1.15**	**2.47**
FF [9]	0.06	0.17	1.07	0.55	1.20	2.54
Ours	**0.04**	**0.09**	**0.88**	0.67	1.36	2.78
NMP [31]	0.04	0.12	*0.87*	*0.53*	*1.25*	*2.67*
P3 [23]	*0.00*	*0.05*	1.03	0.59	1.34	2.82

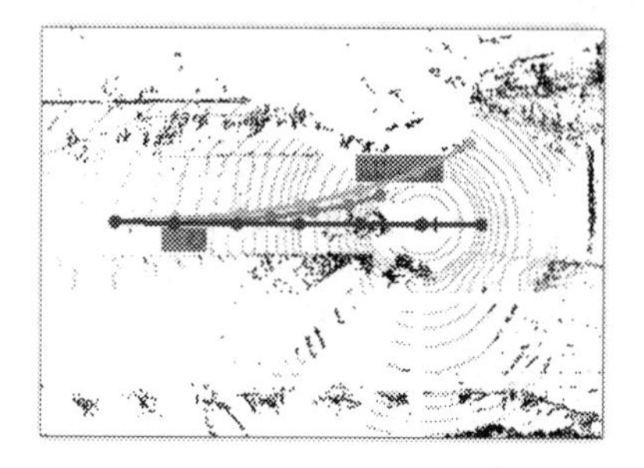

Fig. 7. A vanilla spacetime trajectory with a lower L2 error wrt. expert, may collide into objects unlike a proposed trajectory with larger L2 error but no collision.

Inspired by [3], we sample a data-driven trajectories to complement the model-driven samples (right). The supplement provides more details on our data-driven sampler.

Planning on nuScenes

Baselines: We compare our proposed approach to four baseline end-to-end motion planners. First, we implement a pure imitation learning (IL) baseline, a max-margin neural motion planner self-supervised by expert trajectories, as described in Eq. (7). Second, we re-implement future-freespace-guided max-margin planner (FF) proposed by Hu *et al.* [9], as captured by Eq. (8). Third, we re-implement a simplified neural motion planner (NMP) without modeling costs related to map information and traffic light status as such information is unavailable on nuScenes. Last, we re-implement a simplified version of perceive, predict, and plan (P3) where we do not distinguish semantic occupancy of different classes. To ensure a fair comparison, we adopt the same neural net architecture for the baselines and our approach.

Table 3. Ablation studies on nuScenes-val. Note that (a) is IL, (b) is FF, and (e) is Ours in Table 2.

	Freespace Guided	Multi Task	Diff. Raycast	Box Collision (%)			Point Collision (%)			L2 Error (m)		
				1*s*	2*s*	3*s*	1*s*	2*s*	3*s*	1*s*	2*s*	3*s*
(a)	–	–	–	0.08	0.27	1.95	**0.00**	**0.00**	0.35	0.44	1.15	2.47
(b)	✓	–	–	0.06	0.17	1.07	**0.00**	0.01	0.04	0.55	1.20	2.54
(c)	–	✓	–	0.08	0.17	1.29	**0.00**	0.02	0.08	**0.42**	**1.06**	**2.30**
(d)	✓	✓	–	**0.02**	0.10	1.10	**0.00**	**0.00**	0.08	0.52	1.22	2.64
(e)	✓	✓	✓	0.04	**0.09**	**0.88**	**0.00**	0.01	**0.03**	0.67	1.36	2.78

Main Results: As Table 2 shows, in terms of collision rates, our self-supervised approach outperforms both self-supervised baselines (IL and FF) by a large margin. Moreover, our approach achieves the same collision rate at 3s as the best of supervised baselines. We also observe a commonly observed trade-off between L2 errors and collision rates [31]. For example, pure imitation learning achieves the lowest L2 errors with the highest collision rates.

Ablation Studies: We perform extensive ablation studies in Table 3 to understand where improvements come from. There are three main observations:

- Differentiable raycasting reduces collision rate at further horizon (3s), as seen in (d) vs. (e), suggesting decoupling motion of the world (space-time occupancy) from ego-motion is helpful when learning long range cost maps.
- Multi-task learning further reduces collision rates, as seen in (a) vs. (c). Training max-margin planners with an auxiliary self-supervised forecasting task significantly reduces the collision rates without hurting L2.
- Freespace-guided cost margin is crucial to lowering collision rates, as seen in (a) vs. (b), (c) vs. (d). However, there is a trade-off: the L2 errors tend to increase as being expert-like (at all costs) is no longer the only objective. In Fig. 7, we show an example result describing why L2 error is a misleading metric that doesn't allow for alternate future plans that are otherwise viable. Additionally, Casas *et al.* [3] show that collision rate is a more consistent metric between evaluation in the open- and closed-loop setups.

Planning on ONCE

Baseline: ONCE offers a massive amount of unlabeled, diverse LiDAR sweeps paired with ego-vehicle trajectories and a small fully labeled subset of about 8K samples. We train a re-implemented neural motion planner as a supervised baseline on the fully labeled subset. We train our self-supervised approach over a wide range of training sizes, from 2K to 94K.

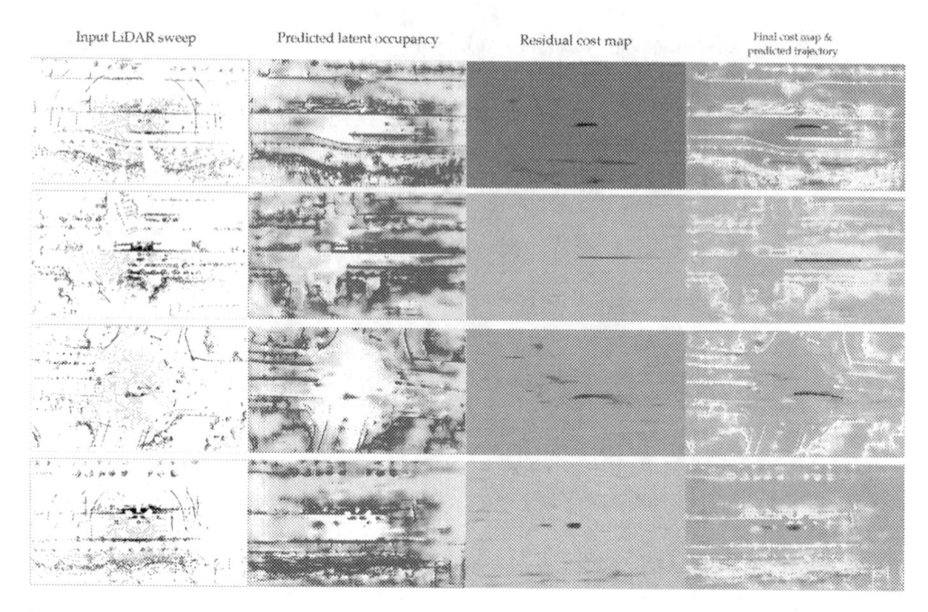

Fig. 8. Qualitative results of our learned model. From top to bottom, we visualize various scenarios, including slowing down, speeding up, navigating an intersection and staying still. All columns after the first one are visualized at future timestamp t = 0.5s. We successfully forecast the motion of surrounding objects, e.g. in third row, which results in safer planned trajectories.

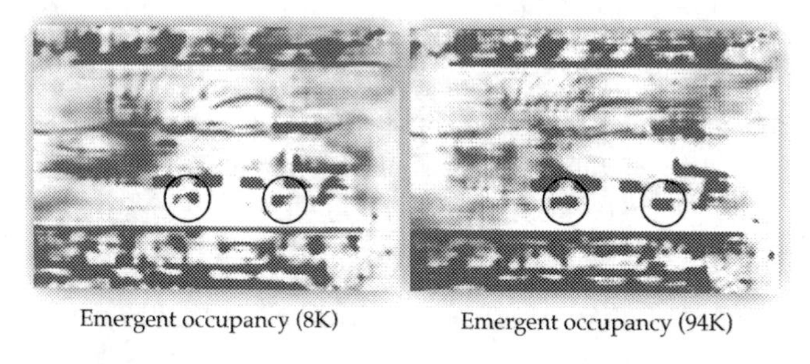

Fig. 9. Evolution of estimated emergent future occupancy.

Main Results: Perhaps unsurprisingly, our first observation is that the metrics on ONCE are inflated as compared to nuScenes, because of the diverse range of environments ONCE features, ranging from straight highways to complex city road structures. To show the scalability of our approach on such a diverse and large dataset, we plot the L2 error and (box) collision rate at 3s as a function of the amount of training data in Fig. 10. Both the L2 error and the collision rate of our approach continue to improve as we increase

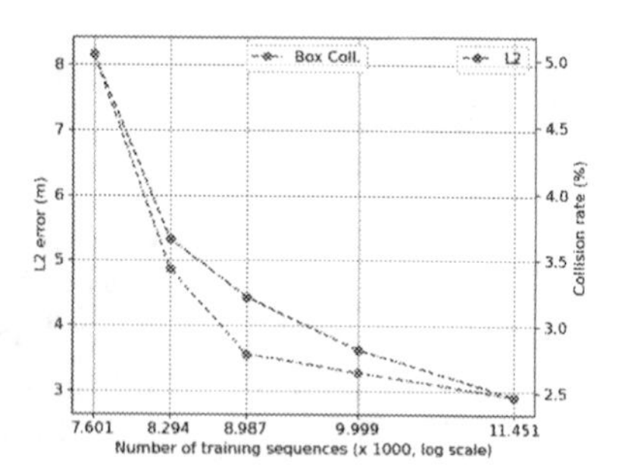

Fig. 10. Planning performance vs. larger ONCE training set size.

the size of the training set. In comparison, the supervised neural motion planner achieves an L2 error of 4.45m and a box collision rate of 2.54% at a training size of 8K.

At 94K training samples, our self-supervised approach achieves a dramatically lower L2 error of 2.9m and a lower collision rate of 2.47%. Importantly, such scalability for motion planning comes for free as our approach is self-supervised. We show some qualitative results on the ONCE dataset in Fig. 8 where our approach is able to deal with a number of varying driving scenarios; decelerate and stop when necessary, predict long trajectories when unoccupied regions are predicted ahead, avoid collisions with other vehicles while navigating an intersection, or stay stationary. Please refer to our supplement for further quantitative evaluation, visualization of future cost maps and more qualitative examples that feature failure cases (e.g., forecasted occupancy diffuses over time).

Evolution of Occupancy Estimates: Our model tends to produce better estimates of emergent occupancy as we increase the amount of training data. The percent of semantic object pixels recalled from the ground-truth semantic object labels in our predicted occupancy map increases from 51% to 59% at t=0s when we increase the amount of training data from 8K to 94K. Qualitatively, this can be seen in Fig. 9 where the shape of two cars in the right lane looks more "space-time complete" for the model trained with increased data.

5 Conclusion

We propose *emergent occupancy* as a self-supervised and explainable representation for motion planning. Our novel differentiable raycasting procedure enables the learning of occupancy forecasting under the self-supervised task of LiDAR sweep forecasting. The raycasting setup also allows us to decouple ego motion from scene motion, making forecasting an easier task for the network to learn. Experimental results suggest that such decoupling is also helpful for downstream motion planning. Such training at scale allows object shape, tracks, and multiple futures to "emerge" in the predicted emergent occupancy.

Acknowledgments. This work was supported by the CMU Argo AI Center for Autonomous Vehicle Research.

References

1. Amanatides, J., Woo, A.: A fast voxel traversal algorithm for ray tracing. In: EG 1987-Technical Papers, Eurographics Association (1987). https://doi.org/10.2312/egtp.19871000
2. Caesar, H., et al.: A multimodal dataset for autonomous driving. arXiv preprint arXiv:1903.11027 (2019)
3. Casas, S., Sadat, A., Urtasun, R.: Mp3: a unified model to map, perceive, predict and plan. In: Proceedings of the IEEE/CVF Conference on Computer Vision and Pattern Recognition, pp. 14403–14412 (2021)

4. Chai, Y., Sapp, B., Bansal, M., Anguelov, D.: Multipath: multiple probabilistic anchor trajectory hypotheses for behavior prediction. arXiv preprint arXiv:1910.05449 (2019)
5. Chen, D., Zhou, B., Koltun, V., Krähenbühl, P.: Learning by cheating. In: Conference on Robot Learning, pp. 66–75. PMLR (2020)
6. Codevilla, F., Miiller, M., López, A., Koltun, V., Dosovitskiy, A.: End-to-end driving via conditional imitation learning. In: 2018 IEEE International Conference on Robotics and Automation, pp. 1–9. IEEE (2018)
7. Codevilla, F., Santana, E., López, A.M., Gaidon, A.: Exploring the limitations of behavior cloning for autonomous driving. In: Proceedings of the IEEE International Conference on Computer Vision, pp. 9329–9338 (2019)
8. Dosovitskiy, A., Ros, G., Codevilla, F., Lopez, A., Koltun, V.: Carla: an open urban driving simulator. arXiv preprint arXiv:1711.03938 (2017)
9. Hu, P., Huang, A., Dolan, J., Held, D., Ramanan, D.: Safe local motion planning with self-supervised freespace forecasting. In: Proceedings of the IEEE/CVF Conference on Computer Vision and Pattern Recognition, pp. 12732–12741 (2021)
10. Hu, P., Ziglar, J., Held, D., Ramanan, D.: What you see is what you get: exploiting visibility for 3d object detection. In: Proceedings of the IEEE/CVF Conference on Computer Vision and Pattern Recognition, pp. 11001–11009 (2020)
11. Lang, A.H., Vora, S., Caesar, H., Zhou, L., Yang, J., Beijbom, O.: PointPillars: fast encoders for object detection from point clouds. In: CVPR (2019)
12. Lombardi, S., Simon, T., Saragih, J., Schwartz, G., Lehrmann, A., Sheikh, Y.: Neural volumes: learning dynamic renderable volumes from images. arXiv preprint arXiv:1906.07751 (2019)
13. Mao, J., et al.: One million scenes for autonomous driving: Once dataset. arXiv preprint arXiv:2106.11037 (2021)
14. Mildenhall, B., Srinivasan, P.P., Tancik, M., Barron, J.T., Ramamoorthi, R., Ng, R.: NERF: representing scenes as neural radiance fields for view synthesis. In: Vedaldi, A., Bischof, H., Brox, T., Frahm, J.-M. (eds.) ECCV 2020. LNCS, vol. 12346, pp. 405–421. Springer, Cham (2020). https://doi.org/10.1007/978-3-030-58452-8_24
15. Mittal, H., Okorn, B., Held, D.: Just go with the flow: self-supervised scene flow estimation. In: Proceedings of the IEEE/CVF Conference on Computer Vision and Pattern Recognition (CVPR) (2020)
16. Montemerlo, M., et al.: Junior: the Stanford entry in the urban challenge. J. Field Robot. **25**(9), 569–597 (2008)
17. Osa, T., Pajarinen, J., Neumann, G., Bagnell, J.A., Abbeel, P., Peters, J.: An algorithmic perspective on imitation learning. arXiv preprint arXiv:1811.06711 (2018)
18. Park, K., et al.: Deformable neural radiance fields. In: Proceedings of the IEEE/CVF International Conference on Computer Vision, pp. 5865–5874 (2021)
19. Pomerleau, D.A.: ALVINN: an autonomous land vehicle in a neural network. In: Advances in neural information processing systems, pp. 305–313 (1989)
20. Qi, C.R., et al.: Offboard 3D object detection from point cloud sequences. In: Proceedings of the IEEE/CVF Conference on Computer Vision and Pattern Recognition, pp. 6134–6144 (2021)
21. Ratliff, N.D., Bagnell, J.A., Zinkevich, M.A.: Maximum margin planning. In: Proceedings of the 23rd international conference on Machine learning, pp. 729–736 (2006)
22. Rhinehart, N., McAllister, R., Levine, S.: Deep imitative models for flexible inference, planning, and control. arXiv preprint arXiv:1810.06544 (2018)

23. Sadat, A., Casas, S., Ren, M., Wu, X., Dhawan, P., Urtasun, R.: Perceive, predict, and plan: safe motion planning through interpretable semantic representations. In: Vedaldi, A., Bischof, H., Brox, T., Frahm, J.-M. (eds.) ECCV 2020. LNCS, vol. 12368, pp. 414–430. Springer, Cham (2020). https://doi.org/10.1007/978-3-030-58592-1_25
24. Sadat, A., Ren, M., Pokrovsky, A., Lin, Y.C., Yumer, E., Urtasun, R.: Jointly learnable behavior and trajectory planning for self-driving vehicles. In: 2019 IEEE/RSJ International Conference on Intelligent Robots and Systems, pp. 3949–3956. IEEE (2019)
25. Urmson, C., Anhalt, J., Bagnell, D., Baker, C., et al.: Autonomous driving in urban environments: boss and the urban challenge. J. Field Robot. **25**(8), 425–466 (2008)
26. Weng, X., Kitani, K.: A baseline for 3D multi-object tracking. arXiv preprint arXiv:1907.03961 1(2), 6 (2019)
27. Weng, X., Wang, J., Levine, S., Kitani, K., Rhinehart, N.: 4D forecasting: sequential forecasting of 100,000 points (2020)
28. Weng, X., Wang, J., Levine, S., Kitani, K., Rhinehart, N.: Inverting the pose forecasting pipeline with spf2: Sequential pointcloud forecasting for sequential pose forecasting. arXiv preprint arXiv:2003.08376 (2020)
29. Wilson, B., et al.: Argoverse 2.0: next generation datasets for self-driving perception and forecasting. In: Thirty-Fifth Conference on Neural Information Processing Systems Datasets and Benchmarks Track (Round 2) (2021)
30. Yen-Chen, L., Florence, P., Barron, J.T., Rodriguez, A., Isola, P., Lin, T.Y.: iNeRF: inverting neural radiance fields for pose estimation. arXiv preprint arXiv:2012.05877 (2020)
31. Zeng, W., Luo, W., Suo, S., Sadat, A., Yang, B., Casas, S., Urtasun, R.: End-to-end interpretable neural motion planner. In: Proceedings of the IEEE Conference on Computer Vision and Pattern Recognition, pp. 8660–8669 (2019)

InAction: Interpretable Action Decision Making for Autonomous Driving

Taotao Jing[1(✉)], Haifeng Xia[1], Renran Tian[2], Haoran Ding[2], Xiao Luo[2], Joshua Domeyer[3], Rini Sherony[3], and Zhengming Ding[1]

[1] Tulane University, New Orleans, LA 70118, USA
{tjing,hxia,zding1}@tulane.edu

[2] Indiana University-Purdue University Indianapolis, Indianapolis, IN 46202, USA
{rtian,luo25}@iupui.edu, hd10@iu.edu

[3] Collaborative Safety Research Center (CSRC), Toyota Motor North America, Ann Arbor, MI 48105, USA
{joshua.domeyer,rini.sherony}@toyota.com

Abstract. Autonomous driving has attracted interest for interpretable action decision models that mimic human cognition. Existing interpretable autonomous driving models explore static human explanations, which ignore the implicit visual semantics that are not explicitly annotated or even consistent across annotators. In this paper, we propose a novel Interpretable Action decision making (**InAction**) model to provide an enriched explanation from both explicit human annotation and implicit visual semantics. First, a proposed visual-semantic module captures the region-based action-inducing components from the visual inputs, which learns the implicit visual semantics to provide a human-understandable explanation in action decision making. Second, an explicit reasoning module is developed by incorporating global visual features and action-inducing visual semantics, which aims to jointly align the human-annotated explanation and action decision making. Experimental results on two autonomous driving benchmarks demonstrate the effectiveness of our **InAction** model for explaining both implicitly and explicitly by comparing it to existing interpretable autonomous driving models. The source code is available at https://github.com/scottjingtt/InAction.git.

Keywords: Interpretable machine learning · Action decision prediction

1 Introduction

Deep learning has recently accelerated the progress of autonomous through remarkable success in computer vision tasks. Existing driving action decision systems can primarily be recognized to be in two major groups, one is the *pipelined*

Supplementary Information The online version contains supplementary material available at https://doi.org/10.1007/978-3-031-19839-7_22.

S. Avidan et al. (Eds.): ECCV 2022, LNCS 13698, pp. 370–387, 2022.
https://doi.org/10.1007/978-3-031-19839-7_22

framework [41] and the other is *end-to-end* system [17,18,33–35,38]. Specifically, pipelined systems decompose the problem into a series of smaller tasks, such as pedestrian trajectory planning and object detection. The final driving action decision is made by relying on the performance of all the modules designed for the sub-tasks. However, pipelined systems are vulnerable to inaccuracies in each sub-task module, which may cause the entire system to perform unreliably if the interactions between modules are ignored. On the contrary, end-to-end systems take advantage of the entire visual scene to directly predict driving action, avoiding the loss of information caused by the intermediate decisions adopted in pipelined systems.

Unfortunately, most end-to-end systems are complex deep neural network models, performing as a black box with opaque reasoning for human interpretation. In safety-critical domains, such as autonomous driving and medical diagnosis, building a transparent and interpretable learning model has recently attracted attention beyond the performance alone [28]. Various interpretation strategies have been explored to explain learning models, e.g., part-based methods [45,47], saliency maps [1,12,46], activation maximization to visualize neurons [23,24], deconvolution/upconvolution to explain layers [10,42]. However, such post-hoc methods give a superficial understanding of the black box models, rather than being a comprehensive explainable system [28]. Alternatively, prototypical visual explanations are incorporated in deep network architecture for intrinsic interpretation and case-based reasoning [2,21,22,29]. Most prior prototype-based work explicitly explores the presence of prototypical parts, which are utilized to recognize objects. However, such strategies ignore the notion of spatial relationships, which is crucial for tasks like driving decision making with complicated context and multiple objects.

For explainable autonomous driving decision making, Xu *et al.* proposed a new paradigm to predict driving action based on finite action-inducing objects, and generated a set of potential explanations in a multi-task fashion [39]. Unfortunately, there are four major limitations of this work from an interpretability perspective. First, although the multi-task framework is supervised by both driving action and a human-defined explanation, the proposed model does not interpret the reasoning process of the prediction for black-box model. Second, the proposed BDD-OIA dataset annotates the reasons of action into 21 explanations; however, it is impractical that the human-defined finite explanation set can cover all possible scenarios considering the complex scene context and objects input for autonomous driving action prediction tasks. For example, the explanation set in the BDD-OIA dataset recognizes "obstacles on the right lane" as a reason "cannot turn right", which is not accurate since different distances and locations of the obstacles could lead to different decisions for drivers. Moreover, the logical reasoning process from the explanation to driving action decision is ambiguous, especially under a multi-label setting that all possible actions are annotated. For instance, we notice that the proposed model predicts two explanations "traffic light is green" and "obstacle: car", but still predicts the action as "forward", without any reasoning about how the predicted explanation results in the action prediction. Last but not least, OIA estimates the driving decision only based on the last frame of observed sequence, ignoring the temporal information.

In this paper, we propose a novel Interpretable Action decision making (**InAction**) to provide reasoning of action prediction from both explicit human annotation and implicit visual semantics (Fig. 1). Generally, we consider the explanation for action decision from two perspectives to compensate for the limitations of each method: existing human-annotated interpretation and AI-based implicit visual hints. To sum up, our contributions are in three areas:

- First, we propose an inherently interpretable reasoning framework for autonomous driving action prediction from both implicit visual semantics and explicit human annotation perspectives.
- Second, the proposed *Implicit Visual-Semantic Interpretation* module interacts with the *Explicit Human-like Reasoning* module by revealing action-inducing concepts, and the learned implicit and explicit explanations compensate for the limitations of each other in predicting the action decision.
- Finally, experimental results on two explainable autonomous driving benchmarks demonstrate the effectiveness of the proposed model by comparing with existing models showing enriched interpretation and reasoning.

2 Related Work

2.1 Autonomous Driving Action Prediction

Existing autonomous driving action prediction solutions can be roughly grouped into two branches, i.e., *end-to-end* and *pipelined*. Generally, pipelined frameworks separate the problem into a series of smaller tasks, such as object detection [3,6,15], pedestrian trajectory planning [5,26,30,31], scene segmentation [11,32,44], and object tracking [7,19,20]. Exploring the performance of each sub-module and assessing its potential contribution to the final prediction can help users understand the reasoning of the final decision. However, because the final decision prediction relies on the performance of all sub-task modules, pipelined systems are vulnerable to failures of individual sub-task modules, leading to unreliable systems.

End-to-end autonomous driving systems have achieved promising progress thanks to the success of computer vision deep leaning algorithms [9,17,18,18,33–38]. However, most end-to-end systems are complex deep neural networks, requiring large-scale datasets to train. Unfortunately, the black box nature of deep neural networks makes the decisions not always trustworthy. Xu *et al.*. design an explainable object-inducing driving action prediction system (OIA) together with a new benchmark consisting of ego-view driving videos annotated with action and static explanations [39]. Through a model that jointly predicts action and explanation, the learned model can explain the decision predicted within the pre-defined reasoning set.

2.2 Interpretable Machine Learning

Building a transparent and interpretable model is crucial for safety-critical problems, such as autonomous driving and medical diagnosis [25,43]. Many efforts have been made to interpret deep neural networks from different perspectives. Typically, researches include part-based methods [45,47], attributes-based methods [13,14], saliency maps [1,12,46], activation maximization [23,24], deconvolution/upconvolution to explain layers [10,42] and have achieved inspiring progress to create human-interpretable black box models. However, such post-hoc solutions have limited capability in enhancing transparency and interpretability. Alternatively, prototype-based frameworks are proposed to build an inherently explainable architecture [2,16,22,28,29].

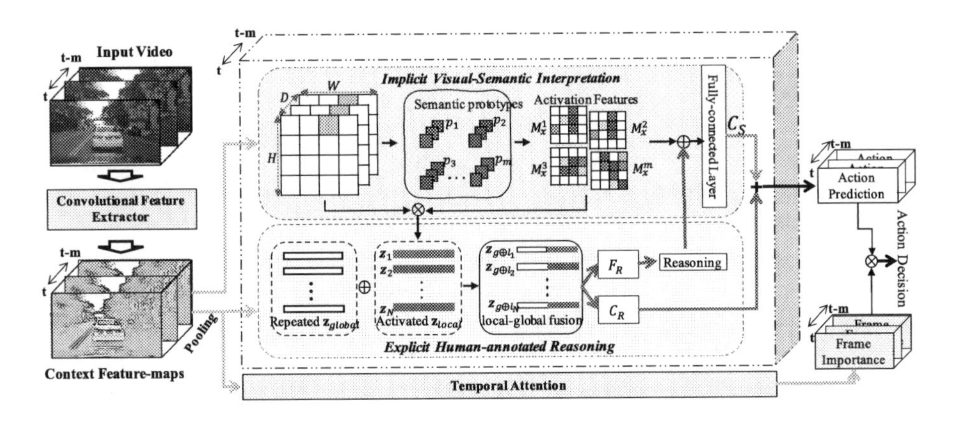

Fig. 1. Illustration of the proposed framework.

For the paradigm proposed by OIA, we argue that a human-defined finite explanation set provides inconsistent and insufficient explanations due to a lack of direct reasoning and failing to leverage temporal knowledge. In this work, we propose a novel prototype-based interpretable action decision making model (**InAction**) from implicit visual-semantic and explicit human-annotation perspectives. Different from prior prototype-based object recognition, our proposed model leverages and integrates action-inducing visual-semantic regions discovery, spatial relationships among objects, and temporal knowledge for driving decision prediction simultaneously and generates enriched explanations.

3 The Proposed Framework

3.1 Motivation

For autonomous driving, beyond pursuing high performance, interpretability is needed for safety-critical domains [25,43]. This aims to imbue autonomous vehicles with reasoning abilities similar to human drivers. Existing efforts mainly adopt human-annotated explanations to guide system learning and generate human-understandable reasoning given the video inputs [39], which skews the model towards human annotation.

Unfortunately, human annotation has some drawbacks like insufficient explanation and inconsistent reasoning. Insufficient explanation means there are always implicit visual semantics not annotated by finite human-defined explanation set, which cannot be easily tracked through an end-to-end system with visual inputs and explanation outputs. Inconsistent reasoning is particularly challenging since different people have different explanations, especially for complicated scenarios, leading to biases and insufficiency of the ground-truth annotation.

Motivated by this, we explore both the implicit visual-semantic interpretation and explicit human annotation jointly, and propose the Interpretable Action decision making model (**InAction**), whose goal is to enhance transparency and interpretability for autonomous driving action decision making.

3.2 Framework Architecture

An overview of the proposed InAction framework is shown as Fig. 1. The model consists of a convolutional backbone $G(\cdot)$, and two interpretable action prediction modules—an implicit visual-semantic module and an explicit human-annotated reasoning module—to predict driving action and reasoning of the decision from different perspectives. Specifically, the implicit visual-semantic module is denoted as $G_S(\cdot)$, which takes the feature map per frame extracted by convolutional backbone as input to discover action-inducing concepts and the presence of learned semantic prototypes as visual cues for following prediction. For the explicit reasoning module, global visual features and the discovered action-inducing local regions are fused and input to two multi-task classifiers, predicting the driving action and human-annotated explanations, denoted as $C_R(\cdot)$ and $F_R(\cdot)$, respectively. Finally, the learned prototypical visual cues and predicted human-annotated explanations are fused and input to a fully-connected layer without bias as the action predictor, denoted as $C_S(\cdot)$. For the input video sequence, such prediction is applied to each frame, with a temporal attention layer employed to explore the contribution of each frame.

Mathematically, given an input video with m frames, $\mathbf{X} = \{\mathbf{x}_i\}_{i=1}^{m}$, whose action label as $\mathbf{y}_a \in \mathbf{A}$ and human annotated explanation $\mathbf{y}_e \in \mathbf{E}$, where $C_{\text{act}} = |\mathbf{A}|$ and $C_{\text{exp}} = |\mathbf{E}|$ are the numbers of categories of actions and human-annotated explanations, respectively. For each frame $\mathbf{x}$, the convolutional backbone extracts the feature map $\mathbf{f} = G(\mathbf{x})$ with shape $H \times W \times D$, where W and H denote the width and height, respectively, and D is the number of channels. For the clarity of description, denoting all the patches in the feature map as $\mathbf{Z}_{\mathbf{x}} = \{\mathbf{z}_i \in \mathbf{f}\}_{i=1}^{HW}$, and the shape of each patch $\mathbf{z}_i$ is $\mathbb{R}^{D\times 1\times 1}$. The implicit visual-semantic module will slide over the whole feature map and calculate the activation scores for all patches in the feature map with respect to the presence of learned semantic prototypes. On the one hand, those regions primarily activated corresponding to specific prototypes are selected as action-inducing semantic regions and being fused with the global features to predict the action and explicit human-annotated explanation. On the other hand, the limitations of the activation map will be compensated by the predicted human-annotated explanations for the action prediction.

Implicit Visual Semantic Interpretation. To explore the action-inducing local regions in the visual input, we assign m_k semantic prototypes for each action class k, resulting in $m = m_k \times C_{\text{act}}$ prototypes in total, making up the visual-semantic layer $\mathbf{P} = \{\mathbf{P}_k\}|_{k=1}^{C_{\text{act}}}$, in which $\mathbf{P}_k = \{\mathbf{p}_j\}|_{j=1}^{m_k}$, and $\mathbf{p}_j$ denotes the semantic visual prototypes to be learned for predicting action class k. Given the convolutional output feature map $\mathbf{Z}_\mathbf{x}$ and prototype $\mathbf{p}_j$, the visual-semantic layer will go though all patches $\mathbf{z}_i \in \mathbf{Z}_\mathbf{x}$ of the feature map to compute the activation score between them:

$$s_{ij} = \log\left(\frac{\|\mathbf{z}_i - \mathbf{p}_j\|^2 + 1}{\|\mathbf{z}_i - \mathbf{p}_j\|^2 + \epsilon}\right), \tag{1}$$

where ϵ is a small positive value, and the activation score s_{ij} represents how strongly a semantic prototype is presented in the specific region of the input frame. The activation scores of all the patches in the feature map produce an activation heat map $\mathbf{M}_\mathbf{x}^j$ with shape $H \times W$, identifying how similar each part of the input frame is to one specific prototype $\mathbf{p}_j$. Calculating activation maps for all prototypes results in an activation feature set $\mathbf{M}_\mathbf{x} = \{\mathbf{M}_\mathbf{x}^j\}_{j=1}^m, \mathbf{M}_\mathbf{x}^j \in \mathbb{R}^{H \times W}$.

Intuitively, the most important patches for making action decision should be clustered around semantically similar prototypes of each specific action category, and the clusters centered at prototypes from different action categories are well separated. Thus, we also adopt a discriminative prototype learning loss as:

$$\mathcal{L}_d = \lambda_1 \mathbb{E}_{\mathbf{x}\in\mathbf{X}} \min_{\mathbf{p}_j \in \mathbf{P}_{\mathbf{y}_a}} \min_{\mathbf{z}\in\mathbf{Z}_\mathbf{x}} \|\mathbf{z} - \mathbf{p}_j\|^2 - \lambda_2 \mathbb{E}_{\mathbf{x}\in\mathbf{X}} \min_{\mathbf{p}_j \notin \mathbf{P}_{\mathbf{y}_a}} \min_{\mathbf{z}\in\mathbf{Z}_\mathbf{x}} \|\mathbf{z} - \mathbf{p}_j\|^2, \tag{2}$$

where λ_1 and λ_2 are two hyper-parameters determining the contributions of the two loss terms. Minimizing $\mathcal{L}_d$ encourages that every input frame at least has one prototype from its own action strongly activated in one of its latent feature map patches, while maximizing the distances between the patches and the prototypes from different classes. Such an optimization objective shapes the latent space into a semantically meaningful clustering structure.

Explicit Human-Annotated Reasoning. Compared to implicit region-based action-inducing prototypes searching, human-annotated reasoning explains the driving decision in a more intuitive and abstract way. Normally natural language annotation involves temporal and spatial knowledge from visual inputs, which provides a more high-level explanation to the decision making. Intuitively, such explanation includes the global scene understanding and corresponding action-inducing objects.

Inspired by OIA [39], we propose an Explicit Human-annotated Reasoning module in a multi-task fashion to jointly generate human-annotated explanations and predict action. Specifically, for all the patches in the extracted feature map, we select top-N patches that activate any one of the prototypes assigned to the same action class as the action-inducing local components, denoted as $\mathbf{Z}_{local} = \{\mathbf{z}_l\}_{l=1}^N$, where $\mathbf{z}_l \in \mathbf{Z}_\mathbf{x}$. The activation scores denote the importance of such patches contributing to the action decision making. It is noteworthy that the action-inducing local components $\mathbf{Z}_{local}$ are the presence of specific learned semantic prototypes, thus are not limited to be objects detected by the pre-trained object detection backbone, which is one of the limitations of OIA [39]. The selected top-N most activated patches can represent various scene contexts, environmental information, in addition to human-defined objects. Furthermore, we consider that the global feature map provides an overall understanding of the visual input and the information like environmental status, e.g., "Road is clear", and agent relationship, e.g., "There is a vehicle parking on the right". In this sense, the local action-inducing components are concatenated with the global features, then input into to the action predictor $C_R(\cdot)$ and human-annotated explanation predictor $F_R(\cdot)$.

Specifically, the global feature map $\mathbf{Z}_\mathbf{x}$ is processed with global average pooling and represented as a feature vector with the same dimension as each local patch $\mathbf{z}_l$, denoted as $\mathbf{z}_{global}$. Every local patch $\mathbf{z}_l$ is concatenated with the global feature $\mathbf{z}_{global}$ producing the local-global fused feature $\mathbf{Z}_{g\oplus l} = \{\mathbf{z}_l \oplus \mathbf{z}_{global}\}_{l=1}^N$, where $\mathbf{z}_l \in \mathbf{Z}_{local}$, and $\oplus$ is concatenation operation. The local-global feature is further vectorized then input to the following action and explanation prediction networks, optimizing the important local components that are highly associated with both action and explanation prediction. Eventually the predicted action and explanation are denoted as $\hat{\mathbf{y}}_a^R$ and $\hat{\mathbf{y}}_e^R$, respectively.

Considering the possible action decisions, we can explore to make a prediction with only one action or more than one action. If more than one action can be made, which is for a multi-label prediction task, the prediction logits are normalized by sigmoid function to the range between 0 and 1. If only one action can be made, which is a multi-class single-label task, the prediction logits are normalized by softmax function. Therefore, we formulate the multi-task learning objective of the explicit reasoning module as:

$$\mathcal{L}_r = L(\mathbf{y}_a, \hat{\mathbf{y}}_a^R) + L(\mathbf{y}_e, \hat{\mathbf{y}}_e^R), \tag{3}$$

where $L(\cdot,\cdot)$ denotes the cross-entropy loss and binary cross-entropy loss for single-label and multi-label prediction tasks, respectively.

Interpretable Decision Prediction. So far, we design two kinds of explanations, i.e., $\mathbf{M}_\mathbf{x}$ and $\hat{\mathbf{y}}_e^R$, for the decision making from two different perspectives. In order for these two explanations to interact and compensate for one another, the concatenated explanation vector $\hat{\mathbf{y}}_e = [\mathbf{M}_\mathbf{x}, \hat{\mathbf{y}}_e^R]$ is exploited to a fully-connected layer $C_S(\cdot)$ to predict the action decision $\hat{\mathbf{y}}_a^S = C_S(\hat{\mathbf{y}}_e)$.

It is noteworthy that driver action decision making has more complicated scene contexts with many different agents, which is different from other prototype-based interpretable object recognition only considering the presence of some specific prototypical parts [2,21,22,29]. Thus, the learned semantically meaningful prototypes that contribute to the final decision could be a part of or a complete object, even a set of objects or an environment region, in the input frame. Moreover, the location of a specific prototype, and the relationships between it with other objects and the environment, play crucial roles in determining the final action. Thus, rather than only choosing the maximum activation score for each prototype in the corresponding activation heat map, the whole activation feature set is considered for the fully-connected layer $C_S(\cdot)$ to integrate the spatial and relationship knowledge for predicting the action decision.

Similarly, we consider single-label and multi-label tasks with different activation functions and the learning objective of action prediction is defined as:

$$\mathcal{L}_s = L(\mathbf{y}_a, \hat{\mathbf{y}}_a^S), \tag{4}$$

where $L(\cdot, \cdot)$ represents cross-entropy loss for multi-class single-label tasks, while it is the binary cross-entropy loss for multi-label prediction tasks.

Cross-module Fusion and Temporal Aggregation. Two action decision predictions $\hat{\mathbf{y}}_a^R$ and $\hat{\mathbf{y}}_a^S$ are obtained with different input knowledge. The former one is based on the visual features, while the latter one is based on explored explicit-and-implicit explanations. Thus, we accept two prediction logits followed by the specific activation function for multi-label or single-label problem, making the final aggregated action prediction, which is denoted as $\hat{\mathbf{y}}_a = \hat{\mathbf{y}}_a^R + \hat{\mathbf{y}}_a^S$.

Moreover, for the video input $\mathbf{X} = \{\mathbf{x}_i\}_{i=1}^m$ with m frames, we make the decision prediction for each frame $\mathbf{x}_i \in \mathbf{X}$, resulting in a sequence of predictions $\{\hat{\mathbf{y}}_a^1, \dots, \hat{\mathbf{y}}_a^m\}$. To find the most relevant information (key frames) in the observed sequence, a temporal attention layer is developed with a fully-connected layer followed by Softmax activation function, generating the importance δ_i for each frame $\mathbf{x}_i$. The objective with a temporal attention layer is defined as:

$$\mathcal{L}_t = L(\mathbf{y}_a, \sum\nolimits_{i=1}^{m} \delta_i \hat{\mathbf{y}}_a^i), \tag{5}$$

wherer $L(\cdot, \cdot)$ is cross-entropy loss or binary cross-entropy loss for single-label and multi-label prediction tasks, respectively.

Overall Objective . To sum up, we integrate two explanation modules into our unified framework and formulate the overall optimization objective as follows:

$$\mathcal{L} = \mathcal{L}_d + \mathcal{L}_r + \mathcal{L}_s + \mathcal{L}_t, \tag{6}$$

which includes two action decision classifiers and one explicit explanation predictor, and these two action decision classifiers will compensate for each other as they are based on different knowledge. In the test stage, we fuse the two predictions of action decision to obtain a more robust output.

4 Experiments

4.1 Experimental Setup

Pedestrian Situated Intent (PSI) Dataset. [4] contains 110 about 15 s long videos with 30 fps, and each is annotated with one of 3 speed change actions ("maintain speed", "slow down", and "stop") on frame level. The reasoning of the action decision is described in natural language, which will be used as explanation knowledge in our experiments. We split all videos into train/validation/test set with the ratio of 75%/5%/20%. We sample the tracks with length of 15 frames, and the overlap ratio is 0.8, while predicting the 16th frame's action and explanation. Samples in PSI dataset are assigned one single label out of three actions, so we evaluate the model by overall prediction accuracy and class-wise average accuracy for action prediction.

Table 1. Statistics of BDD-OIA and PSI dataset.

Dataset	Action	# Frame	# Reasoning
BDD-OIA [39]	Forward	12,491	21 [Human-defined]
	Stop/Slow Down	10,432	
	Turn Left	5,902	
	Turn Right	6,541	
PSI [4]	Maintain Speed	5,800	29 [k-means clustered]
	Slow Down	4,925	
	Stop	1,177	

The original explanations are sentence-based, and each sentence contains descriptions of environmental context and human behaviors. We first split the original sentences into segments reflecting the environmental context or human behaviors. A syntactic dependency tree is applied to generate the dependency tagging of words, and then a set of heuristic rules are adopted to group each sentence into segments. Afterwards, the pre-trained BERT [8] is used to generate embeddings for all segments. The embedding of each segment is generated by averaging the embeddings of the words within the sentence segment. Consequently, we apply k-means clustering to obtain k semantic categories ($k = 29$ in our experiment). Given an explanation, since it is split into multiple segments and each might belong to different semantic categories, we generate k binary labels for each explanation to represent its semantics. For the human-annotated explanation, we report the overall F1 score and class-wise mean F1 score.

BDD-OIA Dataset. [39] is a subset of BDD100K [40] consisting of 22,924 5-second video clips, which were annotated with 4 action decisions ("move forward", "stop/slow down", "left turn", and "right turn") and 21 human-defined explanations. Specifically, each video contains at least 5 pedestrians or bicycle riders and more than 5 vehicles. The videos are collected with complex driving scenes to increase the scene diversity. Following the setting of [39], only the final frame of each video clip is used thus the temporal attention layer is neglected. As there are multiple possible action choices for each sample, we evaluate the performance by F1 score for each specific action, overall F1 score, and the class-wise average F1 score for both action and explanation prediction.

More statistics of the benchmarks are shown in Table 1.

Table 2. Single-label action and multi-label explanation prediction on PSI dataset.

Method	Maintain	Slow	Stop	act. Acc_{all}	act. mAcc	exp. $F1_{all}$	exp. mF1
OIA-global [39]	0.540	<u>0.774</u>	0.537	0.635	0.617	0.178	0.119
OIA [39]	0.693	0.622	0.463	0.643	0.593	0.189	0.110
Ours-f	<u>0.703</u>	0.771	<u>0.641</u>	<u>0.719</u>	<u>0.704</u>	<u>0.277</u>	<u>0.203</u>
Ours-v	**0.717**	**0.776**	**0.672**	**0.734**	**0.722**	**0.285**	**0.223**

Table 3. Multi-label action and explanation prediction on BDD-OIA dataset.

Method	F	S	L	R	act. $F1_{all}$	act. mF1	exp. $F1_{all}$	exp. mF1
Res-101 [39]	0.755	0.607	0.098	0.108	0.601	0.392	0.331	0.180
OIA [39]	**0.829**	**0.781**	**0.630**	**0.634**	**0.734**	**0.718**	0.422	0.208
OIA* [39]	0.792	0.742	0.594	<u>0.627</u>	0.705	0.689	0.501	0.293
Ours(proposals)	0.795	0.743	0.597	0.613	0.706	0.687	<u>0.558</u>	<u>0.332</u>
Ours(global)	<u>0.800</u>	<u>0.747</u>	<u>0.612</u>	0.619	<u>0.714</u>	<u>0.694</u>	**0.565**	**0.347**

Implementation Details . The Faster R-CNN [27] is pre-trained on the annotated images from BDD100K [40] and set as the backbone, which is followed by two 3×3 convolutional layers generating the global feature map with shape $7 \times 7 \times 256$ for each input frame. For implicit visual semantic interpretation module, we assign $m_k = 6$ prototypes with dimension 128 for each action class, resulting in $m = 24$ prototypes for BDD-OIA dataset, and $m = 18$ prototypes in total for PSI dataset. For our InAction model, we set $N = 10$ thus the $top - 10$ patches from the input feature map with the smallest distances compared to all semantic prototypes are selected to be fused with the global features for explicit human-annontated explanation and action prediction. The feature map is input to two additional 1×1 convolutional layers to reduce the channel dimension to be same as the prototypes dimension and normalized by sigmoid function following [2] before calculating the activation scores. The action predictor $C_S(\cdot)$ based on the fused explanation vector is one fully-connected layer without bias. We follow the same strategy of [2] to initialize and train the model. For the

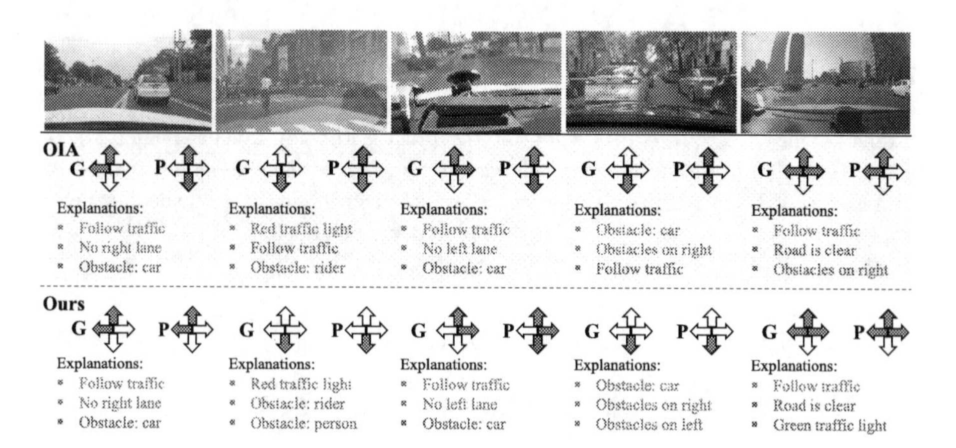

Fig. 2. Selected comparison examples of action and explicit explanation prediction between OIA and InAction on BDD-OIA dataset. **G** denotes the ground-truth annotation, and **P** shows the predicted result from OIA/Ours. green predictions are True Positive, red are False Positive, and gray are False Negative. (Color figure online)

explicit human-annotated reasoning module, the action decision predictor $C_R(\cdot)$ is a three-layer fully-connected neural network, and the explanation predictor $F_R(\cdot)$ is two-layer fully-connected neural network. ReLU activation is used for all hidden layers. The model is optimized by Adam optimizer with learning rate initialized as 10^{-3}, and decayed by 0.1 every 10 epochs. For simplicity, we set $\lambda_1 = 0.1$ and $\lambda_2 = 0.01$ by default for all experiments. We empirically fix $m_k = 6$, and we observe the results are not sensitive to it if $m_k > 3$ on validation set.

4.2 Comparison Results

We compare our proposed InAction model with the OIA method [39] on the PSI and BDD-OIA datasets, and the results are reported in Tables 2 and 3. OIA model only adopts the last frame of a sequence as input, thus we report two results produced by our model with only the last frame or the whole observed video sequence as input, denoted as Ours-f and Ours-v in Table 2, respectively. For experiments on BDD-OIA in Table 3, we reproduce the OIA model based on the official implementation released by the author, denoted as OIA*, in addition to the results reported by OIA [39]. The reproduced results of OIA on BDD-OIA are lower in action decision while better in explanation in term of F-1 score, compared with the reported OIA. Note that OIA adopts the detected proposals generated by the backbone as local features. We utilize the implicit visual-semantic prototypes learned from the global feature map and from the detected proposals, and report the results as Ours(global) and Ours(proposals), respectively. Specifically, to obtain Ours(proposals), we extract the top-100 detected proposals features after average pooling process into the same size as the learned prototypes, then follow the same fusing strategy as aforementioned.

For the PSI results (Table 2), we notice that our proposed InAction model with only the last frame as input outperforms OIA around **0.07** and **0.01** for the overall and class-wise mean action prediction accuracy, respectively. When the whole video sequence is input to our model, the performance is improved further by 0.015 and 0.018, respectively, demonstrating our model can benefit from the temporal knowledge from the input sequence. The PSI dataset has an imbalanced distribution and there are much fewer samples belonging to the category "Stop", thus both OIA-global and OIA* obtain worse performance on this category compared to "Main speed" and "Slow down". Surprisingly, our model is able to achieve better performance on this decision. Moreover, as OIA adopts both global and local detection proposals as input for prediction, while InAction only uses the global feature map, so that we compare our model with another baseline OIA-global, which has the same architecture with OIA excluding the local proposal branch. From the results, we observe that OIA-global obtains worse overall performance compared to OIA and InAction.

From the BDD-OIA results (Table 3), we observe InAction can improve the action prediction performance compared to the reproduced OIA. For the reason prediction, we notice that the reproduced results outperform the numbers reported in the OIA paper around 0.08, and our proposed method can further improve the overall F1 and class-wise mean F1 both over **0.5**. This demonstrates that our model works well in both action prediction and explanation reasoning. Moreover, we observe that the results produced with prototypes learned on the global feature maps are better than based on the detected proposals. We argue that relying on the detected proposals will make the model fail, and constrain the representative capabilities of learned semantic prototypes, compared to exploring the implicit visual-semantic knowledge based on the whole input image.

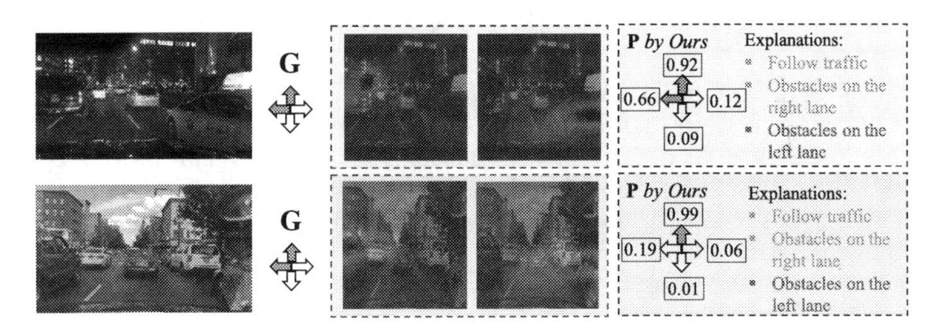

Fig. 3. Comparison of explanations produced by the implicit visual-semantic module and the explicit human-annotated reasoning module for examples on BDD-OIA.

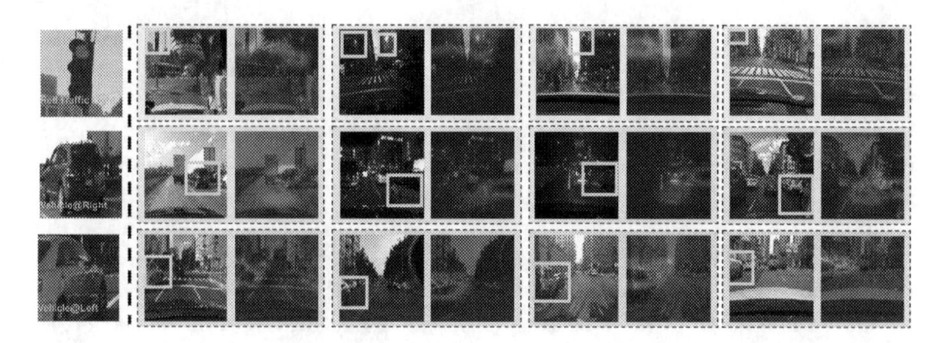

Fig. 4. Visualizing prototypes by selecting the most similar patches from the training samples, where each row shows one explanation.

4.3 Interpretability Analysis

Comparison with OIA. We present qualitative results in Fig. 2 to demonstrate the interpretability and transparency of the propose InAction model. For the same visual input, we compare both action and explanation prediction of OIA and our InAction. From the selected examples, we notice that OIA made wrong action predictions while InAction can achieve correct results in some cases. The only wrong prediction in the 3rd example is that both OIA and the explicit human-annotated reasoning module in InAction recognize the white vehicle in front and predict the explanation as "Obstacle: car", then make the "Stop/Slow down" decision. However, the ground-truth action annotation does not contain this label. Such an observation demonstrates that insufficient explanation and inconsistency reasoning always exists in the human-defined annotations, especially on single-frame based prediction tasks.

Compensation between Implicit and Explicit Interpretation. In Fig. 3, we compare the generated explanations from the implicit and explicit modules for the same task. We notice that some human-annotated reasoning are also captured by the implicit semantic prototypes, e.g., "Obstacles on the right lane". However, some explanations discovered by the implicit visual prototypes compensate the lack of human annotation. For example, the vehicle on the left lane in the second row example is quite close but not annotated, and the ground-truth label is "forward" and "turn left", while fortunately, our model notices the obstacle on the left lane and predict "forward" only.

Implicit Visual Semantics Analysis. To illustrate the learned implicit visual semantic prototypes in an intuitive way, we visualize the prototypes via the most similar patches of images in the BDD-OIA dataset [2]. Figure 4 shows the selected examples with patches highly activated by specific semantic prototypes from action decision "Stop", "Turn left", and "Turn right". The most activated patch of the given input for selected prototypes are marked by bounding boxes in the original input, which represent the image patches that InAction considers to focus on corresponding to specific prototypes. From the results, we observe that when the implicit visual-semantic reasoning module slides over the whole input to obtain activation map, these three prototypes are represented as "Red traffic light", "Vehicle at right", and "Vehicle at left", respectively. Any region is strongly activated by one of the specific prototypes, or, in other words, one of the prototypes presents strongly in the input frame, will play a crucial role in the final prediction.

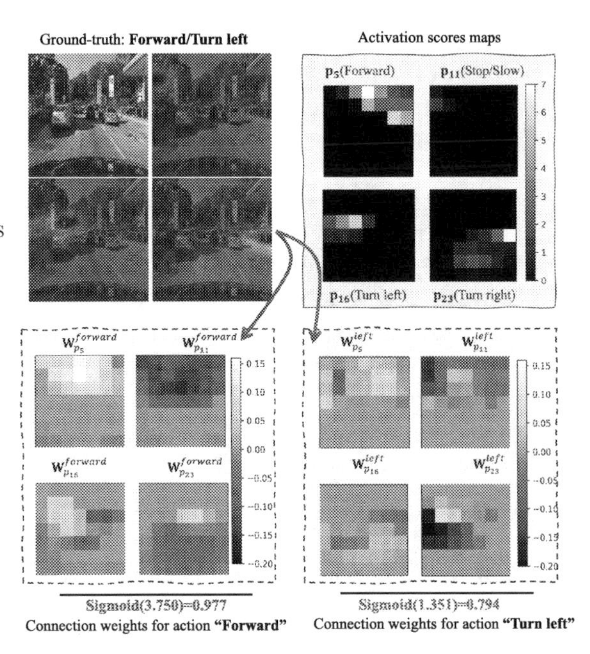

Fig. 5. Visualization of reasoning of selected instance.

Reasoning Process of InAction. Prior prototype-based models only observe the most strongly activated region. However, driving action prediction has much more complicated scene context and multiple objects involved as hints, so the spatial location of each prototype presence and the relationships among different components make crucial influence for the final decision prediction. Figure 5 shows the reasoning process of our InAction predicting the action decision for a test sample, which is annotated as "Forward/Turn left". Given the input frame, the implicit visual semantic interpretation module compares every patch in the feature map against the learned prototypes, producing the activation score maps. The activation maps that are most strongly activated by prototypes are shown as the top-right heatmaps in Fig. 5, where $\mathbf{p}_5, \mathbf{p}_{11}, \mathbf{p}_{16}, \mathbf{p}_{23}$ are assigned to action classes "Forward", "Stop/Slow down", "Turn left", and "Turn right", respectively. Although m prototypes are assigned to C action decision resulting in m_k prototypes per class during training, all activation scores produced by m prototypes over the feature map of the input frame are multiplied by the weight matrix in the last fully-connected layer $C_S(\cdot)$ to generate the output prediction. The weights in the fully-connected layer represent the connections between prototypes and the predicted classes. In Fig. 5, we select the weights ($\mathbf{W}$) for class "Forward" and "Turn left" corresponding to the selected prototypes, and show them after reshaped into the same shape as the activation map. From the weights over different regions of the feature map/activation map, we observe that the same prototype plays different roles for different action decisions. For example, components similar to prototype $\mathbf{p}_{11}$ appearing in the top area of the view will make negative contribution to the prediction of "Forward", while for the prediction of class "Turn left", it will reduce the probability of "Turn left" only when it appears at the top-left corner, otherwise, this prototype is comparably neutral. Interestingly, the prototype shown in the first row of Fig. 4 is prototype $\mathbf{p}_{11}$, which represents "Red traffic light".

5 Conclusion

In this paper, we developed a novel Interpretable Action (**InAction**) decision making model to provide enriched explanations from both explicit human annotation and implicit visual semantics perspectives. To implement this, two interpretable modules were proposed including a visual semantic module and an explicit reasoning module. Specifically, the first module aimed to capture the region-based action-inducing semantic concepts from the visual inputs, so that our model could automatically learn the implicit visual cues to provide a human-understandable explanation. The second module attempted to benefit from the human-annotated reasoning for action decision making so that our model was able to provide a more high-level interpretation by aligning visual inputs to human annotations. Experimental results on two autonomous driving benchmarks demonstrated the effectiveness of our **InAction** model.

Acknowledgment. We thank the Toyota Collaborative Safety Research Center for funding support.

References

1. Bach, S., Binder, A., Montavon, G., Klauschen, F., Müller, K.R., Samek, W.: On pixel-wise explanations for non-linear classifier decisions by layer-wise relevance propagation. PLoS ONE **10**(7), e0130140 (2015)
2. Chen, C., Li, O., Tao, D., Barnett, A., Rudin, C., Su, J.: This looks like that: deep learning for interpretable image recognition. In: Advances in Neural Information Processing Systems, pp. 8928–8939 (2019)
3. Chen, L., Yang, T., Zhang, X., Zhang, W., Sun, J.: Points as queries: weakly semi-supervised object detection by points. In: Proceedings of the IEEE/CVF Conference on Computer Vision and Pattern Recognition (CVPR), pp. 8823–8832 (2021)
4. Chen, T., et al.: PSI: a pedestrian behavior dataset for socially intelligent autonomous car. arXiv preprint arXiv:2112.02604 (2021)

5. Choi, C., Choi, J.H., Li, J., Malla, S.: Shared cross-modal trajectory prediction for autonomous driving. In: Proceedings of the IEEE/CVF Conference on Computer Vision and Pattern Recognition (CVPR), pp. 244–253 (2021)
6. Dai, Z., Cai, B., Lin, Y., Chen, J.: UP-DETR: unsupervised pre-training for object detection with transformers. In: Proceedings of the IEEE/CVF Conference on Computer Vision and Pattern Recognition (CVPR), pp. 1601–1610 (2021)
7. Darms, M., Rybski, P., Urmson, C.: Classification and tracking of dynamic objects with multiple sensors for autonomous driving in urban environments. In: 2008 IEEE Intelligent Vehicles Symposium, pp. 1197–1202 (2008). https://doi.org/10.1109/IVS.2008.4621259
8. Devlin, J., Chang, M.W., Lee, K., Toutanova, K.: BERT: pre-training of deep bidirectional transformers for language understanding. In: Proceedings of the 2019 Conference of the North American Chapter of the Association for Computational Linguistics: Human Language Technologies, Volume 1 (Long and Short Papers), pp. 4171–4186. Association for Computational Linguistics (2019)
9. Dong, J., Chen, S., Zong, S., Chen, T., Labi, S.: Image transformer for explainable autonomous driving system. In: 2021 IEEE International Intelligent Transportation Systems Conference (ITSC), pp. 2732–2737 (2021)
10. Dosovitskiy, A., Brox, T.: Inverting visual representations with convolutional networks. In: Proceedings of the IEEE conference on computer vision and pattern recognition, pp. 4829–4837 (2016)
11. Feng, D., et al.: Deep multi-modal object detection and semantic segmentation for autonomous driving: Datasets, methods, and challenges. In: IEEE Transactions on Intelligent Transportation Systems (2020)
12. Fong, R.C., Vedaldi, A.: Interpretable explanations of black boxes by meaningful perturbation. In: Proceedings of the IEEE International Conference on Computer Vision, pp. 3429–3437 (2017)
13. Jing, T., Liu, H., Ding, Z.: Towards novel target discovery through open-set domain adaptation. In: Proceedings of the IEEE/CVF International Conference on Computer Vision, pp. 9322–9331 (2021)
14. Jing, T., Xia, H., Hamm, J., Ding, Z.: Augmented multi-modality fusion for generalized zero-shot sketch-based visual retrieval. In: IEEE Transactions on Image Processing (2022)
15. Joseph, K.J., Khan, S., Khan, F.S., Balasubramanian, V.N.: Towards open world object detection. In: Proceedings of the IEEE/CVF Conference on Computer Vision and Pattern Recognition (CVPR), pp. 5830–5840 (2021)
16. Kim, E., Kim, S., Seo, M., Yoon, S.: XprotoNet: diagnosis in chest radiography with global and local explanations. In: Proceedings of the IEEE/CVF Conference on Computer Vision and Pattern Recognition, pp. 15719–15728 (2021)
17. Kim, J., Canny, J.: Interpretable learning for self-driving cars by visualizing causal attention. In: Proceedings of the IEEE international conference on computer vision, pp. 2942–2950 (2017)
18. Kim, J., Rohrbach, A., Darrell, T., Canny, J., Akata, Z.: Textual explanations for self-driving vehicles. In: Ferrari, V., Hebert, M., Sminchisescu, C., Weiss, Y. (eds.) ECCV 2018. LNCS, vol. 11206, pp. 577–593. Springer, Cham (2018). https://doi.org/10.1007/978-3-030-01216-8_35
19. Li, J., Zhan, W., Hu, Y., Tomizuka, M.: Generic tracking and probabilistic prediction framework and its application in autonomous driving. IEEE Trans. Intell. Transp. Syst. **21**, 3634–3649 (2020)

20. Li, P., Qin, T., Shen, S.: Stereo vision-based semantic 3d object and ego-motion tracking for autonomous driving. In: Ferrari, V., Hebert, M., Sminchisescu, C., Weiss, Y. (eds.) ECCV 2018. LNCS, vol. 11206, pp. 664–679. Springer, Cham (2018). https://doi.org/10.1007/978-3-030-01216-8_40
21. Ming, Y., Xu, P., Qu, H., Ren, L.: Interpretable and steerable sequence learning via prototypes. In: Proceedings of the 25th ACM SIGKDD International Conference on Knowledge Discovery & Data Mining, pp. 903–913 (2019)
22. Nauta, M., van Bree, R., Seifert, C.: Neural prototype trees for interpretable fine-grained image recognition. In: Proceedings of the IEEE/CVF Conference on Computer Vision and Pattern Recognition, pp. 14933–14943 (2021)
23. Nguyen, A., Dosovitskiy, A., Yosinski, J., Brox, T., Clune, J.: Synthesizing the preferred inputs for neurons in neural networks via deep generator networks. Adv. Neural. Inf. Process. Syst. **29**, 3387–3395 (2016)
24. Olah, C., Mordvintsev, A., Schubert, L.: Feature visualization. Distill **2**(11), e7 (2017)
25. Omeiza, D., Webb, H., Jirotka, M., Kunze, L.: Explanations in autonomous driving: a survey. IEEE Transactions on Intelligent Transportation Systems, pp. 1–21 (2021)
26. Pang, B., Zhao, T., Xie, X., Wu, Y.N.: Trajectory prediction with latent belief energy-based model. In: Proceedings of the IEEE/CVF Conference on Computer Vision and Pattern Recognition (CVPR), pp. 11814–11824 (2021)
27. Ren, S., He, K., Girshick, R., Sun, J.: Faster R-CNN: towards real-time object detection with region proposal networks. Adv. Neural. Inf. Process. Syst. **28**, 91–99 (2015)
28. Rudin, C.: Stop explaining black box machine learning models for high stakes decisions and use interpretable models instead. Nat. Mach. Intell. **1**(5), 206–215 (2019)
29. Rymarczyk, D., Struski, Ł., Tabor, J., Zieliński, B.: Protopshare: prototypical parts sharing for similarity discovery in interpretable image classification. In: Proceedings of the 27th ACM SIGKDD Conference on Knowledge Discovery & Data Mining, pp. 1420–1430 (2021)
30. Shafiee, N., Padir, T., Elhamifar, E.: Introvert: human trajectory prediction via conditional 3d attention. In: Proceedings of the IEEE/CVF Conference on Computer Vision and Pattern Recognition (CVPR), pp. 16815–16825 (2021)
31. Shi, L., et al.: SGCN: sparse graph convolution network for pedestrian trajectory prediction. In: Proceedings of the IEEE/CVF Conference on Computer Vision and Pattern Recognition (CVPR), pp. 8994–9003 (2021)
32. Siam, M., Gamal, M., Abdel-Razek, M., Yogamani, S., Jagersand, M., Zhang, H.: A comparative study of real-time semantic segmentation for autonomous driving. In: Proceedings of the IEEE Conference on Computer Vision and Pattern Recognition (CVPR) Workshops (2018)
33. Tampuu, A., Matiisen, T., Semikin, M., Fishman, D., Muhammad, N.: A survey of end-to-end driving: architectures and training methods. In: IEEE Transactions on Neural Networks and Learning Systems (2020)
34. Wang, D., Devin, C., Cai, Q.Z., Krähenbühl, P., Darrell, T.: Monocular plan view networks for autonomous driving. In: 2019 IEEE/RSJ International Conference on Intelligent Robots and Systems (IROS), pp. 2876–2883. IEEE (2019)
35. Wang, D., Devin, C., Cai, Q.Z., Yu, F., Darrell, T.: Deep object-centric policies for autonomous driving. In: 2019 International Conference on Robotics and Automation (ICRA), pp. 8853–8859. IEEE (2019)

36. Xia, H., Ding, Z.: HGNet: hybrid generative network for zero-shot domain adaptation. In: Vedaldi, A., Bischof, H., Brox, T., Frahm, J.-M. (eds.) ECCV 2020. LNCS, vol. 12372, pp. 55–70. Springer, Cham (2020). https://doi.org/10.1007/978-3-030-58583-9_4
37. Xia, H., Ding, Z.: Structure preserving generative cross-domain learning. In: Proceedings of the IEEE/CVF Conference on Computer Vision and Pattern Recognition, pp. 4364–4373 (2020)
38. Xu, H., Gao, Y., Yu, F., Darrell, T.: End-to-end learning of driving models from large-scale video datasets. In: Proceedings of the IEEE conference on computer vision and pattern recognition, pp. 2174–2182 (2017)
39. Xu, Y., Yang, X., Gong, L., Lin, H.C., Wu, T.Y., Li, Y., Vasconcelos, N.: Explainable object-induced action decision for autonomous vehicles. In: Proceedings of the IEEE/CVF Conference on Computer Vision and Pattern Recognition, pp. 9523–9532 (2020)
40. Yu, F., et al.: BDD100k: a diverse driving dataset for heterogeneous multitask learning. In: Proceedings of the IEEE/CVF Conference on Computer Vision and Pattern Recognition (CVPR) (2020)
41. Yurtsever, E., Lambert, J., Carballo, A., Takeda, K.: A survey of autonomous driving: common practices and emerging technologies. IEEE Access **8**, 58443–58469 (2020)
42. Zeiler, M.D., Fergus, R.: Visualizing and understanding convolutional networks. In: Fleet, D., Pajdla, T., Schiele, B., Tuytelaars, T. (eds.) ECCV 2014. LNCS, vol. 8689, pp. 818–833. Springer, Cham (2014). https://doi.org/10.1007/978-3-319-10590-1_53
43. Zhang, Y., Tiňo, P., Leonardis, A., Tang, K.: A survey on neural network interpretability. IEEE Transactions on Emerging Topics in Computational Intelligence (2021)
44. Zhang, Z., Fidler, S., Urtasun, R.: Instance-level segmentation for autonomous driving with deep densely connected MRFs. In: 2016 IEEE Conference on Computer Vision and Pattern Recognition (CVPR), pp. 669–677 (2016)
45. Zheng, H., Fu, J., Mei, T., Luo, J.: Learning multi-attention convolutional neural network for fine-grained image recognition. In: Proceedings of the IEEE international conference on computer vision, pp. 5209–5217 (2017)
46. Zhou, B., Khosla, A., Lapedriza, A., Oliva, A., Torralba, A.: Learning deep features for discriminative localization. In: Proceedings of the IEEE conference on computer vision and pattern recognition, pp. 2921–2929 (2016)
47. Zhou, B., Sun, Y., Bau, D., Torralba, A.: Interpretable basis decomposition for visual explanation. In: Ferrari, V., Hebert, M., Sminchisescu, C., Weiss, Y. (eds.) ECCV 2018. LNCS, vol. 11212, pp. 119–134. Springer, Cham (2018). https://doi.org/10.1007/978-3-030-01237-3_8

CramNet: Camera-Radar Fusion with Ray-Constrained Cross-Attention for Robust 3D Object Detection

Jyh-Jing Hwang(✉), Henrik Kretzschmar, Joshua Manela, Sean Rafferty, Nicholas Armstrong-Crews, Tiffany Chen, and Dragomir Anguelov

Waymo, Mountain View, USA
jyhjinghwang@gmail.com

Abstract. Robust 3D object detection is critical for safe autonomous driving. Camera and radar sensors are synergistic as they capture complementary information and work well under different environmental conditions. Fusing camera and radar data is challenging, however, as each of the sensors lacks information along a perpendicular axis, that is, depth is unknown to camera and elevation is unknown to radar. We propose the camera-radar matching network CramNet, an efficient approach to fuse the sensor readings from camera and radar in a joint 3D space. To leverage radar range measurements for better camera depth predictions, we propose a novel ray-constrained cross-attention mechanism that resolves the ambiguity in the geometric correspondences between camera features and radar features. Our method supports training with sensor modality dropout, which leads to robust 3D object detection, even when a camera or radar sensor suddenly malfunctions on a vehicle. We demonstrate the effectiveness of our fusion approach through extensive experiments on the RADIATE dataset, one of the few large-scale datasets that provide radar radio frequency imagery. A camera-only variant of our method achieves competitive performance in monocular 3D object detection on the Waymo Open Dataset.

Keywords: Sensor fusion · Cross attention · Robust 3D object detection

1 Introduction

3D object detection that is robust to different weather conditions and sensor failures is critical for safe autonomous driving. Fusion between camera and radar sensors stands out as they are both relatively resistant to various weather conditions [2] compared to the popular lidar sensor [3]. A fusion design that naturally accepts single-sensor failures (lidar, radar, or camera or radar) is thus desired and boosts safety in an autonomous driving system (Fig. 1).

Supplementary Information The online version contains supplementary material available at https://doi.org/10.1007/978-3-031-19839-7_23.

S. Avidan et al. (Eds.): ECCV 2022, LNCS 13698, pp. 388–405, 2022.
https://doi.org/10.1007/978-3-031-19839-7_23

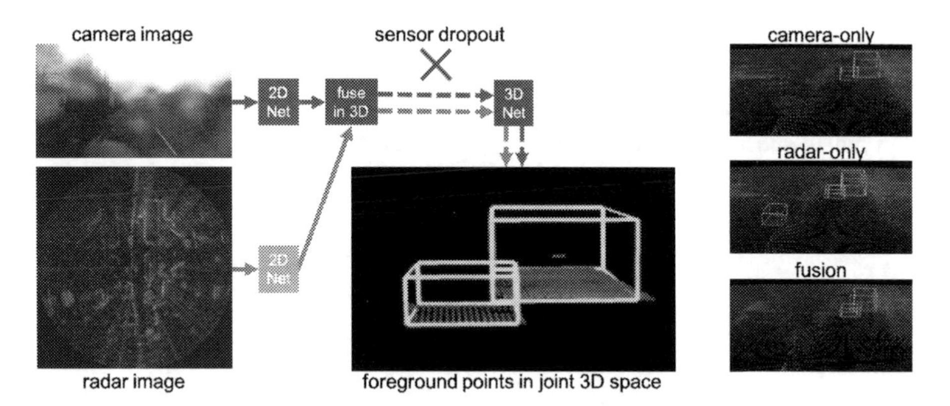

Fig. 1. Our approach takes as input a camera image (top left) and a radar RF image (bottom left). The model then predicts foreground segmentation for both native 2D representations before projecting the foreground points with features into a joint 3D space (middle bottom) for sensor fusion. Finally, the method runs sparse convolutions in the joint space for 3D object detection. The network architecture naturally supports training with sensor dropout. This allows the resulting model to cope with sensor failures at inference time as it can run on camera only and radar only input depending on which sensors are available.

Most sensor fusion research has focused on fusion between lidar and another sensor [7,11,11,19,31,32,39,50,51,54,57] because lidar provides complete geometric information, i.e., azimuth, range, and elevation. Sparse correspondences between lidar and another sensor is thus well defined, making lidar an ideal carrier for fusion. On the other hand, even though camera and radar sensors are lighter and cheaper, consume less power, and endure longer than lidar, camera-radar fusion is understudied. Camera-radar fusion is especially challenging as each sensor lacks information along one perpendicular axis: depth unknown for camera and elevation unknown for emerging imaging radar, as summarized in Table 1. Radar produces radio frequency (RF) imagery that encodes the environment approximately in the bird's-eye view (BEV) with various noise patterns, an example shown in Fig. 1. As a result, camera data (in perspective view) and radar data (in BEV) form many-to-many mappings and the exact matching is unclear from geometry alone.

To solve the matching problem, we consider three possible schemes for fusion: **(1) Perspective view primary** [32]: This scheme implies we trust the depth reasoning from the perspective view. One can project camera pixels to their 3D locations with depth estimates and find their vertical nearest neighbors of corresponding radar points. If depth is unknown, one can project a pixel along a ray in 3D and perform matching. **(2) Bird's-eye view primary** [50]: This scheme implies we trust the elevation reasoning from the bird's-eye view. However, since it's difficult to predict elevation from radar imagery directly, one might borrow elevation information from the map. Hence, the inferred elevation for radar is sometimes inaccurate, resulting in rare usage unless LiDAR is available. **(3) Cross-view matching** [13]: This scheme implies we perform matching in a joint 3D space. For example, one can use supplementary information (map or camera depth estimation) to upgrade camera and radar 2D image pixels to 3D point clouds (with some uncertainty) and perform matching between point clouds directly. This is supposedly the most powerful scheme if we can properly handle uncertainties. Our architecture is designed to enable this matching scheme, hence we name it CramNet (Camera and RAdar Matching Network).

Table 1. Characteristics of major sensors commonly used for autonomous driving. Both camera and radar tend to be less affected by inclement weather compared to lidar scanners. However, whereas regular camera does not directly measure range, radar does not measure elevation. This poses a unique challenge for fusing camera and radar readings as the geometric correspondences between the two sensors are underconstrained. Overall, camera-radar fusion is still underexplored in the literature. *Although there exists radars with elevation, this paper focuses on planar radar which, at the moment, is more common for automotive radar.

Sensor	Azimuth	Range	Elevation	Resistance to weather	3D detection literature
Camera	✓	x	✓	Medium	Abundant
Radar	✓	✓	x*	High	Scarce
Lidar	✓	✓	✓	Low	Abundant

Since the effectiveness of projecting into 3D space heavily relies on accurate camera depth estimates, we propose a ray-constrained cross-attention mechanism to leverage radar for better depth estimation. The idea is to match radar responses along each camera ray emitted from a pixel. The correct projection should be the locations where radar senses reflections. Our architecture is further designed to accept sensor failures naturally. As shown in Fig. 1, the model is able to operate even when one of the modalities is corrupted during inference. To this end, we incorporate sensor dropout [7,52] in the point cloud fusion stage during training to boost the sensor robustness.

We summarize the contributions of this paper as follows:

1. We present a camera-radar fusion architecture for 3D object detection that is flexible enough to fall back to a single sensor modality in the event of a sensor failure.
2. We demonstrate that the sensor fusion model effectively leverages data from both sensors as the model outperforms both the camera-only and the radar-only variants significantly.
3. We propose a ray-constrained cross-attention mechanism that leverages the range measurements from radar to improve camera depth estimates, leading to improved detection performance.
4. We incorporate sensor dropout during training to further improve the accuracy and the robustness of camera-radar 3D object detection.
5. We demonstrate state-of-the-art radar-only and camera-radar detection performance on the RADIATE dataset [40] and competitive camera-only detection performance on the Waymo Open Dataset [47].

2 Related Work

Camera-based 3D object detection. Monocular camera 3D object detection is first approached by directly extending 2D detection architectures and incorporating geometric relationships between the 2D perspective view and 3D space [4,6,8,16,23,23,27,43,44]. Utilizing pixel-wise depth maps as an additional input shows improved results, either for lifting detected boxes [26,42] or projecting image pixels into 3D point clouds [9,24,53,55,58] (also known as Pseudo-LiDAR [53]). More recently, another camp of methods emerge to be promising, i.e., projecting intermediate features into BEV grid features along the projection ray without explicitly forming 3D point clouds [18,34,36,46].

The BEV grid methods benefit from naturally expressing the 3D projection uncertainty along the depth dimension. However, these methods suffer from significantly increased compute requirements as the detection range expands. In contrast, we model the depth uncertainty through sampling along the projection ray and consulting radar features for more accurate range signals. This also enables the adoption of foreground extraction that allows a balanced trade-off between detection range and computation.

Radar-Based 3D Object Detection. Frequency modulated continuous wave (FMCW) radar is usually presented by two kinds of data representations, i.e., radio frequency (RF) images and radar points. The RF images are generated from the raw radar signals using a series of fast Fourier transforms that encode a wide variety of sensing context whereas the radar points are derived from these RF images through a peak detection algorithm, such as Constant False Alarm Rate (CFAR) algorithm [35]. The downside of the radar points is that recall is imperfect and the contextual information of radar returns is lost, with only the range, azimuth and doppler information retained. As a result, radar points are not suitable for effective single modality object detection [33,38], which is why most works use this data format only to foster fusion [2,13,28,29]. On the other hand, the RF images maintain rich environmental context information and even complete object motion information to enable a deep learning model to understand the semantic meaning of a scene [25,40]. Our work is therefore built upon radar RF images and can produce reasonable 3D object detection predictions with radar-only inputs.

Sensor Fusion for 3D Object Detection. Sensor fusion for 3D object detection has been studied extensively using lidar and camera. The reasons are twofold: 1) Lidar scans provide comprehensive representations in 3D for inferring correspondences between sensors, and 2) camera images contain more semantic information to further boost the recognition ability. Various directions have been explored, such as image detection in 2D before projecting into frustums [32,54], two-stage frameworks with object-centric modality fusion [7,11,17], image feature-based lidar point decoration [50,51], or multi-level fusion [11,19,31]. Since sparse correspondences between camera and lidar

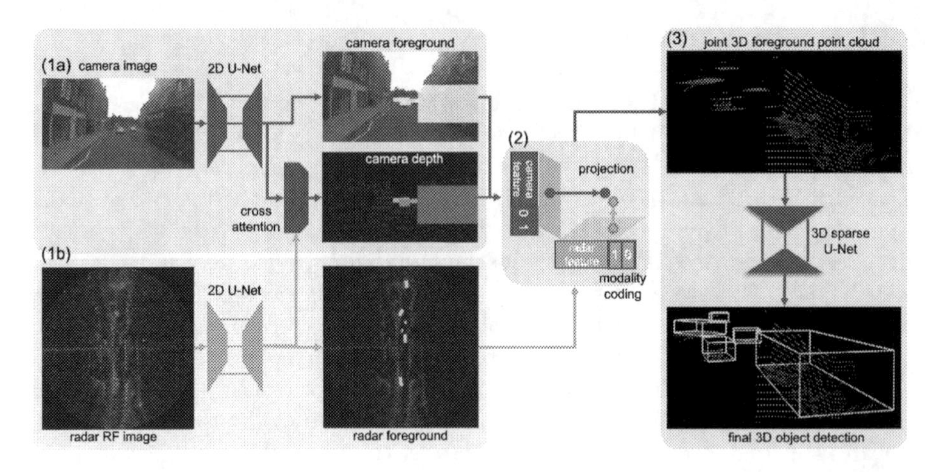

Fig. 2. Architecture overview. Our method can be partitioned into three stages: (1a) camera 2D foreground segmentation and depth estimation, (1b) radar 2D foreground segmentation, (2) projection from 2D to 3D and subsequent point cloud fusion, and (3) 3D foreground point cloud object detection. The cross-attention mechanism modifies the camera depth estimation by consulting radar features, as further illustrated in Fig. 3. The modality coding module appends a camera or radar binary code to the features that are fed into the 3D stage, enabling sensor dropout and enhancing robustness. We depict the camera stream in blue, the radar stream in green, and the fused stream in red. (Color figure online)

are well defined, fusion is mostly focused on integrating information rather than matching points from different sensors.

As a result, these fusion techniques are not directly applicable to camera-radar fusion where associations are underconstrained. Early work, Lim et al. [20], applies feature fusion directly between camera and radar features without any geometric considerations. Recently, more works tend to leverage camera models and geometry for association. For example, CenterFusion [28] creates camera object proposal frustums to associate radar features and GRIF Net [13] projects 3D RoI to camera perspective and radar BEV to associate features. Our model, on the other hand, fuses camera-radar data in a joint 3D space with the flexibility to perform 3D detection with either single modality, leading to increased robustness.

3 CramNet for Robust 3D Object Detection

We describe the overall architecture for camera-radar fusion in Sect. 3.1. In Sect. 3.2, We then introduce a ray-constrained cross-attention mechanism to leverage radar for better camera 3D point localization. Finally, we propose sensor dropout that can be integrated seamlessly into the architecture in Sect. 3.3 to further improve the robustness of 3D object detection.

3.1 Overall Architecture

Our model architecture, in Fig. 2, is inspired by Range Sparse Net (RSN) [48], which is an efficient two-stage lidar-based object detection framework. The RSN framework takes input of perspective range images, segments perspective foreground pixels, extracts 3D (BEV) features on foreground regions using sparse convolution [56], and performs CenterNet-style [60] detection. We adapt the framework for camera-radar fusion and the overall architecture can be partitioned into three stages: (1) 2D foreground segmentation, (2) 2D to 3D projection and point cloud fusion, and (3) 3D foreground point cloud detection.

Stage 1: 2D Foreground Segmentation. The goal of this stage is to perform efficient foreground segmentation for native dense representations from two modalities. This allows us to restrict the expensive 3D operations to foreground points. The network takes as input a pair of camera images $\boldsymbol{I}_C$ and radar RF images $\boldsymbol{I}_R$. We then employ two identical lightweight U-Nets [37] to extract 2D features and predict foreground segmentation masks for each modality, $\boldsymbol{F}_C$ and $\boldsymbol{F}_R$, respectively. For camera image feature extraction, one can also adopt a more powerful, multi-scale feature extractor, such as a feature pyramid network [21]. The detailed design of the U-Net can be found in the supplementary.

To train such a segmentation network (for both camera and radar), we use the 2D projection of 3D bounding box labels as ground truth – a pixel belongs to the foreground class if it falls inside any of the projected 2D boxes. This might introduce some noise as background pixels sometimes fall within a box, but we find that this noise is insignificant in practice. We then apply a pixel-wise focal loss [22] to classify each pixel:

$$L_{\text{seg}} = \frac{-1}{N}\Big(\sum_{i \in \boldsymbol{F}}(1-p_i)^{\gamma_s}\log(p_i) + \sum_{i \in \boldsymbol{B}} p_i^{\gamma_s}\log(1-p_i)\Big), \tag{1}$$

where N is the total number of pixels, $\boldsymbol{F}$ and $\boldsymbol{B}$ are the sets of foreground and background pixels, and p_i is the model's estimated probability of foreground for pixel i. The hyperparameter γ_s controls the penalty reduction. A pixel with foreground score higher than τ will be selected. Since the 3D stage can resolve false positives, whereas false negatives cannot be recovered, we typically set a low value for τ to attain high recall.

Stage 2: 2D to 3D Projection and Point Cloud Fusion. Once we obtain the foreground pixels, we project them into 3D for the following 3D stage. For the camera projection, we predict a depth value for each pixel from the same U-Net with additional convolutional layers. The depth ground truth is obtained by projecting lidar points to the camera view and overlaying them with depth values from projected ground truth 3D boxes. The use of depth from ground truth boxes is to enable 3D detection where lidar data alone is insufficient. This is especially true outside of lidar range, as well as when lidar points are deteriorated due to weather. We train the depth estimation using pixelwise L2 losses on valid regions, or L_{depth}. The camera projection relies on the camera model, i.e., the intrinsics and extrinsics, with depth to infer the 3D location of each pixel.

For radar projection, we use the radar model to transform radar BEV points to 3D using the sensor height as elevation. If map is available, the road elevation can be used to offset this value to handle non-planar scenes like hills.

There are several options to combine the camera and radar 3D point clouds. One plausible choice is to select one modality as a major sensor and gather features from the other modality. This is usually how researchers fuse lidar with other sensors [50]. However, the drawback is obvious: the major sensor is a single point of failure. Instead, we directly place two point clouds in a joint 3D space. We align the feature dimensions of both modalities and append a modality code to the feature so that the 3D network can leverage the multi-modality information easily. The major benefit is to enable robust detection especially when one modality fails to perform.

Stage 3: 3D Foreground Point Cloud Detection. We apply dynamic voxelization [61] on the fused foreground point cloud, whose features are then encoded into sparse voxel features. A 2D or 3D sparse convolution network [10] (for pillar style [15], or 3D voxelization, respectively) is applied on the sparse voxels. The network details can be found in the supplementary.

We follow RSN [48] for CenterNet-style [60] 3D box regression. We calculate a ground truth objectness heatmap for every point $x \in \mathbb{R}^3$: $h(x) = \max\{\exp(-\frac{||x-c||-||x_c-c||}{\sigma^2}) \mid c \in C(x)\}$ where $C(x)$ is the set of centers of boxes containing x, x_c is the closest point to box center c, and σ is a constant. In other words, the objectness of a point is inversely related to its distance to the closest box center. We train the network to predict a heatmap using a focal loss [22]:

$$\begin{aligned} L_{\text{hm}} = & \frac{-1}{N} \sum_x \Big(\big(1 - \tilde{h}(x)\big)^{\gamma_h} \log\big(\tilde{h}(x)\big) \mathbb{1}(h > 1 - \epsilon_h) + \\ & \big(1 - h(x)\big)_h^{\alpha} \tilde{h}(x)^{\gamma_h} \log\big(1 - \tilde{h}(x)\big) \mathbb{1}(h \leq 1 - \epsilon_h) \Big), \end{aligned} \tag{2}$$

where $\mathbb{1}(\cdot)$ is the indicator function, h and $\tilde{h}$ are the ground truth and predicted heat map, $(1 - \epsilon_h)$ decides the threshold for ground truth objectness, and α_h and γ_h are hyperparameters in the focal loss.

The 3D boxes are parameterized as $\boldsymbol{b} = (b_x, b_y, b_z, l, w, h, \theta)$ where b_x, b_y, b_z are the offsets of a 3D box center relative to a voxel center, and l, w, h, θ are the length, width, height, and heading of a box. All the box parameters are trained with smooth L1 losses except for the heading that is trained with a bin loss [41]. An additional IoU loss [59] is employed for better accuracy. The box regression loss is as follows:

$$L_{\text{box}} = \frac{1}{B} \sum_i \Big(L_{\text{SmoothL1}}(\boldsymbol{b}_i \backslash \theta_i - \tilde{\boldsymbol{b}}_i \backslash \tilde{\theta}_i) + L_{\text{bin}}(\theta_i, \tilde{\theta}_i) + L_{\text{IoU}i} \Big), \tag{3}$$

where B is the total number of boxes with ground truth heatmap value greater than a threshold τ_{hm}, and $\boldsymbol{b}_i$ and $\tilde{\boldsymbol{b}}_i$ denote the ground truth and prediction for box $\boldsymbol{b}_i$, respectively; the same for θ_i. For more details on heatmap and box regression, we refer interested readers to RSN [48].

We train the fusion network end-to-end with losses summarized as:

$$L = \lambda_{\text{seg}} L_{\text{seg}} + \lambda_{\text{depth}} L_{\text{depth}} + \lambda_{\text{hm}} L_{\text{hm}} + L_{\text{box}}, \tag{4}$$

where λ_* are hyperparameters for the respective loss weighting.

3.2 Ray-Constrained Cross-Attention

It is widely known [30,53] that camera-based 3D object detection relies heavily on accurate depth estimation, either explicitly or implicitly through BEV representations. Luckily, for camera-radar fusion, we don't have to rely solely on camera to infer depth as radar provides relatively accurate range estimates. To utilize the complementary sensing directions, we propose a ray-constrained cross attention mechanism to leverage radar for improving camera 3D point localization, illustrated in Fig. 3.

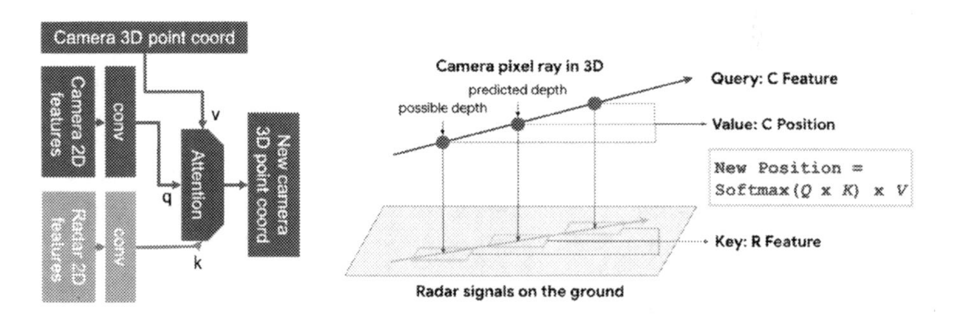

Fig. 3. The proposed ray-constrained cross-attention mechanism resolves the ambiguity in the geometric correspondences between camera features and radar features. Following the Transformer [49], we take camera features as queries and radar features as keys to transform 3D camera points as values.

Our observation is that an optimal 3D location for each foreground camera pixel usually accompanies a corresponding peak response from radar. Thus we propose to consult radar features along a camera 3D ray, emitted from each pixel, to rectify the camera 3D point location after projection. Since there is infinite possible locations, we perform sampling along the ray, centered at the initial depth estimation. The final 3D location is decided by matching between camera and radar features among these sampled locations.

We denote the projected camera 3D location from a depth estimate $\tilde{d}_i$ for pixel i as $M(\tilde{d}_i)$. We sample s points farther and closer around the estimated location respectively, or $M(\tilde{d}_i \pm \epsilon \times k)$, where ϵ is a hyperparameter for depth error, and k ranges from 1 to s. We denote this set of 3D locations as $\tilde{\boldsymbol{M}}_i \in \mathbb{R}^{(2s+1)\times 3}$. We gather closest radar features for every sampled location, denoted as $\boldsymbol{\psi}_{Ri} \in \mathbb{R}^{(2s+1)\times d}$. Likewise, we denote the camera feature for a pixel i as $\boldsymbol{\psi}_{Ci} \in \mathbb{R}^{1\times d}$. The final camera 3D point location $M_i \in \mathbb{R}^{1\times 3}$ for pixel i can thus be obtained using a cross-attention formulation [49]:

$$M_i = \text{softmax}(\frac{\boldsymbol{\psi}_{Ci}\boldsymbol{\psi}_{Ri}^T}{\sqrt{d}})\tilde{\boldsymbol{M}}_i. \tag{5}$$

To relate to the naming convention in attention [49], we use the camera feature $\boldsymbol{\psi}_{Ci}$ as a query, a set of radar features $\boldsymbol{\psi}_{Ri}$ as keys, and the sampled 3D locations $\tilde{\boldsymbol{M}}_i$ as values. Therefore, the final location is calculated by matching the query with the most active keys, associated with the respective values.

Notably, this design is computationally efficient. The time complexity is asymptotically proportional to $(N \times d \times s)$, where N is the number of (foreground) pixels. Since s is a small constant, this operation is as cheap as a conv layer.

3.3 Sensor Dropout

One appealing property of this architecture is the independence of each sensor. We can perform camera-only or radar-only 3D object detection with the same architecture when one modality is unavailable. This is desired in practice as one never knows when a sensor might be unavailable due to various situations, e.g., occlusions, weather, or sensor failure.

To enhance the model ability to handle sensor failures, we incorporate a sensor dropout mechanism [7,52] during training. With a probability P_{drop}, we randomly drop out the entire set of point features of camera $\boldsymbol{\psi}_C$ or radar $\boldsymbol{\psi}_R$, or

$$\begin{aligned} \boldsymbol{X}_C &= \mathbb{1}(r_1 \geq P_{\text{drop}})\, \boldsymbol{X}_C + \mathbb{1}(r_1 < P_{\text{drop}} \wedge r_2 \geq 0.5)\, \mathbf{0} \\ \boldsymbol{X}_R &= \mathbb{1}(r_1 \geq P_{\text{drop}})\, \boldsymbol{X}_R + \mathbb{1}(r_1 < P_{\text{drop}} \wedge r_2 < 0.5)\, \mathbf{0} \end{aligned} \tag{6}$$

where $\mathbf{0}$ is a zero matrix and r_1 and r_2 are uniform random numbers in [0, 1]. Note that camera and radar features won't be dropped out at the same time. We use $p_{\text{drop}} = 0.2$ in our experiments.

The reason why we choose to mask out 3D point features instead of input data directly is that we can still train the cross-attention with proper 2D features normally. If radar sensor is corrupted during inference and produces noisy 2D features, it results in a uniform attention map inside cross-attention and little effect on 3D camera point locations.

4 Experiments

We present experiments on the RADIATE [40] and Waymo Open [47] datasets to verify the efficacy of our proposed CramNet model. We introduce the settings in Sects. 4.1 and 4.2. We include the main results, ablation studies, robustness tests, and visualization on the RADIATE dataset in Sects. 4.3, 4.4, 4.5, and 4.7, respectively. We also present our camera-only results on the Waymo Open dataset in Sect. 4.6. More ablation studies can be found in the supplementary.

4.1 Dataset and Evaluation

RADIATE Dataset. We evaluate our method on the challenging RADIATE dataset [40]. This dataset features radar sensor data collected for scene understanding for safe autonomous driving in various weather conditions, including sunny, night, rainy, foggy, and snowy. The dataset includes 3 h of annotated radar imagery with more than 200k labeled objects for 8 categories. These properties make the RADIATE dataset one of the few public datasets that contain high-resolution radar data along with a large number of ground truth labels for road actors. While the dataset provides high-quality radar data, the quality of its camera and LiDAR data is not comparable to that of other autonomous driving datasets, such as the Waymo Open Dataset [47]. This shortcoming, however, makes the evaluation of the robustness of our proposed sensor fusion algorithm even more compelling. In all of our experiments, we train the models on the training set that contains both good and bad weather conditions, and we evaluate the resulting models on the standard validation set.

Table 2. Main results evaluated in BEV AP (%) on the RADIATE dataset [40]. CramNet-C (*notes evaluation on camera/lidar-specific labels), CramNet-R, and CramNet denote our camera-only, radar-only, and fusion models, respectively. Our final model outperforms the baseline Faster R-CNN [40] by 16 percentage points, the camera-only variant by 38 points, and the radar-only variant by 6 points. These large gains validate the efficacy of our proposed sensor fusion model.

Method	Overall	Sunny (Parked)	Overcast (Motorway)	Sun/OC (Urban)	Night (Motorway)	Rain (Suburban)	Fog (Suburban)	Snow (Surburban)
Baseline [40]	46.55	79.72	44.23	35.45	64.29	31.96	51.22	8.14
CramNet-C*	23.66	67.98	6.50	23.43	2.24	17.69	9.50	0.12
CramNet-R	56.19	83.58	37.65	48.33	60.38	42.86	71.11	**15.84**
CramNet	**62.07**	**96.68**	**50.49**	**52.25**	**79.56**	**57.90**	**85.26**	8.89

Evaluation. The (pseudo) 3D labels in the RADIATE dataset are 2D BEV labels with assumed heights for each category. We therefore report our 3D detection results in terms of BEV AP to align with the baselines, unless otherwise noted. We follow the proposed evaluation in the dataset and define the category "vehicle" to encompass the six categories "car", "van", "truck", "bus", "motorbike", and "bicycle". The final BEV/3D AP numbers are therefore weighted sums of the objects from these categories. For all radar and fusion experiments, we evaluate the performance on the region that is captured by both the cameras and the radar sensors, up to the radar range of 100 m. For all camera-only experiments, we exclude labels that do not contain any LiDAR points. The motivation for this is that camera depth estimates beyond the LiDAR supervision (up to 70 m) tend to be inaccurate.

4.2 Implementation Details

Hyperparameters. CramNet follows the implementation of RSN [48]. The sparse convolution implementation is also similar to [56]. The input camera and radar RF images are both normalized to be in [0, 1]. The foreground score cutoff is set to 0.15, the segmentation loss weight is set to 400, and the depth loss weight is set to 20. For cross-attention, we sample 1 point closer and farther around the predicted depth location, with 10% error. The voxelization region is [-100 m, 100 m] × [-100 m, 100 m] × [-5 m, 5 m] with 0.2 m voxel sizes. In the heatmap computation, σ_h is set to 1.0, the heatmap loss weight is set to 4 and threshold ϵ_h are set to 0.2. We use 12 bins in the heading bin loss for heading regression.

Training and Inference. We train CramNet from scratch end-to-end using the Adam optimizer [14] on Tesla V100 GPUs. The models are trained with 5 batches on 8 GPUs. We use a cosine learning rate decay, where we set the initial learning rate to 0.006, with 1k warm-up steps starting at 0.003 and 50k steps in total. We use layer normalization [1] instead of batch normalization [12] in the 3D network for the number of foreground points varies among different scenes. We do not perform 2D data augmentation but adopt two 3D data augmentation strategies, namely, random flipping along the x-axis and a global rotation around the z-axis, with a random angle from $[-\pi/4, \pi/4]$ on the selected foreground points.

4.3 Performance on the RADIATE Dataset

We evaluate the performance of our method on the RADIATE dataset [40] and summarize the results in Table 2. We report the BEV AP at a 0.5 IoU threshold to align with the baseline proposed in the RADIATE dataset [40]. The baseline runs a Faster R-CNN detector with a ResNet-101 backbone on radar RF images.

Our radar-only variant, CramNet-R, outperforms the baseline by a large margin, $\sim$ 10 percentage points in AP. Our two-stage framework effectively filters out radar noise in the segmentation stage to focus inference on the remaining radar signals in subsequent stages. Our camera-only variant, CramNet-C, performs the worst. Several factors may contribute to the poor performance. First, adverse weather affects the cameras more than the radar sensors, which is exacerbated by the lack of wipers mounted on the vehicles. Second, the effective range of the LiDAR sensors, which we use for camera depth supervision at training time, tends to drop from 70 m in clear weather to about 40 m in adverse weather, whereas we have labeled ground truth boxes within a range of 100 m. Overall, we observe that the short sensing range and the sparsity of the points prevent the model from learning accurate camera depth estimation, resulting in poor camera-only 3D detection performance.

Our proposed fusion model, CramNet, equipped with ray-constrained cross attention and sensor dropout, outperforms the baseline BEV AP by 16 percentage points, the camera-only variant by 38 points, and the radar-only variant by 6 points. These large gains validate the efficacy of our proposed sensor fusion model. In the next sections, we study the performance of our method in more detail.

4.4 Ablation Study

Table 3. Ablation study on CramNet on the RADIATE dataset [40]. Left: The cross-attention and sensor dropout both improve over the vanilla fusion model by 4 to 5 points in AP. Putting them together yields the final fusion model with the best performance. Right: We simulate the radar sparse signals by setting the intensity thresholds to 0.25 or 0.5, resulting in ∼70K or 2K points, respectively. As a result, our model performance is degraded relatively by 15% to 70%. This confirms radar RF imagery contains critical information for 3D detection.

Attention	Dropout	BEV AP
		56.19
✓		60.20
	✓	61.23
✓	✓	62.07

Radar Intensity Threshold	# of Points	BEV AP	Degradation
None	–	60.20	–
0.25	70K	50.90	−15.45%
0.5	2K	17.81	−70.42%

Dropout Location	BEV AP
Normal	57.00
Input	55.78
Point Cloud	58.64
Point Feature	61.23

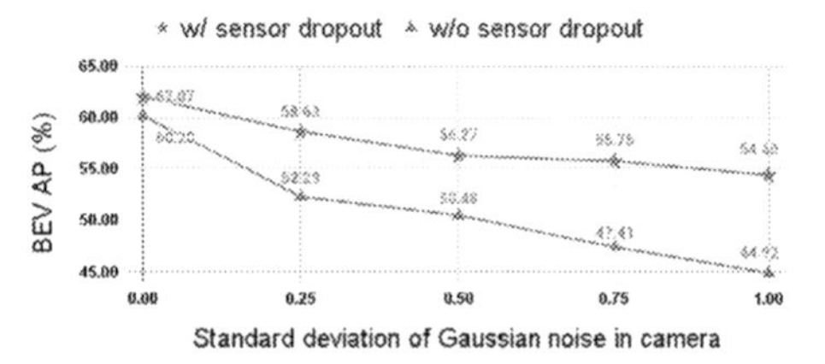

Fig. 4. Left: Analysis on different sensor dropout strategies. Masking out point features yields the best performance. Two possible benefits for this dropout location: 1) Reduce the 3D network reliance on features, which are disrupted the most given sensor noise. 2) Remain smooth training of 2D feature extractors and cross-attention. **Right**: Analysis on model performance degradation on corrupted data. We add varying degrees of Gaussian white noises to corrupt camera images and evaluate the performance. Our fusion model trained with sensor dropout greatly outperforms the one without by 2 to 10% points in BEV AP. This demonstrates that sensor dropout can drastically enhance sensor robustness.

Effects of Ray-Constrained Cross-Attention and Dropout. We experiment the fusion model with different settings to enable/disable ray-constrained cross-attention and sensor dropout mechanisms. The experimental results are summarized in Table 3 (left). The cross-attention and sensor dropout both improve over the vanilla fusion model by 4 to 5 points in BEV AP. Putting them together yields our final fusion model, achieving the performance of 62.07% AP.

Effects of Sampling Radar Points. Most of the camera-radar fusion methods, such as GRIF Net [13] and CenterFusion [28], perform experiments on the nuScenes dataset [5] that contains only sparse radar points, at the scale of hundreds of points in a scene. The resulted radar-only model usually performs poorly, such as 25.5% AP reported in [13]. On the other hand, our model is specifically designed to perform either single modality effectively by taking as input the RF images instead of sparse points.

To quantitatively study how the sparsity of radar signals affects the performance, we filter RF images with varying intensity thresholds, as summarized in Table 3 (right). We set the intensity thresholds to 0.25 or 0.5, resulting in ∼70K or 2K points, respectively, which are already denser than sparse radar points available on nuScenes [5] or SeeingThroughFog [2] datasets. As a result, our model performance is degraded relatively by 15% to 70%. We conclude that radar RF imagery contains critical information for effective 3D object detection.

4.5 Detection Robustness

Detection robustness against sensor deterioration is critical for safe autonomous driving. In this section, we study the effects of our proposed sensor dropout with ablation study and corrupted sensor data.

Where to Drop Out Sensor Data? Dropout is a popular technique for training neural network models [45]. It is usually applied on a layer to randomly mask out neuron activations. We experiment on various places to drop out sensor data and summarize them in Table 4 (left). The 'normal' dropout applies the conventional dropout on point cloud features, regardless of which sensor the points are from. This conventional dropout does not provide benefits in either overall performance or in bad weather conditions. The 'input' dropout randomly masks out a sensor (radar or camera) entirely. The 'point cloud' dropout randomly masks out the 3D points from one sensor entirely. The 'point feature' dropout randomly masks out the initial point cloud features from one sensor entirely but leaves the point cloud positions intact. As the numbers dictate, masking out point features yields the best performance. Two possible benefits for this dropout location: (1) Reduce the 3D network reliance on features, which are disrupted the most due to sensor noise. (2) Remain training of 2D feature extractors and cross-attention. As such, we conclude dropping out sensor point features randomly is the most effective.

Sensor Dropout Improves Robustness Against Input Corruption. We study how the corruption of sensors will affect the performance with and without sensor dropout and summarize the experiments in Table 4 (right). For this purpose, we add random Gaussian white noise with varying standard deviation to corrupt the camera images to different degrees. We evaluate the fusion model on the corrupted data, with or without sensor dropout during training. The experimental results show that our fusion model trained with sensor dropout greatly outperforms the one without by 2 to 10% points in BEV AP. This study demonstrates that sensor dropout can drastically enhance sensor robustness.

4.6 Camera-only CramNet on Waymo Open Dataset

Our radar-only and camera-radar fusion models perform strongly on the RADIATE dataset [40]. However, the camera-only model suffers from the poor image quality and adverse weather conditions in the dataset. Since we do not have access to another public dataset that contains radar RF imagery, we evaluate our camera-only model performance on the Waymo Open Dataset [47], as summarized in Table 4. We report 3D AP/APH with 0.7 IoU threshold on the LEVEL_1 difficulty in Table 4. Our camera-only model, CramNet-C, achieves competitive performance. More details can be found in the supplementary.

4.7 Visualization

We present the visual comparisons between our camera-only, radar-only, and fusion models in Fig. 5. Since the camera visibility is severely reduced due to either underexposure or adverse weather, the camera-only model tends to miss

Table 4. Camera-only 3D detection results on the Waymo Open Dataset [47] validation set on the vehicle class, evaluated in terms of 3D AP/APH at 0.7 IoU on the LEVEL_1 difficulty. Baseline numbers are from [34]. Our camera-only model, CramNet-C, achieves competitive performance among state-of-the-art.

Method	3D AP	0 - 30 m	30 - 50 m	50 m - *infty*	3D APH	0 - 30 m	30 - 50 m	50 m - *infty*
M3D-RPN [4]	0.35	1.12	0.18	0.02	0.34	1.10	0.18	0.02
CaDDN [34]	**5.03**	14.54	**1.47**	0.10	**4.99**	14.43	**1.45**	0.10
CramNet-C	4.14	**15.46**	1.20	**0.15**	4.10	**15.31**	1.19	**0.13**

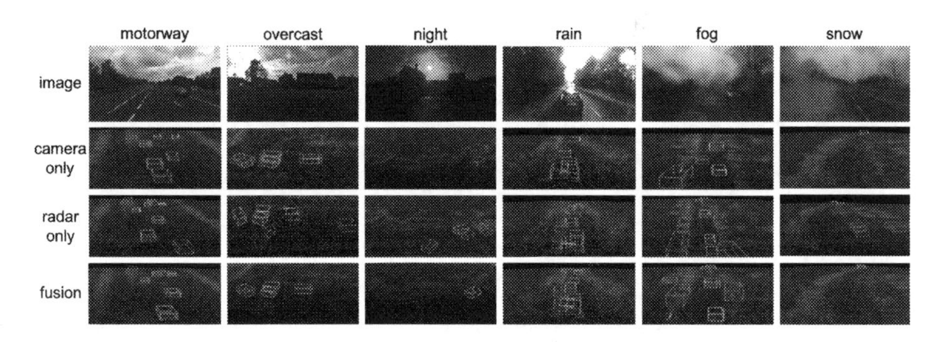

Fig. 5. Visual comparison between CramNet-C, CramNet-R, and CramNet from 6 scenarios. We visualize the predicted boxes in red and the ground truth boxes in yellow with projected radar and camera pixels. Whereas the camera-only model tends to miss detections and predict inaccurate localization, the radar-only model suffers from false positives. Our camera-radar fusion model combines the advantages of the two and produces the most accurate predictions.

detection and the predicted localization tends to be inaccurate. In contrast, the radar-only model suffers from false positives due to lack of appearance features from RF images. Overall, our camera-radar fusion model combines the advantages from the two and produces the most accurate predictions.

5 Conclusion

We introduced a camera-radar sensor fusion approach for robust 3D object detection for autonomous driving. The method relies on a ray-constrained cross-attention mechanism to leverage the range measurements from radar to improve camera depth estimates. Training with sensor dropout allows the method to fall back to a single modality when one of the sensors malfunctions. We present experiments on the RADIATE dataset and the Waymo Open Dataset.

Limitations. Whereas a camera pixel corresponds to a ray, a (range, azimuth) radar reading corresponds to an arc in 3D space. Intersecting a camera ray and a radar arc yields their correspondence. We approximate the radar arc as a pillar, that is, we assume that the radar points are at the same elevation as the sensor. This assumption works well in practice when most objects are at a similar elevation as the sensor. We currently use the RF images in Cartesian coordinates, which may be suboptimal as the radar natively operates in polar coordinates. We will explore a polar convolutional network design and radar-specific spherical voxelization in future work.

References

1. Ba, J.L., Kiros, J.R., Hinton, G.E.: Layer normalization. arXiv preprint arXiv:1607.06450 (2016)
2. Bijelic, M., et al.: Seeing through fog without seeing fog: deep multimodal sensor fusion in unseen adverse weather. In: CVPR (2020)
3. Bijelic, M., Gruber, T., Ritter, W.: A benchmark for lidar sensors in fog: is detection breaking down? In: 2018 IEEE Intelligent Vehicles Symposium (IV) (2018)
4. Brazil, G., Liu, X.: M3D-RPN: Monocular 3D region proposal network for object detection. In: ICCV (2019)
5. Caesar, H., et al.: A multimodal dataset for autonomous driving. In: CVPR (2020)
6. Chen, X., Kundu, K., Zhang, Z., Ma, H., Fidler, S., Urtasun, R.: Monocular 3D object detection for autonomous driving. In: CVPR (2016)
7. Chen, X., Ma, H., Wan, J., Li, B., Xia, T.: Multi-view 3D object detection network for autonomous driving. In: CVPR (2017)
8. Chen, Y., Tai, L., Sun, K., Li, M.: MonoPair: monocular 3D object detection using pairwise spatial relationships. In: CVPR (2020)
9. Ding, M., et al.: Learning depth-guided convolutions for monocular 3D object detection. In: CVPR workshops (2020)
10. Graham, B., van der Maaten, L.: Submanifold sparse convolutional networks. arXiv preprint arXiv:1706.01307 (2017)

11. Huang, T., Liu, Z., Chen, X., Bai, X.: EPNet: enhancing point features with image semantics for 3D object detection. In: Vedaldi, A., Bischof, H., Brox, T., Frahm, J.-M. (eds.) ECCV 2020. LNCS, vol. 12360, pp. 35–52. Springer, Cham (2020). https://doi.org/10.1007/978-3-030-58555-6_3
12. Ioffe, S., Szegedy, C.: Batch normalization: accelerating deep network training by reducing internal covariate shift. In: ICML (2015)
13. Kim, Y., Choi, J.W., Kum, D.: GRIF Net: gated region of interest fusion network for robust 3D object detection from radar point cloud and monocular image. In: IROS (2020)
14. Kingma, D.P., Ba, J.: Adam: a method for stochastic optimization. arXiv preprint arXiv:1412.6980 (2014)
15. Lang, A.H., Vora, S., Caesar, H., Zhou, L., Yang, J., Beijbom, O.: PointPillars: fast encoders for object detection from point clouds. In: CVPR (2019)
16. Li, P., Zhao, H., Liu, P., Cao, F.: RTM3D: real-time monocular 3D detection from object keypoints for autonomous driving. In: Vedaldi, A., Bischof, H., Brox, T., Frahm, J.-M. (eds.) ECCV 2020. LNCS, vol. 12348, pp. 644–660. Springer, Cham (2020). https://doi.org/10.1007/978-3-030-58580-8_38
17. Li, Y., et al.: DeepFusion: lidar-camera deep fusion for multi-modal 3D object detection. In: Proceedings of the IEEE/CVF Conference on Computer Vision and Pattern Recognition, pp. 17182–17191 (2022)
18. Li, Z., Wang, W., Li, H., Xie, E., Sima, C., Lu, T., Yu, Q., Dai, J.: BEVFormer: learning bird's-eye-view representation from multi-camera images via spatiotemporal transformers. In: ECCV (2022)
19. Liang, M., Yang, B., Wang, S., Urtasun, R.: Deep continuous fusion for multi-sensor 3D object detection. In: Ferrari, V., Hebert, M., Sminchisescu, C., Weiss, Y. (eds.) ECCV 2018. LNCS, vol. 11220, pp. 663–678. Springer, Cham (2018). https://doi.org/10.1007/978-3-030-01270-0_39
20. Lim, T.Y., et al.: Radar and camera early fusion for vehicle detection in advanced driver assistance systems. In: Machine Learning for Autonomous Driving Workshop at the 33rd Conference on Neural Information Processing Systems (2019)
21. Lin, T.Y., Dollár, P., Girshick, R., He, K., Hariharan, B., Belongie, S.: Feature pyramid networks for object detection. In: CVPR (2017)
22. Lin, T.Y., Goyal, P., Girshick, R., He, K., Dollár, P.: Focal loss for dense object detection. In: ICCV (2017)
23. Liu, L., Wu, C., Lu, J., Xie, L., Zhou, J., Tian, Q.: Reinforced axial refinement network for monocular 3D object detection. In: Vedaldi, A., Bischof, H., Brox, T., Frahm, J.-M. (eds.) ECCV 2020. LNCS, vol. 12362, pp. 540–556. Springer, Cham (2020). https://doi.org/10.1007/978-3-030-58520-4_32
24. Ma, X., Liu, S., Xia, Z., Zhang, H., Zeng, X., Ouyang, W.: Rethinking pseudo-LiDAR representation. In: Vedaldi, A., Bischof, H., Brox, T., Frahm, J.-M. (eds.) ECCV 2020. LNCS, vol. 12358, pp. 311–327. Springer, Cham (2020). https://doi.org/10.1007/978-3-030-58601-0_19
25. Major, B., et al.: Vehicle detection with automotive radar using deep learning on range-azimuth-doppler tensors. In: ICCV Workshops (2019)
26. Manhardt, F., Kehl, W., Gaidon, A.: ROI-10D: monocular lifting of 2D detection to 6D pose and metric shape. In: CVPR (2019)
27. Mousavian, A., Anguelov, D., Flynn, J., Kosecka, J.: 3D bounding box estimation using deep learning and geometry. In: CVPR (2017)
28. Nabati, R., Qi, H.: CenterFusion: center-based radar and camera fusion for 3D object detection. In: Proceedings of the IEEE/CVF Winter Conference on Applications of Computer Vision (2021)

29. Nobis, F., Shafiei, E., Karle, P., Betz, J., Lienkamp, M.: Radar voxel fusion for 3D object detection. Appl. Sci. **11**(12), 5598 (2021)
30. Park, D., Ambrus, R., Guizilini, V., Li, J., Gaidon, A.: Is pseudo-lidar needed for monocular 3D object detection? In: ICCV (2021)
31. Piergiovanni, A., Casser, V., Ryoo, M.S., Angelova, A.: 4D-Net for learned multi-modal alignment. In: ICCV (2021)
32. Qi, C.R., Liu, W., Wu, C., Su, H., Guibas, L.J.: Frustum pointNets for 3D object detection from RGB-D data. In: CVPR (2018)
33. Qi, C.R., Yi, L., Su, H., Guibas, L.J.: PointNet++: deep hierarchical feature learning on point sets in a metric space. In: NeurIPS (2017)
34. Reading, C., Harakeh, A., Chae, J., Waslander, S.L.: Categorical depth distribution network for monocular 3D object detection. In: CVPR (2021)
35. Richards, M.A., Scheer, J., Holm, W.A., Melvin, W.L.: Principles of modern radar (2010)
36. Roddick, T., Kendall, A., Cipolla, R.: Orthographic feature transform for monocular 3D object detection. arXiv preprint arXiv:1811.08188 (2018)
37. Ronneberger, O., Fischer, P., Brox, T.: U-Net: convolutional networks for biomedical image segmentation. In: MICCAI (2015)
38. Schumann, O., Hahn, M., Dickmann, J., Wöhler, C.: Semantic segmentation on radar point clouds. In: 2018 21st International Conference on Information Fusion (FUSION) (2018)
39. Shah, M., et al.: End-to-end trajectory prediction using spatio-temporal radar fusion. In: CoRL (2020)
40. Sheeny, M., De Pellegrin, E., Mukherjee, S., Ahrabian, A., Wang, S., Wallace, A.: Radiate: a radar dataset for automotive perception in bad weather. In: ICRA (2021)
41. Shi, S., Wang, X., Li, H.: PointRCNN: 3D object proposal generation and detection from point cloud. In: CVPR (2019)
42. Shi, X., Chen, Z., Kim, T.-K.: Distance-normalized unified representation for monocular 3D object detection. In: Vedaldi, A., Bischof, H., Brox, T., Frahm, J.-M. (eds.) ECCV 2020. LNCS, vol. 12374, pp. 91–107. Springer, Cham (2020). https://doi.org/10.1007/978-3-030-58526-6_6
43. Simonelli, A., Bulo, S.R., Porzi, L., López-Antequera, M., Kontschieder, P.: Disentangling monocular 3D object detection. In: ICCV (2019)
44. Simonelli, A., Buló, S.R., Porzi, L., Ricci, E., Kontschieder, P.: Towards generalization across depth for monocular 3D object detection. In: Vedaldi, A., Bischof, H., Brox, T., Frahm, J.-M. (eds.) ECCV 2020. LNCS, vol. 12367, pp. 767–782. Springer, Cham (2020). https://doi.org/10.1007/978-3-030-58542-6_46
45. Srivastava, N., Hinton, G., Krizhevsky, A., Sutskever, I., Salakhutdinov, R.: Dropout: a simple way to prevent neural networks from overfitting. J. Mach. Learn. Res. **15**(1), 1929–1958 (2014)
46. Srivastava, S., Jurie, F., Sharma, G.: Learning 2D to 3D lifting for object detection in 3D for autonomous vehicles. In: IROS (2019)
47. Sun, P., et al.: Scalability in perception for autonomous driving: Waymo Open Dataset. In: CVPR (2020)
48. Sun, P., et al.: RSN: range sparse net for efficient, accurate lidar 3D object detection. In: CVPR (2021)
49. Vaswani, A., et al.: Attention is all you need. In: NeurIPS (2017)
50. Vora, S., Lang, A.H., Helou, B., Beijbom, O.: Pointpainting: sequential fusion for 3D object detection. In: CVPR (2020)

51. Wang, C., Ma, C., Zhu, M., Yang, X.: Pointaugmenting: cross-modal augmentation for 3D object detection. In: CVPR (2021)
52. Wang, W., Tran, D., Feiszli, M.: What makes training multi-modal classification networks hard? In: CVPR (2020)
53. Wang, Y., Chao, W.L., Garg, D., Hariharan, B., Campbell, M., Weinberger, K.Q.: Pseudo-lidar from visual depth estimation: Bridging the gap in 3D object detection for autonomous driving. In: CVPR (2019)
54. Wang, Z., Jia, K.: Frustum convnet: sliding frustums to aggregate local point-wise features for amodal 3D object detection. In: IROS (2019)
55. Weng, X., Kitani, K.: Monocular 3D object detection with pseudo-lidar point cloud. In: ICCV Workshops (2019)
56. Yan, Y., Mao, Y., Li, B.: Second: sparsely embedded convolutional detection. Sensors **18**(10), 3337 (2018)
57. Yang, B., Guo, R., Liang, M., Casas, S., Urtasun, R.: RadarNet: exploiting radar for robust perception of dynamic objects. In: Vedaldi, A., Bischof, H., Brox, T., Frahm, J.-M. (eds.) ECCV 2020. LNCS, vol. 12363, pp. 496–512. Springer, Cham (2020). https://doi.org/10.1007/978-3-030-58523-5_29
58. You, Y., et al.: Pseudo-LIDAR++: accurate depth for 3D object detection in autonomous driving. arXiv preprint arXiv:1906.06310 (2019)
59. Zhou, D., et al.: IoU loss for 2D/3D object detection. In: 2019 International Conference on 3D Vision (3DV) (2019)
60. Zhou, X., Wang, D., Krähenbühl, P.: Objects as points. arXiv preprint arXiv:1904.07850 (2019)
61. Zhou, Y., et al.: End-to-end multi-view fusion for 3D object detection in lidar point clouds. In: CoRL (2020)

CODA: A Real-World Road Corner Case Dataset for Object Detection in Autonomous Driving

Kaican Li[1], Kai Chen[3], Haoyu Wang[1], Lanqing Hong[1(✉)], Chaoqiang Ye[1], Jianhua Han[1], Yukuai Chen[2], Wei Zhang[1], Chunjing Xu[1], Dit-Yan Yeung[3], Xiaodan Liang[4], Zhenguo Li[1], and Hang Xu[1]

[1] Huawei Noah's Ark Lab, Shenzhen, China
honglanqing@huawei.com
[2] Huawei Intelligent Automotive Solution BU, Shenzhen, China
[3] Hong Kong University of Science and Technology, Hong Kong, China
[4] Sun Yat-sen University, Guangzhou, China

Abstract. Contemporary deep-learning object detection methods for autonomous driving usually presume fixed categories of common traffic participants, such as pedestrians and cars. Most existing detectors are unable to detect uncommon objects and corner cases (*e.g.*, a dog crossing a street), which may lead to severe accidents in some situations, making the timeline for the real-world application of reliable autonomous driving uncertain. One main reason that impedes the development of truly reliably self-driving systems is the lack of public datasets for evaluating the performance of object detectors on corner cases. Hence, we introduce a challenging dataset named CODA that exposes this critical problem of vision-based detectors. The dataset consists of 1500 carefully selected real-world driving scenes, each containing four object-level corner cases (on average), spanning more than 30 object categories. On CODA, the performance of standard object detectors trained on large-scale autonomous driving datasets significantly drops to no more than 12.8% in mAR. Moreover, we experiment with the state-of-the-art open-world object detector and find that it also fails to reliably identify the novel objects in CODA, suggesting that a robust perception system for autonomous driving is probably still far from reach. We expect our CODA dataset to facilitate further research in reliable detection for real-world autonomous driving. Our dataset is available at https://coda-dataset.github.io.

Keywords: Autonomous driving · Object detection · Corner case

K. Li, K. Chen and H. Wang—Equal contribution.

Supplementary Information The online version contains supplementary material available at https://doi.org/10.1007/978-3-031-19839-7_24.

S. Avidan et al. (Eds.): ECCV 2022, LNCS 13698, pp. 406–423, 2022.
https://doi.org/10.1007/978-3-031-19839-7_24

1 Introduction

Deep learning has achieved prominent success in object detection for autonomous driving in the wild [5,17,38,47]. The success is mainly attributed to deep neural networks trained on an extensive amount of data extracted from real-life driving scenarios, which have become an indispensable component of existing autonomous driving systems [6,13,28]. Though such models are proficient in detecting common traffic participants (*e.g.*, cars, pedestrians, and cyclists), they are generally incapable of detecting novel objects that are not seen or rarely seen in the training process, i.e., the out-of-distribution samples [43,45,46]. For instance, a vehicle equipped with state-of-the-art detectors galloping on the highway may fail to detect a runaway tire or an overturned truck straight ahead of the road. These failure cases of object detection in autonomous driving may result in severe consequences, putting lives at risk.

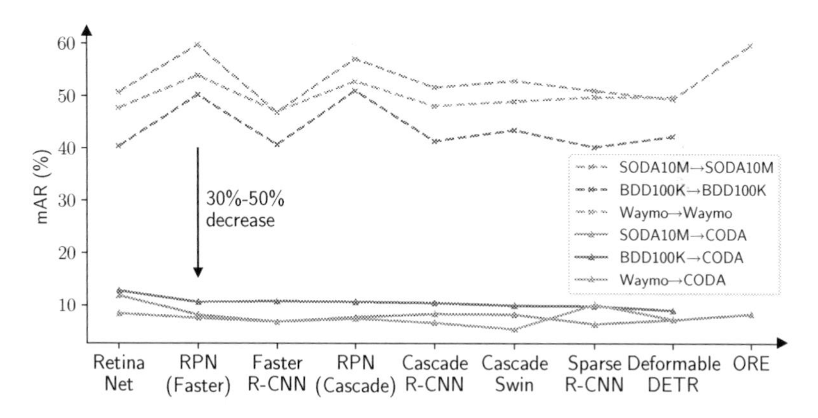

Fig. 1. Detection results on CODA compared with common autonomous driving datasets. All detectors suffer from a significant 30%-50% performance drop, with the best achieved at 12.8% mAR, which is definitely far from solved. Here $A \rightarrow B$ represents that the detector is trained on dataset A and evaluated on dataset B.

To address the problem, we introduce CODA, a novel dataset of *object-level corner cases*[1] in real-world driving scenes. CODA is constructed from three major object detection benchmarks for autonomous driving—KITTI [11], nuScenes [4], and ONCE [28]. In Fig. 2, the examples from CODA exhibit a diverse set of scenes and a great variety of novel objects. In total, 1500 scenes (images) are selected from the combined dataset of over one million scenes, leading to nearly 6000 high-quality annotated road corner cases. The selection process of CODA consists of two stages: a fully-automated generation of proposals on potential corner cases followed by manual inspections and corrections on the proposals. Our approach for corner-case proposal generation, COPG, which

[1] We adopt the definition of object-level corner case proposed in [3].

significantly reduces the amount of human labor in the second stage, is a generic pipeline that only requires raw sensory data from camera and lidar sensor, *i.e.*, no annotation is needed. We believe that the approach can be utilized to efficiently produce more corner case datasets in the future.

Fig. 2. Examples from CODA. Corner cases are indicated by the bounding boxes, while each color stands for a different object class. CODA contains both *instances of novel classes* (e.g., the dog in the top-left image) and *novel instances of common classes* (e.g., the cyclist in the top-middle image).

On CODA, we have evaluated various kinds of object detection methods including standard (closed-world) detectors such as Faster R-CNN [33]; a recently-proposed open-world detector, ORE [18], which is capable of detecting certain objects of unseen classes; and two anomaly detection methods [12,42] which are also in some sense suited to the task. Our experiment results show that none of the methods can consistently detect the novel objects in CODA, demonstrating how challenging CODA is. In general, there is no clear winner among the methods, even though ORE shows some improvements over the closed-world detectors. Finally, we hope that CODA can serve as an effective means for evaluating the robustness of machine perception in autonomous driving, and in turn, facilitate the development of truly reliably self-driving systems. The main contribution of this work can be summarized as follows:

- We propose CODA, the first real-world road corner case dataset, serving as a benchmark for the development of fully reliable self-driving vehicles.
- We evaluate various state-of-the-art object detectors (*e.g.*, Cascade R-CNN [5], Deformable DETR [47], and Sparse R-CNN [38]), suggesting that truly reliably self-driving systems are probably still far from reach.
- We introduce COPG, a generic pipeline for corner-case discovery, reducing human labeling effort by nearly 90% on a large-scale dataset.

2 Related Work

Road Anomaly and Corner Case Dataset. One of the pioneering datasets in road anomaly and corner case detection is the Lost and Found dataset [29] which features small objects in artificial scenes. Later introduced datasets mainly focus on semantic segmentation. Notable ones include the road anomaly dataset of Lis *et al.* [23] containing 60 real-world scenes, and Fishyscapes [1], a synthesized dataset created by overlaying objects crawled from the web onto the scenes of Cityscapes [9] and the Lost and Found dataset. StreetHazards [16] is another synthesized dataset where the scenes are simulated by computer graphics. In the same paper, the authors also introduced BDD-Anomaly, a subset of BDD100K [44], treating trains and motorcycles as anomalous objects.

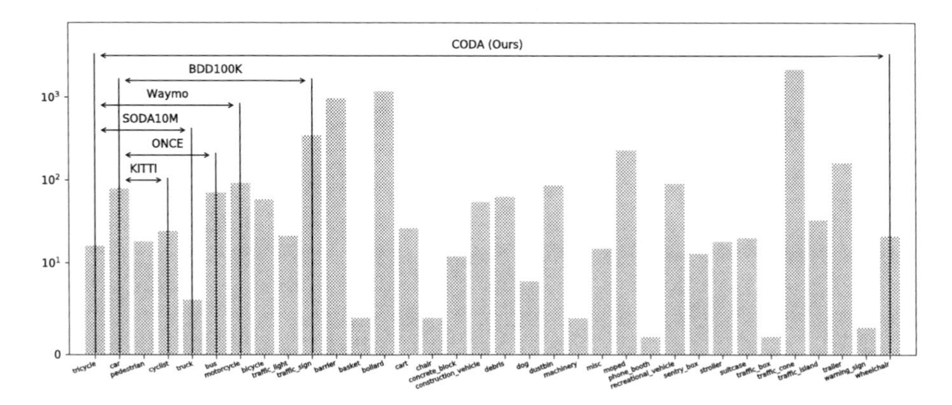

Fig. 3. Class distribution of CODA and annotation coverage of common large-scale autonomous driving benchmarks in comparison with ours. The distribution is inherently long-tailed as suggested by Zipf's law. Class *tram* in SODA10M and class *train* in BDD100K are omitted because CODA does not contain such instance.

Object Detection. Existing methods can be generally categorized into one-stage and two-stage based on how the proposals are generated. One-stage detectors [21,24,32] densely predict class distributions and box coordinates on each position of a given image, while two-stage detectors [5,20,33] utilize the Region Proposal Network (RPN) to generate regions of interest (RoI), which are then fed into multi-head networks for class and coordinate offset prediction. Cascade R-CNN [5] further improves by adding a sequence of heads trained with increasing IoU thresholds. ImageNet-supervised pre-training is adopd to accelerate training, while self-supervised pre-training [6,14,27] has recently demonstrated better transfer performance. Previous detectors are mostly trained in the closed-world setting, which can only detect objects belonging to a pre-defined semantic class set. To build a real-world perception system, open-world detection [18] has raised more attention, which can explicitly detect objects of unseen classes as *unknown*.

3 Properties of CODA

Composition. The scenes in CODA are carefully selected from three large-scale autonomous driving datasets: KITTI [11], nuScenes [4], and ONCE [28]. Together, they contribute 1500 diverse scenes to CODA, each containing at least one object-level corner case that is hazardous to self-driving vehicles or their surrounding lives and assets. The corner cases can be generally grouped into 7 super-classes: *vehicle*, *pedestrian*, *cyclist*, *animal*, *traffic facility*, *obstruction*, and *misc*, governing the 34 fine-grained classes listed in Fig. 3. Moreover, these classes can be divided into *novel classes* and *common classes*. Common classes stand for common object categories (e.g., cars and pedestrians) of existing autonomous driving benchmarks; whereas novel classes stand for the opposites, such as dogs and strollers. More than 90% of the instances in CODA are of novel classes. On one hand, instances of novel classes are inherently undetectable by (closed-world) object detectors that are trained on the common classes. On the other hand, the detectors ought to correctly identify novel instances of common classes, but often fail in doing so. Detailed definitions of common/novel classes will be introduced in Sect. 5, which is important to the evaluation of prevalent object detectors.

Table 1. Comparison with other datasets. CODA is the largest dataset of its kind in multiple aspects. Here we do not compare with the Fishyscapes Web dataset [1], which is neither publicly available nor with detailed statistics. "†" means rough estimates.

Dataset	#Scenes	Real	Weather	Period	#Classes	#Instances
Lis *et al.* [23]	60	✓	✗	✗	2	$300^{\dagger}$
Fishyscapes L&F [1]	375	✓	✗	✗	3	$500^{\dagger}$
Fishyscapes Static [1]	1030	✗	✗	✗	3	$1200^{\dagger}$
StreetHazards [16]	**1500**	✗	✗	✗	1	$1500^{\dagger}$
BDD-Anomaly (v1) [16]	361	✓	✗	✗	2	4476
CODA-KITTI (Ours)	309	✓	✓	✓	6	399
CODA-nuScenes (Ours)	134	✓	✓	✓	17	1125
CODA-ONCE (Ours)	1057	✓	✓	✓	32	4413
CODA (Ours)	**1500**	✓	✓	✓	**34**	**5937**

Diversity. The data diversity of CODA can be seen from both object level and scene level. On the object level, CODA comprises a wide range of object classes, most of which are neglected by the existing benchmarks (see Fig. 3). Though some class only has several instances (due to the natural scarcity of corner cases), they constitute a nontrivial portion of real-world driving environments. Notably, traffic facilities such as *traffic cone* and *barrier* take up a majority of the corner cases because they are indeed more common and often appear in large quantities.

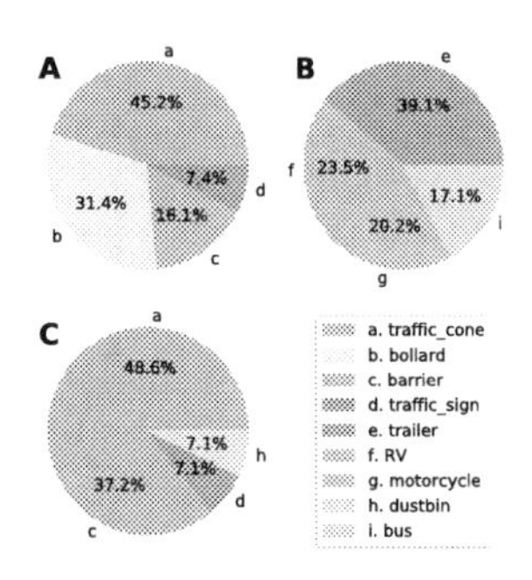

Fig. 4. Distribution of the top-4 classes in the three domains of CODA: **A** ONCE, **B** KITTI, and **C** nuScenes. The distribution largely differs across the domains.

On the scene level, CODA contains scenes from three different countries[2], which are distinct from one another as shown by the examples in Fig. 2. As a result, they introduce more novelty to the corner cases as the difference in object appearance is also a part of the domain shift of the scenes. The disparity between the domains can be seen from Fig. 4, where the distribution of top-4 common classes largely differs. In addition, the scenes in CODA exhibit different weather conditions, of which 75% are clear, 22% are cloudy, and 4% are rainy. Lastly, 9% of the scenes are night scenes apart from the daytime scenes.

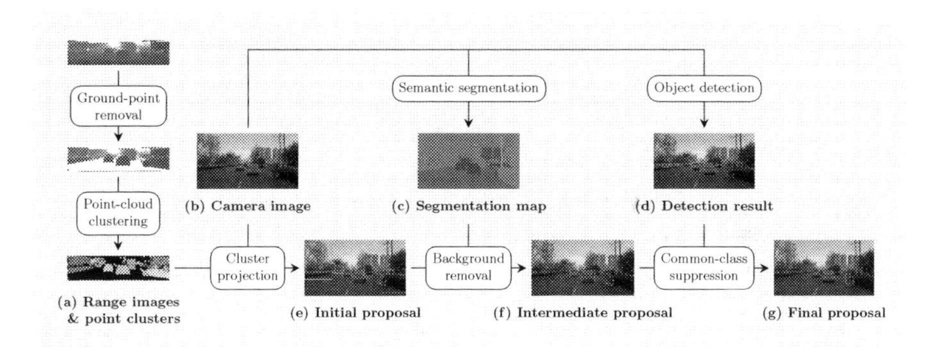

Fig. 5. Pipeline for generating proposals of corner case (COPG). The input to the pipeline is the point cloud and the camera image of a given scene. The point cloud is used to compute **(a)**, whereas the camera image **(b)** is used to produce **(c)** and **(d)**, which then help remove invalid proposals. The output **(g)** is a set of bounding boxes indicating the proposed corner cases in the camera image.

[2] KITTI are captured in a mid-size city of Germany, nuScenes are captured in Singapore, and ONCE are captured in various cities of China.

Comparison with Road Anomaly Datasets. In Table 1, we compare CODA against several prominent road anomaly datasets that also have object-level annotations. In contrast to CODA, the datasets are either synthetic or small in scale. The largest one of real-world road anomalies, BDD-Anomaly (v1) [16] only contains two object classes, albeit it is comparable to CODA on the number of instances.

4 Construction of CODA

As mentioned earlier, CODA is constructed from three autonomous driving benchmarks, of which most scenes are captured in well-regulated urban areas and therefore contain very few corner cases. To identify them in the large pools of data, we must first define what "corner cases" are in a clearer sense. The main criteria we use for determining whether an object is a corner case are as follows:

- **Risk:** The object blocks or is about to block a potential path of the self-driving vehicle mounted with the camera. Static objects not on the road such as trees and buildings are not considered to block the vehicle.
- **Novelty:** The object does not belong to any of the common classes of autonomous driving benchmarks, or it is a novel instance of the common classes. For simplicity, we take the classes of SODA10M [13] as the common classes.

If an object satisfies both criteria then it is a corner case. The first criterion suggests that the object could be hit by the vehicle and the second criterion suggests that the object is difficult to detect.

4.1 Overview

Adhering to the high-level criteria above, the construction of CODA is carried out in two main stages. The first stage is an automatic generation of proposals that identifies potential corner cases from initial data, followed by the second stage, a manual selection and labeling process that eliminates the false positives of the proposals, and then classifies the remaining true positives while adjusting their bounding boxes to be more precise.

For ONCE [28] consisting of a million scenes, the first stage helps filter out nearly 90% of scenes that are unlikely to contain any corner case, significantly reducing human efforts in the subsequent stage. For KITTI [11] and nuScenes [4], which are considerably smaller than ONCE, we skip the first stage by adopting the ground-truth annotations of uncommon objects that are already provided by the datasets as proposals.

Next, we introduce COPG, our pipeline for corner-case proposal generation (illustrated in Fig. 5). It only requires raw sensory data from a camera and a lidar sensor, *i.e.*, 2D images and 3D point clouds, to identify potential corner cases in any given dataset.

4.2 Identifying Potential Corner Cases

Unsupervised Point-Cloud Clustering. To reliably identify objects satisfying the first criterion, the first step is to learn the location of nearby objects that could obstruct the road. Hence, we turn to lidar point clouds. Since we do not assume any annotation on the points, we start by clustering them so as to separate the objects in the cloud. But before that, we remove all ground-level points by RANSAC [10] to avoid ground points being then clustered as parts of other objects and to suppress the noise from insignificant objects (*e.g.*, tin cans and small branches) on the ground.

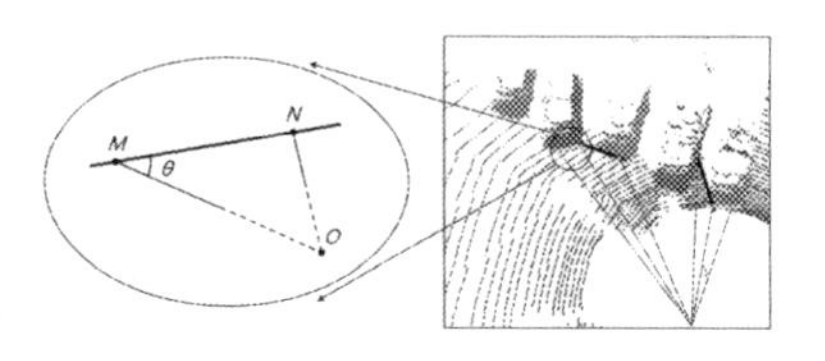

Fig. 6. Abstraction of the point-cloud clustering algorithm [2]. The right figure is a top-view example separating five cars. In the left figure, O denotes the location of the lidar sensor, M and N denote two points in the cloud, while OM and ON denote two lidar beams (OM is the longer beam). If the angle θ is greater than a fixed threshold, then the algorithm labels M and N as points belonging to the same object. The rule is based on the observation that in most cases, if M and N are from the same object, θ is relatively large; however, for those from different objects, θ turns out to be substantially smaller.

Given a point cloud with ground-level points removed, we adopt the algorithm proposed by Bogoslavskyi and Stachniss [2] to cluster the remaining points. The algorithm operates on the range image of the point cloud. A range image is a 2D image showing the distance to points in a scene from a specific point (which is the location of the lidar sensor in our case) and the image has pixel values that correspond to the distance. Given a range image, the algorithm conducts a breadth-first search over the pixels of the image, and eventually assigns every pixel to a cluster. Specifically, the algorithm compares each pixel p with its four neighboring pixels during the search. If a neighbor p' is sufficiently *close* to p, then they are given the same cluster label. The closeness between pixels is determined by the geometric relationship between the underlying points (in the 3D cloud) of these pixels. See Fig. 6 for a detailed explanation.

After separating points of different objects apart by the clustering algorithm, the points are projected onto the camera images. 2D bounding boxes are then generated for each of the clusters, except those that are too small or too far away from the lidar sensor. These bounding boxes are our initial proposals of corner cases, which is a superset of the final proposals. Next, we apply two other techniques to remove the proposals that violate the predefined criteria.

Background Removal. Not all objects found by the point-cloud clustering algorithm satisfy the first criterion since most of the objects are usually off the road. Static objects in the background (*e.g.*, vegetation and buildings) are the most common ones in this category. Discerning these objects from the others requires a semantic understanding of the scene, which could not be derived from merely point-cloud data. Instead, we find semantic segmentation on camera images particularly useful. We utilize a DeepLabv3+ [8] model pre-trained on Cityscapes [9] to produce fine segmentation maps, and then filter out the proposals that has a large overlap (over some threshold) with background regions in the corresponding segmentation map. The following classes are considered as backgrounds: *road*, *sidewalk*, *building*, *wall*, *fence*, *pole*, *vegetation*, *terrain*, and *sky*. After removing the backgrounds, we obtain a set of objects that mostly agrees with our first criterion.

Common-class Suppression. To meet the second criterion, the one on the novelty of corner case, we make use of object detectors used in autonomous driving systems to filter out objects that are not considered novel by our standard. Specifically, we utilize Cascade R-CNN [5] with SP-Net backbone [17] trained on a private dataset that is similar to ONCE and consists of millions of scenes to detect common-class objects, producing a set of bounding boxes for each scene. The bounding boxes are subsequently compared with the proposals from the previous step, and those proposals that have IoUs over a threshold with any of the detected common objects are removed.

Note that we do not use ground-truth annotations to suppress the common-class proposals. Our approach has two important advantages: 1) it applies to unlabeled data as long as there is a working detector trained on a similar dataset of the same task; and 2) it keeps some novel instances of the common classes, *i.e.*the "hard cases" in object detection, that would otherwise be suppressed by the ground truth. The effectiveness of COPG is demonstrated in Sect. 6.

4.3 Further Examination

In the previous subsection, we have discussed how to extract potential corner cases from an abundance of unlabeled data. On ONCE [28], the process leaves only around 10% of the scenes for further examination. It is perhaps worth noting that by increasing the thresholds of background removal and common-class suppression, one can further reduce the number of candidate scenes, but it would also cause more corner cases to be neglected. In some sense, the thresholds control the trade-off between the final amount of true positives and the amount of human labor required to pick them out (see Appendix B for relevant ablations).

Selection. Given the generated proposals of ONCE, we start by examining the scenes containing these proposals. Those that do not contain any valid corner case (according to our criteria) are discarded. After the process, we finally arrive at the 1057 scenes of CODA-ONCE, roughly 0.1% of the one million scenes in the original dataset. This shows that corner cases are indeed rare in real-world data. As for KITTI [11] and nuScenes [4], without undergoing the automatic generation of proposals, all data are manually selected, resulting in 309 and 134 scenes respectively.

Labeling. To ease the labeling process, we use CLIP [31] to pre-label the objects in CODA. After that, we use the toolkit [39] inspired by LabelMe [35] to label the class of each corner case and to revise the bounding boxes since the proposals in each scene may not all be valid and the projection from point clouds to camera images is often inaccurate. Meanwhile, some bounding boxes are also added to corner-case objects missed by the proposals in the selected scenes. For quality assurance, the output of each annotator is verified by two other annotators. In the end, most of the corner cases are given a label of a specific class, except the ones that are either unrecognizable or difficult to categorize, which are placed under the *misc* class.

5 Experiment

5.1 Implementation Details

Baselines. Four categories of baselines are evaluated on CODA: 1) for *closed-world object detectors*, state-of-the-art detectors of both one-stage (*e.g.*, RetinaNet [21]) and two-stage (*e.g.*, Faster [33] and Cascade R-CNN [5]) pre-trained on SODA10M [13], BDD100K [44] and Waymo [37] are selected; 2) *region proposal network (RPN)* [33] can recognize foreground objects in a class-agnostic manner, which might learn a more generalizable representation, so we further report the performance of the RPN of Faster R-CNN and Cascade R-CNN; 3) for *open-world object detectors*, we adopt the state-of-the-art ORE model [18] but without incremental learning; and 4) for *anomaly detection*, we modify the synthesize then compare [42] and memory-based OOD detection [12] to generate anomaly bounding boxes based on the proposals of a pre-trained RPN.

Table 2. Detection results (%) on CODA. The best performance is achieved at 12.8% AR, suggesting that truly reliable object detection is probably still far from reach. Definitions of ORIGIN, CORNER, COMMON, and NOVEL are provided in "class separation" of Sect. 5.1. "D-DETR" is short for Deformable DETR and "Cascade Swin" stands for Swin-Tiny-based Cascade R-CNN. **Bold** values highlight the best performance among detectors pre-trained on the same dataset, and "†" means official checkpoints are adopted. "*" indicates that AR is the primary evaluation metric on CODA, while "-" suggests that the detector cannot report the corresponding values, with reasons explained in "evaluation" of Sect. 5.1. See more results in Appendix D.

CODA		ORIGIN		CORNER				COMMON				NOVEL			
Method	Dataset	AP	AR	AR*	AR_{50}	AR_{75}	AR^{10}	AR*	AR_{50}	AR_{75}	AR^{10}	AR*	AR_{50}	AR_{75}	AR^{10}
RetinaNet† [21]		34.0	50.7	**11.9**	25.2	9.5	5.4	28.7	58.9	23.5	23.9	–	–	–	–
Faster R-CNN† [33]		36.7	46.9	6.8	13.0	6.4	4.9	23.9	46.8	20.1	23.1	–	–	–	–
Cascade R-CNN† [5]		39.4	51.6	8.3	15.5	7.6	5.5	27.2	47.0	29.4	25.3	–	–	–	–
D-DETR [47]		31.8	49.4	7.2	16.7	4.9	3.6	**34.6**	60.2	36.5	29.6	–	–	–	–
Sparse R-CNN [38]	SODA10M [13]	31.2	51.0	6.4	13.2	5.4	3.9	26.4	47.1	25.6	23.0	–	–	–	–
Cascade Swin [26]		41.1	52.9	8.2	15.5	7.6	5.7	30.4	51.3	32.2	29.3	–	–	–	–
RPN (Faster)† [33]		–	59.7	8.1	16.2	7.4	3.1	–	–	–	–	–	–	–	–
RPN (Cascade)† [5]		–	57.1	7.7	16.0	6.8	2.8	–	–	–	–	–	–	–	–
ORE [18]		**49.2**	**59.7**	8.3	16.4	7.4	5.6	18.5	35.5	18.2	18.1	**3.4**	7.6	2.8	2.9
RetinaNet† [21]		28.6	40.4	**12.8**	23.2	11.9	4.8	27.5	58.1	21.5	23.6	**9.7**	17.7	9.1	5.9
Faster R-CNN† [33]		31.0	40.7	10.7	19.2	10.2	4.3	24.4	48.1	20.9	22.0	7.2	13.3	6.8	5.9
Cascade R-CNN† [5]		32.4	41.4	10.4	18.5	9.7	4.5	25.7	48.4	23.3	23.6	6.9	12.5	6.5	5.7
D-DETR [47]	BDD100K [44]	28.5	42.3	9.0	22.2	5.6	2.8	28.5	63.0	22.3	26.2	7.0	17.3	4.3	3.9
Sparse R-CNN† [38]		26.7	40.2	9.8	19.0	8.9	4.5	27.4	51.7	25.8	24.3	8.0	15.4	7.4	5.1
Cascade Swin [26]		**34.5**	43.5	9.9	17.2	9.7	4.9	**31.0**	55.0	29.9	29.4	6.5	11.4	6.4	5.9
RPN (Faster)† [33]		–	50.2	10.6	20.0	10.2	3.7	–	–	–	–	–	–	–	–
RPN (Cascade)† [5]		–	**51.0**	10.6	20.0	10.2	3.9	–	–	–	–	–	–	–	–
RetinaNet [21]		39.7	47.7	8.4	15.6	7.7	5.1	24.5	43.2	24.4	22.2	6.7	11.9	6.4	4.6
Faster R-CNN [33]		40.9	47.0	6.8	12.4	6.4	4.8	20.9	36.0	19.6	19.1	5.5	9.6	5.2	4.3
Cascade R-CNN [5]		42.6	48.1	6.6	11.4	6.6	5.0	18.9	32.6	20.1	17.6	5.3	8.7	5.5	4.4
D-DETR [47]	Waymo [37]	40.4	49.8	7.3	15.8	5.4	3.6	28.5	49.4	24.6	22.5	5.2	11.5	4.0	3.0
Sparse R-CNN [38]		38.8	49.8	**10.1**	19.6	9.0	4.7	**29.5**	51.8	27.0	22.1	**7.6**	14.3	7.1	4.2
Cascade Swin [26]		**44.2**	49.0	5.4	8.7	5.5	4.4	21.8	38.1	18.8	21.3	4.3	6.7	4.6	3.7
RPN (Faster) [33]		–	**53.9**	7.5	13.7	7.5	3.6	–	–	–	–	–	–	–	–
RPN (Cascade) [5]		–	52.8	7.4	13.8	7.3	3.9	–	–	–	–	–	–	–	–

Optimization. We adopt ResNet-50 [15] initialized with ImageNet-supervised pre-trained weights as the backbone for all baselines except Swin Transformer [26] based Cascade R-CNN, denoted as *Cascade Swin* in Table 2. We utilize the officially released checkpoints of closed-world detectors pre-trained on SODA10M and BDD100K, while re-implementing all selected baselines on Waymo, whose official checkpoints are not available, using the MMDetection [7] toolbox. All the BDD100K and Waymo baselines are trained with a batch size of 16 for 12 epochs with an 1000-iteration warmup using the SGD optimizer. The learning rate is set as 0.02, decreased by a factor of 10 at the 8th and 11th epoch. Lastly, we construct ORE based on Faster R-CNN using Detectron2 [41] following the original paper, which is then trained on SODA10M with a batch size of 8 for 24 epochs, the same with the closed-world counterparts. More optimization details are provided in Appendix A.

Class Separation. Considering the fact that the semantic class sets of SODA10M, BDD100K, and Waymo differ from each other, all of which are just subsets of the CODA class set, a unified separation of common and novel classes is necessary for a fair comparison of different detectors. Without loss of generality, we define: **1) COMMON** classes as the class set of SODA10M (*i.e.*, *pedestrian*, *cyclist*, *car*, *truck*, *tram*, and *tricycle*), since ORE is trained on SODA10M; **2) NOVEL** classes as the remaining classes of CODA beyond COMMON; **3) CORNER** combines all COMMON and NOVEL classes to match detector predictions in a class-agnostic manner since it is more important to detect an obstacle before distinguishing its semantic class; and **4) ORIGIN** reports detector performance on their pre-trained datasets for reference (*i.e.*, SODA10M test set for SODA10M detectors and the corresponding validation sets for BDD100K and Waymo) since robustness to corner cases should not come at a high cost of detection precision.

Evaluation. By the class separation described above, we divide detector predictions according to the corresponding semantic classes. Specifically, we treat all predictions but the ORE *unknown*, which should be considered as predictions for NOVEL objects, of SODA10M detectors as predictions for COMMON objects. Predictions of *pedestrian*, *rider*, *car*, *truck*, and *train* of BDD100K detectors are considered as COMMON, while the remaining ones are considered as NOVEL. Such a disjoint division, however, is not applicable for Waymo. According to the official document, all recognizable vehicles are annotated as *vehicle* uniformly, suggesting that Waymo baselines can only detect vehicles in a class-agnostic manner. So here, the *vehicle* predictions of Waymo detectors are not only considered as COMMON (along with *pedestrian* and *cyclist*), but also considered as NOVEL, which might put Waymo detectors at advantage, especially for the recall-based evaluation described below, but it does not affect our conclusion. We further project all detected COMMON vehicles to a unified *vehicle* class so that detectors of different datasets have the same COMMON class set, *i.e.pedestrian*, *cyclist*, and *vehicle*; while we combine all NOVEL objects to evaluate in a class-agnostic manner since detectors cannot discriminate unseen classes.

Note that under two circumstances, detectors cannot be evaluated (marked as "-" in Table 2), including: 1) RPNs can only perform class-agnostic detection, which are only evaluated under ORIGIN and CORNER; and 2) closed-world detectors pre-trained on SODA10M cannot recognize any NOVEL objects, whose semantic class set is considered as CODA COMMON class set.

We utilize the COCO-style Average Recall (AR) as the evaluation metrics instead of Average Precision (AP) since the annotated objects are the most challenging **subset** of all CODA foreground objects. A model that can detect all foreground objects, including those not obstructing the road, would in fact have low AP on CODA. Hence, AR is much more informative than AP. We also consider 1) AR_{50} and AR_{75} for IoU thresholds of 0.5 and 0.75; 2) AR^1 and AR^{10} for at most 1 and 10 boxes per image; and 3) AR^s, AR^m and AR^l for different box scales following COCO definition [22].

5.2 Results

Significance of CODA. Experiment results are reported in Table 2. As summarized in Fig. 1, detectors suffer from a significant performance drop of 30%-50% AR when deployed on CODA (*e.g.*, 43.3% decrease for SODA10M Cascade R-CNN). Even for COMMON classes, the average decrease has also exceeded 21%. The best performance is achieved at 12.8% AR, which is still far from solved even considering the domain gap between CODA and pre-trained datasets. See more complete performance statistics in Appendix D.

Detectors. As shown in Table 2, Cascade R-CNN outperforms Faster R-CNN on CODA in general, not only for COMMON classes but also in the setting of CORNER class with a consistent improvement on the ORIGIN datasets, demonstrating the possibility to achieve higher AR on CODA without a decrease of AP on common datasets for more powerful detectors. On the contrary, RetinaNet exceeds Cascade R-CNN at the expense of AP drop, probably due to the dense prediction design. Note that Cascade R-CNN performs comparably or even better than RetinaNet referring to AR^{10} (*e.g.*, 5.5% vs. 5.4% pre-trained on SODA10M), suggesting that the AR improvement might come from more box predictions (*e.g.*, averaged 86 and 21 boxes/image for SODA10M RetinaNet and Cascade R-CNN). RPN brings minor improvement but is significantly surpassed by RetinaNet even though RPN generates more box predictions (*e.g.*, 1000 vs. 86 boxes/image on SODA10M), showing that class-aware training might be beneficial to learn a more discriminative and robust detector. Surprisingly, we observe that ORE, the open-world detector, brings improvement on both CODA and SODA10M test set, about which more analyses are provided in Sect. 6.

Pre-train Datasets. BDD100K detectors perform the best among three datasets, especially for the NOVEL class since BDD100K has the largest annotated semantic class set, which is definitely beneficial to detect more complicated objects and learn a more discriminative representation as previously discussed. However, it is impossible to annotate all possible semantic classes due to the complexity of real-world road scenes. So we hope CODA can motivate researchers to consider more scalable and effective solutions to build a robust perception system.

6 Discussion

Effectiveness of COPG. The examples in Appendix E qualitatively demonstrates the effectiveness of COPG, reliably identifying nearby objects and retaining corner cases in a progressive manner. We further quantitatively study the effectiveness of COPG by considering it as a *corner-case detector*, instead of a *corner-case proposal generator*. The evaluation result is shown in Table 3, where COPG is compared with other object and anomaly detectors on detecting the corner cases in CODA-KITTI. Note that CODA-KITTI is curated by manually examining all the "misc"-category annotation of KITTI [11]. In other words, the construction of CODA-KITTI does not involve COPG. As reported, COPG shows significant improvements and is much more comparable to human than the baselines.

Moreover, comparing the baseline performances on CODA (Table 2) with those on CODA-ONCE (Table 8 in Appendix D), we notice all detectors generally achieve higher AR on CODA than CODA-ONCE (*e.g.*, 12.8% vs 10.2% AR for BDD100K-trained RetinaNet), suggesting that CODA-ONCE constructed based on the proposals of COPG is much harder than the corner cases of the other two subsets whose construction does not involve COPG.

Table 3. Evaluation of COPG and other object/anomaly detectors on detecting corner cases. The experiments are conducted on CODA-KITTI whose construction does not involve COPG (whereas the construction of CODA-ONCE does). Here AR^m_{50} and AR^l_{50} represent AR_{50} for medium and large objects, since no small corner cases are included in CODA-KITTI, with the same definition for AR^m_{30} and AR^l_{30} under 0.3 IoU threshold.

Method	AR^m_{50}	AR^l_{50}	AR^m_{30}	AR^l_{30}
Faster R-CNN [33]	6.7	8.3	26.4	28.8
Memory-based OOD [12]	2.2	21.8	6.6	39.5
Synthesize then Compare [42]	9.0	17.7	12.3	33.3
COPG (Ours)	**23.8**	**44.9**	**39.6**	**63.9**

Comparison Between Closed-World and Open-World object Detection. We visualize and compare the detection results of Faster R-CNN, ORE and CODA ground truth in Fig. 7. Considering that *unknown* objects are usually trained as background for object detection, ORE utilizes the SODA10M validation set to estimate the known and unknown energy functions based on EBM [19]. As shown in Fig. 7, by using an extra data source, ORE can successfully deal with the corner cases of both common and novel classes, which is consistent with the experiment results in Table 2. The usage of an extra data source might put ORE at advantage, but the improvement is still impressive since the extra data is only used for the energy function estimation without updating the parameters of the detector at training time.

The performance of ORE does remind us that it is possible to build a more robust perception system by utilizing an additional data source to separate background and *unknown* objects. However, for ORE, the extra data source is required to be labeled. Considering the annotation cost, it is more desirable to build a system requiring unlabeled data only (*e.g.*, SODA10M large-scale unlabeled set, which has demonstrated to improve cross-domain performance [13]), of which CODA would be a great help in the evaluation.

Evaluation of Few-Shot Object Detection (FSOD). The main goal of CODA is to evaluate the generalization ability of object detectors in self-driving systems *without* model adaptation. Nevertheless, it can also be used to evaluate adaptation methods like FSOD. So, apart from the typical baselines included in Table 2, we have also evaluated two of the state-of-the-art FSOD methods, FsDet [40] and DeFRCN [30], on CODA in a 34-way-1-shot setting with **5-time** repeated experiments (see Table 4). Neither method demonstrates satisfying performance.

Fig. 7. Visualization of Faster R-CNN (left), ORE (middle) detection results and corner case ground truth (right) on CODA. We annotate the *unknown* predictions of ORE and CODA ground truth with *red* boxes, while the common-class predictions are annotated by *blue* boxes. ORE solves the corner cases of both common (top, cyclist) and novel (bottom, traffic cone & sign) classes. (Color figure online)

Table 4. Evaluation of FSOD on CODA. "††" suggests that the reported values are evaluated in a class-agnostic manner, same as the CORNER setting adopted in Table 2.

Method	34-way (class-wise)			1-way†† (class-agnostic)		
	AR	AR_{50}	AR_{75}	AR	AR_{50}	AR_{75}
FsDet [40]	$4.9_{\pm0.8}$	$9.4_{\pm1.9}$	$4.4_{\pm0.8}$	$4.2_{\pm0.4}$	$7.7_{\pm0.7}$	$4.0_{\pm0.3}$
DeFRCN [30]	$6.7_{\pm1.2}$	$12.1_{\pm1.6}$	$6.6_{\pm1.6}$	$4.5_{\pm0.5}$	$8.9_{\pm0.9}$	$4.2_{\pm0.5}$

Limitation and Potential Negative Societal Impact. We would continue to enlarge CODA by exploring: 1) Use COPG on more real-world road scenes. 2) Since CODA is collected in the real world with high-quality annotation, we can generate more synthesized images following [1,16], or mine large-scale unlabeled road scene images in a semi-supervised manner [25,30,34,36]. Further discussion about potential negative societal impact of CODA are provided in Appendix C.

7 Conclusion

In this paper, we propose CODA, a real-world road corner case dataset for object detection in autonomous driving, constructed by ground truth class separation and automatic proposal. We observe a significant performance drop for state-of-the-art detectors when deployed on CODA. We further provide a thorough comparison of different methods and shed light on potential solutions to a more robust perception system. We hope that CODA can motivate further research in reliable detection for real-world autonomous driving.

Acknowledgement. We gratefully acknowledge the support of MindSpore, CANN (Compute Architecture for Neural Networks) and Ascend AI Processor used for this research.

References

1. Blum, H., Sarlin, P.E., Nieto, J., Siegwart, R., Cadena, C.: The fishyscapes benchmark: Measuring blind spots in semantic segmentation. arXiv preprint arXiv:1904.03215 (2019)
2. Bogoslavskyi, I., Stachniss, C.: Fast range image-based segmentation of sparse 3D laser scans for online operation. In: IROS (2016)
3. Breitenstein, J., Termöhlen, J.A., Lipinski, D., Fingscheidt, T.: Corner cases for visual perception in automated driving: some guidance on detection approaches. arXiv preprint arXiv:2102.05897 (2021)
4. Caesar, H., et al.: A multimodal dataset for autonomous driving. arXiv preprint arXiv:1903.11027 (2019)
5. Cai, Z., Vasconcelos, N.: Cascade R-CNN: delving into high quality object detection. In: CVPR (2018)
6. Chen, K., Hong, L., Xu, H., Li, Z., Yeung, D.Y.: Multisiam: self-supervised multi-instance Siamese representation learning for autonomous driving. In: ICCV (2021)
7. Chen, K., et al.: MMDetection: Open MMLab detection toolbox and benchmark. arXiv preprint arXiv:1906.07155 (2019)
8. Chen, L.-C., Zhu, Y., Papandreou, G., Schroff, F., Adam, H.: Encoder-decoder with atrous separable convolution for semantic image segmentation. In: Ferrari, V., Hebert, M., Sminchisescu, C., Weiss, Y. (eds.) ECCV 2018. LNCS, vol. 11211, pp. 833–851. Springer, Cham (2018). https://doi.org/10.1007/978-3-030-01234-2_49
9. Cordts, M., et al.: The cityscapes dataset for semantic urban scene understanding. In: CVPR (2016)
10. Fischler, M.A., Bolles, R.C.: Random sample consensus: a paradigm for model fitting with applications to image analysis and automated cartography. Commun. ACM **24**(6), 381–395 (1981)
11. Geiger, A., Lenz, P., Urtasun, R.: Are we ready for autonomous driving? the kitti vision benchmark suite. In: CVPR (2012)
12. Gong, D., et al.: Memorizing normality to detect anomaly: memory-augmented deep autoencoder for unsupervised anomaly detection. In: ICCV (2019)
13. Han, J., et al.: SODA10M: a large-scale 2D self/semi-supervised object detection dataset for autonomous driving. arXiv preprint arXiv:2106.11118 (2021)
14. He, K., Fan, H., Wu, Y., Xie, S., Girshick, R.: Momentum contrast for unsupervised visual representation learning. In: CVPR (2020)
15. He, K., Zhang, X., Ren, S., Sun, J.: Deep residual learning for image recognition. In: CVPR (2016)
16. Hendrycks, D., Basart, S., Mazeika, M., Mostajabi, M., Steinhardt, J., Song, D.: A benchmark for anomaly segmentation. arXiv preprint arXiv:1911.11132 (2019)

17. Jiang, C., Xu, H., Zhang, W., Liang, X., Li, Z.: SP-NAS: serial-to-parallel backbone search for object detection. In: CVPR (2020)
18. Joseph, K., Khan, S., Khan, F.S., Balasubramanian, V.N.: Towards open world object detection. In: CVPR (2021)
19. LeCun, Y., Chopra, S., Hadsell, R., Ranzato, M., Huang, F.: A tutorial on energy-based learning. Predicting Structured Data 1 (2006)
20. Lin, T.Y., Dollár, P., Girshick, R., He, K., Hariharan, B., Belongie, S.: Feature pyramid networks for object detection. In: CVPR (2017)
21. Lin, T.Y., Goyal, P., Girshick, R., He, K., Dollár, P.: Focal loss for dense object detection. In: ICCV (2017)
22. Lin, T.-Y., Maire, M., Belongie, S., Hays, J., Perona, P., Ramanan, D., Dollár, P., Zitnick, C.L.: Microsoft COCO: common objects in context. In: Fleet, D., Pajdla, T., Schiele, B., Tuytelaars, T. (eds.) ECCV 2014. LNCS, vol. 8693, pp. 740–755. Springer, Cham (2014). https://doi.org/10.1007/978-3-319-10602-1_48
23. Lis, K., Nakka, K., Fua, P., Salzmann, M.: Detecting the unexpected via image resynthesis. In: ICCV (2019)
24. Liu, W., et al.: SSD: single shot multibox detector. In: Leibe, B., Matas, J., Sebe, N., Welling, M. (eds.) ECCV 2016. LNCS, vol. 9905, pp. 21–37. Springer, Cham (2016). https://doi.org/10.1007/978-3-319-46448-0_2
25. Liu, Y.C., et al.: Unbiased teacher for semi-supervised object detection. In: ICLR (2021)
26. Liu, Z., et al.: Swin transformer: hierarchical vision transformer using shifted windows. In: ICCV (2021)
27. Liu, Z., et al.: Task-customized self-supervised pre-training with scalable dynamic routing. In: AAAI (2022)
28. Mao, J., et al.: One million scenes for autonomous driving: ONCE dataset. arXiv preprint arXiv:2106.11037 (2021)
29. Pinggera, P., Ramos, S., Gehrig, S., Franke, U., Rother, C., Mester, R.: Lost and found: detecting small road hazards for self-driving vehicles. In: IROS (2016)
30. Qiao, L., Zhao, Y., Li, Z., Qiu, X., Wu, J., Zhang, C.: DeFRCN: decoupled faster R-CNN for few-shot object detection. In: ICCV (2021)
31. Radford, A., et al.: Learning transferable visual models from natural language supervision. In: ICML (2021)
32. Redmon, J., Divvala, S., Girshick, R., Farhadi, A.: You only look once: unified, real-time object detection. In: CVPR (2016)
33. Ren, S., He, K., Girshick, R., Sun, J.: Faster R-CNN: towards real-time object detection with region proposal networks. In: NeurIPS (2015)
34. Reza, M.A., Naik, A.U., Chen, K., Crandall, D.J.: Automatic annotation for semantic segmentation in indoor scenes. In: IROS (2019)
35. Russell, B.C., Torralba, A., Murphy, K.P., Freeman, W.T.: LabelMe: a database and web-based tool for image annotation. IJCV **77**(1–3), 157–173 (2008)
36. Sohn, K., Zhang, Z., Li, C.L., Zhang, H., Lee, C.Y., Pfister, T.: A simple semi-supervised learning framework for object detection. arXiv:2005.04757 (2020)
37. Sun, P., et al.: Scalability in perception for autonomous driving: Waymo open dataset. In: CVPR (2020)
38. Sun, P., et al.: Sparse R-CNN: end-to-end object detection with learnable proposals. In: CVPR (2021)
39. Wada, K.: LabelMe: image polygonal annotation with python. https://github.com/wkentaro/labelme (2016)
40. Wang, X., Huang, T., Gonzalez, J., Darrell, T., Yu, F.: Frustratingly simple few-shot object detection. In: ICML (2020)

41. Wu, Y., Kirillov, A., Massa, F., Lo, W.Y., Girshick, R.: Detectron2. https://github.com/facebookresearch/detectron2 (2019)
42. Xia, Y., Zhang, Y., Liu, F., Shen, W., Yuille, A.L.: Synthesize then compare: detecting failures and anomalies for semantic segmentation. In: Vedaldi, A., Bischof, H., Brox, T., Frahm, J.-M. (eds.) ECCV 2020. LNCS, vol. 12346, pp. 145–161. Springer, Cham (2020). https://doi.org/10.1007/978-3-030-58452-8_9
43. Ye, N., et al.: OoD-bench: quantifying and understanding two dimensions of out-of-distribution generalization. In: CVPR (2022)
44. Yu, F., Chen, H., Wang, X., Xian, W., Chen, Y., Liu, F., Madhavan, V., Darrell, T.: BDD100k: a diverse driving dataset for heterogeneous multitask learning. In: CVPR (2020)
45. Zhou, X., et al.: Model agnostic sample reweighting for out-of-distribution learning. In: ICML (2022)
46. Zhou, X., Lin, Y., Zhang, W., Zhang, T.: Sparse invariant risk minimization. In: ICML (2022)
47. Zhu, X., Su, W., Lu, L., Li, B., Wang, X., Dai, J.: Deformable DETR: deformable transformers for end-to-end object detection. arXiv preprint arXiv:2010.04159 (2020)

Motion Inspired Unsupervised Perception and Prediction in Autonomous Driving

Mahyar Najibi, Jingwei Ji, Yin Zhou(✉), Charles R. Qi, Xinchen Yan, Scott Ettinger, and Dragomir Anguelov

Waymo LLC, Mountain View, USA
{najibi,jingweij,yinzhou,rqi,xcyan,settinger,dragomir}@waymo.com

Abstract. Learning-based perception and prediction modules in modern autonomous driving systems typically rely on expensive human annotation and are designed to perceive only a handful of predefined object categories. This closed-set paradigm is insufficient for the safety-critical autonomous driving task, where the autonomous vehicle needs to process arbitrarily many types of traffic participants and their motion behaviors in a highly dynamic world. To address this difficulty, this paper pioneers a novel and challenging direction, *i.e.,* training perception and prediction models to understand *open-set* moving objects, with no human supervision. Our proposed framework uses self-learned flow to trigger an automated meta labeling pipeline to achieve automatic supervision. 3D detection experiments on the Waymo Open Dataset show that our method significantly outperforms classical unsupervised approaches and is even competitive to the counterpart with supervised scene flow. We further show that our approach generates highly promising results in open-set 3D detection and trajectory prediction, confirming its potential in closing the safety gap of fully supervised systems.

Keywords: Autonomous driving · Unsupervised learning · Generalization · Detection · Motion prediction · Scene understanding

1 Introduction

Modern 3D object detection [61,68,102,112] and trajectory prediction models [10,32,51,104] are often designed to handle a predefined set of object types and rely on costly human annotated datasets for their training. While such paradigm has achieved great success in pushing the capability of autonomy systems, it has difficulty in generalizing to the *open-set* environment that includes a long-tail distribution of object types far beyond the predefined taxonomy.

M. Najibi and J. Ji—Equal contribution.

Supplementary Information The online version contains supplementary material available at https://doi.org/10.1007/978-3-031-19839-7_25.

S. Avidan et al. (Eds.): ECCV 2022, LNCS 13698, pp. 424–443, 2022.
https://doi.org/10.1007/978-3-031-19839-7_25

Towards solving the 3D object detection and behavior prediction of those open-set objects, an alternative and potentially more scalable approach to supervised training is unsupervised perception and prediction.

One central problem in autonomous driving is perceiving the amodal shape of moving objects in space and forecasting their future trajectories, such that the planner and control systems can maneuver safely. As motion estimation (also known as the scene flow problem) is a fundamental task agnostic to the scene semantics [50], it provides an opportunity to address the problem of perception and prediction of open-set moving objects, without any human labels. This leads to our motion-inspired unsupervised perception and prediction system.

Using only LiDAR, our system decomposes the unsupervised, open-set learning task to two steps, as shown in Fig. 1: (1) Auto Meta Labeling (AML) assisted by scene flow estimation and temporal aggregation, which generates pseudo labels of any moving objects in the scene; (2) Training detection and trajectory prediction models based on the auto meta labels. Realizing such an automatic supervision, we transform the challenging open-set learning task to a known, well-studied task of supervised detection or behavior prediction model training.

To derive high-quality auto meta labels, we propose two key technologies: an unsupervised scene flow estimation model and a flow-based object proposal and concept construction approach. Most prior works on unsupervised scene flow estimation [45,55,59,96] optimize for the overall flow quality without specifically focusing on the moving objects or considering the usage of scene flow for onboard perception and prediction tasks. For example, the recently proposed Neural Scene Flow Prior (NSFP) [45] achieved state-of-the-art performance in overall scene flow metrics by learning to estimate scene flow through run-time optimization, without any labels. However, there are too many false positive flows generated for the background, which makes it not directly useful for flow-based object discovery. To tackle its limitations, we extend NSFP to a novel, more accurate and scalable version termed NSFP++. Based on the estimated flow, we propose an automatic pipeline to generate proposals for all moving objects and reconstruct the object shapes (represented as amodal 3D bounding boxes) through tracking, shape registration and refinement. The end product of the process is a set of 3D bounding boxes and tracklets. Given the auto labels, we can train top-performing 3D detection models to localize the open-set moving objects and train behavior prediction models to forecast their trajectories.

Evaluated on the Waymo Open Dataset [75], we show that our unsupervised and data-driven method significantly outperforms non-parametric clustering based approaches and is even competitive to supervised counterparts (using ground truth scene flow). More importantly, our method substantially extends the capability of handling open-set moving objects for 3D detection and trajectory prediction models, leading to a safety improved autonomy system.

2 Related Works

LiDAR-Based 3D Object Detection: Supervised 3D detection based on point clouds has been extensively studied. Based on their input representation, these detectors can be categorized as those operating directly on the points [46, 54,61,68,69,102], on a voxelized space [21,43,56,70,73,87,89,100,103,109,112], a perspective projection [5,27,53], or a combination of these representations [12, 34,67,76,111]. Semi-supervised 3D detection with a smaller labeled training set or under the annotator-in-the-loop setting has also been considered [7,62,99]. However, unsupervised 3D detection has been mostly unexplored. Recently, Tian *et al.* [78] proposed to use 3D clues to perform unsupervised 2D detection in images. In contrast, we propose a novel method for 3D detection of moving objects in an unsupervised manner.

Scene Flow Estimation: Most previous learning-based works for 3D point cloud scene flow estimation were supervised [33,47,60,91]. More recently, the unsupervised setting has been also studied. [55] used self-supervised cycle consistency and nearest-neighbour losses to train a flow prediction network. In contrast, [45] took an inference-time optimization approach and trained a network per scene. We follow [45] to build our scene flow module given its unsupervised nature and relatively better performance. However, our analysis reveals the limitations of this method in handling complex scenes, making its direct adaptation for proposing high-quality auto labels challenging. In our paper, we noticeably improve the performance of this method by proposing novel techniques to better capture the locality constraints of the scene and to reduce its false predictions.

Unsupervised Object Detection: Existing efforts have been concentrated in the image and video domain, mostly evaluated on object-centric datasets or datasets with a handful of objects per frame. These include statistic-based methods [65,71], visual similarity-based clustering [26,29,40], linkage analysis [42] with appearance and geometric consistency [15,84–86], visual saliency [39,105], and generative unsupervised feature learning methods [3,44,63,72]. In contrast, unsupervised object detection from LiDAR sequences is fairly underexplored [18,48,78,94]. [18,57] proposed to sequentially update the detections and perform tracking based on motion cues. Cen *et al.* [9] used predictions of a supervised detector to yield proposals of unknown categories, making it inapplicable to fully unsupervised settings. Wong *et al.* [94] introduced a bottom-up approach to segment known and unknown objects by clustering and aggregating points based on their embedding similarities. In contrast, our work leverages both motion cues and point locations for clustering, which puts more emphasis on detecting motion coherent objects and can generate amodal bounding boxes.

Shape Registration: Shape registration has been an important topic in vision and graphics community for decades, spanning from classical methods including Iterative Closest Point (ICP) [4,13,30,64] and Structure-from-Motion (SfM) [1, 38,66,79] to their deep learning variants [37,77,80–82,90,92,97,98,101,110,113]. These methods usually work under the assumption that the object or scene to register is mostly static or at least non-deformable. In autonomous driving, shape registration has gained increasing attentions where offline processing is required [20,22,23,31,56,74,88]. The shape registration outcome can further support downstream applications such as offboard auto labeling [62,99,107], and perception simulation [14,52]. In this work, we use sequential ICP with motion-inspired initializations to aggregate partial views of objects and produce the auto-labeled bounding boxes.

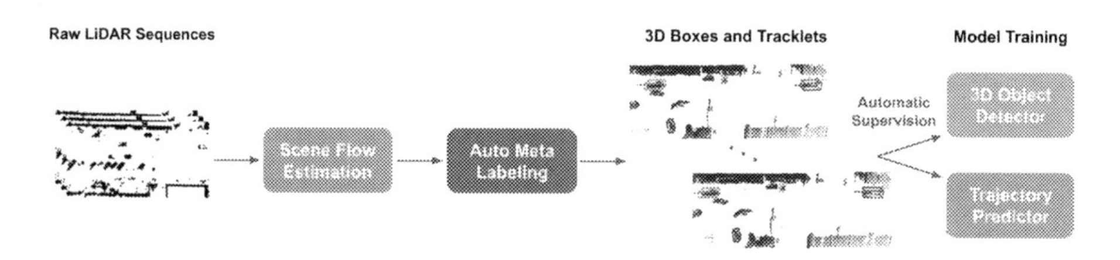

Fig. 1. Proposed framework. Taking as input LiDAR sequences (after ground removal), our approach first reasons about per point motion status (static or dynamic) and predicts accurate scene flow. Based on the motion signal, Auto Meta Labeling clusters points into semantic concepts, connects them across frames and estimates object amodal shapes (3D bounding boxes). The derived amodal boxes and tracklets will serve as automatic supervision to train 3D detection and trajectory prediction models.

Trajectory Prediction: The recent introduction of the large-scale trajectory prediction datasets [6,11,25,36] helped deep learning based methods to demonstrate new state-of-the-art performance. From a problem formulation standpoint, these methods can be categorized into uni-modal and multi-modal. Uni-modal approaches [8,19,28,51] predict a single trajectory per agent. Multi-modal methods [2,10,16,35,49,58,104,106,108] take into account the possibility of having multiple plausible trajectories per agent. However, all these methods rely on fully labeled datasets. Unsupervised or open-set settings, although practically important for autonomous driving, have so far remained unexplored. Our method enables existing behavior prediction models to generalize to all moving objects, without the need for predefining an object taxonomy.

3 Method

Figure 1 illustrates an overview of our proposed method, which primarily relies on motion cues for recognizing moving objects in an unsupervised manner. The pipeline has two main modules: unsupervised scene flow estimation (Sect. 3.1) and Auto Meta Labeling (Sect. 3.2).

3.1 Neural Scene Flow Prior++

Background. Many prior works [33,41,47,95] on scene-flow estimation only considered the supervised scenario where human annotations are available for training. However, these methods cannot generalize well to new environments or to newly seen categories [45]. Recently, Li *et al.* [45] propose neural scene flow

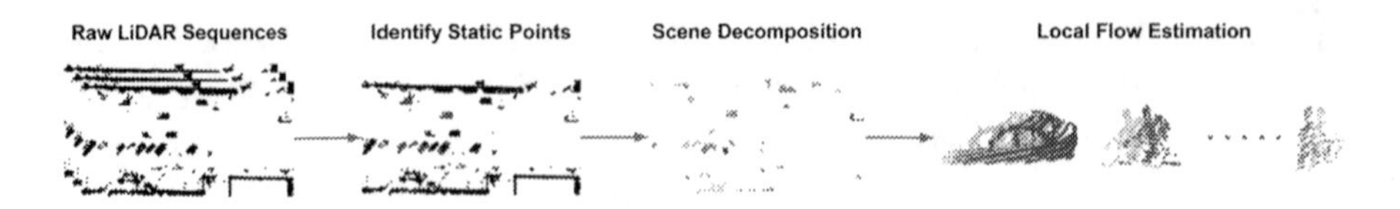

Fig. 2. Proposed NSFP++. Taking as input raw LiDAR sequences (after ground removal), our approach first reasons about the motion status of each point, decomposes the scene into connected components and predicts local flows accurately for each semantically meaningful component.

prior (NSFP), which can learn point-wise 3D flow vectors by solving an optimization problem at run-time without the need of human annotation. Thanks to its unsupervised nature, NSFP can generalize to new environments. It also achieved state-of-the-art performance in 3D scene flow estimation. Still, our study shows that it has notable limitations in handling complex scenes when a mixture of low and high speed objects are present. For example, as illustrated in Fig. 3, NSFP suffers from underestimating the velocity of moving objects, *i.e.*, false negative flows over pedestrians and inaccurate estimation of fast-moving vehicles. It also introduces excessive false positive flows over static objects (e.g., buildings). We hypothesize that such issues are due to the fact that NSFP applies global optimization to the entire point cloud and the highly diverse velocities of different objects set contradictory learning targets for the network to learn properly.

Overview. Our goal is to realize an unsupervised 3D scene flow estimation algorithm that can adapt to various driving scenarios. Here, we present our neural scene flow prior++ (NSFP++) method. As illustrated in Fig. 2, our method features three key innovations: 1) robustly identifying static points; 2) divide-and-conquer strategy to handle different objects by decomposing a scene into semantically meaningful connected components and targetedly estimating local flow for each of the them; 3) flow consistency among points in each component.

Problem Formulation. Let $\mathbf{S}_t \in \mathbb{R}^{N_1 \times 3}$ and $\mathbf{S}_{t+1} \in \mathbb{R}^{N_2 \times 3}$ be two sets of points captured by the LiDAR sensor of an autonomous vehicle at time t and time $t+1$, where N_1 and N_2 denote the number of points in each set. We denote $\mathbf{F}_t \in \mathbb{R}^{N_1 \times 3}$ as the scene flow, a set of flow vectors corresponding to each point in $\mathbf{S}_t$. Given a point $\mathbf{p} \in \mathbf{S}_t$, we define $\mathbf{f} \in \mathbf{F}_t$ be the corresponding flow vector such that $\hat{\mathbf{p}} = \mathbf{p} + \mathbf{f}$ represents the future position of $\mathbf{p}$ at $t+1$. Typically, points in $\mathbf{S}_t$ and $\mathbf{S}_{t+1}$ have no correspondence and N_1 differs from N_2.

As in Li *et al.* [45], we model the flow vector $\mathbf{f} = h(\mathbf{p}; \boldsymbol{\Theta})$ as the output of a neural network h, containing a set of learnable parameters as $\boldsymbol{\Theta}$. To estimate $\mathbf{F}_t$, we solve for $\boldsymbol{\Theta}$ by minimizing the following objective function:

$$\boldsymbol{\Theta}^*, \boldsymbol{\Theta}^*_{\text{bwd}} = \arg \min_{\boldsymbol{\Theta}, \boldsymbol{\Theta}_{\text{bwd}}} \sum_{\mathbf{p} \in \mathbf{S}_t} \mathcal{L}(\mathbf{p} + \mathbf{f}, \mathbf{S}_{t+1}) + \sum_{\hat{\mathbf{p}} \in \hat{\mathbf{S}}_t} \mathcal{L}(\hat{\mathbf{p}} + \mathbf{f}_{\text{bwd}}, \mathbf{S}_t) \tag{1}$$

where $\mathbf{f} = h(\mathbf{p}; \boldsymbol{\Theta})$ is the forward flow, $\mathbf{f}_{\text{bwd}} = h(\hat{\mathbf{p}}; \boldsymbol{\Theta}_{\text{bwd}})$ is the backward flow, $\hat{\mathbf{S}}_t$ is the set of predicted future positions for points in $\mathbf{S}_t$ and $\mathcal{L}$ is Chamfer distance function. Here we have the forward and backward flow models share the same network architecture but parameterized by $\boldsymbol{\Theta}$ and $\boldsymbol{\Theta}_{\text{bwd}}$ respectively. The model parameters, $\boldsymbol{\Theta}$ and $\boldsymbol{\Theta}_{\text{bwd}}$ are initialized and optimized for each time stamp t. Although we only take the forward flow into the next-step processing, learning the flows bidirectionally help improve the scene flow quality [45,47].

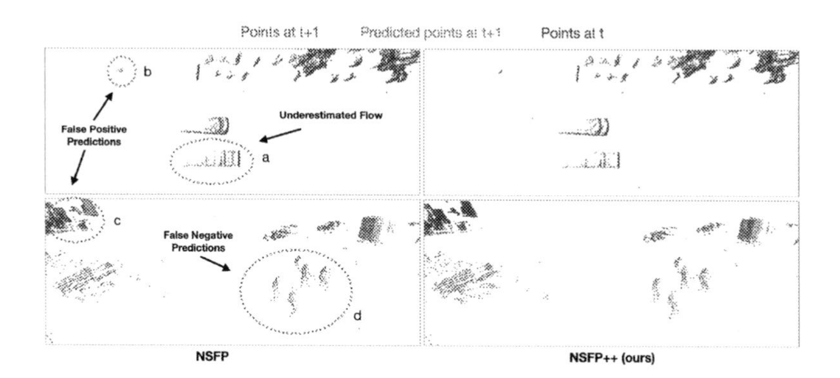

Fig. 3. Flow quality comparison between NSFP [45] and our NSFP++ over the Waymo Open Dataset. Dashed circles in orange color highlight the major shortcomings suffered by NSFP, *i.e.,* (a) underestimated flow for a fast-moving vehicle, (b)(c) false positive predictions at the background and (d) false negative predictions at pedestrians with subtle motion. In contrast, NSFP++ generates accurate predictions in all these cases.

Identifying Static Points. Since our focus is moving objects, we start by strategically removing static points to reduce computational complexity and benefit scene flow estimation. In autonomous driving datasets, one large body of static points is ground. Ground is usually captured as a flat surface for which predicting local motion is not possible due to the aperture problem. We follow [45,47] and remove ground points prior to motion estimation. This is achieved by a RANSAC-based algorithm in which a parameterized close-to-horizontal plane is fitted to the points and points in its vicinity are marked as static. However, ground is not the only static part of the scene and unsupervised flow predictions in these static regions (*e.g.*walls, buildings, trees, *etc.*) introduce noise, reducing the quality of our final auto labels. As a result, we further propose to identify more static regions in the scene prior to scene flow estimation. This is achieved by comparing the Chamfer distance between the points in the current frame with those in earlier frames. We mark points as static if the computed Chamfer distance is less than a threshold. We set a small threshold to have a high precision in this step (*i.e.*20 cm/s in our experiments).

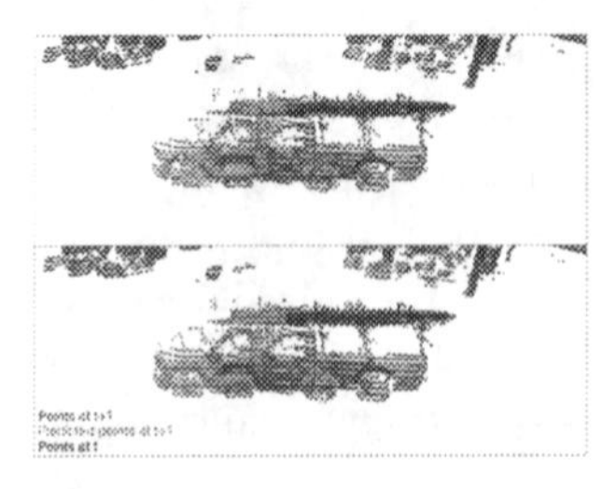

Fig. 4. Illustration of the effectiveness of box query with expansion in more accurately estimating flow over the object shape. Top and bottom show predictions without and with expansion.

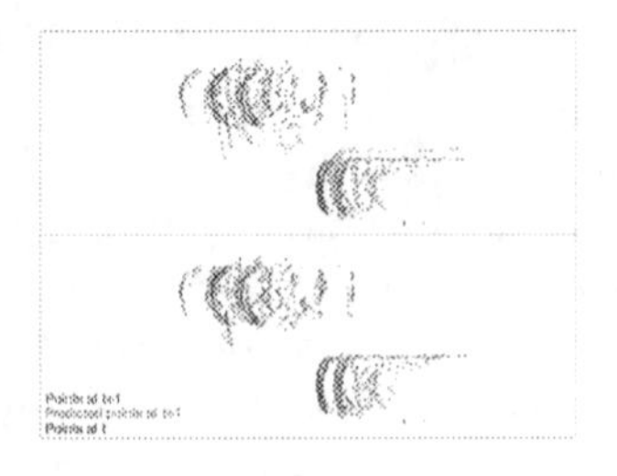

Fig. 5. Illustration of the effectiveness of box query followed by pruning in preserving accurately the local flow for nearby objects. Top and bottom show predictions without and with pruning.

Estimate Local Flow via Scene Decomposition. Inspired by the fact that objects in outdoor scenes are often well-separated after detecting and isolating the ground points, we propose to further decompose the dynamic part of the scene into connected components. This strategy allows us to solve for local flows for each cluster targetedly, which can greatly improve the accuracy of flow estimation for various traffic participants, e.g., vehicles, pedestrians, cyclists, travelling at highly different velocities. Figure 2 gives an overview of our method.

More precisely, given the identified static points, we split the point sets as $\mathbf{S}_t = \mathbf{S}_t^s \cup \mathbf{S}_t^d$ and $\mathbf{S}_{t+1} = \mathbf{S}_{t+1}^s \cup \mathbf{S}_{t+1}^d$, where $\mathbf{S}_t^s$ and $\mathbf{S}_{t+1}^s$ contain static points while $\mathbf{S}_t^d$ and $\mathbf{S}_{t+1}^d$ store dynamic points. This separation, not only helps decompose the scene into semantically meaningful connected components, but also substantially reduces false positive flow predictions on static objects. We then further break down the dynamic points into $\mathbf{S}_t^d = \bigcup_{i=1}^{K} \mathbf{C}_t^i$, where $\mathbf{C}_t^i \in \mathbb{R}^{m_i \times 3}$ is one disjoint cluster of m_i points (the number of clusters K can vary as the scene changes). In the rest of this section, we omit index i for brevity and let $\mathbf{C}_t$ to represent one of the clusters. For every $\mathbf{C}_t \subseteq \mathbf{S}_t^d$ at time t, we solve for model parameters to derive local flows, by minimizing the objective function as:

$$\begin{aligned}\boldsymbol{\Theta}^*, \boldsymbol{\Theta}_{\text{bwd}}^* = \arg \min_{\boldsymbol{\Theta}, \boldsymbol{\Theta}_{\text{bwd}}} \sum_{\mathbf{p} \in \mathbf{C}_t \subseteq \mathbf{S}_t^d} \mathcal{L}(\mathbf{p} + \mathbf{f}, \mathbf{C}_{t+1}) + \sum_{\hat{\mathbf{p}} \in \hat{\mathbf{C}}_t \subseteq \hat{\mathbf{S}}_t^d} \mathcal{L}(\hat{\mathbf{p}} + \mathbf{f}_{\text{bwd}}, \mathbf{C}_t) \\ + \frac{\alpha}{|\mathbf{C}_t|} \sum_{\substack{\mathbf{f}_i, \mathbf{f}_j \in \mathbf{F}_{\mathbf{C}_t} \\ i \neq j}} \|\mathbf{f}_i - \mathbf{f}_j\|_2^2 \end{aligned} \tag{2}$$

where the last term is the newly introduced local consistency regularizer with α set to 0.1, $\mathbf{F}_{\mathbf{C}_t}$ consists of flow vectors for each point in $\mathbf{C}_t$, $\hat{\mathbf{S}}_t^d$ contains predicted future positions of all points residing in $\mathbf{S}_t^d$, $\hat{\mathbf{C}}_t$ is a subset of $\hat{\mathbf{S}}_t^d$ only storing future positions of points in $\mathbf{C}_t \subseteq \mathbf{S}_t^d$ and $\mathbf{C}_{t+1}$ is a subset of $\mathbf{S}_{t+1}^d$, derived based on box query within a neighborhood of $\mathbf{C}_t$. Next we will present our box query strategy: expansion with pruning.

Box Query Strategy. Considering that some objects (vehicles) may move at a high speed, we need to expand the field of view to find match points in the next frame. Given a cluster $\mathbf{C}_t$, we find the axis-aligned (along X and Y axes) bounding box tightly covering $\mathbf{C}_t$, in the bird's eye view (BEV). The box is represented as $\mathbf{b} = [x_{\text{min}}, y_{\text{min}}, x_{\text{max}}, y_{\text{max}}]$. Note that fast-moving objects, *e.g.*, vehicles, can travel multiple meters between two LiDAR scans. To satisfactorily capture the points of such objects at time $t+1$, we propose to expand the box query with axis-aligned buffer distances δ_x, δ_y and use $\mathbf{b}' = [x_{\text{min}} - \delta_x, y_{\text{min}} - \delta_y, x_{\text{max}} + \delta_x, y_{\text{max}} + \delta_y]$ to retrieve points from $\mathbf{S}^d_{t+1}$, resulting in $\mathbf{C}_{t+1}$. We set the buffer distances according to the aspect ratio of the box $\mathbf{b}$, *i.e.*, $\frac{\delta_y}{\delta_x} = \frac{y_{\text{max}} - y_{\text{min}}}{x_{\text{max}} - x_{\text{min}}}$. We empirically set $\max\{\delta_x, \delta_y\} = 2.5\,\text{m}$. Figure 4 illustrates that expanding box query captures the full shape of a fast-moving truck, resulting in accurate prediction of the future position of the entire object point cloud (*i.e.*, predicted future positions align nicely with the next frame).

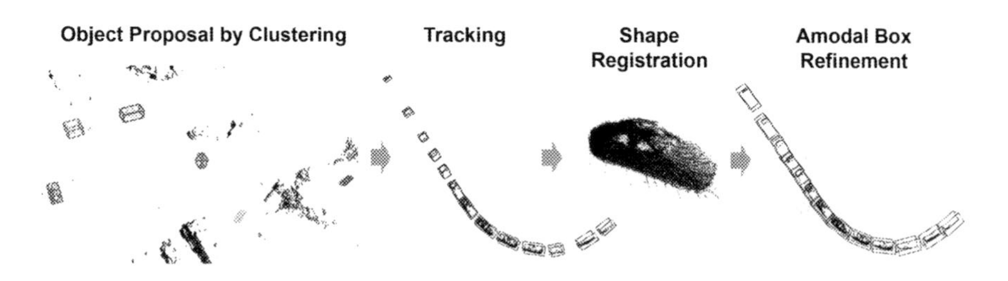

Fig. 6. Auto Meta Labeling pipeline. Given point locations and scene flows on each scene, our Auto Meta Labeling pipeline first proposes objects by spatio-temporal clustering, connects visible bounding boxes of proposals into tracks, then performs shape registration on each track to obtain 3D amodal bounding boxes on each scene.

In crowded areas of the scene, retrieved points with $\mathbf{b}'$ may include irrelevant points into the optimization process, causing flow to drift erroneously. See Fig. 5 as an example, where two vehicles are moving fast and close to each other. Box query with $\mathbf{b}'$ can include points from the other vehicle and lead to flow drifting. To address this challenge, we propose to prune retrieved points based on the statistics of $\mathbf{C}_t$. Formally, let $\mathbf{\Omega}$ be the set of retrieved points by $\mathbf{b}'$ from $\mathbf{S}^d_{t+1}$. We select $n = \min\{|\mathbf{\Omega}|, |\mathbf{C}_t|\}$ nearest points from $\mathbf{\Omega}$ with respect to the first moment of $\mathbf{C}_t$ and store them in set $\mathbf{C}_{t+1} \in \mathbb{R}^{n \times 3}$. The effectiveness of pruning in keeping relevant points and thus preserving local flow is shown in Fig. 5.

3.2 Auto Meta Labeling

With the motion signals provided by the unsupervised scene flow module, we are able to generate 3D proposals for moving objects without any manual labels. We propose an Auto Meta Labeling pipeline, which takes point clouds and scene flows as inputs and generates high quality 3D auto labels (Fig. 6). The Auto Meta

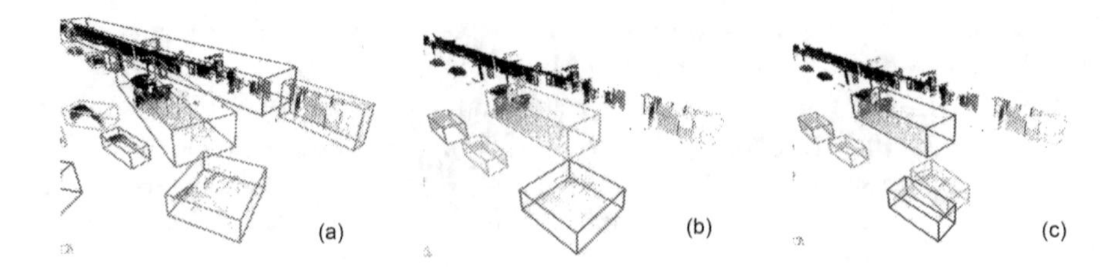

Fig. 7. Comparison between object proposals by different clustering approaches. Points are colored by scene flow magnitudes and directions. Dark for static points. (a) Clustering by location only. (b) Filter by flow magnitude and then cluster based on location (c) Filter by flow and cluster based on both location and motion (ours).

Algorithm 1. Object proposal by spatio-temporal clustering on each scene.

Input: point locations $\mathbf{S} = \{\mathbf{p}_n\}_{n=1}^N$; point-wise scene flows $\mathbf{F} = \{\mathbf{f_n}\}_{n=1}^N$
Hyperparams: neighborhood thresholds ϵ_p, ϵ_f; minimum flow magnitude $|\mathbf{f}|_{min}$
Output: 3D bounding boxes $\mathbf{B}_{vis} = \{\mathbf{b}_k\}_{k=1}^{M_{pf}}$ of the visible parts of proposed objects
function FlowBasedClustering($\mathbf{S}$, $\mathbf{F}$; ϵ_p, ϵ_f, $|\mathbf{f}|_{min}$):
  $\mathbf{S}', \mathbf{F}' \leftarrow$ FilterByFlowMagnitude($\mathbf{S}, \mathbf{F}; |\mathbf{f}|_{min}$)
  $\mathbf{C_p} \leftarrow$ DBSCAN($\mathbf{S}'; \epsilon_p$) ▷ $\mathbf{C_p} = \{\mathbf{c}_i\}_{i=1}^{M_p}$, point sets clustered by locations
  $\mathbf{C_f} \leftarrow$ DBSCAN($\mathbf{F}'; \epsilon_f$) ▷ $\mathbf{C_f} = \{\mathbf{c}_j\}_{j=1}^{M_f}$, point sets clustered by flows
  for $\mathbf{c}_i$ in $\mathbf{C_p}$ **do**
    for $\mathbf{c}_j$ in $\mathbf{C_f}$ **do**
      $\mathbf{c}_k \leftarrow \mathbf{c}_i \cap \mathbf{c}_j$
      $\bar{\mathbf{f}}_k \leftarrow$ Average($\{\mathbf{f}_l \mid \forall l : \mathbf{p}_l \in \mathbf{c}_k\}$)
      $\mathbf{b}_k \leftarrow$ MinAreaBBoxAlongDirection($\mathbf{c}_k, \bar{\mathbf{f}}_k$)
  return $\{\mathbf{b}_k\}_{k=1}^{M_{pf}}$

Labeling pipeline has four components: (a) object proposal by clustering, which leverages spatio-temporal information to cluster points into visible boxes (tight boxes covering visible points), forming the concept of objects in each scene; (b) tracking, which connects visible boxes of objects across frames into tracklets; (c) shape registration, which aggregates points of each track to complete the shape for the object; (d) amodal box refinement, which transforms visible boxes into amodal boxes. See supplementary materials for implementation details.

Object Proposal by Clustering. On each scene, given the point cloud locations $\mathbf{S} = \{\mathbf{p}_n \mid \mathbf{p}_n \in \mathbb{R}^3\}_{n=1}^N$ and the corresponding point-wise scene flows $\mathbf{F} = \{\mathbf{f}_n \mid \mathbf{f}_n \in \mathbb{R}^3\}_{n=1}^N$, the clustering module segments points into subsets where each subset represents an object proposal. We further compute a bounding box of each subset as an object representation. Traditional clustering methods on point cloud often consider 3D point locations $\mathbf{S}$ as the only feature. In the autonomous driving data, with a large portion of points belonging to the background, such methods generate many irrelevant clusters (Fig. 7a). As we focus on moving objects, we leverage the motion signals to reduce false positives. Hence, a clustering method based on both point locations and scene flows is desired.

Algorithm 2. Sequential shape registration and box refinement.

Input: An object track with point locations $\{\mathbf{X}_l\}_{l=1}^L$, bounding boxes $\{\mathbf{b}_l\}_{l=1}^L$, headings $\{\theta_l\}_{l=1}^L$. All in world coordinate system.
Output: Refined boxes $\{\mathbf{b}'_l\}_{l=1}^L$.
function ShapeRegistrationAndBoxRefinement($\{\mathbf{X}_l\}_{l=1}^L$, $\{\mathbf{b}_l\}_{l=1}^L$, $\{\theta_l\}_{l=1}^L$):
  $\mathbf{X}'_l = \mathbf{X}_l - \bar{\mathbf{X}}_l, \forall l \in \{1, ..., L\}$ ▷ Normalize points to object-centered
  $\mathbf{X}'_{tgt} \leftarrow \mathbf{X}'_{\hat{i}} : \hat{i} = \text{argmax}_i |\mathbf{X}'_i|$ ▷ Init target as the most dense point cloud
  $I = \{\hat{i}+1, \hat{i}+2, ..., L, \hat{i}-1, \hat{i}-2, ..., 1\}$ ▷ Shape registration ordering
  for i in I **do**
    for $\mathbf{T}_j$ in SearchGrid($\mathbf{b}_t$) **do**
      $\mathcal{T}_{\text{init}} \leftarrow [\mathbf{R}_{\theta_{tgt}-\theta_i} \mid \mathbf{T}_j]$
      $\mathbf{X}'_{tgt,j}, \mathcal{T}_{i\to tgt,j}, \epsilon_j \leftarrow \text{ICP}(\mathbf{X}'_i, \mathbf{X}'_{tgt}, \mathcal{T}_{\text{init}})$
    $\mathbf{X}'_{tgt}, \mathcal{T}_{i\to tgt} \leftarrow \mathbf{X}'_{tgt,\hat{j}}, \mathcal{T}_{i\to tgt,\hat{j}} : \hat{j} = \text{argmin}_j \epsilon_j$ ▷ Registration w/ least error
  $\mathbf{b}'_{tgt} = \text{MinAreaBBoxAlongDirection}(\mathbf{X}'_{tgt} + \bar{\mathbf{X}}_{tgt}, \theta_{tgt})$
  for i in I **do**
    $\mathbf{b}'_i = \text{Transform}(\mathbf{b}'_{tgt}, \mathcal{T}^{-1}_{i\to tgt})$
  return $\{\mathbf{b}'_l\}_{l=1}^L$

One simple yet effective strategy can be filtering point cloud by scene flows before object proposal: we only keep points with a flow magnitude larger than a threshold. We then apply the DBSCAN. [24] clustering algorithm on the filtered point sets. This filtering can largely reduce the false positives (Fig. 7b).

However, there is still a common case where the aforementioned approach cannot handle well: close-by objects tend to be under-segmented into a single cluster. To solve this issue, we propose clustering by both spatial locations and scene flows (Algorithm 1). After removing points with flow magnitudes smaller than a threshold $|\mathbf{f}|_{min}$, we obtain the filtered point locations $\mathbf{S}'$ and point-wise scene flows $\mathbf{F}'$. Then we apply DBSCAN. to $\mathbf{S}'$ and $\mathbf{F}'$ separately, resulting in two sets of clusters. Based on its location and motion, a point may fall into different subsets based on these two clusterings. We then intersect the subsets obtained by the location-based and the flow-based clusterings to formulate the final clusters. In this way, two points are clustered together only if they are close with respect to both their location and motion (Fig. 7c).

Having the cluster label for each point, we form the concept of an object via a bounding box covering each cluster. Given the partial observation of objects within a single frame, we only generate boxes tightly covering the visible part in this stage, $\mathbf{B}_{vis} = \{\mathbf{b}_k\}$. Without object semantics, we use motion information to decide the heading of each box. We compute the average flow $\bar{\mathbf{f}}_k$ of each cluster $\mathbf{c}_k$. Then we find the 7 DoF bounding box $\mathbf{b}_k$ surrounding $\mathbf{c}_k$ which has the minimum area on the xy-plane along the chosen heading direction parallel to $\bar{\mathbf{f}}_k$.

Multi-object Tracking. The tracking module connects visible boxes $\mathbf{B}_{vis}$ into object tracks. Following the tracking-by-detection paradigm [62,93], we use $\mathbf{B}_{vis}$ for data associations and Kalman filter for state updates. However, rather than relying on the Kalman filter to estimate object speeds, our tracking module leverages our estimated scene flows in the associations. In each step of the association, we advance previously tracked boxes using scene flows and match the advanced boxes with those in the next frame.

Shape Registration and Amodal Box Refinement. In the unsupervised setting, human annotations of object shapes are unavailable. It is hard to infer the amodal shapes of occluded objects purely based on sensor data from one timestamp. However, the observed views of an object often change across time as the autonomous driving car or the object moves. This enables temporal data aggregation to achieve more complete amodal perception of each object.

For temporal aggregation, we propose a shape registration method built upon sequentially applying ICP [4,13,64] (Algorithm 2). ICP performance is sensitive to the transformation initialization. In clustering, we have obtained the headings $\{\theta_l\}_{l=1}^{L}$ of all visible boxes in each track. The difference in headings of each source and target point set constructs a rotation initialization $R_{\theta_{tgt}-\theta_{src}}$ for ICP.

In autonomous driving scenarios, shape registration among a sequence of observations poses special challenges: (a) objects are moving with large displacements in the world coordinate system; (b) many observations of objects are very sparse due to their far distance from the sensor and/or heavy occlusions. These two challenges make it hard to register points from different frames. To tackle this problem, we search in a grid to obtain the best translation for aligning the source (from frame A) and target (from frame B) point sets. The grid, or the search range, is defined by the size of the target frame bounding box. We initialize the translation $\mathbf{T_j}$ corresponding to different grid points and find the best registration results out of them.

Sequentially, partial views of an object in a track are aggregated into a more complete point set, whose size is often close to amodal boxes. We then compute a bounding box around the target point set similar to the last step in object proposal. During registration, we have estimated the transformation from each source point set to the target, and we can propagate the target bounding box back to each scene by inversing each transformation matrix. Finally, we obtain 3D amodal bounding boxes of detected objects.

4 Experiments

We evaluate our framework using the challenging Waymo Open Dataset (WOD) [75], as it provides a large collection of LiDAR sequences with 3D labels for each frame (we only use labels for evaluation unless noted otherwise). In our experiments, objects with speed > 1 m/s are regarded moving. Hyperparameters and ablation studies are presented in the supplementary material.

4.1 Scene Flow

Metrics. We employ the widely adopted metrics as [45,96], which are 3D end-point error (EPE3D) computed as the mean L2 distance between the prediction and the ground truth for all points; Acc_5 denoting the percentage of points with EPE3D <5 cm or relative error <5%; Acc_{10} denoting the percentage of points with EPE3D <10 cm or relative error <10%; and θ, the mean angle error between predictions and ground truths. In addition, we evaluate our approach based on fine grained speed breakdowns. We assign each point to one speed class (*e.g.*, 0–3 m/s, 3–6 m/s, *etc.*) and employ the Intersection-over-Union (IoU) metric to measure the performance in terms of class-wise IoU and mean IoU. IoU is computed as $\frac{\mathrm{TP}}{\mathrm{TP}+\mathrm{FP}+\mathrm{FN}}$, same as in 3D semantic segmentation [6].

Results. We evaluate our NSFP++ over all frames of the WOD [75] validation set and compare it with the previous state-of-the-art scene flow estimator, NSFP [45]. Following [41], we use the provided vehicle pose to compensate for the ego motion, such that our metrics is independent from the autonomous vehicle motion and can better reflect the flow quality on the moving objects. Figure 3 visualizes the improvement of the proposed NSFP++ compared to NSFP. Our approach accurately predicts flows for both high- and low-speed objects (a, d). In addition, NSFP++ not only is highly reliable in detecting the subtle motion of vulnerable road users (d) but can also robustly distinguish all moving objects from the static background (b, c). Finally, our approach outperforms NSFP substantially across all quantitative metrics, as listed in Table 1.

Table 1. Comparison of scene flow methods on the WOD validation set.

Method	EPE3D (m)↓	Acc_5 (%)↑	Acc_{10} (%)↑	θ (rad)↓	IoU per Speed Breakdown (m/s)						mIOU
					0–3	3–6	6–9	9–12	12-15	15+	
NSFP [45]	0.455	23.65	43.06	0.9190	0.657	0.152	0.216	0.166	0.130	0.140	0.244
NSFP++	**0.017**	**95.05**	**96.45**	**0.4737**	**0.989**	**0.474**	**0.522**	**0.479**	**0.442**	**0.608**	**0.586**

4.2 Unsupervised 3D Object Detection

Our method aims at generating auto labels for training downstream autonomous driving tasks in a fully unsupervised manner. 3D object detection is a core component in autonomous driving systems. In this section, we evaluate the effectiveness of our unsupervised AML pseudo labels by training a 3D object detector. We adopt the PointPillars [43] detector for our experiments. All models are trained and evaluated on WOD [75] training and validation sets. Since there is no category information during training, we use a single-class detector to detect any moving objects. We train and evaluate the detectors on a 100 m × 40 m rectangular region around the ego vehicle to reflect the egocentric importance of

Table 2. Comparisons between 3D detectors trained with autolabels generated by AML with supervised flow and unsupervised flow.

Method	Supervision	3D mAP		2D mAP	
		L1	L2	L1	L2
Sup Flow [41] + Clustering	Supervised	30.8	29.7	42.7	41.2
Sup Flow [41] + AML		49.9	48.0	56.8	54.8
No flow + Clustering	Unsupervised	4.7	4.5	5.8	5.6
No flow + AML		9.6	9.4	11.0	10.8
Unsup Flow + Clustering		30.4	29.2	36.7	35.3
Unsup Flow + AML		**42.1**	**40.4**	**49.1**	**47.4**

the predictions [17]. We set a 3D IoU of 0.4 during evaluation to count for the large variation in size of the class-agnostic moving objects, e.g., vehicles, pedestrians, cyclists. We employ a top-performing flow model [41] as the supervised counterpart to our unsupervised flow model NSFP++.

Table 2 compares performance of detectors trained with auto labels generated by our pipelines and several baselines. The first two rows show detection results when a fully supervised flow model [41] (flow supervision derived from human box labels) is deployed for generating the auto labels. The first row represents a baseline where our hybrid clustering method is used to form the auto labels based on motion cues [18]. The second row shows the performance when the same supervised flow predictions are used in combination with our AML pipeline. Clearly, our AML pipeline greatly outperforms the clustering baseline, verifying the high-quality auto labels generated by our method. The last four rows consider the unsupervised setting. *No flow + Clustering* is a baseline where DBSCAN. is applied to the point locations to form the auto labels. *No flow + AML* is our pipeline when purely relying on a regular tracker without using any flow information. *Unsup Flow + Clustering* uses our proposed hybrid clustering technique on the outputs of our NSFP++ scene flow estimator without connecting with our AML. *Usup Flow + AML* is our full unsupervised pipeline. Notably, not only does it outperforms other unsupervised baselines by a large margin, but it also achieves a comparable performance with the supervised *Sup Flow + AML* counterpart. Moreover, comparing it with other unsupervised baselines by removing parts of our pipeline validates the importance of all components in our design (please see the supplementary for more ablations). Most importantly, our approach is a fully unsupervised 3D pipeline, capable of detecting moving objects in the open-set environment. This new feature is cost efficient and safety critical for the autonomous vehicle to reliably detect arbitrary moving objects, removing the need of human annotation and the constraint of predefined taxonomy.

Table 3. Open-set 3D object detection and trajectory prediction results.

	Human labeled		Object detection			Trajectory prediction	
			3D AP		3D mAP	minADE	minFDE
	Vehicle	VRU	Vehicle	VRU			
Supervised method	✓		97.5	0.0	48.8	2.12	5.93
		✓	0.0	88.7	44.4	9.53	22.31
Ours (Sup. + AML)	✓		97.5	20.8	**59.2**	**1.89**	**4.79**
		✓	65.4	88.7	**77.1**	**2.15**	**5.55**

4.3 Open-set 3D Object Detection

In this section, we turn our attention to the open-set setting where only a subset of categories are annotated. Since there is no public 3D dataset designed for this purpose, we perform experiments in a leave-one-out manner on WOD [75]. WOD has three categories, namely vehicle, pedestrian, and cyclist. Considering the similar appearances and safety requirements, we combine pedestrian and cyclist into a larger category called VRU (vulnerable road user), resulting in a data size comparable with the vehicle category. We then assume to only have access to human annotations for one of the two categories, leaving the other one out for our auto meta label pipeline to pseudo label.

The middle part in Table 3 represents the open-set 3D detection results. The first two rows show the performance of a fully supervised point pillars detector. As expected, when the detector is trained on one of the categories, it can not generalize to the other. In the last two rows, when human annotations are not available, we rely on our auto labels to fill in for the unknown category. When no vehicle label is available, our pipeline helps the detector to generalize and consequently improves the mAP from 48.8 to 77.1. Although generalizing to VRUs without any human labels is a more challenging scenario, our pipeline still improves the mAP by a noticeable margin, showing its effectiveness in the open-set settings.

4.4 Open-set Trajectory Prediction

For trajectory prediction, we have extracted road graph information for a subset of WOD (consisting of 625 training and 172 validation sequences). We use those WOD run segments with road graph information for our trajectory prediction experiments. Following [25], a trajectory prediction model is required to forecast the future positions for surrounding agents for 8 s into the future, based on the observation of 1 s history. We use the MultiPath++ [83] model for our study. The model predicts 6 different trajectories for each object and a probability for each trajectory. To evaluate the impact of open-set moving objects on the behavior prediction task, we train models using perception labels derived via different strategies as the ground truth data and then evaluate the behavior prediction metrics of the trained models on a manually labeled validation set. We use the minADE and minFDE metrics as described in [25].

The last two columns in Table 3 show the trajectory prediction results. While the supervised method achieves a reasonable result when the vehicle class is labeled, its performance is poor when trained only on the VRU class. This is expected, as the motion learned from slow vehicles can be generalized to VRUs to some extent, but predicting the trajectory of the fast moving vehicles is out of reach for a model trained on only VRUs. The last two rows show the performance of the same model when AML is deployed for auto-labeling the missing category. Consistent with our observation in 3D detection, our method can bridge the gap in the open-set setting. Namely, our approach significantly remedies the generalization problem from VRUs to vehicles and achieves the best performance when combining human labels of the vehicle class with our auto labels for VRUs.

5 Conclusion

In this paper, we proposed a novel unsupervised framework for training onboard 3D detection and prediction models to understand open-set moving objects. Extensive experiments show that our unsupervised approach is competitive in regular detection tasks to the counterpart which uses supervised scene flow. With promising results, it demonstrates great potential in enabling perception and prediction systems to handle open-set moving objects. We hope our findings encourage more research toward solving autonomy in an open-set environment.

References

1. Agarwal, S., Snavely, N., Seitz, S.M., Szeliski, R.: Bundle adjustment in the large. In: Daniilidis, K., Maragos, P., Paragios, N. (eds.) ECCV 2010. LNCS, vol. 6312, pp. 29–42. Springer, Heidelberg (2010). https://doi.org/10.1007/978-3-642-15552-9_3
2. Bansal, M., Krizhevsky, A., Ogale, A.: ChauffeurNet: learning to drive by imitating the best and synthesizing the worst. arXiv preprint arXiv:1812.03079 (2018)
3. Bau, D., et al.: GAN dissection: visualizing and understanding generative adversarial networks. In: ICLR (2019)
4. Besl, P.J., McKay, N.D.: Method for registration of 3-D shapes. In: Schenker, P.S. (eds.) Sensor Fusion IV: Ccontrol Paradigms and data Structures. vol. 1611, pp. 586–606. SPIE, Bellingham Wash (1992)
5. Bewley, A., Sun, P., Mensink, T., Anguelov, D., Sminchisescu, C.: Range conditioned dilated convolutions for scale invariant 3D object detection (2020)
6. Caesar, H., et al.: Nuscenes: a multimodal dataset for autonomous driving. In: CVPR (2020)
7. Caine, B., et al.: Pseudo-labeling for scalable 3D object detection. arXiv preprint arXiv:2103.02093 (2021)
8. Casas, S., Luo, W., Urtasun, R.: IntentNet: learning to predict intention from raw sensor data. In: CoRL (2018)
9. Cen, J., Yun, P., Cai, J., Wang, M.Y., Liu, M.: Open-set 3D object detection. In: 3DV (2021)

10. Chai, Y., Sapp, B., Bansal, M., Anguelov, D.: Multipath: Multiple probabilistic anchor trajectory hypotheses for behavior prediction. In: CoRL (2019)
11. Chang, M.F., et al.: Argoverse: 3D tracking and forecasting with rich maps. In: CVPR (2019)
12. Chen, Y., Liu, S., Shen, X., Jia, J.: Fast point r-CNN. In: ICCV (2019)
13. Chen, Y., Medioni, G.: Object modelling by registration of multiple range images. Image Vis. Comput. **10**(3), 145–155 (1992)
14. Chen, Y., et al.: GeoSim: realistic video simulation via geometry-aware composition for self-driving. In: CVPR (2021)
15. Cho, M., Kwak, S., Schmid, C., Ponce, J.: Unsupervised object discovery and localization in the wild: Part-based matching with bottom-up region proposals. In: CVPR (2015)
16. Cui, H., et al.: Multimodal trajectory predictions for autonomous driving using deep convolutional networks. In: ICRA (2019)
17. Deng, B., Qi, C.R., Najibi, M., Funkhouser, T., Zhou, Y., Anguelov, D.: Revisiting 3D object detection from an egocentric perspective. In: NeurIPS (2021)
18. Dewan, A., Caselitz, T., Tipaldi, G.D., Burgard, W.: Motion-based detection and tracking in 3D lidar scans. In: ICRA (2016)
19. Djuric, N., et al.: Short-term motion prediction of traffic actors for autonomous driving using deep convolutional networks (2018)
20. Duggal, S., et al.: Mending neural implicit modeling for 3D vehicle reconstruction in the wild. In: WACV (2022)
21. Engelcke, M., Rao, D., Wang, D.Z., Tong, C.H., Posner, I.: Vote3deep: fast object detection in 3D point clouds using efficient convolutional neural networks. In: ICRA (2017)
22. Engelmann, F., Stückler, J., Leibe, B.: Joint object pose estimation and shape reconstruction in urban street scenes using 3D shape priors. In: Rosenhahn, B., Andres, B. (eds.) GCPR 2016. LNCS, vol. 9796, pp. 219–230. Springer, Cham (2016). https://doi.org/10.1007/978-3-319-45886-1_18
23. Engelmann, F., Stückler, J., Leibe, B.: SAMP: shape and motion priors for 4d vehicle reconstruction. In: WACV (2017)
24. Ester, M., et al.: A density-based algorithm for discovering clusters in large spatial databases with noise. In: KDD (1996)
25. Ettinger, S., et al.: Large scale interactive motion forecasting for autonomous driving: the Waymo open motion dataset. In: ICCV (2021)
26. Faktor, A., Irani, M.: "Clustering by composition"-unsupervised discovery of image categories. In: ECCV (2012)
27. Fan, L., Xiong, X., Wang, F., Wang, N., Zhang, Z.: RangeDet: in defense of range view for lidar-based 3D object detection. In: ICCV (2021)
28. Gao, J., et al.: Encoding HD maps and agent dynamics from vectorized representation. In: CVPR (2020)
29. Grauman, K., Darrell, T.: Unsupervised learning of categories from sets of partially matching image features. In: CVPR (2006)
30. Groß, J., Ošep, A., Leibe, B.: AlignNet-3D: fast point cloud registration of partially observed objects. In: 3DV (2019)
31. Gu, J., et al.: Weakly-supervised 3D shape completion in the wild. In: Vedaldi, A., Bischof, H., Brox, T., Frahm, J.-M. (eds.) ECCV 2020. LNCS, vol. 12350, pp. 283–299. Springer, Cham (2020). https://doi.org/10.1007/978-3-030-58558-7_17
32. Gu, J., Sun, C., Zhao, H.: DenseTNT: end-to-end trajectory prediction from dense goal sets. In: ICCV (2021)

33. Gu, X., Wang, Y., Wu, C., Lee, Y.J., Wang, P.: HplflowNet: hierarchical permutohedral lattice flowNet for scene flow estimation on large-scale point clouds. In: CVPR (2019)
34. He, C., Zeng, H., Huang, J., Hua, X.S., Zhang, L.: Structure aware single-stage 3D object detection from point cloud. In: CVPR, June 2020
35. Hong, J., Sapp, B., Philbin, J.: Rules of the road: Predicting driving behavior with a convolutional model of semantic interactions. In: CVPR (2019)
36. Houston, J., et al.: One thousand and one hours: Self-driving motion prediction dataset. arXiv preprint arXiv:2006.14480 (2020)
37. Insafutdinov, E., Dosovitskiy, A.: Unsupervised learning of shape and pose with differentiable point clouds. In: NeurIPS (2018)
38. Izadi, S., et al.: KinectFusion: real-time 3D reconstruction and interaction using a moving depth camera. In: Proceedings of the 24th Annual ACM Symposium on User Interface Software and Technology. pp. 559–568 (2011)
39. Jerripothula, K.R., Cai, J., Yuan, J.: CATS: co-saliency activated tracklet selection for video co-localization. In: Leibe, B., Matas, J., Sebe, N., Welling, M. (eds.) ECCV 2016. LNCS, vol. 9911, pp. 187–202. Springer, Cham (2016). https://doi.org/10.1007/978-3-319-46478-7_12
40. Joulin, A., Bach, F., Ponce, J.: Discriminative clustering for image co-segmentation. In: CVPR (2010)
41. Jund, P., Sweeney, C., Abdo, N., Chen, Z., Shlens, J.: Scalable scene flow from point clouds in the real world. IEEE Rob. Autom. Lett. **7**(2), 1589–1596 (2022). https://doi.org/10.1109/LRA.2021.3139542
42. Kim, G., Torralba, A.: Unsupervised detection of regions of interest using iterative link analysis. In: NIPS (2009)
43. Lang, A.H., Vora, S., Caesar, H., Zhou, L., Yang, J., Beijbom, O.: Pointpillars: fast encoders for object detection from point clouds. In: CVPR (2019)
44. Lee, H., Grosse, R., Ranganath, R., Ng, A.Y.: Convolutional deep belief networks for scalable unsupervised learning of hierarchical representations. In: ICML (2009)
45. Li, X., Pontes, J.K., Lucey, S.: Neural scene flow prior. In: NeurIPS (2021)
46. Li, Z., Wang, F., Wang, N.: Lidar r-CNN: An efficient and universal 3d object detector. In: CVPR (2021)
47. Liu, X., Qi, C.R., Guibas, L.J.: Flownet3d: learning scene flow in 3d point clouds. In: CVPR (2019)
48. Liu, Y., et al.: Opening up open-world tracking. In: CVPR (2022)
49. Liu, Y., Zhang, J., Fang, L., Jiang, Q., Zhou, B.: Multimodal motion prediction with stacked transformers. In: CVPR (2021)
50. Luo, C., Yang, X., Yuille, A.: Self-supervised pillar motion learning for autonomous driving. In: CVPR (2021)
51. Luo, W., Yang, B., Urtasun, R.: Fast and furious: Real time end-to-end 3d detection, tracking and motion forecasting with a single convolutional net. In: CVPR (2018)
52. Manivasagam, S., et al.: LiDARSim: realistic lidar simulation by leveraging the real world. In: CVPR (2020)
53. Meyer, G.P., Laddha, A., Kee, E., Vallespi-Gonzalez, C., Wellington, C.K.: LaserNet: an efficient probabilistic 3D object detector for autonomous driving. In: CVPR (2019)
54. Misra, I., Girdhar, R., Joulin, A.: An end-to-end transformer model for 3D object detection. In: ICCV (2021)
55. Mittal, H., Okorn, B., Held, D.: Just go with the flow: self-supervised scene flow estimation. In: CVPR (2020)

56. Najibi, M., et al.: DOPS: learning to detect 3D objects and predict their 3D shapes. In: CVPR (2020)
57. Pang, Z., Li, Z., Wang, N.: Model-free vehicle tracking and state estimation in point cloud sequences. In: IROS (2021)
58. Phan-Minh, T., Grigore, E.C., Boulton, F.A., Beijbom, O., Wolff, E.M.: CoverNet: Multimodal behavior prediction using trajectory sets. In: CVPR (2020)
59. Pontes, J.K., Hays, J., Lucey, S.: Scene flow from point clouds with or without learning. In: 2020 International Conference on 3D Vision (3DV). pp. 261–270 (2020). https://doi.org/10.1109/3DV50981.2020.00036
60. Puy, G., Boulch, A., Marlet, R.: FLOT: scene flow on point clouds guided by optimal transport. In: Vedaldi, A., Bischof, H., Brox, T., Frahm, J.-M. (eds.) ECCV 2020. LNCS, vol. 12373, pp. 527–544. Springer, Cham (2020). https://doi.org/10.1007/978-3-030-58604-1_32
61. Qi, C.R., Litany, O., He, K., Guibas, L.J.: Deep Hough voting for 3D object detection in point clouds. In: ICCV (2019)
62. Qi, C.R., et al.: Offboard 3D object detection from point cloud sequences. In: CVPR (2021)
63. Radford, A., Metz, L., Chintala, S.: Unsupervised representation learning with deep convolutional generative adversarial networks. In: ICLR (2015)
64. Rusinkiewicz, S., Levoy, M.: Efficient variants of the ICP algorithm. In: Proceedings Third International Conference on 3-D Digital Imaging and Modeling, pp. 145–152. IEEE (2001)
65. Russell, B.C., Freeman, W.T., Efros, A.A., Sivic, J., Zisserman, A.: Using multiple segmentations to discover objects and their extent in image collections. In: CVPR (2006)
66. Schönberger, J.L., Frahm, J.M.: Structure-from-motion revisited. In: CVPR (2016)
67. Shi, S., et al.: PV-RCNN: point-voxel feature set abstraction for 3D object detection. In: CVPR (2020)
68. Shi, S., Wang, X., Li, H.: PointRCNN : 3D object proposal generation and detection from point cloud. In: CVPR (2019)
69. Shi, W., Rajkumar, R.R.: Point-GNN: graph neural network for 3D object detection in a point cloud. In: CVPR (2020)
70. Simon, M., Milz, S., Amende, K., Gross, H.-M.: Complex-YOLO: an Euler-region-proposal for real-time 3D object detection on point clouds. In: Leal-Taixé, L., Roth, S. (eds.) ECCV 2018. LNCS, vol. 11129, pp. 197–209. Springer, Cham (2019). https://doi.org/10.1007/978-3-030-11009-3_11
71. Xia, S., Hancock, E.R.: Graph-based object class discovery. In: Jiang, X., Petkov, N. (eds.) CAIP 2009. LNCS, vol. 5702, pp. 385–393. Springer, Heidelberg (2009). https://doi.org/10.1007/978-3-642-03767-2_47
72. Sohn, K., Zhou, G., Lee, C., Lee, H.: Learning and selecting features jointly with point-wise gated boltzmann machines. In: ICML (2013)
73. Song, S., Xiao, J.: Deep sliding shapes for Amodal 3D object detection in RGB-D images images. In: CVPR (2016)
74. Stutz, D., Geiger, A.: Learning 3d shape completion from laser scan data with weak supervision. In: CVPR (2018)
75. Sun, P., et al.: Scalability in perception for autonomous driving: Waymo open dataset. In: CVPR (2020)
76. Sun, P., et al.: RSN: range sparse net for efficient, accurate lidar 3d object detection. In: CVPR, pp. 5725–5734 (2021)

77. Tang, C., Tan, P.: Ba-Net: dense bundle adjustment network. In: ICLR (2019)
78. Tian, H., Chen, Y., Dai, J., Zhang, Z., Zhu, X.: Unsupervised object detection with lidar clues. In: CVPR (2021)
79. Triggs, B., McLauchlan, P.F., Hartley, R.I., Fitzgibbon, A.W.: Bundle adjustment — a modern synthesis. In: Triggs, B., Zisserman, A., Szeliski, R. (eds.) IWVA 1999. LNCS, vol. 1883, pp. 298–372. Springer, Heidelberg (2000). https://doi.org/10.1007/3-540-44480-7_21
80. Tulsiani, S., Efros, A.A., Malik, J.: Multi-view consistency as supervisory signal for learning shape and pose prediction. In: CVPR (2018)
81. Tulsiani, S., Zhou, T., Efros, A.A., Malik, J.: Multi-view supervision for single-view reconstruction via differentiable ray consistency. In: CVPR (2017)
82. Ummenhofer, B., et al.: Demon: depth and motion network for learning monocular stereo. In: CVPR (2017)
83. Varadarajan, B., et al.: Multipath++: efficient information fusion and trajectory aggregation for behavior prediction. CoRR arXiv:2111.14973 (2021)
84. Vo, H.V., et al.: Unsupervised image matching and object discovery as optimization. In: CVPR (2019)
85. Vo, H.V., Pérez, P., Ponce, J.: Toward unsupervised, multi-object discovery in large-scale image collections. In: Vedaldi, A., Bischof, H., Brox, T., Frahm, J.-M. (eds.) ECCV 2020. LNCS, vol. 12368, pp. 779–795. Springer, Cham (2020). https://doi.org/10.1007/978-3-030-58592-1_46
86. Vo, V.H., Sizikova, E., Schmid, C., Pérez, P., Ponce, J.: Large-scale unsupervised object discovery. In: NeurIPS (2021)
87. Wang, D.Z., Posner, I.: Voting for voting in online point cloud object detection. In: Proceedings of Robotics: Science and Systems. Rome, Italy, July 2015
88. Wang, R., Yang, N., Stückler, J., Cremers, D.: Directshape: direct photometric alignment of shape priors for visual vehicle pose and shape estimation. In: ICRA (2020)
89. Wang, Y., et al.: Pillar-based object detection for autonomous driving. In: Vedaldi, A., Bischof, H., Brox, T., Frahm, J.-M. (eds.) ECCV 2020. LNCS, vol. 12367, pp. 18–34. Springer, Cham (2020). https://doi.org/10.1007/978-3-030-58542-6_2
90. Wang, Y., Solomon, J.M.: Deep closest point: Learning representations for point cloud registration. In: ICCV (2019)
91. Wang, Z., Li, S., Howard-Jenkins, H., Prisacariu, V., Chen, M.: Flownet3d++: Geometric losses for deep scene flow estimation. In: WACV (2020)
92. Wei, X., Zhang, Y., Li, Z., Fu, Y., Xue, X.: DeepSFM: structure from motion via deep bundle adjustment. In: Vedaldi, A., Bischof, H., Brox, T., Frahm, J.-M. (eds.) ECCV 2020. LNCS, vol. 12346, pp. 230–247. Springer, Cham (2020). https://doi.org/10.1007/978-3-030-58452-8_14
93. Weng, X., Kitani, K.: A baseline for 3D multi-object tracking. arXiv preprint arXiv:1907.03961 (2019)
94. Wong, K., Wang, S., Ren, M., Liang, M., Urtasun, R.: Identifying unknown instances for autonomous driving. In: CoRL. PMLR (2020)
95. Wu, P., Chen, S., Metaxas, D.N.: MotionNet: joint perception and motion prediction for autonomous driving based on bird's eye view maps. In: CVPR (2020)
96. Wu, W., Wang, Z.Y., Li, Z., Liu, W., Fuxin, L.: PointPWC-Net: cost volume on point clouds for (Self-)supervised scene flow estimation. In: Vedaldi, A., Bischof, H., Brox, T., Frahm, J.-M. (eds.) ECCV 2020. LNCS, vol. 12350, pp. 88–107. Springer, Cham (2020). https://doi.org/10.1007/978-3-030-58558-7_6
97. Yan, X., et al.: Learning 6-DOF grasping interaction via deep geometry-aware 3D representations. In: ICRA (2018)

98. Yan, X., Yang, J., Yumer, E., Guo, Y., Lee, H.: Perspective transformer nets: Learning single-view 3D object reconstruction without 3D supervision. In: NIPS (2016)
99. Yang, B., Bai, M., Liang, M., Zeng, W., Urtasun, R.: Auto4d: learning to label 4D objects from sequential point clouds. arXiv preprint arXiv:2101.06586 (2021)
100. Yang, B., Luo, W., Urtasun, R.: PIXOR: real-time 3D object detection from point clouds. In: CVPR (2018)
101. Yang, H., Shi, J., Carlone, L.: Teaser: fast and certifiable point cloud registration. IEEE Trans. Rob. **37**(2), 314–333 (2020)
102. Yang, Z., Sun, Y., Liu, S., Jia, J.: 3DSSD: point-based 3D single stage object detector. In: CVPR (2020)
103. Ye, M., Xu, S., Cao, T.: HvNet: hybrid voxel network for lidar based 3d object detection. In: CVPR (2020)
104. Ye, M., Cao, T., Chen, Q.: TPCN: temporal point cloud networks for motion forecasting. In: CVPR (2021)
105. Yuan, J., Liu, Z., Wu, Y.: Discriminative subvolume search for efficient action detection. In: CVPR (2009)
106. Yuan, Y., Weng, X., Ou, Y., Kitani, K.M.: AgentFormer: agent-aware transformers for socio-temporal multi-agent forecasting. In: ICCV (2021)
107. Zakharov, S., Kehl, W., Bhargava, A., Gaidon, A.: Autolabeling 3D objects with differentiable rendering of SDF shape priors. In: CVPR (2020)
108. Zeng, W., et al.: End-to-end interpretable neural motion planner. In: CVPR (2019)
109. Zheng, W., Tang, W., Jiang, L., Fu, C.W.: SE-SSD: self-ensembling single-stage object detector from point cloud. In: CVPR (2021)
110. Zhou, T., Brown, M., Snavely, N., Lowe, D.G.: Unsupervised learning of depth and ego-motion from video. In: CVPR (2017)
111. Zhou, Y., et al.: End-to-end multi-view fusion for 3D object detection in lidar point clouds. In: CoRL (2020)
112. Zhou, Y., Tuzel, O.: VoxelNet: end-to-end learning for point cloud based 3D object detection. In: CVPR (2018)
113. Zhu, R., Kiani Galoogahi, H., Wang, C., Lucey, S.: Rethinking reprojection: closing the loop for pose-aware shape reconstruction from a single image. In: ICCV (2017)

StretchBEV: Stretching Future Instance Prediction Spatially and Temporally

Adil Kaan Akan(✉) and Fatma Güney

KUIS AI Center, Koc University, Istanbul, Turkey
{kakan20,fguney}@ku.edu.tr
https://kuis-ai.github.io/stretchbev

Abstract. In self-driving, predicting future in terms of location and motion of all the agents around the vehicle is a crucial requirement for planning. Recently, a new joint formulation of perception and prediction has emerged by fusing rich sensory information perceived from multiple cameras into a compact bird's-eye view representation to perform prediction. However, the quality of future predictions degrades over time while extending to longer time horizons due to multiple plausible predictions. In this work, we address this inherent uncertainty in future predictions with a stochastic temporal model. Our model learns temporal dynamics in a latent space through stochastic residual updates at each time step. By sampling from a learned distribution at each time step, we obtain more diverse future predictions that are also more accurate compared to previous work, especially stretching both spatially further regions in the scene and temporally over longer time horizons. Despite separate processing of each time step, our model is still efficient through decoupling of the learning of dynamics and the generation of future predictions.

1 Introduction

Future prediction with sequential visual data has been studied from different perspectives. Stochastic video prediction methods operate in the pixel space and learn to predict future frames conditioned on the previous frames. These methods achieve impressive results by modelling the uncertainty of the future with stochasticity, however, long-term predictions in real-world sequences tend to be quite blurry due to the complexity of directly predicting pixels. A more practical approach is to consider a compact representation that is tightly connected to the modalities required for the downstream application. In self-driving, understanding the 3D properties of the scene and the motion of other agents plays a key role in future predictions. The bird's-eye view (BEV) representation meets these requirements by first fusing information from multiple cameras into a 3D point

Supplementary Information The online version contains supplementary material available at https://doi.org/10.1007/978-3-031-19839-7_26.

S. Avidan et al. (Eds.): ECCV 2022, LNCS 13698, pp. 444–460, 2022.
https://doi.org/10.1007/978-3-031-19839-7_26

cloud and then projecting the points to the ground plane [30]. This leads to a compact representation where the future location and motion of multiple agents in the scene can be reliably predicted. In this paper, we explore the potential of stochastic future prediction for self-driving to produce admissible and diverse results in long sequences with an efficient and compact BEV representation.

Future prediction from the BEV representation has been recently proposed in FIERY [21]. The BEV representations of past frames are first related in a temporal model and then used to learn two distributions representing the present and the future. Based on sampling from one of these distributions depending on train or test time, various future modalities are predicted with a recurrent model. For planning, long term multiple future predictions are crucial, however, the predictions of FIERY degrade over longer time spans due to the limited representation capability of a single distribution for increasing diversity in longer predictions. Following the official implementation, the predictions do not differ significantly based on random samples but converge to the mean, therefore lack diversity. We start from the same BEV representation and predict the same output modalities to be comparable. Differently, instead of two distributions for the present and the future, we propose to learn time-dependent distributions by predicting a residual change at each time step to better capture long-term dynamics. Furthermore, we show that by sampling a random variable at each time step, we can increase the diversity of future predictions while still being accurate and efficient. For efficiency, we use a state-space model [29] instead of costly auto-regressive models.

The main idea behind the efficiency of the state-space model is to decouple the learning of dynamics and the generation of predictions [14]. We first learn a low dimensional latent space from the BEV representation to capture the dynamics and then learn to decode some output modalities from the predictions in that latent space. These output modalities show the location and the motion of the agents in the scene. We can train our dynamics model independently by learning to match the BEV representations of future frames that are predicted by our model to the ones that are extracted from a pre-trained BEV segmentation model [30]. Through residual updates to the latent space, our model can capture the changes to the BEV representation over time. This way, the only information we use from the future is the BEV representation of future frames. Another option is to encode the future modalities to predict and provide the model with this encoded representation to learn a future distribution [21]. We experiment with both options in this paper. While providing labels in the future distribution leads to better performance, learning the dynamics becomes dependent on the labels. From the BEV predictions, we train a decoder to predict the location and the motion of future instances in a supervised manner. These output modalities increase the interpretability of the predicted BEV representations that can be used for learning a driving policy in future work.

We present a stochastic future instance prediction method in BEV from multiple cameras. We formulate the prediction task as learning temporal dynamics through stochastic residual updates to a latent representation of the state at each time step. Our model can generate diverse predictions which are interpretable through supervised decoding of the predictions. Our proposed approach clearly outperforms the state of the art in various metrics evaluating the decoded predictions, especially by large margins in challenging cases of spatially far regions and temporally long spans. We also show increased diversity in the predictions.

2 Related Work

2.1 Stochastic Future Prediction

Stochastic future prediction has been mostly explored in the context of next frame prediction in videos. In stochastic video generation, the goal is to predict future frames conditioned on a few initial frames. Typically, the main focus is the diversity of future predictions with a large number of samples for future and the number of frames to be predicted is at least twice as many as the initial conditioning frames. Our work fits into the stochastic future prediction framework, producing long term, diverse predictions, however, we predict future in the BEV space instead of the noisy pixel space.

Most of the stochastic video prediction methods [3,12] use a recurrent neural network in an auto-regressive manner by feeding the generated predictions back to the model to predict future. The performance of auto-regressive methods can be improved by increasing the network capacity [35] or introducing a hierarchy into the latent variables [7], which also increase the complexity of these methods. Due to complexity of predicting pixels, another group of work moves away from the pixel space to the keypoints [28] or to the motion space by incorporating motion history [1]. Our proposed approach follows a similar strategy by performing future prediction in the BEV representation, but more efficiently by avoiding auto-regressive predictions.

Auto-regressive strategy requires encoding the predictions, leading to high computational cost and creates a tight coupling between the temporal dynamics and the generation process [16,31]. The state-space models (SSM) break this coupling by separating the learning of dynamics from the generation process, resulting in a computationally more efficient approach. Low-dimensional states still depend on previous states but not on predictions. Furthermore, learned states can be directly used in reinforcement learning [16] and can be interpreted [31], making SSMs particularly appealing for self-driving. Earlier SSMs with variational deep inference suffer from complicated inference schemes and typically target low-dimensional data [23,25]. An efficient training strategy with a temporal model based on residual updates is proposed for high-dimensional video prediction in the state of the art SSM [14]. We adapt a similar residual update strategy for predicting future BEV representations. We also experimentally show that the content variable for the static part of the scene is not as effective in the BEV space as it is in pixel space [14].

2.2 Future Prediction in Driving

The typical approach to the prediction problem in self-driving is to first perform detection and tracking, and then do the trajectory prediction [8,19]. In these methods, errors are propagated at each step. There are some recent methods [5, 6,13,27] which directly address the prediction problem from the sensory input including LiDAR, radar, and other sensors. These methods also typically rely on an HD map of the environment. Due to high performance and efficiency of end-to-end approach, we follow a similar approach for future prediction but using cameras only and without relying on HD maps.

Despite their efficiency, most of the previous work focus on the most likely prediction [6] or only models the uncertainty regarding the ego-vehicle's trajectory [5,13]. The motion prediction methods which consider the behavior of all the agents in the scene typically assume a top-down rasterised representation as input, e.g.Argoverse setting [9]. Even then, multiple future prediction problem is typically addressed by generating a fixed number of predictions [2,15,26], for example by estimating the likelihood of multiple target locations [17,36]. There are some exceptions [22,32,34] which directly address the diversity aspect with a probabilistic framework. These works, especially the ones using a deep variational framework [32,34] are similar to our approach in spirit, however, they operate in the coordinate space by assuming the availability of a top-down map where locations of agents are marked. We aim to learn this top-down BEV representation from multiple cameras by also segmenting the agents in the scene.

FIERY [21] is the first to address probabilistic future prediction from multiple cameras. However, future predictions are limited both in terms of diversity and length considering the typical video prediction setting. We propose a probabilistic future prediction method that can generate diverse predictions extending to different temporal horizons with a stochastic temporal dynamics model.

3 Methodology

3.1 A Compact Representation for Future Prediction

Modern self-driving vehicles are typically equipped with multiple cameras observing the scene from multiple viewpoints. Placing cameras on the vehicle is cheap but processing information even from a single camera can be quite expensive. The traditional approach in computer vision is to extract low-level and semantic cues from these cameras and then fuse them into a holistic scene representation to perform prediction and planning. Recent success of end-to-end methods in driving has led to a rethinking of this approach. Furthermore, building and maintaining HD maps require a significant effort which is expensive and hard to scale. A better approach is to learn a geometrically consistent scene representation which can also mark the location, the motion, and even the semantics of the dynamic objects in the scene.

The bird's-eye view (BEV) representation initially proposed in [30], takes image $\mathbf{x}_t^i$ at time t from each camera $i \in \{1, \cdots, 6\}$ and fuses them into a compact BEV representation $\mathbf{s}_t$. This is achieved by encoding each image and also

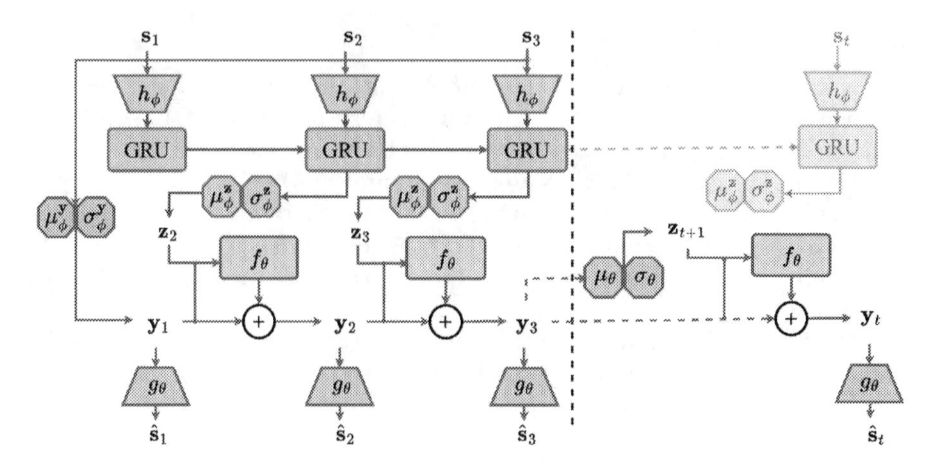

Fig. 1. Architecture for learning temporal dynamics. This figure shows the inference procedure of our model StretchBEV. We start with the first $k = 3$ conditioning frames where we sample the stochastic latent variables from the posterior distribution (purple). On the right, we show the prediction at a step t after the conditioning frames where we sample from the learned future distribution (red). The dashed vertical line marks the conditioning frames.

by predicting a distribution over the possible depth values. The BEV features are obtained by weighting the encoded image features according to depth probabilities predicted. These features are first lifted to 3D by using known camera intrinsic and extrinsic parameters and then the height dimension is pooled over to project the features into the bird's-eye view. This results in the state representation $\mathbf{s}_t$ that we use for future prediction as explained next.

3.2 Learning Temporal Dynamics in BEV

Notation. In our formulation, $\mathbf{s}_{1:T}$ denotes a sequence of BEV feature maps representing the state of the vehicle and its environment for T time steps. In stochastic future prediction, the goal is to predict possible future states $\hat{\mathbf{s}}_{k+1:T}$ conditioned on the state in the first k time steps. Precisely, we condition on the first $k = 3$ steps and predict future in varying lengths from 4 to 12 steps ahead.

Differently from previous work on stochastic video prediction [12,14], the BEV state $\mathbf{s}_t$ that is input to our stochastic prediction framework is the intermediate representations in a high dimensional space rather than a video frame in the pixel space as explained in Sect. 3.1. Similarly, the predicted output $\hat{\mathbf{s}}_t$ represents the predictions of future in the same high dimensional space. We decode these high dimensional future predictions into various output modalities $\hat{\mathbf{o}}_t$ such as future instance segmentation and motion as explained in Sect. 3.4. While these modalities need to be trained in a supervised manner in contrast to typical self-supervised stochastic video prediction frameworks [12,14], they provide interpretability which is critical in self-driving. Furthermore, using these modalities in the posterior in addition to the future state representations improves the results significantly as we show in our experiments.

Stochastic Residual Dynamics. Following [14], we learn the changes in the states through time with stochastic residual updates to a sequence of latent variables. For each state $\mathbf{s}_t$, there is a corresponding latent variable $\mathbf{y}_t$ generating it, independent of the previous states (Fig. 1). Each $\mathbf{y}_{t+1}$ only depends on the previous $\mathbf{y}_t$ and a stochastic variable $\mathbf{z}_{t+1}$. The randomness is introduced by the stochastic latent variable $\mathbf{z}_{t+1}$ which is sampled from a normal distribution learned from the previous state's latent variable only:

$$\mathbf{z}_{t+1} \sim \mathcal{N}\left(\mu_\theta(\mathbf{y}_t), \sigma_\theta(\mathbf{y}_t)\ \mathbf{I}\right) \tag{1}$$

Given $\mathbf{z}_{t+1}$, the dependency between the latent variables $\mathbf{y}_t$ and $\mathbf{y}_{t+1}$ is deterministic through the residual update:

$$\mathbf{y}_{t+1} = \mathbf{y}_t + f_\theta\left(\mathbf{y}_t, \mathbf{z}_{t+1}\right) \tag{2}$$

where f_θ is a small CNN to learn the residual updates to $\mathbf{y}_t$. We learn the distribution of future states from the corresponding latent variable as a normal distribution with constant diagonal variance: $\hat{\mathbf{s}}_t \sim \mathcal{N}(g_\theta(\mathbf{y}_t))$. The first latent variable is inferred from the conditioning frames by assuming a standard Gaussian prior: $\mathbf{y}_1 \sim \mathcal{N}\left(\mathbf{0}, \mathbf{I}\right)$.

On the Content Variable. In video prediction, a common practice is to represent the static parts of the scene with a content variable which allows the model to focus on the moving parts. On the contrary to the state of the art in video prediction [14], the content variable does not improve the results in our case (see Supplementary), therefore we omit it in our formulation here. This can be attributed to the details in the background that are confusing for learning dynamics while operating in the pixel space, but in our case, most of these details are already suppressed in the BEV representation.

On Diversity. In contrast to a present and a future distribution in FIERY [20, 21], there is a distribution learned at each time step in our model. This corresponds to sampling stochastic random variables at each time step as opposed to sampling once to represent all the future frames. This is the key property which allows our model to produce diverse predictions in long sequences. By sampling from a learned distribution at each time step, our model learns to represent the complex dynamics of future frames, even for predictions further away from the conditioning frames. Furthermore, our model does not need a separate temporal block for learning the dynamics prior to learning these distributions. The dynamics are learned through the temporal evolution of latent variables by also considering the randomness of the future predictions with stochastic random variables at each time step. This not only increases the diversity of predictions but also alleviates the need for a separate temporal block, e.g.with 3D convolutions. Note that our formulation is still efficient, almost the same inference time as FIERY (see Supplementary), because the latent variables are low dimensional and each state is generated independently.

Moreover, FIERY uses only a single vector of latent variables which is expanded to the spatial grid to generate futures states probabilistically. Therefore, it uses the same stochastic noise in all the coordinates of the grid. However, in our model, we have a separate random variable at each coordinate of the grid to model the uncertainty spatially as well. Training FIERY with a separate random variable at each location of the grid results in diverging loss values.

3.3 Variational Inference and Architecture

Following the generative process in [14], the joint probability of the BEV states $\mathbf{s}_{1:T}$, the output modalities $\mathbf{o}_t$, and the latent variables $\mathbf{z}_{1:T}$ and $\mathbf{y}_{1:T}$ is as follows:

$$p(\mathbf{s}_{1:T}, \mathbf{o}_{1:T}, \mathbf{z}_{2:T}, \mathbf{y}_{1:T}) = p(\mathbf{y}_1) \prod_{t=2}^{T} p(\mathbf{z}_t, \mathbf{y}_t | \mathbf{y}_{t-1}) \prod_{t=1}^{T} p(\mathbf{o}_t | \mathbf{s}_t)\, p(\mathbf{s}_t | \mathbf{y}_t) \quad (3)$$

$$p(\mathbf{z}_t, \mathbf{y}_t | \mathbf{y}_{t-1}) = p(\mathbf{y}_t | \mathbf{y}_{t-1}, \mathbf{z}_t)\, p(\mathbf{z}_t | \mathbf{y}_{t-1}) \quad (4)$$

The relationship between $\mathbf{y}_t$ and $\mathbf{y}_{t-1}$ in $p(\mathbf{y}_t | \mathbf{y}_{t-1}, \mathbf{z}_t)$ (4) is deterministic through the stochastic latent residual as formulated in (2). Similarly for the term $p(\mathbf{o}_t | \mathbf{s}_t)$ in (3), the output modalities $\mathbf{o}_t$ is learned from $\mathbf{s}_t$ with a deterministic decoder in a supervised manner (Sect. 3.4).

Our goal is to maximize the likelihood of the BEV states extracted from the frames (Sect. 3.1) and the corresponding output modalities $p(\mathbf{s}_{1:T}, \mathbf{o}_{1:T})$. For that purpose, we learn a deep variational inference model q parametrized by ϕ which is factorized as follows:

$$\begin{aligned} q_{Z,Y} &\triangleq q(\mathbf{z}_{2:T}, \mathbf{y}_{1:T} | \mathbf{s}_{1:T}, \mathbf{o}_{2:T}) \\ &= q(\mathbf{y}_1 | \mathbf{s}_{1:k}) \prod_{t=2}^{T} q(\mathbf{z}_t | \mathbf{s}_{1:t}, \mathbf{o}_{2:t})\, q(\mathbf{y}_t | \mathbf{y}_{t-1}, \mathbf{z}_t) \end{aligned} \quad (5)$$

where $k = 3$ is the number of conditioning frames and $q(\mathbf{y}_t | \mathbf{y}_{t-1}, \mathbf{z}_t)$ is equal to $p(\mathbf{y}_t | \mathbf{y}_{t-1}, \mathbf{z}_t)$ with the residual update as explained above. We obtain two versions of our model by keeping (StretchBEV-P) or removing (StretchBEV) the dependency of $\mathbf{z}_t$ on $\mathbf{o}_{1:t}$ in the posterior in (5). We refer the reader to Supplementary for the derivation of the following evidence lower bound (ELBO):

$$\begin{aligned} \log p(\mathbf{s}_{1:T}, \mathbf{o}_{1:T}) &\geq \mathcal{L}(\mathbf{s}_{1:T}, \mathbf{o}_{1:T}; \theta, \phi) \\ &\triangleq - D_{\mathrm{KL}}(q(\mathbf{y}_1 | \mathbf{s}_{1:k}) \,||\, p(\mathbf{y}_1)) \\ &\quad + \mathbb{E}_{(\tilde{\mathbf{z}}_{2:T}, \tilde{\mathbf{y}}_{1:T}) \sim q_{Z,Y}} \Bigg[\sum_{t=1}^{T} \log p(\mathbf{s}_t | \tilde{\mathbf{y}}_t)\, p(\mathbf{o}_t | \mathbf{s}_t) \\ &\quad - \sum_{t=2}^{T} D_{\mathrm{KL}}\big(q(\mathbf{z}_t | \mathbf{s}_{1:t}, \mathbf{o}_{1:t}) \,||\, p(\mathbf{z}_t | \tilde{\mathbf{y}}_{t-1})\big) \Bigg] \end{aligned} \quad (6)$$

where D_{KL} denotes the Kullback-Leibler (KL) divergence, θ and ϕ represent model and variational parameters, respectively. Following the common practice, we choose $q(\mathbf{y}_1|\mathbf{s}_{1:k})$ and $q(\mathbf{z}_t|\mathbf{s}_{1:t}, \mathbf{o}_{1:t})$ to be factorized Gaussian for analytically computing the KL divergences and use the re-parametrization trick [24] to compute gradients through the inferred variables. Then, the resulting objective function is maximizing the ELBO as defined in (6) and minimizing the supervised losses for the output modalities (Sect. 3.4).

We provide a summary of the steps in our temporal model as shown in Fig. 1. We start by fusing the images $\mathbf{x}_t^i$ at time t from each camera i into the BEV state $\mathbf{s}_t$ at each time step as explained in Sect. 3.1.

1. The resulting BEV states are still in high resolution, $\mathbf{s}_t \in \mathbb{R}^{C \times H \times W}$ where $(H, W) = (200, 200)$. Therefore, we first process them with an encoder h_ϕ to reduce the spatial resolution to 50×50.
2. The first latent variable $\mathbf{y}_1$ is inferred using a convolutional neural network on the first three encoded states.
3. The stochastic latent variable $\mathbf{z}_t$ is inferred at each time step from the respective encoded state, using a recurrent neural network which is a combination of a ConvGRU and convolutional blocks.
4. The residual change in the dynamics is predicted with f_θ based on both the previous state dynamics $\mathbf{y}_t$ and the stochastic latent variable $\mathbf{z}_{t+1}$ and added to $\mathbf{y}_t$ to obtain $\mathbf{y}_{t+1}$.
5. From each $\mathbf{y}_t$, the state $\hat{\mathbf{s}}_t$ is predicted in the original resolution with g_θ.
6. Finally, the output modalities $\hat{\mathbf{o}}_t$ are decoded from the state prediction $\hat{\mathbf{s}}_t$.

3.4 Decoding Future Predictions

Based on the predictions of the future states at each time step, we train a supervised decoder to output semantic segmentation, instance center, offsets, and future optical flow, similar to the previous work [11,21]. The decoding function is a deterministic neural network that can be trained either jointly with the dynamics or independently, e.g.later for interpretability. The output modalities show both the location and the motion of instances at each time step. The motion predicted as future flow is used to track instances. We use the same supervised loss functions for each modality as the FIERY [21]. Although the decoding of future predictions is not necessary for planning and control, for example when training a driving agent to act on predictions, these predictions provide interpretability and allow to compare the methods in terms of various metrics evaluating each modality.

4 Experiments

4.1 Dataset and Evaluation Setting

We evaluate the performance of the proposed approach and compare to the state of the art method, FIERY [21], on the nuScenes dataset [4]. On the nuScenes, there are 6 cameras with overlapping views which provide the ego vehicle with a complete view of its surroundings. The nuScenes dataset consists of 1000 scenes with 20 s long at 2 frames per second.

We first follow the training and the evaluation setting proposed in FIERY [21] for comparison by using 1.0 s of past context to predict 2.0 s of future context. Given the sampling rate of 2 frames per second, this setting corresponds to predicting 4 future frames conditioned on 3 past frames. We call this setting *short* in terms of temporal length and define two more settings for longer temporal predictions. All the models are trained to predict 2.0 s into the future and only the evaluation is changed to predict longer time steps. In the *mid* and *long* settings, we double and triple the number of future frames to predict, i.e.8 and 12, respectively, that corresponds to 4.0 and 6.0 s into the future. These settings are closer to the stochastic video prediction setup [1,12,14] where there are typically many more frames to predict than the conditioning frames for measuring diversity and the performance of the models further away from the conditioning frames. Note that *short* and *long* refer to temporal length in our evaluations as opposed spatial coverage as defined in the previous work [21]. We also evaluate in terms of spatial coverage but call it *near* ($30\text{m} \times 30\text{m}$) and *far* ($100\text{m} \times 100\text{m}$) for clarity.

4.2 Training Details

Our models follow the input and output setting proposed in the previous work [21]. We process 6 camera images at a resolution of 224×480 pixels for each frame and construct the BEV state of size $200 \times 200 \times 64$. We further process the states into a smaller spatial resolution (50×50) for efficiency before learning the temporal dynamics but increase it back to the initial resolution afterwards. Given the predicted states, we use the same decoder architecture as the FIERY to decode the object centers, the segmentation masks, the instance offsets, and the future optical flow at a resolution of 200×200 pixels. We provide the details of the architecture in Supplementary and we will release the code upon publication.

In our approach, learning temporal dynamics and decoding output modalities are separated from each other. Therefore, we can pre-train the temporal dynamics part without using the labels for the output modalities. In pre-training, our objective is to learn to match the future states that are extracted using a pre-trained BEV segmentation model [30], conditioned on the past states. This approach is more similar to self-supervised stochastic video prediction methods [12,14]. Furthermore, this way, learning of temporal dynamics can be improved by using camera sequences only as input which can be easily collected in large quantities. Then, we fine-tune the temporal dynamics with a smaller learning rate (see Supplementary for details) while learning to decode the output modalities in a supervised manner. The alternative is to jointly train the temporal dynamics and supervised decoding without pre-training. We present the results of our model StretchBEV with and without pre-training.

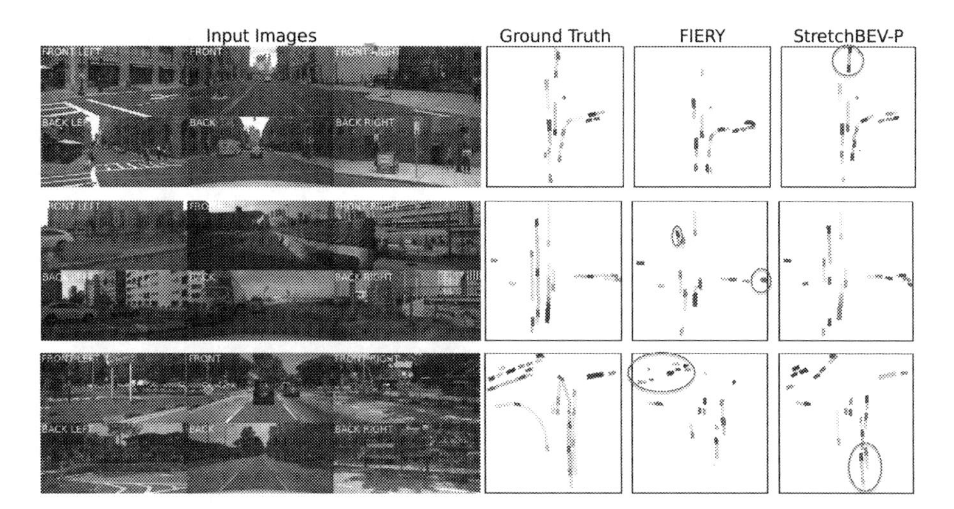

Fig. 2. Qualitative comparison over different temporal horizons. In this figure, we qualitatively compare the results of our model StretchBEV-P **(right)** to the ground truth **(left)** and FIERY [21] **(middle)** over short **(top)**, mid **(middle)**, and long **(bottom)** temporal horizons. Each color represents an instance of a vehicle with its trajectory trailing in the same color transparently.

StretchBEV does not use the labels ($\mathbf{o}_t$) for learning the temporal dynamics, it only uses them in the supervised loss to decode the output modalities. In our full model StretchBEV-P, we encode output modalities following FIERY and use them in the posterior for learning the temporal dynamics. During training, we sample the stochastic latent variables from the posterior and learn to minimize the difference between the posterior and the future distribution. During inference, we sample from the posterior in the conditioning frames and sample from the learned future distribution in the following steps as shown in Fig. 1.

4.3 Metrics

We use two different metrics for evaluating the decoded modalities, one frame level and another video level, that are also used in the previous work [21]. The first is Intersection over Union (IoU) to measure the quality of the segmentation at each frame. The second is Video Panoptic Quality (VPQ) to measure the quality of the segmentation and consistency of the instances through the video.

We evaluate the diversity quantitatively in terms of Generalized Energy Distance (D_{GED}) [33] by using ($1 - \text{VPQ}$) as the distance as proposed in FIERY [21].

Table 1. Ablation Study. In this table, we present the results for the two versions of our model with (StretchBEV-P) and without (StretchBEV) using the labels for the output modalities in the posterior while learning the temporal dynamics and also show the effect of pre-training for the latter in comparison to FIERY [21] and our reproduced version of their results (Reproduced).

	Pre-training	Posterior w/labels	IoU (↑)		VPQ (↑)	
			Near	Far	Near	Far
StretchBEV	—	—	53.3	35.8	41.7	26.0
	✓		55.5	37.1	46.0	29.0
FIERY [21]	—	✓	**59.4**	36.7	50.2	29.4
Reproduced			58.8	35.8	50.5	29.0
StretchBEV-P	—	✓	58.1	**52.5**	**53.0**	**47.5**

4.4 Ablation Study

In Table 1, we evaluate the effect of different versions of our model using IoU and VPQ metrics in the short temporal setting to be comparable to the previous work FIERY [21]. We reproduced their results as shown in the row *Reproduced.* In the first part of the table (StretchBEV), we show the results without explicitly using the labels for future prediction. In that case, labels are only used for decoding the output modalities and back-propagated to future prediction through decoding. Although this introduces a two-stage training, we believe that reporting results using this separation is important for future work to focus on future prediction with more unlabelled data. We measure the effect of pre-training by learning to match our future predictions to the results of a pre-trained model [30] in terms of the BEV state representation. Pre-training allows our model to learn the dynamics before decoding and improves the results significantly in each metric.

In the second half of the Table 1, we report the results using the labels in future prediction by explicitly feeding their encoding to the posterior distribution with the same encoding used in [21] to learn the future distribution. The difference between StretchBEV and StretchBEV-P is that the first has access to the BEV encoding of future predictions while the latter has access to both the BEV encoding and the encoding of the output modalities to predict in the posterior distribution. As can be seen from the results, both FIERY and our model using the labels in the future distribution perform better. This shows the importance of using a more direct and accurate information about future while learning the posterior. Compared to FIERY, our model can use the labels in the conditioning frames during inference and improves the results, especially in spatially far regions and in terms of VPQ, which point to a higher quality in our predictions stretching spatially and temporally over the video. Please see Supplementary for an extended version of Table 1 with multiple samples including standard deviation as an indication of uncertainty.

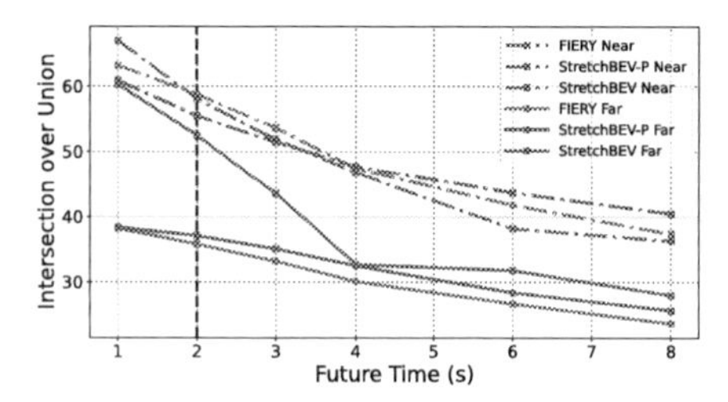

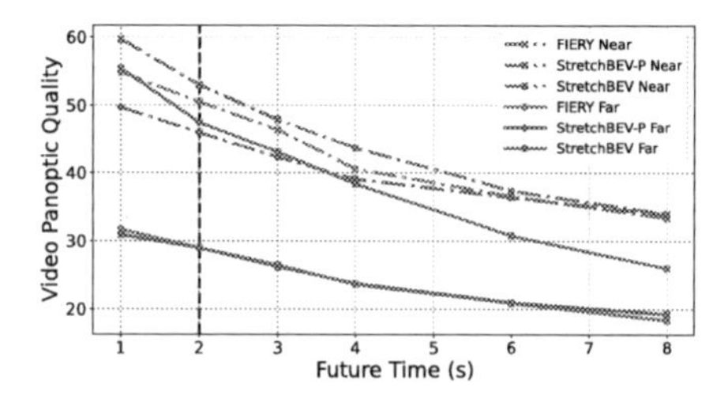

Fig. 3. Evaluation over different temporal horizons. We plot the performance of our models StretchBEV and StretchBEV-P in comparison to FIERY [21] over a range of temporal horizons from 1 s to 8 s in terms of IoU **(left)** and VPQ **(right)** for spatially far **(solid)** and near **(dashed)** regions separately. The vertical dashed line marks the training horizon.

4.5 Temporally Long Predictions

In longer temporal horizons, future prediction becomes increasingly difficult. This is mainly due to increasing uncertainty of future further away from conditioning frames. In Fig. 3, we present the results over different temporal horizons for our model with pre-training without using the labels in the posterior (StretchBEV), FIERY [21], and our model by using the labels in the posterior (StretchBEV-P). There is a separate plot for IoU on the left and for VPQ on the right with respect to the future time steps predicted, ranging from 1 s to 8 s. The vertical line in 2 s marks the training horizon. In Supplementary, we provide a table for the results over short, mid, and long temporal spans.

The negative effect of uncertain futures on each metric can be observed from the results of all the methods degrading from shorter to longer temporal spans. Our models perform better than FIERY in longer temporal spans. This is due to better handling of uncertainty with stochastic latent residual variables. Our method StretchBEV-P outperforms FIERY by significant margins, especially in terms of far VPQ in longer temporal horizons, showing consistent predictions in the overall scene throughout the video. This can be attributed to the difficulty of locating small vehicles in spatially far regions. Since StretchBEV-P has access to the labels via the posterior in the conditioning frames, it learns the temporal dynamics to correctly propagate them to the future frames, while StretchBEV and FIERY struggle to locate the instances in the first place. FIERY learns a single distribution for present and future each, therefore we cannot utilize the labels in the conditioning frames with FIERY. The results of StretchBEV outperforming the other two methods in terms of near IOU in longer temporal spans is promising for future prediction methods with less supervision.

In Fig. 2, we qualitatively compare the performance of our model StretchBEV-P on the right to FIERY in the middle over short, mid, and long temporal horizons in each row. In the first row, our model predicts the future trajectories that are more similar to the ground truth shown on the left. For example, FIERY fails to predict the trajectory of the vehicle in front (marked with red circle). In the second row, our model correctly segments the vehicles,

Table 2. Comparison of semantic segmentation prediction. In this table, we compare the predictions of our models, StretchBEV and StretchBEV-P for semantic segmentation to other BEV prediction methods in terms of IoU using the setting proposed in [18], i.e.32.0 m × 19.2 m at 10 cm resolution over 2s future.

Fishing-Cam [18]	Fishing-LiDAR [18]	FIERY [21]	StretchBEV	StretchBEV-P
30.0	44.3	57.3	<u>58.8</u>	**65.7**

whereas FIERY misses several vehicles far on the right and also, predicts a vehicle that does not exist (in purple on the top left). In the third row, our model predicts the future trajectories of the vehicles correctly while FIERY misses some of the vehicles (marked with red circles). The challenging case of a vehicle turning on the left (green in ground truth) is missed by both models. Some of the vehicles are not visible on the input images, e.g.the back camera in the long temporal horizon. We provide a gif version of these results and more examples in Supplementary.

4.6 Segmentation

The previous work on bird's-eye view segmentation typically focuses on single image segmentation task with a couple of exceptions focusing on prediction. In Table 2, we compare to two BEV segmentation prediction methods [18,21] using their setting with 32.0 m × 19.2 m at 10 cm resolution. Both methods predict 2 s into the future which corresponds to our short temporal setting. FIERY [21] outperforms the previous method [18] even when using LiDAR, and our method significantly outperforms both methods.

4.7 Diversity

We quantitatively evaluate diversity by computing D_{GED} over 10 samples and show the results in Table 3 for our model StretchBEV-P and FIERY (our reproduced version). Our model outperforms FIERY with lower distance scores, demonstrating higher levels of diversity in the samples quantitatively. The difference is especially apparent in spatially far regions. For qualitative comparison, in Fig. 4, we visualize three samples from FIERY (left) and our model (right) over short, mid, and long temporal spans from top to bottom. While FIERY generates almost the same predictions in all three samples, our model can generate diverse predictions of future (marked with red). In the first row, our model can predict the turning behavior of the green vehicle at different speeds. In the second row, our model learns to adjust the speed of nearby vehicles proportionally, as in the case of purple, blue, and gray vehicles in the middle. Similarly, in the third row, our model can generate different predictions for the moving vehicles in the middle. Please see Supplementary for the gif results with more examples.

Table 3. Quantitative evaluation of diversity. This table compares the results of our models to the reproduced results of FIERY [21] in terms of Generalized Energy Distance based on VPQ (lower better) for evaluating diversity.

	Generalized Energy Distance (↓)					
	Short		Mid		Long	
	Near	Far	Near	Far	Near	Far
FIERY [21]	106.09	140.36	118.74	147.26	127.18	152.38
StretchBEV	<u>103.97</u>	<u>132.38</u>	<u>114.11</u>	<u>138.15</u>	<u>119.01</u>	<u>142.51</u>
StretchBEV-P	**82.04**	**85.51**	**94.02**	**98.45**	**101.90**	**109.12**

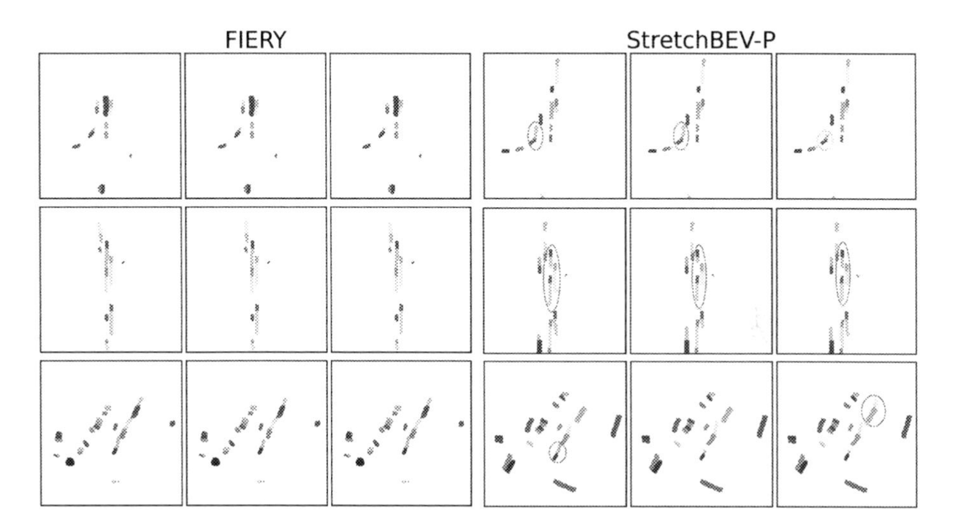

Fig. 4. Qualitative comparison of diversity. In this figure, we visualize random samples from FIERY [21] **(left)** and our model StretchBEV-P **(right)** over short **(top)**, mid **(middle)**, and long **(bottom)** temporal horizons.

5 Conclusion and Future Work

We introduced StretchBEV, a stochastic future instance prediction method that improve over the state of the art, especially in challenging cases, with more diverse predictions. We proposed two versions of our method with and without the labels for output modalities explicitly in the posterior while learning the dynamics. Both models improve the state of the art in spatially far regions and over temporally long horizons. Using labels in the posterior significantly improves the results in almost all metrics but introduces a dependency on the availability of labels in the conditioning frames during inference. Future work on learning dynamics should focus on closing the gap between the two approaches, for example with scheduled sampling.

Our temporal dynamics model can be interpreted as a Neural-ODE [10] because of its residual update dynamics. In our model, we use only one update in between time steps but in future, we plan to explore increasing the number of updates in between time step as done in the previous work [14]. We showed that our model increases the diversity of predictions due to improved modeling of stochasticity with sampling at every time step. In future, we plan to explore driving policies that can utilize stochastic future predictions. Learned latent states at each time step can be directly fed into a policy learning algorithm, e.g.as states in deep reinforcement learning. Furthermore, these states can be interpreted via supervised decoding into various future modalities that we predict.

Acknowledgments. K. Akan was supported by KUIS AI Center fellowship, F. Güney by TUBITAK 2232 International Fellowship for Outstanding Researchers.

References

1. Akan, A.K., Erdem, E., Erdem, A., Güney, F.: SLAMP: stochastic latent appearance and motion prediction. In: Proceedings of the IEEE International Conference on Computer Vision (ICCV) (2021)
2. Aydemir, G., Akan, A.K., Güney, F.: Trajectory forecasting on temporal graphs. arXiv preprint arXiv:2207.00255 (2022)
3. Babaeizadeh, M., Finn, C., Erhan, D., Campbell, R.H., Levine, S.: Stochastic variational video prediction. In: Proceedings of the International Conference on Learning Representations (ICLR) (2018)
4. Caesar, H., et al.: Uuscenes: a multimodal dataset for autonomous driving. In: Proceedings of IEEE Conference on Computer Vision and Pattern Recognition (CVPR) (2020)
5. Casas, S., Gulino, C., Liao, R., Urtasun, R.: SpaGNN: spatially-aware graph neural networks for relational behavior forecasting from sensor data. In: Proceedings of IEEE International Conference on Robotics and Automation (ICRA) (2020)
6. Casas, S., Luo, W., Urtasun, R.: IntentNet: learning to predict intention from raw sensor data. In: Proceedings Conference on Robot Learning (CoRL) (2018)
7. Castrejon, L., Ballas, N., Courville, A.: Improved conditional VRNNs for video prediction. In: Proceedings of the IEEE International Conference on Computer Vision (ICCV) (2019)
8. Chai, Y., Sapp, B., Bansal, M., Anguelov, D.: Multipath: Multiple probabilistic anchor trajectory hypotheses for behavior prediction. In: Proceedings Conference on Robot Learning (CoRL) (2020)
9. Chang, M.F., et al.: Argoverse: 3D tracking and forecasting with rich maps. In: Proceedings of IEEE Conference on Computer Vision and Pattern Recognition (CVPR) (2019)
10. Chen, R.T., Rubanova, Y., Bettencourt, J., Duvenaud, D.K.: Neural ordinary differential equations. In: Advances in Neural Information Processing Systems (NeurIPS) (2018)
11. Cheng, B., et al.: Panoptic-DeepLab: a simple, strong, and fast baseline for bottom-up panoptic segmentation. In: CVPR (2020)
12. Denton, E., Fergus, R.: Stochastic video generation with a learned prior. In: Proceedings of the International Conference on Machine learning (ICML) (2018)

13. Djuric, N., et al.: MultixNet: multiclass multistage multimodal motion prediction. In: Proceedings of IEEE Intelligent Vehicles Symposium (IV) (2021)
14. Franceschi, J.Y., Delasalles, E., Chen, M., Lamprier, S., Gallinari, P.: Stochastic latent residual video prediction. In: Proceedings of the International Conference on Machine Learning (ICML) (2020)
15. Gao, J., et al.: VectorNet: encoding HD maps and agent dynamics from vectorized representation. In: Proceedings IEEE Conference on Computer Vision and Pattern Recognition (CVPR) (2020)
16. Gregor, K., Besse, F.: Temporal difference variational auto-encoder. In: Proceedings of the International Conference on Learning Representations (ICLR) (2019)
17. Gu, J., Sun, C., Zhao, H.: DenseTNT: end-to-end trajectory prediction from dense goal sets. In: Proceedings of the IEEE International Conference on Computer Vision (ICCV) (2021)
18. Hendy, N., et al.: FISHING net: future inference of semantic heatmaps in grids. In: Proceedings of the IEEE Conference on Computer Vision and Pattern Recognition (CVPR) Workshops (2020)
19. Hong, J., Sapp, B., Philbin, J.: Rules of the road: predicting driving behavior with a convolutional model of semantic interactions. In: Proceedings of IEEE Conference on Computer Vision and Pattern Recognition (CVPR) (2019)
20. Hu, A., Cotter, F., Mohan, N., Gurau, C., Kendall, A.: Probabilistic future prediction for video scene understanding. In: Vedaldi, A., Bischof, H., Brox, T., Frahm, J.-M. (eds.) ECCV 2020. LNCS, vol. 12361, pp. 767–785. Springer, Cham (2020). https://doi.org/10.1007/978-3-030-58517-4_45
21. Hu, A., et al.: FIERY: future instance segmentation in bird's-eye view from surround monocular cameras. In: Proceedings of the IEEE International Conference on Computer Vision (ICCV) (2021)
22. Huang, X., et al.: Diversitygan: diversity-aware vehicle motion prediction via latent semantic sampling. IEEE Rob. Autom. Lett. (RA-L) **5** (2020)
23. Karl, M., Soelch, M., Bayer, J., van der Smagt, P.: Deep variational bayes filters: Unsupervised learning of state space models from raw data. In: Proceedings of the International Conference on Learning Representations (ICLR) (2017)
24. Kingma, D.P., Welling, M.: Auto-encoding variational bayes. In: Proceedings of the International Conference on Learning Representations (ICLR) (2014)
25. Krishnan, R.G., Shalit, U., Sontag, D.: Structured inference networks for nonlinear state space models. In: Proceedings of the Conference on Artificial Intelligence (AAAI) (2017)
26. Liang, M., et al.: Learning lane graph representations for motion forecasting. In: Vedaldi, A., Bischof, H., Brox, T., Frahm, J.-M. (eds.) ECCV 2020. LNCS, vol. 12347, pp. 541–556. Springer, Cham (2020). https://doi.org/10.1007/978-3-030-58536-5_32
27. Luo, W., Yang, B., Urtasun, R.: Fast and furious: real time end-to-end 3D detection, tracking and motion forecasting with a single convolutional net. In: Proceedings of IEEE Conference on Computer Vision and Pattern Recognition (CVPR) (2018)
28. Minderer, M., Sun, C., Villegas, R., Cole, F., Murphy, K.P., Lee, H.: Unsupervised learning of object structure and dynamics from videos. In: Advances in Neural Information Processing Systems (NeurIPS) (2019)
29. Murphy, K.P.: Probabilistic Machine Learning: Advanced Topics. MIT Press, Cambridge (2023)

30. Philion, J., Fidler, S.: Lift, splat, shoot: encoding images from arbitrary camera rigs by implicitly unprojecting to 3D. In: Proceedings of the European Conference on Computer Vision (ECCV) (2020)
31. Rubanova, Y., Chen, R.T.Q., Duvenaud, D.K.: Latent ordinary differential equations for irregularly-sampled time series. In: Advances in Neural Information Processing Systems (NeurIPS) (2019)
32. Sriram, N.N., Liu, B., Pittaluga, F., Chandraker, M.: SMART: simultaneous multi-agent recurrent trajectory prediction. In: Proceedings of the European Conference on Computer Vision (ECCV) (2020)
33. Székely, G.J., Rizzo, M.L.: The energy of data. Ann. Rev. Stat. Appl. **4**(1), 447–479 (2017)
34. Tang, Y.C., Salakhutdinov, R.: Multiple futures prediction. In: Advances in Neural Information Processing Systems (NeurIPS) (2019)
35. Villegas, R., Pathak, A., Kannan, H., Erhan, D., Le, Q.V., Lee, H.: High fidelity video prediction with large stochastic recurrent neural networks. In: Advances in Neural Information Processing Systems (NeurIPS) (2019)
36. Zhao, H., et al.: TNT: target-driven trajectory prediction. In: Proceedings Conference on Robot Learning (CoRL) (2020)

RCLane: Relay Chain Prediction for Lane Detection

Shenghua Xu[1], Xinyue Cai[2], Bin Zhao[1], Li Zhang[3(✉)], Hang Xu[2], Yanwei Fu[1], and Xiangyang Xue[1]

[1] Fudan University, Shanghai, China
[2] Huawei Noah's Ark Lab, Shanghai, China
[3] School of Data Science, Fudan University, Shanghai, China
lizhangfd@fudan.edu.cn

Abstract. Lane detection is an important component of many real-world autonomous systems. Despite a wide variety of lane detection approaches have been proposed, reporting steady benchmark improvements over time, lane detection remains a largely unsolved problem. This is because most of the existing lane detection methods either treat the lane detection as a dense prediction or a detection task, few of them consider the unique topologies (*Y-shape, Fork-shape, nearly horizontal lane*) of the lane markers, which leads to sub-optimal solution. In this paper, we present a new method for lane detection based on *relay chain* prediction. Specifically, our model predicts a segmentation map to classify the foreground and background region. For each pixel point in the foreground region, we go through the forward branch and backward branch to recover the whole lane. Each branch decodes a transfer map and a distance map to produce the direction moving to the next point, and how many steps to progressively predict a relay station (next point). As such, our model is able to capture the keypoints along the lanes. Despite its simplicity, our strategy allows us to establish new state-of-the-art on four major benchmarks including *TuSimple, CULane, CurveLanes* and *LLAMAS*.

Keywords: Lane detection · Relay chain

1 Introduction

Lane detection, the process of identifying lanes as approximated curves, is a fundamental step in developing advanced autonomous driving system and plays a vital role in applications such as driving route planning, lane keeping, real-time positioning and adaptive cruise control.

S. Xu and X. Cai—Equal contribution.

Supplementary Information The online version contains supplementary material available at https://doi.org/10.1007/978-3-031-19839-7_27.

S. Avidan et al. (Eds.): ECCV 2022, LNCS 13698, pp. 461–477, 2022.
https://doi.org/10.1007/978-3-031-19839-7_27

Fig. 1. Challenging scenes (curve lanes, Y-shape lanes). The first row of (a) shows the ground truth while the second row is our predictions. The first row of (b) shows the result of segmentation-based methods that global shape of lane is not well fitted. While the second row of (b) shows proposal-based methods, can not depict local locations of Y-shape and curve lanes.

Early lane detection methods [3,8–11,14,28,34] usually extract hand-crafted features and cluster foreground points on lanes through post-processing. However, traditional methods can not detect diverse lanes correctly for so many complicated scenes in driving scenarios. Thanks to the development of deep-learning, a wide variety of lane detection approaches based on convolution neural network(CNN) have been proposed, such as segmentation-based methods and proposal-based methods, reporting steady benchmark improvements over time.

Proposal-based methods initialize a fixed number of anchors directly and model global information focusing on the optimization of proposal coordinates regression. LaneATT [26] designs slender anchors according to long and thin characteristic of lanes. However, line proposals fail to generalize local locations of all lane points for curve lanes or lanes with more complex topologies. While segmentation-based methods treat lane detection as dense prediction tasks to capture local location information of lanes. LaneAF [1] focuses on local geometry to integrate into global results. However, this bottom-up manner can not capture the global geometry of lanes directly. In some cases such as occlusion or resolution reduction for points on the far side of lane, model performance will be affected due to the loss of lane shape information. Visualization results in Fig. 1(b) of these methods show their shortcomings. Lanes always span half or almost all of the image, these methods neglect this long and thin characteristic of lanes which requires networks to focus on the global shape message and local location information simultaneously. In addition, complex lanes such as Y-shape lanes and Fork-shape lanes are common in the current autonomous driving scenario, while existing methods often fail at these challenging scenes which are shown in Fig. 1(a).

To address this important limitation of current algorithms, we propose a more accurate lane detection solution in the unconstrained driving scenarios, which is called *RCLane* inspired by the idea of ***Relay Chain*** for focusing on local location and global shape information of lanes at the meanwhile. Each foreground point on the lane can be treated as a relay station for recovering the whole lane sequentially in a chain mode. Relay station construction is proposed for strengthening the model's ability of learning local message that is fundamental to describe flexible shapes of lanes. To be specific, we construct a transfer map representing the relative location from current pixel to its two neighbors on the same lane. Furthermore, we apply bilateral prediction strategy aiming to improve generalization ability for lanes with complex topologies. Finally, we design global shape message learning module. Concretely, this module predicts the distance map describing the distance from each foreground point to the two end points on the same lane. The contributions of this work are as follows:

- We propose novel relay chain representation for lanes to model global geometry shape and local location information of lanes simultaneously.
- We introduce a novel pair of lane encoding and decoding algorithms to facilitate the process of lane detection with relay chain representation.
- Extensive experiments on four major lane detection benchmarks show that our approach beats the state-of-the-art alternatives, often by a clear margin and achieves real-time performance.

2 Related Work

Existing methods for lane detection can be categorized into: segmentation-based methods, proposal-based methods, row-wise methods and polynomial regression methods.

Segmentation-Based Methods. Segmentation-based methods [7,12,13,20,21], typically make predictions based on pixel-wise classification. Each pixel will be classified as either on lane or background to generate a binary segmentation mask. Then a post-processing step is used to decode it into a set of lanes. But it is still challenging to assign different points to their corresponding lane instances. A common solution is to predict the instance segmentation mask. However, the number of lanes has to be predefined and fixed when using this strategy, which is not robust for real driving scenarios.

Proposal-Based Methods. Proposal-based methods [4,26,32], take a top-to-down pipeline that directly regresses the relative coordinates of lane shapes. Nevertheless, they always struggle in lanes with complex topologies such as curve lanes and Y-shaped lanes. The fixed anchor shape has a major flaw when regressing the variable lane shapes in some hard scenes.

Row-Wise Methods. Based on the grid division of the input image, row-wise detection approaches [6,15,22,23,33] have achieved great progress in terms of accuracy and efficiency. Generally, row-wise detection methods directly predict the lane position for each row and construct the set of lanes through post-processing. However, detecting nearly horizontal lanes which fall at small vertical intervals is still a major problem.

Polynomial Regression Methods. Polynomial regression methods [16,27] directly outputs polynomials representing each lane. The deep network is firstly used in [27] to predict the lane curve equation, along with the domains for these polynomials and confidence scores for each lane. [16] uses a transformer [30] to learn richer structures and context, and reframes the lane detection output as parameters of a lane shape model. However, despite of the fast speed polynomial regression methods achieve, there is still some distance from the state of the art results.

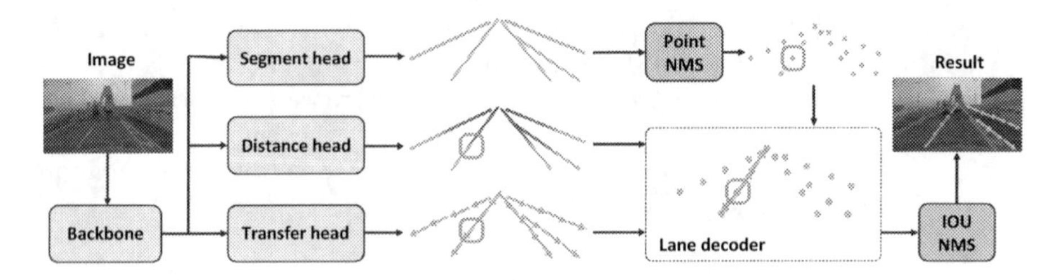

Fig. 2. Schematic illustration of proposed RCLane. Standard Segformer [31] is used as backbone. The output head consists of three branches. The segment head predicts segmentation map (S). The distance head and the transfer head predict distance map (D) and transfer map (T) respectively. Both kinds of maps contain forward and backward parts. Then, Point-NMS is used for sparse segmentation results. All predictions are fed into the lane decoder (Fig. 5), to get final results.

3 Method

Given an input image $I \in \mathbb{R}^{H \times W \times C}$, the goal of RCLane is to predict a collection of lanes $L = \{l_1, l_2, \cdots, l_N\}$, where N is the total number of lanes. Generally, each lane l_k is represented as follows:

$$l_k = \{(x_1, y_1), (x_2, y_2), \cdots, (x_{N_k}, y_{N_k})\}, \tag{1}$$

The overall structure of our RCLane is shown in Fig. 2. This section will first present the concept of lane detection with relay chain, then introduce the lane encoder for relay station construction, followed by a lane decoder to attain curve lanes. Finally, the network architecture and losses we adopt is detailed.

3.1 Lane Detection with Relay Chain

Focusing on the combination of local location and global shape information to detect lanes with complex topologies, we propose a novel lane detection method RCLane with the idea of relay chain. Relay chain is a structure composed of relay stations which are connected in a chain mode. Relay station is responsible for data processing and transmitting it to adjacent stations, while chain is a kind of structure that organizes these stations from an overall perspective. All stations are associated to corresponding lane points respectively.

We design the structure of relay chain which is appropriate for combining local location and global geometry message in lane detection and propose RCLane in this work. To be specific, each foreground point on the lane is treated as a relay station and can extend to the neighbor points iteratively to decode the lane in a chain mode. All foreground points are supervised by two kinds of message mentioned above. Moreover, the structure of chain has high flexibility to fit lanes with complex topologies.

Next, we will introduce the relay station construction and propose bilateral predictions for complex topologies and global shape message learning to explain how to detect lanes with the idea of *Relay Chain* progressively.

Relay Station Construction. Segmentation-based approaches normally predict all foreground points on lanes and cluster them via post-processing. [1] predicts horizontal and vertical affinity fields for clustering and associating pixels belonging to the same lane. [24] regresses a vector describing the local geometry of the curve that current pixel belongs to and refines shape further in the decoding algorithm. Nevertheless, they both fix the vertical intervals between adjacent points and decode lanes row-by-row from bottom to top. In fact, horizontal offsets are used for refining the position of current points while vertical offsets are for exploring the vertical neighbors of them. And the fixed vertical offsets can not adapt to the high degree of freedom for lanes. For example, they can only detect a fraction of the nearly horizontal lanes. Thus, we propose relay station construction module to establish relationships between neighboring points on the lane. Each relay station $p = (p_x, p_y)$ predicts offsets to its neighboring point $p^{next} = (p_x^{next}, p_y^{next})$ on the same lane with a fixed step length d as is shown in Eq. 2, 3 in two directions. And the deformation trend of lanes can be fitted considerably by eliminating vertical constraints. All relay stations are then connected to form a chain which is the lane exactly.

$$(p_x^{next}, p_y^{next}) = (p_x, p_y) + (\Delta x, \Delta y), \tag{2}$$

$$\Delta x^2 + \Delta y^2 = d^2. \tag{3}$$

Bilateral Predictions for Complex Topologies. The current autonomous driving scenario contains lanes with complex topologies such as Y-shape and Fork-shape lanes, which can be regarded as that two lanes merges as the stem. One-way prediction can only detect one of lanes because it can only extend to one limb when starting from the stem of these lanes. We adopt a two-way detection strategy that splits the next neighboring point p^{next} into the forward point p^f and the backward point p^b. Points on different limbs can recover lanes they belong to respectively and compose the final Y-shape or fork-shape lanes as is illustrated in Fig. 3(b). Let F denotes the output feature map from the backbone whose resolution drops by a factor of 4 compared to the original image. We design a transfer output head and pick F as input. F goes through convolution-based transfer head to get the transfer map T which consists of forward and backward components $T_f, T_b \in \mathbb{R}^{H \times W \times 2}$. Each location in T_f is a 2D vector, which represents the offsets between the forward neighboring point p^f and the current pixel p. The definition of T_b is similar as T_f. Consequently, we can detect the forward and backward neighboring points p^f, p^b of p guided by T.

$$p^f = p + T_f(p), \quad p^b = p + T_b(p). \tag{4}$$

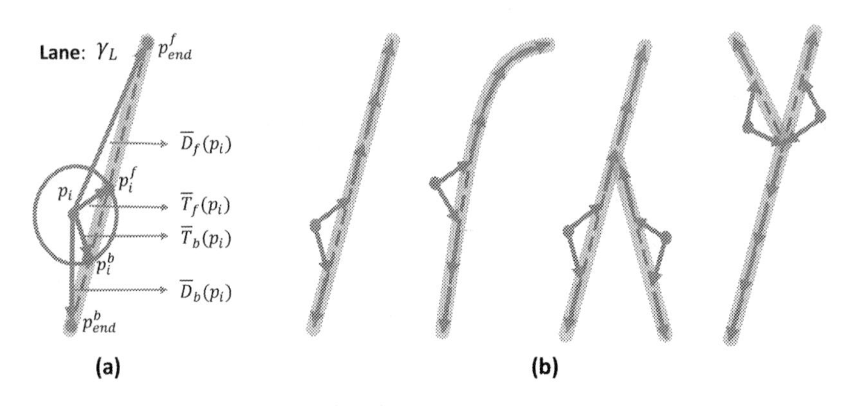

Fig. 3. (a) is an illustration of the transfer vectors and distance scalars for p_i. $\overline{T}_{f,b}(p_i)$ are the forward and backward transfer vectors. $\overline{D}_{f,b}(p_i)$ are the forward and backward distance scalars. (b) shows our bilateral predictions can not only decode Y-shape or fork-shape lanes, but also fit simple structures, like straight lanes and curved lanes.

With the guidance of local location information in transfer map T, the whole lane can be detected iteratively via bilateral strategy.

Global Shape Message Learning. Previous works predict positions of end points for lanes to guide decoding process. FastDraw [22] predicts end tokens to encode the global geometry while CondLaneNet [15] recovers the row-wise shape through the vertical range prediction. These methods actually ignores the relation between the end points and other points on the same lane. We make every relay station learns the global shape message transmitted in the chain by utilizing the relation mentioned above. In detail, we design a distance head to predict the distance map D that consists of the forward and backward components $D_f, D_b \in \mathbb{R}^{H \times W \times 1}$. Each location in D_f is a scalar, which represents the distance from the current pixel p to the forward end point p^f_{end} on the lane. With this global shape information, we can know when to stop the lane decoding process. Specifically speaking, the iterations for decoding the forward branch of p is $\frac{D_f}{d}$. The definition of D_b is similar as D_f as well. With the combination

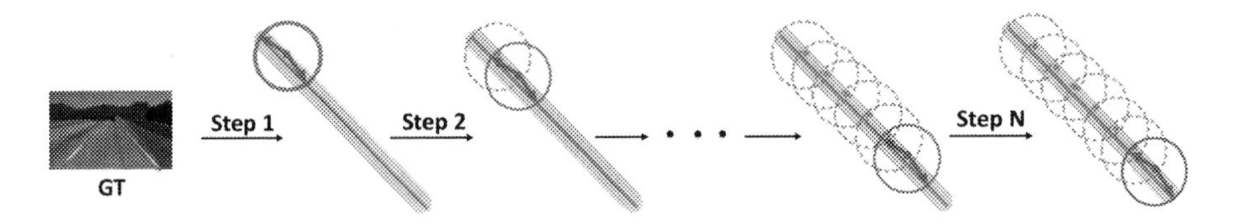

Fig. 4. Lane encoder. All foreground points are matched with the nearest lanes. The arrows in a circle indicate transfer vectors of a foreground point to its two neighbors on lane. The distance scalars represent distances between the current point and two end points of the lane. All results are generated with point-wise traversal.

of local location and global geometry information, our relay chain prediction strategy performs considerably well even in complex scenarios. Next, we will introduce the novel pair of lane encoding and decoding algorithms designed for lane detection.

3.2 Lane Encoder for Relay Station Construction

The lane encoder is to create the supervision of transfer and distance maps for training. Given an image $I \in \mathbb{R}^{H \times W \times 3}$ and its segmentation mask $\overline{S} \in \mathbb{R}^{H \times W \times 1}$, for any foreground point $p_i = (x_i, y_i) \in \overline{S}$ we denote its corresponding lane as γ_L. The two forward and backward end points of γ_L are denoted as $p^f_{end} = (x^f_{end}, y^f_{end})$ and $p^b_{end} = (x^b_{end}, y^b_{end})$, which have the minimum and maximum y-coordinates respectively. The forward distance scalar $\overline{D}_f(p_i)$ and backward distance scalar $\overline{D}_b(p_i)$ of p_i are formulated as the following:

$$\overline{D}_f(p_i) = \sqrt{(x_i - x^f_{end})^2 + (y_i - y^f_{end})^2}, \tag{5}$$

$$\overline{D}_b(p_i) = \sqrt{(x_i - x^b_{end})^2 + (y_i - y^b_{end})^2}. \tag{6}$$

To generate the forward transfer vector and backward transfer vector for pixel p_i, we first find the two neighbors on γ_L of it with the fixed distance d. They are denoted as $p^f_i = (x^f_i, y^f_i)$ and $p^b_i = (x^b_i, y^b_i)$ and represent the forward neighbor and backward neighbor respectively. Then the forward transfer vector $\overline{T}_f(p_i)$ and the backward transfer vector $\overline{T}_b(p_i)$ for pixel p_i are defined :

$$\overline{T}_f(p_i) = (x^f_i - x_i, y^f_i - y_i), \tag{7}$$

$$\overline{T}_b(p_i) = (x^b_i - x_i, y^b_i - y_i), \tag{8}$$

$$||\overline{T}_f(p_i)||_2 = ||\overline{T}_b(p_i)||_2 = d. \tag{9}$$

The details are shown in Fig. 3(a). In addition, for two separate parts of one Y-shape lane: $l_1 = \{(x_1, y_1), \cdots, (x_m, y_m), (x^1_{m+1}, y^1_{m+1}), \cdots, (x^1_{n_1}, y^1_{n_1})\}$, $l_2 =$

$\{(x_1, y_1), \cdots, (x_m, y_m), (x^2_{m+1}, y^2_{m+1}), \cdots, (x^2_{n_2}, y^2_{n_2})\}$. $\{(x_1, y_1), \cdots, (x_m, y_m)\}$ is the shared stem. We randomly choose one point from (x^1_{m+1}, y^1_{m+1}) and (x^2_{m+1}, y^2_{m+1}) as the forward neighboring point of (x_m, y_m) while (x_m, y_m) is the common backward neighboring point of (x^1_{m+1}, y^1_{m+1}) and (x^2_{m+1}, y^2_{m+1}). All foreground pixels on the $\overline{S}$ are processed following the same formula and then $\overline{T}_{f,b}$ and $\overline{D}_{f,b}$ can be generated. The process is shown in Fig. 4.

3.3 Lane Decoder with Transfer and Distance Map

With the predictions of local location and global geometry, we propose a novel lane decoding algorithm to detect all curves in a given image.

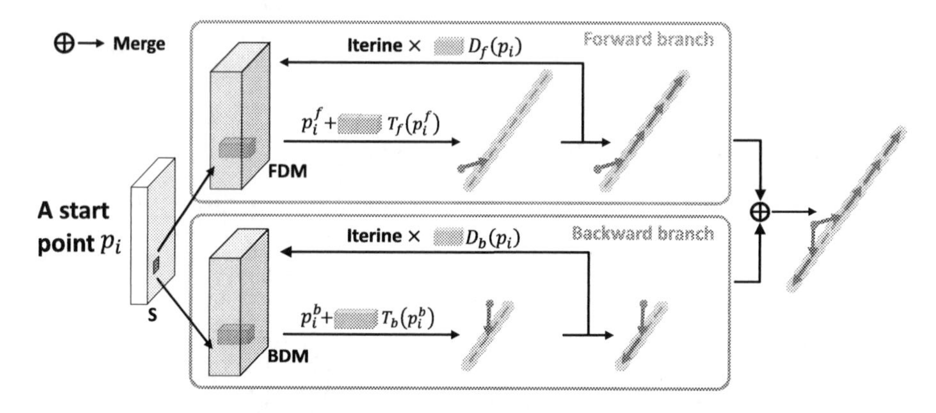

Fig. 5. The illustration of the lane decoder. The forward branch predicts the forward part of the lane via forward transfer map T_f and forward distance map D_f. The backward part can be decoded from the backward branch similarly.

Given the predicted binary segmentation mask S, transfer map T and distance map D, we collect all the foreground points of S and use a Point-NMS to get a sparse set of key points K. Every key point $p \in K$ serves as a start point to recover one global curve.

Step1: Find the forward transfer vector $T_f(p)$ and forward distance scalar $D_f(p)$ for p. The moving steps we should extend the neighbors for the forward branch can be defined as $M^f = \frac{D_f(p)}{d}$. In other words, we can infer the location of the forward end point of p with $D_f(p)$ on the same lane.

Here d is the step length. Then the forward neighbor pixel p^f_{i+1} of p^f_i can be calculated iteratively by:

$$p^f_{i+1} = p^f_i + T_f(p^f_i),\ i \in \{0, 1, 2, \cdots, M^f - 1\},\ p_0 = p. \tag{10}$$

The forward branch of the curve can be recovered by connecting $\{p, p^f_1, \cdots, p^f_{M^f}\}$ sequentially. The detail is shown on the top of Fig. 5.

Step2: We calculate the point set $\{p, p_1^b, p_2^b, \cdots, p_{M^b}^b\}$ following Eq. 10 via T_b and D_b and connect them sequentially to recover the backward branch.
Step3: We then merge the backward and forward curve branches together to get the global curve:

$$\gamma_L = \{p_{M^b}^b, \cdots, p_2^b, p_1^b, p, p_1^f, p_2^f, \cdots, p_{M^f}^f\}. \tag{11}$$

Finally, the non-maximum suppression [19] is performed on all the predicted curves to get the final results.

3.4 Network Architecture

The overall framework is shown in Fig. 2. SegFormer [31] is utilized as our network backbone, aiming to extract global contextual information and learn the long and thin structures of lanes. SegFormer-B0, B1 and B2 are used as small, medium and large backbones in our experiments respectively. Given an image $I \in R^{H \times W \times 3}$, the segmentation head predicts the binary segmentation mask $S \in R^{H \times W \times 1}$, the transfer head predicts the transfer map T which consists of the forward and backward parts $T_f, T_b \in \mathbb{R}^{H \times W \times 2}$, and the distance head predicts the distance map D that consists of $D_f, D_b \in \mathbb{R}^{H \times W \times 1}$.

3.5 Loss Function

To train our proposed model, we adopt different losses for predictions. For the binary segmentation mask, we adopt the OHEM loss [25] to train it in order to solve class imbalance problem due to the sparsity of lane segmentation points. The OHEM loss is formulated as follows:

$$L_{seg} = \frac{1}{N_{pos} + N_{neg}} \Big(\sum_{i \in S_{pos}} y_i log(p_i) + \sum_{i \in S_{neg}} (1 - y_i) log(1 - p_i) \Big). \tag{12}$$

where S_{pos} is the set of positive points and S_{neg} is the set of hard negative points which is most likely to be misclassified as positive. N_{pos} and N_{neg} denote the number of points in S_{pos} and S_{neg} respectively. The ratio of N_{neg} to N_{pos} is a hyperparmeter μ. As for the per-pixel transfer and distance maps, we simply adopt the smooth L_1 loss, which are denoted as L_T and L_D, to train them.

$$L_D = \frac{1}{N_{pos}} \sum_{i \in S_{pos}} L_{smooth_{L_1}}(D(p_i), \overline{D}(p_i)), \tag{13}$$

$$L_T = \frac{1}{N_{pos}} \sum_{i \in S_{pos}} L_{smooth_{L_1}}(T(p_i), \overline{T}(p_i)). \tag{14}$$

In the training phase, the total loss is defined as follows:

$$L_{total} = L_{seg} + L_T + L_D. \tag{15}$$

4 Experiment

4.1 Experimental Setting

Dataset. We conduct experiments on four widely used lane detection benchmark datasets: CULane [21], TuSimple [29], LLAMAS [2] and CurveLanes [32]. CULane consists of 55 h of videos which comprises nine different scenarios, including normal, crowd, dazzle night, shadow, no line, arrow, curve, cross and night. The TuSimple dataset is collected with stable lighting conditions on highways. LLAMAS is a large lane detection dataset obtained on highway scenes with annotations auto-generated by using high-definition maps. CurveLanes is a recently proposed benchmark with cases of complex topologies such as Y-shape lanes and dense lanes. The details of four datasets are shown in Table 1.

Table 1. Lane detection datasets.

Dataset	Train	Val	Test	Road type
CULane [21]	88.9k	9.7k	34.7k	Urban, highway
CurveLanes [32]	100K	20K	30K	Urban, highway
TuSimple [29]	3.3k	0.4k	2.8k	Highway
LLAMAS [2]	58.3k	20.8k	20.9k	Highway

Table 2. State-of-the-art comparison on CULane. Even the small version of our RCLane achieves the state-of-art performance with only 6.3M parameters.

Method	Total	Normal	Crowded	Dazzle	Shadow	No line	Arrow	Curve	Cross	Night	Params(M)	FPS
SCNN [21]	71.60	90.60	69.70	58.50	66.90	43.40	84.10	64.40	1990	66.10	–	7.5
CurveLanes-NAS-S [32]	71.40	88.30	68.60	63.20	68.00	47.90	82.50	66.00	2817	66.20	–	–
CurveLanes-NAS-M [32]	73.50	90.20	70.50	65.90	69.30	48.80	85.70	67.50	2359	68.20	–	–
CurveLanes-NAS-L [32]	74.80	90.70	72.30	67.70	70.10	49.40	85.80	68.40	1746	68.90	–	–
LaneATT-S [26]	75.13	91.17	72.71	65.82	68.03	49.13	87.82	63.75	1020	68.58	13.3	250
LaneATT-M [26]	76.68	92.14	75.03	66.47	78.15	49.39	88.38	67.72	1330	70.72	23.4	171
LaneATT-L [26]	77.02	91.74	76.16	69.47	76.31	50.46	86.29	64.05	1264	70.81	18.8	26
LaneAF (DLA-34) [1]	77.41	91.80	75.61	71.78	79.12	51.38	86.88	72.70	1360	73.03	20.2	–
FOLO [24]	78.80	92.70	77.80	75.20	79.30	52.10	89.00	69.40	1569	74.50	–	–
CondLaneNet-S [15]	78.14	92.87	75.79	70.72	80.01	52.39	89.37	72.40	1364	73.23	12.1	220
CondLaneNet-M [15]	78.74	93.38	77.14	71.17	79.93	51.85	89.89	73.88	1387	73.92	22.2	152
CondLaneNet-L [15]	79.48	93.47	77.44	70.93	80.91	**54.13**	90.16	75.21	1201	74.80	49.9	58
RCLane-S (Ours)	79.52	93.41	77.93	**73.32**	80.31	53.84	89.04	75.66	1298	74.33	6.3	45.6
RCLane-M (Ours)	80.03	93.59	78.77	72.44	**84.37**	52.77	90.31	78.39	**907**	73.96	17.2	43.8
RCLane-L (Ours)	**80.50**	**94.01**	**79.13**	72.92	81.16	53.94	**90.51**	**79.66**	931	**75.10**	30.9	24.5

Evaluation Metrics. For CULane, CurveLanes and LLAMAS, we utilize F1-measure as the evaluation metric. While for TuSimple, accuracy is presented as the official indicator. And we also report the F1-measure for TuSimple. The calculation method follows the same formula as in CondLaneNet [15].

Implementation Details. The small, medium and large versions of our RCLane-Det are used on all four datasets. Except when explicitly indicated, the input resolution is set to 320×800 during training and testing. For all training sessions, we use AdamW optimizer [17] to train 20 epochs on CULane, CurveLanes and LLAMAS, 70 epochs on TuSimple respectively with a batch size of 32. The learning rate is initialized as 6*e*-4 with a "poly" LR schedule. We set η for calculating IOU between lines as 15, the ratio of N_{neg} to N_{pos} μ as 15, the minimum distance between any two foreground pixels of in Point-NMS τ as 2. We implement our method using the Mindspore [18] on Ascend 910.

4.2 Results

CULane. As illustrated in Table 2, RCLane achieves a new state-of-the-art result on the CULane testing set with an 80.50% F1-measure. Compared with the best model as far as we know, CondLaneNet [15], although our method performs better only 1.02% of F1-measure compared with the best model before CondLaneNet since CULane is a simpler dataset with may straight lines, it has an considerable improvements in crowded and curve scenes, which demonstrates that *Relay Chain* can strengthen local location connectivity through global shape learning for local occlusions and complex line topologies.

Table 3. Performance of different methods on CurveLanes.

Method	F1 (%)	Precision (%)	Recall (%)
SCNN [21]	65.02	76.13	56.74
Enet-SAD [7]	50.31	63.60	41.60
PointLaneNet [4]	78.47	86.33	72.91
CurveLane-S [32]	81.12	93.58	71.59
CurveLane-M [32]	81.80	93.49	72.71
CurveLane-L [32]	82.29	91.11	75.03
CondLaneNet-S [15]	85.09	87.75	82.58
CondLaneNet-M [15]	85.92	88.29	83.68
CondLaneNet-L [15]	86.10	88.98	83.41
RCLane-S (Ours)	90.47	93.33	87.78
RCLane-M (Ours)	90.96	93.47	88.58
RCLane-L (Ours)	**91.43**	**93.96**	**89.03**

CurveLanes. CurveLanes [32] is a challenging benchmark with many hard scenarios. The evaluation results are shown in Table 3. We can see that our largest model (with SegFormer-B2) surpasses CondLaneNet-L by 5.33% in F1-measure, which is more pronounced than it on CULane. Due to that CurveLanes is more complex with Fork-shape, Y-shape and other curve lanes, improvements both in recall rate and accuracy prove that RCLane has generalization ability on lanes.

TuSimple. The results on TuSimple are shown in Table 4. As Tusimple is a small dataset and scenes are more simple with accurate annotations, the gap between all methods is small. Moreover, our method also achieves a new state-of-the-art F1 score of 97.64%.

LLAMAS. LLAMAS [2] is a new dataset with more than $100K$ images from highway scenarios. The results of our RCLane on LLAMAS is shown in Table 5. The best result of our method is 96.13% F1 score with RCLane-L.

4.3 Ablation Study

Different Modules. In this section, we perform the ablation study to evaluate the impact of the proposed relay station construction, bilateral predictions and global shape message learning on CurveLanes. The results is shown in Table 6. The first row shows the baseline result, which only uses binary segmentation plus post processing named DBSCAN [5] to detect lanes. In the second row, the lane is recovered from bottom to top gradually with the guidance of the forward transfer map and forward distance map. While the third row detect lanes from top to bottom. In the fourth row, we only use the forward and backward transfer maps to predict the lane. And we present our full version of RCLane in the last row, which attains a new state-of-art result 91.43% on CurveLanes.

Table 4. Performance of different methods on TuSimple.

Method	F1 (%)	Acc (%)	FP (%)	FN (%)
SCNN [21]	95.97	96.53	6.17	**1.80**
PointLaneNet [4]	95.07	96.34	4.67	5.18
LaneATT-ResNet18 [26]	96.71	95.57	3.56	3.01
LaneATT-ResNet34 [26]	96.77	95.63	3.53	2.92
LaneATT-ResNet122 [26]	96.06	96.10	5.64	2.17
CondLaneNet-S [15]	97.01	95.48	2.18	3.80
CondLaneNet-M [15]	96.98	95.37	2.20	3.82
CondLaneNet-L [15]	97.24	96.54	**2.01**	3.50
LaneAF(DLA-34) [1]	96.49	95.62	2.80	4.18
FOLO [24]	–	**96.92**	4.47	2.28
RCLane-S (Ours)	97.52	96.49	2.21	2.57
RCLane-M (Ours)	97.61	96.51	2.24	2.36
RCLane-L (Ours)	**97.64**	96.58	2.28	2.27

Table 5. Performance of different methods on LLAMAS.

Method	F1 (%)	Precision (%)	Recall (%)
PolyLaneNet [27]	88.40	88.87	87.93
LaneATT-ResNet-18 [26]	93.46	**96.92**	90.24
LaneATT-ResNet-34 [26]	93.74	96.79	90.88
LaneATT-ResNet-122 [26]	93.54	96.82	90.47
LaneAF(DLA-34) [1]	96.07	96.91	95.26
RCLane-S (Ours)	96.05	96.70	95.42
RCLane-M (Ours)	96.03	96.62	95.45
RCLane-L (Ours)	**96.13**	96.79	**95.48**

Comparing the first two rows, we can see that the proposed relay station construction has greatly improved the performance. Then, we add global shape information learning with distance map which can improve the performance from 88.19% to 91.43%. While we do additional two experiments in the second and third lines, the lane is detected by transfer and distance maps from one-way direction and there is a certain gap with the highest F1-score. It proves that our bilateral prediction has generalization in depicting topologies of lanes. In addition, there exists a gap between the forward the backward models. As the near lanes (the bottom region of the image) are usually occluded by the ego car, the corresponding lane points get low confidence scores from the segmentation results. Therefore the starting points are usually outside of the occluded area and the forward counterpart eventually has no chance back to cover the lanes at the bottom of the image. In contrast, the backward model detects lanes more completely with the help of the distance map when decoding from the top, including the occluded area.

Table 6. Comparison of different components on CurveLanes. The T_f, T_b, D_f, D_b represent the forward transfer map, backward transfer map, forward distance map and backward distance map respectively.

Baseline	T_f	T_b	D_f	D_b	F1 (%)
√					51.22
	√		√		$75.06^{+23.84}$
		√		√	$83.78^{+32.56}$
	√	√			$88.19^{+36.97}$
	√	√	√	√	$91.43^{+40.21}$

Comparisons with Other Methods Using the Same Backbone. We additionally use Segformer-B2 [31] as backbone to train CondLaneNet [15] and

LaneAF [1] respectively and show their results on Table 7 below. Without changing the parameters of their models, our model still outperforms LaneAF and CondLaneNet by a margin on CULane [21] dataset due to its superior precision, which demonstrates the high quality of lanes detected by RCLane. It further fairly verifies the superiority of our proposed relay chain prediction method, which can process local location and global geometry information simultaneously to improve the capacity of the model.

Table 7. Comparisons with other methods using the same backbone Segformer-B2.

Method	Precision (%)	Recall (%)	F1 (%)
LaneAF [1]	80.89	71.71	76.02
CondLaneNet [15]	82.58	**76.01**	79.16
RCLane (Ours)	**88.52**	73.82	**80.50**

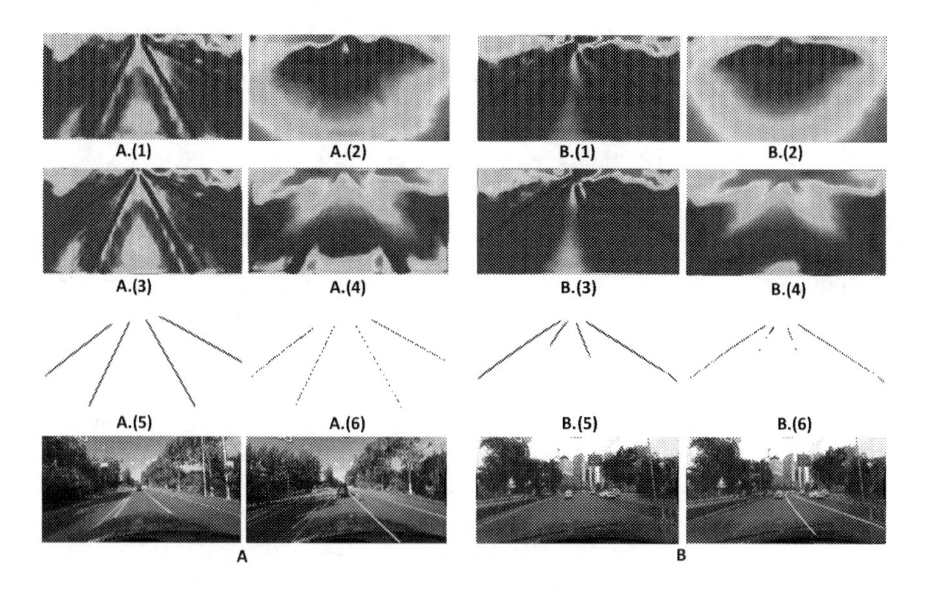

Fig. 6. Visualization of network outputs. A.(1, 3) are features of D_f and D_b, while A.(2, 4) are features of T_f and T_b. A.(5) is the segmentation result and becomes sparse map A.(6) via Point-NMS. B is a harder frame compared to A.

Local Location and Global Shape Message Modeling. In Fig. 6 $A.(1,3)$, the transfer map can capture local location information depicting topology of the lane precisely, while the distance map in Fig. 6 $A.(2,4)$ models global shape information with large receptive field. Furthermore, in some driving scenarios, there occurs loss of lane information due to the disappearance of trace for lanes as is shown in Fig. 6(B). However, lanes are still captured faintly in the transfer map with the global shape information learning. The results show the robustness of our RCLane with local location and global shape message modeling.

5 Conclusion

In this paper, we have proposed to solve lane detection problem by learning a novel relay chain prediction model. Compared with existing lane detection methods, our model is able to capture global geometry and local information progressively with the novel relay station construction and global shape message learning. Furthermore, bilateral predictions can adapt to hard topologies, such as Fork-shape and Y-shape. Extensive experiments on four benchmarks including CULane, CurveLanes, Tusimple and LLAMAS demonstrate state-of-the-art performance and generalization ability of our *RCLane*.

Acknowledgments. This work was supported in part by National Natural Science Foundation of China (Grant No. 6210020439), Lingang Laboratory (Grant No. LG-QS-202202-07), Natural Science Foundation of Shanghai (Grant No. 22ZR1407500), Shanghai Municipal Science and Technology Major Project (Grant No. 2018SHZDZX01 and 2021SHZDZX0103), Science and Technology Innovation 2030 - Brain Science and Brain-Inspired Intelligence Project (Grant No. 2021ZD0200204), MindSpore and CAAI-Huawei MindSpore Open Fund.

References

1. Abualsaud, H., Liu, S., Lu, D., Situ, K., Rangesh, A., Trivedi, M.M.: LaneAF: robust multi-lane detection with affinity fields. arXiv preprint (2021)
2. Behrendt, K., Soussan, R.: Unsupervised labeled lane markers using maps. In: ICCV Workshops (2019)
3. Borkar, A., Hayes, M., Smith, M.T.: Robust lane detection and tracking with ransac and Kalman filter. In: ICIP (2009)
4. Chen, Z., Liu, Q., Lian, C.: PointLaneNet: efficient end-to-end CNNs for accurate real-time lane detection. In: IEEE Intelligent Vehicles Symposium (2019)
5. Ester, M., Kriegel, H.P., Sander, J., Xu, X., et al.: A density-based algorithm for discovering clusters in large spatial databases with noise. In: KDD (1996)
6. Hou, Y., Ma, Z., Liu, C., Hui, T.W., Loy, C.C.: Inter-region affinity distillation for road marking segmentation. In: CVPR (2020)
7. Hou, Y., Ma, Z., Liu, C., Loy, C.C.: Learning lightweight lane detection CNNs by self attention distillation. In: ICCV (2019)
8. Hur, J., Kang, S.N., Seo, S.W.: Multi-lane detection in urban driving environments using conditional random fields. In: IV (2013)
9. Jiang, R., Klette, R., Vaudrey, T., Wang, S.: New lane model and distance transform for lane detection and tracking. In: International Conference on Computer Analysis of Images and Patterns (2009)

10. Jiang, Y., Gao, F., Xu, G.: Computer vision-based multiple-lane detection on straight road and in a curve. In: 2010 International Conference on Image Analysis and Signal Processing (2010)
11. Kim, Z.: Robust lane detection and tracking in challenging scenarios. IEEE Trans. Intell. Transp. Syst. **9**(1), 16–26 (2008)
12. Ko, Y., Lee, Y., Azam, S., Munir, F., Jeon, M., Pedrycz, W.: Key points estimation and point instance segmentation approach for lane detection. IEEE Trans. Intell. Transp. Syst. **23**(7), 8949–8958 (2021)
13. Lee, S., et al.: VPGNet: vanishing point guided network for lane and road marking detection and recognition. In: ICCV (2017)
14. Liu, G., Wörgötter, F., Markelić, I.: Combining statistical Hough transform and particle filter for robust lane detection and tracking. In: IEEE Intelligent Vehicles Symposium (2010)
15. Liu, L., Chen, X., Zhu, S., Tan, P.: CondLaneNet: a top-to-down lane detection framework based on conditional convolution. In: ICCV (2021)
16. Liu, R., Yuan, Z., Liu, T., Xiong, Z.: End-to-end lane shape prediction with transformers. In: WACV (2021)
17. Loshchilov, I., Hutter, F.: Fixing weight decay regularization in Adam (2018)
18. Mindspore (2020). www.mindspore.cn/
19. Neubeck, A., Van Gool, L.: Efficient non-maximum suppression. In: ICPR (2006)
20. Neven, D., De Brabandere, B., Georgoulis, S., Proesmans, M., Van Gool, L.: Towards end-to-end lane detection: an instance segmentation approach. In: IEEE Intelligent Vehicles Symposium (2018)
21. Pan, X., Shi, J., Luo, P., Wang, X., Tang, X.: Spatial As Deep: spatial CNN for traffic scene understanding. In: AAAI (2018)
22. Philion, J.: FastDraw: addressing the long tail of lane detection by adapting a sequential prediction network. In: CVPR (2019)
23. Qin, Z., Wang, H., Li, X.: Ultra fast structure-aware deep lane detection. In: Vedaldi, A., Bischof, H., Brox, T., Frahm, J.-M. (eds.) ECCV 2020. LNCS, vol. 12369, pp. 276–291. Springer, Cham (2020). https://doi.org/10.1007/978-3-030-58586-0_17
24. Qu, Z., Jin, H., Zhou, Y., Yang, Z., Zhang, W.: Focus on local: detecting lane marker from bottom up via key point. In: CVPR (2021)
25. Shrivastava, A., Gupta, A., Girshick, R.: Training region-based object detectors with online hard example mining. In: CVPR (2016)
26. Tabelini, L., Berriel, R., Paixao, T.M., Badue, C., De Souza, A.F., Oliveira-Santos, T.: Keep your eyes on the lane: real-time attention-guided lane detection. In: CVPR (2021)
27. Tabelini, L., Berriel, R., Paixao, T.M., Badue, C., De Souza, A.F., Oliveira-Santos, T.: PolyLaneNet: lane estimation via deep polynomial regression. In: ICPR (2021)
28. Tan, H., Zhou, Y., Zhu, Y., Yao, D., Li, K.: A novel curve lane detection based on improved River Flow and RANSA. In: IEEE Conference on Intelligent Transportation Systems (ITSC) (2014)
29. TuSimple: Tusimple benchmark (2019). https://github.com/TuSimple/tusimple-benchmark
30. Vaswani, A., et al.: Attention is all you need. In: NeurIPS (2017)
31. Xie, E., Wang, W., Yu, Z., Anandkumar, A., Alvarez, J.M., Luo, P.: SegFormer: simple and efficient design for semantic segmentation with transformers. arXiv preprint (2021)

32. Xu, H., Wang, S., Cai, X., Zhang, W., Liang, X., Li, Z.: CurveLane-NAS: unifying lane-sensitive architecture search and adaptive point blending. In: Vedaldi, A., Bischof, H., Brox, T., Frahm, J.-M. (eds.) ECCV 2020. LNCS, vol. 12360, pp. 689–704. Springer, Cham (2020). https://doi.org/10.1007/978-3-030-58555-6_41
33. Yoo, S., et al.: End-to-end lane marker detection via row-wise classification. In: CVPR Workshops (2020)
34. Zhou, S., Jiang, Y., Xi, J., Gong, J., Xiong, G., Chen, H.: A novel lane detection based on geometrical model and Gabor filter. In: IEEE Intelligent Vehicles Symposium (2010)

Drive&Segment: Unsupervised Semantic Segmentation of Urban Scenes via Cross-Modal Distillation

Antonin Vobecky[1,2(✉)], David Hurych[2], Oriane Siméoni[2], Spyros Gidaris[2], Andrei Bursuc[2], Patrick Pérez[2], and Josef Sivic[1]

[1] Czech Institute of Informatics, Robotics and Cybernetics, Czech Technical University in Prague, Prague, Czechia
[2] valeo.ai, Paris, France
antonin.vobecky@cvut.cz
https://vobecant.github.io/DriveAndSegment

Abstract. This work investigates learning pixel-wise semantic image segmentation in urban scenes without any manual annotation, just from the raw non-curated data collected by cars which, equipped with cameras and LiDAR sensors, drive around a city. Our contributions are threefold. First, we propose a novel method for cross-modal unsupervised learning of semantic image segmentation by leveraging synchronized LiDAR and image data. The key ingredient of our method is the use of an object proposal module that analyzes the LiDAR point cloud to obtain proposals for spatially consistent objects. Second, we show that these 3D object proposals can be aligned with the input images and reliably clustered into semantically meaningful pseudo-classes. Finally, we develop a cross-modal distillation approach that leverages image data partially annotated with the resulting pseudo-classes to train a transformer-based model for image semantic segmentation. We show the generalization capabilities of our method by testing on four different testing datasets (Cityscapes, Dark Zurich, Nighttime Driving and ACDC) without any finetuning, and demonstrate significant improvements compared to the current state of the art on this problem.

Keywords: Autonomous driving · Unsupervised semantic segmentation

1 Introduction

In this work, we investigate whether it is possible to learn pixel-wise semantic image segmentation of urban scenes without the need for any manual annotation, just from the raw non-curated data collected by cars equipped with cameras and LiDAR sensors while driving in town. This topic is important as current methods

Supplementary Information The online version contains supplementary material available at https://doi.org/10.1007/978-3-031-19839-7_28.

S. Avidan et al. (Eds.): ECCV 2022, LNCS 13698, pp. 478–495, 2022.
https://doi.org/10.1007/978-3-031-19839-7_28

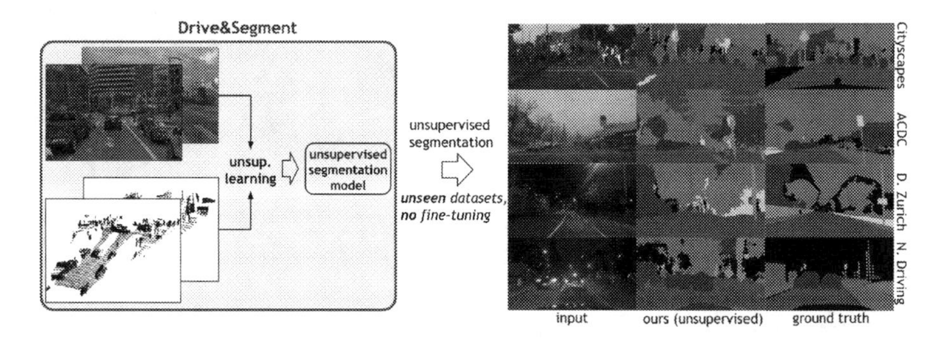

Fig. 1. Proposed fully-unsupervised approach. From uncurated images and LiDAR data, Drive&Segment learns a semantic image segmentation model with no manual annotations. The resulting model performs unsupervised semantic segmentation of new unseen datasets without any human labeling. It can segment complex scenes with many objects, including thin structures such as people, bicycles, poles or traffic lights. Black color denotes the ignored/missing label. (Color figure online)

require large amounts of pixel-wise annotations over various driving conditions and situations. Such a manual segmentation of images on a large scale is very expensive, time-consuming, and prone to biases.

Currently, the best methods for unsupervised learning of semantic segmentation assume that images contain centered objects [50] rather than full scenes or use spatial self-supervision available in the image domain [15]. They do not leverage additional modalities, such as the LiDAR data, available for urban scenes in the autonomous driving set-ups. In this work, we develop an approach for unsupervised semantic segmentation that learns to segment complex scenes containing many objects, including thin structures such as pedestrians or traffic lights, without the need for any manual annotation. Instead, it leverages cross-modal information available in (aligned) LiDAR point clouds and images, see Fig. 1. Exploiting point clouds as a form of supervision is not straightforward: data from LiDAR and camera are rarely perfectly synchronized; moreover, point clouds are unstructured and of much lower resolution compared to images; finally, extracting useful semantic information from LiDAR is still a very hard problem. In this work, we overcome these issues and show that it is nevertheless possible to extract useful pixel-wise semantic supervision from LiDAR data.

The contributions of our work are threefold. First, we propose a novel method for unsupervised cross-modal learning of semantic image segmentation by leveraging synchronized LiDAR and image data. The key ingredient is a module that analyzes the LiDAR point cloud to obtain proposals for spatially consistent objects that can be clearly separated from each other and the ground plane in the 3D scene. Second, we show that these 3D object proposals can be aligned with input images and reliably clustered into semantically meaningful pseudo-classes by using image features from a network trained without supervision. We demonstrate that this approach is robust to noise in point clouds and delivers,

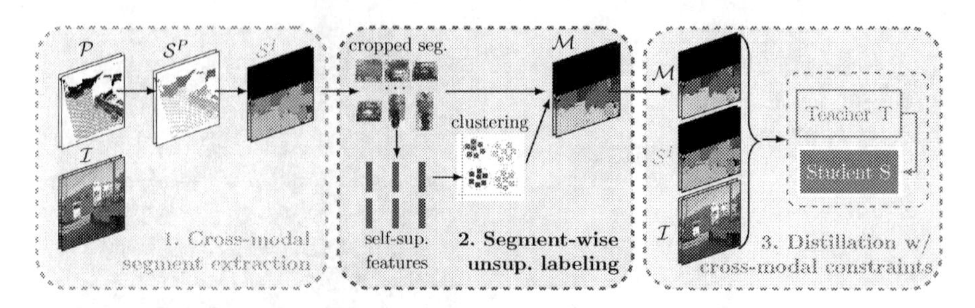

Fig. 2. Overview of Drive&Segment. We first perform cross-modal segment extraction on training dataset by exploiting raw *LiDAR* point clouds $\mathcal{P}$ and raw *images* $\mathcal{I}$. This yields segments $\mathcal{S}^I$ projected onto the image space (Sect. 3.1). By clustering their self-supervised features, we obtain an unsupervised labeling of these segments (Sect. 3.2) and, as a consequence, of their pixels. This provides pixel-wise *pseudo ground truth* for the next learning step. Finally, given the pseudo-labels and the segments, we perform distillation with cross-modal constraints (Sect. 3.3) that conjugates information of the LiDAR and the images to learn a final segmentation model using a teacher-student architecture. The learnt segmentation model S –highlighted in the figure– is used for inference on unseen datasets, yielding compelling results (Sect. 4).

without the need for any manual annotation, pseudo-classes with pixel-wise segmentation for various objects present in driving scenes. These classes include objects such as pedestrians or traffic lights that are notoriously hard to segment automatically in the image domain. Third, we develop a novel cross-modal distillation approach that first trains a teacher network with the available partial pseudo labels and then exploits its predictions for training the student with pixel-wise pseudo annotations that cover the whole image. Additionally, our approach exploits geometric constraints extracted from the LiDAR point cloud during the teacher-student learning process to refine teacher predictions that are distilled into the student network. Implemented with transformer-based networks, this cross-modal distillation approach results in a trained student model that performs well in various challenging conditions such as day, night, fog, or rain, outside the domain of the original training dataset, as shown in Fig. 1.

We train our proposed unsupervised semantic segmentation method on two datasets, Waymo Open [47] and nuScenes [8] (nuScenes results are in the appendix available in the extended version of the paper [52]), and test it on four different datasets in the autonomous driving domain, Cityscapes [16], Dark-Zurich [44], Nighttime driving [17] and ACDC [45] dataset. We demonstrate significant improvements compared to the current state of the art on this problem, improving the current best published unsupervised semantic segmentation results on Cityscapes from 15.8 to 21.8 and from 4.6 to 14.2 on Dark Zurich, measured by mean intersection over union.

2 Related Work

Image Semantic Segmentation. Semantic segmentation is a challenging key visual perception task, especially for autonomous driving [16,39,45,51,58]. Current top-performing models are based on fully convolutional networks [36] with encoder-decoder structures and a large diversity of designs [12,14,35,43,54,63]. Recent progress in vision transformers (ViT) [19] opened the door to a new wave of decoders [46,56,59,64] with appealing predictive performance. These methods attain impressive performance by exploiting large amounts of pixel-wise labeled data. Yet, urban scenes are expensive to annotate manually (1.5 h–3.3 h per image [16,45]). This motivates recent works to rely less on pixel-wise supervision.

Reducing Supervision for Semantic Segmentation. A popular strategy when dealing with limited labeled data is to pre-train some of the blocks of the architecture on related auxiliary tasks with plentiful labels [18,60]. Pre-training encoder for ImageNet [18] classification has been shown to be a successful recipe for both convnets [12] and ViT-based models [46]. Pre-training can be conducted even without any human annotations on artificially-designed self-supervised pretext tasks [9,22–24,26,28] with impressive results on a variety of downstream tasks. Some works also make use of synthetically generated data for pre-training [20,34,53]. Fully unsupervised semantic segmentation [6,13,15,29,31,32,40,50,61] has been recently addressed via generative models to generate object masks [6,13,40] or self-supervised clustering [15,31]. Prior methods are limited to segmenting foreground objects of a single class [6,13] or to *stuff* pixels that far outnumber *things* pixels [31,40]. Others assume that images contain centered objects [50], rely on weak spatial cues from the image domain [13,15,31] or require instance masks during pre-training and annotated data at test time [29]. On the contrary, our approach exploits cross-modal supervision from aligned LiDAR point clouds and images. We show that leveraging this information can considerably improve segmentation performance in complex autonomous driving scenes with multiple classes and strong class imbalance, outperforming PiCIE [15], the current state of the art in unsupervised segmentation. Concurrent work STEGO [25] develops a contrastive formulation for unsupervised semantic segmentation but does not use LiDAR during training.

Cross-Modal Self-supervised Learning. Leveraging language, vision, and/or audio, self-supervised representation learning has seen tremendous progress in recent years [2–4,38,41,42,62]. Besides learning useful representations, these approaches show that signals from one modality can help train object detectors in the other, e.g., detecting instruments that sound in a scene [11,41,62], and even other object types [1]. In autonomous driving, a vehicle is equipped with diverse sensors (e.g., camera, LiDAR, radar), and cross-modal self-supervision is often used to generate labels from a sensor for augmenting the perception of another [5,30,48,55]. LiDAR clues [48] have been recently shown to boost unsupervised object detection (for *things* classes). Both our work and [48] use the same prior method [7] to extract object proposals from LiDAR scans. However, we consider a different problem of dense pixel-wise unsupervised semantic segmentation (for both *things* and *stuff*) and design a new approach for both extraction and learning with pixel-level pseudo labels.

3 Proposed Unsupervised Semantic Segmentation

Our goal is to train an *image segmentation model* with *no human annotation*, by exploiting easily-available aligned *LiDAR* and *image* data. To that end, we propose a novel method, Drive&Segment, that consists of three major steps and is illustrated in Fig. 2. First, as discussed in Sect. 3.1, we extract *segment* proposals for the objects of interest from 3D LiDAR point clouds and project them to the aligned RGB images. In the second step (Sect. 3.2), we build *pseudo-labels* by clustering *self-supervised* image features corresponding to these segments. Finally, in Sect. 3.3, we propose a new teacher-student training scheme that incorporates *spatial constraints* from the LiDAR data and learns an unsupervised segmentation model from the noisy and partial pseudo-annotations generated in the previous two steps.

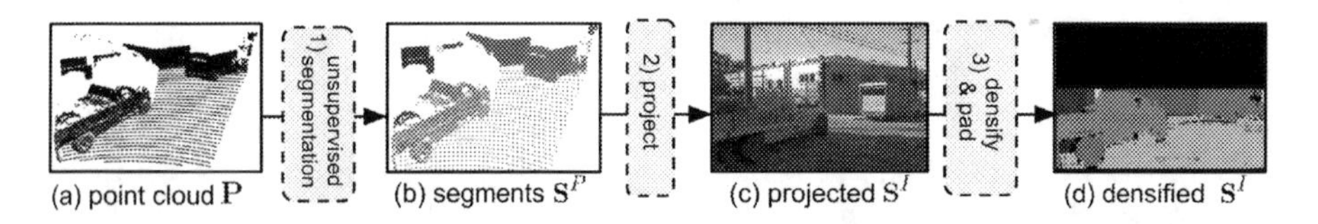

Fig. 3. Cross-modal segment extraction. Input raw point cloud (a) is first segmented with [7] into object segment candidates (b), which are then projected into the image (c); Projected segments are densified to get pixel-level pseudo-labels, with missed pixels being labeled as "ignore", as shown in black (d). (Color figure online)

3.1 Cross-Modal Segment Extraction

Throughout the next sections, we consider a dataset composed of a set $\mathcal{P}$ of 3D point clouds and a set $\mathcal{I}$ of images aligned with the point clouds. In this section, we detail the process of extracting segments of interest in an image $\mathbf{I} \in \mathcal{I}$ using the corresponding aligned LiDAR point cloud $\mathbf{P} \in \mathcal{P}$. The process, illustrated in Fig. 3, consists of three steps. We start by segmenting the LiDAR point cloud $\mathbf{P}$ using its geometrical properties. Then, we project the resulting 3D segments into the image $\mathbf{I}$, and densify the output to obtain pixel-level segments.

Geometric Point Cloud Segmentation. We first extract J non-overlapping object segmentation proposals (*segments*), from the LiDAR point cloud $\mathbf{P}$. Let $\mathbf{S}^P = \{s_j^P\}_{j=1}^J$ be this set, where each segment s_j^P is a subset of the 3D point cloud $\mathbf{P}$ and $\forall j \neq j'$, $s_j^P \cap s_{j'}^P = \varnothing$. Additionally, we refer to the set of segments

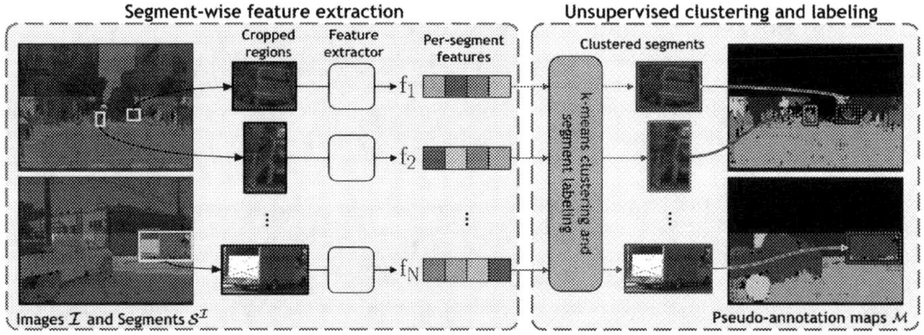

Fig. 4. Segment-wise unsupervised pseudo-labeling. First, given object segments $\mathcal{S}^{\mathcal{I}}$ obtained in the segment extraction stage (left), we take crops around all N objects and feed them to a feature extractor to get a set of N feature vectors. Then, we use the k-means algorithm to cluster the feature vectors into k clusters. Finally, we assign pixel-wise *pseudo-labels* to all pixels belonging to each segment based on the corresponding cluster id. Pixels not covered by a segment are assigned the label "ignore" (black).

across the entire data set as $\mathcal{S}^P$, with $\mathbf{S}^P \subset \mathcal{S}^P$. The J segments detected in one point cloud should ideally correspond to J individual objects in the scene. To get them, we use the unsupervised 3D point cloud segmentation proposed in [7], which exploits the geometrical properties of point clouds and range images.[1] It is a two-stage process that segments the ground plane and objects using greedy labeling by breadth-first search in the range image domain. Urban scenes are particularly suited to this purely geometry-based method as most objects are spatially well separated and the ground plane is relatively easy to segment out.

Point-Cloud-to-Image Transfer. The next step of the segment extraction is to transfer the set $\mathbf{S}^P$ of point cloud segments to the image $\mathbf{I}$, producing the set $\mathbf{S}^I$. Although LiDAR data and camera images are captured at the same time, one-to-one matching is not straightforward. Indeed, among other difficulties, LiDAR data only covers a fraction of the image plane because of its different field of view, its lower density, and its lack of usable measurements on far away objects or on the sky for instance. To overcome the mismatch between the two modalities, we proceed as follows. First, we project the points from the point cloud to the image using the known sensors' calibration. This gives us the locations of 3D points from the point cloud in the image. We also identify locations with invalid measurements in the LiDAR range image, e.g., reflective surfaces or the sky, and assign an "ignore" label to the respective locations.

Densify and Pad. Next, we perform nearest-neighbor interpolation to propagate the $J+1$ segment labels to all pixels, where J is the number of segments (ideally corresponding to objects) and $+1$ denotes the additional "ignore" label. Last, we pad the image with "ignore" label to the input image size.

[1] Range images are depth maps corresponding to the raw LiDAR measurements. Valid measurements are back-projected to the 3D space to form a point cloud.

3.2 Segment-Wise Unsupervised Labeling

Next, we produce *pseudo-labels* for all extracted segments in the image space without using any supervision. To that end, we leverage the recent ViT [19] model pre-trained in a fully unsupervised fashion [10] which has shown impressive results on various downstream tasks. We use this representation for unsupervised learning of pseudo-labels as described next and illustrated in Fig. 4.

Considering the image **I**, we crop a tight rectangular region in the image around each segment $s_j^I \in \mathbf{S}^I$ obtained using the proposal mechanism described in the previous section. We resize it and feed it to the ViT model to extract the feature $\mathbf{f}_j$ corresponding to the output features of the `CLS` token. To limit the influence of pixels outside the object segment, which may correspond to other objects or the background, we mask out these pixels before computing the features. We repeat this operation for all segments in each image **I** in the training dataset and cluster the `CLS` token features using k-means algorithm, thus discovering k clusters of visually similar segments. Therefore, each feature $\mathbf{f}_j$ and its corresponding segment s_j^I, is assigned a cluster id l_j in $[\![1, k]\!]$.

To obtain a dense *segmentation* map **M** corresponding to the image **I**, we assign discovered cluster ids to each pixel belonging to a segment in the image. Additionally, we assign a predefined *ignore* label to pixels not covered by segments, which correspond to missing annotations. This allows us to construct a set $\mathcal{M}$ of dense *maps of pseudo-annotations*, which we later use as a pseudo-ground-truth. Examples of resulting segmentation maps are shown in Fig. 4.

3.3 Distillation with Cross-Modal Spatial Constraints

After previous steps have a set of pseudo-annotated segmentation maps $\mathbf{M} \in \mathcal{M}$, one for every image **I** in the training dataset. However, as explained above, the pseudo-annotations are **only partial**, since the segments that were used to construct them do not cover all pixels of an image. Furthermore, due to imperfections in the segment extraction process or the segment clustering step, these annotations are noisy. Using them to train an image segmentation model directly might be sub-optimal. Instead, we propose a new teacher-student training approach with cross-modal distillation, which is able to learn more accurate unsupervised segmentation models under such partial and noisy pseudo-annotations.

Training the Teacher. The first step of our teacher-student approach is to train the teacher T to make pixel-wise predictions only on the pixels for which pseudo-annotations are available, i.e., only for the pixels that belong to a segment. We denote $\mathbf{Y}_\mathrm{T} = \mathrm{T}\left(\mathbf{I}\right) \in \mathbb{R}^{H \times W}$ the segmentation predictions made by the teacher model on image **I** with a resolution of $H \times W$ pixels. We train the teacher T using loss $\mathcal{L}_\mathrm{T}(\mathbf{I})$ and image **I**:

$$\mathcal{L}_\mathrm{T}(\mathbf{I}) = \frac{1}{\sum_{h,w} B_{(h,w)}} \sum_{h,w} \mathrm{CE}\left(\mathbf{Y}_{\mathrm{T}(h,w)}, \mathbf{M}_{(h,w)}\right) B_{(h,w)}, \tag{1}$$

where CE is the cross-entropy loss measuring the discrepancy between the predicted labels $\mathbf{Y}_\mathrm{T}$ and target pseudo-labels **M** for each pixel (h, w), and B is a

$H \times W$ binary mask for filtering out pixels without pseudo-annotations. The loss is normalized by the number of pseudo-labeled pixels in the image. The trained teacher T is then able to predict pixel-wise segmentation for all pixels in an image, even if they do not belong to a segment. Moreover, since the teacher T is trained on a large set of pseudo-annotated segments, it learns to smooth out some of the noise in the raw pseudo-annotations.

Integrating Spatial Constraints. Considering this smoothing property, we can exploit the trained teacher T for generating new, complete (instead of partial) and smooth pseudo-segmentation maps for training images. In addition, we propose to refine these teacher-generated pseudo-segmentation maps by using the projected LiDAR segments; indeed, these segments encode useful 3D spatial constraints as they often correspond to complete 3D objects, thus respecting the depth discontinuities and occlusion boundaries. In particular, for each image segment s_j^I in image $\mathbf{I}$, we apply majority voting to pixel-wise teacher predictions $\mathbf{Y}_{\mathrm{T}}$ inside the segment. Then we annotate each pixel belonging to the segment with its most frequently predicted label, giving us a new refined segmentation map $\hat{\mathbf{Y}}_{\mathrm{T}} \in \mathbb{R}^{H \times W}$. This procedure is illustrated in Fig. 5.

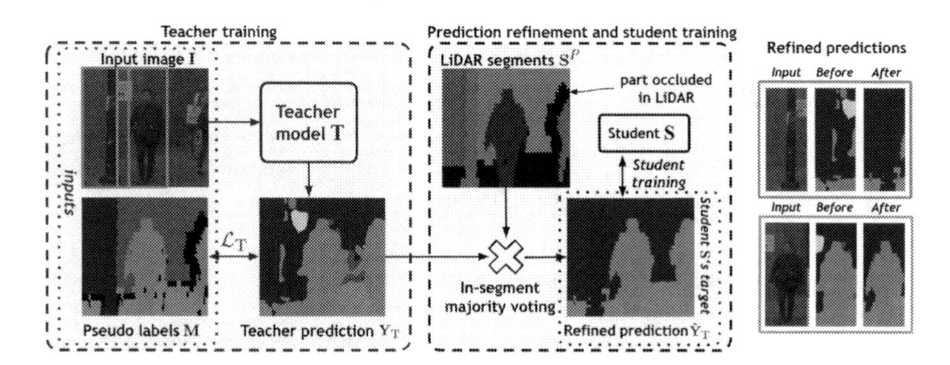

Fig. 5. Teacher prediction refinement using spatial constraints. First, the teacher T is trained using loss $\mathcal{L}_{\mathrm{T}}$ on images in $\mathcal{I}$ together with segmentation maps in $\mathcal{M}$ obtained from segment-wise unsupervised pseudo-labeling. The teacher predictions $\mathbf{Y}_{\mathrm{T}}$ are refined, using LiDAR segments $\mathbf{S}^P$, into maps $\hat{\mathbf{Y}}_{\mathrm{T}}$ that are then used to train the student. Note that teacher's predictions span the whole image, producing outputs even in areas where LiDAR segments $\mathbf{S}^P$ are not available.

Training the Student. Having computed these complete, teacher-generated, and spatially refined pseudo-segmentation maps $\hat{\mathbf{Y}}_{\mathrm{T}}$, we train a student network S using the following loss

$$\mathcal{L}_{\mathrm{distill}}(\mathbf{I}) = \frac{1}{HW} \sum_{h,w} \mathrm{CE}\left(\hat{\mathbf{Y}}_{\mathrm{T},(h,w)}, \mathbf{Y}_{\mathrm{S},(h,w)}\right), \tag{2}$$

where the cross-entropy is computed between $\hat{\mathbf{Y}}_{\mathrm{T}}$ and the segmentation map $\mathbf{Y}_{\mathrm{S}} \in \mathbb{R}^{H \times W}$ predicted by the student at the same resolution as the teacher. The outputs of the trained student are our final unsupervised image segmentation predictions. Further details about our training can be found in Sect. 4.1.

4 Experiments

In this section, we give the implementation details, compare our results with the state-of-the-art unsupervised semantic segmentation methods on four different datasets, and ablate the key components of our approach.

Methods and Architectures. We investigate the benefits of our approach using two different semantic segmentation models to demonstrate the generality of our method. We implement Drive&Segment with both a classical convolutional model and a transformer-based architecture. For the convolutional architecture, we follow [15] and use a ResNet18 [27] backbone followed by an FPN [35] decoder. For the transformer-based architecture, we use the state-of-the-art Segmenter [46] model. We use the ViT-S/16 [19] model as the Segmenter's encoder and use a single layer of the mask transformer [46] as a decoder. We compare our method to three recent unsupervised segmentation models: IIC [31], modified version of DeepCluster [9] (DC), and PiCIE [15]. Please refer to [15] for implementation details of IIC and DC.

Training. In the following, we first discuss how we obtain segment labels by k-means clustering, then we talk about details of pre-training the model backbones, which is followed by the discussion of the datasets for actual training of the models. Finally, we give details of the training procedure.

K-means. We use $k = 30$ in the k-means algorithm (the ablation of the value of k is in the appendix [52]. To extract segment-wise features used for k-means clustering, we use `CLS` token features of the DINO-trained [10] ViT-S [19] model. Obtained segment-wise labels serve as pseudo-annotations for training the ResNet18-FPN and Segmenter models, as discussed in Sect. 3.2.

Pre-training Data and Networks. To be comparable to [15], in our experiments with the ResNet18+FPN model, we initialize its backbone with a ResNet18 trained with supervision on the ImageNet-1k [18] classification task, exactly as all the compared prior methods (PiCIE, DC, and IIC). However, as we aim for a completely unsupervised setup, we initialize the ViT-S backbone of the

Segmenter model with weights the self-supervised approach DINO [10] on ImageNet-1k [18]. The decoders of our models are randomly initialized.

Training Datasets. We train our models on about 7k images from the Waymo Open [47] dataset, which has both image and LiDAR data available. For the baseline methods (IIC [31], modified DC [9] and PiCIE [15]), we take models from PiCIE [15] codebase, i.e., models that are trained on all available images of Cityscapes [16], meaning the 24.5k images from the *train*, *test*, and *train_extra* splits. Note that those models then do not face the problem of domain gap when evaluated on the Cityscapes [16] dataset. To be directly comparable with our approach, we also train a variant of modified DC [9] and PiCIE [15] on the same subset of the Waymo Open [47] dataset as used in our approach. Furthermore, to test the generalizability of our method to other training datasets, we provide results when training on the nuScenes [8] dataset in the appendix [52].

Optimization. To train IIC [31], modified DC [9], and PiCIE [15], we use the setup provided in [15]. For our Drive&Segment, we train the teacher and student models with batches of size 32 and with a learning rate of $2e{-}4$ with a polynomial schedule on a single V100 GPU. During training, we perform data augmentation consisting of random image resizing in the $(0.5, 2.0)$ range, random cropping to 512×512 pixels, random horizontal flipping, and photometric distortions.

Evaluation Protocol. *Mapping.* To evaluate our models in the unsupervised setup, we run trained models on every image, thus getting segmentation predictions with values from 1 to k. Then, we compute the confusion matrix between the C ground-truth classes of the target dataset and the $k \geq C$ pseudo-classes. We map the C ground-truth classes to C out of the k pseudo-classes using Hungarian matching [33]. The pixel predictions for the $k - C$ unmapped pseudo-classes are considered as false negatives.

Test datasets. We evaluate our fully-unsupervised models on the *full-resolution images* of Cityscapes [16], Dark Zurich [44], Nighttime driving [17] and ACDC [45] datasets, *without any finetuning* (no samples from these datasets are ever seen during training). Cityscapes [16] is a well-established dataset with 500 validation images that we use for evaluation. Dark Zurich [44] and Nighttime driving [17] are two nighttime datasets, each with 50 validation images annotated for semantic segmentation that we use for evaluation. ACDC [45] is a recent dataset providing four different adverse weather conditions with 400 training and 100 validation samples per weather condition. We test our approach on the validation images annotated for semantic segmentation. The Cityscapes dataset defines 30 different semantic classes for the pixel-wise semantic segmentation task. Unless stated otherwise, we follow prior work and evaluate our approach on the pre-defined subset of 19 classes [16] for all datasets.

Metrics. Using the mapping, we evaluate the results using two standard metrics for the semantic segmentation task, the mean Intersection over Union, mIoU, and the pixel accuracy, PA, as done in prior work [15]. The mIoU is the mean intersection over union averaged over all classes, while PA defines the percentage of pixels in the image that are segmented correctly, averaged over all images.

4.1 Comparison to State of the Art

Here we evaluate our trained models in the unsupervised setup using the evaluation protocol described above. We compare our method using both the Segmenter [46] and ResNet18+FPN models to three recent unsupervised segmentation models: IIC [31], modified version of DeepCluster [9] (mod. DC), and PiCIE [15]. In the appendix [52], we assess the utility of the features learned by our model in other settings, such as k-NN pixel-wise classification, and linear probing and fine-tuning for semantic segmentation.

We provide results on the Cityscapes, Dark Zurich, and Nighttime Driving datasets in Table 1, and show qualitative results in Fig. 6. As shown in the first two columns of Table 1, our approach (D&S) outperforms [15] on the Cityscapes dataset by a large margin in both the 19-class and 27-class set-ups. Improvements are visible for both architectures, but the best results are usually obtained with the distilled Segmenter architecture using the ViT-S/16 backbone. Our models again outperform [15] in all setups. In addition, we observe a better performance of our models compared to [15] when evaluating on the nighttime scenes. For example, on the Dark Zurich [44] dataset, the mIoU of PiCIE [15] decreases by 71% compared to the results on Cityscapes (15.8 → 4.6), while the mIoU of our Segmenter-based model decreases only by 35% (21.8 → 14.2). This suggests that our models generalize significantly better to out-of-distribution scenes. These findings hold for PiCIE models trained on both Cityscapes and on Waymo Open.

Finally, Table 2 shows results on the ACDC dataset in four different weather conditions. Results follow a similar trend as in Table 1 and show the superiority of our approach measured by mIoU compared to the current SoTA unsupervised

Table 1. Comparison to the state of the art for unsupervised semantic segmentation on Cityscapes [16] (CS), DarkZurich [44] (DZ) and Nighttime driving [17] (ND) datasets measured by the mean IoU (mIoU). The colored differences are reported w.r.t. the SoTA approach of [15] denoted by ⚓. The *sup. init.* abbreviation stands for supervised initialization of the *encoder*, and the column *train. data* indicates whether Cityscapes (CS) or Waymo Open (WO) dataset was used for training.

architecture, method	sup. init.	train. data	CS19 [16] mIoU		CS27 [16] mIoU		DZ [44] mIoU		ND [17] mIoU	
RN18+FPN										
IIC† [31]	yes	CS	-		6.4	(−4.8)	-		-	
Modified DC‡ [9]	yes	CS	11.3	(−4.5)	7.9	(−3.3)	7.5	(+2.9)	8.2	(−1.3)
⚓ PiCIE‡ [15]	yes	CS	15.8		11.2		4.6		9.5	
Modified DC*	yes	WO	11.4	(−4.4)	7.0	(−4.1)	5.9	(+1.3)	8.2	(−1.3)
PiCIE*	yes	WO	13.7	(−2.1)	9.7	(−1.5)	4.9	(+0.3)	9.3	(−0.2)
D&S (Ours, S)	yes	WO	19.5	(+3.7)	**16.2**	(+5.1)	10.9	(+6.3)	14.4	(+4.9)
Segmenter, ViT-S/16										
D&S (Ours, S)	no	WO	**21.8**	(+6.0)	15.3	(+4.1)	**14.2**	(+9.6)	**18.9**	(+9.3)

† Results reported in [15]. ‡ Models provided by the PiCIE [15] authors.
* Trained by PiCIE code base.

Table 2. Comparison to the state-of-the-art for unsupervised semantic segmentation on the ACDC [45] dataset. Please refer to Table 1 for the used symbols.

arch., Method	sup. Init.	train. Data	night mIoU	fog mIoU	rain mIoU	snow mIoU	average mIoU
RN18+FPN							
mod. DC‡ [9]	yes	CS	8.1 (+3.7)	8.3 (-4.0)	6.9 (-5.6)	7.4 (-4.7)	7.7 (-2.6)
⚓PiCIE‡ [15]	yes	CS	4.4	12.2	12.5	12.1	10.3
mod. DC⋆	yes	WO	5.9 (+1.5)	11.7 (-0.5)	9.6 (-2.9)	9.8 (-2.3)	9.2 (-1.0)
PiCIE⋆	yes	WO	4.7 (+0.3)	14.4 (+2.1)	13.7 (+1.2)	14.3 (+2.2)	11.7 (+1.5)
D&S (Ours,S)	yes	WO	11.2 (+6.8)	14.5 (+2.3)	14.9 (+2.5)	14.6 (+2.6)	13.8 (+3.5)
Segmenter, ViT-S/16							
D&S (Ours,S)	no	WO	**13.8** (+9.4)	**18.1** (+5.9)	**16.4** (+3.9)	**18.7** (+6.6)	**16.7** (+6.5)

semantic segmentation method of [15] on images out of the training distribution, such as images at night or in snow. Please see the appendix available in the extended version of the paper [52] for the complete set of results, including results using nuScenes, pixel accuracy, per-class results, and confusion matrices.

4.2 Ablations

In this section, we ablate the main components of our approach, which we present in Table 3, and discuss them in more detail below.

Segment Extraction Approach. To evaluate the benefits of our cross-modal segment extraction module, we investigate using segment proposals generated with a purely image-based segmentation approach by Felzenszwalb and Huttenlocher (FH) [21]. It groups pixels into segments based on similar color and texture properties. We use the same set of hyperparameters as [28]. The results are shown in Table 3a and demonstrate clear benefits of our LiDAR-based cross-modal segment extraction method despite the difficulties of using LiDAR data discussed in Sect. 3.1. We attribute the better results of our approach to the fact that LiDAR data segmentation operates with range information, which is much stronger at separating objects from the background and from each other compared to the purely image-based approach of FH [21]. Indeed, FH relies only on color/texture and is therefore much more likely to join multiple objects into one segment or separate a single object into multiple segments. The benefits of our cross-modal segment extraction are observed for both studied architectures.

Distillation with Cross-Modal Spatial Constraints. To evaluate the benefits of our teacher-student distillation method with cross-modal spatial constraints (Sect. 3.3), we compare the predictions of the teacher T (before distillation) and the student S (after distillation). Table 3b presents results on the Cityscapes dataset using both convolutional- and transformer-based architectures. The results show consistent improvements using our distillation technique, particularly regarding the pixel accuracy metric. We believe that this could be attributed to improvements in predictions for classes such as vegetation and

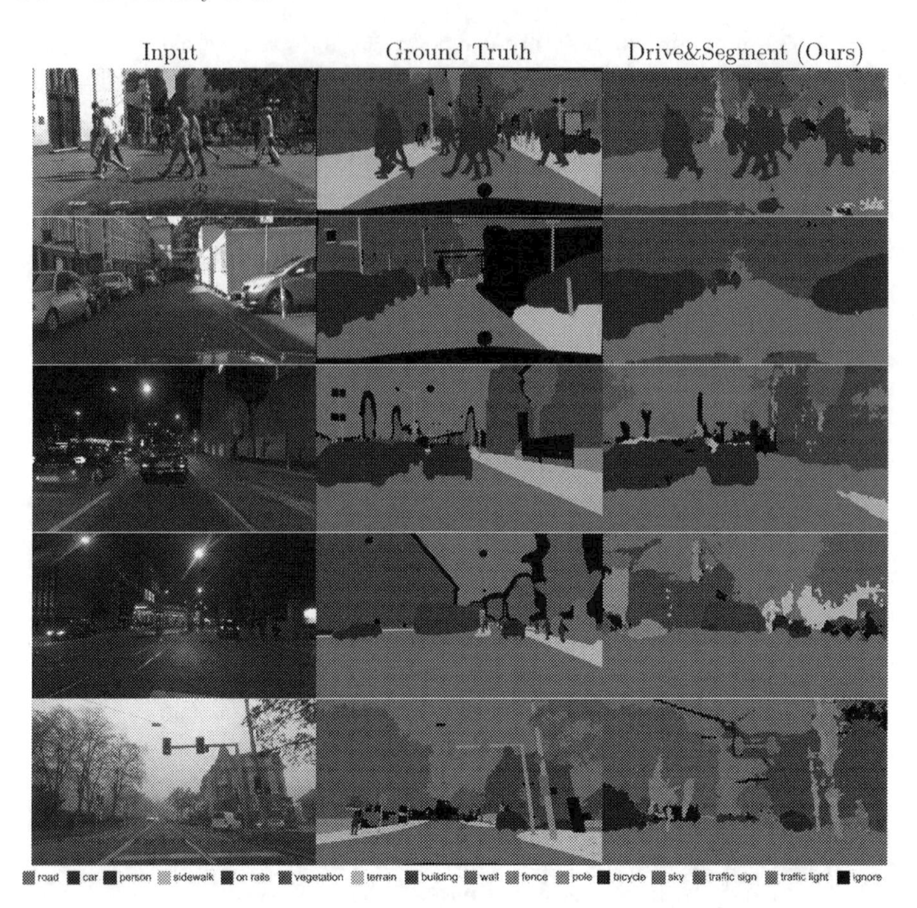

Fig. 6. Qualitative results for *unsupervised* semantic segmentation using our Drive&Segment approach. To obtain the matching between our pseudo-classes and the set of ground-truth classes, we use the Hungarian algorithm. The first two rows show samples from Cityscapes [16], and the three bottom rows show samples from the night and fog splits of the ACDC [45] dataset. See appendix in [52] for more results.

buildings. They often occupy large areas of the image and benefit most from the distillation as they are usually not well covered by the LiDAR scans. Furthermore, the results show clear benefits of using this distillation step both with and without cross-modal spatial constraints (LiD) by Student S outperforming Teacher T in both scenarios. Please also note that our distillation technique works well even in combination with another training approach (PiCIE [15]).

Table 3. Ablations on the Cityscapes dataset. (a) Benefits of our segment extraction method over segment proposals from [21]. **(b)** Benefits of our distillation approach showing an improvement of the student (S) over the teacher (T) and benefits of our LiDAR cross-modal spatial constraints (LiD). **(c)** Ablation of different feature extractors for the k-means clustering.

(a) Segment extraction

arch. / seg. prop.	mIoU	PA
RN18+FPN		
FH [21]	15.5	52.8
Ours	**17.4** (+1.9)	**55.9** (+3.1)
Segmenter		
FH [21]	15.8	51.8
Ours	**20.4** (+4.6)	**65.4** (+13.6)

(b) Distillation

model	LiD.	mIoU	PA
RN18+FPN			
PiCIE (T)		13.7	48.6
PiCIE (S)		14.8 (+1.1)	64.1 (+15.5)
PiCIE (S)	✓	15.1 (+1.4)	**68.4** (+19.8)
Ours (T)		17.4	55.9
Ours (S)		18.8 (+1.4)	63.4 (+7.5)
Ours (S)	✓	**19.5** (+2.1)	**66.4** (+10.5)
Segmenter			
Ours (T)		20.4	65.4
Ours (S)		20.8 (+0.4)	68.5 (+3.1)
Ours (S)	✓	**21.8** (+1.4)	**69.5** (+4.1)

(l) Feature extractors

arch. / method	mIoU	PA
ViT-S/16		
DeiT [49]	21.7	73.0
DINO [10]	20.2	64.4
ResNet18		
supervised [27]	19.6	70.0
ResNet50		
supervised [27]	21.3	67.6
OBOW [22]	20.7	65.9
PixPro [57]	20.7	65.9
MaskCon. [50]	19.1	68.0

Sensitivity to the Initialization. To study the influence of initialization, we take the features extracted by DINO [10] and run the k-means clustering (Sect. 3.2) four times. For each k-means clustering outcome, we run the segmentation model training four times with different initializations. The variance over all k-means and training runs is only 0.5 for mIoU and 1.5 for pixel accuracy (i.e., $20.4 \pm 0.5/65.4 \pm 1.5$). These results clearly show that our method is not very sensitive to k-means initialization or to the network initialization.

Feature Extractors. An ablation of different convolutional and ViT feature extractors for the task of segment-wise unsupervised labelling is shown in Table 3c. The results on the Cityscapes [16] dataset using our Segmenter model demonstrate that our approach works well with feature extractors.

LiDAR Resolution and Number of Clusters. Ablation of the influence of LiDAR resolution is in the appendix [52] and demonstrates that our method is robust to LiDAR's sparsity. Furthermore, we study the choice of the number of clusters for the k-means clustering in the appendix [52].

4.3 Limitations and Failure Modes

Our approach has the following three main limitations. First, LiDAR point clouds do not provide information about very distant or even infinitely distant objects, e.g., the sky, which our approach cannot learn to segment. Second, LiDAR point clouds paired with geometric segmentation can not correctly distinguish road from sidewalk or grass, when all surfaces are similarly flat. Both the above limitations might be possibly tackled by pairing our LiDAR-based segment proposals with an unsupervised image-based method such as [21], or by introducing simple heuristics. Also, the LiDAR points must not be too sparse (e.g., only 4 beams), since otherwise the LiDAR-based segments would be of poor quality. However, this is not an overly restricting requirement as it is common to use LiDAR sensors with sufficient beam resolution, e.g., as in the recent Waymo Open [47] or ONCE [37] datasets. Finally, we encounter semantically similar objects appearing in multiple pseudo-classes, a natural side effect of clustering. This issue may be mitigated by using different feature clustering methods that would allow the measurement of similarities on manifolds in the feature space.

5 Conclusion

We have developed Drive&Segment, a fully unsupervised approach for semantic image segmentation in urban scenes. The approach relies on novel modules for (i) cross-modal segment extraction and (ii) distillation with cross-modal constraints that leverage LiDAR point clouds aligned with images. We evaluate our approach on four different autonomous driving datasets in challenging weather and illumination conditions and demonstrate major gains over prior work. This work opens up the possibility of large-scale autonomous learning of embodied perception models without explicit human supervision.

Acknowledgments. This work was supported by the European Regional Development Fund under the project IMPACT (no. CZ.02.1.010.00.015_0030000468), by the Ministry of Education, Youth and Sports of the Czech Republic through the e-INFRA CZ (ID:90140), and by CTU Student Grant SGS21184OHK33T37.

References

1. Afouras, T., Asano, Y.M., Fagan, F., Vedaldi, A., Metze, F.: Self-supervised object detection from audio-visual correspondence. arXiv (2021)
2. Alayrac, J.B., et al.: Self-supervised multimodal versatile networks. In: NeurIPS (2020)
3. Alwassel, H., Mahajan, D., Korbar, B., Torresani, L., Ghanem, B., Tran, D.: Self-supervised learning by cross-modal audio-video clustering. In: NeurIPS (2020)
4. Arandjelovic, R., Zisserman, A.: Look, listen and learn. In: ICCV (2017)
5. Bartoccioni, F., Zablocki, É., Pérez, P., Cord, M., Alahari, K.: Lidartouch: Monocular metric depth estimation with a few-beam lidar. In: arXiv (2021)
6. Bielski, A., Favaro, P.: Emergence of object segmentation in perturbed generative models. In: NeurIPS (2019)
7. Bogoslavskyi, I., Stachniss, C.: Efficient online segmentation for sparse 3d laser scans. In: PFG (2017)
8. Caesar, H., et al.: nuscenes: A multimodal dataset for autonomous driving. In: CVPR (2020)
9. Caron, M., Bojanowski, P., Joulin, A., Douze, M.: Deep clustering for unsupervised learning of visual features. In: Ferrari, V., Hebert, M., Sminchisescu, C., Weiss, Y. (eds.) Computer Vision – ECCV 2018. LNCS, vol. 11218, pp. 139–156. Springer, Cham (2018). https://doi.org/10.1007/978-3-030-01264-9_9

10. Caron, M., et al.: Emerging Properties in Self-Supervised Vision Transformers. In: ICCV (2021)
11. Chen, H., Xie, W., Afouras, T., Nagrani, A., Vedaldi, A., Zisserman, A.: Localizing visual sounds the hard way. In: CVPR (2021)
12. Chen, L.C., Zhu, Y., Papandreou, G., Schroff, F., Adam, H.: Encoder-decoder with atrous separable convolution for semantic image segmentation. In: ECCV (2018)
13. Chen, M., Artières, T., Denoyer, L.: Unsupervised object segmentation by redrawing. In: NeurIPS (2019)
14. Cheng, B., et al.: Panoptic-deeplab: A simple, strong, and fast baseline for bottom-up panoptic segmentation. In: CVPR (2020)
15. Cho, J.H., Mall, U., Bala, K., Hariharan, B.: PiCIE: Unsupervised semantic segmentation using invariance and equivariance in clustering. In: CVPR (2021)
16. Cordts, M., et al.: The cityscapes dataset for semantic urban scene understanding. In: CVPR (2016)
17. Dai, D., Van Gool, L.: Dark model adaptation: Semantic image segmentation from daytime to nighttime. In: IEEE ITSC (2018)
18. Deng, J., Dong, W., Socher, R., Li, L.J., Li, K., Fei-Fei, L.: ImageNet: A Large-Scale Hierarchical Image Database. In: CVPR (2009)
19. Dosovitskiy, A., et al.: An image is worth 16×16 words: Transformers for image recognition at scale. In: ICLR (2021)
20. Dosovitskiy, A., et al.: Flownet: Learning optical flow with convolutional networks. In: ICCV (2015)
21. Felzenszwalb, P.F., Huttenlocher, D.P.: Efficient graph-based image segmentation. In: IJCV (2004)
22. Gidaris, S., Bursuc, A., Puy, G., Komodakis, N., Cord, M., Pérez, P.: Obow: Online bag-of-visual-words generation for self-supervised learning. In: CVPR (2021)
23. Gidaris, S., Singh, P., Komodakis, N.: Unsupervised representation learning by predicting image rotations. In: ICLR (2018)
24. Grill, J., et al.: Bootstrap your own latent - A new approach to self-supervised learning. In: NeurIPS (2020)
25. Hamilton, M., et al.: Unsupervised semantic segmentation by distilling feature correspondences. In: ICLR (2022)
26. He, K., Fan, H., Wu, Y., Xie, S., Girshick, R.B.: Momentum contrast for unsupervised visual representation learning. In: CVPR (2020)
27. He, K., Zhang, X., Ren, S., Sun, J.: Deep residual learning for image recognition. In: CVPR (2016)
28. Hénaff, O.J., Koppula, S., Alayrac, J.B., Oord, A.v.d., Vinyals, O., Carreira, J.: Efficient visual pretraining with contrastive detection. In: ICCV (2021)
29. Hwang, J.J., et al.: Segsort: Segmentation by discriminative sorting of segments. In: ICCV, pp. 7334–7344 (2019)
30. Jaritz, M., Vu, T.H., Charette, R.d., Wirbel, E., Pérez, P.: xmuda: Cross-modal unsupervised domain adaptation for 3d semantic segmentation. In: CVPR (2020)
31. Ji, X., Henriques, J.F., Vedaldi, A.: Invariant information clustering for unsupervised image classification and segmentation. In: ICCV (2019)
32. Kanezaki, A.: Unsupervised image segmentation by backpropagation. In: ICASSP (2018)
33. Kuhn, H.W., Yaw, B.: The hungarian method for the assignment problem. NRLQ (1955)
34. Li, D., Yang, J., Kreis, K., Torralba, A., Fidler, S.: Semantic segmentation with generative models: Semi-supervised learning and strong out-of-domain generalization. In: CVPR, pp. 8300–8311 (2021)

35. Lin, T.Y., Dollár, P., Girshick, R., He, K., Hariharan, B., Belongie, S.: Feature pyramid networks for object detection. In: CVPR (2017)
36. Long, J., Shelhamer, E., Darrell, T.: Fully convolutional networks for semantic segmentation. In: CVPR (2015)
37. Mao, J., et al.: One million scenes for autonomous driving: Once dataset. In: NeurIPS (2021)
38. Miech, A., Alayrac, J.B., Smaira, L., Laptev, I., Sivic, J., Zisserman, A.: End-to-end learning of visual representations from uncurated instructional videos. In: CVPR (2020)
39. Neuhold, G., Ollmann, T., Rota Bulo, S., Kontschieder, P.: The mapillary vistas dataset for semantic understanding of street scenes. In: ICCV (2017)
40. Ouali, Y., Hudelot, C., Tami, M.: Autoregressive unsupervised image segmentation. In: Vedaldi, A., Bischof, H., Brox, T., Frahm, J.-M. (eds.) ECCV 2020. LNCS, vol. 12352, pp. 142–158. Springer, Cham (2020). https://doi.org/10.1007/978-3-030-58571-6_9
41. Owens, A., Efros, A.A.: Audio-visual scene analysis with self-supervised multisensory features. In: Ferrari, V., Hebert, M., Sminchisescu, C., Weiss, Y. (eds.) ECCV 2018. LNCS, vol. 11210, pp. 639–658. Springer, Cham (2018). https://doi.org/10.1007/978-3-030-01231-1_39
42. Recasens, A., et al.: Broaden your views for self-supervised video learning. In: ICCV (2021)
43. Ronneberger, O., Fischer, P., Brox, T.: U-Net: Convolutional networks for biomedical image segmentation. In: Navab, N., Hornegger, J., Wells, W.M., Frangi, A.F. (eds.) MICCAI 2015. LNCS, vol. 9351, pp. 234–241. Springer, Cham (2015). https://doi.org/10.1007/978-3-319-24574-4_28
44. Sakaridis, C., Dai, D., Van Gool, L.: Map-guided curriculum domain adaptation and uncertainty-aware evaluation for semantic nighttime image segmentation. In: IEEE TPAMI (2020)
45. Sakaridis, C., Dai, D., Van Gool, L.: ACDC: The adverse conditions dataset with correspondences for semantic driving scene understanding. In: ICCV (2021)
46. Strudel, R., Garcia, R., Laptev, I., Schmid, C.: Segmenter: Transformer for semantic segmentation. In: ICCV (2021)
47. Sun, P., et al.: Scalability in perception for autonomous driving: Waymo open dataset. In: CVPR (2020)
48. Tian, H., Chen, Y., Dai, J., Zhang, Z., Zhu, X.: Unsupervised object detection with lidar clues. In: CVPR (2021)
49. Touvron, H., Cord, M., Douze, M., Massa, F., Sablayrolles, A., Jégou, H.: Training data-efficient image transformers & distillation through attention. In: ICML (2021)
50. Van Gansbeke, W., Vandenhende, S., Georgoulis, S., Van Gool, L.: Unsupervised semantic segmentation by contrasting object mask proposals. In: ICCV (2021)
51. Varma, G., Subramanian, A., Namboodiri, A., Chandraker, M., Jawahar, C.: Idd: A dataset for exploring problems of autonomous navigation in unconstrained environments. In: WACV (2019)
52. Vobecky, A., et al.: Drive&segment: Unsupervised semantic segmentation of urban scenes via cross-modal distillation. arxiv.org/abs/2203.11160 (2022)
53. Vobecky, A., Hurych, D., Uřičář, M., Pérez, P., Sivic, J.: Artificial dummies for urban dataset augmentation. In: AAAI, vol. 35, pp. 2692–2700 (2021)
54. Wang, J., et al.: Deep high-resolution representation learning for visual recognition. In: IEEE TPAMI (2020)
55. Weston, R., Cen, S., Newman, P., Posner, I.: Probably unknown: Deep inverse sensor modelling radar. In: ICRA (2019)

56. Xie, E., Wang, W., Yu, Z., Anandkumar, A., Alvarez, J.M., Luo, P.: Segformer: Simple and efficient design for semantic segmentation with transformers. arXiv preprint arXiv:2105.15203 (2021)
57. Xie, Z., Lin, Y., Zhang, Z., Cao, Y., Lin, S., Hu, H.: Propagate yourself: Exploring pixel-level consistency for unsupervised visual representation learning. In: CVPR (2021)
58. Yu, F., et al.: Bdd100k: A diverse driving dataset for heterogeneous multitask learning. In: CVPR (2020)
59. Yuan, Y., Chen, X., Wang, J.: Object-contextual representations for semantic segmentation. In: Vedaldi, A., Bischof, H., Brox, T., Frahm, J.-M. (eds.) ECCV 2020. LNCS, vol. 12351, pp. 173–190. Springer, Cham (2020). https://doi.org/10.1007/978-3-030-58539-6_11
60. Zamir, A.R., Sax, A., Shen, W., Guibas, L.J., Malik, J., Savarese, S.: Taskonomy: Disentangling task transfer learning. In: CVPR (2018)
61. Zhang, X., Maire, M.: Self-supervised visual representation learning from hierarchical grouping. Adv. Neural. Inf. Process. Syst. **33**, 16579–16590 (2020)
62. Zhao, H., Gan, C., Rouditchenko, A., Vondrick, C., McDermott, J., Torralba, A.: The sound of pixels. In: Ferrari, V., Hebert, M., Sminchisescu, C., Weiss, Y. (eds.) ECCV 2018. LNCS, vol. 11205, pp. 587–604. Springer, Cham (2018). https://doi.org/10.1007/978-3-030-01246-5_35
63. Zhao, H., Shi, J., Qi, X., Wang, X., Jia, J.: Pyramid scene parsing network. In: CVPR (2017)
64. Zheng, S., et al.: Rethinking semantic segmentation from a sequence-to-sequence perspective with transformers. In: CVPR (2021)

CenterFormer: Center-Based Transformer for 3D Object Detection

Zixiang Zhou[1,2(✉)], Xiangchen Zhao[1], Yu Wang[1], Panqu Wang[1], and Hassan Foroosh[2]

[1] TuSimple, San Diego, USA
{xiangchen.zhao,yu.wang,panqu.wang}@tusimple.ai
[2] Computational Imaging Lab., University of Central Florida, Orlando, USA
zhouzixiang@knights.ucf.edu, hassan.foroosh@ucf.edu

Abstract. Query-based transformer has shown great potential in constructing long-range attention in many image-domain tasks, but has rarely been considered in LiDAR-based 3D object detection due to the overwhelming size of the point cloud data. In this paper, we propose **CenterFormer**, a center-based transformer network for 3D object detection. CenterFormer first uses a center heatmap to select center candidates on top of a standard voxel-based point cloud encoder. It then uses the feature of the center candidate as the query embedding in the transformer. To further aggregate features from multiple frames, we design an approach to fuse features through cross-attention. Lastly, regression heads are added to predict the bounding box on the output center feature representation. Our design reduces the convergence difficulty and computational complexity of the transformer structure. The results show significant improvements over the strong baseline of anchor-free object detection networks. CenterFormer achieves state-of-the-art performance for a single model on the Waymo Open Dataset, with 73.7% mAPH on the validation set and 75.6% mAPH on the test set, significantly outperforming all previously published CNN and transformer-based methods. Our code is publicly available at https://github.com/TuSimple/centerformer

Keywords: LiDAR point cloud · 3D object detection · Transformer · Multi-frame fusion

1 Introduction

LiDAR is an important sensing and perception tool in autonomous driving due to its ability to provide highly accurate 3D point cloud data of the scanned

Z. Zhou– Work done during an internship at TuSimple.

Supplementary Information The online version contains supplementary material available at https://doi.org/10.1007/978-3-031-19839-7_29.

S. Avidan et al. (Eds.): ECCV 2022, LNCS 13698, pp. 496–513, 2022.
https://doi.org/10.1007/978-3-031-19839-7_29

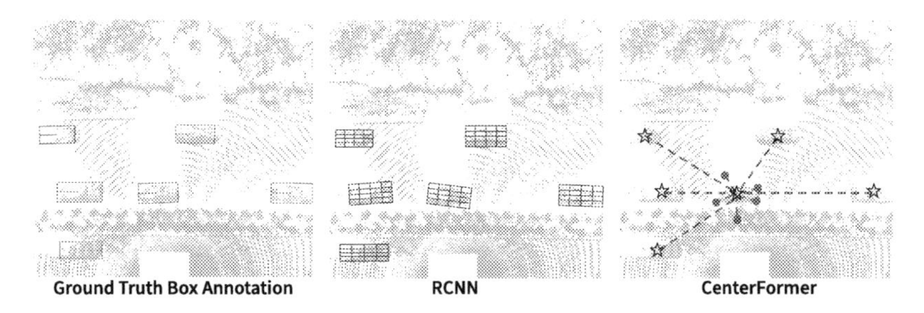

Fig. 1. The comparison of CenterFormer with RCNN-style detector. RCNN aggregates point or grid features in RoI, while CenterFormer can learn object-level contextual information and long range features through an attention mechanism.

environment. LiDAR-based 3D object detection aims to detect the bounding boxes of the objects in the LiDAR point cloud. Compared to image-domain object detection, the scanned points in LiDAR data may be sparse and irregularly spaced depending on the distance from the sensor. Most recent methods rely on discretizing the point clouds into voxels [47,58] or projected bird's eye view (BEV) feature maps [18] to use 2D or 3D convolution networks. Sometimes, it requires a second stage RCNN [11]-style refinement network to compensate for the information loss in the voxelization. However, current two-stage networks [34,51] lack contextual and global information learning. They only use the local features of the proposal (RoI) to refine the results. The features in other boxes or neighboring positions that could also be beneficial to the refinement are neglected. Moreover, the environment of the autonomous driving scene is not stationary. The local feature learning has more limitations when using a sequence of scans.

In the image domain, transformer encoder-decoder structure has become a competitive method for the detection [4,62] and segmentation [2,42] tasks. The transformer is able to capture long-range contextual information in the whole feature map and different feature domains. One of the most representative methods is DETR [4], which uses the parametric query to directly learn object information from an encoder-decoder transformer. DETR is trained end-to-end as a set matching problem to avoid any handcrafting processes like non-maximum suppression (NMS). However, there are two major problems in the DETR-style encoder-decoder transformer network: First, the computational complexity grows quadratically as the input size increases. This limits the transformer to take only low-dimensional features as input which leads to low performance on small objects. Second, the query embedding is learned through the network so that the training is hard to converge.

Can we design a transformer encoder-decoder network for the LiDAR point cloud in order to better perceive the global connection of point cloud data? Considering the sheer size of LiDAR point cloud data, and the relatively small sizes of objects to be detected, voxel or BEV feature map representations need to be large enough to keep the features for such objects to be separable. As a result, it is impractical to use the transformer encoder structure on the feature map due to the large input size. In addition, if we use a large feature map for the transformer decoder, the query embedding is also difficult to focus on meaningful attention during training. To mitigate these converging problems, one solution is to provide the transformer with a good initial query embedding and confine the attention learning region to a smaller range. In the center-based 3D object detection network [51], the feature at the center of an object is used to capture all object information, hence the center feature is a good substitute for the object feature embedding. Multi-scale image pyramid and deformable convolution [16] are two common methods to increase the receptive field of the feature learning without significantly increasing the complexity. Some recent works [15,62] apply these two methods in the transformer networks.

Taking the aforementioned aspects into consideration, we propose a center-based transformer network, named **Center Transformer (CenterFormer)**, for 3D object detection. Specifically, we first use a standard voxel-based backbone network to encode the point cloud into a BEV feature representation. Next, we employ a multi-scale center proposal network to convert the feature into different scales and predict the initial center locations. The feature at the proposed center is fed into a transformer decoder as the query embedding. In each transformer module, we use a deformable cross attention layer to efficiently aggregate the features from the multi-scale feature map. The output object representation then regresses to other object properties to create the final object prediction. As shown in Fig. 1, our method can model object-level connection and long-range feature attention. To further explore the ability of the transformer, we also propose a multi-frame design to fuse features from different frames through cross-attention. We test CenterFormer on the large scale Waymo Open Dataset [38] and the nuScenes dataset [3]. Our method outperforms the popular center-based 3D object detection networks which are dominant on public benchmarks by a large margin, achieving state-of-the-art performance, with 73.7% and 75.6% mAPH on the waymo validation and test sets, respectively. The contributions of our method can be summarised as follows:

- We introduce a center-based transformer network for 3D object detection.
- We use the center feature as the initial query embedding to facilitate learning of the transformer.
- We propose a multi-scale cross-attention layer to efficiently aggregate neighboring features without significantly increasing the computational complexity.
- We propose using the cross-attention transformer to fuse object features from different frames.
- Our method outperforms all previously published methods by a large margin, setting a new state-of-the-art performance on the Waymo Open Dataset.

2 Related Work

2.1 LiDAR-Based 3D Object Detection

Compared to the well-established point cloud processing networks like PointNet [31] and PointNet++ [32], most recent LiDAR detection and segmentation methods [18,54,58,59,61] voxelize the point cloud in a fixed 3D space into a BEV/voxel representation and use conventional 2D/3D convolutional networks to predict the 3D bounding boxes. Other methods [1,9,39,44] detect the objects on a projected range image. There are also some methods that use hybrid features along with the voxel network [30,34,49], and combine multi-view features in the voxel feature representation [34]. VoxelNet [58] uses a PointNet inside each voxel to encode all points into a voxel feature. This feature encoder later became an essential method in voxel-based point cloud networks. PointPillar [18] proposes the pillar feature encoder to directly encode the point cloud into the BEV feature map so that only 2D convolution is needed in the network.

Similar to image object detection, 3D object detection methods can be divided into anchor-based [18,34,36,47] and anchor-free [10,51] methods. Anchor-based methods detect objects through a classification of all predefined object anchors, while anchor-free methods generally consider objects as keypoints and find those keypoints at the local heatmap maxima. Even though anchor-based methods can achieve a good performance, they rely heavily on hyper-parameter tuning. On the other hand, as the anchor-free methods become more prevailing in image-domain tasks, many 3D and LiDAR works have adopted the same design and show a more efficient and competitive performance. Many works [6,20,24,34] also require an RCNN-style second stage refinement. Feature maps for each bounding box proposal are aggregated through RoIAlign or RoIPool. CenterPoint [51] detects objects using center heatmap and regresses other bounding box information using center feature representation.

Most methods directly concatenate points from different frames based on the ego-motion estimation to use the multi-frame information. This assumes the model can align object features from different frames. However, independently moving objects cause misalignment of features across frames. Recent multi-frame methods [14,50] use an LSTM or a GNN module to fuse the previous state feature with the current feature map. 3D-MAN [48] uses a multi-frame alignment and aggregation module to learn the temporal attention of predictions from multiple frames. The feature of each box is generated from the RoI pooling.

2.2 Vision Transformer

Originally proposed in the Natural Language Processing (NLP) community, transformer [40] is becoming a competitive feature learning module in computer vision. Compared to traditional CNN, the transformer has a bigger receptive field, and feature aggregation is based on the response learned directly from pairwise features. A transformer encoder [7,15,52] usually serves as a replacement for the convolution layer in the backbone network. Meanwhile, the transformer decoder uses high-level query feature embedding as the input and extracts features from feature encoding through cross-attention, which is more common in detection and segmentation tasks [4,42,57,62]. DETR [4] uses a transformer encoder-decoder structure to predict objects from learned query embedding. Deformable DETR [62] improves the DETR training through a deformable attention layer. Some recent methods [22,26,53] show that DETR is easier to converge using guidance like anchor boxes.

2.3 3D Transformer

An important design in transformer structure is the position embedding due to permutation invariance of the transformer input. However, 3D point clouds already have position information in them, which leads to deviance in the design of 3D transformers. Point transformer [56] proposes a point transformer layer in the PointNet structure, where the position embedding in the transformer is the pairwise point distances. 3DETR [27] and [23] use a DETR-style transformer decoder in the point cloud except that the query embedding in the decoder is sampled from Farthest Point Sampling (FPS) and learned through classification. Voxel Transformer [25] introduces a voxel transformer layer to replace the sparse convolution layer in the voxel-based point cloud backbone network. SST [8] uses a Single-stride Sparse Transformer as the backbone network to prevent information loss in downsampling of the previous 3D object detector. CT3D [33] uses a transformer to learn a refinement of the initial prediction from local points. In contrast to the above methods, our CenterFormer tailors the DETR to work on LiDAR point clouds with lower memory usage and faster convergence. Moreover, CenterFormer can learn both object-level self-attentions and local cross-attentions without requiring a first-stage bounding box prediction.

3 Method

3.1 Preliminaries

Center-Based 3D Object Detection is motivated by the recent anchor-free image-domain object detection methods [17,19]. It detects each object as a center keypoint by predicting a heatmap on the BEV feature map. Given the output of a common voxel point cloud feature encoder $M \in R^{h*w*c}$, where h and w are the BEV map size and c is the feature dimension, center-based LiDAR object detection predicts both a center heatmap $H \in R^{h*w*l}$ and the box regression $B \in R^{h*w*8}$ through two separated heads. Center heatmap H has l channels, one for each object class. In training, the ground truth is generated from the Gaussian heatmap of the annotated box center. Box regression B contains 8 object properties: the grid offset from the predicted grid center to the real box center, the height of the object, the 3D size, and the yaw rotation angle. During the evaluation, it takes the class and regression predictions at the top N highest heatmap scores and uses NMS to predict the final bounding box.

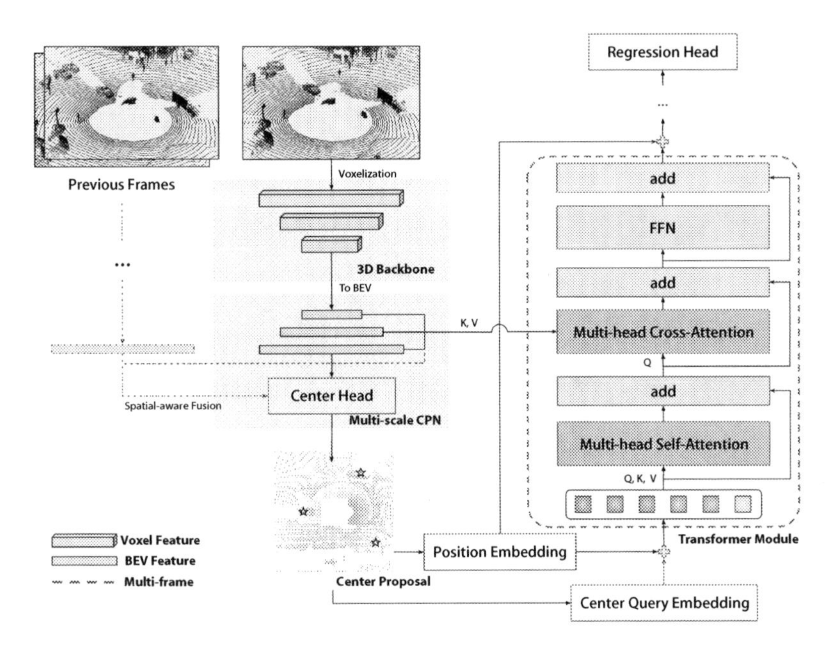

Fig. 2. The overall architecture of CenterFormer. The network consists of four parts: a voxel feature encoder that encodes the raw point cloud into a BEV feature representation, a multi-scale center proposal network (CPN), the center-based transformer decoder, and a regression head to predict the bounding box.

Transformer Decoder aggregates features from the source representation to each query based on the query-key pairwise attention. Each transformer module consists of three layers: A multi-head self-attention layer, a multi-head cross-attention layer, and a feed-forward layer. In each layer, there is also a skip connection that connects the input and the output features and layer normalization. Let f^q and f^k be the query feature and key feature. The multi-head attention can be formulated as:

$$f_i^{out} = \sum_{m=1}^{M} W_m[\sum_{j\in\Omega_j} \sigma(\frac{Q_i K_j}{\sqrt{d}}) \cdot V_j] \tag{1}$$

$$Q_i = f_i^q W_q + E_i^{pos}, K_j = f_j^k W_k + E_j^{pos}, V_j = f_j^k W_v \tag{2}$$

where i and j are the indices of query feature and source feature respectively, m is the head index, Ω_j is the set of attending key features, σ is the softmax function, d is the feature dimension, E^{pos} is the position embedding and W is the learnable weight. In the self-attention layer, the query feature and the key feature come from the same set of query feature embedding, while in the cross-attention layer, the set of key features is the source feature representation.

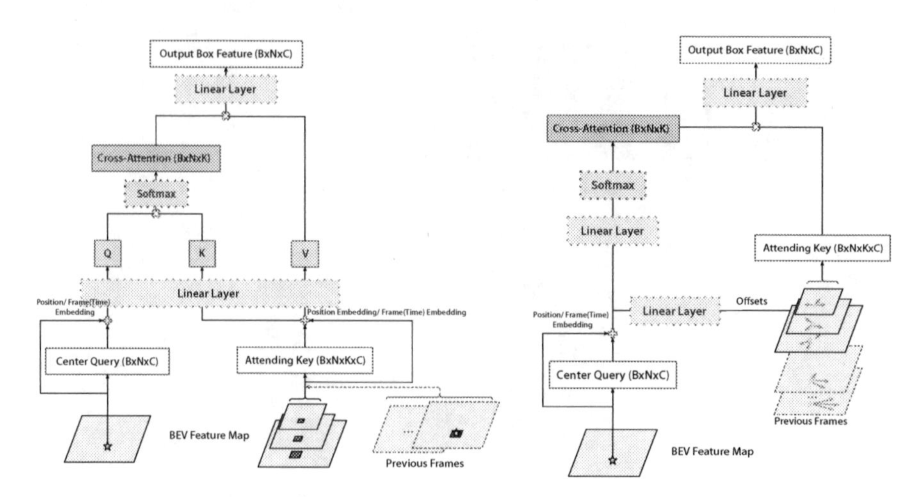

Fig. 3. Illustration of the cross-attention layer. (Left) Multi-scale cross-attention. (Right) Multi-scale deformable cross-attention.

3.2 Center Transformer

The architecture of our model is illustrated in Fig. 2. We use a standard sparse voxel-based backbone network [51] to process each point cloud into a BEV feature representation. We then encode the BEV feature into a multi-scale feature map and predict the center proposals. The proposed centers are then used as the query feature embedding in a transformer decoder to aggregate features from other centers and from multi-scale feature maps. Finally, we use a regression head to predict the bounding box at each enhanced center feature. In our multi-frame CenterFormer, the last BEV features of frames are fused together in both the center prediction stage and the cross-attention transformer.

Multi-scale Center Proposal Network. A DETR-style transformer encoder requires the feature map to be compressed into a small size so that the computation cost is acceptable. This makes the network lose fine-grained features that are crucial for the detection of small objects, which typically occupy $< 1\%$ of the space in the BEV map. Therefore, we propose a multi-scale center proposal network (CPN) to replace the transformer encoder for the BEV feature. In order to prepare a multi-scale feature map, we use a feature pyramid network to process the BEV feature representation into three different scales. At the end of each scale, we add a convolutional block attention module (CBAM) [43] to enhance the feature via channel-wise and spatial attention.

We use a center head on the highest scale feature map $\mathcal{C}$ to predict an l-channel heatmap of object centers. Each channel contains the heatmap score of one class. The location of the top N heatmap scores will be taken out as the center proposals. We used $N = 500$ in our experiments empirically.

Multi-scale Center Transformer Decoder. We extract the features at the proposed center locations as the query embedding for the transformer decoder. We use a linear layer to encode the location of the centers into a position embedding. Traditional DETR decoder initializes the query with a learnable parameter. Consequently, the attention weights acquired in the decoder are almost the same among all features. By using the center feature as the initial query embedding, we can guide the training to focus on the feature that contains meaningful object information. We use the same self-attention layer in the vanilla Transformer decoder to learn contextual attention between objects. The complexity of computing the cross-attention of a center query to all multi-scale BEV features is $O(\sum_{s=1}^{S} H_s W_s K)$. Since the BEV map resolution needs to be relatively large to maintain the fine-grained features for small objects, it is impractical to use all BEV features as the attending keypoints. Alternatively, we confine the attending keypoints to a small 3×3 window near the center location at each scale, as illustrated in Fig. 3. The complexity of this cross-attention is $O(9SK)$, which is more efficient than the normal implementation. Because of multi-scale features, we are able to capture a wide range of features around proposed centers. The Multi-scale cross-attention can be formulated as:

$$\mathrm{MSCA}(p) = \sum_{m=1}^{M} W_m [\sum_{s=1}^{S} \sum_{j \in \Omega_j} \sigma(\frac{Q_i K_j^s}{\sqrt{d}}) \cdot V_j^s], \tag{3}$$

where p denotes the center proposal, Ω_j here is the window around the center, and s is the index of the scale. The feed-forward layer is also kept unchanged.

Multi-scale Deformable Cross-Attention Layer. Inspired by [62], we also used a deformable cross-attention layer to sample the attending keypoints automatically. Figure 3 shows the structure of the deformable cross-attention layer. Compared to the normal multi-head cross-attention layer, deformable cross-attention uses a linear layer to learn 2D offsets Δp of the reference center location p at all heads and scales. The feature at $p + \Delta p$ will be extracted as the cross-attention attending feature through bilinear sampling. We use a linear layer to directly learn the attention scores from the query embedding. Features from multiple scales are aggregated together to form the cross-attention layer output:

$$\mathrm{MSDCA}(p) = \sum_{m=1}^{M} W_m [\sum_{s=1}^{S} \sum_{k=1}^{K} \sigma(W_{msk} \mathcal{C}(p)) x^s (p + \Delta p_{msk})], \tag{4}$$

where x^s is the multi-scale BEV feature, $\mathcal{C}(p)$ is the center feature, and $\sigma(W_{msk}$ $\mathcal{C}(p))$ is the attention weight. We used $K = 15$ in our experiments.

3.3 Multi-frame CenterFormer

Multi-frame is commonly used in 3D detection to improve performance. Current CNN-based detectors cannot effectively fuse features from a fast-moving object,

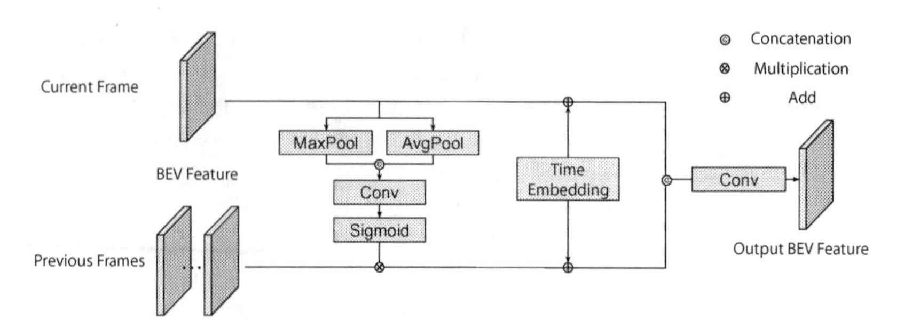

Fig. 4. The network structure of spatial-aware fusion. To focus on the current centers, we use current BEV feature as the reference to learn attention.

while the transformer structure is more suitable for the fusion due to the attention mechanism. To further explore the potential of CenterFormer, we propose a multi-frame feature fusing method using the cross-attention transformer. As shown in Fig. 2, we process each frame individually using the same backbone network. The last BEV feature of the previous frames is transformed to current coordinates and fused with the current BEV feature in both the center head and cross-attention layer.

Due to object movements, the center of an object may shift in different frames. Since we only need to predict the center in the current frame, we use a spatial-aware fusion in the center head to alleviate the misalignment error. As shown in Fig. 4, the spatial-aware module uses a similar spatial attention layer as CBAM [43] to calculate pixel-wise attention based on the current BEV feature. We concatenate the current BEV feature and weighted previous BEV feature and use an additional convolution layer to fuse them together. We also add the time embedding to the BEV features based on their relative time. Finally, we feed the output fused features to the center head to predict the center candidates.

In the cross-attention layer, we use the location of the center proposal to find the corresponding features in the aligned previous frames. The extracted features will be added to the attending keys. Since our normal cross-attention design uses features in a small window close to the center location, it has limited learnability if the object was out of the window area due to fast movement. Meanwhile, our deformable cross-attention is able to model any level of movement, and is more suitable for the long time-range case. Because our multi-frame model only needs the final BEV feature of the previous frame, it is easy to be deployed to the online prediction by saving the BEV feature in a memory bank.

3.4 Loss Functions

Besides the general classification and regression loss functions, we add two additional loss functions to better account for the center-based object detection. First, Inspired by the design in CIA-SSD [45], we move the IoU-aware confidence rectification module from the second stage of other methods to the regression head. More specifically, we predict an IoU score iou for each bounding box proposal, which is supervised with the highest IoU between the prediction and all ground truth annotations in a smooth L1 loss. During the evaluation, we rectify the confidence score with the predicted IoU score using $\alpha' = \alpha * iou^{\beta}$, where α is the confidence score and β is a hyperparameter controlling the degree of rectification. Second, similar to [13,55], we also added a corner heatmap head alongside the center heatmap head as auxiliary supervision. For each box, we generate the corner heatmap of four bounding box edge centers and the object center using the same methods to draw the center heatmap except that the Gaussian radius is half size. During training, we supervise the corner prediction with an MSE loss on the region where the ground truth heatmap score is above 0.

The final loss used in our model is the weighted combination of the following four parts: $\mathcal{L} = w_{hm}\mathcal{L}_{hm} + w_{reg}\mathcal{L}_{reg} + w_{iou}\mathcal{L}_{iou} + w_{cor}\mathcal{L}_{cor}$. We use focal loss [21] and L1 loss for the heatmap classification and box regression. The weights of heatmap classification loss, box regression loss, IoU rectification loss, and corner classification loss are [1, 2, 1, 1], respectively, in our experiment.

4 Experiments

We present our experimental results on two large-scale LiDAR object detection benchmarks: Waymo Open Dataset [38] and nuScenes dataset [3]. Due to page limitation, the experimental results on the nuScenes dataset, more analysis as well as the details on the choice of network parameters, and additional visualizations are included in the supplementary material.

4.1 Dataset

Waymo Open Dataset (WOD) is a large-scale LiDAR point cloud dataset for the autonomous driving environment. It contains 700, 150, and 150 sequences for training, validation, and testing, respectively. Each sequence is 20-second long, captured by a 10 FPS LiDAR sensor with 64 lines in 360°. WOD provides bounding box annotations for three classes: vehicles, pedestrians, and cyclists. The evaluation metrics used in WOD are mean average precision (mAP) and mAP weighted by heading accuracy (mAPH). The evaluation is split into two levels of difficulty, where LEVEL_1 denotes that there are at least 5 points on the object and LEVEL_2 denotes that there is at least 1 point on the object. We set the range of the 3D voxel space as $[-75.2\text{m}, 75.2\text{m}]$ for the X and Y axes, and $[-2\text{m}, 4\text{m}]$ for the Z axis. The size of each voxel is set to $(0.1\text{m}, 0.1\text{m}, 0.15\text{m})$.

4.2 Implementation Details

We follow the same VoxelNet backbone network design as [34,51,60]. In our center proposal network, we process the output of the BEV feature into three

Table 1. The detection results on WOD validation set. †: Deformable CenterFormer. ‡ Reported by PVRCNN++.

Method	Frame	Vehicle L1 (mAP/APH)	Vehicle L2 (mAP/APH)	Pedestrain L1 (mAP/APH)	Pedestrain L2 (mAP/APH)	Cyclist L1 (mAP/APH)	Cyclist L2 (mAP/APH)	Mean L2 mAPH
StarNet [28]	1	55.1/54.6	48.7/48.3	68.3/60.9	59.3/52.8	-/-	-/-	-
SECOND‡ [47]	1	72.3/71.7	63.9/63.3	68.7/58.2	60.7/51.3	60.6/59.3	58.3/57.1	57.2
LiDAR R-CNN [20]	1	73.5/73.0	64.7/64.2	71.2/58.7	63.1/51.7	68.6/66.9	66.1/64.4	60.1
Part-A^2 [37]	1	77.1/76.5	68.5/68.0	75.2/66.9	66.2/58.6	68.6/67.4	66.1/64.9	63.8
3D-MAN [48]	16	74.5/74.0	67.6/67.2	71.7/67.7	62.6/59.9	-/-	-/-	-
PVRCNN++ [35]	1	78.8/78.2	70.3/69.7	76.7/67.2	68.5/59.7	69.0/67.6	66.5/65.2	64.9
CenterPoint [51]	1	-/-	-/67.9	-/-	-/65.6	-/-	-/68.6	67.4
CenterPoint [51]	2	-/-	-/69.7	-/-	-/70.3	-/-	-/70.9	70.3
CenterFormer	1	75.0/74.4	69.9/69.4	78.0/72.4	73.1/67.7	73.8/72.7	71.3/70.2	69.1
CenterFormer†	1	75.2/74.7	70.2/69.7	78.6/73.0	73.6/68.3	72.3/71.3	69.8/68.8	69.0
CenterFormer	2	77.1/76.6	72.2/71.7	80.9/77.6	76.2/73.0	76.0/75.1	73.6/72.7	72.5
CenterFormer†	2	77.0/76.5	72.1/71.6	81.4/78.0	76.7/73.4	**76.6/75.7**	**74.2/73.3**	72.8
CenterFormer†	4	78.1/77.6	73.4/72.9	81.7/78.6	77.2/74.2	75.6/74.8	73.4/72.6	73.2
CenterFormer†	8	**78.8/78.3**	**74.3/73.8**	**82.1/79.3**	**77.8/75.0**	75.2/74.4	73.2/72.3	**73.7**

scales through one upsample layer and one downsample layer. We set the transformer layer/head numbers to 3/4 when using the normal cross-attention, and to 2/6 when using the deformable cross-attention. During evaluation, we use the NMS IoU threshold of [0.8, 0.55, 0.55] and $\beta = [1, 1, 4]$ for vehicle, pedestrian and cyclist. For our 8 frames model, we use $\beta = [1, 1, 1]$ to get a better result for the cyclist. We also increase the center proposal number N to 1000 in evaluation. We trained our model using AdamW optimizer with the one-cycle policy. We trained the single-frame and multi-frame models on 8 Nvidia A100 GPUs with batch sizes 32 and 16. Due to the memory limitation, 4 frames and 8 frames are first split into two 2 frames and 4 frames mini-batch. The points from frames in each mini-batch will be first concatenated together. Hence our multi-frame model only needs to fuse two BEV features together. We apply the object copy & paste augmentation during training with the same fade strategy in [41]. More details are included in the supplementary material.

4.3 Object Detection Results

In Table 1, we show the results on the validation set. Anchor-free center-based method CenterPoint achieves better performance than the anchor-based methods. PVRCNN++ is the best anchor-based method so far, yet shows weak performance on small objects. This demonstrates the limitation of using manually designed anchors for the detection of objects that have large size variations. Our single frame model outperforms CenterPoint by 1.7%. Using a multi-frame model can significantly increase the mAPH performance. Our 2/4/8 frames model reaches the mAPH of 72.8%, 73.2% and 73.7%, respectively, becoming the new state-of-the-art. The pedestrian class benefits most from the multi-frame since the pedestrian point cloud suffers most from the occlusion and noise, as well as its small size. The overall better performance verifies the effectiveness of our proposed transformer model.

Table 2. The single-model detection result on WOD testing set.

Method	Vehicle L1 (mAP/APH)	Vehicle L2 (mAP/APH)	Pedestrain L1 (mAP/APH)	Pedestrain L2 (mAP/APH)	Cyclist L1 (mAP/APH)	Cyclist L2 (mAP/APH)	Mean L2 mAPH
StarNet [28]	61.5/61.0	54.9/54.5	67.8/59.9	61.1/54.0	-/-	-/-	-
PPBA [5]	64.6/64.1	56.2/55.8	69.7/61.7	63.0/55.8	-/-	-/-	-
M3DETR [12]	77.7/77.1	70.5/70.0	68.2/58.5	60.6/52.0	67.3/65.7	65.3/63.8	61.9
3D-MAN [48]	78.7/78.2	70.4/70.0	70.0/66.0	64.0/60.3	-/-	-/-	-
RSN [1]	80.7/80.3	71.9/71.6	78.9/75.6	70.7/67.8	-/-	-/-	-
PVRCNN++ [35]	81.1/80.6	73.7/73.2	80.3/76.3	74.0/70.2	75.1/73.8	72.4/71.2	71.5
CenterPoint [51]	81.1/80.6	73.4/73.0	80.5/77.3	74.6/71.5	74.6/73.7	72.2/71.3	71.9
SST_3f [8]	81.0/80.6	73.1/72.7	83.3/79.7	76.9/73.5	75.7/74.6	73.2/72.2	72.8
AFDetV2 [13]	81.7/81.2	74.3/73.9	81.3/78.1	75.5/72.4	**76.4/75.4**	**74.1/73.0**	73.1
CenterFormer	**84.7/84.4**	**78.1/77.7**	**84.6/81.8**	**79.4/76.6**	75.5/74.5	73.3/72.4	**75.6**

Table 3. The comparison of single-frame CenterFormer (trained on vehicle only) with other methods on WOD validation set for vehicle class.

	Method	Ref	L2 mAP	L2 mAPH
CNN	RSN [1]	CVPR 2021	66.0	65.5
	Voxel R-CNN [25]	AAAI 2021	66.6	–
	Pyramid R-CNN [24]	ICCV 2021	67.2	66.7
	LiDAR R-CNN [20]	CVPR 2021	68.3	67.9
	CenterPoint [51]	CVPR 2021	–	67.9
	AFDetV2 [13]	AAAI 2022	69.7	69.2
	BtcDet [46]	AAAI 2022	70.1	69.6
	PVRCNN++ [35]	arXiv 2021	70.3	69.7
Transformer	BoxeR-3D [29]	CVPR 2022	63.9	63.7
	Voxel transformer [25]	ICCV 2021	65.9	65.3
	M3DETR [12]	WACV 2022	66.6	66.0
	3D-MAN [48]	CVPR 2021	67.6	67.1
	SST_1f [8]	CVPR 2022	68.0	67.6
	CT3D [33]	ICCV 2021	69.0	–
	CenterFormer	ECCV 2022	**70.5**	**70.0**

In Table 2, we show the single model results on the test set. The prediction result is submitted to the online server for evaluation. Our method outperforms all the previous methods by a large margin. The result on vehicle and pedestrian classes have significant improvements (+3.8% and +3.1% on L2 mAPH) as a result of the long-range contextual information learning of the transformer.

To fairly compare with more recent methods [6,24,25,33] that train the model and report the result on only the vehicle class, we train the single frame CenterFormer only on Vehicle class too. We show the results in Table 3. As shown in the table, even with the simplest design, CenterFormer outperforms both the recent transformer-based methods and CNN-based baselines.

Table 4. The ablation of the LEVEL_2 mAPH result improvement of each component on the validation set using single frame. SA, CA and DCA denote the self-attention layer, cross-attention layer and deformable cross-attention layer. IoU and Corner denote IoU rectification and Corner auxiliary supervision. Fade denotes the fade augmentation strategy.

CPN	SA	CA	Corner	DCA	IoU	Fade	Vehicle	Pedstrain	Cyclist	Mean
							66.4	63.4	67.8	65.9
✓							68.5	62.7	67.0	66.1
✓	✓						68.7	64.5	67.3	66.8
✓		✓					68.7	64.2	66.7	66.5
✓	✓	✓					69.3	64.8	66.8	67.0
✓	✓	✓	✓				69.5	65.3	66.4	67.1
✓	✓	✓	✓		✓		69.5	66.9	69.7	68.7
✓	✓	✓	✓		✓	✓	69.4	67.7	70.2	69.1
✓	✓		✓	✓			69.3	65.1	67.5	67.3
✓	✓		✓	✓	✓		69.2	66.7	69.1	68.3
✓	✓		✓	✓	✓	✓	69.7	68.3	68.8	69.0

4.4 Ablation Study

In Table 4, we investigate the effect of each added component in our method on a single frame. We use the previous center-based method as the baseline. After changing the RPN to multi-scale CPN and separating the detection into the center proposal and box regression, our method reaches a similar performance despite we flatten the regression head to 1D. The transformer self-attention layer and cross-attention layer both can improve the results, and when used together, the result reaches 67.0%. This indicates that the self-attention layer and the cross-attention layer learn features separately. Corner auxiliary supervision can additionally improve the result by 0.1%. On the other hand, deformable cross-attention achieves a better result of 67.3%. When trained with IoU rectification, the results get a significant boost to 68.7% and 68.3% for the models using cross-attention and deformable cross-attention. The fade augmentation strategy can further improve the result by 0.4% and 0.7%. This is because the model can adjust to the real data distribution at the end of the training.

4.5 Analysis

Comparison with Deformable DETR. Deformable DETR [62] aims to speed up the learning speed and reduce the computation cost in the DETR structure. Compared to Deformable DETR, our method has three major differences. First, we completely remove the transformer encoder to enable a larger encoded feature map. Second, we use the center feature rather than the learnable parameter as the query embedding for the transformer decoder. Experiment shows that using the center feature as the initial query embedding outperforms the parametric embedding by 1.5% mAPH. This is because the center feature already contains object-level information, which makes it easier to learn pairwise attentions. Third, we use a similar training strategy as [51] rather than the end-to-end set matching training strategy. The set matching training is known for being hard to converge. Since we already have an initial center proposal, we can limit the network to learn only when the proposal is close to the ground truth annotation to speed up the training. Experiment shows that if we use the same set matching training strategy in DETR, the mAPH result is 46.3%, which is more than 20% lower than our current training method.

Visualization of the Learned Attention. The visualizations of the learned self- and cross-attention are illustrated in Fig. 5. The self-attention learning is mainly focusing on the feature of the same class or nearby objects that have the same attributes. For instance, the vehicles on the same line or same parking area will have higher attention weight. The offsets learned by the deformable attention layer vary among different scales. The offsets in two lower scales mostly lead to the keypoints inside or around the object, whereas the offsets in the higher scale can sample far-range features.

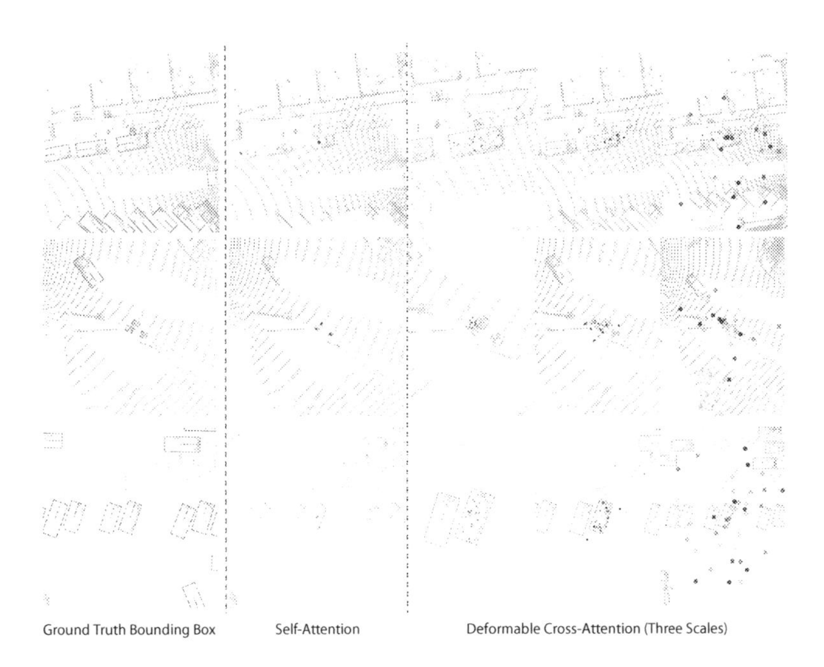

Fig. 5. The visualization of the learned self- and cross-attention weight. The lightness or darkness of the color represents the value of attention weight. The red box denotes the ground truth box and blue box denotes the predicted box. In cross-attention, the sampled keypoints are drawn with red, green and blue for the scale from low to high. Best viewed in color. (Color figure online)

5 Conclusion

In this paper, we propose a novel center-based transformer for 3D object detection. Our method provides a solution to improve the anchor-free 3D object detection network through object-level attention learning. Compared to the DETR-style transformer networks, we use the center feature as the initial query embedding in the transformer decoder to speed up the convergence. We also avoid high computational complexity by focusing the cross-attention learning of each query in a small multi-scale window or a deformable region. Results show that the proposed method outperforms the strong baseline in the Waymo Open Dataset, and reaches state-of-the-art performance when extended to multi-frame. We hope our design will inspire more future work in query-based transformers for LiDAR point cloud analysis.

Acknowledgements. We thank Yufei Xie for his help with refactoring the code for release.

References

1. Bewley, A., Sun, P., Mensink, T., Anguelov, D., Sminchisescu, C.: Range conditioned dilated convolutions for scale invariant 3d object detection. In: CoRL (2021)
2. Bowen, C., Alexander, G.S., Alexander, K.: Per-pixel classification is not all you need for semantic segmentation. In: NeurIPS (2021)
3. Caesar, H., et al.: nuscenes: A multimodal dataset for autonomous driving. In: CVPR (2020)
4. Carion, N., Massa, F., Synnaeve, G., Usunier, N., Kirillov, A., Zagoruyko, S.: End-to-end object detection with transformers. In: Vedaldi, A., Bischof, H., Brox, T., Frahm, J.-M. (eds.) ECCV 2020. LNCS, vol. 12346, pp. 213–229. Springer, Cham (2020). https://doi.org/10.1007/978-3-030-58452-8_13
5. Cheng, S., et al.: Improving 3d object detection through progressive population based augmentation. In: Vedaldi, A., Bischof, H., Brox, T., Frahm, J.-M. (eds.) ECCV 2020. LNCS, vol. 12366, pp. 279–294. Springer, Cham (2020). https://doi.org/10.1007/978-3-030-58589-1_17
6. Deng, J., Shi, S., Li, P., Zhou, W., Zhang, Y., Li, H.: Voxel r-cnn: Towards high performance voxel-based 3d object detection. In: AAAI (2021)
7. Dosovitskiy, A., et al.: An image is worth 16×16 words: Transformers for image recognition at scale. In: ICLR (2021)
8. Fan, L., et al.: Embracing single stride 3d object detector with sparse transformer. In: CVPR (2022)
9. Fan, L., Xiong, X., Wang, F., Wang, N., Zhang, Z.: Rangedet: In defense of range view for lidar-based 3d object detection. In: ICCV (2021)
10. Ge, R., et al.: Afdet: Anchor free one stage 3d object detection. In: CVPRW (2020)
11. Girshick, R., Donahue, J., Darrell, T., Malik, J.: Rich feature hierarchies for accurate object detection and semantic segmentation. In: CVPR (2014)

12. Guan, T., et al.: M3detr: Multi-representation, multi-scale, mutual-relation 3d object detection with transformers. In: WACV (2022)
13. Hu, Y., et al.: Afdetv2: Rethinking the necessity of the second stage for object detection from point clouds. In: AAAI (2022)
14. Huang, R., et al.: An LSTM approach to temporal 3d object detection in lidar point clouds. In: Vedaldi, A., Bischof, H., Brox, T., Frahm, J.-M. (eds.) ECCV 2020. LNCS, vol. 12363, pp. 266–282. Springer, Cham (2020). https://doi.org/10.1007/978-3-030-58523-5_16
15. Jianwei, Y., et al.: Focal self-attention for local-global interactions in vision transformers. In: NeurIPS (2021)
16. Jifeng, D., et al.: Deformable convolutional networks. In: ICCV (2017)
17. Kaiwen, D., Song, B., Lingxi, X., Honggang, Q., Qingming, H., Qi, T.: Centernet: Keypoint triplets for object detection. In: ICCV (2019)
18. Lang, A.H., Vora, S., Caesar, H., Zhou, L., Yang, J., Beijbom, O.: PointPillars: Fast encoders for object detection from point clouds. In: CVPR (2019)
19. Law, H., Deng, J.: CornerNet: Detecting objects as paired keypoints. In: Ferrari, V., Hebert, M., Sminchisescu, C., Weiss, Y. (eds.) Computer Vision – ECCV 2018. LNCS, vol. 11218, pp. 765–781. Springer, Cham (2018). https://doi.org/10.1007/978-3-030-01264-9_45
20. Li, Z., Wang, F., Wang, N.: Lidar r-cnn: An efficient and universal 3d object detector. In: CVPR (2021)
21. Lin, T.Y., Goyal, P., Girshick, R., He, K., Dollár, P.: Focal loss for dense object detection. In: CVPR (2017)
22. Liu, S., et al.: Dab-detr: Dynamic anchor boxes are better queries for detr. In: ICLR (2021)
23. Liu, Z., Zhang, Z., Cao, Y., Hu, H., Tong, X.: Group-free 3d object detection via transformers. In: ICCV (2021)
24. Mao, J., Niu, M., Bai, H., Liang, X., Xu, H., Xu, C.: Pyramid r-cnn: Towards better performance and adaptability for 3d object detection. In: CVPR (2021)
25. Mao, J., et al.: Voxel transformer for 3d object detection. In: CVPR (2021)
26. Meng, D., et al.: Conditional detr for fast training convergence. In: ICCV (2021)
27. Misra, I., Girdhar, R., Joulin, A.: An end-to-end transformer model for 3d object detection. In: CVPR (2021)
28. Ngiam, J., et al.: Starnet: Targeted computation for object detection in point clouds. arXiv (2019)
29. Nguyen, D.K., Ju, J., Booji, O., Oswald, M.R., Snoek, C.G.: Boxer: Box-attention for 2d and 3d transformers. In: CVPR (2022)
30. Noh, J., Lee, S., Ham, B.: Hvpr: Hybrid voxel-point representation for single-stage 3d object detection. In: CVPR (2021)
31. Qi, C.R., Su, H., Mo, K., Guibas, L.J.: Pointnet: Deep learning on point sets for 3d classification and segmentation. In: CVPR (2017)
32. Qi, C.R., Yi, L., Su, H., Guibas, L.J.: Pointnet++: Deep hierarchical feature learning on point sets in a metric space. In: NeurIPS (2017)
33. Sheng, H., et al.: Improving 3d object detection with channel-wise transformer. In: ICCV (2021)
34. Shi, S., et al.: Pv-rcnn: Point-voxel feature set abstraction for 3d object detection. In: CVPR (2020)
35. Shi, S., et al.: Pv-rcnn++: Point-voxel feature set abstraction with local vector representation for 3d object detection. arXiv (2021)
36. Shi, S., Wang, X., Li, H.: Pointrcnn: 3d object progposal generation and detection from point cloud. In: CVPR (2019)

37. Shi, S., Wang, Z., Shi, J., Wang, X., Li, H.: From points to parts: 3d object detection from point cloud with part-aware and part-aggregation network. In: TPAMI (2020)
38. Sun, P., ., et al.: Scalability in perception for autonomous driving: Waymo open dataset. In: CVPR (2020)
39. Sun, P., et al.: Rsn: Range sparse net for efficient, accurate lidar 3d object detection. In: CVPR (2021)
40. Vaswani, A., et al.: Attention is all you need. In: NeurIPS (2017)
41. Wang, C., Ma, C., Zhu, M., Yang, X.: Pointaugmenting: Cross-modal augmentation for 3d object detection. In: CVPR (2021)
42. Wang, H., Zhu, Y., Adam, H., Yuille, A., Chen, L.C.: Max-deeplab: End-to-end panoptic segmentation with mask transformers. In: CVPR (2021)
43. Woo, S., Park, J., Lee, J.-Y., Kweon, I.S.: CBAM: Convolutional block attention module. In: Ferrari, V., Hebert, M., Sminchisescu, C., Weiss, Y. (eds.) ECCV 2018. LNCS, vol. 11211, pp. 3–19. Springer, Cham (2018). https://doi.org/10.1007/978-3-030-01234-2_1
44. Wu, B., Wan, A., Yue, X., Keutzer, K.: Squeezeseg: Convolutional neural nets with recurrent crf for real-time road-object segmentation from 3d lidar point cloud. In: ICRA (2018)
45. Wu, Z., Weiliang, T., Sijin, C., Li, J., Chi-Wing, F.: Cia-ssd: Confident iou-aware single-stage object detector from point cloud. In: AAAI (2021)
46. Xu, Q., Zhong, Y., Neumann, U.: Behind the curtain: Learning occluded shapes for 3d object detection. In: AAAI (2022)
47. Yan, Y., Mao, Y., Li, B.: Second: Sparsely embedded convolutional detection. In: Sensors (2018)
48. Yang, Z., Zhou, Y., Chen, Z., Ngiam, J.: 3d-man: 3d multi-frame attention network for object detection. In: CVPR (2021)
49. Ye, M., Xu, S., Cao, T.: Hvnet: Hybrid voxel network for lidar based 3d object detection. In: CVPR (2020)
50. Yin, J., Shen, J., Guan, C., Zhou, D., Yang, R.: Lidar-based online 3d video object detection with graph-based message passing and spatiotemporal transformer attention. In: CVPR (2020)
51. Yin, T., Zhou, X., Krähenbühl, P.: Center-based 3d object detection and tracking. In: CVPR (2021)
52. Ze, L., et al.: Swin transformer: Hierarchical vision transformer using shifted windows. In: ICCV (2021)
53. Zhang, H., et al.: Dino: Detr with improved denoising anchor boxes for end-to-end object detection. arXiv (2022)
54. Zhang, Y., et al.: Polarnet: An improved grid representation for online lidar point clouds semantic segmentation. In: CVPR (2020)
55. Zhang, Z., Sun, B., Yang, H., Huang, Q.: H3DNet: 3D object detection using hybrid geometric primitives. In: Vedaldi, A., Bischof, H., Brox, T., Frahm, J.-M. (eds.) ECCV 2020. LNCS, vol. 12357, pp. 311–329. Springer, Cham (2020). https://doi.org/10.1007/978-3-030-58610-2_19
56. Zhao, H., Jiang, L., Jia, J., Torr, P.H., Koltun, V.: Point transformer. In: CVPR (2021)
57. Zheng, S., et al.: Rethinking semantic segmentation from a sequence-to-sequence perspective with transformers. In: CVPR (2021)
58. Zhou, Y., Tuzel, O.: Voxelnet: End-to-end learning for point cloud based 3d object detection. In: CVPR (2018)
59. Zhou, Z., Zhang, Y., Foroosh, H.: Panoptic-polarnet: Proposal-free lidar point cloud panoptic segmentation. In: CVPR (2021)

60. Zhu, B., Jiang, Z., Zhou, X., Li, Z., Yu, G.: Class-balanced grouping and sampling for point cloud 3d object detection. arXiv (2019)
61. Zhu, X., et al.: Cylindrical and asymmetrical 3d convolution networks for lidar segmentation. In: CVPR (2021)
62. Zhu, X., Su, W., Lu, L., Li, B., Wang, X., Dai, J.: Deformable detr: Deformable transformers for end-to-end object detection. In: ICLR (2020)

Physical Attack on Monocular Depth Estimation with Optimal Adversarial Patches

Zhiyuan Cheng[1], James Liang[2], Hongjun Choi[1], Guanhong Tao[1], Zhiwen Cao[1], Dongfang Liu[2](✉), and Xiangyu Zhang[1](✉)

[1] Purdue University, West Lafayette, USA
{cheng443,choi293,taog,cao270,xyzhang}@cs.purdue.edu
[2] Rochester Institute of Technology, Rochester, USA
{jcl3689,dongfang.liu}@rit.edu

Abstract. Deep learning has substantially boosted the performance of Monocular Depth Estimation (MDE), a critical component in fully vision-based autonomous driving (AD) systems (*e.g.*, Tesla and Toyota). In this work, we develop an attack against learning-based MDE. In particular, we use an optimization-based method to systematically generate stealthy physical-object-oriented adversarial patches to attack depth estimation. We balance the stealth and effectiveness of our attack with object-oriented adversarial design, sensitive region localization, and natural style camouflage. Using real-world driving scenarios, we evaluate our attack on concurrent MDE models and a representative downstream task for AD (*i.e.*, 3D object detection). Experimental results show that our method can generate stealthy, effective, and robust adversarial patches for different target objects and models and achieves more than 6 m mean depth estimation error and 93% attack success rate (ASR) in object detection with a patch of 1/9 of the vehicle's rear area. Field tests on three different driving routes with a real vehicle indicate that we cause over 6 m mean depth estimation error and reduce the object detection rate from 90.70% to 5.16% in continuous video frames.

Keywords: Physical adversarial attack · Monocular depth estimation · Autonomous driving

1 Introduction

Monocular Depth Estimation (MDE) is a technique for estimating the distance between an object and the camera from RGB image inputs. It is a critical vision task for autonomous driving (AD) because it bridges the gap between Lidar sensors and RGB cameras [52] and its measurement has an effect on a variety of

Supplementary Information The online version contains supplementary material available at https://doi.org/10.1007/978-3-031-19839-7_30.

S. Avidan et al. (Eds.): ECCV 2022, LNCS 13698, pp. 514–532, 2022.
https://doi.org/10.1007/978-3-031-19839-7_30

downstream perception tasks (*e.g.*, object detection [12,26], visual SLAM [55], and visual relocalization [27]). For its importance, Tesla has integrated MDE into its production-grade Autopilot system [2,3], and other AD companies such a Toyota [20] and Huawei [5] are also actively investigating this technique. With the increasing popularity of MDE, ensuring its security becomes a critical challenge.

Existing adversarial attacks against MDE are implemented in digital- [56,66] or physical-world platforms [63]. Compared to digital-world attacks, attacks in the physical world are more challenging because they require robust perturbations to overcome various photometric and geometric changes [6], reducing their stealth. Prior efforts for physical-world adversarial attacks [7,23,45,63] generally employ an unnatural-looking adversarial patch and sacrifice stealth for attack effectiveness, leaving plenty of room for improvement. Additionally, with MDE's rapid development, many downstream tasks that previously require expensive Lidar sensors or depth cameras can now be performed entirely with MDE's measurement and achieve competitive performance. However, the investigation of the impact of compromised MDE on these downstream tasks remains largely unknown.

To address the aforementioned problems, in this paper, we investigate the *stealth of physical-world attack against MDE* and present a physical-object-oriented adversarial patch optimization framework to generate *stealthy, effective and robust adversarial patches* for target objects (*e.g.*, vehicles and pedestrians), which, to our best knowledge, is the **FIRST** work in the community. In particular, we are able to achieve the followings: ❶ we design a physical-object-oriented adversarial optimization, which binds the patch and the target object together regarding attack effects and physical-world transformations (Subsect. 3.2); ❷ we optimize the patch region on the target object with a differentiable patch mask representation, which automatically locates the highly effective area for attack on the target object and improves attack performance with a small patch size (Subsect. 3.3); ❸ we camouflage the adversarial pattern with natural styles (*e.g.*, rusty and dirty) with deep photo style transfer [29], resulting in stealthier patch for the attack (Subsect. 3.4); ❹ we investigate the impact of compromised MDE on a representative downstream task in AD — 3D object detection (Subsect. 4.4). Our attack causes over 6 m of mean depth estimation error for a real vehicle, with a patch only 1/9 of the vehicle's rear area, and achieves more than 90% attack success rate in 3D object detection (Fig. 1).

2 Related Work

AD Systems Security. In AD, sensor security and autonomy software security are the two important challenges. For sensor security, prior works focus on spoofing/jamming on camera [32,35,64], LiDAR [10,44], RADAR [64], ultrasonic [64], GPS [43] and IMU [48,50]. For autonomy software security, some prior works study regression tasks (*e.g.*, depth estimation [63] and optical flow estimation [37]), and others focus on classification tasks (*e.g.*, 2D object detection and classification [7,45], tracking [21], lane detection [40,41], and traffic

Fig. 1. Attack MDE and 3D object detection with a natural adversarial patch. The left is a benign scenario and the right is the corresponding adversarial scenario. 3D object detection takes the pseudo-Lidar (*i.e.*, point cloud projected from 2D depth map) as input and outputs bounding boxes of recognized objects. Observe in the adversarial scenario (b) that our optimized adversarial patch can disturb the depth estimation of the target vehicle significantly and the effect propagates to an area larger than the patch itself. Pseudo-Lidar of the vehicle is thus distorted and it cannot be detected in the downstream task.

light detection [46]). This work focuses on autonomy software security, that is, compromising MDE and its related downstream tasks.

Physical-World Adversarial Attacks. Many prior efforts in adversarial attacks have been directed toward generating patches or perturbations in the digital space [19,33,34,36,49,57–60]. In comparison, we conduct extensive experiments on adversarial attacks in the physical world. Although existing physical-world attacks have addressed tasks such as image classification [7,45], object detection [11,47,61], face recognition [23,42], the domain of depth estimation attack has received scant attention. Moreover, the correlations between stealth and attack effectiveness are largely understudied in the literature. In this paper, we make an attempt to close the aforementioned knowledge gap.

MDE Attacks. Zhang [66] proposes a multi-task attack strategy to improve the performance in the universal attack scenario. Wong [56] proposes a way to generate targeted adversarial perturbation on images and alter the depth map arbitrarily. These two attacks focus on digital-space perturbations thus are not directly applicable in the physical world. Yamanaka [63] proposes a method to generate printable adversarial patch for MDE but it does not consider stealth of the patch. Different from prior efforts, we focus on the stealth and to the best of our knowledge, we are the **FIRST** work to examine the stealth of adversarial patches for physical-world attack against MDE.

3 Method

3.1 Physical-Object-Oriented MDE Attack

Motivation. Compared with unconstrained adversarial patches (see Fig. 2a) which often look suspicious, stealthy patches may draw less attention and hence can stay on the target vehicle for an extended period of time, posing a greater threat. We divide the challenge of achieving stealth into two sub-problems: patch size minimization and achieving natural appearance. To minimize patch size, we

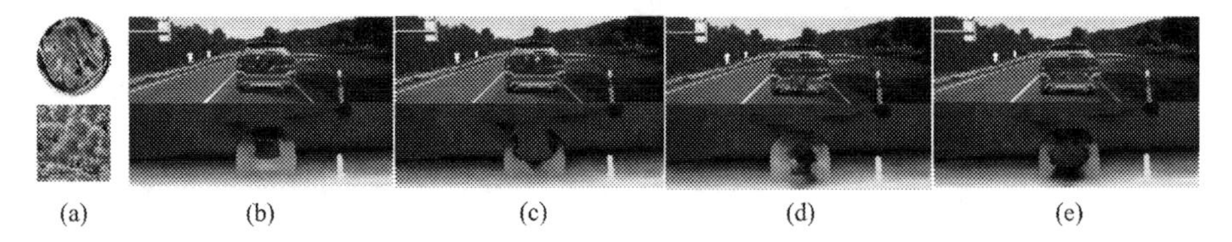

Fig. 2. (a): Unconstrained adversarial patches in [30,63] are easy to be identified; traditional patch-oriented attack in (b) affects smaller area than our object-oriented attack in (c); (c), (d) and (e): different regions on the target object have different sensitivity regarding attack effect even with the same total area.

investigate how to maximize the attack effect with smaller patches and propose two approaches: ❶ enlarging the patch's affected area (see comparison in Fig. 2b and c), and ❷ locating the adversarial patch in a more sensitive region of the target vehicle (see Fig. 2c, d and e). In terms of naturalness, as the magnitude of perturbations required to launch attack in the physical world is much more substantial, we cannot simply bound the adversarial noise to a human unnoticeable level via various L_p-norms as in digital-world attacks, which provides little physical-world robustness. Instead, we use style transfer to disguise the adversarial pattern as natural styles (*e.g.*, dirty or rusty).

Attack Pipeline. We use an optimization-based method to generate adversarial patches and there are three main optimization goals: ❶ increasing the estimated distance of target object (Subsect. 3.2); ❷ minimizing the patch to locate a sensitive (*i.e.*, most effective) region for attack (Subsect. 3.3), and ❸ camouflaging the adversarial patch with natural styles (Subsect. 3.4). The optimization is conducted in the digital world. Figure 3 shows the overview of our attack. From the top left, we start with style transfer on the patch content image. Next, we crop the style-transferred patch with an optimizable patch mask (m_p^{Θ}) and paste it onto a target object (O) (*e.g.*, a vehicle) creating an adversarial one (O'). Then, we synthesize adversarial scenarios (R_t') by placing the adversarial object into random scenes with physical transformations (t) and estimate scenarios' depth ($\mathcal{D}(R_t')$). We define an adversarial loss ($\mathcal{L}_a$) to increase depth of the target object. Together with a style transfer loss ($\mathcal{L}_{st}$) maintaining the naturalness and a patch size loss ($\mathcal{L}_m$) minimizing the patch, we perform back propagation and update the patch content and the mask iteratively to address the three optimization goals. The solid lines denote data flow and the dashed lines represent back propagation paths. Each component is explained in details in the following sections.

3.2 Adversarial Perturbation Generation

In preparation, we take a photo of the target object (O) and select a patch content image (x) and a style image. Given the patch mask (m_p), we create an adversarial object (O') by applying the style-transferred patch (x') on the

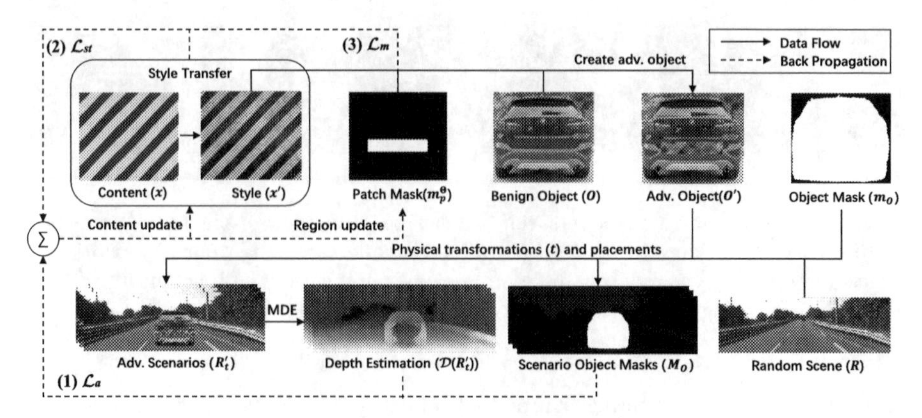

Fig. 3. Overview of the physical-object-oriented framework to generate a stealthy adversarial patch.

benign object in the following way:

$$O' = O \odot (1 - m_p) + x' \odot m_p, \tag{1}$$

where $\odot$ denotes the element-wise multiplication and O, m_p, x' have the same width and height. We explain the patch mask definition and style transfer later in Subsect. 3.3 and Subsect. 3.4. We evaluate the depth of the target object inside a scene because the camera on the victim vehicle captures scene frames as input instead of independent objects. Specifically, in each optimization iteration, we randomly sample a scene from the dataset and paste the adversarial object into the scene to create an adversarial scenario. Unlike previous attacks against autonomous driving systems [9,39] that aim at a particular scene or a road section, our attack is universal and scene-independent.

To improve the robustness of our attack in the physical world, we apply Expectation of Transformation (EoT) [6] by randomly transforming the object in size, rotation, brightness, saturation, etc., before pasting. The horizontal position of pasting is random, while the vertical position is calculated according to the size of the object considering physical constraints. Specifically, Fig. 4 shows the perspective model of a vehicle in a side view and we assume the camera is facing straight forward without tilt. H is the height of the target vehicle; h is the height of the camera with respect to the victim vehicle; f is the focus length of the camera and α relates to the camera's view angle. On the image, the vertical position of the vehicle (d) is calculated from the height of the vehicle (s) with Eq. 2. Intuitively, objects farther away appear smaller in perspective so a smaller object after transformation is pasted to a higher vertical position on the image, which is closer to *the vanishing point* (of the camera), which denotes the furthest physical point in the camera view, and has further depth estimation.

$$d = -\frac{h}{H}s + \frac{f}{\tan \alpha} \tag{2}$$

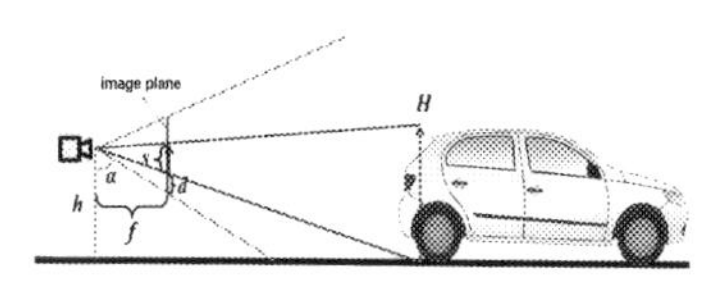

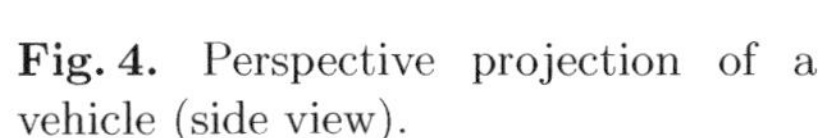

Fig. 4. Perspective projection of a vehicle (side view).

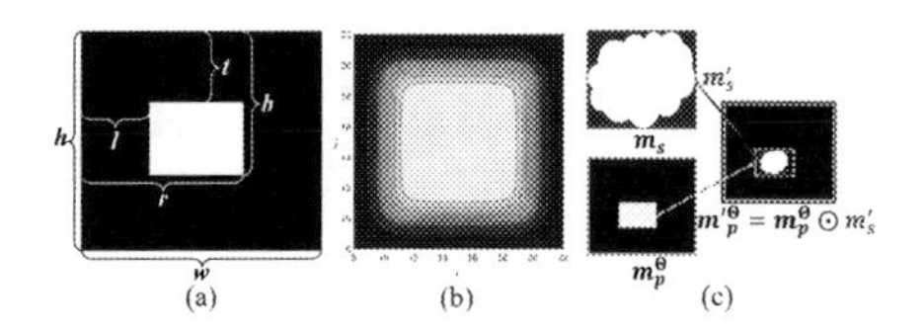

Fig. 5. Patch region definitions.

Formally, the adversarial scenario R'_t is described as $R'_t = \Lambda_t\left(t(O' \odot m_o), R\right)$, where t is the random transformation applied on the target object; m_o is the object mask used to extract the object from the image; R is the randomly sampled scene from database and $\Lambda(\cdot,\cdot)$ is the paste operation to combine an adversarial object and a scene following the physical constraint in Eq. 2. Since our goal is to make the target object further away, we want to maximize the object's depth estimation (*i.e.*, minimize the reciprocal). Hence, we define the adversarial loss in Eq. 3, where T is a set of transformations; D_R is a set of scenes; $MSE(\cdot,\cdot)$ is the mean square error between two variables; $\mathcal{D}$ is the depth estimation model and M_o is the object mask in the scenario.

$$\mathcal{L}_a = \mathbf{E}_{t\sim T, R\sim D_R}\left[MSE\left(\mathcal{D}\left(R'_t\right)^{-1} \odot M_o, 0\right)\right] \tag{3}$$

3.3 Sensitive Region Localization

As described in Subsect. 3.2, we apply the style transferred patch x' onto the target object by a patch mask m_p which defines the patch region on the target object. Prior works [25,28,51] optimizing masks treat each pixel of the mask as a parameter and the generated mask suffers from low deployability due to sparse and scattered mask regions (See Fig. 11b). Instead we design a novel rectangular patch region optimization method (we call it *regional optimization*) to locate a sensitive region automatically. Although we define the patch region as rectangular, the final patch is not necessarily rectangular but have an arbitrary predefined shape. Details are explained later.

A typical rectangular patch mask has ones within the rectangular borders and zeros otherwise. However, this mask is not differentiable regarding the border parameters because the mask values are not continuous across the borders and border information is not encoded into each mask values, which means that the region cannot be optimized via gradient descent and back propagation. To solve this problem, we design a differentiable soft version of the rectangular mask making it optimizable with respect to four border parameters. Specifically, we define border parameters $\Theta = [l, r, t, b]$ as shown in Fig. 5a. l and r are the left and right borders' column indices and t and b are the top and bottom borders' row indices. Let w and h be the width and height of the mask respectively and we have $0 \le l \le r \le w$ and $0 \le t \le b \le h$.

$$\begin{aligned} m_p^{\Theta} &= \{m_p^{\Theta}[i,j] \mid i \in 1...w, j \in 1...h\} \\ m_p^{\Theta}[i,j] &= \frac{1}{4}(-sign(i-t) \cdot sign(i-b)+1) \\ &\quad \cdot(-sign(j-l) \cdot sign(j-r)+1), \end{aligned} \tag{4}$$

Typically, a mask is defined by Eq. 4 with Θ as parameters, where $m_p^{\Theta} \in \{0,1\}^{w \times h}$ is the patch mask and $[i,j]$ is index of the pixel at i-th row and j-th column; $sign(x)$ outputs 1 when $x \ge 0$ and -1 when $x < 0$; and $m_p^{\Theta}[i,j]$ evaluates to one if and only if the pixel is within the four borders defined by Θ and zero otherwise. To make each mask value differentiable regarding border parameters and maintain the property of original definition, we approximate $sign(\cdot)$ by $\tanh(\cdot)$ and define the patch mask with Eq. 5.

$$\begin{aligned} m_p^{\Theta}[i,j] &= \frac{1}{4}(-\tanh(i-t) \cdot \tanh(i-b)+1) \\ &\quad \cdot(-\tanh(j-l) \cdot \tanh(j-r)+1) \end{aligned} \tag{5}$$

Fig. 5b is an example of the mask defined by us. In this example, w and h are 30, l and t are 10, and r and b are 20. Observe that the borders of the rectangular region change gradually. Each pixel value is encoded with border parameters Θ.

In the beginning, the patch mask is initialized to cover the whole image, ($i.e.$, $l = t = 0$, $b = h$ and $r = w$). One of our optimization goal is to minimize the mask area, thus we define a mask loss term (Eq. 6) to penalize the area of mask.

$$\mathcal{L}_m = \frac{r-l+b-t}{w+h} \tag{6}$$

We use a linear combination of the width and height of the rectangular region to avoid bias in the update of edges. Otherwise, if we use the ratio of area ($i.e.$, $(r-l) \times (b-t)/(w \times h)$) as the mask loss, parameters of the longer edge ($e.g.$, b and t when $(b-t) < (r-l)$) would have larger gradients and tend to change faster than the shorter edges, which leads to a bias towards updating the longer-edge parameters. Using a linear combination avoids this problem and each mask parameter has the same weight.

Although we define a rectangular patch region, the final patch mask can be an arbitrary shape within the region. As shown in Fig. 5c, given a predefined patch shape mask m_s ($m_s[i,j] \in \{0,1\}$), the final patch mask $m_p'^{\Theta}$ is calculated by element-wise multiplying the scaled shape mask m_s' with the region mask m_p^{Θ} inside the rectangular region. Specifically, in each iteration, given border parameters Θ, we can scale and fit the predefined shape mask m_s into the center of the rectangular region getting mask m_s' , which is denoted by the red color in Fig. 5c. The final patch mask is calculated with Eq. 7 by multiplying the region mask and the shape mask within the rectangular region. Without loss of generality, we focus on rectangular shapes ($i.e.$, $m_s \equiv 1$) in our evaluation.

$$m_p'^{\Theta}[i,j] = \begin{cases} m_p^{\Theta}[i,j] * m_s'[i,j] & i \in l...r, j \in t...b \\ m_p^{\Theta}[i,j] & others \end{cases} \tag{7}$$

In addition, our mask definition also supports optimizing with multiple patches. The key point is to take the union of several regions and optimize them together.

3.4 Attack Camouflage

Patches generated in existing adversarial attacks against depth estimation models have obvious perturbations as shown in Fig. 2a. Unlike them, we use style transfer to camouflage the attack with natural styles. There have been works using style transfer [14] in attacking classification models but we are the first to combine style transfer with the more challenging depth estimation attack. We use deep photo style transfer [29] as our style transfer method. This method is a kind of neural style transfer which has demonstrated remarkable results for image stylization [16]. It uses a convolutional neural network (CNN) to extract the deep features of an image and separate the content and style information in the deep feature representations. The source image will be updated iteratively to approach the style information extracted from the style image and keep the content information of the source image. Specifically, as defined in deep photo style transfer [29], there are four terms regarding the style transfer components in the loss function. They are style loss ($\mathcal{L}_s$), content loss ($\mathcal{L}_c$), smoothness loss ($\mathcal{L}_t$) and photorealism regularization loss ($\mathcal{L}_r$). We refer the readers to [29] for more detailed explanation on each term. The style transfer loss therefore is:

$$\mathcal{L}_{st} = \mathcal{L}_s + \mathcal{L}_c + \mathcal{L}_t + \mathcal{L}_r \tag{8}$$

In summary, our adversarial patch generation process can be formulated by the following optimization problem:

$$\begin{aligned} \min_{x',\Theta} \quad & \mathcal{L}_a + \mathcal{L}_m + \lambda \mathcal{L}_{st} \\ s.t. \quad & x' \in [0, 255]^{3 \times w \times h}, \Theta = \{l, r, t, b\} \\ & 0 \le l \le r \le w,\ 0 \le t \le b \le h, \end{aligned} \tag{9}$$

where λ is an adjustable weight parameter to balance the style transfer naturalness and attack performance. The weights of other terms are fixed in our experiments. In each iteration, we calculate gradients of x' and Θ with back propagation and, same as in deep photo style transfer [29], we use LBFGS [8] to update the patch x'. We update border parameters Θ with Adam [22] and we only update the edge with the maximum absolute gradient instead of four, which avoids the constraint of compressing the region from all directions in each iteration and provides more flexibility. We set a target ratio of the patch region in advance (*i.e.*, the area of the patch region relative to the object) as the stopping criteria of mask optimization. In other words, the mask will stop updating when it is smaller than the predefined target ratio.

Fig. 6. Target objects and fixed regions.

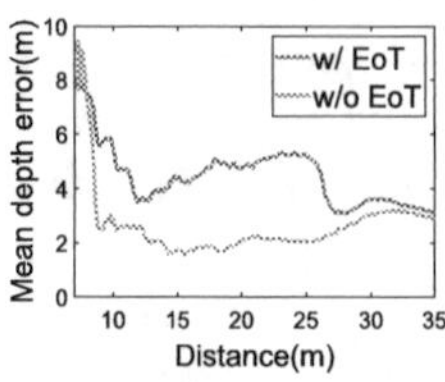

Fig. 7. $\mathcal{E}_d$ of the target vehicle at **different distance**.

Table 1. Mean depth estimation error ($\mathcal{E}_d$) in attacking **fixed regions and optimized regions**.

	Mono			DH			Many		
	V	TB	P	V	TB	P	V	TB	P
Ours	**16.84**	**8.26**	**14.06**	**15.23**	**4.54**	**13.17**	**6.31**	**3.57**	**10.15**
LO	13.90	5.21	11.53	2.51	1.63	10.79	3.03	2.94	8.93
R1	3.70	2.35	10.20	2.25	1.50	11.78	1.12	2.77	9.21
R2	7.41	2.67	11.28	4.66	1.40	10.52	4.23	1.40	8.66
R3	5.20	4.96	5.05	3.92	1.45	4.08	1.33	3.05	5.06
R4	7.31	1.59	-	5.58	1.59	-	4.89	1.59	-
R5	14.95	2.39	-	7.70	0.90	-	5.66	2.43	-
R6	9.69	2.59	-	2.37	0.49	-	1.36	1.15	-
R7	3.23	-	-	2.62	-	-	1.67	-	-
R8	7.74	-	-	4.44	-	-	4.91	-	-
R9	5.36	-	-	1.38	-	-	1.32	-	-

Mono: Monodepth2, DH: DepthHints, Many: Manydepth
V: Vehicle, TB: Traffic Barrier, P: Pedestrian
LO: Location Optimize in [38], R: Region

4 Experiments

4.1 Experimental Setup

MDE Model Selection. In our evaluation, we use three widely known, representative monocular depth estimation models: Monodepth2 [18], Depthhints [54], and Manydepth [53].

Target Object Selection. Our attack is generic so it can be applied to any class of objects on public roads[1]. This paper focuses on three representative types of objects to attack: vehicles, traffic barriers, and pedestrians as shown in Fig. 6. We choose them because they are most common on public roads in regular driving scenarios, and a failure in detecting them could lead to life-threatening consequences. Vehicles are the most attractive objects for attackers since they are the main targets of perception systems on autonomous driving cars. We mainly focus on vehicles in our experiments.

Evaluation Scene Selection. We select 100 real-world driving scenes from KITTI dataset [17] to evaluate the attack performance of the generated patch on each object in the digital-world. These scenes cover a wide range of roads (*e.g.*, high-way, local, and rural roads) and background objects (*e.g.*, trucks, traffic lights, and cars). Physical-world experiments use three driving routes with various lighting conditions.

Evaluation Metrics. We use mean depth estimation error ($\mathcal{E}_d$) of the target object and ratio of affected region ($\mathcal{R}_a$) as our evaluation metrics. We use depth

[1] Code can be found at https://github.com/Bob-cheng/MDE_Attack.

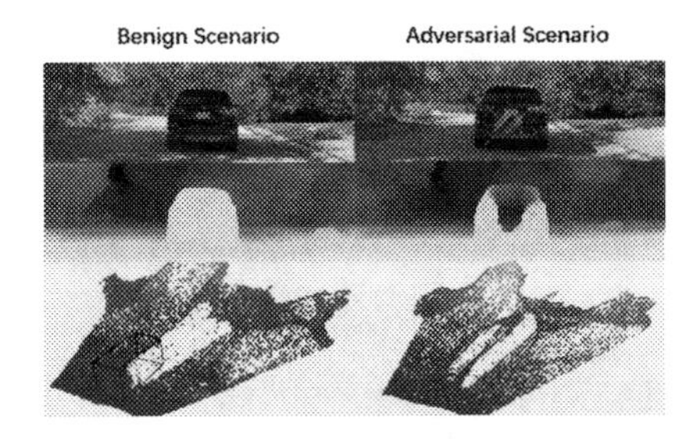

Fig. 8. Physical world attack example.

Table 2. Physical world attack result.

	Time (s)	Frames	$\mathcal{E}_d$	Detected	Detection Rate
Route 1 Benign	95	477	0.52	469	98.32%
Route 2 Benign	82	412	0.77	354	85.92%
Route 3 Benign	80	402	0.62	348	86.57%
Total Benign	**257**	**1291**	**0.64**	**1171**	**90.70%**
Route 1 Adv.	94	468	6.73	45	9.62%
Route 2 Adv.	82	408	8.92	11	2.70%
Route 3 Adv.	80	402	7.68	10	2.49%
Total Adv.	**256**	**1278**	**7.77**	**66**	**5.16%**

estimation of the original object as the ground truth and compare with depth estimation of the adversarial object. The mean depth estimation error denotes the attack effectiveness of our adversarial patch. The larger it is, the better the performance. Equation 10 is the formal definition. Meanings of the symbols are the same as those in Sect. 3.

$$\mathcal{E}_d = \frac{\mathrm{sum}\left(\left|\mathcal{D}\left(\Lambda(O,R)\right) - \mathcal{D}\left(\Lambda(O',R)\right)\right| \odot M_o\right)}{\mathrm{sum}(M_o)} \tag{10}$$

The ratio of affected region $\mathcal{R}_a$ is defined as:

$$\mathcal{R}_a = \frac{\mathrm{sum}\left(\mathbf{I}\left(\left|\mathcal{D}\left(\Lambda(O,R)\right) - \mathcal{D}\left(\Lambda(O',R)\right)\right| \odot M_o \geq 10\right)\right)}{\mathrm{sum}(M_o)}, \tag{11}$$

where $\mathbf{I}(x)$ is the indicator function that evaluates to 1 only when x is true. We define ≥ 10 m error of depth estimation for a pixel as a valid attack and this pixel will be included in the affected region. $\mathcal{R}_a$ is the ratio between the number of affected pixels and all pixels of the object.

4.2 Main Results

We present our main results regarding effectiveness, robustness and stealth.

Attack Effectiveness. We run our attack with the three MDE models and we target the three types of objects for each model. For each object, we split it into several regions with equal size as shown in Fig. 6 and attack these fixed regions respectively (*i.e.*, optimize the patch on each region.), then we compare with two patch region optimization techniques: our sensitive region localization (Subsect. 3.3) and the location-optimized patch [38]. In [38], the authors update the location of a fixed-size patch after each optimization iteration. They tentatively move the patch towards four directions with a predefined stride and select the direction with the least adversarial loss as the next patch location. For a fair comparison, we set the target ratio of patch region the same as that of those

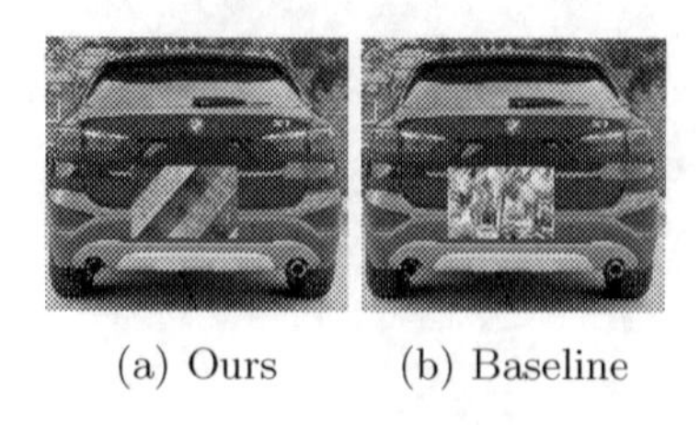

(a) Ours (b) Baseline

Fig. 9. Naturalness comparison.

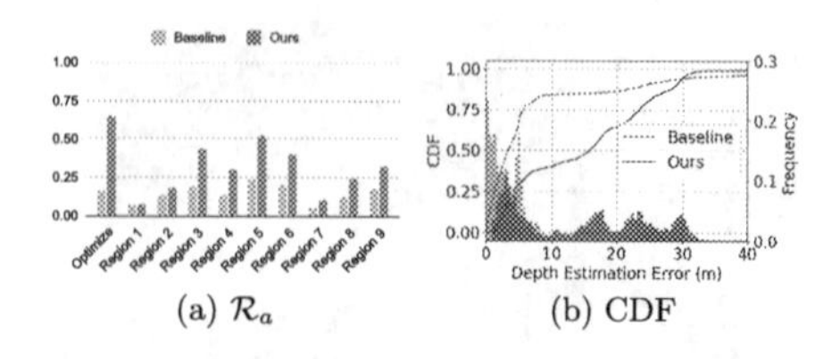

(a) $\mathcal{R}_a$ (b) CDF

Fig. 10. Comparing **patch-oriented attack** (baseline) with our **object-oriented attack**.

fixed regions (*e.g.*, 1/9 of the vehicle's read area). Our regional optimization stops when the patch ratio is smaller than the target ratio. In each test, we evaluate the mean depth estimation error ($\mathcal{E}_d$) of the target object in 100 scenes and take the average of them as the result. In each scene, the object is placed at 7 m away from the victim's camera. We choose 7 m since it is the breaking distance [1] while driving at a speed of 25 mph, which is almost the lowest in normal driving. In other words, it is the smallest distance at which the object has to be detected by the victim to avoid a crash in normal driving scenarios [9].

Table 1 reports the effectiveness evaluation result. As shown, our attack is generic and effective on different depth estimation models and objects. With our sensitive region localization, an adversarial patch with 1/9 of the vehicle's rear area causes at least 6 m $\mathcal{E}_d$ across different depth estimation models. Observe that attack performance differs with patch regions. Our sensitive region localization can locate an optimal place that outperforms all those fixed regions and the location optimized regions in [38]. For the physical world experiments, Fig. 8 presents an example. As shown, the adversarial patch on the vehicle fools the vehicle's depth estimation, and the effect is not limited to the patch area but propagates to a broader area. After being projected to 3D space, it is more obvious that the point cloud of the adversarial vehicle is distorted comparing with the benign one. Table 2 reports the physical world attack performance. The first column in the table denotes different drives. The second column shows the time of each drive in seconds. The third column shows the total frames evaluated from the video, and we evaluate frames at a frequency 5 Hz. The fourth column reports the mean depth estimation error ($\mathcal{E}_d$) of the vehicle. As shown, in benign scenarios, the error is under 1 m while the error in adversarial scenarios is over 7 m, which justifies our attack in the physical world.

Attack Robustness. Relative to the victim vehicle, we place the adversarial object at places with longitudinal distances (*i.e.*, forward and back) ranging from 7 m to 35 m and lateral distances (*i.e.*, left and right) ranging from -1 m to 1 m. The 7 m to 35 m longitudinal distance corresponds to the brake distance for driving speed from about 25 to 55 mph [4]. We consider the victim vehicle at the center of the lane, and -1 m to 1 m of lateral deviation from the lane center covers most driving scenarios of the vehicle ahead [13]. We use a vehicle as the target object and Monodepth2 as the depth estimation network. We use the regional optimization and set the target patch size to 1/9 of the vehicle's rear area. We test our attack with and without EoT [6] (see Sect. 3.2) during optimization.

Figure 7 shows the result of the robustness evaluation. We report the mean depth estimation error of the target object under different longitude distances with the victim vehicle. Observe that our attack is robust and causes more than 3 m of mean depth estimation error in different victim approaching positions. EoT increases the attack performance by 40.63% and makes our attack more robust in different distances. As shown, the closer the target object, the larger the error in depth estimation, which makes the victim vehicle harder to detect the object from the distorted pseudo-Lidar and continue approaching it until collision. In the physical world experiments, our attack is conducted with real driving scenarios. Compared to evaluating with a single image from a specific position in prior work, continuous and dynamic movement is more challenging and practical. Our attack is shown to be robust under different lighting conditions (*e.g.*, shadows and different light directions), driving operations (*e.g.*, moving straight and turning) and background scenes. The dynamic moving video of our physical world attack is at https://youtu.be/L-SyoAsAM0Y.

Stealth. As we discussed in our motivation, we consider the stealth in two directions: the naturalness of appearance and the patch size. In terms of naturalness, we compare the adversarial patch generated by our method with the baseline method proposed by Yamanaka et al. [63]. As shown in Fig. 9, our method with style transfer-based camouflage generates more natural patches and is less likely to be identified as adversarial but just a normal sticker. Human studies conducted in [14,29] also justify the naturalness of style-transfer-based image processing. As for the patch size, a smaller size suggests more stealth and less effectiveness. We hence investigate maximizing the attack effect with small patches. We compare the $\mathcal{R}_a$ caused by our object-oriented attack and the patch-oriented attack in [63] which only attacks the patch area in their adversarial loss design instead of considering the whole object. For a fair comparison, we use style-transfer-based camouflage in both methods and we test with fixed regions and the regional optimization. This experiment is conducted on Monodepth2 [18] targeting the vehicle and other settings are the same as the previous setup.

As shown in Fig. 10a, our method (object-oriented) has over 2.5 times higher $\mathcal{R}_a$ on the vehicle than the baseline (patch-oriented) in all cases, and our method in the regional optimization case outperforms all other fixed-region cases. Hence, with the same total patch area, our object-oriented attack with regional optimization affects a broader area than the baseline. In other words, *to achieve similar attack effect, using our method requires a smaller patch and is thus stealthier.* Fig. 10b additionally shows the CDF and histogram of depth estimation error in the case with our regional optimization. As shown, more than 80% errors caused by the baseline method are below 10 m, which corresponds to our observation in Fig. 2c that the patch-oriented attack mainly affects the limited patch area and the effect of our method propagates to a broader area causing larger errors.

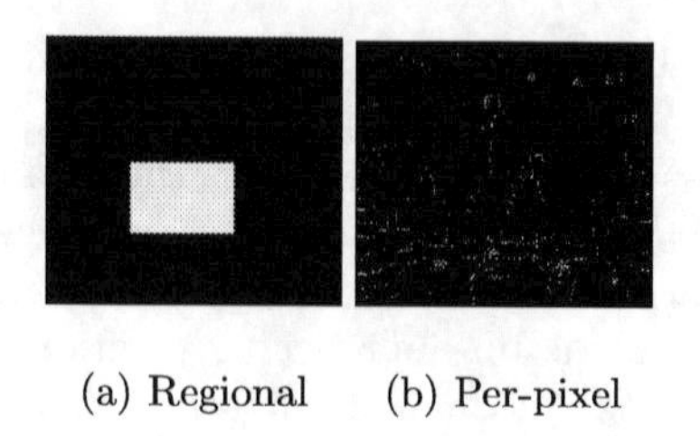

(a) Regional (b) Per-pixel

Fig. 11. Different **mask optimization methods.**

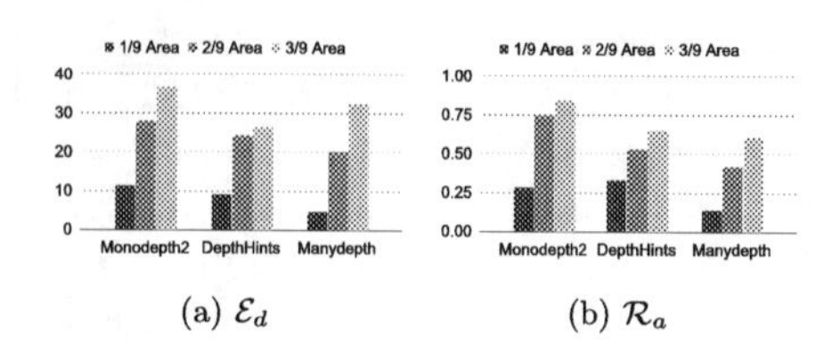

(a) $\mathcal{E}_d$ (b) $\mathcal{R}_a$

Fig. 12. Attack performance of regional optimization with **different target sizes.**

4.3 Ablation Study

We investigate our method through a set of ablation studies.

Combinations. As detailed in Sect. 3, we use the object-oriented adversarial loss design and the regional optimization of the patch mask to maximize the attack effect with a small patch. We conduct ablations on these techniques to see how each component contributes. Table 3 shows the result. We attack Monodepth2 and use the vehicle as the target object and report $\mathcal{E}_d$ and $\mathcal{R}_a$. For those tests without regional optimization, we use #5 fixed region because its attack performance is the best among all the fixed regions in previous evaluations. As shown, the object-oriented adversarial loss itself can improve the attack performance while the regional optimization cannot. The regional optimization is useful only when object-oriented adversarial loss is applied together. The regional optimization has to consider the whole object to find an optimal place regarding the target object. Since the patch-oriented design does not encode the global information, our regional optimization cannot converge to the most effective region.

Mask Optimization Methods. We compare our regional optimization with another commonly used mask optimization technique which treats pixels of the patch mask m_p as optimizable parameters instead of the four borders. This method has been used in many backdoor scanning works such as Neural Cleanse [51] and ABS [28] to find a trigger that modifies a limited portion of image and causes misclassification (see Fig. 11). Note that the patch mask generated by the baseline method is more sparse and scattered. The patch unit is tiny. Compared with our method, it is not suitable as a physical world attack vector because it is hard to print and deploy these scattered tiny patches.

Table 3. Ablation study .

OA	RO	$\mathcal{E}_d$	$\mathcal{R}_a$
		8.47	0.23
	✓	6.38	0.16
✓		14.95	0.52
✓	✓	**16.84**	**0.65**

OA: Object-oriented Adv. Loss
RO: Regional Optimization

Patch Sizes. Larger patches have more effect on depth estimation but are less stealthy. We evaluate our attack on a vehicle object with three different target

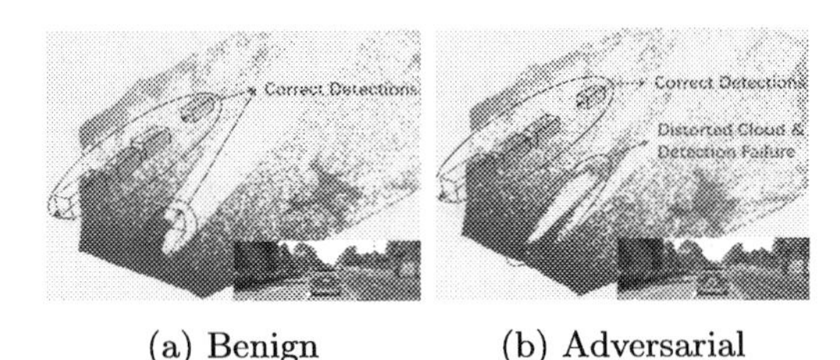

(a) Benign (b) Adversarial

Fig. 13. Attack against **3D object detection.**

Table 4. Attack success rate of different adversarial patches.

	Monodepth2	DepthHints	Manydepth
1/9 Area	95%	93%	98%
2/9 Area	98%	97%	100%
1/3 Area	100%	100%	100%

patch sizes and use three depth estimation models (see Fig. 12). Note that the mean depth estimation error $\mathcal{E}_d$ and the ratio of affected region $\mathcal{R}_a$ increase with the size of patch for all three target networks.

4.4 Downstream Task Impact

We evaluate the impact of our attack on a point cloud based 3D object detection model – PointPillars [24] and use attack success rate (ASR) as the metric to evaluate our method on 3D object detection. We consider the attack is successful when the benign vehicle can be detected by PointPillar while the adversarial object cannot. Figure 13 gives an example of a successful attack. Figure 13a presents a benign scenario where the benign vehicle can be correctly detected with a 3D bounding box. Figure 13b shows the corresponding adversarial scenario where the pseudo-Lidar point cloud of the adversarial vehicle is severely distorted by the patch, and thus the vehicle is not detected. The PointPillar network can correctly detect the benign vehicle in all the 100 scenes and the attack success rate (ASR) of different adversarial patches are reported in Table 4. The first column denotes different patch sizes and columns 2–4 refer to the three different target networks. As shown, the ASR is over 90% with all the patch sizes and target networks. Even when the patch size is just 1/9 of the vehicle's rear area, it can still achieve at least 93% ASR, which shows that our attack is an effective method in fooling the 3D object detection model. In the physical world experiments, the fifth column of Table 2 denotes the number of frames in which the vehicle is detected from the pseudo-Lidar point cloud, and the sixth column reports the object detection rate. For benign cases, the rate of successful object detection is 90.70% in 1291 data frames. The rate drops to 5.16% in adversarial cases with 1278 data frames.

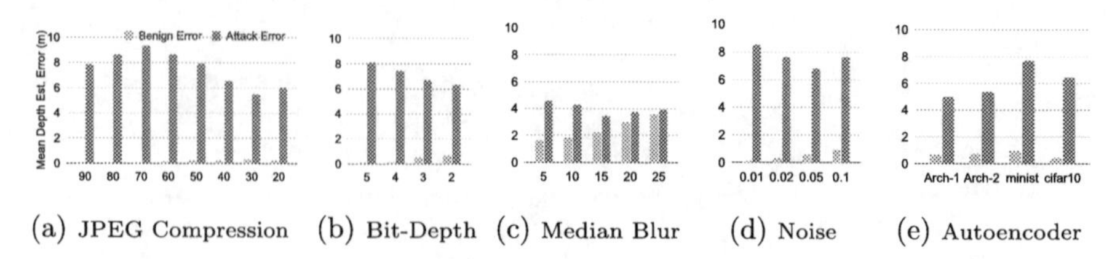

(a) JPEG Compression (b) Bit-Depth (c) Median Blur (d) Noise (e) Autoencoder

Fig. 14. Five directly-applicable defence methods. *Benign Error*: Error caused by the defence in benign cases. *Attack Error*: Error caused by our attack.

4.5 Defence Discussion

Although many defense techniques against adversarial examples have been proposed, none of them focuses on MDE to the best of our knowledge. As a best effort to understand the performance of our attack under different defences, we apply five popular defence techniques which perform input transformations without retraining the victim network. They are JPEG compression [15], bit-depth reduction [62], median blurring [62], adding Gaussian noise [65] and autoencoder reformation [31]. Figure 14 presents our results. We report the $\mathcal{E}_d$ of the benign vehicle and the adversarial vehicle under different input transformations. An ideal defence should minimize both errors. As shown, our attack can still cause over 5 m $\mathcal{E}_d$ in all methods except median blur. In median blur, the attack is mitigated but the benign performance also drops a lot. This shows that these techniques cannot effectively defend our attack. We argue that these defenses are mainly for attacks in digital space [39] instead of physical world settings.

5 Conclusion

In this paper, we investigate stealthy physical-world adversarial patch attack against MDE in the AD scenario. We design a novel physical-object-oriented optimization framework to generate stealthy and effective adversarial patches for attack via an object-oriented adversarial loss design. Experimental results show that our attack is effective, stealthy and robust against different target objects, state-of-the-art models and a representative downstream task (*i.e.*, 3D object detection) in AD.

Acknowledgments. This research was supported, in part by IARPA TrojAI W911NF-19-S-0012, NSF 1901242 and 1910300, ONR N000141712045, N00014-1410468 and N000141712947.

References

1. Break distance. www.csgnetwork.com/stopdistcalc.html
2. Tesla AI day. youtu.be/j0z4FweCy4M?t=5295

3. Tesla use per-pixel depth estimation with self-supervised learning. youtu.be/hx7BXih7zx8?t=1334
4. Vehicle stopping distance. www.csgnetwork.com/stopdistcalc.html
5. Aich, S., Vianney, J.M.U., Islam, M.A., Kaur, M., Liu, B.: Bidirectional attention network for monocular depth estimation. arXiv preprint arXiv:2009.00743 (2020)
6. Athalye, A., Engstrom, L., Ilyas, A., Kwok, K.: Synthesizing robust adversarial examples. In: International Conference on Machine Learning, pp. 284–293. PMLR (2018)
7. Brown, T.B., Mané, D., Roy, A., Abadi, M., Gilmer, J.: Adversarial patch. arXiv preprint arXiv:1712.09665 (2017)
8. Byrd, R.H., Lu, P., Nocedal, J., Zhu, C.: A limited memory algorithm for bound constrained optimization. SIAM J. Sci. Comput. **16**(5), 1190–1208 (1995)
9. Cao, Y., et al.: Invisible for both camera and lidar: Security of multi-sensor fusion based perception in autonomous driving under physical-world attacks. In: 2021 IEEE Symposium on Security and Privacy (SP), pp. 176–194. IEEE (2021)
10. Cao, Y., et al.: Adversarial sensor attack on lidar-based perception in autonomous driving. In: Proceedings of the 2019 ACM SIGSAC Conference on Computer and Communications Security, pp. 2267–2281 (2019)
11. Chen, S.-T., Cornelius, C., Martin, J., Chau, D.H.P.: ShapeShifter: robust physical adversarial attack on faster R-CNN object detector. In: Berlingerio, M., Bonchi, F., Gärtner, T., Hurley, N., Ifrim, G. (eds.) ECML PKDD 2018. LNCS (LNAI), vol. 11051, pp. 52–68. Springer, Cham (2019). https://doi.org/10.1007/978-3-030-10925-7_4
12. Cui, Y., Yan, L., Cao, Z., Liu, D.: TF-blender: temporal feature blender for video object detection. In: Proceedings of the IEEE/CVF International Conference on Computer Vision, pp. 8138–8147 (2021)
13. Dominguez, S., Ali, A., Garcia, G., Martinet, P.: Comparison of lateral controllers for autonomous vehicle: experimental results. In: 2016 IEEE 19th International Conference on Intelligent Transportation Systems (ITSC), pp. 1418–1423. IEEE (2016)
14. Duan, R., Ma, X., Wang, Y., Bailey, J., Qin, A.K., Yang, Y.: Adversarial camouflage: hiding physical-world attacks with natural styles. In: Proceedings of the IEEE/CVF Conference on Computer Vision and Pattern Recognition, pp. 1000–1008 (2020)
15. Dziugaite, G.K., Ghahramani, Z., Roy, D.M.: A study of the effect of jpg compression on adversarial images. arXiv preprint arXiv:1608.00853 (2016)
16. Gatys, L.A., Ecker, A.S., Bethge, M.: Image style transfer using convolutional neural networks. In: Proceedings of the IEEE Conference on Computer Vision and Pattern Recognition, pp. 2414–2423 (2016)
17. Geiger, A., Lenz, P., Urtasun, R.: Are we ready for autonomous driving? the kitti vision benchmark suite. In: Conference on Computer Vision and Pattern Recognition (CVPR) (2012)
18. Godard, C., Mac Aodha, O., Firman, M., Brostow, G.J.: Digging into self-supervised monocular depth prediction, October 2019
19. Goodfellow, I.J., Shlens, J., Szegedy, C.: Explaining and harnessing adversarial examples. arXiv preprint arXiv:1412.6572 (2014)
20. Guizilini, V., Ambrus, R., Pillai, S., Raventos, A., Gaidon, A.: 3d packing for self-supervised monocular depth estimation. In: IEEE Conference on Computer Vision and Pattern Recognition (CVPR) (2020)

21. Jia, Y., Lu, Y., Shen, J., Chen, Q.A., Zhong, Z., Wei, T.: Fooling detection alone is not enough: first adversarial attack against multiple object tracking. In: International Conference on Learning Representations (ICLR) (2020)
22. Kingma, D.P., Ba, J.: Adam: a method for stochastic optimization. arXiv preprint arXiv:1412.6980 (2014)
23. Komkov, S., Petiushko, A.: Advhat: Real-world adversarial attack on arcface face id system. In: 2020 25th International Conference on Pattern Recognition (ICPR), pp. 819–826. IEEE (2021)
24. Lang, A.H., Vora, S., Caesar, H., Zhou, L., Yang, J., Beijbom, O.: Pointpillars: fast encoders for object detection from point clouds. In: Proceedings of the IEEE/CVF Conference on Computer Vision and Pattern Recognition, pp. 12697–12705 (2019)
25. Lee, J., Yi, J., Shin, C., Yoon, S.: BBAM: bounding box attribution map for weakly supervised semantic and instance segmentation. In: Proceedings of the IEEE/CVF Conference on Computer Vision and Pattern Recognition, pp. 2643–2652 (2021)
26. Liu, D., Cui, Y., Tan, W., Chen, Y.: SG-Net: Spatial granularity network for one-stage video instance segmentation. In: Proceedings of the IEEE/CVF Conference on Computer Vision and Pattern Recognition, pp. 9816–9825 (2021)
27. Liu, D., Cui, Y., Yan, L., Mousas, C., Yang, B., Chen, Y.: DenserNet: weakly supervised visual localization using multi-scale feature aggregation. In: Proceedings of the AAAI Conference on Artificial Intelligence, vol. 35, pp. 6101–6109 (2021)
28. Liu, Y., Lee, W.C., Tao, G., Ma, S., Aafer, Y., Zhang, X.: ABS: scanning neural networks for back-doors by artificial brain stimulation. In: Proceedings of the 2019 ACM SIGSAC Conference on Computer and Communications Security, pp. 1265–1282 (2019)
29. Luan, F., Paris, S., Shechtman, E., Bala, K.: Deep photo style transfer. In: Proceedings of the IEEE Conference on Computer Vision and Pattern Recognition, pp. 4990–4998 (2017)
30. Mathew, A., Patra, A.P., Mathew, J.: Monocular depth estimators: vulnerabilities and attacks. arXiv preprint arXiv:2005.14302 (2020)
31. Meng, D., Chen, H.: MagNet: a two-pronged defense against adversarial examples. In: Proceedings of the 2017 ACM SIGSAC Conference on Computer and Communications Security, pp. 135–147 (2017)
32. Nassi, B., Nassi, D., Ben-Netanel, R., Mirsky, Y., Drokin, O., Elovici, Y.: Phantom of the ADAS: phantom attacks on driver-assistance systems. IACR Cryptol. ePrint Arch. **2020**, 85 (2020)
33. Papernot, N., McDaniel, P., Jha, S., Fredrikson, M., Celik, Z.B., Swami, A.: The limitations of deep learning in adversarial settings. In: 2016 IEEE European Symposium on Security and Privacy (EuroS&P), pp. 372–387. IEEE (2016)
34. Pei, K., Cao, Y., Yang, J., Jana, S.: DeepXplore: automated whitebox testing of deep learning systems. In: proceedings of the 26th Symposium on Operating Systems Principles, pp. 1–18 (2017)
35. Petit, J., Stottelaar, B., Feiri, M., Kargl, F.: Remote attacks on automated vehicles sensors: experiments on camera and lidar. Black Hat Eur. **11**(2015), 995 (2015)
36. Qiu, H., Xiao, C., Yang, L., Yan, X., Lee, H., Li, B.: SemanticAdv: generating adversarial examples via attribute-conditioned image editing. In: Vedaldi, A., Bischof, H., Brox, T., Frahm, J.-M. (eds.) ECCV 2020. LNCS, vol. 12359, pp. 19–37. Springer, Cham (2020). https://doi.org/10.1007/978-3-030-58568-6_2
37. Ranjan, A., Janai, J., Geiger, A., Black, M.J.: Attacking optical flow. In: Proceedings of the IEEE/CVF International Conference on Computer Vision, pp. 2404–2413 (2019)

38. Rao, S., Stutz, D., Schiele, B.: Adversarial training against location-optimized adversarial patches. In: Bartoli, A., Fusiello, A. (eds.) ECCV 2020. LNCS, vol. 12539, pp. 429–448. Springer, Cham (2020). https://doi.org/10.1007/978-3-030-68238-5_32
39. Sato, T., Shen, J., Wang, N., Jia, Y., Lin, X., Chen, Q.A.: Dirty road can attack: security of deep learning based automated lane centering under physical-world attack. In: 30th {USENIX} Security Symposium ({USENIX} Security 21), pp. 3309–3326 (2021)
40. Sato, T., Shen, J., Wang, N., Jia, Y.J., Lin, X., Chen, Q.A.: Hold tight and never let go: security of deep learning based automated lane centering under physical-world attack. arXiv preprint arXiv:2009.06701 (2020)
41. Sato, T., Shen, J., Wang, N., Jia, Y.J., Lin, X., Chen, Q.A.: WIP: Deployability improvement, stealthiness user study, and safety impact assessment on real vehicle for dirty road patch attack. In: Workshop on Automotive and Autonomous Vehicle Security (AutoSec), vol. 2021, p. 25 (2021)
42. Sharif, M., Bhagavatula, S., Bauer, L., Reiter, M.K.: Accessorize to a crime: real and stealthy attacks on state-of-the-art face recognition. In: Proceedings of the 2016 ACM SIGSAC Conference on Computer and Communications Security, pp. 1528–1540 (2016)
43. Shen, J., Won, J.Y., Chen, Z., Chen, Q.A.: Drift with devil: security of multi-sensor fusion based localization in high-level autonomous driving under {GPS} spoofing. In: 29th {USENIX} Security Symposium ({USENIX} Security 20), pp. 931–948 (2020)
44. Shin, H., Kim, D., Kwon, Y., Kim, Y.: Illusion and dazzle: adversarial optical channel exploits against lidars for automotive applications. In: Fischer, W., Homma, N. (eds.) CHES 2017. LNCS, vol. 10529, pp. 445–467. Springer, Cham (2017). https://doi.org/10.1007/978-3-319-66787-4_22
45. Song, D., et al.: Physical adversarial examples for object detectors. In: 12th {USENIX} Workshop on Offensive Technologies ({WOOT} 18) (2018)
46. Tang, K., Shen, J.S., Chen, Q.A.: Fooling perception via location: a case of region-of-interest attacks on traffic light detection in autonomous driving. In: NDSS Workshop on Automotive and Autonomous Vehicle Security (AutoSec) (2021)
47. Thys, S., Van Ranst, W., Goedemé, T.: Fooling automated surveillance cameras: adversarial patches to attack person detection. In: Proceedings of the IEEE/CVF Conference on Computer Vision and Pattern Recognition Workshops (2019)
48. Trippel, T., Weisse, O., Xu, W., Honeyman, P., Fu, K.: Walnut: waging doubt on the integrity of mems accelerometers with acoustic injection attacks. In: 2017 IEEE European Symposium on Security and Privacy (EuroS&P), pp. 3–18. IEEE (2017)
49. Tsai, T., Yang, K., Ho, T.Y., Jin, Y.: Robust adversarial objects against deep learning models. In: Proceedings of the AAAI Conference on Artificial Intelligence, vol. 34, pp. 954–962 (2020)
50. Tu, Y., Lin, Z., Lee, I., Hei, X.: Injected and delivered: fabricating implicit control over actuation systems by spoofing inertial sensors. In: 27th {USENIX} Security Symposium ({USENIX} Security 18), pp. 1545–1562 (2018)
51. Wang, B., et al.: Neural cleanse: identifying and mitigating backdoor attacks in neural networks. In: 2019 IEEE Symposium on Security and Privacy (SP), pp. 707–723. IEEE (2019)
52. Wang, Y., Chao, W.L., Garg, D., Hariharan, B., Campbell, M., Weinberger, K.: Pseudo-lidar from visual depth estimation: bridging the gap in 3d object detection for autonomous driving. In: CVPR (2019)

53. Watson, J., Aodha, O.M., Prisacariu, V., Brostow, G., Firman, M.: The temporal opportunist: self-supervised multi-frame monocular depth. In: Computer Vision and Pattern Recognition (CVPR) (2021)
54. Watson, J., Firman, M., Brostow, G.J., Turmukhambetov, D.: Self-supervised monocular depth hints. In: The International Conference on Computer Vision (ICCV), October 2019
55. Wimbauer, F., Yang, N., von Stumberg, L., Zeller, N., Cremers, D.: MonoRec: semi-supervised dense reconstruction in dynamic environments from a single moving camera. In: IEEE Conference on Computer Vision and Pattern Recognition (CVPR) (2021)
56. Wong, A., Cicek, S., Soatto, S.: Targeted adversarial perturbations for monocular depth prediction. arXiv preprint arXiv:2006.08602 (2020)
57. Xiao, C., Li, B., Zhu, J.Y., He, W., Liu, M., Song, D.: Generating adversarial examples with adversarial networks. arXiv preprint arXiv:1801.02610 (2018)
58. Xiao, C., et al.: Characterizing attacks on deep reinforcement learning. arXiv preprint arXiv:1907.09470 (2019)
59. Xiao, C., Yang, D., Li, B., Deng, J., Liu, M.: MeshAdv: adversarial meshes for visual recognition. In: Proceedings of the IEEE/CVF Conference on Computer Vision and Pattern Recognition, pp. 6898–6907 (2019)
60. Xiao, C., Zhu, J.Y., Li, B., He, W., Liu, M., Song, D.: Spatially transformed adversarial examples. arXiv preprint arXiv:1801.02612 (2018)
61. Xu, K., et al.: Adversarial t-shirt! evading person detectors in a physical world. In: Vedaldi, A., Bischof, H., Brox, T., Frahm, J.-M. (eds.) ECCV 2020. LNCS, vol. 12350, pp. 665–681. Springer, Cham (2020). https://doi.org/10.1007/978-3-030-58558-7_39
62. Xu, W., Evans, D., Qi, Y.: Feature squeezing: detecting adversarial examples in deep neural networks. arXiv preprint arXiv:1704.01155 (2017)
63. Yamanaka, K., Matsumoto, R., Takahashi, K., Fujii, T.: Adversarial patch attacks on monocular depth estimation networks. IEEE Access **8**, 179094–179104 (2020)
64. Yan, C., Xu, W., Liu, J.: Can you trust autonomous vehicles: contactless attacks against sensors of self-driving vehicle. Def Con **24**(8), 109 (2016)
65. Zhang, Y., Liang, P.: Defending against whitebox adversarial attacks via randomized discretization. In: The 22nd International Conference on Artificial Intelligence and Statistics, pp. 684–693. PMLR (2019)
66. Zhang, Z., Zhu, X., Li, Y., Chen, X., Guo, Y.: Adversarial attacks on monocular depth estimation. arXiv preprint arXiv:2003.10315 (2020)

ST-P3: End-to-End Vision-Based Autonomous Driving via Spatial-Temporal Feature Learning

Shengchao Hu[1], Li Chen[2(✉)], Penghao Wu[1,3], Hongyang Li[1,2], Junchi Yan[1,2], and Dacheng Tao[4]

[1] MoE Key Lab of Artificial Intelligence, Shanghai Jiao Tong University, Shanghai, China
charles-hu@sjtu.edu.cn
[2] Shanghai AI Laboratory, Shanghai, China
lichen@pjlab.org.cn
[3] The University of California, San Diego, CA, USA
[4] JD Explore Academy, JD.com Inc., Beijing, China

Abstract. Many existing autonomous driving paradigms involve a multi-stage discrete pipeline of tasks. To better predict the control signals and enhance user safety, an end-to-end approach that benefits from joint spatial-temporal feature learning is desirable. While there are some pioneering works on LiDAR-based input or implicit design, in this paper we formulate the problem in an interpretable vision-based setting. In particular, we propose a spatial-temporal feature learning scheme towards a set of more representative features for perception, prediction and planning tasks simultaneously, which is called ST-P3. Specifically, an egocentric-aligned accumulation technique is proposed to preserve geometry information in 3D space before the bird's eye view transformation for perception; a dual pathway modeling is devised to take past motion variations into account for future prediction; a temporal-based refinement unit is introduced to compensate for recognizing vision-based elements for planning. To the best of our knowledge, we are the first to systematically investigate each part of an interpretable end-to-end vision-based autonomous driving system. We benchmark our approach against previous state-of-the-arts on both open-loop nuScenes dataset as well as closed-loop CARLA simulation. The results show the effectiveness of our method. Source code, model and protocol details are made publicly available at https://github.com/OpenPerceptionX/ST-P3.

1 Introduction

A classical paradigm design for autonomous driving systems often adopts a modular based spirit [2,49], where the input of a planning or controlling unit is based

S. Hu and P. Wu—Work done during internship at Shanghai AI Laboratory.

Supplementary Information The online version contains supplementary material available at https://doi.org/10.1007/978-3-031-19839-7_31.

S. Avidan et al. (Eds.): ECCV 2022, LNCS 13698, pp. 533–549, 2022.
https://doi.org/10.1007/978-3-031-19839-7_31

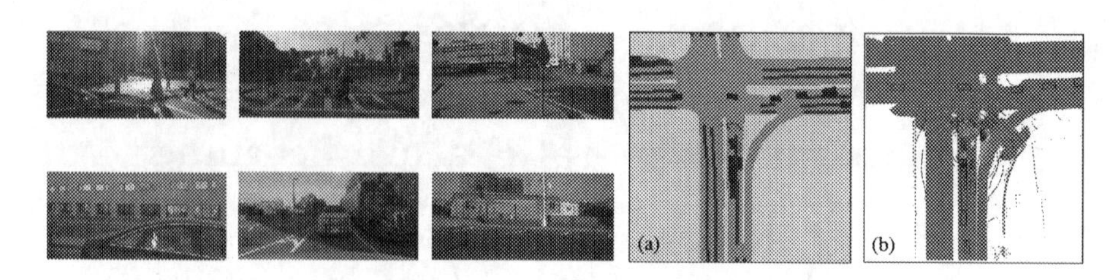

Fig. 1. Problem setup whereby an interpretable vision-based end-to-end framework in (a) is devised, parallel to the LiDAR-based counterpart by aid of HD maps in (b)

on the outputs from preceding modules in perception. As we witness the blossom of end-to-end algorithms and success applications into various domains [22,37], there are some attempt implementing such a philosophy in autonomous driving as well [1,3,7,8,12,15,40,41,52]. Rather than an isolated staged pipeline, we aim for a framework to directly take raw sensor data as inputs and generate the planning routes or control signals. A straightforward incentive to do so is that feature representations can thus be optimized simultaneously *within* one network towards the ultimate goal of the system (*e.g.*, acceleration, steering).

One direction of end-to-end pipelines is to focus on the ultimate planning task mainly without explicit design of the intermediate representation [9,12–14,25,41]. Reinforcement learning (RL) fits well as a feasible resolution since the planned routes are not unique and each action should be rewarded correspondingly based on the environment. RL algorithms are applied to mimic experienced human experts and guide the behavior learning of driving agents [6,52]. Besides RL approaches, some propose to generate cost map with a trajectory sampler with knowledge of the perception environment [12,25] or fusion of sensors in an attention manner [19,41]. These work aforementioned achieve impressive performance on challenging scenarios in closed-loop simulation [18]. The plausible transfer from synthetic setting to realistic application remains an open question.

Another direction is to explicitly design the intermediate representations in the network, provide convincing interpretability of each module and thus enhance safety towards a stable and robust system. Based on the input type, explicit approaches are divided into LiDAR-based [5,16,50,51] and vision-based [24,38,46] respectively. LiDAR-based methods bundled with high-definition (HD) maps exhibit good performance on various benchmarks, and they investigate the effect of each module of the system exhaustively. However, the inherent defects of LiDAR, such as recognition of traffic lights and short-range detection of objects, might confine their applications.

In this paper, concurrent to the interpretable LiDAR-based pipelines [5,51], we propose to investigate the potential of *vision-based* end-to-end framework (Fig. 1). If each module is exquisitely designed, to which extent the performance of each task as in perception, prediction and planning should be improved?

To answer the above question, the first key challenge for vision-based methods is to appropriately transform feature representations from perspective views to the bird's eye view (BEV) space. The pioneering LSS approach [38] extracts perspective features from multi-view cameras, lifts them into 3D with depth estimation and fuses into the BEV space. It is observed that latent depth prediction for the feature transformation between two views is crucial [24,42,46]. On a theoretical analysis, this is true since lifting 2D planar information to 3D requires an additional dimension, which is the depth that fits into 3D geometric autonomous driving tasks. To further improve feature representation, it is natural to incorporate temporal information into the framework as most scenarios are tasked with video sources. Descending from [38], the follow-up literature project past frames' feature onto current coordinate view either by ego-motion of the self-driving vehicle (SDV) provided by dataset [24] or learned mapping from optical flow [46]. These approaches project features in the past frame-by-frame in isolation and feed them into the temporal unit; instead we accumulate all past aligned features in 3D space before transforming to BEV, preserving geometry information at best and compensating for more robust feature representations of the current state, which is empirically proven as a better design choice.

Equipped with representative features in BEV space, we formulate prediction task as the future instance segmentation, which is the same as does in FIERY [24]. The common practice includes generating uncertainty from data distribution for current state and feeding it in a temporal model to infer predictions under a window of future horizons [24,46]. A natural incentive to boost future predictions, which is missing in current literature, is to take into account the motion variations in the *past*. To do so, an additional temporal model with fusion unit could be introduced to reason about the probabilistic nature of both past and future motions. A stronger version of scene representations could therefore be obtained, which serves as recipe towards the ultimate planning task.

The general idea for motion planners is to output the most likely trajectory, given a sampling of possible candidates and semantic results from preceding modules [5,25,41,51]. In light of an interpretable spirit, most previous work construct cost volumes, learning-based [5,12] and/or rule-based [6,50,51], to indicate the confidence of trajectories with a certain form of trajectory modelling in a sampler. We follow such a philosophy to indicate the most possible candidate with the help of a high-level command, without HD map as guidance. The outcome trajectory is further refined with the features from the front-view camera (front-view features) to consider vision-based elements (*e.g.*, traffic lights). This is inspired by MP3 [5], where they also remove HD maps and feed the network with a high-level command. However, we argue that the vision recognition module in [5] is off-the-shelve; in this work, we integrate vision information in form of a lightweight GRU unit within the same network. Such a refinement process serves as a complementary feature boosting towards the final outcome.

To this end, we propose an interpretable vision-based end-to-end system that improves feature learning for perception, prediction and planning altogether, namely **ST-P3**. Figure 2 describes the overall framework. Specifically, given a set of surrounding camera videos, we feed them into the backbone to generate preliminary front-view features. An auxiliary depth estimation is performed to transform 2D features onto 3D space. An egocentric aligned accumulation scheme first aligned past features to the current view coordinate system. The current and

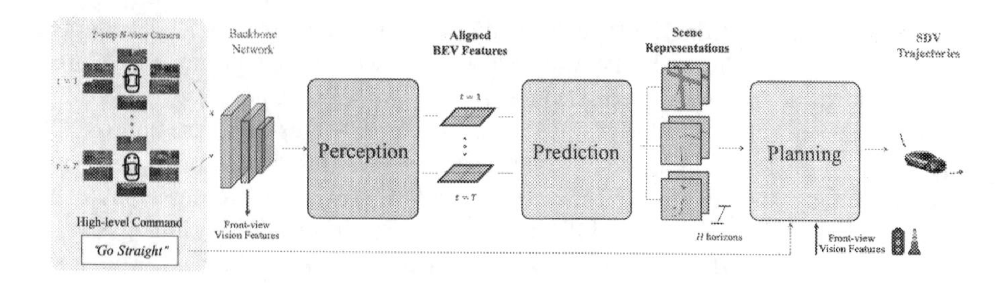

Fig. 2. We present **ST-P3**, an interpretable end-to-end vision-based framework. For **perception**, the egocentric aligned accumulation guarantees features (past and present) aligned and aggregated in 3D space to preserve geometry information before BEV transformation. For **prediction**, a dual pathway scheme is introduced to bring in past variations in pursuit of future predictions. For **planning**, the prior knowledge is fed into a refinement unit to generate the final trajectory with an integrated cost volume and sampler from high-level commands

past features are then aggregated in 3D space to preserve geometric information before the BEV representation. Apart from a commonly used temporal model for prediction, this module is further boosted by constructing a second pathway to account for motion variations in the past. Such a dual pathway modelling ensures stronger feature representations to infer future semantic outcomes. Towards the ultimate goal of trajectory planning, we integrate prior knowledge from features in the early stage of the network. A refinement module is devised to generate the final trajectory with help of high-level commands and no presence of HD maps. We benchmark our approach against previous state-of-the-arts on both open-loop nuScenes dataset as well as closed-loop simulator CARLA environment. To sum up, ST-P3 owns the following contributions baked into it:

1. For better spatial-temporal feature learning, we propose three novel improvements, *i.e.*, the egocentric aligned accumulation, the dual pathway modelling, and the prior-knowledge refinement for perception, prediction and planning modules respectively. The resulting new end-to-end vision-based network for autonomous driving is called ST-P3.
2. We investigate each part of an interpretable end-to-end system for autonomous driving tasks systematically. As a vision-based counterpart to the study of LiDAR-based approaches [51], to our best knowledge, we provide the first detailed analysis and comparison for a vision-based pipeline.
3. ST-P3 achieves state-of-the-art performance on benchmarks from the popular nuScenes dataset and simulator CARLA. The full suite of codebase, as well as protocols are made publicly available.

2 Related Work

We briefly discuss the related works in four aspects.

Interpretable End-to-End Framework. We review popular approaches [5, 16, 44, 50, 51] that adopt an explicit design to have clear interpretability of the system and hence prompts safety. These are LiDAR-based approaches mainly as there are few vision-based solutions in the wild (compared in the previous section already). The pioneering NMP [50] takes as input LiDAR and HD maps to first predict the bounding boxes of actors in the future, and then learns a cost volume to choose the best planned trajectory. The subsequent P3 [44] work further achieves consistency between planning and perception by a differentiable occupancy representation, which is explicitly used as cost by planning to produce safe maneuvers. To avoid heavy reliance on HD maps, Casas *et al.* (MP3) [5] constructs an online map from segmentation as well as the current and future states of other agents; these results are then fed into a sampler-based planner to obtain a safe and comfortable trajectory. LookOut [16] predicts a diverse set of futures of how the scene might unroll and estimates the trajectory by optimizing a set of contingency plans. DSDNet [51] considers the interactions between actors and produces socially consistent multimodal future predictions. It explicitly exploits the predicted future distributions of actors to plan a safe maneuver by using a structured planning cost. In general, LiDAR based approaches demonstrate good performance on challenging urban scenarios. Unfortunately, the datasets and baselines in these work are not released to compare.

Bird's eye view (BEV) Representation is a natural and perfect fit for planning and control tasks [1,12,34,36,52], since it avoids problems such as occlusion, scale distortion, *etc.*, and preserves 3D scene layout. Although information in LiDAR and HD maps can be easily represented in BEV, how to project vision inputs from camera view to BEV space is a non-trivial problem. Some learning based methods [12,36] implicitly project image input into BEV, but the quality can not be guaranteed since usually we do not have ground truth in BEV to supervise the projection process. Loukkal *et al.* [31] explicitly projects image into BEV using homography between image and BEV plane. [10,29] aquires BEV features through spatial cross-attention with pre-defined BEV queries. LSS [38] and FIERY [24] perform the projection with estimated depth and image intrinsic, which have shown impressive performance. We predict depth and project images in a similar fashion. Different from FIERY [24] which transforms past features to current timestamp frame-by-frame correspondingly, we append all past 3D features to the current ego view (egocentric) and accumulate the aligned features, providing better representation for subsequent tasks.

Future Prediction. Current motion prediction methods [21,26,30,35] usually takes ground truth perception information and HD map as input, but they are susceptible to cumulative error when the perception input comes from other modules in real-life application. Taking raw sensor data as input, end-to-end methods which focus on future trajectory prediction or take it as an intermediate step usually rely on LiDAR and HD map [16,33,47,50,51] to detect and predict. Recently, future prediction in the form of BEV semantic segmentation using only camera input [24,46] has been proposed and shown great performance. However, the evolution process of the past is not well captured and exploited [24,38,46]. Inspired by video future prediction [23], we combine the probabilistic uncertainty with the dynamics of past to predict diverse and plausible future scenes.

Motion Planning. We cover previous learning-based motion planning methods and omit traditional approaches in this part. For implicit methods [14,40,41], the network directly generates planned trajectory or control commands. Although such design is direct and simple, it suffers from robustness and lack of interpretability. On the contrary, explicit methods usually build a cost map with a trajectory sampler to generate the desired trajectory by choosing the optimal candidate with the lowest cost. The cost map can be constructed with hand-crafted rules [5,16,44] based on intermediate representations such as segmentations and HD maps; or it can be learned directly from the network [50]. DSDNet [51] combines hand-crafted and learning-based costs to obtain an integrated cost volume. We also adopt this combination to choose the best trajectory. However, we modify the pipeline by adding an additional GRU refinement unit with navigation signal to further adjust and optimize the chosen trajectory.

3 Methodology

An overview of ST-P3 is given in Fig. 2. ST-P3 first extracts features of a sequence of surrounding images and lift them to 3D with depth prediction. They are fused in both spatial and temporal domains with the egocentric aligned accumulation (see Sect. 3.1). We show the dual pathway mechanism in Sect. 3.2, a novel uncertainty modeling to incorporate history information. Sect. 3.3 elaborates on how we utilize the intermediate representations to plan a safe and comfortable trajectory.

3.1 Perception: Egocentric Aligned Accumulation

In this stage we need to build a spatiotemporal BEV feature from multi-view camera inputs in past t timestamps. As discussed in Sect. 1, the direct concatenation [46] has the alignment issue while FIERY [24] suffers from losing height information. Towards these problems, here we introduce our accumulative ego-centric alignment method which incorporates two steps, *i.e.*, spatial and temporal fusion. In the spatial fusion part, multi-view images in all timestamps are processed and transformed to the current ego-centric 3D space. While in the temporal fusion step, we enhance the feature discrimination of static elements and objects in motion in an accumulative way and adopt a temporal model to achieve the final fusion. An illustration is depicted in Fig. 3.

Spatial Fusion. On one hand, features from multi-view images should be transformed to a common frame. Inspired by [38], we extract the feature and predict the corresponding depth for each image and then lift it into the global 3D frame. In particular, each camera image $I_t^n \in \mathbb{R}^{3\times H\times W}$ is passed through a backbone network to obtain features $f_i^k \in \mathbb{R}^{C\times H_e\times W_e}$ and depth estimation $d_i^k \in \mathbb{R}^{D\times H_e\times W_e}$ respectively, where $i \in \{1,\dots,t\}$, $n \in \{1,\dots,6\}$, C is the number of feature channels, D denotes the number of discrete depths and (H_e, W_e) indicates the spatial size. Since the exact depth information is not available, we spread the feature across the entire ray of space according to the predicted depth distribution by taking the outer product of the matrices:

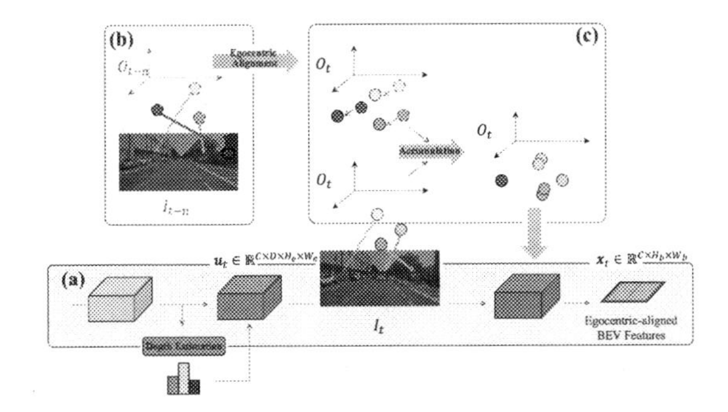

Fig. 3. Egocentric aligned accumulation for Perception. **(a)** Feature at current timestamp is lifted to 3D with depth estimation, and pooled to BEV features x_t after alignment; **(b–c)** 3D features in previous frames are aligned to current view and fused with all past and current states, so that the feature representation get enhanced

$$u_i^k = f_i^k \otimes d_i^k, \tag{1}$$

with $u_i^k \in \mathbb{R}^{C \times D \times H_e \times W_e}$. Then we use the camera extrinsics and intrinsics to transform the camera feature frustums $u_i \in \{u_i^1, \ldots, u_i^n\}$ to the global 3D coordinate whose origin is at the inertial center of the ego-vehicle at time i.

On the other hand, the spatial fusion needs to align past features to the current frame for the downstream prediction and planning tasks. With the ego-motion from time $t-1$ to t, we can transform the cube obtained at time $t-1$ into the coordinate system centered on the SDV at time t. The same process could be applied to the past frames as well, resulting in ego-centric features $\{u_i^{'}\}$ for all previous timestamps. Ultimately, the BEV feature maps $b_i \in \mathbb{R}^{C \times H_b \times W_b}$ could be sum pooled from $\{u_i^{'}\}$ along the vertical dimension.

Temporal Fusion. Classical temporal fusion methods directly exploit 3D convolutions with stacked BEV features. However, it is noted that the corresponding features at the same location of various cubes should be similar if there are objects that are stationary on the ground, such as lanes and static vehicles. Due to this property, we perform a self-attention to boost the perceptual ability of static objects by adding the previous BEV feature maps to the current, which can be formulated as follows (where the discount $\alpha = 0.5$ and $\tilde{x}_1 = b_1$):

$$\tilde{x}_t = b_t + \sum_{i=1}^{t-1} \alpha^i \times \tilde{x}_{t-i}, \tag{2}$$

In order to perceive dynamic objects more accurately, we then feed these features into a temporal fusion network achieved by 3D convolution. To compensate for deviations caused by ego-vehicle motion, we add the motion matrix to the features by concatenating it in the spatial channel:

$$x_{1 \sim t} = \mathcal{C}(\tilde{x}_{1 \sim t}, m_{1 \sim t}), \tag{3}$$

where $m_{1 \sim t}$ denotes the ego-motion matrices and $\mathcal{C}$ denotes the 3D conv network.

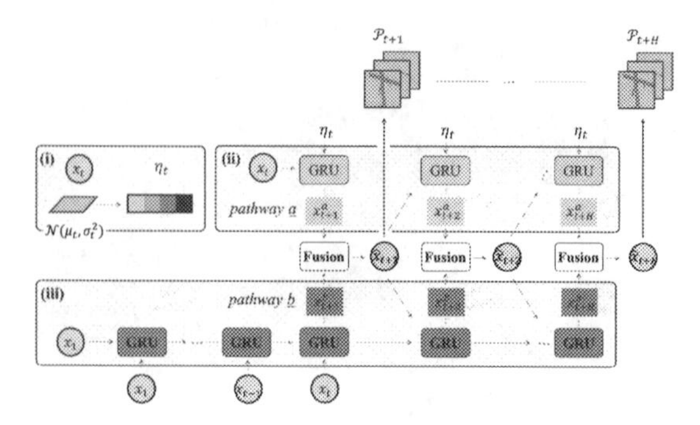

Fig. 4. Dual pathway modelling for prediction. **(i)** the latent code is from distribution of feature maps; **(ii–iii)** pathway $\underline{a}$ incorporates uncertainty distribution to indicate the multi-modal of future, while $\underline{b}$ learns from variations in the past, which is beneficial to compensate for information in $\underline{a}$

3.2 Prediction: Dual Pathway Probabilistic Future Modelling

In a dynamic driving environment, traditional motion prediction algorithms [17, 20, 48] often predict future trajectories as a deterministic or multi-modal results. However, a finite number of probabilities modelling could not cover the complex situation of future due to the interaction among agents, traffic elements and road environments. In order to deal with the stochasticity of the future, we wish to reason about the conditional uncertainty in the prediction features. Motivated by Hu *et al.* [23], we model the future uncertainty as diagonal Gaussians with mean $\mu \in \mathbb{R}^L$ and variance $\sigma^2 \in \mathbb{R}^L$, where L is the latent channel. Samples from the distribution could serve as a hidden state feature for future use. It is noted that during training, we use samples $\eta_t \sim \mathcal{N}(\mu_{\mathrm{t}}, \sigma_{\mathrm{t}}^2)$ while it will be sampled from $\eta_t \sim \mathcal{N}(\mu_{\mathrm{t}}, 0)$ during inference time.

The architecture of the prediction module is depicted in Fig. 4. We integrate the BEV features till current timestamp and the future uncertainty distribution into our prediction model, corresponding to two pathways in the dual modelling respectively. One uses historical features $(x_1, \ldots, x_t)$ as GRU inputs, and x_1 as the initial hidden state for prediction. The other uses samples from η_t as the GRU input and x_t as the initial hidden state. When predicting the feature at time $t+1$, we combine the predicted features in the form of a mixed Gaussian:

$$\hat{x}_{t+1} = \mathcal{G}(x_t, \eta_t) \oplus \mathcal{G}(x_{0:t}), \tag{4}$$

where $\mathcal{G}$ represents the process of GRU. And we use the mixture prediction as the base for the following prediction progress. Through this method, Dual Modelling recursively predicts future states $(\hat{x}_{t+1}, \ldots, \hat{x}_{t+H})$.

All the features $(x_1, \ldots, x_t), (\hat{x}_{t+1}, \ldots, \hat{x}_{t+H})$ are fed into the decoder $\mathcal{D}$ which has multiple output heads to generate different interpretable intermediate representations. The outcome is shown in Fig. 6. For the instance segmentation task, we follow the evaluation metrics in [24], where the heads output the instance

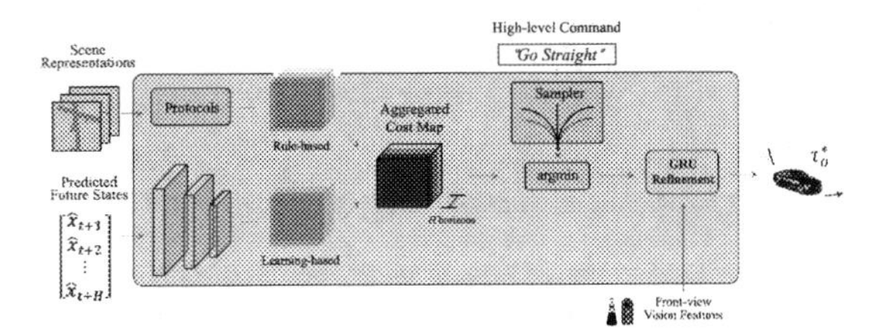

Fig. 5. Prior knowledge integration and refinement for Planning. The overall cost map includes two subcosts. The min-cost trajectory is further refined with front-view features to aggregate vision-based information from camera inputs

centerness, offset and future flow. Meanwhile, the semantic segmentation head mainly focuses on the vehicle and pedestrian which are the main actors in a driving setting. Furthermore, as HD map plays a vital role in autonomous driving [44,45,50], we explicitly generate two elements - drivable area and lanes, to provide an interpretable map representation. A cost volume head is designed for representing the *expense* of each possible location that the SDV can take within the planning horizon. More detailed information on cost volume is illustrated in Sect. 3.3. Note that we also decode features in the past frames to boost historical features' accuracy, which is required by Dual Modelling. As demonstrated in Sect. 4.1, more accurate feature information could lead to better prediction.

3.3 Planning: Prior Knowledge Incorporation and Refinement

As the ultimate goal, a motion planner needs to plan a safe and comfortable trajectory towards the target point. Given the occupancy predictions o and perceptions of map representations m, we design the motion planner which samples a diverse set of trajectories, and picks the one minimizing the learned cost function, inspired by [5,44,50]. However, we differentiate from them with an additional optimization step with a temporal model to integrate information of target point and traffic lights. The overall module is illustrated in Fig. 5.

The cost function makes full use of the learned occupancy probability field (segmentation maps in Prediction) and rich pre-knowledge as well, such as traffic rules, to ensure the safety and smoothness of the final trajectory. Formally, given the SDV's dynamic state, we adopt the bicycle model [39] to sample a set of trajectories τ. The objective cost function f is composed of subcosts, f_o that evaluates the predicted trajectory at every timestamp t according to the prior knowledge, f_v from prediction decoder (learning-based), and f_r that relates to the overall performances of the trajectory such as the comfort, progress. Thus the overall cost function can be an equally weighted combination as:

$$f(\tau, o, m; w) = f_o(\tau, o, m; w_o) + f_v(\tau; w_v) + f_r(\tau; w_r), \tag{5}$$

with $w = (w_o, w_v, w_r)$ being the vector of all learnable parameters. We briefly describe the subcosts below and refer readers to the Appendix for details.

Safety Cost. The SDV should not collide with other objects on the road and need to consider their future motion when planning its trajectory. The planned trajectories cannot overlap the grids occupied by other agents or road elements and need to maintain a certain safe distance at the high-velocity motion.

Cost Volume. Due to the complexity of road information, we cannot manually enumerate all possible cases or planning cost, thus we introduce a learned Cost Volume generated by the head in Sect. 3.2. In order to balance the cost scale, we clip the value so that it does not take a dominant role in evaluating trajectories.

Comfort and Progress. We penalize trajectories with large lateral acceleration, jerk or curvature. We wish the trajectory to be efficient towards the destination, hence the trajectory progressing forward would be awarded.

However, the above cost does not contain the target information which is often provided by a routed map; yet, this is not available in our setting. Thus we adopt the high level command including forward, turn left and turn right, and only evaluate the trajectories according to the corresponding command. To sum up, the motion planner now outputs one with the minimum cost as:

$$\tau^* = \arg\min_{\tau_h} f(\tau_h, c) = \arg\min_{\tau_h} f(\tau_h, o, m; w), \tag{6}$$

where τ^* is the selected trajectory, τ_h is the trajectory set under the corresponding high level command and c is the overall cost map. Furthermore, traffic lights are critical for the SDV while it is not explicitly utilized in previous modules. We take it into account through a GRU network to refine the trajectory. We initialize the hidden state with front camera features from the encoder module, and use each sample point in τ^* as the input. It is demonstrated in Table 5 that the front camera features indeed capture the information of traffic lights, which helps the SDV to launch or stop at intersections.

3.4 Overall Loss for End-to-End Learning

We optimize our model with perception, prediction, planning in an end-to-end manner by exploiting the following loss functions:

$$\mathcal{L} = \mathcal{L}_{per} + \alpha\mathcal{L}_{pre} + \beta\mathcal{L}_{pla}. \tag{7}$$

Note that we follow the protocol in [11,27] where the weights α, β are learnable rather than fixed, to balance the scale in different tasks according to gradients of the corresponding task loss.

Perception Loss. This loss includes the segmentation loss for current and past frames, the mapping loss and an auxiliary depth loss. For semantic segmentation, we use a top-k cross-entropy loss as in [24] since the BEV image is largely dominated by the background. For instance segmentation, the l_2 distance loss is adopted for the centerness supervision while l_1 distance loss for the offset and flow tasks. We use a cross-entropy loss for the lane and drivable area prediction. Current methods [38,46] utilize downstream tasks to implicitly optimize the

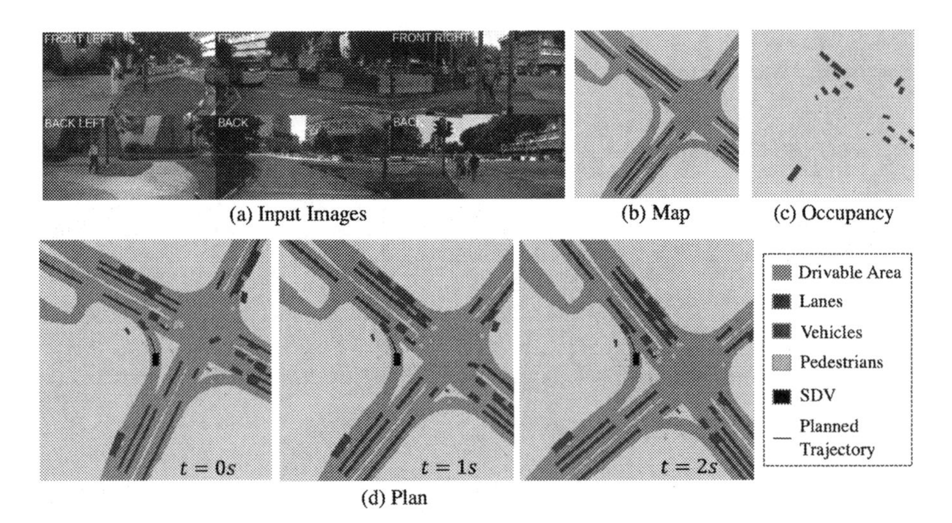

Fig. 6. Qualitative results on nuScenes. (b) shows predicted map representation including drivable area and lanes. (c) depicts the segmentation of vehicles and pedestrians. (a–c) are at $t = 1$ s. (d) represents the overall results from our model - perception, prediction, planning. The intermediate scene representations are robust in the time period, and the SDV successfully generate a safe trajectory to do left-turn without collision with curbsides or the front car

depth prediction rather than a direct supervision, yet this approach is remarkably affected by the design of the final loss function without clear explainability. Therefore we generate the depth value with other networks beforehand, then use it to direct the prediction. More details are in the Appendix.

Prediction Loss. As our prediction module infers the future semantic segmentation and instance segmentation, we keep the same top-k cross-entropy loss as in Perception. Nonetheless, loss in future timestamps would be discounted exponentially due to the uncertainty of the prediction.

Planning Loss. Our planning module first select the *best* trajectory τ^* from the sampled trajectory set τ as in Eq. (6), then a GRU-based network is applied to further refine the trajectory to obtain final output τ_o^*. Thus the planning loss contains two parts: a max-margin loss, which treats expert behavior τ_h as a positive example while trajectories sampled as negative ones, and a naive l_1 distance loss between the output and the expert trajectory. In particular, we set

$$\mathcal{L}_{pla} = \max_{\tau} \left[f(\tau_h, c) - f(\tau, c) + d(\tau_h, \tau) \right]_+ + d(\tau_h, \tau_o^*), \tag{8}$$

where $[\cdot]_+$ denotes ReLU and d is the l_1 distance between input trajectories.

4 Experiments

We evaluate ST-P3 in both open-loop and closed-loop environments. We adopt nuScenes dataset [4] for the open-loop setting, and CARLA simulator [18] for

Table 1. Perception results. We report the BEV segmentation IoU (%) of intermediate representations and their mean value. ST-P3 outperforms in most cases

Method	Mean Value	Drivable Area	Lane	Vehicle	Pedestrian
VED [32]	28.19	60.82	16.74	23.28	11.93
VPN [36]	30.36	65.97	17.05	28.17	10.26
PON [43]	30.52	63.05	17.19	27.91	13.93
Lift-Splat [38]	34.61	72.23	19.98	31.22	15.02
IVMP [46]	36.76	74.70	20.94	34.03	**17.38**
FIERY [24]	40.18	71.97	33.58	38.00	17.15
ST-P3	**42.69**	**75.97**	**40.20**	**40.10**	14.48

the closed-loop demonstration. By default we take the 1.0 s of past context and predict the future 2.0 s contexts, corresponding to 3 frames in the past and 4 frames in the future. More details on protocols are provided in the Appendix.

4.1 Open-Loop Experimental Results on NuScenes

Perception. We evaluate the models on map representation and semantic segmentation. The perceived map includes drivable area and lanes - two most critical elements for the driving behavior, since the SDV is assumed to be driving in the drivable areas and keeping itself in the center of lanes. The semantic segmentation focuses on vehicles and pedestrians, both of which are the main agents in a driving environment. We use the Intersection-over-Union (IoU) as the metric, modeling the perception module as a BEV segmentation task. As shown in Table 1, ST-P3 gets the highest mean value on the nuScenes validation set, surpassing previous SOTA by **2.51%** with our Egocentric Aligned Accumulation algorithm.

Prediction. Predicting future segmentation in BEV is first proposed in [24], thus we select it as our baseline. We evaluate our model by the metric of IoU, existing panoptic quality (PQ), recognition quality (RQ), and segmentation quality (SQ) following metrics in video prediction area [28]. Note that we predict vehicles which have been shown in the past frames only as we cannot predict those out of nowhere. Results are shown in Table 2. Our model achieves the state-of-the-art in *all* metrics as a consequence of the novel design of the prediction module. Though the Gaus. version performs a little worse than the Ber. one, we finally choose it for relatively smaller memory usage.

Planning. For the open-loop planning, we focus on two evaluation metrics: L2 error and collision rate, and adjust the planning horizon to 3.0 s for a fair comparison. We compute the L2 error between the planned trajectory and the human driving trajectory, and evaluate how often the SDV would collide with other agents on the road. Detailed comparison with previous methods is shown in Table 3. Note that the Vanilla algorithm is penalized based on how much the

Table 2. Prediction results. We report semantic segmentation IoU (%) and instance segmentation metrics from video prediction area. The *static* method assumes all obstacles static in the prediction horizon. "Ber." denotes modeling the future uncertainty as Bernoulli distribution, while "Gaus." means Gaussian distribution. ST-P3 achieves best performance on all temporal segmentation metrics

Method	Future semantic seg	Future instance seg.		
	IoU ↑	PQ ↑	SQ ↑	RQ ↑
Static	32.20	27.64	70.05	39.08
FIERY [24]	37.00	30.20	70.20	42.90
ST-P3 Gaus.	38.63	31.72	70.15	45.22
ST-P3 Ber.	**38.87**	**32.09**	**70.39**	**45.59**

Table 3. Open-loop planning results. *vanilla* approach is supervised with ground truth trajectories only. ST-P3 achieves the lowest collision rate in all time intervals

Method	L2 (m) ↓			Collision (%) ↓		
	1 s	2 s	3 s	1 s	2 s	3 s
Vanilla	**0.50**	**1.25**	**2.80**	0.68	0.98	2.76
NMP [50]	0.61	1.44	3.18	0.66	0.90	2.34
Freespace [25]	0.56	1.27	3.08	0.65	0.86	1.64
ST-P3	1.33	2.11	2.90	**0.23**	**0.62**	**1.27**

Table 4. Closed-loop simulation results. ST-P3 outperforms vision-based baselines in all scenarios, and achieves better route completion performance in long-range tests compared to LiDAR-based method. *: LiDAR-based method

Method	Town05 short		Town05 long	
	DS ↑	RC ↑	DS ↑	RC ↑
CILRS [14]	7.47	13.40	3.68	7.19
LBC [9]	30.97	55.01	7.05	32.09
Transfuser* [41]	54.52	78.41	**33.15**	56.36
ST-P3	**55.14**	**86.74**	11.45	**83.15**

trajectory deviates from the ground truth, thus it achieves the lowest L2 error but largest collision rates in all prediction horizons. ST-P3 obtains the lowest collision rate, implying the superior safety of our planned trajectory.

4.2 Closed-Loop Planning Results on CARLA Simulator

We conduct closed-loop experiments in CARLA simulator to demonstrate the applicability and robustness of ST-P3. It is far more challenging since the driving errors would stack up and lead to dangerous crashes. Following [41], we adopt

Table 5. Ablation on nuScenes validation set. Exp. 1–3 explore the effectiveness of depth supervision (Depth) and egocentric aligned accumulation module (EAA.) for perception. Exp. 4–6 is on prediction module, with "Dual." representing Dual Modelling and "LFA." indicating loss for all timestamps. Exp. 7–9 show the superiority of combining the sampler (S.) and refinement (R.) units in planning. "V.IoU" is the IoU metric of vehicles. "V.PQ" is the panoptic quality of vehicles. "Col." denotes the collision rate

Exp	EAA	Depth	Dual	LFA	S	R	V. IoU ↑	V. PQ ↑	L2 ↓	Col. ↓
1							38.00	–	–	–
2	✓						38.79	–	–	–
3	✓	✓					40.10	–	–	–
4	✓	✓					37.09	28.63	–	–
5	✓	✓	✓				38.16	31.35	–	-
6	✓	✓	✓	✓			38.63	31.72	–	–
7	✓	✓	✓	✓	✓		–	–	2.128	0.850
8	✓	✓	✓	✓		✓	–	–	2.321	1.089
9	✓	✓	✓	✓	✓	✓	-	-	1.890	0.513

the Route Completion (RC) - the percentage of route distance completed, and the Driving Score (DS) - RC weighted by an penalty factor that accounts for collisions with pedestrians, vehicles, *etc.* Table 4 shows the comparison with two camera-based algorithms and a LiDAR-based SOTA method. ST-P3 outperforms the camera-based methods on all metrics and is comparable with the LiDAR-based method. Tansfuser [41] has a higher driving score in long routes mainly due to the lower penalty resulting from the shorter traveling distance. ST-P3 obtains impressive route completion performance which indicates the ability of recovering from collisions, with help of the front-view vision refinement.

4.3 Ablation Study

Table 5 shows the effectiveness of different modules in ST-P3. We report the vehicle IoU and vehicle PQ for perception and prediction tasks, L2 and collision rate for planning evaluation. For Exp. 1–3 in Table 5 we present the impact of depth supervision and Egocentric Aligned Accumulation (EAA.) in perception. Note that Exp.1 is identical to FIERY [24]. Our module improves **0.79%** by adopting EAA. algorithem (Exp. 2), and supervising depth explicitly brings an improvement to **1.31%** (Exp. 3). Exp. 4–6 demonstrate the impact of Dual Modelling and corresponding training method - loss for all states (LFA.) on prediction task. Since Dual Modelling considers both uncertainty and historical continuity, the correctness of past features plays a vital role in it. As the results show, these two strategies improves **1.54%** and **3.09%** to V.IoU and V.PQ respectively. Exp. 7–9 is on the sampler and GRU refinement unit in planning. A sampler without front-view vision refinement (Exp. 7) or an implicit model without prior sampling knowledge (Exp.8) both get a high L2 error and collision rate. Our design remarkably improves the safety and accuracy of the planned trajectory.

5 Conclusions

In this paper, we have proposed an interpretable end-to-end vision-based framework for autonomous driving tasks. The motivation behind the improved design is to boost feature representations both in spatial and temporal domains. An egocentric aligned accumulation to aggregate features in 3D space and preserve geometry information is proposed; a dual pathway modelling to reason about the probabilistic character of semantic representations across frames is devised; a prior knowledge refinement unit to take into account road elements is introduced. Together with these improvements within the ST-P3 pipeline, we achieve impressive performance compared to previous state-of-the-arts.

Acknowledgements. The project is partially supported by the Shanghai Committee of Science and Technology (Grant No. 21DZ1100100). This work is also supported in part by National Key Research and Development Program of China (2020AAA0107600), Shanghai Municipal Science and Technology Major Project (2021SHZDZX0102) and NSFC (61972250, 72061127003).

References

1. Bansal, M., Krizhevsky, A., Ogale, A.: ChauffeurNet: Learning to drive by imitating the best and synthesizing the worst. arXiv preprint arXiv:1812.03079 (2018)
2. Behere, S., Torngren, M.: A functional architecture for autonomous driving. In: WASA (2015)
3. Bojarski, M., et al.: End to end learning for self-driving cars. arXiv preprint arXiv:1604.07316 (2016)
4. Caesar, H., et al.: nuScenes: a multimodal dataset for autonomous driving. In: CVPR (2020)
5. Casas, S., Sadat, A., Urtasun, R.: Mp3: a unified model to map, perceive, predict and plan. In: CVPR (2021)
6. Chekroun, R., Toromanoff, M., Hornauer, S., Moutarde, F.: GRI: general reinforced imitation and its application to vision-based autonomous driving. arXiv preprint 2111.08575 (2021)
7. Chen, D., Koltun, V., Krähenbühl, P.: Learning to drive from a world on rails. In: ICCV (2021)
8. Chen, D., Krähenbühl, P.: Learning from all vehicles. In: CVPR (2022)
9. Chen, D., Zhou, B., Koltun, V., Krähenbühl, P.: Learning by cheating. In: CoRL (2020)
10. Chen, L., et al.: PersFormer: 3d lane detection via perspective transformer and the OpenLane benchmark. In: ECCV (2022)
11. Chen, Z., Badrinarayanan, V., Lee, C.Y., Rabinovich, A.: GradNorm: gradient normalization for adaptive loss balancing in deep multitask networks. In: International Conference on Machine Learning, pp. 794–803. PMLR (2018)

12. Chitta, K., Prakash, A., Geiger, A.: Neat: neural attention fields for end-to-end autonomous driving. In: ICCV (2021)
13. Codevilla, F., Müller, M., López, A., Koltun, V., Dosovitskiy, A.: End-to-end driving via conditional imitation learning. In: ICRA (2018)
14. Codevilla, F., Santana, E., López, A.M., Gaidon, A.: Exploring the limitations of behavior cloning for autonomous driving. In: ICCV (2019)
15. commaai: Openpilot. https://github.com/commaai/openpilot (2022)
16. Cui, A., Casas, S., Sadat, A., Liao, R., Urtasun, R.: Lookout: diverse multi-future prediction and planning for self-driving. In: ICCV (2021)
17. Djuric, N., et al.: Uncertainty-aware short-term motion prediction of traffic actors for autonomous driving. In: WACV (2020)
18. Dosovitskiy, A., Ros, G., Codevilla, F., Lopez, A., Koltun, V.: Carla: an open urban driving simulator. In: CoRL (2017)
19. Fadadu, S., et al.: Multi-view fusion of sensor data for improved perception and prediction in autonomous driving. In: Proceedings of the IEEE/CVF Winter Conference on Applications of Computer Vision (2022)
20. Fang, L., Jiang, Q., Shi, J., Zhou, B.: TPNet: trajectory proposal network for motion prediction. In: CVPR (2020)
21. Gu, J., Sun, C., Zhao, H.: DenseTNT: end-to-end trajectory prediction from dense goal sets. In: ICCV (2021)
22. Hämäläinen, A., Arndt, K., Ghadirzadeh, A., Kyrki, V.: Affordance learning for end-to-end visuomotor robot control. In: IROS (2019)
23. Hu, A., Cotter, F., Mohan, N., Gurau, C., Kendall, A.: Probabilistic future prediction for video scene understanding. In: ECCV (2020)
24. Hu, A., et al.: Fiery: future instance prediction in bird's-eye view from surround monocular cameras. In: ICCV (2021)
25. Hu, P., Huang, A., Dolan, J., Held, D., Ramanan, D.: Safe local motion planning with self-supervised freespace forecasting. In: CVPR (2021)
26. Jia, X., Sun, L., Zhao, H., Tomizuka, M., Zhan, W.: Multi-agent trajectory prediction by combining egocentric and allocentric views. In: CoRL (2022)
27. Kendall, A., Gal, Y., Cipolla, R.: Multi-task learning using uncertainty to weigh losses for scene geometry and semantics. In: CVPR (2018)
28. Kim, D., Woo, S., Lee, J.Y., Kweon, I.S.: Video panoptic segmentation. In: CVPR (2020)
29. Li, Z., et al.: BEVFormer: learning bird's-eye-view representation from multi-camera images via spatiotemporal transformers. In: ECCV (2022)
30. Liao, W., et al.: Trajectory prediction from ego view: a coordinate transform and tail-light event driven approach. In: ICME (2022)
31. Loukkal, A., Grandvalet, Y., Drummond, T., Li, Y.: Driving among flatmobiles: Bird-eye-view occupancy grids from a monocular camera for holistic trajectory planning. In: WACV (2021)
32. Lu, C., van de Molengraft, M.J.G., Dubbelman, G.: Monocular semantic occupancy grid mapping with convolutional variational encoder-decoder networks. In: RAL (2019)
33. Luo, W., Yang, B., Urtasun, R.: Fast and furious: real time end-to-end 3d detection, tracking and motion forecasting with a single convolutional net. In: CVPR (2018)
34. Ng, M.H., et al.: BEV-Seg: bird's eye view semantic segmentation using geometry and semantic point cloud. arXiv preprint arXiv:2006.11436 (2020)
35. Ngiam, J., et al.: Scene transformer: a unified multi-task model for behavior prediction and planning. arXiv e-prints pp. arXiv-2106 (2021)

36. Pan, B., Sun, J., Leung, H.Y.T., Andonian, A., Zhou, B.: Cross-view semantic segmentation for sensing surroundings. IEEE Robot. Autom. Lett. **5**(3), 4867–4873 (2020)
37. Pfeiffer, M., Schaeuble, M., Nieto, J., Siegwart, R., Cadena, C.: From perception to decision: a data-driven approach to end-to-end motion planning for autonomous ground robots. In: ICRA (2017)
38. Philion, J., Fidler, S.: Lift, splat, shoot: encoding images from arbitrary camera rigs by implicitly unprojecting to 3d. In: ECCV (2020)
39. Polack, P., Altché, F., d'Andréa Novel, B., de La Fortelle, A.: The kinematic bicycle model: a consistent model for planning feasible trajectories for autonomous vehicles? In: IV (2017)
40. Pomerleau, D.A.: ALVINN: an autonomous land vehicle in a neural network. Adv. Neural Inf. Process. Syst. (1988)
41. Prakash, A., Chitta, K., Geiger, A.: Multi-modal fusion transformer for end-to-end autonomous driving. In: CVPR (2021)
42. Reading, C., Harakeh, A., Chae, J., Waslander, S.L.: Categorical depth distribution network for monocular 3d object detection. In: CVPR (2021)
43. Roddick, T., Cipolla, R.: Predicting semantic map representations from images using pyramid occupancy networks. In: CVPR (2020)
44. Sadat, A., Casas, S., Ren, M., Wu, X., Dhawan, P., Urtasun, R.: Perceive, predict, and plan: safe motion planning through interpretable semantic representations. In: ECCV (2020)
45. Sadat, A., Ren, M., Pokrovsky, A., Lin, Y.C., Yumer, E., Urtasun, R.: Jointly learnable behavior and trajectory planning for self-driving vehicles. In: IROS (2019)
46. Wang, H., Cai, P., Sun, Y., Wang, L., Liu, M.: Learning interpretable end-to-end vision-based motion planning for autonomous driving with optical flow distillation. In: ICRA (2021)
47. Wei, B., Ren, M., Zeng, W., Liang, M., Yang, B., Urtasun, R.: Perceive, attend, and drive: learning spatial attention for safe self-driving. In: ICRA (2021)
48. Wu, P., Chen, S., Metaxas, D.N.: MotionNet: joint perception and motion prediction for autonomous driving based on bird's eye view maps. In: CVPR (2020)
49. Yurtsever, E., Lambert, J., Carballo, A., Takeda, K.: A survey of autonomous driving: common practices and emerging technologies. IEEE Access **8**, 58443–58469 (2020)
50. Zeng, W., Luo, W., Suo, S., Sadat, A., Yang, B., Casas, S., Urtasun, R.: End-to-end interpretable neural motion planner. In: CVPR (2019)
51. Zeng, W., Wang, S., Liao, R., Chen, Y., Yang, B., Urtasun, R.: DSDNet: deep structured self-driving network. In: Vedaldi, A., Bischof, H., Brox, T., Frahm, J.-M. (eds.) ECCV 2020. LNCS, vol. 12366, pp. 156–172. Springer, Cham (2020). https://doi.org/10.1007/978-3-030-58589-1_10
52. Zhang, Z., Liniger, A., Dai, D., Yu, F., Van Gool, L.: End-to-end urban driving by imitating a reinforcement learning coach. In: ICCV (2021)

PersFormer: 3D Lane Detection via Perspective Transformer and the OpenLane Benchmark

Li Chen[1], Chonghao Sima[1], Yang Li[1], Zehan Zheng[1], Jiajie Xu[1], Xiangwei Geng[1], Hongyang Li[1,2(✉)], Conghui He[1], Jianping Shi[3], Yu Qiao[1], and Junchi Yan[1,2]

[1] Shanghai AI Laboratory, Shanghai, China
{lichen,simachonghao,liyang,lihongyang}@pjlab.org.cn
[2] Shanghai Jiao Tong University, Shanghai, China
yanjunchi@sjtu.edu.cn
[3] SenseTime Research, Hong Kong, China

Abstract. Methods for 3D lane detection have been recently proposed to address the issue of inaccurate lane layouts in many autonomous driving scenarios (uphill/downhill, bump, *etc.*). Previous work struggled in complex cases due to their simple designs of the spatial transformation between front view and bird's eye view (BEV) and the lack of a realistic dataset. Towards these issues, we present PersFormer: an end-to-end monocular 3D lane detector with a novel Transformer-based spatial feature transformation module. Our model generates BEV features by attending to related front-view local regions with camera parameters as a reference. PersFormer adopts a unified 2D/3D anchor design and an auxiliary task to detect 2D/3D lanes simultaneously, enhancing the feature consistency and sharing the benefits of multi-task learning. Moreover, we release one of the first large-scale real-world 3D lane datasets: OpenLane, with high-quality annotation and scenario diversity. OpenLane contains 200,000 frames, over 880,000 instance-level lanes, 14 lane categories, along with scene tags and the closed-in-path object annotations to encourage the development of lane detection and more industrial-related autonomous driving methods. We show that PersFormer significantly outperforms competitive baselines in the 3D lane detection task on our new OpenLane dataset as well as Apollo 3D Lane Synthetic dataset, and is also on par with state-of-the-art algorithms in the 2D task on OpenLane. The project page is available at https://github.com/OpenPerceptionX/PersFormer_3DLane and OpenLane dataset is provided at https://github.com/OpenPerceptionX/OpenLane.

L. Chen, C. Sima and Y. Li—Equal contribution.

Supplementary Information The online version contains supplementary material available at https://doi.org/10.1007/978-3-031-19839-7_32.

S. Avidan et al. (Eds.): ECCV 2022, LNCS 13698, pp. 550–567, 2022.
https://doi.org/10.1007/978-3-031-19839-7_32

1 Introduction

Autonomous driving is one of the most successful applications for AI algorithms to deploy in recent years. Modern Advanced Driver Assistance Systems (ADAS) for either L2 or L4 routes provide functionalities such as Automated Lane Centering (ALC) and Lane Departure Warning (LDW), where the essential need for perception is a lane detector to generate robust and generalizable lane lines [12]. With the prosperity of deep learning, lane detection algorithms in the 2D image space has achieved impressive results [28,39,46], where the task is formulated as a 2D segmentation problem given front view (perspective) image as input [1,25,34,37]. However, such a framework to perform lane detection in the perspective view is not applicable for industry-level products where complicated scenarios dominate.

On one side, downstream modules as in planning and control *often* require the lane location to be in the form of the orthographic bird's eye view (BEV) instead of a front view representation. Representation in BEV is for better task alignment with interactive agents (vehicle, road marker, traffic light, *etc.*.) in the environment and multi-modal compatibility with other sensors such as LiDAR and Radar. The conventional approaches to address such a demand are either to simply project perspective lanes to ones in the BEV space [32,52], or more elegantly to cast perspective features to BEV by aid of camera in/extrinsic matrices [15,19,60]. The latter solution is inspired by the spatial transformer network (STN) [22] to generate a one-to-one correspondence from the image to BEV feature grids. By doing so, the quality of features in BEV depends solely on the quality of the *corresponding* feature in the front view. The predictions using these outcome features are not adorable as the blemish of scale variance in the front view, which inherits from the camera's pinhole model, remains.

On the other side, the height[1] of lane lines has to be considered when we project perspective lanes into BEV space. As illustrated in Fig. 1, the lanes would diverge/converge in case of uphill/downhill if the height is ignored, leading to improper action decisions as in the planning and control module. Previous literature [34,44,52] inevitably hypothesize that lanes in the BEV space lie on a flat ground, *i.e.*, the height of lanes is zero. The planar assumption does not hold true in most autonomous driving scenarios, *e.g.*, uphill/downhill, bump, crush turn, *etc.*. Since the height information is unavailable on public benchmarks or complicated to acquire accurate ground truth, 3D lane detection is ill-posed. There are some attempts to address this issue by creating 3D synthetic benchmarks [15,19]. Their performance still needs improvement in complex, realistic scenarios nonetheless (c.f. (b–c) in Fig. 1). Moreover, the domain adaption between simulation and real data is not well-studied [16].

To address these bottlenecks aforementioned, we propose Perspective Transformer, shortened as **PersFormer**, which has a spatial feature transformation

[1] We define the height of lane line z to be the relative height concerning the zero point in the ego vehicle coordinate system (x, y, z) in BEV 3D space. The coordinate of the perspective (front view) 2D space in the image plane is referred to as (u, v).

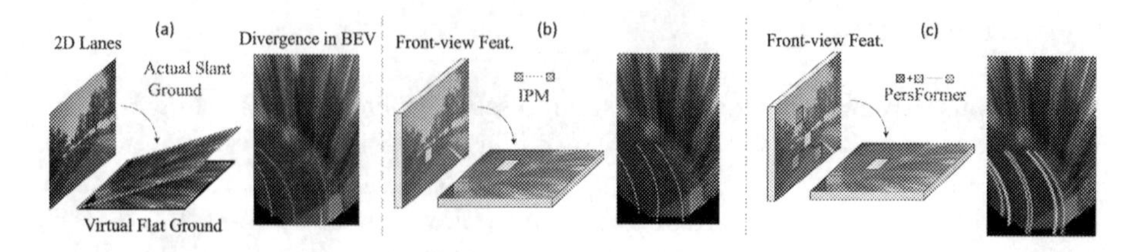

Fig. 1. Motivation of performing lane detection from 2D in (a) to BEV in (b); and the superiority of our method in (c) versus (b). Lanes would diverge/converge in projected BEV on planar assumption, and a 3D solution with height to be considered can accurately predict the parallel topology in this case

module to generate better BEV representations for the task. The proposed framework unifies 2D/3D lane detection tasks, and substantiates performance on the proposed large-scale realistic 3D lane dataset, **OpenLane**.

First, we model the spatial feature transformation as a learning procedure that has an attention mechanism to capture the interaction both among local region in the front view feature and between two views (front view to BEV), consequently being able to generate a fine-grained BEV feature representation. Inspired by [8,51], we construct a Transformer-based module to realize this, while the deformable attention mechanism [62] is adopted to remarkably reduce the computational memory requirement and dynamically adjust keys through the cross-attention module to capture prominent feature among the local region. Compared with direct 1–1 transformation via Inverse Perspective Mapping (IPM), the resultant features would be more representative and robust as it attends to the surrounding local context and aggregates relevant information. We further aim at unifying 2D and 3D lane detection tasks to benefit from the co-learning optimization. Second, we release the first real-world, large-scale 3D lane dataset and corresponding benchmark, OpenLane, to support research into the problem. OpenLane contains 200,000 annotated frames and over 880,000 lanes - each with one of 14 category labels (single white dash, double yellow solid, left/right curbside, *etc.*.), which exceeds all of the existing lane datasets. It also has some distinguishing elements such as scenes, weather, and closed-in-path-object (CIPO) for other research topics in autonomous driving.

The Main Contributions of our Work are Three-fold: 1) Perspective Transformer, a novel Transformer-based architecture to realize spatial transformation of features; **2)** An architecture to simultaneously unify 2D and 3D lane detection, which is feasibly needed in the application. Experiments show that our PersFormer outperforms state-of-the-art 3D lane detection algorithms; **3)** The OpenLane dataset, the first large-scale realistic 3D lane dataset with high-quality labeling and vast diversity. The dataset, baselines, as well as the whole suite of codebase, is released to facilitate the research in this area.

2 Related Work

Vision Transformers in Bird's-Eye-View (BEV). Projecting features to BEV and performing downstream tasks in it has become more dominant and ensured better performance recently [29]. Compared with conventional CNN structure, the cross attention scheme in Vision Transformers [8,13,31,51,62] is naturally introduced to serve as a learnable transformation of features across different views in an elegant spirit [29]. Instead of simply projecting features via IPM, the successful application of Transformers in view transformation has demonstrated great success in various domains, including 3D object detection [18,26,53,58], prediction [14,17,36], planning [11,38], *etc.*

Previous work [7,15,41,53,57] bring the BEV philosophy into pipeline, and yet they do not consider attention mechanism and/or 3D vision geometry (in this case, camera parameters). For instance, 3D-LaneNet [15] is set up with camera in/extrinsic matrices; the IPM process generates a virtual BEV representation from front view features. DETR3D [53] also considers camera geometry and formulates a learnable 3D-to-2D query search with attention scheme. However, there is no explicit BEV modelling for robust feature representation; the aggregated features might not be properly represented in 3D space. To address these shortcomings, our proposed PersFormer takes into account both the effect of camera parameters to generate BEV features and the convenience of cross-attention mechanism to model view transformation, achieving better feature representation in the end.

Lane Detection Benchmarks. A large-scale, diverse dataset with high-quality annotation is a pivot for lane detection. Along with the progress of lane detection approaches, numerous datasets have been proposed [4,10,21,25,37,49,55,59]. However, they usually fit into one or the other lane detection scenario. Table 1 depicts more details of the existing benchmarks and their comparison with our proposed OpenLane dataset. OpenLane is the first large-scale, realistic 3D lane dataset. It equips with a wide span of diversity in both data distribution and task applicability.

3D Lane Detection. As discussed in Sect. 1, planar assumption does not always reserve in some cases, *i.e.*, uphill/downhill, bump. Several approaches [3,5,33] utilize multi-modal or multi-view sensors, such as a stereo camera or LiDAR, to get the 3D ground topology. However, these sensors have shortages of high cost in hardware and computation resources, confining their practical applications. Recently, some monocular methods [15,19,23,30] take a single image and employ IPM to predict lanes in 3D space. 3D-LaneNet [15] is the pioneering work in this domain with one simple end-to-end neural network, which adopts STN [22] to accomplish the spatial projection of features. Gen-LaneNet [19] builds on top of 3D-LaneNet and designs a two-stage network for decoupling the segmentation encoder and 3D lane prediction head. These two approaches [15,19] suffer from improper feature transformation and unsatisfying performance in curving or crush turn cases. Confronted with the issues above, we bring in PersFormer to provide better feature representation and optimize anchor design to unify 2D and 3D lane detection simultaneously.

3 Methodology

In this section, we propose PersFormer, a unified 2D/3D lane detection framework with Transformer. We first describe the problem formulation, followed by an introduction to the overall structure in Sect. 3.1. In Sect. 3.2, we present Perspective Transformer, an explicit feature transformation module from front view to BEV space by the aid of camera parameters. In Sect. 3.3, we give details on the anchor design to unify 2D/3D tasks and in Sect. 3.4 we further elaborate on the auxiliary task and loss function to finalize our training strategy.

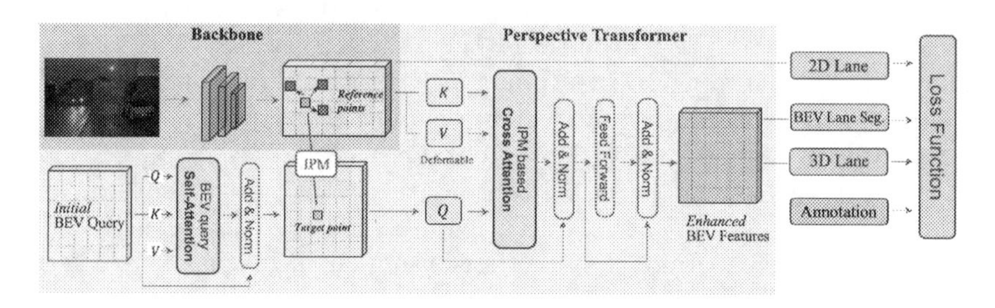

Fig. 2. Our proposed PersFormer pipeline. The core is to learn a spatial feature transformation from front view to BEV space so that the generated BEV features at target point would be more representative by attending local context around reference point. PersFormer consists of the self-attention module to interact with its own BEV queries; the cross-attention module that takes the key-value pair from the IPM-based front view features to generate fine-grained BEV feature

Problem Formulation. Given an input image $I_{org} \in \mathbb{R}^{H_{org} \times W_{org}}$, the goal of PersFormer is to predict a collection of 3D lanes $L_{3D} = \{l_1, l_2, \ldots, l_{N_{3D}}\}$ and 2D lanes $L_{2D} = \{l_1, l_2, \ldots, l_{N_{2D}}\}$, where N_{3D}, N_{2D} are the total number of 3D lanes in the pre-defined BEV range and 2D lanes in the original image space (front view) respectively. Mathematically, each 3D lane l_d is represented by an ordered set of 3D coordinates:

$$l_d = \left[(x_1, y_1, z_1), (x_2, y_2, z_2), \ldots, (x_{N_d}, y_{N_d}, z_{N_d})\right], \tag{1}$$

where d is the lane index, and N_d is the max number of sample points of this lane. The form of 2D lane is represented similarly with 2D coordinate (u, v) accordingly. Each lane has a categorical attribute $c_{\mathrm{3D/2D}}$, indicating the type of this lane (*e.g.*, single-white dash line). Also, for each point in a single 2D/3D lane, there exists an attribute property indicating whether the point is visible or not, denoted by $\mathbf{vis}_{\mathrm{fv/bev}}$ as a vector for the lane.

3.1 Approach Overview

The overall structure, as illustrated in Fig. 2, consists of three parts: the backbone, the Perspective Transformer, and lane detection heads. The backbone takes the resized image as input and generates multi-scale front view features, where the popular ResNet variant [47] is adopted. Note that these features might suffer from the defect of scale variance, occlusion, *etc.* - residing from the inherent feature extraction in the front view space. The Perspective Transformer takes the front view features as input and generates BEV features by the aid of camera intrinsic and extrinsic parameters. Instead of simply projecting the one-to-one feature correspondence from the front view to BEV, we introduce Transformer to attend local context and aggregate surrounding features to form a robust representation in BEV. By doing so, we learn the inverse perspective mapping from front view to BEV in an elegant manner with Transformer. Finally, the lane detection heads are responsible for predicting 2D/3D coordinates as well as lane types. The 2D/3D detection heads are referred to as LaneATT [46] and 3D-LaneNet [15], with modification on the structure and anchor design.

3.2 Proposed Perspective Transformer

We present Perspective Transformer, a spatial transformation method that combines camera parameters and data-driven learning procedures. The general idea of Perspective Transformer is to use the coordinates transformation matrix from IPM as a reference to generate BEV feature representation, by attending related region (local context) in front view feature. On the assumption that the ground is flat and the camera parameters are given, a classical IPM approach calculates a set of coordinate mapping from front-view to BEV, where the BEV space is defined on the flat ground (see [20], Sect. 8.1.1). Given a point p_{fv} with its coordinate (u, v) in the front-view feature $F_{\text{fv}} \in \mathbb{R}^{H_{\text{fv}} \times W_{\text{fv}} \times C}$, IPM maps the point p_{fv} to the corresponding point p_{bev} in BEV, where (x, y) is the coordinate in the BEV space $\mathbb{R}^{H_{\text{bev}} \times W_{\text{bev}} \times C}$. The transform is achieved with camera in/extrinsic and can be represented mathematically as:

$$\begin{pmatrix} x \\ y \\ 0 \end{pmatrix} = \alpha_{f2b} \cdot R_{\theta} \cdot K^{-1} \cdot \begin{pmatrix} u \\ v \\ 1 \end{pmatrix} + \begin{pmatrix} 0 \\ 0 \\ -h \end{pmatrix}, \tag{2}$$

where α_{f2b} implies the scale factor between front-view and BEV, R_{θ} denotes the pitch rotation matrix from extrinsic, K is the intrinsic matrix, and h stands for camera height. Such a transformation in Eqn.(2) enframes a strong prior on the attention unit in PerFormer to generate more representative BEV features.

The architecture of Perspective Transformer is inspired by popular approaches such as DETR [8], and consists of the self-attention module and cross-attention module (see Fig. 2). We differentiate from them in that the queries are not implicitly updated. However, instead, they are piloted by an explicit meaning - the physical location to detect objects or lanes in BEV. In the **self-attention** module, the output Q_{bev} descends from the triplet (key, value, query) input through their interaction. The formulation of such a self-attention can be described as:

$$Q_{\text{bev}} = \texttt{softmax}\left(\frac{QK^{\top}}{\sqrt{d_k}}\right)V, \tag{3}$$

where $K, Q, V \in \mathbb{R}^{(H_{\text{bev}} \times W_{\text{bev}} \times C)}$ are the same query that is pre-defined in BEV, $\sqrt{d_k}$ is the dimensional normalized factor.

In the **cross-attention** module, the input query $Q^{'}_{\text{bev}}$ is the outcome of several additional layers feeding the self-attention output Q_{bev} as input. Note that $Q^{'}_{\text{bev}}$ is an explicit feature representation as to which part in BEV should be paid more attention since the generation of queries is location-sensitive in BEV. This is quite different compared with queries that do not consider view transformation in most Vision Transformers [18,53,62]. Furthermore, the intuition behind employing Transformer to map features from front view to BEV is that such an attention mechanism would automatically attend which part of features contribute *most* towards the target point (query) in the destination view. The direct feature transformation would suffer from camera parameter noise or scale variance issues, as discussed and illustrated in Sect. 1. Note that the naive Transformer cannot be applied directly since the number of key-value pairs is huge and thus be confined by computational burden. Inspired by Deformable DETR [62], we attend partial key-value pairs around the local region in a learnable manner to save cost and improve efficiency.

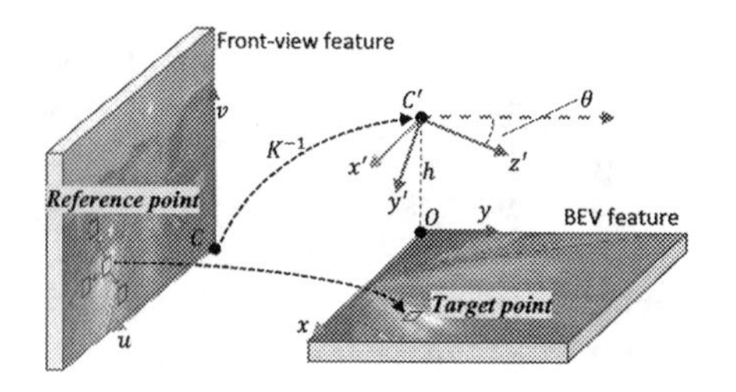

Fig. 3. Generation of keys in the cross attention. Point (x, y) in BEV space casts the corresponding point (u, v) in front view through intermediate state $(x^{'}, y^{'})$; by learning offsets, the network learns target-reference points mapping from green rectangles to yellow and related blue rectangles as keys to Transformer

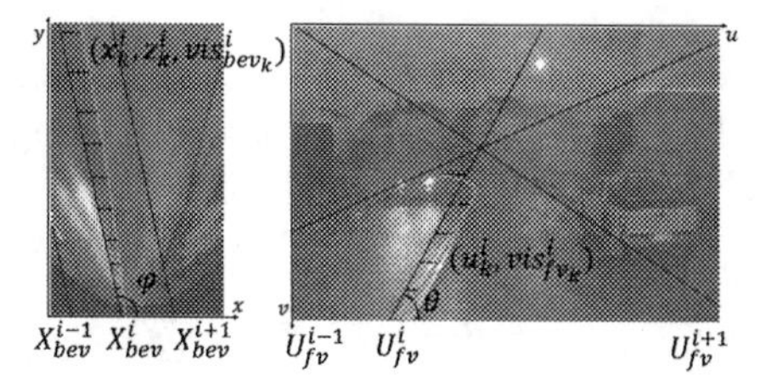

Fig. 4. Unifying anchor design in 2D and 3D. We first put curated anchors (red) in the BEV space (left), then project them to the front view (right). Offset x_k^i and u_k^i (dashed line) are predicted to match ground truth (yellow and green) to anchors. The correspondence is thus built, and features are optimized together (Color figure online)

Figure 3 depicts the feature transformation process and the generation of key-value pairs in cross-attention. Specifically, given a query point (x, y) in the target BEV map Q'_{bev}, we project it to the corresponding point (u, v) in the front view via Eqn.(2). As does similarly in [62], we learn some offsets based on point (u, v) to generate a set of most related points around it. These learned points, together with (u, v) are defined as *reference points*. They contribute most to the query point (x, y), defined as *target point*, in BEV-space. The reference points serve as the surrounding context in the local region that contributes most to the feature representation from perspective view to BEV space. They are the desired keys we try to find, and their features are values for the cross attention module. Note that the initial locations of reference points from IPM are used as preliminary locations for the coordinate mapping; the location are adjusted gradually during the learning procedure, which is the core role of Deformable Attention.

As a result, the output of the cross-attention module can be formulated as:

$$F_{\text{bev}} = \texttt{DeformAttn}(Q'_{\text{bev}}, F_{\text{fv}}, p_{\text{fv2bev}}), \tag{4}$$

where $F_{\text{bev}} \in \mathbb{R}^{(H_{\text{bev}} \times W_{\text{bev}} \times C)}$ is the final desired features for the subsequent 3D head to get lane predictions, Q'_{bev} denotes the input queries, $F_{\text{fv}} \in \mathbb{R}^{(H_{\text{fv}} \times W_{\text{fv}} \times C)}$ indicates the front view features from backbone, and p_{fv2bev} is the IPM-inited coordinate mapping from front view to BEV space. Considering F_{fv} and p_{fv2bev} with the deformable unit, we get the explicit transformed BEV feature F_{bev}.

To sum up, Perspective Transformer extracts front-view features among the reference points to construct representative BEV features. As demonstrated in Sect. 5, such a feature transformation in an aggregation spirit via Transformer is proven to perform better than a direct IPM-based projection across views.

3.3 Simultaneous 2D and 3D Lane Detection

Although the main focus in this paper lies in 3D detection, we formulate the PersFormer framework to detect 2D and 3D lanes in one shot. On one side, 2D lane detection in the perspective view still draws interest in the community as part of the general high-level vision problems [1,28,39,46]; on the other side, unifying 2D and 3D tasks are naturally feasible since the BEV features to predict 3D outputs descend from the counterpart in the 2D branch. An end-to-end unified framework would leverage features and benefit from the co-learning optimization process as proven in most multi-task literature [24,27,50].

Unified Anchor Design. Since our method is anchor-based detection, the core issue to achieve the unified framework is to integrate anchors in both 2D and 3D. Unfortunately, anchors in these two domains usually do not share similar distribution. For example, the popular 2D approach LaneATT [46] settles too many anchors, spanning different directions in the image; while the recent 3D work Gen-LaneNet [19] puts too few anchors, which are parallel and sparse in BEV. Based on these observations, we thereby design anchors such that the redesigned anchors could leverage the network to optimize shared features across *two* domains. We start with several groups of anchors (here, the group number is set to 7) sampled with different incline angles in the BEV space and then projected to the front view. Figure 4 elaborates on the integration of 2D and 3D anchors. Below we describe how the lane line is modeled via anchors.

3D Anchor Design. To match ground truth lanes tightly, the anchors are placed approximately longitudinal along x-axis, with an incline angle φ. As denoted in Fig. 4(left), the initial line (equally spaced) with staring position along x-axis is denoted by X_{bev}^i for each anchor i. Similar to anchor regression in object detection, the network predicts the relative offset $\mathbf{x}^i$ w.r.t. the initial position X_{bev}^i; hence the resultant lane prediction along x-axis is $(\mathbf{x}^i + X_{\text{bev}}^i)$. As indicated in Eqn.(1), each lane is represented as a number of N_d points. The prediction head generates three vectors related to lane shape as follows:

$$(\mathbf{x}^i, \mathbf{z}^i, \mathbf{vis}_{\text{bev}}^i) = \{(x^{(i,k)}, z^{(i,k)}, \text{vis}_{\text{bev}}^{(i,k)})\}_{k=1}^{N_d} \tag{5}$$

where $\mathbf{z}^i$ is the lane height in 3D sense, the binary $\text{vis}_{\text{bev}}^{(i,k)}$ denotes the visibility of each location k in lane i, which controls the endpoint or length of a lane. Note that the lane position along y-axis does not need to be predicted since each y value of the N_d samples in a lane is pre-defined - we predict the $x^{(i,k)}$ value at the corresponding (fixed) y location. To sum up, the description of a lane's location in the world coordinate system is denoted as $(\mathbf{x}^i + X_{\text{bev}}^i, \mathbf{y}, \mathbf{z}^i)$.

2D Anchor Design. The anchor description and prediction are similar to those defined in 3D view, except that the (u, v) is in 2D space and there is no height (see Fig. 4(right)). We omit the detailed notations for brevity. It is worth mentioning that each 3D anchor X_{bev}^i with an incline angle φ corresponds to a specific 2D anchor U_{fv}^i with the incline angle θ; the connection is built via the projection in Eqn.(2). We achieve the goal of unifying 2D and 3D tasks simultaneously by setting the *same* set of anchors. Such a design would optimize features together and features being more aligned and representative across views.

3.4 Prediction Loss

Binary Segmentation under BEV. As do in many preceding work [21,35,54], adding more intermediate supervision into the network training would boost the performance of network. Since lane detection belongs to image segmentation and requires general large resolution, we concatenate a U-Net structure [40] head on top of the generated BEV features. Such an auxiliary task is to predict lanes in BEV, but instead in a conventional 2D segmentation manner, aiming for better feature representation for the main task. The ground truth S_{gt} is a binary segmentation map projected from 3D lane ground truth to the BEV space. The prediction output is denoted by S_{pred} and owns the same size as S_{gt}.

Loss Function. Equipped with the anchor representation and segmentation head aforementioned, we summarize the overall loss. Given an image input and its ground truth labels, it finally computes a sum of all anchors' loss; the loss is a combination of the 2D lane detection, 3D lane detection and intermediate segmentation with learnable weights (α, β, γ) accordingly:

$$\mathcal{L} = \sum_i \alpha \mathcal{L}_{2\text{D}}(c^i_{2\text{D}}, \mathbf{u}^i, \mathbf{vis}^i_{\text{fv}}) + \beta \mathcal{L}_{3\text{D}}(c^i_{3\text{D}}, \mathbf{x}^i, \mathbf{z}^i, \mathbf{vis}^i_{\text{bev}}) + \gamma \mathcal{L}_{\text{seg}}(S_{\text{pred}}), \quad (6)$$

where $c^i_{(\cdot)}$ is the predicted lane category in 2D and 3D domain respectively. The loss input above shows the prediction part only; we omit the ground truth notation for brevity. The loss of lane category classification for the 2D/3D task is the cross-entropy; the loss of lane shape regression is the l_1 norm; the loss of lane visibility prediction is the binary cross-entropy loss. The loss of the auxiliary task is a binary cross-entropy loss between two segmentation maps.

4 OpenLane: A Large-scale Realistic 3D Lane Benchmark

4.1 Highlights over Previous Benchmarks

OpenLane is the *first* real world 3D lane dataset and the *largest* scale to date compared with existing benchmarks. We construct OpenLane on top of the influential Waymo Open dataset [45], following the same data format and evaluation pipeline - leveraging existent practice in the community so that users would not handle additional rules for a new benchmark. Table 1 compares OpenLane with existing counterparts in various aspects. In short, OpenLane owns 200K frames

Table 1. Comparison of OpenLane with existing benchmarks. "Avg. Length" denotes the average time duration of segments. "Inst. Anno." indicates whether lanes are annotated instance-wise (c.f. semantic-wise). "Track. Anno." implies if a lane has a unique tracking ID. Numbers in '#Frames' are the number of annotated frames/total frames respectively. Details of "Scenario" can be found in Appendix

Dataset	#Segments	#Frames	Avg.Length	Inst.Anno.	Track.Anno.	Max#Lanes	LineCategory	Scenario
Caltech Lanes [2]	4	1224/1224	-	✓	✗	4	-	Easy
TuSimple [49]	6.4K	6.4K/128K	1 s	✓	✗	5	-	Easy
3D Synthetic [19]	-	10K/10K	-	✓	-	6	-	Easy
VIL-100 [61]	100	10K/10K	10 s	✓	✗	6	10	Medium
VPG [25]	-	20K/20K	-	✗	-	-	7	Medium
OpenDenseLane [10]	1.7K	57K/57K	-	✓	✗	-	4	Medium
LLAMAS [4]	14	79K/100K	-	✓	✗	4	-	Easy
ApolloScape [21]	235	115K/115K	16 s	✗	✗	-	13	Medium
BDD100K [59]	100K	100K/120M	40 s	✗	✗	-	11	Medium
CULane [37]	-	133K/133K	-	✓	-	4	-	Medium
CurveLanes [55]	-	150K/150K	-	✓	-	9	-	Medium
ONCE-3DLanes [56]	-	211K/211K	-	✓	-	8	-	Medium
OpenLane	1K	**200K/200K**	20 s	✓	✓	**24**	**14**	**Hard**

and over 880K carefully annotated lanes, 33% and 35% more compared with existing largest lane dataset CurveLanes [55] respectively, with rich annotations.

We annotate all the lanes in each frame, including those in the *opposite* direction if no curbside exists in the middle. Due to the complicated lane topology, *e.g.*, intersection/roundabout, one frame could contain as many as **24** lanes in OpenLane. Statistically, about 25% frames of OpenLane have more than 6 lanes, which exceeds the maximum number in most lane datasets. **14** lane categories are annotated alongside to cover a wide range of lane types in most scenarios, including road edges. Double yellow solid lanes, single white solid and dash lanes take up almost 90% of total lanes. This is imbalanced, and yet it falls into a long-tail distribution problem, which is common in realistic scenarios. In addition to the lane detection task, we also annotate: (a) scene tags, such as weather and locations; (b) the closest-in-path object (CIPO), which is defined as the most concerned target w.r.t. ego vehicle; such a tag is quite pragmatic for subsequent modules as in planning/control, besides a whole set of objects from perception. An annotation example is provided in Fig. 5(d), along with some typical samples in existing 2D lane datasets in Fig. 5(a–c). The detailed statistics, annotation criterion and visualization can be found in Appendix.

Fig. 5. Annotation samples of OpenLane compared with other lane datasets. OpenLane is challenging with more lane categories per frame in average and has rich labels including scene, weather, hours, CIPO

4.2 Generation of High-Quality Annotation

Building a real-world 3D lane dataset has challenges mainly in an accurate localization system and occlusions. We compare several popular sensor datasets [6,9,45] by projecting 3D object annotations to image planes and constructing 3D scene maps using both learning-based [48] or SLAM algorithms [42,43]. The reconstruction precision and scalability of Waymo Open Dataset [45] outperforms other candidates, leading to employing it as our basis.

Primarily, we generate the necessary high-quality 2D lane labels. They contain the final annotations of tracking ID, category, and 2D points ground truth. Then for each frame, the point clouds are first filtered with the original 3D object bounding boxes and then projected back into the corresponding image. We further keep those points related to 2D lanes only with a certain threshold. However, the output directly after a static threshold filtering could lead to an unsatisfying ground truth due to the perspective scaling issue. To solve this and keep the slender shape of lanes, we use the filtered point clouds to interpolate the 3D position for each point in 2D annotations. Afterward, with the help of the localization system, 3D lane points in frames within a segment could be spliced into long, high-density lanes. This process could bring some unreasonable parts into the current frame; thus, points in one lane whose 2D projections are higher than the ending position of its 2D annotation are labeled as invisible. A smoothing step is ultimately deployed to filtrate any outliers and generate the 3D labeling results. We omit some technical details, such as how to deal with a large U-turn during smoothing, and we refer the audience to Appendix.

5 Experiments

We examine PersFormer on two 3D lane benchmarks, the newly proposed real-world OpenLane dataset, and the synthetic Apollo dataset. For both 3D lane datasets, we follow the evaluation metrics designed by Gen-LaneNet [19], with additional category accuracy on OpenLane dataset. For the 2D task, the classical metric in CULane [37] is adopted. We put correlated details in Appendix.

5.1 Results on OpenLane

We provide 3D and 2D evaluation results on the proposed OpenLane dataset. In order to evaluate the models thoroughly, we report F-Score on the entire validation set and different scenario sets. The scenario sets are selected from the entire validation set based on the scene tags of each frame. In Table 2, PersFormer gets the highest F-Score on the entire validation set and every scenario set, surpassing previous SOTA methods in varying degrees. In Table 3, PersFormer outperforms LaneATT [46], which is our baseline 2D method, by **11%**. Detailed comparison with previous 3D SOTAs is presented in Table 4. PersFormer outperforms the previous best method in F-Score by **6.4%**, realizes satisfying accuracy on the classification of lane type, and presents the first baseline result. Note that PersFormer is not satisfying on the metric of near error on x-axis. This is probably because the unified anchor design is more suitable in fitting the main body of a

Table 2. Comparison with other open-sourced 3D methods on OpenLane. PersFormer achieves the best F-Score on the entire validation set and every scenario set

Method	All	Up & Down	Curve	Extreme Weather	Night	Intersection	Merge & Split
3D-LaneNet [15]	44.1	40.8	46.5	47.5	41.5	32.1	41.7
Gen-LaneNet [19]	32.3	25.4	33.5	28.1	18.7	21.4	31.0
PersFormer (ours)	**50.5**	**42.4**	**55.6**	**48.6**	**46.6**	**40.0**	**50.7**

Table 3. Comparison with state-of-the-art 2D method on OpenLane. The result from the 2D head of PersFormer also achieves competitive performance

Method	All	Up & Down	Curve	Extreme Weather	Night	Intersection	Merge & Split
LaneATT-S [46]	28.3	25.3	25.8	32.0	27.6	14.0	24.3
LaneATT-M [46]	31.0	28.3	27.4	34.7	30.2	17.0	26.5
CondLaneNet-S [28]	52.3	55.3	57.5	45.8	46.6	48.4	45.5
CondLaneNet-M [28]	55.0	58.5	59.4	49.2	48.6	50.7	47.8
CondLaneNet-L [28]	**59.1**	**62.1**	**62.9**	**54.7**	**51.0**	**55.7**	**52.3**
PersFormer (ours)	42.0	40.7	46.3	43.7	36.1	28.9	41.2

Table 4. Comprehensive 3D Lane evaluation under different metrics. On the strength of unified anchor design, PersFormer outperforms previous 3D methods on the metrics of far error while retains comparable results on near error (m). * denotes projecting 2D lane results from CondLaneNet [28] to BEV using IPM

Method	F-Score	Category Accuracy	X error near	X error far	Z error near	Z error far
3D-LaneNet [15]	44.1	–	**0.479**	0.572	0.367	0.443
Gen-LaneNet [19]	32.3	–	0.591	0.684	0.411	0.521
Cond-IPM*	36.6	–	0.563	1.080	0.421	0.892
PersFormer (ours)	**50.5**	**92.3**	0.485	**0.553**	**0.364**	**0.431**

lane rather than the starting point. Qualitative results are shown in Fig. 6, indicating that PersFormer is good at catching dense and unapparent lanes in usual autonomous driving scenes. Overall, PersFormer reaches the best performance on 3D lane detection and gains remarkable improvement in 2D on OpenLane.

5.2 Results on Apollo 3D Synthetic

We evaluate PersFormer on Apollo 3D Lane Synthetic dataset [19]. In Table 5, while limited by the scale of the dataset (10K frames), our PersFormer still

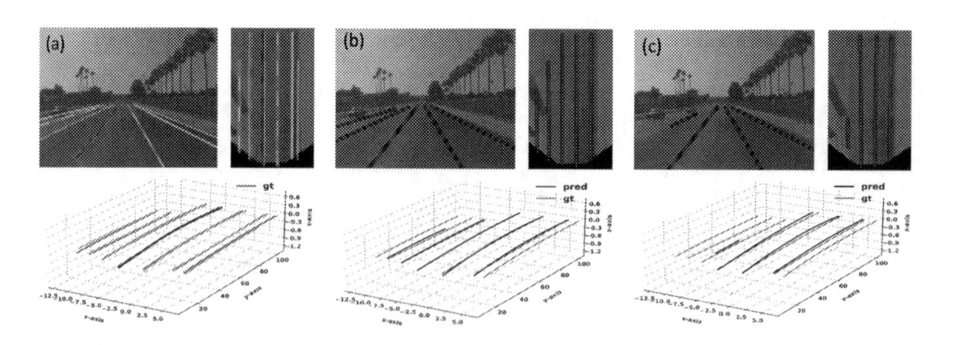

Fig. 6. Qualitative results of PersFormer (a), 3D-LaneNet (b) [15], and Gen-LaneNet (c) [19]. Under a straight road scenario, PersFormer can provide lane-type information and even detect subtle curbside while other methods are missing it

Table 5. Comparison with previous 3D methods on Apollo 3D Lane Synthetic. PersFormer achieves best F-Score on every scene set with comparable X/Z error (m)

Scene	Method	F-Score	X error near	X error far	Z error near	Z error far
Balanced Scenes	3D-LaneNet [15]	86.4	0.068	0.477	0.015	**0.202**
	Gen-LaneNet [19]	88.1	0.061	0.496	0.012	0.214
	3D-LaneNet(1/att) [23]	91.0	0.082	0.439	0.011	0.242
	Gen-LaneNet(1/att) [23]	90.3	0.080	0.473	0.011	0.247
	CLGo [30]	91.9	0.061	0.361	0.029	0.250
	PersFormer (ours)	**92.9**	**0.054**	**0.356**	**0.010**	0.234
Rarely Observed	3D-LaneNet [15]	72.0	0.166	0.855	0.039	**0.521**
	Gen-LaneNet [19]	78.0	0.139	0.903	0.030	0.539
	3D-LaneNet(1/att) [23]	84.1	0.289	0.925	0.025	0.625
	Gen-LaneNet(1/att) [23]	81.7	0.283	0.915	0.028	0.653
	CLGo [30]	86.1	0.147	**0.735**	0.071	0.609
	PersFormer (ours)	**87.5**	**0.107**	0.782	**0.024**	0.602
Vivual Variants	3D-LaneNet [15]	72.5	0.115	0.601	0.032	**0.230**
	Gen-LaneNet [19]	85.3	0.074	0.538	0.015	0.232
	3D-LaneNet(1/att) [23]	85.4	0.118	0.559	0.018	0.290
	Gen-LaneNet(1/att) [23]	86.8	0.104	0.544	0.016	0.294
	CLGo [30]	87.3	0.084	0.464	0.045	0.312
	PersFormer (ours)	**89.6**	**0.074**	**0.430**	**0.015**	0.266

achieves the best F-Score on every scene set. In terms of X/Z error, our model gets comparable results compared to previous methods.

5.3 Ablation Study

We present ablation studies on the anchor design, multi-task strategy, transformer-based view transformation, and auxiliary segmentation task. We mainly report the improvement on 3D lane detection and provide related results on 2D task.

Anchor Design and Multi-task. Starting with a pure 3D lane detection framework (similar to 3D-LaneNet [15]), PersFormer gains **1.7%** by adopting multi-task scheme (Exp.2) and **0.98%** with new anchor design (Exp.4) respectively. By jointly using the new anchor and multi-task trick, PersFormer acquires an improvement of **2.5%** in 3D task and **2.6%** in 2D task (Exp.5).

Spatial Feature Transformation. By using Perspective Transformer with the new anchor design, the improvement increases to **4.9%** (Exp.6), almost doubling the previous improvement. Adding auxiliary binary segmentation task further brings an improvement to **6.02%** (Exp.7), which is our complete model. These ablations support our assumption that PersFormer indeed generates a fine-grained BEV feature, and the spatial feature transformation does illustrate its importance in 3D lane detection task. Surprisingly, a better BEV feature helps 2D task a lot as well, improving **9.7%** (Exp.7).

Table 6. Ablative Study on a 300 segments subset of OpenLane. Exp. 1 is the baseline 3D method, growing with anchor design and multi-task learning (Exp. 2–5). The performance culminates with our spatial feature transformation module and explicit BEV supervision (Exp. 6, 7)

Exp	Unified Anchor	3D Det	2D Det	Perspective Transformer	Binary Seg	3D F-Score	2D F-Score
1		✓				41.77	–
2		✓	✓			43.49	32.33
3	✓		✓			–	34.90
4	✓	✓				42.75	–
5	✓	✓	✓			44.29	34.98
6	✓	✓	✓	✓		46.62	37.00
7	✓	✓	✓	✓	✓	47.79	42.00

6 Conclusions

In this paper, we have proposed Persformer, a novel Transformer-based 2D/3D lane detector, along with OpenLane, a large-scale realistic 3D lane dataset. We demonstrate experimentally that a fine-grained BEV feature with explicit prior and supervision can significantly improve the performance of lane detection. Meanwhile, a large-scale real-world 3D lane dataset effectively align the demand from both the academic and the industrial side.

Acknowledgements. The project is partially supported by the Shanghai Committee of Science and Technology (Grant No. 21DZ1100100). This work was supported in part by National Key Research and Development Program of China (2020AAA0107600), Shanghai Municipal Science and Technology Major Project (2021SHZDZX0102). We would like to acknowledge the great support from SenseBee labelling team at SenseTime Research, constructive contribution from Zihan Ding at BUAA, and the fruitful discussions and comments for this project from Zhiqi Li, Yuenan Hou, Yu Liu, Jing Shao, Jifeng Dai.

References

1. Abualsaud, H., Liu, S., Lu, D.B., Situ, K., Rangesh, A., Trivedi, M.M.: Laneaf: Robust multi-lane detection with affinity fields. IEEE Robot. Autom. Lett. RA-L **6**(4), 7477–7484 (2021)
2. Aly, M.: Real time detection of lane markers in urban streets. In: IV IEEE Intelligent Vehicles Symposium (2008)
3. Bai, M., Mattyus, G., Homayounfar, N., Wang, S., Lakshmikanth, S.K., Urtasun, R.: Deep multi-sensor lane detection. In: IROS (2018)
4. Behrendt, K., Soussan, R.: Unsupervised labeled lane markers using maps. In: ICCV, pp. 832–839 (2019)
5. Benmansour, N., Labayrade, R., Aubert, D., Glaser, S.: Stereovision-based 3d lane detection system: a model driven approach. In: ITSC (2008)
6. Caesar, H., et al.: A multimodal dataset for autonomous driving. In: CVPR (2020)

7. Can, Y.B., Liniger, A., Paudel, D.P., Van Gool, L.: Structured bird's-eye-view traffic scene understanding from onboard images. In: ICCV (2021)
8. Carion, N., Massa, F., Synnaeve, G., Usunier, N., Kirillov, A., Zagoruyko, S.: End-to-end object detection with transformers. In: Vedaldi, A., Bischof, H., Brox, T., Frahm, J.-M. (eds.) ECCV 2020. LNCS, vol. 12346, pp. 213–229. Springer, Cham (2020). https://doi.org/10.1007/978-3-030-58452-8_13
9. Chang, M.F., et al.: Argoverse: 3d tracking and forecasting with rich maps. In: CVPR (2019)
10. Chen, X., Liao, W., Liu, B., Yan, J., He, T.: Opendenselane: a new lidar-based dataset for hd map construction. In: ICME (2022)
11. Chitta, K., Prakash, A., Geiger, A.: Neat: Neural attention fields for end-to-end autonomous driving. In: ICCV (2021)
12. Comma.ai: Openpilot. https://github.com/commaai/openpilot
13. Dosovitskiy, A., et al.: An image is worth 16x16 words: Transformers for image recognition at scale. In: ICLR (2021)
14. Gao, J., et al.: Vectornet: Encoding hd maps and agent dynamics from vectorized representation. In: CVPR (2020)
15. Garnett, N., Cohen, R., Pe'er, T., Lahav, R., Levi, D.: 3D-lanenet: End-to-end 3D multiple lane detection. In: ICCV (2019)
16. Garnett, N., Uziel, R., Efrat, N., Levi, D.: Synthetic-to-real domain adaptation for lane detection. In: ACCV (2020)
17. Gu, J., Sun, C., Zhao, H.: Densetnt: End-to-end trajectory prediction from dense goal sets. In: ICCV (2021)
18. Guan, T., Wang, J., Lan, S., Chandra, R., Wu, Z., Davis, L., Manocha, D.: M3detr: Multi-representation, multi-scale, mutual-relation 3d object detection with transformers. In: WACV (2022)
19. Guo, Y., et al.: Gen-LaneNet: a generalized and scalable approach for 3D lane detection. In: Vedaldi, A., Bischof, H., Brox, T., Frahm, J.-M. (eds.) ECCV 2020. LNCS, vol. 12366, pp. 666–681. Springer, Cham (2020). https://doi.org/10.1007/978-3-030-58589-1_40
20. Hartley, R.I., Zisserman, A.: Multiple View Geometry in Computer Vision. Cambridge University Press, ISBN: 0521540518, second edn. (2004)
21. Huang, X., Wang, P., Cheng, X., Zhou, D., Geng, Q., Yang, R.: The apolloscape open dataset for autonomous driving and its application. TPAMI (2019)
22. Jaderberg, M., Simonyan, K., Zisserman, A., et al.: Spatial transformer networks. In: NeurIPS (2015)
23. Jin, Y., Ren, X., Chen, F., Zhang, W.: Robust monocular 3d lane detection with dual attention. In: ICIP (2021)
24. Kumar, V.R., Yogamani, S., Rashed, H., Sitsu, G., Witt, C., Leang, I., Milz, S., Mäder, P.: Omnidet: Surround view cameras based multi-task visual perception network for autonomous driving. RA-L (2021) 8
25. Lee, S., Kim, J., Shin Yoon, J., Shin, S., Bailo, O., Kim, N., Lee, T.H., Seok Hong, H., Han, S.H., So Kweon, I.: Vpgnet: Vanishing point guided network for lane and road marking detection and recognition. In: ICCV (2017) 2, 4, 10
26. Li, Z., et al.: Bevformer: Learning bird's-eye-view representation from multi-camera images via spatiotemporal transformers. arXiv preprint arXiv:2203.17270 (2022)
27. Liang, M., Yang, B., Chen, Y., Hu, R., Urtasun, R.: Multi-task multi-sensor fusion for 3d object detection. In: CVPR (2019)
28. Liu, L., Chen, X., Zhu, S., Tan, P.: Condlanenet: a top-to-down lane detection framework based on conditional convolution. In: CVPR (2021)

29. Liu, P.L.: Monocular bev perception with transformers in autonomous driving. https://towardsdatascience.com/monocular-bev-perception-with-transformers-in-autonomous-driving-c41e4a893944
30. Liu, R., Chen, D., Liu, T., Xiong, Z., Yuan, Z.: Learning to predict 3d lane shape and camera pose from a single image via geometry constraints. In: AAAI (2022)
31. Liu, Z., et al.: Swin transformer: Hierarchical vision transformer using shifted windows. In: ICCV (2021)
32. Meyer, A., Salscheider, N.O., Orzechowski, P.F., Stiller, C.: Deep semantic lane segmentation for mapless driving. In: IROS (2018)
33. Nedevschi, S., et al.: 3d lane detection system based on stereovision. In: ITSC (2004)
34. Neven, D., De Brabandere, B., Georgoulis, S., Proesmans, M., Van Gool, L.: Towards end-to-end lane detection: an instance segmentation approach. In: IV (2018)
35. Newell, A., Yang, K., Deng, J.: Stacked hourglass networks for human pose estimation. In: Leibe, B., Matas, J., Sebe, N., Welling, M. (eds.) ECCV 2016. LNCS, vol. 9912, pp. 483–499. Springer, Cham (2016). https://doi.org/10.1007/978-3-319-46484-8_29
36. Ngiam, J., et al.: Scene transformer: A unified architecture for predicting future trajectories of multiple agents. In: ICLR (2021)
37. Pan, X., Shi, J., Luo, P., Wang, X., Tang, X.: Spatial as deep: Spatial cnn for traffic scene understanding. In: AAAI (2018)
38. Prakash, A., Chitta, K., Geiger, A.: Multi-modal fusion transformer for end-to-end autonomous driving. In: CVPR (2021) 4
39. Qu, Z., Jin, H., Zhou, Y., Yang, Z., Zhang, W.: Focus on local: Detecting lane marker from bottom up via key point. In: CVPR (2021)
40. Ronneberger, O., Fischer, P., Brox, T.: U-net: Convolutional networks for biomedical image segmentation. In: MICCAI (2015)
41. Saha, A., Maldonado, O.M., Russell, C., Bowden, R.: Translating images into maps. In: ICRA (2022)
42. Shan, T., Englot, B., Meyers, D., Wang, W., Ratti, C., Daniela, R.: Lio-sam: Tightly-coupled lidar inertial odometry via smoothing and mapping. In: IROS (2020)
43. Shan, T., Englot, B., Ratti, C., Daniela, R.: Lvi-sam: Tightly-coupled lidar-visual-inertial odometry via smoothing and mapping. In: ICRA (2021)
44. Su, J., Chen, C., Zhang, K., Luo, J., Wei, X., Wei, X.: Structure guided lane detection. In: IJCAI-21 (2021)
45. Sun, P., et al.: Scalability in perception for autonomous driving: Waymo open dataset. In: CVPR (2020)
46. Tabelini, L., Berriel, R., Paixao, T.M., Badue, C., De Souza, A.F., Oliveira-Santos, T.: Keep your eyes on the lane: Real-time attention-guided lane detection. In: CVPR (2021)
47. Tan, M., Le, Q.: Efficientnet: Rethinking model scaling for convolutional neural networks. In: ICML (2019)
48. Teed, Z., Deng, J.: Droid-slam: Deep visual slam for monocular, stereo, and rgb-d cameras. In: NeurIPS (2021)
49. TuSimple: https://github.com/TuSimple/tusimple-benchmark (2017)
50. Vandenhende, S., et al.: Multi-task learning for dense prediction tasks: A survey. In: TPAMI (2021)
51. Vaswani, A., et al.: Attention is all you need. In: NeurIPS (2017)

52. Wang, J., Mei, T., Kong, B., Wei, H.: An approach of lane detection based on inverse perspective mapping. In: ITSC (2014)
53. Wang, Y., Guizilini, V.C., Zhang, T., Wang, Y., Zhao, H., Solomon, J.: Detr3d: 3d object detection from multi-view images via 3d-to-2d queries. In: CoRL (2022)
54. Wei, S.E., Ramakrishna, V., Kanade, T., Sheikh, Y.: Convolutional pose machines. In: CVPR (2016)
55. Xu, H., Wang, S., Cai, X., Zhang, W., Liang, X., Li, Z.: CurveLane-NAS: unifying lane-sensitive architecture search and adaptive point blending. In: Vedaldi, A., Bischof, H., Brox, T., Frahm, J.-M. (eds.) ECCV 2020. LNCS, vol. 12360, pp. 689–704. Springer, Cham (2020). https://doi.org/10.1007/978-3-030-58555-6_41
56. Yan, F., et al.: Once-3dlanes: Building monocular 3d lane detection. In: CVPR (2022)
57. Yang, W., et al.: Projecting your view attentively: Monocular road scene layout estimation via cross-view transformation. In: CVPR (2021)
58. Yin, J., Shen, J., Guan, C., Zhou, D., Yang, R.: Lidar-based online 3d video object detection with graph-based message passing and spatiotemporal transformer attention. In: CVPR (2020)
59. Yu, F., et al.: Bdd100k: A diverse driving dataset for heterogeneous multitask learning. In: CVPR (2020)
60. Yu, Z., Ren, X., Huang, Y., Tian, W., Zhao, J.: Detecting lane and road markings at a distance with perspective transformer layers. In: ITSC (2020)
61. Zhang, Y., et al.: Vil-100: A new dataset and a baseline model for video instance lane detection. In: ICCV (2021)
62. Zhu, X., Su, W., Lu, L., Li, B., Wang, X., Dai, J.: Deformable DETR: Deformable transformers for end-to-end object detection. In: ICLR (2021)

PointFix: Learning to Fix Domain Bias for Robust Online Stereo Adaptation

Kwonyoung Kim[1], Jungin Park[1], Jiyoung Lee[2], Dongbo Min[3], and Kwanghoon Sohn[1](✉)

[1] Yonsei University, Seoul, South Korea
{kyk12,newrun,khsohn}@yonsei.ac.kr
[2] NAVER AI Lab, Seongnam, South Korea
lee.j@navercorp.com
[3] Ewha Womans University, Seoul, South Korea
dbmin@ewha.ac.kr

Abstract. Online stereo adaptation tackles the domain shift problem, caused by different environments between synthetic (training) and real (test) datasets, to promptly adapt stereo models in dynamic real-world applications such as autonomous driving. However, previous methods often fail to counteract particular regions related to dynamic objects with more severe environmental changes. To mitigate this issue, we propose to incorporate an auxiliary point-selective network into a meta-learning framework, called *PointFix*, to provide a robust initialization of stereo models for online stereo adaptation. In a nutshell, our auxiliary network learns to fix local variants intensively by effectively back-propagating local information through the meta-gradient for the robust initialization of the baseline model. This network is model-agnostic, so can be used in any kind of architectures in a plug-and-play manner. We conduct extensive experiments to verify the effectiveness of our method under three adaptation settings such as short-, mid-, and long-term sequences. Experimental results show that the proper initialization of the base stereo model by the auxiliary network enables our learning paradigm to achieve state-of-the-art performance at inference.

Keywords: Online adaptation · Stereo depth estimation · Meta-learning

This work was supported by the National Research Foundation of Korea (NRF) grant funded by the Korea government (MSIP) (NRF-2021R1A2C2006703) and the Yonsei University Research Fund of 2021 (2021-22-0001).

Supplementary Information The online version contains supplementary material available at https://doi.org/10.1007/978-3-031-19839-7_33.

S. Avidan et al. (Eds.): ECCV 2022, LNCS 13698, pp. 568–585, 2022.
https://doi.org/10.1007/978-3-031-19839-7_33

1 Introduction

Stereo depth estimation to predict 3D geometry for practical real-world applications such as autonomous driving [1] has been developed by handcrafted methods [9,11,13,43] and deep stereo models based on supervised learning [2,15,23,34] that leverage the excellent representation power of deep neural networks. In general, given that the high performance of deep networks is guaranteed when test and training data are derived from a similar underlying distribution [4,5,12,20,22], they demand a huge amount of annotated training data to reflect a real-world distribution. Acquiring groundtruth disparity maps, but unfortunately, is laborious and impractical [36]. Especially for autonomous driving, constructing datasets from all possible different conditions (*e.g.* weather and road conditions) is impossible while it is a very fatal problem [37]. To mitigate the aforementioned issues, an intuitive solution is to fine-tune the stereo model trained on a large-scale synthetic dataset that is easier to collect groundtruth. However, despite the help of large-scale synthetic datasets, most recent works [26,35,44,46] have pointed out the limitation of fine-tuning that is incapable of collecting sufficient data in advance when running the stereo models in the *open world.* While domain generalization methods [32,42] have shown promising results without real images, they require high computations to provide generalized stereo models and often fail to respond to continuously changing environments.

As an alternative solution, online stereo adaptation [19,29,36,37,39,45] is proposed to incorporate unsupervised domain adaptation [6,10] into a continual learning process [31]. Formally, a baseline network is trained offline using a large

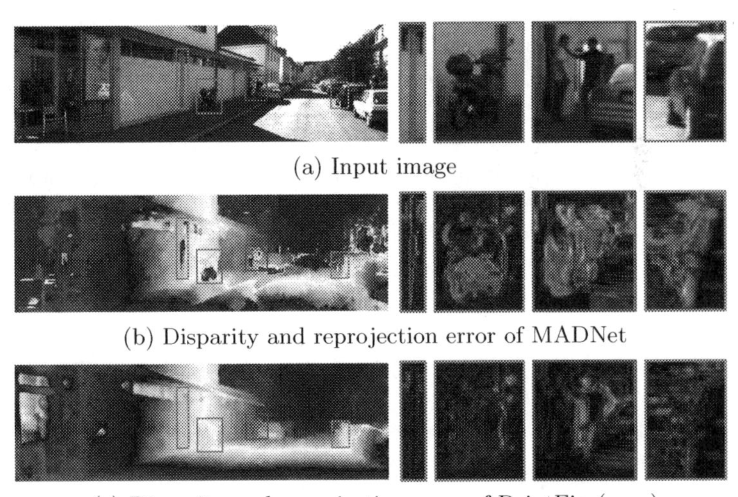

(a) Input image

(b) Disparity and reprojection error of MADNet

(c) Disparity and reprojection error of PointFix (ours)

Fig. 1. Estimated disparity after online adaptation from MADNet and our PointFix. Our method has a much stronger adaptation ability, especially in local detail.

number of the labeled synthetic datasets (*e.g.* Synthia [30], FlyingThings3D [23]) and continually adapted to unlabeled unseen scenarios at test time in an unsupervised manner. To demonstrate a faster inference speed for real-world applications, MADNet [29,37] proposed a lightweight network and modular adaptation framework to rely on self-supervision via reprojection loss [10]. Meanwhile, Learning-to-adapt (L2A) [36] introduced a new learning framework based on model agnostic meta-learning (MAML) [4] for the improved adaptation ability of the network by well-suited base parameters. It shows that the meta-learning framework has great potential in making the network parameters in learning process to make the parameters into a very adaptable state. Despite their great progress, most online adaptation methods [36,37] have merely attempted to only impose a global average errors from the whole prediction as a learning objective during an offline training without attention to a domain gap in local, showing poor initial performance.

In particular, we observe that given stereo images from a novel environment, incorrectly estimated disparities are concentrated on specific *local* regions, as depicted in Fig. 1(b). The domain shift [27] issue arises because the local context of test data (*e.g.* appearance deformations of objects, occlusion type, or the form of a shadow etc.) is significantly diverse from those deployed throughout the training process. This means that without taking such locally varying discrepancies between training and test data into account, the global adaptation strategy used in the existing methods [19,29,36,37,39,45] has fundamental limitations in improving the adaptation performance. In addition, a plug-and-play algorithm to easily combine with evolving deep stereo networks is also needed.

In this paper, we propose a novel model-agnostic training method for robust online stereo adaptation, called *PointFix*, that can be flexibly built on the top of existing stereo models and learns the base stereo network on a meta-learning framework. Unlike the existing methods [3,19,29,39] that focus on a new online adaptation strategy, we leverage the meta-learning strategy for *learning-to-fix* a base stereo network offline so that it can have generalized initial model parameters and respond to novel environments more robustly. Specifically, we incorporate an auxiliary point-selective network, termed PointFixNet into meta-learning to rectify the local detriment of the base network and alternately fix the base network by an additional update as in the online meta-learning methods [5,20]. As a result, the parameters of the base network are updated to grasp and utilize the incoming local context and can be robust to the local variants at test time by preventing the network from being biased to global domain dependencies only.

In the experiment, we learn two base stereo models, DispNetC [23], MADNet [37], together with our framework on the synthetic data in the offline training, and then update the whole models (full adaptation) or the sub-module of the models (MAD adaptation) in the online adaptation using the unsupervised reprojection loss [6,10] on the real-world dataset. Note that the proposed auxiliary network is not used in the online adaptation during inference, maintaining an original inference speed of the base stereo models. Given that our PointFix is a general and synergistic strategy that can be adopted with any kinds of stereo

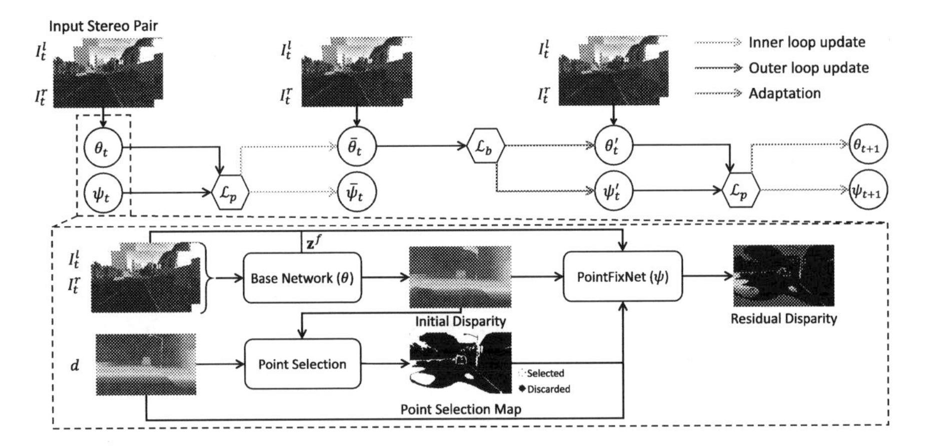

Fig. 2. The overall framework of the proposed *PointFix* and PointFixNet. We alternatively update the base network and PointFixNet underlying meta-learning. The detail of main flowchart is illustrated in the dotted box.

networks, it improves a generalization capability of the base stereo model to novel environments through the robust parameter initialization. Extensive experimental results show that PointFix outperforms recent state-of-the-art results by a significant margin on various adaptation scenarios including short-, mid-, and long-term adaptation. In addition, comparison with domain generalization methods [21,32,42] demonstrates the superiority of our PointFix in terms of both accuracy and speed.

2 Related Work

For depth estimation from stereo images, there is an extensive literature, but here we briefly introduce related work in the application of convolutional neural networks (CNNs). Modern approaches using CNNs are mostly categorized by matching-based approaches [15,33,41] that learn how to match corresponding points, and regression-based methods [2,14,23] that learn to directly regress sub-pixel disparities. To further enhance the performance, some works [16,17,28] consider exploiting auxiliary network or module to assist base network by estimating confidence of prediction map from stereo inputs, prediction or cost volume of the base network. Although their results are promising, they have a limitation in retaining the superb accuracy in new domains [29].

Pointing out the domain shift issue, recent works [3,19,29,36,37,39,45] have proposed online stereo adaptation methods to consider more practical solution for real-world applications. They have argued that we should consider a new *open-world* scenario in which input frames are sequentially provided to the model with certain time intervals. By continually updating parameters at test time in an unsupervised manner, they observe the adaptability of the models in changing environments. As one of the pioneering works, [29,37] has proposed a light weight model (MADNet) and its modular updating frameworks (MAD, MAD++) to improve the adaptation speed drastically. Following this idea, [39] has picked up the speed even more by implementing 'Adapt or Hold' mechanism based on deep Q-learning network [25].

Closely related work to ours is meta-learning based online adaptation approaches [36,45] that train to learn proper model parameters to be better suitable for online adaptation. L2A [36] has directly incorporated online adaptation process into the inner loop updating process, and also learn a confidence measure to use adaptive weighted loss. However, L2A is inherently unstable during training due to multiple adaptation steps, especially when coupled with the lightweight model (*e.g.* MADNet [29]). Also, it is worth noting a difference in purpose between confidence-weighted adaptation of L2A and our PointFixNet. The role of the auxiliary network in L2A is to eliminate uncertain errors from inherently noisy reprojection loss, whereas ours is to stabilize the generality of the base model by generating a proper point-wise learning objective for particular bad pixels. We hereby deliver the gradient of the point-wise loss to the base stereo model to prevent the model from being learned by minimizing only the global errors and to remedy domain-invariant representations in local regions.

3 Problem Statement and Preliminaries

Online Stereo Adaptation. Here we first formulate online stereo adaptation. Given stereo image pairs from source domain $\mathcal{D}_s$ with available ground-truth maps, online stereo adaptation aims to learn a stereo model capable of adapting itself dynamically in a novel unseen domain $\mathcal{D}_u$. In the inference, given a set of parameters θ from the base stereo model trained on $\mathcal{D}_s$, the base network parameters are updated in a single iteration step t to adapt the stereo model with respect to continuous input sequence without ground-truth disparity maps:

$$\theta_{t+1} \leftarrow \theta_t - \alpha \nabla_{\theta_t} \mathcal{L}_u(\theta_t, I_t^l, I_t^r), \tag{1}$$

where $\mathcal{L}_u$ is an unsupervised loss function, α is the learning rate, I_t^l, and I_t^r are left and right images of the t-th stereo pair from $\mathcal{D}_u$. We evaluate the performance of the online adaptation under the short-term (sequence-level), mid-term (environment-level), and long-term (full) settings. Note that following previous approaches [36,37], we use a reprojection error [10] as the unsupervised learning objective $\mathcal{L}_u$.

Model Agnostic Meta-learning. MAML [4] has proposed to learn initial base parameters that are suited to adapt to new domain with only few updates. This is attained by implementing a nested optimization which consists of an inner loop and an outer loop. Specifically, the inner loop updates the base parameters for each sample in batch separately in the standard gradient descent way. The outer loop performs the update of the base parameters using the sum of sample-specific gradients (meta-gradient) which are computed by the parameters updated in the inner loop. Formally, for a set of tasks $\mathcal{T}$, let the meta-training and meta-testing sets be $\mathcal{D}_{\tau}^{\text{train}}$ and $\mathcal{D}_{\tau}^{\text{test}}$ respectively. A set of parameters θ^* can be obtained for a specific task $\tau \in \mathcal{T}$ with a single gradient step in the inner loop:

$$\theta^* = \min_{\theta} \sum_{\tau \in \mathcal{T}} \mathcal{L}(\theta - \alpha \nabla_{\theta} \mathcal{L}(\theta, \mathcal{D}_{\tau}^{\text{train}}), \mathcal{D}_{\tau}^{\text{test}}), \tag{2}$$

where $\mathcal{L}$ is a objective function for the task and α is the learning rate. They carries out meta-update, including inner and outer updates, during training and then adaptation for the target task after the whole meta-learning processes are completely over. In this paper, we treat the disparity prediction for each stereo pair as a single task.

Recent works [5,20] have introduced two stage meta-learning procedure (a.k.a online meta-learning) that implements meta-update and adaptation stages alternatively during training to achieve more efficient gradient path and better performance. Inspired by their alternative updating scheme, we propose a novel *learning-to-fix* strategy. Our objective is to get base network and PointFixNet parameters that enable the base network to quickly adapt regardless of local variants in unseen domains. The details are described in the following section.

Algorithm 1. Parameter update with PointFix loss

Input: $\mathcal{I} = \{(I_n^l, I_n^r, d_n)\}_{n=0}^{N-1}$; learning rate α; base model parameters θ; PointFix network parameters ψ

Output: updated parameters $\bar{\theta}$, $\bar{\psi}$

1: **function** PointUpdate($\mathcal{I}, \alpha, \theta, \psi$)
2: $\quad \mathcal{L}_p^{\tau} \leftarrow 0$ ▷ Initialize loss
3: $\quad$ **for** $n = 0$ **to** $N-1$ **do**
4: $\quad\quad \hat{d}, \mathbf{z}^b = \mathcal{F}(I_n^l, I_n^r|\theta)$
5: $\quad\quad \mathbf{p}(\theta) = \{(i,j)|\ |(\hat{d}_{ij} - d_{n,ij})|_1 > 3\}$ ▷ Select points
6: $\quad\quad \mathbf{z}^c = \mathcal{F}(I_n^l, \hat{d}, d_n|\psi^c)$
7: $\quad\quad$ **for** $(i,j) \in \mathbf{p}(\theta)$ **do**
8: $\quad\quad\quad r_{ij} = \mathcal{F}(\mathbf{z}_{ij}^b, \mathbf{z}_{ij}^c|\psi^p)$ ▷ Residual disparity
9: $\quad\quad\quad \mathcal{L}_p^{\tau} \leftarrow \mathcal{L}_p^{\tau} + \mathcal{L}_p(r_{ij} + \hat{d}_{ij}, d_{n,ij})$ ▷ PointFix loss
10: $\quad\quad$ **end for**
11: $\quad$ **end for**
12: $\quad \bar{\theta} \leftarrow \theta - \alpha \nabla_{\theta} \mathcal{L}_p^{\tau}$ ▷ Update base network
13: $\quad \bar{\psi} \leftarrow \psi - \alpha \nabla_{\psi} \mathcal{L}_p^{\tau}$ ▷ Update PointFixNet
14: $\quad$ **return** $\bar{\theta}, \bar{\psi}$
15: **end function**

4 PointFix: Learning to Fix

We design a novel meta-learning framework for online stereo adaptation to learn good base model and quickly adapt to novel environments (*i.e.*, unseen domain), especially concentrating on erroneous pixels. As illustrated in Fig. 2, we leverage

Algorithm 2. Overall training procedure

Hyperparameters: batch size N; max iteration K; learning rate of inner and outer loop α, β
Input: pre-trained base model parameters θ; source training dataset $\mathcal{S}$
Output: optimized base model parameters θ^*

1: **function** TRAINING($\theta, \mathcal{S}$)
2: Initialize θ and ψ.
3: **for** $k = 0$ **to** $K - 1$ **do**
4: $\mathcal{L}_k \leftarrow 0$ ▷ Initialize loss
5: $\mathcal{I}_k \sim \mathcal{S}$ ▷ Sample a batch of size N
6: **for** $n = 0$ **to** $N - 1$ **do**
7: $\theta_n \leftarrow \theta, \psi_n \leftarrow \psi$ ▷ Copy parameters
8: $\bar{\theta}_n, \bar{\psi}_n, \leftarrow \text{PointUpdate}(\mathcal{I}_{k,n}, \alpha, \theta_n, \psi_n)$ ▷ Inner loop update
9: $\hat{d}_n, \mathbf{z}^b_{k,n} = \mathcal{F}(I^l_{k,n}, I^r_{k,n}|\bar{\theta}_n)$
10: $\mathcal{L}_k \leftarrow \mathcal{L}_k + \mathcal{L}_b(\hat{d}_n, d_{k,n})$ ▷ Base loss
11: **end for**
12: $\theta' \leftarrow \theta - \beta \nabla_\theta \mathcal{L}_k$ ▷ Outer loop update
13: $\psi' \leftarrow \psi - \beta \nabla_\psi \mathcal{L}_k$ ▷ Outer loop update
14: $\theta, \psi \leftarrow \text{PointUpdate}(\mathcal{I}_k, \alpha, \theta', \psi')$ ▷ Adaptation
15: **end for**
16: **return** $\theta^* \leftarrow \theta$
17: **end function**

off-the-shelf deep stereo model as a base network and incorporate an auxiliary network, PointFixNet, to make parameters of the base network robust to the local distortion.

4.1 Base Stereo Models

Our goal is to train base models offline to be more suitable for the online adaptation by correcting bias to the seen domain. We have employed two stereo networks as a base model: 1) DispNet-Corr1D [23] (shortened as DispNetC) and 2) MADNet [37]. Besides taking the initial disparity $\hat{d}$ estimated from the base model, we extract intermediate features that is useful to exploit the fine-grained information. The intermediate features consist of the matching cost $\mathbf{c}$ (same as correlation layer in DispnetC) and its corresponding left feature $\mathbf{f}^l$. They are concatenated and taken as a base feature $\mathbf{z}^b = \Pi(\mathbf{c}, \mathbf{f}^l)$, where $\Pi(\cdot, \cdot)$ is a concatenation operation. We note that in MADNet, the matching cost calculation is similar with the one in DispNetC, but before calculation their right features are warped with a disparity map on a coarse resolution to reduce search range and computation. However, to extract base features, we apply the same matching cost computation scheme used in DispNetC regardless of the base network to ensure the generality of our method.

In each inner loop of our method, we select a pixel i, j to fix local deformations by computing ℓ_1 loss between $\hat{d}$ and the ground-truth d, such that the set of points $\mathbf{p}(\theta)$ with the base parameter θ can be derived as:

$$\mathbf{p}(\theta) = \{(i,j)| \ |\hat{d}_{ij} - d_{ij}|_1 > 3\}. \tag{3}$$

Note that we represent $\mathbf{p}$ as a function of θ to indicate that the selected point varies depending on θ updated in the learning procedure. In next section, we describe a way to leverage the set of points for correcting local distortions caused by the seen domain bias, depicted in Fig. 1.

4.2 PointFixNet

To mitigate the seen domain bias of the base network, we deploy an additional auxiliary network, called PointFixNet, which individually repairs a disparity by incurring a proper point-wise gradient. The PointFixNet consists of two modules: a feature extraction module (parameterized by ψ^c) that extracts feature $\mathbf{z}^c$ from heterogeneous inputs; and a point-wise prediction module (parameterized by ψ^p) that generates residual disparity value of each point and back-propagates the point-wise errors. Specifically, the feature extraction module consists of three convolution layers and takes the left image, I^l, the initial disparity $\hat{d}$, and ground-truth d as inputs to integrate context around each erroneous pixel. Therefore, the feature $\mathbf{z}^c$ can be obtained as follows:

$$\mathbf{z}^c = \mathcal{F}(I^l, \hat{d}, d|\psi^c), \tag{4}$$

where $\mathcal{F}$ is a feed-forward process.

Then, the base feature $\mathbf{z}^b$ from the base network and feature $\mathbf{z}^c$ from the feature extraction module are concatenated and fed into the point-wise prediction module to generate the residual disparity value r_{ij}. Inspired by structure in [18], the module consists of four fully-connected (FC) layers to produce a single value for each pixel, such that:

$$r_{ij} = \mathcal{F}(\mathbf{z}^b_{ij}, \mathbf{z}^c_{ij}|\psi^p), \tag{5}$$

where $(i.j) \in \mathbf{p}(\theta)$. The final disparity for (i,j)-th pixel is obtained by adding r_{ij} to $\hat{d}_{ij}$. We note that FC layers share weights across all selected points.

4.3 Learning to Fix

The key idea underlying our framework is iterating *learning how to fix* first and then *fixing* alternatively. First, the parameters of the PointFixNet learn how to generate a proper gradient to the base model in a point-wise manner such that the base model can be improved with less domain bias and then secondly the base model is updated by the learned PointFixNet. This strategy is essential because if we keep network training only with point loss, the performance of prediction after PointFixNet can be guaranteed but the one after the base model may not. Thus, to enhance maximal performance of the base network, it is necessary to employ the alternative meta-learning structure. To this end, we deploy two loss functions: a base loss, $\mathcal{L}_b$, derived from the whole disparity map predicted by the base network and a point loss, $\mathcal{L}_p$, applied to the final disparity values.

They are alternatively optimized to update θ and ψ by relying on the two-stage meta-learning scheme [5,20], as described in Fig. 2 and Algorithm 2.

In the inner loop, parameters are copied for each sample in batch, $\theta_n \leftarrow \theta$ and $\psi_n \leftarrow \psi$. Then we calculate the PointFix loss, to evaluate of the current parameters:

$$\bar{\theta}_n \leftarrow \theta_n - \alpha\nabla_{\theta_n} \sum\nolimits_{(i,j)\in\mathbf{p}(\theta)} \mathcal{L}_p(\hat{d}_{ij} + r_{ij}, d_{ij}), \tag{6}$$

$$\bar{\psi}_n \leftarrow \psi_n - \alpha\nabla_{\psi_n} \sum\nolimits_{(i,j)\in\mathbf{p}(\theta)} \mathcal{L}_p(\hat{d}_{ij} + r_{ij}, d_{ij}), \tag{7}$$

where α is a learning rate, $\mathbf{p}(\theta)$ is a set of selected points, and $\mathcal{L}_p$ is a point-wise ℓ_1 loss between the final disparity and its corresponding ground-truth. Given in Algorithm 1, since the PointFix loss is imposed on the local distortion of the erroneous pixels selected on the initial prediction, we can update base parameters that refer to the fine-grained details.

In the outer loop, we evaluate the performance of the updated base parameter after the inner loop. To measure the performance, we apply the conventional supervised loss between the initial disparity map and ground-truth. Following the procedure of [4], the parameters θ and ψ are updated based on sum of $\mathcal{L}_k$ as follows:

$$\theta' \leftarrow \theta - \beta\nabla_\theta\mathcal{L}_k, \quad \psi' \leftarrow \psi - \beta\nabla_\psi\mathcal{L}_k, \tag{8}$$

where $\mathcal{L}_k = \sum_N \mathcal{L}_b(\hat{d}_n, d_{k,n})$ and k is the current iteration step. Note that the gradients are computed along with the parameters before being updated in the inner loop.

Unlike traditional MAML where the parameters are optimized via meta-update only, we deploy an additional update inspired by online meta-learning. θ' and ψ' are updated in the same way as the inner loop so that the parameters after final update, θ and ψ can be written as:

$$\theta \leftarrow \theta' - \alpha\nabla_{\theta'} \sum\nolimits_{(i,j)\in\mathbf{p}(\theta')} \mathcal{L}_p(\hat{d}_{ij} + r_{ij}, d_{ij}), \tag{9}$$

$$\psi \leftarrow \psi' - \alpha\nabla_{\psi'} \sum\nolimits_{(i,j)\in\mathbf{p}(\theta')} \mathcal{L}_p(\hat{d}_{ij} + r_{ij}, d_{ij}), \tag{10}$$

where $r_{ij} = \mathcal{F}(\mathbf{z}^b_{ij}, \mathbf{z}^c_{ij}, d_{ij}|\bar{\psi}^p)$ for $(i,j) \in \mathbf{p}(\theta')$. Finally, the networks are updated with PointFix loss which are, at the first stage, trained to generate proper back-propagation to enhance the performance of the base network in the next training step. At test time, we use the final parameters of the base network θ^* and perform adaptation according to Eq. (1).

5 Experiments

5.1 Experimental Settings

Datasets. In order to evaluate our method on realistic scenario, we use synthetic dataset for offline training and real dataset for the test. Therefore, the training and test data exist in completely different data distributions. Following the previous work [36], we train our networks using the Synthia [30] dataset and evaluate each model on the KITTI-raw [7] dataset and the subset of the DrivingStereo [40] dataset. All datasets are recorded in driving scene but the Synthia [30] is synthetic data, the KITTI and DrivingStereo are obtained from real world. The Synthia [30] dataset contains 50 sequences which have different combination of weathers, seasons and locations. To set similar disparity ranges in training and test, we resized Synthia [30] dataset images to half resolution as in [36]. We exploit stereo images from front direction only and there are 45,591 total number of stereo frames. The KITTI [7] dataset consists of 71 sequences and total 42,917 frames of stereo images and sparse depth maps. Different from the scenario on the KITTI [7] dataset, we present an additional adaptation scenario that the models adapt to various unseen weather conditions using the subset of the DrivingStereo [40] dataset. The DrivingStereo contains four different weather sequences (*i.e.*, cloudy, foggy, rainy, and sunny) that each sequence includes 500 stereo images with high quality labels obtained from multi-frame LiDAR points.

Metrics. We evaluate the performance using two popular evaluation metrics, the percentage of pixels with disparity outliers larger than 3 (D1-all) and average end point error (EPE). Following the scheme of [36,37], we average each score from all the frames which belong to the same sequence and reset the model to the base parameters at the next sequence, based on the definition of a sequence for different evaluation protocols.

Evaluation Protocols. We perform online stereo adaptation under three different settings according to the definition of the sequence, including short-, mid-, and long-term adaptation. For **short-term adaptation**, each sequence is defined as a distinct sequence provided by the dataset (*e.g. 2011_09_30_drive_0028_sync*). This setting is appeared in [36]. The sequences in **mid-term adaptation** are divided according to the environment (*i.e.*, *City*, *Residential*, *Campus*, *Road*). In **long-term adaptation**, we perform adaptation for all frames by concatenating all mid-term sequences. The mid- and long-term adaptation settings are shown in [37] as short- and long-term adaptation. The implementation details are provided in supplementary material.

Table 1. Mid-term adaptation: Performance comparison for several methods on the KITTI [7] dataset.

Method	Training	Adapt.	City		Residential		Campus		Road		Avg.	
			D1-all	EPE	D1-all	EPE	D1-all	EPE	D1-all	EPE	D1-all	EPE
DispNetC-GT	KITTI	No	1.94	0.68	2.43	0.77	5.43	1.10	1.67	0.69	2.87	0.81
MADNet-GT	KITTI	No	2.05	0.65	2.67	0.82	6.87	1.24	1.57	0.66	3.29	0.84
L2A-Disp	Synthia	No	12.78	1.67	12.80	1.72	17.57	2.06	12.34	1.59	13.87	1.76
MADNet	Synthia	No	38.78	8.36	35.73	7.89	40.59	7.68	38.31	8.77	38.35	8.18
Ours-Disp	Synthia	No	9.98	1.47	10.99	1.62	17.01	2.06	7.98	1.33	11.49	1.62
Ours-MAD	Synthia	No	15.51	1.82	14.24	1.78	22.40	3.04	15.61	1.84	16.94	2.12
L2A-Disp	Synthia	Full	2.05	0.78	2.57	0.86	4.43	1.07	1.63	0.77	2.67	0.87
MADNet	Synthia	Full	2.11	0.81	2.79	0.90	6.24	1.41	1.60	0.72	3.19	0.96
Ours-Disp	Synthia	Full	2.03	0.99	2.46	0.83	4.21	**1.02**	1.58	0.74	2.57	0.90
Ours-MAD	Synthia	Full	**1.55**	**0.72**	**1.55**	**0.70**	**3.84**	1.08	**1.15**	**0.67**	**2.02**	**0.79**
MADNet	Synthia	MAD	2.36	0.84	1.94	0.77	10.03	1.70	2.27	0.83	4.15	1.04
MADNet	Synthia	MAD++	1.95	0.80	1.86	0.76	8.57	1.65	1.94	0.80	3.56	0.99
Ours-MAD	Synthia	MAD	1.63	0.74	1.62	0.73	4.16	1.12	1.23	0.69	2.16	0.82

5.2 Synthetic to Real Adaptation

We evaluate the performance corresponding to the different adaptation methods: *No adaptation (No)*, which measures the performance for all sequences without performing adaptation from the base parameters to estimate the capacity of the initial parameters; *Full adaptation (Full)*, which updates parameters of whole network; *MAD adaptation (MAD)*, which performs faster modular adaptation on a prediction of certain resolution selected at every iteration by their own handcrafted method as proposed in [37]; *MAD++ adaptation (MAD++)* is an extension from MAD and utilizes predictions obtained by handcrafted methods (*e.g.* SGM [11], WILD [38]) as proxy supervision. The cases of *-GT* are regarded as supervised learning that is fine-tuned on KITTI 2012 [8] and 2015 [24] datasets. In the experiment, (L2A, Ours)-Disp. and Ours-MAD. employ DispNetC and MADNet as base networks, respectively.

Mid-Term Adaptation. Table 1 shows the results according to the adaptation methods under mid-term adaptation setting. From the performance with *No adaptation* (row 3–6), we observe that our method helps to learn better base parameters compared to other methods using the same base network. Especially before the adaptation, with MADNet (row 4 and 6), our PointFix significantly improves the performance of the base network by 20.19% and 4.39 in terms of D1-all and EPE respectively, that demonstrates PointFix is effective in a generalization capability.

The results of *Full adaptation* (row 7–10) and *MAD adaptation* (row 11–13) show the adaptation capability of the base network. Our PointFix outperforms previous methods with large margin in both metrics, achieving state-of-the-art performance on all domains. The PointFix with MAD adaptation (row 13) outperforms MADNet with MAD++ adaptation that leverages the additional supervision and even L2A and MADNet with full adaptation (row 7 and

Table 2. Short-term and Long-term adaptation: Performance comparison for several methods on the KITTI [7] dataset.

Method	Training	Adapt.	Short-term		Long-term	
			D1-all	EPE	D1-all	EPE
DispNetC-GT	KITTI	No	2.38	0.77	2.32	0.75
MADNet-GT	KITTI	No	2.57	0.75	2.52	0.78
L2A-Disp	Synthia	No	12.99	1.73	12.86	1.70
MADNet	Synthia	No	27.63	3.59	36.77	8.09
Ours-Disp	Synthia	No	9.47	1.21	10.56	1.57
Ours-MAD	Synthia	No	11.59	1.67	23.50	3.23
L2A-Disp	Synthia	Full	2.64	0.84	2.37	0.84
MADNet	Synthia	Full	6.68	1.31	2.86	0.93
Ours-Disp	Synthia	Full	2.62	0.81	2.30	0.82
Ours-MAD	Synthia	Full	**2.00**	**0.74**	1.56	0.71
MADNet	Synthia	MAD	11.82	1.90	1.92	0.75
MADNet	Synthia	MAD++	9.56	1.61	1.70	0.75
Ours-MAD	Synthia	MAD	2.64	0.87	**1.47**	**0.70**

8) while enabling fast inference by adapting only a few parameters. As pointed out in [37], all adapted models perform worse on *Campus* domain that has a small number of frames (1149) compared to the other domain (5674, 28067, 8027). The results on *Campus* domain show that PointFix adapt better than previous works with a small number of frames in all adaptation methods.

Short-Term Adaptation. In Table 2, to examine the adaptability in short sequences, we evaluate models on each sequence independently as represented in [36]. For each sequence, parameters are initialized at every beginning of sequences. Measured performance is first averaged in each sequence and then they are averaged out. Thanks to fast adaptation speed and inherent robustness of our framework, we surpass the performance than previous works with a large margin. Especially, due to its light weight structure, inherent weakness of MADNet is maximized in short-term environment (row 4 and 8) because they requires a number of frames to be adapted. Nevertheless, MADNet with our framework shows superior results. This suggests that our framework is worthy to be developed with light weight networks.

We conduct additional experiments on the DrivingStereo [40] dataset to hypothesize more difficult scenarios under various weather conditions. Specifically, we train all models on Synthia [30] dataset and evaluate the performance on each sequence including four types of weather conditions in the short-term adaptation setting. As shown in Table 3, our model outperforms MADNet [37] with a large margin for all novel weather conditions. In particular, our method

Table 3. Short-term adaptation: Performance comparison on the subset of DrivingStereo [40] under different weather conditions.

Method	Adapt.	Cloudy		Foggy		Rainy		Sunny		Avg.	
		D1-all	EPE	D1-all	EPE	D1-all	EPE	D1-all	EPE	D1-all	EPE
MADNet	No	56.83	19.16	70.14	23.85	54.20	19.20	51.30	16.08	58.12	19.57
Ours-MAD	No	32.76	5.34	37.25	6.55	34.06	4.78	30.00	4.58	33.52	5.31
MADNet	Full	15.71	3.11	18.09	3.28	18.37	2.86	14.71	2.63	16.72	2.97
Ours-MAD	Full	**7.00**	**1.28**	**7.25**	**1.39**	15.31	2.52	**8.33**	**1.49**	**9.47**	1.67
MADNet	MAD	28.46	6.65	33.56	6.13	31.34	5.79	27.10	6.15	30.11	6.18
Ours-MAD	MAD	8.46	1.46	8.57	1.46	**11.99**	**1.69**	8.90	1.52	9.48	**1.53**

represents error rates of about half those of MADNet [37]. The implementation details and additional experimental results are provided in supplementary material.

Long-Term Adaptation. The adaptation on a long sequence followed by various environments without network resets can be regarded as the most practical scenario in the real world. To simulate this scenario, we report the results evaluated on the concatenation of four environments of the KITTI [7] database ($\sim$ 43000 frames) in Table 2. As analyzed in [37], the results show much smaller average errors than the mid- and short-term adaptation for all adaptation methods, as the length of the sequence increased. Among them, our PointFix shows drastically improved performance and significantly outperforms previous works. Therefore, PointFix framework can be further improved, continually adapting to the real world environment.

5.3 Analysis

Ablation Study. To investigate the effectiveness of the components of within our model, we conduct ablation experiments on the KITTI [7] dataset according to PointFixNet and the meta-learning framework (ML), as shown in Table 4. Note that we use MADNet as the base network and evaluate the performance using the full adaptation under the short-term adaptation setting for all experiments in this section. As a baseline, we remove all components of the proposed method such that the first row in Table 4 corresponds to MADNet [37].

Table 4. Ablation studies for PointFixNet and the meta-learning framework (ML) evaluated on the KITTI [7] dataset under short-term adaptation setting.

PointFixNet	ML	Adapt	D1-all	EPE
✗	✗	Full	6.68	1.31
✓	✗	Full	8.06	1.47
✗	✓	Full	3.12	0.96
✓	✓	Full	**2.00**	**0.74**

Effectiveness of PointFixNet. To validate the effectiveness of the point-wise backpropagation, we ablate PointFixNet and apply ℓ_1 loss between the initial disparity $\hat{d}$ and groundtruth d instead of the point loss in Algorithm 1. The

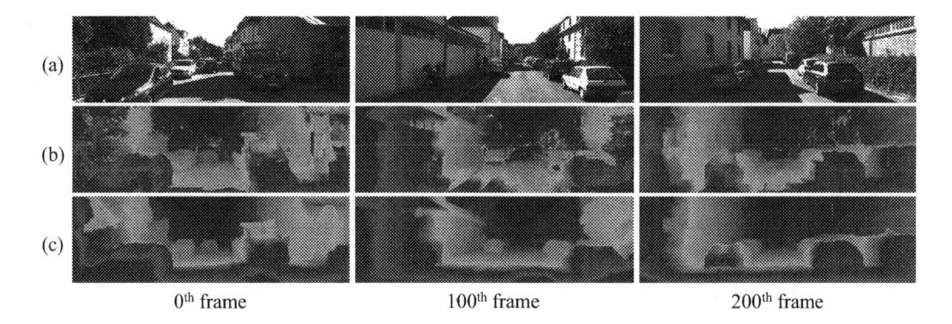

Fig. 3. Disparity maps predicted using MADNet as the base network on the KITTI [7] sequence. (a) Left images, (b) MADNet with MAD adaptation [37], (c) Ours-MAD. with MAD adaptation. Red pixel values indicate closer objects. (Color figure online)

comparison between the third row of Table 4 and the full use of components shows PointFixNet contributes 1.12% and 0.22 in terms of D1-all error and EPE and demonstrates fixing local detriments is simple yet effective to improve the robustness of the stereo model.

Effectiveness of ML. As described in 4.3, the performance improvement of the base network using the point loss is not guaranteed without meta-learning. The results in the second row of Table 4 show poor performance even than the baseline. The comparison between the first and third rows of Table 4 further validates the effectiveness of the ML framework, showing significant performance improvements. Finally, the state-of-the-art performance is shown by demonstrating the advantage of the full use of all components of learning to fix the base network through meta-learning.

Convergence. In Fig. 3, we evaluate the qualitative results related to the convergence analysis. The results contain good initialized parameters (first column), fast adaptation (second column), and convergence to low errors (last column) as analyzed above. As the adaptation proceeds, the MADNet (row 2) estimates better prediction, yet still shows a high error while Ours-MAD (row 3) shows not only robust initial performance but also faster convergence to low errors.

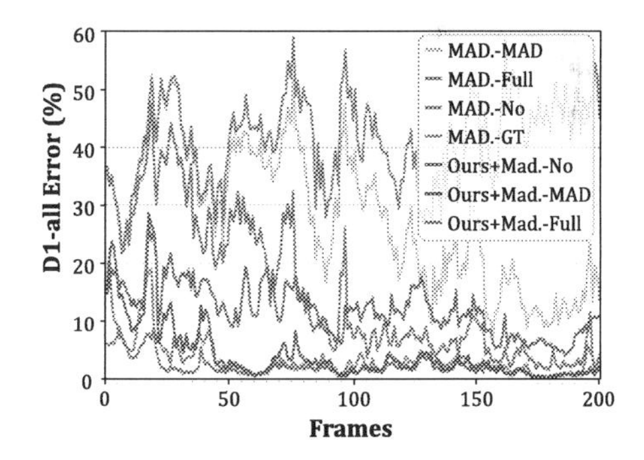

Fig. 4. D1-all error (%) across frames in sequence from the KITTI [7] dataset with respect to the adaptation methods. (Color figure online)

To analyze and compare the adaptation cost corresponding to the methods, we visualize the adaptation performance over frames of the sequence from the KITTI [7] dataset in Fig. 4[1]. In overall view, the results show that our PointFix adapts faster than [37] and converges with lower errors regardless of the adaptation method. Furthermore, the comparison between **MAD.-No** (green) and **Ours+Mad.-No** (brown) shows the effectiveness of the initial base parameters. The comparison between **MAD.-MAD** (yellow) and **Ours+Mad.-MAD** (purple) shows that the performance is improved by PointFix with MAD adaptation, while **MAD.-Full** (red) overtakes from the about 100-th frame. Finally, **Ours+Mad.-Full** (blue) adapts faster than all the other methods and converges to the low D1-all error, showing comparable performance with **MAD.-GT** (gray) fine-tuned with ground-truth. The additional experimental results and analysis are shown in supplementary material.

Comparison with Domain Generalization Methods. To argue the practicality of online stereo adaptation, we compare our model with the state-of-the-art domain generalization (DG) methods [21,32,42] that the stereo models are trained on the synthetic dataset and evaluated on the unseen real dataset without the additional adaptation.

For a fair comparison, the models are pretrained on the SceneFlow [23] dataset and evaluated on the KITTI [7] dataset[2]. Note that we report the performance of our model measured under the long-term adaptation setting. As shown in Table 5[3], our model not only outperforms the generalization approaches in terms of D1-all error but also shows about ×27, ×8.3, and ×1.6 faster inference speed than [42], [32], and [21], respectively, despite of additional adaptation steps. While the domain generalization approaches [21,32,42] estimate the depth maps without the adaptation, they require a large number of parameters to obtain a generalized stereo model, making them impractical. This is worth noting that our method has high applicability to the practical application such as autonomous driving in terms of accuracy and inference speed.

Table 5. Comparisons of domain generalization methods and our method evaluated on the KITTI [7] dataset.

Method	Adapt	D1-all	EPE	FPS
DSMNet [42]	N/A	1.59	0.68	1.30
CFNet [32]	N/A	1.93	0.97	4.27
Raft [21]	N/A	1.66	0.71	22.44
MADNet [37]	MAD	1.95	0.82	35.7
Ours-MAD	MAD	1.47	0.70	35.7

6 Conclusion

In this paper, we proposed PointFix, a novel meta-learning framework to effectively adapt any deep stereo models in online setting. Compared with previ-

[1] The results with DispNetC are shown in supplementary materials.
[2] We conducted a custom experiment using a publicly available code for each paper.
[3] For Raft-stereo [21], a real-time version was employed which shows a much faster inference speed.

ous online stereo adaptation approaches facing global domain bias problem to synthetic data, our model can induce maximal performance of the base stereo networks by the proposed the auxiliary network PointFixNet and learning-to-fix strategy, that can adapt well to the fine-grained domain gap. Our extensive experiments show PointFix achieves state-of-the-art results, outperforming several online stereo adaptation methods in a wide variety of environments. In addition, the results demonstrate that PointFix is capable of improving the generalization ability of the stereo models.

References

1. Achtelik, M., Bachrach, A., He, R., Prentice, S., Roy, N.: Stereo vision and laser odometry for autonomous helicopters in GPS-denied indoor environments. In: SPIE Defense, Security, and Sensing, pp. 733219–733219 (2009)
2. Chang, J.R., Chen, Y.S.: Pyramid stereo matching network. In: CVPR (2018)
3. Chen, Y., Van Gool, L., Schmid, C., Sminchisescu, C.: Consistency guided scene flow estimation. In: Vedaldi, A., Bischof, H., Brox, T., Frahm, J.-M. (eds.) ECCV 2020. LNCS, vol. 12352, pp. 125–141. Springer, Cham (2020). https://doi.org/10.1007/978-3-030-58571-6_8
4. Finn, C., Abbeel, P., Levine, S.: Model-agnostic meta-learning for fast adaptation of deep networks. In: PMLR (2017)
5. Finn, C., Rajeswaran, A., Kakade, S., Levine, S.: Online meta-learning. In: ICML (2019)
6. Garg, R., B.G., V.K., Carneiro, G., Reid, I.: Unsupervised CNN for single view depth estimation: geometry to the rescue. In: Leibe, B., Matas, J., Sebe, N., Welling, M. (eds.) ECCV 2016. LNCS, vol. 9912, pp. 740–756. Springer, Cham (2016). https://doi.org/10.1007/978-3-319-46484-8_45
7. Geiger, A., Lenz, P., Stiller, C., Urtasun, R.: Vision meets robotics: the Kitti dataset. Int. J. Rob. Res. **32**(11), 1231–1237 (2013)
8. Geiger, A., Lenz, P., Urtasun, R.: Are we ready for autonomous driving? the Kitti vision benchmark suite. In: CVPR (2012)
9. Geiger, A., Roser, M., Urtasun, R.: Efficient large-scale stereo matching. In: Kimmel, R., Klette, R., Sugimoto, A. (eds.) ACCV 2010. LNCS, vol. 6492, pp. 25–38. Springer, Heidelberg (2011). https://doi.org/10.1007/978-3-642-19315-6_3
10. Godard, C., Mac Aodha, O., Brostow, G.J.: Unsupervised monocular depth estimation with left-right consistency. In: CVPR (2017)
11. Hirschmuller, H.: Stereo processing by semiglobal matching and mutual information. IEEE Trans. Pattern Anal. Mach. Intell. **30**(2), 328–341 (2008)
12. Hoffman, J., Wang, D., Yu, F., Darrell, T.: FCNS in the wild: pixel-level adversarial and constraint-based adaptation. arXiv preprint arXiv:1612.02649 (2016)
13. Hu, X., Mordohai, P.: A quantitative evaluation of confidence measures for stereo vision. IEEE Trans. Pattern Anal. Mach. Intell. **34**(11), 2121–2133 (2012)
14. Kendall, A., Martirosyan, H., Dasgupta, S., Henry, P., Kennedy, R., Bachrach, A., Bry, A.: End-to-end learning of geometry and context for deep stereo regression. In: ICCV (2017)
15. Khamis, S., et al.: StereoNet: guided hierarchical refinement for real-time edge-aware depth prediction. In: Ferrari, V., Hebert, M., Sminchisescu, C., Weiss, Y. (eds.) ECCV 2018. LNCS, vol. 11219, pp. 596–613. Springer, Cham (2018). https://doi.org/10.1007/978-3-030-01267-0_35

16. Kim, S., Kim, S., Min, D., Sohn, K.: LAF-Net: locally adaptive fusion networks for stereo confidence estimation. In: CVPR (2019)
17. Kim, S., Min, D., Kim, S., Sohn, K.: Unified confidence estimation networks for robust stereo matching. IEEE Trans. Image Process. **28**(3), 1299–1313 (2018)
18. Kirillov, A., Wu, Y., He, K., Girshick, R.: PointRend: image segmentation as rendering. In: CVPR (2020)
19. Knowles, M., Peretroukhin, V., Greene, W.N., Roy, N.: Toward robust and efficient online adaptation for deep stereo depth estimation. In: ICRA (2021)
20. Li, D., Hospedales, T.: Online meta-learning for multi-source and semi-supervised domain adaptation. In: Vedaldi, A., Bischof, H., Brox, T., Frahm, J.-M. (eds.) ECCV 2020. LNCS, vol. 12361, pp. 382–403. Springer, Cham (2020). https://doi.org/10.1007/978-3-030-58517-4_23
21. Lipson, L., Teed, Z., Deng, J.: Raft-stereo: multilevel recurrent field transforms for stereo matching. In: 3DV (2021)
22. Liu, M.Y., Tuzel, O.: Coupled generative adversarial networks. In: NIPS (2016)
23. Mayer, N., et al.: A large dataset to train convolutional networks for disparity, optical flow, and scene flow estimation. In: CVPR (2016)
24. Menze, M., Geiger, A.: Object scene flow for autonomous vehicles. In: CVPR (2015)
25. Mnih, V., et al.: Human-level control through deep reinforcement learning. Nature **518**(7540), 529–533 (2015)
26. Pang, J., et al.: Zoom and learn: generalizing deep stereo matching to novel domains. In: CVPR (2018)
27. Patricia, N., Carlucci, F.M., Caputo, B.: Deep depth domain adaptation: a case study. In: ICCV (2017)
28. Poggi, M., et al.: On the confidence of stereo matching in a deep-learning era: a quantitative evaluation. IEEE Trans. Pattern Anal. Mach. Intell. (2021)
29. Poggi, M., Tonioni, A., Tosi, F., Mattoccia, S., Di Stefano, L.: Continual adaptation for deep stereo. IEEE Trans. Pattern Anal. Mach. Intell. 1 (2021)
30. Ros, G., Sellart, L., Materzynska, J., Vazquez, D., Lopez, A.M.: The Synthia dataset: a large collection of synthetic images for semantic segmentation of urban scenes. In: CVPR (2016)
31. Rusu, A.A., et al.: Progressive neural networks. arXiv preprint arXiv:1606.04671 (2016)
32. Shen, Z., Dai, Y., Rao, Z.: CFNet: cascade and fused cost volume for robust stereo matching. In: CVPR (2021)
33. Sun, D., Yang, X., Liu, M.Y., Kautz, J.: PWC-Net: CNNs for optical flow using pyramid, warping, and cost volume. In: CVPR (2018)
34. Tankovich, V., Hane, C., Zhang, Y., Kowdle, A., Fanello, S., Bouaziz, S.: Hitnet: Hierarchical iterative tile refinement network for real-time stereo matching. In: CVPR (2021)
35. Tonioni, A., Poggi, M., Mattoccia, S., Di Stefano, L.: Unsupervised adaptation for deep stereo. In: ICCV (2017)
36. Tonioni, A., Rahnama, O., Joy, T., Stefano, L.D., Ajanthan, T., Torr, P.H.: Learning to adapt for stereo. In: CVPR (2019)
37. Tonioni, A., Tosi, F., Poggi, M., Mattoccia, S., Stefano, L.D.: Real-time self-adaptive deep stereo. In: CVPR (2019)
38. Tosi, F., Poggi, M., Mattoccia, S., Tonioni, A., di Stefano, L.: Learning confidence measures in the wild. In: BMVC (2017)
39. Wang, H., Wang, X., Song, J., Lei, J., Song, M.: Faster self-adaptive deep stereo. In: Ishikawa, H., Liu, C.-L., Pajdla, T., Shi, J. (eds.) ACCV 2020. LNCS, vol.

12622, pp. 175–191. Springer, Cham (2021). https://doi.org/10.1007/978-3-030-69525-5_11
40. Yang, G., Song, X., Huang, C., Deng, Z., Shi, J., Zhou, B.: DrivingStereo: a large-scale dataset for stereo matching in autonomous driving scenarios. In: CVPR (2019)
41. Zbontar, J., LeCun, Y., et al.: Stereo matching by training a convolutional neural network to compare image patches. J. Mach. Learn. Res. **17**(1), 2287–2318 (2016)
42. Zhang, F., Qi, X., Yang, R., Prisacariu, V., Wah, B., Torr, P.: Domain-invariant stereo matching networks. In: Vedaldi, A., Bischof, H., Brox, T., Frahm, J.-M. (eds.) ECCV 2020. LNCS, vol. 12347, pp. 420–439. Springer, Cham (2020). https://doi.org/10.1007/978-3-030-58536-5_25
43. Zhang, K., Lu, J., Lafruit, G.: Cross-based local stereo matching using orthogonal integral images. IEEE Trans. Circuit Syst. Video Technol. **19**(7), 1073–1079 (2009)
44. Zhang, Y., et al.: ActiveStereoNet: end-to-end self-supervised learning for active stereo systems. In: Ferrari, V., Hebert, M., Sminchisescu, C., Weiss, Y. (eds.) ECCV 2018. LNCS, vol. 11212, pp. 802–819. Springer, Cham (2018). https://doi.org/10.1007/978-3-030-01237-3_48
45. Zhang, Z., Lathuilière, S., Pilzer, A., Sebe, N., Ricci, E., Yang, J.: Online adaptation through meta-learning for stereo depth estimation. arXiv preprint arXiv:1904.08462 (2019)
46. Zhou, C., Zhang, H., Shen, X., Jia, J.: Unsupervised learning of stereo matching. In: ICCV (2017)

BRNet: Exploring Comprehensive Features for Monocular Depth Estimation

Wencheng Han[1], Junbo Yin[2], Xiaogang Jin[3], Xiangdong Dai[4], and Jianbing Shen[1](✉)

[1] SKL-IOTSC, Computer and Information Science, University of Macau, Zhuhai, China
shenjianbingcg@email.com
[2] School of Computer Science, Beijing Institute of Technology, Beijing, China
[3] State Key Lab of CAD&CG, Zhejiang University, Hangzhou 310058, China
[4] Guangdong OPPO Mobile Telecommunications Corp., Ltd., Dongguan, China

Abstract. Self-supervised monocular depth estimation has achieved encouraging performance recently. A consensus is that high-resolution inputs often yield better results. However, we find that the performance gap between high and low resolutions in this task mainly lies in the inappropriate feature representation of the widely used U-Net backbone rather than the information difference. In this paper, we address the comprehensive feature representation problem for self-supervised depth estimation by paying attention to both local and global feature representation. Specifically, we first provide an in-depth analysis of the influence of different input resolutions and find out that the receptive fields play a more crucial role than the information disparity between inputs. To this end, we propose a bilateral depth encoder that can fully exploit detailed and global information. It benefits from more broad receptive fields and thus achieves substantial improvements. Furthermore, we propose a residual decoder to facilitate depth regression as well as save computations by focusing on the information difference between different layers. We named our new depth estimation model Bilateral Residual Depth Network (BRNet). Experimental results show that BRNet achieves new state-of-the-art performance on the KITTI benchmark with three types of self-supervision. Codes are available at: https://github.com/wencheng256/BRNet.

1 Introduction

Depth estimation is a fundamental problem in many applications, which aims to estimate the depth for each pixel in a 2D image. Traditional methods formulate this task as a stereo matching problem [29,38], but the exhausted matching process often limits their deployment. In recent years, deep learning based monocular depth estimation [7] has drawn much attention, which predicts depth only

Supplementary Information The online version contains supplementary material available at https://doi.org/10.1007/978-3-031-19839-7_34.

S. Avidan et al. (Eds.): ECCV 2022, LNCS 13698, pp. 586–602, 2022.
https://doi.org/10.1007/978-3-031-19839-7_34

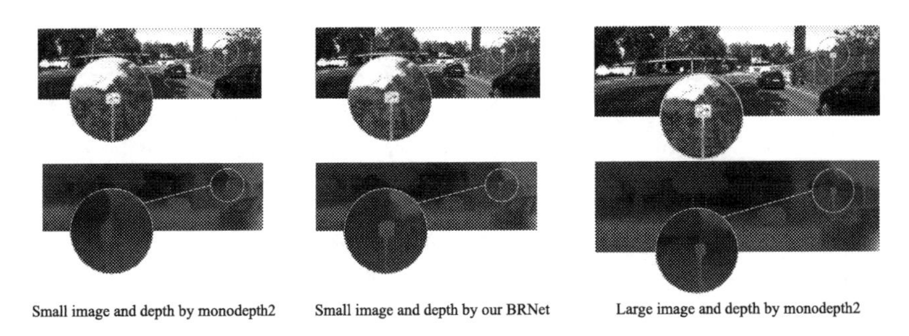

Fig. 1. Predictions from monodepth2 [8] and our BRNet with different input resolutions. The board in the images shows no obvious difference in the two resolutions, but the results of the monodepth2 differ substantially. This motivates us to explore the reason for the different predictions and devise a more suitable model of fully utilizing information in the images for depth estimation.

relying on a single-view input image. However, such a method requires large-scale samples with accurate annotation for fully-supervised training, leading to expensive and elaborate manual work [5,28]. An alternative is to apply self-supervised learning [7] for monocular depth estimation, which requires only video sequences or stereo images to provide supervision signal. This largely decreases the heavy annotation burden and still achieves competitive performance.

Most of the self-supervised learning works, e.g., Monodepth2 [8], HRDepth [17] and PackNet-SfM [9], adopt the U-Net-like [26] architectures. U-Net [26] contains an encoder for extracting hierarchical feature maps and a decoder for predicting depth based on encoded feature maps. It is worth mentioning that almost all these models achieve notably better results when taking higher resolution images as input. Despite this, little work has given deeper attention to how the input resolution influences the performance. Intuitive thought is that the more detailed information in larger images may be the reason. However, as shown in Fig. 1, the boards in the red circles are equally clear in both resolutions, but the model still gives different predictions for them. This observation motivates us to find out the real mechanism behind it.

In this paper, we argue that the performance gap between the high and low-resolution inputs mainly lies in the different receptive fields rather than the information disparity between the inputs. Features extracted from high-resolution images have smaller receptive fields than those extracted from low-resolution ones. The smaller receptive enforces the network to focus more on the detailed information of the images, which is of crucial importance for a depth estimation model. To prove this point, we conduct a heuristic experiment by interpolating the images from a lower resolution to a higher resolution, e.g., the large fake image in Fig. 2, and perform depth estimation based on this fake large image. It turns out that there is little performance decrease in comparison to the input with real large images, suggesting the information disparity is not the main reason for the performance gap. In addition, we find that only smaller receptive fields could not always come with better results, so we should also integrate the feature with large receptive fields, which provides global information such as perspective and object relations [7,20]. To this end, we propose a bilateral encoder. One branch is designed to encode detailed information with small receptive fields, and the other accounts for global information with large receptive fields. Combining these comprehensive and complementary features, our model significantly outperforms existing U-Net-based encoders [26].

Since the input resolution determines the output resolution, we further investigate how much the output depth map sizes of the decoder affect the performance. There are five decoder layers in U-Net, where each layer upsamples the current feature map to a larger one and fuses it with the corresponding feature map from the encoder. We evaluate the output depth from each decoder layer and surprisingly find that there is no obvious performance gap between them. This is contrary to the intuition that larger outputs should encode more details and be preciser than smaller ones. We infer that the incremental information can not be fully exploited as the decoder layers go deeper. To address this, we propose to add the depth obtained from the previous layers to the input of the current layer and enforce the network to predict residual information between different depth sizes. We call the proposed decoder as residual decoder, which is inspired by the residual operation in ResNet. Our residual decoder effectively focuses on the difference between features maps from different layers and thus facilitates depth regression.

Our main contributions can be summarized as follows:

- We provide a deeper insight into the performance gaps between different input resolutions by thorough and exhaustive heuristic experiments. We find that appropriate receptive fields are critical to the performance of the depth estimation model.
- We propose a new Bilateral Depth Encoder that can fully exploit details and global information simultaneously, providing a more comprehensive feature representation for depth estimation.
- A Residual Decoder is introduced for dense depth regression. It makes full use of the output depth from each decoder layer, focusing on the information difference between features maps of different scales.
- Our depth estimator achieves a new state-of-the-art performance on the KITTI benchmark. The approach remarkably outperforms other approaches with all the types of self-supervision, e.g., achieving 0.097 and 4.378 in terms of Abs Rel and RMSE, respectively, with MS training.

2 Related Work

Current monocular depth estimation approaches can be roughly categorized into two groups. One group applies the fully-supervised learning to regress the ground truth depth maps generated from the LiDAR sensor. Another group provides supervision by leveraging synchronized monocular videos or stereo pairs in order to optimize the depth estimation model in a self-supervised fashion.

2.1 Fully-Supervised Monocular Depth Estimation

The first method for monocular depth estimation based on deep learning is developed by Eigen *et al.* [5], which directly regresses the depth by two components. One is for estimating the global structure of the scene, and the other is used to refine it using local information. The two components are trained separately, and this increases the training expenses. Later this architecture is replaced by a fully convolutional network [14] designed for semantic segmentation by Evan *et al.* [30]. This work enables the depth estimation task to be trained end-to-end with a deeper network, and it performs comparably with the depth sensor. And then, many methods are exploited to improve the performance [2,25,31].

Besides point clouds, some networks also try to exploit extra labels to improve the performance. Klodt *et al.* [12] employ sparse depths and poses from the traditional SLAM system, and Ramirez *et al.* [24] demonstrate that jointly predicting depth and semantic labels can improve the performance of depth estimation. Tosi *et al.* [32] exploit proxy ground truth labels, which are generated by a traditional stereo matching method.

Although these works achieve notable success, they rely on ground truth labels that are expensive to obtain, which limits the training data's scale. In consequence, many works focus on self-supervised methods.

2.2 Self-Supervised Monocular Depth Estimation

For self-supervised approaches in monocular depth estimation, there are mainly two categories. One is based on stereo pairs, and the other uses consecutive video frames during the training phase.

Stereo Training. Garg *et al.* [6] first propose the self-supervised training approach for monocular depth estimation with a photometric-consistency loss between stereo pairs. Specifically, they take the images from one of the views and construct pseudo images for the other view based on the depth predicted and the relative position between the two views. Then a L_2 loss is employed to measure the difference between generated images and the real ones. Godard *et al.* [7] improve the method by replacing the L_2 loss with a L_1 loss and introducing the structural similarity index measure to generate sharper results. Godard *et al.* [8] show that computing projection loss at a higher resolution will improve the depth map quality, and Pillai *et al.* [21] introduce differentiable flip augmentation and subpixel convolutions for increasing the fidelity of the depth map. Watson *et al.* [35] introduce Depth Hints to alleviate the effects of ambiguous reprojections in depth-prediction.

Although some works have achieved satisfying performance [1,13,16,22,39], stereo image pairs require specialized equipment. Therefore, some other works like [10,18,34,36,37] turn to investigate methods based on video sequences.

Video Training. Zhou *et al.* [41] design an additional network to predict the camera pose between adjacent frames and construct the pseudo frame based on the previous and subsequent frames, respectively. To deal with non-rigid scene motion, they employ a motion explanation mask so that it allows the network to ignore specific regions. Godard *et al.* [8] then propose a new strategy to replace this mask and achieve better performance. To fully exploit the view difference between frames, they did not average the loss from the future and past frames but took the minimum value of the loss instead. They also propose a simple auto-masking method to filter out pixels that do not change appearance in a sequence from one frame to the next. Lyu *et al.* [17] prove that predicting more accurate boundaries can improve performance and redesign the skip connection generating high-resolution feature maps to get sharper edges. Zhou *et al.* [42] proposed a lightweight architecture that performs better in a more efficient way. Johnston *et al.* [11] and Bakhanova *et al.* [19] introduce attention mechanism into depth estimation area to improve the performance of models.

These methods show better performances when taking large inputs in their experiments, but little work investigates the reason.

3 Methods

This section first analyses the reason behind the performance gap between the different input resolutions and reveals several limitations of the prevalent depth estimation backbone U-Net [26]. Then, based on our observations, we develop a new depth estimation network, BRNet, which effectively mines the detail and global information of the input image by the Bilateral Encoder and fully exploits the incremental data of different layers by the Residual Decoder. Notably, even taking small inputs, our model significantly outperforms the baseline method taking large inputs.

3.1 Analysis of Current Depth Estimator

As shown in previous works, models with larger inputs always outperform those with small inputs, which is a consensus of the community. An intuitive reason is that larger images keep more real-world information while small ones can not. However, as shown in the Fig. 1, although some objects can be seen clearly in both resolutions, the model still gives different results. Stemming from these observations, we believe that information difference between inputs may not be the main reason leading to the performance gap.

To explore this, we conduct an experiment as shown in Fig. 2. We first downsample a large input image into a small resolution and then interpolate it back to the original size. The resultant image, called a fake-large image, has the same information as the small image except for a larger size. Afterward, we apply monodepth2 [8] as the baseline model to train and evaluate on fake-large, small and large images, respectively. If the performance gap comes from the information difference, the model's performance with fake-large inputs will be the same as that of small inputs. Otherwise, the result will be similar to the one by taking large inputs. As shown in the Table 1, the model's performance with fake-large inputs is comparable with that of large inputs and is better than the model of

Table 1. Experiments about different input resolutions and different architecture of encoder and decoder. We adopt Abs Rel, RMSE and δ as our metrics. For Abs Rel and RMSE, lower values are better, and higher values are better for $\delta > 1.25$.

Experiments	Metrics		
	Abs Rel ↓	RMSE ↓	$\delta < 1.25$ ↑
Input resolution			
Small	0.115	4.863	0.877
Large (1024×320)	0.115	4.701	0.879
Fake large (1024×320)	0.115	4.708	0.880
Encoder			
1/2 Receptive field	0.111	4.672	0.880
1/4 Receptive field	0.116	4.761	0.875
1/8 Receptive field	0.129	5.098	0.844
Decoder			
Extra layer	0.118	4.826	0.869
Disp0 (full size)	0.115	4.863	0.877
Disp1 (1/2 size)	0.114	4.858	0.877
Disp2 (1/4 size)	0.114	4.834	0.875
Disp3 (1/8 size)	0.116	4.869	0.868

small inputs. This experiment proves that most of the performance gap between large and small inputs in the depth-estimation model does not come from the information difference between them.

As input information is not the reason, it may be in the process of feature extraction in the encoder or result generation in the decoder. So we design a series of experiments to find it.

Encoder. We find that the bad cases from small inputs are mainly objects that are far from the camera. The prediction for these objects heavily depends on the detailed occlusion relationship with other elements. However, the radical down-sampling strategy of the encoder provides a large receptive field, making it hard to focus on these details and thus neglecting this information. Larger inputs counterbalance the down-sampling and provide relatively smaller receptive fields, which can extract more information, leading to the performance gap between different resolutions. To validate this, we adjust the stride of the convolutions in the encoder and feed small input into it so that the produced feature maps have $\frac{1}{2}$ receptive fields of the origin network and are similar to those of large input. As shown in Table 1, it generates even better results than that of large inputs. The results support our theory that the receptive fields of the model may be the main reason of the performance gap.

On the other hand, as mentioned in Miangole *et al.* [20], higher resolution inputs do not always come with better results. As the resolution gets higher, the network starts losing the overall structure of the scene. As a result, the relative depth between objects is better predicted, but their absolute depth in the whole image is less precise than that of smaller inputs. To verify this conclusion in monocular depth estimation, we prepare another experiment by adjusting more convolutions in the decoder and trying smaller receptive fields. As shown in the Table 1, when reducing the receptive fields into $\frac{1}{4}$ and $\frac{1}{8}$ of the original ones, the performance of this model will degenerate gradually.

In summary, relatively smaller receptive fields can help provide more detailed information, contributing to the better performance of the depth estimator. Larger inputs also lead to smaller receptive fields, which may be the main reason behind the influence of the input resolution. Before we make a conclusion, we still need to clarify the role of the decoder.

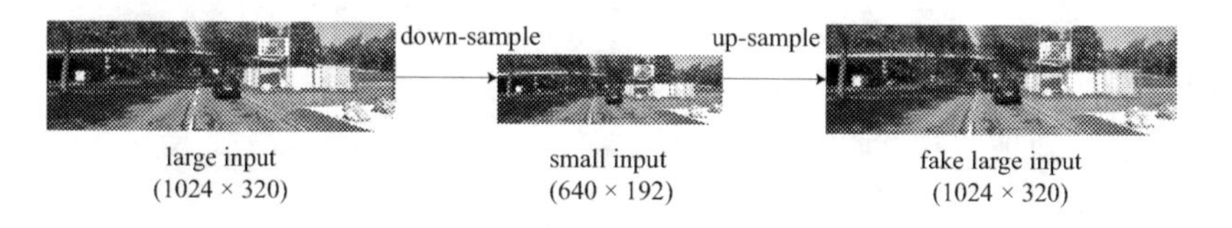

Fig. 2. Experiments in our analysis. An example of our larger inputs, small inputs and fake-large inputs.

Decoder. Larger inputs lead to smaller receptive fields, but they also induce larger depth maps. Here is another question, how much does the result size affect the performance. To figure out this question, we prepare another experiment. In this experiment, we feed the network large inputs but down-sample the first feature map into small resolution and send it to the rest of the encoder. The feature maps of the encoder keep the same receptive fields as it takes small inputs. As for the decoder, we insert an extra decoder layer, which upsamples the decoder feature map and fuses it with the large feature map, and we finally obtain the results with the same size as larger inputs as shown in the Fig. 2. As a result, it has similar results with smaller inputs which confirms that the larger output size can obviously improve the performance.

Since a larger result resolution does not affect the performance markedly, does a smaller result affect the performance a lot? To form multi-scale disparity loss, monodepth2 [8] modifies the decoder to generate four results with different sizes. So we evaluate the results produced by different layers of the decoder, which has a full, $\frac{1}{2}$, $\frac{1}{4}$, and $\frac{1}{8}$ resolution, respectively. As shown in the Table 1, they all perform comparably. The result with the full size is even worse than that with $\frac{1}{4}$ size. This observation is counterintuitive because larger depth maps are supposed to keep more details, so they should be preciser than small results. The experiment shows that the decoder does not perform as it is designed. We attribute this problem to that the U-Net decoder designed for semantic segmentation is not well suited to depth estimation and will discuss it in Sect. 3.3.

Summaries. The analysis above leads us to the following conclusion: The receptive field's difference contributes most to the performance gap between different input resolutions rather than the information disparity. Therefore, adjusting the convolutions' stride in the encoder providing more detailed information can lead to better performance. Besides, global information is also indispensable. To achieve better performance, we must fully exploit both detailed and global features. Finally, we find that U-Net decoder can not achieve this as it is designed in the depth estimation, so we need a new decoder that is more suitable for this task.

3.2 Bilateral Depth Encoder

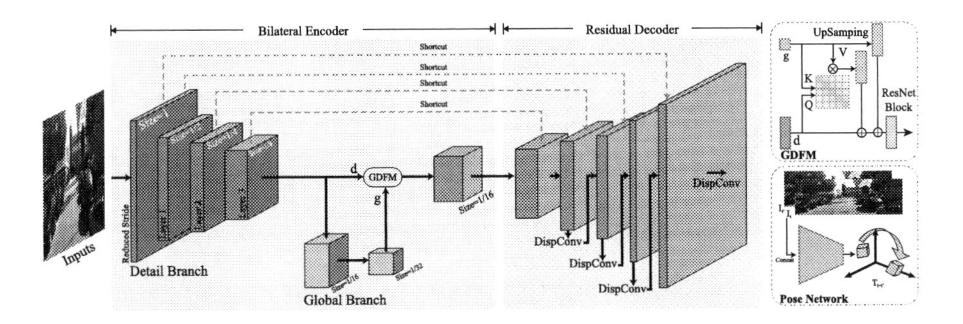

Fig. 3. Overview of the proposed network which mainly contains two parts. One is the bilateral encoder the other is the residual decoder. There is a *detail branch* for extracting detail features for the encoder, a *global branch* to extract high-level information, and finally, a global-detail fuse module (GDFM) that fuses the two feature maps. The pose network is employed to give the relative pose between two frames and is only used during the training phase.

According to the conclusions drawn above, a better depth encoder should properly deal with global and detailed information. However, there is a paradox. Detailed information needs the encoder to have relatively smaller receptive fields to focus on occlusion relationships. On the other hand, only larger receptive fields can provide an objective overview. To meet their needs simultaneously, we design a two-branch architecture to extract both global and detailed features and then fuse them into more comprehensive ones.

As shown in the Fig. 3, we reduce the stride of the first convolution so that the whole network downsamples the input more slowly, concentrating more on the details. After several blocks, information such as occlusion relationships will be well encoded. Still, as the receptive fields are reduced, they cannot get enough global information. Thus it will lead to inconsistent predictions, as discussed above. To alleviate this problem, we develop a new branch, i.e., *Global Branch*, after Layer-3. For convenience, we call the original one as *Detail Branch.*

The global branch further down-samples the feature maps to get larger receptive fields while the detail-branch keeps high resolution of the feature maps maintaining detail features. Then the two branches are joined together by GDFM (Global-Detail Fusion Module) to generate comprehensive feature maps, as shown in Fig. 3. Some global information is critical to every location in the image, such as the perspective point and view point. To share this information to every place, we employ a multi-head attention [33] to fuse global and detailed information. Specifically, the detailed features are mapped into query features, and the global features are mapped into key and value features. Then the query and key features are multiplied to generate an affinity map, and the affinity map is used to aggregate information from value features. In this way, every pixel in the detail feature map can aggregate corresponding global features as they need, even though they are far away from current location. Finally, we add this feature map into a detailed feature map to fuse them. Also, we add a residual shortcut of the global features by upsampling and merging it into the final feature map to facilitate optimization.

Bilateral Depth Encoder can fully exploit hidden information from input images by taking advantage of global and detailed features. With these comprehensive feature maps, our model is able to achieve significant improvement.

3.3 Residual Decoder

Intuitively, larger outputs contain more details and are supposed to perform better than smaller ones. On the contrary, as discussed in Sect. 3.1, the results of different layers in the current depth decoder perform closely, which means that the decoder does not ideally perform as designed. Decoders in most depth estimators are adopted from U-Net [26], which is intended for semantic segmentation. However, semantic segmentation is a dense classification task, while depth estimation is a dense regression task so they are different.

In the classification task, as types are discrete, every pixel needs to be judged independently to generate sharp borders. But in depth estimation, depth values are changed smoothly, and every pixel in the depth map is close to its' neighbors. Thus, we can generate depth for large regions and fine-tune the depth for every pixel in a region.

In this work, we regard the outputs of higher layers as residuals:

$$Disp(l) = UpSample(Disp(l-1)) + R(l). \tag{1}$$

If there is no extra information for some regions in the view, their residuals $R(l)$ are close to 0. On the contrary, if there is valuable information, the residuals are used to adjust the predictions made by the former layer. As shown in the Fig. 4, different from the original decoder layer, which upsamples the feature maps from the previous layer, our residual layer upsamples the one-channel depth map to a larger size. Then the depth map is modulated by a convolution, and then we concatenate it with the shortcut feature map extracting the residual using another convolution as:

$$R(l) = f(Disp(l-1), F(l); \theta), \quad (2)$$

where θ is the parameters and $F(l)$ is the shortcut features of layer l. Finally, the residual is added to the previous depth map. Compared to independent decoder layers, residuals can focus on the differences between layers, which helps the current layer to generate better predictions than their predecessors. Benefited from the multi-scale loss [8], every layer in the decoder can be fully optimized during the training phase. Finally, with this residual decoder, our model yields significant improvements compared to the U-Net decoder.

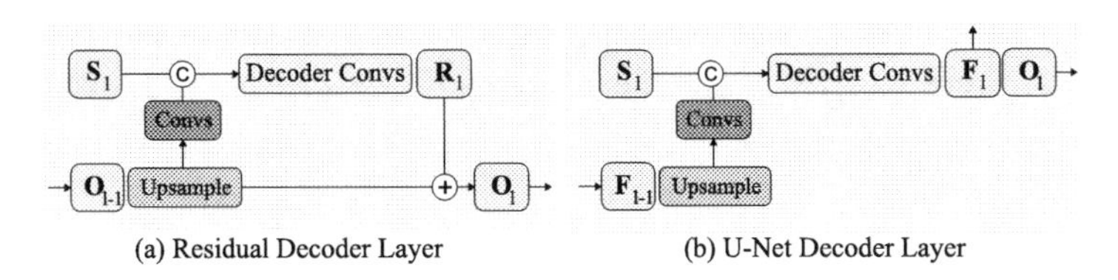

Fig. 4. Comparison between Residual Decoder layer and U-Net Decoder layer. In a U-Net decoder layer, the feature map from the previous layer F_{l-1} is upsampled and concatenated with the shortcut feature map S_l. Then the feature map is used to generate fused featuremaps F_l and a depth map O_l. In a residual decoder layer, the depth map from the previous layer (O_{l-1}) is upsampled and concatenated with S_l. And the output of this layer is regarded as residual (R_l) and should be added to the previous depth map generating O_l.

3.4 Loss Function

To optimize the proposed model, we formulate our problem as minimising the photometric reprojection error during the training phase. As described in monodepth2 [8], at each pixel, minimization is used to merge results from different sources:

$$L_p = \min_{t'} pe\left(I_t, I_{t' \to t}\right), \quad (3)$$

where I_t is the target image and $I_{t' \to t}$ is the pseudo target image which is generated by:

$$I_{t' \to t} = I_{t'} \left\langle \operatorname{proj}\left(D_t, T_{t \to t'}, K\right)\right\rangle \quad (4)$$

where $I_{t'}$ is the source image and D_t is the depth map generated by our model. K is the pre-computed intrinsics and $T_{t \to t'}$ is the camera pose between $I_{t'}$ and I_t. When training with video sequences, $T_{t \to t'}$ is the result of the pose model or when training with stereo image pairs, it is the pose between two cameras. $pe(I_a, I_b)$ is our photometric error function, which is designed as:

$$\frac{\alpha}{2}(1 - \text{SSIM}(I_a, I_b)) + (1 - \alpha)\|I_a - I_b\|_1. \tag{5}$$

In our experiments, α is set to 0.85. Also, the edge-aware smoothness [7] is employed to encourage the disparities to be locally smooth:

$$L_s = |\partial_x d_t^*| e^{-|\partial_x I_t|} + |\partial_y d_t^*| e^{-|\partial_y I_t|}. \tag{6}$$

Then the final loss is:

$$L = \mu L_p + \lambda L_s, \tag{7}$$

where μ and λ are the auto-mask introduced by [8]. When training with stereo image pairs, $I_{t'}$ is the image from the other view. For video sequences, $I_{t'} \in \{I_{t-1}, I_{t+1}\}$ where I_{t-1} is the previous frame and I_{t+1} is the subsequent frame.

4 Experiments

We implement our network using Pytorch, and all the training and evaluation are performed on a workstation with an Intel E5-2698 v4CPU, 512G memory, and a single V100 GPU.

4.1 KITTI

The KITTI benchmark is the most widely used dataset in depth estimation, consisting of calibrated videos registered to LiDAR measurements of city scenarios. The depth evaluation is done on the LiDAR point cloud. We adopt the data split like [4], followed by pre-processing as [41] to remove static frames. Finally, 39,810 triplets are used for training and 4,424 for validation. We use the same intrinsics for all images and set the camera's principal point to the image center and the focal length to the average of all the focal lengths in KITTI.

4.2 State-of-the-Art Comparison

Followed by [8], we compare the results of several variants of the proposed model, which is trained on different types of self-supervision. M, S and MS mean monocular video, stereo image pairs only and both, respectively. For the metrics, following [4] we adopt Abs Rel, Sq Rel, RMSE, RMSE log for which lower is better and $\delta < 1.25$, $\delta < 1.25^2$, $\delta < 1.25^3$ for which higher is better.

We compare our model to several state-of-the-art methods. As shown in Table 2, our approach outperforms all existing self-supervised networks with all the three types of self-supervision and two input resolutions. Compared to our baseline model monodepth2 [8], our model with monocular video training and small inputs achieves 0.010, 0.401 and 0.013 improvement in terms of AbsRel, RMSE and $\delta < 1.25$, respectively. As discussed in Sect. 3.1, the original U-Net

Table 2. Comparison with other SOTA networks on KITTI Eigen split test set. The train column refers to the training data of the models. M means monocular videos only and S means stereo image pairs, and MS means both. The best two results are shown in bold and underlined, respectively.

Method	Resolution	Trian	Lower is better				Higher is better		
			Abs Rel	Sq Rel	RMSE	RMSE log	$\delta < 1.25$	$\delta < 1.25^2$	$\delta < 1.25^3$
Low resolution									
EPC++ [15]	640×192	M	0.141	1.029	5.350	0.216	0.816	0.941	0.976
Struct2depth [3]	416×128	M	0.141	1.026	5.291	0.215	0.816	0.945	0.979
Monodepth2 [8]	640×192	M	0.115	0.903	4.863	0.193	0.877	0.959	0.981
Monodepth2 R50 [8]	640×192	M	0.110	0.831	4.642	0.187	0.883	0.962	0.982
Johnston [11]	640×192	M	0.106	0.861	4.699	0.185	0.889	0.962	0.982
PackNet-SfM [9]	640×192	M	0.111	0.785	4.601	0.189	0.878	0.960	0.982
HR-Depth [17]	640×192	M	0.109	0.792	4.632	0.185	0.884	0.962	0.983
R-MSFM6 [42]	640×192	M	0.112	0.806	4.704	0.191	0.878	0.960	0.981
BRNet (ours)	640×192	M	**0.105**	**0.698**	**4.462**	**0.179**	**0.890**	**0.965**	**0.984**
Monodepth R50 [7]	512×256	S	0.133	1.142	5.533	0.230	0.83	0.936	0.970
3Net (R50) [23]	512×256	S	0.129	0.996	5.281	0.223	0.831	0.939	0.974
3Net (VGG) [23]	512×256	S	0.119	1.201	5.888	0.208	0.844	0.941	0.978
Monodepth2 [8]	640×192	S	0.109	0.873	4.960	0.209	0.864	0.948	0.975
BRNet (Ours)	640×192	S	**0.103**	**0.792**	**4.716**	**0.197**	**0.876**	**0.954**	**0.978**
EPC++ [15]	640×192	MS	0.128	0.935	5.011	0.209	0.831	0.945	0.979
Monodepth2 [8]	640×192	MS	0.106	0.818	4.750	0.196	0.874	0.957	0.979
HR-Depth [17]	640×192	MS	0.107	0.785	4.612	0.185	**0.887**	**0.962**	0.982
R-MSFM6 [42]	640×192	MS	0.111	0.787	4.625	0.189	0.882	0.961	0.981
BRNet (ours)	640×192	MS	**0.099**	**0.685**	**4.453**	**0.183**	0.885	**0.962**	**0.983**
High resolution									
Monodepth2 [8]	1024×320	M	0.115	0.882	4.701	0.190	0.879	0.961	0.982
Zhao *et al.* [40]	832×256	M	0.113	0.704	4.581	0.184	0.871	0.961	0.984
PackNet-SfM [9]	1280×384	M	0.107	0.802	4.538	0.186	0.889	0.962	0.981
HR-Depth [17]	1024×320	M	0.106	0.755	4.472	0.181	**0.892**	**0.966**	0.984
R-MSFM6 [42]	1024×320	M	0.108	0.748	4.470	0.185	0.889	0.963	0.982
BRNet (ours)	1024×320	M	**0.103**	**0.684**	**4.385**	**0.175**	0.889	0.965	**0.985**
SuperDepth + pp [21]	1024×382	S	0.112	0.875	4.958	0.207	0.852	0.947	0.977
Monodepth2 [8]	1024×320	S	0.107	0.849	4.764	0.201	0.874	0.953	0.977
BRNet (ours)	1024×320	S	**0.097**	**0.729**	**4.510**	**0.191**	**0.886**	**0.958**	**0.979**
Monodepth2 [8]	1024×320	MS	0.106	0.806	4.630	0.193	0.876	0.958	0.980
HR-Depth [17]	1024×320	MS	0.101	0.716	4.395	**0.179**	**0.899**	0.966	0.983
R-MSFM6 [42]	1024×320	MS	0.108	0.753	4.469	0.185	0.888	0.963	0.982
BRNet (ours)	1024×320	MS	**0.097**	**0.677**	**4.378**	**0.179**	0.888	0.965	**0.984**

Table 3. Make3D results with monocular training and 640×192 inputs.

Architecture	Abs Rel $\downarrow$	Sq Rel $\downarrow$	RMSE$\downarrow$	log_{10} $\downarrow$
Monodepth	0.544	10.94	11.760	0.193
Monodepth2	0.322	3.589	7.414	0.163
BRNet	**0.302**	**3.133**	**7.068**	**0.156**

cannot fully exploit the detailed information in small images. Therefore, they cannot fully use the more detailed information carried by the large information either. Different from it, BRNet can extract abundant details from the inputs. When taking larger images, BRNet can extract more information than smaller ones and achieve better results. As a result, our model significantly outperforms our baseline when taking large inputs. We achieve 0.252 improvements on RMSE with 1024×320 inputs and MS training. To show the generality of our model, we follow the setting of [7] and evaluate BRNet on Make3D [27] dataset. According to the results shown in Table.3, BRNet significantly outperform our baseline with monocular training and 640×192 input.

We also compare the qualitative performance with other representative networks. As shown in Fig. 5, our model performs better than other networks, especially for objects far from the camera (marked by the red circles).

4.3 Ablation Study

We perform ablation studies to understand how the proposed components contribute to the overall performance improvements. All the experiments are evaluated on the KITTI Eigen split [4]. There are three components need evaluation:

Detail Branch (DB). The detail branch is responsible for extracting comprehensive features from inputs. As shown in the Table 4, the model taking small inputs with the detail branch even performs better with the baseline accepting large inputs gaining 0.004 improvement in terms of AbsRel and 0.192 in RMSE.

Global Branch (GB). The global branch is incorporated to extract perspective and other global information. In default, GDFM is a part of the global branch

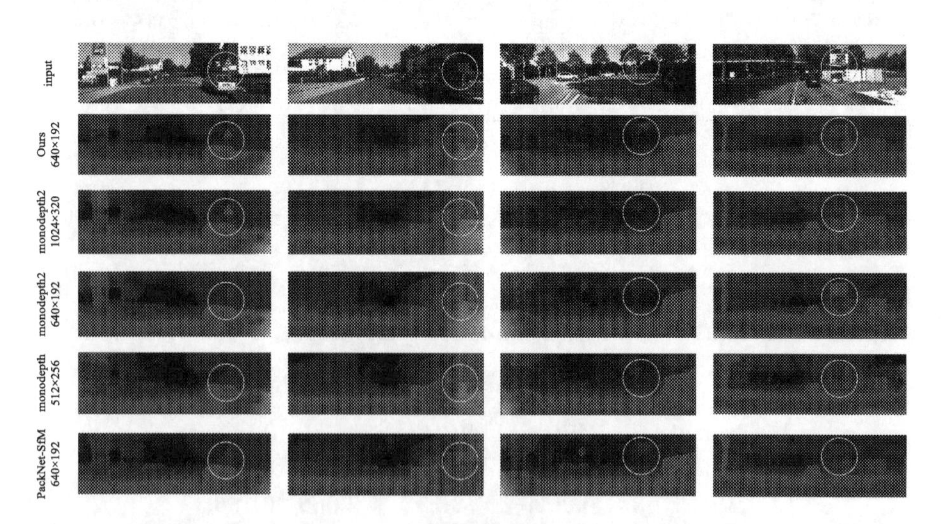

Fig. 5. Qualitative results on the KITTI Eigen split test set. Our model can fully exploit global and detailed information and outperforms previous SOTA networks. (Color figure online)

Table 4. Ablation study of the proposed components We compare the improvements from DB (Detail Branch), GB (Global Branch) RD (Residual Decoder) and Global Branch without GDFM, and the model with all the three components and GDFM achieves our best performance.

Components				Abs Rel ↓	RMSE ↓	$\delta < 1.25$ ↑
DB	GB	GB w/o GDFM	RD			
				0.115	4.863	0.877
✓				0.111	4.672	0.880
✓		✓		0.109	4.579	0.889
✓	✓			0.108	4.538	0.890
✓	✓		✓	0.105	4.462	0.890
			✓	0.113	4.780	0.877

and is used to fuse the features from the two branches. To show the efficiency of GDFM, we also show a result without GDFM by replacing GDFM with an element-wise summation. As shown in Table 4, with global branch, the model achieves 0.108 in AbsRel while GDFM provide an obvious improvement.

Residual Decoder (RD). We demonstrate the performance improvement from our decoder. As shown in Table 4, the residual decoder outperforms the original decoder by 0.076 in terms of RMSE. As shown in the table, RD alone can improve the AbsRel by 0.002 and about 0.1 in RMSE.

Efficiency. To extract abundant details from the inputs, BRNet introduces more computation than our baseline. To verify the complexity of our model, we compare its efficiency with two representative works in Table 5. As shown in it, BRNet is only slightly slower than its baseline, running at 14 ms (71 fps), a speed far beyond real-time requirement. Compared with PackNet-SfM, BRNet largely outperform it with 15% computation, which can prove the improvement of our model does not come from the increasing computation.

Distance Comparison. As discussed above, our method mainly improves the performance on objects far from the camera. To verify this, we split the validation dataset into three groups according to the depth of each pixel and the illustration and proportion of each group are shown in Fig. 6. We then com-

Table 5. Comparison of speed, computation. All the speeds are reported on the same device with a Tesla V100 GPU.

Architecture	Abs Rel ↓	FLOPs (B)	Params (M)	Speed (ms)
Monodepth2	0.115	8	14	10
BRNet	0.105	31	19	14
PackNet-SfM	0.111	205	128	154

Table 6. Far objects comparison on different parts over the whole dataset.

Method	Abs Rel		
	Near	Middle	Far
BRNet	0.059	0.073	0.147
Monodepth2	0.061	0.082	0.159

(a) Near 0-8m (25%) (b) Middle 8-14m (35%) (c) Far 14-∞m (40%)

Fig. 6. Illustration and the proportion of different depth groups. We split the prediction and ground truth into three groups, Near, Middle and Far, according to the depth of each pixel.

pare our model and the baseline model on the groups. In Table 6, two methods perform comparably on the Near part but BRNet outperforms Monodepth2 [8] significantly on the Middle and Far parts, which indicates that for more than 75% pixels in the dataset, our model can improve the performance from our baseline.

5 Conclusion

In this paper, we first make a deep study on the mechanism behind the influence of input resolutions and find out that the receptive field affects the performance rather than information difference between inputs. Indeed, depth estimators need a wide range of receptive fields. Thus they can extract detailed information, such as occlusion and global information like perspective. Based on these findings, we designed a bilateral depth encoder that simultaneously extracts details and global features and fuses them. Finally, we propose a residual decoder focusing on the difference between layers in the decoder, which is able to generate better predictions.

Acknowledgements. This work was supported in part by the National Natural Science Foundation of China (Grant Nos. 62036010, 61972344), and the Start-up Research Grant (SRG) of University of Macau.

References

1. Aleotti, F., Tosi, F., Poggi, M., Mattoccia, S.: Generative adversarial networks for unsupervised monocular depth prediction. In: Leal-Taixé, L., Roth, S. (eds.) ECCV 2018. LNCS, vol. 11129, pp. 337–354. Springer, Cham (2019). https://doi.org/10.1007/978-3-030-11009-3_20
2. Bhat, S.F., Alhashim, I., Wonka, P.: AdaBins: depth estimation using adaptive bins. In: CVPR (2021)

3. Casser, V., Pirk, S., Mahjourian, R., Angelova, A.: Depth prediction without the sensors: leveraging structure for unsupervised learning from monocular videos. In: AAAI (2019)
4. Eigen, D., Fergus, R.: Predicting depth, surface normals and semantic labels with a common multi-scale convolutional architecture. In: ICCV (2015)
5. Eigen, D., Puhrsch, C., Fergus, R.: Depth map prediction from a single image using a multi-scale deep network. arXiv preprint arXiv:1406.2283 (2014)
6. Garg, R., B.G., V.K., Carneiro, G., Reid, I.: Unsupervised CNN for single view depth estimation: geometry to the rescue. In: Leibe, B., Matas, J., Sebe, N., Welling, M. (eds.) ECCV 2016. LNCS, vol. 9912, pp. 740–756. Springer, Cham (2016). https://doi.org/10.1007/978-3-319-46484-8_45
7. Godard, C., Aodha, O.M., Brostow, G.J.: Unsupervised monocular depth estimation with left-right consistency. In: CVPR (2017)
8. Godard, C., Aodha, O.M., Firman, M., Brostow, G.J.: Digging into self-supervised monocular depth estimation. In: ICCV (2019)
9. Guizilini, V., Ambrus, R., Pillai, S., Raventos, A., Gaidon, A.: 3D packing for self-supervised monocular depth estimation. In: CVPR (2020)
10. Janai, J., Güney, F., Ranjan, A., Black, M., Geiger, A.: Unsupervised learning of multi-frame optical flow with occlusions. In: Ferrari, V., Hebert, M., Sminchisescu, C., Weiss, Y. (eds.) ECCV 2018. LNCS, vol. 11220, pp. 713–731. Springer, Cham (2018). https://doi.org/10.1007/978-3-030-01270-0_42
11. Johnston, A., Carneiro, G.: Self-supervised monocular trained depth estimation using self-attention and discrete disparity volume. In: CVPR (2020)
12. Klodt, M., Vedaldi, A.: Supervising the new with the old: learning SFM from SFM. In: Ferrari, V., Hebert, M., Sminchisescu, C., Weiss, Y. (eds.) ECCV 2018. LNCS, vol. 11214, pp. 713–728. Springer, Cham (2018). https://doi.org/10.1007/978-3-030-01249-6_43
13. Kuznietsov, Y., Stuckler, J., Leibe, B.: Semi-supervised deep learning for monocular depth map prediction. In: CVPR (2017)
14. Long, J., Shelhamer, E., Darrell, T.: Fully convolutional networks for semantic segmentation. In: CVPR (2015)
15. Luo, C., et al.: Every pixel counts++: joint learning of geometry and motion with 3D holistic understanding. PAMI **42**(10), 2624–2641 (2019)
16. Luo, Y., et al.: Single view stereo matching. In: CVPR (2018)
17. Lyu, X., et al.: HR-depth: high resolution self-supervised monocular depth estimation. CoRR abs/2012.07356 (2020)
18. Mahjourian, R., Wicke, M., Angelova, A.: Unsupervised learning of depth and ego-motion from monocular video using 3d geometric constraints. In: CVPR (2018)
19. Makarov, I., Bakhanova, M., Nikolenko, S., Gerasimova, O.: Self-supervised recurrent depth estimation with attention mechanisms. PeerJ Comput. Sci. **8**, e865 (2022)
20. Mahdi, S., Miangoleh, H., Dille, S., Mai, L., Paris, S., Aksoy, Y.: Boosting monocular depth estimation models to high-resolution via content-adaptive multi-resolution merging. In: CVPR (2021)
21. Pillai, S., Ambruş, R., Gaidon, A.: SuperDepth: self-supervised, super-resolved monocular depth estimation. In: ICRA (2019)
22. Poggi, M., Aleotti, F., Tosi, F., Mattoccia, S.: Towards real-time unsupervised monocular depth estimation on CPU. In: IROS (2018)
23. Poggi, M., Tosi, F., Mattoccia, S.: Learning monocular depth estimation with unsupervised trinocular assumptions. In: 3DV (2018)

24. Zama Ramirez, P., Poggi, M., Tosi, F., Mattoccia, S., Di Stefano, L.: Geometry meets semantics for semi-supervised monocular depth estimation. In: Jawahar, C.V., Li, H., Mori, G., Schindler, K. (eds.) ACCV 2018. LNCS, vol. 11363, pp. 298–313. Springer, Cham (2019). https://doi.org/10.1007/978-3-030-20893-6_19
25. Ranftl, R., Bochkovskiy, A., Koltun, V.: Vision transformers for dense prediction. In: CVPR (2021)
26. Ronneberger, O., Fischer, P., Brox, T.: U-net: convolutional networks for biomedical image segmentation. In: International Conference on Medical Image Computing and Computer-Assisted Intervention (2015)
27. Saxena, A., Sun, M., Ng, A.Y.: Make3D: depth perception from a single still image. In: AAAI (2008)
28. Saxena, A., Sun, M., Ng, A.Y.: Make3D: learning 3D scene structure from a single still image. PAMI **31**(5), 824–840 (2008)
29. Scharstein, D., Szeliski, R.: High-accuracy stereo depth maps using structured light. In: CVPR (2003)
30. Shelhamer, E., Barron, J.T., Darrell, T.: Scene intrinsics and depth from a single image. In: ICCV Workshops (2015)
31. Song, M., Lim, S., Kim, W.: Monocular depth estimation using Laplacian pyramid-based depth residuals. IEEE Trans. Circuits Syst. Video Technol. (2021)
32. Tosi, F., Aleotti, F., Poggi, M., Mattoccia, S.: Learning monocular depth estimation infusing traditional stereo knowledge. In: CVPR (2019)
33. Vaswani, A., et al.: Attention is all you need. In: Advances in Neural Information Processing Systems, pp. 5998–6008 (2017)
34. Vijayanarasimhan, S., Ricco, S., Schmid, C., Sukthankar, R., Fragkiadaki, K., SFM-Net: learning of structure and motion from video. arXiv preprint arXiv:1704.07804 (2017)
35. Watson, J., Firman, M., Brostow, G.J., Turmukhambetov, D.: Self-supervised monocular depth hints. In: ICCV (2019)
36. Yang, Z., Wang, P., Wang, Y., Xu, W., Nevatia, R.: Lego: learning edge with geometry all at once by watching videos. In: CVPR (2018)
37. Yang, Z., Wang, P., Xu, W., Zhao, L., Nevatia, R.: Unsupervised learning of geometry with edge-aware depth-normal consistency. arXiv preprint arXiv:1711.03665 (2017)
38. Zbontar, J., LeCun, Y., et al.: Stereo matching by training a convolutional neural network to compare image patches. J. Mach. Learn. Res. **17**(1), 2287–2318 (2016)
39. Zhan, H., Garg, R., Weerasekera, C.S., Li, K., Agarwal, H., Reid, I.: Unsupervised learning of monocular depth estimation and visual odometry with deep feature reconstruction. In CVPR (2018)
40. Zhao, W., Liu, S., Shu, Y., Liu, Y.-J.: Towards better generalization: joint depth-pose learning without posenet. In: CVPR (2020)
41. Zhou, T., Brown, M., Snavely, N., Lowe, D.G.: Unsupervised learning of depth and ego-motion from video. In: CVPR (2017)
42. Zhou, Z., Fan, X., Shi, P., Xin, Y.: R-MSFM: recurrent multi-scale feature modulation for monocular depth estimating. In: ICCV (2021)

SiamDoGe: Domain Generalizable Semantic Segmentation Using Siamese Network

Zhenyao Wu[1], Xinyi Wu[1], Xiaoping Zhang[2], Lili Ju[1(✉)], and Song Wang[1(✉)]

[1] University of South Carolina, Columbia, USA
{zhenyao,xinyiw}@email.sc.edu, ju@math.sc.edu, songwang@cec.sc.edu
[2] Wuhan University, Wuhan, China
xpzhang.math@whu.edu.cn

Abstract. Deep learning-based approaches usually suffer from performance drop on out-of-distribution samples, therefore domain generalization is often introduced to improve the robustness of deep models. Domain randomization (DR) is a common strategy to improve the generalization capability of semantic segmentation networks, however, existing DR-based algorithms require collecting auxiliary domain images to stylize the training samples. In this paper, we propose a novel domain generalizable semantic segmentation method, "**SiamDoGe**", which builds upon a DR approach without using auxiliary domains and employs a Siamese architecture to learn domain-agnostic features from the training dataset. Particularly, the proposed method takes two augmented versions of each training sample as input and produces the corresponding predictions in parallel. Throughout this process, the features from each branch are randomized by those from the other to enhance the feature diversity of training samples. Then the predictions produced from the two branches are enforced to be consistent conditioned on feature sensitivity. Extensive experiment results demonstrate the proposed method exhibits better generalization ability than existing state-of-the-arts across various unseen target domains.

Keywords: Domain generalization · Semantic segmentation · Siamese network · Domain randomization

1 Introduction

Semantic image segmentation associates each pixel to a semantic label and has a wide range of applications in real world, such as autonomous driving [5,65], robotic navigation [41,53] and medical image diagnostic [40,68]. Current

Code is available at https://github.com/W-zx-Y/SiamDoGe.

Supplementary Information The online version contains supplementary material available at https://doi.org/10.1007/978-3-031-19839-7_35.

S. Avidan et al. (Eds.): ECCV 2022, LNCS 13698, pp. 603–620, 2022.
https://doi.org/10.1007/978-3-031-19839-7_35

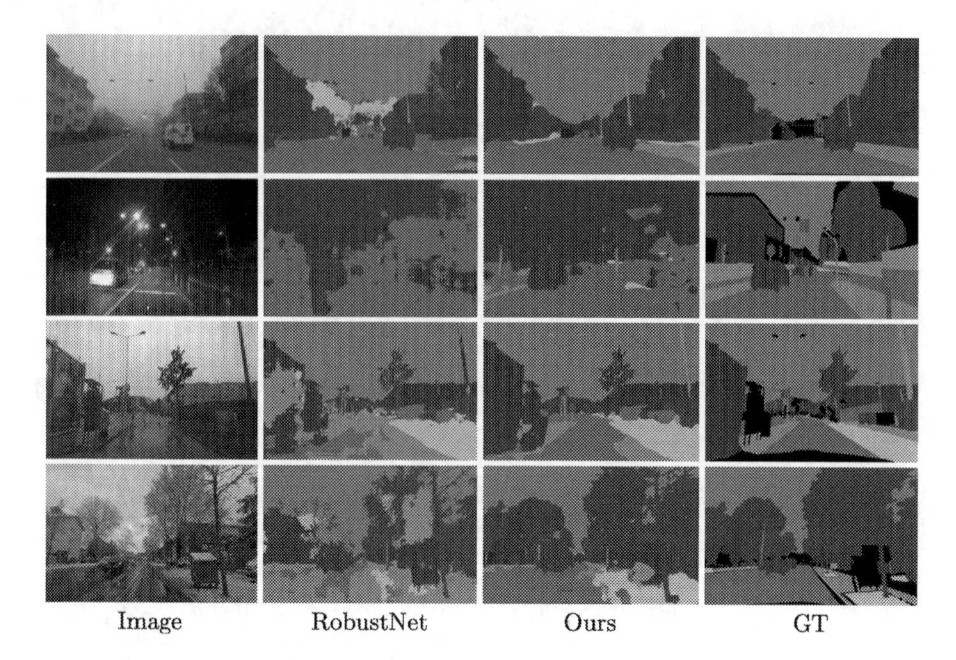

Fig. 1. Some visual comparisons of the domain generalizable semantic segmentation results produced by the proposed method and RobustNet [10]. Both models are trained on the GTAV dataset and evaluated using the ACDC dataset with four conditions of Fog, Night, Rain and Snow, as shown in rows 1 to 4 respectively.

deep learning-based approaches [5,34,65] have achieved very promising results through training on large-scale labeled datasets [1,11,13], but these datasets are usually very laborious to collect and annotate. Another well-known phenomenon is that a deep model trained on one dataset often fits well on its own test split (in-domain) but suffers from a huge performance drop on other datasets (cross-domain), and this phenomenon is usually called *domain shift*.

In recent years, domain adaptive (DA) semantic segmentation [8,22,44,55, 57,60,63] was proposed to bridge domain gaps so that a model trained on one domain (source) also works well on other domains (target). This is achieved by leveraging multiple unlabeled target samples as references while training the source with supervision. However, this approach has two limitations: 1) target samples are always required even if it can be as few as one [36]; and 2) multiple times of adaptation are required to be performed when there are multiple desired target domains. Compared with DA, domain generalization (DG) is a more universal solution to handle arbitrary domain shifts thus does not have a preference towards a particular target domain. It aims to reduce the model sensitivity to the change of data distribution via domain-agnostic feature learning. DG has been typically studied with two different settings: multi-source DG [14,29–31,42,67] and single-source DG [46,58,66]. In this paper, we study single-source DG for semantic segmentation.

Existing domain generalizable semantic segmentation approaches are mainly based on feature normalization [10,45] and domain randomization [23,27,62]

(DR). It is observed that the domain randomization-based approaches usually can achieve better generalization capacity than the domain normalization-based ones due to the use of auxiliary real-world domains, *e.g.*, ImageNet [12] or web-crawled images, for source image stylization. On the other hand, the DR-based methods also have the following drawbacks: 1) their DG performances highly depend on the choice of auxiliary domains and it takes a lot of time to carefully collect data in the domains related to the task in order to avoid impure DG[1] [27]; and 2) most of them lack enough control and could undesirably alter the semantic structures (domain invariant features) of images [23].

With this observation, we propose in this paper a novel domain generalizable semantic segmentation method, "SiamDoGe", which is based on domain randomization but does not use other auxiliary domains. Our work is partially inspired by SimSiam [7], a Siamese network for unsupervised representation learning by comparing two views of one image. In the proposed method, two augmented versions of a source sample are first generated, then a Siamese network is employed to find the crucial shared invariant representations from the two branches for domain generalization. Specifically, the features from the two branches are randomized interdependently during training. There are two natural advantages for such design over existing DR-based algorithms: 1) collecting extra data in auxiliary domains is no longer needed; and 2) more controllability is obtained since the randomization is performed by using two images that share common content. Besides, we also study the feature sensitivity by comparing low-level features from the two branches. The prediction consistency of the two branches is then enforced with more attention being paid to more sensitive regions since it is usually difficult to obtain domain-agnostic features in those regions. Extensive experimental results verify the effectiveness of our approach and show that the proposed SiamDoGe generalizes very well to multiple unseen domains both qualitatively (see Fig. 1 for sample visual comparisons) and quantitatively.

The main contributions of our work are summarized as follows:

- A new domain generalizable semantic segmentation approach, "SiamDoGe", is developed for domain-invariant feature learning by employing Siamese structure.
- Our method achieves a more controllable self-guided randomization without using auxiliary domains, and better domain-agnostic features by taking account of feature sensitivity when ensuring the prediction consistency.
- We evaluate the performance on various unseen domains and the results show that our method exhibits better generalization capacity than existing state-of-the-arts.

2 Related Works

In this section, we discuss the related works on DA- and DG-based semantic segmentation and the background of Siamese networks.

[1] The auxiliary domains should not share common data samples with the unseen target domain according to the definition of DG.

Domain Adaptive Semantic Segmentation. Domain adaptation (DA) is related to our work since its goal is to minimize domain discrepancy. There are two commonly-used strategies for domain adaptive semantic segmentation: adversarial training [21,22,35,44,55,57] and self-training [17,26,33,38,60,63,69, 70]. The former usually trains a discriminator and a segmentation network alternatively to align source and target domains and the latter leverages the confident pseudo labels to achieve more performance gains via multiple rounds of retraining on the target domain. A lot of DA scenarios have been studied for semantic segmentation such as synthetic-to-real [44,55,57,60,63,69,70], cross-time of day [49–51,59], cross weathers [51], cross cities [22,55], and many of them benefit autonomous driving. However, retraining is required whenever a new DA scenario (target) appears. Differently, in this paper, we explore domain generalization, which is a more universal solution than DA for handling the domain discrepancy since it does not require to specify a target domain.

Domain Generalizable Semantic Segmentation. Existing DG semantic segmentation is usually achieved by specific designs of feature normalization [10,45], knowledge distillation [6], or domain randomization [23,27,62].

IBN-Net is the first DG semantic segmentation approach [45] integrating instance normalization [56] and batch normalization [25], where the former learns appearance-invariant features and the latter preserves content information. This work was recently extended by incorporating an instance selective whitening loss to selectively remove the feature co-variances that respond sensitively to the domain shift in [10]. In [6], Chen *et al.*formulate the synthetic-to-real generalization as a lifelong learning problem by enforcing the representation similarity between synthetically trained models and the ImageNet pre-trained model via a distillation loss.

Domain randomization (DR) is a more frequently used strategy for DG. Yue *et al.*[62] first explored the DR for semantic segmentation where auxiliary-domain images are carefully picked from ImageNet [12] to stylize the training images and the prediction consistency is enforced across all stylized images of one training sample. However, DR is not controllable and might hurt the domain invariant features [23] crucial to DG. Huang *et al.*[23] further refined the DR strategy by transferring images into the frequency domain and only randomizing the domain-variant frequency components with the domain-invariant ones unchanged. Very recently, Kim *et al.*[27] proposed a non-parametric style injection module to randomize the training images on-the-fly using a large amount of web-crawled images which are real and related to the application of autonomous driving. In general, our proposed method falls in the group of DR but does not use auxiliary domains for source image stylization and thus is more controllable.

Siamese Network. Siamese neural networks [4] were proposed to learn semantic similarity and have been shown to work well on various vision tasks such as object-tracking [3,9,18,54,64], image co-segmentation [2,32], one-shot learning [28] and unsupervised visual representation learning [7,16,19], *etc.*. By

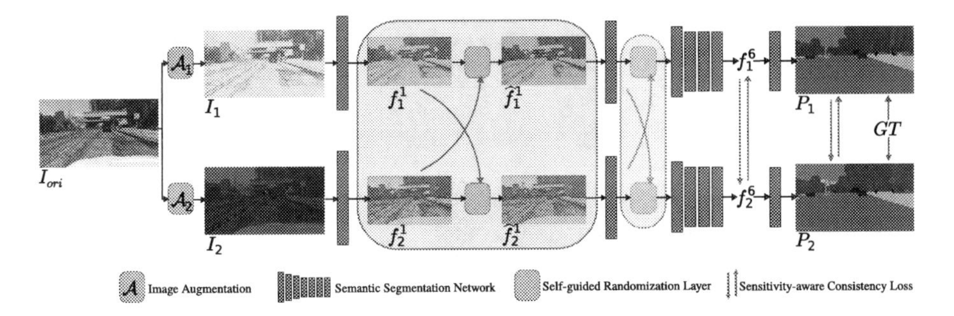

Fig. 2. The overall architecture and training pipeline of the proposed SiamDoGe. A training sample I_{ori} is first taken as input to obtain two augmented versions I_1 and I_2, respectively. Then each branch i, $(i = 1, 2)$ is fed with an augmented image I_i and produces a prediction P_i with a group of intermediate features $\{f_i^j\}_{j=1}^6$. The proposed self-guided randomization is particularly applied to the features obtained from the first and second layers of the segmentation network, *i.e.*, f_i^1 and f_i^2, to produce the corresponding randomized features $\widehat{f_i^1}$ and $\widehat{f_i^2}$. The semantic segmentation network shares weights across the two branches and the whole pipeline is trained via the standard cross-entropy loss and regularized by a novel sensitivity-guided consistency loss.

definition, Siamese networks are weight-sharing neural networks applied to two or more inputs for comparing the entities [4,7]. It has been employed by many of existing DA approaches [22,39,55,57] to bridge the domain gap between source and target images. In this paper, we propose a novel DG semantic segmentation approach based on the Siamese architecture.

3 Proposed Method

3.1 Overview

Our model improves the performance and robustness of semantic segmentation networks via two specially designed components: a *self-guided randomization* and a *sensitivity-guided consistency training*. The former allows to perform domain randomization with the training sample itself, and the latter encourages to find domain-invariant features based on feature sensitivity by comparing the two branches.

The overall architecture and training pipeline of the proposed SiamDoGe is illustrated in Fig. 2. Given a training sample image I_{ori}, we first obtain two augmented views I_1 and I_2 for it with two random simple color jittering transformations $\mathcal{A}_1$ and $\mathcal{A}_2$, respectively. The two views which share the same content but different visual styles are then fed into a weight-sharing Siamese network (two branches) for semantic segmentation. During feature extraction, the proposed self-guided randomization is particularly applied to the features obtained from the first and second layers of the segmentation network, *i.e.*, f_i^1 and f_i^2, to produce the corresponding randomized features $\widehat{f_i^1}$ and $\widehat{f_i^2}$ for each branch

$i \in \{1, 2\}$. The whole pipeline is trained under the supervision of source domain ground truths for both branches via the standard cross-entropy loss plus a novel sensitivity-guided consistency loss.

3.2 The Self-Guided Randomization

To achieve more controllable domain randomization without accessing to auxiliary domains, we propose to randomize the source image in a self-guided way, which is implemented via feature normalization inspired by [67]. The details of this operation are shown in Fig. 3.

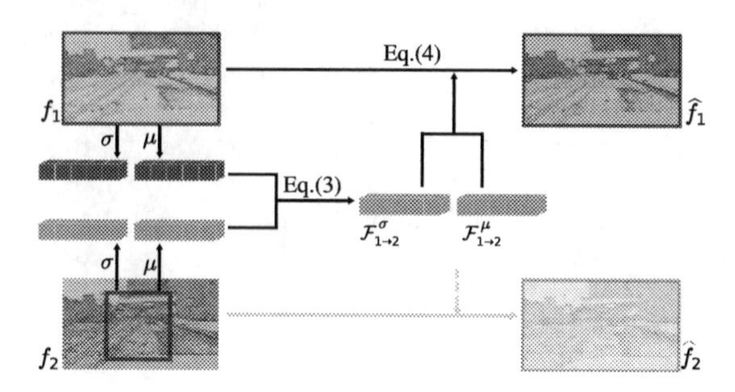

Fig. 3. An illustration of the proposed self-guided randomization process.

Specifically, the inputs of the self-guided randomization layer are two intermediate feature maps f_1 and $f_2 \in \mathbb{R}^{C\times H\times W}$ from the two branches with C, H and W representing channel, height and width, respectively. Following [24], we first denote the spatial feature statistics $\mu(.)$ and $\sigma(.) \in \mathbb{R}^C$ of a feature f by

$$\mu(f)_{(c)} = \frac{1}{HW}\sum_{h=1}^{H}\sum_{w=1}^{W} f_{(c,h,w)}, \tag{1}$$

and

$$\sigma(f)_{(c)} = \sqrt{\frac{1}{HW}\sum_{h=1}^{H}\sum_{w=1}^{W}(f_{(c,h,w)} - \mu(f)_{(c)})^2 + \epsilon}, \tag{2}$$

where ϵ is set to 10^{-10}.

A naive randomization can be easily obtained via adding Gaussian noise to the source feature statistics, however, such perturbation is also lacking in control and still might destroy the domain-invariant features. Based on the concept of "domain flow" introduced in [15] to describe intermediate domains between the source and target domains for domain adaptation, we propose a concept of "intra-domain flow" which represents intermediate intra-domains within the source domain for domain generalization. We first define the intra-domain flow from f_1 to f_2 based on their feature statistics as:

$$\mathcal{F}^{\mu}_{1\to 2} = \mu(\mathcal{C}(f_2)) - \mu(f_1); \quad \mathcal{F}^{\sigma}_{1\to 2} = \sigma(\mathcal{C}(f_2)) - \sigma(f_1), \tag{3}$$

where the function $\mathcal{C}$ stands for the random cropping operation with a size of 64×64 used to help improve the diversity of the flow. Then the computed intra-domain flow $\{\mathcal{F}^{\mu}_{1\to 2}, \mathcal{F}^{\sigma}_{1\to 2}\}$ is adopted to randomize f_1 as follows:

$$\begin{aligned}\hat{f}_1 &= (\sigma(f_1) + \lambda \mathcal{F}^{\sigma}_{1\to 2}) \left(\frac{f_1 - \mu(f_1)}{\sigma(f_1)}\right) \\ &\quad + (\mu(f_1) + \lambda \mathcal{F}^{\mu}_{1\to 2}) \\ &= f_1 + \lambda \left(\mathcal{F}^{\sigma}_{1\to 2} \frac{f_1 - \mu(f_1)}{\sigma(f_1)} + \mathcal{F}^{\mu}_{1\to 2}\right),\end{aligned} \tag{4}$$

where $\lambda \in [0, 1]$ is a randomly generated hyper-parameter used to control the randomization. Similarly, we also compute the flow $\{\mathcal{F}^{\mu}_{2\to 1}, \mathcal{F}^{\sigma}_{2\to 1}\}$ and then randomize f_2 via Eqs. (3) and (4) by switching f_1 and f_2, and replacing $1 \to 2$ with $2 \to 1$.

3.3 The Sensitivity-Guided Consistency Training

Consistency training was explored in [62] for domain generalizable semantic segmentation, where various stylized source images of one training sample in auxiliary domains are enforced to have consistent predictions. In this paper, we make one further step to propose a sensitivity-guided consistency training to ensure the consistency of the two branches as illustrated in Fig. 4. Our insight is that the difference between low-level features from the two augmented versions can well describe how the features are sensitive to the "domain" shift, *i.e.*, a small difference means the feature is robust while a large one means it is variant to the shift. Therefore, the proposed loss function will pay more attention to the sensitive regions and less attention to the insensitive regions that are already generalized well.

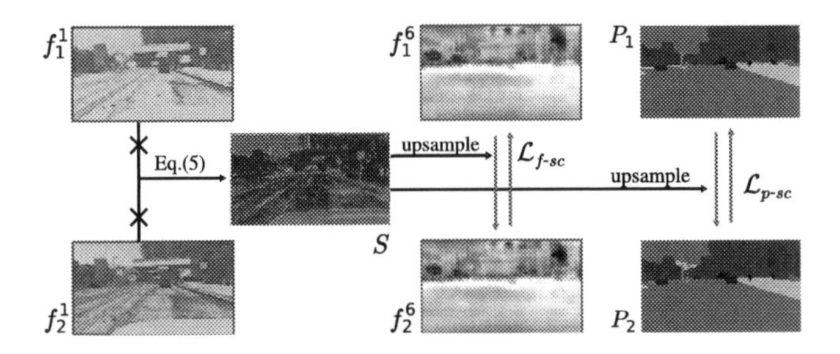

Fig. 4. An illustration of the proposed sensitivity-guided consistency training. In a sensitivity map S, the darker blue regions are less sensitive than the light blue ones. (Color figure online)

We first measure the distance between the low-level features f_1^1 and f_2^1 from the two branches and obtain the feature sensitivity map $S \in \mathbb{R}^{H \times W}$ by

$$S_{(h,w)} = \zeta\left(\frac{1}{C}\sum_{c=1}^{C}|f^1_{1,(c,h,w)} - f^1_{2,(c,h,w)}|\right), \tag{5}$$

where ζ is the stop-gradient operation that is borrowed from [7]. S then is involved in consistency loss computation. Specifically, we build the sensitivity-guided consistency loss in both the feature level ($\mathcal{L}_{f-sc}$) and the prediction level ($\mathcal{L}_{p-sc}$), which are formulated as:

$$\mathcal{L}_{f-sc} = \frac{1}{CHW}\sum_{h,w}\left(S_{(h,w)}\sum_{c=1}^{C}|f^6_{1,(c,h,w)} - f^6_{2,(c,h,w)}|\right), \tag{6}$$

$$\mathcal{L}_{p-sc} = \frac{1}{CHW}\sum_{h,w}\left(S_{(h,w)}\sum_{c=1}^{C}|P_{1,(c,h,w)} - P_{2,(c,h,w)}|\right), \tag{7}$$

where the feature f_i^6 and the prediction P_i are upsampled to the same resolution as S.

3.4 Implementation Details

Architecture. Following [10], we use the DeeplabV3+ [5] as our semantic segmentation network with Resnet-50 [20] as the backbone of the proposed SiamDoGe. It contains seven layers in total with the first one being a combination of Conv-BN-ReLU-MAX POOLING operations, the 2–5th ones being the residual blocks, the 6th one being the ASPP layer with the output stride of 16 and the last one is the classifier layer.

Loss Function. The training objective of SiamDoGe is a combination of semantic segmentation loss and the sensitivity-guided consistency loss defined as:

$$\mathcal{L}_{total} = \mathcal{L}_{CE} + \alpha(\mathcal{L}_{f-sc} + \mathcal{L}_{p-sc}), \tag{8}$$

where α is a weighting parameter for balancing the two loss terms.

Augmentation. Our SiamDoGe only uses the random color jittering for input image augmentation, *i.e.*, $\mathcal{A}_1$ and $\mathcal{A}_2$ in Fig. 2, which produces two different views of the same image. Following [10], the parameters for color jittering are set to 0.8, 0.8, 0.8 and 0.3 for brightness, contrast, saturation and hue, respectively.

Optimization. The SGD optimizer is used with an initial learning rate of 10^{-2} and a momentum of 0.9. The polynomial learning rate scheduler [65] with the power of 0.9 is also applied to stabilize the training. The whole network is trained using two Nvidia Tesla V100 GPU cards with a batch size of 8 for 40K iterations in total for all experiments and each experiment requires around 22 h for training. The hyper-parameter α is set to 0 for the first 10K iterations and 10 for the rest iterations during training.

4 Experiments

In this section, we first introduce all the datasets that are involved in the experiments. Then we demonstrate the excellent performance of our SiamDoGe by comparing it with existing state-of-the-arts, and also empirically study the effects of its key components via ablation studies.

4.1 Datasets

We evaluate the proposed SiamDoGe on two DG settings based on different domains, including GTAV(G) $\rightarrow$ {Cityscapes(C), SYNTHIA(S), Mapillary(M), BDD-100K(B), ACDC(A)} and Cityscapes(C) $\rightarrow$ {GTAV(G), SYNTHIA(S), Mapillary(M), BDD-100K(B), ACDC(A)}. We adopt the mean intersection-over-union (mIoU) as the evaluation metric (the higher the better) and all the datasets are evaluated based on 19 classes defined by Cityscapes.

GTAV. [47] is a large-scale self-annotated synthetic dataset collected from commercial video games. It contains images with a resolution of 1,914 $\times$ 1,052 and is divided into 12,403/6,382/6,181 for training, validation and testing purposes. Here we use the training and validation images with their labels for training when it serves as the seen domain and only use its validation set when it is treated as an unseen domain.

Cityscapes. [11] captures real-world urban street scenes from different cities and it provides 5,000 high quality manually-annotated images in pixel level with a resolution of 2,048 $\times$ 1,024. The images are split into subsets of 2,975/500/1,525 images for training, validation and testing, respectively. Here we use the training set when it serves as the seen domain and the validation set when it serves as an unseen domain.

SYNTHIA. [48] is a synthetic dataset consisting of 9,400 self-labeled images with a resolution of 960 $\times$ 720. Here we use the 2,820 images split by [10] for evaluation.

Mapillary. [43] is a real-world dataset that is designed to capture scenes in the wild variations across season/weather conditions, viewing perspectives, time and resolution (at least 1,920$\times$1,080), *etc.* Here we use its original validation split (2,000 images) for evaluation.

BDD-100K. [61] is an another real-world dataset recorded in diverse weather conditions at different times of the day. The resolution of the images is 1,280$\times$720. Here we use its original validation split (1,000 images) for evaluation.

ACDC. [51] is the largest adverse condition dataset for semantic segmentation to date. Different from Mapillary and BDD-100K which also contain normal condition scenes, ACDC purely consists of images with four common adverse conditions of fog, nighttime, rain and snow. The resolution of the images is 1,920$\times$1,080. Here we use its validation set (406 images) for evaluation, *i.e.*, 100/100/100/106 for fog/rain/snow/nighttime conditions, respectively.

4.2 Comparison with State-of-the-Arts

We first compare the performance of our SiamDoGe with several existing DG approaches for semantic segmentation under the setting G → {C, B, M, S, A}, including the feature normalization-based ones (not using auxiliary domains): IBN-Net [45] and RobustNet [10], and the domain randomization-based ones (using auxiliary domains): DRPC [62] and WEDGE [27]. Table 1 reports all the quantitative comparison results. Note that only the results on G → {C, B, M} are available for DRPC and WEDGE in the literature. It can be observed that our SiamDoGe achieves the best performance across all unseen target domains among the methods of not using auxiliary domains and surpasses the second best by a large margin on Cityscapes (6.38 mIoU) and ACDC (3.79 mIoU). More surprisingly, our SiamDoGe also outperforms the methods of using auxiliary domains, *e.g.*, DRPC significantly on all the three unseen target domains and WEDGE slightly on Cityscapes and BDD-100K. It is worth mentioning that WEDGE uses around 5K auxiliary domain images for randomization which is

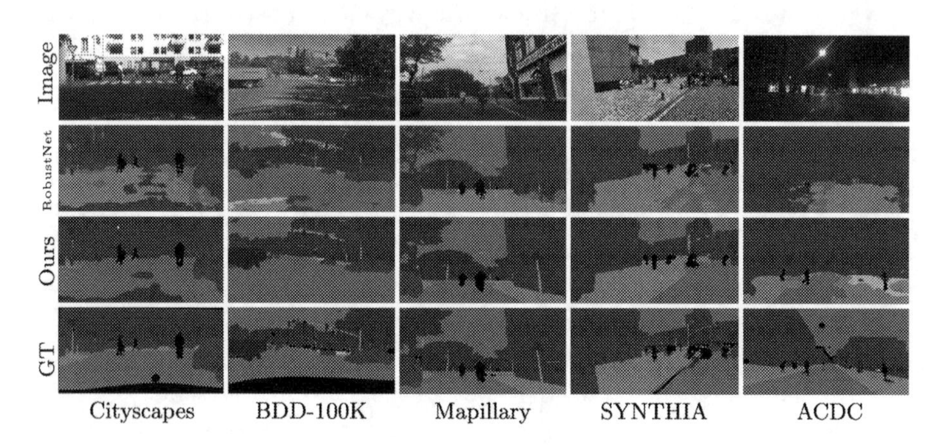

Fig. 5. Qualitative results of SiamDoGe and RobustNet [10] under the setting G → $\{C, B, M, S, A\}$.

Table 1. Quantitative comparison results of our SiamDoGe and the existing state-of-the-art DG approaches for semantic segmentation under the setting G → $\{C, B, M, S, A\}$. All the methods use ResNet-50 as backbone. The *avg.* represents the average mIoU (%) over the five datasets. †indicates that auxiliary domains are required. The best results are highlighted with **bold**.

Methods	C	B	M	S	A	*Avg.*
DRPC †[62]	37.41	32.14	34.12	–	–	–
WEDGE †[27]	38.15	36.14	**43.21**	–	-	–
IBN-Net [45]	33.85	32.30	37.75	27.90	22.55	30.87
RobustNet [10]	36.58	35.20	40.33	28.30	25.46	33.14
SiamDoGe (ours)	**42.96**	**37.54**	40.64	**28.34**	**29.25**	**35.75**

even larger than those in the unseen target domains. Some qualitative comparison results with RobustNet under this setting are provided in Fig. 5, where we observe that our method achieves better results visually for the truck in the sample from BDD-100K, the traffic sign in the sample from Mapillary and the person in the sample from ACDC.

Table 2. Quantitative comparison results of SiamDoGe and the existing state-of-the-art DG approaches without using auxiliary domains for semantic segmentation under the setting C $\rightarrow \{G, B, M, S, A\}$. All the methods use ResNet-50 as backbone.

Methods	G	B	M	S	A	*Avg.*
IBN-Net [45]	45.06	48.56	57.04	26.14	44.05	44.17
RobustNet [10]	45.00	50.73	58.64	26.20	46.91	45.50
SiamDoGe	**45.08**	**51.53**	**59.00**	**26.67**	**52.34**	**46.92**

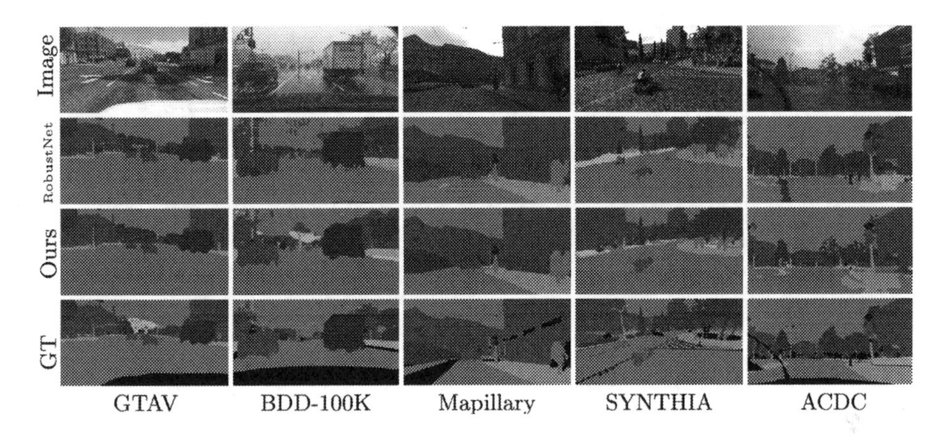

Fig. 6. Qualitative comparison results of SiamDoGe and RobustNet [10] under the setting C $\rightarrow \{G, B, M, S, A\}$.

Similarly, we compare the performance of our SiamDoGe with the state-of-the-arts [10,45] under the setting C $\rightarrow$ {G, B, M, S, A}. The quantitative results are reported in Table 2 with some visualization results shown in Fig. 6. Consistently, our SiamDoGe wins across all unseen domains again and surpasses the second best [10] on ACDC dataset by 5.44 mIoU. In addition, from Fig. 6 we find that our method obtains better predictions especially for the classes of track, car, motor, train and *etc.*.

For a fair comparison, we also report the comparison of computational costs of the proposed SiamDoGe with RobustNet [10] in Table 3. It is observed that our method is more efficient than RobustNet.

Table 3. Comparison of computational costs with ResNet-50 as backbone.

Methods	# of Params	GFLOPs	Inference Time (ms)
RobustNet	40.35	60.69	7.58
SiamDoGe	40.23	43.00	5.71

Table 4. Quantitative comparison results of our SiamDoGe and the existing state-of-the-art DG approaches without using auxiliary domains for semantic segmentation under the setting G → {C, B, M, S, A}. All the methods use ShuffleNetV2 as backbone.

Methods	C	B	M	S	A	*Avg.*
IBN-Net [45]	27.10	31.82	34.89	**25.56**	22.33	28.34
RobustNet [10]	30.98	32.06	35.31	24.31	21.27	28.79
SiamDoGe	**34.40**	**34.23**	**35.87**	21.95	**25.22**	**30.33**

Table 5. Quantitative comparison results of our SiamDoGe and the existing state-of-the-art DG approaches without using auxiliary domains for semantic segmentation under the setting G → {C, B, M, S, A}. All the methods use MobileNetV2 as backbone.

Methods	C	B	M	S	A	*Avg.*
IBN-Net [45]	30.14	27.66	27.07	**24.98**	20.30	26.03
RobustNet [10]	30.86	30.05	30.67	24.43	23.26	27.85
SiamDoGe	**34.15**	**34.50**	**32.34**	23.53	**24.17**	**29.74**

Other Backbones. Following [10], we also employ ShuffleNetV2 [37] and MobileNetV2 [52] as additional backbones for performance evaluation. The models are compared under the setting G → {C, B, M, S, A} with corresponding results reported in Tables 4 and 5, respectively. We observe that our method still achieves the best performance on average over all the unseen domains. Among all five test domains, our SiamDoGe achieves the best on the four real-world domains for both backbones.

4.3 Ablation Studies

On Main Model Components. We first examine how each of our model components impacts the DG performance for semantic segmentation by testing several model variants. The numerical results obtained under both the settings G → {C, B, M, S, A} and C → {G, B, M, S, A} are shown in Table 6. "Single branch" serves as the baseline of our method by feeding the randomly augmented images (using $\mathcal{A}$) into only one branch to produce the segmentation results. "Siamese Network" means two branches take two augmented views of a sample as input and the predictions from the two branches are supervised by the ground-truth independently. The third one further models the relationship between the branches with the proposed sensitivity-guided consistency. The last one is our full model by adding the self-guided randomization. By comparing the former three variants, we can find that the Siamese structure is meaningless without modeling the consistency, *i.e.*, just doubles the batch size of the "Single branch". For the two settings, the sensitivity-guided consistency brings 2.12/1.17 mIoU gains on average and the self-guided randomization further brings 1.50/1.75 mIoU gains.

Table 6. Ablation study for main components of our SiamDoGe, including the Siamese network, the sensitivity-guided consistency (*SC*) and the self-guided randomization (*SR*).

	Trained on Cityscapes						Trained on GTAV					
Variants	G	B	M	S	A	*Avg.*	C	B	M	S	A	*Avg.*
Single branch	42.27	47.02	55.05	24.66	45.78	42.96	40.75	33.69	37.09	29.42	23.67	32.92
+ Siamese Network	40.98	46.88	56.84	24.36	47.45	43.30	39.81	34.93	37.36	**29.51**	22.54	32.83
+ *SC*	43.71	48.97	58.68	25.37	50.36	45.42	40.75	35.47	38.54	28.86	26.38	34.00
+ *SR*	**45.08**	**51.53**	**59.00**	**26.67**	**52.34**	**46.92**	**42.96**	**37.54**	**40.64**	28.34	**29.25**	**35.75**

On the Self-Guided Randomization. We then study the choice of layers to perform the self-guided randomization. As shown in Table 7, for both DG settings, we achieve the best performance when randomizing features from both layer1 and layer2, *i.e.*, f^1 and f^2. Besides, for all the test domains, the best performance is always located in the last three rows which also verifies the effectiveness of the self-guided randomization. From the Table 7, we also observe that the croppping operation $\mathcal{C}$ and randomly generated hyper-parameter λ are effective in both generalization scenarios. Our proposed self-guided randomization is also outperform MixStyle [67] on the semantic segmentation task.

Table 7. Ablation study on the choice of layers for the self-guided randomization in our SiamDoGe.

	Trained on Cityscapes						Trained on GTAV					
Variants	G	B	M	S	A	*Avg.*	C	B	M	S	A	*Avg.*
w/o randomization	43.71	48.97	58.68	25.37	50.36	45.42	40.75	35.47	38.54	28.86	26.38	34.00
Using MixStyle [67]	44.40	50.87	57.30	24.39	48.85	45.16	40.68	36.00	38.55	27.81	27.69	34.15
w/o $\mathcal{C}$	45.01	51.44	58.62	25.57	52.83	46.69	40.15	37.94	38.18	27.50	28.80	34.51
$\lambda = 0.5$	44.58	50.87	58.67	25.62	51.54	46.26	39.89	38.21	38.24	27.92	28.72	34.60
Layer 1 only	**45.53**	**51.85**	**59.13**	25.20	52.01	46.74	41.18	**37.73**	40.34	27.28	**29.60**	35.23
Layer 2 only	43.29	50.18	57.49	25.78	49.77	45.30	41.51	37.30	38.65	**29.22**	27.95	34.93
Layers 1 and 2	45.08	51.53	59.00	**26.67**	**52.34**	**46.92**	**42.96**	37.54	**40.64**	28.34	29.25	**35.75**

On the Sensitivity-Guided Consistency Training. Next, we study several variants of the proposed sensitivity-guided consistency training. We can observe from the top part of Table 8 (Rows 1–3) that both the feature-level consistency loss $\mathcal{L}_{f-cs}$ and the prediction-level consistency loss $\mathcal{L}_{p-cs}$ can boost the performance of domain generalization on most test domains and the feature-level one seems even more important. We also find (from Row 4 and Row 5) that the performance of each column is improved except for BDD-100K and ACDC (trained on GTAV), which verifies the importance of sensitivity guidance for consistency training.

Hyper-Parameter Tuning. Finally, we tune the values of the hyper-parameter α and the number of iterations before launching the consistency loss during the training of our full model. The results are reported in the bottom part of Table 8 and we observe that $\alpha = 10$ and $iter = 10$k gives the best performance.

Table 8. Ablation study on each factor of sensitivity-guided consistency loss in our SiamDoGe. "iter" represents the iterations required before launching the sensitivity-guided consistent loss.

Variants				Trained on Cityscapes						Trained on GTAV					
$\mathcal{L}_{f-cs}$	$\mathcal{L}_{p-cs}$	α	iter	G	B	M	S	A	*Avg.*	C	B	M	S	A	*Avg.*
		–	–	43.28	49.40	57.93	24.67	47.88	44.63	40.08	34.52	37.94	**28.62**	25.71	33.37
✓		10.0	10k	43.54	48.99	57.40	26.13	48.21	44.85	39.39	34.13	36.44	26.88	27.14	32.80
	✓	10.0	10k	**45.23**	51.21	58.92	26.13	**52.44**	46.79	40.85	37.31	37.99	26.94	27.72	34.12
w/o S	w/o S	10.0	10k	44.26	51.07	57.22	24.55	51.70	45.76	41.86	**39.09**	40.04	26.96	**29.72**	35.53
✓	✓	10.0	10k	45.08	**51.53**	**59.00**	**26.67**	52.34	**46.92**	**42.96**	37.54	**40.64**	28.34	29.25	**35.75**
✓	✓	1.0	10k	44.62	50.17	58.50	25.59	50.41	45.86	41.42	34.96	38.94	27.41	25.61	33.67
✓	✓	50.0	10k	43.84	50.13	56.88	23.75	52.13	45.35	41.07	37.53	40.52	25.52	28.08	34.54
✓	✓	10.0	0	44.06	51.49	58.37	25.18	51.75	46.17	40.28	35.75	38.82	26.43	28.00	33.86
✓	✓	10.0	20k	44.17	51.24	58.01	25.11	51.69	46.04	40.71	36.47	39.86	25.88	27.00	33.98

5 Conclusion

In this paper, we explored a novel domain generalizable semantic segmentation approach with a more controllable domain randomization strategy. The proposed method, "SiamDoGe", is built upon a Siamese network with two branches performing semantic segmentation. It is integrated with two novel designs: one is the self-guided randomization which randomizes each training sample without using auxiliary domain images (different from other existing DR-based alternatives), the other is the sensitivity-guided consistency training which helps learn domain-agnostic features from two views of each training sample. Comprehensive numerical experiments demonstrated that our SiamDoGe generalizes well on several unseen target domains by training on a single domain and achieves a new state-of-the-art performance.

Acknowledgments. Dr. Lili Ju's work is partially supported by U.S. Department of Energy, Office of Advanced Scientific Computing Research through Applied Mathematics program under grant DE-SC0022254. This work used GPUs provided by the NSF MRI-2018966.

References

1. Alhaija, H., Mustikovela, S., Mescheder, L., Geiger, A., Rother, C.: Augmented reality meets computer vision: efficient data generation for urban driving scenes. In: IJCV (2018)
2. Banerjee, S., Hati, A., Chaudhuri, S., Velmurugan, R.: Cosegnet: image co-segmentation using a conditional Siamese convolutional network. In: IJCAI, pp. 673–679 (2019)
3. Bertinetto, L., Valmadre, J., Henriques, J.F., Vedaldi, A., Torr, P.H.S.: Fully-convolutional Siamese networks for object tracking. In: Hua, G., Jégou, H. (eds.) ECCV 2016. LNCS, vol. 9914, pp. 850–865. Springer, Cham (2016). https://doi.org/10.1007/978-3-319-48881-3_56
4. Bromley, J., et al.: Signature verification using a "Siamese" time delay neural network. Int. J. Pattern Recognit Artif Intell. **7**(04), 669–688 (1993)
5. Chen, L.C., Papandreou, G., Kokkinos, I., Murphy, K., Yuille, A.L.: DeepLab: semantic image segmentation with deep convolutional nets, atrous convolution, and fully connected CRFs. In: IEEE TPAMI (2018)
6. Chen, W., Yu, Z., Wang, Z., Anandkumar, A.: Automated synthetic-to-real generalization. In: International Conference on Machine Learning, pp. 1746–1756. In: PMLR (2020)
7. Chen, X., He, K.: Exploring simple Siamese representation learning. In: CVPR, pp. 15750–15758 (2021)
8. Chen, Y.H., Chen, W.Y., Chen, Y.T., Tsai, B.C., Frank Wang, Y.C., Sun, M.: No more discrimination: cross city adaptation of road scene Segmenters. In: ICCV, pp. 1992–2001 (2017)
9. Cheng, S., et al.: Learning to filter: Siamese relation network for robust tracking. In: CVPR, pp. 4421–4431 (2021)
10. Choi, S., Jung, S., Yun, H., Kim, J.T., Kim, S., Choo, J.: Robustnet: improving domain generalization in urban-scene segmentation via instance selective whitening. In: CVPR, pp. 11580–11590 (2021)
11. Cordts, M., et al.: The cityscapes dataset for semantic urban scene understanding. In: CVPR, pp. 3213–3223 (2016)
12. Deng, J., Dong, W., Socher, R., Li, L.J., Li, K., Fei-Fei, L.: ImageNet: a large-scale hierarchical image database. In: CVPR, pp. 248–255. IEEE (2009)
13. Everingham, M., Van Gool, L., Williams, C.K., Winn, J., Zisserman, A.: The pascal visual object classes (voc) challenge. IJCV **88**(2), 303–338 (2010)
14. Gan, C., Yang, T., Gong, B.: Learning attributes equals multi-source domain generalization. In: CVPR, pp. 87–97 (2016)
15. Gong, R., Li, W., Chen, Y., Gool, L.V.: Dlow: domain flow for adaptation and generalization. In: CVPR, pp. 2477–2486 (2019)
16. Grill, J.B., Strub, F., et al.: Bootstrap your own latent: a new approach to self-supervised learning. arXiv preprint arXiv:2006.07733 (2020)
17. Guo, X., Yang, C., Li, B., Yuan, Y.: Metacorrection: domain-aware meta loss correction for unsupervised domain adaptation in semantic segmentation. In: CVPR, pp. 3927–3936 (2021)

18. He, A., Luo, C., Tian, X., Zeng, W.: A twofold Siamese network for real-time object tracking. In: CVPR, pp. 4834–4843 (2018)
19. He, K., Fan, H., Wu, Y., Xie, S., Girshick, R.: Momentum contrast for unsupervised visual representation learning. In: CVPR, pp. 9729–9738 (2020)
20. He, K., Zhang, X., Ren, S., Sun, J.: Deep residual learning for image recognition. In: CVPR, pp. 770–778 (2016)
21. Hoffman, J., et al.: Cycada: cycle-consistent adversarial domain adaptation. In: International Conference on Machine Learning, pp. 1989–1998. PMLR (2018)
22. Hoffman, J., Wang, D., Yu, F., Darrell, T.: FCNs in the wild: pixel-level adversarial and constraint-based adaptation. arXiv preprint arXiv:1612.02649 (2016)
23. Huang, J., Guan, D., Xiao, A., Lu, S.: FSDR: frequency space domain randomization for domain generalization. In: CVPR, pp. 6891–6902 (2021)
24. Huang, X., Belongie, S.: Arbitrary style transfer in real-time with adaptive instance normalization. In: ICCV, pp. 1501–1510 (2017)
25. Ioffe, S., Szegedy, C.: Batch normalization: accelerating deep network training by reducing internal covariate shift. In: International Conference on Machine Learning (2015)
26. Kim, M., Byun, H.: Learning texture invariant representation for domain adaptation of semantic segmentation. In: CVPR, pp. 12975–12984 (2020)
27. Kim, N., Son, T., Lan, C., Zeng, W., Kwak, S.: Wedge: web-image assisted domain generalization for semantic segmentation. arXiv preprint arXiv:2109.14196 (2021)
28. Koch, G., et al.: Siamese neural networks for one-shot image recognition. In: ICML deep learning workshop, vol. 2. Lille (2015)
29. Li, D., Yang, Y., Song, Y.Z., Hospedales, T.M.: Deeper, broader and artier domain generalization. In: ICCV, pp. 5542–5550 (2017)
30. Li, D., Zhang, J., Yang, Y., Liu, C., Song, Y.Z., Hospedales, T.M.: Episodic training for domain generalization. In: ICCV, pp. 1446–1455 (2019)
31. Li, H., Pan, S.J., Wang, S., Kot, A.C.: Domain generalization with adversarial feature learning. In: CVPR, pp. 5400–5409 (2018)
32. Li, W., Hosseini Jafari, O., Rother, C.: Deep object co-segmentation. In: Jawahar, C.V., Li, H., Mori, G., Schindler, K. (eds.) ACCV 2018. LNCS, vol. 11363, pp. 638–653. Springer, Cham (2019). https://doi.org/10.1007/978-3-030-20893-6_40
33. Lian, Q., Lv, F., Duan, L., Gong, B.: Constructing self-motivated pyramid curriculums for cross-domain semantic segmentation: a non-adversarial approach. In: ICCV, pp. 6758–6767 (2019)
34. Long, J., Shelhamer, E., Darrell, T.: Fully convolutional networks for semantic segmentation. In: CVPR (2015)
35. Luo, Y., Liu, P., Guan, T., Yu, J., Yang, Y.: Significance-aware information bottleneck for domain adaptive semantic segmentation. In: ICCV, pp. 6778–6787 (2019)
36. Luo, Y., Liu, P., Guan, T., Yu, J., Yang, Y.: Adversarial style mining for one-shot unsupervised domain adaptation. In: NeurIPS (2020)
37. Ma, N., Zhang, X., Zheng, H.-T., Sun, J.: ShuffleNet V2: practical guidelines for efficient CNN architecture design. In: Ferrari, V., Hebert, M., Sminchisescu, C., Weiss, Y. (eds.) Computer Vision – ECCV 2018. LNCS, vol. 11218, pp. 122–138. Springer, Cham (2018). https://doi.org/10.1007/978-3-030-01264-9_8
38. Mei, K., Zhu, C., Zou, J., Zhang, S.: Instance adaptive self-training for unsupervised domain adaptation. In: Vedaldi, A., Bischof, H., Brox, T., Frahm, J.-M. (eds.) ECCV 2020. LNCS, vol. 12371, pp. 415–430. Springer, Cham (2020). https://doi.org/10.1007/978-3-030-58574-7_25
39. Melas-Kyriazi, L., Manrai, A.K.: Pixmatch: unsupervised domain adaptation via pixelwise consistency training. In: CVPR, pp. 12435–12445 (2021)

40. Milletari, F., Navab, N., Ahmadi, S.A.: V-net: fully convolutional neural networks for volumetric medical image segmentation. In: 2016 fourth international conference on 3D vision (3DV), pp. 565–571. IEEE (2016)
41. Mousavian, A., Toshev, A., Fišer, M., Košecká, J., Wahid, A., Davidson, J.: Visual representations for semantic target driven navigation. In: 2019 International Conference on Robotics and Automation (ICRA), pp. 8846–8852. IEEE (2019)
42. Muandet, K., Balduzzi, D., Schölkopf, B.: Domain generalization via invariant feature representation. In: International Conference on Machine Learning, pp. 10–18. PMLR (2013)
43. Neuhold, G., Ollmann, T., Rota Bulo, S., Kontschieder, P.: The mapillary vistas dataset for semantic understanding of street scenes. In: ICCV (2017)
44. Pan, F., Shin, I., Rameau, F., Lee, S., Kweon, I.S.: Unsupervised intra-domain adaptation for semantic segmentation through self-supervision. In: CVPR, pp. 3764–3773 (2020)
45. Pan, X., Luo, P., Shi, J., Tang, X.: Two at once: enhancing learning and generalization capacities via IBN-Net. In: Ferrari, V., Hebert, M., Sminchisescu, C., Weiss, Y. (eds.) ECCV 2018. LNCS, vol. 11208, pp. 484–500. Springer, Cham (2018). https://doi.org/10.1007/978-3-030-01225-0_29
46. Qiao, F., Zhao, L., Peng, X.: Learning to learn single domain generalization. In: CVPR, pp. 12556–12565 (2020)
47. Richter, S.R., Vineet, V., Roth, S., Koltun, V.: Playing for data: ground truth from computer games. In: Leibe, B., Matas, J., Sebe, N., Welling, M. (eds.) ECCV 2016. LNCS, vol. 9906, pp. 102–118. Springer, Cham (2016). https://doi.org/10.1007/978-3-319-46475-6_7
48. Ros, G., Sellart, L., Materzynska, J., Vazquez, D., Lopez, A.M.: The synthia dataset: a large collection of synthetic images for semantic segmentation of urban scenes. In: CVPR, pp. 3234–3243 (2016)
49. Sakaridis, C., Dai, D., Gool, L.V.: Guided curriculum model adaptation and uncertainty-aware evaluation for semantic nighttime image segmentation. In: ICCV, pp. 7374–7383 (2019)
50. Sakaridis, C., Dai, D., Van Gool, L.: Map-guided curriculum domain adaptation and uncertainty-aware evaluation for semantic nighttime image segmentation. IEEE TPAMI (2020). https://doi.org/10.1109/TPAMI.2020.3045882
51. Sakaridis, C., Dai, D., Van Gool, L.: ACDC: the adverse conditions dataset with correspondences for semantic driving scene understanding. arXiv preprint arXiv:2104.13395 (2021)
52. Sandler, M., Howard, A., Zhu, M., Zhmoginov, A., Chen, L.C.: Mobilenetv 2: inverted residuals and linear bottlenecks. In: CVPR, pp. 4510–4520 (2018)
53. Silberman, N., Hoiem, D., Kohli, P., Fergus, R.: Indoor segmentation and support inference from RGDB images. In: Fitzgibbon, A., Lazebnik, S., Perona, P., Sato, Y., Schmid, C. (eds.) ECCV 2012. LNCS, vol. 7576, pp. 746–760. Springer, Heidelberg (2012). https://doi.org/10.1007/978-3-642-33715-4_54
54. Tao, R., Gavves, E., Smeulders, A.W.: Siamese instance search for tracking. In: CVPR, pp. 1420–1429 (2016)
55. Tsai, Y.H., Hung, W.C., Schulter, S., Sohn, K., Yang, M.H., Chandraker, M.: Learning to adapt structured output space for semantic segmentation. In: CVPR, pp. 7472–7481 (2018)
56. Ulyanov, D., Vedaldi, A., Lempitsky, V.: Instance normalization: the missing ingredient for fast stylization. arXiv preprint arXiv:1607.08022 (2016)

57. Vu, T.H., Jain, H., Bucher, M., Cord, M., Pérez, P.: Advent: adversarial entropy minimization for domain adaptation in semantic segmentation. In: CVPR, pp. 2517–2526 (2019)
58. Wang, Z., Luo, Y., Qiu, R., Huang, Z., Baktashmotlagh, M.: Learning to diversify for single domain generalization. In: ICCV, pp. 834–843 (2021)
59. Wu, X., Wu, Z., Guo, H., Ju, L., Wang, S.: DANNet: a one-stage domain adaptation network for unsupervised nighttime semantic segmentation. In: CVPR, pp. 15769–15778 (2021)
60. Yang, Y., Soatto, S.: FDA: Fourier domain adaptation for semantic segmentation. In: CVPR, pp. 4085–4095 (2020)
61. Yu, F., et al.: BDD100K: a diverse driving dataset for heterogeneous multitask learning. In: CVPR (2020)
62. Yue, X., Zhang, Y., Zhao, S., Sangiovanni-Vincentelli, A., Keutzer, K., Gong, B.: Domain randomization and pyramid consistency: simulation-to-real generalization without accessing target domain data. In: CVPR, pp. 2100–2110 (2019)
63. Zhang, P., Zhang, B., Zhang, T., Chen, D., Wang, Y., Wen, F.: Prototypical pseudo label denoising and target structure learning for domain adaptive semantic segmentation. In: Proceedings of the IEEE/CVF Conference on Computer Vision and Pattern Recognition, pp. 12414–12424 (2021)
64. Zhang, Z., Peng, H.: Deeper and wider Siamese networks for real-time visual tracking. In: CVPR, pp. 4591–4600 (2019)
65. Zhao, H., Shi, J., Qi, X., Wang, X., Jia, J.: Pyramid scene parsing network. In: CVPR, pp. 2881–2890 (2017)
66. Zhao, L., Liu, T., Peng, X., Metaxas, D.: Maximum-entropy adversarial data augmentation for improved generalization and robustness. In: NeurIPS (2020)
67. Zhou, K., Yang, Y., Qiao, Y., Xiang, T.: Domain generalization with mixstyle. In: ICLR (2021)
68. Zhou, Z., Rahman Siddiquee, M.M., Tajbakhsh, N., Liang, J.: UNet++: a nested u-net architecture for medical image segmentation. In: Stoyanov, D., et al. (eds.) DLMIA/ML-CDS -2018. LNCS, vol. 11045, pp. 3–11. Springer, Cham (2018). https://doi.org/10.1007/978-3-030-00889-5_1
69. Zou, Y., Yu, Z., Vijaya Kumar, B.V.K., Wang, J.: Unsupervised domain adaptation for semantic segmentation via class-balanced self-training. In: Ferrari, V., Hebert, M., Sminchisescu, C., Weiss, Y. (eds.) ECCV 2018. LNCS, vol. 11207, pp. 297–313. Springer, Cham (2018). https://doi.org/10.1007/978-3-030-01219-9_18
70. Zou, Y., Yu, Z., Liu, X., Kumar, B., Wang, J.: Confidence regularized self-training. In: ICCV, pp. 5982–5991 (2019)

Context-Aware Streaming Perception in Dynamic Environments

Gur-Eyal Sela[1(✉)], Ionel Gog[1], Justin Wong[1], Kumar Krishna Agrawal[1], Xiangxi Mo[1], Sukrit Kalra[1], Peter Schafhalter[1], Eric Leong[1], Xin Wang[2], Bharathan Balaji[3], Joseph Gonzalez[1], and Ion Stoica[1]

[1] University of California, Berkeley, Berkeley, USA
ges@berkeley.edu
[2] Microsoft Research, Redmond, USA
[3] Amazon, Seattle, USA

Abstract. Efficient vision works maximize accuracy under a latency budget. These works evaluate accuracy offline, one image at a time. However, real-time vision applications like autonomous driving operate in streaming settings, where ground truth changes between inference start and finish. This results in a significant accuracy drop. Therefore, a recent work proposed to maximize accuracy in streaming settings on average. In this paper, we propose to maximize streaming accuracy for every environment context. We posit that scenario difficulty influences the initial (offline) accuracy difference, while obstacle displacement in the scene affects the subsequent accuracy degradation. Our method, Octopus, uses these scenario properties to select configurations that maximize streaming accuracy at test time. Our method improves tracking performance (S-MOTA) by 7.4% over the conventional static approach. Further, performance improvement using our method comes in addition to, and not instead of, advances in offline accuracy.

1 Introduction

Recent works like EfficientDet [26], YOLO [3], and SSD [16] were designed for real-time computer vision applications that require high accuracy in the presence of latency constraints. However, these solutions are evaluated offline, one image at a time, and do not consider the impact of increase in inference latency on the application performance. In real-time systems such as autonomous vehicles, the models are deployed in an online streaming setting where the ground truth changes during inference time as shown in Fig. 1a. To evaluate performance in

I. Gog—Now at Google Research.
B. Balaji—Work unrelated to Amazon.

Supplementary Information The online version contains supplementary material available at https://doi.org/10.1007/978-3-031-19839-7_36.

S. Avidan et al. (Eds.): ECCV 2022, LNCS 13698, pp. 621–638, 2022.
https://doi.org/10.1007/978-3-031-19839-7_36

streaming settings, Li et al. [12] proposed a modified metric that measures the model performance against the ground truth at the end of inference. They evaluated object detection models in a streaming fashion, and found that the average precision of the best performing model drops from 38.0 to 6.2, and picking the model that maximizes streaming average precision reduces the drop to 17.8.

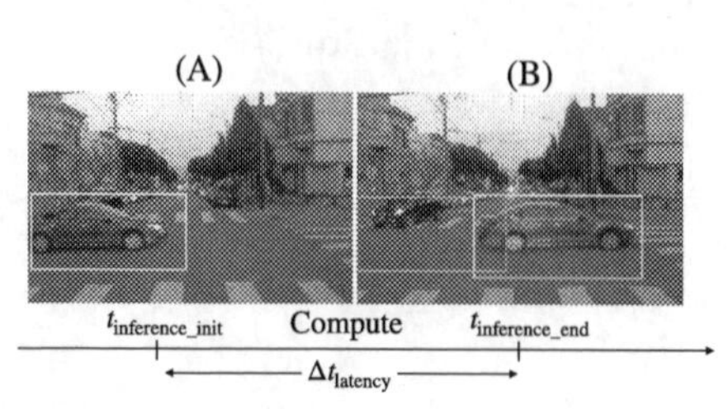

(a) In online streaming settings, the environment changes during inference. Streaming accuracy is computed by evaluating the prediction run on ground truth A against ground truth B.

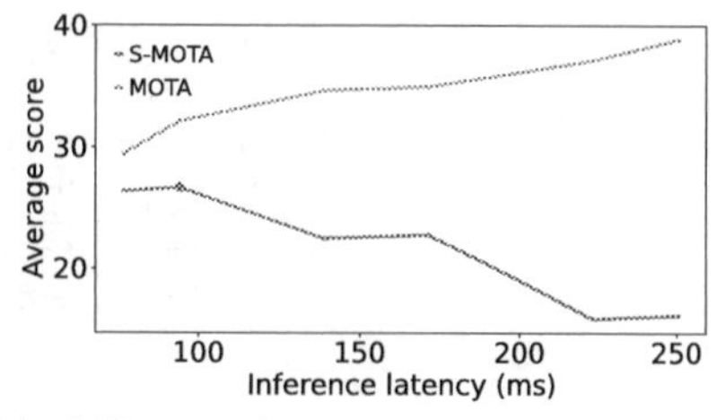

(b) Offline and streaming accuracy of a tracker of increasing model size. While the offline MOTA (orange, y-axis) of the model increases as inference latency (x-axis) increases, its streaming MOTA (S-MOTA in blue, y-axis) decreases.

Fig. 1. Streaming accuracy deviates from offline accuracy because the ground truth changes during inference.

We confirm the findings of Li et al. [12], and extend their analysis to object tracking. The standard metric for object tracking is MOTA (multiple object tracking accuracy) [11,18], and we refer to its streaming counterpart as S-MOTA. Figure 1b shows that larger models with higher MOTA deteriorate in S-MOTA for the Waymo dataset [28] as the higher latency widens the gap between ground truth between inference start and finish. The MOTA of the largest model (EfficientDet-D7x) is 38.9 while the S-MOTA is 16.2. The model that maximizes S-MOTA is EfficientDet-D4 with MOTA of 32.1 and S-MOTA of 26.7.

Since environment context varies, to further analyze the tradeoffs between latency and accuracy, we identify the model that maximizes the S-MOTA of 1-second video segment scenarios in the Waymo dataset. We observe that the best performing model varies widely from scenario to scenario (Fig. 2). Scenarios that are difficult (e.g., sun glare, drops on camera and reflection) and still (e.g., standing cars in intersection) show (Fig. 2 center) benefit from stronger perception while incurring marginal penalty from the latency increase. On the other hand, simple and fast scenes with rapid movement (e.g., turning, in Fig. 2 left), behave the opposite, where performance degrades sharply with latency. As a result, at the video segment-level the optimization landscape of streaming accuracy looks vastly different than on aggregate (Fig. 1b).

In this paper, we propose leveraging contextual cues to optimize S-MOTA dynamically at test time. The object detection model is just one of several choices that we refer to as *metaparameters* to consider in an object tracking system.

Our method, called Octopus, optimizes S-MOTA at test time by dynamically tuning the metaparameters. Concretely, we train a light-weight second-order model to switch between the metaparameters using a battery of environment features extracted from video segments, like obstacle movement speed, obstacle proximity, and time of day.

Our contributions in this work can be summarized as:

- We are the first to analyze object tracking in a streaming setting. We show that the models that maximize S-MOTA change per scenario and propose that the optimization tradeoffs are a result of scene difficulty as well as obstacle displacement.
- We present a novel method of S-MOTA optimization that leverages contextual features to switch the object tracking configuration at test-time.
- Our policy improves tracking performance (S-MOTA) by 7.4% over a static approach by evaluating it on the Waymo dataset [28]. We improve S-MOTA by 3.4% when we apply this approach on the Argoverse dataset [5].

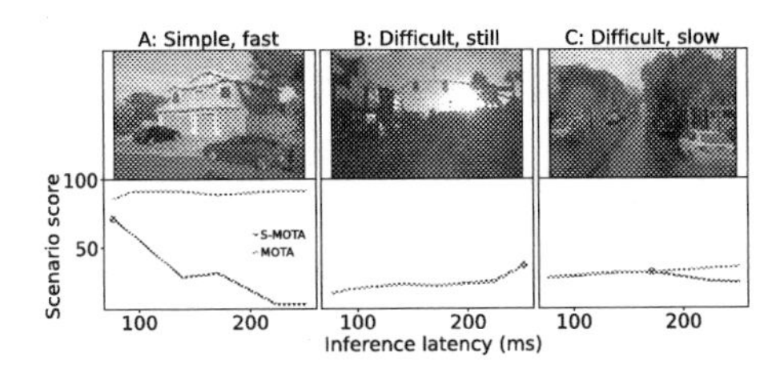

Fig. 2. Offline and streaming accuracies in three different scenarios. While offline MOTA (orange, y-axis) of the EfficientDet [26] models increases as inference latency (x-axis) increases, streaming MOTA (S-MOTA in blue, y-axis), responds depending on the scenario context (frames). Compare these plots to Fig. 1b, which shows the same on average over the entire dataset. (Color figure online)

2 Related Works

Latency *vs.* Accuracy. Prior works have recognized the tension between latency and accuracy in perception models [3,7,9,26], and examined facets of the tradeoff between decision speed *vs.* accuracy [24]. However, these works study this tradeoff in offline static settings. Li et al. [12] examine this tradeoff in streaming settings, where the ground truth of the world changes continuously. They show that conventional accuracy misrepresents perception performance in such settings, and proposed streaming accuracy. Their results were shown in detection (as well as semantic segmentation [6]), so we first verified that the idea also holds in tracking. Further, while Li et al. maximize streaming accuracy on average, we expose how the environment context plays a crucial role on its behavior. We leverage the context to dynamically maximize streaming accuracy at test time.

Model Serving Optimization and Dynamic Test-Time Adaptation. To reduce the inference latency, techniques such as model pruning [8,13,17] and quantization [23,29,31] have been proposed. While these techniques focus on effectively reducing the size of the models, our focus is on the policy: when and where to do it, in order to optimize accuracy in streaming settings. These techniques could be applied to reduce the latency of the models we use. Most similar to our setting are works that leverage context to dynamically adjust model architecture at test time to reduce resource consumption [10,27] or to improve throughput [25]. This work employs a similar approach in leveraging insight about latency *vs.* accuracy tradeoff in AV (autonomous vehicle) perception.

Inferring Configuration Performance Without Running. Several existing works achieve resource savings by modeling from data how a candidate configuration would perform without actually running it. Hyperstar [19] learns to approximate how a candidate hyperparameter configuration would perform on a dataset without training. Chameleon [10] reduces profiling cost of configurations for surveillance camera detection by assuming temporal locality among profiles. One key differentiating factor with our work is that the prior works were designed for a hybrid setting where some profiling is allowed. In our case, the strict time and compute constraints in the AV setting [14] restrict this approach, requiring an approximation-only method like Octopus.

3 Problem Setup

We first lay out the problem formulation (Sect. 3.1). Next, after an introduction of the Octopus dataset (Sect. 3.2), we measure the accuracy opportunity gap between the global best policy [12] and the optimal dynamic policy (Sect. 3.3). Finally, we perform a breakdown analysis of the components needed in order to optimize streaming accuracy at test time (Sect. 3.4).

3.1 Problem Formulation

Given real-time video stream as a series of images, we consider the inference of an AV pipeline (obstacle detection and tracking) on this stream. Let $\mathbb{S}$ denote the mean S-MOTA score of the tracking model, which depends on the values of the metaparameters $\mathcal{H}$, such as object detection model architecture and maximum age of tracked objects. Currently, the metaparameters $h \in \mathcal{H}$ are chosen using offline datasets, and are kept constant during deployment [12]. We refer to this method as the *global best* approach for statically choosing global metaparameters (h_{global}), which are expected to be best across all driving scenarios.

In contrast, in this work we study whether the S-MOTA score $\mathbb{S}$ can be improved by dynamically changing h every $\Delta\tau$ at test time. The metaparameters we choose at each time period $[\tau, \tau + \Delta\tau)$ is h_τ, and the corresponding score-optimal values is h^*_τ.

3.2 The Octopus Dataset

To generate the Octopus dataset ($\mathcal{D}$), we divide each video of a driving dataset into consecutive segments of duration $\Delta\tau$. We run the perception pipeline with a range of values of metaparameters $\mathcal{H}$, and record the S-MOTA score for each segment s^h_τ. We assign the optimal h^*_τ to the metaparameters that achieve the highest S-MOTA score (i.e., the optimal S-MOTA score s^*_τ). The segment duration $\Delta\tau$ is chosen to be short. This allows more accurately studying the performance potential of dynamic streaming accuracy optimization, because decision-making over smaller intervals generally performs better.

We generate the Octopus dataset by recording metaparameter values h_τ and the corresponding S-MOTA scores s^h_τ of the Pylot AV pipeline [7] for the Argoverse and Waymo datasets [5,28]. We execute Pylot's perception consisting of a suite of 2D object detection models from the EfficientDet model family [26] followed by the Simple, Online, and Real-Time tracker [2]. For each video scenario, we explore the following metaparameters:

Table 1. Dynamically changing metaparameters creates an accuracy opportunity gap. The *streaming* accuracy (S-MOTA) of the *global best* metaparameters h_{global} is 6.1 points lower on average than that of the *optimal* metaparameters h^*_τ. Similarly, there is a 3.2 gap in MOTA (top).

Method	Dataset	MOTA↑	MOTP↑	FP↓	FN↓	ID_{sw}↓
Global best	Waymo	37.3	78.1	21515	543193	11615
Optimal	Waymo	40.5	77.6	15738	532678	9137
Global best	Argoverse	63.0	82.1	6721	43376	1284
Optimal	Argoverse	70.8	81.0	5772	34210	828
Method	Dataset	S-MOTA↑	S-MOTP↑	S-FP↓	S-FN↓	S-ID_{sw}↓
Global best	Waymo	25.1	72.2	33616	633159	11212
Optimal	Waymo	31.2	71.0	28907	590847	6997
Global best	Argoverse	49.4	75.2	13485	55484	1092
Optimal	Argoverse	57.9	74.1	9562	48354	708

- **Detection model architecture**: selects the model from the EfficientDet family of models, which offers different latency *vs.* accuracy tradeoff points.
- **Tracked obstacles' maximum age**: limits the duration for which the tracker continues modeling the motion of previously-detected obstacles, under the assumption of temporary occlusion or low detection confidence (flickering).

Other metaparameters had limited effect on performance (Appendix A).

We run every metaparameter configuration in a Cartesian product of selected values for each metaparameter, and record latency metrics, detected objects, and tracked objects. The resulting 18 metaparameter configurations yield $\approx 18,000$ trials. We make this dataset public (https://github.com/EyalSel/Contextual-Streaming-Perception).

3.3 Accuracy Opportunity Gap

In order to study if the *global best* metaparameters h_{global} offer the best accuracy in all driving scenarios, we split the Octopus dataset ($\mathcal{D}$) into train ($\mathcal{D}_{train}$) and test ($\mathcal{D}_{test}$) sets. Next, we compute global best metaparameters (h_{global}) as the configuration that yields the highest mean S-MOTA score across all video segments in $\mathcal{D}_{train}$. We denote h_{global}'s mean S-MOTA scores on the train and test set as s_{train}^{global} and s_{test}^{global}, respectively. Similarly, we denote the mean S-MOTA scores of $h_{\mathcal{T}}^{*}$ (i.e., optimally changing metaparameters) as s_{train}^{*} and s_{test}^{*}.

We define the *S-MOTA opportunity gap* between the optimal dynamic metaparameters and the global best metaparameters as the upper bound of $s_{test}^{*} - s_{test}^{global}$. We repeat the same calculation for MOTA. In Table 1, we show the opportunity gap for the Argoverse and Waymo datasets [5,28] using $\Delta\mathcal{T} = 1$ s. We conclude that optimally choosing the metaparameters at test time offers a 6.1 (Waymo) and 8.5 (Argoverse) S-MOTA improvement on average, along with reductions in streaming false positives/negatives and streaming ID switches.

Of note, if offline accuracy were to increase uniformly across all configurations and scenarios, the performance improvement of the dynamic approach over the static baseline is expected to persist. This applies to the opportunity gap shown above (the optimal improvement), as well as for any dynamic policy improvement in this space. This means that performance improvement achieved by dynamic optimization come in addition to, and not instead of, further advances in conventional (offline) tracking.

3.4 Streaming Accuracy Analysis

We approach dynamic configuration optimization as a ranking problem [15], and solve it by learning to predict the difference in score of configuration pairs in a given scenario context [30]. We decompose this learning task into predicting the difference in (*i*) MOTA, and (*ii*) accuracy degradation during inference. To our knowledge we're the first to perform this analysis.

Decomposition. Streaming accuracy (S-MOTA) is tracking accuracy against ground truth at the end of inference, instead of the beginning (MOTA) [12]. Offline accuracy (MOTA) degrades as a result of change in ground truth during inference. The gap between MOTA and S-MOTA is defined here as the "degradation". Therefore, S-MOTA is expressed as $S - D$ where S is MOTA and D is the degradation. The difference in S-MOTA between two configurations is $(S_1 - S_2) - (D_1 - D_2)$. This decomposition separates the difference in S-MOTA of two configurations into two parts:

- $(S_1 - S_2)$ is the difference in offline accuracy. This difference originates from the MOTA boost, which is affected by *(i)* the added modeling capacity influenced by the scene difficulty for detection, a specific case of the more general

example difficulty [1], and *(ii)* the max-age choice which depends on the scene obstacle displacement [32].
- $(D_1 - D_2)$ is the difference in accuracy degradation. This may be derived from: a. the difference configuration latencies and b. scene obstacle displacement.

Predicting Both Components to Optimize Streaming Accuracy. At test-time, $S_1 - S_2$ and $D_1 - D_2$ are predicted for each scenario. We find that accurately predicting both components per environment context is necessary to realize the *opportunity gap* (see Sect. 3.3). First, we show that perfectly predicting $D_1 - D_2$ (ΔD^*) and $S_1 - S_2$ (ΔS^*) (Table 2, top left) yields 31.2 S-MOTA, the same as the optimal policy h^*_τ on Waymo (Table 1 bottom panel, row 2). Then, we predict $D_1 - D_2$ and $S_1 - S_2$ on average across all scenarios for each configuration ($\overline{\Delta D}$ and $\overline{\Delta S}$ respectively in Table 2). Combining $\overline{\Delta D}$ and $\overline{\Delta S}$ (Table 2, bottom-right) yields 25.1 S-MOTA, the same as the global best policy h_{global} score on Waymo (Table 1 bottom panel, row 1). The hybrid-optimal policies (Table 2 top-right and bottom-left) achieve 1.8 (26.9 – 25.1) and 2.7 (27.8 – 25.1) of the 6.1 (31.2 – 25.1) optimal policy opportunity gap. Taken together, these results demonstrate the need to accurately predict both the change in MOTA $(S_1 - S_2)$ and in degradation $(D_1 - D_2)$ per scenario in order to optimize S-MOTA at test time.

Table 2. Both streaming accuracy components must be predicted per scenario in order to realize the full dynamic policy's opportunity gap. ΔD^* is optimal degradation prediction, ΔS^* is optimal offline gap prediction. $\overline{\Delta D}$ and $\overline{\Delta S}$ are the corresponding global-static policies.

	ΔD^*	$\overline{\Delta D}$
ΔS^*	31.2	26.9
$\overline{\Delta S}$	27.8	25.1

4 Octopus: Environment-Driven Perception

We propose leveraging properties of the AV environment context (e.g., ego speed, number of agents, time of day) that can be perceived from sensors in order to dynamically change metaparameters at test time. We first formally present our approach for choosing metaparameters, which uses regression to infer a ranking of metaparameter configurations (Sect. 4.1). Then, we describe the environment representation that allows to effectively infer each component of the streaming accuracy (Sect. 4.2).

4.1 Configuration Ranking via Regression

In order to find the metaparameters h_τ that maximize the S-MOTA score s_τ^h for each video segment, Octopus first learns a regression model M. The model predicts s_τ^h given the metaparameters h_τ and the representation of the environment e_τ for the period τ. Following, Octopus considers all metaparameter values, and picks the metaparameters that give the highest predicted S-MOTA.

Executing the model M at the beginning of each segment τ requires an up-to-date representation of the environment. However, building a representation requires the output of the perception pipeline (e.g., number of obstacles). Octopus could run the current perception configuration in order to update the environment before executing the model M, but this approach would greatly increase the response time of the AV. Instead, Octopus makes a Markovian assumption, and inputs the environment representation of the previous segment $e_{\tau-1}$ to the model M.

$$M(h_\tau, e_{\tau-1}) = \hat{s}_\tau^h \tag{1}$$

Thus, it is important to limit the segment length $\Delta\tau$ as the longer a segment is, the more challenging it is to accurately predict the scores due to using an older environment representation (see Sect. 5.1 for our methodology for choosing $\Delta\tau$).

The model is trained using the mean squared error loss. We choose the metaparameters that the model predicts as the highest score. Let s_τ^{global} denote the S-MOTA score obtained with h^{global} metaparameters, which by definition is a lower bound of the optimal S-MOTA score s_τ^* (i.e., $s_\tau^{global} \leq s_\tau^*$). Thus, in order to pick the best metaparameters, Octopus can only predict s_τ^h relative to s_τ^{global}. As a result, Octopus utilizes the following as the final loss:

$$L = \frac{1}{N} \sum_{i=0}^{N} \left(\hat{r}_i^h - clip((s_i^h - s_i^{global}), \epsilon) \right)^2 \tag{2}$$

where N is the size of the training data and $\hat{r}_i^h$ is the relative S-MOTA score predicted by the model M, $\hat{r}_i^h = \hat{s}_i^h - s_i^{global}$.

Finally, Octopus clips by lower bounding the predictions by $s_i^{global} - \epsilon$ as the predictions significantly below s_i^{global} are irrelevant to the optimization problem. Moreover, by clipping the predictions, Octopus reduces the dynamic range of the regressor and makes it easier to predict the higher S-MOTA scores.

4.2 Environment Representation

We can represent the environment context e_τ by capturing the characteristics of the video segment. In order to keep the decision-making latency small, we eschew more complex learned representation designs. Instead, we developed hand-engineered features from sensors and the outputs of the object detection model following prior work by Nishi et al. [20], where the features were used to predict human driving behavior. We collect the features per frame and then aggregate by averaging across frames in the video segment. Unless otherwise specified, we use the 10th percentile, mean, and 90th percentile of the following features:

1. **Bounding box speed** is the distance traveled by an object in pixel space across two frames. This feature infers the speed of objects, and thus it is important for capturing obstacle displacement to predict the streaming accuracy degradation.
2. **Bounding box self IoU** is the Intersection-over-Union (IoU) of an object's bounding box in the current frame relative to the previous frame. Along with the bounding speed, this feature helps isolate the change in size of an obstacle, as it moves towards or away from the ego vehicle.
3. **Number of objects** is measured per frame, and indicates the complexity of a scene as the more objects in a scene, the more likely the AV is to encounter object path crossings and occlusions (scene difficulty). Therefore, this feature signals when to prioritize for high offline accuracy configurations that are robust to object occlusions.
4. **Obstacle longevity** is the number of frames for which an obstacle has been tracked. Lower obstacle longevity implies more occlusion as obstacles enter and leave the scene, making perception more difficult. This feature also guides the choice of tracking metaparameters as low obstacle longevities correlate with lower tracking maximum age giving better performance.
5. **Ego driving speed** is the speed at which the AV is traveling, and indicates the environment in which the AV is driving (e.g., highway vs. city).
6. **Ego turning speed** is the angle change of the AV's direction between consecutive frames. This feature helps differentiate the source of apparent obstacle displacement between obstacle movement and ego movement.
7. **Time of day** hints if a high-accuracy configuration is required to handle challenging scenes (e.g., night driving, sun glare at sunset).

We compute these feature statistics for different bounding box size ranges in order to reason about obstacle behavior depending on the detection model strength needed for their accurate detection. For example, higher offline accuracy models do not confer better streaming accuracy if the obstacles in the detectable size range move too rapidly for the longer detection inference time to keep up with.

Note that the choice of metaparameters in each video segment changes the configuration, and hence the outputs of the tracking pipeline. Therefore, the resulting environment representation is no longer independently and identically distributed (i.i.d), making it challenging to use traditional supervision techniques. To keep the data distribution stationary during train time, we use the object detection model as given by the global best metaparameters h_{global}. While this training objective is biased considering that the features may be derived from any configuration, we find that it reduces variance during training, and works better than using features derived from the ground truth.

We concatenate the metaparameters h and the environment representation vector e_{τ}, and then use a supervised regression model to predict the score, optimizing using the loss given in Eq. (2).

In addition, we compare our method to a conventional CNN model approach (ignoring its resource and runtime requirements) in Appendix I.

5 Experiments

Next, we analyze our proposed approach. The subsections discuss the following:

1. **Setup** (Sect. 5.1): Description of Datasets and Model Details
2. **Main results** (Sect. 5.2): What is the performance improvement conferred by the dynamic policy over the static baseline?
3. **Explainability** (Sect. 5.3): **I.** Does the learned policy behavior match human understanding of the driving scenario? **II.** Where is the learned policy similar to and different from the optimal policy? **III.** How are scenarios clustered by metaparameter score? **IV.** What is the relative importance of the hand-picked features and of the metaparameters towards S-MOTA optimization?
4. **Ablation study (Appendix H)**: How do various ranking implementation choices affect the final performance?

5.1 Methodology

We evaluate on the Argoverse and Waymo datasets [5,28], covering a variety of environments, traffic, and weather conditions. We do not use the private Waymo test set as it does not support streaming metrics. However, we treat the Waymo validation set as the test set, and we perform cross validation on the training dataset (798 videos for training, and 202 videos for validation). Similarly, we divide the Argoverse videos into 75 videos for training, and 24 for validation. In addition, we follow the methodology from Li et al. [12] in order to create ground truth 2D bounding boxes and tracking IDs, which are not present in the Argoverse dataset. We generate these labels using QDTrack [22] trained on the BDD100k dataset, which is the highest offline accuracy model available.

In our experiments, we run object detectors from the EfficientDet architecture [26] on NVIDIA V100 GPUs, and the SORT tracker [2] on a CPU (simulated evaluation on faster hardware is in Appendix J). We chose to use the EfficientDet models because they are especially optimized for trading off between latency and offline accuracy, and because they are close to the state-of-the-art. The EfficientDet models were further optimized using Tensor RT [21], which both reduces inference latencies to less than 250 ms (the latency of the largest model, EfficientDet-D7x) and decreases resource requirements to at most 3 GPUs. We use these efficient models to investigate the effect of optimizing two metaparameters: detection model and tracking max age (see Sect. 3.2 for details). We compile 18 configurations of the AV perception pipeline by exploring the Cartesian product of the values of the two metaparameters (see Listing 1.1).

```
1 # Metaparameters
2 detection-model = {EfficientDet: 3, 4, 5, 6, 7, 7x}
3 tracking-maximum-age = {1, 3, 7}
```

Listing 1.1. Values for detection and tracking metaparameters.

We implement Octopus's policy regression model as a Random Forest [4] using Eq. (2) to choose metaparameters for video segment length $\Delta\tau = 1\,\mathrm{s}$. We describe training details in Appendix B.

Policy Runtime Overhead. Every step $\Delta\tau$, the Octopus policy applies Random Forest regression for all the 18 perception pipeline configurations in order to predict the best one to apply in the next step. Due to the lightweight policy design, inference on all 18 configurations has a latency of at most 6 ms using a single Intel Xeon Platinum 8000 core. As a result, the policy decision finishes before the sensor data of the next segment arrives, and thus does not affect latency of the perception pipeline. Moreover, Octopus pre-loads the perception model weights (13.86 GB in total in our experiments) and forward pass activations in the GPU memory (32 GB), and thus actuates metaparameter changes quickly.

5.2 Main Results

We compare the Octopus policy with the optimal policy (h^*_τ) and the global best policy (h^{global}) as proposed by Li et al. [12]. In addition, following the discussion in Sect. 4.2, we show ablations of the Octopus policy in order to highlight the relative contribution in both the setting of predictive and close-loop metaparameter optimization. Concretely, we include the following setups that Octopus can use to optimize the metaparameters at time t:

- *Ground truth from current segment*: features are derived from the sensor data and the labels at time t.
- *Ground truth from previous segment*: features are derived from the sensor data and the labels at time $t - \Delta\tau$.
- *Closed-loop prediction from previous segment*: features are derived from the sensor data and the output of the perception pipeline for the previous segment (i.e., at time $t - \Delta\tau$).

In Table 3 we show the tracking accuracy results for the Waymo and Argoverse datasets. The results show that the Octopus policy with closed-loop prediction outperforms the global best policy by 1.9 S-MOTA (Waymo) and 1.7 S-MOTA (Argoverse). Both the optimal and Octopus policies achieve further accuracy increases using features derived from the current segment, further illustrating how rapidly configuration score changes over time. The consistent accuracy improvements across the two datasets show that leveraging environment context to dynamically optimize streaming accuracy provides substantial improvements over the state-of-the-art. Moreover, as discussed in Sect. 3.3, streaming accuracy improvement over the global best policy will likely persist independently of innovation in offline accuracy of the underlying perception models.

Table 3. Streaming tracking accuracy results on two datasets.

(a) Waymo

Method	S-MOTA↑	S-MOTP↑	S-FP↓	S-FN↓	S-ID$_{sw}$↓
Global best	25.1	72.2	33616	633159	11212
Optimal	31.2	71.0	28907	590847	6997
Optimal from the prev. segment	27.2	71.2	38631	603777	8092
Octopus with:					
Ground truth from current segment	27.9	72.3	31489	608870	8966
Ground truth from prev. segment	27.3	72.7	31056	612862	9103
Prediction from prev. segment	**27.0**	**72.8**	**30272**	**615780**	**9511**

(b) Argoverse

Method	S-MOTA↑	S-MOTP↑	S-FP↓	S-FN↓	S-ID$_{sw}$↓
Global best	49.4	75.2	13485	55484	1092
Optimal	57.9	74.1	9562	48354	708
Optimal from the prev. segment	51.9	74.0	12652	52804	810
Octopus with:					
Ground truth from current segment	53.2	74.9	11447	52638	917
Ground truth from prev. segment	51.6	75.1	11348	54502	1010
Prediction from prev. segment	**51.1**	**74.9**	**12062**	**54469**	**1008**

S-MOTA *vs.* S-MOTP. Table 3 highlights that both Octopus and optimal policy occasionally deteriorate S-MOTP, inversely to the improvement in S-MOTA. This result reflects on a broader pattern where in streaming settings bounding boxes in general lag after the ground truth, even if by a small enough margin to be counted as true positives. S-MOTP, which is weighted by the IOU between correct predictions and the ground truth is especially hurt as a result. S-MOTA, which just counts the number of false positives, is less affected by this. We provide a longer analysis of this tradeoff/pareto-frontier between S-MOTA and S-MOTP in Appendix D.

5.3 Explainability

We survey various aspects of the metaparameter optimization problem, and qualitatively compare the global best (baseline), the optimal, and the Octopus (learned) policies.

I. Case study: Busy Intersection. Figure 3 shows a scenario where the ego vehicle enters a busy intersection. The scenario is divided into three phases, where the learned (Octopus) policy adaptively tunes the metaparameter to varying road conditions similarly to the optimal policy plan. First, the ego vehicle approaches the intersection with oncoming traffic on the left. The vehicles in the distance cannot be picked up by any of the candidate models. The learned policy chooses D4, the same model as the global best policy. Then, as the light turns yellow and the ego vehicle comes to a stop first in the queue, the learned policy adapts by increasing the strength of the model to D7x. The perception pipeline is now able to detect the smaller vehicles in the opposing lane and adjacent to the road. The optimal policy makes a similar decision while the global best policy remains with the D4 choice, incurring a net performance loss. Finally, as vehicles in the cross-traffic start passing and occluding the vehicles in the background, the learned policy returns to the lower latency D3 model. This scenario illustrates how the learned policy can select the best model in each phase of the scenario, adapting to the change in driving environment.

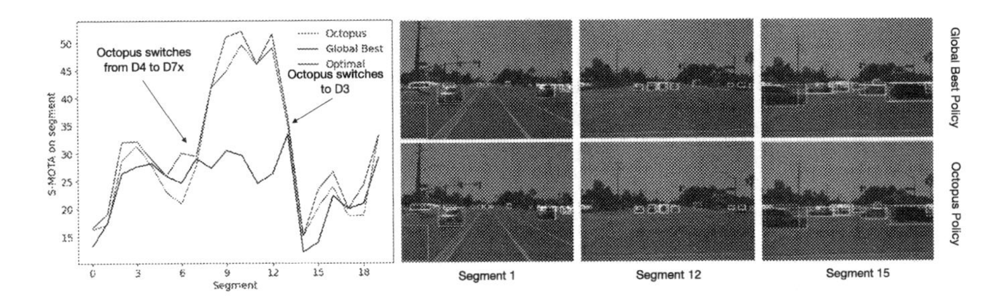

Fig. 3. Busy Intersection Scenario. Left: The S-MOTA score of each $\tau = 1s$ segment for the Octopus (learned), global best, and optimal policies. Right: The front-facing camera feed. Red bounding-boxes represent the ground truth, and orange represent the pipeline's predictions using the policy's configuration choice.

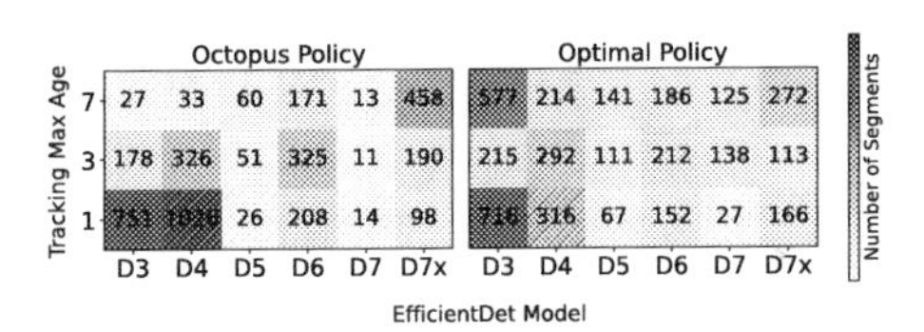

Fig. 4. Policy decision frequency. The configuration choice frequency (color intensity) of the Octopus policy (left) and of the optimal policy (right). The global best configuration is EfficientDet-D4 with tracking max age of 1, emphasized with gray stripes. (Color figure online)

II. Policy Action Heatmap. In Fig. 4, we visualize the learned policy in comparison to the optimal policy. As we expect, the learned policy often selects the global best configuration (D4-1), but expands out similarly to the optimal policy when performance can be improved by changing configurations. This result illustrates how, contrary to common belief that onboard perception must have low latency (e.g. under 100 ms [28]), higher perception latencies are tolerable and even preferable in certain environments. The Octopus policy learns to take advantage of this when it opts for higher offline accuracy models.

III. Scenarios are Clustered by metaparameter Score. Here, we evaluate whether scenarios are grouped by common metaparameter score behaviors. To this end, we first visualize the scenario score space (explained below) as a t-SNE plot and then perform more formal centroid analysis.

To this end, each video segment of length $\Delta\tau = 1s$ is vectorized by computing the MOTA score difference $(S_c - S_{h_{global}})$ and degradation difference $(D_c - D_{h_{global}})$ for every metaparameter configuration c and the global best configuration h_{global} (see Sect. 3.4). These values are then concatenated and normalized (z-score) across each scenario.

Cluster Visualization. We visualize this space in a t-SNE plot in Fig. 5. The scenario segment points are colored according to the model that optimizes MOTA (left) and S-MOTA (right). We observe a non-uniform impact of accuracy degradation on the optimal model choice in different video segment regions. This varying effect illustrates that S-MOTA deviates from MOTA setting on an environment context-dependent basis. For case-study analysis of points in the t-SNE please see Appendix F.

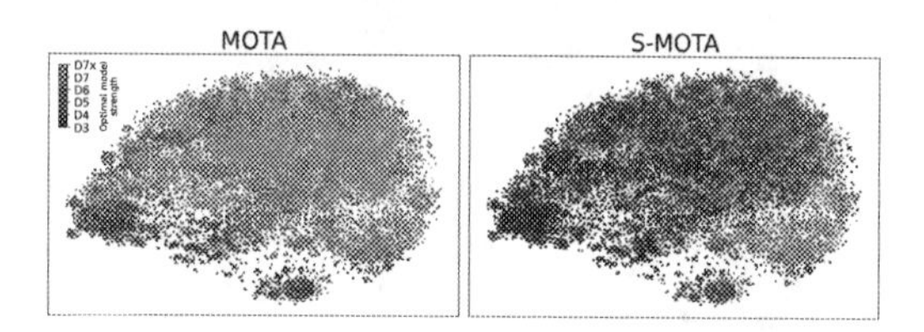

Fig. 5. The impact of the accuracy degradation incurred in online streaming context is scenario dependent. Each point is a 1 s driving scenario. Its color is the model that maximizes offline accuracy (left) and streaming accuracy (right).

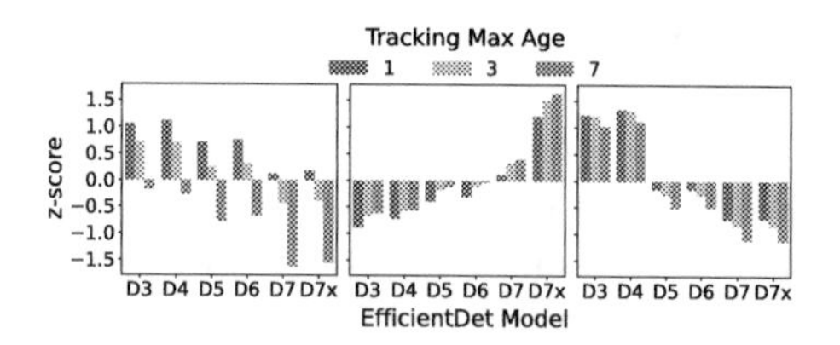

Fig. 6. Three centroids of the configuration score distribution: (i) preference for low max age (left), (ii) preference for offline accuracy (middle), and (iii) preference for low latency (right). The three clusters account for 34% of the dataset.

Centroid Analysis. We now perform a more formal clustering analysis that reveals modes of metaparameter optimality.

We perform K-means clustering, with k = 8 on the z-score space. Figure 6 shows centroids of three representative scenario clusters (right to left): (i) scenarios that benefit from lower latency detection. (ii) scenarios that benefit from higher accuracy detection, and (iii) scenarios that primarily benefit from a lower

Table 4. Importance study

(a) Feature importance (Regression)

Feature	Importance Score	Feature	Importance Score
Mean BBox self IOU	0.304	BBox bin [665, 1024)	0.087
Mean BBox Speed	0.200	BBox bin [1024, 1480)	0.071
Num. BBoxes	0.063	BBox bin [1480, 2000)	0.055
BBox Longevity	0.076	BBox bin [2000, 2565)	0.040
Ego movement	0.096	BBox bin [2565, ∞)	0.417
Time of Day	0.008		

(b) Metaparameter importance (Regression)

Metaparameter	S-MOTA$\uparrow$	S-MOTP$\uparrow$	S-FP$\downarrow$	S-FN$\downarrow$	S-ID$_{sw}\downarrow$
Global best	25.1	72.2	33616	633159	11212
Detection-model	27.6	72.4	26116	615771	11254
Tracking-max-age	25.6	71.9	39400	625184	8973
Both	27.9	72.3	31489	608870	8966

tracking max age, and to a lesser extent from lower latency detection. These distinct modes reflect that scenarios are grouped around similar metaparameter behaviors, corroborating with variation in scene difficulty and obstacle displacement. The rest of the scenario cluster visualizations are in Appendix G.

IV. Feature and Configuration Metaparameter Importance.

Feature Importance. To study the relative importance of the environment features used in our solution (described in Sect. 4.2), we show the feature importance scores derived from the trained Random Forest regressor in Table 4a. The mean bounding box self-IOU and speed together constitute over half of the normalized importance score, as they are predictive of the accuracy degradation that higher-latency detection models would incur. Aggregate bounding box statistics that capture scene difficulty, such as the average number and longevity of the bounding boxes per frame, are also useful for prediction.

Configuration metaparameter Importance. In Table 4b, we evaluate the gains attributed to each metaparameter by only optimizing one metaparameter at a time, fixing the other parameter to the value in the global best configuration. We do this to ablate the performance shown in Table 3, where they are optimized together. In both cases, the models were trained on the ground truth and the present. Although the detection model choice is the primary contributor, the additional choice of occlusion tolerance (max age) further contributes to the achieved performance. We also observe how the performance gain when optimizing the tracker's maximum occlusion tolerance (see Sect. 3 for details) does not combine additively with the improvement in the detection model optimization.

6 Conclusions

Streaming accuracy is a much more accurate representation of tracking for real-time vision systems because it uses ground truth at the end of inference to measure performance. In this study we show the varying impact that environment context has on the deviation of streaming accuracy from offline accuracy. We propose a new method, Octopus, to leverage environment context to maximize streaming accuracy at test time. Further, we decompose streaming accuracy into two components: difference in offline accuracy MOTA and the degradation, and show that both must be inferred in every scenario to achieve optimal performance. Octopus improves streaming accuracy over the global best policy in multiple autonomous vehicle datasets.

Acknowledgements. We thank Daniel Rothchild and Horia Mania for helpful discussions.

References

1. Baldock, R., Maennel, H., Neyshabur, B.: Deep learning through the lens of example difficulty. Adv. Neural. Inf. Process. Syst. **34**, 10876–10889 (2021)
2. Bewley, A., Ge, Z., Ott, L., Ramos, F., Upcroft, B.: Simple online and realtime tracking. In: Proceedings of the 23th IEEE International Conference on Image Processing (ICIP), pp. 3464–3468 (2016)
3. Bochkovskiy, A., Wang, C.Y., Liao, H.Y.M.: YOLOv4: optimal speed and accuracy of object detection. arXiv preprint arXiv:2004.10934 (2020)
4. Breiman, L.: Random forests. Mach. Learn. **45**(1), 5–32 (2001)
5. Chang, M.F., et al.: Argoverse: 3D tracking and forecasting with rich maps (2019)
6. Courdier, E., Fleuret, F.: Real-time segmentation networks should be latency aware. In: Proceedings of the Asian Conference on Computer Vision (2020)
7. Gog, I., Kalra, S., Schafhalter, P., Wright, M.A., Gonzalez, J.E., Stoica, I.: Pylot: a Modular platform for exploring latency-accuracy tradeoffs in autonomous vehicles. In: Proceedings of IEEE International Conference on Robotics and Automation (ICRA) (2021)
8. Han, S., Pool, J., Tran, J., Dally, W.J.: Learning both weights and connections for efficient neural networks. In: Proceedings of the 28th International Conference on Neural Information Processing (NeurIPS), pp. 1135–1143 (2015)
9. Huang, J., et al.: Speed/accuracy trade-offs for modern convolutional object detectors. In: Proceedings of the IEEE Conference on Computer Vision and Pattern Recognition (CVPR) (2017)
10. Jiang, J., Ananthanarayanan, G., Bodik, P., Sen, S., Stoica, I.: Chameleon: scalable adaptation of video analytics. In: Proceedings of the 2018 Conference of the ACM Special Interest Group on Data Communication (SIGCOMM), pp. 253–266 (2018)
11. Leal-Taixé, L., Milan, A., Schindler, K., Cremers, D., Reid, I., Roth, S.: Tracking the trackers: an analysis of the state of the art in multiple object tracking. arXiv preprint arXiv:1704.02781 (2017)
12. Li, M., Wang, Y.-X., Ramanan, D.: Towards streaming perception. In: Vedaldi, A., Bischof, H., Brox, T., Frahm, J.-M. (eds.) ECCV 2020. LNCS, vol. 12347, pp. 473–488. Springer, Cham (2020). https://doi.org/10.1007/978-3-030-58536-5_28

13. Lin, J., Rao, Y., Lu, J., Zhou, J.: Runtime Neural Pruning. In: Proceedings of the 31st International Conference on Neural Information Processing Systems (NeurIPS), pp. 2178–2188 (2017)
14. Lin, S.C., et al.: The architectural implications of autonomous driving: constraints and acceleration. In: Proceedings of the 23rd International Conference on Architectural Support for Programming Languages and Operating Systems (ASPLOS), pp. 751–766 (2018)
15. Liu, T.Y., et al.: Learning to rank for information retrieval. Found. Trends® Inf. Retrieval **3**(3), 225–331 (2009)
16. Liu, W., et al.: SSD: single shot MultiBox detector. In: Leibe, B., Matas, J., Sebe, N., Welling, M. (eds.) ECCV 2016. LNCS, vol. 9905, pp. 21–37. Springer, Cham (2016). https://doi.org/10.1007/978-3-319-46448-0_2
17. Luo, J.H., Wu, J., Lin, W.: Thinet: a filter level pruning method for deep neural network compression. In: Proceedings of the IEEE/CVF Conference on Computer Vision and Pattern Recognition (CVPR), pp. 5058–5066 (2017)
18. Milan, A., Leal-Taixé, L., Reid, I., Roth, S., Schindler, K.: Mot16: a benchmark for multi-object tracking. arXiv preprint arXiv:1603.00831 (2016)
19. Mittal, G., Liu, C., Karianakis, N., Fragoso, V., Chen, M., Fu, Y.: HyperSTAR: task-aware hyperparameters for deep networks. In: Proceedings of the IEEE/CVF Conference on Computer Vision and Pattern Recognition (CVPR), pp. 8736–8745 (2020)
20. Nishi, K., Shimosaka, M.: Fine-grained driving behavior prediction via context-aware multi-task inverse reinforcement learning. In: 2020 IEEE International Conference on Robotics and Automation (ICRA), pp. 2281–2287. IEEE (2020)
21. NVIDIA: Tensor RT. https://developer.nvidia.com/tensorrt
22. Pang, J., et al.: Quasi-dense similarity learning for multiple object tracking. In: Proceedings of the IEEE/CVF Conference on Computer Vision and Pattern Recognition, pp. 164–173 (2021)
23. Park, E., Ahn, J., Yoo, S.: Weighted-entropy-based quantization for deep neural networks. In: Proceedings of the IEEE/CVF Conference on Computer Vision and Pattern Recognition (CVPR), pp. 5456–5464 (2017)
24. Pleskac, T.J., Busemeyer, J.R.: Two-stage dynamic signal detection: a theory of choice, decision time, and confidence. Psychol. Rev. **117**(3), 864 (2010)
25. Shen, H., Han, S., Philipose, M., Krishnamurthy, A.: Fast video classification via adaptive cascading of deep models. In: Proceedings of the IEEE Conference on Computer Vision and Pattern Recognition (CVPR), pp. 3646–3654 (2017)
26. Tan, M., Pang, R., Le, Q.V.: EfficientDet: scalable and efficient object detection. In: Proceedings of the IEEE Conference on Computer Vision and Pattern Recognition (CVPR) (2020)
27. Wang, X., Yu, F., Dou, Z.-Y., Darrell, T., Gonzalez, J.E.: SkipNet: learning dynamic routing in convolutional networks. In: Ferrari, V., Hebert, M., Sminchisescu, C., Weiss, Y. (eds.) ECCV 2018. LNCS, vol. 11217, pp. 420–436. Springer, Cham (2018). https://doi.org/10.1007/978-3-030-01261-8_25
28. Waymo Inc.: Waymo Open Dataset. https://waymo.com/open/
29. Xu, Y., Wang, Y., Zhou, A., Lin, W., Xiong, H.: Deep neural network compression with single and multiple level quantization. In: Proceedings of the AAAI Conference on Artificial Intelligence, vol. 32 (2018)
30. Yogatama, D., Mann, G.: Efficient transfer learning method for automatic hyperparameter tuning. In: Artificial Intelligence and Statistics, pp. 1077–1085. PMLR (2014)

31. Zhao, R., Hu, Y., Dotzel, J., De Sa, C., Zhang, Z.: Improving neural network quantization without retraining using outlier channel splitting. In: International Conference on Machine Learning (ICML), pp. 7543–7552 (2019)
32. Zhou, X., Koltun, V., Krähenbühl, P.: Tracking objects as points. In: Vedaldi, A., Bischof, H., Brox, T., Frahm, J.-M. (eds.) ECCV 2020. LNCS, vol. 12349, pp. 474–490. Springer, Cham (2020). https://doi.org/10.1007/978-3-030-58548-8_28

SpOT: Spatiotemporal Modeling for 3D Object Tracking

Colton Stearns[1(✉)], Davis Rempe[1], Jie Li[2], Rareş Ambruş[2], Sergey Zakharov[2], Vitor Guizilini[2], Yanchao Yang[1], and Leonidas J. Guibas[1]

[1] Stanford University, Stanford, USA
coltongs@stanford.edu
[2] Toyota Research Institute, Los Altos, USA

Abstract. 3D multi-object tracking aims to uniquely and consistently identify all mobile entities through time. Despite the rich spatiotemporal information available in this setting, current 3D tracking methods primarily rely on abstracted information and limited history, *e.g.* single-frame object bounding boxes. In this work, we develop a holistic representation of traffic scenes that leverages both spatial and temporal information of the actors in the scene. Specifically, we reformulate tracking as a spatiotemporal problem by representing tracked objects as sequences of time-stamped points and bounding boxes over a long temporal history. At each timestamp, we improve the location and motion estimates of our tracked objects through learned refinement over the full sequence of object history. By considering time and space jointly, our representation naturally encodes fundamental physical priors such as object permanence and consistency across time. Our spatiotemporal tracking framework achieves state-of-the-art performance on the Waymo and nuScenes benchmarks.

Keywords: 3D object detection · 3D object tracking · point clouds · LiDAR · Autonomous driving · NuScenes Dataset

1 Introduction

3D multi-object tracking (MOT) is an essential task for modern robotic systems designed to operate in the real world. It is a core capability to ensure safe navigation of autonomous platforms in dynamic environments, connecting object detection with downstream tasks such as path-planning and trajectory forecasting. In recent years, new large scale 3D scene understanding datasets of driving scenarios [3,33] have catalyzed research around 3D MOT [6,12,35,36]. Nevertheless, establishing high-fidelity object tracks for this safety-critical application remains a challenge. Notably, recent literature suggests that even small errors in 3D tracking can lead to significant failures in downstream tasks [34,40].

Supplementary Information The online version contains supplementary material available at https://doi.org/10.1007/978-3-031-19839-7_37.

S. Avidan et al. (Eds.): ECCV 2022, LNCS 13698, pp. 639–656, 2022.
https://doi.org/10.1007/978-3-031-19839-7_37

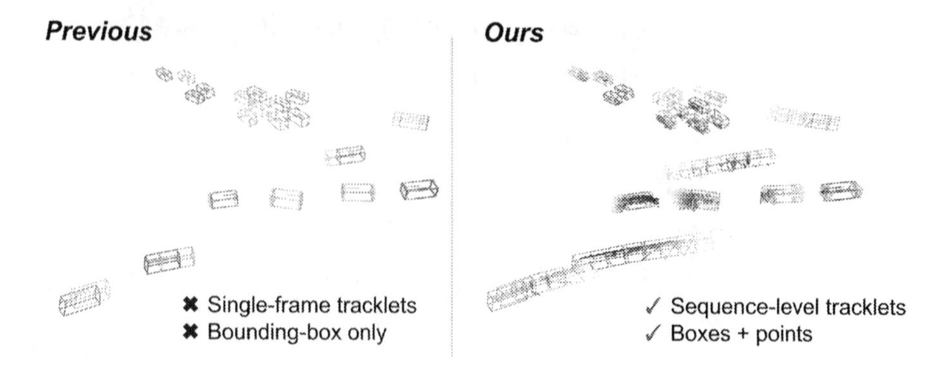

Fig. 1. Previous works use a highly abstracted tracklet representation (*e.g.* bounding boxes) and a compressed motion model (Kalman filter or constant velocity). We efficiently maintain an active history of object-level point clouds and bounding boxes for each tracklet.

A distinct challenge faced by 3D MOT is that of data association when using LIDAR data as the main source of observation, due to the sparse and irregular scanning patterns inherent in time-of-flight sensors designed for outdoor use. Established works in 2D use appearance-based association [1,44], however, these cannot be directly adapted to 3D MOT. Sensor fusion methods combine camera and LIDAR in an effort to provide appearance-based cues in 3D association [5, 12,36]. However, this comes at the cost of additional hardware requirements and increased system complexity.

Most of the recent works in 3D MOT from LIDAR data address the association problem by matching single-frame tracks to current detection results with close 3D proximity. Single-frame detection results are modeled as bounding boxes [35] or center-points [39] and compared to the same representation of the tracked objects from the last visible frame. Although it touts simplicity, this strategy does not fully leverage the spatiotemporal nature of the 3D tracking problem: temporal context is often over-compressed into a simplified motion model such as a Kalman filter [6,35] or a constant-velocity assumption [39]. Moreover, these approaches largely ignore the low-level information from sensor data in favor of abstracted detection entities, making them vulnerable to crowded scenes and occlusions.

However, improving spatiotemporal context by integrating scene-level LIDAR data over time is challenging due to the large quantity of sampled points along with sparse and irregular scanning patterns. Some methods aggregate LIDAR to improve 3D detection over short time horizons and in static scenes [3,9], as well as over longer time horizons in an *offline* manner [23]. There is also recent work for *single*-object tracking that leverages low-level features in building object representations [19,38], while object-centric 4D canonical representations [18,22,28] have demonstrated the power of spatiotemporal information in object reconstruction. However, these methods are restricted to object-centric datasets, require clean data (*i.e.* low levels of noise), and run on heavy architectures that are not suitable for real time.

In this work, we propose a **spatiotemporal representation** for object tracklets (see Fig. 1). Our method, **SpOT** (**Sp**atiotemporal **O**bject **T**racking), actively maintains the history of *both* object-level point clouds and bounding boxes for each tracked object. At each frame, new object detections are associated with these maintained past sequences, as shown in Fig. 2; the sequences are then updated using a novel 4D backbone to *refine* the entire sequence of bounding boxes and to predict the current velocity, both of which are used to forecast the object into the next frame. This refinement step improves the quality of bounding box and motion estimates by ensuring spatiotemporal consistency, allowing tracklet association to benefit from low-level geometric context over a long time horizon. We perform extensive evaluations on both nuScenes [3] and Waymo Open [33] datasets to demonstrate that maintaining and refining sequences of tracked objects has several advantages, which together enable state-of-the-art tracking performance. Our method is particularly helpful in tracking sparse and occluded objects such as pedestrians, which can especially benefit from temporal priors.

In summary, we contribute: (**i**) a novel tracking algorithm that leverages spatiotemporal object context by storing and updating object bounding boxes and object-level point cloud sequences, (**ii**) a new 4D point cloud architecture for refining object tracks, and (**iii**) state-of-the-art results for 3D multi-object tracking on the standard nuScenes [3] and Waymo [33] benchmark datasets.

2 Related Work

2.1 3D Object Detection on LIDAR Point Clouds

3D detection is one of the most important modules for most 3D tracking frameworks. While 3D detectors from camera data have seen great improvement in recent years [11,15,21,31], LIDAR-based 3D detection offers significantly better performance, especially in driving scenes [7,8,13,30,39,47]. The majority of works on LIDAR-based detection have centered around improving feature extraction from unorganized point clouds. VoxelNet [47] groups points by 3D voxels and extracts voxel-level features using PointNet [25]. PointPillar [13] organizes point clouds into vertical columns (pillars) to achieve higher efficiency. PV-RCNN [30] aggregates voxel-level and point-level features to achieve better accuracy. On the other hand, CenterPoint [39] improves the 3D detector by looking at the output representation, proposing a point-based object representation at the decoding stage. While our proposed approach is not constrained to a specific input detector, we use CenterPoint in our experiments due to its popularity and to facilitate comparison with other state-of-the-art tracking algorithms.

In addition to improving the 3D detector architecture, aggregating temporal information has also been shown to improve 3D detection results as temporal information compensates for the sparsity of LIDAR sensor inputs. nuScenes [3] uses a simple motion compensation scheme to accumulate multiple LIDAR sweeps, providing a richer point cloud with an added temporal dimension. In addition, multiple detector architectures [13,48] use accumulated point clouds to improve performance and reason over object visibility [9]. Nevertheless, limited by the static scene assumption of motion compensation, temporal aggregation for the detection task is primarily done over short intervals.

Recently in offline perception, Qi et al. [23] address the use of a longer time horizon as a post-processing step. After running an offline detection and tracking algorithm, Qi et al. apply a spatiotemporal sliding window refinement on each tracked object at each frame.

In our work, we explore the utilization of temporal information over a longer time horizon in the task of multi-object tracking. Distinct from Qi et al., we utilize a long-term time horizon *within* our tracking algorithm, we operate in the noisier online setting, and we maintain and refine full object sequences.

2.2 3D Multi-Object Tracking

Thanks to the advances in 3D object detection discussed above, most state-of-the-art 3D MOT algorithms follow the *tracking-by-detection* paradigm. The performance of 3D MOT algorithms is mainly affected by three factors other than detection results: the motion prediction model, the association metric, and the life-cycle management of the tracklets.

CenterPoint [39] proposes a simple, yet effective, approach that gives reliable detection results and estimates velocities to propagate detections between sequential frames. The distance between object centers is used as the association metric. However, CenterPoint's constant velocity can be less robust to missing detections and long-term occlusions (as we demonstrate later in Fig. 4).

The most popular category of 3D MOT algorithms leverages Kalman Filters to estimate the location of tracked objects and their dynamics, providing predictions for future association. AB3DMOT [35] provides the prior baseline in this direction and leverages 3D Intersection-over-Union (IoU) as the association metric. Following this line, Chiu et al. [6] proposes to replace 3D IoU with Mahalanobis distance to better capture the uncertainty of the tracklets. SimpleTrack [20] conducts an analysis on different components of a tracking-by-detection pipeline and proposes corresponding enhancements to different modules.

Other works jointly train detection and tracking in a more data-driven fashion. FaF [16] proposes to jointly solve detection, tracking, and prediction using a multi-frame architecture on a voxel representation. However, the architecture is hard to scale up to a long history. PnPNet [14] extends the idea of FaF and proposes a more general framework with an explicit tracking model. Zaech et al. [41] combines detection and association in a graph structure and employs neural message passing.

Another line of work explores feature learning for data association in sensor fusion scenarios [5,12,36,43]. In this work, we focus on the application scenario of the LIDAR sensor only.

2.3 3D Single-Object Tracking

Given an initial *template* ground-truth bounding box of an object, the goal of single-object tracking (SOT) is to track the template object through all future frames. Unlike MOT, it is common for SOT methods to aggregate object-level information over a large temporal interval.

Many works use a Siamese network to compare the template encoding with a surrounding region of interest. P2B [27] uses a PointNet++ to directly propose seeds within the surrounding region and avoid an exhaustive search. BAT [45] improves the template object representation with a box-aware coordinate space. PTTR [46] uses cross-attention to improve feature comparison.

Recent works propose SOT without direct supervision. Pang et al. [19] perform template matching by optimizing hand-crafted shape and motion terms. Ye et al. [38] extend the work of Pang et al. with a deep SDF matching term. In contrast to SOT methods, we operate in the MOT setting on sequences originally generated from an imperfect 3D detector, and we maintain object sequence histories to avoid propagating error over time.

3 Multi-object Tracking Using Sequence Refinement

In this section, we provide an overview of our basic notation and SpOT tracking pipeline in Sect. 3.1. We then introduce our novel spatiotemporal sequence refinement module in detail in Sect. 3.2.

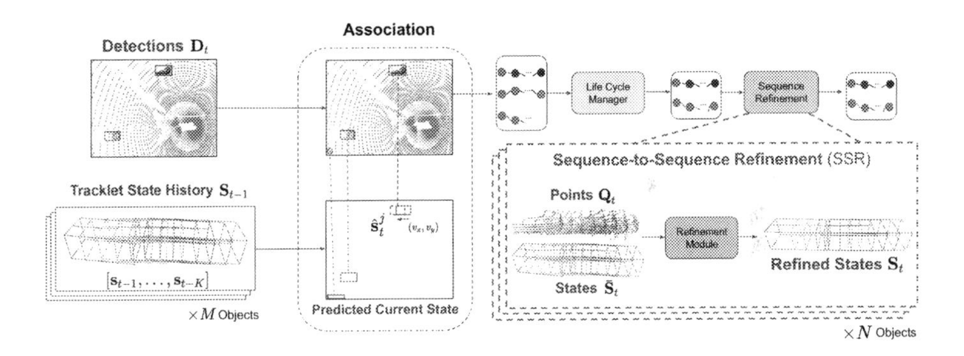

Fig. 2. Tracking algorithm overview. SpOT maintains sequence-level tracklets containing both object bounding box states and object point clouds from the last K timesteps. At each step, tracklets are associated to current detections using the predicted current state, and then the updated tracklets are refined by the learned SSR module to improve spatiotemporal consistency.

3.1 Tracking Pipeline

In this work, we address the problem of 3D multi-object tracking (MOT) from LIDAR sensor input. The goal of this task is to uniquely and consistently identify every object in the scene in the form of tracklets $\mathbf{O}_t = \{\mathbf{T}_t\}$ at each frame of input t. As input, the current LIDAR point cloud $\mathbf{P}_t$ is given along with detection results $\mathbf{D}_t = \{\mathbf{b}_t\}$, in the form of 7-DoF amodal bounding boxes $\mathbf{b}_i = (x, y, z, l, w, h, \theta)_i$ and confidences s_i, from a given detector. In contrast to previous works that maintain only single-frame tracklets, we model our tracklet as $\mathbf{T}_t = \{\mathbf{S}_t, \mathbf{Q}_t\}$ to include both a low-level history of *object* points $\mathbf{Q}_t$ and the corresponding sequence of detections $\mathbf{S}_t$. In each tracklet, $\mathbf{S}_t$ includes the estimated state trajectory of the tracked object within a history window:

$$\mathbf{S}_t = \{\mathbf{s}_i = (\mathbf{b}, v_x, v_y, c, s)_i\}, \quad t - K \leq i \leq t \tag{1}$$

where K is a pre-defined length of maximum history. Tracklet state has 11 elements and includes a 7-DoF bounding box $\mathbf{b}$, a birds-eye-view velocity (v_x, v_y), an object class c (*e.g.* "car"), and a confidence score $s \in [0, 1]$. On the other hand, $\mathbf{Q}_t$ encodes the spatiotemporal information from raw sensor observations in the form of time-stamped points:

$$\mathbf{Q}_t = \{\hat{\mathbf{P}}_i = \{(x, y, z, i)\}\}, \quad t - K \leq i \leq t \tag{2}$$

where $\hat{\mathbf{P}}_i$ is the cropped point cloud region from $\mathbf{P}_i$ according to the associated detected bounding box $\mathbf{b}_i$ at time i. We enlarge the cropping region by a factor of 1.25 to ensure that $\hat{\mathbf{P}}_i$ is robust through imperfect detection results.

As depicted in Fig. 2, we propose a tracking framework that follows the tracking-by-detection paradigm while leveraging low-level sensory information. At each timestep t, we first predict the current tracklets $\hat{\mathbf{T}}_t$ based on the stored previous ones $\mathbf{T}_{t-1}$:

$$\hat{\mathbf{T}}_t = \mathbf{Predict}(\mathbf{T}_{t-1}) = \{\hat{\mathbf{S}}_t, \mathbf{Q}_{t-1}\} \tag{3}$$

$$\hat{\mathbf{S}}_t = \{\hat{\mathbf{s}}_i = (x + v_x, y + v_y, z, l, w, h, \theta, v_x, v_y, c, s)_{i-1}\}, \quad t - K \leq i \leq t. \tag{4}$$

We compare the *last* state of the predicted tracklets $\hat{\mathbf{s}}_t$ to off-the-shelf detection results $\mathbf{D}_t$ to arrive at associated tracklets:

$$\bar{\mathbf{T}}_t = \mathbf{Association}(\hat{\mathbf{S}}_t, \mathbf{D}_t, \mathbf{P}_t) = \{\bar{\mathbf{S}}_t, \mathbf{Q}_t\} \tag{5}$$

where $\bar{\mathbf{S}}_t$ is the previous state history $\mathbf{S}_{t-1}$ concatenated with its associated detection and $\mathbf{Q}_t$ is the updated spatiotemporal history. Without loss of generality, we follow CenterPoint's [39] association strategy in our experiments. Finally, we conduct the posterior tracklet update using a novel sequence-to-sequence refinement (SSR) module:

$$\mathbf{S}_t = \mathbf{SSR}(\bar{\mathbf{S}}_t, \mathbf{Q}_t), \tag{6}$$

which provides the final updated tracklet estimation $\mathbf{T}_t = \{\mathbf{S}_t, \mathbf{Q}_t\}$. In the following section, we will provide technical details of the SSR module.

3.2 Sequence-to-Sequence Refinement (SSR) Module

We propose a novel algorithm to update a full tracklet history of estimated states by accounting for its spatiotemporal context, including raw sensor observations. Figure 3 displays our spatiotemporal sequence-to-sequence refinement (SSR) module, which takes the *associated* tracklet states $\bar{\mathbf{S}}_t$ and the time-stamped object point cloud segments $\mathbf{Q}_t$ as input and outputs refined final tracklet states $\mathbf{S}_t$. SSR first processes the sequential information with a 4D backbone to extract per-point context features. In the decoding stage, it predicts a global object size across all frames, as well as per-frame time-relevant object attributes including center, pose, velocity, and confidence.

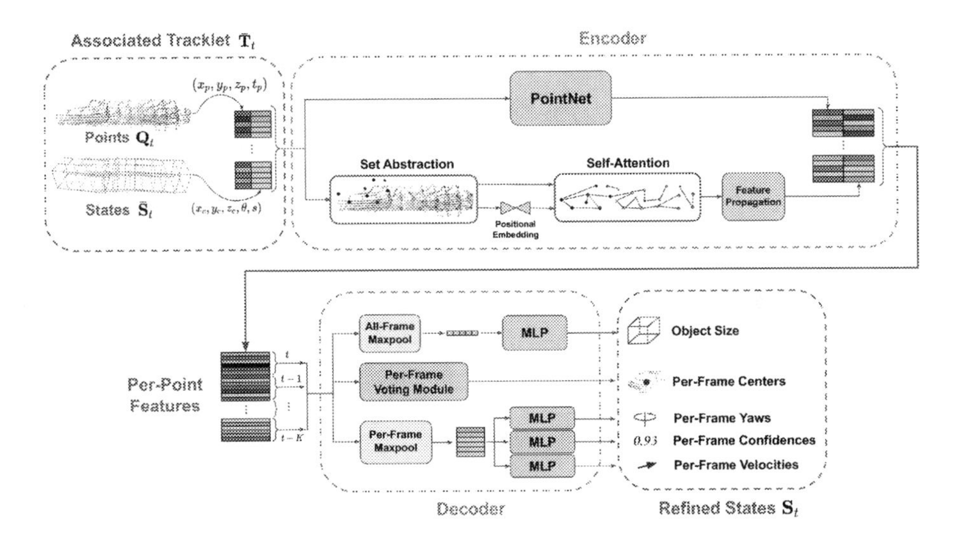

Fig. 3. Architecture of the sequence-to-sequence refinement (SSR) network. Given a tracklet containing object points and bounding boxes after association to a detection, the encoder first extracts spatiotemporal features corresponding to each input point. The features are given to the decoder which predicts a refined state trajectory and velocities to be used for subsequent association.

Split Self-Attention Encoder. The top part of Fig. 3 illustrates the encoding backbone, which processes each associated tracklet independently. Since the inputs contain two streams of information $\bar{\mathbf{S}}_t, \mathbf{Q}_t$, which are at different levels of abstraction (object vs. point), we first append the bounding-box-level information as an additional dimension to each point in $\mathbf{Q}_t$. This yields a set of object-aware features:

$$\mathbf{f}_p = [x_p, y_p, z_p, t_p, x_c, y_c, z_c, \sin(\theta), \cos(\theta), s], \tag{7}$$

where (x_p, y_p, z_p, t_p) denotes the 4D geometric point and $(x_c, y_c, z_c, \theta, s)$ is the center location, yaw, and confidence score of the corresponding bounding box at frame t_p.

Similar to previous works on spatiotemporal representation learning [28], the encoder is a two-branch point cloud backbone as depicted in Fig. 3. In the first branch, we apply a PointNet [25] to directly encode the high-dimensional inputs into per-point features. For the second branch, we apply a novel self-attention architecture inspired by the encoder of 3Detr [17]. First, we apply a per-frame PointNet++ [26] set abstraction layer, so that at each frame i we have a sub-sampled set of anchor-point features $\{\mathbf{a}_i^k\}_{k=1}^A$ where A is a hyperparameter for the number of anchor points. For each anchor point, a 4D positional embedding is generated using a 3-layer multi-layer perceptron (MLP):

$$\mathbf{pos}_{\mathbf{a}_i^k} = \mathbf{MLP}(\mathbf{a}_i^k). \tag{8}$$

The anchor features and positional embedding are concatenated as $[\mathbf{a}_i^k, \mathbf{pos}_{\mathbf{a}_i^k}]$ before applying four layers of self-attention across *all* anchor features. Notably, this self-attention allows information flow across both space and time. Finally, updated anchor features are propagated back to the full resolution point cloud via a feature propagation layer [26]. Layer normalization is applied to the features from each branch before concatenating to get the final per-point features.

Each branch of the encoder uses a 256-dim feature, yielding a concatenated 512-dim feature at the output. Set abstraction uses $A = 10$ anchor points per frame and a feature radius of 1.5 m for cars/vehicles and 0.6 m for pedestrians. Additional architectural details are provided in the supplemental material.

Sequence Decoder. The SSR decoder outputs a refined, ordered sequence of object states $\mathbf{S}_t$ that is amenable to association in subsequent frames. To output object state trajectories, some recent works use explicit priors on temporal continuity, such as anchor-based trajectories [4] or an autoregressive motion rollout [29]. In contrast, we choose a decoder without an explicit prior: the decoder directly predicts the ordered sequence of bounding boxes in one forward pass. This choice allows the model to learn temporal priors where needed through training. Our design is motivated by the discontinuous nature of many sequences that SSR operates on, which contain identity switches, false-positives, and occlusions.

As depicted in the bottom portion of Fig. 3, given an encoded set of per-point spatiotemporal features, we group features by their time of acquisition (*i.e.* by frame). We pass our time-grouped point features into 5 decoding heads. The first decoding head performs a max-pool on the entire feature set to regress a single object size (l, w, h), which is used for every output frame. The second head applies a voting module [24] to each set of time-grouped features; this outputs per-timestep object center predictions (x_c, y_c, z_c). The remaining heads perform a max-pool on each set of time-grouped features to obtain a single feature per timestep. This feature is passed through 2-layer MLPs to regress a yaw, confidence, and velocity (θ, s, v_x, v_y) for each frame.

Training Losses. Our sequence refinement module balances two loss terms: a bounding-box loss and a confidence-score loss. Our total loss is as follows:

$$L = w_{\text{conf}} L_{\text{conf}} + L_{\text{box}}, \tag{9}$$

where w_{conf} is a hyperparameter that balances the two losses.

We formulate our bounding-box loss similar to standard 3D detection works [10,13,23,37,39]. We apply an L1 loss on 3D box center $[x, y, z]$. We apply a cross-entropy loss on the predicted size bin and an L1 loss on the predicted size residual. We apply an L1 loss on the polar angle representation $\sin(\theta), \cos(\theta)$. This yields a bounding box loss of:

$$L_{\text{box}} = w_{\text{c}} L_{\text{c}} + w_{\theta} L_{\theta} + w_{\text{vel}} L_{\text{vel}} + w_{\text{wlh-cls}} L_{\text{wlh-cls}} + w_{\text{wlh-res}} L_{\text{wlh-res}}, \quad (10)$$

where $w_{\text{c}}, w_{\theta}, w_{\text{vel}}, w_{\text{wlh-cls}}$, and $w_{\text{wlh-res}}$ balance losses for the bounding box center, yaw, velocity, size-bin, and size-residual, respectively.

We desire the model's predicted confidence to match the prediction's quality. To achieve this, we set a target confidence score $\bar{s}$ in a manner proportional to the accuracy of the bounding-box estimate. Concretely, if a bounding box is not close to a ground-truth object, we assign the target confidence to 0. Otherwise, we follow [32] and assign the target confidence to be proportional to the L2 distance from the closest ground-truth object as $\bar{s} = e^{-\alpha \mathbf{b}_{\text{err}}}$, where α is a temperature hyperparameter and $\mathbf{b}_{\text{err}}$ is the L2 box center error. Our confidence loss is then a binary cross-entropy loss, $L_{\text{conf}} = \text{BCE}(s, \bar{s})$.

Training Data. During online tracking, the refinement module must robustly handle noisy inputs from the 3D detector, which may contain false-positives, identity switches, occlusions, and more. Therefore, we must use a set of suitable training sequences that faithfully capture these challenging test-time phenomena. To achieve this, we use the outputs of previous tracking methods to generate object tracks that are used as training sequences. For all experiments, we generate data using the CenterPoint [39] tracker with varied track-birth confidence threshold, $c_{thresh} \in \{0.0, 0.3, 0.45, 0.6\}$, and varied track-kill age, $t_{kill} \in \{1, 2, 3\}$. We additionally augment these tracks with transformations, noise, and random frame dropping. Our final training set averages 750k sequences per object class. These augmentation methods are detailed in the supplementary material.

4 Experimental Evaluation

4.1 Datasets and Evaluation Metrics

nuScenes Dataset The nuScenes dataset contains 1000 sequences of driving data, each 20 s in length. 32-beam LIDAR data is provided 20 Hz, but 3D labels are only given 2 Hz. The relatively sparse LIDAR data and low temporal sampling rate make our proposed method particularly suitable for data like that in nuScenes, where leveraging spatiotemporal history provides much-needed additional context. We follow the official nuScenes benchmark protocol for tracking, which uses the AMOTA and AMOTP metrics [35]. For a thorough definition of AMOTA and AMOTP, we refer the reader to the supplementary material. We evaluate on the two most observed classes: car and pedestrian.

Waymo Open Dataset. The Waymo Open Dataset [33] contains 1150 sequences, each with 20 s of contiguous driving data. Unlike nuScenes, the Waymo Open Dataset provides sensor data for four short-range LIDARs and one long-range LIDAR. Each LIDAR is sampled 10 Hz, and the long-range LIDAR is significantly denser than the nuScenes 32-beam device. 3D labels are provided for every frame 10 Hz. We follow official Waymo Open Dataset benchmark protocol, which uses MOTA and MOTP [2] to evaluate tracking. For a thorough definition of MOTA and MOTP, we refer the reader to the supplementary material. We evaluate on the two most observed classes: vehicle and pedestrian.

Note that AMOTA averages MOTA at different recall thresholds. In our experiments, different sets of parameters were used between nuScenes and Waymo. For nuScenes, lower thresholds were used to balance the recall.

4.2 Implementation Details

In this section, we highlight the most important implementation details, and refer the reader to the supplementary material for additional information.

SSR Training Details. We train a different network for each object class using sequences of length $K = 40$ for nuScenes and $K = 15$ for Waymo (we investigate the effect sequence length has on tracking performance in Table 3). To improve robustness and mimic test time when stored tracklets are iteratively refined each time a new frame is observed, during training we refine a sequence a random number of times before backpropagation, *i.e.* the network sees its own output as input. In practice, we find that *not* refining bounding-box size is beneficial for Waymo vehicles, due to the large variance in sizes.

Table 1. Tracking performance on the nuScenes dataset validation split. An asterisk* denotes a preprint.

Method	Car		Pedestrian	
	AMOTA↑	AMOTP↓	AMOTA↑	AMOTP↓
AB3DMOT [35]	72.5	0.638	58.1	0.769
Centerpoint [39]	84.2	**0.380**	77.3	0.392
ProbabalisticTracking [6]	84.2	–	75.2	–
MultimodalTracking [5]	84.3	–	76.6	–
SimpleTrack-2Hz* [20]	83.8	0.396	79.4	0.418
SpOT-No-SSR (Ours)	84.5	**0.380**	81.1	0.391
SpOT (Ours)	**85.1**	0.390	**82.5**	**0.386**

Test-Time Tracking Details. Similar to prior works [5,20,42], we use off-the-shelf detections from CenterPoint [39] as input at each frame of tracking. For nuScenes, CenterPoint provides detections 2 Hz so we upsample 20 Hz by backtracking the estimated velocities to match the LIDAR sampling rate. CenterPoint detections are pre-processed with a birds-eye-view non-maximal-suppression using thresholds of 0.3 IoU for nuScenes and 0.5 IoU for Waymo.

Detections are associated with the last frame of object tracklets using a greedy bipartite matching algorithm over L2 center-distances that uses detection confidence [39]. We set a maximum matching distance of 1 m and 4 m for nuScenes pedestrians and cars, respectively, and 0.4 m and 0.8 m for Waymo Open pedestrians and vehicles. We use a track-birth confidence threshold of 0.0 for nuScenes. For Waymo Open, this is 0.6 for pedestrians and 0.7 for vehicles. The track-kill age is 3 for both datasets. For nuScenes, we start refining tracklets at a minimum age of 30 frames with a maximum temporal context of 40 frames. For Waymo, the minimum refinement age is 5 for vehicles and 2 for pedestrians and the maximum temporal context is 10 frames.

As discussed in Sect. 3.1 the tracklet state trajectory $\mathbf{S}_t$ stores the history of bounding boxes for each object. On nuScenes, these boxes are the output of our refinement network such that all boxes continue to be refined at each new frame. On Waymo, we instead directly store the given CenterPoint detections.

Table 2. Tracking performance on the Waymo Open dataset validation split. An asterisk* denotes a preprint.

Method	MOTA↑	FP%↓	Miss%↓	Mismatch%↓
	Vehicle			
AB3DMOT [35]	55.7	–	–	0.40
CenterPoint [39]	55.1	10.8	33.9	0.26
SimpleTrack* [20]	**56.1**	**10.4**	33.4	**0.08**
SpOT-No-SSR (Ours)	55.1	10.8	33.9	0.21
SpOT (Ours)	55.7	11.0	**33.2**	0.18
	Pedestrian			
AB3DMOT [35]	52.2	–	–	2.74
CenterPoint [39]	54.9	**10.0**	34.0	1.13
SimpleTrack* [20]	57.8	10.9	30.9	**0.42**
SpOT-No-SSR (Ours)	56.5	11.4	31.5	0.61
SpOT (Ours)	**60.5**	11.3	**27.6**	0.56

4.3 Comparison with State-of-the-Art Tracking

In this section, all reported tracking results are obtained with CenterPoint detections [39]. We build our tracking pipeline based on CenterPoint and adopt NMS pre-processing to detection results as suggested in SimpleTrack [20]. We denote this baseline version of CenterPoint tracking with NMS pre-processing as *SpOT-No-SSR* and report its performance for a fair comparison.

In Table 1, we compare SpOT to various tracking methods on the nuScenes dataset. SpOT significantly outperforms all previous methods in correctly tracking objects (AMOTA) and is on-par with previous methods in estimating high-quality object tracklets (AMOTP). In Table 2, we compare SpOT to various tracking methods on the Waymo Open dataset. For pedestrians, SpOT significantly outperforms all previous methods. For vehicles, SpOT notably improves tracking over the CenterPoint baseline. Examining metric breakdowns, it becomes clear that SpOT is able to robustly track objects in more cluttered environments compared to previous methods. That is, we remove fewer detections in pre-processing to yield fewer *Misses*, yet we still maintain a very low *Mismatch* score. Additionally, we note that many contributions of the competitive method SimpleTrack [20], such as a 2-stage association strategy and a generalized-IoU similarity metric, could be seamlessly integrated into SpOT.

The SSR module of SpOT exhibits greater improvements on the nuScenes dataset and on the pedestrian class in general. This is unsurprising because sparser LIDAR frames and smaller objects are expected to benefit disproportionately from increased temporal context. Figure 5 illustrates examples of our refined sequences compared to tracklets composed of off-the-shelf CenterPoint detections. We observe the greatest improvement in sequence quality when individual frames are sparse. Furthermore, we can qualitatively observe improved temporal consistency within sequences.

Additionally, we observe that our SSR module can handle noisy input detections and/or associations by learning *when* to make use of physical priors on object permanence and consistency. We provide some examples illustrating this property in Fig. 4. The first row displays an example when both CenterPoint and SpOT encounter an ID-switch error in the tracklet. For CenterPoint, this error will be propagated to future prediction and association. For SpOT, even though we can not retroactively correct the misassociation, the SSR module still refines the sequence bounding boxes in a manner that accurately reflects two disjoint objects; this accurate update will help to avoid future tracking errors. The second row shows a discontinuous sequence due to *occlusion* where different parts of an object are observed. Our SSR module refines the occluded region in a manner that reflects temporal continuity of a single object.

4.4 Ablative Analysis

Length of Maximum History. Table 3 reports how the length of maximum history affects tracking performance. The table emphasizes the significant advantage of using a large history. Because nuScenes is sampled 20 Hz and uses a sparse 32-beam LIDAR, we observe tracking performance monotonically improve up to a 40-frame history (2 s). In contrast, Waymo tracking performance peaks at a 10-frame history (1 s) and declines beyond 10 frames.

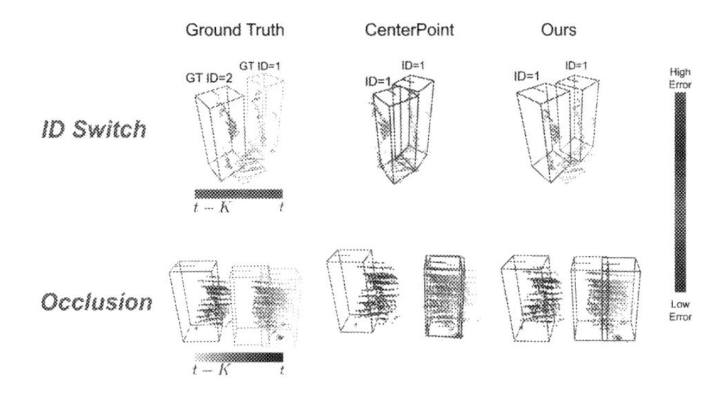

Fig. 4. Examples of discontinuous object tracklets. We visualize every 10th prediction for clarity, and predicted boxes are colored according to L2 center error. Our refinement is robust to different types of input sequence discontinuities. In the first row, our refinement immediately recovers after an identity-switch by correctly updating the bounding boxes to reflect the existence of two disjoint objects. In the second row, it correctly updates bounding boxes to reflect single-object continuity through occlusion.

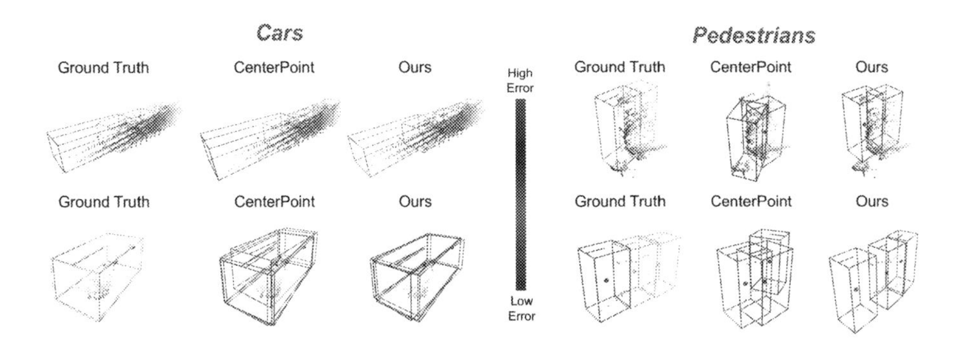

Fig. 5. Qualitative results of our spatiotemporal sequence refinement. We visualize every 10th prediction for clarity. Predicted bounding boxes are colored according to their L2 center error. Refinement improves the temporal consistency of detections, especially for sparse sequences.

SSR Update Components. Recall that for each object tracklet, our SSR module predicts per-timestep refinements consisting of a bounding box (with velocity) and a confidence score. Table 4a displays an ablation study comparing these two SSR refinements. All reported values are the AMOTA metric on the nuScenes dataset. As observed, both bounding box and confidence refinements contribute to tracking, and we achieve the best performance when we refine both.

SSR Backbone Architecture. Table 4b displays an ablation analysis of our SSR backbone on the nuScenes dataset. All reported values are the AMOTA metric. The first row shows tracking metrics with only a PointNet backbone.

The second row corresponds to a two-branch backbone where the second branch consists of set abstraction and feature propagation. The third row corresponds to our full backbone. As observed, each part of our backbone improves refinement.

4.5 Runtime Analysis

All components of SpOT except our SSR module are benchmarked in previous real-time tracking algorithms [20,35,39]. We benchmark the real-time performance of our SSR module on an Nvidia RTX3090 GPU. On the nuScenes dataset, our SSR module averages 51 Hz for pedestrians 28 Hz for cars. On the Waymo dataset, it averages 26 Hz for pedestrian 17 Hz for vehicles.

Table 3. Ablation on how the length of maximum history (number of input LIDAR sweeps to SSR) affects the quality of tracking. Tracking performance peaks at a 2 s history for nuScenes and a 1 s history for Waymo.

nuScenes (20 Hz): AMOTA↑			**Waymo** (10 Hz): MOTA↑		
Num Sweeps	*Car*	*Pedestrian*	Num Sweeps	*Vehicle*	*Pedestrian*
10	84.7	81.1	1	53.4	58.4
20	85.0	81.4	5	54.6	59.8
30	85.0	81.7	10	**55.7**	**60.4**
40	**85.1**	**82.5**	15	54.5	60.2

Table 4. Ablation experiments on the nuScenes dataset. All reported values are the AMOTA tracking metric. (a) Ablation holding out box and confidence-score refinements of our SSR module. (b) Ablation holding out parts of our refinement backbone. We denote set abstraction as SA and feature propagation as FP.

SSR Refinement Components

Refinement Type	*Car*	*Pedestrian*
No Refinement	84.5	81.1
Boxes Only	84.9	81.9
Confidences Only	84.9	81.6
Boxes and Confidences	**85.1**	**82.5**

SSR Encoder

Architecture	*Car*	*Pedestrian*
PointNet only	84.8	81.6
+ SA and FP	84.8	81.7
+ Self-attention	**85.1**	**82.5**

5 Discussion

We have introduced SpOT, a method for 3D multi-object tracking in LIDAR data that leverages a spatiotemporal tracklet representation in the form of object bounding boxes and point cloud history. Furthermore, we have proposed a 4D refinement network to iteratively update stored object sequences after associating new detections at each frame. Through evaluations on standard tracking benchmarks, SpOT compares favorably to prior works that use only single-frame tracks, thanks to the ability to leverage larger spatiotemporal context and use low-level geometry cues to improve bounding box and motion estimates. Our method particularly excels given longer temporal history and when operating on pedestrians due to naturally increased occlusions and sparsity.

Although our results indicate a promising first step to improving 3D tracking with spatiotemporal representations, we believe there are many future directions to explore, such as building end-to-end association in lieu of a 2-stage strategy and more explicitly utilizing aggregated tracklet geometry for downstream tasks.

Acknowledgments. This work was supported by grants from the Toyota Research Institute (TRI) University 2.0 program, a Vannevar Bush Faculty Fellowship, and a gift from the Amazon Research Awards program. Toyota Research Institute ("TRI") provided funds to assist the authors with their research but this article solely reflects the opinions and conclusions of its authors and not TRI or any other Toyota entity.

References

1. Bergmann, P., Meinhardt, T., Leal-Taixé, L.: Tracking without bells and whistles. In: The IEEE International Conference on Computer Vision (ICCV), October 2019
2. Bernardin, K., Stiefelhagen, R.: Evaluating multiple object tracking performance: the clear mot metrics. EURASIP J. Image Video Process. (2008)
3. Caesar, H., et al.: nuscenes: a multimodal dataset for autonomous driving. In: CVPR (2020)
4. Chai, Y., Sapp, B., Bansal, M., Anguelov, D.: Multipath: multiple probabilistic anchor trajectory hypotheses for behavior prediction. CoRR abs/1910.05449 (2019). arxiv:1910.05449
5. Chiu, H., Li, J., Ambrus, R., Bohg, J.: Probabilistic 3D multi-modal, multi-object tracking for autonomous driving. CoRR abs/2012.13755 (2020). arxiv:2012.13755
6. Chiu, H., Prioletti, A., Li, J., Bohg, J.: Probabilistic 3D multi-object tracking for autonomous driving. In: 2021 IEEE International Conference on Robotics and Automation (ICRA) (2020). arxiv:2001.05673
7. Engelcke, M., Rao, D., Wang, D.Z., Tong, C.H., Posner, I.: Vote3deep: fast object detection in 3D point clouds using efficient convolutional neural networks. In: 2017 IEEE International Conference on Robotics and Automation (ICRA), pp. 1355–1361. IEEE (2017)
8. Ge, R., et al.: Afdet: anchor free one stage 3D object detection. CoRR abs/2006.12671 (2020). arxiv:2006.12671
9. Hu, P., Ziglar, J., Held, D., Ramanan, D.: What you see is what you get: exploiting visibility for 3D object detection. In: Proceedings of the IEEE/CVF Conference on Computer Vision and Pattern Recognition, pp. 11001–11009 (2020)
10. Huang, R., et al.: An LSTM approach to temporal 3D object detection in LiDAR point clouds. In: Vedaldi, A., Bischof, H., Brox, T., Frahm, J.-M. (eds.) ECCV 2020. LNCS, vol. 12363, pp. 266–282. Springer, Cham (2020). https://doi.org/10.1007/978-3-030-58523-5_16

11. Kehl, W., Manhardt, F., Tombari, F., Ilic, S., Navab, N.: SSD-6D: Making RGB-based 3d detection and 6d pose estimation great again. In: Proceedings of the IEEE International Conference on Computer Vision, pp. 1521–1529 (2017)
12. Kim, A., Ošep, A., Leal-Taixé, L.: Eagermot: real-time 3D multi-object tracking and segmentation via sensor fusion. In: CVPR-Workshops, vol. 1, p. 3 (2020)
13. Lang, A.H., Vora, S., Caesar, H., Zhou, L., Yang, J., Beijbom, O.: Pointpillars: fast encoders for object detection from point clouds. In: 2019 IEEE/CVF Conference on Computer Vision and Pattern Recognition (CVPR), pp. 12689–12697 (2019). https://doi.org/10.1109/CVPR.2019.01298
14. Liang, M., et al.: PNPNet: end-to-end perception and prediction with tracking in the loop. In: 2020 IEEE/CVF Conference on Computer Vision and Pattern Recognition (CVPR), pp. 11550–11559 (2020). https://doi.org/10.1109/CVPR42600.2020.01157
15. Liu, Z., Wu, Z., Tóth, R.: Smoke: single-stage monocular 3D object detection via keypoint estimation. In: Proceedings of the IEEE/CVF Conference on Computer Vision and Pattern Recognition Workshops, pp. 996–997 (2020)
16. Luo, W., Yang, B., Urtasun, R.: Fast and furious: Real time end-to-end 3D detection, tracking and motion forecasting with a single convolutional net. In: Proceedings of the IEEE Conference on Computer Vision and Pattern Recognition, pp. 3569–3577 (2018)
17. Misra, I., Girdhar, R., Joulin, A.: An end-to-end transformer model for 3D object detection. In: ICCV (2021)
18. Niemeyer, M., Mescheder, L., Oechsle, M., Geiger, A.: Occupancy flow: 4D reconstruction by learning particle dynamics. In: Proceedings of the IEEE/CVF International Conference on Computer Vision, pp. 5379–5389 (2019)
19. Pang, Z., Li, Z., Wang, N.: Model-free vehicle tracking and state estimation in point cloud sequences. In: IROS (2021)
20. Pang, Z., Li, Z., Wang, N.: Simpletrack: understanding and rethinking 3D multi-object tracking. arXiv preprint arXiv:2111.09621 (2021)
21. Park, D., Ambrus, R., Guizilini, V., Li, J., Gaidon, A.: Is pseudo-lidar needed for monocular 3D object detection? In: IEEE/CVF International Conference on Computer Vision (ICCV) (2021)
22. Pumarola, A., Corona, E., Pons-Moll, G., Moreno-Noguer, F.: D-nerf: neural radiance fields for dynamic scenes. In: Proceedings of the IEEE/CVF Conference on Computer Vision and Pattern Recognition, pp. 10318–10327 (2021)
23. Qi, C., et al.: Offboard 3D object detection from point cloud sequences. In: 2021 IEEE/CVF Conference on Computer Vision and Pattern Recognition (CVPR), pp. 6130–6140 (2021)
24. Qi, C.R., Litany, O., He, K., Guibas, L.J.: Deep hough voting for 3D object detection in point clouds. In: Proceedings of the IEEE International Conference on Computer Vision (2019)
25. Qi, C.R., Su, H., Mo, K., Guibas, L.J.: Pointnet: deep learning on point sets for 3D classification and segmentation. In: CVPR (2017)
26. Qi, C.R., Yi, L., Su, H., Guibas, L.J.: Pointnet++: deep hierarchical feature learning on point sets in a metric space. arXiv preprint arXiv:1706.02413 (2017)
27. Qi, H., Feng, C., Cao, Z.G., Zhao, F., Xiao, Y.: P2b: Point-to-box network for 3D object tracking in point clouds, pp. 6328–6337, June 2020. https://doi.org/10.1109/CVPR42600.2020.00636
28. Rempe, D., Birdal, T., Zhao, Y., Gojcic, Z., Sridhar, S., Guibas, L.J.: CASPR: learning canonical spatiotemporal point cloud representations. In: Advances in Neural Information Processing Systems (NeurIPS) (2020)

29. Rempe, D., Philion, J., Guibas, L.J., Fidler, S., Litany, O.: Generating useful accident-prone driving scenarios via a learned traffic prior. In: Conference on Computer Vision and Pattern Recognition (CVPR) (2022)
30. Shi, S., et al.: PV-RCNN: point-voxel feature set abstraction for 3D object detection. In: Proceedings of the IEEE Conference on Computer Vision and Pattern Recognition (2020)
31. Simonelli, A., Bulò, S.R., Porzi, L., Kontschieder, P., Ricci, E.: Demystifying pseudo-lidar for monocular 3D object detection. arXiv preprint arXiv:2012.05796 (2020)
32. Simonelli, A., Bulò, S.R., Porzi, L., Antequera, M.L., Kontschieder, P.: Disentangling monocular 3D object detection: from single to multi-class recognition. IEEE Trans. Pattern Anal. Mach. Intell. **44**(3), 1219–1231 (2022). https://doi.org/10.1109/TPAMI.2020.3025077
33. Sun, P., et al.: Scalability in perception for autonomous driving: Waymo open dataset. In: Proceedings of the IEEE/CVF Conference on Computer Vision and Pattern Recognition, pp. 2446–2454 (2020)
34. Weng, X., Ivanovic, B., Pavone, M.: MTP: multi-hypothesis tracking and prediction for reduced error propagation. CoRR abs/2110.09481 (2021)
35. Weng, X., Wang, J., Held, D., Kitani, K.: AB3DMOT: a baseline for 3d multi-object tracking and new evaluation metrics. In:ECCVW (2020)
36. Weng, X., Wang, Y., Man, Y., Kitani, K.M.: GNN3DMOT: graph neural network for 3D multi-object tracking with 2d–3d multi-feature learning. In: 2020 IEEE/CVF Conference on Computer Vision and Pattern Recognition, CVPR 2020, Seattle, WA, USA, June 13–19, 2020. pp. 6498–6507. Computer Vision Foundation/IEEE (2020). https://doi.org/10.1109/CVPR42600.2020.00653
37. Yan, Y., Mao, Y., Li, B.: Second: sparsely embedded convolutional detection. Sensors 18(10) (2018). https://www.mdpi.com/1424-8220/18/10/3337 https://doi.org/10.3390/s18103337
38. Ye, J., Chen, Y., Wang, N., Wang, X.: Online adaptation for implicit object tracking and shape reconstruction in the wild. IEEE Robot. Autom. Lett. 1–8 (2022). https://doi.org/10.1109/LRA.2022.3189185
39. Yin, T., Zhou, X., Krähenbühl, P.: Center-based 3d object detection and tracking. In: CVPR (2021)
40. Yu, C., Ma, X., Ren, J., Zhao, H., Yi, S.: Spatio-temporal graph transformer networks for pedestrian trajectory prediction. In: Vedaldi, A., Bischof, H., Brox, T., Frahm, J.-M. (eds.) ECCV 2020. LNCS, vol. 12357, pp. 507–523. Springer, Cham (2020). https://doi.org/10.1007/978-3-030-58610-2_30
41. Zaech, J.N., Liniger, A., Dai, D., Danelljan, M., Van Gool, L.: Learnable online graph representations for 3D multi-object tracking. IEEE Robot. Autom. Lett. **7**, 5103–5110 (2022)
42. Zaech, J.N., Liniger, A., Dai, D., Danelljan, M., Van Gool, L.: Learnable online graph representations for 3D multi-object tracking. IEEE Robot. Autom. Lett. **7**(2), 5103–5110 (2022). https://doi.org/10.1109/LRA.2022.3145952
43. Zhang, W., Zhou, H., Sun, S., Wang, Z., Shi, J., Loy, C.C.: Robust multi-modality multi-object tracking. In: Proceedings of the IEEE/CVF International Conference on Computer Vision, pp. 2365–2374 (2019)
44. Zhang, Y., Wang, C., Wang, X., Zeng, W., Liu, W.: Fairmot: on the fairness of detection and re-identification in multiple object tracking. Int. J. Comput. Vis. **129**(11), 3069–3087 (2021)
45. Zheng, C., et al.: Box-aware feature enhancement for single object tracking on point clouds. CoRR abs/2108.04728 (2021). arxiv:2108.04728

46. Zhou, C., Luo, Z., Luo, Y., Liu, T., Pan, L., Cai, Z., Zhao, H., Lu, S.: PTTR: relational 3D point cloud object tracking with transformer. In: Proceedings of the IEEE/CVF Conference on Computer Vision and Pattern Recognition (CVPR) (2022)
47. Zhou, Y., Tuzel, O.: Voxelnet: end-to-end learning for point cloud based 3D object detection. In: Proceedings of the IEEE Conference on Computer Vision and Pattern Recognition, pp. 4490–4499 (2018)
48. Zhu, B., Jiang, Z., Zhou, X., Li, Z., Yu, G.: Class-balanced grouping and sampling for point cloud 3D object detection. arXiv e-prints arXiv:1908.09492 (2019)

Multimodal Transformer for Automatic 3D Annotation and Object Detection

Chang Liu(✉), Xiaoyan Qian, Binxiao Huang, Xiaojuan Qi, Edmund Lam, Siew-Chong Tan, and Ngai Wong

The University of Hong Kong, Pokfulam, Hong Kong
{lcon7,qianxy10,huangbx7}@connect.hku.hk,
{xjqi,elam,sctan,nwong}@eee.hku.hk

Abstract. Despite a growing number of datasets being collected for training 3D object detection models, significant human effort is still required to annotate 3D boxes on LiDAR scans. To automate the annotation and facilitate the production of various customized datasets, we propose an end-to-end multimodal transformer (MTrans) autolabeler, which leverages both LiDAR scans and images to generate precise 3D box annotations from weak 2D bounding boxes. To alleviate the pervasive sparsity problem that hinders existing autolabelers, MTrans densifies the sparse point clouds by generating new 3D points based on 2D image information. With a multi-task design, MTrans segments the foreground/background, densifies LiDAR point clouds, and regresses 3D boxes simultaneously. Experimental results verify the effectiveness of the MTrans for improving the quality of the generated labels. By enriching the sparse point clouds, our method achieves 4.48% and 4.03% better 3D AP on KITTI moderate and hard samples, respectively, versus the state-of-the-art autolabeler. MTrans can also be extended to improve the accuracy for 3D object detection, resulting in a remarkable 89.45% AP on KITTI hard samples. Codes are at https://github.com/Cliu2/MTrans.

Keywords: 3d Autolabeler · 3d Object detection · Multimodal vision · Self-attention · Self-supervision · Transformer

1 Introduction

3D detection technologies have seen rapid development in recent years, with considerable importance for tasks such as autonomous driving and robotics. Large-scale datasets can often improve the performance and generality of deep neural networks, hence new datasets such as nuScenes [1] and Waymo Open Dataset [20] include millions of annotated objects and hundreds of thousands of LiDAR scans. Nonetheless, the annotation procedure is extremely labor-intensive, especially for

Supplementary Information The online version contains supplementary material available at https://doi.org/10.1007/978-3-031-19839-7_38.

S. Avidan et al. (Eds.): ECCV 2022, LNCS 13698, pp. 657–673, 2022.
https://doi.org/10.1007/978-3-031-19839-7_38

3D point clouds [12,26]. Therefore, it is imperative to explore ways to accelerate the annotation procedure.

Compared with 3D annotations, 2D bounding boxes are much easier to obtain. Hence, we investigate the generation of 3D annotations from weak 2D boxes. Despite its substantial practical value, only a few existing works study the automatic annotation problem. Taking point clouds and images as the multimodal inputs, FGR [26] converts 2D boxes to 3D boxes with a non-learning algorithm. Alternatively, WS3D [11] leverages center clicks as weak annotations, and directly regresses 3D boxes from cylindrical proposals centered by the clicks. Both methods leverage the image information to reduce the problem space and achieves state-of-the-art performances on the KITTI dataset [5]. Nevertheless, their methods overlook the pervasive sparsity issue of point clouds. As shown on the left of Fig. 1, even though the problem domain is narrowed down by the weak 2D annotations, it is still challenging to generate a 3D box if the points are too sparse, due to the ambiguity in orientation and scale. Consequently, although these methods generate human-comparable annotations for easy samples, they suffer obvious accuracy drops for hard samples that are usually sparse, by 9.13% and 6.47% in 3D AP, respectively.

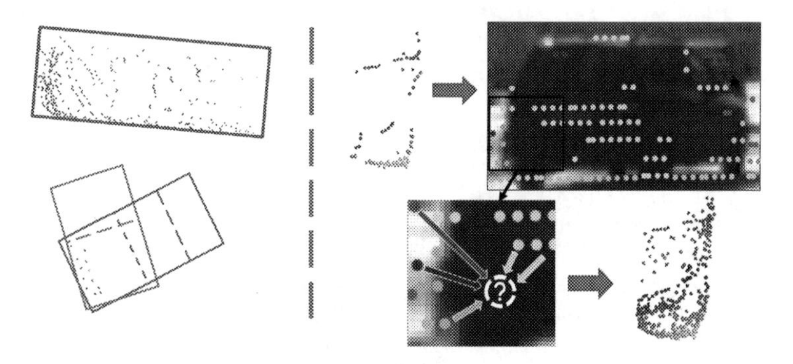

Fig. 1. Left: The sparsity issue of point cloud for 3D box annotation. Compared with the top object, the bottom is ambiguous in orientation and scale. Both are from the Bird's Eye View of LiDAR scans. **Right:** Interpolating 3D points by referring to context points based on 2D semantic and geometric relationships. Point's depth shown in color. (Color figure online)

The sparsity problem remains a fundamental challenge for automatic annotations. Besides the sparsity issue, point clouds lack color information. Therefore, similar-shaped objects (e.g., traffic light post and pedestrian) are difficult to distinguish, which further complicates the problem [23]. Fortunately, images provide dense fine-grid RGB information, which is an effective complement to the sparse point clouds. Motivated by this, we investigate point cloud enrichment with image information. As shown on the right of Fig. 1, 3D LiDAR points can be projected onto the image, and therefore an image pixel's 3D coordinates can be estimated by referring to the context LiDAR points, based on their 2D spatial and semantic relationships. This way, new 3D points can be generated from 2D pixels, and hence densify the sparse point cloud.

To this end, we propose the end-to-end multimodal transformer (MTrans) which alleviates the sparsity problem and outputs accurate 3D box annotations given weak 2D boxes. In order to leverage the dense image information to enrich sparse point clouds, we present the multimodal self-attention module, which efficiently extracts multimodal 2D and 3D features, modeling points' geometric and semantic relationships. To train MTrans, we also introduce a multi-task design, so that it learns to generate new 3D points to reduce the sparsity of point clouds while predicting 3D boxes. The multi-task design also includes a foreground segmentation task to encourage MTrans to uncover point-wise semantics. The multimodal self-attention and multi-task design work collectively and enable the MTrans to efficiently exploit multimodal information to address the sparsity issue, and hence facilitate more accurate 3D box prediction. Moreover, to exploit unlabelled data for the point generation task, a *mask-and-predict* self-supervision strategy can be carried out. With a small amount of annotated data (e.g., 500 frames) and optional unlabeled data, our method can generate high-quality 3D box annotations for training other 3D object detectors. Comprehensive experiments are conducted, and our MTrans outperforms all the other baseline methods significantly. Using only 500 annotated frames, PointRCNN trained with MTrans can achieve 99.62% of its original performance trained with full human annotations, which is 3.65% higher in mAP than the state-of-the-art autolabeler. Furthermore, our point generation approach for handling the sparsity issue also has direct value for 3D object detection tasks. Extended for object detection, MTrans surpasses similar 2D-box-aided methods of F-PointNet [14] and F-ConvNet [25] by 19.6% and 10.8% mAP, respectively. Our contributions are threefold:

1. We propose a novel autolabeler, MTrans, which generates high-quality 3D boxes from weak 2D bounding boxes, thus greatly saving manual work.
2. MTrans effectively alleviates the pervasive sparsity issue for 3D point clouds, and it can also be extended for 3D object detection tasks.
3. Comprehensive experiments are conducted to justify our method. MTrans outperforms existing state-of-the-art autolabelers by large margins. Visualization results are also provided for interpretability.

2 Related Work

2.1 Multimodal 3D Object Detection

Several previous studies investigate multimodal networks, utilizing images and LiDAR point clouds for a more comprehensive perception. Most of them extract features from images and point clouds in different branches, followed by fusing the information for final decision [2,6,27,33]. MV3D [2] proposes the deep fusion manner to repetitively exchange information from the different views including the image, LiDAR's front view and Bird's Eye View (BEV), during the inference. Later, EPNet [6] adopts Feature Pyramid Network to encode LiDAR's front view and RGB images, and fuses the two feature pyramids with the gating mechanism. Similarly, ACMNet [33] and 3D-CVF [31] are based on gating fusion as well. PI-RCNN [27] presents a hybrid PACF fusion module. To align different views of the image and point clouds during fusion, AVOD [7] introduces the 3D anchor grid to generate 3D proposals and applies fusion only for the proposed region. Alternatively, Pi-RCNN [27] and PointPainting [23] directly extract point cloud features with point-based backbones (e.g., PointNet [15] and PointNet++ [16]), and augment them with image features through LiDAR-image calibration.

Rather than fusing different modalities, some other research employs images to narrow down the problem space. PointPainting [23] performs image semantic segmentation and highlights the points corresponding to the interested 2D region. Similarly, Ref. [10] performs instance segmentation on images to filter the 3D point clouds. F-PointNet [14] and F-Convnet [25] leverage 2D bounding boxes on images to locate a frustum region of the point cloud, and hence lowers the difficulty and improves accuracy. CLOCs [13] generates 2D box proposals from images to filter 3D proposals, exploiting the consistency of the two modalities to improve the overall detection accuracy.

The approaches above demonstrated the value of multiple modalities. The dense image data can complement the sparse point clouds. Nevertheless, the number of points remains unchanged, hence the sparsity problem is not substantially solved. Also, the multi-branch design introduces extra computation and memory overheads, leaving it still an open question on how to fully utilize multimodal information [24].

2.2 Single-Modal Detectors Addressing the Sparsity Problem

Noticing the sparsity problem, some researchers try to enrich the sparse clouds to improve detection accuracy. SPG [29] firstly locates foreground regions, where an object is likely to exist, and then generates semantic points with a voxel-based point generation module. Specifically, the network predicts whether each voxel is in the foreground or background, and then generates a new point for each foreground voxel. BtcDet [28] densifies the point clouds within voxel-based proposals as well. Instead of directly generating new points, BtcDet finds a best matched prior object for each proposal, and densifies the proposal by mirroring the original points as well as aggregating the points from the prior object.

Both SPG and BtcDet alleviate the sparsity problem, resulting in state-of-the-art performances on the KITTI detection dataset. The two works demonstrate the value of enriching the sparse point cloud for the 3D object detection task. Nonetheless, they are all trained with a large amount of annotated data, which are not accessible for the automatic annotation task. Additionally, they only leverage single modality, omitting the dense RGB information from images.

2.3 Autolabelers for 3D Object Detection

Although autolabelers have considerable potential for real-world applications, only a few works have investigated this problem. FGR [26] is a non-learning approach that takes 2D bounding boxes to locate frustum sub-point-clouds and removes ground points with the RANSAC algorithm. After that, they heuristically calculate the tightest 3D bounding box that can wrap all remaining points for each object. In contrast, other deep-learning approaches train their networks with a small amount of human annotations, and generate 3D boxes from inexpensive weak annotations.

CC [21] extracts frustum sub-clouds in the same way as FGR, but adopts PointNets [15] for segmenting and regressing 3D boxes. SDF [32] uses predefined CAD models to estimate the 3D geometry of cars detected in 2D images. VS3D [17] generates 3D proposals based on cloud density with an unsupervised UPM module. Points in the proposals together with cropped images are fed into a classification/regression network for the final 3D boxes. Later, WS3D [12] and WS3D (2021) [11] utilize center clicks as weak 2D annotations. Human annotators are asked to click on the objects' centroids on images, followed by adjusting the position in BEV of LiDAR scans. Points within a cylindrical proposal centered by each click are processed by a backbone network for the 3D pseudo box.

The above autolabelers utilize image modality and weak annotations to automatically generate 3D boxes for point clouds. However, they mainly use images to coarsely locate objects but do not address the sparsity problem of point clouds. Consequently, their performances are still sub-optimal compared with human annotations, having noticeable accuracy drops, especially for moderate and hard samples, which are usually far and sparse.

3 MTrans for Automatic Annotation

In this section, we first define the automatic annotation problem, and then introduce the proposed MTrans as an autolabeler, including multimodal self-attention mechanism, multi-task design, and self-supervised point generation.

Automatic Annotation. Given the LiDAR scan and its corresponding image, this research aims to generate 3D bounding box annotations for interested objects (i.e., Cars) from weak annotations of 2D bounding boxes. LiDAR points and image pixels are connected by the LiDAR-image projection based on the calibration parameters. Each point in the LiDAR point cloud can be projected to a pixel on the corresponding image. The autolabeler is trained with a small amount of human annotated data (e.g., 500 frames) and then used to label the remaining data. Another object detection network can be trained with the auto-generated 3D labels, saving manual work significantly.

Overview. As shown in Fig. 2, to address the sparsity problem and generate high quality 3D annotations, our MTrans takes multimodal inputs of LiDAR point clouds and their corresponding images (Sect. 3.1). The multimodal inputs are encoded and fused as point-level embedding vectors for further processing (Sect. 3.2). With the proposed multimodal self-attention module, object features are extracted based on the points' semantic and geometric relationships (Sect. 3.3). Then the extracted features are used for multiple tasks of foreground segmentation, point generation, and 3D box regression, encouraging the MTrans to comprehensively utilize the multimodal information and alleviate the sparsity

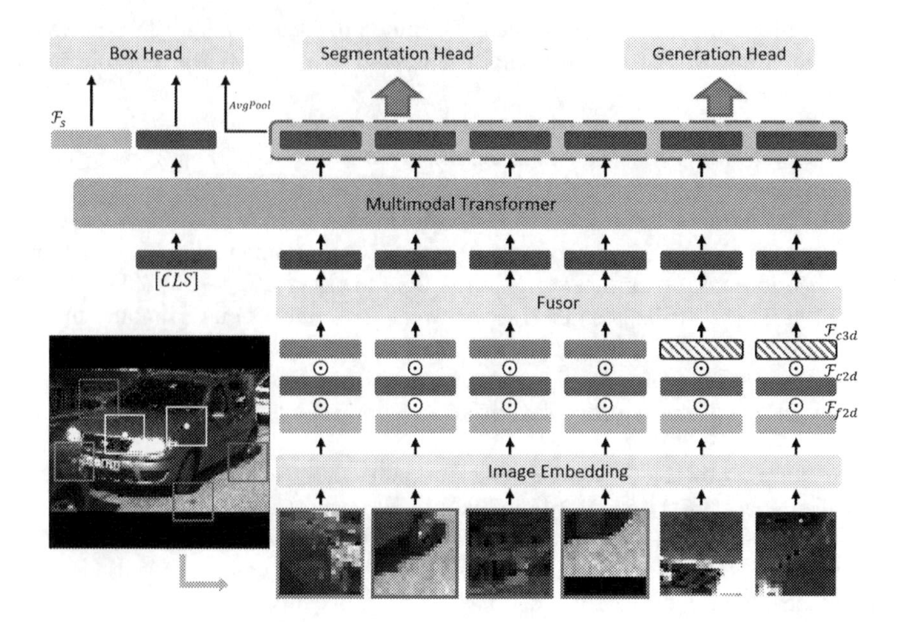

Fig. 2. Workflow of the MTrans. The network models both 3D and 2D information from the LiDAR scans and images. The frustum points and the image are fed into the network for simultaneous foreground segmentation, point generation, and box regression. $\mathcal{F}_{f2d}$, $\mathcal{F}_{c2d}$ and $\mathcal{F}_{c3d}$ are the embeddings for image patches, 2D coordinates and 3D coordinates. $\mathcal{F}_s$ is the global image feature derived by averaging $\mathcal{F}_{f2d}$.

problem (Sect. 3.4). Additionally, a self-supervised training strategy is proposed to leverage unlabeled data for the point generation task (Sect. 3.5).

3.1 Data Preparation

Our MTrans generates one 3D bounding box for each object in a weak 2D box. Using the LiDAR-image calibration parameters, a 3D point (x, y, z) can be projected onto the image plane (u, v) by the mapping function f_{cal}. Therefore, we can extract the frustum sub-cloud $\mathcal{P}_F$ corresponding to a 2D box:

$$\mathcal{P}_F = \{(x, y, z) \mid f_{cal}(x, y, z) \in \mathcal{B}_{2D}\}, \tag{1}$$

where $\mathcal{B}_{2D}$ is the region cropped by the 2D box. We use n to represent the frustum cloud size (i.e., the number of points). The points' 2D projections are defined as:

$$\mathcal{C}_{2D} = \{(u, v) \mid (u, v) = f_{cal}(x, y, z),\ (x, y, z) \in \mathcal{P}_F\}, \tag{2}$$

A shown in Fig. 2, centered by the 2D projections, image patches of shape $k \times k$ are extracted, denoted as $\mathcal{I}_p$. Thus, every LiDAR point has three categories of features, $\mathcal{P}_F^{(i)}$, $\mathcal{C}_{2D}^{(i)}$ and $\mathcal{I}_p^{(i)}$.

As mentioned in Sect. 1, in order to alleviate the sparsity problem, we generate 3D points from the image pixels, by referring to the n real LiDAR points as *context*. Since the number of real LiDAR points, n, varies from sample to sample, we randomly drop points or sample pixels from images as *padding*, resulting in fixed n' elements. Besides the padding points, we further sample another m pixels uniformly from the image, called *target* points. The padding and target points only have 2D features ($\mathcal{C}_{2D}^{(i)}$ and $\mathcal{I}_p^{(i)}$) without known 3D coordinates ($\mathcal{P}_F^{(i)}$).

3.2 Feature Embeddings and Object Representation

Same as vanilla transformers [4,22], we encode the object features as embedding vectors. For the 2D position, we use sinusoidal position embedding [22] to encode u and v into a half-length vector respectively, and then concatenate them for the 2D position embedding $\mathcal{F}_{c2d}$. As the 3D coordinates are continuous, a multilayer perceptron (MLP) is used to encode the (x, y, z) coordinates into 3D position embeddings $\mathcal{F}_{c3d}$. Lastly, we extract image patch features $\mathcal{F}_{f2d}$ with a 3-layer convolutional sub-network from $\mathcal{I}_p$. Specifically, we empirically set the patch size as 7×7 and kernel size as 3×3. All of $\mathcal{F}_{c2d}$, $\mathcal{F}_{c3d}$ and $\mathcal{F}_{f2d}$ have shapes of $((n' + m), d)$, where d is the embedding vector length, for the $n' + m$ points. Note that for the padding and target points, they have no known 3D coordinates. Therefore, their $\mathcal{F}_{c3d}$ features are replaced by a trainable embedding vector $\mathcal{E}$.

Afterwards, we fuse the three kinds of features by concatenating them along the d-axis, and decrease the channel from $3d$ back to d by a fully-connected layer. As shown in Fig. 2, we introduce an object-level token $[CLS]$, which is a randomly-initialized trainable embedding vector. The total $n' + m + 1$ element representations of the object, $\mathcal{F} \in \mathbb{R}^{(n'+m+1)\times d}$, are then input into the MTrans for feature extraction with the proposed multimodal self-attention mechanism.

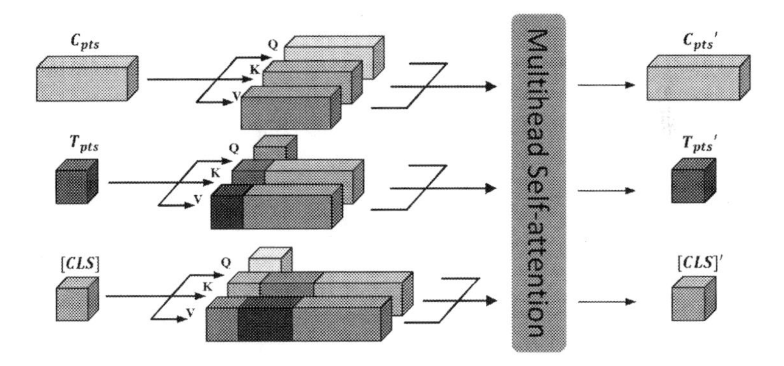

Fig. 3. Multimodal Self-attention. C-pts with known 3D coordinates attend to each other, while each of the T-pts attends to itself as well as the C-pts. $[CLS]$ attends to all available C-pts and T-pts for object-level representation. For better visual effect, we show one T-pt doing self-attention in the middle row. Best viewed in color. (Color figure online)

3.3 Multimodal Self-attention

Following the vanilla transformer [4,22], we manipulate the element representations with the multi-head self-attention mechanism. The element representations are transformed into query Q, key K, and value V. The values V are recombined based on attention scores calculated by multiplying Q and K. Owing to the space limitation, we do not repeat the technical details of the multi-head self-attention mechanism, but only stress on the modifications for our MTrans.

Different from the original transformer, under our problem setting, some elements contain incomplete information (i.e., the padding and target points have only 2D information but no 3D information). Therefore, we treat the original frustum sub-cloud points with full information as the context points (*C-pts* in Fig. 3), and recover 3D coordinates for the padding and target elements (*T-pts*). Specifically, the T-pts look up C-pts based on their position and semantic relationships, and interpolate their 3D coordinates to enrich the sparse point cloud. During our multimodal self-attention, the C-pts only attend to each other but not T-pts, while each T-pt can also attend to itself, and $[CLS]$ attends to all available points to aggregate object-level information.

As shown in Fig. 3, the C-pts perform vanilla self-attention, and have representations updated as C-pts$'$. For each T-pts, a one-element Q is built from its element representation, while K and V are from the concatenation of itself and the C-pts. Similarly, $[CLS]$ has Q from itself, but K and V from the concatenation of $[CLS]$, T-pts, and C-pts. With this asymmetric design, T-pts update themselves by absorbing 3D information from C-pts, and $[CLS]$ encodes a global representation of the object. For our MTrans, we stack 4 multimodal self-attention layers, and vanilla feed-forward-layer [22] is adopted between each pair of the self-attention layers.

3.4 Multi-task Prediction Heads

After the self-attention layers, the transformed element representations are fed into multiple prediction heads for the multi-task supervision. Firstly, a 2-layer MLP $\mathcal{H}_{seg}$ performs binary segmentation for each point. Points within the ground-truth 3D bounding box are regarded as foreground, otherwise background. The segmentation loss $\mathcal{L}_{seg}$ is the summation of Cross-Entropy loss and Dice loss. Since T-pts have no 3D coordinates, only C-pts are supervised.

Another 2-layer MLP head $\mathcal{H}_{xyz}$ predicts the 3D coordinates for the point generation task. Since there are no ground-truth values for the padding and target points, we adopt a mask-and-predict strategy to train this task with self-supervision, which will be introduced in detail in the next section. Smooth L_1 loss is used for the point generation task as $\mathcal{L}_{xyz}$. Intuitively, the foreground is more critical than background, so we only calculate $\mathcal{L}_{xyz}$ for the foreground points predicted by $\mathcal{H}_{seg}$. Moreover, we multiply the loss with a weight of 0.1 if the point is a context point due to the leaked 3D coordinates, otherwise 1.

The third prediction head $\mathcal{H}_{box}$ directly regresses 3D bounding boxes from the object-level representations, which is derived by concatenating the $[CLS]$

representation, the averaged element representations, and the averaged image feature of $\mathcal{F}_s$, as shown in Fig. 2. The output box is a 7-element vector for the location ((x, y, z) coordinates), dimension (length, width, height), and rotation along the z-axis. We use the dIoU loss [9,34,35] for the box loss $\mathcal{L}_{box}$.

As the IoU (intersection over union) metric is direction-invariant (i.e., rotating a box for 180 °C does not change the IoU), we further introduce a direction head for the binary direction classification. Specifically, orientation within $[-\pi/2, \pi/2)$ is regarded as the front, and $[-\pi, -\pi/2) \cup [\pi/2, \pi]$ as the back. Cross Entropy loss is used for the direction loss $\mathcal{L}_{dir}$.

All the heads are trained simultaneously, and the overall loss is given by:

$$\mathcal{L} = \mathcal{L}_{seg} + \mathcal{L}_{xyz} + \lambda_{box}\mathcal{L}_{box} + \mathcal{L}_{dir}, \tag{3}$$

where the λ_{box} is a coefficient and we empirically set it as 5.

3.5 Self-supervised Point Generation

One major motivation of the proposed MTrans is to enrich the sparse point clouds with image information. As mentioned above, beyond the context points, some extra T-pts (i.e., padding and target points) are appended with unknown 3D information. The point generation task, supervised by the loss $\mathcal{L}_{xyz}$, aims to uncover the 3D coordinates for these T-pts based on their 2D geometric and semantic relationships with the context points. Due to the lack of ground truths, we adopt a self-supervised strategy for this task. Specifically, we randomly mask a portion ($0 \sim 95\%$) of the context points, by replacing their $\mathcal{F}_{c3d}$ with the trainable embedding vector $\mathcal{E}$. The masked points are then treated as T-pts and supervised by the $\mathcal{L}_{xyz}$. With this partial supervision, we managed to train the MTrans for the point generation task. During the inference, no context points are masked and the model recovers the 3D coordinates for the real T-pts.

4 Extension of MTrans to 3D Object Detection

Inspired by F-PointNet [14] and F-ConvNet [25], we can also convert the MTrans into a 3D object detector. Similarly, our model lifts 2D detection results up to 3D bounding boxes. The architecture and workflow remain the same as in Sect. 3. However, to be compatible for the detection metric of Average Precision (AP), we add another prediction head $\mathcal{H}_c$ for the confidence of the generated 3D box. The confidence is indicated by the estimated IoU score of each generated 3D box. Instead of using a small amount of data for the automatic annotation task, the detector version of the MTrans is trained with full data. The overall loss is also updated as:

$$\mathcal{L} = \mathcal{L}_{seg} + \mathcal{L}_{xyz} + \lambda_{box}\mathcal{L}_{iou} + \mathcal{L}_{dir} + \mathcal{L}_c, \tag{4}$$

where the new term $\mathcal{L}_c$ is the Smooth L_1 loss for confidence head $\mathcal{H}_c$.

5 Experimental Evaluations

The proposed MTrans is evaluated for two tasks, the automatic annotation and 3D object detection. We also conduct ablation studies to justify the contributions of each module.

KITTI Dataset. KITTI [5] is a benchmark dataset for 3D detection for autonomous driving. We follow the official practice to split the dataset into training and validation sets with 3,712 and 3,769 frames. The standard evaluation metric of AP with the IoU threshold being 0.7 is adopted. Same as FGR [26], we focus on the Car class and filter out objects with less than 5 foreground LiDAR points, due to some samples having too few points to possibly draw a 3D box.

5.1 Implementation Details

For the MTrans, 4 multimodal self-attention layers with the hidden size of 768 and 12 heads are used. As an autolabeler, the model is trained with 500 or 125 annotated frames for 300 epochs. The remaining frames are also used as unlabeled data for self-supervision on the point generation task. Adam optimizer is employed with a learning rate of 10^{-4}. We also use standard image augmentation methods of auto contrast, random sharpness and color jittering, as well as point cloud augmentation methods of random translation, scaling and mirroring. Due to some 2D boxes having overlaps, an overlap mask of shape is added as a new channel of the image, where the overlapped areas have values of 0, and otherwise 1. For each object, the cloud size n' is set to 512 for context plus padding points, and m is 784 for the number of image-sampled target points.

5.2 Automatic Annotation Results

To evaluate our MTrans as an autolabeler, we employ two popular backbone object detection networks, PointRCNN [19] and PointPillars [18], and retrain them with the auto-generated 3D annotations. To examine the influence of different data scales, we also train them with all frames, 500 frames, and 125 frames of human-annotated data, respectively. As shown in Table 1, the accuracy drops dramatically when the data are insufficient, especially for the moderate and hard samples. With 125 frames of annotated data, PointPillars and PointRCNN suffer mean accuracy drops of 16.9% and 6.23% in AP.

Meanwhile, the proposed MTrans is also trained with 500 frames and 125 frames of annotated data, and then re-annotates the whole KITTI *train* set. As shown in Table 1, trained with the MTrans-generated annotations, the two networks both significantly outperform their counterparts. For PointPillars, our method brings 7.06% and 12.42% absolute average increases in AP on the *val* set for 500 and 125 frames of data, respectively. Impressively, using only 500 and 125 frames, PointRCNN trained with MTrans can achieve 99.62% and 97.24% of the original model trained with full data on the *val* set. On the *test* set,

Table 1. Automatic annotation results on KITTI *val* set and *test* set.

Method	Fully supervised	AP_{3D} Val			AP_{3D} Test		
		Easy	Mod	Hard	Easy	Mod	Hard
PointPillars [8]	✓	86.10	76.58	72.79	82.58	74.31	68.99
PointPillars [8]	500f	80.65	67.65	66.14	–	–	–
PointPillars [8]	125f	71.36	58.29	55.11	–	–	–
Ours + PointPillars	500f	86.69	76.56	72.38	77.65	67.48	62.38
Ours + PointPillars	125f	83.70	71.66	66.67	–	–	–
PointRCNN [19]	✓	88.99	78.71	78.21	86.96	75.64	70.70
PointRCNN [19]	500f	88.54	76.85	70.12	–	–	–
PointRCNN [19]	125f	85.63	73.60	67.98	–	–	–
Ours + PointRCNN	500f	88.72	78.84	77.43	83.42	75.07	68.26
Ours + PointRCNN	125f	87.64	77.31	74.32	–	–	–
Compare with other autolabelers							
WS3D [12]	BEV Centroid	84.04	75.10	73.29	80.15	69.64	63.71
WS3D(2021) [11]	BEV Centroid	85.04	*75.94*	*74.38*	*80.99*	*70.59*	*64.23*
FGR [26]	2D Box	*86.67*	73.55	67.90	80.26	68.47	61.57
Ours (PointRCNN)	2D Box	**88.72**	**78.84**	**77.43**	**83.42**	**75.07**	**68.26**

MTrans-trained PointRCNN also yields 97.19% of the original model. The above results demonstrate that our MTrans can generate high-quality 3D annotations, producing object detection networks comparable to their counterparts trained with human annotations. Automatically raising 2D boxes into 3D annotations, our method greatly saves human efforts for the annotation procedure.

We also compare our method with existing state-of-the-art autolabelers, FGR [26] and WS3D [11,12]. Although they also generate 3D boxes from weak 2D annotations, we want to stress the difference in task design for the three approaches. FGR uses 2D boxes, same as in MTrans, but is a non-learning algorithm, while WS3D uses center-clicks on the image and BEV, which might provide extra 3D information. Using the same PointRCNN backbone, our method surpasses all the above baselines significantly. Especially for the moderate and hard samples, MTrans improves the 3D AP by 4.48% and 4.03% respectively. This observation aligns with the motivation of enriching the sparse point cloud for higher quality of generated labels, since moderate and hard samples usually suffer more from the sparsity issue than easy samples.

5.3 Detection Results

As mentioned in Sect. 4, with an extra prediction head for the box confidence, the proposed MTrans can also be applied for 3D object detection task. Same as F-PointNet [14] and F-ConvNet [25], we train our MTrans with full data, and use ground-truth 2D boxes for frustum proposals. We evaluate our method on the KITTI *val* set, using metrics of AP_{3D} and AP_{3D} R40 (40 recall thresholds).

Table 2. Detection results on KITTI *val* set.

Method	Modality	Extra Ref. Signal	AP_{3D}			AP_{3D} R40		
			Easy	Mod	Hard	Easy	Mod	Hard
MV3D [2]	LiDAR+RGB	✗	71.29	62.68	56.56	–	–	–
F-PointNet [14]	LiDAR+RGB	2D box	83.76	70.92	63.65	–	–	–
F-ConvNet [25]	LiDAR+RGB	2D box	89.02	78.80	77.09	–	–	–
VPFNet [36]	LiDAR+RGB	✗	–	–	–	93.42	88.76	86.05
SECOND [30]	LiDAR	✗	87.43	76.48	69.10	90.97	79.94	77.09
Voxel-RCNN [3]	LiDAR	✗	89.41	84.52	78.93	92.38	85.29	82.86
PV-RCNN [18]	LiDAR	✗	89.35	83.69	78.70	92.57	84.83	82.69
SPG [29]	LiDAR	✗	89.81	84.45	79.14	92.53	85.31	82.82
BtcDet [28]	LiDAR	✗	–	–	–	93.15	86.28	83.86
Ours	LiDAR+RGB	2D box	**97.82**	**89.89**	**89.45**	**98.83**	**93.57**	**90.95**

Shown in Table 2, our method achieves impressive performances of 92.38% and 94.46% for mAP and mAP(R40). The detection accuracy is significantly higher than other state-of-the-art baselines. However, same as F-PointNet [14] and F-ConvNet [25], our method utilizes the ground-truth 2D boxes as extra reference signals, which reduces the problem difficulty. Nonetheless, in the fair comparison with F-PointNet and F-ConvNet under identical task and experiment settings, our method still boosts the mAP by 19.61% and 10.75%, respectively.

Table 3. Ablation results.

	MM	Seg	Gen	ExtraPts	SelfSup	Metric			
						mIoU	Recall	mAP	mAP_{R40}
Vanilla	✗	✗	✗	✗	✗	60.07	41.53	48.57	46.89
Multimodal	✓	✗	✗	✗	✗	65.57	53.30	58.59	57.33
Seg Loss	✓	✓	✗	✗	✗	69.45	61.53	67.84	69.42
XYZ Loss	✓	✗	✓	✗	✗	67.61	57.76	63.62	63.82
Self-supervise	✓	✗	✓	✗	✓	68.42	59.82	65.25	66.89
Densify	✓	✗	✓	✓	✗	75.37	74.95	80.82	83.47
Full generate	✓	✗	✓	✓	✓	76.84	78.48	83.98	85.72
Ours	✓	✓	✓	✓	✓	**77.50**	**81.22**	**86.24**	**90.30**
Ours (all frames)	✓	✓	✓	✓	✓	81.01	87.65	92.36	94.43

5.4 Ablation Study

Ablation studies are conducted to verify the technical contributions of the proposed method. We train the MTrans with 500 annotated frames of the KITTI *train* set, and evaluate the model on the *val* set. The generated 3D boxes are compared with the ground-truth human annotations, and three metrics are employed, namely, the mean IoU (mIoU), recall with IoU threshold of 0.7, and mean average precision (mAP, mAP_{R40}) over the Car class.

As shown in Table 3, the vanilla transformer, which directly processes 3D LiDAR points $\mathcal{F}_{c3d}$, cannot model the problem well with the small amount of data with only 500 frames. The *Multimodal* variant introduces extra modality of images, and improves the mIoU and recall significantly by 5.5% and 11.77%, respectively. *Seg Loss* and *XYZ Loss* introduce the multi-task design, and again improve the accuracy significantly by 12.09 and 6.49% mAP_{R40}, demonstrating the contribution of the auxiliary losses. For *Densify*, extra target points are evenly sampled from the image, and their 3D coordinates are estimated to densify the point cloud a step further. Alternatively, *Self-supervise* utilize unlabeled data for the point generation task, also improving the accuracy, resulting in 68.42% mIoU. Combining all the point cloud densification techniques, *Full Generate*'s mIoU and recall both surpass 70%. Leveraging all the modules, our MTrans archives impressive 81.22% Recall and 86.24% mAP using only 500 frames of annotated data, which is even comparable with state-of-the-art object detectors trained with full annotations.

The ablation studies justify the contributions of our four innovations, namely, the multimodal network, auxiliary multiple tasks, point generation, and self-supervision. The multimodal inputs can effectively boost accuracy with negligible modification on the vanilla transformer (*Multimodal* variant). Moreover, the introduction of the auxiliary multi-task design (*Seg Loss* and *XYZ Loss*) raises the mIoU by 2.04% and 3.88%, showing that the model is encouraged to understand the problem more comprehensively from multiple aspects. The largest gain, however, comes from adding extra target points from the image, which increases the mIoU by 7.76% and mAP by 20.35% (*Densify*). This observation well supports our motivation that image-guided interpolation can enrich the sparse point cloud for better annotations. Combining all the techniques, our final MTrans yields a surprising 77.5% mIoU and 81.22% recall, with only 500 annotated frames. Nonetheless, the gains from these techniques tend to be saturated when combining all of the modules. Compared with the mAP accuracy for all-data-trained MTrans at the bottom of Table 1, we hypothesize that the main constraint is the amount of training data.

6 Visualization of Attention on Context Points

In Fig. 4 we visualize the attention distribution over context points. Specifically, for each context point, we average the attention scores it received from all other points. The attention scores of all context points sum to 1. As shown in the figure, points that belong to the car tends to receive more attention than background points, which aligns with our intuition, since background points provide little reference value of labeling 3D boxes. Moreover, the attention distribution also demonstrates a coherent spatial pattern in both the 2D image contents and the 3D space. This observation suggests that multimodal self-attention extracts object features based on 3D & 2D geometric, and 2D semantic relationships.

Another interesting phenomenon is that the foreground points closer to object edges tend to receive more attention. As shown in the figures, yellow points are

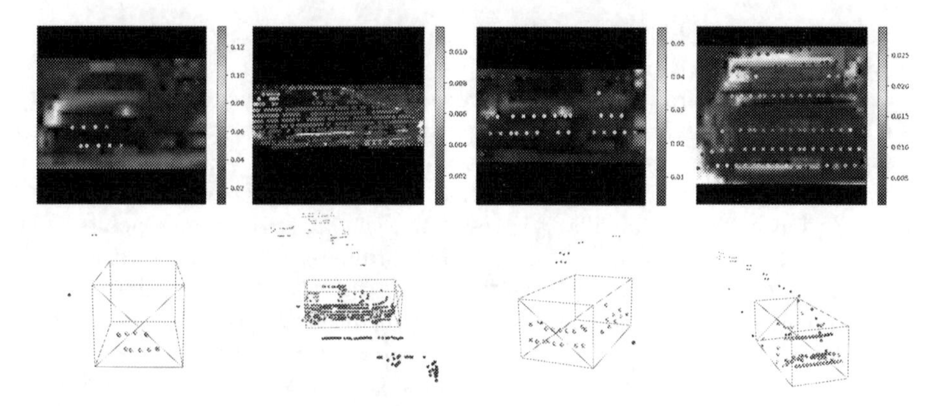

Fig. 4. Attention score visualization results. Context points and their average attention scores are plotted. Each point's color represents the amount of attention it receives from other points. The attention scores sum to 1 for each data sample. It is observed that the foreground points tend to receive more attention than background points, which provides an intuitive interpretation for the multimodal attention MTrans.

presented near box corners for all four samples. This is also reasonable since corner points usually provide more critical information for 3D box prediction. The observation provides good interpretability for our MTrans.

7 Conclusion

This work has proposed the MTrans, an autolabeler that generates 3D boxes from weak 2D box annotations, given the LiDAR point clouds and corresponding images. Our method effectively alleviates the pervasive sparsity problem of point clouds. Specifically, we introduce the multimodal self-attention mechanism, together with auxiliary multi-task and self-supervision design, which leverage image information to generate extra 3D points. Experimental results demonstrate that our method can outperform existing autolabelers significantly, generating 3D box annotations comparable to human annotations. Moreover, the proposed MTrans can also be applied for the 3D object detection task, achieving state-of-the-art performances. Beyond LiDAR + monocular image, an interesting future direction would be to incorporate more modalities to improve the detection accuracy and robustness, such as radar and/or stereo images. It would also be valuable to investigate the domain transfer problem for the autolabelers, which could save human effort further. We hope this study can inspire future works to solve the sparsity problem as well as to develop automatic annotation workflows to reduce manual workloads.

References

1. Caesar, H., et al.: nuScenes: a multimodal dataset for autonomous driving. In: Proceedings of the IEEE/CVF Conference on Computer Vision and Pattern Recognition, pp. 11621–11631 (2020)
2. Chen, X., Ma, H., Wan, J., Li, B., Xia, T.: Multi-view 3D object detection network for autonomous driving. In: Proceedings of the IEEE Conference on Computer Vision and Pattern Recognition, pp. 1907–1915 (2017)
3. Deng, J., Shi, S., Li, P., Zhou, W., Zhang, Y., Li, H.: Voxel R-CNN: towards high performance voxel-based 3D object detection. In: Proceedings of the AAAI Conference on Artificial Intelligence, vol. 35, pp. 1201–1209 (2021)
4. Devlin, J., Chang, M.W., Lee, K., Toutanova, K.: Bert: pre-training of deep bidirectional transformers for language understanding. arXiv preprint arXiv:1810.04805 (2018)
5. Geiger, A., Lenz, P., Urtasun, R.: Are we ready for autonomous driving? The Kitti vision benchmark suite. In: Conference on Computer Vision and Pattern Recognition (CVPR) (2012)
6. Huang, T., Liu, Z., Chen, X., Bai, X.: EPNet: enhancing point features with image semantics for 3D object detection. In: Vedaldi, A., Bischof, H., Brox, T., Frahm, J.-M. (eds.) ECCV 2020. LNCS, vol. 12360, pp. 35–52. Springer, Cham (2020). https://doi.org/10.1007/978-3-030-58555-6_3
7. Ku, J., Mozifian, M., Lee, J., Harakeh, A., Waslander, S.L.: Joint 3D proposal generation and object detection from view aggregation. In: 2018 IEEE/RSJ International Conference on Intelligent Robots and Systems (IROS), pp. 1–8. IEEE (2018)
8. Lang, A.H., Vora, S., Caesar, H., Zhou, L., Yang, J., Beijbom, O.: PointPillars: fast encoders for object detection from point clouds. In: Proceedings of the IEEE/CVF Conference on Computer Vision and Pattern Recognition, pp. 12697–12705 (2019)
9. lilanxiao: Differentiable iou of oriented boxes. https://github.com/lilanxiao/Rotated_IoU (2021)
10. McCraith, R., Insafutdinov, E., Neumann, L., Vedaldi, A.: Lifting 2D object locations to 3D by discounting lidar outliers across objects and views. arXiv preprint arXiv:2109.07945 (2021)
11. Meng, Q., Wang, W., Zhou, T., Shen, J., Jia, Y., Van Gool, L.: Towards a weakly supervised framework for 3D point cloud object detection and annotation. IEEE Trans. Pattern Anal. Mach. Intell. **44**(8), 4454–4468 (2021)
12. Meng, Q., Wang, W., Zhou, T., Shen, J., Van Gool, L., Dai, D.: Weakly supervised 3D object detection from lidar point cloud. In: Vedaldi, A., Bischof, H., Brox, T., Frahm, J.-M. (eds.) ECCV 2020. LNCS, vol. 12358, pp. 515–531. Springer, Cham (2020). https://doi.org/10.1007/978-3-030-58601-0_31
13. Pang, S., Morris, D., Radha, H.: CLOCs: camera-lidar object candidates fusion for 3D object detection. In: 2020 IEEE/RSJ International Conference on Intelligent Robots and Systems (IROS), pp. 10386–10393. IEEE (2020)
14. Qi, C.R., Liu, W., Wu, C., Su, H., Guibas, L.J.: Frustum pointnets for 3D object detection from RGB-D data. In: Proceedings of the IEEE Conference on Computer Vision and Pattern Recognition, pp. 918–927 (2018)
15. Qi, C.R., Su, H., Mo, K., Guibas, L.J.: PointNet: deep learning on point sets for 3D classification and segmentation. In: Proceedings of the IEEE Conference on Computer Vision and Pattern Recognition, pp. 652–660 (2017)

16. Qi, C.R., Yi, L., Su, H., Guibas, L.J.: PointNet++: deep hierarchical feature learning on point sets in a metric space. arXiv preprint arXiv:1706.02413 (2017)
17. Qin, Z., Wang, J., Lu, Y.: Weakly supervised 3D object detection from point clouds. In: Proceedings of the 28th ACM International Conference on Multimedia, pp. 4144–4152 (2020)
18. Shi, S., et al.: PV-RCNN: point-voxel feature set abstraction for 3D object detection. In: Proceedings of the IEEE/CVF Conference on Computer Vision and Pattern Recognition, pp. 10529–10538 (2020)
19. Shi, S., Wang, X., Li, H.: PointRCNN: 3D object proposal generation and detection from point cloud. In: Proceedings of the IEEE/CVF Conference on Computer Vision and Pattern Recognition, pp. 770–779 (2019)
20. Sun, P., et al.: Scalability in perception for autonomous driving: Waymo open dataset. In: Proceedings of the IEEE/CVF Conference on Computer Vision and Pattern Recognition, pp. 2446–2454 (2020)
21. Tang, Y.S., Lee, G.H.: Transferable semi-supervised 3D object detection from RGB-D data. In: Proceedings of the IEEE/CVF International Conference on Computer Vision, pp. 1931–1940 (2019)
22. Vaswani, A., et al.: Attention is all you need. In: Advances in Neural Information Processing Systems, pp. 5998–6008 (2017)
23. Vora, S., Lang, A.H., Helou, B., Beijbom, O.: Pointpainting: sequential fusion for 3D object detection. In: Proceedings of the IEEE/CVF Conference on Computer Vision and Pattern Recognition, pp. 4604–4612 (2020)
24. Wang, W., Tran, D., Feiszli, M.: What makes training multi-modal classification networks hard? In: Proceedings of the IEEE/CVF Conference on Computer Vision and Pattern Recognition, pp. 12695–12705 (2020)
25. Wang, Z., Jia, K.: Frustum convnet: sliding frustums to aggregate local point-wise features for amodal 3D object detection. In: 2019 IEEE/RSJ International Conference on Intelligent Robots and Systems (IROS), pp. 1742–1749. IEEE (2019)
26. Wei, Y., Su, S., Lu, J., Zhou, J.: FGR: frustum-aware geometric reasoning for weakly supervised 3D vehicle detection. In: 2021 IEEE International Conference on Robotics and Automation (ICRA), pp. 4348–4354. IEEE (2021)
27. Xie, L., et al.: PI-RCNN: an efficient multi-sensor 3D object detector with point-based attentive Cont-Conv fusion module. In: Proceedings of the AAAI Conference on Artificial Intelligence, vol. 34, pp. 12460–12467 (2020)
28. Xu, Q., Zhong, Y., Neumann, U.: Behind the curtain: learning occluded shapes for 3D object detection. arXiv preprint arXiv:2112.02205 (2021)
29. Xu, Q., Zhou, Y., Wang, W., Qi, C.R., Anguelov, D.: SPG: unsupervised domain adaptation for 3D object detection via semantic point generation. In: Proceedings of the IEEE/CVF International Conference on Computer Vision, pp. 15446–15456 (2021)
30. Yan, Y., Mao, Y., Li, B.: Second: sparsely embedded convolutional detection. Sensors **18**(10), 3337 (2018)
31. Yoo, J.H., Kim, Y., Kim, J., Choi, J.W.: 3D-CVF: generating joint camera and LiDAR features using cross-view spatial feature fusion for 3D object detection. In: Vedaldi, A., Bischof, H., Brox, T., Frahm, J.-M. (eds.) ECCV 2020. LNCS, vol. 12372, pp. 720–736. Springer, Cham (2020). https://doi.org/10.1007/978-3-030-58583-9_43
32. Zakharov, S., Kehl, W., Bhargava, A., Gaidon, A.: Autolabeling 3D objects with differentiable rendering of SDF shape priors. In: Proceedings of the IEEE/CVF Conference on Computer Vision and Pattern Recognition, pp. 12224–12233 (2020)

33. Zhao, S., Gong, M., Fu, H., Tao, D.: Adaptive context-aware multi-modal network for depth completion. IEEE Trans. Image Process. **30**, 5264–5276 (2021)
34. Zheng, Z., Wang, P., Liu, W., Li, J., Ye, R., Ren, D.: Distance-IoU loss: faster and better learning for bounding box regression. In: Proceedings of the AAAI Conference on Artificial Intelligence, vol. 34, pp. 12993–13000 (2020)
35. Zhou, D., et al.: IoU loss for 2D/3D object detection. In: 2019 International Conference on 3D Vision (3DV), pp. 85–94 (2019)
36. Zhu, H., et al.: VPFNet: improving 3D object detection with virtual point based lidar and stereo data fusion. arXiv preprint arXiv:2111.14382 (2021)

Dynamic 3D Scene Analysis by Point Cloud Accumulation

Shengyu Huang[1(✉)], Zan Gojcic[2], Jiahui Huang[3], Andreas Wieser[1], and Konrad Schindler[1]

[1] ETH Zürich, Zurich, Switzerland
shenhuan@ethz.ch
[2] NVIDIA, Zurich, Switzerland
[3] BRCist, Beijing, China

Abstract. Multi-beam LiDAR sensors, as used on autonomous vehicles and mobile robots, acquire sequences of 3D range scans ("frames"). Each frame covers the scene sparsely, due to limited angular scanning resolution and occlusion. The sparsity restricts the performance of downstream processes like semantic segmentation or surface reconstruction. Luckily, when the sensor moves, frames are captured from a sequence of different viewpoints. This provides complementary information and, when accumulated in a common scene coordinate frame, yields a denser sampling and a more complete coverage of the underlying 3D scene. However, often the scanned scenes contain moving objects. Points on those objects are not correctly aligned by just undoing the scanner's ego-motion. In the present paper, we explore multi-frame point cloud accumulation as a mid-level representation of 3D scan sequences, and develop a method that exploits inductive biases of outdoor street scenes, including their geometric layout and object-level rigidity. Compared to state-of-the-art scene flow estimators, our proposed approach aims to align all 3D points in a common reference frame correctly accumulating the points on the individual objects. Our approach greatly reduces the alignment errors on several benchmark datasets. Moreover, the accumulated point clouds benefit high-level tasks like surface reconstruction. [project page]

1 Introduction

LiDAR point clouds are a primary data source for robot perception in dynamically changing 3D scenes. They play a crucial role in mobile robotics and autonomous driving. To ensure awareness of a large field of view at any point in time, 3D point measurements are acquired as a sequence of sparse scans that each cover a large field-of-view—typically, the full 360°. In each individual scan *(i)* the point density is low, and *(ii)* some scene parts are occluded. Both issues complicate downstream processing. One way to mitigate the problem is to assume

Supplementary Information The online version contains supplementary material available at https://doi.org/10.1007/978-3-031-19839-7_39.

S. Avidan et al. (Eds.): ECCV 2022, LNCS 13698, pp. 674–690, 2022.
https://doi.org/10.1007/978-3-031-19839-7_39

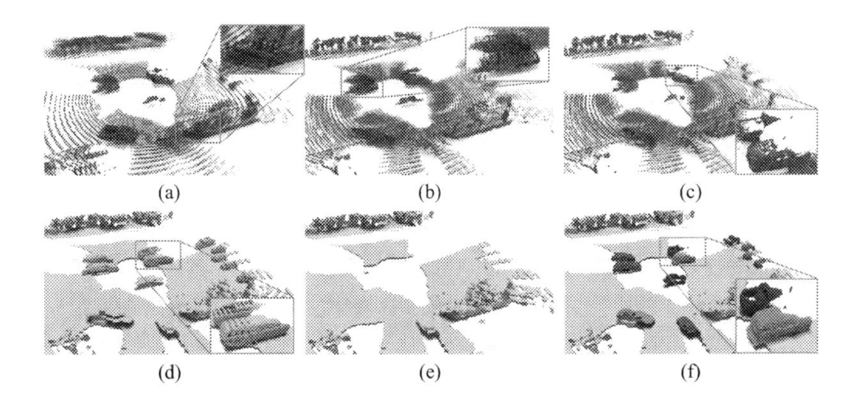

Fig. 1. Points in LiDAR frames acquired over time are not aligned due to the motion of the sensor and of other agents in the scene (a, d). Static background points can be aligned using ego-motion, but this smears the dynamic points across their trajectories (b). While motion segmentation only enables removing the moving points from the scene (e), our method properly disentangles individual moving objects from the static part and accumulates both correctly (c, f).

the sensor ego-motion is known and to align multiple consecutive scans into a common scene coordinate frame, thus accumulating them into a denser and more complete point cloud. This simple accumulation strategy already boosts performance for perception tasks like object detection [7] and semantic segmentation [4], but it also highlights important problems. First, to obtain sufficiently accurate sensor poses the ego-motion is typically computed in post-processing to enable sensor fusion and loop closures—meaning that it would actually not be available for online perception. Second, compensating the sensor ego-motion only aligns scan points of the static background, while moving foreground objects are smeared out along their motion trajectories (Fig. 1b).

To properly accumulate 3D points across multiple frames, one must disentangle the individual moving objects from the static background and reason about their spatio-temporal properties. Since the grouping of the 3D points into moving objects itself depends on their motion, the task becomes a form of multi-frame 3D scene flow estimation. Traditional scene flow methods [32,40,58] model dynamics in the form of a free-form velocity field from one frame to the next, only constrained by some form of (piece-wise) smoothing [13,51,52].

While this is a very flexible and general approach, it also has two disadvantages in the context of autonomous driving scenes: *(i)* it ignores the geometric structure of the scene, which normally consists of a dominant, static background and a small number of discrete objects that, at least locally, move rigidly; *(ii)* it also ignores the temporal structure and only looks at the minimal setup of two frames. Consequently, one risks physically implausible scene flow estimates [19], and does not benefit from the dense temporal sequence of scans.

Starting from these observations, we propose a novel point cloud accumulation scheme tailored to the autonomous driving setting. To that end, we aim to accumulate point clouds over time while abstracting the scene into a collection of rigidly moving agents [3,19,47] and reasoning about each agent's motion on the basis of a longer sequence of frames [22,23]. Along with the accumulated point cloud (Fig. 1c), our method provides more holistic scene understanding, including foreground/background segmentation, motion segmentation, and per-object parametric motion compensation. As a result, our method can conveniently serve as a common, low-latency preprocessing step for perception tasks including surface reconstruction (Fig. 1f) and semantic segmentation [4].

We carry out extensive evaluations on two autonomous driving datasets *Waymo* [46] and *nuScenes* [7], where our method greatly outperforms prior art. For example, on *Waymo* we reduce the average endpoint error from 12.9 cm to 1.8 cm for the static part and from 23.7 cm to 17.3 cm for the dynamic objects. We observe similar performance gains also on *nuScenes*.

In summary, we present a novel, learnable model for temporal accumulation of 3D point cloud sequences over multiple frames, which disentangles the background from dynamic foreground objects. By decomposing the scene into agents that move rigidly over time, our model is able to learn multi-frame motion and reason about motion in context over longer time sequences. Moreover, our method allows for low-latency processing, as it operates on raw point clouds and requires only their sequence order as further input. It is therefore suitable for use in online scene understanding, for instance as a low-level preprocessor for semantic segmentation or surface reconstruction.

2 Related Work

Temporal Point Cloud Processing. Modeling a sequence of point clouds usually starts with estimating accurate correspondences between frames, for which scene flow emerged as a popular representation. Originating from [49], scene flow estimation builds an intuitive and effective dynamic scene representation by computing a flow vector for each source point. While traditional scene flow methods [50–52,55] leverage motion smoothness as regularizer within their optimization frameworks, modern learning-based methods learn the preference for smooth motions directly from large-scale datasets [32,36,40,58]. Moreover, manually designed scene priors proved beneficial for structured scenes, for instance supervoxel-based local rigidity constraints [29], or object-level shape priors learned in fully supervised [3] or weakly supervised [19] fashion. Methods like SLIM [2] take a decoupled approach, where they first run motion segmentation before deriving scene flows for each segment separately. Treating the entire point cloud sequence as 4D data and applying a spatio-temporal backbone [10,16,33] demonstrates superior performance and efficiency. Subsequent works enhance such backbones by employing long-range modeling techniques such as Transformers [15,48,60], or by coupling downstream tasks like semantic segmentation [1], object detection [42,59] and multi-modal fusion [38]. While our method employs the representation of multi-frame scene flow, we explicitly model individual dynamic objects, which not only provides a high-level scene decomposition, but also yields markedly higher accuracy.

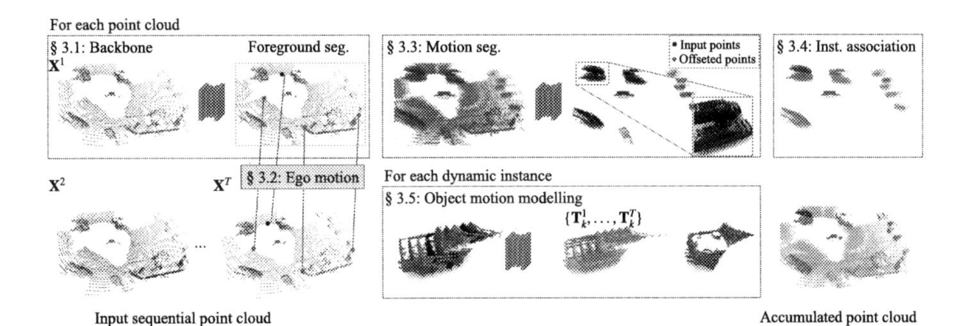

Fig. 2. Overview. Our method takes in a point cloud sequence of T frames and starts by extracting foreground points (marked yellow) for each frame. To obtain ego motion, $T-1$ pairwise registrations are performed. Next, points belonging to dynamic foreground object are extracted using our motion segmentation module (marked orange). To boost subsequent spatio-temporal instance association, we additionally predict per-point offset vectors. After instance association, we finally compute the rigid motion separately for each segmented dynamic object. (Color figure online)

Motion Segmentation. Classification of the points into static and dynamic scene parts serves as an essential component in our pipeline. Conventional geometric approaches either rely on ray casting [8,44] over dense terrestrial laser scans to build clean static maps, or on visibility [26,39] to determine the dynamics of the query point by checking its occlusion state in a dense map. Removert [26] iteratively recovers falsely removed static points from multi-scale range images. Most recently, learning-based methods formulate and solve the segmentation task in a data-driven way: Chen *et al.* [9] propose a deep model over multiple range image residuals and show SoTA results on a newly-established motion segmentation benchmark [4]. Any Motion Detector [17] first extracts per-frame features from bird's-eye-view projections and then aggregates temporal information from ego-motion compensated per-frame features (in their case with a with convolutional RNN). Our work is similar in spirit, but additionally leverages information from the foreground segmentation task and object clustering in an end-to-end framework.

Dynamic Object Reconstruction. Given sequential observations of a rigid object, dynamic object reconstruction aims to recover the 3D geometric shape as well as its rigid pose over time. Such a task can be handled either by directly hallucinating the full shape or by registering and accumulating partial observations. Approaches of the former type usually squash partial observations into a global feature vector [18,30,62] and ignore the local geometric structure. [21] go one step further by disentangling shape and pose with a novel supervised loss. However, there is still no guarantee for the fidelity of the completed shape. We instead rely on registration and accumulation. Related works include AlignNet-3D [20] that directly regresses the relative transformation matrix from concatenated global features of the two point clouds. NOCS [53] proposes a category-aware canonical representation that can be used to estimate instance pose w.r.t.its canonical pose. Caspr [43] implicitly accumulates the shapes by mapping a sequence of partial observations to a continuous latent space. [23] and [22] respectively propose multi-way registration methods that accumulate multi-body and non-rigid dynamic point clouds, but do not scale well to large scenes.

3 Method

The network architecture of our multitask model is schematically depicted in Fig. 2. To accumulate the points over time, we make use of the inductive bias that scenes can be decomposed into agents that move as rigid bodies [19]. We start by extracting the latent base features of each individual frame (Sect. 3.1), which we then use as input to the task-specific heads. To estimate the ego-motion, we employ a differentiable registration module (Sect. 3.2). Instead of using the ego-motion only to align the static scene parts, we also use it to spatially align the base features, which are reused in the later stages. To explain the motion of the dynamic foreground, we utilize the aligned base features and perform motion segmentation (Sect. 3.3) as well as spatio-temporal association of dynamic foreground objects (Sect. 3.4). Finally, we decode the rigid body motion of each foreground object from its spatio-temporal features (Sect. 3.5). We train the entire model end-to-end with a loss $\mathcal{L}$ composed of five terms:

$$\mathcal{L} = \mathcal{L}_{\text{ego}}^{\bullet} + \mathcal{L}_{\text{FG}}^{\bullet} + \mathcal{L}_{\text{motion}}^{\bullet} + \lambda_{\text{offset}}\mathcal{L}_{\text{offset}}^{\bullet} + \lambda_{\text{obj}}\mathcal{L}_{\text{obj}}^{\bullet} \,. \tag{1}$$

In the following, we provide a high-level description of each module. Detailed network architectures, including parameters and loss formulations, are given in the supplementary material (unless already elaborated).

Problem Setting. Consider an *ordered* point cloud sequence $\mathcal{X} = \{\mathbf{X}^t\}_{t=1}^T$ consisting of T frames $\mathbf{X}^t = [\mathbf{x}_1^t, ..., \mathbf{x}_i^t, ..., \mathbf{x}_{n_t}^t] \in \mathbb{R}^{3\times n_t}$ of variable size, captured by a *single moving observer* at constant time intervals Δt. We denote the first frame $\mathbf{X}^1$ the *target* frame, while the remaining frames $\{\mathbf{X}^t \mid t > 1\}$ are termed *source* frames. Our goal is then to estimate the flow vectors $\{\mathbf{V}^t \in \mathbb{R}^{3\times n_t} | t > 1\}$ that align each of the source frames to the target frame, and hence *accumulate* the point clouds. Instead of predicting unconstrained pointwise flow vectors, we make use of the inductive bias that each frame can be decomposed into a *static* part $\mathbf{X}_{\text{static}}^t$ and K_t rigidly-moving *dynamic* parts $\mathcal{X}_{\text{dynamic}}^t = \{\mathbf{X}_k^t\}_{k=1}^{K_t}$ [19]. For an individual frame, the scene flow vectors $\mathbf{V}_{\text{static}}^t$ of the static part and $\mathbf{V}_k^t$ of the k-th dynamic object can be explained by the rigid ego-motion $\mathbf{T}_{\text{ego}}^t \in \text{SE}(3)$ and the motion of the dynamic object relative to the static background $\mathbf{T}_k^t \in \text{SE}(3)$, respectively as:

$$\mathbf{V}_{\text{static}}^t = \mathbf{T}_{\text{ego}}^t \circ \mathbf{X}_{\text{static}}^t - \mathbf{X}_{\text{static}}^t, \qquad \mathbf{V}_k^t = \mathbf{T}_k^t \mathbf{T}_{\text{ego}}^t \circ \mathbf{X}_k^t - \mathbf{X}_k^t, \tag{2}$$

where $\mathbf{T} \circ \mathbf{X}$ (or $\mathbf{T} \circ \mathbf{x}$) denotes applying the transformation to the point set $\mathbf{X}$ (or point $\mathbf{x}$).

3.1 Backbone Network

Similar to [2,24,57], our backbone network converts the 3D point cloud of a single frame into a bird's eye view (BEV) latent feature image. Specifically, we lift the point coordinates to a higher-dimensional latent space using a point-wise MLP and then scatter them into a $H \times W$ feature grid aligned with the gravity axis. The features per grid cell ("pillar") are aggregated with max-pooling, then fed through a 2D UNet [34] to enlarge their receptive field and strengthen the local context. The output of the backbone network is a 2D latent *base* feature map $\mathbf{F}^t_{\text{base}}$ for each of the T frames.

3.2 Ego-motion Estimation

We estimate the ego-motion $\mathbf{T}^t_{\text{ego}}$ using a correspondence-based registration module separately for each source frame. Points belonging to dynamic objects can bias the estimate of ego-motion, especially when using a correspondence-based approach, and should therefore be discarded. However, at an early stage of the pipeline it is challenging to reason about scene dynamics, so we rather follow the conservative approach and classify points into background and foreground, where foreground contains all the *movable* objects (*e.g.*, cars and pedestrians), irrespective of their actual dynamics [19]. The predicted foreground mask is later used to guide motion segmentation in Sect. 3.3.

We start by extracting ego-motion features $\mathbf{F}^t_{\text{ego}}$ and foreground scores $\mathbf{s}^t_{\text{FG}}$ from each $\mathbf{F}^t_{\text{base}}$ using two dedicated heads, each consisting of two convolutional layers separated by a ReLU activation and batch normalization. We then randomly sample N_{ego} background pillars whose $\mathbf{s}^t_{\text{FG}} < \tau$ and compute the pillar centroid coordinates $\mathbf{P}^t = \{\mathbf{p}^t_l\}$. The ego motion $\mathbf{T}^t_{\text{ego}}$ is estimated as:

$$\mathbf{T}^t_{\text{ego}} = \arg\min_{\mathbf{T}} \sum_{l=1}^{N_{\text{ego}}} w^t_l \left\| \mathbf{T} \circ \mathbf{p}^t_l - \phi(\mathbf{p}^t_l, \mathbf{P}^1) \right\|^2 . \quad (3)$$

Here, $\phi(\mathbf{p}^t_l, \mathbf{P}^1)$ finds the *soft correspondence* of $\mathbf{p}^t_l$ in $\mathbf{P}^1$, and w^t_l is the weight of the correspondence pair $\left(\mathbf{p}^t_l, \phi(\mathbf{p}^t_l, \mathbf{P}^1)\right)$. Both $\phi(\mathbf{p}^t_l, \mathbf{P}^1)$ and w^t_l are estimated using an entropy-regularized Sinkhorn algorithm from $\mathbf{F}^t_{\text{ego}}$ with slack row/column padded [11,61] and the optimal $\mathbf{T}^t_{\text{ego}}$ is computed in closed form via the differentiable Kabsch algorithm [25].

We supervise the ego-motion with an L_1 loss over the transformed pillars $\mathcal{L}_{\text{trans}} = \frac{1}{|\mathbf{P}^t|} \sum_{l=1}^{|\mathbf{P}^t|} \|\mathbf{T} \circ \mathbf{p}^t_l - \overline{\mathbf{T}} \circ \mathbf{p}^t_l\|_1$ and an inlier loss $\mathcal{L}_{\text{inlier}}$ [61] that regularizes the Sinkhorn algorithm, $\mathcal{L}^{\circledast}_{\text{ego}} = \mathcal{L}_{\text{trans}} + \mathcal{L}_{\text{inlier}}$. The foreground score $\mathbf{s}^t_{\text{FG}}$ is supervised using a combination of weighted binary cross-entropy (BCE) loss $\mathcal{L}_{\text{bce}}$ and Lovasz-Softmax loss $\mathcal{L}_{\text{ls}}$ [5]: $\mathcal{L}^{\circledast}_{\text{FG}} = \mathcal{L}_{\text{bce}}(\mathbf{s}^t_{\text{FG}}, \overline{\mathbf{s}}^t_{\text{FG}}) + \mathcal{L}_{\text{ls}}(\mathbf{s}^t_{\text{FG}}, \overline{\mathbf{s}}^t_{\text{FG}})$, with $\overline{\mathbf{s}}^t_{\text{FG}}$ the binary ground truth. The weights in $\mathcal{L}_{\text{bce}}$ are inversely proportional to the square root of elements in each class.

3.3 Motion Segmentation

To separate the *moving* objects from the *static* ones we perform motion segmentation, reusing the per-frame base features $\{\mathbf{F}^t_{\text{base}}\}$. Specifically, we apply a differentiable feature warping scheme [45] that warps each $\mathbf{F}^t_{\text{base}}$ using the predicted ego-motion $\mathbf{T}^t_{\text{ego}}$, and obtain a spatio-temporal 3D feature tensor of size $C \times T \times H \times W$ by stacking the warped feature maps along the channel dimension. This feature tensor is then fed through a series of 3D convolutional layers, followed by max-pooling across the temporal dimension T. Finally, we apply a small 2D UNet to obtain the 2D motion feature map $\mathbf{F}_{\text{motion}}$.

To mitigate discretization error, we bilinearly interpolate grid motion features to all foreground points in each frame.[1] The point-level motion feature for $\mathbf{x}^t_i$ is computed as:

$$\mathbf{f}^t_{\text{motion},i} = \text{MLP}\big(\text{cat}[\psi(\mathbf{x}^t_i, \mathbf{F}_{\text{motion}}), \text{MLP}(\mathbf{x}^t_i)]\big), \tag{4}$$

where MLP$(\cdot)$ denotes a multi-layer perceptron, cat$[\cdot]$ concatenation, and $\psi(\mathbf{x}, \mathbf{F})$ a bilinear interpolation from $\mathbf{F}$ to $\mathbf{x}$. The dynamic score $\mathbf{s}^t_i$ of the point $\mathbf{x}^t_i$ is then decoded from the motion feature $\mathbf{f}^t_{\text{motion},i}$ using another MLP, and supervised similar to the foreground segmentation, with a loss $\mathcal{L}^{\circledast}_{\text{motion}} = \mathcal{L}_{\text{bce}}(\mathbf{s}^t_i, \overline{\mathbf{s}}^t_i) + \mathcal{L}_{\text{ls}}(\mathbf{s}^t_i, \overline{\mathbf{s}}^t_i)$, where $\overline{\mathbf{s}}^t_i$ denotes the ground-truth motion label of point $\mathbf{x}^t_i$.

3.4 Spatio-temporal Instance Association

To segment the dynamic points (extracted by thresholding the $\mathbf{s}^t_i$) into individual objects and associate them over time, we perform spatio-temporal instance association. Different from the common tracking-by-detection [12,56] paradigm, we propose to directly *cluster the spatio-temporal point cloud*, which simultaneously provides instance masks and the corresponding associations. However, naive clustering of the ego-motion aligned point clouds often fails due to LiDAR sparsity and fast object motions, hence we predict a per-point offset vector $\boldsymbol{\delta}^t_i$ pointing towards the (motion-compensated) instance center:

$$\boldsymbol{\delta}^t_i = \text{MLP}\big(\text{cat}[\psi(\mathbf{x}^t_i, \mathbf{F}_{\text{motion}}), \text{MLP}(\mathbf{x}^t_i)]\big). \tag{5}$$

The DBSCAN [14] algorithm is subsequently applied over the deformed point set $\{\mathbf{x}^t_i + \boldsymbol{\delta}^t_i \mid \forall i, \forall t\}$ to obtain an instance index for each point. This association scheme is simple yet robust, and can seamlessly handle occlusions and misdetections. Similar to 3DIS [28] we supervise the offset predictions $\boldsymbol{\delta}^t_i$ with both an L_1-distance loss and a directional loss:

$$\mathcal{L}^{\circledast}_{\text{offset}} = \frac{1}{n} \sum_{\{i,t\}}^{n} \left(\left\| \boldsymbol{\delta}^t_i - \overline{\boldsymbol{\delta}}^t_i \right\|_1 + 1 - \langle \frac{\boldsymbol{\delta}^t_i}{\|\boldsymbol{\delta}^t_i\|}, \frac{\overline{\boldsymbol{\delta}}^t_i}{\|\overline{\boldsymbol{\delta}}^t_i\|} \rangle \right), \tag{6}$$

where $\overline{\boldsymbol{\delta}}$ is the ground truth offset $\mathbf{o} - \mathbf{x}$ from the associated instance centroid $\mathbf{o}$ in the target frame, and $\langle \cdot \rangle$ is the inner product.

[1] We predict motion labels only for foreground and treat background points as static.

3.5 Dynamic Object Motion Modelling

Once we have spatio-temporally segmented objects, we must recover their motions at each frame. As LiDAR points belonging to a single object are sparse and explicit inter-frame correspondences are hard to find, we take a different approach from the one used in the ego-motion head and construct a novel TubeNet to directly regress the transformations. Specifically, TubeNet takes T frames of the same instance $\mathbf{X}_k$ as input, and regresses its rigid motion parameters $\mathbf{T}_k^t$ as:

$$\mathbf{T}_k^t = \text{MLP}\left(\text{cat}[\tilde{\mathbf{f}}_{\text{motion}}, \tilde{\mathbf{f}}_{\text{ego}}, \tilde{\mathbf{f}}_{\text{pos}}^t, \tilde{\mathbf{f}}_{\text{pos}}^1]\right), \tag{7}$$

where $\tilde{\mathbf{f}}_{\text{motion}}$ and $\tilde{\mathbf{f}}_{\text{ego}}$ are instance-level global features obtained by applying PointNet [41] to the respective point-level features of that instance, $\tilde{\mathbf{f}}_* = \text{PN}(\{\mathbf{f}_{*,i}^t \mid \mathbf{x}_i^t \in \mathbf{X}_k^t\})$. Recall that point-level features $\mathbf{f}_{*,i}^t$ are computed from $\mathbf{F}_*$ via the interpolation scheme described in Eq (4). Here, $\tilde{\mathbf{f}}_{\text{motion}}$ encodes the overall instance motion while $\tilde{\mathbf{f}}_{\text{ego}}$ supplements additional geometric cues. The feature $\tilde{\mathbf{f}}_{\text{pos}}^t = \text{PN}(\mathbf{X}_k^t)$ is a summarized encoding over individual frames and provides direct positional information for accurate transformation estimation. The transformations are initialised to identity and TubeNet is applied in iterative fashion to regress residual transformations relative to the last iteration, similar to RPMNet [61].

For the loss function, we choose to parameterise each $\mathbf{T}$ as an un-normalised quaternion $\mathbf{q} \in \mathbb{R}^4$ and translation vector $\mathbf{t} \in \mathbb{R}^3$, and supervise it with:

$$\mathcal{L}_{\text{obj}}^{\circledast} = \mathcal{L}_{\text{trans}} + \frac{1}{T-1}\sum_{t=2}^{T}\left(\left\|\bar{\mathbf{t}}_k^t - \mathbf{t}_k^t\right\|_2 + \lambda\left\|\bar{\mathbf{q}}_k^t - \frac{\mathbf{q}_k^t}{\|\mathbf{q}_k^t\|}\right\|_2\right), \tag{8}$$

where $\bar{\mathbf{t}}_k^t$ and $\bar{\mathbf{q}}_k^t$ are the ground truth transformation, and λ is a constant weight, set to 50 in our experiments. $\mathcal{L}_{\text{trans}}$ is the same as in the ego-motion (Sect. 3.2).

3.6 Comparison to Related Work

WsRSF [19]. Our proposed method differs from WsRSF in several ways: *(i)* WsRSF is a pair-wise scene flow estimation method, while we can handle multiple frames; *(ii)* unlike WsRSF we perform motion segmentation for a more complete understanding of the scene dynamics; *(iii)* our method outputs instance-level associations, while WsRSF simply connects each instance to the complete foreground of the other point cloud.

MotionNet [57]. Similar to our method, MotionNet also deals with sequential point clouds and uses a BEV representation. However, MotionNet *(i)* assumes that ground truth ego-motion is available, while we estimate it within our network; and *(ii)* does not provide object-level understanding, rather it only separates the scene into a static and a dynamic part.

3.7 Implementation Details

Our model is implemented in pytorch [37] and can be trained on a single RTX 3090 GPU. During training we minimize Eq. (1) with the Adam [27] optimiser, with an initial learning rate 0.0005 that exponentially decays at a rate of 0.98 per epoch. For both *Waymo* and *nuScenes*, the size of the pillars is $(\delta_x, \delta_y, \delta_z) = (0.25, 0.25, 8)$ m. We sample $N_{\text{ego}} = 1024$ points for ego-motion estimation and set $\tau = 0.5$ for foreground/background segmentation. The feature dimensions of $\mathbf{F}^t_{\text{base}}$, $\mathbf{F}^t_{\text{ego}}$, $\mathbf{F}_{\text{motion}}$ are 32, 64, 64 respectively. During inference we additionally use ICP [6] to perform test-time optimisation of the ego-motion as well as the transformation parameters of each dynamic object. ICP thresholds for ego-motion and dynamic object motion are 0.1/0.2 and 0.15/0.25 m for *Waymo*/*nuScenes*.

4 Experimental Evaluation

In this section, we first describe the datasets and the evaluation setting for our experiments (Sect. 4.1). We then proceed with a quantitative evaluation of our method and showcase its applicability to downstream tasks with qualitative results for surface reconstruction (Sect. 4.2). Finally, we validate our design choices in an ablation study (Sect. 4.3).

4.1 Datasets and Evaluation Setting

Waymo. The Waymo Open Dataset [46] includes 798/202 scenes for training/validation, where each scene is a 20-second clip captured by a 64-beam LiDAR 10 Hz. We randomly sample 573/201 scenes for training/validation from the training split, and treat the whole validation split as a held-out test set. We consider every 5 consecutive frames as a *sample* and extract 20 *samples* from each clip, for a total of 11440/4013/4031 samples for training/validation/test.

nuScenes. The nuScenes dataset [7] consists of 700 training and 150 validation scenes, where each scene is a 20-second clip captured by a 32-beam LiDAR at 20 Hz. We use all validation scenes for testing and randomly hold out 150 training scenes for validation. We consider every 11 consecutive frames as a *sample*, resulting in a total of 10921/2973/2973 samples for training/validation/testing.

Ground Truth. We follow [24] to construct pseudo ground-truth from the detection and tracking annotations. Specifically, the flow vectors of the background part are obtained from ground truth ego-motion. For foreground objects, we use the unique instance IDs of the tracking annotations and recover their rigid motion parameters by registering the bounding boxes. Notably, for *nuScenes* the bounding boxes are only annotated every 10 frames. To obtain pseudo ground truth for the remaining frames, we linearly interpolate the boxes, which may introduce a small amount of label noise especially for fast-moving objects.

Table 1. Scene flow results on *Waymo* and *nuScenes* datasets.

Dataset	Method	Static part				Dynamic foreground				
		EPE avg.↓	AccS↑	AccR↑	ROutlier↓	EPE avg. ↓	EPE med.↓	AccS↑	AccR↑	ROutliers ↓
Waymo	PPWC-Net [58]	0.414	17.6	40.2	12.1	0.475	0.201	9.0	29.3	22.4
	FLOT [40]	0.129	65.2	85.0	2.8	0.625	0.231	9.8	27.4	33.8
	WsRSF [19]	0.063	87.3	96.6	0.6	0.381	0.094	31.3	64.0	10.1
	NSFPrior [31]	0.187	79.8	89.1	4.7	0.237	0.077	44.7	68.6	11.5
	Ours	0.028	97.5	99.5	0.1	0.197	0.062	53.3	77.5	5.9
	Ours+	**0.018**	**99.0**	**99.7**	**0.1**	**0.173**	**0.043**	**69.1**	**86.9**	**5.1**
	Ours (w. ground)	0.042	91.9	98.8	0.1	0.219	0.071	47.1	72.8	8.5
nuScenes	PPWC-Net [58]	0.316	16.1	37.0	8.7	0.661	0.307	7.6	24.2	31.9
	FLOT [40]	0.153	51.7	78.3	4.3	1.216	0.710	3.0	10.3	63.9
	WsRSF [19]	0.195	57.4	82.6	4.8	0.539	0.204	17.9	37.4	22.9
	NSFPrior [31]	0.584	38.9	56.7	26.9	0.707	0.222	19.3	37.8	32.0
	Ours	0.111	65.4	88.6	1.1	0.301	0.146	26.6	53.4	**12.1**
	Ours+	**0.091**	**72.8**	**91.9**	**0.9**	**0.301**	**0.135**	**32.7**	**56.7**	13.7
	Ours (w. ground)	0.134	55.3	83.8	1.9	0.37	0.182	18.2	43.8	17.5

Metrics. We use standard *scene flow* evaluation metrics [2,32] to compare the performance of our approach to the selected baselines. These metric include: *(i)* 3D end-point-error (*EPE* [m]) which denotes the mean L_2-error of all flow vectors averaged over all frames; *(ii)* strict/relaxed accuracy (*AccS* [%] / *AccR* [%]). *i.e.*, the fraction of points with *EPE* $< 0.05/0.10$m or relative error $< 0.05/0.10$; *(iii) Outliers [%]* which denotes the ratio of points with *EPE* > 0.30m or relative error > 0.10; and *(iv) ROutliers* [%], the fraction of points whose *EPE* > 0.30m and relative error > 0.30. We evaluate these metrics for the static and dynamic parts of the scene separately.[2] Following [35,57], we evaluate the performance of all methods only on the points that lie within the square of size 64×64 m^2 centered at the ego-car location in the target frame. Additionally we remove ground points by thresholding along the z-axis.[3] Ablations studies additionally report the quality of *motion segmentation* in terms of recall and precision of *dynamic* parts, and the quality of *spatio-temporal instance association* in terms of mean weighted coverage (*mWCov*, the *IoU* of recovered instances [54]). For further details, see the supplementary material.

Baselines. We compare our method to 4 baseline methods. PPWC [58] and FLOT [40] are based on dense matching and are trained in a fully supervised manner; WsRSF [19] assumes a *multi-body* scene and can be trained with weak supervision; NSFPrior [31] is an optimisation-based method without pre-training. For PPWC [58], FLOT [40] and WsRSF [19], we sample at most 8192 points from each frame due to memory constraints, and use the k-nn graph to up-sample flow vectors to full resolution at inference time. For NSFPrior [31], we use the full point clouds and take the default hyper-parameter settings given by the authors, except for the early-stopping patience, which we set to 50 to make it computationally tractable on our large-scale dataset. For all baseline methods, we directly estimate flow vectors from any *source* frame to the *target* frame.

[2] A point is labelled as *dynamic* if its ground-truth velocity is > 0.5 m/s.

[3] This setting turns out to better suit the baseline methods [31,40,58]. However, we keep ground points in our dynamic point cloud accumulation task, as thresholding could falsely remove points that are of interest for reconstruction or mapping.

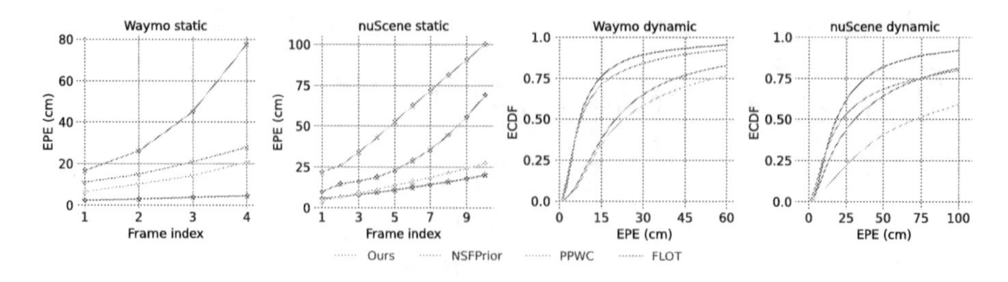

Fig. 3. Our method scales better to more frames. For the challenging dynamic parts, it also has smaller errors as well as fewer extreme outliers.

4.2 Main Results

The detailed comparison on the *Waymo* and *nuScenes* data is given in Tab. 1. *Ours* denotes the direct output of our model, while *Ours+* describes the results after test-time refinement with ICP. Many downstream tasks (e.g., surface reconstruction) rely on the accumulation of full point clouds and require also ground points. We therefore also train a variant of our method on point clouds with ground points and denote it *Ours (w. ground)*. To facilitate a fair comparison, we use full point clouds as input during training and inference, but only compute the evaluation metrics on points that do not belong to ground.

Comparison to State of the Art. On the static part of the ***Waymo*** dataset, FLOT [40] reaches the best performance among the baselines, but still has an average EPE error of 12.9 cm, which is more than 4 times larger than that of *Ours* (2.8 cm). This result corroborates our motivation for decomposing the scene into a static background and dynamic foreground. Modeling the motion of the background with a single set of transformation parameters also enables us to run ICP at test time (*Ours+*), which further reduces the EPE of the static part to 1.8 cm. On the dynamic foreground points, NSFPrior [31] reaches a comparable performance to *Ours*. However, based on our spatio-temporal association of foreground points we can again run ICP at test-time, which reduces the median EPE error to 4.3 cm, ≈40 % lower than that of NSFPrior. Furthermore, NSFPrior is an optimization method and not amenable to online processing (see Tab. 3). The results for ***nuScenes*** follow a similar trend as the ones for *Waymo*, but are larger in absolute terms due to the lower point density. Our method achieves the best performance on both static and dynamic parts. The gap to the closest competitors is even larger, which auggests that our method is more robust to low point density.

To further understand the error distribution of the dynamic parts, and the evolution of the errors of the static part as the gap between the *source* and *target* frames increases, we plot detailed results in Fig. 3. On both *Waymo* and *nuScenes* our method degrades gracefully, and slower than the baselines. The ECDF curve of the EPE error for foreground points also shows that our method performs best at all thresholds.

Table 2. Scene flow results on *Waymo* dataset w.r.t. input length.

	3	4	5	6	7	8	9	10
static EPE avg	**0.022**	0.025	0.028	0.032	0.037	0.044	0.054	0.066
dynamic EPE avg	0.199	**0.168**	0.190	0.218	0.250	0.294	0.348	0.412

Table 3. Runtimes for the *Waymo* and *nuScenes* datasets. All numbers are seconds per 5-frame (*Waymo*), respectively 11-frame (*nuScenes*) *sample.*

	Waymo	*nuScenes*
PPWC-Net [58]	0.608	0.990
FLOT [40]	1.028	2.010
WsRSF [19]	1.252	1.460
NSFPrior [31]	212.256	63.460
Ours	**0.174**	**0.250**

	Waymo	*nuScenes*
ego-motion estimation	0.100	0.188
motion segmentation	0.024	0.040
instance association	0.036	0.009
TubeNet	0.014	0.013

Breakdown of the Performance Gain. Overall, our method improves over baseline methods by a large margin on two datasets. The gains are a direct result of our design choices: *(i)* by modelling the flow of the background as rigid motion instead of unconstrained flow, we can greatly reduce *ROutlier* and improve the accuracy; *(ii)* different from [19] we perform motion segmentation, and can thus assign ego-motion flow to points on movable, but *static* objects ($\approx 75\%$ of the foreground). This further improves the results (see EPE of static FG in Tab. 4); *(iii)* reasoning on the object level, combined with spatio-temporal association and modelling, improves flow estimates for the *dynamic* foreground.

Generalisation to Variable Input Length. T When trained with a fixed input length ($T = 5$ on *Waymo*), our model is able to generalize to different input lengths (see Tab. 2). The performance degrades moderately with increasing T, as the motions become larger than seen during training. Also, larger displacements make the correspondence problem inherently harder.

Runtime. We report runtimes for our model and several baseline methods on both datasets in Table 3 (left). Our method is significantly faster than all baselines under the multi-frame scene flow setting. We also report detailed runtimes of our model for individual steps in Table 3 (right). As we can see, backbone feature extraction and pairwise registration (*ego-motion estimation*) account for the majority of the runtime, 57.5% on *Waymo* and 75.2% on *nuScenes*. Note that this runtime is calculated over all frames, while under a data streaming setting, we only need to run the first part for a single incoming frame, then re-use the features of the previous frames at later stages, which will greatly reduce runtime: after initialisation, the runtime for every new sample decreases to around 0.094 s for *Waymo*, respectively 0.079 s on *nuScenes*.

Qualitative Results. In Fig. 4 we show qualitative examples of scene and object reconstruction with our approach. By jointly estimating the ego-motion of the static part and the moving object motions, our method accumulates the corresponding points into a common, motion-compensated frame. It thus provides an excellent basis for 3D surface reconstruction.

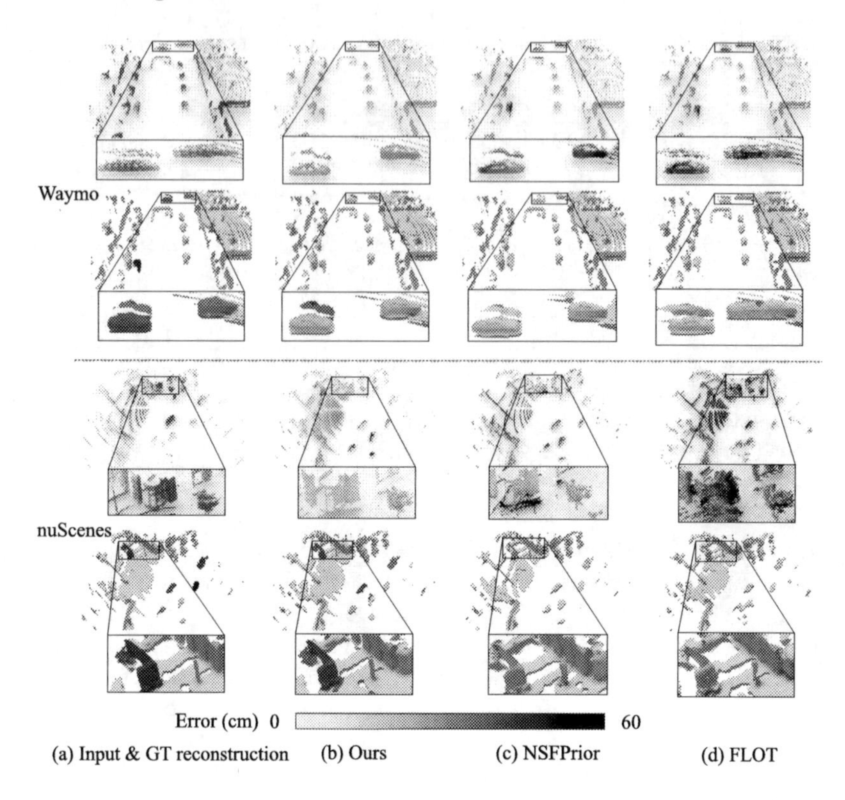

Fig. 4. Qualitative results showing scene flow estimation (top) and surface reconstruction (bottom) for two example scenes from *Waymo* and *nuScenes*.

4.3 Ablation Study

Sequential Model. We evaluate the individual modules of our sequential model, namely the foreground segmentation (*FG*), motion segmentation (*MOS*) and offset compensation (*Offset*). We train two models with and without foreground segmentation. For variants without *MOS* or *Offset*, we take the trained full model but remove *MOS* or *Offset* at inference time. The detailed results are summarised in Table 4. *FG* enables us to exclude dynamic foreground objects during pairwise registration. On *Waymo*, this reduces EPE of the static parts by 30% from 4.1 to 2.9 cm, and as a result also reduces EPE of dynamic parts from 28.6 to 19.7 cm. By additionally extracting the static foreground parts with *MOS*, the model can recover more accurate ego-motion for them, which reduces EPE from 19.0/19.9 to 2.1/7.4 cm on *Waymo*/*nuScenes*. *Offset* robustifies the instance association against low point density and fast object motion (+3.1 *pp* in *WCov* on *nuScenes*), this further reduces the EPE of dynamic parts by 3.7 cm.

Ego-motion Estimation Strategy. By default, we directly estimate the ego-motion from any *source* frame to the *target* frame. We compare to an alternative which estimates the ego-motion relative to the previous frame. Although that

Table 4. Ablation studies on *Waymo* and *nuScenes* datasets.

	Modules			Motion seg.		Association	Static BG	Static FG	Dynamic FG				
	FG	MOS	Offset	recall↑	precision↑	WCov↑	EPE avg.↓	EPE avg.↓	EPE avg.↓	EPE med.↓	AccS↑	AccR↑	ROutliers↓
Waymo		✓	✓	<u>92.7</u>	94.6	82.2	0.041	0.028	0.286	0.071	45.5	71.3	10.8
	✓		✓	–	–	**83.0**	0.031	0.190	0.198	0.062	**53.5**	77.4	6.3
	✓	✓		92.2	96.5	79.1	0.029	0.021	0.202	0.064	52.5	76.1	6.1
	✓	✓	✓	92.2	**96.8**	80.4	**0.029**	**0.021**	0.197	**0.062**	53.3	**77.5**	5.9
nuScenes		✓	✓	87.8	**92.5**	65.1	0.115	0.076	0.333	0.153	25.0	50.9	13.8
	✓		✓	–	–	**66.8**	0.118	0.199	**0.296**	**0.143**	**26.9**	**53.9**	**11.7**
	✓	✓		89.2	90.7	60.2	**0.113**	0.075	0.348	0.149	25.8	51.9	13.9
	✓	✓	✓	**89.3**	90.8	63.2	0.114	**0.074**	0.301	0.146	26.6	53.4	12.1
Waymo	Kalman tracker			92.3	96.6	77.1	0.030	0.027	0.586	0.099	36.3	61.6	11.7
	chained poses			**93.2**	94.9	81.9	0.044	0.030	**0.171**	0.068	48.0	77.0	**4.8**
nuScenes	Kalman tracker			89.4	90.8	42.9	0.114	0.092	1.238	0.364	10.7	25.1	44.0
	chained poses			88.1	90.2	62.1	0.225	0.151	0.315	0.155	23.4	51.8	13.4

achieves smaller pairwise errors, after chaining the estimated poses the errors w.r.t. the *target* frame explode, resulting in inferior scene flow estimates (*chained poses* in Table 4).

Comparison to Tracking-Based Method. Instead of running spatio-temporal association followed by TubeNet to model the motion of each moving object, an alternative would be to apply Kalman tracker so as to simultaneously solve association and motion. We compare to the modified AB3DMOT [56], which is based on a constant velocity model. That method first clusters moving points for each frame independently to obtain instances, then associates instances by greedy matching of instance centroids based on L_2 distances. The results in Table 4 (*Kalman tracker*) show clearly weaker performance, due to the less robust proximity metric based on distances between noisy centroid estimates.

5 Conclusion

We have looked at the analysis of 3D point cloud sequences from a fresh viewpoint, as point cloud accumulation across time. In that view we integrate point cloud registration, motion segmentation, instance segmentation, and piecewise rigid scene flow estimation into a complete multi-frame 4D scene analysis method. By jointly considering sequences of frames, our model is able to disentangle scene dynamics and recover accurate instance-level rigid-body motions. The model processes (ordered) raw point clouds, and can operate online with low latency. A major *limitation* is that our approach is fully supervised and heavily relies on annotated data: it requires instance-level segmentations as well as ground truth motions, although we demonstrate some robustness to label noise from interpolated pseudo ground truth. Also, our system consists of multiple processing stages and cannot fully recover from mistakes in early stages, like incorrect motion segmentation.

In *future work* we hope to explore our method's potential for downstream scene understanding tasks. We also plan to extend it to an incremental setting, where longer sequences of frames can be summarized into our holistic, dynamic scene representation in an online fashion.

References

1. Aygun, M., et al.: 4D panoptic LiDAR segmentation. In: Proceedings of the CVPR (2021)
2. Baur, S.A., Emmerichs, D.J., Moosmann, F., Pinggera, P., Ommer, B., Geiger, A.: SLIM: self-supervised LiDAR scene flow and motion segmentation. In: Proceedings of the ICCV (2021)
3. Behl, A., Paschalidou, D., Donné, S., Geiger, A.: PointFlowNet: learning representations for rigid motion estimation from point clouds. In: Proceedings of the CVPR (2019)
4. Behley, J., et al.: SemanticKITTI: a dataset for semantic scene understanding of lidar sequences. In: Proceedings of the ICCV (2019)
5. Berman, M., Triki, A.R., Blaschko, M.B.: The Lovász-Softmax loss: a tractable surrogate for the optimization of the intersection-over-union measure in neural networks. In: Proceedings of the CVPR (2018)
6. Besl, P.J., McKay, N.D.: Method for registration of 3-D shapes. In: Sensor Fusion IV: Control Paradigms and Data Structures, vol. 1611, pp. 586–606. International Society for Optics and Photonics (1992)
7. Caesar, H., et al.: nuScenes: a multimodal dataset for autonomous driving. In: Proceedings of the CVPR (2020)
8. Chen, C., Yang, B.: Dynamic occlusion detection and inpainting of in situ captured terrestrial laser scanning point clouds sequence. ISPRS J. Photogrammetry Remote Sens. **119**, 90–107 (2016)
9. Chen, X., et al.: Moving object segmentation in 3D LiDAR data: a learning-based approach exploiting sequential data. In: IEEE RA-L (2021)
10. Choy, C., Gwak, J., Savarese, S.: 4d spatio-temporal convnets: Minkowski convolutional neural networks. In: Proceedings of the CVPR (2019)
11. Cuturi, M.: Sinkhorn distances: Lightspeed computation of optimal transport. In: Proceedings of the NeurIPS (2013)
12. Dendorfer, P., et al.: MOTchallenge: a benchmark for single-camera multiple target tracking. IJCV (2021)
13. Dewan, A., Caselitz, T., Tipaldi, G.D., Burgard, W.: Rigid scene flow for 3D lidar scans. In: Proceedings of the IROS (2016)
14. Ester, M., Kriegel, H.P., Sander, J., Xu, X., et al.: A density-based algorithm for discovering clusters in large spatial databases with noise. In: Proceedings of the KDD (1996)
15. Fan, H., Yang, Y., Kankanhalli, M.: Point 4D transformer networks for spatio-temporal modeling in point cloud videos. In: Proceedings of the CVPR (2021)
16. Fan, H., Yu, X., Ding, Y., Yang, Y., Kankanhalli, M.: PSTNet: point spatio-temporal convolution on point cloud sequences. In: Proceedings of the ICLR (2020)
17. Filatov, A., Rykov, A., Murashkin, V.: Any motion detector: learning class-agnostic scene dynamics from a sequence of lidar point clouds. In: Proceeding of the ICRA (2020)
18. Giancola, S., Zarzar, J., Ghanem, B.: Leveraging shape completion for 3D siamese tracking. In: Proceedings of the CVPR (2019)
19. Gojcic, Z., Litany, O., Wieser, A., Guibas, L.J., Birdal, T.: Weakly supervised learning of rigid 3D scene flow. In: Proceedings of the CVPR (2021)
20. Groß, J., Ošep, A., Leibe, B.: AlignNet-3D: fast point cloud registration of partially observed objects. In: Proceedings of the 3DV (2019)

21. Gu, J., et al.: Weakly-supervised 3D shape completion in the wild. In: Vedaldi, A., Bischof, H., Brox, T., Frahm, J.-M. (eds.) ECCV 2020. LNCS, vol. 12350, pp. 283–299. Springer, Cham (2020). https://doi.org/10.1007/978-3-030-58558-7_17
22. Huang, J., Birdal, T., Gojcic, Z., Guibas, L.J., Hu, S.M.: Multiway non-rigid point cloud registration via learned functional map synchronization. IEEE T-PAMI (2022)
23. Huang, J., et al.: MultiBodySync: multi-body segmentation and motion estimation via 3d scan synchronization. In: Proceedings of the CVPR (2021)
24. Jund, P., Sweeney, C., Abdo, N., Chen, Z., Shlens, J.: Scalable scene flow from point clouds in the real world. arXiv preprint arXiv:2103.01306 (2021)
25. Kabsch, W.: A solution for the best rotation to relate two sets of vectors. Acta Crystallogr. Sect. Crystal Phys. Diffr. Theoret. Gen. Crystallogr. **32**(5), 922–923 (1976)
26. Kim, G., Kim, A.: Remove, then revert: Static point cloud map construction using multiresolution range images. In: Proceedings of the IROS (2020)
27. Kingma, D.P., Ba, J.: Adam: A method for stochastic optimization. arXiv preprint arXiv:1412.6980 (2014)
28. Lahoud, J., Ghanem, B., Pollefeys, M., Oswald, M.R.: 3D instance segmentation via multi-task metric learning. In: Proceedings of the ICCV (2019)
29. Li, R., Lin, G., He, T., Liu, F., Shen, C.: HCRF-Flow: scene flow from point clouds with continuous high-order CRFs and position-aware flow embedding. In: Proceedings of the CVPR (2021)
30. Li, R., Li, X., Fu, C.W., Cohen-Or, D., Heng, P.A.: PU-GAN: a point cloud upsampling adversarial network. In: Proceedings of the ICCV (2019)
31. Li, X., Kaesemodel Pontes, J., Lucey, S.: Neural scene flow prior. In: Proceedings of the NeurIPS (2021)
32. Liu, X., Qi, C.R., Guibas, L.J.: FlowNet3D: learning scene flow in 3D point clouds. In: Proceedings of the CVPR (2019)
33. Liu, X., Yan, M., Bohg, J.: Meteornet: deep learning on dynamic 3D point cloud sequences. In: Proceedings of the ICCV (2019)
34. Long, J., Shelhamer, E., Darrell, T.: Fully convolutional networks for semantic segmentation. In: Proceedings of the CVPR (2015)
35. Luo, C., Yang, X., Yuille, A.: Self-supervised pillar motion learning for autonomous driving. In: Proceedings of the CVPR (2021)
36. Ouyang, B., Raviv, D.: Occlusion guided self-supervised scene flow estimation on 3D point clouds. arXiv preprint arXiv:2104.04724 (2021)
37. Paszke, A., et al.: Pytorch: an imperative style, high-performance deep learning library. In: Proceedings of the NeurIPS (2021)
38. Piergiovanni, A., Casser, V., Ryoo, M.S., Angelova, A.: 4D-net for learned multimodal alignment. arXiv preprint arXiv:2109.01066 (2021)
39. Pomerleau, F., Krüsi, P., Colas, F., Furgale, P., Siegwart, R.: Long-term 3D map maintenance in dynamic environments. In: Proceedings of the ICRA (2014)
40. Puy, G., Boulch, A., Marlet, R.: FLOT: scene flow on point clouds guided by optimal transport. In: Vedaldi, A., Bischof, H., Brox, T., Frahm, J.-M. (eds.) ECCV 2020. LNCS, vol. 12373, pp. 527–544. Springer, Cham (2020). https://doi.org/10.1007/978-3-030-58604-1_32
41. Qi, C.R., Su, H., Mo, K., Guibas, L.J.: Pointnet: deep learning on point sets for 3D classification and segmentation. In: Proceedings of the CVPR (2017)
42. Qi, C.R., et al.: Offboard 3D object detection from point cloud sequences. In: Proceedings of the CVPR (2021)

43. Rempe, D., Birdal, T., Zhao, Y., Gojcic, Z., Sridhar, S., Guibas, L.J.: CaSPR: learning canonical spatiotemporal point cloud representations. In: Proceedings of the NeurIPS (2020)
44. Schauer, J., Nüchter, A.: The peopleremover–removing dynamic objects from 3-D point cloud data by traversing a voxel occupancy grid. IEEE RA-L **3**(3), 1679–1686 (2018)
45. Sun, D., Yang, X., Liu, M.Y., Kautz, J.: PWC-net: CNNs for optical flow using pyramid, warping, and cost volume. In: Proceedings of the CVPR (2018)
46. Sun, P., et al.: Scalability in perception for autonomous driving: waymo open dataset. In: Proceedings of the CVPR (2020)
47. Teed, Z., Deng, J.: RAFT-3D: scene flow using rigid-motion embeddings. In: Proceedings of the CVPR (2021)
48. Vaswani, A., et al.: Attention is all you need. In: Proceedings of the NeurIPS (2017)
49. Vedula, S., Baker, S., Rander, P., Collins, R., Kanade, T.: Three-dimensional scene flow. In: Proceedings of the ICCV (1999)
50. Vogel, C., Schindler, K., Roth, S.: 3D scene flow estimation with a rigid motion prior. In: Proceedings of the ICCV (2011)
51. Vogel, C., Schindler, K., Roth, S.: Piecewise rigid scene flow. In: Proceedings of the ICCV (2013)
52. Vogel, C., Schindler, K., Roth, S.: 3D scene flow estimation with a piecewise rigid scene model. Int. J. Comput. Vision **115**(1), 1–28 (2015)
53. Wang, H., Sridhar, S., Huang, J., Valentin, J., Song, S., Guibas, L.J.: Normalized object coordinate space for category-level 6d object pose and size estimation. In: Proceedings of the CVPR (2019)
54. Wang, X., Liu, S., Shen, X., Shen, C., Jia, J.: Associatively segmenting instances and semantics in point clouds. In: Proceedings of the CVPR (2019)
55. Wedel, A., Rabe, C., Vaudrey, T., Brox, T., Franke, U., Cremers, D.: Efficient dense scene flow from sparse or dense stereo data. In: Forsyth, D., Torr, P., Zisserman, A. (eds.) ECCV 2008. LNCS, vol. 5302, pp. 739–751. Springer, Heidelberg (2008). https://doi.org/10.1007/978-3-540-88682-2_56
56. Weng, X., Wang, J., Held, D., Kitani, K.: 3D multi-object tracking: a baseline and new evaluation metrics. In: Proceedings of the IROS (2020)
57. Wu, P., Chen, S., Metaxas, D.N.: MotionNet: Joint perception and motion prediction for autonomous driving based on bird's eye view maps. In: Proceedings of the CVPR (2020)
58. Wu, W., Wang, Z.Y., Li, Z., Liu, W., Fuxin, L.: PointPWC-Net: cost volume on point clouds for (self-)supervised scene flow estimation. In: Vedaldi, A., Bischof, H., Brox, T., Frahm, J.-M. (eds.) ECCV 2020. LNCS, vol. 12350, pp. 88–107. Springer, Cham (2020). https://doi.org/10.1007/978-3-030-58558-7_6
59. Yang, B., Bai, M., Liang, M., Zeng, W., Urtasun, R.: Auto4d: learning to label 4D objects from sequential point clouds. arXiv preprint arXiv:2101.06586 (2021)
60. Yang, Z., Zhou, Y., Chen, Z., Ngiam, J.: 3D-MAN: 3d multi-frame attention network for object detection. In: Proceedings of the CVPR (2021)
61. Yew, Z.J., Lee, G.H.: RPM-Net: robust point matching using learned features. In: Proceedings of the CVPR (2020)
62. Yuan, W., Khot, T., Held, D., Mertz, C., Hebert, M.: PCN: point completion network. In: Proceedings of the 3DV (2018)

Homogeneous Multi-modal Feature Fusion and Interaction for 3D Object Detection

Xin Li[1], Botian Shi[2], Yuenan Hou[2], Xingjiao Wu[1,3], Tianlong Ma[1,3], Yikang Li[2(✉)], and Liang He[1(✉)]

[1] East China Normal University, Shanghai, China
{tlma,lhe}@cs.ecnu.edu.cn
[2] Shanghai AI Lab, Shanghai, China
{shibotian,houyuenan,liyikang}@pjlab.org.cn
[3] Fudan University, Shanghai, China

Abstract. Multi-modal 3D object detection has been an active research topic in autonomous driving. Nevertheless, it is non-trivial to explore the cross-modal feature fusion between sparse 3D points and dense 2D pixels. Recent approaches either fuse the image features with the point cloud features that are projected onto the 2D image plane or combine the sparse point cloud with dense image pixels. These fusion approaches often suffer from severe information loss, thus causing sub-optimal performance. To address these problems, we construct the homogeneous structure between the point cloud and images to avoid projective information loss by transforming the camera features into the LiDAR 3D space. In this paper, we propose a homogeneous multi-modal feature fusion and interaction method (HMFI) for 3D object detection. Specifically, we first design an image voxel lifter module (IVLM) to lift 2D image features into the 3D space and generate homogeneous image voxel features. Then, we fuse the voxelized point cloud features with the image features from different regions by introducing the self-attention based query fusion mechanism (QFM). Next, we propose a voxel feature interaction module (VFIM) to enforce the consistency of semantic information from identical objects in the homogeneous point cloud and image voxel representations, which can provide object-level alignment guidance for cross-modal feature fusion and strengthen the discriminative ability in complex backgrounds. We conduct extensive experiments on the KITTI and Waymo Open Dataset, and the proposed HMFI achieves better performance compared with the state-of-the-art multi-modal methods. Particularly, for the 3D detection of cyclist on the KITTI benchmark, HMFI surpasses all the published algorithms by a large margin.

Keywords: 3D object detection · Multi-modal · Feature-level fusion · Self-attention

Supplementary Information The online version contains supplementary material available at https://doi.org/10.1007/978-3-031-19839-7_40.

S. Avidan et al. (Eds.): ECCV 2022, LNCS 13698, pp. 691–707, 2022.
https://doi.org/10.1007/978-3-031-19839-7_40

1 Introduction

3D object detection is an important task that aims to precisely localize and classify each object in the 3D space, thus allowing vehicles to perceive and understand their surrounding environment comprehensively. So far, various LiDAR-based and image-based 3D detection approaches [3,6,12,17,19,24–26,29–31] have been proposed.

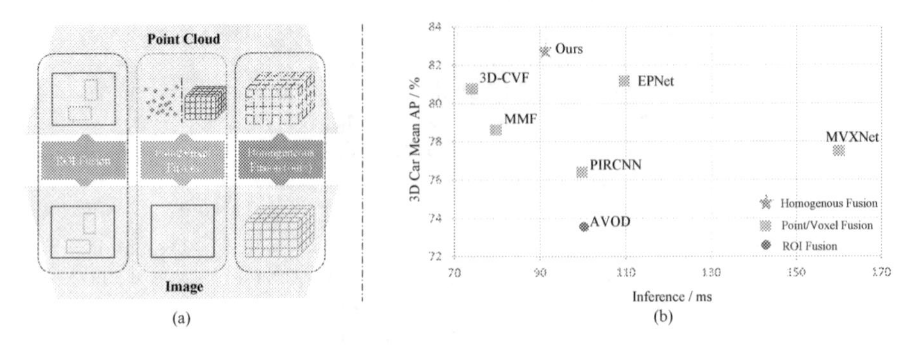

Fig. 1. (a) Schematic comparison between different feature-level fusion based methods. (b) Quantitative comparison with competitive multi-modal feature-level fusion methods. Our method achieves good performance-efficiency trade-off for the car category (Mean AP of all difficulty levels) on the KITTI [4] benchmark. (Color figure online)

LiDAR-based methods can achieve superior performance over image-based approaches as point cloud contains precise spatial information. However, LiDAR points are usually sparse and do not have enough color and texture information. As to image-based approaches, they perform better in capturing semantic information while suffering from the lack of depth signal. Therefore, multi-modal 3D object detection is a promising direction that can fully utilize the complementary information of images and point cloud.

Recent multi-modal approaches can be generally categorized into two types: decision-level fusion and feature-level fusion. Decision-level fusion methods ensemble the detected objects in respective modalities and their performance is bounded by each stage [21]. Feature level fusion is more prevalent as they fuse the rich informative features of two modalities. Three representative feature-level fusion methods are depicted in Fig. 1(a). The first one is fusing multi-modal features at the regions of interest (RoI). However, these methods have severe spatial information loss when projecting 3D points onto the bird's eye view (BEV) or front view (FV) in 2D plane, while 3D information plays a key role in accurate 3D object localization. Another line of work conducts fusion on the point/voxel-level [9,14,15,33,39,40,44,47], which can achieve complementary fusion at a much finer granularity and involve the combination of low-level multi-modal features at 3D points or 2D pixels. However, they can only approximately establish a relatively coarse correspondence between the point/voxel features and image features. Moreover, these two schemes of feature fusion usually suffer from severe information loss due to the mismatched projection between 2D dense image pixels and 3D sparse LiDAR points.

To address the aforementioned problems, we propose a homogeneous fusion scheme that lifts image features from 2D plane to 3D dense voxel structure. In our homogeneous fusion scheme, we propose the Homogeneous Multi-modal Feature Fusion and Interaction method (HMFI), which exploits the complementary information in multi-modal features and alleviates severe information loss caused by the dimensional reduction mapping. Furthermore, we build the cross-modal feature interaction between the point cloud features and image features at object-level based on the homogeneous 3D structure to strengthen the model's ability to fuse image semantic information with the point cloud.

Specifically, we design an image voxel lifter module (IVLM) to lift the 2D image features to the 3D space first and construct a homogeneous voxel structure of 2D images for multi-modal feature fusion, which is guided by the point cloud as depth hint. It will not cause information loss for fusing these two multi-modal data. We also notice that the homogeneous voxel structure of cross-modal data can help in feature fusion and interaction. Thus, we introduce the query fusion mechanism (QFM) that introduces a self-attention based operation that can adaptively combine point cloud and image features. Each point cloud voxel will query all image voxels to achieve homogeneous feature fusion and combine with the original point cloud voxel features to form the joint camera-LiDAR features. QFM enables each point cloud voxel to perceive image features in the common 3D space adaptively and fuse these two homogeneous representations effectively.

Besides, we explore building a feature interaction between the homogeneous point cloud and image voxel features instead of refining in regions of interest (RoI) based pooling which is applied to fuse low-level LiDAR and camera features with the joint camera-LiDAR features. We consider that, although point cloud and image representations are in different modalities, the object-level semantic properties should be similar in the homogeneous structure. Therefore, to strengthen the abstract representation of point cloud and images in a shared 3D space and exploit the similarity of identical objects' properties in two modalities, we propose a voxel feature interaction module (VFIM) at the object-level to improve the consistency of point cloud and image homogeneous representations in the 3D RoI. To be specific, we use the voxel RoI pooling [3] to extract features in these two homogeneous features according to the predicted proposals and produce the paired RoI feature set. Then we adopt the cosine similarity loss [2] between each pair of RoI features and enforce the consistency of object-level properties in point cloud and images. In VFIM, building the feature interaction in these homogeneous paired RoI features improves the object-level semantic consistency between two homogeneous representations and enhances the model's ability to achieve cross-modal feature fusion. Extensive experiments conducted on KITTI and Waymo Open Dataset demonstrate that the proposed method can achieve better performance compared to the state-of-the-art multi-modal methods. Our **contributions** are summarized as below:

1. We propose an image voxel lifter module (IVLM) to lift 2D image features into the 3D space and construct two homogeneous features for multi-modal fusion, which retains original information of image and point cloud.
2. We introduce the query fusion mechanism (QFM) to fuse two homogeneous representations of the point cloud voxel features and image voxel features effectively, which enables the fused voxels to perceive objects in a unified 3D space for each frame adaptively.
3. We propose a voxel feature interaction module (VFIM) to improve the consistency of identical objects' semantic information in the homogeneous point cloud and image voxel features which can guide the cross-modal feature fusion and greatly improve the detection performance.
4. Extensive experiments demonstrate the effectiveness of the proposed HMFI and achieve competitive performance on KITTI and Waymo Open Dataset. Notably, on the KITTI benchmark, HMFI surpasses all the published competitive methods by a large margin on detecting cyclist.

2 Related Works

2.1 LiDAR-Based 3D Object Detection

Point-Based Methods: These methods [24,25,30,32] take the raw point cloud as input and employ stacked MLP layers to extract point features. PointR-CNN [30] uses the PointNets [24,25] as point cloud encoder, then generates proposals based on the extracted semantic and geometric features, and refines these coarse proposals via 3D ROI pooling operation. Point-GNN [32] designs a graph neural network to detect 3D objects and encodes the point clouds in a fixed radius near the neighbors' graph. Since the point clouds are unordered and large in number, point-based methods typically suffer from high computational costs.

Voxel-Based Methods: These voxel-based approaches [3,13,20,29,31,46,50] tend to convert the point cloud into voxels and utilize voxel encoding layers to extract voxel features. SECOND [46] proposes a novel sparse convolution layer to replace the original computation-intensive 3D convolution. PointPillars [12] converts the point cloud to a pseudo-image and applies 2D CNN to produce the final detection results. Some other works [3,13,19,29,31] follow [46] to utilize the 3D sparse convolutional operations to encode the voxel features and obtain more accurate detection results in the coarse-to-refine two-stage manner. The more recent CT3D [28] designs a channel-wise transformer architecture to constitute 3D object detection framework with minimal hand-crafted design.

2.2 Image-Based 3D Object Detection

Many researchers are also very concerned about how to use camera images to perform 3D detection [6,17,18,26,48]. Specifically, CaDDN [26] designs a Frustum

Feature Network to project image information into 3D space. We directly introduce depth bins through point cloud projection and use a non-parametrical module to lift image features into 3D space. LIGA-Stereo [6] utilizes the LiDAR-based model to guide the training of stereo-based 3D detection model and achieves the state-of-the-art stereo detection performance. Although cameras are the most common sensors and inexpensive, the performance of image-based methods is still inferior to the LiDAR-based approaches due to the lack of accurate depth information.

2.3 Multi-modal 3D Object Detection

Multi-modal 3D object detection has received more and more attention [41] as it can utilize the complementary information of each single modality to the maximum extent. There are two levels of fusion: decision-level fusion [1,11,21,23, 43]and feature-level fusion [9,14,15,33,39,44,45,47]. The former fusion methods [21] ensemble the detection results of each modality directly. Their performance is limited by each stage.

As for feature-level fusion methods which fuse multi-modal data in a much finer granularity, AVOD [11] utilizes point clouds BEV as well as images features and feeds the features into region proposal network (RPN) for improving detection performance. F-ConvNet [43] follows [23] to utilize frustum point clouds and front view images for 3D object detection. PointFusion [45] and PointPainting [39] enhance raw point cloud with the corresponding class prediction scores through a well pre-trained image semantic segmentation network [7]. EPNet [9] projects the point cloud into image plane to retrieve semantic information at multi-level resolutions in a point-wise manner. MVXNet [33] utilizes pre-trained 2D detectors [27] to produce semantic image features to strengthen the voxel feature representations in the early stage. These methods only exploit part of the rich information contained in an image and suffer from severe information loss [42]. 3D-CVF [47] lifts image features to the dense 3D voxel space but fuses the multi-modal feature in BEV via a cross-view spatial feature fusion strategy and it causes feature overlap in 3D space when constructing image voxel features.

Although many multi-modal networks have been proposed, they do not easily outperform state-of-the-art LiDAR-only based detectors. These fusion methods establish a coarse relationship between the point cloud features and semantic image features. Besides, they suffer from severe information loss by perspective projection. Moreover, existing fusion methods do not exploit the similarity of object-level semantic information in the cross-modal fusion. Our approach is designed to overcome these challenges and achieve better 3D detection performance.

2.4 Methodology

2.5 Framework Overview

The overall architecture of the proposed homogeneous multi-modal fusion and interaction (HMFI) method is illustrated in Fig. 2. We first leverage a point

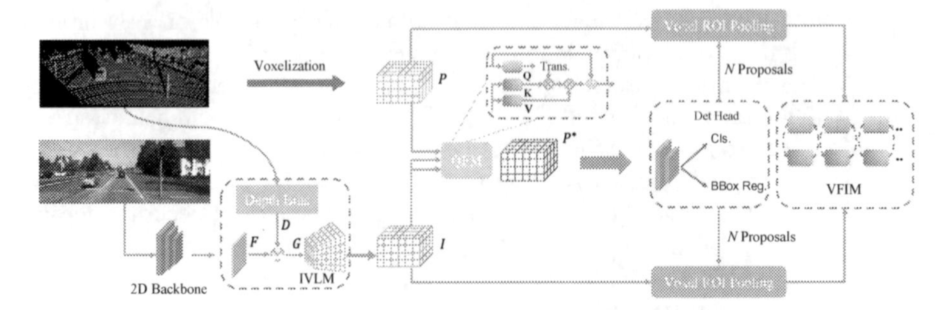

Fig. 2. The architecture of HMFI. Each image is processed by a 2D backbone network and fed into an image voxel lifter module (IVLM) to produce a homogeneous structure based on the depth bins transformed by the point cloud. Then, the processed homogeneous image and point cloud features are fused by the query fusion mechanism (QFM). Next, a voxel-based object detector is employed on fused features to produce 3D detection results. Finally, the voxel feature interaction module (VFIM) conducts feature interaction at object-level based on the detection results to improve semantic consistency in these two homogeneous cross-modal features.

encoding network to extract the features of the point cloud and then pool them to obtain the voxel features $P \in \mathbb{R}^{X_P \times Y_P \times Z_P \times C_F}$ [50] where the C_F is the number of channels of the voxel feature and the (X_P, Y_P, Z_P) is the grid size. The image $\tilde{I} \in \mathbb{R}^{W_{\tilde{I}} \times H_{\tilde{I}} \times 3}$ is fed into a ResNet-50 [8] backbone to extract image features $F \in \mathbb{R}^{W_F \times H_F \times C_F}$, where $W_{\tilde{I}}$ and $H_{\tilde{I}}$ are the width and height of the image and W_F, H_F and C_F are the width, height and number of channels of the image features.

In order to fuse point cloud features and image features in 3D space, we propose an image voxel lifter module (IVLM) to project the image feature F into 3D homogeneous image voxel space as the $I \in \mathbb{R}^{X_I \times Y_I \times Z_I \times C_F}$. Then we use the query fusion mechanism (QFM) to fuse the homogeneous point voxel P and image voxel I to generate the fused representation $P^* \in \mathbb{R}^{X_P \times Y_P \times Z_P \times C_F}$. Afterward, we use the detection module to generate the classification and 3D bounding box of each object based on P^*. Meanwhile, a voxel feature interaction module (VFIM) is proposed to conduct the feature interaction at object-level based on the detection results to improve semantic consistency in these two homogeneous cross-modal features. We introduce the details in the following sections.

2.6 Image Voxel Lifter Module

To encode perceptual depth information in the image effectively and construct a homogeneous structure for multi-modal feature fusion and interaction, we propose the image voxel lifter module (IVLM) to lift 2D image features into 3D space by associating image features and discretized depth maps. The procedure is shown in Fig. 3.

To construct an image feature voxel, we follow [22,26] and convert the image plane features into frustum features G which can encode depth information in image features. Thus, we scatter the vector $F_{m,n} \in \mathbb{R}^{C_F}$ of each pixel (m,n) in the image feature map F into the 3D space determined by the depth bin $D_{m,n}$ along the ray of image frustum perspective projection. The depth bins D are produced by discretizing the depth map with a linear-increasing depth discretization (LID) method [26,36]. The $D \in \mathbb{R}^{W_F \times H_F \times R}$ consists of $W_F \times H_F$ one-hot discretized depth bins in $\mathbb{R}^R$. In order to associate image features with discretized depth information, we utilize the outer product to process the image features F and depth bins D to generate a frustum feature $G \in \mathbb{R}^{W_F \times H_F \times R \times C_F}$. Each $G_{m,n} \in \mathbb{R}^{R \times C_F}$ on pixel (m,n) can be calculated by:

$$G_{m,n} = F_{m,n} \otimes D_{m,n} \tag{1}$$

where $\otimes$ represents the outer product, (m,n) is the index of the each feature pixel.

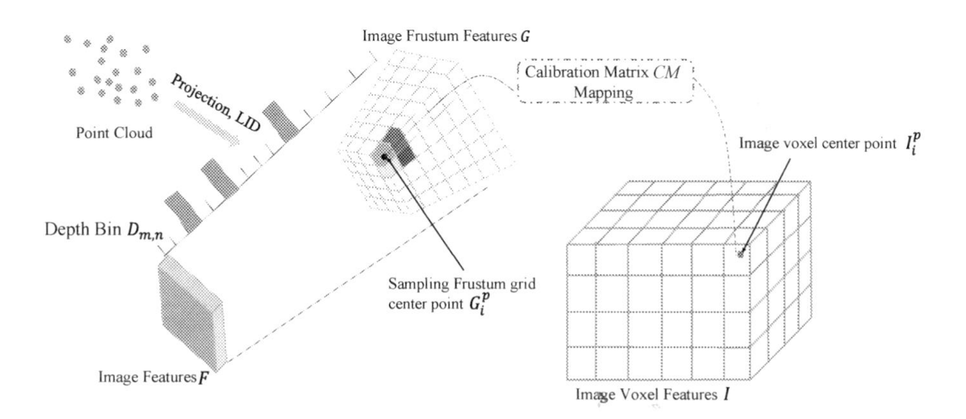

Fig. 3. The image voxel lifter module. Each feature pixel $F_{m,n}$ along the ray is determined by the discrete depth bins D to generate frustum features $G_{m,n}$. Then sampling grid center points in image voxel are projected into the frustum grid based on the calibration matrix CM. The neighboring sampled voxel grids (shown as blue in the image frustum features G) are combined using trilinear interpolation and assigned to the corresponding voxel in I. (Color figure online)

Next, we transform the features from the frustum space $G \in \mathbb{R}^{W_F \times H_F \times R \times C_F}$ into the 3D space $I \in \mathbb{R}^{X_I \times Y_I \times Z_I \times C_F}$ by the trilinear interpolation. Specifically, to acquire the i-th image voxel feature $I_i \in \mathbb{R}^{C_F}$, we sample the corresponding centroid in image frustum features G by a transformation based on calibration matrix CM as $G_i^p = CM \cdot I_i^p$, where the $G_i^p, I_i^p \in \mathbb{R}^3$ indicates the 3D position of the i-th grid in G, I. After that, we conduct the trilinear interpolation around the neighborhood of G_i^p to form the I_i^p. Finally, the image voxel features I is constructed by this process on each spatial index i.

2.7 Query Fusion Mechanism

To exploit the complementary information from point cloud and images, we introduce the query fusion mechanism (QFM) that enables each point cloud voxel feature to perceive the whole image and selectively combines image voxel features. Instead of simply fusing the cross-modal voxel pairs, we consider that the LiDAR voxel can perceive the whole image voxel feature. In order to aggregate these complementary information of two modalities effectively, we propose to use a self-attention [38] module which regards each voxel feature vector of image and point cloud as a homogeneous token.

To be more specific, we use the point cloud voxel features F_P as the queries, the image voxel features F_I as the keys as well as values to conduct the fusion and form the fused voxel features P^*. The construction of F_P and F_I is described as follows.

Considering that most of LiDAR voxels are empty, we produce $F_P \in \mathbb{R}^{M \times C_F}$ by selecting all M non-empty voxels within the homogeneous point cloud voxel features P. However, the image voxel features I is much denser than point cloud voxels. In order to reduce the computational cost, we adopt the 3D max-pooling on I with a scale factor λ to obtain the most informative features $I^* \in \mathbb{R}^{\frac{X_I}{\lambda} \times \frac{Y_I}{\lambda} \times \frac{Z_I}{\lambda} \times C_F}$. Then, we flatten I^* along the first three dimensions to make $F_I \in \mathbb{R}^{L \times C_F}$ where $L = \frac{X_I}{\lambda} * \frac{Y_I}{\lambda} * \frac{Z_I}{\lambda}$.

After constructing the point voxel F_P and the image voxel F_I, we utilize a multi-head self-attention [38] layer as the query fusion mechanism (QFM). We adopt three learnable linear transformation for each head i on the query F_P, key F_I and value F_I, denoted as $Q_i \in \mathbb{R}^{M \times d_k}$, $K_i \in \mathbb{R}^{L \times d_k}$ and $V_i \in \mathbb{R}^{L \times d_v}$ respectively:

$$Q_i = F_P \cdot W_i^Q, \quad K_i = F_I \cdot W_i^K, \quad V_i = F_I \cdot W_i^V \tag{2}$$

where $W_i^Q \in \mathbb{R}^{C_F \times d_k}$, $W_i^K \in \mathbb{R}^{C_F \times d_k}$ and $W_i^V \in \mathbb{R}^{C_F \times d_v}$.

Then we perform the multi-head self-attention with r heads:

$$\begin{aligned} A_M &= \text{Concat}(\text{head}_1, \text{head}_2, \cdots, \text{head}_r) W^O \\ \text{head}_i &= \text{softmax}\left(\frac{Q_i K_i^T}{\sqrt{d_k}}\right) V_i \end{aligned} \tag{3}$$

where $A_M \in \mathbb{R}^{M \times C_F}$ is the output of multi-head attention module, $W_O \in \mathbb{R}^{r * d_v \times C_F}$ is a linear transformation matrix to project the concatenation of r attention heads into the homogeneous point voxel space. Then we concatenate the A_M and the non-empty point voxel features F_P to acquire the fused voxel features $F_P^* \in \mathbb{R}^{M \times (2 * C_F)}$. Finally, we restore F_P^* into the homogeneous voxel space as $P^* \in \mathbb{R}^{X_P \times Y_P \times Z_P \times (2 * C_F)}$ as the input of the downstream 3D object detection module.

2.8 Voxel Feature Interaction Module

LiDARs and cameras have different representations of identical objects in the scene. Though the modalities are different from each other, the object-level representations should be similar. Motivated by this observation, we design a voxel

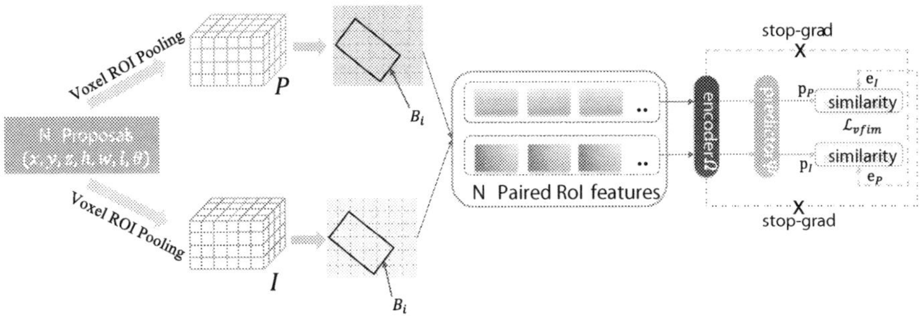

Fig. 4. The voxel feature interaction module. We use the voxel RoI pooling to extract features in these two homogeneous features according to the predicted proposals and form the paired RoI feature set. Then we adopt feature interaction between each pair of RoI features based on the symmetry similarity constraint loss to improve the object-level semantic consistency between two homogeneous representations

feature interaction module (VFIM) to build the feature interaction in these two cross-modal features based on the consistency of object-level properties in point cloud and images. And we can fully utilize the similarity constraint between homogeneous features P and I with the object-level guidance to achieve better cross-modal feature fusion.

As shown in Fig. 4, we sample the N 3D detection proposals from 3D detection head as $B = \{B_1, B_2, , ..., B_N\}$. Then, we introduce the voxel RoI pooling [3] on homogeneous point voxel features P and image voxel features I to obtain the respective RoI features including $P_B = \{P_{B_1}, P_{B_2}, \ldots, P_{B_N}\}$ and $I_B = \{I_{B_1}, I_{B_2}, \ldots, I_{B_N}\}$.

Finally, inspired by [2], to improve the similarity between the output vectors from the paired RoI features P_{B_i} and I_{B_i}, we feed both of them into an encoder Ω and use a MLP-based predictor Ψ to transform the output of these encoded RoI features into the metric space: $p_P = \Psi(\Omega(P_{B_i}))$, $e_I = \Omega(I_{B_i})$, $p_I = \Psi(\Omega(I_{B_i}))$, $e_P = \Omega(P_{B_i})$.

We minimize the paired feature distance by using cosine similarity:

$$CosSim(p, e) = -\frac{p}{\|p\|_2} * \frac{e}{\|e\|_2} \tag{4}$$

where $\|\cdot\|_2$ means l_2 normalization.

Meanwhile, the stop-gradient operation is also adopted for better modeling the similarity constraint, then we utilize the symmetry similarity constraint loss $\mathcal{L}_{vfim}$ as:

$$\mathcal{L}_{vfim} = \frac{1}{2}CosSim(p_P, stop\_grad(e_I)) + \frac{1}{2}CosSim(p_I, stop\_grad(e_P)) \tag{5}$$

2.9 Loss Function

In previous methods, the image backbone is directly initialized with the fixed pre-trained weights from other external datasets such as ImageNet. On the contrary, our HMFI is trained via two-stage training process in an end-to-end manner. We utilize a multi-task loss function for jointly optimizing the whole network. The total loss $\mathcal{L}_{total}$ can be formulated as:

$$\mathcal{L}_{\text{total}}=\mathcal{L}_{\text{rpn}} + \mathcal{L}_{rcnn} + \gamma\mathcal{L}_{vfim} \tag{6}$$

where γ is set to 0.1. $\mathcal{L}_{rpn}$ and $\mathcal{L}_{rcnn}$ denote the training objectives for the region proposal network (RPN) and the refinement network, We follow [3,46] to devise the loss of the RPN as:

$$\mathcal{L}_{\text{rpn}}=\omega_1\mathcal{L}_{\text{cls}} + \omega_2\mathcal{L}_{reg} \tag{7}$$

where ω_1 and ω_2 are set to 1 and 2, respectively. We adopt the focal loss [16] to balance the positive and negative samples in classification loss with default hyperparameters and the $smooth_{L1}$ loss is utilized for the box regression. The proposal refinement loss $\mathcal{L}_{rcnn}$ includes the IoU-guided confidence prediction loss $\mathcal{L}_{iou}$ and the box refinement loss $\mathcal{L}_{refine}$ as

$$\mathcal{L}_{\text{rcnn}}=\mathcal{L}_{\text{iou}} + \mathcal{L}_{refine} \tag{8}$$

3 Experiments

In this section, we evaluate the performance of the proposed HMFI on the KITTI [4] and Waymo Open Dataset [35].

3.1 Datasets

KITTI is a widely used dataset. It consists of 7,481 training frames and 7,518 testing frames, with 2D and 3D annotations of cars, pedestrians and cyclists on the streets. Objects are divided into three difficulty levels: easy, moderate and hard, according to their size, occlusion level and truncation level. For validation, training samples are commonly divided into a train set with 3,712 samples and a val set with 3,769 samples.

Waymo Open Dataset (WOD) is a large-scale dataset for autonomous driving. There are totally 798 scenes for training and 202 scenes for validation. Each scene is a sequential segment that has around 20 s of sensor data. Note that cameras in WOD only cover around 250° field of view (FOV), which is different from LiDAR points and 3D labels in full 360°. To follow the same setting of KITTI, we only select LiDAR points and ground-truth in the FOV of front camera for training and evaluating. We sample every 5th frames from all the training samples to form the new training set (~32k frames) due to the large dataset size and high frame rate.

3.2 Implementation Details

Experimental Settings. On the KITTI benchmark, we set the range of point cloud to [0, 70.4], [−40, 40], [−3, 1]m in the (x, y, z) axis. The LiDAR voxel structure is divided by a voxel size (0.05, 0.05, 0.1)m, while each image voxel size is set to (0.2, 0.2, 0.4)m to fit with the feature size of the point cloud branch. As for Waymo, we use [0, 75.2], [−75.2, 75.2], [−2, 4]m for the point cloud range, (0.1, 0.1, 0.15)m for the voxel size. And each image voxel size is set to (0.4, 0.4, 0.6)m to fit the point cloud feature size. In the QFM, the scale factor λ is set as 4, the count r and hidden units of attention heads are set to 4 and 64, respectively. In the VFIM, the settings of the voxel RoI pooling operation are the same as Voxel-RCNN [3] and we sample $N = 128$ proposals, half of them are positive samples that have $IoU > 0.55$ with the corresponding ground truth boxes. The number of hidden units of the encoder Ω and predictor Ψ are both set to 256.

Table 1. Quantitative comparison with the state-of-the-art 3D object detection methods on KITTI *test* set.

Method	Modality	Car 3D AP			Pedestrian 3D AP			Cyclist 3D AP		
		Easy	Mod.	Hard	Easy	Mod.	Hard	Easy	Mod.	Hard
PointPillars [12]	LiDAR	82.58	74.31	68.99	51.45	41.92	38.89	77.10	58.65	51.92
SECOND [46]	LiDAR	87.44	79.46	73.97	–	–	–	–	–	–
PointRCNN [30]	LiDAR	86.96	75.64	70.70	47.98	39.37	36.01	74.96	58.82	52.53
PointGNN [32]	LiDAR	88.33	79.47	72.29	51.92	**43.77**	40.14	78.60	63.48	57.08
Part A^2 [31]	LiDAR	87.81	78.49	73.51	**53.10**	43.35	40.06	79.17	63.52	56.93
PV-RCNN [29]	LiDAR	90.25	81.43	76.82	52.17	43.29	**40.29**	78.60	63.71	57.65
Voxel-RCNN [3]	LiDAR	**90.90**	81.62	77.06	–	–	–	–	–	–
M3DETR [5]	LiDAR	90.28	81.73	76.96	–	–	–	**83.83**	**66.74**	**59.03**
CT3D [28]	LiDAR	87.83	81.77	77.16	–	–	–	–	–	–
Pyramid RCNN-V [19]	LiDAR	87.06	81.28	76.85	–	–	–	–	–	–
Pyramid RCNN-PV [19]	LiDAR	88.39	**82.08**	**77.49**	–	–	–	–	–	–
MV3D [1]	LiDAR+RGB	74.97	63.63	54.00	–	–	–	–	–	–
Confuse [3]	LiDAR+RGB	83.68	68.78	61.67	–	–	–	–	–	–
F-PointNet [23]	LiDAR+RGB	82.19	69.79	60.59	50.53	42.15	38.08	72.27	56.12	49.01
MVXNet [33]	LiDAR+RGB	83.20	72.70	65.20	–	–	–	–	–	–
PointPainting [39]	LiDAR+RGB	82.11	71.70	67.08	50.32	40.97	37.77	77.63	63.78	55.89
AVOD-FPN [11]	LiDAR+RGB	83.07	71.76	65.73	50.46	42.27	39.04	63.76	50.55	44.93
MAFF [49]	LiDAR+RGB	85.52	75.04	67.60	–	–	–	–	–	–
PI-RCNN [44]	LiDAR+RGB	84.37	74.82	70.03	–	–	–	–	–	–
F-Convnet [43]	LiDAR+RGB	87.36	76.39	66.69	**52.16**	**43.38**	38.80	81.98	65.07	56.54
MMF [14]	LiDAR+RGB	88.40	77.43	70.22	–	–	–	–	–	–
CLOCs_PVCas [21]	LiDAR+RGB	88.94	80.67	77.15	47.30	39.42	36.97	77.33	62.02	55.52
EPNet [9]	LiDAR+RGB	**89.81**	79.28	74.59	–	–	–	–	–	–
3D-CVF [47]	LiDAR+RGB	89.20	80.05	73.11	–	–	–	–	–	–
HMFI (ours)	LiDAR+RGB	88.90	**81.93**	**77.30**	50.88	42.65	**39.78**	**84.02**	**70.37**	**62.57**

Training. To validate the effectiveness of our HMFI, we select the Voxel-RCNN [3] as the baseline. Our HMFI is trained via the two-stage training process. We adopt OpenPCDet [37] as our codebase, and a pre-trained ResNet50 [8] is adopted as the 2D backbone to produce image features F for the image voxel lifter module. We train the model with the Adam [10] optimizer, which uses the one-cycle policy [34] with the initial learning rate being 0.0005. Batch size is set as 2. The total number of training epochs is set as 80 for KITTI [4] and 30 epochs for WOD [35].

3.3 Results on KITTI Dataset

KITTI Test set. Experiments on the KITTI test split [4] are evaluated using average precision (AP) via 40 recall positions. We compare our HMFI with other state-of-the-art approaches by submitting the detection results to the KITTI server for evaluation. Table 1 presents the quantitative comparison with state-of-the-art 3D object detection methods on the KITTI test set. It is apparent that the HMFI achieves better or comparable performance over the state-of-the-art methods on car and cyclist for all difficulty levels, respectively. The HMFI achieves up to 1.88% gains (for moderate difficulty) over 3D-CVF [47] which is the best feature-level fusion based method. The HMFI outperforms most of the LiDAR-based 3D object detectors except for the Pyramid RCNN-PV [19] which introduces the raw point features to achieve a better result but with a worse efficiency. By contrast, our method outperforms the Pyramid RCNN-V [19] in the same settings. Especially, our HMFI surpasses all the published algorithms by a large margin for the 3D detection of cyclist. Note that none of the models in Table 1 can achieve superior performance to our model on car and cyclist simultaneously.

KITTI Val set. In addition, we also report the performance on the KITTI val set with AP calculated by 11 recall positions. As shown in Table 2, our HMFI achieves the state-of-the-art performance on moderate level on the val set, even better than the LiDAR-based method [19].

To sum up, the results on both val set and test set consistently demonstrate that our proposed HMFI achieves superior 3D detection performance. Specifically, we achieve satisfactory performance on pedestrian and cyclist which usually have very few points in LiDAR measurements. As shown in Fig. 1 (b), we also report the inference time per frame of some feature-level fusion methods, and our HMFI achieves the best balance between the accuracy and efficiency among all methods.

3.4 Results on Waymo Open Dataset

To further validate the effectiveness of the proposed HMFI, we also conduct experiments on the large-scale Waymo Open Dataset. Two difficulty levels are also introduced, where the LEVEL_1 mAP is calculated on objects that have more than 5 points and the LEVEL_2 mAP is measured on objects that have 1∼5 points. Table 3 summarizes the performance of our method and baselines. It is obvious that our HMFI performs superbly over all the object classes and two

Table 2. Performance comparison on the moderate level of KITTI val split with AP calculated by 11 recall positions, † means the our re-implementation results. $\text{Car}_{Mod.}$, $\text{Pedestrian}_{Mod.}$ and $\text{Cyclist}_{Mod.}$ donate the performance of Car, Pedestrian and Cyclist on moderate level respectively.

Method	Modality	$\text{Car}_{Mod.}$	$\text{Pedestrian}_{Mod.}$	$\text{Cyclist}_{Mod.}$
SECOND [46]	LiDAR	76.48	59.84	64.89
PointRCNN [30]	LiDAR	76.05	51.59	66.67
PV-RCNN [29]	LiDAR	83.69	57.90	70.47
Voxel-RCNN [3]	LiDAR	84.52	–	–
Voxel-RCNN† [3]	LiDAR	84.27	60.11	72.07
Pyramid RCNN-PV [19]	LiDAR	84.38	–	–
PointPainting [30]	LiDAR+RGB	77.74	61.67	71.62
CLOCs_SecCas [21]	LiDAR+RGB	79.31	56.20	67.92
3D-CVF [47]	LiDAR+RGB	79.88	–	–
HMFI (ours)	LiDAR+RGB	**85.14**	**62.41**	**74.11**

difficulty levels. In particular, we achieve remarkable gains on pedestrian and cyclist with +2.17% and +1.86% mAP on LEVEL 2, which demonstrates the outstanding performance of our method on detecting objects with fewer than 5 LiDAR points. The results on the Waymo Open Dataset further validate both the effectiveness and generalization of the HMFI.

Table 3. Performance comparison on the Waymo Open Dataset with 202 validation sequences (~40k samples)

Method	Vehicle L1/L2		Pedestrian L1/L2		Cyclist L1/L2	
	AP	APH	AP	APH	AP	APH
Baseline	66.46/64.21	64.81/62.40	64.35/62.74	58.11/55.43	62.34/59.85	59.64/57.67
Ours	68.34/65.66	66.84/64.57	66.62/64.91	59.76/57.24	64.25/61.71	61.23/59.21
Improvements	+1.88/+1.45	+2.03/+2.17	+2.27/+2.17	+1.65/+1.81	+1.91/+1.86	+1.59/+1.54

3.5 Ablation Study

In this section, we present an ablation study for validating the effect of each component in the HMFI method. The ablation study is conducted on the KITTI validation set. We adopt the mean average precision (mAP) on easy, moderate and hard difficulty levels via 11 recall positions for evaluation. As shown in Table 4, our HMFI can bring over 1.8% AP performance gain on all difficulty levels of three objects.

Effect of Query Fusion Mechanism. The query fusion mechanism (QFM) combines the image and point cloud features selectively depending on their relevance according to the attention map between the image features and point cloud features. In Table 4, we observe that QFM can generate the enhanced joint camera-LiDAR features and lead to 0.83%, 0.58%, and 0.62% performance gains in AP_{Easy}, $AP_{Mod.}$, AP_{Hard}, respectively.

Table 4. Effect of each component of our HMFI on KITTI val set. AP_{Easy}, $AP_{Mod.}$, and AP_{Hard} are the mAP performance of easy, moderate, and hard levels respectively.

Method	QFM	IVLM	VFIM	AP_{Easy}	$AP_{Mod.}$	AP_{Hard}
Baseline [3]	–	–	–	81.34	71.76	67.09
Ours	✓	–	–	82.17	72.34	67.73
	✓	✓	–	82.52	72.94	68.45
	✓	✓	✓	83.36	73.89	68.98
Improvements				+2.02	+2.13	+1.89

Effect of Multi-modal Feature Structure. In Table 4, We observe that the IVLM can bring 0.35%, 0.60%, and 0.72% performance gains in AP_{Easy}, $AP_{Mod.}$, AP_{Hard}. IVLM lifts image features to the homogeneous space with point cloud voxel features, which not only facilitates feature fusion, but also enables object-level semantic consistency modeling between two homogeneous features.

Effect of Voxel Feature Interaction. We observe that the voxel feature interaction module (VFIM) improves the baseline by 0.84%, 0.95%, and 0.53% in AP_{Easy}, $AP_{Mod.}$, AP_{Hard}, respectively. It indicates that our VFIM plays a pivotal role in our multi-modal detection framework. It can improve object-level semantic consistency between two homogeneous features and enables the detector to aggregate paired features across homogenous representations based on object-level semantic similarity.

4 Conclusions

In this paper, we propose the homogeneous multi-modal feature fusion and interaction (HMFI) method for 3D detection which fuses image and point cloud features in a homogeneous structure and enforces the consistency of object-level semantic information between two homogeneous features. We propose an image voxel lifter module (IVLM) to lift 2D image features to the 3D space and generate homogeneous image voxel features with point cloud voxel features. Then, image and point cloud features are selectively combined by the query fusion mechanism (QFM). Besides, we build the feature interaction in the homogeneous image and point cloud voxel features based on the similarity of object-level semantic information. Extensive experiments conducted on KITTI and Waymo Open Dataset show that significant performance gains can be obtained by our proposed HMFI. Particularly, for the detection of cyclist on the KITTI benchmark, HMFI surpasses all published algorithms by a large margin.

Acknowledgments. This research is funded by the Science and Technology Commission of Shanghai Municipality (19511120200), The computation is performed in ECNU Multifunctional Platform for Innovation (001).

References

1. Chen, X., Ma, H., Wan, J., Li, B., Xia, T.: Multi-view 3D object detection network for autonomous driving. In: CVPR, pp. 1907–1915 (2017)
2. Chen, X., He, K.: Exploring simple siamese representation learning. In: Proceedings of the IEEE/CVF Conference on Computer Vision and Pattern Recognition, pp. 15750–15758 (2021)
3. Deng, J., Shi, S., Li, P., Zhou, W., Zhang, Y., Li, H.: Voxel R-CNN: towards high performance voxel-based 3d object detection. In: AAAI, pp. 1201–1209 (2021)
4. Geiger, A., Lenz, P., Urtasun, R.: Are we ready for autonomous driving? The kitti vision benchmark suite. In: CVPR, pp. 3354–3361 (2012)
5. Guan, T., Wang, J., Lan, S., Chandra, R., Wu, Z., Davis, L., Manocha, D.: M3detr: multi-representation, multi-scale, mutual-relation 3d object detection with transformers. In: WACV, pp. 772–782 (2022)
6. Guo, X., Shi, S., Wang, X., Li, H.: Liga-stereo: learning lidar geometry aware representations for stereo-based 3D detector. In: CVPR, pp. 3153–3163 (2021)
7. He, K., Gkioxari, G., Dollár, P., Girshick, R.: Mask R-CNN. In: ICCV, pp. 2961–2969 (2017)
8. He, K., Zhang, X., Ren, S., Sun, J.: Deep residual learning for image recognition. In: Proceedings of the IEEE Conference on Computer Vision and Pattern Recognition, pp. 770–778 (2016)
9. Huang, T., Liu, Z., Chen, X., Bai, X.: EPNet: enhancing point features with image semantics for 3D object detection. In: Vedaldi, A., Bischof, H., Brox, T., Frahm, J.-M. (eds.) ECCV 2020. LNCS, vol. 12360, pp. 35–52. Springer, Cham (2020). https://doi.org/10.1007/978-3-030-58555-6_3
10. Kingma, D.P., Ba, J.: Adam: a method for stochastic optimization. ICLR (2015)
11. Ku, J., Mozifian, M., Lee, J., Harakeh, A., Waslander, S.L.: Joint 3D proposal generation and object detection from view aggregation. In: IROS, pp. 1–8. IEEE (2018)
12. Lang, A.H., Vora, S., Caesar, H., Zhou, L., Yang, J., Beijbom, O.: Pointpillars: fast encoders for object detection from point clouds. In: CVPR, pp. 12697–12705 (2019)
13. Li, Z., Wang, F., Wang, N.: Lidar R-CNN: an efficient and universal 3D object detector. In: CVPR, pp. 7546–7555 (2021)
14. Liang, M., Yang, B., Chen, Y., Hu, R., Urtasun, R.: Multi-task multi-sensor fusion for 3D object detection. In: CVPR, pp. 7345–7353 (2019)
15. Liang, M., Yang, B., Wang, S., Urtasun, R.: Deep continuous fusion for multi-sensor 3D object detection. In: Ferrari, V., Hebert, M., Sminchisescu, C., Weiss, Y. (eds.) ECCV 2018. LNCS, vol. 11220, pp. 663–678. Springer, Cham (2018). https://doi.org/10.1007/978-3-030-01270-0_39

16. Lin, T.Y., Goyal, P., Girshick, R., He, K., Dollár, P.: Focal loss for dense object detection. In: ICCV, pp. 2980–2988 (2017)
17. Liu, Z., Wu, Z., Tóth, R.: Smoke: Single-stage monocular 3D object detection via keypoint estimation. In: CVPRW, pp. 996–997 (2020)
18. Lu, Y., et al.: Geometry uncertainty projection network for monocular 3D object detection. In: ICCV, pp. 3111–3121 (2021)
19. Mao, J., Niu, M., Bai, H., Liang, X., Xu, H., Xu, C.: Pyramid R-CNN: Towards better performance and adaptability for 3D object detection. In: Proceedings of the IEEE/CVF International Conference on Computer Vision, pp. 2723–2732 (2021)
20. Mao, J., et al.: Voxel transformer for 3D object detection. In: ICCV, pp. 3164–3173 (2021)
21. Pang, S., Morris, D., Radha, H.: Clocs: camera-lidar object candidates fusion for 3D object detection. In: IROS, pp. 10386–10393. IEEE (2020)
22. Philion, J., Fidler, S.: Lift, splat, shoot: encoding images from arbitrary camera rigs by implicitly unprojecting to 3D. In: Vedaldi, A., Bischof, H., Brox, T., Frahm, J.-M. (eds.) ECCV 2020. LNCS, vol. 12359, pp. 194–210. Springer, Cham (2020). https://doi.org/10.1007/978-3-030-58568-6_12
23. Qi, C.R., Liu, W., Wu, C., Su, H., Guibas, L.J.: Frustum pointnets for 3D object detection from RGB-D data. In: CVPR, pp. 918–927 (2018)
24. Qi, C.R., Su, H., Mo, K., Guibas, L.J.: Pointnet: deep learning on point sets for 3D classification and segmentation. In: CVPR, pp. 652–660 (2017)
25. Qi, C.R., Yi, L., Su, H., Guibas, L.J.: Pointnet++: deep hierarchical feature learning on point sets in a metric space. In: NeurIPS, vol. 30 (2017)
26. Reading, C., Harakeh, A., Chae, J., Waslander, S.L.: Categorical depth distribution network for monocular 3D object detection. In: CVPR, pp. 8555–8564 (2021)
27. Ren, S., He, K., Girshick, R., Sun, J.: Faster R-CNN: towards real-time object detection with region proposal networks. In: NIPS, vol. 28 (2015)
28. Sheng, H., et al.: Improving 3D object detection with channel-wise transformer. In: ICCV, pp. 2743–2752 (2021)
29. Shi, S., et al.: PV-RCNN: point-voxel feature set abstraction for 3d object detection. In: CVPR, pp. 10529–10538 (2020)
30. Shi, S., Wang, X., Li, H.: PointrCNN: 3D object proposal generation and detection from point cloud. In: CVPR, pp. 770–779 (2019)
31. Shi, S., Wang, Z., Shi, J., Wang, X., Li, H.: From points to parts: 3D object detection from point cloud with part-aware and part-aggregation network. PAMI **43**(8), 2647–2664 (2020)
32. Shi, W., Rajkumar, R.: Point-GNN: graph neural network for 3D object detection in a point cloud. In: CVPR, pp. 1711–1719 (2020)
33. Sindagi, V.A., Zhou, Y., Tuzel, O.: MVX-Net: multimodal voxelnet for 3D object detection. In: 2019 International Conference on Robotics and Automation (ICRA), pp. 7276–7282. IEEE (2019)
34. Smith, L.N.: A disciplined approach to neural network hyper-parameters: Part 1-learning rate, batch size, momentum, and weight decay. arXiv preprint arXiv:1803.09820 (2018)
35. Sun, P., et al.: Scalability in perception for autonomous driving: Waymo open dataset. In: Proceedings of the IEEE/CVF Conference on Computer Vision and Pattern Recognition, pp. 2446–2454 (2020)
36. Tang, Y., Dorn, S., Savani, C.: Center3D: center-based monocular 3D object detection with joint depth understanding. In: Akata, Z., Geiger, A., Sattler, T. (eds.) DAGM GCPR 2020. LNCS, vol. 12544, pp. 289–302. Springer, Cham (2021). https://doi.org/10.1007/978-3-030-71278-5_21

37. Team, O.D.: Openpcdet: an open-source toolbox for 3D object detection from point clouds (2020). https://github.com/open-mmlab/OpenPCDet
38. Vaswani, A., et al.: Attention is all you need. In: NIPS, vol. 30 (2017)
39. Vora, S., Lang, A.H., Helou, B., Beijbom, O.: Pointpainting: sequential fusion for 3D object detection. In: CVPR, pp. 4604–4612 (2020)
40. Wang, C., Ma, C., Zhu, M., Yang, X.: Pointaugmenting: cross-modal augmentation for 3D object detection. In: CVPR, pp. 11794–11803 (2021)
41. Wang, Y., Mao, Q., Zhu, H., Zhang, Y., Ji, J., Zhang, Y.: Multi-modal 3D object detection in autonomous driving: a survey. CoRR (2021)
42. Wang, Y., Mao, Q., Zhu, H., Zhang, Y., Ji, J., Zhang, Y.: Multi-modal 3D object detection in autonomous driving: a survey. arXiv preprint arXiv:2106.12735 (2021)
43. Wang, Z., Jia, K.: Frustum convnet: sliding frustums to aggregate local point-wise features for amodal 3D object detection. In: IROS, pp. 1742–1749. IEEE (2019)
44. Xie, L., Xiang, C., Yu, Z., Xu, G., Yang, Z., Cai, D., He, X.: PI-RCNN: an efficient multi-sensor 3D object detector with point-based attentive cont-conv fusion module. In: AAAI, pp. 12460–12467 (2020)
45. Xu, D., Anguelov, D., Jain, A.: Pointfusion: deep sensor fusion for 3D bounding box estimation. In: CVPR, pp. 244–253 (2018)
46. Yan, Y., Mao, Y., Li, B.: Second: sparsely embedded convolutional detection. Sensors **18**(10), 3337 (2018)
47. Yoo, J.H., Kim, Y., Kim, J., Choi, J.W.: 3D-CVF: generating joint camera and lidar features using cross-view spatial feature fusion for 3D object detection. In: Vedaldi, A., Bischof, H., Brox, T., Frahm, J.-M. (eds.) ECCV 2020. LNCS, vol. 12372, pp. 720–736. Springer, Cham (2020). https://doi.org/10.1007/978-3-030-58583-9_43
48. You, Y., et al.: Pseudo-lidar++: accurate depth for 3D object detection in autonomous driving. In: ICLR (2020)
49. Zhang, Z., et al.: Maff-net: filter false positive for 3D vehicle detection with multi-modal adaptive feature fusion. arXiv preprint arXiv:2009.10945 (2020)
50. Zhou, Y., Tuzel, O.: Voxelnet: end-to-end learning for point cloud based 3D object detection. In: CVPR, pp. 4490–4499 (2018)

JPerceiver: Joint Perception Network for Depth, Pose and Layout Estimation in Driving Scenes

Haimei Zhao[1], Jing Zhang[1(✉)], Sen Zhang[1], and Dacheng Tao[1,2]

[1] The University of Sydney, 6 Cleveland Street, Darlington NSW 2008, Australia
{hzha7798,szha2609}@uni.sydney.edu.au, jing.zhang1@sydney.edu.au
[2] JD Explore Academy, Beijing, China

Abstract. Depth estimation, visual odometry (VO), and bird's-eye-view (BEV) scene layout estimation present three critical tasks for driving scene perception, which is fundamental for motion planning and navigation in autonomous driving. Though they are complementary to each other, prior works usually focus on each individual task and rarely deal with all three tasks together. A naive way is to accomplish them independently in a sequential or parallel manner, but there are three drawbacks, i.e., 1) the depth and VO results suffer from the inherent scale ambiguity issue; 2) the BEV layout is usually estimated separately for roads and vehicles, while the explicit overlay-underlay relations between them are ignored; and 3) the BEV layout is directly predicted from the front-view image without using any depth-related information, although the depth map contains useful geometry clues for inferring scene layouts. In this paper, we address these issues by proposing a novel joint perception framework named JPerceiver, which can simultaneously estimate scale-aware depth and VO as well as BEV layout from a monocular video sequence. It exploits the cross-view geometric transformation (CGT) to propagate the absolute scale from the road layout to depth and VO based on a carefully-designed scale loss. Meanwhile, a cross-view and cross-modal transfer (CCT) module is devised to leverage the depth clues for reasoning road and vehicle layout through an attention mechanism. JPerceiver can be trained in an end-to-end multi-task learning way, where the CGT scale loss and CCT module promote inter-task knowledge transfer to benefit feature learning of each task. Experiments on Argoverse, Nuscenes and KITTI show the superiority of JPerceiver over existing methods on all the above three tasks in terms of accuracy, model size, and inference speed. The code and models are available at https://github.com/sunnyHelen/JPerceiver.

Keywords: Depth estimation · Visual odometry · Layout estimation

---

Supplementary Information The online version contains supplementary material available at https://doi.org/10.1007/978-3-031-19839-7_41.

S. Avidan et al. (Eds.): ECCV 2022, LNCS 13698, pp. 708–726, 2022.
https://doi.org/10.1007/978-3-031-19839-7_41

1 Introduction

Autonomous driving has witnessed great progress in recent years, where deep learning is playing an increasing role in perception [27,37,61], planning [6,48], navigation [2,44], and decision making [18,21]. Among them, scene perception is the basis for other subsequent procedures in autonomous driving [56], which includes various sub-tasks for different perception purposes, e.g., depth estimation for 3D measurement [14,17], ego motion estimation for localization and visual odometry (VO) [25,49,58,62], as well as bird's-eye-view (BEV) or front-view (FV) layout estimation for detecting obstacles and drivable areas [10,31,36,54]. Although these tasks have underlying relations to each other intuitively, they are usually tackled separately in prior works [17,54]. The joint estimation for all these tasks has not drawn enough attention so far, and the benefits and challenges of doing so remain unclear, which is the focus of this paper.

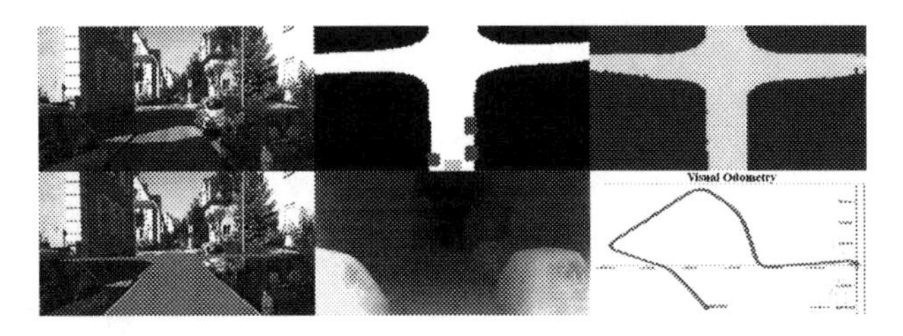

Fig. 1. The proposed JPerceiver can predict BEV semantic layout (middle top, white for roads, blue for cars and cyan for the ego car), scale-aware depth map (middle bottom) and VO result (bottom right) simultaneously from a monocular video sequence (top left). The drivable area is visualized in green (bottom left) by projecting the BEV road layout onto the image plane. The ground truth BEV layout is shown in top right. (Color figure online)

Depth and VO estimation are two closely related computer vision tasks that have been studied for decades [33,45]. Recent self-supervised learning methods use the photometric consistency between consecutive frames to achieve the simultaneous estimation of scene depth and VO from monocular video sequences, where no ground truth depth labels are required [17,42,60]. On the other hand, BEV scene layout estimation refers to the task of estimating the semantic occupancy of roads and vehicles in the metric-scale BEV plane directly from FV images [20,31,36,54]. Though significant progress has been made in each individual task, they still suffer from some inherent problems: i.e., (1) the scale ambiguity in monocular depth and VO estimation since the photometric error between corresponding pixels is equivalent up to an arbitrary scaling factor w.r.t. depth and translation, and (2) the lack of geometry priors for predicting complex BEV layout. Consequently, monocular depth and VO predictions need to be rescaled with a scaling ratio derived from ground truth [4,17], which is not appealing in real-world applications. And previous BEV methods [31,36,54] usually predict the BEV layout of different semantic categories separately and ignore potentially useful geometry clues such as the depth order between cars.

In this paper, we propose to handle these three tasks simultaneously and provide complementary information for each other to address the aforementioned issues. We are inspired by the following two key observations. First, we note that the BEV road layout can provide absolute scale under the weak assumption that the road is flat, which allows us to exploit the cross-view geometric transformation and obtain a depth map with an absolute scale corresponding to the distance field that existed in the layout. As a result, the absolute scale can be introduced into our depth and VO predictions, resolving the scale ambiguity problem. Second, the learned depth predictions can provide useful priors about scene geometry (e.g., the relationship between near and far as well as overlay and underlay between objects and roads in the scene) to help solve the challenges (e.g., occlusions) in BEV layout estimation.

To this end, we propose a novel joint perception network named JPerceiver that can estimate scale-aware depth and VO as well as BEV layout of roads and vehicles simultaneously, as shown in Fig. 1. JPerceiver follows the multi-task learning framework, consisting of three networks for depth, pose and layout, respectively, which can be efficiently trained in an end-to-end manner. Specifically, we design a cross-view geometric transformation-based (CGT) scale loss to propagate the absolute scale from the road layout to depth and VO. Meanwhile, a cross-view and cross-modal transfer (CCT) module is devised to leverage the depth clues for inferring the road and vehicle layouts through an attention mechanism. Our proposed scale loss and CCT module not only promote inter-task knowledge transfer but also benefit the feature learning of each task via network forward computation and gradient back-propagation.

The contributions of this paper are summarized as follows: 1) we propose the first joint perception framework JPerceiver for depth, VO and BEV layout estimation simultaneously; 2) we design a CGT scale loss to leverage the absolute scale information from the BEV layout to achieve scare-aware depth and VO; 3) we devise a CCT module that leverages the depth clues to help reason the spatial relationships between roads and vehicles implicitly, and facilitates the feature learning for BEV layout estimation; and 4) we conduct extensive experiments on public benchmarks and show that JPerceiver outperforms the state-of-the-art methods on the above three tasks by a large margin.

2 Related Work

Self-supervised Depth Estimation and VO. SfMLearner [60] is one of the first works that propose to optimize depth and pose jointly in a self-supervised manner, utilizing the photometric consistency among continuous frames. Though this self-supervised learning scheme has drawn great attention from researchers and achieved promising results [17,17,29,42,47], current monocular unsupervised methods still suffer from the scale ambiguity problem. McCraith et al. [32] fit sample points to get the road plane estimation in the 3D world during test to obtain scale hint, but the hard formulation limits its general applicability. DNet [52] proposed to recover the scale by calculating the ratio of the estimated camera height and a given one, which requires a visible ground plane to be detected during inference. Wagstaff and Kelly [46] also use the camera height as the scale hint by training a plane segmentation network, which is the most similar work to ours. However, they use a three-stage training strategy to train networks separately which is much more complex than our end-to-end method.

BEV-Based Environment Perception. Due to the limited field of view (FOV) of FV cameras, BEV representation is commonly used in environment perception and motion planning for autonomous driving [39,50]. Traditional methods usually predict depth and segmentation from front images, and then warp them to BEV through inverse perspective mapping (IPM) [30,43], which loses a large amount of information and cause distortions due to potential occlusions. Recently deep learning-based methods have been developed to estimate the road and vehicle layout in the orthographic BEV plane, taking the advantage of the strong hallucination ability of CNN [1,20,26,31,36,54]. The newly released self-driving datasets like Argoverse [7] and Nuscenes [5] provide a large number of BEV maps that contain annotations of drivable areas, which makes it possible to train models for BEV perception using real-world data. Compared with prior methods, we explore the incorporation of self-supervised depth learning explicitly, which provides an important perception output with useful geometric clues for BEV layout estimation. Dwivedi et al. [13] also conduct explicit depth estimation but just take it as an intermediate process to model 3D geometry rather than a joint learning perception task. Besides, prior works usually predict different semantic categories separately, while JPerceiver exploits the synergy of different semantics and predicts the layouts of all categories simultaneously.

MTL-Based Environment Perception. Recently, some multi-task learning works [11,12,24,38,41,55,59,63] propose to combine related perception tasks with depth estimation and VO to exploit complementary information such as segmentation [24,38,41] and optical flow [12,55,63], which effectively boost the network performance. However none of them tackles the scale ambiguity problem of monocular depth and VO via multi-task learning, which is one of our key purposes in this paper.

3 Method

3.1 Overview of JPerciver

As shown in Fig. 2, JPerceiver consists of three networks for depth, pose and layout, respectively, which are all based on the encoder-decoder architecture. The depth network aims to predict the depth map D_t of the current frame I_t, where each depth value indicates the distance between a 3D point and the camera. And the goal of the pose network is to predict the pose transformations $T_{t\to t+m}$ between the current frame and its adjacent frames I_{t+m}. The layout network targets to estimate the BEV layout L_t of the current frame, i.e. semantic occupancy of roads and vehicles in the top-view Cartesian plane. The three

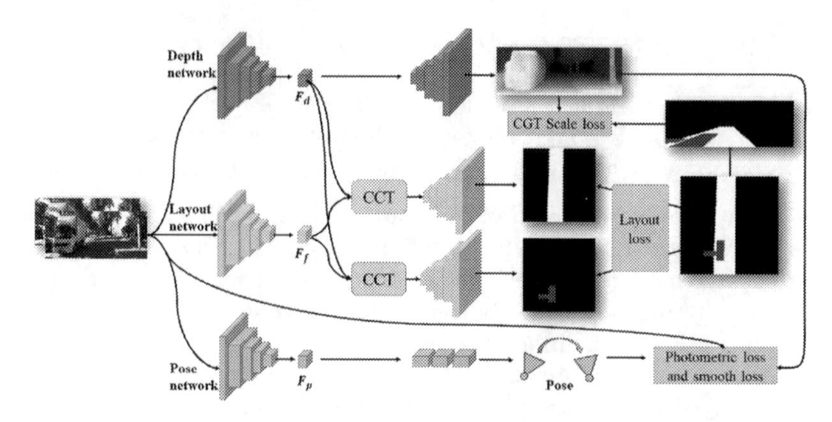

Fig. 2. JPerceiver consists of three networks for depth, pose and layout estimation, and is trained in the end-to-end manner. F_d, F_f, F_p represent the feature learned for three tasks and CCT denotes the cross-view and cross-modal transfer module.

networks are jointly optimized during training. The overall objective function consists of the loss items of all the three tasks and can be formulated as:

$$\ell_{total} = \ell_{dp} + \ell_{layout}, \tag{1}$$

where ℓ_{dp} is the loss of depth and VO estimation in the self-supervised learning scheme, and ℓ_{layout} is the loss of the layout estimation task. We explain the details of ℓ_{dp} and ℓ_{layout} in Sec. 3.2 and Sec. 3.3, respectively.

3.2 Self-supervised Depth Estimation and VO

We adopt two networks to predict depth and pose, respectively, which are jointly optimized using the photometric loss and the smoothness loss in a self-supervised manner, following the baseline method [17]. We additionally devise a CGT scale loss to address the scale ambiguity problem of monocular depth and VO estimation. We describe the loss items of depth and pose networks in this section.

Self-supervised monocular depth and pose estimation is achieved by leveraging the geometry consistency among continuous frames. During training, the depth D_t of current frame I_t and the poses $\{T_{t\to t-1}, T_{t\to t+1}\}$ between I_t and its adjacent frames $\{I_{t-1}, I_{t+1}\}$ are used to obtain the reconstructed current frames $\{\hat{I}_{t-1\to t}, \hat{I}_{t+1\to t}\}$ via the differentiable warping function ω from $\{I_{t-1}, I_{t+1}\}$:

$$\hat{I}_{t+m\to t} = \omega(KT_{t\to t+m}D_tK^{-1}I_{t+m}),\ m \in \{-1, 1\}. \tag{2}$$

Then, the photometric differences between I_t and its reconstructed counterparts $\hat{I}_{t-1\to t}, \hat{I}_{t+1\to t}$ are minimized to train the depth and pose networks. We quantify the photometric differences using the SSIM and L1 losses:

$$\ell_{ph} = \min_{m\in\{-1,1\}} \frac{\alpha(1 - SSIM(I_t, \hat{I}_{t+m\to t}))}{2} + (1-\alpha)|I_t - \hat{I}_{t+m\to t}|, \tag{3}$$

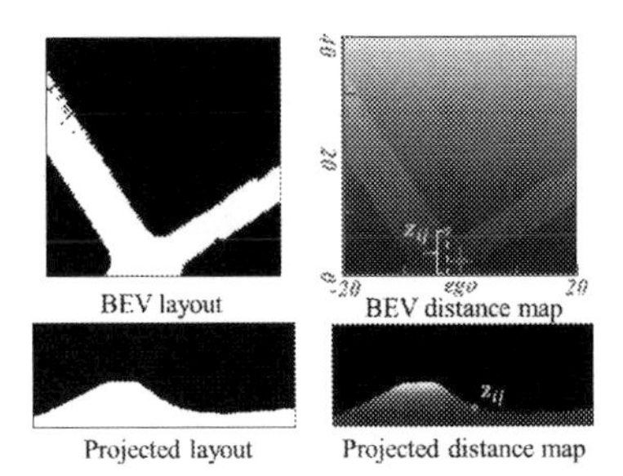

Fig. 3. The demonstration of CGT scale loss.

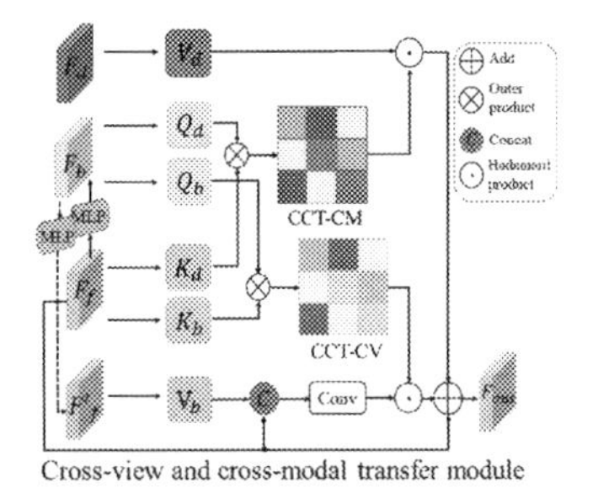

Fig. 4. The structure of cross-view and cross-modal transfer (CCT) module.

where α is set to 0.85. Following our baseline method [17], we also take the per-pixel minimum of the photometric loss and adopt the auto-masking strategy.

To overcome the discontinuity of the predicted depth map, a smoothness loss [16,17] is adopted based on the gradient of I_t:

$$\ell_{sm} = |\partial_x \mu_{D_t}| e^{-|\partial_x I_t|} + |\partial_y \mu_{D_t}| e^{-|\partial_y I_t|}, \tag{4}$$

where μ_{D_t} denotes the normalized inverse depth. By minimizing the above losses, the depth network and the pose network can be optimized simultaneously.

CGT Scale Loss. To accomplish scale-aware environment perception, we propose the cross-view geometric transformation-based (CGT) scale loss for depth estimation and VO by utilizing the scale information in the BEV layout. Since BEV layouts demonstrate the semantic occupancy in the BEV Cartesian plane, covering the range of Z meters in front of the ego vehicle and horizontally covering $\frac{Z}{2}$ meters to the left and right, respectively. It provides a natural distance field z with each pixel having a metric distance value z_{ij} with respect to the ego vehicle, as shown in Fig. 3. By assuming that the BEV plane is a flat plane on the ground with its origin just below the origin of the ego vehicle coordinate system, the BEV plane can then be projected to FV using the camera extrinsic parameters via a homography transformation:

$$H_{bev}^{cam} = K\, T_{ego}^{cam}\, T_{bev}^{ego}, \quad T_{ego}^{cam} \; and \; T_{bev}^{ego} \in \mathcal{SE}(3), \tag{5}$$

where cam, ego, bev represent the camera coordinate system, the ego vehicle coordinate system and the BEV ground system, respectively. T_{bev}^{ego} and T_{ego}^{cam} are the $\mathcal{SE}(3)$ transformations that transform the BEV plane coordinate system to the ego-vehicle coordinate system and then transform to the camera coordinate system, respectively. Therefore, the BEV distance field z can be projected into FV as z^{fv} as shown in Fig. 3, which is then utilized to regulate the predicted depth d, leading to our proposed CGT scale loss:

$$\ell_{CGT} = \frac{1}{h_d w_d} \sum_{j=1}^{h_d} \sum_{i=1}^{w_d} \frac{|z_{ij}^{fv} - d_{ij}|}{z_{ij}^{fv}}, \quad if \; z_{ij}^{fv} \neq 0. \tag{6}$$

To learn the scale-aware depth and pose in a self-supervised manner, we take the weighted sum of ℓ_{ph}, ℓ_{sm}, and ℓ_{CGT} as the final depth-and-pose objective:

$$\ell_{dp} = \ell_{ph} + \ell_{sm} + \beta \cdot \ell_{CGT}, \tag{7}$$

where β is a hyper-parameter and set to 0.1 empirically.

3.3 BEV Layout Estimation

For layout estimation, an encoder-decoder network structure is adopted, following the prior work [54]. It is noteworthy that we use one shared encoder as the feature extractor and different decoders to learn BEV layouts of different semantic categories simultaneously instead of training networks for each category individually as in prior works [31,36,54]. In addition, a CCT module is designed to strengthen feature interaction and knowledge transfer between tasks, and impose 3D geometry information for the spatial reasoning in BEV. To regularize the layout network, we combine various loss items to form a hybrid loss and achieve a balanced optimization for different categories.

CCT Module. To enhance feature interaction and impose 3D geometry information for BEV perception, we devise CCT to investigate the correlation between the FV feature F_f, the BEV layout feature F_b, the retransformed front feature F_f' and the FV depth feature F_d, and refine the layout feature accordingly, as shown in Fig. 4. We describe CCT in two parts, i.e. CCT-CV and CCT-CM for the cross-view module and the cross-modal module, respectively. CCT is inspired by prior work [54], but uses different structures and modal information. In CCT, F_f and F_d are extracted by the encoders of the corresponding perception branches, while F_b is obtained by transforming F_f to BEV with a view projection MLP, which is then re-transformed to F_f' using the same MLP constrained by a cycle loss, following prior work [54]. All the features are set to the same size, i.e., $F_f, F_d, F_b, F_f' \in \mathbb{R}^{H \times W \times D}$. In CCT-CV, a cross-attention mechanism is used to discover the geometry correspondence between FV and BEV features, which is then utilized to guide the FV information refinement and prepared for BEV reasoning. To fully exploit the FV image features, F_b and F_f are projected to patches $Q_{b_i} \in Q_b (i \in [1, ..., HW])$ and $K_{bi} \in K_b (i \in [1, ..., HW])$, acting as the Query and Key respectively. Then each location in the FV will retrieve the correlation from every location in BEV to form a correlation matrix:

$$C_b = \frac{Q_b K_b^T}{\sqrt{D}} \in \mathbb{R}^{H \times W \times H \times W}, \tag{8}$$

where the normalization factor $\frac{1}{\sqrt{D}}$ is used to restrict the value range. The cross-view correspondence can be identified by finding the location with the largest correlation value in C_b, which is differentiable by using the softmax operation:

$$M_b = softmax(C_b) \in \mathbb{R}^{H \times W \times H \times W}. \tag{9}$$

Since F'_f is obtained by first transforming the FV features to BEV and then back to FV, it contains both FV and BEV information. We thus project F'_f to patches $V_{bi} \in V_b (i \in [1, ..., HW])$ and concatenate it with F_f to provide the Value for the cross-view correlation after a convolution layer:

$$F_{cv} = Conv(Concat(F'_f, V_b)) \odot M_b. \tag{10}$$

Except for utilizing the FV features, we also deploy CCT-CM to impose 3D geometry information from F_d. Since F_d is extracted from FV images, it is reasonable to use F_f as the bridge to reduce the cross-modal gap and learn the correspondence between F_d and F_b. Thus, similar to CCT-CV, F_b and F_f are regarded as the Query Q_d and the Key K_d to calculate the correlation matrix:

$$C_d = \frac{Q_d K_d^T}{\sqrt{D}} \in \mathbb{R}^{H \times W \times H \times W}. \tag{11}$$

F_d plays the role of Value so that we can acquire the valuable 3D geometry information correlated to the BEV information and further improve the accuracy of layout estimation. The final CCT-CM feature F_{cm} is then derived as:

$$\begin{aligned} M_d &= softmax(C_d) \in \mathbb{R}^{H \times W \times H \times W}, \\ F_{cm} &= F_d \odot M_d. \end{aligned} \tag{12}$$

In the end, the original FV feature F_f, the cross-view correlated feature F_{cv} and the cross-modal correlated feature F_{cm} are summed up as the input of the layout decoder branch to conduct subsequent learning: $F_{out} = F_f + F_{cv} + F_{cm}$.

Hybrid Loss. BEV layout estimation is a binary classification problem for each semantic category in the BEV grid to determine whether an area belongs to roads, cars, or backgrounds. Thus, this task is usually regularized by minimizing the difference $Diff(\cdot)$ between the predictions L_{pred} and the ground truth L_{gt} using Cross-Entropy (CE) or L2 loss in prior works [20,54]:

$$\ell_{layout}^c = Diff(L_{pred} - L_{gt}), \; c \in \{c_{road}, c_{vehicle}\}. \tag{13}$$

In the process of exploring our joint learning framework to predict different layouts simultaneously, we observe that a great difference exists in the characteristics and distributions of different semantic categories. For characteristics, the road layout in driving scenes usually needs to be connected, while different vehicles instead must be separated. And for distributions, more scenes with straight roads are observed than scenes with turns, which is reasonable in real-world datasets. Such difference and imbalance increase the difficulty of BEV layout learning, especially for predicting different categories jointly, due to the failure of the simple CE or L1 losses in such circumstances. Thus, we incorporate several kinds of segmentation losses including the distribution-based CE loss, the region-based IoU loss, and the boundary loss into a hybrid loss to predict the layout for each category. First, the Weighted Binary Cross-Entropy Loss is adopted, which is most commonly used in semantic segmentation tasks:

$$\ell_{WBCE} = \frac{1}{h_l w_l} \sum_{n=1}^{h_l w_l} \sum_{m=1}^{M} -w_m [y_n^m \cdot log x_n^m + (1 - y_n^m) \cdot log(1 - x_n^m)], \quad (14)$$

where x_n and y_n denote the n-th predicted category value and the counterpart ground truth in a layout of size $h_l w_l$, respectively. $M = 2$ means whether a pixel belongs to foreground roads or cars, while w_m is the hyperparameter for tackling the issue of sampling imbalance between different labels. We set w_m to 5 and 15 for roads and vehicles respectively following [54].

Since the CE loss treats each pixel as an independent sample and thus neglects the interactions between nearby pixels, we then adopt the Soft IoU Loss ℓ_{IoU} to ensure the connectivity within the region:

$$\ell_{IoU} = -\frac{1}{M} \sum_{m=1}^{M} \frac{\sum_{n=1}^{h_l w_l} x_n^m y_n^m}{\sum_{n=1}^{h_l w_l} (x_n^m + y_n^m + x_n^m y_n^m)}, \quad M = 2. \quad (15)$$

For the integrity of the region edges, we further use the Boundary Loss ℓ_{Bound} to constrain the learning of the boundary predictions, which has proven effective for mitigating issues related to regional losses in highly unbalanced segmentation problems such as medical image segmentation [22,28,53]. It is calculated as the Hadamard product of the signed distance (SDF) map of the ground truth layout:

$$\ell_{Bound} = M_{SDF}(L_{gt}) \odot L_{pred}, \quad (16)$$

$$M_{SDF}(L_{gt}) = \begin{cases} -\inf\limits_{y \in L_{gt}^b} ||x - y||_2, x \in L_{gt}^{in}, \\ +\inf\limits_{y \in L_{gt}^b} ||x - y||_2, x \in L_{gt}^{out}, \\ 0, x \in L_{gt}^b, \end{cases}$$

where $L_{gt}^{in}, L_{gt}^{out}$, and L_{gt}^b represent regions inside, outside and at the foreground object boundaries in the ground truth, respectively. $||x - y||_2$ is the Euclidian distance between x and y.

The final loss of the layout estimation for each category then reads:

$$\ell_{layout}^{c_i} = \ell_{WBCE}^{c_i} + \lambda \cdot \ell_{IoU}^{c_i} + \lambda \cdot \ell_{Bound}^{c_i}, \quad (17)$$

where $\lambda = 20$ and $c_i \in \{c_{road}, c_{vehicle}\}$.

Different from prior works [36,54], our joint learning framework predict all semantic categories simultaneously instead of training a network separately for each category. The final optimization loss for our layout network reads:

$$\ell_{layout} = \sum_{c_i} \ell_{layout}^{c_i}, \; c_i \in \{c_{road}, c_{vehicle}\}. \quad (18)$$

4 Experiments

Since there is no previous work that accomplishes depth estimation, visual odometry and BEV layout estimation simultaneously, we evaluate the three tasks on their corresponding benchmarks and compared our method with the SOTA methods of each task. In addition, extensive ablation studies are performed to verify the effectiveness of our joint learning network architecture and loss items.

4.1 Datasets

We evaluate our JPerceiver on three driving scene datasets, i.e., Argoverse [8], Nuscenes [5] and KITTI [15]. Argoverse and Nuscenes are relatively newly published autonomous driving datasets that provide high-resolution BEV semantic occupancy labels for roads and vehicles. We evaluate the performance of BEV layout estimation on Argoverse with 6,723 training images and 2,418 validation images within the range of 40 m $\times$ 40 m, and on Nuscenes with 28,130 training samples and 6,019 validation samples under two settings (in Supplementary). The ablation study for layout estimation is performed on Argoverse. Two semantic categories are included in the evaluated BEV layouts, i.e. roads and vehicles. We adopt the mean of Intersection-over-Union (mIoU) and Average Precision (mAP) as the evaluation metrics, following prior works [31,54]. Due to the insufficient annotations in KITTI, we follow prior works to use three splits for the tasks, i.e., the KITTI Odometry split (15,806 and 6,636 items for training and validation) for depth, VO, as well as road layout estimation, the KITTI Raw split (10,156 and 5,074 items for training and validation) for road layout estimation, and the KITTI 3D Object split (3,712 and 3,769 items for training and validation) for vehicle layout estimation, all within the range of 40 m $\times$ 40 m.

Implementation Details. We adopt the encoder-decoder structure for the three networks, all using pre-trained ResNet18 [19] as the encoder backbone except that we modify the pose network to take two-frame pairs as input. Following the prior method [54], the input sizes of the depth and layout networks are both set to 1024 $\times$ 1024, while a smaller input size 192 $\times$ 640 is used for the pose network to save computation resources since its outputs are not pixel-wise. Our model is implemented in PyTorch [35] and trained for 80 epochs using Adam [23], with a learning rate of 10^{-4} for the first 50 epochs and 10^{-5} for the remaining epochs. The details of the network structure are presented in Supplementary. And there is a potential promising improvement using more powerful network backbones (e.g. transformer [51,57]) in our method (Fig. 2).

4.2 Layout Estimation

Argoverse. We first quantitatively compare JPerceiver with SOTA methods on Argoverse [8], including VPN [34], Monolay [31] and PYVA [54]. We compare two variants of our method, i.e., (1) JPerceiver("1-1") that trains two layout estimation networks separately for each category, which is also the common practice in the compared methods, and (2) JPerceiver("1-2") that predicts the

Table 1. Quantitative comparisons (top part) and ablation study (bottom part) on Argoverse [8]. "CCT-CV" and "CCT-CM" denote the cross-view and cross-modal part in CCT.

Methods	Argoverse Road		Argoverse Vehicle	
	mIoU(%)	mAP(%)	mIoU(%)	mAP(%)
VED [26]	72.84	78.11	24.16	36.83
VPN [34]	71.07	86.83	16.58	39.73
Monolay [31]	73.25	84.56	32.58	51.06
PYVA [54]	76.51	87.21	48.48	64.04
JPeceiver("1-1")	**77.86**	**90.59**	**49.94**	**65.44**
JPerceiver("1-2")	**77.50**	**90.21**	**49.45**	**65.84**
Baseline("1-1")	76.66	87.17	46.97	63.36
+CCT-CV	77.76	88.42	49.33	64.05
+CCT-CV+CCT-CM	77.80	89.00	49.39	64.86
Ours("1-1"+CCT+HLoss)	**77.86**	**90.59**	**49.94**	**65.44**
Baseline("1-2")	76.52	86.54	42.77	59.39
+CCT-CV	76.91	87.19	46.46	61.02
+CCT-CV+CCT-CM	77.38	88.40	47.19	61.43
Ours("1-2"+CCT(S)+HLoss)	76.81	89.39	48.06	63.61
Ours("1-2"+CCT+HLoss)	**77.50**	**90.21**	**49.45**	**65.84**

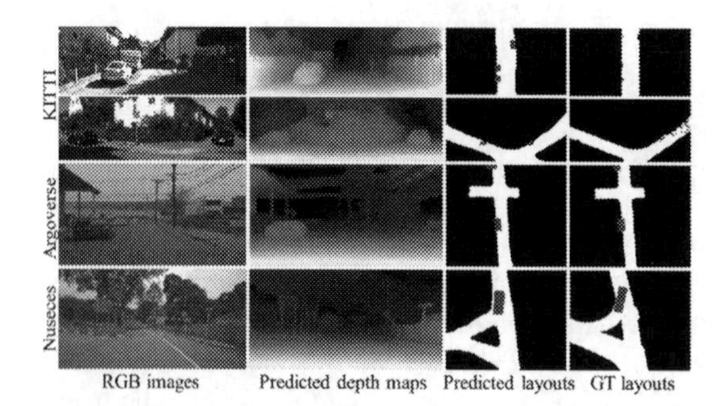

Fig. 5. The qualitative results of depth and layout on the three datasets. White and blue regions indicate road and vehicle layouts. (Color figure online)

Fig. 6. Visualization of features and attention maps aligned with the corresponding views.

two kinds of layouts jointly with one shared encoder. As shown in the top part of Table 1, the two variants both show superiority over other approaches in the road and vehicle layout estimation. Moreover, JPerceiver("1-2") achieves comparable performance with JPerceiver("1-1"), even if JPerceiver("1–2") only uses a shared encoder and thus is more efficient in terms of memory and computation.

Ablation. To investigate the effect of the component and the hybrid loss in our model, we conduct ablation studies on Argoverse and report the results in the bottom part of Table 1. We take a basic "1-1" structure as our baseline, i.e., an encoder-decoder structure trained for each semantic category with a CE loss. We then use CCT and jointly train the depth network, pose network, as well as layout network together. Specifically, we ablate the cross-view module CCT-CV and the cross-modal module CCT-CM, respectively. As can be seen, both the cross-view and the cross-modal modules improve the performance of the baseline, and the complete CCT brings a gain of 1.14% mIoU and 1.83% mAP for road layout estimation and 2.42% mIoU and 1.5% mAP for vehicle layout estimation, respectively. After using the hybrid loss (HLoss), the performance reaches the best, i.e., 77.86% mIoU and 90.59% mAP for roads as well as 49.94% mIoU and 65.44% mAP for vehicles, respectively. We then conduct the same ablation study on the "1-2" structures. It is observed that using one encoder to learn representations for both two semantic layouts significantly decreases the performance of baseline models, especially for the vehicle layout, i.e., from 46.97% mIoU and 63.36% mAP to 42.77% mIoU and 59.39% mAP. After using the proposed CCT module and hybrid loss, the performance drop can be recovered, where the final model achieves comparable results as the "1-1" structure. We further investigate the performance of using a shared CCT module for the two semantic layouts, denoted as "CCT(S)" in Table 1. As can be seen, its results are much worse than using separate CCT for each category. It is probably due to the distinct geometry characteristics of road and vehicle, where different regions should be paid attention to as illustrated in Fig. 6.

KITTI. We train our layout network for roads on the KITTI Odometry and the KITTI Raw splits, and the layout for vehicles on the KITTI Object split. As shown in Table 2 (left), our performance on KITTI Odometry is superior to other methods w.r.t. both mIoU and mAP. The evaluation results on KITTI Raw are listed in Table 2 (middle). We report our reproduced result of PYVA [54] within parentheses because their reported results are trained with processed ground truth. Even so, our method still outperforms their reported results with a gain of 5.39% in mAP. We observe a performance degradation from KITTI Odometry to KITTI Raw, potentially because the ground truth of the latter comes from registered semantic segmentation of Lidar scans while the former obtains the ground truth from the more accurate Semantic KITTI dataset [3], both collected by [31]. For vehicle layout estimation, we show the quantitative results in Table 2 (right). Our method exceeds other works by a large margin, i.e., 2.06% and 6.97% w.r.t. mIoU and mAP.

4.3 Depth Estimation and Visual Odometry

KITTI presents the most commonly used dataset for depth estimation and VO. And we report our scale-aware depth and VO results on KITTI Odometry.

Table 2. Quantitative comparisons of BEV layout estimation results on KITTI Odometry, KITTI Raw and KITTI 3D Object.

Methods	KITTI Odometry road		KITTI raw road		KITTI 3D object	
	mIoU(%)	mAP(%)	mIoU(%)	mAP(%)	mIoU(%)	mAP(%)
VED [26]	65.74	67.84	58.41	66.01	20.45	22.5
Mono3D [9]	–	–	–	–	17.11	26.62
OFT [40]	–	–	–	–	25.24	34.69
VPN [34]	66.81	81.79	59.58	79.07	16.80	35.54
Monolay [31]	76.81	85.25	66.02	75.73	30.18	45.91
PYVA [54]	77.49	86.69	68.34 (65.52)	80.78 (79.52)	38.79	50.26
JPeceiver	**78.13**	**89.57**	66.39	**86.17**	**40.85**	**57.23**

Table 3. Quantitative comparisons and ablation study for depth estimation. "w" and "w/o" denote evaluation results with or without rescaling by the scale factor, which is calculated during inference.

Methods	Resolution	Scaling	Abs Rel (↓)	Sq Rel(↓)	RMSE(↓)	RMSE log(↓)	Scale factor
Monodepth2 [17]	1024×1024	w	**0.113**	0.526	3.656	0.181	42.044 ± 0.076
		w/o	0.976	13.687	17.128	3.754	–
DNet [52]	1024×1024	w	0.121	0.582	3.762	0.192	34.393 ± 0.077
		w/o	0.970	13.528	17.028	3.545	–
JPerceiver	1024×1024	w	0.116	**0.517**	**3.573**	**0.180**	$\mathbf{1.065 \pm 0.071}$
		w/o	**0.112**	**0.559**	**3.817**	**0.196**	–
Baseline	512×512	w	0.120	0.550	3.670	0.184	39.452 ± 0.077
		w/o	0.974	13.616	17.073	0.179	–
+CCT	512×512	w	0.108	0.505	3.574	0.179	37.711 ± 0.067
		w/o	0.973	13.616	17.083	3.645	–
+scale loss	512×512	w	0.135	0.633	3.860	0.194	1.088 ± 0.093
		w/o	0.125	0.643	4.092	0.211	–
Ours	512×512	w	0.128	0.574	3.739	0.189	1.099 ± 0.085
		w/o	0.122	0.628	3.952	0.205	–

Depth Estimation. We compare with several self-supervised depth estimation methods on the KITTI Odometry test set, shown in Table 3. The scaling factor is calculated as the average of all depth map scale ratios, which is the ratio of the median of depth values and the median of ground truth values. We use Monodepth2 [17] as the baseline and DNet [52] as a representative competitor of the scale-aware methods. Though Monodepth2 [17] achieves good up-to-scale accuracy, however, without the scaling factor, its performance significantly degrades. DNet [52] predicts a camera height during inference and calculates the ratio between the ground truth camera height and the predicted one to get the scaling factor. However, its output depths still need to be scaled. Differently, thanks to the CGT scale loss, our depth prediction naturally contains the absolute metric scale and does not require any scaling operation. As shown in the top part

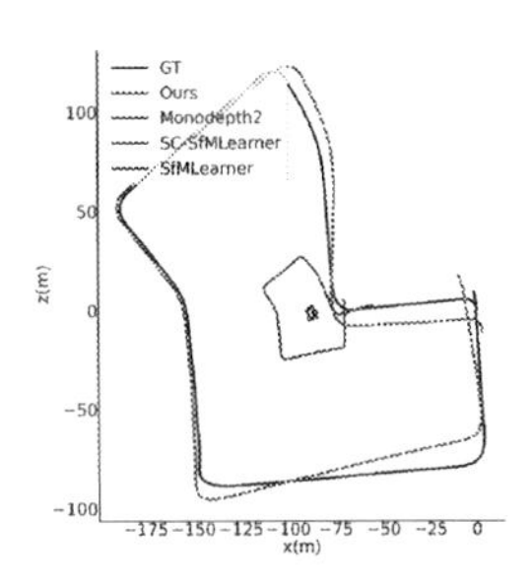

Fig. 7. The comparison of VO trajectories on sequence 07.

Table 4. The comparison of Visual Odometry. t_{err} is the average translational RMSE drift (%) on length from 100, 200 to 800 m, and r_{err} is average rotational RMSE drift (°/100 m) on length from 100, 200 to 800 m.

Methods	Scaling	Sequence 07		Sequence 10	
		t_{err}	r_{err}	t_{err}	r_{err}
SfMLearner [60]	GT	12.61	6.31	15.25	4.06
GeoNet [55]	GT	8.27	5.93	20.73	9.04
Monodepth2 [17]	GT	8.85	5.32	11.60	5.72
SC-Sfmlearner [4]	GT	8.29	4.53	10.74	4.58
Dnet [52]	Camera height	–	–	13.98	4.07
LSR [46]	None	–	–	10.54	4.03
JPerveiver	None	**4.57**	**2.94**	**7.52**	**3.83**

of Table 3, our scale factor computed during inference is 1.065 with a variance of 0.071, while a comparable precision is also achieved.

Ablation. We further conduct the ablation study for depth and report the results in the bottom part of Table 3 with the input resolution of 512 × 512. Introducing CCT on the baseline structure boosts the depth estimation results no matter with or without scaling, and the scaling factor is similar to the baseline counterpart. We then add the CGT scale loss to the baseline to validate its feasibility, resulting in a nearly perfect scaling ratio. However, since our CGT scale loss only takes regional pixels in ground areas into account instead of using all pixels and is based on an assumption that the ground plane is flat, a less accurate result is observed for the overall prediction. Our full model achieves comparable depth performance with the baseline but up to a metric scale. Of note is that all variants with our scale loss obtain lower Abs Rel error without scaling while worse results in other metrics compared with the results with scaling. It may be because the calculation of our Scale loss is the same as the Abs Rel metric. Different ways of calculating and utilizing the scale loss on other baselines might be explored to further improve the estimation accuracy in future work.

Visual Odometry. We train Perceiver on the KITTI Odometry sequences 01–06 and 08–09, and use the sequences 07 and 10 as our test set for evaluating our model for VO. We compare with several self-supervised visual odometry methods in Table 4, including SfMLearner [60], GeoNet [55], Monodepth2 [17], SC-Sfmlearner [4], which are trained on sequence 00–08. "Scaling" means the scaling method is used during inference. "GT" means the scaling factor for correcting the predictions comes from the ground truth. Dnet [52], LSR [46] and Ours all borrow information from the road plane to recover the scale but in different ways. Of note is that Dnet [52] needs invisible ground plane to predict camera height during inference for scale recovery, which is not required by our method. While LSR [46] incorporates a front-view ground plane estimation task, their networks are trained in a serial way, i.e. unscaled depth and VO network, plane segmentation network, and then scaled depth and VO network. In comparison,

Table 5. The Analysis of model complexity with input resolution 512×512 using one single GPU.

Methods	Task	Params(M)	Flops(G)	FPS(BS=1)	FPS(BS=6)
Monodepth2 [17]	depth & pose	39.33	26.93	30.3	36.9
PYVA [54]	layout	29.73	20.42	65.2	72
Monodepth2 [17] +2×PYVA [54]	depth & pose & layout	69.06	47.35	15.6	18.6
JPerceiver	depth & pose & layout	57.15	37.69	19.8	26.9

our method can produce scaled VO results during inference without any hint by using our CGT scale loss. In addition, our method is superior to other competitors w.r.t. the average translational and rotational RMSE drift metrics. The comparison of VO trajectories with other methods without rescaling is shown in Fig. 7, which further proves the effectiveness of our scale loss.

4.4 Model Complexity

We compare the complexity of our JPerciever with the single-task competitors and the simply combined model in Table 5. Since our method can predict the depth, VO, and BEV layouts of the two semantic categories simultaneously, it is less complex in terms of parameters (M), computations (FLOPs), and inference speed (FPS) compared with the combined model while producing better prediction in all the three tasks. Besides, our JPerciever benefits more from parallel acceleration as shown in the last column, where a batch size of 6 is used.

5 Conclusion

In this paper, we propose a joint perception framework named JPerceiver for the autonomous driving scenarios, which accomplishes scale-aware depth estimation, visual odometry, and also BEV layout estimation of multiple semantic categories simultaneously. To realize the joint learning of multiple tasks, we introduce a cross-view and cross-modal transfer module and fully make use of the metric scale from the BEV layout to devise a cross-view geometry transformation-based scale loss to obtain scale-aware predictions. Our method achieves better performance towards all the above three perception tasks using less computation resource and training time. We hope our work can provide valuable insight to the future study of designing more effective joint environment perception model.

Acknowledgement. Ms. Haimei Zhao, Dr. Jing Zhang, and Mr. Sen Zhang are supported by ARC FL-170100117, DP-180103424, IC-190100031, and LE-200100049.

References

1. Translating images into maps. arXiv preprint arXiv:2110.00966 (2021)
2. Badki, A., Gallo, O., Kautz, J., Sen, P.: Binary TTC: a temporal geofence for autonomous navigation. In: Proceedings of the IEEE/CVF Conference on Computer Vision and Pattern Recognition (CVPR), pp. 12946–12955, June 2021
3. Behley, J., et al.: Semantickitti: a dataset for semantic scene understanding of lidar sequences. In: Proceedings of the IEEE/CVF International Conference on Computer Vision, pp. 9297–9307 (2019)
4. Bian, J.W., et al.: Unsupervised scale-consistent depth and ego-motion learning from monocular video. In: Advances in Neural Information Processing Systems (2019)
5. Caesar, H., et al.: nuscenes: a multimodal dataset for autonomous driving. In: Proceedings of the IEEE/CVF Conference on Computer Vision and Pattern Recognition, pp. 11621–11631 (2020)
6. Casas, S., Sadat, A., Urtasun, R.: Mp3: a unified model to map, perceive, predict and plan. In: Proceedings of the IEEE/CVF Conference on Computer Vision and Pattern Recognition (CVPR), pp. 14403–14412, June 2021
7. Chang, M.F., et al.: Argoverse: 3D tracking and forecasting with rich maps. In: Proceedings of the IEEE/CVF Conference on Computer Vision and Pattern Recognition (CVPR), June 2019
8. Chang, M.F., et al.: Argoverse: 3D tracking and forecasting with rich maps. In: Proceedings of the IEEE/CVF Conference on Computer Vision and Pattern Recognition, pp. 8748–8757 (2019)
9. Chen, X., Kundu, K., Zhang, Z., Ma, H., Fidler, S., Urtasun, R.: Monocular 3D object detection for autonomous driving. In: Proceedings of the IEEE Conference on Computer Vision and Pattern Recognition, pp. 2147–2156 (2016)
10. Chen, Z., Zhang, J., Tao, D.: Progressive lidar adaptation for road detection. IEEE/CAA J. Automatica Sinica **6**(3), 693–702 (2019)
11. Chen, Z., Wang, C., Yuan, B., Tao, D.: Puppeteergan: arbitrary portrait animation with semantic-aware appearance transformation. In: Proceedings of the IEEE/CVF Conference on Computer Vision and Pattern Recognition, pp. 13518–13527 (2020)
12. Chi, C., Wang, Q., Hao, T., Guo, P., Yang, X.: Feature-level collaboration: joint unsupervised learning of optical flow, stereo depth and camera motion. In: Proceedings of the IEEE/CVF Conference on Computer Vision and Pattern Recognition (CVPR), pp. 2463–2473, June 2021
13. Dwivedi, I., Malla, S., Chen, Y.T., Dariush, B.: Bird's eye view segmentation using lifted 2D semantic features. In: British Machine Vision Conference (BMVC), pp. 6985–6994 (2021)
14. Fu, H., Gong, M., Wang, C., Batmanghelich, K., Tao, D.: Deep ordinal regression network for monocular depth estimation. In: Proceedings of the IEEE Conference on Computer Vision and Pattern Recognition, pp. 2002–2011 (2018)
15. Geiger, A., Lenz, P., Urtasun, R.: Are we ready for autonomous driving? The kitti vision benchmark suite. In: Proceedings of the IEEE Conference on Computer Vision and Pattern Recognition, pp. 3354–3361. IEEE (2012)
16. Godard, C., Mac Aodha, O., Brostow, G.J.: Unsupervised monocular depth estimation with left-right consistency. In: Proceedings of the IEEE Conference on Computer Vision and Pattern Recognition, pp. 270–279 (2017)
17. Godard, C., Mac Aodha, O., Firman, M., Brostow, G.J.: Digging into self-supervised monocular depth prediction, October 2019

18. Hang, P., Lv, C., Xing, Y., Huang, C., Hu, Z.: Human-like decision making for autonomous driving: a noncooperative game theoretic approach. IEEE Trans. Intell. Transp. Syst. **22**(4), 2076–2087 (2020)
19. He, K., Zhang, X., Ren, S., Sun, J.: Deep residual learning for image recognition. In: Proceedings of the IEEE Conference on Computer Vision and Pattern Recognition, pp. 770–778 (2016)
20. Hu, A., et al.: FIERY: future instance segmentation in bird's-eye view from surround monocular cameras. In: Proceedings of the International Conference on Computer Vision (ICCV) (2021)
21. Huang, C., Lv, C., Hang, P., Xing, Y.: Toward safe and personalized autonomous driving: decision-making and motion control with DPF and CDT techniques. IEEE/ASME Trans. Mechatron. **26**(2), 611–620 (2021)
22. Kervadec, H., Bouchtiba, J., Desrosiers, C., Granger, E., Dolz, J., Ayed, I.B.: Boundary loss for highly unbalanced segmentation. In: International Conference on Medical Imaging with Deep Learning, pp. 285–296. PMLR (2019)
23. Kingma, D.P., Ba, J.: Adam: a method for stochastic optimization. arXiv preprint arXiv:1412.6980 (2014)
24. Klingner, M., Termöhlen, J.-A., Mikolajczyk, J., Fingscheidt, T.: Self-supervised monocular depth estimation: solving the dynamic object problem by semantic guidance. In: Vedaldi, A., Bischof, H., Brox, T., Frahm, J.-M. (eds.) ECCV 2020. LNCS, vol. 12365, pp. 582–600. Springer, Cham (2020). https://doi.org/10.1007/978-3-030-58565-5_35
25. Li, R., Wang, S., Long, Z., Gu, D.: Undeepvo: monocular visual odometry through unsupervised deep learning. In: IEEE International Conference on Robotics and Automation, pp. 7286–7291. IEEE (2018)
26. Lu, C., van de Molengraft, M.J.G., Dubbelman, G.: Monocular semantic occupancy grid mapping with convolutional variational encoder-decoder networks. IEEE Robot. Autom. Lett. **4**(2), 445–452 (2019)
27. Luo, C., Yang, X., Yuille, A.: Self-supervised pillar motion learning for autonomous driving. In: Proceedings of the IEEE/CVF Conference on Computer Vision and Pattern Recognition, pp. 3183–3192 (2021)
28. Ma, J., et al.: How distance transform maps boost segmentation CNNs: an empirical study. In: Medical Imaging with Deep Learning, pp. 479–492. PMLR (2020)
29. Mahjourian, R., Wicke, M., Angelova, A.: Unsupervised learning of depth and egomotion from monocular video using 3D geometric constraints. In: Proceedings of the IEEE Conference on Computer Vision and Pattern Recognition, pp. 5667–5675 (2018)
30. Mallot, H.A., Bülthoff, H.H., Little, J., Bohrer, S.: Inverse perspective mapping simplifies optical flow computation and obstacle detection. Biol. Cybern. **64**(3), 177–185 (1991)
31. Mani, K., Daga, S., Garg, S., Narasimhan, S.S., Krishna, M., Jatavallabhula, K.M.: Monolayout: Amodal scene layout from a single image. In: The IEEE Winter Conference on Applications of Computer Vision, pp. 1689–1697 (2020)
32. McCraith, R., Neumann, L., Vedaldi, A.: Calibrating self-supervised monocular depth estimation. arXiv preprint arXiv:2009.07714 (2020)
33. Nistér, D., Naroditsky, O., Bergen, J.: Visual odometry. In: Proceedings of the 2004 IEEE Computer Society Conference on Computer Vision and Pattern Recognition. CVPR 2004. vol. 1, pp. I-I. IEEE (2004)
34. Pan, B., Sun, J., Leung, H.Y.T., Andonian, A., Zhou, B.: Cross-view semantic segmentation for sensing surroundings. IEEE Robot. Autom.tion Lett. **5**(3), 4867–4873 (2020)

35. Paszke, A., et al.: Automatic differentiation in pytorch (2017)
36. Philion, J., Fidler, S.: Lift, splat, shoot: encoding images from arbitrary camera rigs by implicitly unprojecting to 3D. In: Vedaldi, A., Bischof, H., Brox, T., Frahm, J.-M. (eds.) ECCV 2020. LNCS, vol. 12359, pp. 194–210. Springer, Cham (2020). https://doi.org/10.1007/978-3-030-58568-6_12
37. Phillips, J., Martinez, J., Barsan, I.A., Casas, S., Sadat, A., Urtasun, R.: Deep multi-task learning for joint localization, perception, and prediction. In: Proceedings of the IEEE/CVF Conference on Computer Vision and Pattern Recognition (CVPR), pp. 4679–4689 (June 2021)
38. Ranjan, A., et al.: Competitive collaboration: joint unsupervised learning of depth, camera motion, optical flow and motion segmentation. In: Proceedings of the IEEE Conference on Computer Vision and Pattern Recognition, pp. 12240–12249 (2019)
39. Reading, C., Harakeh, A., Chae, J., Waslander, S.L.: Categorical depth distribution network for monocular 3D object detection. In: Proceedings of the IEEE/CVF Conference on Computer Vision and Pattern Recognition, pp. 8555–8564 (2021)
40. Roddick, T., Kendall, A., Cipolla, R.: Orthographic feature transform for monocular 3D object detection. arXiv preprint arXiv:1811.08188 (2018)
41. Schön, M., Buchholz, M., Dietmayer, K.: MGNet: monocular geometric scene understanding for autonomous driving. In: Proceedings of the IEEE/CVF International Conference on Computer Vision, pp. 15804–15815 (2021)
42. Shu, C., Yu, K., Duan, Z., Yang, K.: Feature-metric loss for self-supervised learning of depth and egomotion. In: Vedaldi, A., Bischof, H., Brox, T., Frahm, J.-M. (eds.) ECCV 2020. LNCS, vol. 12364, pp. 572–588. Springer, Cham (2020). https://doi.org/10.1007/978-3-030-58529-7_34
43. Simond, N., Parent, M.: Obstacle detection from IPM and super-homography. In: 2007 IEEE/RSJ International Conference on Intelligent Robots and Systems, pp. 4283–4288. IEEE (2007)
44. Thavamani, C., Li, M., Cebron, N., Ramanan, D.: Fovea: foveated image magnification for autonomous navigation. In: Proceedings of the IEEE/CVF International Conference on Computer Vision (ICCV), pp. 15539–15548, October 2021
45. Torralba, A., Oliva, A.: Depth estimation from image structure. IEEE Trans. Pattern Anal. Mach. Intell. **24**(9), 1226–1238 (2002)
46. Wagstaff, B., Kelly, J.: Self-supervised scale recovery for monocular depth and egomotion estimation. arXiv preprint arXiv:2009.03787 (2020)
47. Wang, C., Miguel Buenaposada, J., Zhu, R., Lucey, S.: Learning depth from monocular videos using direct methods. In: Proceedings of the IEEE Conference on Computer Vision and Pattern Recognition, pp. 2022–2030 (2018)
48. Wang, H., Cai, P., Fan, R., Sun, Y., Liu, M.: End-to-end interactive prediction and planning with optical flow distillation for autonomous driving. In: Proceedings of the IEEE/CVF Conference on Computer Vision and Pattern Recognition (CVPR) Workshops, pp. 2229–2238 (June 2021)
49. Wang, S., Clark, R., Wen, H., Trigoni, N.: Deepvo: towards end-to-end visual odometry with deep recurrent convolutional neural networks. In: 2017 IEEE International Conference on Robotics and Automation (ICRA), pp. 2043–2050. IEEE (2017)
50. Wang, Y., Mao, Q., Zhu, H., Zhang, Y., Ji, J., Zhang, Y.: Multi-modal 3D object detection in autonomous driving: a survey. arXiv preprint arXiv:2106.12735 (2021)
51. Xu, Y., Zhang, Q., Zhang, J., Tao, D.: Vitae: vision transformer advanced by exploring intrinsic inductive bias. Adv. Neural. Inf. Process. Syst. **34**, 28522–28535 (2021)

52. Xue, F., Zhuo, G., Huang, Z., Fu, W., Wu, Z., Ang, M.H.: Toward hierarchical self-supervised monocular absolute depth estimation for autonomous driving applications. In: 2020 IEEE/RSJ International Conference on Intelligent Robots and Systems (IROS), pp. 2330–2337. IEEE (2020)
53. Xue, Y., et al.: Shape-aware organ segmentation by predicting signed distance maps. In: Proceedings of the AAAI Conference on Artificial Intelligence, vol. 34, pp. 12565–12572 (2020)
54. Yang, W., et al.: Projecting your view attentively: monocular road scene layout estimation via cross-view transformation. In: Proceedings of the IEEE/CVF Conference on Computer Vision and Pattern Recognition, pp. 15536–15545 (2021)
55. Yin, Z., Shi, J.: Geonet: unsupervised learning of dense depth, optical flow and camera pose. In: Proceedings of the IEEE Conference on Computer Vision and Pattern Recognition, pp. 1983–1992 (2018)
56. Zhang, J., Tao, D.: Empowering things with intelligence: a survey of the progress, challenges, and opportunities in artificial intelligence of things. IEEE Internet Things J. **8**(10), 7789–7817 (2020)
57. Zhang, Q., Xu, Y., Zhang, J., Tao, D.: Vitaev2: vision transformer advanced by exploring inductive bias for image recognition and beyond. arXiv preprint arXiv:2202.10108 (2022)
58. Zhang, S., Zhang, J., Tao, D.: Towards scale consistent monocular visual odometry by learning from the virtual world. In: 2022 IEEE International Conference on Robotics and Automation (ICRA) (2022)
59. Zhao, H., Bian, W., Yuan, B., Tao, D.: Collaborative learning of depth estimation, visual odometry and camera relocalization from monocular videos. In: IJCAI, pp. 488–494 (2020)
60. Zhou, T., Brown, M., Snavely, N., Lowe, D.G.: Unsupervised learning of depth and ego-motion from video. In: Proceedings of the IEEE Conference on Computer Vision and Pattern Recognition, pp. 1851–1858 (2017)
61. Zhuang, Z., Li, R., Jia, K., Wang, Q., Li, Y., Tan, M.: Perception-aware multi-sensor fusion for 3D lidar semantic segmentation. In: Proceedings of the IEEE/CVF International Conference on Computer Vision (ICCV), pp. 16280–16290, October 2021
62. Zou, Y., Ji, P., Tran, Q.-H., Huang, J.-B., Chandraker, M.: Learning monocular visual odometry via self-supervised long-term modeling. In: Vedaldi, A., Bischof, H., Brox, T., Frahm, J.-M. (eds.) ECCV 2020. LNCS, vol. 12359, pp. 710–727. Springer, Cham (2020). https://doi.org/10.1007/978-3-030-58568-6_42
63. Zou, Y., Luo, Z., Huang, J.-B.: DF-Net: unsupervised joint learning of depth and flow using cross-task consistency. In: Ferrari, V., Hebert, M., Sminchisescu, C., Weiss, Y. (eds.) ECCV 2018. LNCS, vol. 11209, pp. 38–55. Springer, Cham (2018). https://doi.org/10.1007/978-3-030-01228-1_3

Semi-supervised 3D Object Detection with Proficient Teachers

Junbo Yin[1], Jin Fang[2,3,4], Dingfu Zhou[2,3], Liangjun Zhang[2,3], Cheng-Zhong Xu[4], Jianbing Shen[4(✉)], and Wenguan Wang[5(✉)]

[1] School of Computer Science, Beijing Institute of Technology, Beijing, China
[2] Baidu Research, Beijing, China
[3] National Engineering Laboratory of Deep Learning Technology and Application, Beijing, China
[4] SKL-IOTSC, CIS, University of Macau, Zhuhai, China
[5] ReLER, AAII, University of Technology Sydney, Ultimo, Australia
wenguanwang.ai@gmail.com
https://github.com/yinjunbo/ProficientTeachers

Abstract. Dominated point cloud-based 3D object detectors in autonomous driving scenarios rely heavily on the huge amount of accurately labeled samples, however, 3D annotation in the point cloud is extremely tedious, expensive and time-consuming. To reduce the dependence on large supervision, semi-supervised learning (SSL) based approaches have been proposed. The Pseudo-Labeling methodology is commonly used for SSL frameworks, however, the low-quality predictions from the teacher model have seriously limited its performance. In this work, we propose a new Pseudo-Labeling framework for semi-supervised 3D object detection, by enhancing the teacher model to a proficient one with several necessary designs. First, to improve the recall of pseudo labels, a Spatial-temporal Ensemble (STE) module is proposed to generate sufficient seed boxes. Second, to improve the precision of recalled boxes, a Clustering-based Box Voting (CBV) module is designed to get aggregated votes from the clustered seed boxes. This also eliminates the necessity of sophisticated thresholds to select pseudo labels. Furthermore, to reduce the negative influence of wrongly pseudo-labeled samples during the training, a soft supervision signal is proposed by considering Box-wise Contrastive Learning (BCL). The effectiveness of our model is verified on both ONCE and Waymo datasets. For example, on ONCE, our approach significantly improves the baseline by 9.51 mAP. Moreover, with half annotations, our model outperforms the oracle model with full annotations on Waymo.

Keywords: 3D object detection · Semi-supervised learning · Point cloud

J. Yin and J. Fang—Equal contribution. Work done when J. Yin was an intern at Baidu Research.

S. Avidan et al. (Eds.): ECCV 2022, LNCS 13698, pp. 727–743, 2022.
https://doi.org/10.1007/978-3-031-19839-7_42

1 Introduction

With the rapid development of range sensors (*e.g.*, LiDAR) and their wide application in the field of robotics and autonomous driving, point cloud-based scene understanding such as 3D object detection has received great attention recently. With the great capabilities of deep neural networks (DNNs) and a huge number of annotated samples, impressive performances have been achieved on different public benchmarks [2,8,9,26]. VoxelNet [41], PointRCNN [22], PointPillars [11], PV-RCNN [21] and CenterPoint [38] are several representative 3D object detection frameworks. Nevertheless, the results highly rely on huge annotations, while the annotation in 3D data is extremely expensive and time-consuming, *e.g.*, a skilled worker may take hundreds of hours to annotate just one hour of driving data [7,12–14].

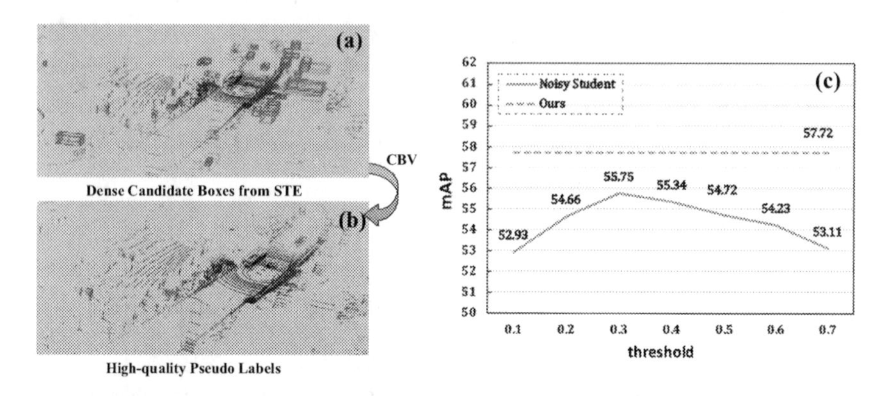

Fig. 1. The main idea of our *ProficientTeachers*. Given an unlabeled point cloud, the spatial-temporal ensemble (STE) module first produces **(a)** seed boxes by combing predictions from multiple augmented views. Then, the clustering-based box voting (CBV) module adaptively aggregates these boxes to get **(b)** the final high-quality pseudo labels. In this way, we not only achieve better detection results, but also remove the necessity of sophisticated thresholds for selecting pseudo labels, as seen in **(c)**.

Semi-supervised learning (SSL) techniques, which train a model with a small number of labeled samples together with an abundance of unlabeled data, are a promising alternative to the fully-supervised learning frameworks. Compared to labeled data, unlabeled data is obtained very conveniently and cheaply. However, due to the inherent difficulty of point cloud (*e.g.*, orderless, textureless, and sparsity), only a few SSL-based 3D object detection frameworks have been proposed. Up to now, SESS [39] and 3DIoUMatch [29] are two pioneers of this domain. To handle the unlabeled data, SESS leverages asymmetric data augmentation and enforces consistency regularization between the predictions of teacher and student models. Although noticeable improvements have been achieved upon a vanilla VoteNet [16] on indoor datasets, other researchers [18,29] found that the consistency regularization is suboptimal if it is uniformly enforced on all the student and teacher predictions because the quality of these predictions may be quite different. To well handle this limitation, 3DIoUMatch seeks a pseudo-labeling approach and applies a confidence-based filtering strategy for pseudo-label selection, where the confidence is defined as a combination of classification score and IoU estimation. Though achieving better performance, it takes tremendous effort to select a suitable confidence score. Moreover, the pseudo labels produced by its plain teacher model limit the final detection performance.

To address these challenges, we propose a new pseudo-labeling SSL framework, *ProficientTeachers*, that not only provides high-quality pseudo labels via an enhanced teacher model, but also reduces the necessity of deliberately selected thresholds. To be specific, false negative (FN) and false positive (FP) in the pseudo labels are two main challenges. The LiDAR point cloud is sparse and noisy, and some street objects are of small sizes (e.g., less than $2\,m$ for pedestrian) and unevenly distributed across a considerably wide range (e.g., $150\,m \times 150\,m$ in Waymo [26]). Thus it is prone to cause FN detections from only a single point cloud view. To handle this, a spatial-temporal ensemble (STE) module is proposed to generate sufficient seed boxes from spatially and temporally augmented views. Aggregating predictions from different views reduces the prediction bias of the teacher model, thus essentially boosting the *recall*. The redundant seed boxes from STE inevitably involve FP detections. To further resolve the FP problem, we propose a clustering-based box voting (CBV) module. Our CBV module groups the seed boxes into different clusters and generates votes (*i.e.*, a refined bounding box) for each box in a cluster. These votes are then aggregated to produce a more accurate box for each cluster. In this way, more clean and precise pseudo labels can be obtained by a simple NMS, removing the need of selecting thresholds. Our CBV module significantly improves the *precision* of pseudo labels produced by the STE module, without losing the *recall*. By equipping the STE and CBV modules, the vanilla teacher model has become to be proficient teachers model. Furthermore, we find that the original pseudo-labeling method enforces a hard training target, where the inaccurate pseudo labels will undermine the performance of the student. To alleviate this problem, a soft training target is proposed by box-wise contrastive learning (BCL), which aims to learn the cross-view feature consistency based on the informative boxes.

To summarize, we propose a new 3D SSL framework, *ProficientTeachers*, for LiDAR-based 3D object detection, which is achieved by promoting the plain teacher model to proficient teachers inspired by ensemble learning. Our framework not only performs better results, but also removes the necessity of confidence-based thresholds for filtering pseudo labels. In our model, spatial-temporal ensemble (STE) and clustering-based box voting (CBV) modules are developed to improve the recall and precision of pseudo labels. Furthermore, a box-wise contrastive learning (BCL) strategy has been advocated to explore representation learning based on expressive 3D boxes. Comprehensive evaluations have been conducted on ONCE and Waymo. Our *ProficientTeachers* can improve the baseline detector by 9.51 mAP on ONCE, and save half annotations on Waymo.

2 Related Work

3D Object Detection. 3D object detection from LiDAR point cloud has been studied for decades with various approaches being proposed. The mainstream methodologies can be divided into two categories: voxel-based [4,35,36,38,41] and point-based [22,23,34,40]. Voxel-based expression is a popular way of processing the point cloud in deep learning, and has been widely applied for 3D object detection with the development of sparse convolution [33]. Voxelnet [41] splits the LiDAR data into voxels and sends the points in each voxel into a voxel feature encoding layer to get voxel-wise features. To accelerate the speed, PointPillars [11] uses a pillar representation to replace the voxel representation. Different from the above approaches, the point-based approaches take the point cloud directly into the DNNs. PointRCNN [22] is representative of this, which employs the Pointnet++ [17] as the backbone for semantic segmentation first, and then regions of interest (RoIs) are generated based on foreground points. Besides this, PV-RCNN [21] proposes to combine both point cloud and multi-scale voxel representation and achieve high performance.

Semi-supervised Learning (SSL). SSL has been studied for a long time and many approaches have been proposed [28,42]. The recently popular SSL approaches such as Temporal Ensembling [19], Mean Teacher [27] and Noisy Student [31] have achieved impressive performance on the 2D tasks. The temporal model, which explores consistency in the prediction level, tries to minimize the difference between the predictions from the current step and the EMA (an exponential moving average) predictions over multiple previous training epochs. The EMA predictions can largely improve the quality of the predictions. Mean Teacher [27] further improves the temporal model by replacing network prediction average with network parameter average. It contains two network branches, *i.e.*, teacher and student, with the same architecture. The parameters of the teacher are the EMA of the student, while the parameters of the student are updated by stochastic gradient descent. The student network is trained to yield consistent predictions with the teacher network. Noisy Student [31] deliberately injects noise to the student model from both input level and model level to strengthen the student. Then it makes the student a new teacher and performs iterative self-training to further improve performance.

SSL on 3D Object Detection. Most of the SSL approaches are proposed for classification tasks [1,24] and a few SSL approaches have been proposed to leverage the object detection task [10,25], especially 3D object detection. SESS [39] and 3DIoUMatch [29] are two typical recently proposed approaches for 3D object detection from point cloud data. The SESS, which is built based on the Mean Teacher paradigm by updating the parameters of the teacher network with an EMA technique, employs asymmetric data augmentation and enforces three kinds of consistency losses between the teacher and student predictions. Different from SESS, 3DIoUMatch proposes a series of handcrafted strategies to achieve better pseudo labels such as joint class, objectness, localization confidences-based pseudo-label filtering, and IoU-guided lower-half suppression for deduplication, etc. The proposed framework works well on PV-RCNN [21]. By contrast, our method benefits from an enhanced teacher model, without sophisticated thresholds selection, and can be flexibly applied on popular LiDAR-based 3D detectors, *e.g.*, SECOND [33], CenterPoint [38] and PV-RCNN [21].

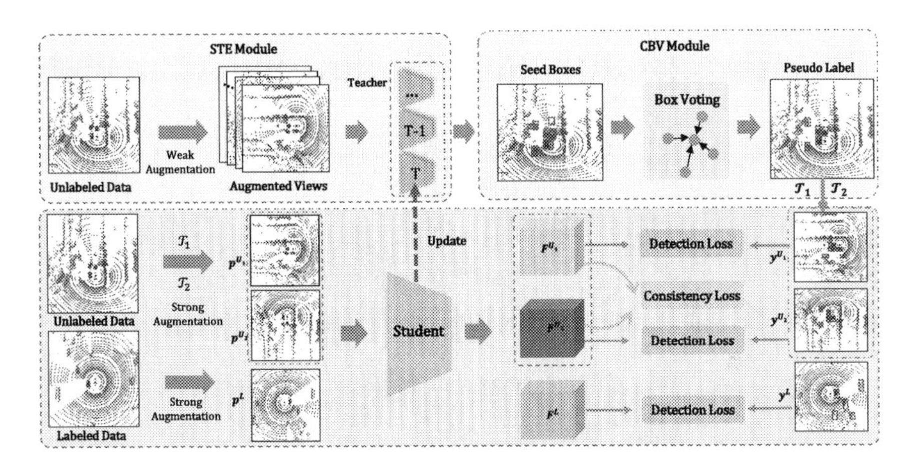

Fig. 2. The framework of our *ProficientTeachers* model. It consists of a spatial-temporal ensemble (STE) module and a clustering-based box voting (CBV) module. STE produces sufficient seed boxes and CBV adaptively fuses them to obtain better pseudo labels. Besides, a consistency loss based on box-wise contrastive learning (BCL) is applied on the student model to explicitly learn from informative box features.

3 Proposed Approach

3.1 SSL Framework

Semi-supervised 3D object detection is very valuable and practical in real self-driving scenarios. Here, we explore the pseudo-labeling framework, which is a popular methodology in semi-supervised learning. Typical pseudo-labeling methods [31] exploit a teacher model to produce pseudo labels which are used to supervise a student model. We argue that the low-quality predictions produced by the vanilla teacher model limit the performance. In this work, we propose to promote the vanilla teacher model to a proficient one, as seen in Fig. 2.

Our *ProficientTeachers* model, as described in Sect. 3.2, contains a spatial-temporal ensemble (STE) module and a clustering-based box voting (CBV) module, which are designed to handle false negative (FN) and false positives (FP) in pseudo labels, respectively. In particular, STE produces multi-group boxes based on augmented and assembled views to recall the missed objects. To further remove the redundant FP boxes, the original pseudo-labeling approaches apply a fixed threshold to filter out boxes with lower confidence scores, which is unstable and inefficient. In contrast, our CBV adaptively aggregates the seed boxes to reduce FP and also improves the precision by voting the boxes within a cluster. Furthermore, we propose a soft supervision signal in Sect. 3.3, *i.e.*, *Contrastive Student* model, by involving box-wise contrastive learning (BCL). Next, we elaborate on each module in subsequent sections.

3.2 Proficient Teachers Model

Assuming that we have total N training samples, including N_l labeled samples $\mathcal{P}^L = \{\mathbf{p}_i^L, \mathbf{y}_i^L\}_{i=1}^{N_l}$ and N_u unlabeled samples $\mathcal{P}^U = \{\mathbf{p}_i^U\}_{i=1}^{N_u}$, where $\mathbf{p}_i \in \mathcal{R}^{n\times\{3+r\}}$ represents a point cloud sample $\mathbf{p}_i$ that has n points with 3-dimensional coordinates and other r-dimensional attributes such as intensity, timestamps, etc. The core idea of our *ProficientTeachers* model is to produce high-quality pseudo labels $\{\mathbf{y}_i^U\}_{i=1}^{N_u}$ for the unlabeled point clouds $\{\mathbf{p}_i^U\}_{i=1}^{N_u}$.

Spatial-Temporal Ensemble Module. As described above, predictions produced by vanilla teacher inevitably encounter the FN problem, which will cause the student model to treat a foreground object as a negative example, thus degenerating the detection performance. Motivated from this observation, we attempt to improve the recall of the teacher model with the STE module, which generates sufficient candidate pseudo boxes with a spatial ensemble module and a temporal ensemble module.

The spatial ensemble in the STE aims to produce multi-group detections based on differently augmented point cloud views, which is inspired by test time augmentation [20]. Detections on a certain view may exist bias and miss objects. By combining detections from multiple views after reverse transformation, less FN will happen, thus improving the recall. Formally, given an unlabeled data $\mathbf{p}_i^U \in \mathcal{P}^U$, we first apply *fixed* data augmentation to spatially transform the input point cloud into different views. This can be deemed as a form of weak augmentation compared with the *random* train-time augmentation policies, which is:

$$\{\mathbf{p}_i^{U_1}, \mathbf{p}_i^{U_2}, \dots, \mathbf{p}_i^{U_K}\} = \mathcal{T}(\mathbf{p}_i^U), \tag{1}$$

where $\mathcal{T}(\cdot)$ is the augmentation function and K is the number of augmented views. More details about our data augmentation can be found in Sect. 4.2.

Then, we use the teacher network $f_{\text{TEA}}(\cdot)$ to give predictions for each view:

$$\{\mathbf{y}_i^{U_1}, \mathbf{y}_i^{U_2}, \dots, \mathbf{y}_i^{U_K}\} = f_{\text{TEA}}(\{\mathbf{p}_i^{U_1}, \mathbf{p}_i^{U_2}, \dots, \mathbf{p}_i^{U_K}\}), \tag{2}$$

With the spatial ensemble module, objects not detected in one view may be detected in another view. All detection results are then projected to the original coordinate with reverse transformations.

Moreover, we can also incorporate a temporal ensemble module. In particular, previous pseudo label methods update weight of teacher model by student model from a certain epoch. We argue that models from different epochs exist bias, and temporal aggregation of models from different epochs can alleviate this problem. This is presented as:

$$\{\mathbf{y}_i^{U_1^T}, \mathbf{y}_i^{U_2^T}, \ldots, \mathbf{y}_i^{U_K^T}\} = f_{\mathrm{TEA}}^{T}(\{\mathbf{p}_i^{U_1}, \mathbf{p}_i^{U_2}, \ldots, \mathbf{p}_i^{U_K}\}), \tag{3}$$

where $f_{\mathrm{TEA}}^{T}(\cdot)$ denotes the teacher model updated by student model from epoch T. We consider the models from recent epochs for temporal ensemble. Afterward, we collect all the predictions produced by both the spatial and temporal ensemble modules. As seen in Table 1, our STE improves the recall by 5.6%, compared with the original teacher model. Here, simple post-processing like Non-maximum Suppression (NMS) can be directly applied to aggregate these predictions.

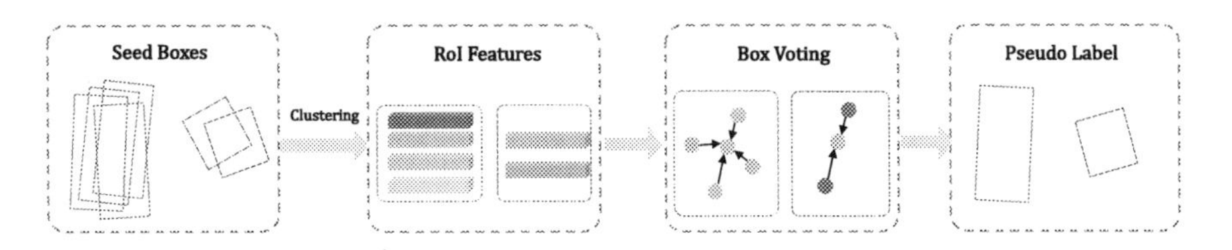

Fig. 3. Illustration of clustering-based box voting (CBV) module, which contains four steps: seed box clustering, RoI feature extraction, box voting and votes aggregation. By adaptively fusing the high-recall candidate boxes output by the STE, CBV significantly improves the precision of pseudo labels.

Clustering-Based Box Voting Module. Though the proposed STE module has ensured a higher recall by combing sufficient pseudo boxes, some inaccurate candidate boxes will also lead to a lower precision (*i.e.*, from 47.2% to 27.4%, as seen in Table. 1), which undermines the quality of pseudo labels. Thus, the main purpose of the CBV module is to address these redundant candidate boxes.

An intuition way is to define a score to formulate the quality of the pseudo boxes, and filter the low-quality boxes under a certain threshold. For instance, FixMatch [24] directly uses the classification score to filter the pseudo boxes. Later, 3DIoUMatch [29] leverages a predicted IoU score for filtering objects. However, there are potential limitations in these threshold-based filtering methods. First of all, it requires manual experience and elaborate experiments to choose a suitable threshold for a class, and different datasets or detectors might not share the same thresholds. Secondly, the predicted box confidence score either formulates class probability, *e.g.*, classification score, or the localization accuracy, *e.g.*, IoU, individually, which requires

Table 1. Recall and precision of "Vehicle" class based on SECOND detector trained with different SSL settings. "c" represents the threshold of the confidence score.

Methods	Recall	Precision
SECOND [33] (Baseline)	78.3%	47.2%
SECOND [33] + STE	83.9%	27.4%
SECOND [33] + STE (c = 0.3)	80.4%	74.6%
SECOND [33] + STE + CBV	83.6%	76.5%

a sophisticated combination from different aspects. Thirdly, they separately consider each individual box, ignoring the relations of boxes in a cluster covering the same ground truth. To this end, we propose a Clustering-based Box Voting (CBV) module to obtain more clean and accurate pseudo labels via a learnable voting process, as well as eliminate the need of sophisticated threshold selection.

The detailed pipeline of the CBV module can be found in Fig. 3. Specifically, given the pseudo labels produced by the STE module, we first cluster these boxes based on the IoU criterion. For example, the boxes are first arranged by the scores. Then, for the box with the highest score, we select other boxes that have a larger IoU with it as the same cluster. This process is iteratively performed for all the boxes to produce all the clusters. Assuming the k-th cluster has $M^{'}$ boxes, *i.e.*, $\{\mathbf{b}_m^k\}_{m=1}^{M^{'}}$, we aim to aggregate these boxes and get a refined box $\mathbf{b}^k$ via box voting and votes aggregation. For box voting, we first obtain features of each box via a pre-trained RoI network [5], *i.e.*, $\tilde{\mathbf{b}}_m^k = f_{\mathrm{RoI}}(\mathbf{b}_m^k)$. Then, we let each box in a cluster predict a vote, based on the context-aware features of the box. The voting network is realized by a shared-weight two-layer MLP, and is trained to regress the offset to the ground-truth box. This can be denoted as $\mathbf{v}_m^k = f_{\mathrm{Vote}}(\tilde{\mathbf{b}}_m^k)$. In this way, we have the vote results $\{\mathbf{v}_m^k\}_{m=1}^{M^{'}}$, where each vote $\mathbf{v}_m^k$ presents a refined bounding box. Furthermore, we also predict an objectness $\mathbf{s}_m^k$ to describe the quality of a box. This is achieved by training another two-layer MLP head for foreground and background classification. Accordingly, we can aggregate these votes to get a more accurate box $\mathbf{b}^k$ for the k-th cluster, by weighting the votes $\mathbf{v}_m^k$ with the objectness $\mathbf{s}_m^k$, which is denoted as:

$$\mathbf{b}^k = \frac{\sum_{m=1}^{M^{'}} \mathbf{s}_m^k \cdot \mathbf{v}_m^k}{\sum_{m=1}^{M^{'}} \mathbf{s}_m^k}, \tag{4}$$

In this way, more clean pseudo labels are produced. The objectness $\mathbf{s}^k$ is also obtained by considering the average scores and the detection number in this cluster. Then, NMS can be directly applied to get the final pseudo labels, *i.e.*, $\mathbf{y}^U = \{\mathbf{b}^k\}_{k=1}^{M}$, where M is the total pseudo box number.

In Table 1, we find that the CBV module significantly improves the precision from 27.4% to 76.5%, and even improves the recall by 5.3%, due to the votes aggregation in clusters. We also implement a heuristic method by selecting better confidence threshold, *i.e.*, $c = 0.3$, to filter boxes as in [24]. As seen in Table. 1, such a method is inferior to ours according to both recall and precision metrics.

3.3 Contrastive Student Model

In this section, we introduce the learning of the student model. Concretely, our student model takes as input both the labeled data $\{\mathbf{p}^L, \mathbf{y}^L\} \in \mathcal{P}^L$ and the unlabeled data $\mathbf{p}^U \in \mathcal{P}^U$. For training the labeled data $\mathbf{p}^L$, we directly obtain the predictions of the student model, and optimize with the detection loss:

$$\mathcal{L}_{\mathrm{det}}^L = \mathcal{L}_{\mathrm{cls}}^L + \mathcal{L}_{\mathrm{smooth}\text{-}\ell_1}^L, \tag{5}$$

For training the unlabeled data $\mathbf{p}^U$, we seek a soft learning target besides the pseudo label-based detection loss. To be specific, though our *ProficientTeachers* model has significantly improved the quality of pseudo labels, it inevitably contains inaccurate predictions. Learning towards these noisy pseudo targets will affect the student performance. To alleviate this problem, we propose to further mine the information in unlabeled data $\mathbf{p}^U$, which is achieved by box-wise contrastive learning (BCL). Though contrastive learning [3,30,32,37] has been explored in 3D point cloud, our BCL is different from those by directly contrasting expressive box-level features based on the pseudo predictions.

In particular, our BCL module enforces the feature consistency of the same box instance from different augmented views. Formally, given an unlabeled point cloud $\mathbf{p}^U$, we first apply two random augmentations to generate different views $\mathbf{p}^{U_1}$ and $\mathbf{p}^{U_2}$. Then, we get the pseudo labels of the two views:

$$\mathbf{y}^{U_1} = f_{\text{STU}}(\mathbf{p}^{U_1}), \mathbf{y}^{U_2} = f_{\text{STU}}(\mathbf{p}^{U_1}), \tag{6}$$

where $\mathbf{y}^{U_1} \in \mathbb{R}^{M_1 \times 7}$, $\mathbf{y}^{U_2} \in \mathbb{R}^{M_2 \times 7}$ and M_1, M_2 are the number of the predicted bounding boxes in each view with 7-d attributes. Next, we transform $\mathbf{y}^{U_1}$ and $\mathbf{y}^{U_2}$ to the same view, and build positive and negative sample pairs by a greedy matching, *i.e.*, box pairs with the smallest distance are treated as positives and other boxes are viewed as negatives. This results in M matched positive box pairs. Afterwards, we extract box features by a point-wise interpolation method, which is inspired by CenterPoint [38]. More precisely, we use the center points features from the six faces of a 3D bounding box to present a box, by applying bilinear interpolation on the bird-eye-view (BEV) feature maps $\mathbf{F}^U$. Due to operating on BEV, the center points of the six faces of a 3D box are equal to the center points of the four sides of a 2D box plus one box center. This essentially simplifies the box feature extraction process. We formulate this as:

$$\mathbf{h}^{U_1} = I(\mathbf{y}^{U_1}, \mathbf{F}^{U_1}), \mathbf{h}^{U_2} = I(\mathbf{y}^{U_2}, \mathbf{F}^{U_2}), \tag{7}$$

where $I(\cdot)$ is the bilinear interpolation function, and $\mathbf{h}^{U_1}$ and $\mathbf{h}^{U_1}$ are resultant box features from different views. These features are then collected and arranged as $\{\mathbf{h}_k\}_{k=1}^{2M}$, where the even indexes denote the boxes from $\mathbf{h}^{U_1}$ and the odd ones present boxes from $\mathbf{h}^{U_2}$. Later, a projection head $\phi(\cdot)$ that contains two 1×1 convolutional layers is used to map the box features to an embedding space, such that $\mathbf{z}_k = \phi(\mathbf{h}_k)$. Then, InfoNCE loss [15] is exploited to build the soft target:

$$\ell(p,q) = -\log \frac{\exp(\mathbf{z}_p \cdot \mathbf{z}_q / \tau)}{\sum_{k=1}^{2M} \mathbb{1}_{[k \neq q]} \exp(\mathbf{z}_p \cdot \mathbf{z}_k / \tau)}, \tag{8}$$

$$\mathcal{L}_{\text{con}}^U = \frac{1}{2M} \sum\nolimits_{k=1}^{M} [\ell(2k-1, 2k) + \ell(2k, 2k-1)], \tag{9}$$

where $p, q \in [1, \dots, 2M]$ and $p \neq q$. τ is a temperature hyper-parameter that is set to 0.1. Furthermore, we leverage the high-quality pseudo labels produced by the *ProficientTeachers* to construct the detection loss, which is denoted as:

$$\mathcal{L}_{\text{det}}^U = \frac{1}{2}(\mathcal{L}_{\text{cls}}^{U_1} + \mathcal{L}_{\text{smooth-}\ell_1}^{U_1} + \mathcal{L}_{\text{cls}}^{U_2} + \mathcal{L}_{\text{smooth-}\ell_1}^{U_2}). \tag{10}$$

Finally, the overall loss of the student model is defined as:

$$\mathcal{L} = \mathcal{L}_{\text{det}}^{L} + \mathcal{L}_{\text{det}}^{U} + \alpha\mathcal{L}_{\text{con}}^{U}, \tag{11}$$

where α is a coefficient and we set it to 0.05 empirically.

4 Experimental Results

In this work, we first introduce the datasets and the implementation details of our model in Sect. 4.1 and Sect. 4.2, respectively. Then, we report the main evaluation results by comparing with other SSL approaches in Sect. 4.3. Finally, the ablation studies are presented in Sect. 4.4.

4.1 Datasets

ONCE Dataset. ONCE [12] is a large-scale autonomous driving dataset with 1 million LiDAR point cloud samples. Only 15,000 samples are with annotations which have been divided into training, validation, and testing split with 5K, 3K, and 8K samples, respectively. In this dataset, five kinds of foreground objects have been annotated i.e., "Car", "Bus", "Truck", "Pedestrian" and "Cyclist", while "Car", "Bus" and "Truck" are merged into one class "Vehicle" during evaluation. In particular, a specific setting is designed for SSL approaches evaluation, *i.e.*, 5K labeled samples and all the unlabeled samples have been divided into 3 subsets: *Small*, *Medium* and *Large* to explore the effects of different data amounts for SSL-based 3D detection. The small unlabeled set *Small* contains 70 sequences (100k samples), the medium set *Medium* contains 321 sequences (500k samples) and the large set *Large* contains 560 sequences (about 1M samples) in total. Similar to other 3D object detection benchmarks, mean AP (average precision) [6] over all the classes is employed for evaluation, based on the 3D IoU thresholds 0.7, 0.3 and 0.5 for "Vehicle", "Pedestrian" and "Cyclist", respectively. In addition, three different perception ranges, '0–30 m", "30–50 m", and "50 m-inf", are specified to well evaluate the performance of 3D detectors.

Waymo Open Dataset. Waymo [26] provides a large-scale LiDAR point cloud dataset that contains 798 sequences (158,361 frames) for training and 202 sequences (40,077 frames) for validation. Since it does not provide additional unlabeled raw data for semi-supervised training. We thus manually tailor a semi-supervised learning dataset following the setting in ONCE [12]. Specifically, we divide the 798 Waymo training sequences equally into two splits, *i.e.*, labeled split $\mathcal{P}^L$ and unlabeled split $\mathcal{P}^U$ (without using the original labels), with each containing 399 sequences. Then, we randomly sample 5%, 10%, 20% and 50% sequences from $\mathcal{P}^L$, which lead to the ratio of labeled data and unlabeled data $\mathcal{P}^L : \mathcal{P}^U$ as 1:20, 1:10, 1:5 and 1:2, respectively. mAP and mAPH under LEVEL_2 metric are used to evaluate the 3D object detection performance on the full validation set, where 3D IoU thresholds for "Vehicle", "Pedestrian" and "Cyclist" are 0.7, 0.5 and 0.5, respectively.

4.2 Implementation Details

For the STE module in the teacher model, we have empirically defined the fixed (weak) augmentation types as rotation and double flip. Grid search is conducted to find the more effective rotation parameters, *i.e.*, $\{0°, +22.5°, -22.5°\}$. The final number of augmented views K is computed by the product of flip times and rotation times, *i.e.*, $4 \times 3 = 12$ in our case. The strong augmentation includes *random* rotation and flip and scaling, which is the same as that in [33]. For the CBV module, we offline train the RoI network [5] for 10 epochs based only on the labeled data, with a stop-gradient operation to detach from the backbones. We define positive and negative box samples according to the boxes classification scores and the IoU between ground truths. For example, boxes with classification scores below 0.1 or IoU below 0.3 are treated as negatives, and other boxes are viewed as positives. The votes objectness is optimized by a sigmoid focal loss and the votes localization is optimized by smooth-ℓ_1 loss. The IoU score for clustering boxes is fixed as 0.5. For the semi-supervised training configuration, we follow the Noisy Student implementation provided by ONCE official benchmark [12], *i.e.*, a model pre-trained on the full training set is used to warm up both the student and teacher model. Then the student is trained for 25 to 75 epochs depending on the amount of unlabeled data with learning rate 0.001, and the teacher is updated every 25 epochs.

Table 2. Evaluation results on ONCE validation set with different amounts of unlabeled samples (*e.g.*, "Small", "Medium" and "Large") following the official implementation in ONCE [12]. For better understanding, the best overall result in each class has been highlighted in **bold** and the relative gains of each SSL method compared to the baseline model (*i.e.*, SECOND [33] trained with only labeled samples) have been illustrated in colors where the positive gains in **blue** and negative gains are in green.

Methods	Vehicle AP (%)				Pedestrian AP (%)				Cyclist AP (%)				mAP (%)
	Overall	0–30 m	30–50 m	50 m-inf	Overall	0–30 m	30–50 m	50 m-inf	Overall	0–30 m	30–50 m	50 m-inf	
Baseline [33]	71.19	84.04	63.02	47.25	26.44	29.33	24.05	18.05	58.04	69.96	52.43	34.61	51.89
Small (100K unlabeled Samples)													
Pseudo Label	72.80	84.46	64.97	51.46	25.50	28.36	22.66	18.51	55.37	65.95	50.34	34.42	51.22 ([illegible])
Noisy Student [31]	73.69	84.69	67.72	53.41	28.81	33.23	23.42	16.93	54.67	65.58	50.43	32.65	52.39 (+ **0.50**)
Mean Teacher [27]	74.46	86.65	68.44	53.59	30.54	34.24	26.31	20.12	61.02	72.51	55.24	39.11	55.34 (+ **3.45**)
SESS [39]	73.33	84.52	66.22	52.83	27.31	31.11	23.94	19.01	59.52	71.03	53.93	36.68	53.39 (+ **1.50**)
3DIoUMatch [29]	73.81	84.61	68.11	54.48	30.86	35.87	25.55	18.30	56.77	68.02	51.80	35.91	53.81 (+ **1.92**)
Our Mehtod	**76.07**	86.78	70.19	56.17	**35.90**	39.98	31.67	24.37	**61.19**	73.97	55.13	36.98	**57.72 (+ 5.83)**
Medium (500K unlabeled Samples)													
Pseudo Label	73.03	86.06	65.96	51.42	24.56	27.28	20.81	17.00	53.61	65.26	48.44	33.58	50.40 ([illegible])
Noisy Student [31]	75.53	86.52	69.78	55.05	31.56	35.80	26.24	21.21	58.93	69.61	53.73	36.94	55.34 (+ **3.45**)
Mean Teacher [27]	76.01	86.47	70.34	55.92	35.58	40.86	30.44	19.82	63.21	74.89	56.77	40.29	58.27 (+ **6.38**)
SESS [39]	72.11	84.06	66.44	53.61	33.44	38.58	28.10	18.67	61.82	73.20	56.60	38.73	55.79 (+ **3.90**)
3DIoUMatch [29]	75.69	86.46	70.22	56.06	34.14	38.84	29.19	19.62	58.93	69.08	54.16	38.87	56.25 (+ **4.36**)
Our Mehtod	**78.07**	87.43	**72.5**	59.51	**38.38**	42.45	34.62	25.58	**63.23**	74.70	58.19	40.73	**59.89 (+ 8.00)**
Large (1M unlabeled Samples)													
Pseudo Label	72.41	84.06	64.54	50.05	23.62	26.80	20.13	16.66	53.25	64.69	48.52	33.47	49.76 ([illegible])
Noisy Student [31]	75.99	86.67	70.48	55.60	33.31	37.81	28.19	21.39	59.81	70.01	55.13	38.33	56.37 (+ **4.48**)
Mean Teacher [27]	76.38	86.45	70.99	57.48	35.95	41.76	29.05	18.81	**65.50**	75.72	60.07	43.66	59.28 (+ **7.39**)
SESS [39]	75.95	86.83	70.45	55.76	34.43	40.00	27.92	19.20	63.58	74.85	58.88	39.51	57.99 (+ **6.10**)
3DIoUMatch [29]	75.81	86.11	71.82	57.84	35.70	40.68	30.34	21.15	59.69	70.69	54.92	39.08	57.07 (+ **5.18**)
Our Mehtod	**78.12**	87.22	72.74	59.58	**41.95**	48.09	35.13	26.01	64.12	75.85	58.04	41.45	**61.40 (+ 9.51)**

Table 3. Generalizability on different detectors with our SSL method.

Methods	mAP(%)	Gain	AP (%)		
			Vehicle	Pedestrian	Cyclist
SECOND [33]	51.89	-	71.19	26.44	58.04
Our Method	59.89	**+8.00**	78.07	38.38	63.23
PV-RCNN [21]	57.24	-	79.35	29.64	62.73
Our Method	63.40	**+6.16**	81.09	41.55	67.57
CenterPoint [38]	62.99	-	75.26	51.65	65.79
Our Method	68.22	**+5.23**	77.77	56.34	70.55

Table 4. Performance on the ONCE test set.

Methods	mAP (%)	AP (%)		
		Vehicle	Pedestrian	Cyclist
SECOND [33]	51.90	69.71	26.09	59.92
Pseudo Label	49.29	70.29	21.85	55.72
Noisy Student [31]	56.61	74.50	33.28	62.05
Mean Teacher [27]	59.99	76.60	36.37	66.99
SESS [39]	58.78	74.52	36.29	65.52
3DIoUMatch [29]	57.43	74.48	35.74	62.06
Our Method	61.44	76.85	41.27	66.19

4.3 Main Results

ONCE Results. First of all, we aim to compare our *ProficientTeachers* with other state-of-the-art SSL approaches on the ONCE dataset [12]. Here, we borrow all the evaluation results from the official benchmark[1] [12], where five typical SSL approaches have been included, *i.e.*, Pseudo Label, Noisy Student [31], Mean Teacher [27], SESS [39], 3DIoUMatch [29]. For a fair comparison, the SECOND [33] detector trained with only the labeled samples has been adopted as the baseline. All the comparison results are given in Table 2. Compared to the baseline model that is trained with only the labeled samples, all these SSL frameworks can obtain positive gains, with the help of large amounts of unlabeled samples except for the Pseudo Label method. This may be because the official implementation of Pseudo Label includes no augmentation when training the student. Interestingly, the improvements increase gradually with the increase of the number of unlabeled samples. Compared to the other SSL methods, our framework achieves the best performance among all the three splits, which obtains 5.83, 8.00, 9.51 mAP improvements, respectively.

Since our method removes the necessity of threshold selection, we compare it with the confidence-based filtering method, *i.e.*, Noisy Student [31] with different pseudo-label threshold c. In our implementation, pseudo boxes with scores above c are viewed as positive samples, meanwhile, we also ignore the pseudo boxes with scores below c to avoid taking potential true positives as negatives. As shown in Fig. 1 (c), $c = 0.3$ achieves better performance (55.75 mAP), which is even stronger than Mean Teacher [27], while our method still surpasses it a lot.

We also verify our SSL method on two more baselines, which are PV-RCNN [21] and CenterPoint [38]. PV-RCNN is a hybrid point-voxel approach that inherits the advantages from both point [17] and voxels features [41]. CenterPoint is a voxel-based anchor-free framework that has achieved SOTA results in several benchmarks. All the methods are trained with the "Medium" unlabeled subset. From Tab. 3, we can see that our method works well on these different detectors, *e.g.*, still yielding 5.23 points improvements over the strong detector CenterPoint. This indicates the good generalizability of our SSL method.

[1] https://once-for-auto-driving.github.io/benchmark.html#benchmark.

Table 5. Semi-supervised 3D Object Detection on Waymo dataset. We train the SECOND [33] baseline with different fractions of labeled data. Then, we compare our *ProficientTeachers* model with a strong competitor, Fixmatch [24]. It shows that our model can consistently improve the detection performance.

Different Label Amounts	Training Paradigm	Performance Gain	3D AP/APH @0.7 (LEVEL 2)			
			Overall	Vehicle	Pedestrian	Cyclist
5% (∼ 4k Labels) $\mathcal{P}^L : \mathcal{P}^U = 1:20$	Baseline [33]	-/-	45.78/40.40	50.03/49.52	45.77/34.98	41.53/36.69
	Fixmatch [24]	+3.02/+2.95	48.80/43.35	51.87/51.27	48.28/36.56	46.26/42.21
	ProficientTeacher	**+5.32/+5.35**	51.10/45.75	53.04/52.54	50.33/38.67	49.92/46.03
10% (∼ 8k Labels) $\mathcal{P}^L : \mathcal{P}^U = 1:10$	Baseline [33]	-/-	50.00/45.83	54.90/54.25	48.45/38.44	46.66/44.79
	Fixmatch [24]	+2.28/+1.90	52.28/47.73	56.60/55.99	51.60/40.63	48.63/46.56
	ProficientTeacher	**+5.01/+4.60**	55.01/50.43	57.59/56.92	54.28/43.19	53.15/51.18
20% (∼ 16k Labels) $\mathcal{P}^L : \mathcal{P}^U = 1:5$	Baseline [33]	-/-	53.09/49.11	57.40/56.81	51.54/41.91	50.33/48.62
	Fixmatch [24]	+2.72/+2.34	55.81/51.45	58.94/58.37	54.37/44.23	54.11/51.75
	ProficientTeacher	**+5.50/+5.05**	58.59/54.16	59.97/59.36	57.88/46.97	57.93/56.15
50% (∼ 40k Labels) $\mathcal{P}^L : \mathcal{P}^U = 1:2$	Baseline [33]	-/-	57.07/53.26	60.93/60.37	55.98/46.68	54.31/52.74
	Fixmatch [24]	+2.72/+2.45	59.79/55.71	61.88/61.34	58.64/49.00	58.85/56.78
	ProficientTeacher	**+4.57/+4.34**	61.64/57.60	63.06/62.50	61.53/51.33	60.33/58.97
100% (∼ 80k Labels) $\mathcal{P}^L : \mathcal{P}^U = 1:1$	Baseline [33]	-/-	59.63/55.94	62.78/62.24	59.45/50.44	56.67/55.13
	Fixmatch [24]	+2.43/+2.02	62.06/57.96	63.50/62.98	62.00/52.52	60.69/58.37
	ProficientTeacher	**+3.33/+3.20**	62.96/59.14	63.56/63.06	62.34/53.19	62.97/61.18
	Oracle Model [33]	+2.80/+2.93	62.43/58.87	65.43/64.91	61.82/52.93	60.03/58.76

The results on the testing split of the ONCE benchmark are presented in Table 4, where the results of other SSL methods are reported in terms of the official benchmark. All the SSL methods are based on the SECOND detector for a fair comparison. Among all the competitors, our method shows remarkable superiority. It outperforms the strong competitor Mean Teacher by 1.45 points, and also exceeds the recently proposed SESS and 3DIoUMatch by a large margin.

Waymo Results. For evaluating on Waymo, we compare our model with the strongest competitor verified on ONCE, *i.e.*, Noisy Student [31] with a carefully selected threshold $c = 0.3$. Since this filtering strategy derives from Fixmatch [24], we name it Fixmatch for simplicity. The SECOND [33] detector trained with different amounts of labeled data is used as the baseline. As shown in Table 5, impressive results are obtained. Both Fixmatch and our *ProficientTeachers* achieve better results than the full-supervised baseline, which proves the advantage of semi-supervised learning. In particular, Fixmatch surpasses the baseline by 1.90 to 2.95 mAPH. By contrast, our method obtains much better results than Fixmatch, *i.e.*, improving the baseline by 4.34 to 5.35 mAPH. Moreover, we also run an oracle model that trains the detector with the full 158,361 labels, while our model with only half labels outperforms it. This further demonstrates the generalization of our method over different datasets.

Qualitative Results. We visualize some examples of pseudo labels on ONCE in Fig. 4. The false positives and false negatives are highlighted in circles. We compare our STE and CBV modules with the threshold-based method FixMatch [24]. It incurs FN when detecting distant objects with sparse points, and

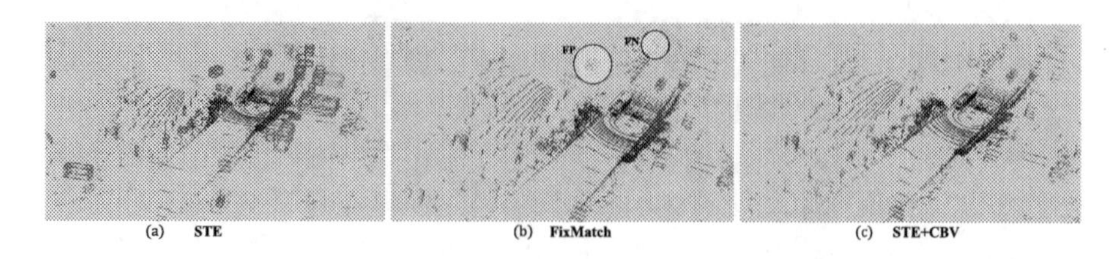

Fig. 4. Visualization of pseudo labels produced by our STE and CBV modules, or by threshold-based FixMatch [24]. Predictions are in red and GTs are in green. (Color figure online)

Table 6. Ablation studies to verify the effect of different modules. The experiments are conducted based on the SECOND with the "Small" unlabeled subset.

Module	Aspect	mAP (%)	Gain	AP (%)		
				Vehicle	Pedestrian	Cyclist
Noisy Student (Baseline)	c = 0.3	55.75	–	74.66	32.25	60.34
BCL Module	c = 0.3, *w/o* STE or CBV	56.54	**+0.79**	75.70	33.65	60.28
STE Module	c = 0.3, *w/o* BCL or CBV	56.21	**+0.46**	75.62	33.14	59.87
STE+CBV Modules	*w/o* BCL	57.17	**+1.42**	75.98	34.86	60.67
Full Model	*w/* STE, CBV and BCL	57.72	**+1.97**	76.07	35.90	61.19

it gives FP when detecting a hard distractor as shown in the circles in (b) (plz zoom in for a better view). In contrast, by gathering boxes from different views, our STE successfully addresses the FN object. As for the FP box, its score will be refined and re-scaled in our CBV such that a box with low detection frequency will be removed accordingly. Moreover, for boxes that are close to GTs, their box coordinates will also be refined after voting in a cluster. All these designs lead to high-quality pseudo labels.

4.4 Ablation Studies

A series of ablation studies are set to verify the effectiveness of different modules and all the results are given in Table 6. All the experiments are conducted based on the SECOND detector with the "Small" unlabeled subset. Our *ProficientTeachers* model is mainly based on the Noisy Student implementation in ONCE, and the default confidence threshold c for filtering the noisy pseudo labels is 0.1. To explore a more effective threshold, we perform grid search and find that $c = 0.3$ gives better detection results, as shown in Fig. 1 (c). Thus we use this threshold for the subsequent experiments. STE, CBV and BCL are the three necessary modules proposed in this work, and we ablate the contribution of each module. First, the STE module has been used to generate more pseudo boxes. We evaluate it by using $c = 0.3$ to filter the boxes. This improves the baseline by 0.46 points. Next, we replace the confidence-based thresholding with the proposed CBV module to adaptively aggregate these pseudo boxes. This exceeds the baseline by 1.42 points. Finally, thanks to the BCL module, our full model achieves 57.72% mAP, improving the baseline by 1.97 points.

5 Discussion and Conclusion

In this work, we proposed a new SSL framework for LiDAR-based 3D object detection. In particular, our work focuses on improving the FP and FN in the pseudo labels produced by the teacher model. First, to address the FN, a spatial-temporal ensemble (STE) module is introduced to produce sufficient seed boxes and ensure a high recall. This is realized by a spatial data augmentation and a temporal model ensemble. Second, to resolve the FP predictions and improve the precision, we developed a clustering-based box voting (CBV) module that performs voting and aggregating based on boxes in a cluster. More importantly, our CBV can yield high-quality pseudo labels without the need of deliberately selecting thresholds. The STE and CBV modules enhance the original teacher to proficient teachers. Finally, we proposed a box-wise contrastive learning (BCL) strategy to optimize the student towards cross-view feature consistency, reducing the effect of inaccurate pseudo labels. Experiments on the large-scale ONCE and Waymo datasets demonstrated the superiority of our method.

Acknowledgements. This work was partially supported by Zhejiang Lab's International Talent Fund for Young Professionals (ZJ2020GZ023), ARC DECRA DE220101390, FDCT under grant 0015/2019/AKP, and the Start-up Research Grant (SRG) of University of Macau.

References

1. Berthelot, D., Carlini, N., Goodfellow, I., Papernot, N., Oliver, A., Raffel, C.A.: Mixmatch: a holistic approach to semi-supervised learning. In: NeurIPS (2019)
2. Caesar, H., et al.: nuscenes: a multimodal dataset for autonomous driving. In: CVPR (2020)
3. Chen, T., Kornblith, S., Norouzi, M., Hinton, G.: A simple framework for contrastive learning of visual representations. In: ICML (2020)
4. Chen, X., Ma, H., Wan, J., Li, B., Xia, T.: Multi-view 3d object detection network for autonomous driving. In: CVPR (2017)
5. Deng, J., Shi, S., Li, P., gang Zhou, W., Zhang, Y., Li, H.: Voxel R-CNN: towards high performance voxel-based 3d object detection. In: AAAI (2021)
6. Everingham, M., Van Gool, L., Williams, C.K., Winn, J., Zisserman, A.: The pascal visual object classes (voc) challenge. IJCV **88**(2), 303–338 (2010)
7. Fang, J., et al.: Augmented lidar simulator for autonomous driving. IEEE Robot. Autom. Lett. **5**(2), 1931–1938 (2020)
8. Geiger, A., Lenz, P., Urtasun, R.: Are we ready for autonomous driving? The kitti vision benchmark suite. In: CVPR (2012)
9. Huang, X., Wang, P., Cheng, X., Zhou, D., Geng, Q., Yang, R.: The apolloscape open dataset for autonomous driving and its application. PAMI **42**(10), 2702–2719 (2019)

10. Jeong, J., Lee, S., Kim, J., Kwak, N.: Consistency-based semi-supervised learning for object detection. In: NeurIPS (2019)
11. Lang, A.H., Vora, S., Caesar, H., Zhou, L., Yang, J., Beijbom, O.: Pointpillars: Fast encoders for object detection from point clouds. In: CVPR (2019)
12. Mao, J., et al.: One million scenes for autonomous driving: once dataset. In: NeurIPS Datasets and Benchmarks (2021)
13. Meng, Q., Wang, W., Zhou, T., Shen, J., Jia, Y., Van Gool, L.: Towards a weakly supervised framework for 3d point cloud object detection and annotation. TPAMI (2021)
14. Meng, Q., Wang, W., Zhou, T., Shen, J., Van Gool, L., Dai, D.: Weakly supervised 3D object detection from lidar point cloud. In: Vedaldi, A., Bischof, H., Brox, T., Frahm, J.-M. (eds.) ECCV 2020. LNCS, vol. 12358, pp. 515–531. Springer, Cham (2020). https://doi.org/10.1007/978-3-030-58601-0_31
15. Oord, A.v.d., Li, Y., Vinyals, O.: Representation learning with contrastive predictive coding. arXiv preprint (2018)
16. Qi, C.R., Litany, O., He, K., Guibas, L.J.: Deep hough voting for 3D object detection in point clouds. In: CVPR (2019)
17. Qi, C.R., Yi, L., Su, H., Guibas, L.J.: Pointnet++: deep hierarchical feature learning on point sets in a metric space. In: NeurIPS (2017)
18. Rizve, M.N., Duarte, K., Rawat, Y.S., Shah, M.: In defense of pseudo-labeling: an uncertainty-aware pseudo-label selection framework for semi-supervised learning. In: ICLR (2021)
19. Samuli, L., Timo, A.: Temporal ensembling for semi-supervised learning. In: ICLR (2017)
20. Shanmugam, D., Blalock, D., Balakrishnan, G., Guttag, J.: Better aggregation in test-time augmentation. In: ICCV (2021)
21. Shi, S., Guo, C., Jiang, L., Wang, Z., Shi, J., Wang, X., Li, H.: Pv-rcnn: Point-voxel feature set abstraction for 3D object detection. In: CVPR (2020)
22. Shi, S., Wang, X., Li, H.: Pointrcnn: 3d object proposal generation and detection from point cloud. In: CVPR (2019)
23. Shi, W., Rajkumar, R.: Point-GNN: Graph neural network for 3d object detection in a point cloud. In: CVPR (2020)
24. Sohn, K., et al.: Fixmatch: simplifying semi-supervised learning with consistency and confidence. In: NeurIPS (2020)
25. Sohn, K., Zhang, Z., Li, C.L., Zhang, H., Lee, C.Y., Pfister, T.: A simple semi-supervised learning framework for object detection. arXiv preprint (2020)
26. Sun, P., et al.: Scalability in perception for autonomous driving: Waymo open dataset. In: CVPR (2020)
27. Tarvainen, A., Valpola, H.: Mean teachers are better role models: weight-averaged consistency targets improve semi-supervised deep learning results. In: NeurIPS (2017)
28. van Engelen, J.E., Hoos, H.H.: A survey on semi-supervised learning. Mach. Learn. **109**(2), 373–440 (2019). https://doi.org/10.1007/s10994-019-05855-6
29. Wang, H., Cong, Y., Litany, O., Gao, Y., Guibas, L.J.: 3dioumatch: leveraging IOU prediction for semi-supervised 3d object detection. In: CVPR (2021)
30. Wang, W., Zhou, T., Yu, F., Dai, J., Konukoglu, E., Van Gool, L.: Exploring cross-image pixel contrast for semantic segmentation. In: ICCV (2021)
31. Xie, Q., Luong, M.T., Hovy, E., Le, Q.V.: Self-training with noisy student improves imagenet classification. In: CVPR (2020)

32. Xie, S., Gu, J., Guo, D., Qi, C.R., Guibas, L., Litany, O.: PointContrast: unsupervised pre-training for 3d point cloud understanding. In: Vedaldi, A., Bischof, H., Brox, T., Frahm, J.-M. (eds.) ECCV 2020. LNCS, vol. 12348, pp. 574–591. Springer, Cham (2020). https://doi.org/10.1007/978-3-030-58580-8_34
33. Yan, Y., Mao, Y., Li, B.: Second: sparsely embedded convolutional detection. Sensors **18**(10), 3337 (2018)
34. Yang, Z., Sun, Y., Liu, S., Jia, J.: 3DSSD: point-based 3D single stage object detector. In: CVPR (2020)
35. Yin, J., Shen, J., Gao, X., Crandall, D., Yang, R.: Graph neural network and spatiotemporal transformer attention for 3d video object detection from point clouds. TPAMI (2021)
36. Yin, J., Shen, J., Guan, C., Zhou, D., Yang, R.: Lidar-based online 3d video object detection with graph-based message passing and spatiotemporal transformer attention. In: CVPR (2020)
37. Yin, J., Zhou, D., Zhang, L., Fang, J., Xu, C.Z., Shen, J., Wang, W.: Proposalcontrast: Unsupervised pre-training for lidar-based 3D object detection. In: ECCV (2022)
38. Yin, T., Zhou, X., Krahenbuhl, P.: Center-based 3D object detection and tracking. In: CVPR (2021)
39. Zhao, N., Chua, T.S., Lee, G.H.: SESS: self-ensembling semi-supervised 3D object detection. In: CVPR (2020)
40. Zhou, D., et al.: Joint 3d instance segmentation and object detection for autonomous driving. In: CVPR (2020)
41. Zhou, Y., Tuzel, O.: VoxelNet: end-to-end learning for point cloud based 3D object detection. In: CVPR (2018)
42. Zhu, X., Goldberg, A.B.: Introduction to semi-supervised learning. Synthesis Lect. Artif. Intell. Mach. Learn. **3**(1), 1–130 (2009)

Point Cloud Compression with Sibling Context and Surface Priors

Zhili Chen, Zian Qian, Sukai Wang, and Qifeng Chen(✉)

The Hong Kong University of Science and Technology, Clear Water Bay, Hong Kong
{zchenei,zqianaa,swangcy,cqf}@ust.hk

Abstract. We present a novel octree-based multi-level framework for large-scale point cloud compression, which can organize sparse and unstructured point clouds in a memory-efficient way. In this framework, we propose a new entropy model that explores the hierarchical dependency in an octree using the context of siblings' children, ancestors, and neighbors to encode the occupancy information of each non-leaf octree node into a bitstream. Moreover, we locally fit quadratic surfaces with a voxel-based geometry-aware module to provide geometric priors in entropy encoding. These strong priors empower our entropy framework to encode the octree into a more compact bitstream. In the decoding stage, we apply a two-step heuristic strategy to restore point clouds with better reconstruction quality. The quantitative evaluation shows that our method outperforms state-of-the-art baselines with a bitrate improvement of 11–16% and 12–14% on the KITTI Odometry and nuScenes datasets, respectively.

Keywords: Point cloud compression · Autonomous driving

1 Introduction

LiDAR is undergoing rapid development and becoming popular in robots and self-driving vehicles. As the eyes of those machines, LiDAR sensors can generate point clouds that provide accurate 3D geometry in diverse environments. However, the large number of point clouds incur a heavy burden on storage and bandwidth usage. For example, a single Velodyne HDL-64E can generate more than 450 GB of data over 30 billion points in just eight hours of driving. Therefore, developing effective algorithms for 3D point cloud compression is imperative.

Unlike image and video compression, point cloud compression is technically challenging due to the sparseness of orderless point clouds. In the early years,

Supplementary Information The online version contains supplementary material available at https://doi.org/10.1007/978-3-031-19839-7_43.

S. Avidan et al. (Eds.): ECCV 2022, LNCS 13698, pp. 744–759, 2022.
https://doi.org/10.1007/978-3-031-19839-7_43

researchers utilize different data structures such as octrees [18] and KD-trees [4] to organize unstructured data and ignore the sparsity of point clouds. However, these methods did not reduce the information redundancy hidden in the data structure representations. Therefore, the potential of reducing the bitrate of a point cloud is still not well explored.

Recent works have shown that learning-based methods have great potential in reducing information redundancy. By encoding point cloud information into an embedded representation, a neural network is used to further reduce the bitrate by predicting a better probability distribution in entropy coding [11,24]. Their approaches can be summarized into three steps: (1) constructing an octree; (2) fusing different features such as ancestor information [11] or neighbor information [24] to construct a contextual feature map; (3) and training a shared weight entropy model to reduce the length of the bitstream. However, their approaches have three limitations. First, although previous approaches largely focus on the spatial dependencies such as ancestor information [11] or neighbor information [24], they do not utilize prior knowledge of decoded siblings' children to reduce the information redundancy during the entropy coding. Second, they ignore local geometric information such as quadratic surface. Third, as illustrated in Fig. 3, since the distributions of occupancy symbols at different levels in an octree are very diverse, training a shared weight entropy model can overfit the octants at deeper levels and lead to low prediction accuracy on octants at shallower levels.

To address these issues, we propose a novel octree-based multi-level framework for point cloud compression. Our framework encodes the point cloud data into the non-leaf octants as eight-bit occupancy symbols. The entropy model encodes each occupancy symbol into a more compact bitstream through entropy coding by accurately estimating symbol occurrence probabilities. Compared to the previous methods, our entropy model is the first to utilize the siblings' children as strong priors when inferring the current octant's symbol probabilities. Our entropy model also incorporates the context of the current octant's neighbor and ancestor to explore the hierarchical contextual information of the octree fully. We also propose a quadratic surface fitting module to provide geometric priors, which is empirically proven to be beneficial in lowering the bitrate. In our compression framework, we train an independent entropy model for each octree level to capture resolution-specific context flow across different levels. At the decoding stage, we further improve the reconstructed point cloud quality by a two-step heuristic strategy that first predicts leaf octants' occupancies to retrieve the missing points by a specific level of entropy model and then apply a refinement module on the aggregated point cloud.

The contributions of this work can be summarized as follows.

- We build a novel octree-based multi-level compression framework for the point cloud. Our approach is the first one that incorporates sibling context for point cloud compression.
- We introduce surface priors into our entropy model, which effectively reduces octree-structured data's overall bitrate.

- In the decoding stage, we propose a two-step heuristic strategy by first predicting the missing points and then refining the aggregated point clouds to achieve a better reconstruction quality.
- Our proposed multi-level compression framework outperforms previous state-of-the-art methods on compression performance and reconstructs high-quality point clouds on two large-scale LiDAR point cloud datasets.

2 Related Work

2.1 Traditional Point Cloud Compression

Traditional point cloud methods usually utilize tree-based data structure to organize unstructured point cloud data, especially on KD-tree [4,9] and Octree [7,10,13–15,25,26]. Meagher et al. [18] directly utilize octree to encode the point cloud without reducing the information redundancy. Huang et al. [13] utilize the neighborhood information for better entropy encoding. To reduce spatial redundancy, Kammerl et al. [15] use XOR to encode sibling octants. In recent years, MPEG also developed an octree-based standard point cloud compression method G-PCC [10]. Moreover, Google utilizes KD-tree in its open-source point cloud compression software Draco [9]. Since all of the methods are hand-crafted, they cannot be implicitly optimized end-to-end. As such, the reduction of information redundancy is likely to be sub-optimal.

2.2 Learning-based Point Cloud Compression

Since point cloud data are in $n \times 3$ unstructured floating points, it is almost impossible to directly apply convolutions to raw point cloud data. Thus, researchers proposed several kinds of methods to organize raw point clouds. Point-based methods [12,35] are usually built based on PointNet framework [22, 23]. Some point-based methods [34] directly downsample a point cloud to a small set of points, then use deconvolution to rebuild the point cloud. However, point-based methods usually suffer from the sparsity of point cloud data and incur high memory costs in processing. Therefore, it cannot handle extremely large LiDAR point cloud data. Range-image-based point cloud compression methods [1,20,27,28,31–33] first transfer point cloud into depth map or range image, then utilize state-of-the-art image compression methods such as [29] for compression. However, such kinds of methods highly depend on the quality of image compression algorithms and could not utilize spatial information. Voxel-based methods [2,11,21,24] ignore the sparsity of the point cloud data, and can utilize diverse geometric and spatial information. However, Huang et al. [11] only utilize the ancestor contexts, Que et al. [24] only use the neighbor contexts, and Fu et al. [6] ignore the local geometric information.

Unlike previous methods, our model is the first to reduce the inter-voxel redundancy by utilizing the context of decoded siblings' children octants. Moreover, local geometric information is also provided as the features for entropy model encoding. We also incorporate both ancestor and neighbor information into our entropy model to enrich the contextual information.

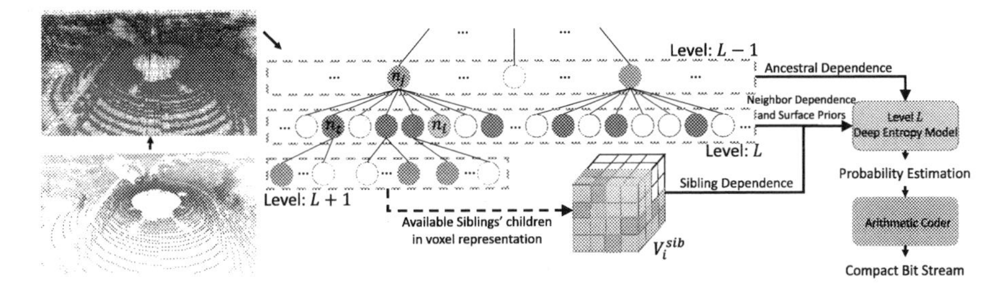

Fig. 1. The overview of our method. The input point cloud is first encoded into an octree. The ancestors, neighbors, and siblings' children octants are represented in orange, blue, and green colors. Our entropy model estimates occupancy probability distribution for each non-empty octant, conditioning on ancestral dependence, neighbor dependence, sibling dependence, and surface priors. Finally, the octree symbols are encoded into a more compact bitstream with arithmetic coding. The voxel block V_i^{sib} is formed by decoded siblings' children. The green and the gray voxels in V_i^{sib} represent the existent and absent siblings' children, respectively. The white voxels represent the unknown occupancies. After encoding the octant n_i in yellow, its occupancy symbol will be filled in the yellow voxels of V_i^{sib}. (Color figure online)

3 Methodology

3.1 Overview

Our compression framework is shown in Fig. 1. Given an input point cloud, we first organize it using an octree with occupancy symbols stored at each non-leaf octant. A corresponding deep entropy model takes the hierarchical contexts as input for each octant to predict its occupancy symbol probabilities with 256 classes. The hierarchical contexts consist of local neighbors, ancestors, and siblings' children. The local neighbor context and sibling context are represented in binary voxel blocks. The ancestor information is an extracted feature map passed from the upper level. Moreover, we also incorporate our entropy model with surface priors by locally fitting quadratic surfaces among the local neighbors. Our entropy model can accurately predict symbol probabilities with these strong priors and further compress the symbols into a more compact bitstream by entropy coding.

In the decoding stage, we first reconstruct an octree with L levels from the compressed bitstream. We then propose a two-step heuristic strategy to produce a point cloud with better quality. As illustrated in Fig. 4, for each leaf octant, we first reuse our deep entropy model at level $L+1$ to retrieve its missing points by decoding the top-1 prediction of the occupancy symbol. Finally, we apply a separately trained refinement module [24] to further refine the aggregated points by adding the predicted offsets.

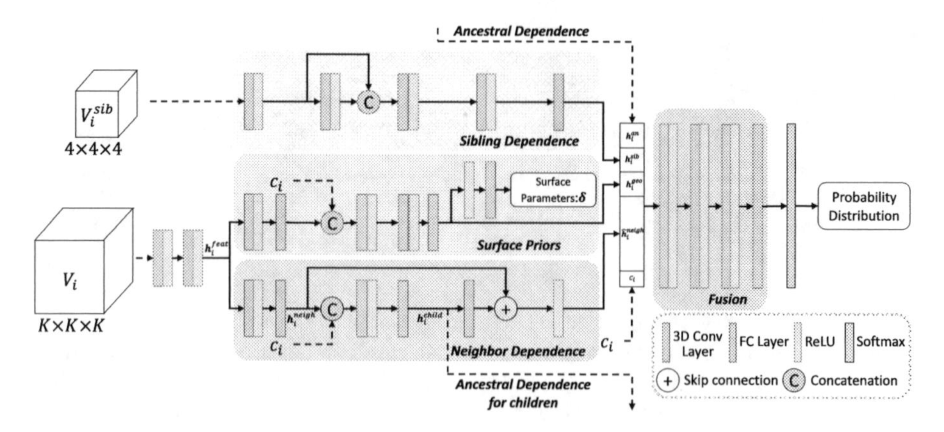

Fig. 2. The network architecture of our deep entropy model. The dash lines indicate the contextual input to our entropy model. The sibling and neighbor contexts are present as V_i^{sib} and V_i in voxel representations, c_i is the octant information (located tree level and corresponding coordinates) and h_i^{an} is the ancestor context. Note that the dimension K of the neighbor context V_i is empirically set to 9 in experiments.

3.2 Octree Construction

Due to the sparseness of the orderless point cloud data, we have to first organize it in a well-structured representation before subsequent processing. Unlike other popular structures, such as voxel-based methods, an octree is more memory efficient as it only partitions the non-empty 3D space and ignores the empty space. Moreover, the tree structure can provide hierarchical spatial information for a better reduction of information redundancy during compression. Therefore, we utilize octree to organize the point clouds.

An octree is built by recursively partitioning the input space into eight non-empty subspaces of the same size until the predefined depth is reached. The octant represents a bounding subspace and uses its center as the point coordinate. Each non-leaf occupied octant consists of an 8-bit occupancy symbol, and each bit represents the existence of a corresponding child (one for existence and zero for the absence). With such representation, a point cloud can be expressed as a serialized 8-bit occupancy symbols stream level-by-level and facilitate further lossless compression by an entropy encoder.

3.3 Hierarchical Context Entropy Model

Given an occupancy symbol stream $\mathbf{s} = [s_1, s_2, ..., s_n]$ with the number of non-leaf octants as its length n, the goal of our entropy model is to minimize the bitstream length. According to information theory, this goal can be reached by minimizing cross-entropy loss $\mathbb{E}_{s\sim p}[-\log q(\mathbf{s})]$, where $q(s)$ is the estimated probability distribution of occupancy symbol s and p is the actual distribution.

The accuracy of the probability distribution $q(s)$ determines the effectiveness of the entropy model. For example, if the actual probability distribution of s is known, no bit is needed to encode the whole point cloud. However, formulating the actual $q(s)$ could be very difficult because it has many complex dependencies such as ancestor or neighbor context. A good formulation of $q(s)$ can provide strong priors for the entropy model and therefore reduce the information redundancy in a bitstream. As illustrated in Fig. 1, our entropy model utilizes hierarchical dependencies from coarse to fine with ancestors, neighbors, and decoded siblings' children. Moreover, we utilize surface information as geometric priors while estimating $q(s)$. Therefore, the formulation of $q(s)$ in our entropy model can be defined as

$$q(\mathbf{s}) = \prod_i q(s_i \mid h_i^{an}, h_i^{neigh}, h_i^{sib}, h_i^{geo}, c_i; \theta), \quad (1)$$

where h_i^{an} is the ancestral dependence, h_i^{neigh} is the neighbor dependence, h_i^{sib} is the siblings' children dependence, h_i^{geo} is the surface priors, c_i is the octant information (located tree level and corresponding coordinates), and θ is the parameters of our entropy model.

3.3.1 Neighbor Dependence

The octree at depth l ($l \in [1, L]$) can be considered as a discretization of the 3D space at the resolution of $2^l \cdot 2^l \cdot 2^l$. Inspired by [24], we construct the neighbor context of an octant n_i by locally forming a $K \times K \times K$ binary voxel block V_i centered at n_i and K is empirically set to 9 in experiments. Each binary value indicates the existence of its corresponding neighbors. The neighbor voxel V_i is transformed by a feature extraction function f_{neigh}:

$$h_i^{neigh} = f_{neigh}(V_i), \quad (2)$$

where h_i^{neigh} is the output feature vector that represents the neighbor contextual information. The feature vector h_i^{neigh} is provided as part of the prior knowledge for inferring the n_i's children distribution.

3.3.2 Ancestral Dependence

The ancestral information brings coarser geometric information from a shallower level to the current octant [11], which can enlarge the entropy model's receptive field. As illustrated in Fig. 2, since the entropy model is trained level-by-level, the neighbor context features h_i^{neigh} of current octant n_i can be further passed to its children as ancestral features.

To avoid ancestral features overwhelming the feature space of its children octants, we first concatenate the h_i^{neigh} with the octant information c_i and then apply an MLP denoted as φ to extract more condensed features h_i^{child}. The processed features h_i^{child} will then be passed to its children as their ancestral dependence. A following linear projection layer γ is applied on h_i^{child} to scale back the features to the original dimension. Then the feature is added with

h_i^{neigh} through skip connection to recover the original neighbor feature $\hat{h}_i^{neigh}$ of the current octant:

$$h_i^{child} = \varphi(\text{Concat}(h_i^{neigh}, c_i)), \hat{h}_i^{neigh} = \gamma(h_i^{child}) + h_i^{neigh}. \quad (3)$$

Note that the ancestral features received from the upper level for the current octant are denoted as h_i^{an}. For the root octant, we initialize its ancestral features with zero values.

3.3.3 Sibling Context

Sibling octants are adjacent in the original 3D space. The decoded children of siblings are represented in a finer voxel size and are even closer to the encoding/decoding octant. Similar to successive pixels in 2D images and videos, siblings' children are strongly correlated with the children of the current octant because the geometric structure of these siblings' children and the current octant's children can be quite similar. Therefore, having a condition on siblings' children can reduce the information redundancy while predicting the probability distribution of the current octant.

As illustrated in Fig. 1, an occupied octant n_j is located at the octree level $L-1$ and represents a space in the 3D world. Until the level of $L+1$, this space is partitioned into $4 \times 4 \times 4$ subspaces. We represent it as a binary voxel V_i^{sib}. Since our entropy model encodes the occupancy symbols of n_j's children sequentially, the previous siblings' occupancy symbols are available while predicting the probability of the current octant n_i. The available occupancy symbols are filled in V_i^{sib}. Note that the first occupied child octant of n_t has no available sibling context, and it will take V_i^{sib} with zero values as input. We parameterize another feature transformation function, f_{sib}, as a 3D convolution network to exploit the sibling context with V_i^{sib} as input. We formulate it as

$$h_i^{sib} = f_{sib}(V_i^{sib}). \quad (4)$$

The h_i^{sib} is the output feature vector that represents the sibling contextual information and will be further provided as a part of prior knowledge for entropy coding.

3.3.4 Geometric Priors

Since LiDAR is a time-of-flight device to estimate the round-trip time of the darting laser beams reflecting from a scene, most of the sampling LiDAR points are densely distributed on large surfaces, such as roads and vehicles. If an octant crosses the boundary of a surface, its children's octants are more likely to be occupied. Therefore, geometric information such as the surface can be a strong prior for reducing the data redundancies.

The neighbor context of n_i is defined as a voxel representation V_i in Sect. 3.3.1. We first transform the neighbor context V_i into a point representation $\{(x_i, y_i, z_i)\}_{i=1}^{N}$ with the coordinate of n_i as the origin. Then we train a module to locally fit a quadratic surface by minimizing the vertical distance from each point to the surface:

$$\mathcal{L}_{sf} = \|\mathbf{z} - \boldsymbol{\delta}\left[\mathbf{x}^2, \mathbf{y}^2, \mathbf{xy}, \mathbf{x}, \mathbf{y}, \mathbf{1}\right]^T\|_2^2, \tag{5}$$

where $\boldsymbol{\delta} \in \mathbb{R}^6$ are six parameters of the quadratic surface. As illustrated in Fig. 2, the module that learns surface priors will take the low-level feature h_i^{feat} from the second layer of f_{neigh} as input. The surface estimation module f_{geo} extract the geometric features from h_i^{feat}, followed by a MLP to estimate the parameters $\boldsymbol{\delta}$ for the quadratic surface. The transformation can be defined as

$$h_i^{geo} = f_{geo}(h_i^{feat}), \tag{6}$$

$$\boldsymbol{\delta} = \mathbf{MLP}(h_i^{geo}), \tag{7}$$

where h_i^{geo} represents the geometric priors provided to the entropy model to infer the probability distribution.

3.3.5 Entropy Model Header

Eventually, the neighbor feature h_i^{neigh}, the ancestral feature h_i^{an}, siblings' children feature h_i^{sib}, the geometric feature h_i^{geo}, and the current octant's information c_i are aggregated to a four-layer MLP followed by a softmax to predict a 256 channels probability $q(s_i)$ for the 8-bit occupancy symbol. Then we minimize the cross-entropy loss on each octant:

$$\mathcal{L}_{CE} = -\sum_i p(s_i) \log q(s_i), \tag{8}$$

where $p(s_i)$ is the ground-truth probability distribution of the occupancy symbol s_i.

3.4 Multi-level Learning Framework

From our statistics on 12-level octrees constructed from the KITTI Odometry [8] dataset, the last level of the non-leaf octants accounts for more than 50% of the total number of all the non-leaf octants. We calculate the occupancy symbols distribution by levels and compute the Jensen-Shannon (JS) divergence among them. As shown in Fig. 3, the greater JS divergence means the larger difference between the two distributions.

Based on these two observations, we separately train entropy models for each level instead of training a single shared-weight entropy model on the entire octree. This multi-level learning framework facilitates our entropy model to better reduce each octant's spatial redundancies in different spatial resolutions.

We use the same objective function for the entropy models at different levels:

$$\mathcal{L} = \mathcal{L}_{CE} + \lambda \mathcal{L}_{sf}, \tag{9}$$

where λ is the weight of the loss and empirically set to 0.2 in experiments.

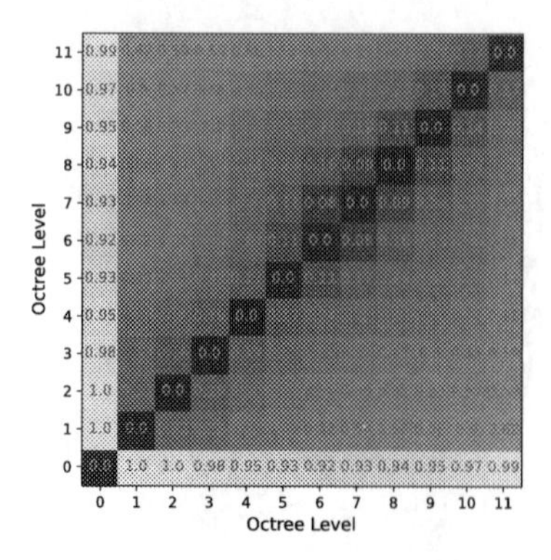

Fig. 3. The heat map of the Jensen-Shannon (JS) divergence of the occupancy symbol distributions among different levels of an octree. The distributions are computed from our training set on KITTI [8]. JS divergence is to measure the difference between two probability distributions. The greater JS divergence means the larger difference between the two distributions.

3.5 Two-step Reconstruction Strategy

The quantization error of our point cloud compression framework comes from the octree construction and depends on the octree level L. Each leaf octant at level L represents a subspace in the large 3D space. Because of the quantization, we can only recover a single point from each leaf octant by taking the center coordinate of its corresponding subspace. VoxelContext [24] introduces a refinement module to reduce the quantization error by adding predicted offsets to these coordinates in the decoding stage. In contrast, we approach this problem by first retrieving missing points with the help of our trained entropy model at level $L+1$ and then further refining the aggregated points. With this strategy, we can reconstruct point clouds with a precision close to level $L+1$ while keeping the bitrate at level L.

As illustrated in Fig. 4, considering a reconstructed octree with L levels, we denote a set of octants at level L as $\mathcal{N}_L$ storing the occupancy symbols. The symbols from $\mathcal{N}_L$ consist of the occupancy information that forms the children's octants at level $L+1$. We first apply our entropy model at level $L+1$ on each children's octant to estimate its occupancy symbol probabilities. We take the predicted occupancy symbols $\mathcal{S}_{L+1}$ with top-1 accuracy to form a new set of octants $\mathcal{N}_{L+1}$ located at the predicted level $L+1$. We denote the corresponding 3D coordinate of $\mathcal{N}_{L+1}$ as $\mathbf{x}^p \in \mathbb{R}^{n\times 3}$. In the second step, we apply the refinement module at level $L+1$ [24] that takes neighbor context V_i and octant information c_i as the input to predict offsets for the coordinates. Our refinement module is predefined to output offsets for every child's octant of $\mathcal{N}_L$, where the offsets are denoted as $\mathbf{x}^o$. Then we obtain the final output of the coordinate by

$$\mathbf{x}^r = \mathbf{x}^p + \mathbf{I}(\mathcal{S}_{L+1}, \mathbf{x}^o), \tag{10}$$

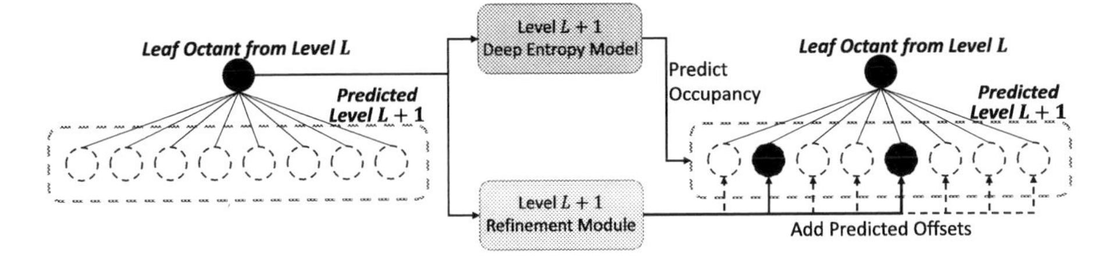

Fig. 4. The illustration of our two-step refinement strategy. The leaf octant of the decoded octree locates at the depth level L, and its occupancies are unknown. We first apply our entropy model on the octant to predict the occupancy symbols and calculate their coordinates. Then the coordinates are added with the corresponding offsets predicted from the refinement module.

where $\mathbf{I}(a, b)$ denote the operation of indexing b according to a. To be specifically, $\mathbf{I}(\mathcal{S}_{L+1}, \mathbf{x}^o)$ means that for those octants whose occupancy codes are zeros, we ignore their offsets. We only add the offsets to the octants whose occupancy codes are ones. $\mathbf{x}^r \in \mathbb{R}^{n\times 3}$ is the resulting refined 3D coordinates.

To obtain the ground-truth coordinate of N_{L+1} when training the entropy model, We build octrees with $L+1$ levels and calculate the coordinates $\mathbf{x}^{gt}$ from $\mathcal{N}_{L+1}$ as the ground truth. Note that the sibling context at level $L+1$ is not available during the encoding and decoding time, and we set V_i^{sib} as zeros voxel block to be the input of our entropy model. We train the refinement module $\mathcal{R}$ by minimizing the Chamfer Distance [5] of the $\mathbf{x}^{gt}$ and $\mathbf{x}^r$:

$$\mathcal{L}_{CD} = \max \left\{ \mathrm{CD}(\mathbf{x}^r, \mathbf{x}^{gt}), \mathrm{CD}(\mathbf{x}^{gt}, \mathbf{x}^r) \right\}, \tag{11}$$

where $\mathcal{L}_{CD}$ is the objective function for the refinement module.

4 Experiments

4.1 Experimental Setup

Dataset. We evaluate the compression performance and reconstruction quality on two real-world datasets with different densities: KITTI Odometry [8] and nuScenes [3], which are collected by a Velodyne HDL-64 sensor and a Velodyne HDL-32 sensor. We also use the KITTI detection dataset [8] for evaluating the 3D object detection as a downstream task using the reconstructed point cloud.

To ensure the diversity of the training and testing scene, we randomly sample 6000 frames from sequences 00 to 10 in the KITTI Odometry dataset for training and 550 frames from sequences 11 to 21 for testing. For nuScenes, we randomly sample 1200 frames from each of the first five data batches (total 6000) for training, and randomly sampled 100 frames from each of the last five data batches (total 500) for testing. Note that we compare our compression performance and reconstruction quality with the original raw point cloud without any pre-processing in the following sections.

Baselines. To evaluate the performance of our framework, we compare our method with three state-of-the-art baselines: hand-drafted Octree-based point cloud compression method MPEG G-PCC [26], KD-tree based method Google Draco [9], and also a learning-based method VoxelContext [24]. VoxelContext has been faithfully self-implemented as the official code has not been released.

Implementation Details. We construct the octree with a maximum level of 13 on both KITTI Odometry and nuScenes datasets. Note that we construct level 13 of the octree only for generating the ground-truth labels for our training. In our experiment section, we evaluate the performance of our framework by truncating octree levels ranging from 9 to 12 on both datasets to vary the compression bitrates. The corresponding spatial quantization error ranges from 14.94 cm to 1.87 cm for KITTI Odometry and 18.29 cm to 2.29 cm for nuScenes.

We implement our framework in PyTorch, and the models are trained on NVIDIA 3090 GPU. The total number of parameters in our deep entropy model is 1.77M. We use Adam optimizer [16] with a learning rate of 1e-4 to train our whole framework. The training epoch for the multi-level entropy model and the refinement module are 20 and 2, respectively. To be memory efficient, we do not backpropagate the gradients of an entropy model from the lower to the upper level during the training of multi-level entropy models. We freeze the weights of the entropy models while training the refinement module.

4.2 Evaluation Metrics

We evaluate our framework from two aspects, compression performance and reconstruction quality. Bits per point (BPP) is the most commonly used metric for evaluating compression performance. Since we only consider the geometric compression of a point cloud in this section, the size of the original point cloud data is calculated by $96 \times N$, where N is the number of points in the raw point cloud, and 96 is the size of the coordinates: x, y and z, where each coordinate is represented in a 32-bit floating-point. BPP is defined as $BPP = |bit|/N$, where the $|bit|$ is the bitstream length.

We utilize four metrics to evaluate the reconstruction quality of the point cloud: point-to-point and point-to-plane PSNR [19,30], F1 score, and maximum Chamfer Distance [5,30]. Same as the evaluation metrics defined in [2], for original input point cloud P and reconstruction point cloud $\hat{P}$, we use $\mathrm{PSNR} = 10\log_{10}\frac{3r^2}{\max\{\mathrm{MSE}(P,\hat{P}),\mathrm{MSE}(\hat{P},P)\}}$, $\mathrm{F1} = \frac{TP}{TP+FP+FN}$, $\mathrm{CD}_{max} = \max\{\mathrm{CD}(P,\hat{P}), \mathrm{CD}(\hat{P},P)\}$, where the peak constant value of r is 59.70 m, and the distance threshold of the F1 score is 2 cm.

4.3 Results

Quantitative Results. To quantitatively evaluate our method, we report BPP versus four reconstruction quality metrics. As illustrated in Fig. 5, our method saves more bitrate and achieves higher reconstruction quality compared with all

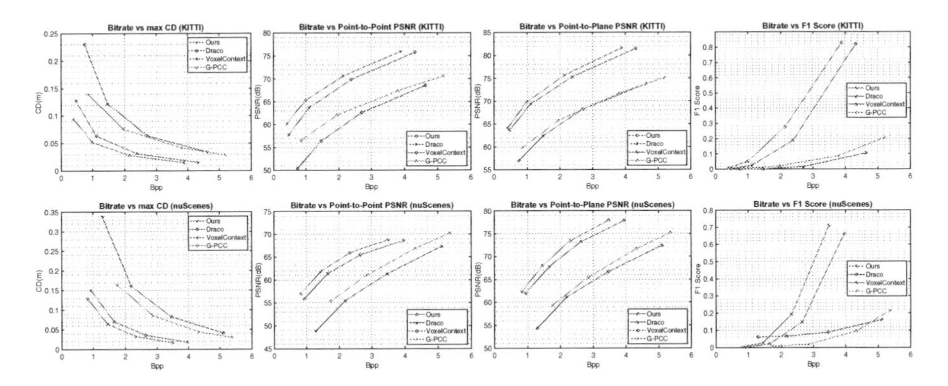

Fig. 5. The quantitative results of our method on the KITTI Odometry dataset (first row) and nuScenes dataset (second raw). From left to right: maximum Chamfer distance (↓), point-to-point PSNR (↑), point-to-plane PSNR (↑), and F1 score (↑). Note that both our method and VoxelContext are applied to the refinement module when calculating the evaluation metrics in this figure.

Table 1. The quantitative results when compared to VoxelContext without any refinement module. The first row is the result of the KITTI Odometry dataset. The second row is the result of the nuScenes dataset. The reconstructed point clouds of the two methods are the same at each level.

Dataset	Method	BPP↓				
		Level 8	Level 9	Level 10	Level 11	Level 12
KITTI	VoxelContext [24]	0.173	0.466	1.123	2.380	4.371
	Ours	**0.149**	**0.409**	**0.999**	**2.137**	**3.878**
nuScenes	VoxelContext [24]	0.474	0.937	1.650	2.642	3.960
	Ours	**0.424**	**0.829**	**1.445**	**2.319**	**3.487**

the state-of-the-art methods on both 64 channel KITTI Odometry dataset and 32 channel nuScenes dataset.

Since we use the same octree construction strategy as VoxelContext, we have the same reconstruction quality at the same octree level without applying any refinement module. As illustrated in Table 1, our method saves 11%–16% bitrate on the KITTI Odometry dataset and 12%–14% on the nuScenes dataset compared to VoxelContext. Our method reaches higher reconstruction quality than VoxelContext at the same octree level despite using the same refinement module, clearly demonstrating the benefit of our newly proposed two-step refinement strategy.

Qualitative Results. We evaluate our method qualitatively by comparing the reconstruction error with state-of-the-art methods on two datasets. As illustrated in Fig. 6, the error between the original point cloud and our reconstructed point

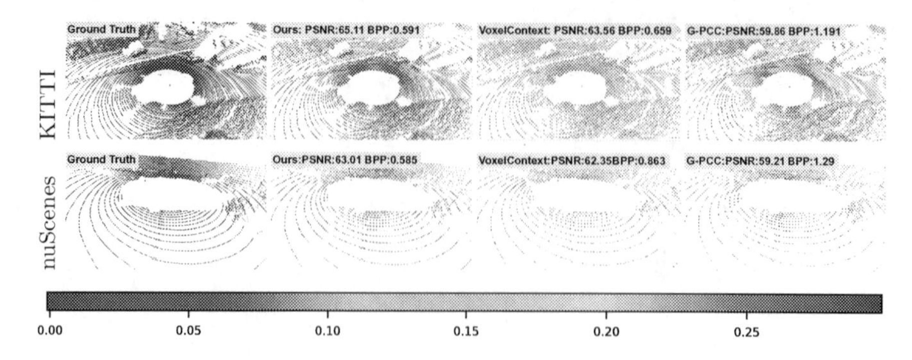

Fig. 6. The qualitative results of our method compared with VoxelContext and G-PCC on the KITTI Odometry dataset (first row) and the nuScenes dataset (second row). The color bar indicates the error in meters between the original and reconstructed point cloud. Our reconstructed point cloud has a lower error at a lower bitrate compared to all baselines.

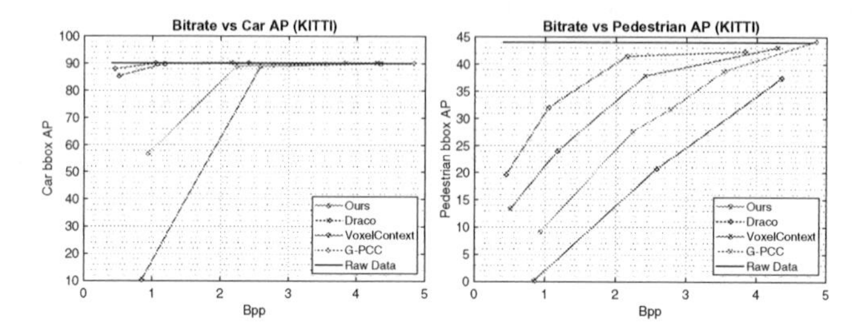

Fig. 7. The qualitative results of downstream tasks on the KITTI detection dataset.

cloud is much smaller at an even lower bitrate as compared to all three baseline methods.

Downstream Tasks. The performance on downstream tasks of the reconstructed point cloud is also a crucial evaluation metric for point cloud compression. We evaluate the 3D object detection method proposed in [17] on reconstructed point cloud data of different compression methods. We report the bitrate versus average precision (AP) at 0.5 IOU as a threshold for pedestrians and 0.7 IOU for cars. As illustrated in Fig. 7, our method outperforms all three baselines in different categories.

4.4 Ablation Study

We perform an ablation study on our deep entropy model to demonstrate the effectiveness of different contextual features. We ablate over the contextual features by training the model without exploiting features from siblings' children, ancestors, or surfaces. The models for ablation studies are trained on the KITTI

Odometry dataset with the same training configurations. We evaluate the models at a level ranging from 8 to 12. From the quantitative results shown in Table 2, we observe that the bitrate increases when not incorporating any dependencies into the entropy model. Moreover, the result also demonstrates that the sibling features are crucial to the compression performance of our entropy model.

Table 2. Ablation study of our entropy model on KITTI [8] without using context from siblings' children, ancestors, or surface priors.

Method	BPP↓				
	Level 8	Level 9	Level 10	Level 11	Level 12
Ours w/o Sibling	0.161 (+8.1%)	0.439 (+7.3%)	1.071 (+7.2%)	2.288 (+7.1%)	4.133 (+6.6%)
Ours w/o Ancestor	0.150 (+0.6%)	0.414 (+1.2%)	1.016 (+1.7%)	2.181 (+2.1%)	3.956 (+2.0%)
Ours w/o Surface	0.151 (+1.3%)	0.415 (+1.5%)	1.019 (+2.0%)	2.180 (+2.0%)	3.953 (+1.9%)
Ours	**0.149**	**0.409**	**0.999**	**2.137**	**3.878**

5 Conclusion

We have presented a novel octree-based point cloud compression method. Our method uses the hierarchical contexts of octree structures and surface priors to reduce the information redundancies in the bitstream. On the decoding side, our method utilizes a two-step heuristic strategy for reconstructing a point cloud with better quality. We have evaluated our method against three state-of-the-art baselines on two datasets. The result demonstrates that our method outperforms previous methods on compression performance, reconstruction quality, and downstream tasks. We hope our work can inspire researchers to further reduce the compression rate and improve the reconstruction quality in future work.

References

1. Ahn, J.K., Lee, K.Y., Sim, J.Y., Kim, C.S.: Large-scale 3D point cloud compression using adaptive radial distance prediction in hybrid coordinate domains. JSTSP **9**(3), 422–434 (2014)
2. Biswas, S., Liu, J., Wong, K., Wang, S., Urtasun, R.: MuSCLE: multi sweep compression of lidar using deep entropy models. In: NIPS (2020)
3. Caesar, H., et al.: nuScenes: a multimodal dataset for autonomous driving. In: CVPR (2020)
4. Devillers, O., Gandoin, P.M.: Geometric compression for interactive transmission. In: VIS 2000 (Cat. No. 00CH37145) (2000)
5. Fan, H., Su, H., Guibas, L.J.: A point set generation network for 3d object reconstruction from a single image. In: CVPR (2017)
6. Fu, C., Li, G., Song, R., Gao, W., Liu, S.: OctAttention: octree-based large-scale contexts model for point cloud compression. In: AAAI (2022)

7. Garcia, D.C., de Queiroz, R.L.: Intra-frame context-based octree coding for point-cloud geometry. In: ICIP (2018)
8. Geiger, A., Lenz, P., Urtasun, R.: Are we ready for autonomous driving? The kitti vision benchmark suite. In: CVPR (2012)
9. Google: Draco (2017). https://github.com/google/draco
10. Graziosi, D., Nakagami, O., Kuma, S., Zaghetto, A., Suzuki, T., Tabatabai, A.: An overview of ongoing point cloud compression standardization activities: video-based (V-PCC) and geometry-based (G-PCC). APSIPA Trans. Signal Inform. Process. **9**, E13 (2020)
11. Huang, L., Wang, S., Wong, K., Liu, J., Urtasun, R.: OctSqueeze: octree-structured entropy model for lidar compression. In: CVPR (2020)
12. Huang, T., Liu, Y.: 3D point cloud geometry compression on deep learning. In: MM (2019)
13. Huang, Y., Peng, J., Kuo, C.C.J., Gopi, M.: A generic scheme for progressive point cloud coding. TVCG **14**(2), 440–453 (2008)
14. Jackins, C.L., Tanimoto, S.L.: Oct-trees and their use in representing three-dimensional objects. CGIP **14**(3), 249–270 (1980)
15. Kammerl, J., Blodow, N., Rusu, R.B., Gedikli, S., Beetz, M., Steinbach, E.: Real-time compression of point cloud streams. In: ICRA (2012)
16. Kingma, D.P., Ba, J.: Adam: a method for stochastic optimization. In: ICLR (2015)
17. Lang, A.H., Vora, S., Caesar, H., Zhou, L., Yang, J., Beijbom, O.: PointPillars: fast encoders for object detection from point clouds. In: CVPR (2019)
18. Meagher, D.: Geometric modeling using octree encoding. CGIP **19**(2), 129–147 (1982)
19. Mekuria, R., Laserre, S., Tulvan, C.: Performance assessment of point cloud compression. In: VCIP (2017)
20. Nenci, F., Spinello, L., Stachniss, C.: Effective compression of range data streams for remote robot operations using h. 264. In: IROS (2014)
21. Nguyen, D.T., Quach, M., Valenzise, G., Duhamel, P.: Multiscale deep context modeling for lossless point cloud geometry compression. In: ICMEW (2021)
22. Qi, C., Yi, L., Su, H., Guibas, L.J.: PointNet++: deep hierarchical feature learning on point sets in a metric space. In: NIPS (2017)
23. Qi, C.R., Su, H., Mo, K., Guibas, L.J.: PointNet: deep learning on point sets for 3D classification and segmentation. In: CVPR (2017)
24. Que, Z., Lu, G., Xu, D.: VoxelContext-Net: an octree based framework for point cloud compression. In: CVPR (2021)
25. Schnabel, R., Klein, R.: Octree-based point-cloud compression. In: PBG@ SIGGRAPH (2006)
26. Schwarz, S., et al.: Emerging MPEG standards for point cloud compression. JETCAS **9**(1), 133–148 (2018)
27. Sun, X., Ma, H., Sun, Y., Liu, M.: A novel point cloud compression algorithm based on clustering. RA-L **4**(2), 2132–2139 (2019)
28. Sun, X., Wang, S., Liu, M.: A novel coding architecture for multi-line lidar point clouds based on clustering and convolutional LSTM network. T-ITS **29**(3), 2190–2201 (2020)
29. Theis, L., Shi, W., Cunningham, A., Huszár, F.: Lossy image compression with compressive autoencoders. In: 5th International Conference on Learning Representations, ICLR (2017)
30. Tian, D., Ochimizu, H., Feng, C., Cohen, R., Vetro, A.: Geometric distortion metrics for point cloud compression. In: ICIP (2017)

31. Tu, C., Takeuchi, E., Carballo, A., Takeda, K.: Point cloud compression for 3D lidar sensor using recurrent neural network with residual blocks. In: ICRA (2019)
32. Wang, J., Zhu, H., Ma, Z., Chen, T., Liu, H., Shen, Q.: Learned point cloud geometry compression. CoRR abs/1909.12037 (2019)
33. Wang, S., Jiao, J., Cai, P., Liu, M.: R-PCC: a baseline for range image-based point cloud compression. CoRR abs/2109.07717 (2021)
34. Wiesmann, L., Milioto, A., Chen, X., Stachniss, C., Behley, J.: Deep compression for dense point cloud maps. RA-L **6**(2), 2060–2067 (2021)
35. Yan, W., Shao, Y., Liu, S., Li, T.H., Li, Z., Li, G.: Deep autoencoder-based lossy geometry compression for point clouds. CoRR abs/1905.03691 (2019)

Author Index

Printed in the United States
by Baker & Taylor Publisher Services